PRINCIPLES AND APPLICATIONS OF ELECTRICAL ENGINEERING

PRINCIPLES AND APPLICATIONS OF ELECTRICAL ENGINEERING

Second Edition

Giorgio Rizzoni
The Ohio State University

IRWIN

Chicago • Bogotá • Boston • Buenos Aires • Caracas
London • Madrid • Mexico City • Sydney • Toronto

This project was supported, in part, by the

National Science Foundation

Opinions expressed are those of the authors and not necessarily those of the Foundation

Publisher: *Tom Casson*
Senior sponsoring editor: *Scott Isenberg*
Senior developmental editor: *Kelley Butcher*
Marketing manager: *Brian Kibby*
Project editor: *Rebecca Dodson*
Production supervisor: *Dina L. Treadaway*
Cover designer: *Terri Ellerbach*
Compositor: *Publication Services*
Typeface: *10/12 Times Roman*
Printer: *Quebecor*

Library of Congress Cataloging-in-Publication Data

Rizzoni, Giorgio.
Principles and applications of electrical engineering / Giorgio Rizzoni. –2nd ed.
p. cm.
Includes index.
ISBN 0-256-17770-8
1. Electric engineering. I. Title.
TK146.R437 1996
621.3—dc20 95–18650
CIP

Printed in the United States of America
4 5 6 7 8 9 0 QF 2 1 0 9 8 7

In memoria di papà

About the Author

Giorgio Rizzoni received his B.S., M.S., and Ph.D. degrees, all in Electrical Engineering, from the University of Michigan, and is currently an Associate Professor in the Department of Mechanical Engineering at the Ohio State University, where he is responsible for teaching in the areas of electromechanics, system dynamics, mechanical measurements, control systems, and signal processing. Upon completing his graduate studies, Dr. Rizzoni served as a Lecturer and Research Scientist in the University of Michigan EECS Department, and as Assistant Director of the Vehicular Electronics Laboratory. As a lecturer at the University of Michigan, he taught the nonmajors electrical engineering lecture and laboratory courses, upgrading the curriculum and designing a new laboratory. For these efforts he received the 1990 Tau Beta Pi College of Engineering Outstanding Teaching Award.

Professor Rizzoni is the recipient of a 1991 NSF Presidential Young Investigator Award and of a 1992 SAE Ralph R. Teetor Educational Award, in addition to several other teaching and research awards. He is the recipient of a curriculum development grant from the National Science Foundation (Division of Undergraduate Education) to support his ongoing development of an interdisciplinary curriculum in the emerging field of Mechatronics. He has developed a number of course modules in various subjects related to Mechatronics, some of which are reflected in this book. His curriculum development efforts at the Ohio State University are also supported by a number of major industrial sponsors, including General Motors, Ford, and Chrysler.

Dr. Rizzoni is an active member of ASME, IEEE, and SAE, and is Associate Editor of the ASME Transactions, *Journal of Dynamic Systems, Measurement and Control,* and of the IEEE *Transactions on Vehicular Technology.* His research has been funded, among others, by NSF, General Motors, Ford, Chrysler, and IBM. He is the author of nearly 100 technical papers published in peer-reviewed journals and in conference proceedings.

Preface

The pervasive presence of electronic devices and instrumentation in all aspects of engineering design and analysis is one of the manifestations of the electronic revolution that has characterized the second half of the 20th century. Every aspect of engineering practice, and even of everyday life, has been affected in some way or another by electrical and electronic devices and instruments. Computers are perhaps the most obvious manifestations of this presence. However, many other areas of electrical engineering are also important to the practicing professional in a range of fields, from mechanical and industrial engineering to chemical, nuclear, and materials engineering, the aerospace and astronautical disciplines, and civil engineering. Engineers today must be able to communicate effectively within the interdisciplinary teams in which they work.

OBJECTIVES

The aim of *Principles and Applications of Electrical Engineering* is to provide the nonelectrical engineering student—typically in the second or third year of a four-year curriculum—with a foundation for understanding the basic principles of electrical and electronic systems. This book should help build a solid foundation for further study, either in the classroom or on the job.

It would be overly ambitious to attempt to summarize all important aspects of electrical engineering in a single book, without its taking on an undesired encyclopedic nature. This book differs in two respects in content and style of presentation from many other textbooks that are currently available in the market. First, no attempt is made to cover all topics in electrical engineering. Rather, the intent is to help the student in building a fundamental understanding of the basic principles, without clouding the main topics with excessive detail. The reader interested in a more in-depth treatment of a particular subject will be able to refer to more advanced books, while the instructor who wishes to cover some particular area in greater depth will find that each chapter includes a number of homework problems designed for more advanced study. Second, a number of topics that are commonly found in books that take a more encyclopedic approach to presenting the subject will not be found here. For example, topics such as control or communication systems are not included, although some discussion of relevant ideas is given when appropriate. The author believes that such subjects are best dealt with in a specially designed course (for example, many undergraduate degree programs require a senior level course in control systems). Further, in attempting to include too many subjects, many authors have been forced to reduce the coverage of those essential topics that form the basis of electrical engineering. An honest effort has been made to avoid this shortcoming in the present book; a multitude of examples and exercises are provided in the early circuit analysis chapters to ensure a solid and complete understanding of the material that forms the basis of the entire discipline. This approach is based on the author's belief that learning the foundations of circuit analysis greatly simplifies the task of understanding the more advanced material.

The emphasis in *Principles and Applications of Electrical Engineering* is thus on teaching relevant electrical engineering concepts to engineers who will be users, not designers, of electrical, electromechanical, and electronic systems. Thus, the book provides a user's perspective of electrical engineering, focusing on those principles and applications that are likely to be of use to tomorrow's engineers. Electrical and electronic subsystems are an integral part of any complex plant or engineering system—from the self-contained power systems aboard a ship or an aircraft, to the control of industrial processes and power plants or the management of propulsion systems for automobiles and spacecrafts. In all of these applications, the common threads are *instrumentation,* which is required to perform measurement for process and plant performance evaluation and control, and the *devices* that are used to implement the controls and which make up the measuring instruments. Thus, this book will focus on the application of devices to specific measurements instrumentation and control tasks, and on the basic building blocks of instrumentation and measurement systems that are likely to be encountered in practice.

ORGANIZATION AND CONTENT

The wide range of requirements that characterize four-year engineering programs in the United States presents a challenge to writing a useful and versatile book for the nonmajor audience. The following paragraphs provide a rationale for the organization and content of this book and suggest various sequences that may be adopted to match the requirements of varied curricula.

The book is structured in three parts: circuit analysis, electronics and computers, and electromechanics. It is difficult to decouple completely one subject from the others. Aside from the fact that circuit analysis forms the foundation of the entire book, the operation of modern electric machines cannot be separated from the solid-state electronic control circuits that drive them. Further, where possible, examples are given of the application of a specific circuit or device to measurements and instrumentation. Chapters 1 through 6 (and part of 11) cover the fundamentals of circuit analysis; Chapters 7 through 14 cover electronics and computers; and Chapters 5, 10, and 15 through 17 cover power systems and electric machines. The material in the first six chapters provides a basic understanding of DC and AC circuit analysis and of transient analysis and frequency response. The instructor who wishes to emphasize the role of operational amplifiers early will find that the first two sections of Chapter 11 can be covered early in conjunction with Chapter 3. Another alternative (employed by the author for several semesters) introduces the op-amp immediately after Chapter 4, to use it as a vehicle for presenting the concept of frequency response early on.

The electronics material in Part II can be covered in a variety of ways, to suit the requirements of different audiences. For example, a very light treatment in a one-quarter course might involve all or only parts of Chapters 7, 8, and 9, while an entire course in electronics might include Chapters 7 through 14. A number of options are available to the instructor between these extremes. If time permits treatment of discrete electronic devices, Chapters 7, 8, and 9 provide substantial amounts of material; if the preferred emphasis is on IC and digital electronics, Chapters 11 through 14 are self-contained and do not require a background in discrete electronics.

The book is designed to be as flexible as possible in tailoring the material on electronics and electromechanics to the needs of a second course in electrical engineering. For example, if emphasis on digital electronics were desired, Chapter 8 would provide sufficient material on transistor technology to, say, move directly to the switching applications of transistors in Section 9.5, and then move on to the digital logic and digital systems material (Chapters 12 and 13). As another example, a course on power systems and electric machinery might include the first half of Chapter 7 (diode fundamentals), Chapter 8 (transistor fundamentals), and Chapter 10 (power electronics), before approaching the section on electromechanics (Chapters 15–17). If the desired emphasis were on electronic instrumentation, the material on discrete electronic devices could be bypassed altogether, and Chapters 11–14 would provide a self-contained unit on modern integrated circuit electronics and instrumentation. Instructors who teach a survey-oriented course will find that the chapters go into progressively more depth in the later sections, so that it is possible to cover only part of each chapter and achieve substantial coverage of different topics. It is conceivable, for example, to touch on part or all of the first 13 chapters in a one-semester course if desired.

Part III, dealing with electric machines, is geared toward providing the student with an understanding of the fundamental principles of operation of the principal electric machines, with some focus on the performance and selection criteria for different applications. In addition to an introduction to the basic properties of DC, synchronous, and induction motors, emphasis is also given to special-purpose machines, such as servos, stepper motors, brushless DC motors, and other devices commonly used in motion control systems, in the hope that the interested student may use this material as a stepping stone to more advanced courses.

The following chart illustrates possible coverage options. Each instructor will be able to tailor the book to the specific needs of his or her course.

Suggested Sequences

First course in Circuit Analysis (with a survey of selected topics in electronics)	
1-quarter course	Part I—Circuits, plus survey of topics in electronics, e.g.: Survey of electronic devices: 7.1–7.3, 8, 10.1–2 Op-amps: 11 Digital Logic Circuits: 12 and parts of 13
1-semester course	Part I—Circuits, plus one topic in electronics, e.g.: Discrete Analog Electronics: 7, 8, 9 Power Electronics: 7.1–7.3, 8, 10 Digital Circuits and Systems: 12, 13 Instrumentation: 11, 14 Electric Machinery: 15, 16

Second course in Analog or Digital Electronics, Instrumentation, Energy Conversion Systems (assumes coverage of Chapters 1–6)	
Analog Electronics	Chapters 7, 8, 9.1–9.4, 10, 11
Digital Electronics	Chapters 7.1–7.3, 8, 9.5, 12, 13
Instrumentation	Chapters 7, 8, 11, 12, 13, 14
Energy Conversion	Chapters 7.1–7.3, 8, 10, 15, 16, 17

PEDAGOGY

Each chapter has an introduction with chapter objectives, an abundance of illustrations (over 1,500 total), many completely worked examples (over 260 overall), drill exercises to reinforce concepts just learned, a chapter summary, and key terms lists. Also, there is a variety of homework problems at the end of each chapter. The problems are intended to assist the student in understanding and applying the basic concepts presented. There are approximately 750 homework problems—an average of 50 per chapter. Answers to selected problems can be found at the back of the book, along with an appendix on linear algebra.

EIT REVIEW

The Engineer-in-Training (EIT) examination is one of the important reasons for the study of electrical engineering in other engineering curricula. This book addresses this important aspect of the education of engineering undergraduates by presenting a brief review of each of the exam's nine subsections at the end of the relevant chapters. In addition, a description of the examination is given in Chapter 1. Each review section includes solved examples, as well as precise references to those examples in the text that are most representative of the material covered in the EIT exam. Further, selected homework problems that are relevant to this exam are specially marked "EIT" in the Homework Problems sections of the relevant chapters; the answers to these problems are supplied in Appendix B.

The unmatched coverage of the EIT examination makes this text uniquely suited to complete a thorough review of the Electrical Circuits part of the EIT exam.

CHANGES TO THE SECOND EDITION

The second edition of this book has been significantly improved in a number of areas.

Pedagogy

Over 260 homework problems have been replaced with new ones. The answers to the Drill Exercises have been moved to the end of each chapter, to permit the student to attempt the exercises without distractions. The index has been significantly expanded.

Content

Material on dynamic circuit equations and on the Laplace Transform method of solutions has been added to Chapters 4 and 6, respectively. The material on Transistors in Chapters 8 and 9 has been reorganized to permit coverage at two different levels: Chapter 8 now introduces the fundamentals of both bipolar and field effect transistors, while Chapter 9 presents more advanced topics, such as small signal amplifiers and switching circuits. The new arrangement allows the instructor greater flexibility. In addition, the material on field effect transistors has been brought up to date and expanded. A new chapter on Power Electronics has been added, which includes some coverage of electric motor drives. This chapter is ideally suited to form an energy systems course together with Chapters 5, 15,

16, and 17. The material on Digital Circuits and Systems has been rearranged, with the addition of a section on sequential logic circuit design. A new chapter on Instrumentation has been added, summarizing important concepts that appeared in a less organized form in the first edition. Finally, the section on electric machines has been supplemented with an overview of motor selection and application criteria.

SUPPLEMENTS

The Instructor's Manual, available to adopters from the publisher, contains complete solutions to the homework problems. Transparency masters of important figures and example problems are also available.

ACKNOWLEDGMENTS

This book has been read critically by the following reviewers. Richard D. Irwin and the author would like to thank these reviewers for their contribution to *Principles and Applications of Electrical Engineering:*

Louis E. Roemer, *Louisiana Tech University*
Ronald L. Klein, *West Virginia University*
M. Paul Murray, *Mississippi State University*
Lippold Hakon, *University of Illinois*
J. D. Aplevich, *University of Waterloo*
Carl L. Epley, *Virginia Polytechnic Institute and State University*
H. Roland Zapp, *Michigan State University*
William E. Bennett, *U.S. Naval Academy*
Richard S. Marleau, *University of Wisconsin*
Richard W. Christiansen, *Brigham Young University*
William H. Dawes, *Kansas State University*
Dennis P. Malone, *University at Buffalo–SUNY*
Khalil Najafi, *University of Michigan*

The author is indebted to these reviewers for their valuable insights and for the patient reviews that have greatly improved the accuracy and readability of this book. Special thanks also go to Gene and Linda Stuffle for their valuable assistance in preparing new homework problems and the solutions manual.

Throughout the development of this edition, great care has been taken to eliminate errors. Thanks to the following individuals for their work in verifying the accuracy of this book:

M. Paul Murray, *Mississippi State University*
Deborah Hicks, *Mississippi State University*

Sam Stone, *West Virginia University*
Richard Marleau, *University of Wisconsin*
Shatil Haque, *Virginia Polytechnic Institute*
Tom Morreale, *North Carolina State University*
Steve Moore, *North Carolina State University*
Yue Yun Wang, *The Ohio State University*

To ensure that future editions are error free, please forward your suggestions or comments to: Dr. Giorgio Rizzoni, c/o IRWIN Editorial-Engineering, Richard D. Irwin, 1333 Burr Ridge Parkway, Burr Ridge, IL 60521.

Finally, the love, patience, and support provided by my wife, Kathryn, have made it possible to endure the long hours of work inevitably associated with such a challenging project, while my son, Alessandro, and my daughter, Maria, gave me inspiration with their smiles.

Giorgio Rizzoni
The Ohio State University

Contents

PART II ELECTRONICS 347

Chapter 7 Semiconductors and Diodes 349

Chapter 8 Transistor Fundamentals 407

Chapter 9 Transistor Amplifiers and Switches 451

Chapter 10 Power Electronics 507

Chapter 11 Operational Amplifiers 541

CHAPTER

1

Introduction to Electrical Engineering

The aim of this chapter is to introduce electrical engineering. The chapter is organized to provide the newcomer with a view of the different specialties making up electrical engineering and to place the intent and organization of the book into perspective. Perhaps the first question that surfaces in the mind of the student approaching the subject is, Why electrical engineering? Since this book is directed at a readership having a mix of engineering backgrounds other than electrical engineering, the question is well justified and deserves some discussion. The chapter begins by defining the various branches of electrical engineering, showing some of the interactions among them, and illustrating by means of a simple example how electrical engineering is intimately connected to many other engineering disciplines. The next section introduces the Engineer-in-Training (EIT) national examination. A brief historical perspective is also provided, to outline the growth and development of this relatively young engineering specialty. Next, the fundamental physical quantities and the system of units are defined, to set the stage for the chapters that follow. Finally, the organization of the book is discussed, to give the student, as well as the teacher, a sense of continuity in the development of the different subjects covered in Chapters 2 through 17.

1.1 ELECTRICAL ENGINEERING

The typical curriculum of an undergraduate electrical engineering student includes the subjects listed in Table 1.1. Although the distinction between some of these subjects is not always clear-cut, the table is sufficiently representative to serve our purposes. Figure 1.1 illustrates a possible interconnection between the disciplines of Table 1.1. The aim of this book is to introduce the non–electrical engineering student to those aspects of electrical engineering that are likely to be most relevant to his or her professional career. Virtually all of the topics of Table 1.1 will be touched on in the book, with varying degrees of emphasis. The following example illustrates the pervasive presence of electrical, electronic, and

Table 1.1 Electrical engineering disciplines

Electrical engineering disciplines
Circuit analysis
Electromagnetics
Solid-state electronics
Electric machines
Electric power systems
Digital logic circuits
Computer systems
Communication systems
Electro-optics
Instrumentation
Control systems

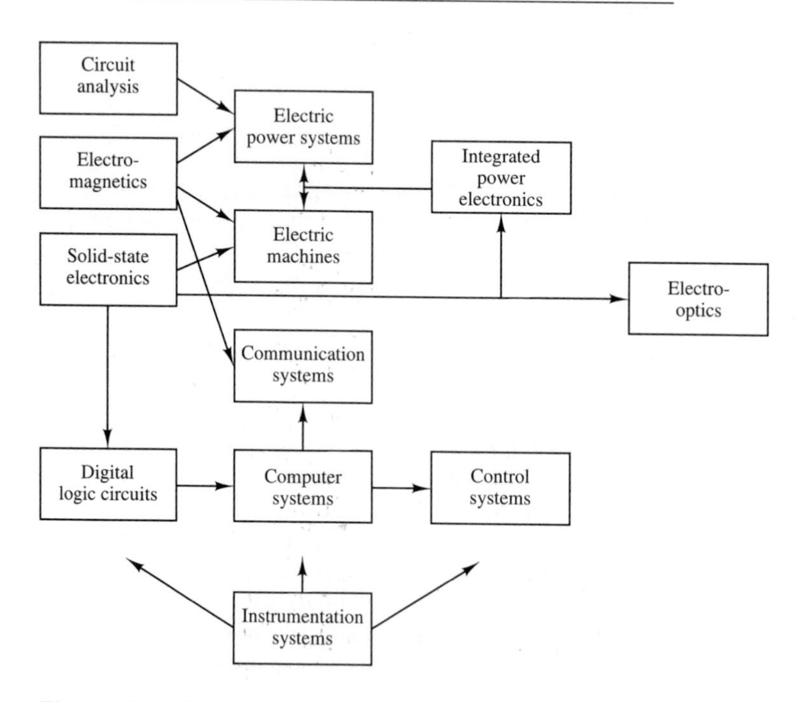

Figure 1.1 Electrical engineering disciplines

electromechanical devices and systems in a very common application: the automobile. As you read through the example, it will be instructive to refer to Figure 1.1 and Table 1.1.

Example 1.1 Electrical Systems in a Passenger Automobile

It may be enlightening to consider a familiar example to illustrate how the seemingly disparate specialties of electrical engineering actually interact to permit the operation of a very familiar engineering system: the automobile. Figure 1.2 presents a view of electrical engineering systems in a modern automobile. Even in older vehicles, the electrical system—in effect, an *electric circuit*—plays a very important part in the overall operation. An inductor coil generates a sufficiently high voltage to allow a spark to form across the spark plug gap, and to ignite the air and fuel mixture; the coil is supplied by a DC voltage provided by a lead-acid battery. In addition to providing the energy for the ignition circuits, the battery also supplies power to many other electrical components, the most obvious of which are the lights, the windshield wipers, and the radio. Electric power is carried from the battery to all of these components by means of a wire harness, which constitutes a

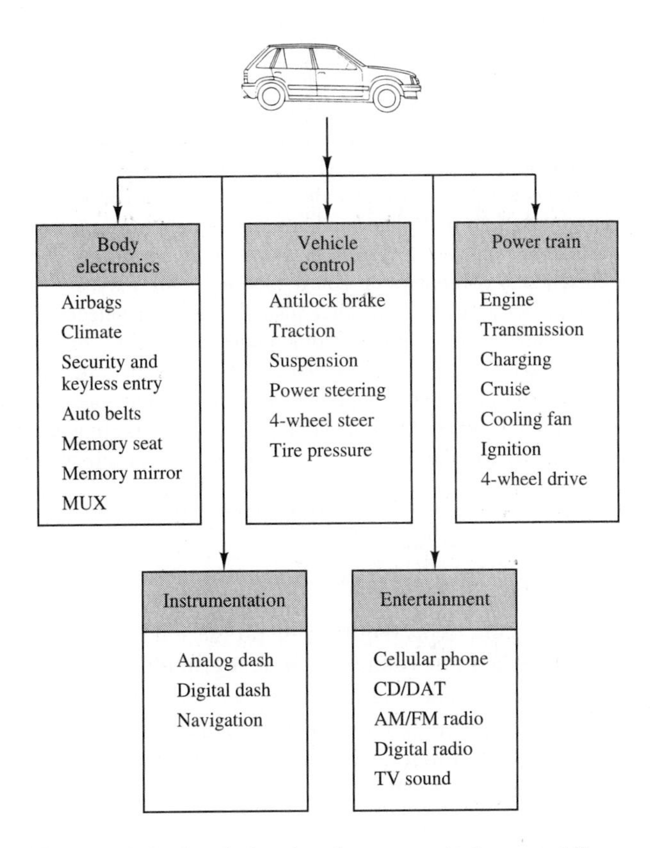

Figure 1.2 Electrical engineering systems in the automobile

rather elaborate electrical circuit. In recent years, the conventional electrical ignition system has been supplanted by *electronic* ignition; that is, solid-state electronic devices called *transistors* have replaced the traditional breaker points. The advantage of transistorized ignition systems over the conventional mechanical ones is their greater reliability, ease of control, and life span (mechanical breaker points are subject to wear).

Other electrical engineering disciplines are fairly obvious in the automobile. The on-board radio receives electromagnetic waves by means of the antenna, and decodes the communication signals to reproduce sounds and speech of remote origin; other common *communication systems* that exploit *electromagnetics* are CB radios and the ever more common cellular phones. But this is not all! The battery is, in effect, a self-contained 12-VDC *electric power system,* providing the energy for all of the aforementioned functions. In order for the battery to have a useful lifetime, a charging system, composed of an alternator and of power electronic devices, is present in every automobile. The alternator is an *electric machine,* as are the motors that drive the power mirrors, power windows, power seats, and other convenience features found in luxury cars. Incidentally, the loudspeakers are also electric machines!

The list does not end here, though. In fact, some of the more interesting applications of electrical engineering to the automobile have not been discussed yet. Consider *computer systems.* You are certainly aware that in the last two decades, environmental concerns related to exhaust emissions from automobiles have led to the introduction of sophisticated engine emission *control systems.* The heart of such control systems is a type of computer called a *microprocessor.* The microprocessor receives signals from devices (called *sensors*) that measure relevant variables—such as the engine speed, the concentration of oxygen in the exhaust gases, the position of the throttle valve (i.e., the driver's demand for engine power), and the amount of air aspirated by the engine—and subsequently computes the optimal amount of fuel and the correct timing of the spark to result in the cleanest combustion possible under the circumstances. The measurement of the aforementioned variables falls under the heading of *instrumentation,* and the interconnection between the sensors and the microprocessor is usually made up of *digital circuits.* Finally, as the presence of computers on board becomes more pervasive—in areas such as antilock braking, electronically controlled suspensions, four-wheel steering systems, and electronic cruise control—communications among the various on-board computers will have to occur at faster and faster rates. Some day in the not-so-distant future, these communications may occur over a fiber optic network, and *electro-optics* will replace the conventional wire harness. It should be noted that electro-optics is already present in some of the more advanced displays that are part of an automotive instrumentation system.

1.2 EIT EXAM REVIEW

Each of the 50 states regulates the engineering profession by requiring individuals who intend to practice the profession to become registered professional engineers (PE). To become a professional engineer, it is necessary to satisfy four requirements. The first is the completion of a B.S. degree in engineering from an accredited college or university (although it is theoretically possible to be registered without having completed a degree). The second is the successful completion of the Engineer-in-Training (Engineering Fundamentals) Examination. This is an eight-hour exam that covers general engineering undergraduate education. The third requirement is two to four years of engineering experience after passing the Engineer-in-Training (EIT) exam. Finally, the fourth requirement is successful completion of the Principles and Practice of Engineering or Professional Engineer (PE) Examination.

The EIT exam is a two-part national examination given twice a year (in April and October). The exam is divided into two four-hour sessions. The morning session consists of 140 multiple choice questions (five possible answers are given); the afternoon session consists of 70 questions. The exam is prepared by the National Council of Examiners for Engineering and Surveying.[1]

One of the aims of this book is to assist you in preparing for one part of the EIT exam, entitled Electrical Circuits. This part of the EIT consists of a total of 14 questions in the morning session and 10 questions in the afternoon session. The examination topics for the electrical circuits part are the following:

DC Circuits
AC Circuits
Three-Phase Circuits
Capacitance and Inductance
Transients
Diode Applications
Operational Amplifiers (Ideal)
Electric and Magnetic Fields
Electric Machinery

Solved sample problems in these areas are given at the ends of many chapters, together with a summary of useful examples you may refer to in the text; all topics are eventually covered. You will find these end-of-chapter reviews invaluable in preparing for the Electrical Circuits part of the EIT exam.

The detailed topics covered in the electrical circuits part of the EIT exam are listed below, along with the relevant chapter in the text.

DC Circuits
- Relationship between current and charge (Chapter 2)
- Current and voltage laws (Chapter 2)
- Power and energy (Chapter 2)
- Equivalent series and parallel elements (Chapter 2)
- Maximum power transfer (Chapter 3)
- Thévenin and Norton theorems (Chapter 3)

AC Circuits
- Algebra of complex numbers (Appendix A)
- Phasor representation (Chapter 4)

[1]P.O. Box 1686 (1820 Seneca Creek Road), Clemson, SC 29633-1686.

- Concept of time domain and frequency domain (Chapters 4 and 6)
- Impedance concepts (Chapter 4)
- Complex power, apparent power, and power factor (Chapter 5)
- Ideal transformers (Chapter 5)

Three-Phase Circuits
- Computation of power (Chapter 5)
- Line and phase currents for delta and wye circuits (Chapter 5)

Capacitance and Inductance
- Energy storage (Chapter 4)
- *RL* or *RC* (first-order) transients (Chapter 6)

Transients (Chapter 6)

Diode Applications (Chapter 7)

Operational Amplifiers (Ideal) (Chapter 11)

Electric and Magnetic Fields
- Work done in moving a charge in an electric field (Chapter 2)
- Relationship between voltage and work (Chapter 2)
- Faraday's and Ampère's laws (Chapter 15)
- Magnetic reluctance concepts (Chapter 15)

Electric Machinery
- AC and DC motor fundamentals (Chapters 16, 17)

1.3 BRIEF HISTORY OF ELECTRICAL ENGINEERING

The historical evolution of electrical engineering can be attributed, in part, to the work and discoveries of the people in the following list. You will find these scientists, mathematicians, and physicists referenced throughout the text.

William Gilbert (1540–1603), English physician, founder of magnetic science, published *De Magnete,* a treatise on magnetism, in 1600.

Charles A. Coulomb (1736–1806), French engineer and physicist, published the laws of electrostatics in seven memoirs to the French Academy of Science between 1785 and 1791. His name is associated with the unit of charge.

James Watt (1736–1819), English inventor, developed the steam engine. His name is used to represent the unit of power.

Alessandro Volta (1745–1827), Italian physicist, discovered the electric pile. The unit of electric potential and the alternate name of this quantity (voltage) are named after him.

Hans Christian Oersted (1777–1851), Danish physicist, discovered the connection between electricity and magnetism in 1820. The unit of magnetic field strength is named after him.

André Marie Ampère (1775–1836), French mathematician, chemist, and physicist, experimentally quantified the relationship between electric current and the magnetic field. His works were summarized in a treatise published in 1827. The unit of electric current is named after him.

Georg Simon Ohm (1789–1854), German mathematician, investigated the relationship between voltage and current and quantified the phenomenon of resistance. His first results were published in 1827. His name is used to represent the unit of resistance.

Michael Faraday (1791–1867), English experimenter, demonstrated electromagnetic induction in 1831. His electrical transformer and electromagnetic generator marked the beginning of the age of electric power. His name is associated with the unit of capacitance.

Joseph Henry (1797–1878), American physicist, discovered self-induction around 1831, and his name has been designated to represent the unit of inductance. He had also recognized the essential structure of the telegraph, which was later perfected by Samuel F. B. Morse.

Carl Friedrich Gauss (1777–1855), German mathematician, and **Wilhelm Eduard Weber** (1804–1891), German physicist, published a treatise in 1833 describing the measurement of the earth's magnetic field. The Gauss is a unit of magnetic field strength, while the Weber is a unit of magnetic flux.

James Clerk Maxwell (1831–1879), Scottish physicist, discovered the electromagnetic theory of light and the laws of electrodynamics. The modern theory of electromagnetics is entirely founded upon Maxwell's equations.

Ernst Werner Siemens (1816–1892) and **Wilhelm Siemens** (1823–1883), German inventors and engineers, contributed to the invention and development of electric machines, as well as to perfecting electrical science. The modern unit of conductance is named after them.

Heinrich Rudolph Hertz (1857–1894), German scientist and experimenter, discovered the nature of electromagnetic waves and published his findings in 1888. His name is associated with the unit of frequency.

Nikola Tesla (1856–1943), Croatian inventor, emigrated to the United States in 1884. He invented polyphase electric power systems and the induction motor and pioneered modern AC electric power systems. His name is used to represent the unit of magnetic flux density.

1.4 SYSTEM OF UNITS

This book employs the International System of Units (also called SI, from the French *S*ystème *I*nternational des Unités). SI units are commonly adhered to by virtually all engineering professional societies. This section summarizes SI units and will serve as a useful reference in reading the book.

SI units are based on six fundamental quantities, listed in Table 1.2. All other units may be derived in terms of the fundamental units of Table 1.2. Since, in practice, one often needs to describe quantities that occur in large multiples or small fractions of a unit, standard prefixes are used to denote powers of 10 of SI (and derived) units. These prefixes are listed in Table 1.3. Note that, in general, engineering units are expressed in powers of 10 that are multiples of 3. For example, 10^{-4} s would be referred to as 100×10^{-6} s, or 100 μs (or, less frequently, 0.1 ms).

Table 1.2 **SI units**

Quantity	Unit	Symbol
Length	Meter	m
Mass	Kilogram	kg
Time	Second	s
Electric current	Ampere	A
Temperature	Kelvin	K
Luminous intensity	Candela	cd

Table 1.3 **Standard prefixes**

Prefix	Symbol	Power
atto	a	10^{-18}
femto	f	10^{-15}
pico	p	10^{-12}
nano	n	10^{-9}
micro	μ	10^{-6}
milli	m	10^{-3}
centi	c	10^{-2}
deci	d	10^{-1}
deka	da	10
kilo	k	10^{3}
mega	M	10^{6}
giga	G	10^{9}
tera	T	10^{12}

1.5 ORGANIZATION OF THIS BOOK

This book is divided into three major sections: Circuits, Electronics and Computers, and Electromechanics. Circuit analysis forms the basis of the entire discipline of electrical engineering; thus, a substantial portion of the book is devoted to this subject. Since circuit analysis forms the core of the first course in electricity for non–electrical engineering majors, you will probably spend a significant amount of time working with Chapters 2 through 6. After completing these chapters, you will be able to analyze simple electric circuits, and will begin to appreciate the inner workings of many devices and systems you may have already encountered in other engineering classes and laboratories. However, circuit analysis does not end with Chapter 6! The material in the introductory chapters is absolutely necessary for understanding the fundamentals of electronic and electromechanical systems. This is true because simple electric circuits provide us with good models (i.e., approximations) of the behavior of more complex electronic and electromechanical devices. Thus, mastering the introductory materials will ensure that the more advanced topics of Chapters 7 through 17 will be more easily understood.

The second and third parts of the book are somewhat different in character from the circuit analysis chapters, in that it is not necessary to follow a prescribed order in covering various topics in electronics and electromechanics. Thus, some instructors will emphasize digital electronics and computer systems over analog electronic amplifiers, while some others may prefer to discuss electric motors and electric power systems. Since the interdependence of topics is relatively small past Chapter 6, you will find that this book will serve you as a useful reference for later classes as well. For example, you may be able to find valuable information to assist you in your senior design project.

To provide as much assistance as possible, this book contains a large number of examples and drill exercises; each important concept is reinforced by one or more examples, and a few drill exercises. Since the answers to many drill exercises are provided in the book, you should use these exercises as a check to verify complete understanding of the material. Using the drill exercises as an aid in reading and understanding the text will also help you in solving the homework problems. Many homework problems are simply extensions of the exercises and examples. Some are somewhat more advanced. As in many other subjects, repetition is essential to learning; this statement is particularly true in circuit analysis, where you need to quickly develop familiarity with many new ideas and build confidence in problem solving. In preparing this book, great effort has been spent in making this task as painless as possible.

HOMEWORK PROBLEMS

1.1 List five applications of electric motors in the common household.

1.2 By analogy with the discussion of electrical systems in the automobile, list examples of applications of the electrical engineering disciplines of Table 1.1 for each of the following engineering systems:

a. A ship.
b. A commercial passenger aircraft.
c. Your household.

1.3 Electric power systems provide energy in a variety of commercial and industrial settings. Make a list of systems and devices that receive electric power in:

a. A large office building.
b. A factory floor.
c. A construction site.

PART I

CIRCUITS

CHAPTER 2

Fundamentals of Electric Circuits

This chapter presents the fundamental laws of circuit analysis and serves as the foundation for the study of electrical circuits. The fundamental concepts developed in these first pages will be called upon throughout the book.

The chapter starts with definitions of charge, current, voltage, and power, and with the introduction of the basic laws of electrical circuit analysis: Kirchhoff's laws. Next, the basic circuit elements are introduced, first in their ideal form, then including the most important physical limitations. The elements discussed in the chapter include voltage and current sources, measuring instruments, and the ideal resistor. Once the basic circuit elements have been presented, the concept of an electrical circuit is introduced, and some simple circuits are analyzed using Kirchhoff's and Ohm's laws. The student should appreciate the fact that, although the material presented at this early stage is strictly introductory, it is already possible to discuss some useful applications of electric circuits to practical engineering problems. To this end, two examples are introduced which discuss simple resistive devices that can measure displacements and forces. The topics introduced in Chapter 2 form the foundations for the remainder of this book and should be mastered thoroughly. By the end of the chapter, you should have accomplished the following learning objectives:

- Application of Kirchhoff's and Ohm's laws to elementary resistive circuits.
- Power computation for a circuit element.
- Use of the passive sign convention in determining voltage and current directions.
- Solution of simple voltage and current divider circuits.
- Assigning node voltages and mesh currents in an electrical circuit.
- Writing the circuit equations for a linear resistive circuit by applying Kirchhoff's voltage law and Kirchhoff's current law.

2.1 CHARGE, CURRENT, AND KIRCHHOFF'S CURRENT LAW

Charles Coulomb (1736–1806). *Photo courtesy of French Embassy, Washington, D.C.*

The earliest accounts of electricity date from about 2,500 years ago, when it was discovered that static charge on a piece of amber was capable of attracting very light objects, such as feathers. The word itself—*electricity*—originated about 600 B.C.; it comes from *elektron,* which was the ancient Greek word for amber. The true nature of electricity was not understood until much later, however. Following the work of Alessandro Volta[1] and his invention of the copper-zinc battery, it was determined that static electricity and the current that flows in metal wires connected to a battery are due to the same fundamental mechanism: the atomic structure of matter, consisting of a nucleus—neutrons and protons—surrounded by electrons. The fundamental electric quantity is **charge,** and the smallest amount of charge that exists is the charge carried by an electron, equal to

$$q_e = -1.602 \times 10^{-19} \text{ coulomb} \qquad \textbf{(2.1)}$$

As you can see, the amount of charge associated with an electron is rather small. This, of course, has to do with the size of the unit we use to measure charge, the **coulomb (C),** named after Charles Coulomb.[2] However, the definition of the coulomb leads to an appropriate unit when we define electric current, since current consists of the flow of very large numbers of charge particles. The other charge-carrying particle in an atom, the proton, is assigned a positive sign, and the same magnitude. The charge of a proton is

$$q_p = +1.602 \times 10^{-19} \text{ coulomb} \qquad \textbf{(2.2)}$$

Electrons and protons are often referred to as **elementary charges.**

Current $i = dq/dt$ is generated by the flow of charge through the cross-sectional area A in a conductor.

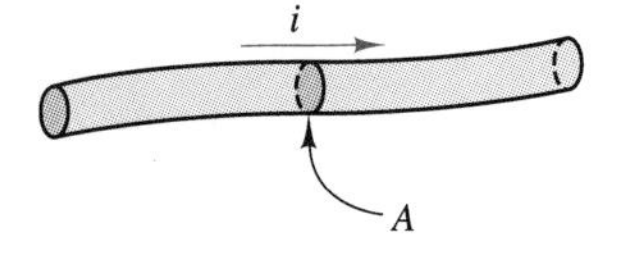

Figure 2.1 Current flow in an electric conductor

Electric current is defined as the time rate of change of charge passing through a predetermined area. Typically, this area is the cross-sectional area of a metal wire; however, there are a number of cases we shall explore later in this book where the current-carrying material is not a conducting wire. Figure 2.1 depicts a macroscopic view of the flow of charge in a wire, where we imagine

[1]See brief biography on page 6.

[2]See brief biography on page 6.

Δq units of charge flowing through the cross-sectional area A in Δt units of time. The resulting current, i, is then given by

$$i = \frac{\Delta q}{\Delta t} \tag{2.3}$$

If we consider the effect of the enormous number of elementary charges actually flowing, we can write this relationship in differential form:

$$i = \frac{dq}{dt} \quad \frac{\text{C}}{\text{s}} \tag{2.4}$$

The units of current are called **amperes (A),** where 1 ampere = 1 coulomb/second. The name of the unit is a tribute to the French scientist André Marie Ampère.[3] The electrical engineering convention states that the positive direction of current flow is that of positive charges. In metallic conductors, however, current is carried by negative charges; these charges are the free electrons in the conduction band, which are only weakly attracted to the atomic structure in metallic elements and are therefore easily displaced in the presence of electric fields.

Example 2.1

A cylindrical conductor is 1 m long and 2 mm in diameter and contains 10^{29} free carriers per cubic meter.

1. Find the total charge of the carriers in this wire.
2. If the wire is used in a circuit, find the current flowing in the wire if the average velocity of the carriers is 19.9×10^{-6} m/s.

Solution:

1. In order to compute the total charge contributed by the electrons, we first need to compute the volume of the conductor.

$$\begin{aligned}\text{Volume} &= \text{Length} \times \text{Cross-sectional area} \\ &= \pi r^2 L = \pi\left(\frac{2 \times 10^{-3}}{2}\right)^2 (1)\end{aligned}$$

Next we compute the charge by determining the total number of charge carriers in the conductor as follows:

$$\text{Charge} = \text{Volume} \times \frac{\text{Charge}}{\text{Unit volume}}$$

$$\begin{aligned}Q &= \pi\left(\frac{2 \times 10^{-3}}{2}\right)^2 (1)(-1.602 \times 10^{-19}\ \text{C})\left(10^{29}\ \frac{\text{carriers}}{\text{m}^3}\right) \\ &= -50.33 \times 10^3\ \text{C}\end{aligned}$$

[3] See brief biography on page 7.

2. If the carriers move with an average velocity of 19.9×10^{-6} m/s, the magnitude of the total current flow in the wire can be computed by considering that current is the flow of charge per unit time:

$$\begin{aligned}\text{Current} &= \text{Charge density per unit length (C/m)} \times \text{Carrier velocity (m/s)} \\ &= \frac{50.33 \times 10^3}{1} \times 19.9 \times 10^{-6} \\ &= 1 \text{ A}\end{aligned}$$

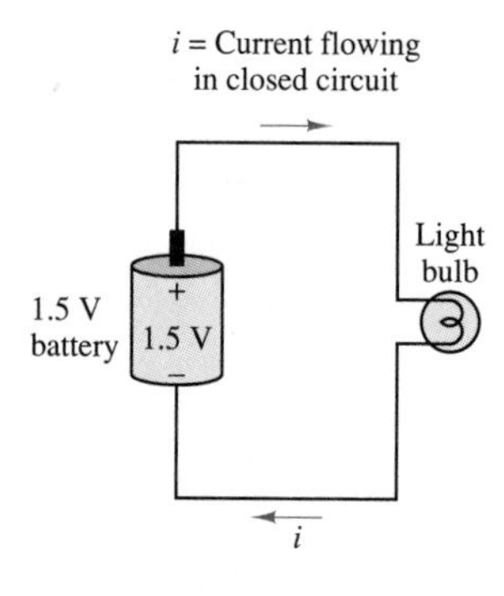

Figure 2.2 A simple electrical circuit

In order for current to flow there must exist a closed circuit. Figure 2.2 depicts a simple circuit, composed of a battery (e.g., a dry-cell or alkaline 1.5-V battery) and a light bulb.

Note that in the circuit of Figure 2.2, the current, i, flowing from the battery to the resistor is equal to the current flowing from the light bulb to the battery. In other words, no current (and therefore no charge) is "lost" around the closed circuit. This principle was observed by the German scientist G. R. Kirchhoff[4] and is now known as **Kirchhoff's current law (KCL).** Kirchhoff's current law states that because charge cannot be created but must be conserved, *the sum of the currents at a node must equal zero* (in an electrical circuit, a **node** is the junction of two or more conductors). Formally:

$$\sum_{n=1}^{N} i_n = 0 \qquad \text{Kirchhoff's current law} \tag{2.5}$$

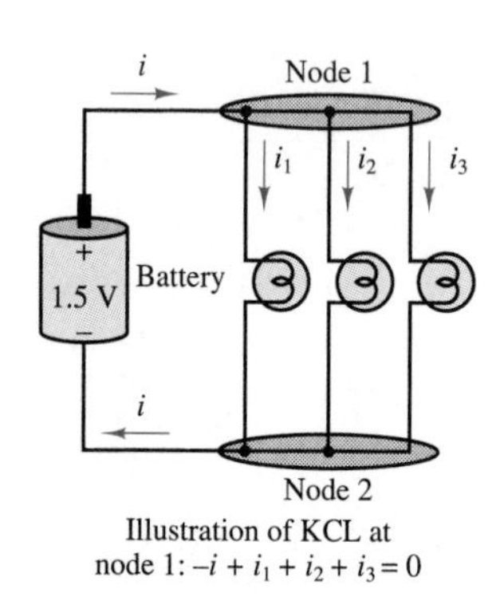

Figure 2.3 Illustration of Kirchhoff's current law

The significance of Kirchhoff's current law is illustrated in Figure 2.3, where the simple circuit of Figure 2.2 has been augmented by the addition of two light bulbs (note how the two nodes that exist in this circuit have been emphasized by the shaded areas). In applying KCL, one usually defines currents entering a node as being negative and currents exiting the node as being positive. Thus, the resulting expression for the circuit of Figure 2.3 is:

$$-i + i_1 + i_2 + i_3 = 0$$

Kirchhoff's current law is one of the fundamental laws of circuit analysis, making it possible to express currents in a circuit in terms of each other; for example, one can express the current leaving a node in terms of all the other currents at the node. The ability to write such equations is a great aid in the systematic solution of large electric circuits. Much of the material presented in Chapter 3 will be an extension of this concept.

[4]Gustav Robert Kirchhoff (1824–1887), a German scientist, who published the first systematic description of the laws of circuit analysis. His contribution—though not original in terms of its scientific content—forms the basis of all circuit analysis.

Example 2.2

An automotive battery provides an excellent intuitive illustration of Kirchhoff's current law. Assume for a moment that the only electrical circuits connected to the car battery are the headlights (Figure 2.4). How does the current provided by the battery divide between headlights?

Solution:

The battery supplies power to the headlights by generating a current that flows through the filaments of the bulbs. Since the headlights are usually of the same type and both have the same power rating, it is reasonable to assume that the currents I_P and I_D must be the same in both magnitude and direction of current flow (incidentally, the power rating of a headlight is usually specified in watts). The battery, then, must supply the current to the headlights, and since the battery is connected so that 12 V is across both of the headlights (this is called a parallel connection), we reason that the total current out of the battery, I_B, is the sum of the two currents I_P and I_D, or:

$$I_B = I_P + I_D$$

If we wanted to consider a more realistic automotive circuit, we would have to account for the tail- and brake-light circuits, the turn signal lights, and a number of other electrical loads all connected in parallel across the battery. The electrical system in an automobile is an excellent example of how circuit analysis applies to real-life problems: the designer of the electrical system of an automobile would have to compute the individual current required by each load, evaluate the total capacity of the battery needed to furnish the current required by the loads under worst-case conditions, and select the wiring scheme and the wire gauges most appropriate for each load.

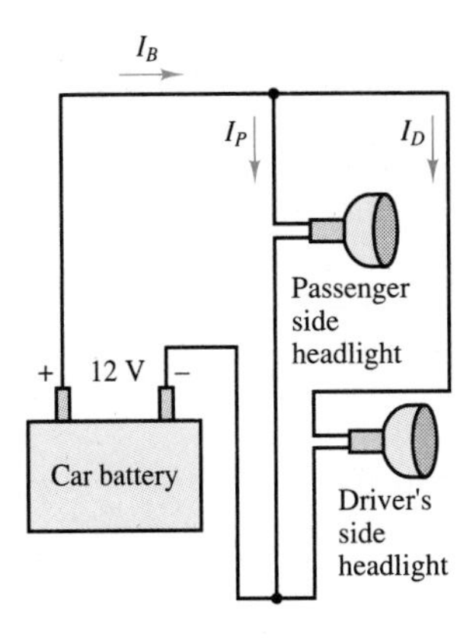

Figure 2.4 KCL in a practical setting

2.2 VOLTAGE AND KIRCHHOFF'S VOLTAGE LAW

Charge moving in an electric circuit gives rise to a current, as stated in the preceding section. Naturally, it must take some work, or energy, for the charge to move between two points in a circuit, say, from point a to point b. The total *work per unit charge* associated with the motion of charge between two points is called **voltage.** Thus, the units of voltage are those of energy per unit charge; they have been called **volts** in honor of Alessandro Volta:

$$1 \text{ volt} = \frac{1 \text{ joule}}{\text{coulomb}} \tag{2.6}$$

The voltage, or **potential difference,** between two points in a circuit indicates the energy required to move charge from one point to the other. As will be presently shown, the direction, or polarity, of the voltage is closely tied to whether energy is being dissipated or generated in the process. The seemingly abstract concept

Gustav Robert Kirchhoff (1824–1887). *Photo courtesy of Deutsches Museum, Munich.*

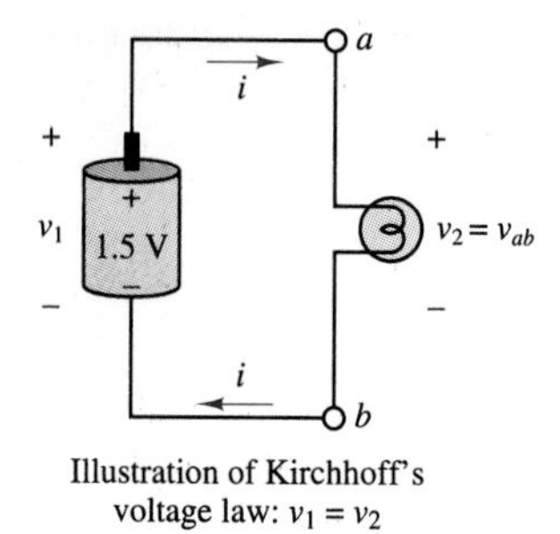

Figure 2.5 Voltages around a circuit

of work being done in moving charges can be directly applied to the analysis of electrical circuits; consider again the simple circuit consisting of a battery and a light bulb. The circuit is drawn again for convenience in Figure 2.5, and nodes are defined by the letters a and b. A series of carefully conducted experimental observations regarding the nature of voltages in an electric circuit led Kirchhoff to the formulation of the second of his laws, **Kirchhoff's voltage law,** or **KVL.** The principle underlying KVL is that no energy is lost or created in an electric circuit; in circuit terms, the sum of all voltages associated with sources must equal the sum of the load voltages, so that *the net voltage around a closed circuit is zero.* If this were not the case, we would need to find a physical explanation for the excess (or missing) energy not accounted for in the voltages around a circuit. Kirchhoff's voltage law may be stated in a form similar to that used for KCL:

$$\sum_{n=1}^{N} v_n = 0 \qquad \text{Kirchhoff's voltage law} \tag{2.7}$$

where the v_n are the individual voltages around the closed circuit. Making reference to Figure 2.5, we see that it must follow from KVL that the work generated by the battery is equal to the energy dissipated in the light bulb to sustain the current flow and to convert the electric energy to heat and light:

$$v_{ab} = -v_{ba}$$

or

$$v_1 = v_2$$

One may think of the work done in moving a charge from point a to point b and the work done moving it back from b to a as corresponding directly to the *voltages across individual circuit elements.* Let Q be the total charge that moves around the circuit per unit time, giving rise to the current i. Then the work done in moving Q from b to a (i.e., across the battery) is

$$W_{ba} = Q \times 1.5 \text{ V} \tag{2.8}$$

Similarly, work is done in moving Q from a to b, that is, across the light bulb. Note that the word *potential* is quite appropriate as a synonym of voltage, in that voltage represents the potential energy between two points in a circuit: if we remove the light bulb from its connections to the battery, there still exists a voltage across the (now disconnected) terminals b and a. This is illustrated in Figure 2.6.

The presence of a voltage, v_2, across the open terminals a and b indicates the potential energy that can enable the motion of charge once a closed circuit is established and allow current to flow.

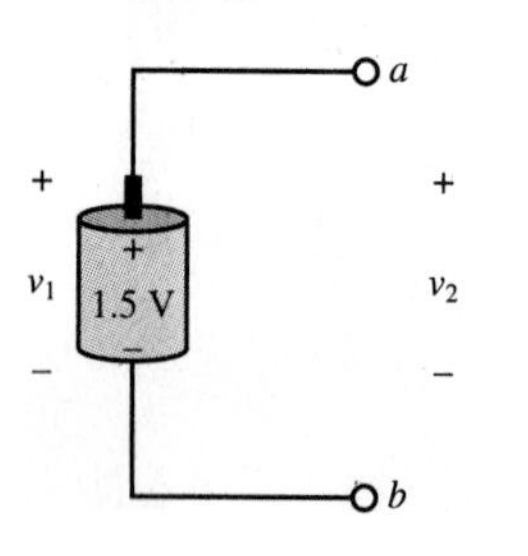

Figure 2.6 Concept of voltage as potential difference

A moment's reflection upon the significance of voltage should suggest that it must be necessary to specify a sign for this quantity. Consider, again, the same dry-cell or alkaline battery, where, by virtue of an electrochemically induced separation of charge, a 1.5-V potential difference is generated. The potential generated by the battery may be used to move charge in a circuit. The rate at which charge is moved once a closed circuit is established (i.e., the current drawn by the circuit connected to the battery) depends now on the circuit element we choose to connect to the battery. Thus, while the voltage across the battery represents the potential for *providing energy* to a circuit, the voltage across the light bulb indicates the amount of work done in *dissipating energy.* In the first case, energy is generated; in the second, it is consumed (note that energy may also be stored, by suitable circuit elements yet to be introduced). This fundamental distinction requires attention in defining the sign (or polarity) of voltages.

We shall, in general, refer to elements that provide energy as **sources,** and to elements that dissipate energy as **loads.** Standard symbols for a generalized source-and-load circuit are shown in Figure 2.7. Formal definitions will be given in a later section.

A symbolic representation of the battery–light bulb circuit of Figure 2.5.

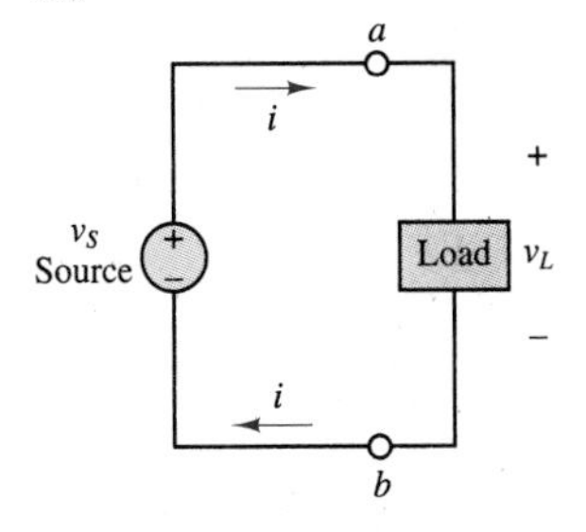

Figure 2.7 Sources and loads in an electrical circuit

Example 2.3

Many years ago, Christmas light manufacturers would manufacture strands of Christmas tree lights with the bulbs connected in a *series circuit,* as depicted in Figure 2.8. Since, by virtue of KVL, the net voltage around the loop is zero, the sum of the voltages across all of the light bulbs must equal the total voltage supplied by the 110-VAC source. If we assume that the bulbs are identical to each other, and if there were, say, 11 bulbs in a strand, then each bulb would have $\frac{1}{11}$ of the outlet's voltage across it, or 10 VAC.

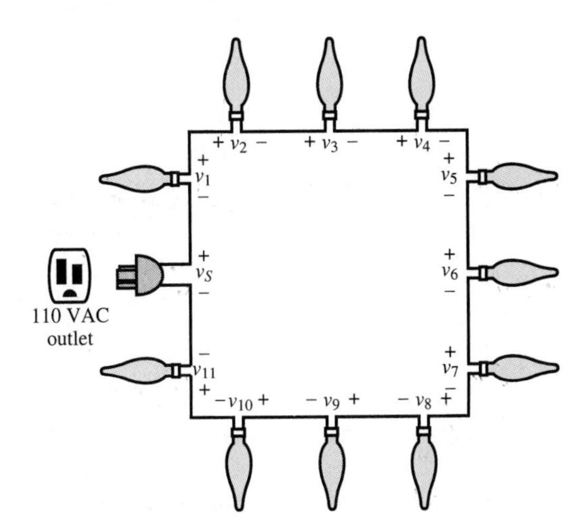

Figure 2.8 Christmas tree lights in a series connection

Although the series connection is very simple to manufacture, it is not used too frequently, in practice. Why is it a poor configuration, from a practical standpoint? (Think what would happen if a light bulb burned out.)

2.3 IDEAL VOLTAGE AND CURRENT SOURCES

In the examples presented in the preceding sections, a battery was used as a source of energy, under the unspoken assumption that the voltage provided by the battery (e.g., 1.5 volts for a dry-cell or alkaline battery, or 12 volts for an automotive lead-acid battery) is fixed. Under such an assumption, we implicitly treat the battery as an ideal source. In this section, we will formally define ideal sources. Intuitively, an ideal source is a source that can provide an arbitrary amount of energy. **Ideal sources** are divided into two types: voltage sources and current sources. Of these, you are probably more familiar with the first, since dry-cell, alkaline, and lead-acid batteries are all voltage sources (they are not ideal, of course). You might have to think harder to come up with a physical example that approximates the behavior of an ideal current source; however, reasonably good approximations of ideal current sources also exist. For instance, a voltage source connected in series with a circuit element that has a large resistance to the flow of current from the source provides a nearly constant—though small—current and therefore acts very nearly like an ideal current source.

Ideal Voltage Sources

An **ideal voltage source** is an electrical device that will generate a prescribed voltage at its terminals. The ability of an ideal voltage source to generate its output voltage is not affected by the current it must supply to the other circuit elements. Another way to phrase the same idea is as follows:

> An ideal voltage source provides a prescribed voltage across its terminals irrespective of the current flowing through it. The amount of current supplied by the source is determined by the circuit connected to it.

Figure 2.9 depicts various symbols for voltage sources that will be employed throughout this book. Note that the output voltage of an ideal source can be

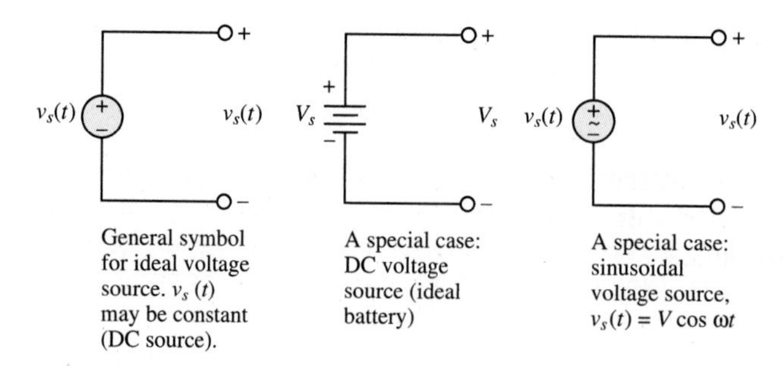

Figure 2.9 Ideal voltage sources

a function of time. In general, the following notation will be employed in this book, unless otherwise noted. A generic voltage source will be denoted by a lowercase v. If it is necessary to emphasize that the source produces a time-varying voltage, then the notation $v(t)$ will be employed. Finally, a constant (DC) voltage source will be denoted by the uppercase character V. Note that by convention the direction of positive current flow out of a voltage source is *out of the positive terminal.*

The notion of an ideal voltage source is best appreciated within the context of the source-load representation of electrical circuits, which will frequently be referred to in the remainder of this book. Figure 2.10 depicts the connection of an energy source with a passive circuit (i.e., a circuit that can absorb and dissipate energy—for example, the headlights and light bulb of our earlier examples). Three different representations are shown to illustrate the conceptual, symbolic, and physical significance of this source-load idea.

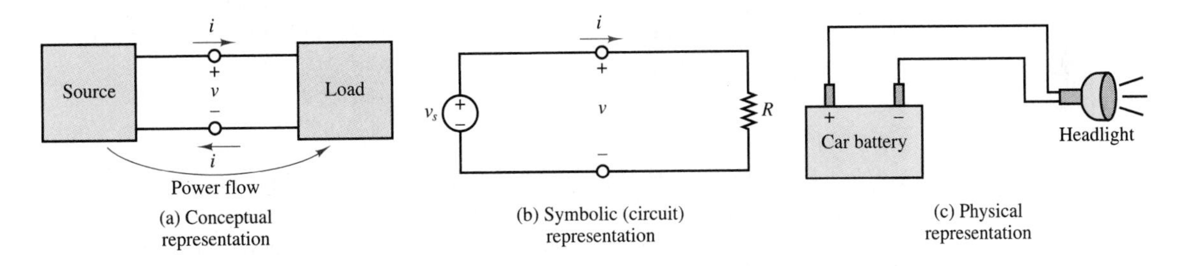

Figure 2.10 Various representations of an electrical system

In the analysis of electrical circuits, we choose to represent the physical reality of Figure 2.10(c) by means of the approximation provided by ideal circuit elements, as depicted in Figure 2.10(b).

Ideal Current Sources

An **ideal current source** is a device that can generate a prescribed current independent of the circuit it is connected to. To do so, it must be able to generate an arbitrary voltage across its terminals. Figure 2.11 depicts the symbol used to represent ideal current sources. By analogy with the definition of the ideal voltage source stated in the previous section, we write:

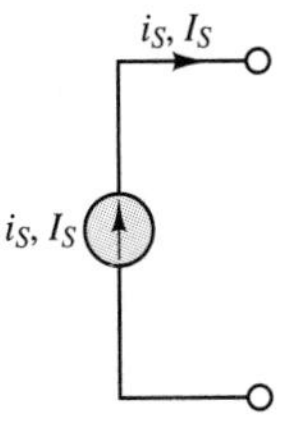

Figure 2.11 Symbol for ideal current source

> An ideal current source provides a prescribed current to any circuit connected to it. The voltage generated by the source is determined by the circuit connected to it.

The same uppercase and lowercase convention used for voltage sources will be employed in denoting current sources.

Dependent (Controlled) Sources

The sources described so far have the capability of generating a prescribed voltage or current independent of any other element within the circuit. Thus, they are termed *independent sources.* There exists another category of sources, however, whose output (current or voltage) is a function of some other voltage or current in a circuit. These are called **dependent** (or **controlled**) **sources.** A different symbol, in the shape of a diamond, is used to represent dependent sources and to distinguish them from independent sources. The symbols typically used to represent dependent sources are depicted in Figure 2.12; the table illustrates the relationship between the source voltage or current and the voltage or current it depends on—v_x or i_x, respectively—which can be any voltage or current in the circuit.

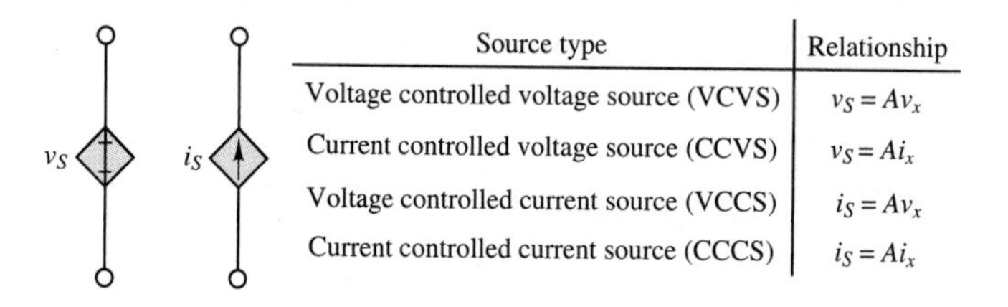

Source type	Relationship
Voltage controlled voltage source (VCVS)	$v_S = Av_x$
Current controlled voltage source (CCVS)	$v_S = Ai_x$
Voltage controlled current source (VCCS)	$i_S = Av_x$
Current controlled current source (CCCS)	$i_S = Ai_x$

Figure 2.12 Symbols for dependent sources

Dependent sources are very useful in describing certain types of electronic circuits. You will encounter dependent sources again in Chapters 8, 9, and 11, when electronic amplifiers are discussed.

2.4 ELECTRIC POWER AND SIGN CONVENTION

The definition of voltage as work per unit charge lends itself very conveniently to the introduction of power. Recall that power is defined as the work done per unit time. Thus, the power, P, either generated or dissipated by a circuit element can be represented by the following relationship:

$$\text{Power} = \frac{\text{Work}}{\text{Time}} = \frac{\text{Work}}{\text{Unit charge}}\frac{\text{Charge}}{\text{Time}} = \text{Voltage} \times \text{Current} \qquad \textbf{(2.9)}$$

Thus,

> The electrical power generated by an active element, or that dissipated or stored by a passive element, is equal to the product of the voltage across the element and the current flowing through it.

$$\boxed{P = VI} \qquad \textbf{(2.10)}$$

It is easy to verify that the units of voltage (joules/coulomb) times current (coulombs/second) are indeed those of power (joules/second, or watts).

It is important to realize that, just like voltage, power is a signed quantity, and that it is necessary to make a distinction between *positive* and *negative power.* This distinction can be understood with reference to Figure 2.13, in which a source and a load are shown side by side. The polarity of the voltage across the source and the direction of the current through it indicate that the voltage source *is doing work in moving charge from a lower potential to a higher potential.* On the other hand, the load is dissipating energy, because the direction of the current indicates that *charge is being displaced from a higher potential to a lower potential.* To avoid confusion with regard to the sign of power, the electrical engineering community uniformly adopts the **passive sign convention,** which simply states that *the power dissipated by a load is a positive quantity* (or, conversely, that the power generated by a source is a positive quantity). Another way of phrasing the same concept is to state that if current flows from a higher to a lower voltage (+ to −), the power dissipated will be a positive quantity.

It is important to note also that the actual numerical values of voltages and currents do not matter: once the proper reference directions have been established and the passive sign convention has been applied consistently, the answer will be correct regardless of the reference direction chosen. The following examples illustrate this point.

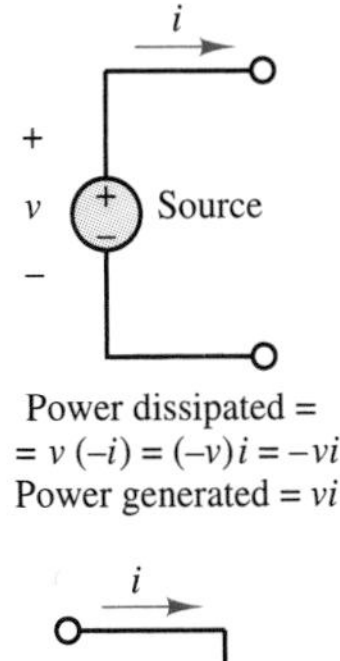

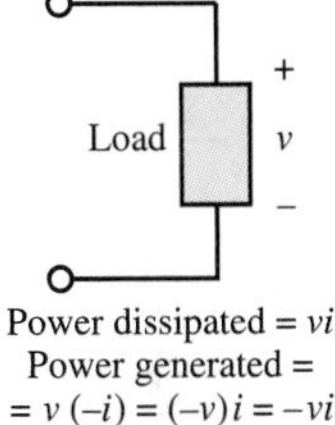

Figure 2.13 The passive sign convention

Example 2.4

This example will show that whichever reference direction is selected in a circuit, the same answer will be obtained, *provided that the passive sign convention is applied consistently.* The circuit of Figure 2.14 consists of one source (the battery) and two loads connected in series. We first arbitrarily select the clockwise direction for the current flowing in the series circuit. Once we have decided that the current flows in the clockwise direction, we can compute the power dissipated by each element. Since current flows from the negative to the positive terminal in the battery, the power dissipated by the battery will be:

$$P_B = (-v_B)(i) = (-12)(0.1) = -1.2 \text{ W}$$

That is, the battery *generates* 1.2 W. By the same convention, the two loads will dissipate power according to the relations

$$P_1 = (v_1)(i) = (8)(0.1) = 0.8 \text{ W}$$

and

$$P_2 = (v_2)(i) = (4)(0.1) = 0.4 \text{ W}$$

Thus, energy is conserved, since the net power dissipation in the circuit is zero.

Now consider the same circuit but with the current direction arbitrarily given as counterclockwise, as shown in Figure 2.15. The voltages resulting from the assumed current direction are also shown in the figure. Note that, assuming a clockwise reference direction, the signs of the voltages are opposite to those of Figure 2.14. Thus, the battery voltage is

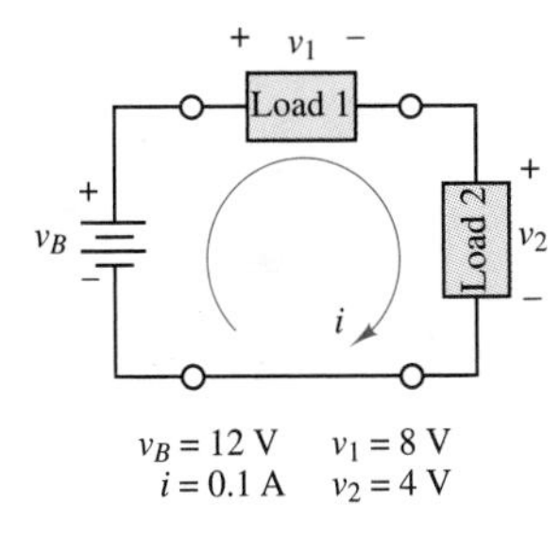

Figure 2.14

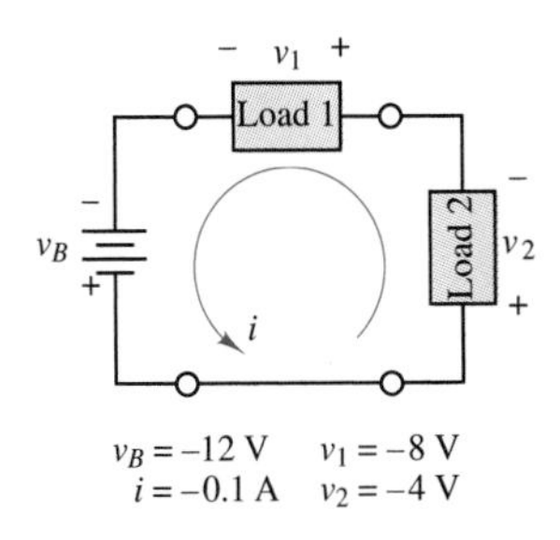

Figure 2.15

now -12 V, and the voltages across the loads are -8 V and -4 V, respectively. Thus, according to the sign convention, the power dissipated by the battery is

$$P_B = (-v_B)(i) = -(-12)(-0.1) = -1.2 \text{ W}$$

while the power dissipated by the loads is given by

$$P_1 = (v_1)(i) = (-8)(-0.1) = 0.8 \text{ W}$$

and

$$P_2 = (v_2)(i) = (-4)(-0.1) = 0.4 \text{ W}$$

As in the previous case, the sign convention tells us that the battery is generating 1.2 W of power, which is dissipated by the two loads.

Example 2.5

This example illustrates how we can determine whether an unknown element is a source or a load simply by identifying voltage polarities and current directions. Assume that devices are available that allow us to measure voltages and currents in the unknown circuit of Figure 2.16(a). Using the diagram, determine which element is supplying power and which is dissipating it.

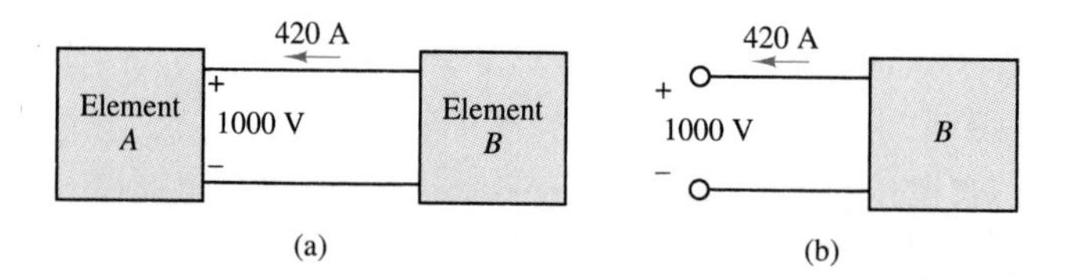

Figure 2.16

Solution:

Looking at element B, we recognize that this circuit element acts as a source, since current flows from a point of lower potential to one of higher potential, as shown in Figure 2.16(b). Therefore, we conclude that the power dissipated by element B is

$$P_B = (1{,}000)(-420) = -420 \text{ kW}$$

This means that element B is delivering 420 kW of power to element A.

DRILL EXERCISES

2.1 Compute the current flowing through each of the headlights of Example 2.2 if each headlight has a power rating of 50 W. How much power is the battery providing?

2.2 Determine which circuit element in the illustration is supplying power and which is dissipating power. Also determine the amount of power dissipated and supplied.

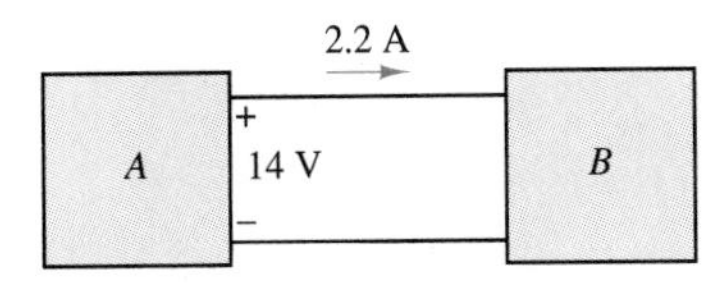

2.3 If the battery in the accompanying diagram supplies a total of 10 mW to the three elements shown and $i_1 = 2$ mA and $i_2 = 1.5$ mA, what is the current i_3? If $i_1 = 1$ mA and $i_3 = 1.5$ mA, what is i_2?

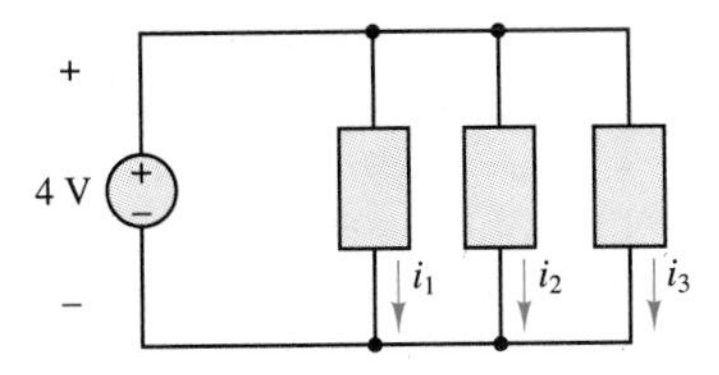

2.5 CIRCUIT ELEMENTS AND THEIR *i-v* CHARACTERISTICS

The relationship between current and voltage at the terminals of a circuit element defines the behavior of that element within the circuit. In this section we shall introduce a graphical means of representing the terminal characteristics of circuit elements. Figure 2.17 depicts the representation that will be employed throughout the chapter to denote a generalized circuit element: the variable i represents the current flowing through the element, while v is the potential difference, or voltage, across the element.

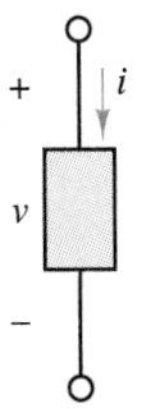

Figure 2.17 Generalized representation of circuit elements

Suppose now that a known voltage were imposed across a circuit element. The current that would flow as a consequence of this voltage, and the voltage itself, form a unique pair of values. If the voltage applied to the element were varied and the resulting current measured, it would be possible to construct a functional relationship between voltage and current known as the ***i-v* characteristic** (or **volt-ampere characteristic**). Such a relationship defines the circuit element, in the sense that if we impose any prescribed voltage (or current), the resulting current (or voltage) is directly obtainable from the *i-v* characteristic. A direct consequence is that the power dissipated (or generated) by the element may also be determined from the *i-v* curve.

Figure 2.18 depicts an experiment for empirically determining the i-v characteristic of a tungsten filament light bulb. A variable voltage source is used to apply various voltages, and the current flowing through the element is measured for each applied voltage.

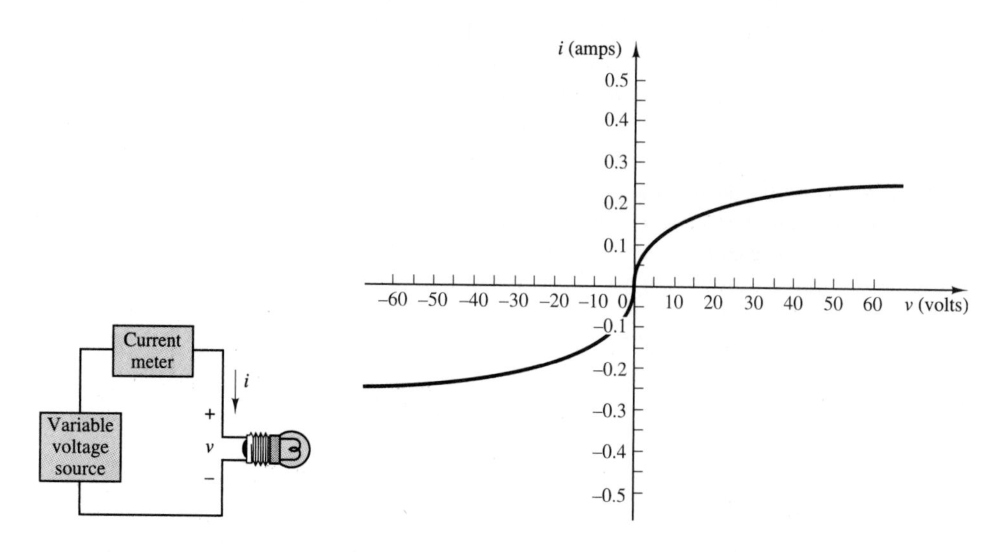

Figure 2.18 Volt-ampere characteristic of a tungsten light bulb

We could certainly express the i-v characteristic of a circuit element in functional form:

$$i = f(v) \qquad v = g(i) \tag{2.11}$$

In some circumstances, however, the graphical representation is more desirable, especially if there is no simple functional form relating voltage to current. The simplest form of the i-v characteristic for a circuit element is a straight line, that is,

$$i = kv \tag{2.12}$$

with k a constant. In the next section we shall see how this simple model of a circuit element is quite useful in practice and can be used to define the most common circuit elements: ideal voltage and current sources and the resistor.

We can also relate the graphical i-v representation of circuit elements to the power dissipated or generated by a circuit element. For example, the graphical representation of the light bulb i-v characteristic of Figure 2.18 illustrates that when a positive current flows through the bulb, the voltage is positive, and that, conversely, a negative current flow corresponds to a negative voltage. In both cases the power dissipated by the device is a positive quantity, as it should be, on the basis of the discussion of the preceding section, since the light bulb is a passive device. Note that the i-v characteristic appears in only two of the four possible quadrants in the i-v plane. In the other two quadrants, the product of voltage and current (i.e., power) is negative, and an i-v curve with a portion in either of these quadrants would therefore correspond to power generated. This is not possible for a passive load such as a light bulb; however, there are electronic devices that can operate, for example, in three of the four quadrants of the i-v characteristic and can therefore act as sources of energy for specific combinations of voltages and currents. An example of this dual behavior will be introduced in Chapter 7, where it is shown that the photodiode can act either in a passive mode (as a light sensor) or in an active mode (as a solar cell).

The i-v characteristics of ideal current and voltage sources can also be useful in visually representing their behavior. An ideal voltage source generates a prescribed voltage independent of the current drawn from the load; thus, its i-v characteristic is a straight vertical line with a voltage axis intercept corresponding to the source voltage. Similarly, the i-v characteristic of an ideal current source is a horizontal line with a current axis intercept corresponding to the source current. Figure 2.19 depicts this behavior.

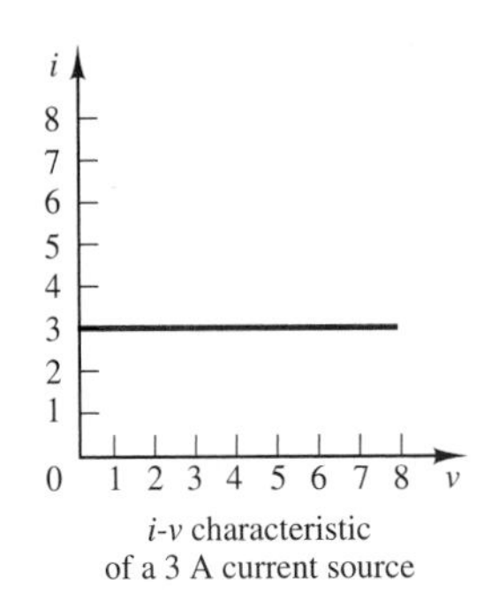

i-v characteristic of a 3 A current source

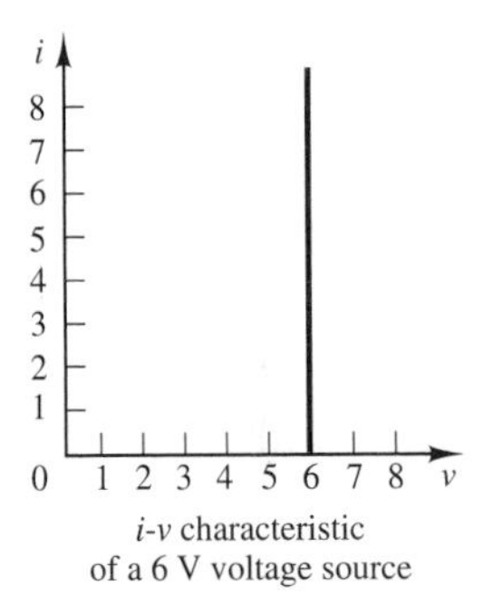

i-v characteristic of a 6 V voltage source

Figure 2.19 i-v characteristics of ideal sources

2.6 RESISTANCE AND OHM'S LAW

When electric current flows through a metal wire or through other circuit elements, it encounters a certain amount of **resistance,** the magnitude of which depends on the electrical properties of the material. Resistance to the flow of current may be undesired—for example, in the case of lead wires and connection cable—or it may be exploited in an electrical circuit in a useful way. Nevertheless, practically all circuit elements exhibit some resistance; as a consequence, current flowing through an element will cause energy to be dissipated in the form of heat. An ideal **resistor** is a device that exhibits linear resistance properties according to **Ohm's law,** which states that

$$V = IR \qquad \text{Ohm's law} \tag{2.13}$$

that is, that the voltage across an element is directly proportional to the current flow through it. R is the value of the resistance in units of **ohms** (**Ω**), where

$$1\,\Omega = 1\text{ V/A} \tag{2.14}$$

The resistance of a material depends on a property called **resistivity,** denoted by the symbol ρ; the inverse of resistivity is called **conductivity** and is denoted by the symbol σ. For a cylindrical resistance element (shown in Figure 2.20), the resistance is proportional to the length of the sample, l, and inversely proportional to its cross-sectional area, A, and conductivity, σ.

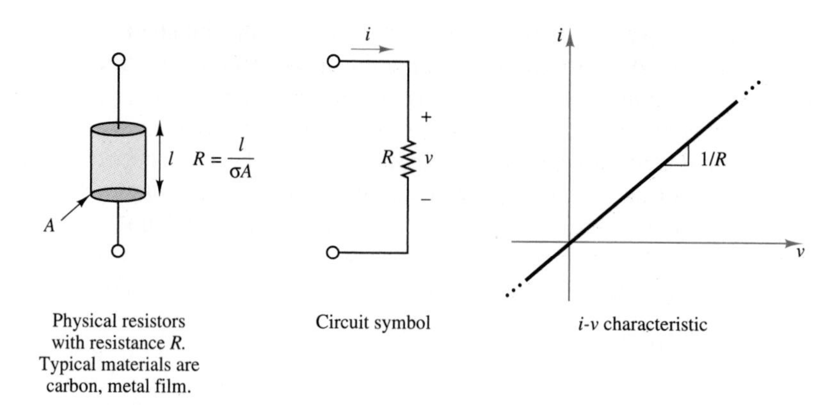

Figure 2.20 The resistance element

$$v = \frac{l}{\sigma A} i \tag{2.15}$$

It is often convenient to define the **conductance** of a circuit element as the inverse of its resistance. The symbol used to denote the conductance of an element is G, where

$$G = \frac{1}{R} \textbf{ siemens (S)} \qquad \text{where} \qquad 1 \text{ S} = 1 \text{ A/V} \tag{2.16}$$

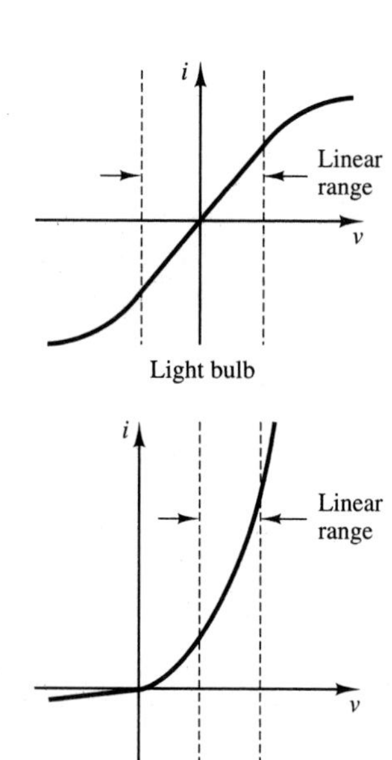

Figure 2.21

Thus, Ohm's law can be restated in terms of conductance as:

$$I = GV \tag{2.17}$$

Ohm's law is an empirical relationship that finds widespread application in electrical engineering, because of its simplicity. It is, however, only an approximation of the physics of electrically conducting materials. Typically, the linear relationship between voltage and current in electrical conductors does not apply at very high voltages and currents. Further, not all electrically conducting materials exhibit linear behavior even for small voltages and currents. It is usually true, however, that for some range of voltages and currents, most elements display a linear *i-v* characteristic. Figure 2.21 illustrates how the linear resistance concept may apply to elements with nonlinear *i-v* characteristics, by graphically defining the linear portion of the *i-v* characteristic of two common electrical devices: the light bulb, which we have already encountered, and the semiconductor diode, which we shall study in greater detail in Chapter 7.

The typical construction and the circuit symbol of the resistor are shown in Figure 2.20. Resistors made of cylindrical sections of carbon (with resistivity $\rho = 3.5 \times 10^{-5}$ Ω-m) are very common and are commercially available in a wide range of values for several power ratings (as will be explained shortly). Another commonly employed construction technique for resistors employs metal film. A common power rating for resistors used in electronic circuits (e.g., in most consumer electronic appliances such as radios and television sets) is ¼ W. Table 2.1 lists the standard values for commonly used resistors and the color code associated with these values (i.e., the common combinations of the digits $b_1 b_2 b_3$ as defined in Figure 2.22). For example, if the first three color bands on a resistor show the colors red ($b_1 = 2$), violet ($b_2 = 7$), and yellow ($b_3 = 4$), the resistance value can be interpreted as follows:

$$R = 27 \times 10^4 = 270{,}000\ \Omega = 270\ \text{k}\Omega$$

In Table 2.1, the leftmost column represents the complete color code; columns to the right of it only show the third color, since this is the only one that changes. For example, a 10-Ω resistor has the code brown-black-*black*, while a 100-Ω resistor has brown-black-*brown*.

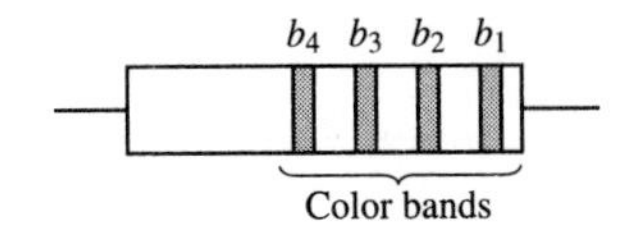

Figure 2.22 Resistor color code

Table 2.1 Common resistor values (⅛-, ¼-, ½-, 1-, 2-W rating)

Ω	Code	Ω	Multiplier	kΩ	Multiplier	kΩ	Multiplier	kΩ	Multiplier
10	Brn-blk-blk	100	Brown	1.0	Red	10	Orange	100	Yellow
12	Brn-red-blk	120	Brown	1.2	Red	12	Orange	120	Yellow
15	Brn-grn-blk	150	Brown	1.5	Red	15	Orange	150	Yellow
18	Brn-gry-blk	180	Brown	1.8	Red	18	Orange	180	Yellow
22	Red-red-blk	220	Brown	2.2	Red	22	Orange	220	Yellow
27	Red-vlt-blk	270	Brown	2.7	Red	27	Orange	270	Yellow
33	Org-org-blk	330	Brown	3.3	Red	33	Orange	330	Yellow
39	Org-wht-blk	390	Brown	3.9	Red	39	Orange	390	Yellow
47	Ylw-vlt-blk	470	Brown	4.7	Red	47	Orange	470	Yellow
56	Grn-blu-blk	560	Brown	5.6	Red	56	Orange	560	Yellow
68	Blu-gry-blk	680	Brown	6.8	Red	68	Orange	680	Yellow
82	Gry-red-blk	820	Brown	8.2	Red	82	Orange	820	Yellow

In addition to the resistance in ohms, the maximum allowable power dissipation (or **power rating**) is typically specified for commercial resistors. Exceeding this power rating leads to overheating and can cause the resistor to literally start on fire. For a resistor R, the power dissipated is given by

$$P = VI = I^2 R = \frac{V^2}{R} \qquad \textbf{(2.18)}$$

That is, the power dissipated by a resistor is proportional to the square of the current flowing through it, as well as the square of the voltage across it. The following example illustrates how one can make use of the power rating to determine whether a given resistor will be suitable for a certain application.

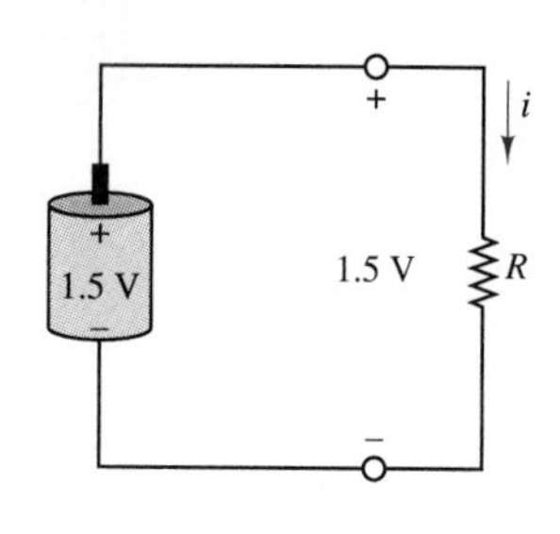

Figure 2.23

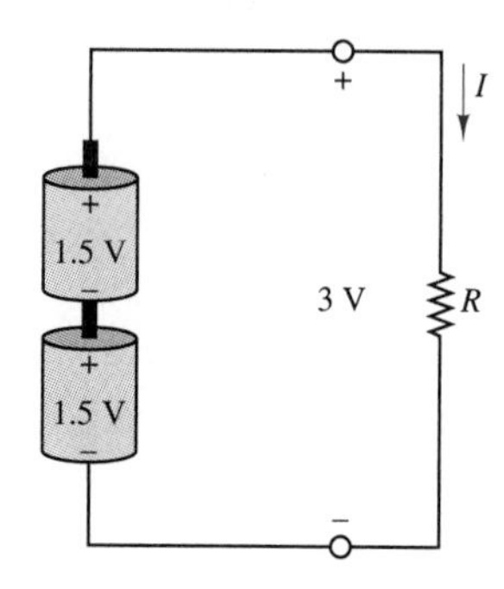

Figure 2.24

Example 2.6

A typical power rating (i.e., maximum power allowable for a device) for carbon resistors employed in low-power electronic circuits is ¼ W. What is the value of the smallest ¼-W resistor we can connect to a 1.5-V dry-cell battery? (See Figure 2.23.)

Solution:

In order to compute the size of the resistor, we first calculate the power dissipated by the resistor as a function of its resistance:

$$P = VI = V\frac{V}{R} = \frac{V^2}{R} = \frac{1.5^2}{R}$$

Since the maximum allowable power dissipation is ¼ W, we can write

$$P_{\text{max}} = 0.25 = \frac{2.25}{R}$$

$$R = \frac{2.25}{0.25} = 9$$

Therefore R must be at least 9 Ω.

How would this result change if the voltage were doubled? We can easily calculate that the power dissipated is now given by $P = 3^2/R$, and therefore that R must be at least 36 Ω in order not to exceed its power rating. This result is shown in Figure 2.24. Note the effect of the square relationship: doubling the voltage increases the minimum allowable value of R by a factor of 4.

Example 2.7 Resistive Displacement Transducer

A simple resistor can serve as an elementary displacement transducer. The principle illustrated in this example actually forms the basis of many commercial displacement transducers. The resistor in Figure 2.25 is a wire-wound resistor—that is, a resistor made of

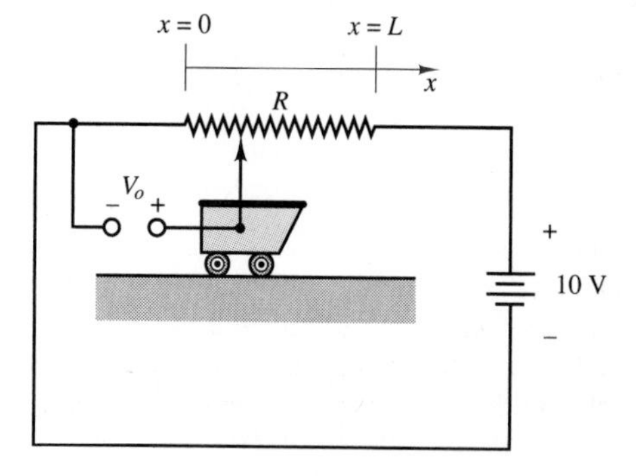

Figure 2.25 A simple resistive displacement transducer

a tightly wound coil of thin wire—that is rated at 0.1 Ω per turn. The pointer attached to the moving cart makes contact with the resistor, thus closing the circuit at a distance x from the origin. Each turn occupies 0.5 mm, and the total length, L, of the resistor is 10 cm. Find the output voltage V_o as a function of the cart displacement, x.

Solution:

The total number of turns in the resistor is given by the following expression:

$$\text{Number of turns} = \frac{\text{Total distance}}{\text{Distance per turn}} = \frac{0.1 \text{ m}}{0.0005 \text{ m/turn}} = 200$$

which means that the total resistance from 0 to L is 20 Ω. Thus, we can determine the relationship between the position of the cart and the resistance between the origin and point x as follows:

$$R(x) = \frac{20}{10}x = 2x \ \Omega$$

where x is the distance traveled in centimeters. The output voltage can then be found from this expression if we consider Ohm's law:

$$I = \frac{10 \text{ V}}{20 \ \Omega} = 0.5 \text{ A}$$

and

$$V_o = IR(x) = (0.5)2x = x$$

Thus, the relative position of the moving cart, in cm, is numerically equal to the voltage V_o. This simple position-sensing concept finds common application in engineering.

Example 2.8 Resistance Strain Gauges

Another common application of the resistance concept to engineering measurements is the resistance **strain gauge.** Strain gauges are devices that are bonded to the surface of an object, and whose resistance varies as a function of the surface strain experienced by the object. Strain gauges may be used to perform measurements of strain, stress, force, torque, and pressure. Recall that the resistance of a cylindrical conductor of cross-sectional area A, length L, and conductivity σ is given by the expression

$$R = \frac{L}{\sigma A}$$

If the conductor is compressed or elongated as a consequence of an external force, its dimensions will change, and with them its resistance. In particular, if the conductor is stretched, its cross-sectional area will decrease and the resistance will increase. If the conductor is compressed, its resistance decreases, since the length, L, will decrease. The relationship between change in resistance and change in length is given by the gauge factor, G, defined by

$$\text{G} = \frac{\Delta R/R}{\Delta L/L}$$

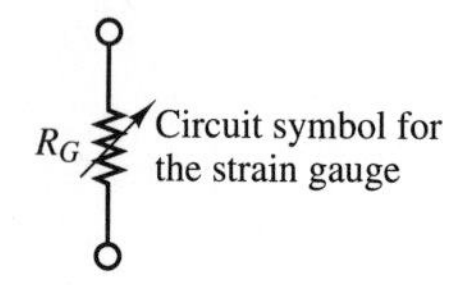

Metal-foil resistance strain gauge. The foil is formed by a photo-etching process and is less than 0.00002 in thick. Typical resistance values are 120, 350, and 1,000 Ω. The wide areas are bonding pads for electrical connections.

Figure 2.26 The resistance strain gauge

and since the strain ε is defined as the fractional change in length of an object, by the formula

$$\varepsilon = \frac{\Delta L}{L}$$

the change in resistance due to an applied strain ε is given by the expression

$$\Delta R = R_0 G\varepsilon$$

where R_0 is the resistance of the strain gauge under no strain and is called the zero strain resistance. The value of G for resistance strain gauges made of metal foil is usually about 2.

Figure 2.26 depicts a typical foil strain gauge. The maximum strain that can be measured by a foil gauge is about 0.4 to 0.5 percent; that is, $\Delta L/L = 0.004$–0.005. For a 120-Ω gauge, this corresponds to a change in resistance of the order of 0.96 to 1.2 Ω. Although this change in resistance is very small, it can be detected by means of suitable circuitry. Resistance strain gauges are usually connected in a circuit called the Wheatstone bridge, which we analyze later in this chapter.

Open and Short Circuits

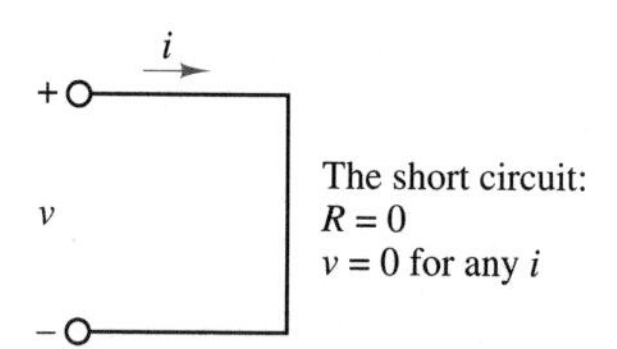

Figure 2.27 The short circuit

Two convenient idealizations of the resistance element are provided by the limiting cases of Ohm's law as the resistance of a circuit element approaches zero or infinity. A circuit element with resistance approaching zero is called a **short circuit.** Intuitively, one would expect a short circuit to allow for unimpeded flow of current. In fact, metallic conductors (e.g., short wires of large diameter) approximate the behavior of a short circuit. Formally, a short circuit is defined as a circuit element across which the voltage is zero, regardless of the current flowing through it. Figure 2.27 depicts the circuit symbol for an ideal short circuit.

Physically, any wire or other metallic conductor will exhibit some resistance, though small. For practical purposes, however, many elements approximate a short circuit quite accurately under certain conditions. For example, a large-diameter copper pipe is effectively a short circuit in the context of a residential electrical power supply, while in a low-power microelectronic circuit (e.g., an FM radio) a short length of 24 gauge wire (refer to Table 2.2 for the resistance of 24 gauge wire) is a more than adequate short circuit.

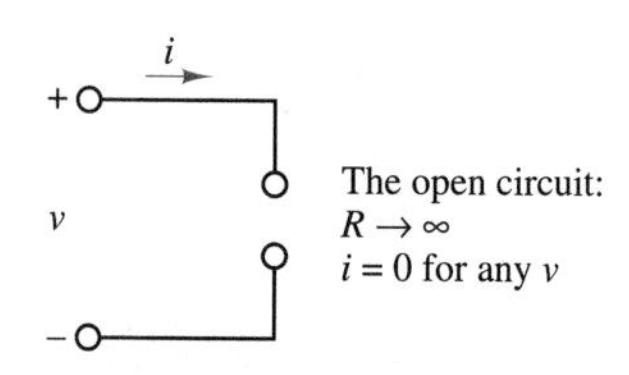

Figure 2.28 The open circuit

A circuit element whose resistance approaches infinity is called an **open circuit.** Intuitively, one would expect no current to flow through an open circuit, since it offers infinite resistance to any current. In an open circuit, we would expect to see zero current regardless of the externally applied voltage. Figure 2.28 illustrates this idea.

In practice, it is not too difficult to approximate an open circuit: any break in continuity in a conducting path amounts to an open circuit. The idealization of the open circuit, as defined in Figure 2.28, does not hold, however, for very

Table 2.2 **Resistance of copper wire**

AWG size	Number of strands	Diameter per strand	Resistance per 1,000 ft (Ω)
24	Solid	0.0201	28.4
24	7	0.0080	28.4
22	Solid	0.0254	18.0
22	7	0.0100	19.0
20	Solid	0.0320	11.3
20	7	0.0126	11.9
18	Solid	0.0403	7.2
18	7	0.0159	7.5
16	Solid	0.0508	4.5
16	19	0.0113	4.7

high voltages. The insulating material between two insulated terminals will break down at a sufficiently high voltage. If the insulator is air, ionized particles in the neighborhood of the two conducting elements may lead to the phenomenon of arcing; in other words, a pulse of current may be generated that momentarily jumps a gap between conductors (thanks to this principle, we are able to ignite the air-fuel mixture in a spark-ignition internal combustion engine by means of spark plugs). The ideal open and short circuits are useful concepts and find extensive use in circuit analysis.

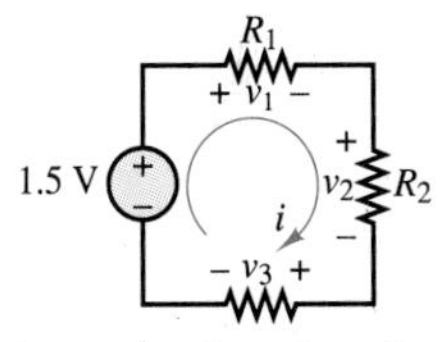

The current i flows through each of the four series elements. Thus, by KVL,

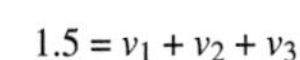

$$1.5 = v_1 + v_2 + v_3$$

Series Resistors and the Voltage Divider Rule

Although electrical circuits can take rather complicated forms, even the most involved circuits can be reduced to combinations of circuit elements *in parallel* and *in series*. Thus, it is important that you become acquainted with parallel and series circuits as early as possible, even before formally approaching the topic of network analysis. Parallel and series circuits have a direct relationship with Kirchhoff's laws. The objective of this section and the next is to illustrate two common circuits based on series and parallel combinations of resistors: the voltage and current dividers. These circuits form the basis of all network analysis; it is therefore important to master these topics as early as possible.

For an example of a series circuit, refer to the circuit of Figure 2.29, where a battery has been connected to resistors R_1, R_2, and R_3. The following definition applies:

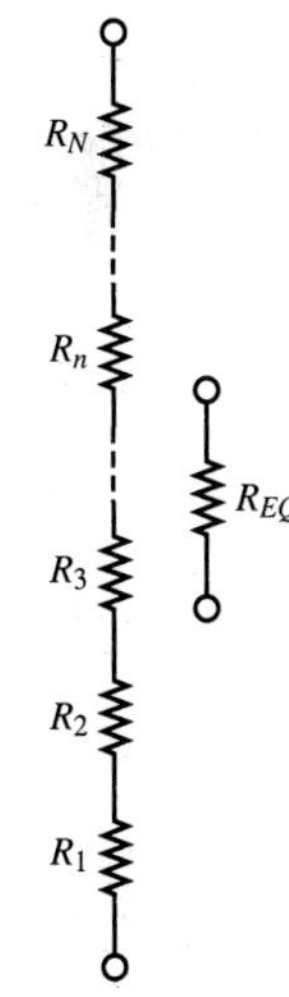

N series resistors are equivalent to a single resistor equal to the sum of the individual resistances.

Figure 2.29

> **Definition**
>
> Two or more circuit elements are said to be **in series** if the same current flows through each of the elements.

By applying KVL, you can verify that the sum of the voltages across the three resistors equals the voltage externally provided by the battery:

$$1.5 = v_1 + v_2 + v_3$$

and since, according to Ohm's law, the separate voltages can be expressed by the relations

$$v_1 = iR_1 \qquad v_2 = iR_2 \qquad v_3 = iR_3$$

we can therefore write

$$1.5 = i(R_1 + R_2 + R_3)$$

This simple result illustrates a very important principle: to the battery, the three series resistors appear as a single equivalent resistance of value R_{EQ}, where

$$R_{EQ} = R_1 + R_2 + R_3$$

The three resistors could thus be replaced by a single resistor of value R_{EQ} without changing the amount of current required of the battery. From this result we may extrapolate to the more general relationship defining the equivalent resistance of N series resistors:

$$R_{EQ} = \sum_{n=1}^{N} R_n \qquad \textbf{(2.19)}$$

which is also illustrated in Figure 2.29. A concept very closely tied to series resistors is that of the **voltage divider.** This terminology originates from the observation that the source voltage in the circuit of Figure 2.29 divides among the three resistors according to KVL. If we now observe that the series current, i, is given by

$$i = \frac{1.5}{R_{EQ}} = \frac{1.5}{R_1 + R_2 + R_3}$$

we can write each of the voltages across the resistors as:

$$v_1 = iR_1 = \frac{R_1}{R_{EQ}}(1.5)$$

$$v_2 = iR_2 = \frac{R_2}{R_{EQ}}(1.5)$$

$$v_3 = iR_3 = \frac{R_3}{R_{EQ}}(1.5)$$

that is,

> the voltage across each resistor in a series circuit is directly proportional to the ratio of its resistance to the total series resistance of the circuit.

An instructive exercise consists of verifying that KVL is still satisfied, by adding the voltage drops around the circuit and equating their sum to the source voltage:

$$v_1 + v_2 + v_3 = \frac{R_1}{R_{EQ}}(1.5) + \frac{R_2}{R_{EQ}}(1.5) + \frac{R_3}{R_{EQ}}(1.5) = 1.5$$

since

$$R_{EQ} = R_1 + R_2 + R_3$$

Therefore, since KVL is satisfied, we are certain that the voltage divider rule is consistent with Kirchhoff's laws. By virtue of the voltage divider rule, then, we can always determine the proportion in which voltage drops are distributed around a circuit. This result will be useful in reducing complicated circuits to simpler forms. The general form of the voltage divider rule for a circuit with N series resistors and a voltage source is:

$$v_n = \frac{R_n}{R_1 + R_2 + \cdots + R_n + \cdots + R_N} v_S \qquad \text{Voltage divider} \qquad \textbf{(2.20)}$$

Example 2.9

This example will illustrate the notion of the voltage divider by means of a simple circuit. We seek to determine the voltage v_x in the circuit of Figure 2.30.

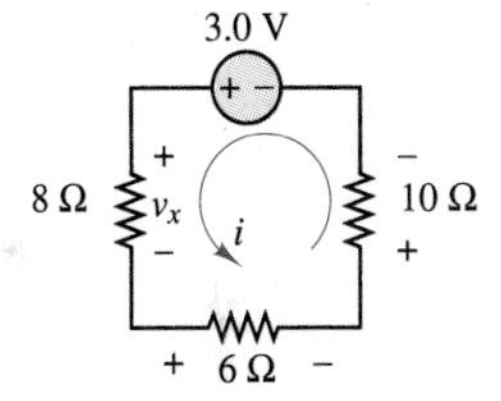

Figure 2.30

Solution:

If we select the reference direction indicated in Figure 2.30 by the direction of the current, we can assign voltage polarities around the circuit as shown. Then, application of KVL yields:

$$3.0 - v_{10} - v_6 - v_8 = 0$$

where v_{10} is the voltage across the 10-Ω resistor, according to the direction of the current indicated in the figure. Note that $v_x = v_8$. According to the voltage divider rule, then,

$$v_x = v_8 = \frac{8}{8 + 6 + 10}(3) = 1 \text{ V}$$

The technique employed in this example is representative of how the voltage divider rule is applied to series circuits. The key factor to be determined before applying this rule is whether the circuit is in fact a series circuit. We shall return to this point shortly.

Parallel Resistors and the Current Divider Rule

A concept analogous to that of the voltage divider may be developed by applying Kirchhoff's current law to a circuit containing only parallel resistances.

Definition

Two or more circuit elements are said to be **in parallel** if the same voltage appears across each of the elements.

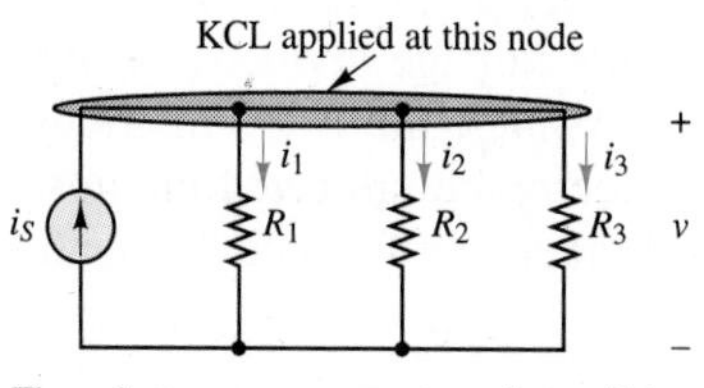

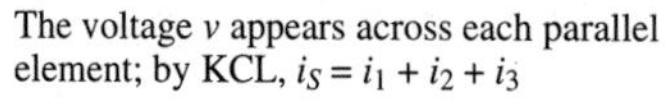
The voltage v appears across each parallel element; by KCL, $i_S = i_1 + i_2 + i_3$

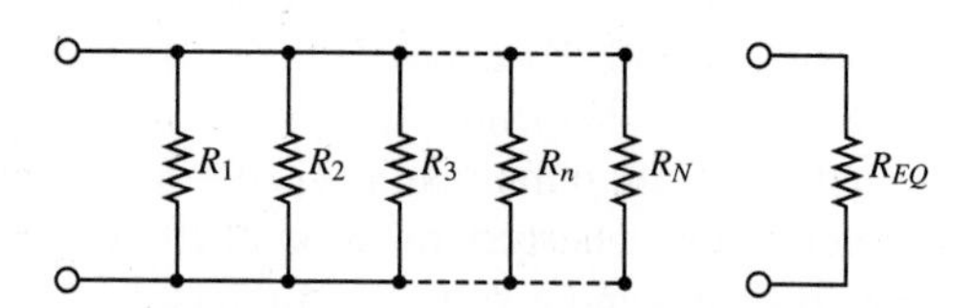

N resistors in parallel are equivalent to a single equivalent resistor with resistance equal to the inverse of the sum of the inverse resistances.

Figure 2.31 Parallel circuits

Figure 2.31 illustrates the notion of parallel resistors connected to an ideal current source. Kirchhoff's current law requires that the sum of the currents into, say, the top terminal of the circuit be zero:

$$i_S = i_1 + i_2 + i_3$$

But by virtue of Ohm's law we may express each current as follows:

$$i_1 = \frac{v}{R_1}$$

$$i_2 = \frac{v}{R_2}$$

$$i_3 = \frac{v}{R_3}$$

since, by definition, the same voltage, v, appears across each element. Kirchhoff's current law may then be restated as follows:

$$i_S = v\left(\frac{1}{R_1} + \frac{1}{R_2} + \frac{1}{R_3}\right)$$

Note that this equation can be also written in terms of a single equivalent resistance:

$$i_S = v\frac{1}{R_{EQ}}$$

where

$$\frac{1}{R_{EQ}} = \frac{1}{R_1} + \frac{1}{R_2} + \frac{1}{R_3}$$

As illustrated in Figure 2.31, one can generalize this result to an arbitrary number of resistors connected in parallel by stating that N resistors in parallel act as a single equivalent resistance, R_{EQ}, given by the expression

$$\frac{1}{R_{EQ}} = \frac{1}{R_1} + \frac{1}{R_2} + \cdots + \frac{1}{R_N} \tag{2.21}$$

or

$$R_{EQ} = \frac{1}{1/R_1 + 1/R_2 + \cdots + 1/R_N} \tag{2.22}$$

Very often in the remainder of this book we shall refer to the parallel combination of two or more resistors with the following notation:

$$R_1 \parallel R_2 \parallel \cdots$$

where the symbol $\parallel$ signifies "in parallel with."

From the results shown in equations 2.21 and 2.22, which were obtained directly from KCL, the **current divider** rule can be easily derived. Consider, again, the three-resistor circuit of Figure 2.31. From the expressions already derived from each of the currents, i_1, i_2, and i_3, we can write:

$$i_1 = \frac{v}{R_1}$$

$$i_2 = \frac{v}{R_2}$$

$$i_3 = \frac{v}{R_3}$$

and since $v = R_{EQ} i_S$, these currents may be expressed by:

$$i_1 = \frac{R_{EQ}}{R_1} i_S = \frac{1/R_1}{1/R_{EQ}} i_S = \frac{1/R_1}{1/R_1 + 1/R_2 + 1/R_3} i_S$$

$$i_2 = \frac{1/R_2}{1/R_1 + 1/R_2 + 1/R_3} i_S$$

$$i_3 = \frac{1/R_3}{1/R_1 + 1/R_2 + 1/R_3} i_S$$

One can easily see that the current in a parallel circuit divides in inverse proportion to the resistances of the individual parallel elements. The general expression for the current divider for a circuit with N parallel resistors is the following:

$$i_n = \frac{1/R_n}{1/R_1 + 1/R_2 + \cdots + 1/R_n + \cdots + 1/R_N} i_S \quad \text{Current divider} \tag{2.23}$$

Example 2.10 illustrates the application of the current divider rule.

Example 2.10

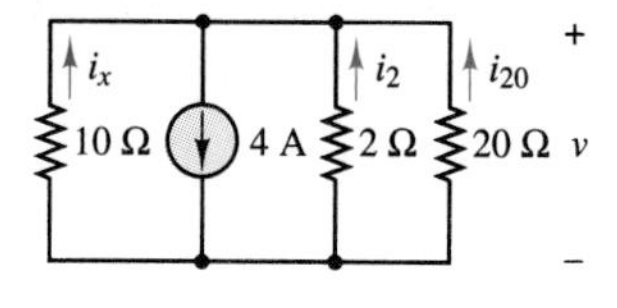

Figure 2.32

Find the current i_x in the circuit shown in Figure 2.32.

Solution:

The most direct solution is obtained by applying the current divider rule, where the 10-Ω resistance is associated with the desired current:

$$i_x = \frac{1/10}{1/10 + 1/2 + 1/20}(4) = (0.1538)4 = 0.6154 \text{ A}$$

An equivalent but less direct way of obtaining the same result would be to compute the equivalent resistance of the parallel circuit, determine the parallel voltage, v, and then compute the unknown current as

$$i_x = \frac{v}{10}$$

It is left as an exercise for you to demonstrate that this method is not as direct as the application of the current divider rule.

Much of the resistive network analysis that will be introduced in Chapter 3 is based on the simple principles of the voltage and current dividers introduced in this section. Unfortunately, practical circuits are rarely composed only of parallel or only of series elements. The following examples and drill exercises illustrate some simple and slightly more advanced circuits that combine parallel and series elements.

Example 2.11

This example illustrates the analysis of a circuit that contains a combination of parallel and series elements. The circuit of Figure 2.33 is neither a series nor a parallel circuit; thus, it is not possible to directly determine the voltage v or the current i by means of the voltage divider or current divider rule introduced earlier. How can v and i be determined?

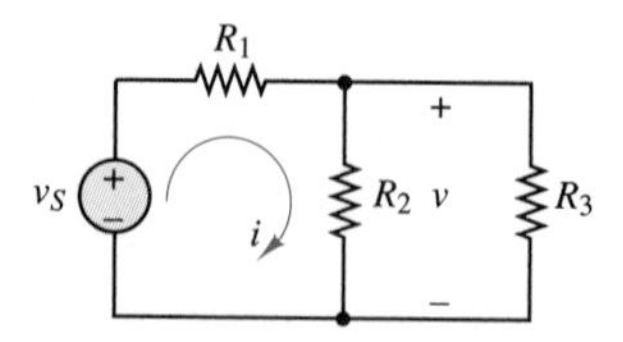

Figure 2.33

Solution:

The circuit takes a much simpler appearance once it becomes evident that the same voltage appears across both R_2 and R_3 and, therefore, that these elements are in parallel. If these two resistors are replaced by a single equivalent resistor according to the procedures described in this section, the circuit of Figure 2.34 is obtained. Note that now the equivalent circuit is a simple series circuit and the voltage divider rule can be applied to determine that:

$$v = \frac{R_2 \parallel R_3}{R_1 + R_2 \parallel R_3} v_S$$

while the current is found to be

$$i = \frac{v_S}{R_1 + R_2 \parallel R_3}$$

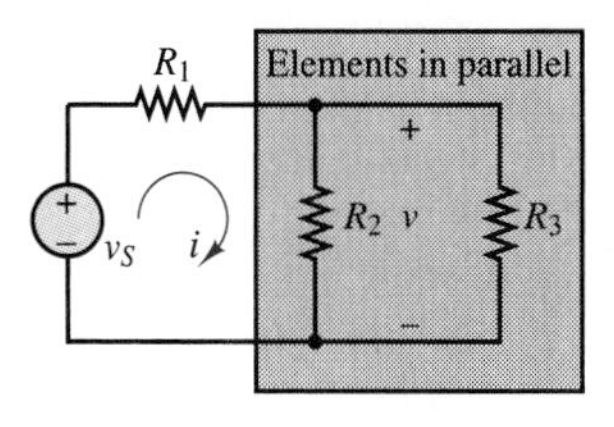

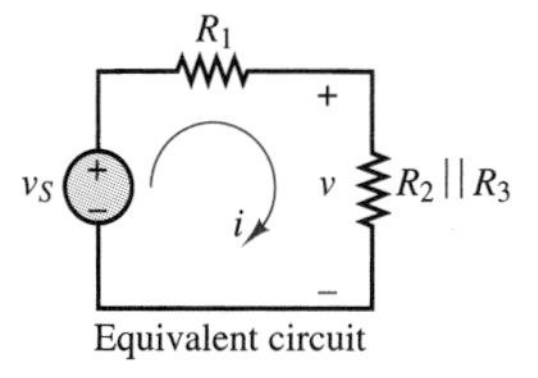

Figure 2.34

Example 2.12 The Wheatstone Bridge

The **Wheatstone bridge** is a resistive circuit that is frequently encountered in a variety of measurement circuits. The general form of the bridge circuit is shown in Figure 2.35(a), where R_1, R_2, and R_3 are known while R_x is an unknown resistance, to be determined. The circuit may also be redrawn as shown in Figure 2.35(b). The latter circuit will be used to demonstrate the use of the voltage divider rule in a mixed series-parallel circuit. The objective is to determine the unknown resistance, R_x.

1. Find the value of the voltage $v_{ab} = v_{ad} - v_{bd}$ in terms of the four resistances and the source voltage, v_S. Note that since the reference point d is the same for both voltages, we can also write $v_{ab} = v_a - v_b$.
2. If $R_1 = R_2 = R_3 = 1\ \text{k}\Omega$, $v_S = 12$ V, and $v_{ab} = 12$ mV, what is the value of R_x?

Solution:

1. First, we observe that the circuit consists of the parallel combination of three subcircuits: the voltage source, the series combination of R_1 and R_2, and the series combination of R_3 and R_x. Since these three subcircuits are in parallel, the same voltage will appear across each of them, namely, the source voltage, v_S.

   Thus, the source voltage divides between each resistor pair, R_1-R_2 and R_3-R_x, according to the voltage divider rule: v_a is the fraction of the source voltage appearing across R_2, while v_b is the voltage appearing across R_x:

$$v_a = v_S \frac{R_2}{R_1 + R_2} \qquad \text{and} \qquad v_b = v_S \frac{R_x}{R_3 + R_x}$$

   Finally, the voltage difference between points a and b is given by:

$$v_{ab} = v_a - v_b = v_S \left(\frac{R_2}{R_1 + R_2} - \frac{R_x}{R_3 + R_x} \right)$$

   This result is very useful and quite general, and it finds application in numerous practical circuits.

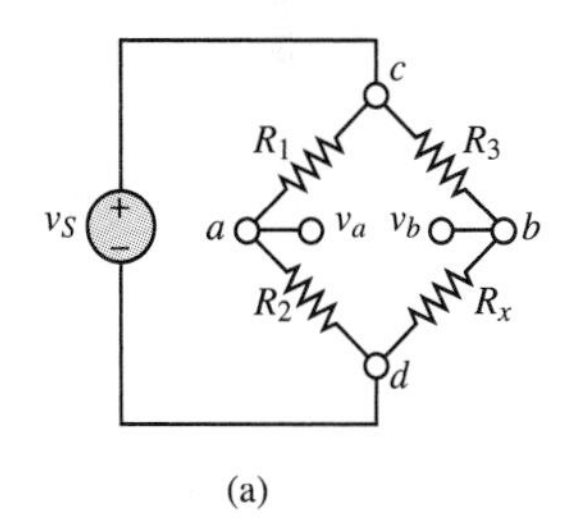

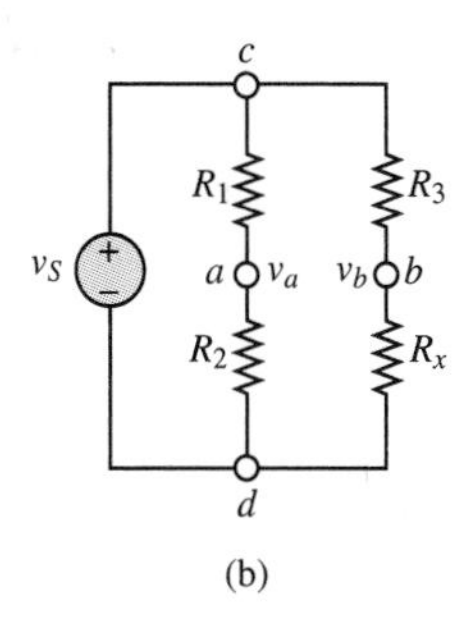

Figure 2.35 Wheatstone bridge circuits

2. In order to solve for the unknown resistance, we substitute the numerical values in the preceding equation to obtain

$$0.012 = 12\left(\frac{1{,}000}{2{,}000} - \frac{R_x}{1{,}000 + R_x}\right)$$

which may be solved for R_x to yield

$$R_x = 996\ \Omega$$

Example 2.13 The Strain Gauge Wheatstone Bridge

Strain gauges, which were introduced in Example 2.8, are frequently employed in the measurement of force. One of the simplest applications of strain gauges is in the measurement of the force applied to a cantilever beam, as illustrated in Figure 2.36. Four strain gauges are employed in this case, of which two are bonded to the upper surface of the beam at a distance L from the point where the external force, F, is applied and two are bonded on the lower surface, also at a distance L. Under the influence of the external force, the beam deforms and causes the upper gauges to extend and the lower gauges to compress. Thus, the resistance of the upper gauges will increase by an amount ΔR, and that of the lower gauges will decrease by an equal amount, assuming that the gauges are symmetrically placed. Let R_1 and R_4 be the upper gauges and R_2 and R_3 the lower gauges. Thus, under the influence of the external force, we have:

$$R_1 = R_4 = R_0 + \Delta R$$

$$R_2 = R_3 = R_0 - \Delta R$$

where R_0 is the zero strain resistance of the gauges (see Example 2.8). It can be shown from elementary statics that the relationship between the strain ε and a force F applied at a distance L for a cantilever beam is:

$$\varepsilon = \frac{6LF}{wh^2Y}$$

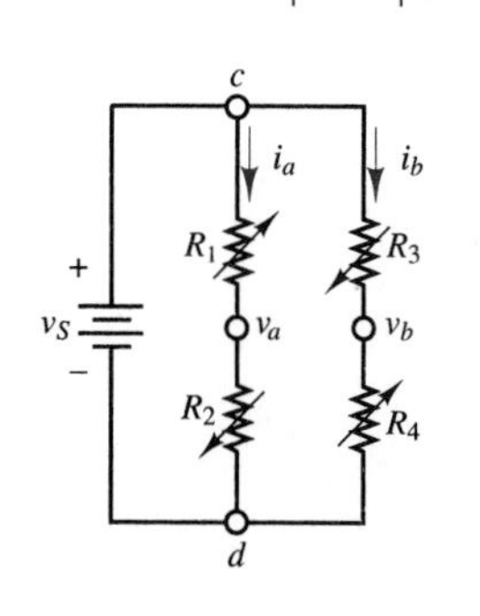

Figure 2.36 A force-measuring instrument

where h and w are as defined in Figure 2.36 and Y is the beam's modulus of elasticity.

In the circuit of Figure 2.36, the currents i_a and i_b are given by

$$i_a = \frac{v_S}{R_1 + R_2} \quad \text{and} \quad i_b = \frac{v_S}{R_3 + R_4}$$

The bridge output voltage is defined by $v_o = v_b - v_a$ and may be found from the following expression:

$$v_o = i_b R_4 - i_a R_2 = \frac{v_S R_4}{R_3 + R_4} - \frac{v_S R_2}{R_1 + R_2}$$

$$= v_S \frac{R_0 + \Delta R}{R_0 + \Delta R + R_0 - \Delta R} - v_S \frac{R_0 - \Delta R}{R_0 + \Delta R + R_0 - \Delta R}$$

$$= v_S \frac{\Delta R}{R_0} = v_S G\varepsilon$$

where the expression for $\Delta R/R_0$ was obtained in Example 2.8. Thus, it is possible to obtain a relationship between the output voltage of the bridge circuit and the force, F, as follows:

$$v_o = v_S G\varepsilon = v_S G\frac{6LF}{wh^2Y} = \frac{6v_S GL}{wh^2Y}F = kF$$

where k is the calibration constant for this force transducer.

Strain gauge Wheatstone bridges are the workhorse of many instruments for the measurement of strain, stress, force, torque, pressure, and related quantities.

DRILL EXERCISES

2.4 Repeat Example 2.9 by reversing the reference direction of the current, to show that the same result is obtained.

2.5 The circuit in the accompanying illustration contains a battery, a resistor, and an unknown circuit element.

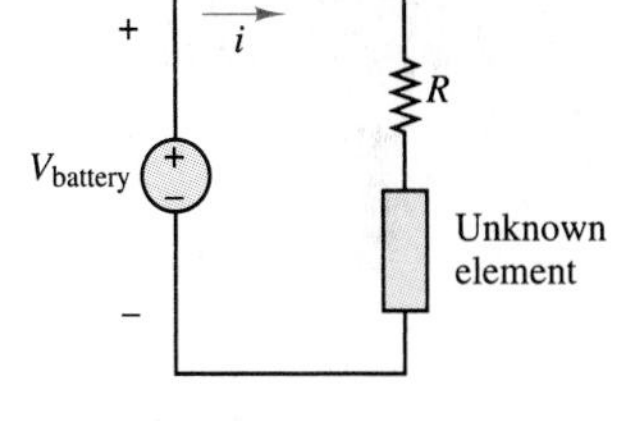

1. If the voltage $v_{battery}$ is 1.45 V and $i = 5$ mA, find power supplied to or by the battery.
2. Repeat part 1 if $i = -2$ mA.

2.6 The battery in the accompanying circuit supplies power to the resistors R_1, R_2, and R_3. Use KCL to determine the current i_B, and find the power supplied by the battery if $V_{battery} = 3$ V.

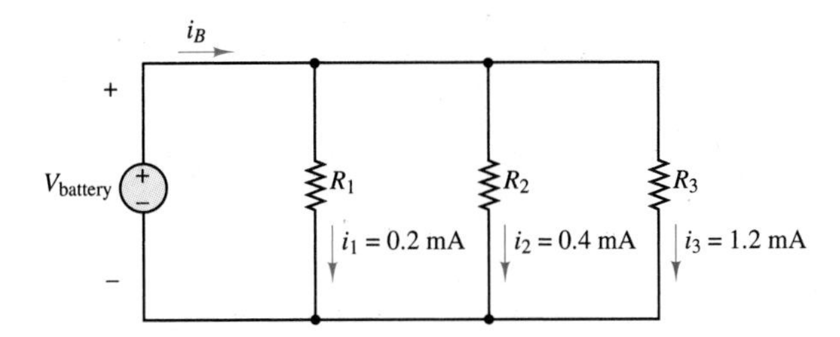

2.7 Use the results of part 1 of Example 2.12 to find the condition for which the voltage $v_{ab} = v_a - v_b$ is equal to zero (this is called the balanced condition for the bridge). Does this result necessarily require that all four resistors be identical? Why?

2.8 Verify that KCL is satisfied by the current divider rule and that the source current i_S divides in inverse proportion to the parallel resistors R_1, R_2, and R_3. (This should not be a surprise, since we would expect to see more current flow through the smaller resistance.)

2.9 Compute the full-scale (i.e., largest) output voltage for the force-measuring apparatus of Example 2.13. Assume that the strain gauge bridge is to measure forces ranging from 0 to 500 N, $L = 0.3$ m, $w = 0.05$ m, $h = 0.01$ m, $G = 2$, and the modulus of elasticity for the beam is 69×10^9 N/m^2 (aluminum). The source voltage is 12 V. What is the calibration constant of this force transducer?

2.10 Repeat the derivation of the current divider law by using conductance elements—that is, by replacing each resistance with its equivalent conductance, $G = 1/R$.

2.7 PRACTICAL VOLTAGE AND CURRENT SOURCES

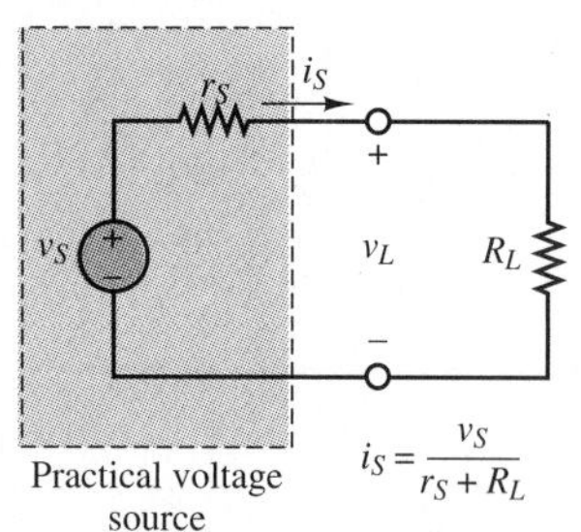

$$i_S = \frac{v_S}{r_S + R_L}$$

$$\lim_{R_L \to 0} i_S = \frac{v_S}{r_S}$$

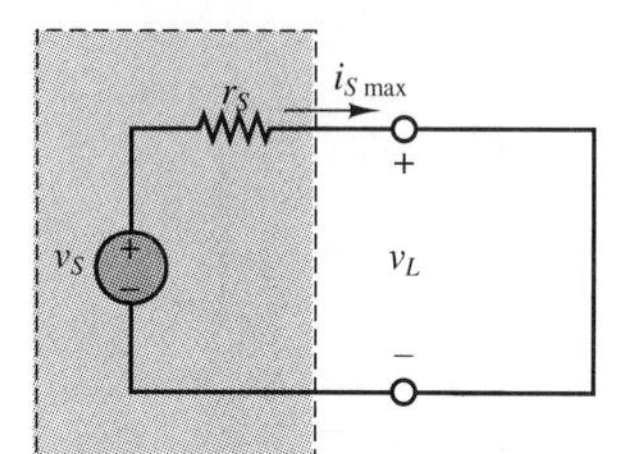

The maximum (short circuit) current which can be supplied by a practical voltage source is

$$i_{S\,\max} = \frac{v_S}{r_S}$$

Figure 2.37 Practical voltage source

The idealized models of voltage and current sources we discussed in Section 2.3 fail to take into consideration the finite-energy nature of practical voltage and current sources. The objective of this section is to extend the ideal models to models that are capable of describing the physical limitations of the voltage and current sources used in practice. Consider, for example, the model of an ideal voltage source shown in Figure 2.9. As the load resistance (R) decreases, the source is required to provide increasing amounts of current to maintain the voltage $v_S(t)$ across its terminals:

$$i(t) = \frac{v_S(t)}{R} \tag{2.24}$$

This circuit suggests that the ideal voltage source is required to provide an infinite amount of current to the load, in the limit as the load resistance approaches zero. Naturally, you can see that this is impossible; for example, think about the ratings of a conventional car battery: 12 V, 450 A-h (amp-hours). This implies that there is a limit (albeit a large one) to the amount of current a practical source can deliver to a load. Fortunately, it will not be necessary to delve too deeply into the physical nature of each type of source in order to describe the behavior of a practical voltage source: the limitations of practical sources can be approximated quite simply by exploiting the notion of the internal resistance of a source. Although the models described in this section are only approximations of the actual behavior of energy sources, they will provide good insight into the limitations of practical voltage and current sources. Figure 2.37 depicts a model for a practical voltage source; this is composed of an ideal voltage source, v_S, in series with a resistance, r_S. The resistance r_S in effect poses a limit to the maximum current the voltage source can provide:

$$i_{S\,\max} = \frac{v_S}{r_S} \tag{2.25}$$

Typically, r_S is small. Note, however, that its presence affects the voltage across the load resistance: now this voltage is no longer equal to the source voltage. Since the current provided by the source is

$$i_S = \frac{v_S}{r_S + R_L} \tag{2.26}$$

the load voltage can be determined to be

$$\begin{aligned} v_L &= i_S R_L \\ &= v_S \frac{R_L}{r_S + R_L} \end{aligned} \tag{2.27}$$

Thus, in the limit as the source internal resistance, r_S, approaches zero, the load voltage, v_L, becomes exactly equal to the source voltage. It should be apparent that a desirable feature of an ideal voltage source is a very small internal resistance, so that the current requirements of an arbitrary load may be satisfied. Often, the effective internal resistance of a voltage source is quoted in the technical specifications for the source, so that the user may take this parameter into account.

A similar modification of the ideal current source model is useful to describe the behavior of a practical current source. The circuit illustrated in Figure 2.38 depicts a simple representation of a practical current source, consisting of an ideal source in parallel with a resistor. Note that as the load resistance approaches infinity (i.e., an open circuit), the output voltage of the current source approaches its limit,

$$v_{S\,\max} = i_S r_S \tag{2.28}$$

A good current source should be able to approximate the behavior of an ideal current source. Therefore, a desirable characteristic for the internal resistance of a current source is that it be as large as possible.

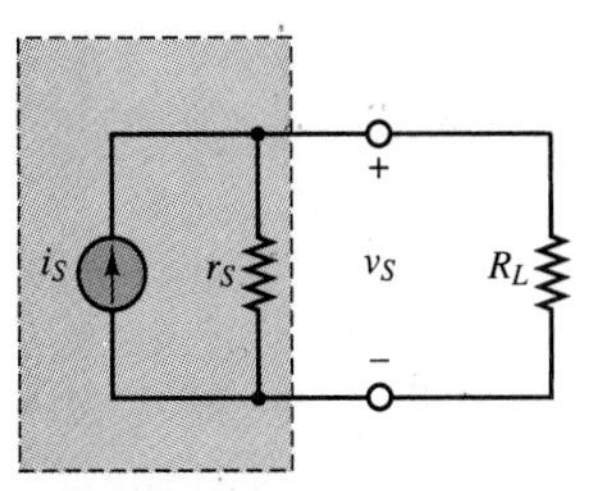

A model for practical current sources consists of an ideal source in *parallel* with an internal resistance.

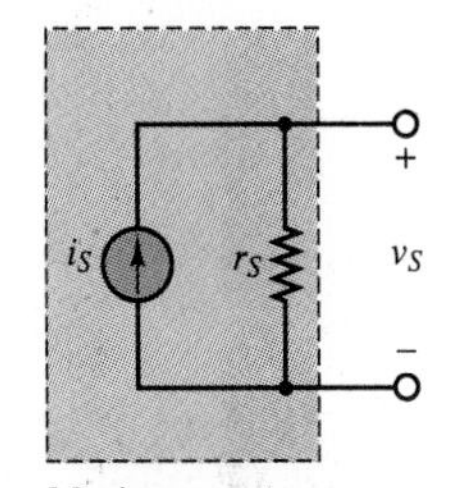

Maximum output voltage for practical current source with open-circuit load:

$$v_{S\,\max} = i_S r_S$$

Figure 2.38 Practical current source

2.8 MEASURING DEVICES

In this section, you should gain a basic understanding of the desirable properties of practical devices for the measurement of electrical parameters. The measurements most often of interest are those of current, voltage, power, and resistance. In analogy with the models we have just developed to describe the nonideal behavior of voltage and current sources, we shall similarly present circuit models for practical measuring instruments suitable for describing the nonideal properties of these devices.

The Ammeter

The **ammeter** is a device that, when connected in series with a circuit element, can measure the current flowing through the element. Figure 2.39 illustrates this idea. From Figure 2.39, two requirements are evident for obtaining a correct measurement of current:

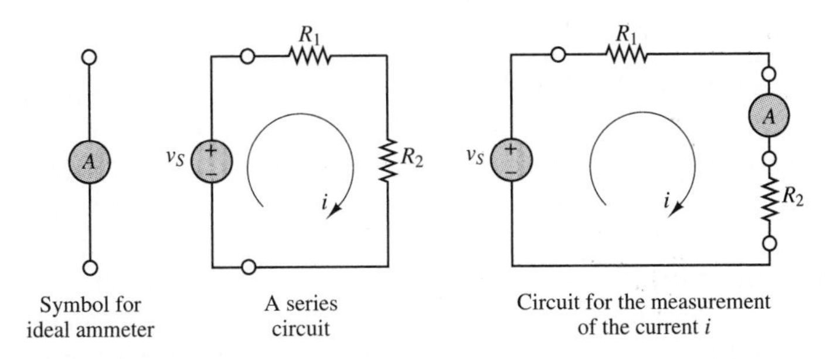

Figure 2.39 Measurement of current

1. The ammeter must be placed in series with the element whose current is to be measured (e.g., resistor R_2).
2. The ammeter should not restrict the flow of current (i.e., cause a voltage drop), or else it will not be measuring the true current flowing in the circuit. *An ideal ammeter has zero internal resistance.*

The Voltmeter

The **voltmeter** is a device that can measure the voltage across a circuit element. Since voltage is the difference in potential between two points in a circuit, the voltmeter needs to be connected across the element whose voltage we wish to measure. A voltmeter must also fulfill two requirements:

1. The voltmeter must be placed in parallel with the element whose voltage it is measuring.
2. The voltmeter should draw no current away from the element whose voltage it is measuring, or else it will not be measuring the true voltage across that element. Thus, *an ideal voltmeter has infinite internal resistance.*

Figure 2.40 illustrates these two points.

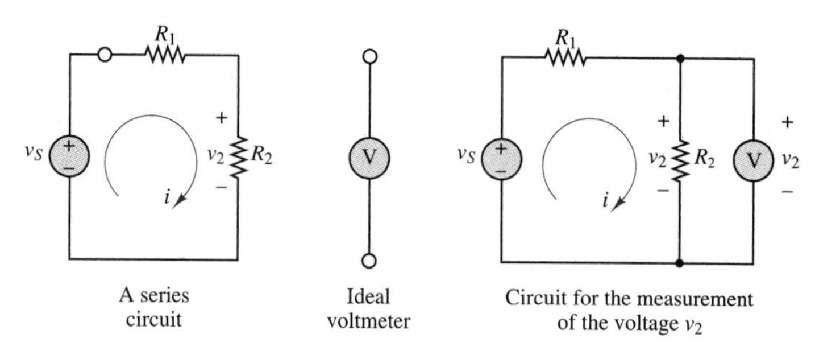

Figure 2.40 Measurement of voltage

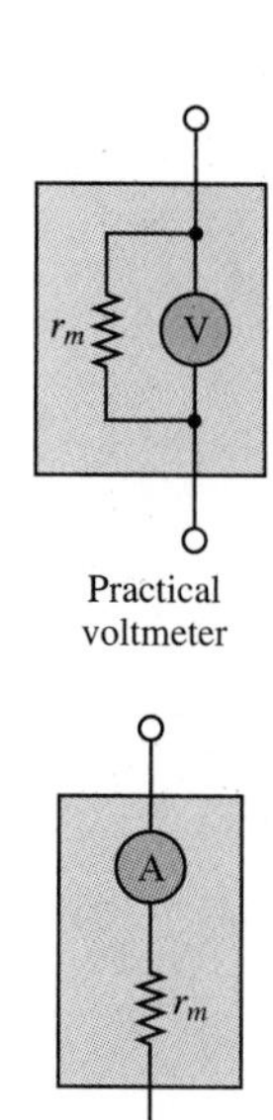

Figure 2.41 Models for practical ammeter and voltmeter

Once again, the definitions just stated for the ideal voltmeter and ammeter need to be augmented by considering the practical limitations of the devices. A practical ammeter will contribute some series resistance to the circuit in which it is measuring current; a practical voltmeter will not act as an ideal open circuit but will always draw some current from the measured circuit. The homework problems verify that these practical restrictions do not necessarily pose a limit to the accuracy of the measurements obtainable with practical measuring devices, as long as the internal resistance of the measuring devices is known. Figure 2.41 depicts the circuit models for the practical ammeter and voltmeter.

All of the considerations that pertain to practical ammeters and voltmeters can be applied to the operation of a **wattmeter,** a measuring instrument that provides a measurement of the power dissipated by a circuit element, since the wattmeter is in effect made up of a combination of a voltmeter and an ammeter.

Figure 2.42 depicts the typical connection of a wattmeter in the same series circuit used in the preceding paragraphs. In effect, the wattmeter measures the current flowing through the load and, simultaneously, the voltage across it and multiplies the two to provide a reading of the power dissipated by the load. The internal power consumption of a practical wattmeter is explored in the homework problems.

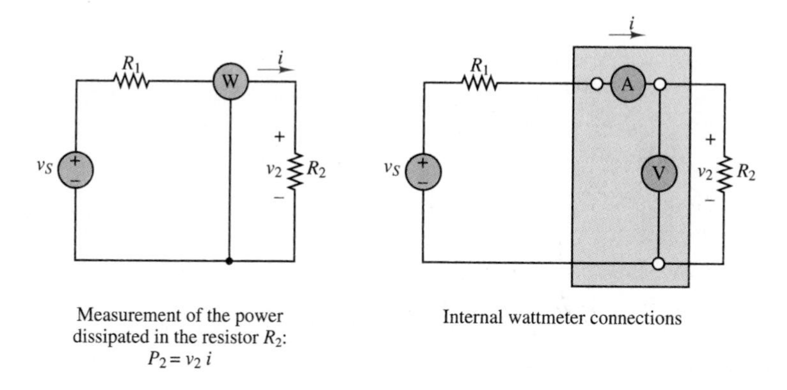

Figure 2.42 Measurement of power

2.9 ELECTRICAL NETWORKS

In the previous sections we have outlined models for the basic circuit elements: sources, resistors, and measuring instruments. We have assembled all the tools and parts we need in order to define an **electrical network.** It is appropriate at this stage to formally define the elements of the electrical circuit; the definitions that follow are part of standard electrical engineering terminology.

Branch

A **branch** is any portion of a circuit with two terminals connected to it. A branch may consist of one or more circuit elements (Figure 2.43). In practice, any circuit element with two terminals connected to it is a branch.

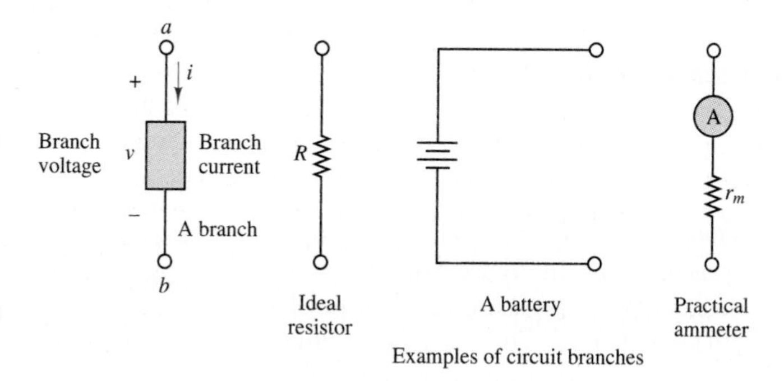

Figure 2.43 Definition of a branch

Node

A node is the junction of two or more branches (one often refers to the junction of only two branches as a *trivial node*). Figure 2.44 illustrates the concept. In effect, any connection that can be accomplished by soldering various terminals together is a node. It is very important to identify nodes properly in the analysis of electrical networks.

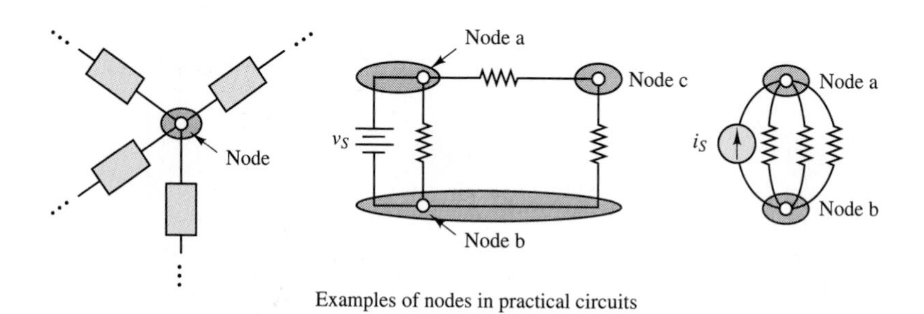

Figure 2.44 Definition of a node

Loop

A **loop** is any closed connection of branches that begins at one terminal of a branch and ends at the opposite terminal of the same branch. Various loop configurations are illustrated in Figure 2.45.

Note how two different loops in the same circuit may include some of the same elements or branches.

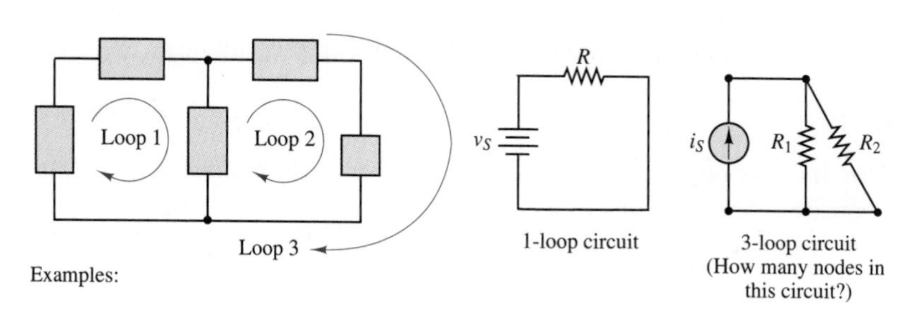

Figure 2.45 Definition of a loop

Mesh

A **mesh** is a loop that does not contain other loops. Meshes are an important aid to certain analysis methods. In Figure 2.45, the circuit with loops 1, 2, and 3 consists of two meshes: loops 1 and 2 are meshes, but loop 3 is not a mesh, because it encircles both loops 1 and 2. The one-loop circuit of Figure 2.45 is also a one-mesh circuit. Figure 2.46 illustrates how meshes are simpler to visualize in complex networks than loops are.

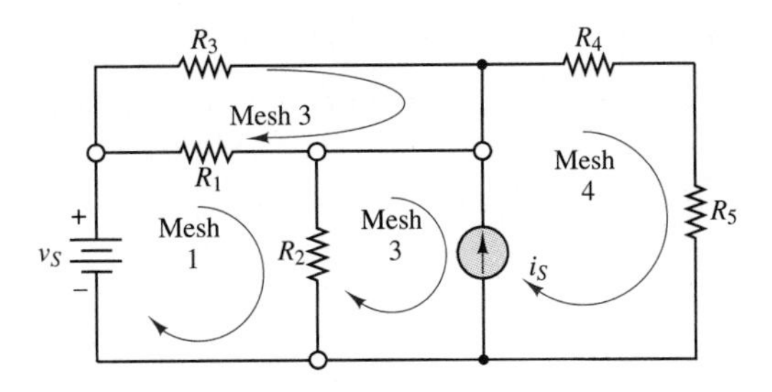

How many loops can you identify in this four-mesh circuit? (Answer: 14)

Figure 2.46 Definition of a mesh

Network Analysis

The analysis of an electrical network consists of determining each of the unknown branch currents and node voltages. It is therefore important to define all of the relevant variables as clearly as possible, and in systematic fashion. Once the known and unknown variables have been identified, a set of equations relating these variables is constructed, and these are solved by means of suitable techniques. The analysis of electrical circuits consists of writing the smallest set of equations sufficient to solve for all of the unknown variables. The procedures required to write these equations are the subject of Chapter 3 and are very well documented and codified in the form of simple rules. The analysis of electrical circuits is greatly simplified if some standard conventions are followed. The objective of this section is precisely to outline the preliminary procedures that will render the task of analyzing an electrical circuit manageable.

Circuit Variables

The first observation to be made is that the relevant variables in network analysis are the node voltages and the branch currents. This fact is really nothing more than a consequence of Ohm's law. Consider the branch depicted in Figure 2.47, consisting of a single resistor. Here, once a voltage v_R is defined across the resistor R, a current i_R will flow through the resistor, according to $v_R = i_R R$. But the voltage v_R, which causes the current to flow, is really the difference in electric potential between nodes a and b:

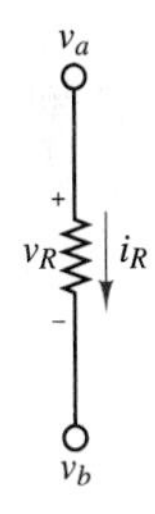

Figure 2.47 Variables in a network analysis problem

$$v_R = v_a - v_b \tag{2.29}$$

What meaning do we assign to the variables v_a and v_b? Was it not stated that voltage is a potential difference? Is it then legitimate to define the voltage at a single point (node) in a circuit? Whenever we reference the voltage at a node in a circuit, we imply an assumption that the voltage at that node is the potential difference between the node itself and a reference node called **ground,** which is located somewhere else in the circuit and which for convenience has been assigned a potential of zero volts. Thus, in Figure 2.47, the expression

$$v_R = v_a - v_b$$

really signifies that v_R is the difference between the voltage differences $v_a - v_c$ and $v_b - v_c$, where v_c is the (arbitrary) ground potential. Note that the equation

$v_R = v_a - v_b$ would hold even if the reference node, c, were not assigned a potential of zero volts, since

$$v_R = v_a - v_b = (v_a - v_c) - (v_b - v_c) \quad \textbf{(2.30)}$$

What, then, is this ground or reference voltage?

Ground

The choice of the word *ground* is not arbitrary. This point can be illustrated by a simple analogy with the physics of fluid motion. Consider a tank of water, as shown in Figure 2.48, located at a certain height above the ground. The potential energy due to gravity will cause water to flow out of the pipe at a certain flow rate. The pressure that forces water out of the pipe is directly related to the head, $(h_1 - h_2)$, in such a way that this pressure is zero when $h_2 = h_1$. Now the point h_3, corresponding to the ground level, is defined as having zero potential energy. It should be apparent that the pressure acting on the fluid in the pipe is really caused by the difference in potential energy, $(h_1 - h_3) - (h_2 - h_3)$. It can be seen, then, that it is not necessary to assign a precise energy level to the height h_3; in fact, it would be extremely cumbersome to do so, since the equations describing the flow of water would then be different, say, in Denver ($h_3 = 1{,}600$ m above sea level) from those that would apply in Miami ($h_3 = 0$ m above sea level). You see, then, that it is the relative difference in potential energy that matters in the water tank problem.

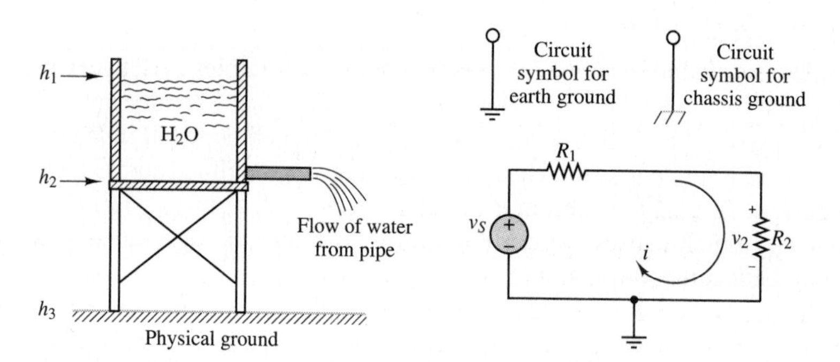

Figure 2.48 Analogy between electrical and earth ground

In analogous fashion, in every circuit a point can be defined that is recognized as "ground" and is assigned the electric potential of zero volts *for convenience.* Note that, unless they are purposely connected together, the grounds in two completely separate circuits are not necessarily at the same potential. This last statement may seem puzzling, but Example 2.14 should clarify the idea.

Example 2.14 Chassis and Earth Grounds

Automobiles and commercial vehicles are equipped with a lead-acid battery (usually 12 V) to provide electrical power to a variety of accessories. The electrical circuits inside the vehicle are all tied to one common ground, which, for convenience, is none other than the metal body of the car (see Figure 2.49). The same notion extends to many electrically powered devices, where the case, enclosure, or chassis becomes a reference point for all electrical circuits. This type of ground connection takes the name of **chassis ground.**

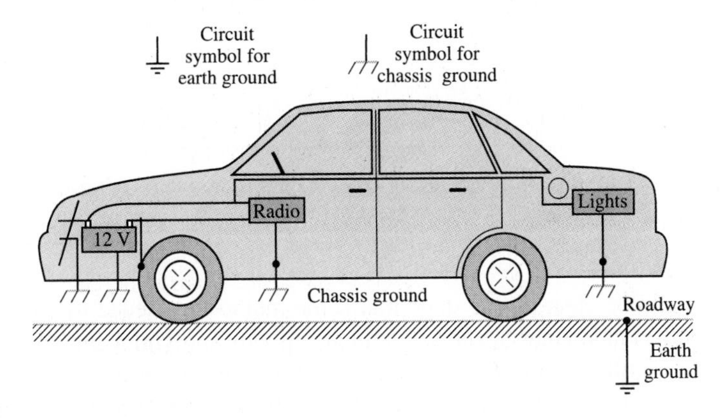

Figure 2.49 Chassis ground versus earth ground

Note that the car sits on an insulating support, that is, its tires. Therefore, the ground potential of the chassis does not necessarily have to be the same as the ground potential of the earth! The latter quantity, in fact, takes a different name: it is commonly referred to as **earth ground.**

Often, chassis ground and earth ground are tied together. For example, a washing machine with internal circuitry tied to the metal enclosure (chassis) is also grounded to earth. This is a very good idea from the standpoint of safety. If the two grounds were not tied together, a potential difference might exist between earth and chassis ground, and the unaware user might risk electric shock by touching the washing machine while standing on a grounded surface, thus closing the loop between chassis and earth ground (Figure 2.50). We shall be more precise with regard to grounding and safety issues when electric power is discussed, in Chapter 5. Note that in this book we shall not make a distinction between chassis and earth ground unless this is dictated by the requirements of a specific problem.

In a washing machine, the chassis ground is also tied to earth for safety reasons.

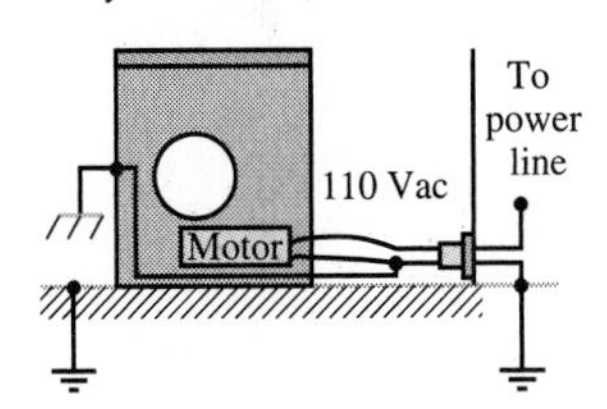

Figure 2.50 Grounding chassis to earth for safety

It is a useful exercise at this point to put the concepts illustrated in this chapter into practice by identifying the relevant variables in a few examples of electrical circuits. In the following examples, we shall illustrate how it is possible to define unknown voltages and currents in a circuit in terms of the source voltages and currents and the resistances in the circuit.

Example 2.15

Identify the branch and node voltages and the loop and mesh currents in the circuit of Figure 2.51.

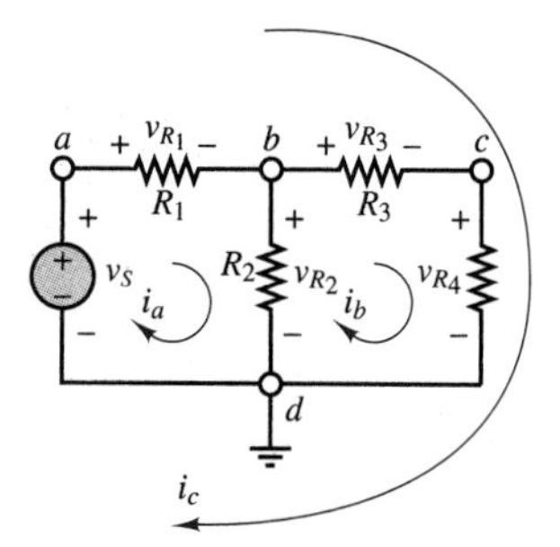

Figure 2.51

Solution:

The following node voltages may be identified:

Node voltages	Branch voltages
$v_a = v_S$ (source voltage)	$v_S = v_a - v_d = v_a$
$v_b = v_{R2}$	$v_{R1} = v_a - v_b$
$v_c = v_{R4}$	$v_{R2} = v_b - v_d = v_b$
$v_d = 0$ (ground)	$v_{R3} = v_b - v_c$
	$v_{R4} = v_c - v_d = v_c$

Currents i_a, i_b, and i_c are loop currents, but only i_a and i_b are mesh currents.

It should be clear at this stage that some method is needed to organize the wealth of information that can be generated simply by applying Ohm's law at each branch in a circuit. What would be desirable at this point is a means of reducing the number of equations needed to solve a circuit to the minimum necessary, that is, a method for obtaining N equations in N unknowns. The next chapter is devoted to the development of systematic circuit analysis methods that will greatly simplify the solution of electrical network problems.

DRILL EXERCISES

2.11 Write expressions for the voltage across each resistor in Example 2.15 in terms of the mesh currents.

2.12 Write expressions for the current through each resistor in Example 2.15 in terms of the node voltages.

CONCLUSION

The objective of this chapter was to introduce the background needed in the following chapters for the analysis of linear resistive networks. The fundamental laws of circuit analysis, Kirchhoff's current law, Kirchhoff's voltage law, and Ohm's law, were introduced, along with the basic circuit elements, and all were used to analyze the most basic circuits: voltage and current dividers. Measuring devices and a few other practical circuits employed in common engineering measurements were also introduced to provide a flavor of the applicability of these basic ideas to practical engineering problems. The remainder of the book draws on the concepts developed in this chapter. Mastery of the principles exposed in these first pages is therefore of fundamental importance.

KEY TERMS

Ammeter *43*
Ampere (A) *15*
Branch *45*
Charge *14*
Chassis Ground *49*
Conductance *28*
Conductivity *27*
Coulomb (C) *14*
Current Divider *37*
Dependent (or Controlled) Source *22*
Earth Ground *49*
Electric Current *14*
Electrical Network *45*
Elementary Charge *14*
Ground *47*
i-v Characteristic (or Volt-Ampere Characteristic) *25*
Ideal Current Source *21*
Ideal Source *20*
Ideal Voltage Source *20*
Kirchhoff's Current Law (KCL) *16*
Kirchhoff's Voltage Law (KVL) *18*
Load *19*
Loop *46*
Mesh *46*
Node *16*
Ohm (Ω) *27*
Ohm's Law *27*
Open Circuit *32*
Parallel Circuit *36*
Passive Sign Convention *23*
Potential Difference *17*
Power Rating *29*
Resistance *27*
Resistivity *27*
Resistor *27*
Series Circuit *33*
Short Circuit *32*
Siemens (S) *28*
Source *19*
Strain Gauge *31*
Volt (V) *17*
Voltage *17*
Voltage Divider *34*
Voltmeter *44*
Wattmeter *44*
Wheatstone Bridge *39*

ANSWERS TO DRILL EXERCISES

2.1 $I_P = I_D = 4.17$ A; 100 W

2.2 *A*, supplying 30.8 W; *B*, dissipating 30.8 W

2.3 $i_3 = -1$ mA; $i_2 = 0$ mA

2.5 $P_1 = 7.25 \times 10^{-3}$ W (supplied by)
$P_2 = 2.9 \times 10^{-3}$ W (supplied to)

2.6 $i_B = 1.8$ mA
$P_B = 5.4$ mW

2.7 $R_1 R_x = R_2 R_3$

2.9 v_o (full scale) $= 62.6$ mV; $k = 0.125$ mV/N

2.11 $v_{R1} = i_a R_1$; $v_{R2} = (i_a - i_b)R_2$; $v_{R3} = i_b R_3$; $v_{R4} = i_b R_4$

2.12 $i_1 = \dfrac{v_a - v_b}{R_1}$; $i_2 = \dfrac{v_b - v_d}{R_2}$; $i_3 = \dfrac{v_b - v_c}{R_3}$; $i_4 = \dfrac{v_c - v_d}{R_4}$

HOMEWORK PROBLEMS

EIT 2.1 An automotive battery is rated at 120 A-h. This means that under certain test conditions it can output 1 A at 12 V for 120 h (under other test conditions, the battery may have other ratings).

a. How much total energy is stored in the battery?
b. If the headlights are left on overnight (8 h), how much energy will still be stored in the battery in the morning? (Assume a 150-W total power rating for both headlights together.)

EIT 2.2 A car battery kept in storage in the basement needs recharging. If the voltage and the current provided by the charger during a charge cycle are as shown in Figure P2.1,

a. Find the total charge transferred to the battery.
b. Find the total energy transferred to the battery.

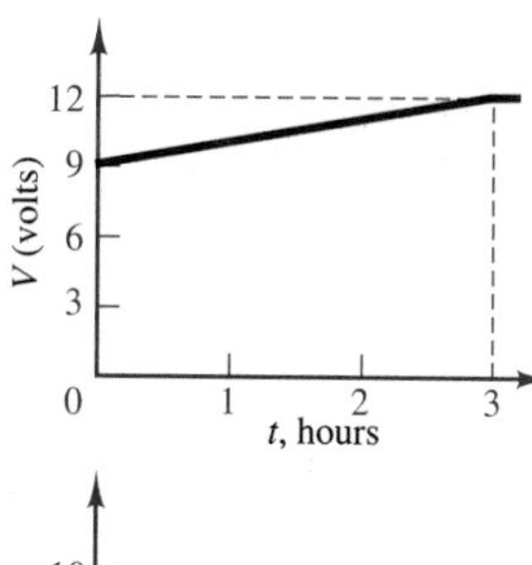

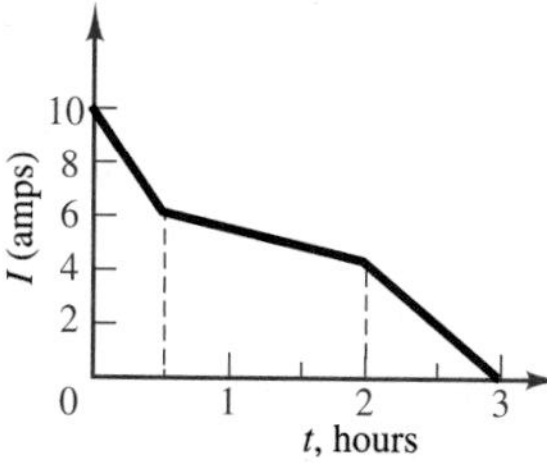

Figure P2.1

2.3 Suppose the current flowing through a wire is given by the curve shown in Figure P2.2.

a. Find the amount of charge, q, that flows through the wire between $t_1 = 0$ and $t_2 = 1$s.
b. Repeat part a for $t_2 = 2, 3, 4, 5, 6, 7, 8, 9,$ and 10s.
c. Sketch $q(t)$ for $0 \leq t \leq 10$s.

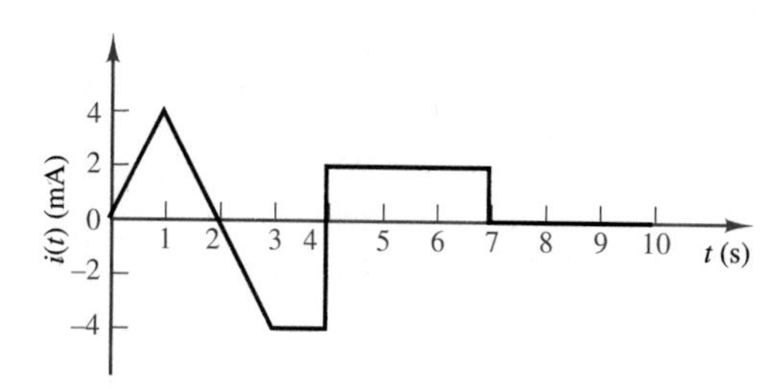

Figure P2.2

2.4 The capacity of a car battery is usually specified in ampere-hours. A battery rated at, say, 100 A-h should be able to supply 100 A for 1 hour, 50 A for 2 hours, 25 A for 4 hours, 1 A for 100 hours, or any other combination yielding a product of 100 A-h.

a. How many coulombs of charge should we be able to draw from a fully charged 100 A-h battery?
b. How many electrons does your answer to part a require?

2.5 The current in a semiconductor device results from the motion of two different kinds of charge carriers: electrons and holes. The holes and electrons have charge of equal magnitude but opposite sign. In a particular device, suppose the electron density is 2×10^{19} electrons/m^3, and the hole density is 5×10^{18} holes/m^3. This device has a cross-sectional area of 50 nm^2. If the electrons are moving to the left at a velocity of 0.5 mm/s, and the holes are moving to the right at a velocity of 0.2 mm/s, what are:

a. The direction of the current in the semiconductor.
b. The magnitude of the current in the device.

2.6 The charge cycle shown in Figure P2.3 is an example of a *two-rate charge.* The current is held constant at 50 mA for 5 h. Then it is switched to 20 mA for the next 5 h. Find:

a. The total charge transferred to the battery.
b. The energy transferred to the battery.

Hint: Recall that energy, w, is the integral of power, or $P = dw/dt$.

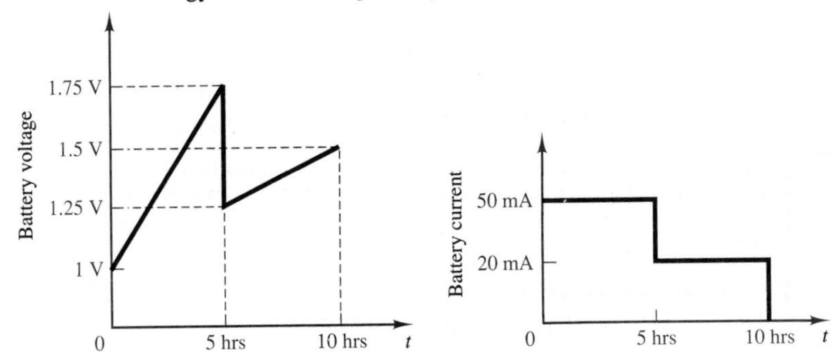

Figure P2.3

EIT 2.7 Apply KCL to find the current i in the circuit of Figure P2.4.

2.8 Apply KCL to find the current i in the circuit of Figure P2.5.

2.9 Apply KVL to find the voltages v_1 and v_2 in Figure P2.6.

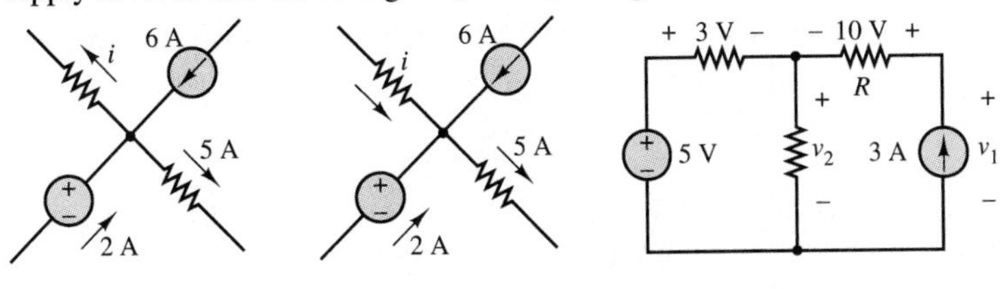

Figure P2.4 **Figure P2.5** **Figure P2.6**

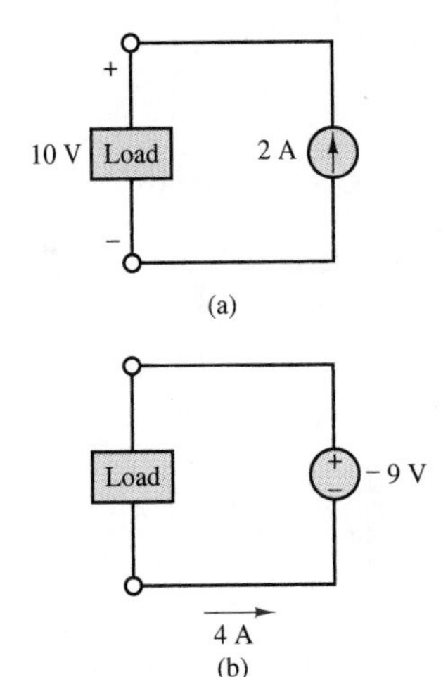

Figure P2.7

EIT 2.10 Find the power delivered by each source in the circuits of Figure P2.7.

2.11 In Figure 2.8 in the text, let v_S = 110 VAC and assume that the circuit wiring has a fixed resistance of 11 Ω. Plot the current i as a function of the resistance of a light bulb.

2.12 Find the power dissipated by the resistor in the circuit of Figure P2.8.

EIT 2.13 Determine which element in the circuit of Figure P2.9 is supplying power and which is dissipating power. Also determine the amount of power dissipated and supplied.

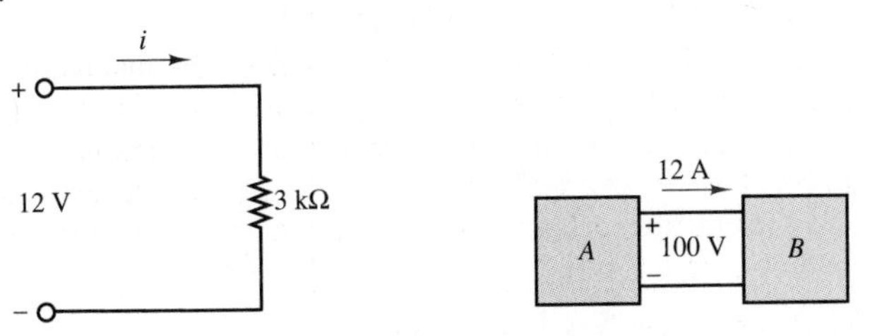

Figure P2.8 **Figure P2.9**

2.14 Determine which elements in the circuit of Figure P2.10 are supplying power and which are dissipating power. Also determine the amount of power dissipated and supplied.

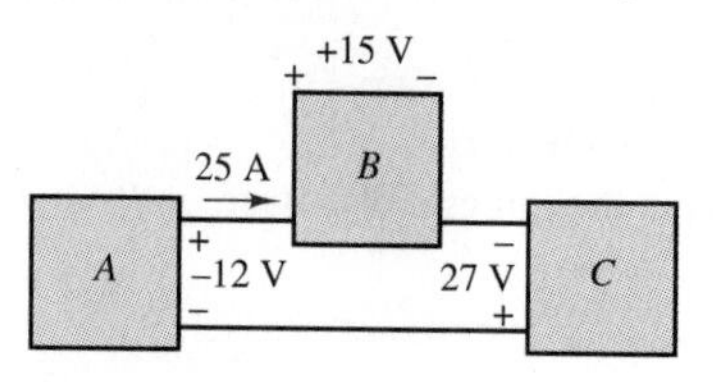

Figure P2.10

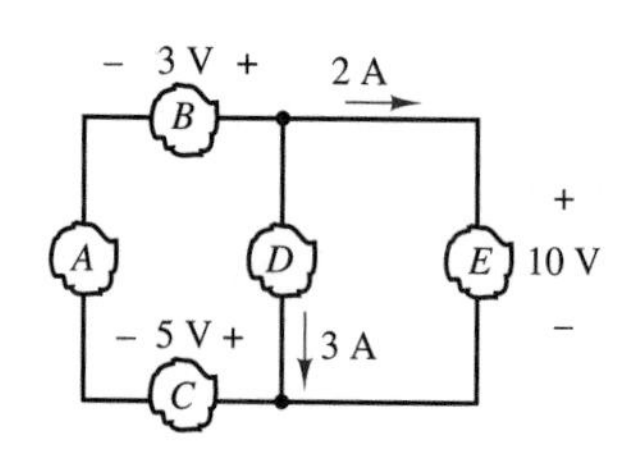

Figure P2.11

2.15 In the circuit of Figure P2.6, determine the power absorbed by the resistor R and the power delivered by the current source.

2.16 For the circuit shown in Figure P2.11:

a. Determine which components are absorbing power and which are delivering power.
b. Is conservation of power satisfied? Explain your answer.

EIT 2.17 Suppose one of the two headlights in Example 2.2 has been replaced with the wrong part and the 12-V battery is now connected to a 75-W and a 50-W headlight. What is the resistance of each headlight, and what is the total resistance seen by the battery?

2.18 What is the equivalent resistance seen by the battery of Example 2.2 if two 10-W taillights are added to the 50-W (each) headlights?

2.19 For the circuit shown in Figure P2.12, determine the power absorbed by the 5 Ω resistor.

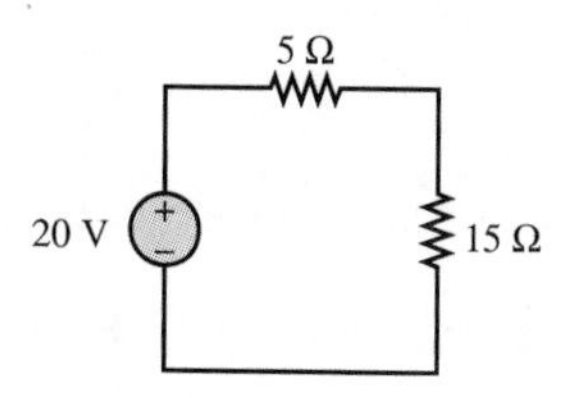

Figure P2.12

2.20 In the circuits of Figure P2.13, the directions of current and polarities of voltage have already been defined. Find the actual values of the indicated currents and voltages.

EIT 2.21 In the circuit of Figure P2.14, if $v_1 = v/8$ and the power delivered by the source is 8 mW, find R, v, v_1, and i.

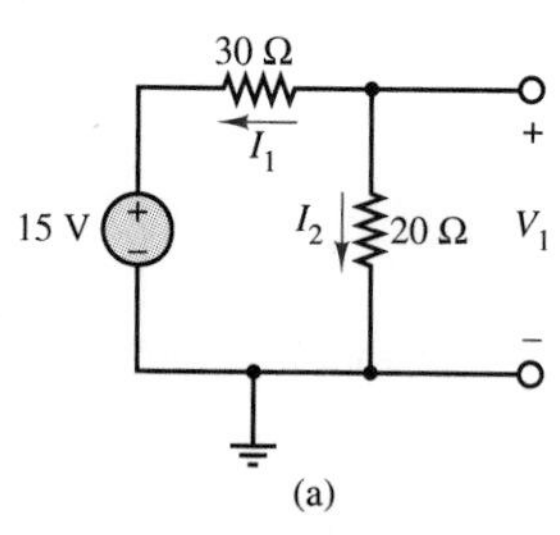

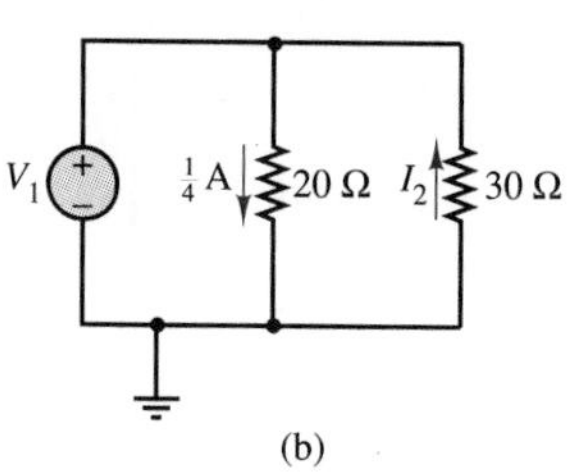

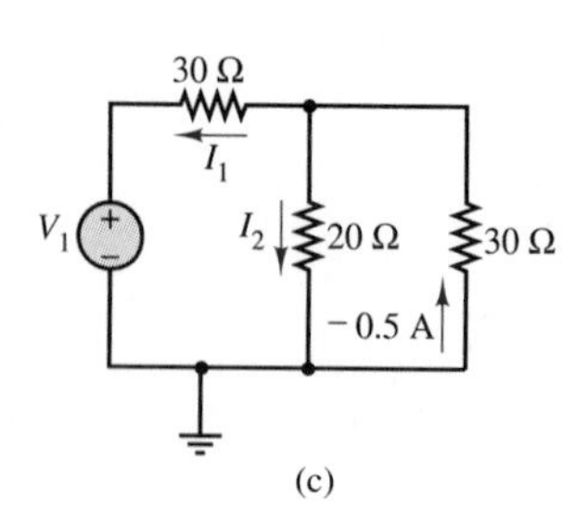

Figure P2.13

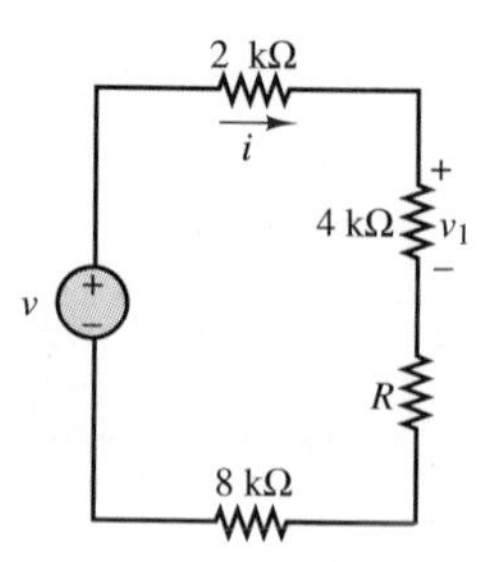

Figure P2.14

2.22 Use Kirchhoff's current law to determine the current in each of the 30-Ω resistors in the circuit of Figure P2.15.

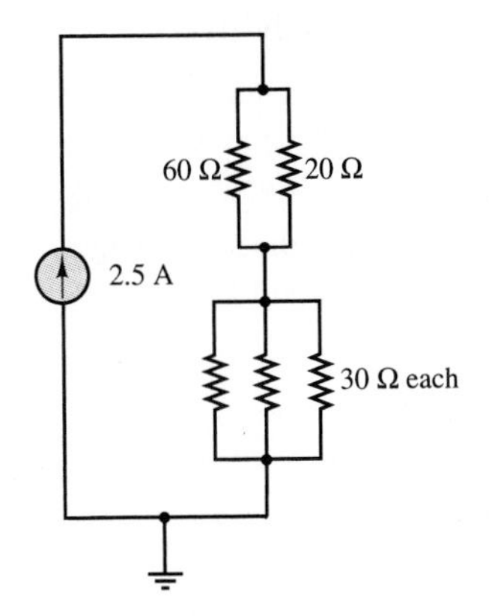

Figure P2.15

2.23 Use KCL and Ohm's Law to determine the current through each resistor in the circuit of Figure P2.16.

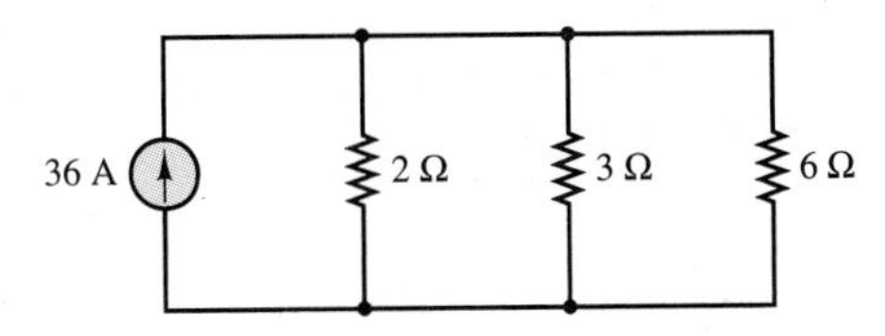

Figure P2.16

2.24 Use Ohm's Law and KVL to determine i_1 and i_2 in the circuit of Figure P2.17.

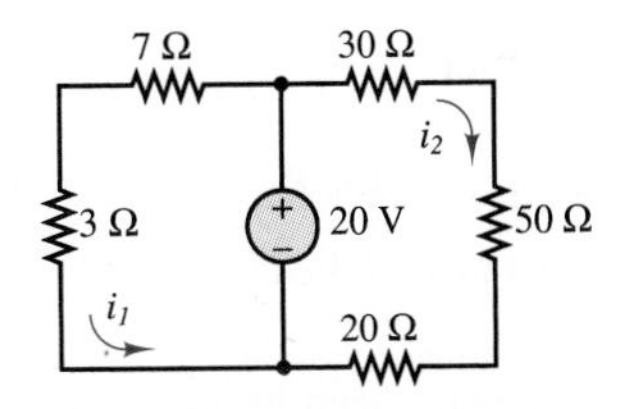

Figure P2.17

EIT 2.25 Use Ohm's Law and KCL to determine the current I_1 in the circuit of Figure P2.18.

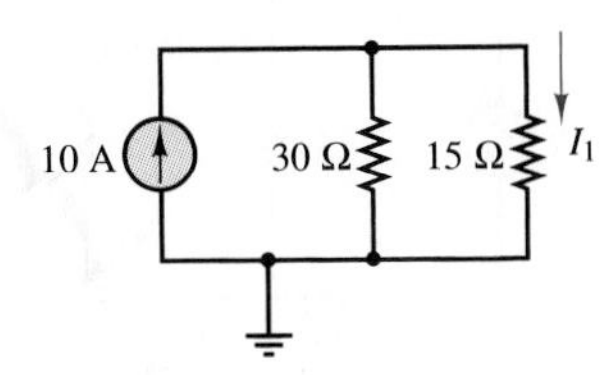

Figure P2.18

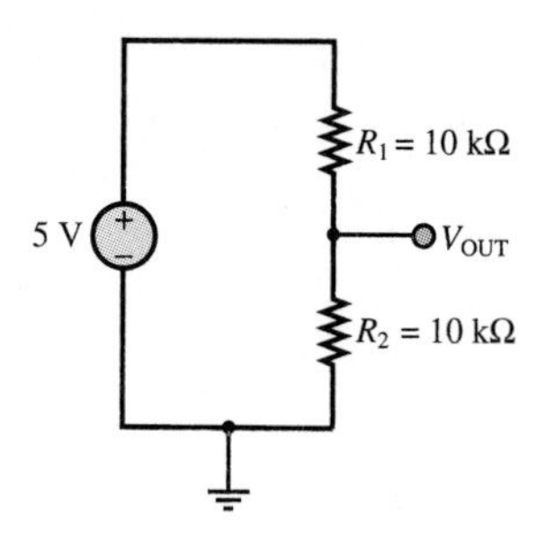

Figure P2.19

2.26 The voltage divider network of Figure P2.19 is expected to provide 2.5 V at the output. The resistors, however, may not be exactly the same; that is, their tolerances are such that the resistances may not be exactly 10 kΩ.

a. If the resistors have ±10 percent tolerance, find the worst-case output voltages.
b. Find these voltages for tolerances of ±5 percent.

2.27 For the circuits of Figure P2.20, determine the resistor values (including the power rating) necessary to achieve the indicated voltages.
Resistors are available in ⅛-, ¼-, ½-, and 1-W ratings.

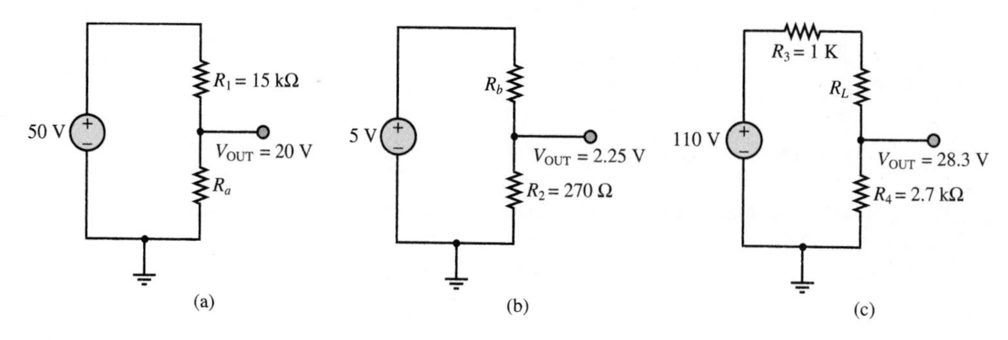

Figure P2.20

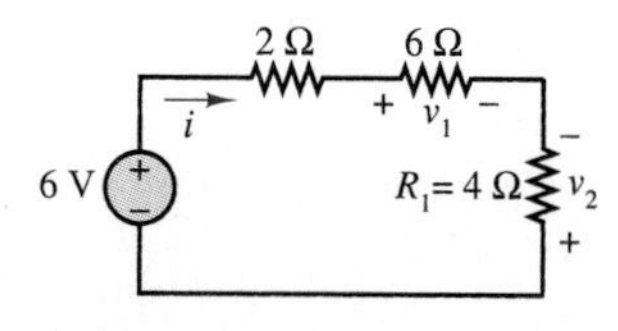

Figure P2.21

EIT 2.28 For the circuit shown in Figure P2.21, find

a. The equivalent resistance seen by the source.
b. The current, i.
c. The power delivered by the source.
d. The voltages, v_1, v_2.
e. The minimum power rating required for R_1.

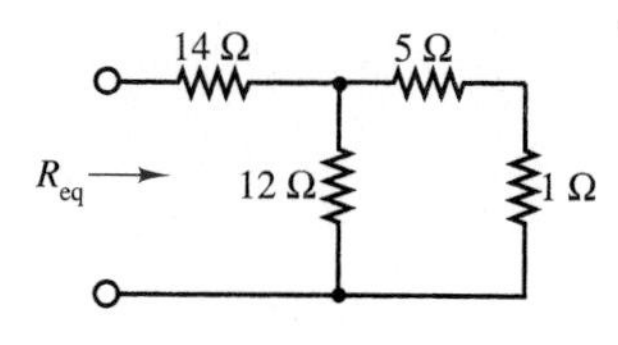

Figure P2.22

2.29 Find the equivalent resistance of the circuit of Figure P2.22 by combining resistors in series and in parallel.

2.30 Find the equivalent resistance seen by the source in Figure P2.23, and use the result to find i, i_1, and v.

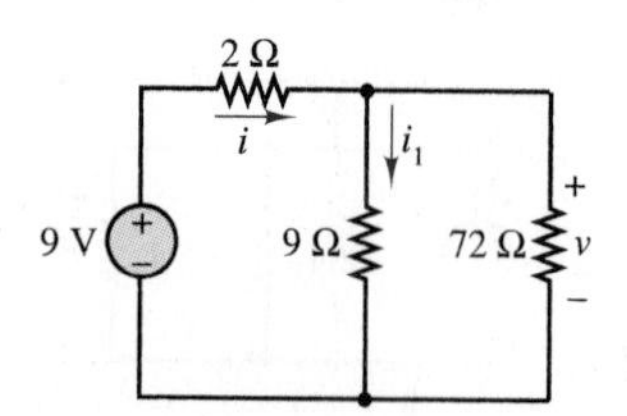

Figure P2.23

2.31 Find the equivalent resistance seen by the source and the current i in the circuit of Figure P2.24.

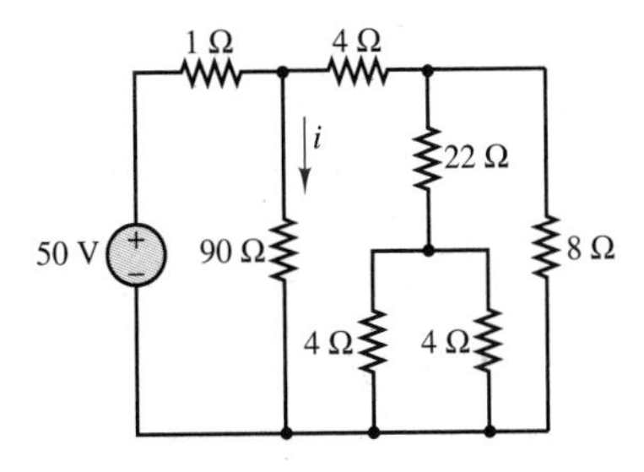

Figure P2.24

2.32 In the circuit of Figure P2.25, the power absorbed by the 15-Ω resistor is 15 W. Find R.

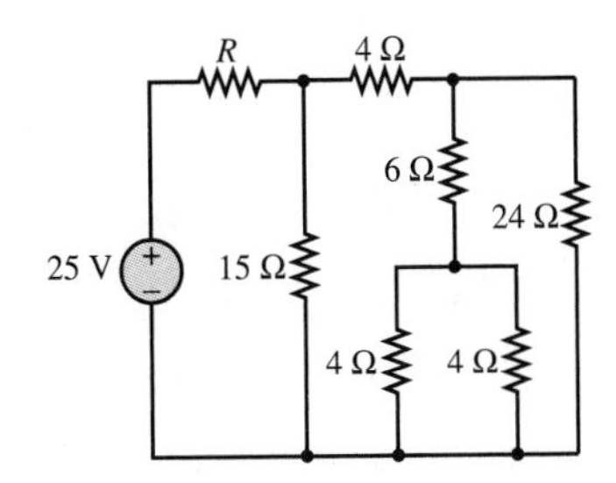

Figure P2.25

2.33 Find the equivalent resistance between terminals a and b in the circuit of Figure P2.26.

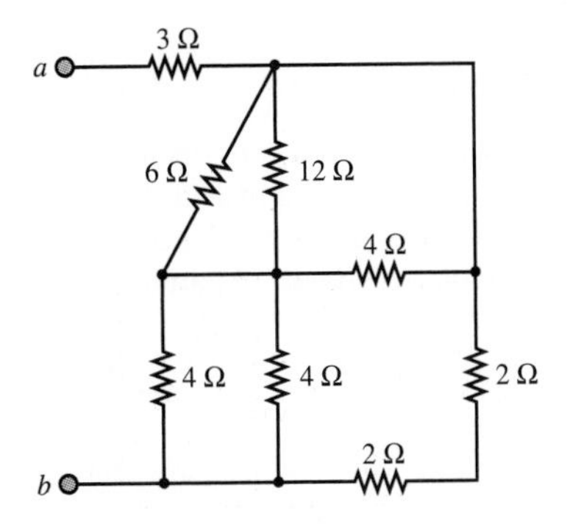

Figure P2.26

2.34 For the circuit shown in Figure P2.27:

a. Find the equivalent resistance seen by the source.
b. How much power is delivered by the source?

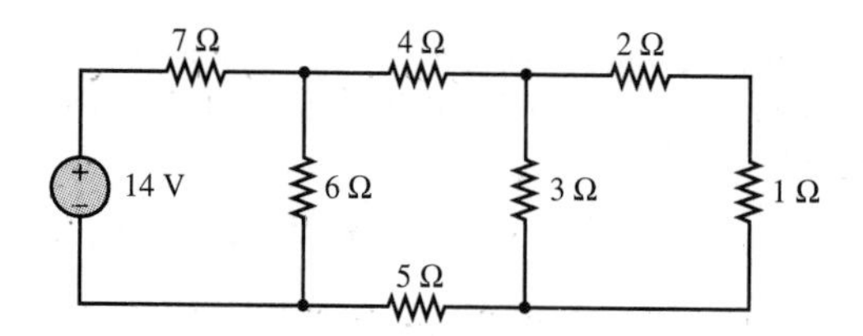

Figure P2.27

2.35 In the circuit of Figure P2.28, find the equivalent resistance looking in at terminals a and b if terminals c and d are open and again if terminals c and d are shorted together. Also, find the equivalent resistance looking in at terminals c and d if terminals a and b are open and if terminals a and b are shorted together.

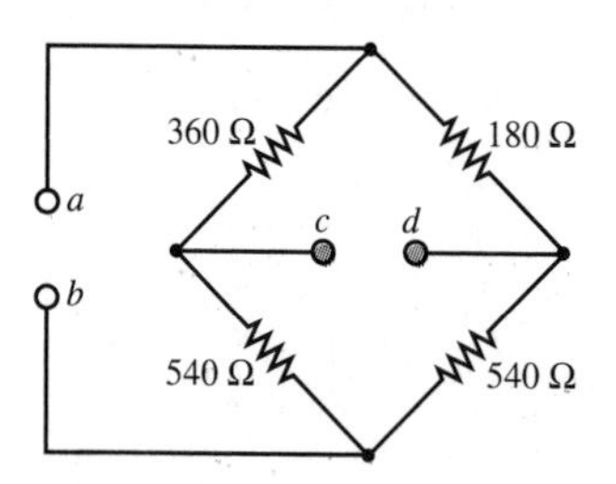

Figure P2.28

2.36 A *thermistor* is a device whose terminal resistance changes with the temperature of its surroundings. Its resistance is an exponential relationship:

$$R_{th}(T) = R_A e^{-\beta T}$$

where R_A is the terminal resistance at $T = 0°C$ and β is a material parameter.

a. If $R_A = 100\ \Omega$ and $\beta = 0.10/C°$, plot $R_{th}(T)$ versus T for $0 \le T \le 100°C$.
b. The thermistor is placed in parallel with a resistor whose value is $100\ \Omega$.
   i. Find an expression for the equivalent resistance.
   ii. Plot $R_{eq}(T)$ on the same plot you made in part a.

EIT 2.37 A certain resistor has the following nonlinear characteristic:

$$R(x) = 100e^{x}$$

where x is a displacement (in cm). The nonlinear resistor is to be used to measure the displacement x in the circuit of Figure P2.29.

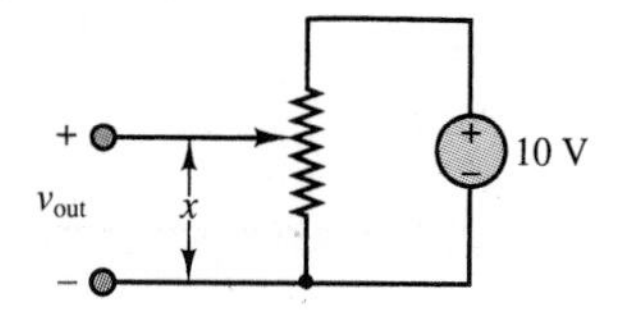

Figure P2.29

a. If the total length of the resistor is 10 cm, find an expression for $v_{out}(x)$.
b. If $v_{out} = 4$ V, what is the distance, x?

2.38 Find the currents i_1 and i_2 and the power delivered by the 2-A current source and the 10-V voltage source and the total power dissipated by the circuit of Figure P2.30. $R_1 = 32\ \Omega$, $R_2 = R_3 = 6\ \Omega$, and $R_4 = 50\ \Omega$.

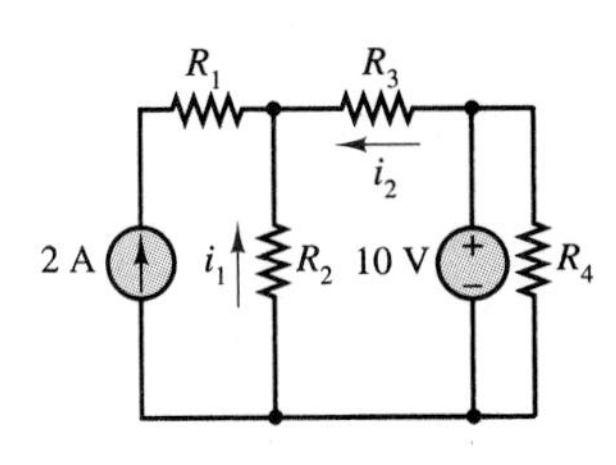

Figure P2.30

2.39 Determine the power delivered by the dependent source in the circuit of Figure P2.31.

2.40 The model shown inside the dashed box in Figure P2.32 is an idealized representation of a device called an operational amplifier. If other components are connected as shown, what is the ratio v_o/v_i?

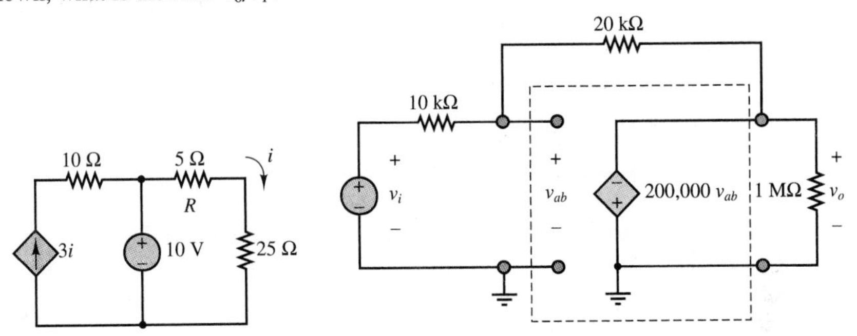

Figure P2.31

Figure P2.32

2.41 Consider the circuit shown in Figure P2.33.

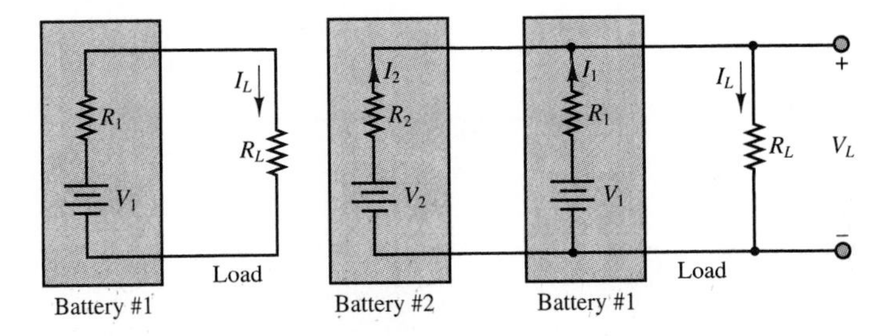

Figure P2.33

a. If $V_1 = 10.0$ V, $R_1 = 0.05\ \Omega$, and $R_L = 0.45\ \Omega$, find the load current I_L and the power dissipated by the load.

b. If we connect a second battery in parallel with battery 1 that has voltage $V_2 = 10$ V and $R_2 = 0.1\ \Omega$, will the load current I_L increase or decrease? Will the power dissipated by the load increase or decrease? By how much?

EIT 2.42 The circuit of Figure P2.34 is used to measure the internal impedance of a battery. The battery being tested is a zinc-carbon dry cell.

a. A fresh battery is being tested, and it is found that the voltage, V_{out}, is 1.64 V with the switch open and 1.63 V with the switch closed. Find the internal resistance of the battery.

b. The same battery is tested one year later, and V_{out} is found to be 1.6 V with the switch open but 0.17 V with the switch closed. Find the internal resistance of the battery.

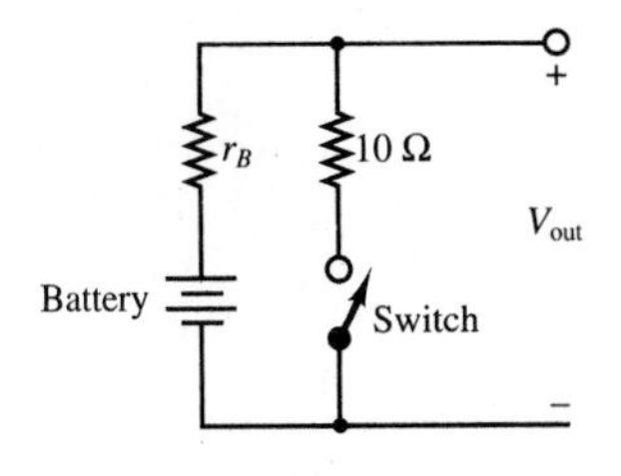

Figure P2.34

2.43 Consider the practical ammeter, diagramed in Figure P2.35, consisting of an ideal ammeter in series with a 2-kΩ resistor. The meter sees a full-scale deflection when the current through it is 50μA. If we wished to construct a multirange ammeter reading full-scale values of 1 mA, 10 mA, or 100 mA, depending on the setting of a rotary switch, what should R_1, R_2, and R_3 be?

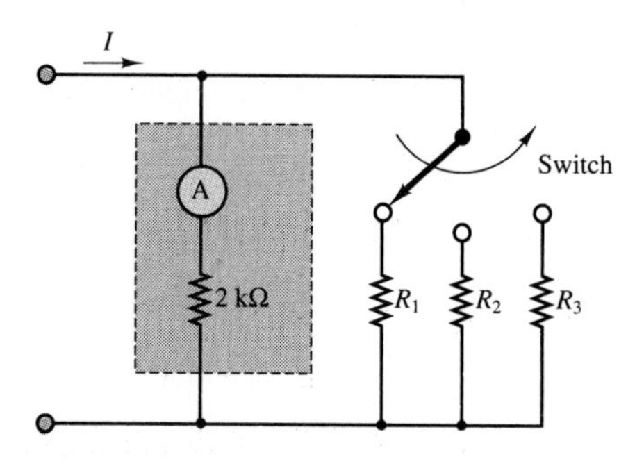

Figure P2.35

EIT 2.44 A circuit that measures the internal resistance of a practical ammeter is shown in Figure P2.36, where $R_S = 10{,}000\ \Omega$, $V_S = 10$ V, and R_p is a variable resistor that can be adjusted at will.

a. Assume that $r_a \ll 10{,}000\ \Omega$. Estimate the current i.
b. If the meter displays a current of 0.43 mA when $R_p = 7\ \Omega$, find the internal resistance of the meter, r_a.

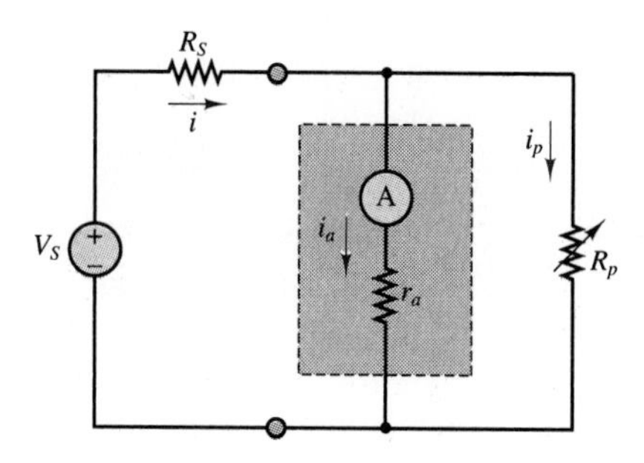

Figure P2.36

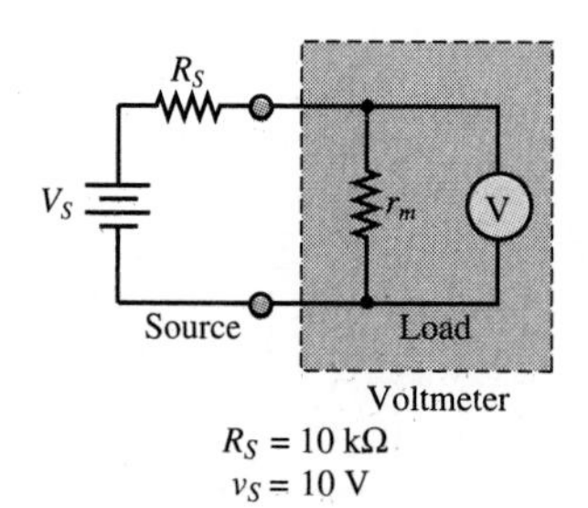

Figure P2.37

2.45 A practical voltmeter has an internal resistance r_m. What is the value of r_m if the meter reads 9.89 V when connected as shown in Figure P2.37?

2.46 Using the circuit of Figure P2.37, find the voltage that the meter reads if $V_S =$ 10 V and R_S has the following values: $R_S = 0.1r_m, 0.3r_m, 0.5r_m, r_m, 3r_m, 5r_m$, and $10r_m$. How large (or small) should the internal resistance of the meter be relative to R_S?

2.47 A voltmeter is used to determine the voltage across a resistive element in the circuit of Figure P2.38. The instrument is modeled by an ideal voltmeter in parallel with a 97-kΩ resistor, as shown. The meter is placed to measure the voltage across R_3. Let $R_1 = 10$ kΩ, $R_S = 100$ kΩ, $R_2 = 40$ kΩ, and $I_S = 90$ mA. Find the voltage across R_3 with and without the voltmeter in the circuit for the following values:

a. $R_3 = 100\ \Omega$
b. $R_3 = 1\ k\Omega$
c. $R_3 = 10\ k\Omega$
d. $R_3 = 100\ k\Omega$

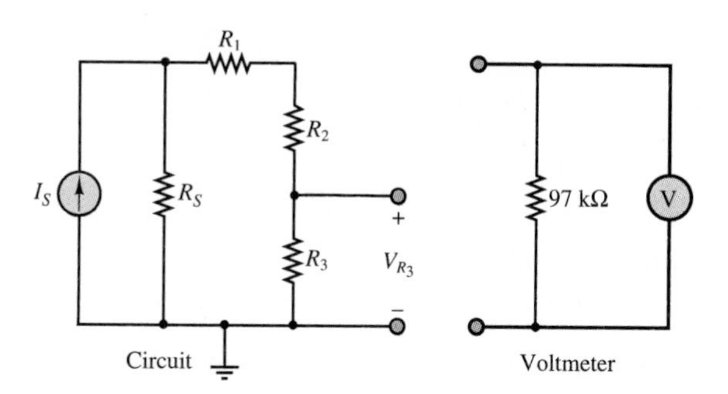

Figure P2.38

EIT 2.48 An ammeter is used as shown in Figure P2.39. The ammeter model consists of an ideal ammeter in series with a resistance. The ammeter model is placed in the branch as shown in the figure. Find the current through R_3 both with and without the ammeter in the circuit for the following values, assuming that $V_S = 10$ V, $R_S = 10\ \Omega$, $R_1 = 1\ k\Omega$, and $R_2 = 100\ \Omega$: (a) $R_3 = 1\ k\Omega$, (b) $R_3 = 100\ \Omega$, (c) $R_3 = 10\ \Omega$, (d) $R_3 = 1\ \Omega$.

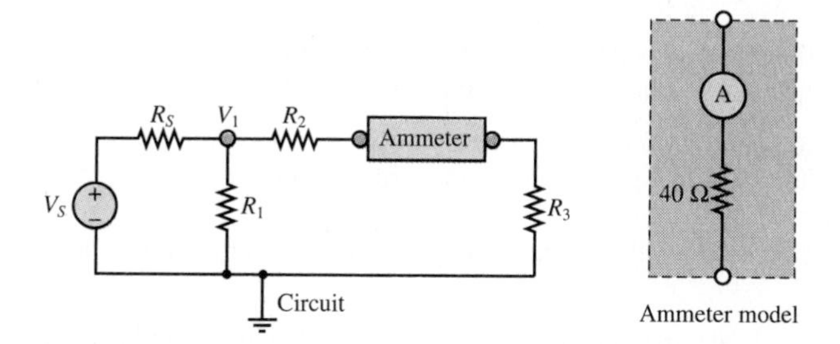

Figure P2.39

2.49 The d'Arsonval movement is a mechanism that moves on the basis of the current flowing through a meter. The movement's resistance is usually given by the voltage at which there is full-scale deflection and by the current at full-scale deflection. This movement can be used to make simple voltmeters when connected as shown in Figure P2.40. R_S is a resistance in series with the movement's internal resistance, R_m. R_S is chosen so that $V_{\text{voltmeter}}$ is equal to the desired full-scale reading when V_m is the movement's actual full-scale deflection voltage. In other words, V_m is a fraction of the external voltage, $V_{\text{voltmeter}}$. Find R_S so that the d'Arsonval movement deflects full scale when $V_{\text{voltmeter}}$ is 20 volts. V_m full scale $= 0.01$ V and I_m full scale $= 0.5$ mA.

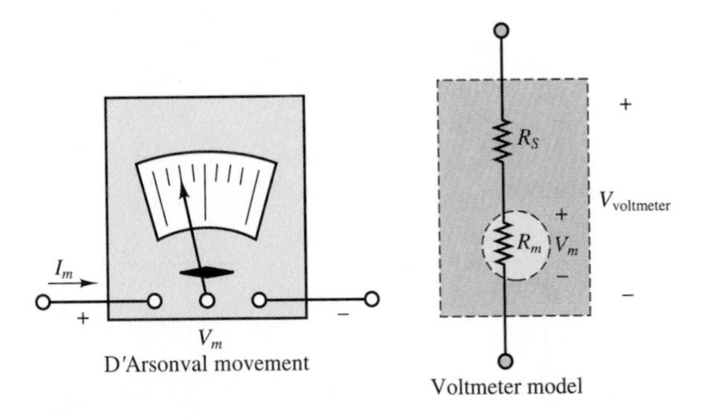

Figure P2.40

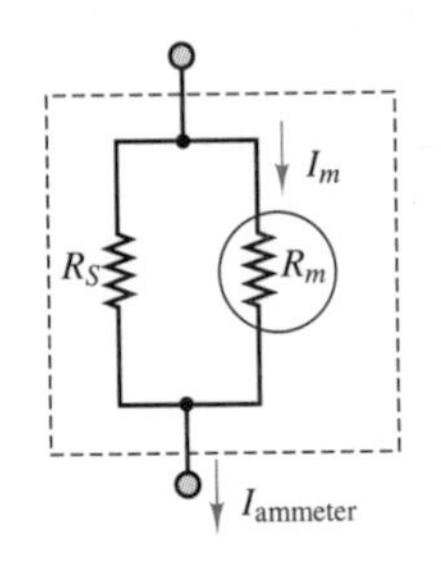

Figure P2.41

EIT 2.50 The d'Arsonval movement can be used to make a simple ammeter as shown in Figure P2.41. R_S is a resistance parallel to the movement's internal resistance, R_m. R_S is shown so that the current through R_m is equal to the full-scale deflection current while $I_{ammeter}$ is the desired full-scale deflection current. Find R_S so that we have full-scale deflection at $I_{ammeter} = 100$ mA. V_m full scale $= 0.01$ V and I_m full scale $= 0.5$ mA.

2.51

a. With reference to Figure 2.42 in the text, find a general expression for the power internally dissipated by the wattmeter as a function of the resistors R_1 and R_2, assuming that the voltmeter may be treated as ideal but the ammeter has an internal resistance r_A.
b. Compute the percent error in the power measurement due to the internal ammeter resistance if $R_1 = 50\ \Omega$, $R_2 = 10\ \text{k}\Omega$, and the internal ammeter resistance is $1.2\ \text{k}\Omega$.

2.52 With no load attached, the voltage at the terminals of a particular power supply is 25.5 V. When a 5 W load is attached, the voltage drops to 25 V.

a. Determine v_S and R_S for this non-ideal source.
b. What voltage would be measured at the terminals in the presence of a 10 Ω load resistor?
c. How much current could be drawn from this power supply under short-circuit conditions?

2.53 A 120 V electric heater has four heating coils which can be switched such that either can be used independently, or the two can be connected in series or parallel, yielding a total of four possible configurations. If the warmest setting corresponds to 1500 W power dissipation and the coolest corresponds to 200 W, determine:

a. The resistance of each of the two coils.
b. The power dissipation for each of the other two possible arrangements.

CHAPTER

3

Resistive Network Analysis

This chapter will illustrate the fundamental techniques for the analysis of resistive circuits. The methods introduced are based on the circuit laws presented in Chapter 2: Kirchhoff's and Ohm's laws. The main thrust of the chapter is to introduce and illustrate various methods of circuit analysis that will be applied throughout the book.

The first topic is the analysis of resistive circuits by the methods of mesh currents and node voltages; these are fundamental techniques, which you should master as early as possible. The second topic is a brief introduction to the principle of superposition. Section 3.5 introduces another fundamental concept in the analysis of electrical circuits: the reduction of an arbitrary circuit to equivalent circuit forms (Thévenin and Norton equivalent circuits). In this section it will be shown that it is generally possible to reduce all linear circuits to one of two equivalent forms, and that any linear circuit analysis problem can be reduced to a simple voltage or current divider problem. The Thévenin and Norton equivalent representations of electrical circuits naturally lead to the description of electrical circuits in terms of sources and loads. This notion, in turn, leads to the analysis of the transfer of power between a source and a load, and of the phenomenon of source loading. Finally, some graphical and numerical techniques are introduced for the analysis of nonlinear circuit elements.

Upon completing this chapter, you should have developed confidence in your ability to compute numerical solutions for a wide range of resistive circuits. Good familiarity with the techniques illustrated in this chapter will greatly simplify the study of AC circuits in Chapter 4. The objective of the chapter is to develop a solid understanding of the following topics:

- Node voltage and mesh current analysis.
- The principle of superposition.
- Thévenin and Norton equivalent circuits.
- Numerical and graphical (load-line) analysis of nonlinear circuit elements.

3.1 THE NODE VOLTAGE METHOD

In the node voltage method, we assign the node voltages v_a and v_b; the branch current flowing from a to b is then expressed in terms of these node voltages.

$$i = \frac{v_a - v_b}{R}$$

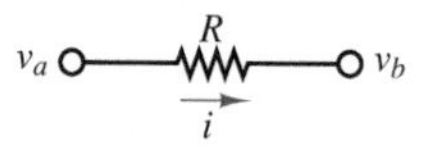

Figure 3.1 Branch current formulation in nodal analysis

Chapter 2 introduced the essential elements of network analysis, paving the way for a systematic treatment of the analysis methods that will be introduced in this chapter. You are by now familiar with the application of the three fundamental laws of network analysis: KCL, KVL, and Ohm's law; these will be employed to develop a variety of solution methods that can be applied to linear resistive circuits. The material presented in the following sections presumes good understanding of Chapter 2. You can resolve many of the doubts and questions that may occasionally arise by reviewing the material presented in the preceding chapter.

Node voltage analysis is the most general method for the analysis of electrical circuits. In this section, its application to linear resistive circuits will be illustrated. The **node voltage method** is based on defining the voltage at each node as an independent variable. One of the nodes is selected as a **reference node** (usually—but not necessarily—ground), and each of the other node voltages is referenced to this node. Once each node voltage is defined, Ohm's law may be applied between any two adjacent nodes in order to determine the current flowing in each branch. In the node voltage method, *each branch current is expressed in terms of one or more node voltages;* thus, currents do not explicitly enter into the equations. Figure 3.1 illustrates how one defines branch currents in this method. You may recall a similar description given in Chapter 2.

By KCL: $i_1 = i_2 + i_3$. In the node voltage method, we express KCL by

$$\frac{v_a - v_b}{R_1} = \frac{v_b - v_c}{R_2} + \frac{v_b - v_d}{R_3}$$

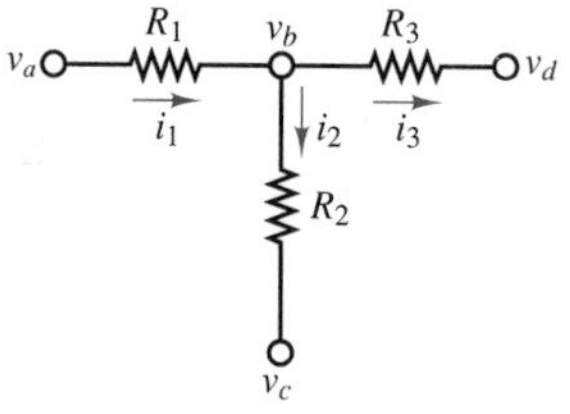

Figure 3.2 Use of KCL in nodal analysis

Once each branch current is defined in terms of the node voltages, Kirchhoff's current law is applied at each node. The particular form of KCL employed in the nodal analysis equates the sum of the currents into the node to the sum of the currents leaving the node:

$$\sum i_{\text{in}} = \sum i_{\text{out}} \tag{3.1}$$

Figure 3.2 illustrates this procedure.

The systematic application of this method to a circuit with n nodes would lead to writing n linear equations. However, one of the node voltages is the reference voltage and is therefore already known, since it is usually assumed to be zero (recall that the choice of reference voltage is dictated mostly by convenience, as explained in Chapter 2). Thus, we can write $n - 1$ *independent linear equations* in the $n - 1$ independent variables (the node voltages). Nodal analysis provides the minimum number of equations required to solve the circuit, since any branch voltage or current may be determined from knowledge of nodal voltages. At this stage, you might wish to review Example 2.15, to verify that, indeed, knowledge of the node voltages is sufficient to solve for any other current or voltage in the circuit.

The nodal analysis method may also be defined as a sequence of steps, as outlined in the following box:

Node Voltage Analysis Method

1. Select a reference node (usually ground). All other node voltages will be referenced to this node.
2. Define the remaining $n - 1$ node voltages as the independent variables.
3. Apply KCL at each of the $n - 1$ nodes, expressing each current in terms of the adjacent node voltages.
4. Solve the linear system of $n - 1$ equations in $n - 1$ unknowns.

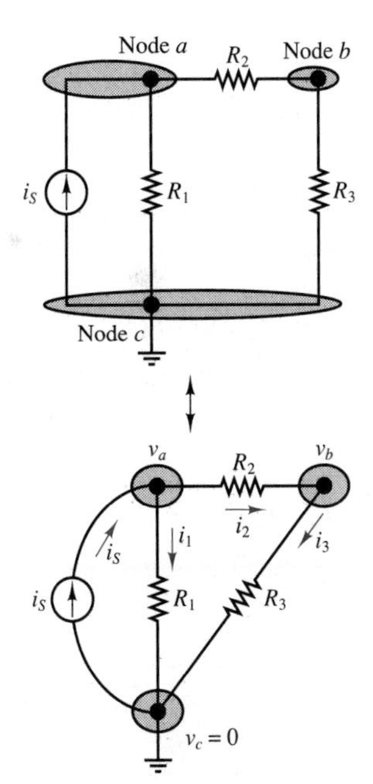

Figure 3.3 Illustration of nodal analysis

Following the procedure outlined in the box guarantees that the correct solution to a given circuit will be found, provided that the nodes are properly identified and KCL is applied consistently. As an illustration of the method, consider the circuit shown in Figure 3.3. The circuit is shown in two different forms to illustrate equivalent graphical representations of the same circuit. The bottom circuit leaves no question where the nodes are. The direction of current flow is selected arbitrarily (assuming that i_S is a positive current). Application of KCL at node a yields:

$$i_S = i_1 + i_2 \tag{3.2}$$

whereas, at node b,

$$i_2 = i_3 \tag{3.3}$$

It is instructive to verify (at least the first time the method is applied) that it is not necessary to apply KCL at the reference node. The equation obtained at node c,

$$i_1 + i_3 = i_S \tag{3.4}$$

is not independent of equations 3.2 and 3.3; in fact, it may be obtained by adding the equations obtained at nodes a and b (verify this, as an exercise). This observation confirms the statement made earlier:

In a circuit containing n nodes, we can write at most $n - 1$ independent equations.

Now, in applying the node voltage method, the currents i_1, i_2, and i_3 are expressed as functions of v_a, v_b, and v_c, the independent variables. Ohm's law requires that i_1, for example, be given by

$$i_1 = \frac{v_a - v_c}{R_1} \tag{3.5}$$

since it is the potential difference, $v_a - v_c$, across R_1 that causes the current i_1 to flow from node a to node c. Similarly,

$$\begin{aligned} i_2 &= \frac{v_a - v_b}{R_2} \\ i_3 &= \frac{v_b - v_c}{R_3} \end{aligned} \tag{3.6}$$

Substituting the expression for the three currents in the nodal equations (equations 3.2 and 3.3), we obtain the following relationships:

$$i_S = \frac{v_a}{R_1} + \frac{v_a - v_b}{R_2} \tag{3.7}$$

$$\frac{v_a - v_b}{R_2} = \frac{v_b}{R_3} \tag{3.8}$$

Equations 3.7 and 3.8 may be obtained directly from the circuit, with a little practice. Note that these equations may be solved for v_a and v_b, assuming that i_S, R_1, R_2, and R_3 are known. The same equations may be reformulated as follows:

$$\begin{aligned} \left(\frac{1}{R_1} + \frac{1}{R_2}\right)v_a + \left(-\frac{1}{R_2}\right)v_b &= i_S \\ \left(-\frac{1}{R_2}\right)v_a + \left(\frac{1}{R_2} + \frac{1}{R_3}\right)v_b &= 0 \end{aligned} \tag{3.9}$$

The following examples further illustrate the application of the method.

Example 3.1

Solve for all unknown currents and voltages in the circuit of Figure 3.4, using nodal analysis.

Solution:

The circuit of Figure 3.4 is shown again in Figure 3.5, with a graphical indication of how KCL will be applied to determine the nodal equations. Note that we have selected to ground the lower part of the circuit, resulting in a reference voltage of zero at that node.

Applying KCL at node 1 yields:

$$0.01 = \frac{v_1 - 0}{1{,}000} + \frac{v_1 - v_2}{10{,}000} + \frac{v_1 - v_2}{2{,}000}$$

whereas at node 2 we have

$$\frac{v_1 - v_2}{10{,}000} + \frac{v_1 - v_2}{2{,}000} = \frac{v_2}{2{,}000} + 0.05$$

Note how each current has been expressed directly in terms of the node voltages. Since the direction of the currents has been (arbitrarily) assigned in Figure 3.5, writing the nodal equations becomes no more than a simple application of KCL and Ohm's law.

The next task is to write the equations in a more systematic fashion. From node 1, we obtain

$$\left(\frac{1}{1{,}000} + \frac{1}{10{,}000} + \frac{1}{2{,}000}\right)v_1 + \left(-\frac{1}{10{,}000} - \frac{1}{2{,}000}\right)v_2 = 0.01$$

while at node 2 we have

$$\left(-\frac{1}{10{,}000} - \frac{1}{2{,}000}\right)v_1 + \left(\frac{1}{10{,}000} + \frac{1}{2{,}000} + \frac{1}{2{,}000}\right)v_2 = -0.05$$

With some manipulation, the equations finally lead to the following form:

$$1.6v_1 - 0.6v_2 = 10$$
$$-0.6v_1 + 1.1v_2 = -50$$

These equations may be solved simultaneously to obtain

$$v_1 = -13.57 \text{ V}$$
$$v_2 = -52.86 \text{ V}$$

Knowing the node voltages, we can determine each of the branch currents and voltages in the circuit. For example, the current through the 10-kΩ resistor is given by:

$$i_{10\text{ k}\Omega} = \frac{v_1 - v_2}{10{,}000} = 3.93 \text{ mA}$$

indicating that the initial (arbitrary) choice of direction for this current was the same as the actual direction of current flow. As another example, consider the current through the 1-kΩ resistor:

$$i_{1\text{ k}\Omega} = \frac{v_1}{1{,}000} = -13.57 \text{ mA}$$

In this case, the current is negative, indicating that current actually flows from ground to node 1, as it should, since the voltage at node 1 is negative with respect to ground. You may continue the branch-by-branch analysis started in this example to verify that the solution obtained in the example is indeed correct.

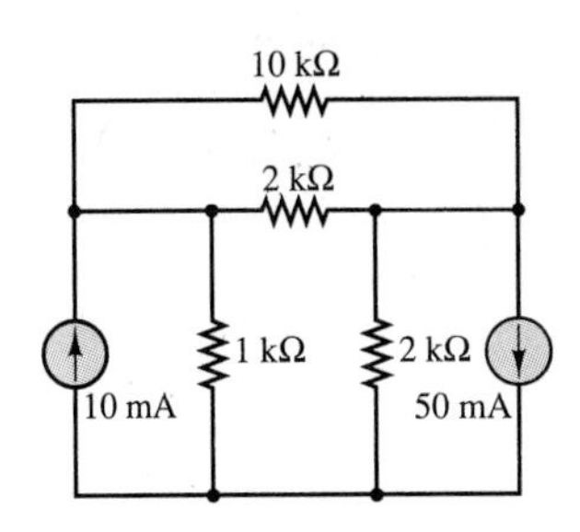

Figure 3.4

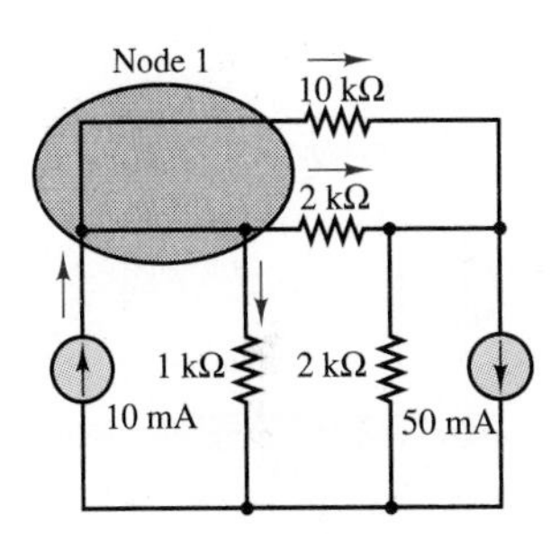

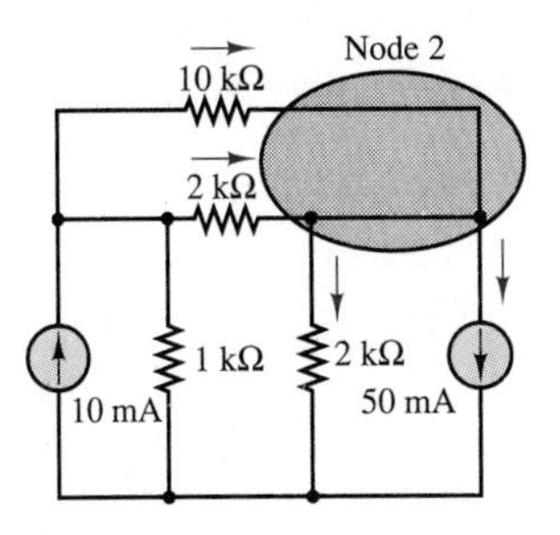

Figure 3.5

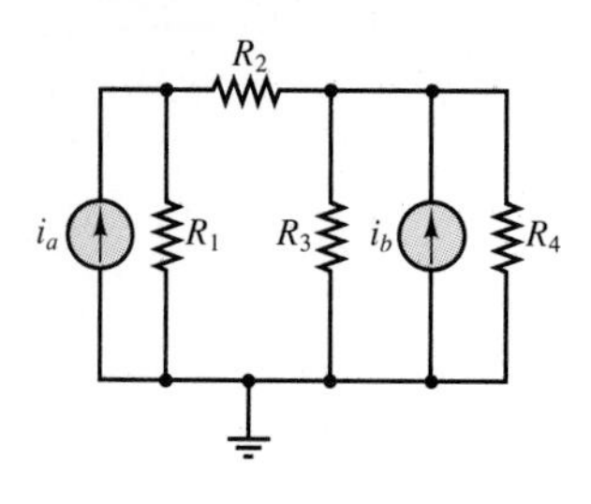

Figure 3.6

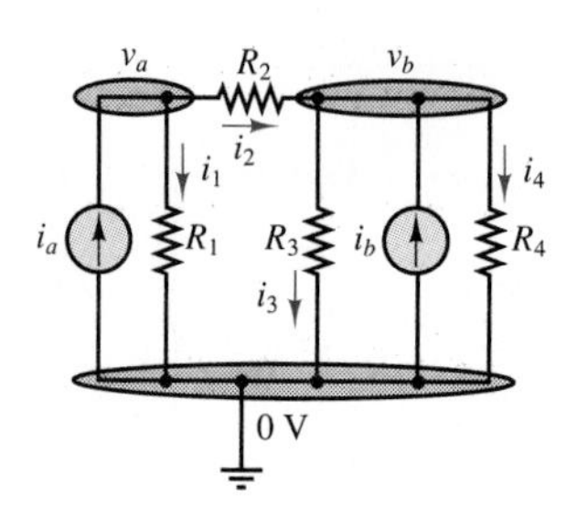

Figure 3.7

Example 3.2

Write the node voltage equations for the circuit shown in Figure 3.6. Assume that $R_1 = 1\ \text{k}\Omega$, $R_2 = 500\ \Omega$, $R_3 = 2.2\ \text{k}\Omega$, $R_4 = 4.7\ \text{k}\Omega$, $i_a = 1$ mA, and $i_b = 2$ mA.

Solution:

To write the node equations, we start by selecting the reference node (step 1). Figure 3.7 illustrates that two nodes remain after the selection of the reference node. Let us label these a and b and define voltages v_a and v_b (step 2).

Next, we apply KCL at each of the nodes, a and b (step 3):

$$i_a = \frac{v_a}{R_1} + \frac{v_a - v_b}{R_2} \qquad \text{(node } a\text{)}$$

$$\frac{v_a - v_b}{R_2} + i_b = \frac{v_b}{R_3} + \frac{v_b}{R_4} \qquad \text{(node } b\text{)}$$

and rewrite the equations to obtain a linear system:

$$\left(\frac{1}{R_1} + \frac{1}{R_2}\right)v_a + \left(-\frac{1}{R_2}\right)v_b = i_a$$

$$\left(-\frac{1}{R_2}\right)v_a + \left(\frac{1}{R_2} + \frac{1}{R_3} + \frac{1}{R_4}\right)v_b = i_b$$

Substituting the numerical values in these equations, we get

$$3 \times 10^{-3} v_a - 2 \times 10^{-3} v_b = 1 \times 10^{-3}$$

$$-2 \times 10^{-3} v_a + 2.67 \times 10^{-3} v_b = 2 \times 10^{-3}$$

or

$$3v_a - 2v_b = 1$$

$$-2v_a + 2.67v_b = 2$$

The solution $v_a = 1.667$ V, $v_b = 2$ V may then be obtained by solving the system of equations.

Example 3.3

The system of equations generated in Example 3.2 may also be solved by using linear algebra methods, by recognizing that the system of equations can be written as:

$$\begin{bmatrix} 3 & -2 \\ -2 & 2.67 \end{bmatrix} \begin{bmatrix} v_a \\ v_b \end{bmatrix} = \begin{bmatrix} 1 \\ 2 \end{bmatrix}$$

By using Cramer's rule (see Appendix A), the solution for the two unknown variables, v_a and v_b, can be written as follows:

$$v_a = \frac{\begin{vmatrix} 1 & -2 \\ 2 & 2.67 \end{vmatrix}}{\begin{vmatrix} 3 & -2 \\ -2 & 2.67 \end{vmatrix}} = \frac{(1)(2.67) - (-2)(2)}{(3)(2.67) - (-2)(-2)} = \frac{6.67}{4} = 1.667$$

$$v_b = \frac{\begin{vmatrix} 3 & 1 \\ -2 & 2 \end{vmatrix}}{\begin{vmatrix} 3 & -2 \\ -2 & 2.67 \end{vmatrix}} = \frac{(3)(2) - (-2)(1)}{(3)(2.67) - (-2)(-2)} = \frac{8}{4} = 2$$

The result is the same as in Example 3.2.

Nodal Analysis with Voltage Sources

It would appear from the examples just shown that the node voltage method is very easily applied when current sources are present in a circuit. This is, in fact, the case, since current sources are directly accounted for by KCL. Some confusion occasionally arises, however, when voltage sources are present in a circuit analyzed by the node voltage method. In fact, the presence of voltage sources actually simplifies the calculations. To further illustrate this point, consider the circuit of Figure 3.8. Note immediately that one of the node voltages is known already! The voltage at node *a* is forced to be equal to that of the voltage source; that is, $v_a = v_S$. Thus, only two nodal equations will be needed, at nodes *b* and *c*:

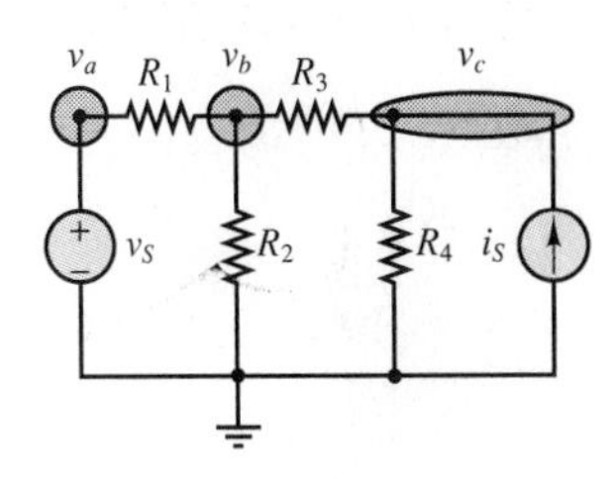

Figure 3.8 Nodal analysis with voltage sources

$$\frac{v_S - v_b}{R_1} = \frac{v_b}{R_2} + \frac{v_b - v_c}{R_3} \quad \text{(node } b\text{)}$$

$$\frac{v_b - v_c}{R_3} + i_S = \frac{v_c}{R_4} \quad \text{(node } c\text{)} \qquad \textbf{(3.10)}$$

Rewriting these equations, we obtain:

$$\left(\frac{1}{R_1} + \frac{1}{R_2} + \frac{1}{R_3}\right)v_b + \left(-\frac{1}{R_3}\right)v_c = \frac{v_S}{R_1}$$

$$\left(-\frac{1}{R_3}\right)v_b + \left(\frac{1}{R_3} + \frac{1}{R_4}\right)v_c = i_S \qquad \textbf{(3.11)}$$

Note how the term v_S/R_1 on the right-hand side of the first equation is really a current, as is dimensionally required by the nature of the node equations.

Example 3.4

For the circuit of Figure 3.9, find the voltages at node *a* and node *b*.

Solution:

Applying KCL at node *a* yields:

$$-0.002 = \frac{v_a - 0}{1{,}000} + \frac{v_a - v_b}{2{,}000}$$

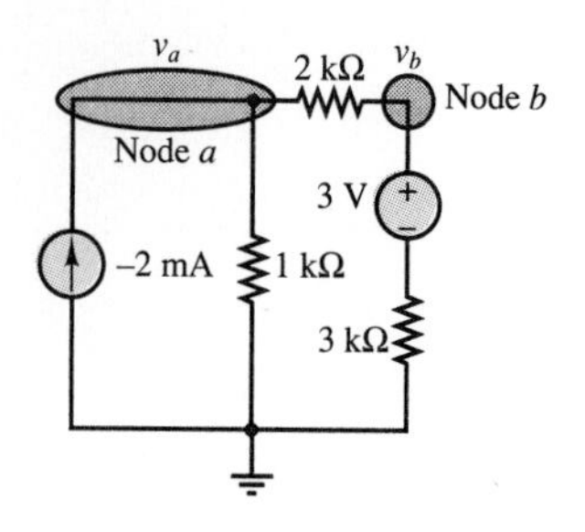

Figure 3.9

whereas at node b, we have

$$\frac{v_a - v_b}{2{,}000} = \frac{v_b - 3}{3{,}000}$$

Reformulating the last two equations, we derive the following system:

$$1.5v_a - 0.5v_b = -2$$
$$-0.5v_a + 0.833v_b = 1$$

Solving the last set of equations, we obtain the following values:

$$v_a = -1.167 \text{ V} \qquad \text{and} \qquad v_b = 0.5 \text{ V}$$

DRILL EXERCISES

3.1 Find the current i_L in the circuit shown, using the node voltage method.

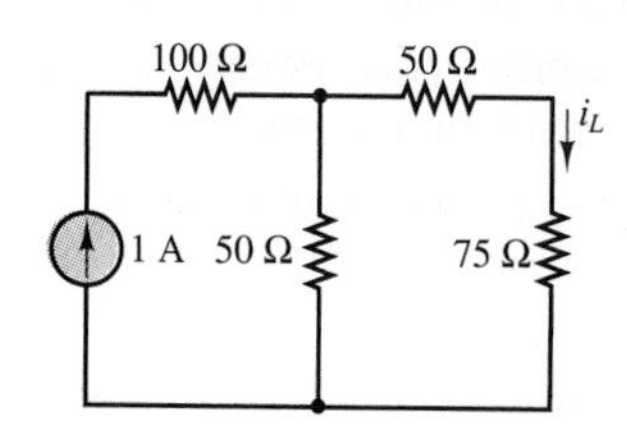

3.2 Find the voltage v_x by the node voltage method for the circuit shown.

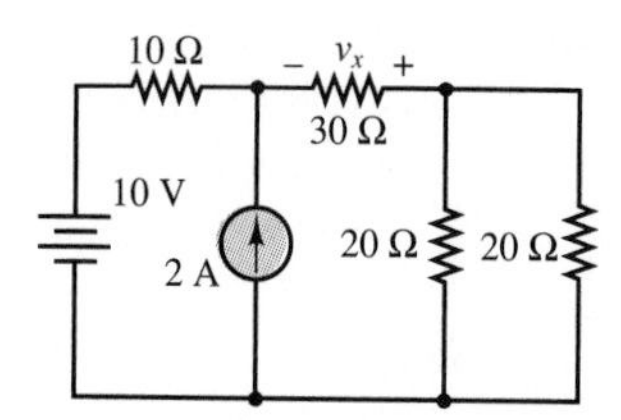

3.3 Show that the answer to Example 3.2 is correct by applying KCL at one or more nodes.

3.2 THE MESH CURRENT METHOD

The second method of circuit analysis discussed in this chapter, which is in many respects analogous to the method of node voltages, employs **mesh currents** as the independent variables. The idea is to write the appropriate number of independent equations, using mesh currents as the independent variables. Analysis by mesh currents consists of defining the currents around the individual meshes as the independent variables. Subsequent application of Kirchhoff's voltage law around each mesh provides the desired system of equations.

In the mesh current method, we observe that a current flowing through a resistor in a specified direction defines the polarity of the voltage across the resistor, as illustrated in Figure 3.10, and that the sum of the voltages around a closed circuit must equal zero, by KVL. Once a convention is established regarding the direction of current flow around a mesh, simple application of KVL provides the desired equation. Figure 3.11 illustrates this point.

The number of equations one obtains by this technique is equal to the number of meshes in the circuit. All branch currents and voltages may subsequently be obtained from the mesh currents, as will presently be shown. Since meshes are easily identified in a circuit, this method provides a very efficient and systematic procedure for the analysis of electrical circuits. The following box outlines the procedure used in applying the mesh current method to a linear circuit.

The current i, defined as flowing from left to right, establishes the polarity of the voltage across R.

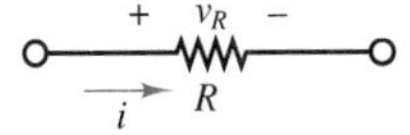

Figure 3.10 Basic principle of mesh analysis

Once the direction of current flow has been selected, KVL requires that $v_1 = v_2 + v_3$.

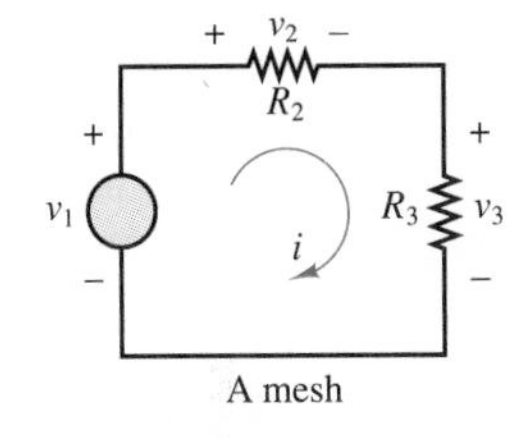

Figure 3.11 Use of KVL in mesh analysis

Mesh Current Analysis Method

1. Define each mesh current consistently. We shall always define mesh currents clockwise, for convenience.
2. Apply KVL around each mesh, expressing each voltage in terms of one or more mesh currents.
3. Solve the resulting linear system of equations with mesh currents as the independent variables.

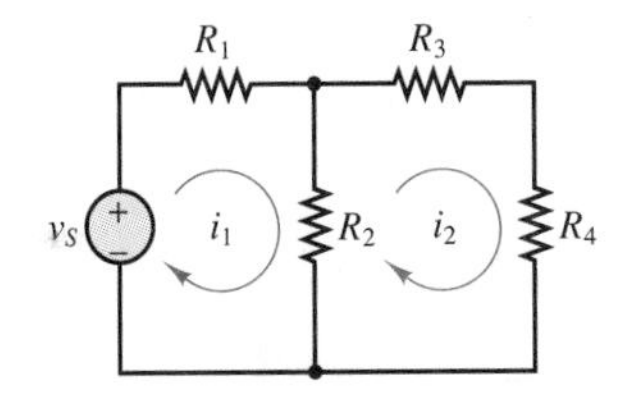

Figure 3.12 A two-mesh circuit

In mesh analysis, it is important to be consistent in choosing the direction of current flow. To avoid confusion in writing the circuit equations, mesh currents will be defined exclusively clockwise when we are using this method. To illustrate the mesh current method, consider the simple two-mesh circuit shown in Figure 3.12. This circuit will be used to generate two equations in the two unknowns, the mesh currents i_1 and i_2. It is instructive to first consider each mesh by itself. Beginning with mesh 1, note that the voltages around the mesh have been assigned in Figure 3.13 according to the direction of the mesh current, i_1. Recall that as long as signs are assigned consistently, an arbitrary direction may be assumed for any current in a circuit; if the resulting numerical answer for the current is negative, then the chosen reference direction is opposite to the direction of actual current flow. Thus, one need not be concerned about the actual direction of current flow in mesh analysis, once the directions of the mesh currents have been assigned. The correct solution will result, eventually.

According to the sign convention, then, the voltages v_1 and v_2 are defined as shown in Figure 3.13. Now, it is important to observe that while mesh current i_1 is equal to the current flowing through resistor R_1 (and is therefore also the branch current through R_1), it is not equal to the current through R_2. The branch

Mesh 1: KVL requires that $v_S = v_1 + v_2$, where $v_1 = i_1 R_1$, $v_2 = (i_1 - i_2)R_1$.

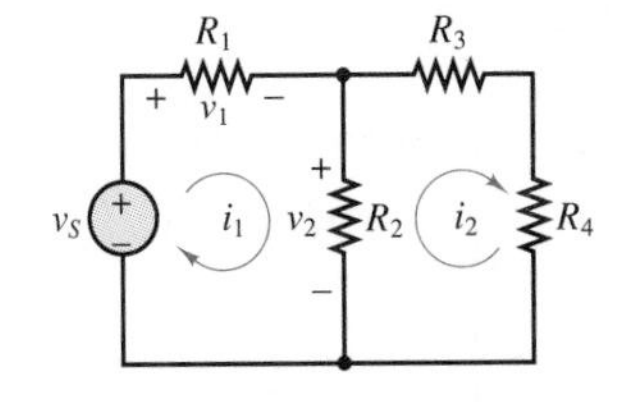

Figure 3.13 Assignment of currents and voltages around mesh 1

current through R_2 is the difference between the two mesh currents, $i_1 - i_2$. Thus, since the polarity of the voltage v_2 has already been assigned, according to the convention discussed in the previous paragraph, it follows that the voltage v_2 is given by:

$$v_2 = (i_1 - i_2)R_2 \tag{3.12}$$

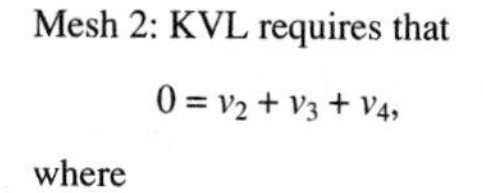

Mesh 2: KVL requires that

$0 = v_2 + v_3 + v_4,$

where

$v_2 = (i_2 - i_1)R_2,$

$v_3 = i_2R_3,$

$v_4 = i_2R_4$

Finally, the complete expression for mesh 1 is:

$$v_S = i_1R_1 + (i_1 - i_2)R_2 \tag{3.13}$$

The same line of reasoning applies to the second mesh. Figure 3.14 depicts the voltage assignment around the second mesh, following the clockwise direction of mesh current i_2. The mesh current i_2 is also the branch current through resistors R_3 and R_4; however, the current through the resistor that is shared by the two meshes, R_2, is now equal to $(i_2 - i_1)$, and the voltage across this resistor is

$$v_2 = (i_2 - i_1)R_2 \tag{3.14}$$

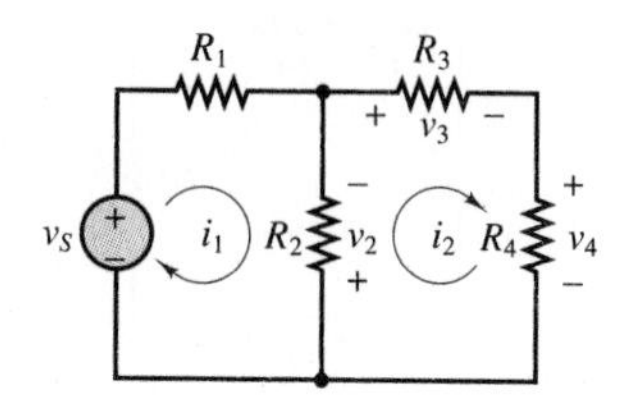

Figure 3.14 Assignment of currents and voltages around mesh 2

and the complete expression for mesh 2 is

$$0 = (i_2 - i_1)R_2 + i_2R_3 + i_2R_4 \tag{3.15}$$

Why is the expression for v_2 obtained in equation 3.14 different from equation 3.12? The reason for this apparent discrepancy is that the voltage assignment for each mesh was dictated by the (clockwise) mesh current. Thus, since the mesh currents flow through R_2 in opposing directions, the voltage assignments for v_2 in the two meshes will also be opposite. This is perhaps a potential source of confusion in applying the mesh current method; you should be very careful to carry out the assignment of the voltages around each mesh separately.

Combining the equations for the two meshes, we obtain the following system of equations:

$$\begin{aligned} (R_1 + R_2)i_1 - R_2i_2 &= v_S \\ -R_2i_1 + (R_2 + R_3 + R_4)i_2 &= 0 \end{aligned} \tag{3.16}$$

These equations may be solved simultaneously to obtain the desired solution, namely, the mesh currents, i_1 and i_2. You should verify that knowledge of the mesh currents permits determination of all the other voltages and currents in the circuit. The following examples further illustrate some of the details of this method.

Example 3.5

In this example, we analyze a simple two-mesh circuit, shown in Figure 3.15, by means of the mesh current method.

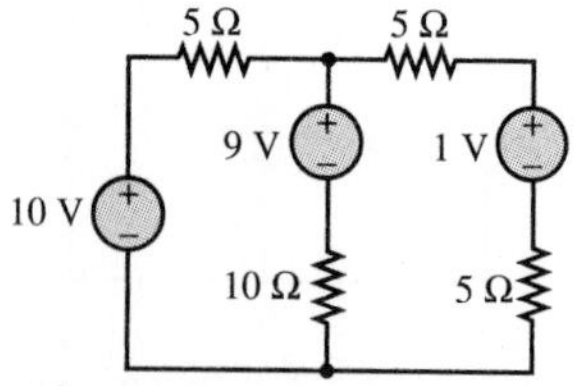

Figure 3.15

Solution:

The circuit of Figure 3.15 will yield two equations in two unknowns, i_1 and i_2. It is instructive to consider each mesh separately in writing the mesh equations; to this end, Figure 3.16 depicts the appropriate voltage assignments around the two meshes, based on the assumed directions of the mesh currents. From Figure 3.16, we write the mesh equations:

$$5i_1 + 10(i_1 - i_2) = 10 - 9$$
$$5i_2 + 5i_2 + 10(i_2 - i_1) = 9 - 1$$

Rearranging the linear system of the equation, we obtain

$$15i_1 - 10i_2 = 1$$
$$-10i_1 + 20i_2 = 8$$

which can be solved to obtain i_1 and i_2:

$$i_1 = 0.5 \text{ A} \quad \text{and} \quad i_2 = 0.65 \text{ A}$$

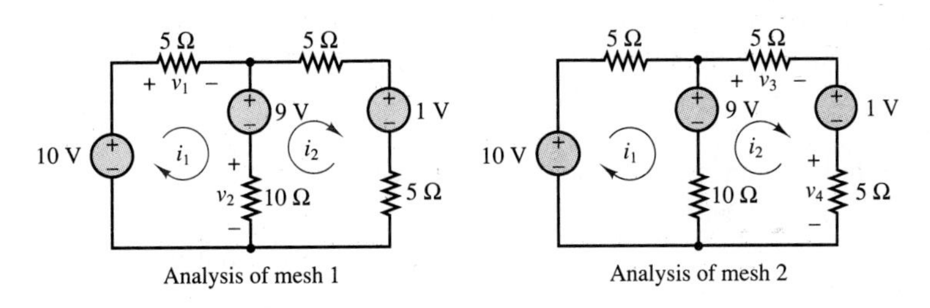

Figure 3.16

Example 3.6

Write the mesh equations for the circuit of Figure 3.17.

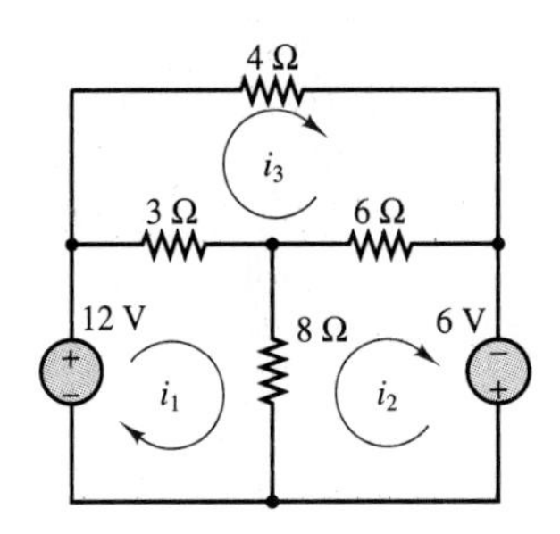

Figure 3.17

Solution:

Starting from mesh 1, we apply KVL using the direction of mesh current i_1 as our reference for assignment of voltages:

$$12 = 3(i_1 - i_3) + 8(i_1 - i_2)$$

For mesh 2, we obtain:

$$6 = 6(i_2 - i_3) + 8(i_2 - i_1)$$

and, finally, for mesh 3:

$$0 = 3(i_3 - i_1) + 6(i_3 - i_2) + 4i_3$$

These equations can be rearranged in standard form to obtain

$$(3 + 8)i_1 - 8i_2 - 3i_3 = 12$$
$$-8i_1 + (6 + 8)i_2 - 6i_3 = 6$$
$$-3i_1 - 6i_2 + (3 + 6 + 4)i_3 = 0$$

You may verify that KVL holds around any one of the meshes, as a test to check that the answer is indeed correct.

A comparison of this result with the analogous result obtained by the node voltage method reveals that we are using Ohm's law in conjunction with KVL (in contrast with the use of KCL in the node voltage method) to determine the minimum set of equations required to solve the circuit.

Mesh Analysis with Current Sources

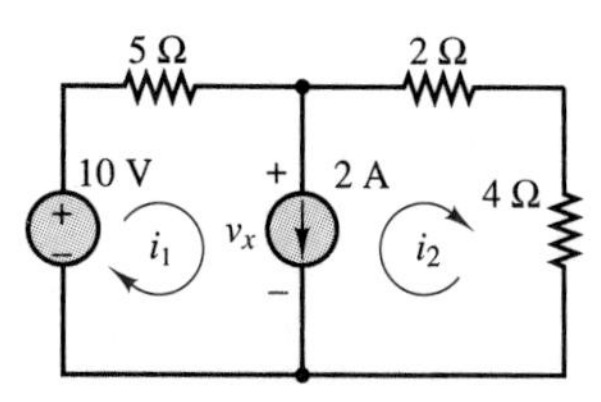

Figure 3.18 Mesh analysis with current sources

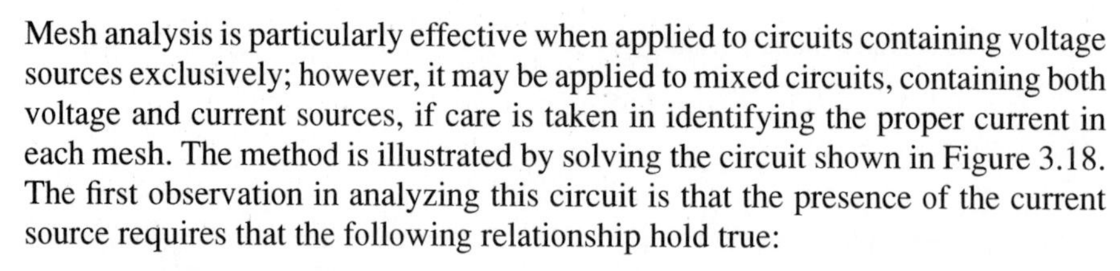

Mesh analysis is particularly effective when applied to circuits containing voltage sources exclusively; however, it may be applied to mixed circuits, containing both voltage and current sources, if care is taken in identifying the proper current in each mesh. The method is illustrated by solving the circuit shown in Figure 3.18. The first observation in analyzing this circuit is that the presence of the current source requires that the following relationship hold true:

$$i_1 - i_2 = 2 \text{ A} \quad \textbf{(3.17)}$$

If the unknown voltage across the current source is labeled v_x, application of KVL around mesh 1 yields:

$$10 = 5i_1 + v_x \quad \textbf{(3.18)}$$

while KVL around mesh 2 dictates that

$$v_x = 2i_2 + 4i_2 \quad \textbf{(3.19)}$$

Substituting equation 3.19 in equation 3.18, and using equation 3.17, we can then obtain the system of equations

$$\begin{aligned} 5i_1 + 6i_2 &= 10 \\ -i_1 + i_2 &= -2 \end{aligned} \quad \textbf{(3.20)}$$

which we can solve to obtain

$$\begin{aligned} i_1 &= 2 \text{ A} \\ i_2 &= 0 \text{ A} \end{aligned} \quad \textbf{(3.21)}$$

Note also that the voltage across the current source may be found by using either equation 3.18 or equation 3.19; for example, using equation 3.19,

$$v_x = 6i_2 = 0 \text{ V} \quad \textbf{(3.22)}$$

The following example further illustrates the solution of this type of circuit.

Example 3.7

Solve the circuit of Figure 3.19, using mesh analysis.

Solution:

Starting from mesh 1, we can immediately see that the current source forces

$$i_1 = 0.5 \text{ A}$$

There is no need to write any further equations around mesh 1, since the value of mesh current i_1 is already known. In this respect, the current source has in fact simplified the problem! Applying KVL around mesh 2, we obtain the following equation:

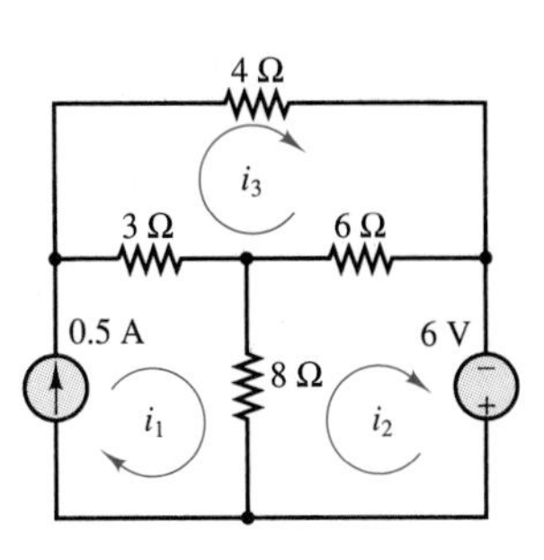

Figure 3.19

$$6(i_2 - i_3) + 8(i_2 - i_1) = 6$$

while around mesh 3 we can write

$$3(i_3 - i_1) + 6(i_3 - i_2) + 4i_3 = 0$$

Rearranging the equations and substituting the known value of i_1, we obtain a system of two equations in two unknowns:

$$14i_2 - 6i_3 = 10$$

$$-6i_2 + 13i_3 = 1.5$$

which can be solved to obtain

$$i_2 = 0.95 \text{ A} \qquad i_3 = 0.55 \text{ A}$$

As usual, you should verify that the solution is correct by applying KVL.

DRILL EXERCISES

3.4 Find the unknown voltage, v_x, by mesh current analysis in the circuit of Figure 3.20.

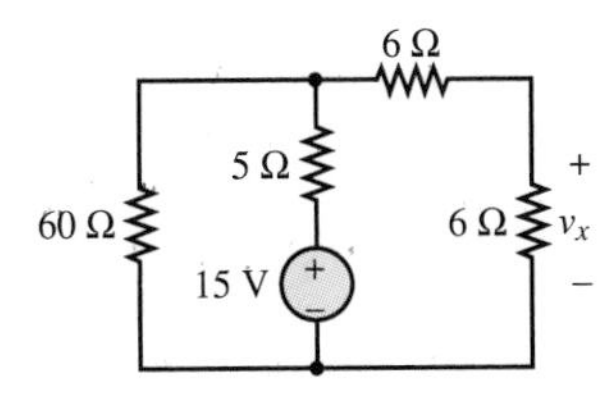

Figure 3.20

3.5 Find the unknown current, I_x, using mesh current methods in the circuit of Figure 3.21.

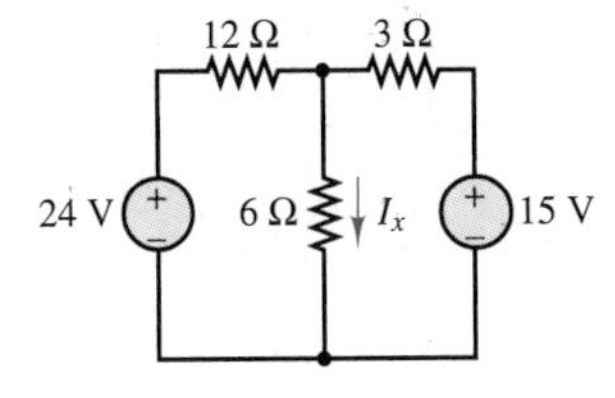

Figure 3.21

3.6 Show that the equations given in Example 3.6 are correct, by applying KCL at each node.

3.3 NODAL AND MESH ANALYSIS WITH CONTROLLED SOURCES

The methods just described also apply, with relatively minor modifications, in the presence of dependent (controlled) sources. Solution methods that allow for the presence of controlled sources will be particularly useful in the study of *transistor amplifiers* in Chapter 8. Recall from the discussion in Section 2.3 that a dependent source is a source that generates a voltage or current that depends on the value of another voltage or current in the circuit. When a dependent source is present in a circuit to be analyzed by node or mesh analysis, one can initially treat it as an ideal source and write the node or mesh equations accordingly. In addition to the equation obtained in this fashion, there will also be an equation relating the dependent source to one of the circuit voltages or currents. This **constraint equation** can then be substituted in the set of equations obtained by the techniques of nodal and mesh analysis, and the equations can subsequently be solved for the unknowns.

It is important to remark that once the constraint equation has been substituted in the initial system of equations, the number of unknowns remains unchanged. Consider, for example, the circuit of Figure 3.22, which is a simplified

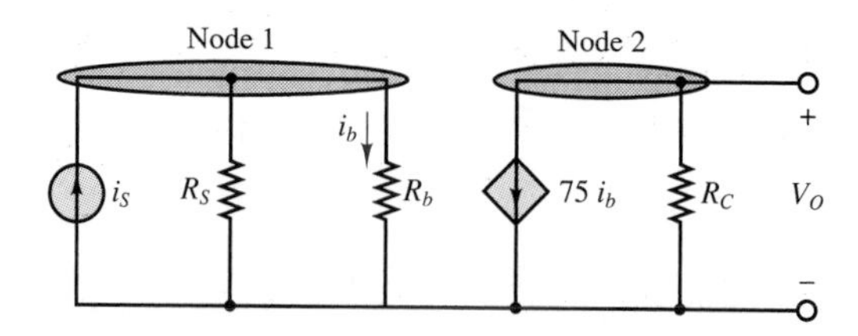

Figure 3.22 Circuit with dependent source

model of a bipolar transistor amplifier (transistors will be introduced in Chapter 8). In the circuit of Figure 3.22, two nodes are easily recognized, and therefore nodal analysis is chosen as the preferred method. Applying KCL at node 1, we obtain the following equation:

$$i_S = v_1\left(\frac{1}{R_S} + \frac{1}{R_b}\right) \qquad \textbf{(3.23)}$$

KCL applied at the second node yields:

$$75 i_b = \frac{v_2}{R_C} \qquad \textbf{(3.24)}$$

Next, it should be observed that the current i_b can be determined by means of a simple current divider:

$$i_b = i_S \frac{1/R_b}{1/R_b + 1/R_S} = i_S \frac{R_S}{R_b + R_S} \qquad \textbf{(3.25)}$$

and therefore a system of two equations is obtained:

$$
\begin{aligned}
i_S &= v_1\left(\frac{1}{R_S} + \frac{1}{R_b}\right) \\
-75 i_S \frac{R_S}{R_b + R_S} &= \frac{v_2}{R_C}
\end{aligned}
\tag{3.26}
$$

which can be used to solve for v_1 and v_2. Note that, in this particular case, the two equations are independent of each other. The following example illustrates a case in which the resulting equations are not independent.

Example 3.8

Solve for the unknown mesh currents in the circuit of Figure 3.23. The dependent source v_x depends on the mesh current i_2 according to the relation $v_x = 0.5i_2$.

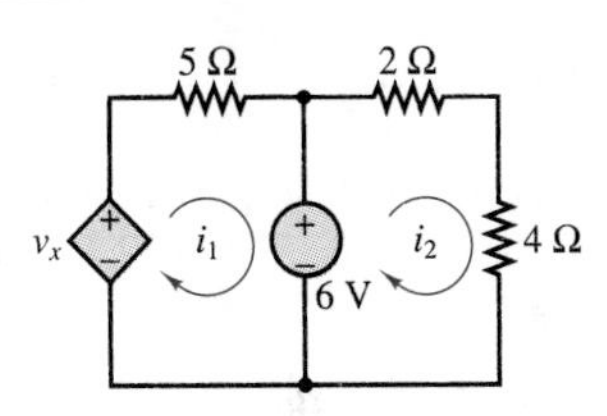

Figure 3.23

Solution:

According to the solution technique proposed in the opening of this section, we first write the conventional mesh equations, treating v_x as an ideal independent source:

$$v_x - 6 = 5i_1 \qquad \text{(mesh 1)}$$
$$6 = (2 + 4)i_2 \qquad \text{(mesh 2)}$$

We then recognize that $v_x = 0.5i_2$ and substitute this quantity in the first equation:

$$0.5i_2 - 6 = 5i_1 \qquad \text{(mesh 1)}$$

From the equation for mesh 2, we obtain

$$i_2 = 1\text{ A}$$

and can then solve for i_1 from the equation for mesh 1:

$$i_1 = \frac{-5.5}{5} = -1.1\text{ A}$$

We may verify that the solution is correct by applying KVL around the first mesh:

$$
\begin{aligned}
v_x - 6 &= 5i_1 \\
0.5i_2 - 6 &= 5i_1 \\
0.5(1) - 6 &= 5(-1.1) \\
-5.5 &= -5.5
\end{aligned}
$$

showing that the solution is consistent.

Remarks on Node Voltage and Mesh Current Methods

The techniques presented in this section and the two preceding sections find use more generally than just in the analysis of resistive circuits. These methods should be viewed as general techniques for the analysis of any linear circuit; they provide systematic and effective means of obtaining the minimum number of equations necessary to solve a network problem. Since these methods are based on the fundamental laws of circuit analysis, KVL and KCL, they also apply to any electrical circuit, even circuits containing nonlinear circuit elements, such as those to be introduced later in this chapter.

You should master both methods as early as possible. Proficiency in these circuit analysis techniques will greatly simplify the learning process for more advanced concepts.

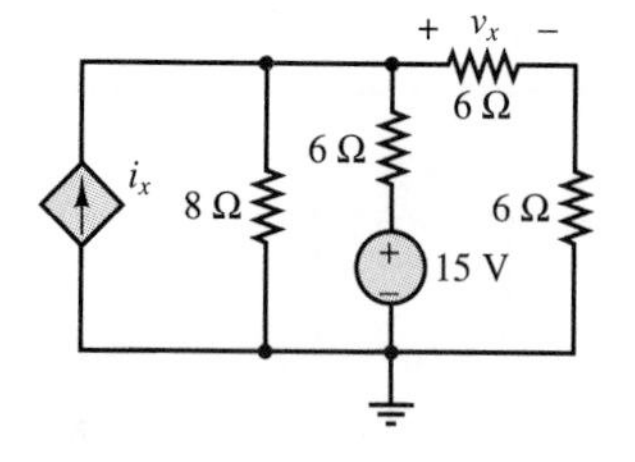

Figure 3.24

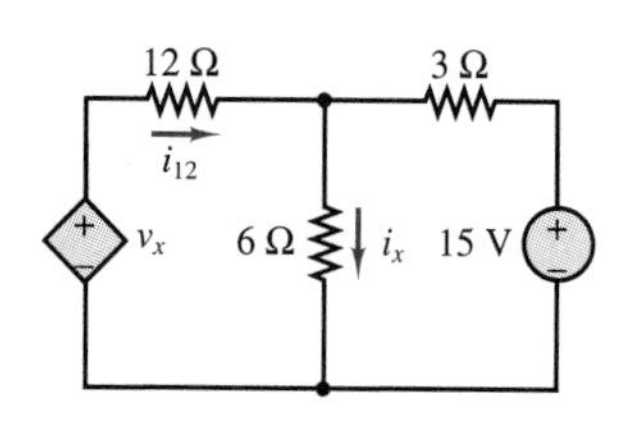

Figure 3.25

DRILL EXERCISES

3.7 The current source i_x is related to the voltage v_x in Figure 3.24 by the relation

$$i_x = \frac{v_x}{3}$$

Find the voltage across the 8-Ω resistor by nodal analysis.

3.8 Find the unknown current i_x in Figure 3.25 using the mesh current method. The dependent voltage source is related to the current i_{12} through the 12-Ω resistor by $v_x = 2i_{12}$.

3.4 THE PRINCIPLE OF SUPERPOSITION

This brief section discusses a concept that is frequently called upon in the analysis of linear circuits. Rather than a precise analysis technique, like the mesh current and node voltage methods, the principle of superposition is a conceptual aid that can be very useful in visualizing the behavior of a circuit containing multiple energy sources. The principle of superposition applies exclusively to linear circuits, and may be stated as follows:

> In a linear circuit containing N sources, each branch voltage and current is the sum of N voltages and currents each of which may be computed by setting all but one source equal to zero and solving the circuit containing that single source.

An elementary illustration of the concept may easily be obtained by simply considering a circuit with two sources connected in series, as shown in Figure 3.26.

The circuit of Figure 3.26 is more formally analyzed as follows. The current, i, flowing in the circuit on the left-hand side of Figure 3.26 may be expressed as:

$$i = \frac{v_{B1} + v_{B2}}{R} = \frac{v_{B1}}{R} + \frac{v_{B2}}{R} = i_{B1} + i_{B2} \tag{3.27}$$

Figure 3.26 also depicts the circuit as being equivalent to the combined effects of two circuits, each containing a single source. In each of the two subcircuits, a

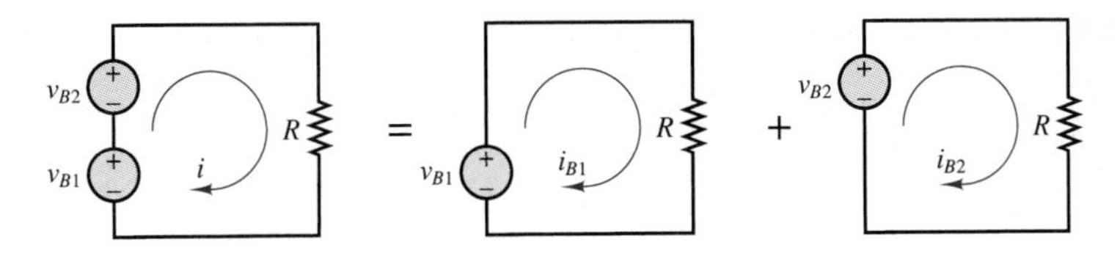

The net current through R is the sum of the individual source currents: $i = i_{B1} + i_{B2}$.

Figure 3.26 The principle of superposition

short circuit has been substituted for the missing battery. This should appear as a sensible procedure, since a short circuit—by definition—will always "see" zero voltage across itself, and therefore this procedure is equivalent to "zeroing" the output of one of the voltage sources.

If, on the other hand, one wished to cancel the effects of a current source, it would stand to reason that an open circuit could be substituted for the current source, since an open circuit is by definition a circuit element through which no current can flow (and which will therefore generate zero current). These basic principles are used frequently in the analysis of circuits, and are summarized in Figure 3.27.

1. In order to set a voltage source equal to zero, we replace it with a short circuit.

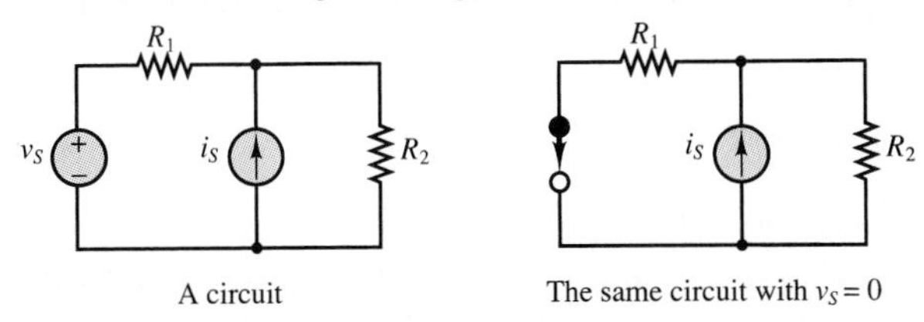

2. In order to set a current source equal to zero, we replace it with an open circuit.

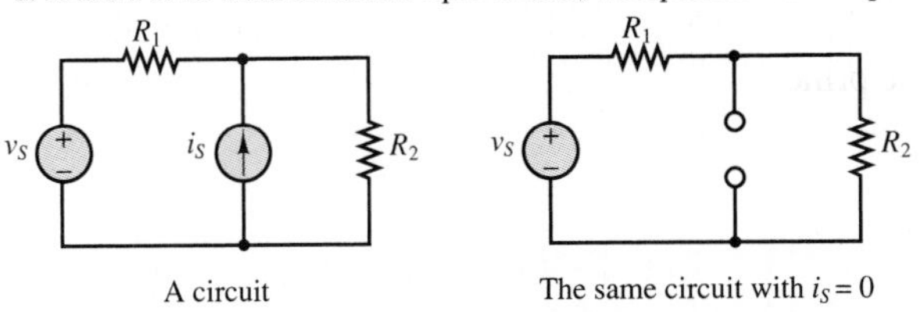

Figure 3.27 Zeroing voltage and current sources

The principle of superposition can easily be applied to circuits containing multiple sources and is sometimes an effective solution technique. More often, however, other methods result in a more efficient solution. Example 3.9 further illustrates the use of superposition to analyze a simple network. The drill exercises at the end of the section illustrate the fact that superposition is often a cumbersome solution method.

Example 3.9

Use the circuits of Figure 3.27 and the principle of superposition to determine the current through resistor R_2. Assume that $v_S = 6$ V, $i_S = 0.5$ A, $R_1 = 50\ \Omega$, and $R_2 = 100\ \Omega$.

Solution:

The current through resistor R_2 in the circuit of Figure 3.27 is due to the superposition of the effects of the voltage and current sources. Let the notation i_{i2} and i_{v2} denote the currents through resistor R_2 due to the current and voltage sources, respectively. Then

$$i_{i2} = 0.5 \times \frac{R_1}{R_1 + R_2} = (0.5)\frac{1}{3} = \frac{1}{6} \text{ A}$$

by the current divider rule, and

$$i_{v2} = \frac{6}{R_1 + R_2} = \frac{6}{150} = \frac{1}{25} \text{ A}$$

by the voltage divider rule and Ohm's law. Finally, by the principle of superposition, the current through resistor R_2, i_2, is given by

$$i_2 = i_{i2} + i_{v2} = \frac{1}{6} + \frac{1}{25} = 0.207 \text{ A}$$

You should verify that this solution is correct by using nodal or mesh analysis.

DRILL EXERCISES

3.9 Find the voltages v_a and v_b for the circuits of Example 3.4 by superposition.

3.10 Repeat Drill Exercise 3.2, using superposition. This exercise illustrates that superposition is not necessarily a computationally efficient solution method.

3.11 Solve Example 3.5, using superposition.

3.12 Solve Example 3.7, using superposition.

3.5 ONE-PORT NETWORKS AND EQUIVALENT CIRCUITS

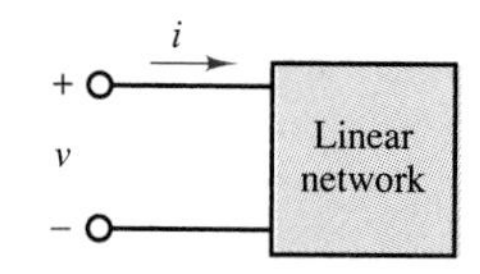

Figure 3.28 One-port network

You may recall that in the discussion of ideal sources in Chapter 2, the flow of energy from a source to a load was described in a very general form, by showing the connection of two "black boxes" labeled source and load (see Figure 2.10). In the same figure, two other descriptions were shown: a symbolic one, depicting an ideal voltage source and an ideal resistor; and a physical representation, in which the load was represented by a headlight and the source by an automotive battery. Whatever the form chosen for source-load representation, each block—source or load—may be viewed as a two-terminal device, described by an *i-v* characteristic. This general circuit representation is shown in Figure 3.28. This configuration is called a **one-port network** and is particularly useful for introducing the notion of equivalent circuits. Note that the network of Figure 3.28 is completely described by its *i-v* characteristic; this point is best illustrated by an example.

Example 3.10

Consider the circuit shown in Figure 3.29. If we wished to determine, say, the current demand placed on the voltage source by the resistive network, we would compute the value of the current provided by the source, i, according to the following expression:

$$i = \frac{v_S}{1/(1/R_1 + 1/R_2 + 1/R_3)}$$

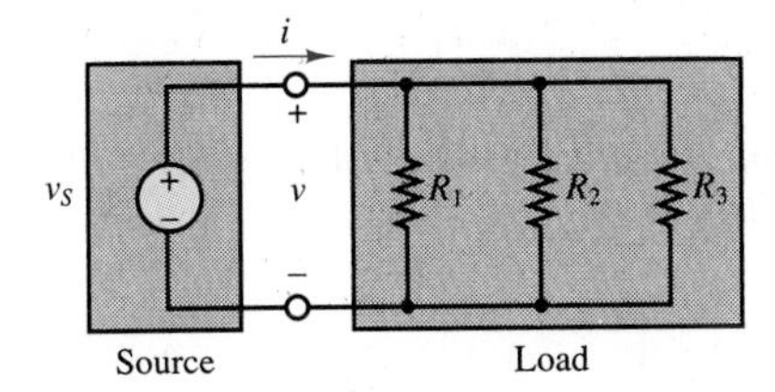

Figure 3.29 Illustration of equivalent-circuit concept

It should be apparent that *insofar as the source is concerned,* it is irrelevant whether the load consists of three resistors in parallel or of a single equivalent resistance given by the expression

$$R_{EQ} = \frac{1}{1/R_1 + 1/R_2 + 1/R_3}$$

In either case, we may represent the load by means of a single resistor of value R_{EQ}, as shown in Figure 3.30.

Similarly, if the voltage source consists of a 6-V battery, *insofar as the load is concerned* it is irrelevant whether the source circuit consists of a single 6-V battery or of four 1.5-V batteries connected in series.

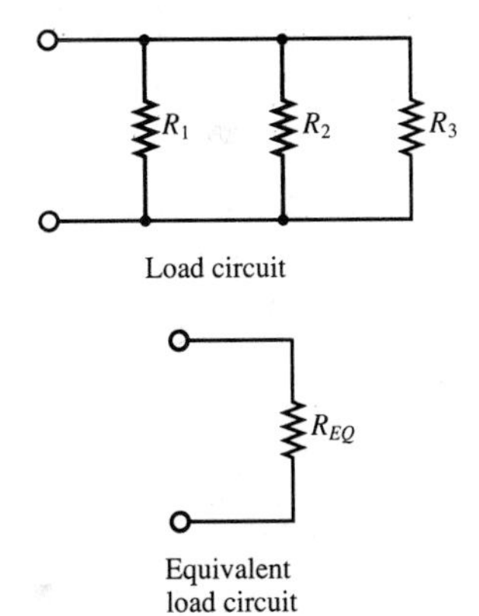

Figure 3.30 Equivalent load resistance concept

For the remainder of this chapter, we shall focus on developing techniques for computing equivalent representations of linear networks. Such representations will be useful in deriving some simple—yet general—results for linear circuits, as well as analyzing simple nonlinear circuits.

Thévenin and Norton Equivalent Circuits

This section discusses one of the most important topics in the analysis of electrical circuits: the concept of an **equivalent circuit.** It will be shown that it is always possible to view even a very complicated circuit in terms of much simpler *equivalent* source and load circuits, and that the transformations leading to equivalent circuits are easily managed, with a little practice. In studying node voltage and mesh current analysis, you may have observed that there is a certain correspondence (called **duality**) between current sources and voltage sources, on the one hand, and parallel and series circuits, on the other. This duality appears again very clearly in the analysis of equivalent circuits: it will shortly be shown that equivalent circuits fall into one of two classes, involving either voltage or current sources and (respectively) either series or parallel resistors, reflecting this same principle of duality. The discussion of equivalent circuits begins with the statement of two very important theorems, summarized in Figures 3.31 and 3.32.

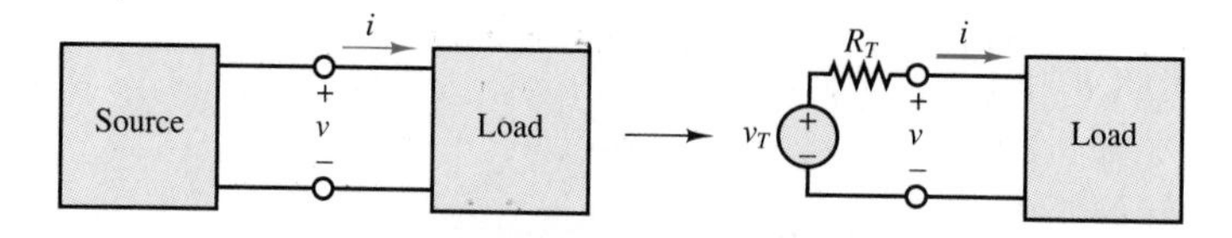

Figure 3.31 Illustration of Thévenin theorem

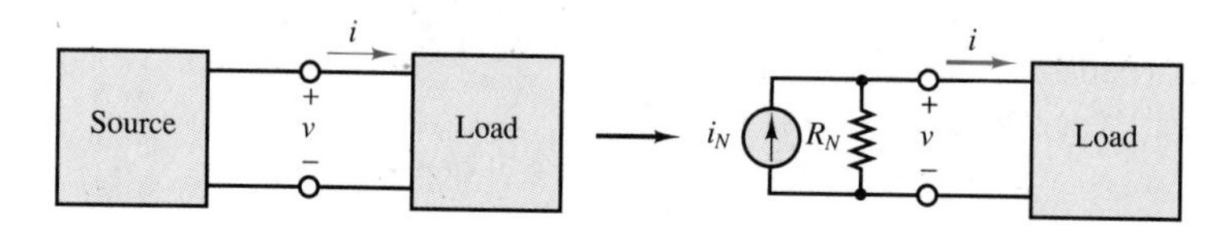

Figure 3.32 Illustration of Norton theorem

> **The Thévenin Theorem**
>
> As far as a load is concerned, any network composed of ideal voltage and current sources, and of linear resistors, may be represented by an equivalent circuit consisting of an ideal voltage source, v_T, in series with an equivalent resistance, R_T.

R_1 R_3 a v_S R_2 R_L b

Complete circuit

> **The Norton Theorem**
>
> As far as a load is concerned, any network composed of ideal voltage and current sources, and of linear resistors, may be represented by an equivalent circuit consisting of an ideal current source, i_N, in parallel with an equivalent resistance, R_N.

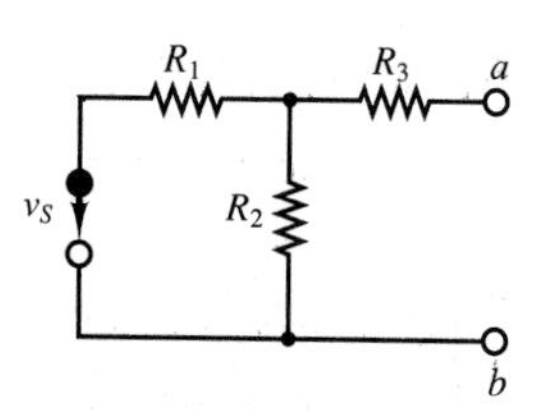

Circuit with load removed for computation of R_T. The voltage source is replaced by a short circuit.

Figure 3.33 Computation of Thévenin resistance

The first obvious question to arise is, how are these equivalent source voltages, currents, and resistances computed? The next few sections illustrate the computation of these equivalent circuit parameters, mostly through examples. A substantial number of drill exercises are also provided, with the following caution: the only way to master the computation of Thévenin and Norton equivalent circuits is by patient repetition.

Determination of Norton or Thévenin Equivalent Resistance

The first step in computing a Thévenin or Norton equivalent circuit consists of finding the equivalent resistance presented by the circuit at its terminals. This is done by setting all sources in the circuit equal to zero and computing the effective resistance between terminals. The voltage and current sources present in the circuit are set to zero by the same technique used with the principle of superposition: voltage sources are replaced by short circuits, current sources by open circuits. To illustrate the procedure, consider the simple circuit of Figure 3.33; the objective is to compute the equivalent resistance the load R_L "sees" at port a-b.

In order to compute the equivalent resistance, we remove the load resistance from the circuit and replace the voltage source, v_S, by a short circuit. At this point—seen from the load terminals—the circuit appears as shown in Figure 3.34. You can see that R_1 and R_2 are in parallel, since they are connected between the same two nodes. If the total resistance between terminals a and b is denoted by R_T, its value can be determined as follows:

$$R_T = R_3 + R_1 \parallel R_2 \tag{3.28}$$

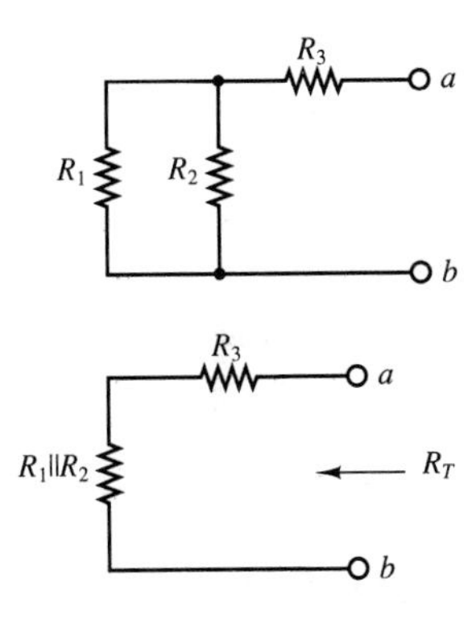

Figure 3.34 Equivalent resistance seen by the load

An alternative way of viewing R_T is depicted in Figure 3.35, where a hypothetical 1-A current source has been connected to the terminals a and b. The voltage v_x appearing across the a-b pair will then be numerically equal to R_T (only because $i_S = 1$ A!). With the 1-A source current flowing in the circuit, it should be apparent that the source current encounters R_3 as a resistor in series with the parallel combination of R_1 and R_2, prior to completing the loop.

Summarizing the procedure, we can produce a set of simple rules as an aid in the computation of the Thévenin (or Norton) equivalent resistance for a linear resistive circuit:

What is the total resistance the current i_S will encounter in flowing around the circuit?

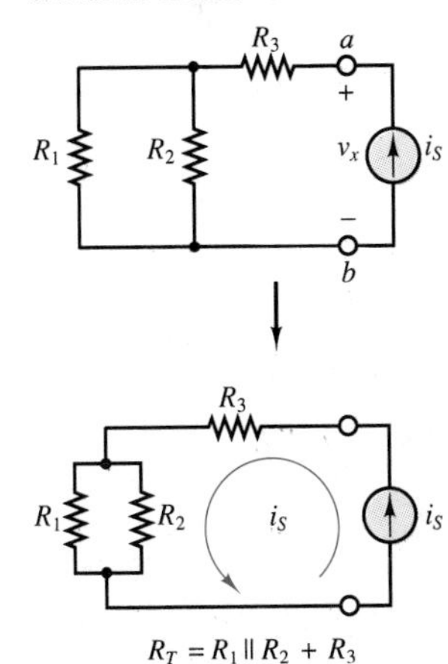

Figure 3.35 An alternative method of determining the Thévenin resistance

Computation of Equivalent Resistance of a One-Port Network

1. Remove the load.
2. Zero all voltage and current sources.
3. Compute the total resistance between load terminals, *with the load removed.* This resistance is equivalent to that which would be encountered by a current source connected to the circuit in place of the load.

We note immediately that this procedure yields a result that is independent of the load. This is a very desirable feature, since once the equivalent resistance has been identified for a source circuit, the equivalent circuit remains unchanged if we connect a different load. The following examples further illustrate the procedure.

Example 3.11

Find the Thévenin equivalent resistance seen by the load resistance, R_L, in the circuit of Figure 3.36.

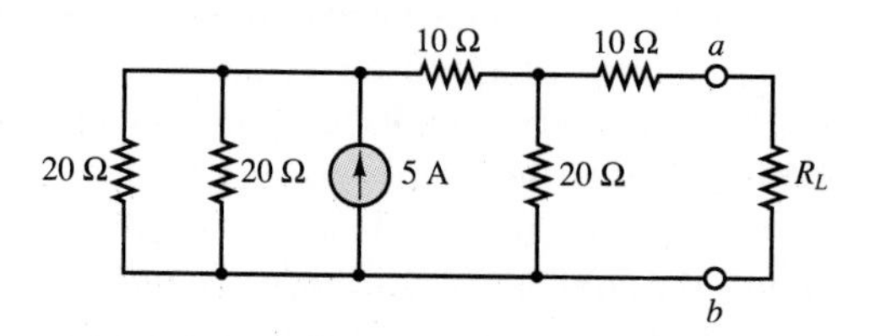

Figure 3.36

Solution:

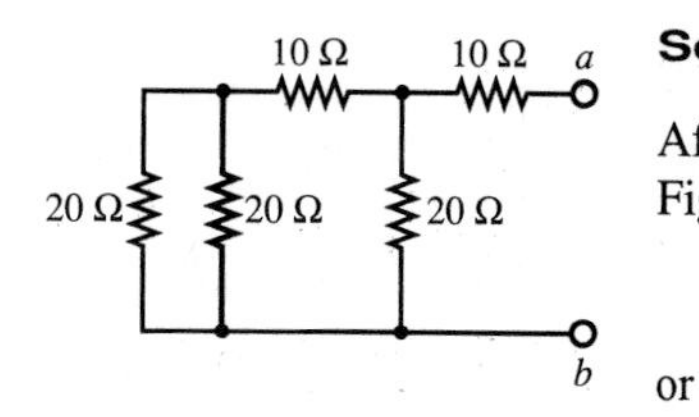

Figure 3.37

After replacing the current source with an open circuit, the circuit appears as shown in Figure 3.37. Looking into terminals a and b, we then give the equivalent resistance by

$$R_T = ((20 \parallel 20) + 10) \parallel 20 + 10$$

or

$$R_T = 20\ \Omega$$

Example 3.12

For the circuit shown in Figure 3.38, find the Thévenin equivalent resistance seen by the load resistance, R_L.

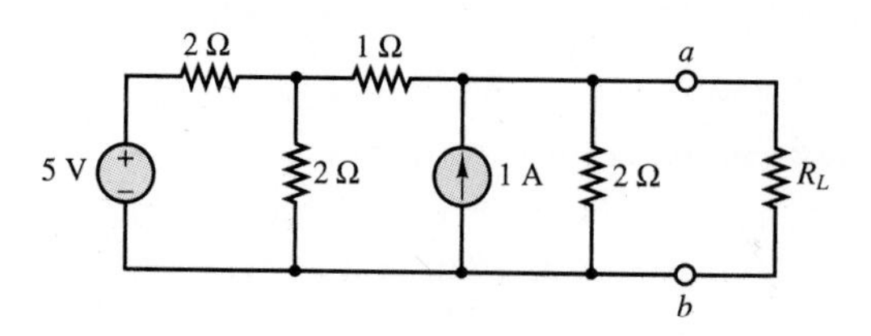

Figure 3.38

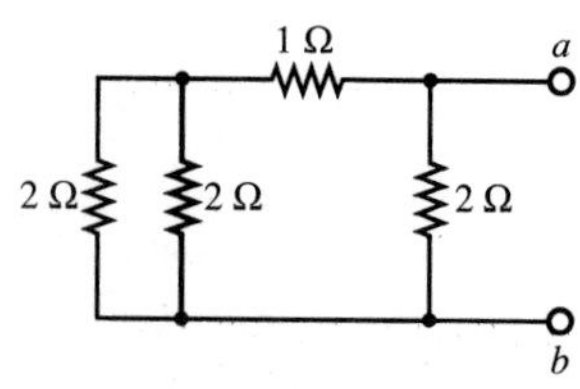

Figure 3.39

Solution:

First we replace the voltage source with a short circuit and the current source with an open circuit. The circuit now has the appearance of Figure 3.39. Looking into terminals a and b, we then give the equivalent resistance by

$$R_T = ((2 \parallel 2) + 1) \parallel 2$$

or

$$R_T = 1\ \Omega$$

As a final note, it should be remarked that the Thévenin and Norton equivalent resistances are one and the same quantity:

$$R_T = R_N \tag{3.29}$$

Therefore, the preceding discussion holds whether we wish to compute a Norton or a Thévenin equivalent circuit. From here on we shall use the notation R_T exclusively, for both Thévenin and Norton equivalents. Drill Exercise 3.13 will give you an opportunity to explain why the two equivalent resistances are one and the same.

DRILL EXERCISES

3.13 Apply the methods described in this section to show that $R_T = R_N$ in the circuits of Figure 3.40.

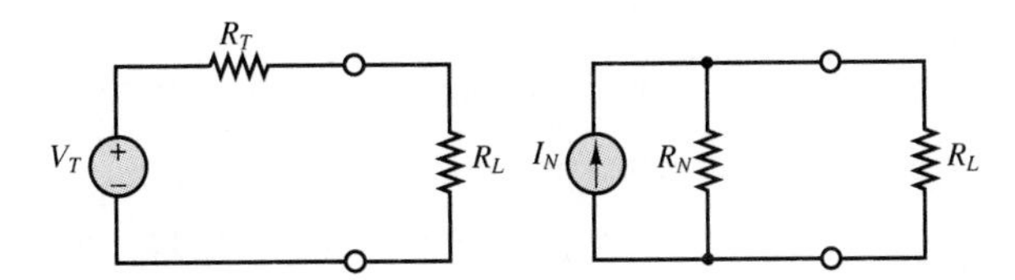

Figure 3.40

3.14 Find the Thévenin resistance of the circuit of Figure 3.41 seen by the load resistor, R_L.

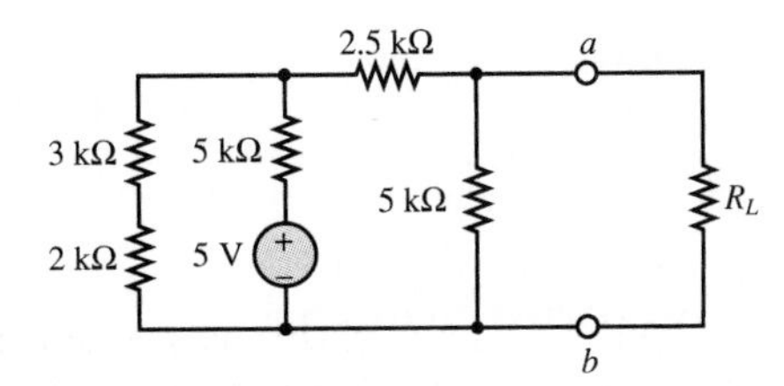

Figure 3.41

3.15 Find the Thévenin resistance seen by the load resistor, R_L, in the circuit of Figure 3.42.

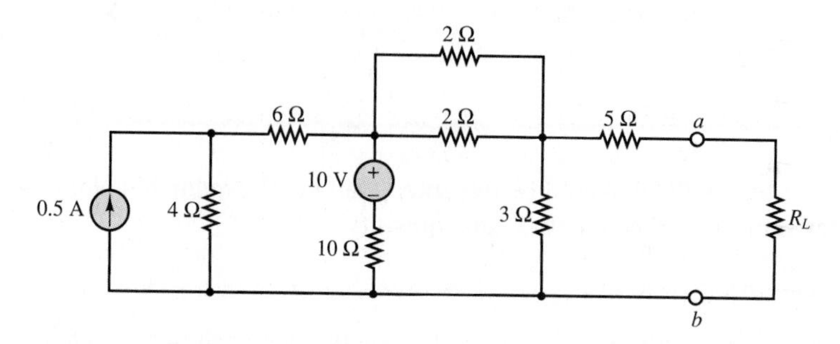

Figure 3.42

3.16 For the circuit of Figure 3.43, find the Thévenin resistance seen by the load resistor, R_L.

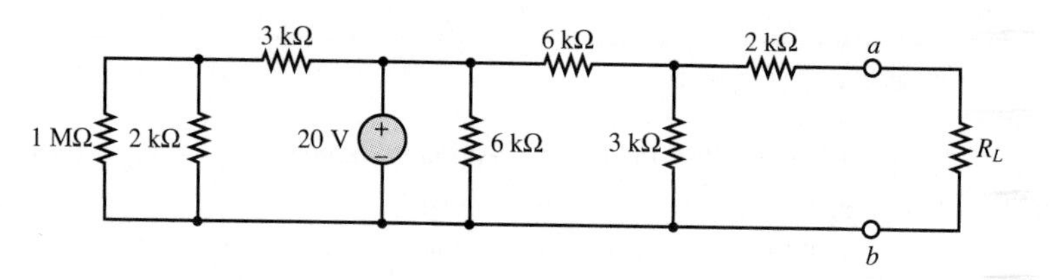

Figure 3.43

3.17 For the circuit of Figure 3.44, find the Thévenin resistance seen by the load resistor, R_L.

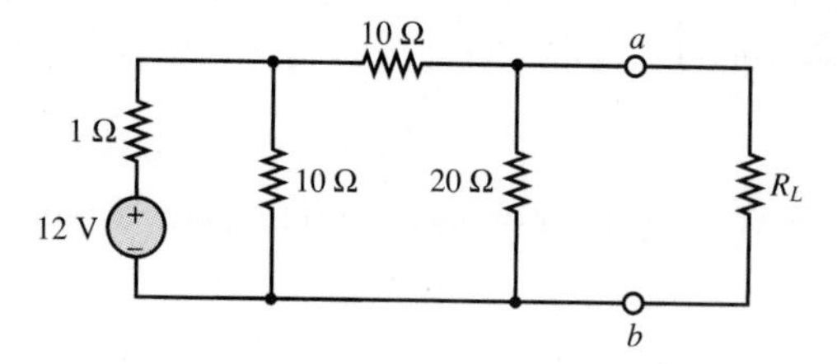

Figure 3.44

Computing the Thévenin Voltage

This section describes the computation of the Thévenin equivalent voltage, v_T, for an arbitrary linear resistive circuit. The Thévenin equivalent voltage is defined as follows:

The equivalent (Thévenin) source voltage is equal to the **open-circuit voltage** present at the load terminals with the load removed.

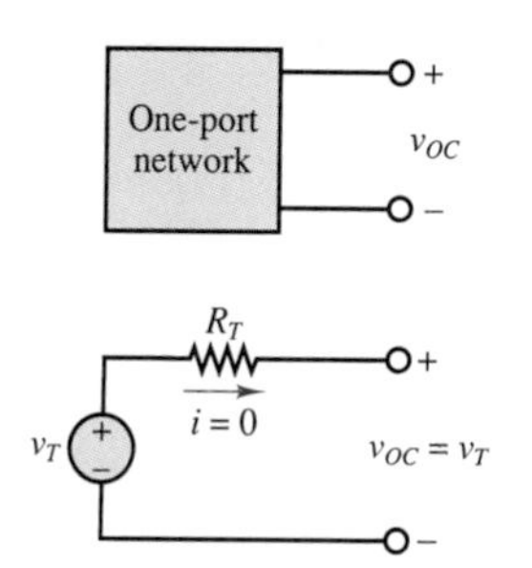

Figure 3.45 Equivalence of open-circuit and Thévenin voltage

This states that in order to compute v_T, it is sufficient to remove the load and to compute the open-circuit voltage at the one-port terminals. Figure 3.45 illustrates that the open-circuit voltage, v_{OC}, and the Thévenin voltage, v_T, must be the same if the Thévenin theorem is to hold. This is true because in the circuit consisting of v_T and R_T, the voltage v_{OC} must equal v_T, since no current flows through R_T and therefore the voltage across R_T is zero. Kirchhoff's voltage law confirms that

$$v_T = R_T(0) + v_{OC} = v_{OC} \tag{3.30}$$

The actual computation of the open-circuit voltage is best illustrated by examples; there is no substitute for practice in becoming familiar with these computations. To summarize the main points in the computation of open-circuit voltages, consider the circuit of Figure 3.33, shown again in Figure 3.46 for convenience. Recall that the equivalent resistance of this circuit was given by $R_T = R_3 + R_1 \| R_2$. To compute v_{OC}, we disconnect the load, as shown in Figure 3.47, and immediately observe that no current flows through R_3, since there is no closed circuit connection at that branch. Therefore, v_{OC} must be equal to the voltage across R_2, as illustrated in Figure 3.48. Since the only closed circuit is the mesh consisting of v_S, R_1, and R_2, the answer we are seeking may be obtained by means of a simple voltage divider:

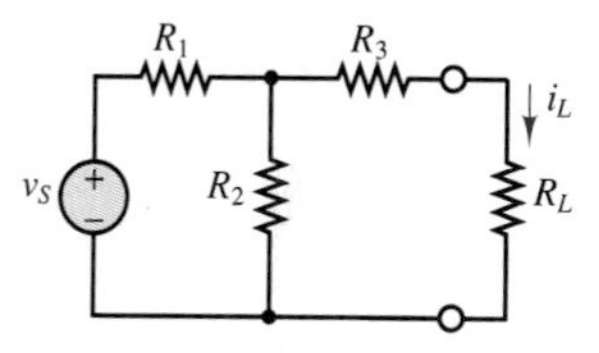

Figure 3.46

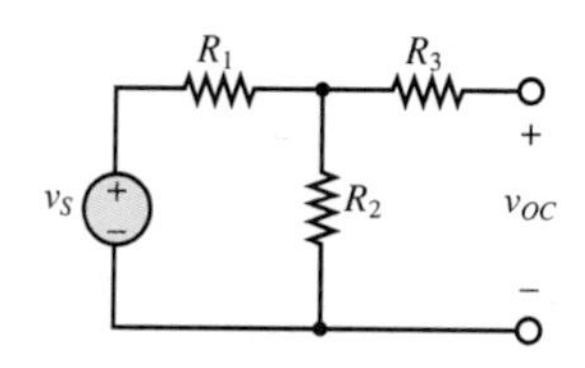

Figure 3.47

$$v_{OC} = v_{R2} = v_S \frac{R_2}{R_1 + R_2}$$

It is instructive to review the basic concepts outlined in the example by considering the original circuit and its Thévenin equivalent side by side, as shown in Figure 3.49. The two circuits of Figure 3.49 are equivalent in the sense that the current drawn by the load, i_L, is the same in both circuits, that current being given by:

$$i_L = v_S \cdot \frac{R_2}{R_1 + R_2} \cdot \frac{1}{(R_3 + R_1 \| R_2) + R_L} = \frac{v_T}{R_T + R_L} \tag{3.31}$$

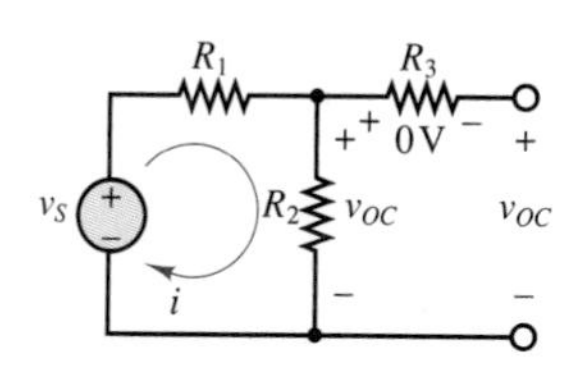

Figure 3.48

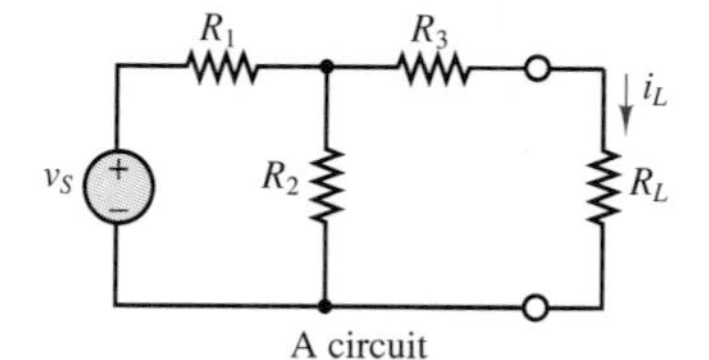

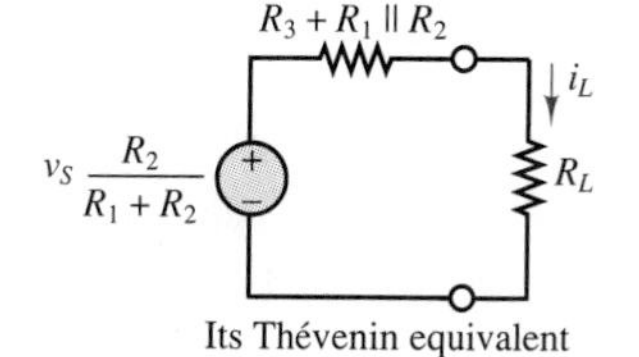

Figure 3.49 A circuit and its Thévenin equivalent

The computation of Thévenin equivalent circuits is further illustrated in the following examples.

Example 3.13

Compute the open-circuit voltage for the circuit of Figure 3.44.

Solution:

With the load open-circuited, the circuit becomes a two-mesh circuit, as shown in Figure 3.50. The open-circuit voltage, v_{OC}, is equal to the voltage across the 20-Ω resistor and is therefore given by the expression

$$v_{OC} = 20i_2$$

Writing the mesh equations for the circuit, we obtain

$$(10 + 1)i_1 - 10i_2 = 12$$
$$-10i_1 + 40i_2 = 0$$

or, in matrix form,

$$\begin{bmatrix} 11 & -10 \\ -10 & 40 \end{bmatrix} \begin{bmatrix} i_1 \\ i_2 \end{bmatrix} = \begin{bmatrix} 12 \\ 0 \end{bmatrix}$$

The system of equations can be solved for i_2 (for example, using Cramer's rule—see Appendix A) to obtain

$$i_2 = 0.3529 \text{ A}$$

or

$$v_{OC} = 20i_2 = 7.06 \text{ V}$$

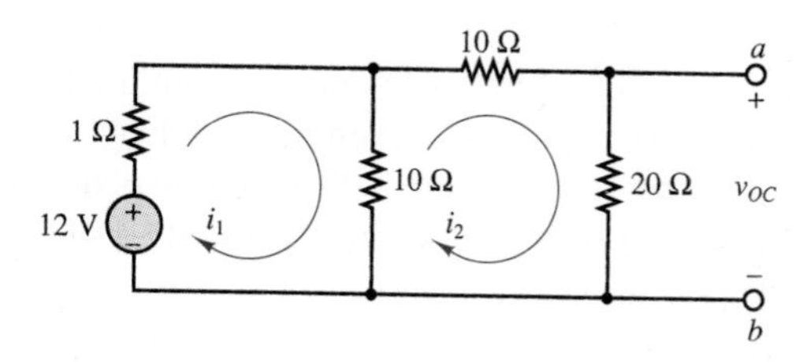

Figure 3.50

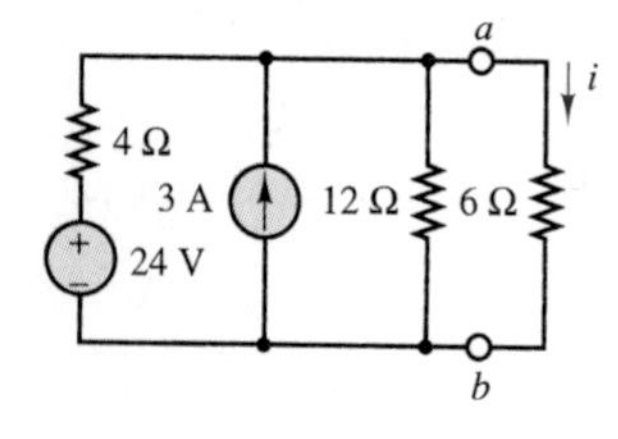

Figure 3.51

Example 3.14

In Figure 3.51, replace the network to the left of the terminals a and b by its Thévenin equivalent circuit, and use the result to find the load current, i.

Solution:

First, we replace the voltage source with a short circuit and the current source by an open circuit. The circuit appears as shown in Figure 3.52. Looking into terminals a and b, we then give the equivalent resistance by

$$R_T = 12 \| 4 = 3 \ \Omega$$

For the open-circuit voltage, we can immediately observe that v_{OC} must be equal to the voltage across the 12-Ω resistor. Mesh analysis can be applied to find v_{OC} as shown in Figure 3.53. Applying KVL to meshes 1 and 2, we obtain

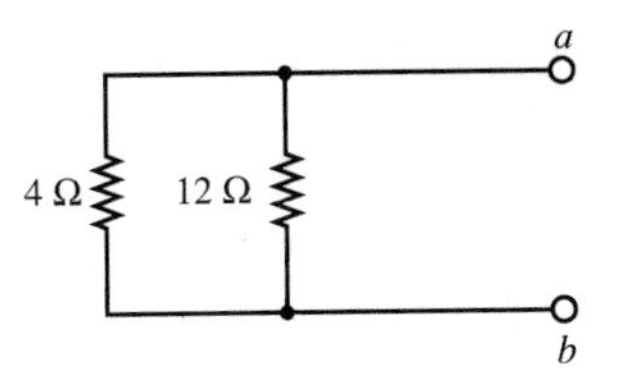

Figure 3.52

$$4i_1 - 24 + v_1 = 0$$
$$12i_2 - v_1 = 0$$

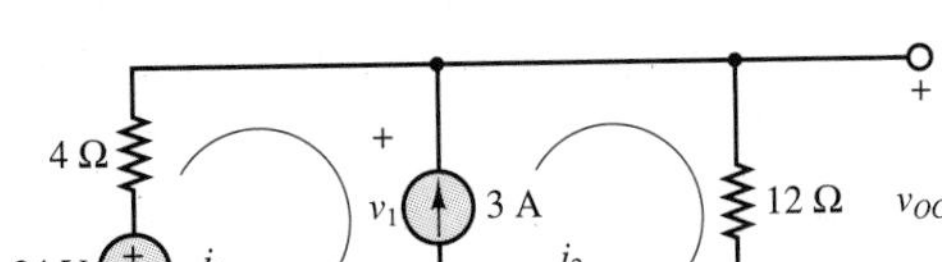

Figure 3.53

Also, we observe that the current source constrains the mesh currents to satisfy the relationship

$$i_2 - i_1 = 3$$

Rearranging the three equations, we can formulate a system of two equations in two unknowns:

$$4i_1 + 12i_2 = 24$$
$$i_2 - i_1 = 3$$

Solving for the two equations, we obtain the following values for the currents i_1 and i_2:

$$i_1 = -\tfrac{3}{4}\ \text{A}$$
$$i_2 = \tfrac{9}{4}\ \text{A}$$

Observing that the open-circuit voltage is actually equal to the voltage across the 12-Ω resistor, we find the Thévenin equivalent voltage to be

$$v_T = v_{OC} = 12i_2 = 27\ \text{V}$$

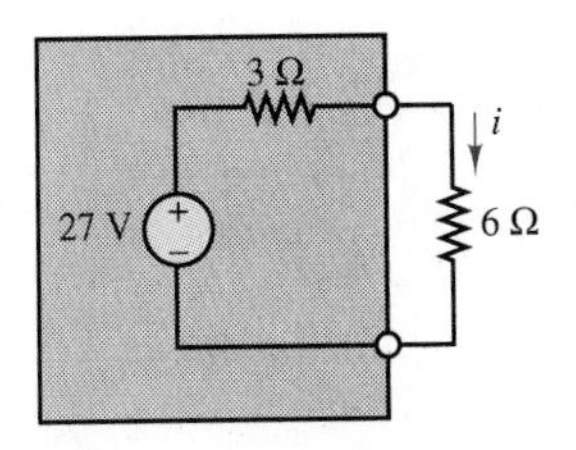

Figure 3.54 Thévenin equivalent

Finally, to determine the load current, i, we connect the load to the Thévenin equivalent circuit, as shown in Figure 3.54, and compute

$$i = \frac{v_{OC}}{R_T + 6} = \frac{27}{3 + 6} = 3\ \text{A}$$

DRILL EXERCISES

3.18 With reference to Figure 3.46, find the load current, i_L, by mesh analysis, if $v_S = 10$ V, $R_1 = R_3 = 50\ \Omega$, $R_2 = 100\ \Omega$, $R_L = 150\ \Omega$.

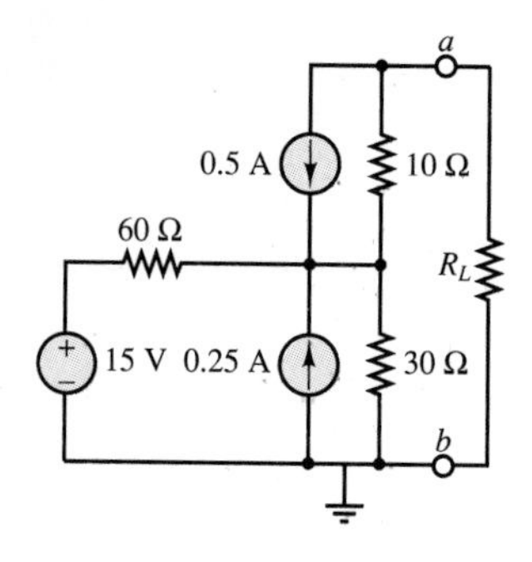

Figure 3.55

3.19 Find the Thévenin equivalent circuit seen by the load resistor, R_L, for the circuit of Figure 3.55.

3.20 Find the Thévenin equivalent circuit for the circuit of Figure 3.56.

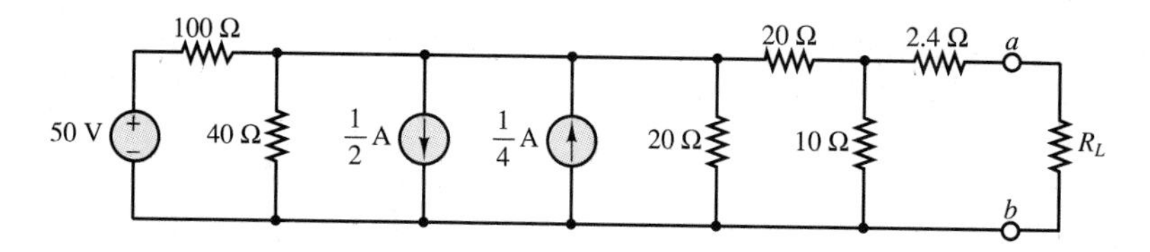

Figure 3.56

Computing the Norton Current

The computation of the Norton equivalent current is very similar in concept to that of the Thévenin voltage. The following definition will serve as a starting point:

Definition

The Norton equivalent current is equal to the **short-circuit current** that would flow were the load replaced by a short circuit.

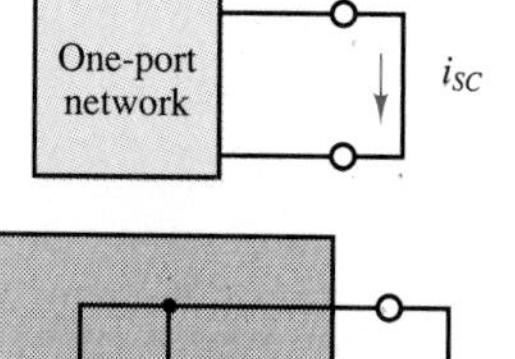

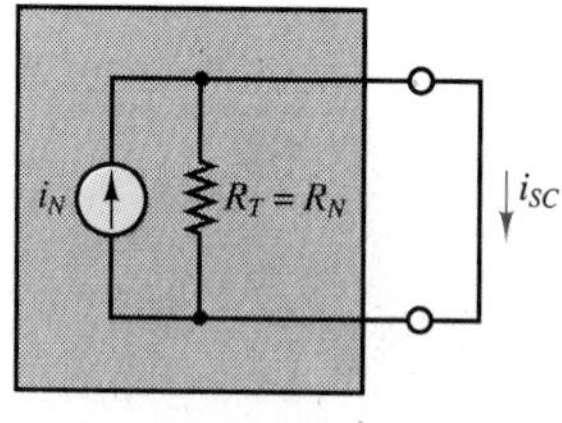

Figure 3.57 Illustration of Norton equivalent circuit

An explanation for the definition of the Norton current is easily found by considering, again, an arbitrary one-port network, as shown in Figure 3.57, where the one-port network is shown together with its Norton equivalent circuit.

It should be clear that the current, i_{SC}, flowing through the short circuit replacing the load is exactly the Norton current, i_N, since all of the source current in the circuit of Figure 3.57 must flow through the short circuit. Consider the circuit of Figure 3.58, shown with a short circuit in place of the load resistance. Any of the techniques presented in this chapter could be employed to determine the current i_{SC}. In this particular case, mesh analysis is a convenient tool, once it is recognized that the short-circuit current is a mesh current. Let i_1 and $i_2 = i_{SC}$ be the mesh currents in the circuit of Figure 3.58. Then, the following mesh equations can be derived and solved for the short-circuit current:

$$\begin{aligned}(R_1 + R_2)i_1 - R_2 i_{SC} &= v_S \\ -R_2 i_1 + (R_2 + R_3)i_{SC} &= 0\end{aligned}$$

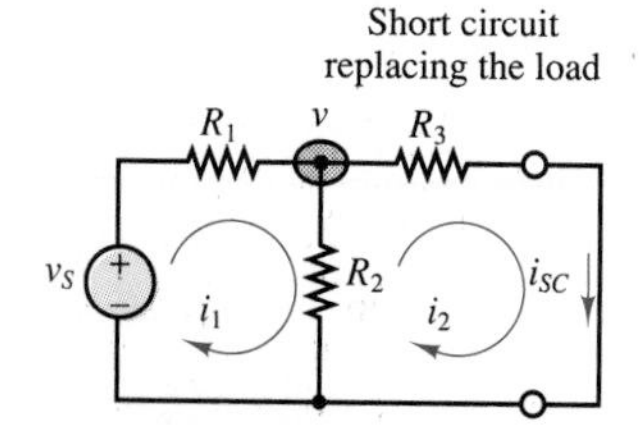

Figure 3.58 Computation of Norton current

An alternative formulation would employ nodal analysis to derive the equation

$$\frac{v_S - v}{R_1} = \frac{v}{R_2} + \frac{v}{R_3}$$

leading to

$$v = v_S \frac{R_2 R_3}{R_1 R_3 + R_2 R_3 + R_1 R_2}$$

Recognizing that $i_{SC} = v/R_3$, we can determine the Norton current to be:

$$i_N = \frac{v}{R_3} = \frac{v_S R_2}{R_1 R_3 + R_2 R_3 + R_1 R_2}$$

Thus, conceptually, the computation of the Norton current simply requires identifying the appropriate short-circuit current. The following example further illustrates this idea.

Example 3.15

Find the Norton equivalent circuit to the left of terminals *a* and *b* for the network of Figure 3.59.

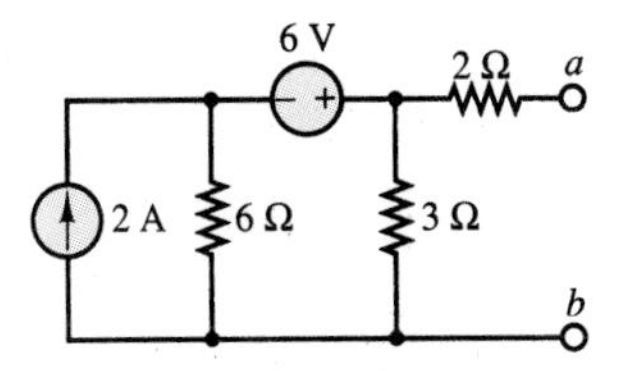

Figure 3.59

Solution:

First, we replace the voltage source with a short circuit, and the current source by an open circuit. The circuit now has the appearance of Figure 3.60 and permits the calculation of the Thévenin resistance:

$$R_T = 6 \| 3 + 2 = 2 + 2 = 4$$

To find the short-circuit, or Norton, current we can short-circuit terminals *a* and *b* and use mesh analysis to calculate i_N, as suggested in Figure 3.61. Using mesh analysis, we may obtain the following equations:

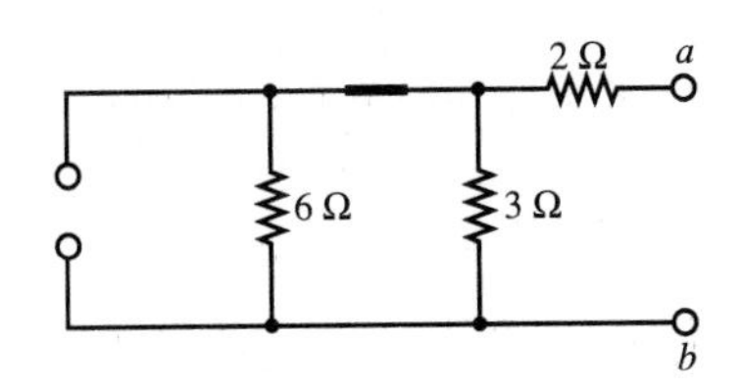

Figure 3.60

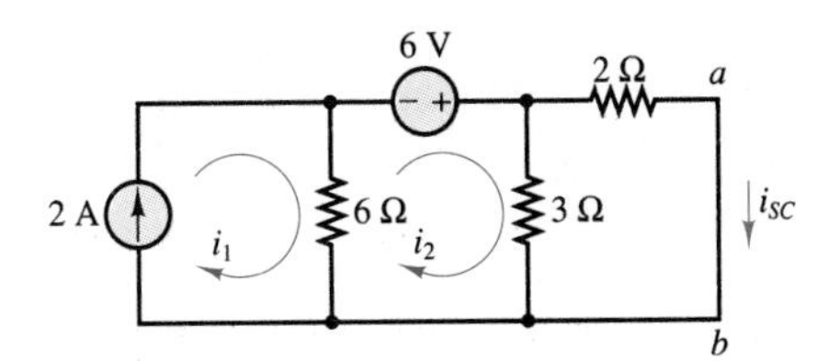

Figure 3.61

$$i_1 = 2 \text{ A}$$
$$9i_2 - 3i_{SC} = 18$$
$$5i_{SC} - 3i_2 = 0$$

We may then solve these to compute

$$i_1 = 2 \text{ A}, \quad i_2 = 2.5 \text{ A}, \quad \text{and} \quad i_{SC} = 1.5 \text{ A}$$

The complete equivalent circuit is shown in Figure 3.62.

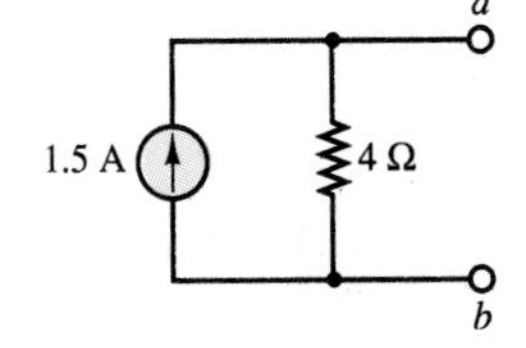

Figure 3.62 Norton equivalent circuit

Source Transformations

This section illustrates **source transformations,** a procedure that may be very useful in the computation of equivalent circuits, permitting, in some circumstances, replacement of current sources with voltage sources, and vice versa. The Norton and Thévenin theorems state that any one-port network can be represented by a voltage source in series with a resistance, or by a current source in parallel with a resistance, and that either of these representations is equivalent to the original circuit, as illustrated in Figure 3.63.

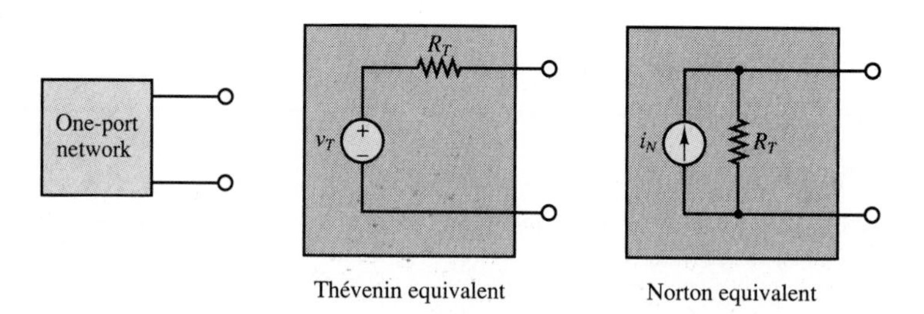

Figure 3.63 Equivalence of Thévenin and Norton representations

An extension of this result is that any circuit in Thévenin equivalent form may be replaced by a circuit in Norton equivalent form, provided that we use the following relationship:

$$v_T = R_T i_N \tag{3.32}$$

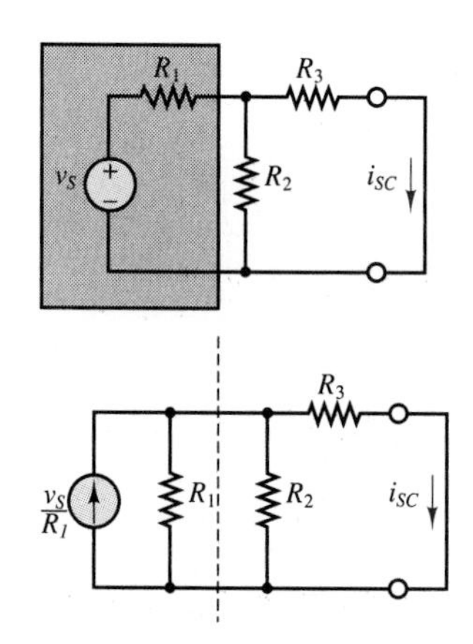

Figure 3.64 Effect of source transformation

Thus, the subcircuit to the left of the dashed line in Figure 3.64 may be replaced by its Norton equivalent, as shown in the figure. Then, the computation of i_{SC} becomes very straightforward, since the three resistors are in parallel with the current source and therefore a simple current divider may be used to compute the short-circuit current. Observe that the short-circuit current is the current flowing through R_3; therefore,

$$i_{SC} = i_N = \frac{1/R_3}{1/R_1 + 1/R_2 + 1/R_3} \frac{v_S}{R_1} = \frac{v_S R_2}{R_1 R_3 + R_2 R_3 + R_1 R_2} \tag{3.33}$$

which is the identical result obtained for the same circuit in the preceding section, as you may easily verify. This source transformation method can be very useful, if employed correctly. Figure 3.65 shows how one can recognize subcircuits

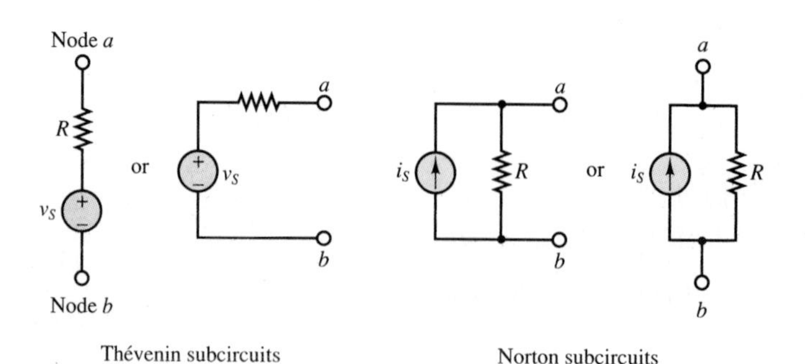

Figure 3.65 Subcircuits amenable to source transformation

amenable to such source transformations. Example 3.16 is a numerical example illustrating the procedure.

Example 3.16

This example will illustrate the connection between Thévenin and Norton source transformations and the computation of Thévenin equivalents. The aim is to determine the Norton equivalent for the circuit shown in Figure 3.66.

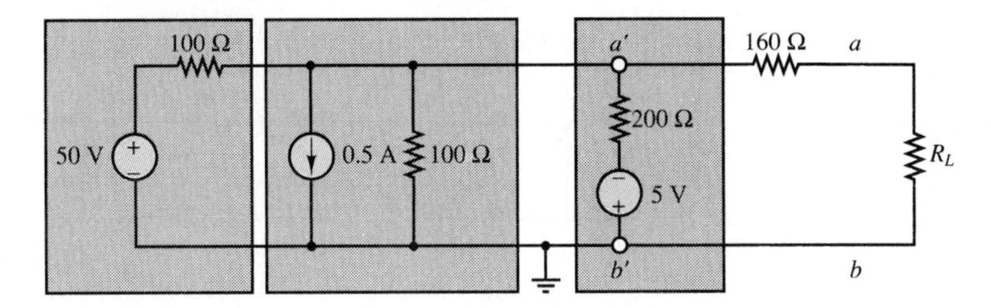

Figure 3.66

Solution:

With the 50-V and 5-V sources replaced by short circuits and the 0.5-A source replaced by an open circuit, we can find the Thévenin resistance to be

$$R_T = 160 + 200 \,\|\, 100 \,\|\, 100 = 200\ \Omega$$

To take advantage of the source transformation technique, it is useful to sketch the circuit again, emphasizing the location of the nodes, as shown in Figure 3.67. Note that we have

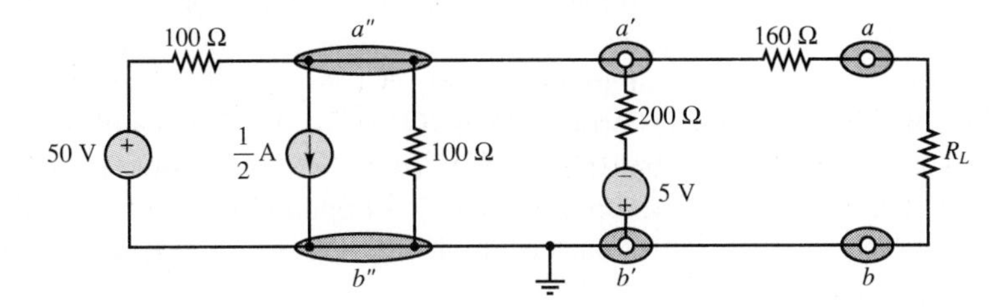

Figure 3.67

purposely separated the points a' and b' from the points a'' and b'' in the circuit, *even though these are the same nodes*. Now, the branch containing the 50-V source in series with the 100-Ω resistor clearly appears between nodes a'' and b'' and can therefore be replaced by an equivalent Norton circuit, with equivalent current source of strength $^{50}/_{100}$, in parallel with a 100-Ω resistor. Similarly, the branch containing the 5-V source in series with the 200-Ω resistor can be manipulated to obtain the circuit shown in Figure 3.68.

Observe how all the elements to the left of the 160-Ω resistor are now in parallel! An additional manipulation, consisting of placing all of the sources next to each other in the diagram, illustrates the simple nature of the original circuit in this new light (Figure 3.69),

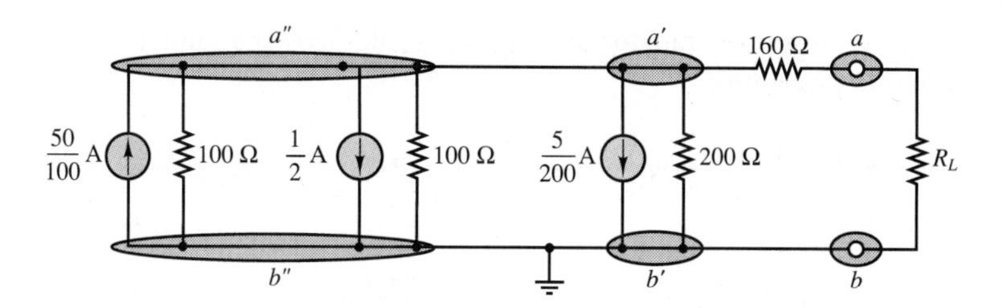

Figure 3.68

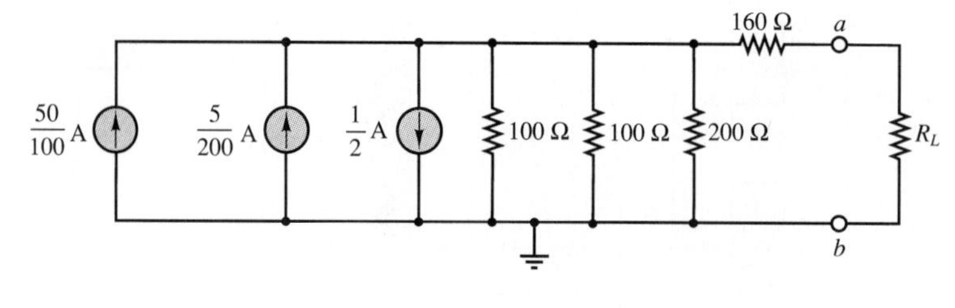

Figure 3.69

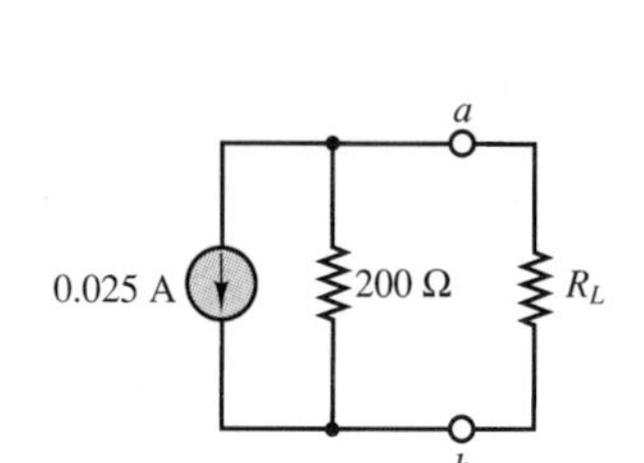

Figure 3.70

revealing that if we simply sum (algebraically) the three current sources and perform a conversion from current to voltage source, combine the resistors, and convert back to a current source, we can finally obtain the Norton equivalent circuit of Figure 3.70.

Experimental Determination of Thévenin and Norton Equivalents

The idea of equivalent circuits as a means of representing complex and sometimes unknown networks is useful not only analytically, but in practical engineering applications as well. It is very useful to have a measure, for example, of the equivalent internal resistance of an instrument, so as to have an idea of its power requirements and limitations. Fortunately, Thévenin and Norton equivalent circuits can also be evaluated experimentally by means of very simple techniques. The basic idea is that the Thévenin voltage is an open-circuit voltage and the Norton current is a short-circuit current. It should therefore be possible to conduct appropriate measurements to determine these quantities. Once v_T and i_N are known, we can determine the Thévenin resistance of the circuit being analyzed according to the relationship

$$R_T = \frac{v_T}{i_N} \tag{3.34}$$

How are v_T and i_N measured, then?

Figure 3.71 illustrates the measurement of the open-circuit voltage and short-circuit current for an arbitrary network connected to any load and also illustrates that the procedure requires some special attention, because of the non-ideal nature of any practical measuring instrument. The figure clearly illustrates that in the presence of finite meter resistance, r_m, one must take this quantity into

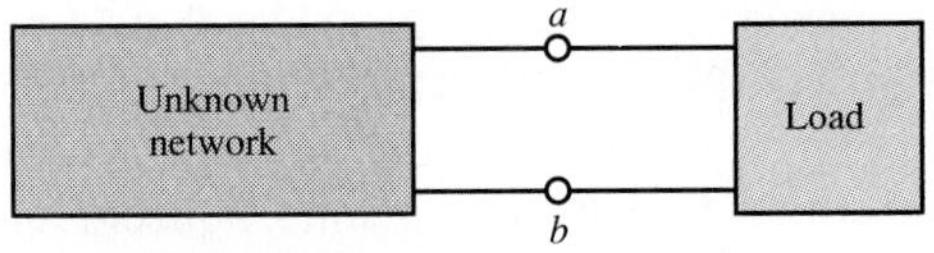

An unknown network connected to a load

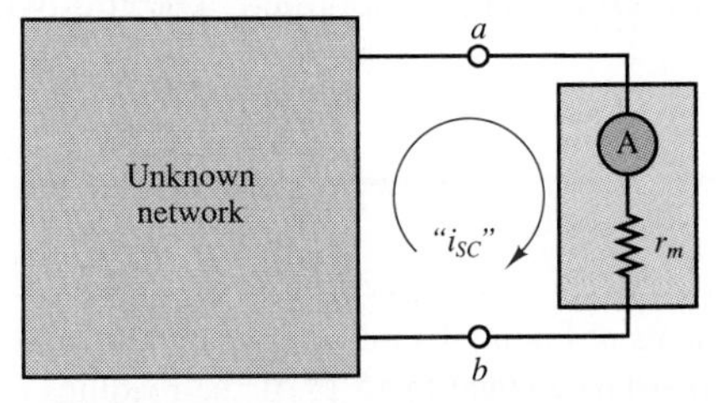

Network connected for measurement of short-circuit current

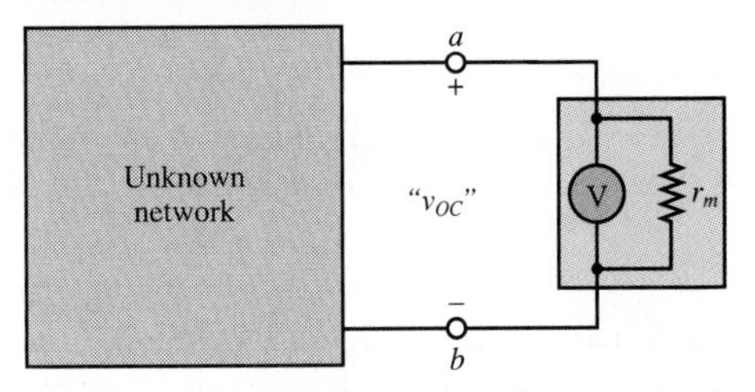

Network connected for measurement of open-circuit voltage

Figure 3.71 Measurement of open-circuit voltage and short-circuit current

account in the computation of the short-circuit current and open-circuit voltage; v_{OC} and i_{SC} appear between quotation marks in the figure specifically to illustrate that the measured "open-circuit voltage" and "short-circuit current" are in fact affected by the internal resistance of the measuring instrument and are not the true quantities.

You should verify that the following expressions for the true short-circuit current and open-circuit voltage apply (see the material on nonideal measuring instruments in Section 2.8):

$$
\begin{aligned}
i_N &= \text{“}i_{SC}\text{”}\left(1 + \frac{r_m}{R_T}\right) \\
v_T &= \text{“}v_{OC}\text{”}\left(1 + \frac{R_T}{r_m}\right)
\end{aligned}
\tag{3.35}
$$

where i_N is the ideal Norton current, v_T the Thévenin voltage, and R_T the true Thévenin resistance. If you recall the earlier discussion of the properties of ideal ammeters and voltmeters, you will recall that for an ideal ammeter, r_m should approach zero, while in an ideal voltmeter, the internal resistance should approach an open circuit (infinity); thus, the two expressions just given permit the determination of the true Thévenin and Norton equivalent sources from an (imperfect) measurement of the open-circuit voltage and short-circuit current, provided that the internal meter resistance, r_m, is known. Note also that, in practice, the internal resistance of voltmeters is sufficiently high to be considered infinite relative to the equivalent resistance of most practical circuits; on the other hand, it is impossible to construct an ammeter that does not have a finite, nonzero internal resistance. If the internal ammeter resistance is known, however, a reasonably accurate measurement of short-circuit current may be obtained. The following example illustrates the point.

Example 3.17

Assume that the measurements of Figure 3.71 are performed with an ideal voltmeter and an ammeter with internal resistance equal to 15 Ω. If the readings obtained are "i_{SC}"= 3.75 mA and "v_{OC}"= 6.5 V, what is the Thévenin resistance of the unknown network?

Solution:

Using the circuit of Figure 3.71 and replacing the unknown source network with its Thévenin equivalent, as shown in Figure 3.72, it should be apparent that

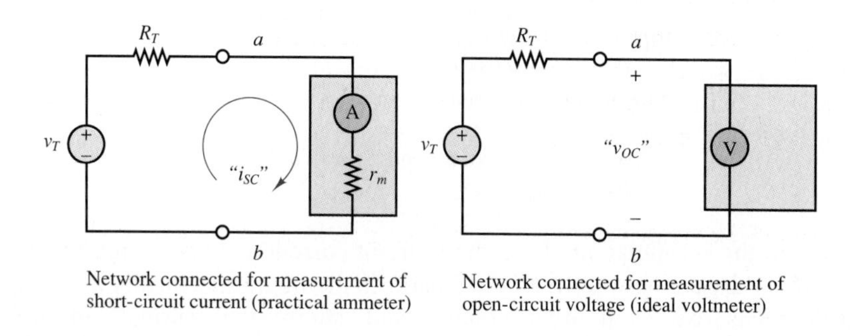

Network connected for measurement of short-circuit current (practical ammeter)

Network connected for measurement of open-circuit voltage (ideal voltmeter)

Figure 3.72

$$\frac{v_{OC}}{i_{SC}} = R_T + r_m$$

Thus,

$$R_T = \frac{\text{"}v_{OC}\text{"}}{\text{"}i_{SC}\text{"}} - r_m = 1{,}733.33 - 15 = 1{,}718.33\ \Omega$$

One last comment is in order concerning the practical measurement of the internal resistance of a network. In most cases, it is not advisable to actually short-circuit a network by inserting a series ammeter as shown in Figure 3.71; permanent damage to the circuit or to the ammeter may be a consequence. For example, imagine that you wanted to estimate the internal resistance of an automotive battery; connecting a laboratory ammeter between the battery terminals would surely result in immediate loss of the instrument. Most ammeters are not designed to withstand currents of such magnitude. Thus, the experimenter should pay attention to the capabilities of the ammeters and voltmeters used in measurements of this type, as well as to the (approximate) power ratings of any sources present. However, there are established techniques especially designed to measure large currents.

3.6 MAXIMUM POWER TRANSFER

The reduction of any linear resistive circuit to its Thévenin or Norton equivalent form is a very convenient conceptualization, as far as the computation of load-related quantities is concerned. One such computation is that of the power absorbed by the load. The Thévenin and Norton models imply that some of the power generated by the source will necessarily be dissipated by the internal circuits within the source. Given this unavoidable power loss, a logical question to ask is, How much power can be transferred to the load from the source under the most ideal conditions? Or, alternatively, What is the value of the load resistance that will absorb maximum power from the source? The answer to these questions is contained in the **maximum power transfer theorem,** which is the subject of the present section.

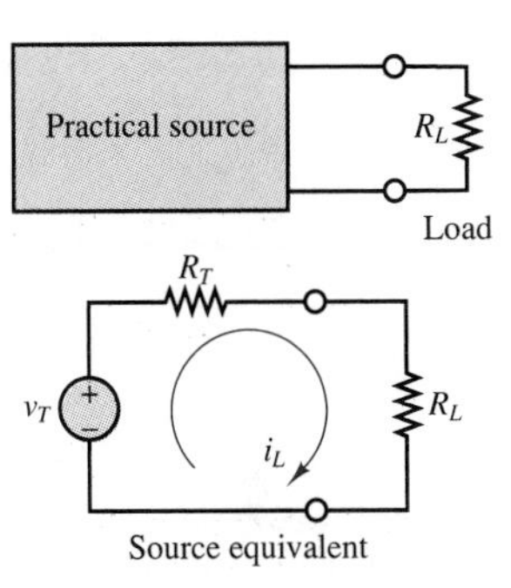

Given v_T and R_T, what value of R_L will allow for maximum power transfer?

Figure 3.73 Power transfer between source and load

The model employed in the discussion of power transfer is illustrated in Figure 3.73, where a practical source is represented by means of its Thévenin equivalent circuit. The maximum power transfer problem is easily formulated if we consider that the power absorbed by the load, P_L, is given by the expression

$$P_L = i_L^2 R_L \tag{3.36}$$

and that the load current is given by the familiar expression

$$i_L = \frac{v_T}{R_L + R_T} \tag{3.37}$$

Combining the two expressions, we can compute the load power as

$$P_L = \frac{v_T^2}{(R_L + R_T)^2} R_L \tag{3.38}$$

To find the value of R_L that maximizes the expression for P_L (assuming that V_T and R_T are fixed), the simple maximization problem

$$\frac{dP_L}{dR_L} = 0 \tag{3.39}$$

must be solved. Computing the derivative, we obtain the following expression:

$$\frac{dP_L}{dR_L} = \frac{v_T^2(R_L + R_T)^2 - 2v_T^2 R_L(R_L + R_T)}{(R_L + R_T)^4} \tag{3.40}$$

which leads to the expression

$$(R_L + R_T)^2 - 2R_L(R_L + R_T) = 0 \tag{3.41}$$

It is easy to verify that the solution of this equation is:

$$R_L = R_T \tag{3.42}$$

Thus, in order to transfer maximum power to a load, the equivalent source and load resistances must be **matched,** that is, equal to each other.

This analysis shows that in order to transfer maximum power to a load, given a fixed equivalent source resistance, the load resistance must match the equivalent source resistance. What if we reversed the problem statement and required that the load resistance be fixed? What would then be the value of source resistance that maximizes the power transfer in this case? The answer to this question can be easily obtained by solving Drill Exercise 3.23.

Example 3.18

A greatly simplified model of a stereo amplifier and a speaker is shown in Figure 3.74. In order to transfer maximum power to the speaker, we should select $R_L = R_T$, as illustrated in the preceding discussion. Suppose, instead, that, not knowing that the internal resistance of the amplifier is $R_T = 8\ \Omega$, we have connected a mismatched speaker, with $R_L = 16\ \Omega$. We would like to determine how much more power we could be delivering to the speaker if its resistance were matched to that of the amplifier.

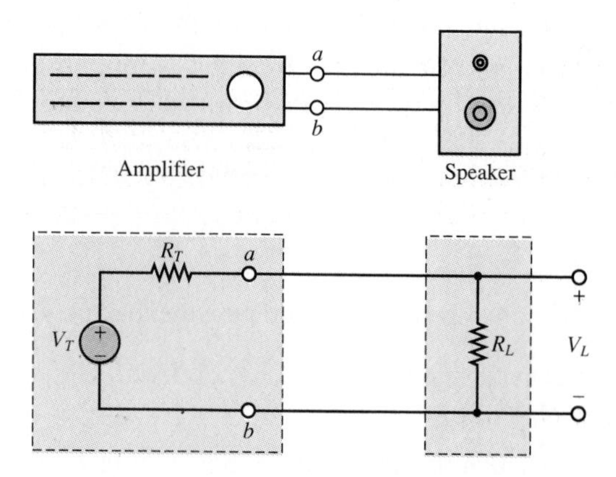

Figure 3.74 A simplified model of an audio system

Solution:

With the 16-Ω speaker, the power delivered by the amplifier is computed as follows:

$$v_L = \frac{R_L}{R_L + R_T} v_T = \frac{2}{3} v_T$$

$$P_L = \frac{v_L^2}{R_L} = \frac{4}{9R_L} v_T^2 = (0.0278) v_T^2$$

If the speaker resistance were matched to the amplifier resistance, that is, equal to 8 Ω, the power delivered to the speaker under matched load conditions—which we will call P'_L—would be greater, according to the following calculation. Since the load resistance is matched to the source, we have

$$v'_L = \frac{v_T}{2}$$

and therefore

$$P'_L = \frac{v'^2_L}{R_L} = \frac{1}{4R_L}v_T^2 = (0.03125)v_T^2$$

with an increase in power equal to $\frac{0.03125-0.0278}{0.0278} \times 100 = 12.5$ percent. In reality, an audio amplifier is not well represented by a single resistance, but requires a somewhat more complex circuit model. The audiophile can find more interesting tidbits concerning loudspeakers and audio circuits in Chapters 6 and 15.

A problem related to power transfer is that of **source loading.** This phenomenon, which is illustrated in Figure 3.75, may be explained as follows: when a practical voltage source is connected to a load, the current that flows from the source to the load will cause a voltage drop across the internal source resistance, v_{int}; as a consequence, the voltage actually seen by the load will be somewhat lower than the *open-circuit voltage* of the source. As stated earlier, the open-circuit voltage is equal to the Thévenin voltage. The extent of the internal voltage drop within the source depends on the amount of current drawn by the load. With reference to Figure 3.75, this internal drop is equal to iR_T, and therefore the load voltage will be:

$$v_L = v_T - iR_T \tag{3.43}$$

It should be apparent that it is desirable to have as small an internal resistance as possible in a practical voltage source.

In the case of a current source, the internal resistance will draw some current away from the load because of the presence of the internal source resistance; this current is denoted by i_{int} in Figure 3.75. Thus the load will receive only part of the *short-circuit current* available from the source (the Norton current):

$$i_L = i_N - \frac{v}{R_T} \tag{3.44}$$

It is therefore desirable to have a very large internal resistance in a practical current source. You may wish to refer back to the discussion of practical sources to verify that the earlier interpretation of practical sources can be expanded in light of the more recent discussion of equivalent circuits.

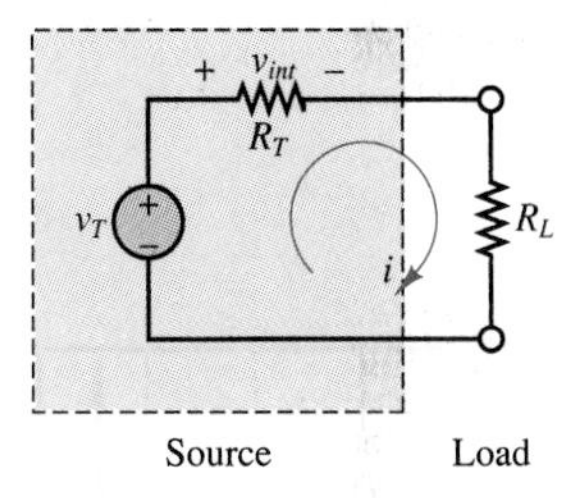

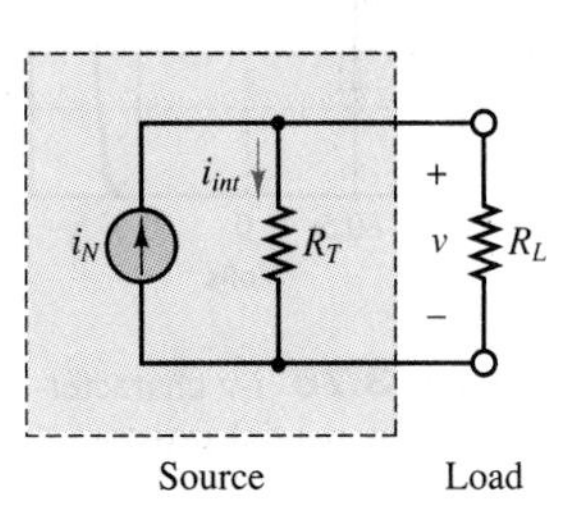

Figure 3.75 Source loading effects

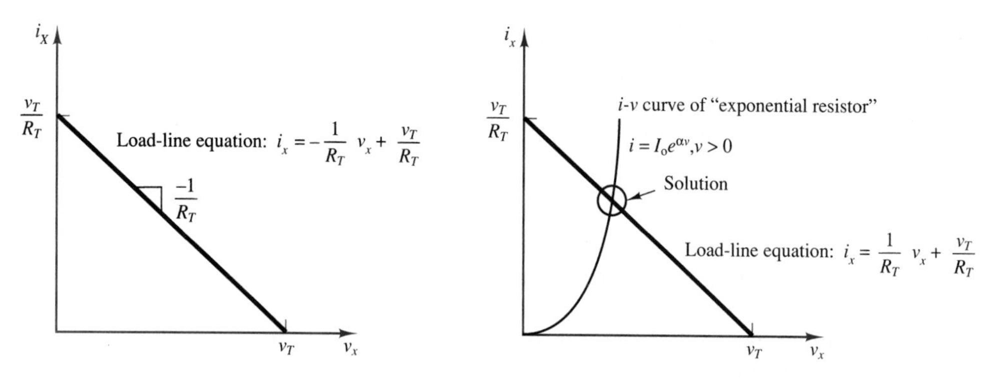

Figure 3.78 Load line

Figure 3.79 Graphical solution of equations 3.48 and 3.49

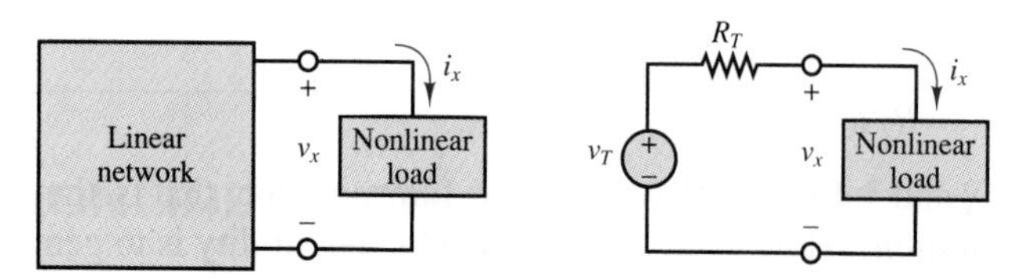

Figure 3.80 Transformation of nonlinear circuit of Thévenin equivalent

developed earlier can be very useful in simplifying problems in which a nonlinear load is present. Example 3.19 illustrates this point.

Example 3.19

The Thévenin equivalent model of a generator has been found experimentally to be $v_T = 15$ V, $R_T = 30\ \Omega$, using the methods described in Section 3.5. The generator is connected to a nonlinear load whose i-v characteristic has also been measured and is available in graphical form, as shown in Figure 3.81. Find the power dissipated by the nonlinear load.

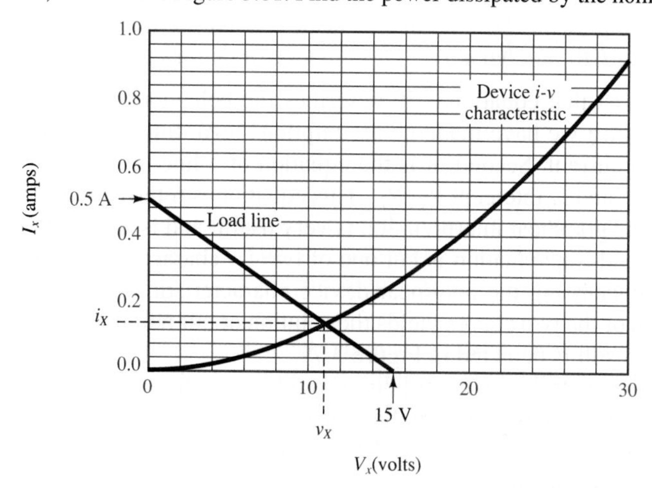

Figure 3.81

Solution:

We can model the circuit as shown in Figure 3.80. The objective is to determine the voltage v_x and the current i_x using graphical methods. The load-line equation for the circuit is given by the expression

$$i_x = -\frac{1}{R_T}v_x + \frac{v_T}{R_T}$$

or

$$i_x = -\frac{1}{30}v_x + \frac{15}{30}$$

This equation represents a line in the i_x-v_x plane, with i_x intercept at 0.5 A and v_x intercept at 15 V. In order to determine the operating point of the circuit, we superimpose the load line on the device *i*-*v* characteristic and determine the solution by finding the intersection of the load line with the device curve. Inspection of the graph reveals that the intersection point is given approximately by

$$i_x = 0.14 \text{ A} \qquad v_x = 11 \text{ V}$$

and therefore the power dissipated by the nonlinear load is

$$P_x = 0.14 \times 11 = 1.54 \text{ W}$$

It is important to observe that the result obtained in this example is, in essence, a description of experimental procedures, indicating that the analytical concepts developed in this chapter also apply to practical measurements.

CONCLUSION

The objective of this chapter was to provide a practical introduction to the analysis of linear resistive circuits. The emphasis on examples is important at this stage, since we believe that familiarity with the basic circuit analysis techniques will greatly ease the task of learning more advanced ideas in circuits and electronics. In particular, your goal at this point should be to have mastered four analysis methods, summarized as follows:

1. *Node voltage and mesh current analysis.* These methods are analogous in concept; the choice of a preferred method depends on the specific circuit. They are generally applicable to the circuits we will analyze in this book and are amenable to solution by matrix methods.
2. *The principle of superposition.* This is primarily a conceptual aid that may simplify the solution of circuits containing multiple sources. It is usually not an efficient method.
3. *Thévenin and Norton equivalents.* The notion of equivalent circuits is at the heart of circuit analysis. Complete mastery of the reduction of linear resistive circuits to either equivalent form is a must.
4. *Numerical and graphical analysis.* These methods apply in the case of nonlinear circuit elements. The load-line analysis method is intuitively appealing and will be employed again in this book to analyze electronic devices.

The material covered in this chapter will be essential to the development of more advanced techniques throughout the remainder of the book.

EIT EXAM REVIEW: DC CIRCUITS

DC circuit analysis is the foundation of electrical engineering. The aim of EIT exam problems in this subject area is to verify the basic understanding of circuit analysis. Chapters 2 and 3 of the text cover all the relevant review topics.

DRILL EXERCISES

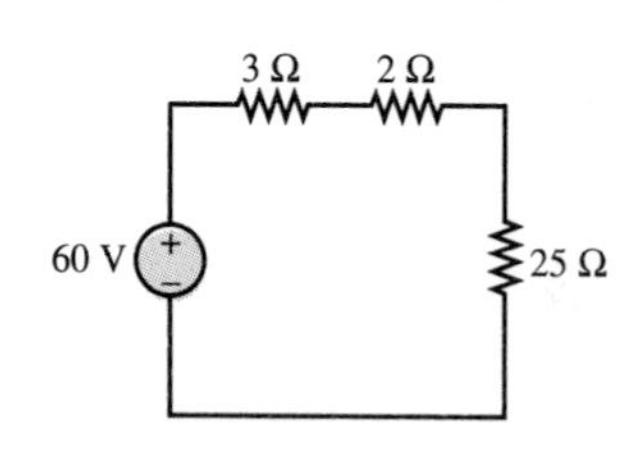

Figure 3.82

3.24 Assuming the connecting wires and the battery have negligible resistance, the voltage across the 25-Ω resistance in Figure 3.82 is

a. 25 V b. 60 V c. 50 V d. 15 V e. 12.5 V

Solution:

This problem calls for application of the voltage divider rule, discussed in Section 2.6. Applying the voltage divider rule to the circuit of Figure 3.82, we have

$$v_{25\ \Omega} = 60\left(\frac{25}{3+2+25}\right) = 50\ \text{V}$$

Thus, the answer is c.

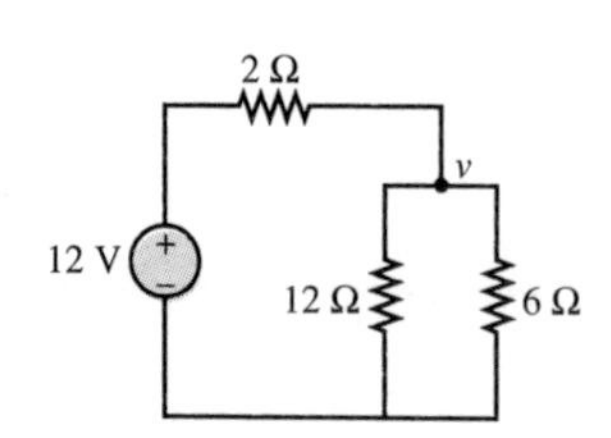

Figure 3.83

3.25 Assuming the connecting wires and the battery have negligible resistance, the voltage across the 6-Ω resistor in Figure 3.83 is

a. 6 V b. 3.5 V c. 12 V d. 4 V e. 3 V

Solution:

This problem can be solved most readily by applying nodal analysis (Section 3.1), since one of the node voltages is already known. Applying KCL at the node v, we obtain

$$\frac{12-v}{2} = \frac{v}{6} + \frac{v}{12}$$

This equation can be solved to show that $v = 4$ V. Note that it is also possible to solve this problem by mesh analysis (Section 3.2). You are encouraged to try this method as well.

3.26 A 125-V battery charger is used to charge a 75-V battery with internal resistance of 1.5 Ω. If the charging current is not to exceed 5 A, the minimum resistance in series with the charger must be

a. 10 Ω b. 5 Ω c. 38.5 Ω d. 41.5 Ω e. 8.5 Ω

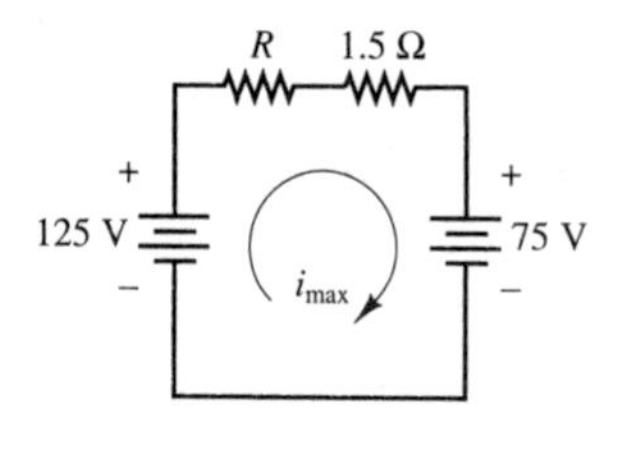

Figure 3.84

Solution:

The circuit of Figure 3.84 describes the charging arrangement. Applying KVL to the circuit of Figure 3.84, we obtain

$$i_{\max}R + 1.5i_{\max} - 125 + 75 = 0$$

and using $i = i_{\max} = 5$ A, we can find R from the following equation:

$$5R + 7.5 - 125 + 75 = 0$$

$$R = 8.5\ \Omega$$

Thus, e is the correct answer.

If you wish to reinforce the concepts illustrated in the sample problems given in this section, you may review some of the examples in Chapters 2 and 3. In particular, the following examples illustrate the fundamental techniques of circuit analysis:

Chapter 2: Examples 2.6, 2.9, 2.10, 2.11
Chapter 3: Examples 3.1, 3.5, 3.11, 3.12, 3.13, 3.14, 3.15

The problems marked "EIT" in the Homework Problems sections of Chapters 2 and 3 will provide additional practice material. Answers to these problems will be found at the end of the book (Appendix B).

KEY TERMS

ANSWERS TO DRILL EXERCISES

3.1 0.2857 A
3.2 −18 V
3.4 5 V
3.5 2 A
3.7 12 V
3.8 1.39 A
3.14 $R_T = 2.5\ \text{k}\Omega$
3.15 $R_T = 7\ \Omega$
3.16 $R_T = 4.0\ \text{k}\Omega$
3.17 $R_T = 7.06\ \Omega$
3.18 $i_L = 0.02857$ A
3.19 $R_T = 30\ \Omega$; $v_{OC} = v_T = 5$ V
3.20 $R_T = 10\ \Omega$; $v_{OC} = v_T = 0.704$ V
3.21 58.8 Ω
3.22 1.64%
3.23 $R_S = 0$ for maximum power transfer to the load

HOMEWORK PROBLEMS

3.1 Use node voltage analysis to find the voltages V_1 and V_2 for the circuit of Figure P3.1.

3.2 Using node voltage analysis, find the voltages V_1 and V_2 for the circuit of Figure P3.2.

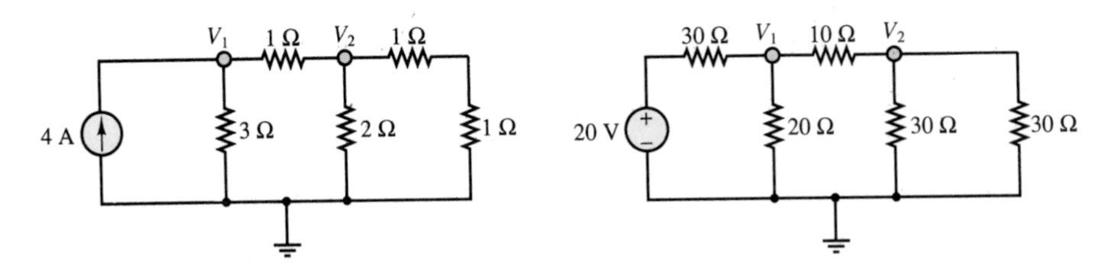

Figure P3.1

Figure P3.2

3.3 Using node voltage analysis in the circuit of Figure P3.3, find the currents i_1 and i_2.

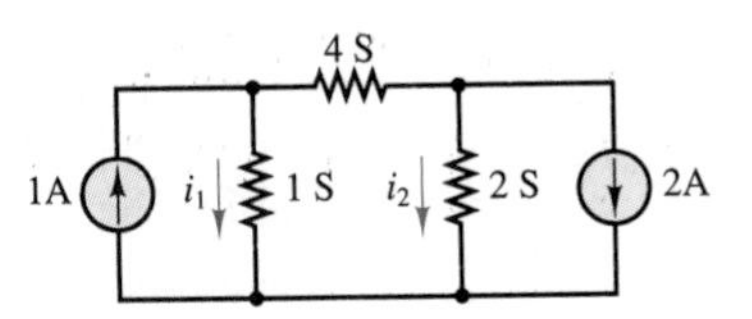

Figure P3.3

3.4 Using node voltage analysis in the circuit of Figure P3.4, find the voltage, v, across the 4-siemens conductance.

3.5 Using node voltage analysis in the circuit of Figure P3.5, find the current, i, through the voltage source.

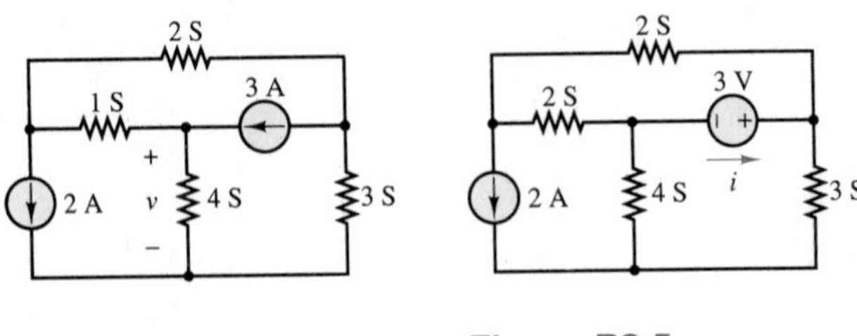

Figure P3.4

Figure P3.5

3.6 Using node voltage analysis in the circuit of Figure P3.6, find the three indicated node voltages.

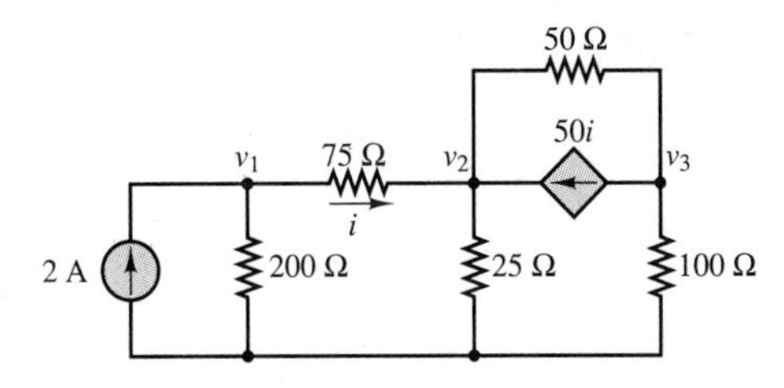

Figure P3.6

3.7 Using node voltage analysis in the circuit of Figure P3.7, find the current, i, drawn from the independent voltage source.

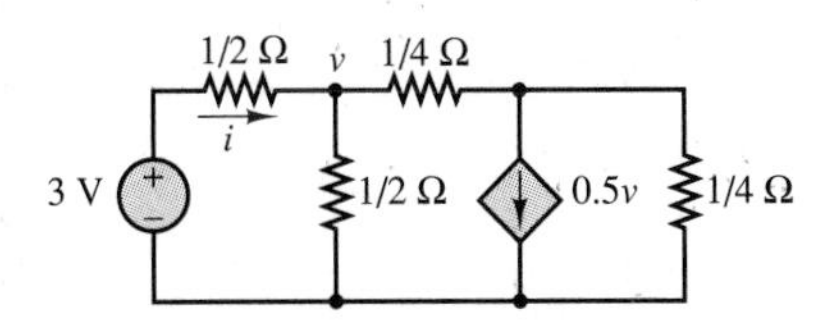

Figure P3.7

EIT 3.8 The circuit shown in Figure P3.8 is a Wheatstone bridge circuit. Use node voltage analysis to determine V_a and V_b, and thus determine $V_a - V_b$.

EIT 3.9 Find the power delivered to the load resistor, R_L, for the circuit of Figure P3.9, using node voltage analysis, given that $R_1 = 2\ \Omega$, $R_V = R_2 = R_L = 4\ \Omega$, $V_S = 4$ V, and $I_S = 0.5$ A.

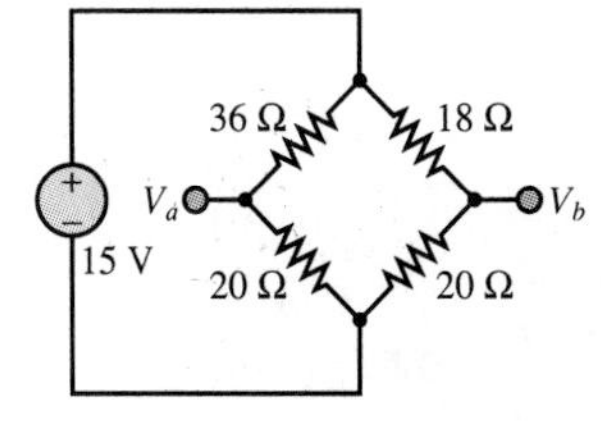

Figure P3.8

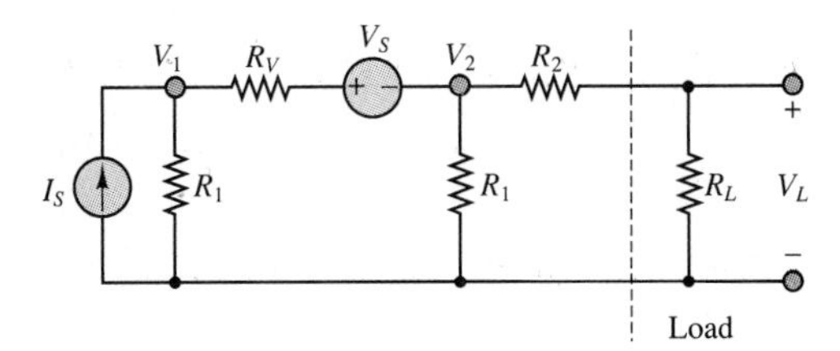

Figure P3.9

***3.10**

a. For the circuit of Figure P3.10, write the node equations necessary to find voltages V_1, V_2, and V_3. Note that $G = 1/R$ = conductance. From the results, note the interesting form that the matrices $[G]$ and $[I]$ have taken in the equation $[G][V] = [I]$, where

$$[G] = \begin{bmatrix} g_{11} & g_{12} & \cdots & g_{1n} \\ g_{21} & g_{22} & & \\ \vdots & & \ddots & \\ g_{n1} & & & g_{nn} \end{bmatrix} \quad \text{and} \quad [I] = \begin{bmatrix} I_1 \\ I_2 \\ \vdots \\ I_n \end{bmatrix}$$

b. Write the matrix form of the node voltage equations again, using the following formulas:

$$g_{ii} = \sum \text{conductances connected to node } i$$
$$g_{ij} = -\sum \text{conductances } \textit{shared} \text{ by nodes } i \text{ and } j$$
$$I_i = \sum \text{all } \textit{source} \text{ currents into node } i$$

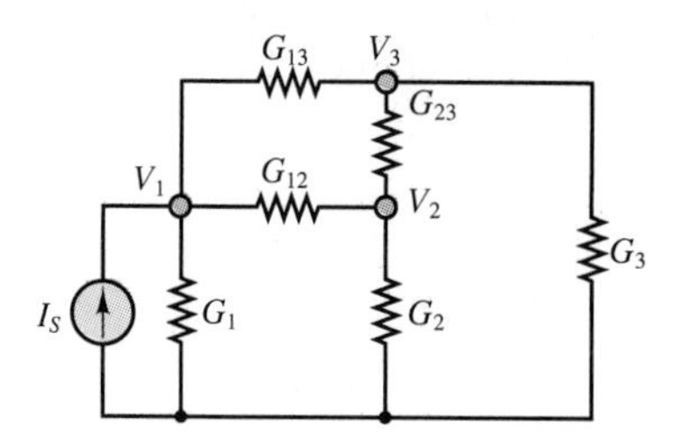

Figure P3.10

3.11 Using mesh current analysis, find the currents i_1 and i_2 for the circuit of Figure P3.11.

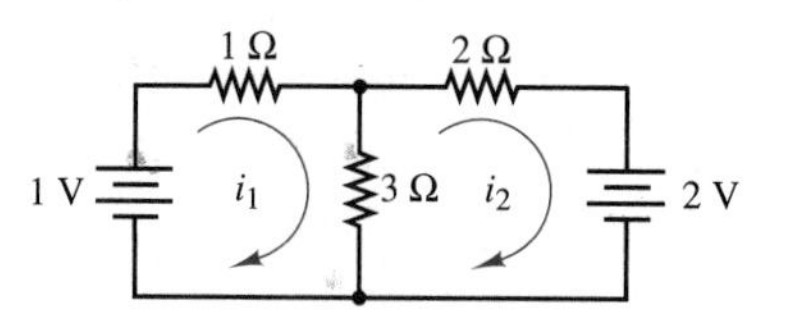

Figure P3.11

EIT 3.12 Using mesh current analysis, find the currents I_1 and I_2 and the voltage across the top 10-Ω resistor in the circuit of Figure P3.12.

3.13 Using mesh current analysis, find the voltage, v, across the 3-Ω resistor in the circuit of Figure P3.13.

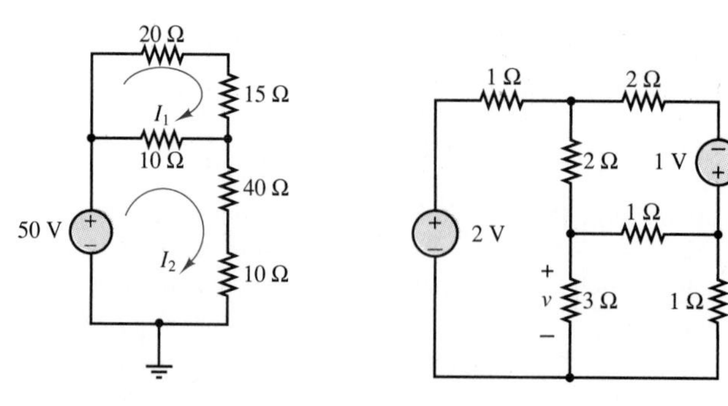

Figure P3.12 **Figure P3.13**

3.14 Using mesh current analysis, find the current, i, through the 2-Ω resistor on the right in the circuit of Figure P3.14.

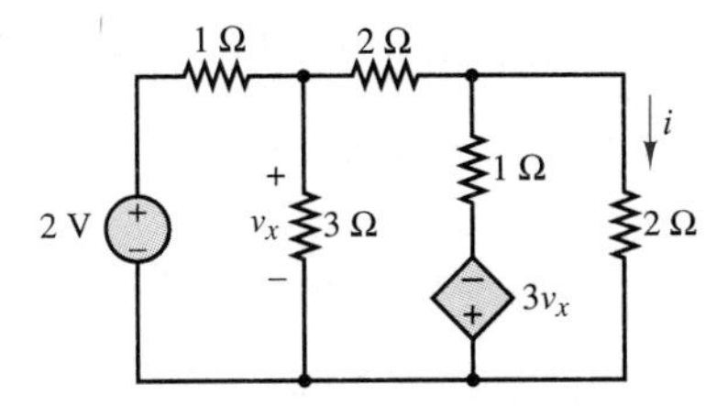

Figure P3.14

3.15 Using mesh current analysis, find the currents I_1, I_2, and I_3 and the voltage across the 40-Ω resistor in the circuit of Figure P3.15 (assume polarity according to I_2).

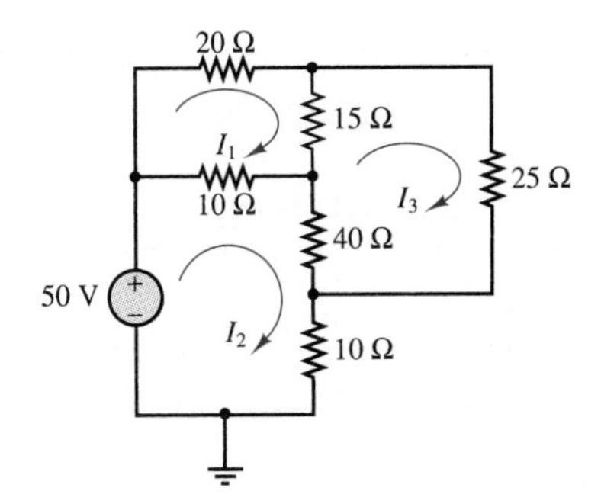

Figure P3.15

3.16 Using mesh current analysis, find the voltage, v, across the current source in the circuit of Figure P3.16.

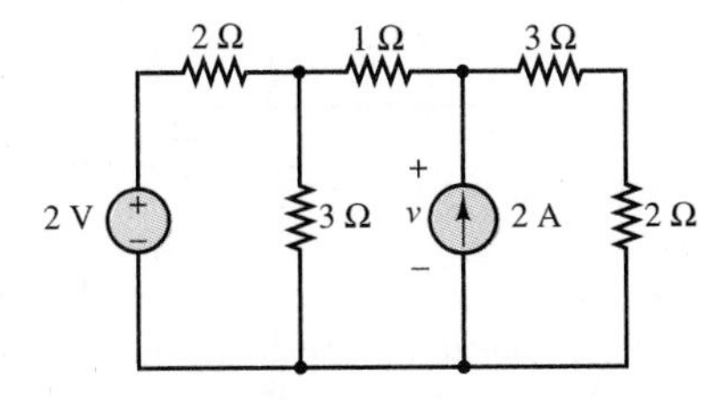

Figure P3.16

3.17 Using mesh current analysis, find the current, i, through the voltage source in the circuit of Figure P3.5.

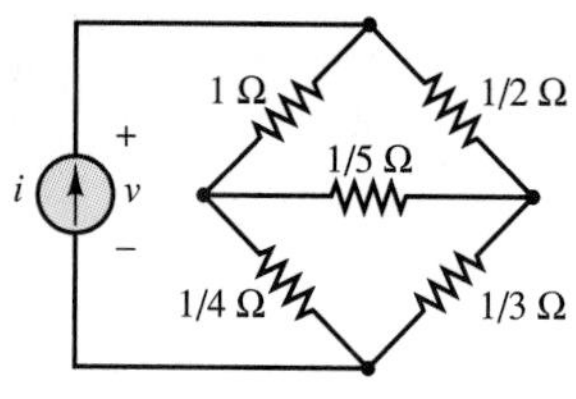

Figure P3.17

3.18 Using mesh current analysis, find the current, i, in the circuit of Figure P3.6.

3.19 Using mesh current analysis, find the equivalent resistance, $R = v/i$, seen by the source of the circuit in Figure P3.17.

3.20 Using mesh current analysis, find the voltage gain, $A_v = v_2/v_1$, in the circuit of Figure P3.18.

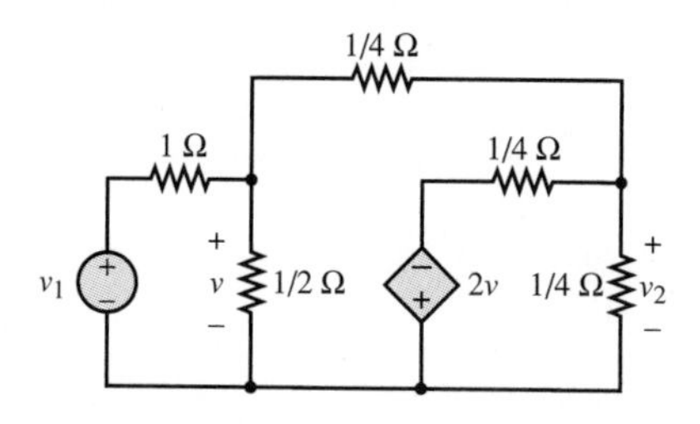

Figure P3.18

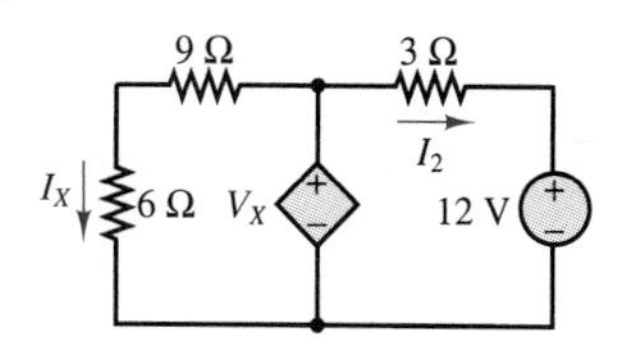

Figure P3.19

EIT 3.21 Using mesh current analysis, find the voltage across the 1-Ω resistor in Figure 3.44, given that $R_L = 10\ \Omega$.

3.22 Find the unknown current I_x and the unknown voltage V_x, using mesh current analysis, for the circuit of Figure P3.19. The dependent voltage source is related to the currents I_x and I_2 by $V_x = 5(I_2 - I_x)$.

EIT 3.23 For the Wheatstone bridge circuit of Figure P3.8, use mesh current methods to determine V_a and V_b and thus to determine $V_a - V_b$.

***3.24**

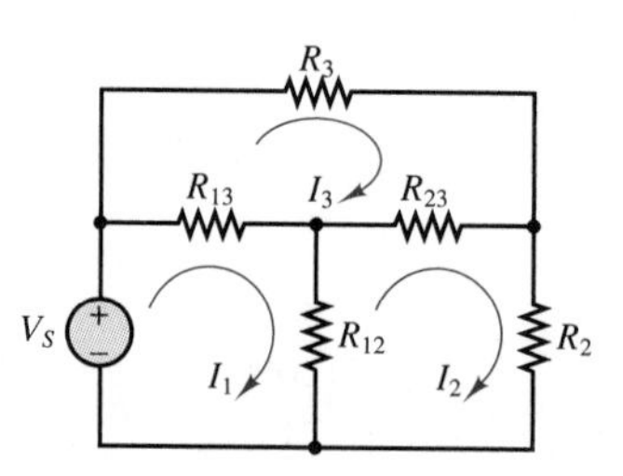

Figure P3.20

a. For the circuit of Figure P3.20, write the mesh equations in matrix form. Notice the form of the [R] and [V] matrices in [R][I] = [V], where

$$[R] = \begin{bmatrix} r_{11} & r_{12} & \cdots & r_{1n} \\ r_{21} & r_{22} & & \\ \vdots & & \ddots & \\ r_{n1} & & & r_{nn} \end{bmatrix} \quad \text{and} \quad [V] = \begin{bmatrix} V_1 \\ V_2 \\ \vdots \\ V_n \end{bmatrix}$$

b. Write the matrix form of the mesh equations again by using the following formulas:

$$r_{ii} = \sum \text{resistances around loop } i$$
$$r_{ij} = -\sum \text{resistances } \textit{shared} \text{ by loops } i \text{ and } j$$
$$V_i = \sum \text{source voltages around loop } i$$

3.25 Find the Thévenin equivalent of the circuit connected to R_L in Figure P3.21.

3.26 Find the Norton equivalent of the circuit connected to R_L in Figure P3.22.

3.27 Find the Thévenin equivalent circuit as seen by the 3-Ω resistor for the circuit of Figure P3.23.

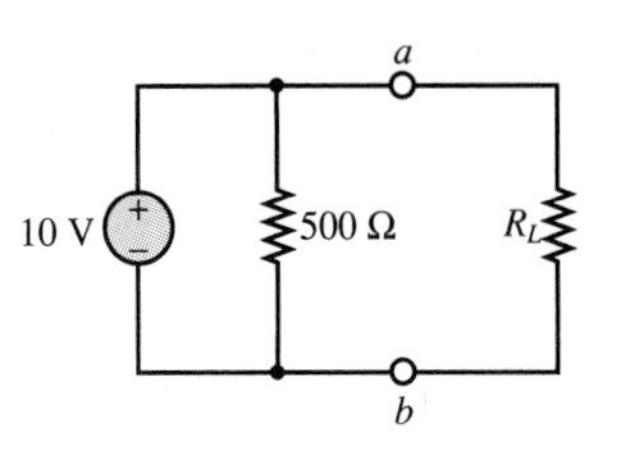

Figure P3.21

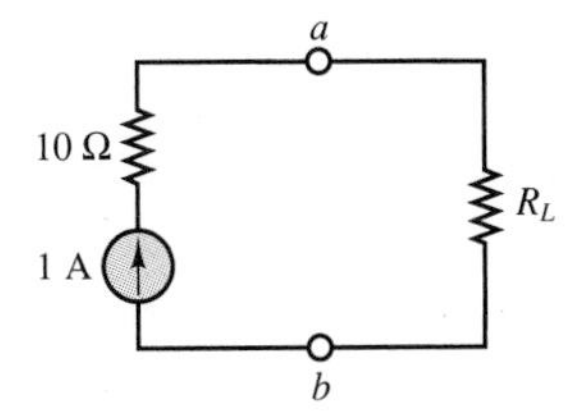

Figure P3.22

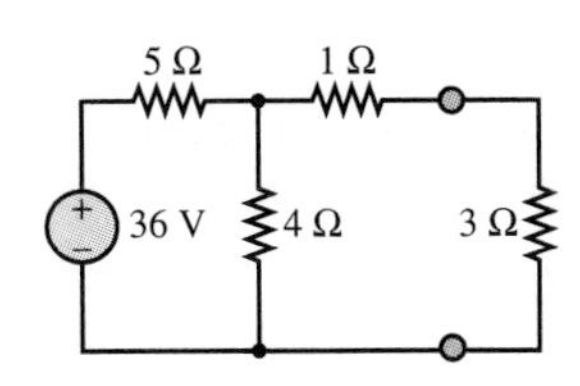

Figure P3.23

3.28 Find the voltage, v, across the 3-Ω resistor in the circuit of Figure P3.24 by replacing the remainder of the circuit with its Thévenin equivalent.

3.29 Find the Thévenin equivalent for the circuit of Figure P3.25.

3.30 Find the Thévenin equivalent for the circuit of Figure P3.26.

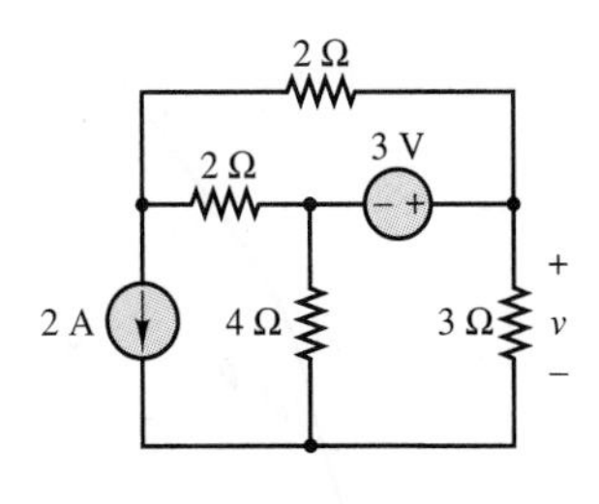

Figure P3.24

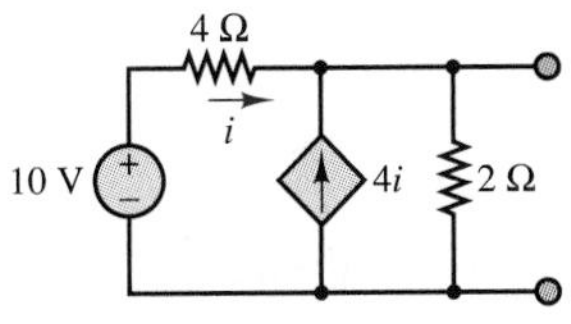

Figure P3.25

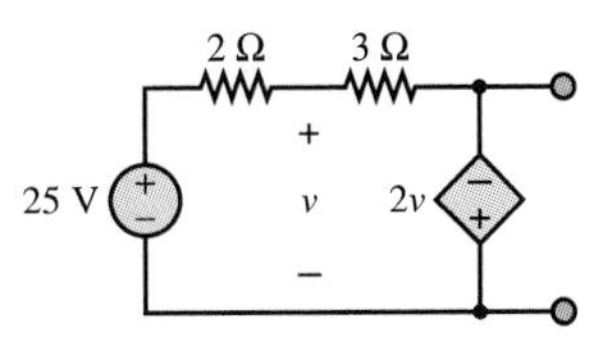

Figure P3.26

3.31 Find the Norton equivalent of the circuit of Figure P3.25.

3.32 Find the Norton equivalent of the circuit of Figure P3.27.

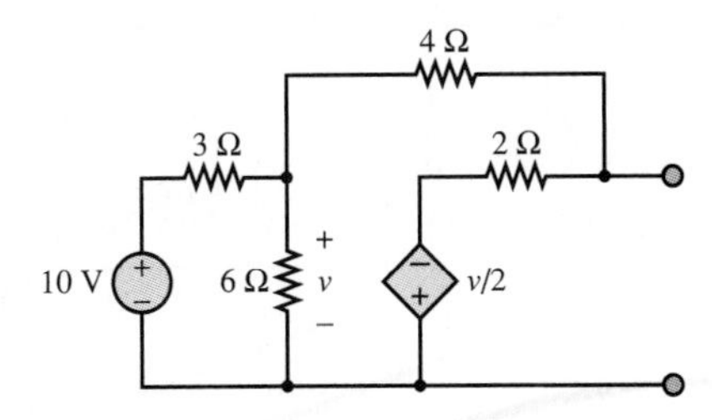

Figure P3.27

3.33 Find the Norton equivalent of the circuit to the left of the 2-Ω resistor in Figure P3.28.

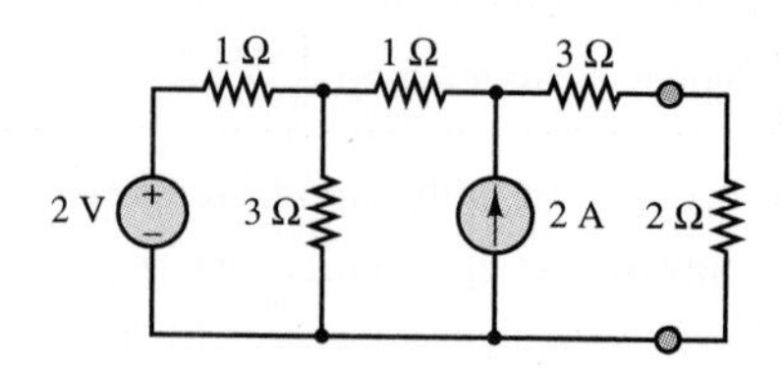

Figure P3.28

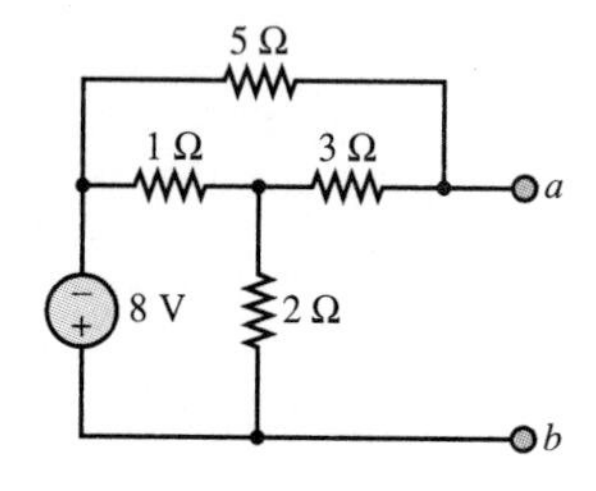

Figure P3.29

3.34 Find the Norton equivalent to the left of terminals *a* and *b* of the circuit shown in Figure P3.29.

EIT 3.35 Find the Thévenin equivalent circuit that the load sees for the circuit of Figure P3.30.

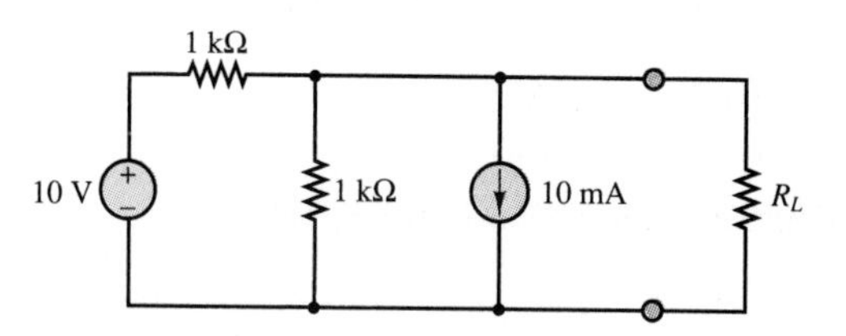

Figure P3.30

3.36 Find the Thévenin equivalent resistance seen by the load resistor, R_L, in the circuit of Figure P3.31.

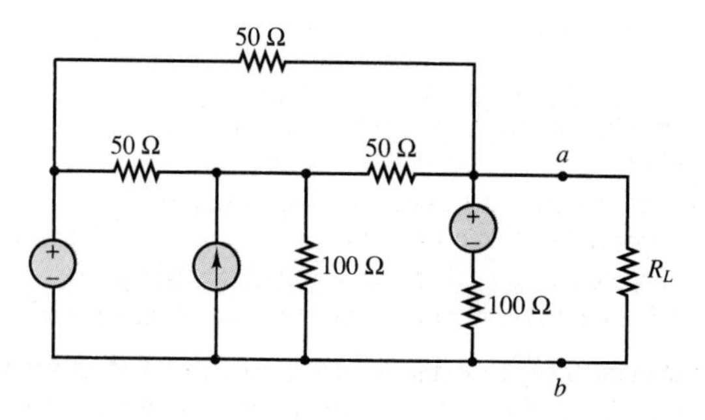

Figure P3.31

EIT 3.37 Find the Thévenin equivalent of the circuit connected to R_L in Figure P3.32.

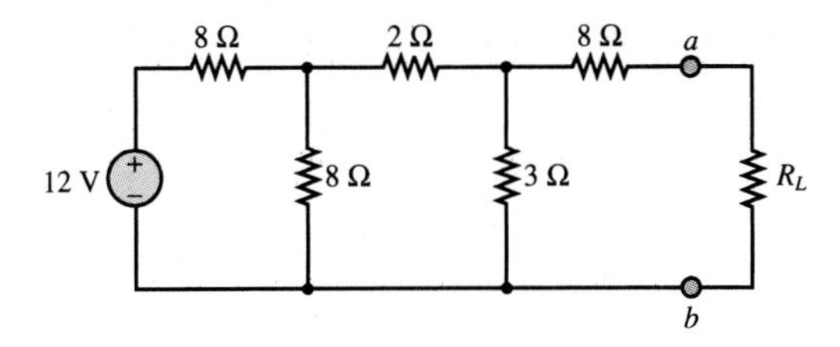

Figure P3.32

3.38 Find the Thévenin equivalent of the circuit connected to R_L in Figure P3.33, where $R_1 = 10\ \Omega$, $R_2 = 20\ \Omega$, $R_g = 0.1\ \Omega$, and $R_p = 1\ \Omega$.

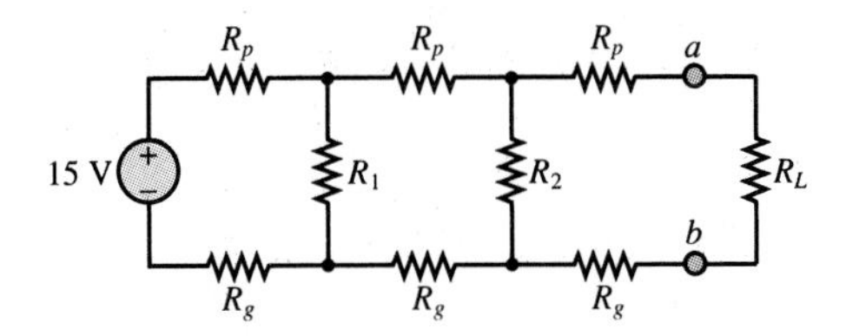

Figure P3.33

EIT **3.39** Consider the circuit of Figure P3.34.

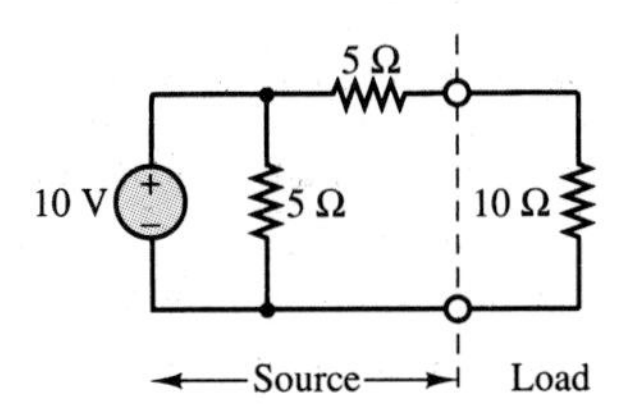

Figure P3.34

a. Find the power delivered to the load resistor, R_L.
b. Find the total power dissipated by the two resistors in the source.
c. Find the Thévenin equivalent circuit of the source.
d. Find the power dissipated by the Thévenin resistance, R_T.
e. Is the power dissipated by the two resistors in the source as found in part b the same as the power dissipated in R_T? Can we say that the Thévenin equivalent circuit is adequate to describe the internal power consumption of the source?

3.40 The Wheatstone bridge circuit shown in Figure P3.35 is used in a number of practical applications. One traditional use is in determining the value of an unknown resistor, R_X.

a. Find the value of the voltage $V_{ab} = V_a - V_b$ in terms of R, R_X, and V_S.
b. If $R = 1\ \text{k}\Omega$, $V_S = 12$ V, and $V_{ab} = 12$ mV, what is the value of R_X?

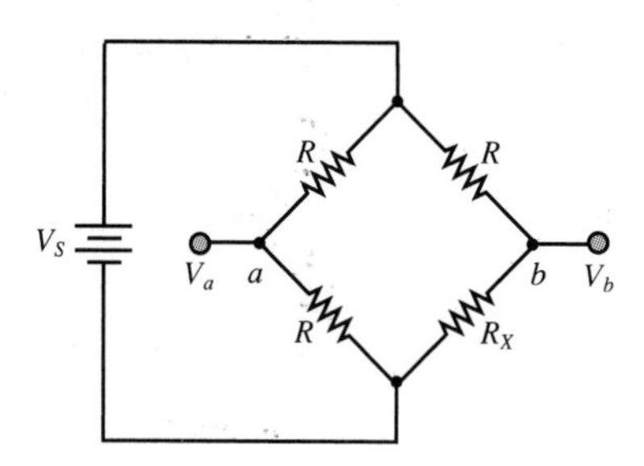

Figure P3.35

EIT 3.41 It is sometimes useful to compute a Thévenin equivalent circuit for a Wheatstone bridge. For the circuit of Figure P3.36,

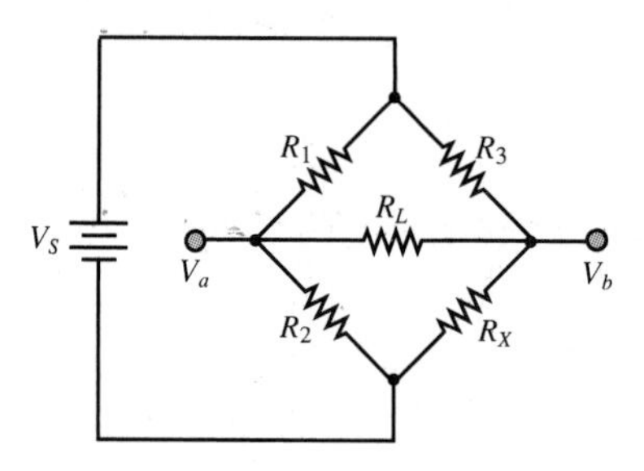

Figure P3.36

a. Find the Thévenin equivalent circuit seen by the load resistor, R_L.
b. If $V_S = 12$ V, $R_1 = R_2 = R_3 = 1$ kΩ, and R_X is the resistance found in part b of the previous problem, use the Thévenin equivalent to compute the power dissipated by R_L, if $R_L = 500\ \Omega$.
c. Find the power dissipated by the Thévenin equivalent resistance, R_T, with R_L included in the circuit.
d. Find the power dissipated by the bridge without the load resistor in the circuit.

3.42 Find the Norton equivalent resistance of the circuit in Figure P3.37 by applying a voltage source v_o and calculating the resulting current i_o.

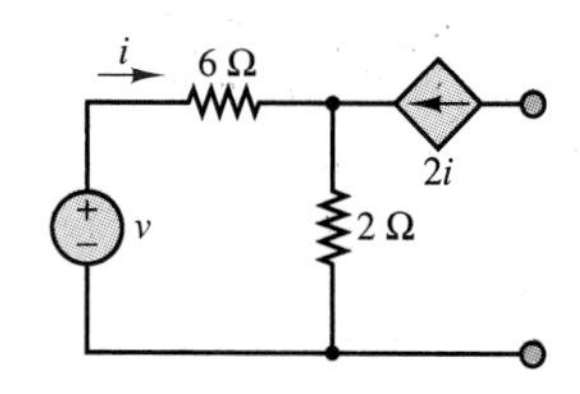

Figure P3.37

***3.43** Find the Thévenin resistance seen by the load, R_L, as well as the open-circuit voltage, v_{OC}, for the circuit of Figure P3.38.

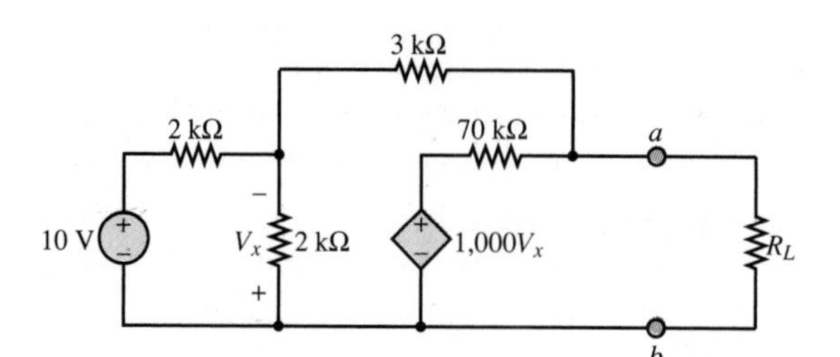

Figure P3.38

3.44 The circuit shown in Figure P3.39 is in the form of what is known as a *differential amplifier.* Find an expression for v_o in terms of v_1 and v_2 using Thévenin's or Norton's theorem.

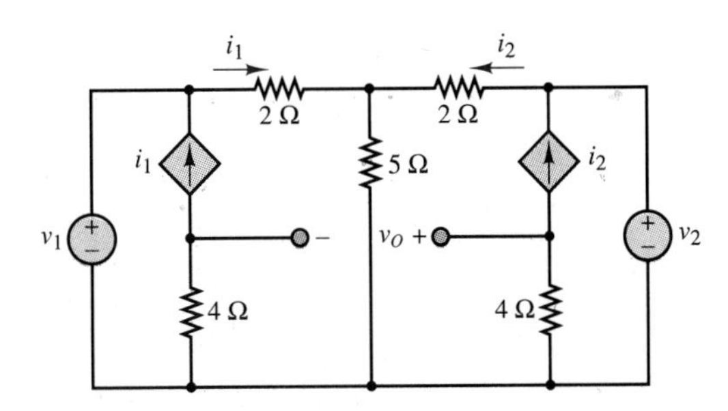

Figure P3.39

3.45 Refer to the circuit of Figure P3.26. Assume the Thévenin voltage is known to be 2 V, positive at the bottom terminal. Find the new source voltage.

***3.46** Solve Example 3.6 using matrix techniques.

3.47 For the circuit of Figure P3.40,

a. Find an expression for the power dissipated by the load as a function of R_L.
b. Plot the power dissipated by the load versus the load resistance. Determine the value of R_L corresponding to maximum load power.

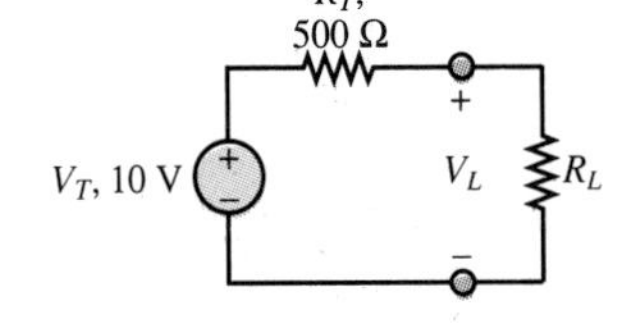

Figure P3.40

EIT 3.48 The circuit of Figure P3.41 has a "square-law resistor" as a load, with *i-v* characteristic given by

$$I^2 = \frac{V}{R'}$$

If $V_T = 12$ V, $R_T = 20\ \Omega$, and $R' = 10\ \Omega$, find the output voltage, V_2, analytically *and* graphically.

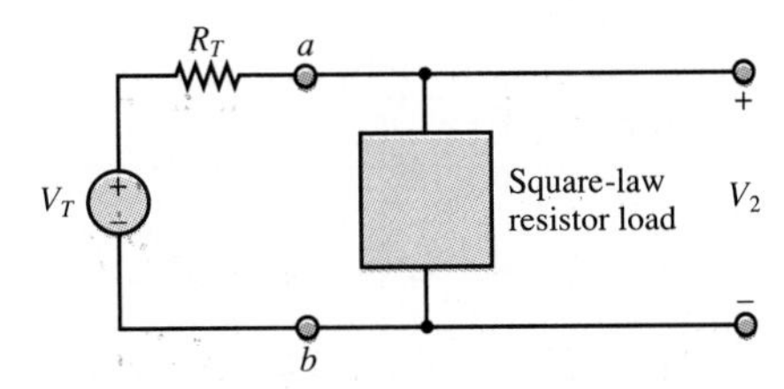

Figure P3.41

3.49 A nonlinear element is placed in the circuit of Figure P3.42. The nonlinear element has the *i-v* characteristics shown in Figure P3.43. Use load-line analysis to determine the operating point of the element and thus V_{out}.

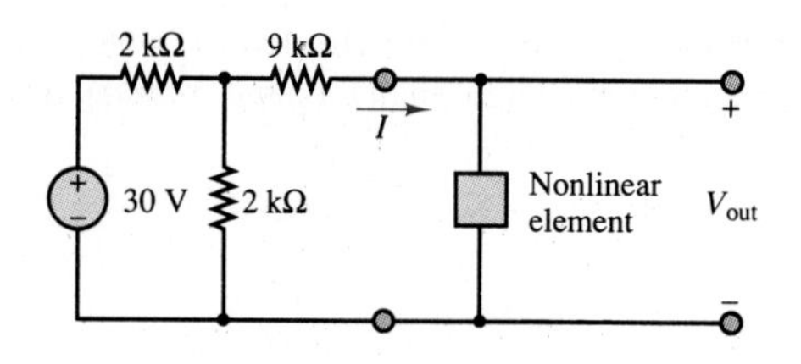

Figure P3.42

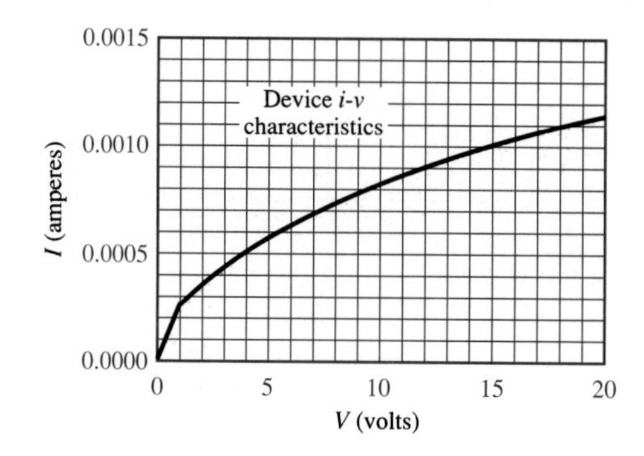

Figure P3.43

EIT 3.50 A nonlinear element is placed in the circuit of Figure P3.44. The nonlinear element has the *i-v* characteristics shown in Figure P3.45. Use load-line analysis to determine the operating point of the element and thus V_{out}.

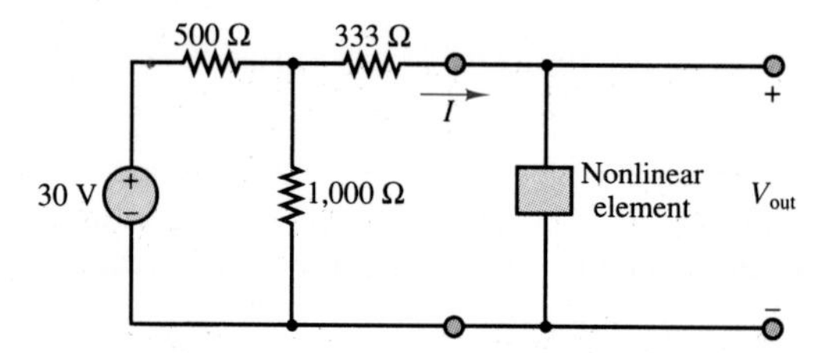

Figure P3.44

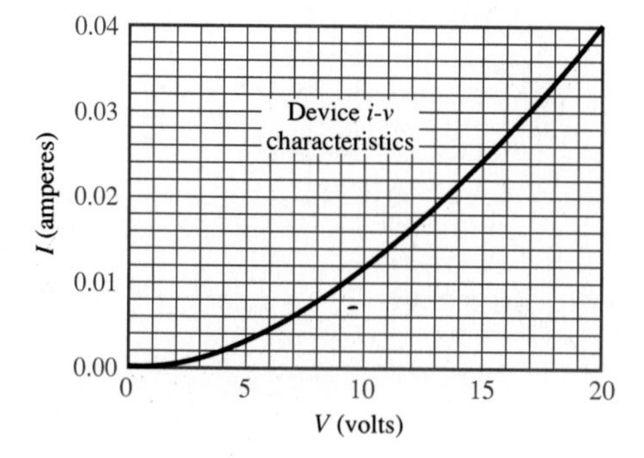

Figure P3.45

3.51 A nonlinear element is placed in the circuit of Figure P3.46. The nonlinear element has the *i-v* characteristics shown in Figure 3.81. Use load-line analysis to determine the operating point of the element and thus V_{out}.

3.52 A nonlinear element is placed in the circuit of Figure P3.47. The nonlinear element has the *i-v* characteristics shown in Figure P3.48. Use load-line analysis to determine the operating point of the element and thus V_{out}.

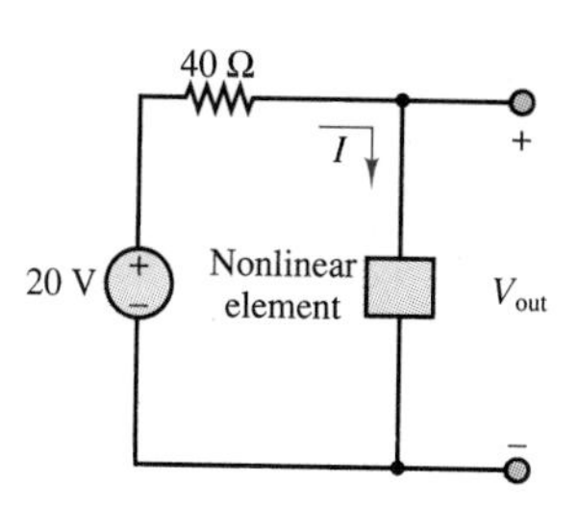

Figure P3.46

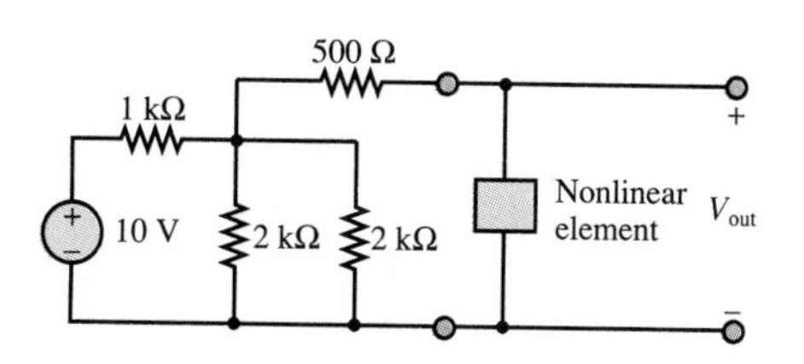

Figure P3.47

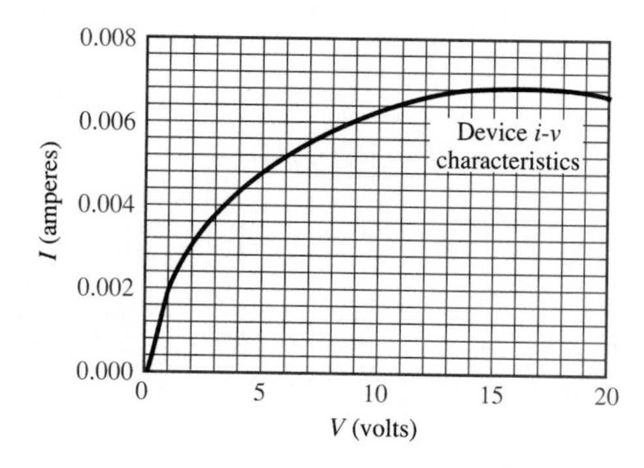

Figure P3.48

EIT 3.53 A thermistor is a device whose resistance and therefore whose *i-v* characteristics change with temperature. Shown in Figure P3.49 is a family of *i-v* characteristic curves for a certain thermistor. The thermistor is connected to the circuit of Figure P3.50.

a. If the voltage V_{out} is 3 V, what is the temperature?
b. If the temperature is $-10°C$, what is the output voltage?
c. Plot V_{out} versus T for the range given in the plot.

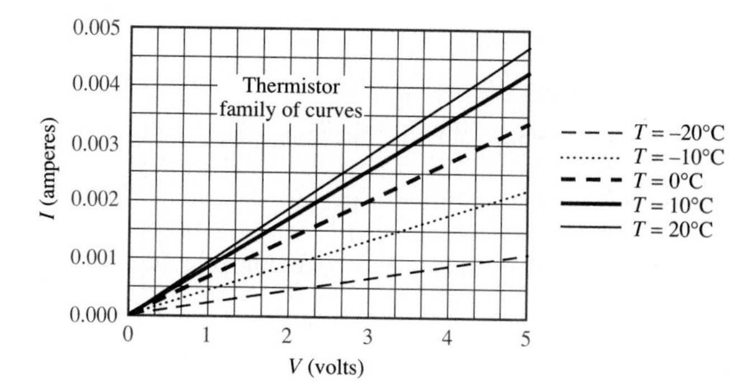

Figure P3.49

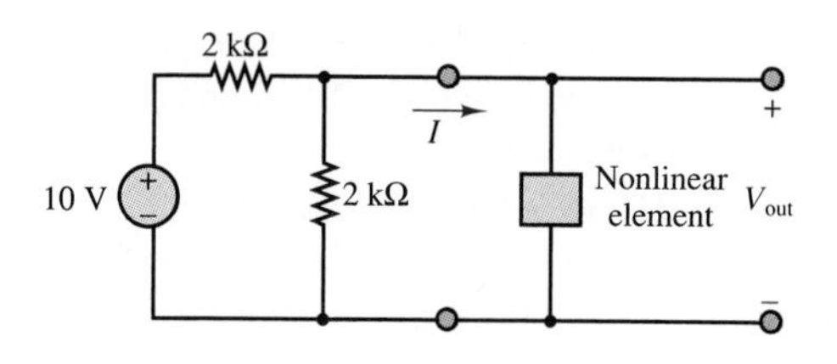

Figure P3.50

3.54 Write the node voltage equations in terms of v_1 and v_2 for the circuit of Figure P3.51. The two nonlinear resistors are characterized by

$$i_a = 2v_a^3$$
$$i_b = v_b^3 + 10v_b$$

Do not solve the resulting equations.

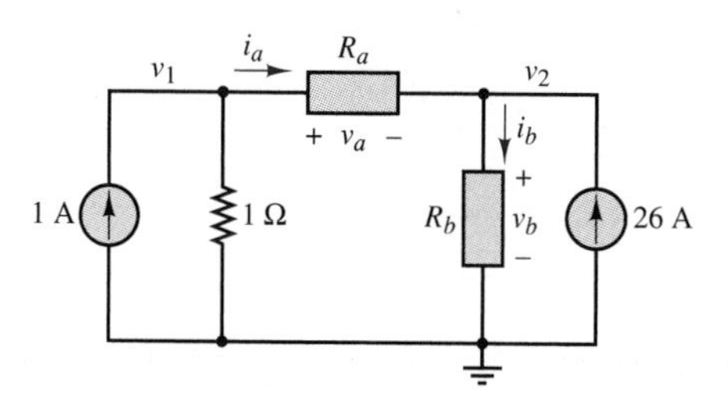

Figure P3.51

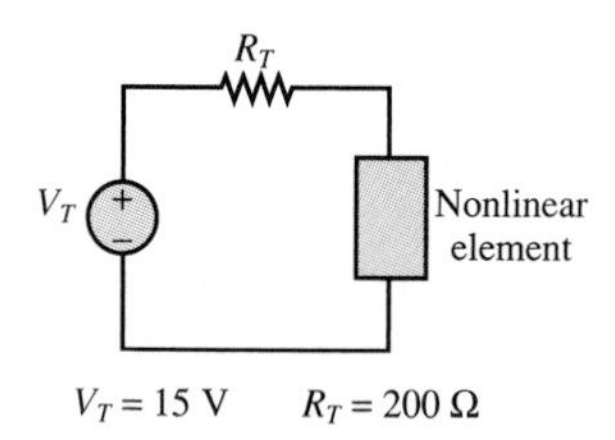

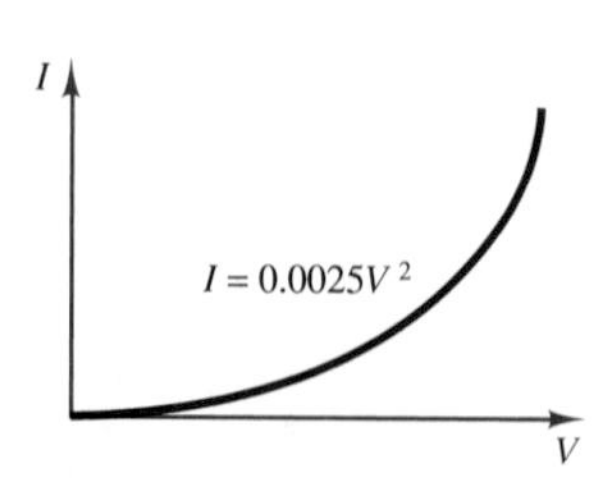

Figure P3.52

3.55 We have seen that some devices do not have a linear current-voltage characteristic for all i and v—that is, R is not constant for all values of current and voltage. For many devices, however, we can estimate the characteristics by piecewise linear approximation. For a portion of the characteristic curve around an operating point, the slope of the curve is relatively constant. The inverse of this slope at the operating point is defined as "incremental resistance," R_{inc}:

$$R_{inc} = \left.\frac{dV}{dI}\right|_{[V_0, I_0]} \approx \left.\frac{\Delta V}{\Delta I}\right|_{[V_0, I_0]}$$

where $[V_0, I_0]$ is the operating point of the circuit.

a. For the circuit of Figure P3.52, find the operating point of the element that has the characteristic curve shown.
b. Find the incremental resistance of the nonlinear element at the operating point of part a.
c. If V_T were increased to 20 V, find the new operating point and the new incremental resistance.

CHAPTER

4

AC Network Analysis

In this chapter we introduce energy-storage elements, dynamic circuits, and the analysis of circuits excited by sinusoidal voltages and currents. Sinusoidal (or AC) signals constitute the most important class of signals in the analysis of electrical circuits. The simplest reason is that virtually all of the electric power used in households and industries comes in the form of sinusoidal voltages and currents.

The chapter is arranged as follows. First, energy-storage elements are introduced, and time-dependent signal sources and the concepts of average and root-mean-square (rms) values are discussed. Next, we analyze the circuit equations that arise when time-dependent signal sources excite circuits containing energy-storage elements; in the course of this discussion, it will become apparent that differential equations are needed to describe the dynamic behavior of these circuits. The remainder of the chapter is devoted to the development of circuit analysis techniques that greatly simplify the solution of dynamic circuits for the special case of sinusoidal signal excitation; the more general analysis of these circuits will be completed in Chapter 6.

By the end of the chapter, you should have mastered a number of concepts that will be used routinely in the remainder of the book; these are summarized as follows:

- Definition of the *i-v* relationship for inductors and capacitors.
- Computation of rms values for periodic waveforms.
- Representation of sinusoidal signals by complex phasors.
- Impedance of common circuit elements.
- AC circuit analysis by Kirchhoff's laws and equivalent-circuit methods.

4.1 ENERGY-STORAGE (DYNAMIC) CIRCUIT ELEMENTS

The ideal resistor was introduced through Ohm's law in Chapter 2 as a useful idealization of many practical electrical devices. However, in addition to resistance to the flow of electric current, which is purely a dissipative (i.e., an energy-loss) phenomenon, electric devices may also exhibit energy-storage properties, much in the same way a spring or a flywheel can store mechanical energy. Two distinct mechanisms for energy storage exist in electric circuits: **capacitance** and **inductance,** both of which lead to the storage of energy in an electromagnetic field. For the purpose of this discussion, it will not be necessary to enter into a detailed electromagnetic analysis of these devices. Rather, two ideal circuit elements will be introduced to represent the ideal properties of capacitive and inductive energy storage: the **ideal capacitor** and the **ideal inductor.** It should be stated clearly that ideal capacitors and inductors do not exist, strictly speaking; however, just like the ideal resistor, these "ideal" elements are very useful for understanding the behavior of physical circuits. In practice, any component of an electric circuit will exhibit some resistance, some inductance, and some capacitance—that is, some energy dissipation and some energy storage.

The situation depicted in this introduction is completely analogous to the description commonly given of mechanical systems, which involves frictional, elastic, and inertial components (namely, dampers, springs, and masses). The analogy with mechanical systems will be developed in parallel with the description of the new circuit elements.

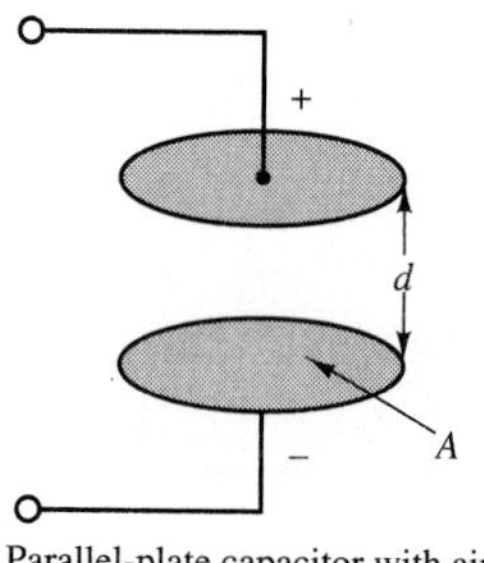

Parallel-plate capacitor with air gap d (air is the dielectric)

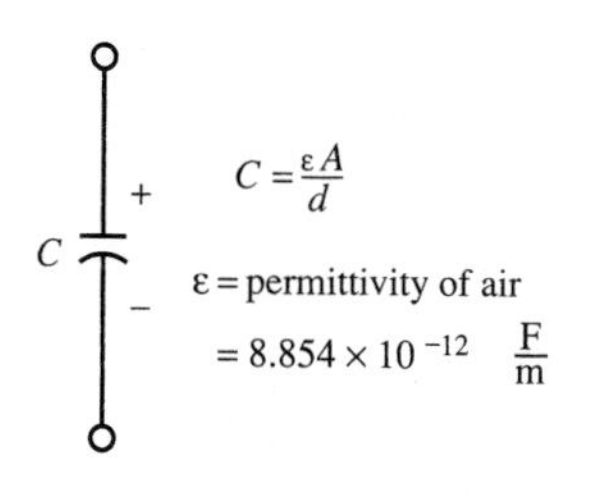

Circuit symbol

Figure 4.1 Structure of parallel-plate capacitor

The Ideal Capacitor

A physical capacitor is a device that can store energy in the form of a charge separation when appropriately polarized by an electric field (i.e., a voltage). The simplest capacitor configuration consists of two parallel conducting plates of cross-sectional area A, separated by air (or another **dielectric**[1] material, such as mica or Teflon). Figure 4.1 depicts a typical configuration and the circuit symbol for a capacitor.

[1] A dielectric material is a material that is not an electrical conductor but contains a large number of electric dipoles, which become polarized in the presence of an electric field.

The presence of an insulating material between the conducting plates does not allow for the flow of DC current; thus, *a capacitor acts as an open circuit in the presence of DC currents*. However, if the voltage present at the capacitor terminals changes as a function of time, so will the charge that has accumulated at the two capacitor plates, since the degree of polarization is a function of the applied electric field, which is time-varying. In a capacitor, the charge separation caused by the polarization of the dielectric is proportional to the external voltage, that is, to the applied electric field:

$$Q = CV \tag{4.1}$$

where the parameter C is called the *capacitance* of the element and is a measure of the ability of the device to accumulate, or store, charge. The unit of capacitance is the coulomb/volt and is called the **farad (F).** The farad is an unpractically large unit; therefore it is common to use microfarads ($1\ \mu\text{F} = 10^{-6}$ F) or picofarads ($1\ \text{pF} = 10^{-12}$ F). From equation 4.1 it becomes apparent that if the external voltage applied to the capacitor plates changes in time, so will the charge that is internally stored by the capacitor:

$$q(t) = Cv(t) \tag{4.2}$$

Thus, although no current can flow through a capacitor if the voltage across it is constant, a time-varying voltage will cause charge to vary in time.

The change with time in the stored charge is analogous to a current. You can easily see this by recalling the definition of current given in Chapter 2, where it was stated that

$$i(t) = \frac{dq(t)}{dt} \tag{4.3}$$

that is, that electric current corresponds to the time rate of change of charge. Differentiating equation 4.2, one can obtain a relationship between the current and voltage in a capacitor:

$$\boxed{i(t) = C\frac{dv(t)}{dt}} \tag{4.4}$$

Equation 4.4 is the defining circuit law for a capacitor. It may be of interest to examine the analogy between equation 4.4 and the equation describing a force acting on a mass. Let $u(t)$ represent the velocity of the mass and $f(t)$ represent the force acting on it. Then Newton's law dictates that

$$f(t) = M\frac{du(t)}{dt} \tag{4.5}$$

This principle is illustrated in Figure 4.2.

If the differential equation that defines the i-v relationship for a capacitor is integrated, one can obtain the following relationship for the voltage across a capacitor:

The defining equation for the capacitance circuit element is analogous to the equation of motion of a mass acted upon by a force.

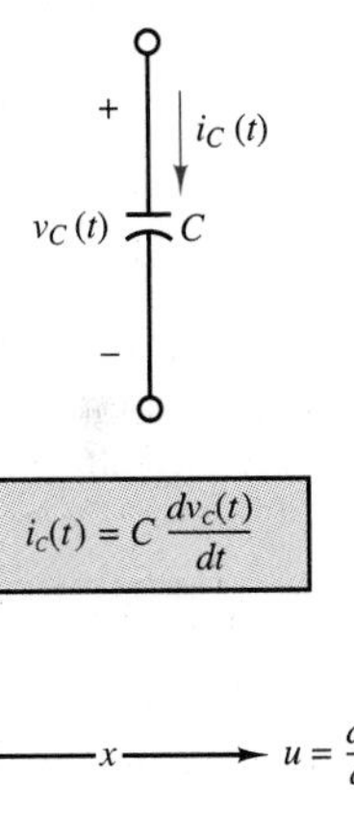

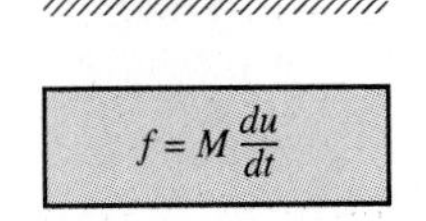

$f = M\frac{du}{dt}$

Figure 4.2 Defining equation for the ideal capacitor, and analogy with force-mass system

$$v_C(t) = \frac{1}{C}\int_{-\infty}^{t} i_C\,dt \tag{4.6}$$

Equation 4.6 indicates that the capacitor voltage depends on the past current through the capacitor, up until the present time, t. Of course, one does not usually have precise information regarding the flow of capacitor current for all past time, and so it is useful to define the initial voltage (or *initial condition*) for the capacitor according to the following, where t_0 is an arbitrary initial time:

$$V_0 = v_C(t = t_0) = \frac{1}{C}\int_{-\infty}^{t_0} i_C\,dt \tag{4.7}$$

The capacitor voltage is now given by the expression

$$v_C(t) = \frac{1}{C}\int_{t_0}^{t} i_C\,dt + V_0 \qquad t \geq t_0 \tag{4.8}$$

The significance of the initial voltage, V_0, is simply that at time t_0 some charge is stored in the capacitor, giving rise to a voltage, $v_C(t_0)$, according to the relationship $Q = CV$. Knowledge of this initial condition is sufficient to account for the entire past history of the capacitor current.

From the standpoint of circuit analysis, it is important to point out that capacitors connected in series and parallel can be combined to yield a single equivalent capacitance. The rule of thumb, which is illustrated in Figure 4.3, is the following:

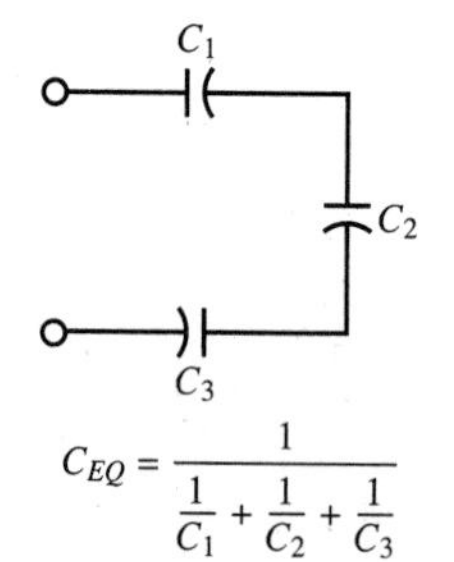

$$C_{EQ} = \frac{1}{\frac{1}{C_1} + \frac{1}{C_2} + \frac{1}{C_3}}$$

Capacitances in series combine like resistors in parallel

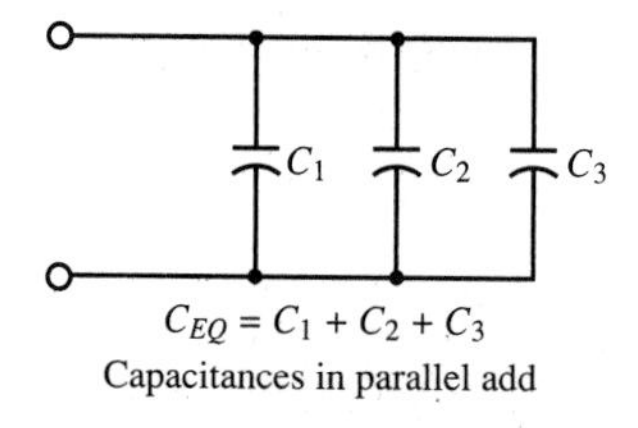

$C_{EQ} = C_1 + C_2 + C_3$

Capacitances in parallel add

Figure 4.3 Combining capacitors in a circuit

> Capacitors in parallel add. Capacitors in series combine according to the same rules used for resistors connected in parallel.

Example 4.1

Calculate the current through a capacitor from knowledge of its terminal voltage. Assume that a 0.1-μF capacitor is charged by an exponential voltage, $v(t)$, shown in Figure 4.4. If $v(t) = 5(1 - e^{-t/10^{-6}})$, with $t \geq 0$, what is the current flowing through the capacitor for $t \geq 0$?

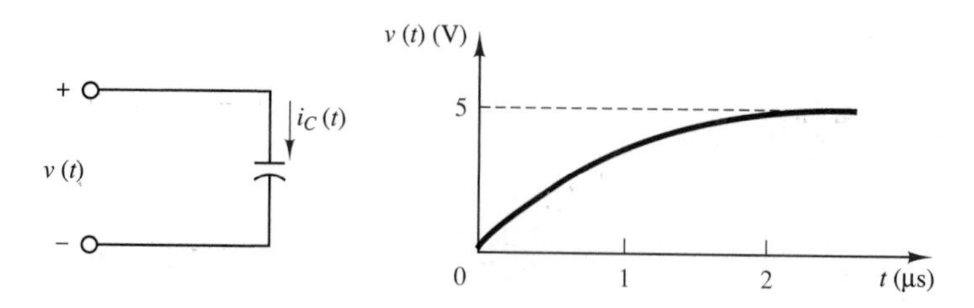

Figure 4.4

Solution

Using the defining relationship for the capacitor, we may obtain the current by differentiating the voltage:

$$\begin{aligned} i_C &= C\frac{dv(t)}{dt} \\ &= \left(10^{-7}\right)\left(\frac{5}{10^{-6}}\right)e^{-t/10^{-6}} \\ &= 0.5e^{-t/10^{-6}} \text{ A} \end{aligned}$$

A sketch of the capacitor current is shown in Figure 4.5. Note how this current jumps instantaneously to 0.5 A as the voltage begins to slowly rise exponentially: the ability of capacitor currents to change suddenly is an important property of capacitors.

It is also worth noting that, as the voltage approaches the constant value 5 V, the capacitor reaches its maximum charge-storage capability for that voltage (since $Q = CV$) and no more current flows through the capacitor. The total charge stored is $Q = 0.5 \times 10^{-6}$ C. This is a fairly small amount of charge, but it can produce a substantial amount of current for a brief period of time. For example, the fully charged capacitor could provide 100 mA of current for a period of time equal to 5 μs:

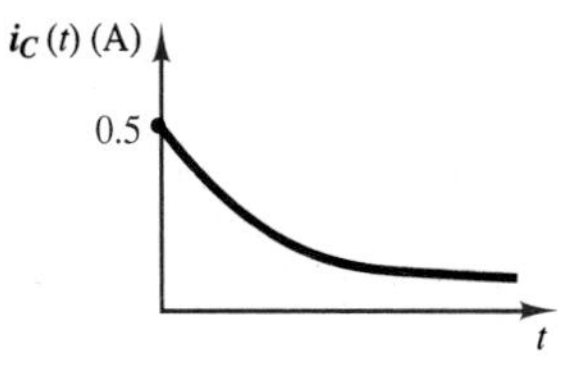

Figure 4.5

$$I = \frac{\Delta Q}{\Delta t} = \frac{0.5 \times 10^{-6}}{5 \times 10^{-6}} = 0.1 \text{ A}$$

There are many useful applications of this energy-storage property of capacitors in practical circuits.

Example 4.2

To illustrate the effect of initial conditions in a capacitor, we provide a simple example in which a 1,000-μF capacitor has been charged to 2 V. At $t = t_0 = 0$, a current of 10 mA is injected into the capacitor; the current is held constant for 1 second. Sketch the capacitor voltage and current waveforms.

Solution:

Integrating the capacitor current from $t = 0$ to $t = 1$ is very simple, since the current is constant in this interval. The general expression for the capacitor voltage is:

$$v_C(t) = \frac{1}{C}\int_{t_0}^{t} i_C \, dt + V_0 \qquad t \geq t_0$$

and therefore,

$$v_C(t) = \frac{1}{C}\int_{0}^{t} I \, dt + 2$$

and

$$\begin{aligned} v_C(t) &= \frac{I}{C}t + 2 \qquad 1 \geq t \geq 0 \\ &= 10t + 2 \end{aligned}$$

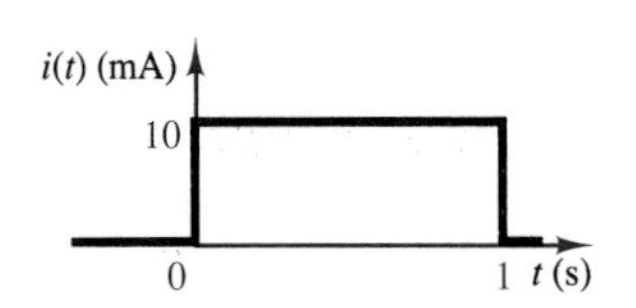

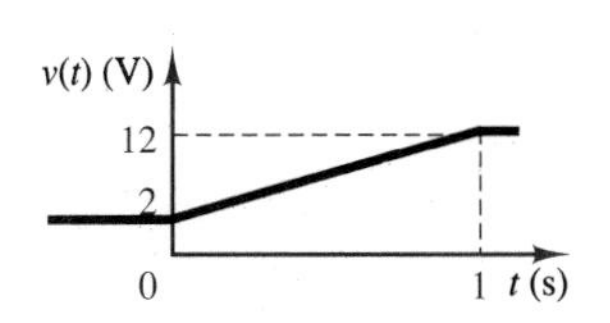

Figure 4.6

Once the current stops, at $t = 1$ s, the capacitor voltage cannot develop any further but remains now at the maximum value it reached at $t = 1$ s: $v_C(t = 1) = 12$ V. Note that this final value of the capacitor voltage after the current source has stopped charging the capacitor depends on two factors: (1) the initial value of the capacitor voltage, and (2) the history of the capacitor current. Figure 4.6 depicts the two waveforms.

Physical capacitors are rarely constructed of two parallel plates separated by air, because this configuration yields very low values of capacitance, unless one is willing to tolerate very large plate areas. In order to increase the capacitance (i.e., the ability to store energy), physical capacitors are often made of tightly rolled sheets of metal film, with a dielectric (paper or Mylar) sandwiched in between. Table 4.1 illustrates typical values, materials, maximum voltage ratings, and useful frequency ranges for various types of capacitors. The voltage rating is particularly important, because any insulator will break down if a sufficiently high voltage is applied across it.

Table 4.1 **Capacitors**

Material	Capacitance range	Maximum voltage (V)	Frequency range (Hz)
Mica	1 pF to 0.1 μF	100–600	10^3–10^{10}
Ceramic	10 pF to 1 μF	50–1,000	10^3–10^{10}
Mylar	0.001 μF to 10 μF	50–500	10^2–10^8
Paper	1,000 pF to 50 μF	100–105	10^2–10^8
Electrolytic	0.1 μF to 0.2 F	3–600	10–10^4

Energy Storage in Capacitors

You may recall that the capacitor was described earlier in this section as an energy-storage element. An expression for the energy stored in the capacitor, $W_C(t)$, may be derived easily if we recall that energy is the integral of power, and that the instantaneous power in a circuit element is equal to the product of voltage and current:

$$\begin{aligned} W_C(t) &= \int P_C(t)\,dt \\ &= \int v_C(t) i_C(t)\,dt \\ &= \int v_C(t) C \frac{dv_C(t)}{dt}\,dt \end{aligned} \tag{4.9}$$

$$W_C(t) = \frac{1}{2} C v_C^2(t) \quad \text{Energy stored in a capacitor (J)}$$

Example 4.3 illustrates the calculation of the energy stored in a capacitor. It will also be observed that the expression for the stored energy in a capacitor is analogous to the expression for the kinetic energy of the mass of Figure 4.2, where

$$W_M(t) = \frac{1}{2}Mu^2(t) \tag{4.10}$$

It should be apparent that these energy equations suggest an analogy in Figure 4.2 between the velocity of the mass and the voltage across the capacitor.

Example 4.3

Consider a 10-μF capacitor that has been charged by a 12-V battery. The charge stored in the capacitor is given by the expression

$$\begin{aligned} Q &= CV = 10^{-5} \times 12 \\ &= 120\ \mu\text{C} \end{aligned}$$

The energy stored in the capacitor is

$$W_C = \frac{1}{2}Cv_C^2 = 720 \times 10^{-6}\ \text{J} = 720\ \mu\text{J}$$

Example 4.4 Capacitive Displacement Transducer and Microphone

As shown in Figure 4.1, the capacitance of a parallel-plate capacitor is given by the expression

$$C = \frac{\varepsilon A}{d}$$

where ε is the **permittivity** of the dielectric material, A the area of each of the plates, and d their separation. The permittivity of air is $\varepsilon_0 = 8.854 \times 10^{-12}$ F/m, so that two parallel plates of area 1 m^2, separated by a distance of 1 mm, would give rise to a capacitance of 8.854×10^{-3} μF, a very small value for a very large plate area. This relative inefficiency makes parallel-plate capacitors impractical for use in electronic circuits. On the other hand, parallel-plate capacitors find application as *motion transducers,* that is, as devices that can measure the motion or displacement of an object. In a capacitive motion transducer, the air gap between the plates is designed to be variable, typically by fixing one plate and connecting the other to an object in motion. Using the capacitance value just derived for a parallel-plate capacitor, one can obtain the expression

$$C = \frac{8.854 \times 10^{-3} A}{x}$$

where C is the capacitance in pF, A is the area of the plates in mm^2, and x is the (variable) distance in mm. It is important to observe that the change in capacitance caused by the displacement of one of the plates is nonlinear, since the capacitance varies as the inverse of the displacement. For small displacements, however, the capacitance varies approximately in a linear fashion.

The *sensitivity, S*, of this motion transducer is defined as the slope of the change in capacitance per change in displacement, x, according to the relation

$$S = \frac{dC}{dx} = -\frac{8.854 \times 10^{-3} A}{2x^2} \frac{\text{pF}}{\text{mm}}$$

Thus, the sensitivity increases for small displacements. This behavior can be verified by plotting the capacitance as a function of x and noting that as x approaches zero, the slope of the nonlinear $C(x)$ curve becomes steeper (thus the greater sensitivity). Figure 4.7 depicts this behavior for a transducer with area equal to 10 mm^2.

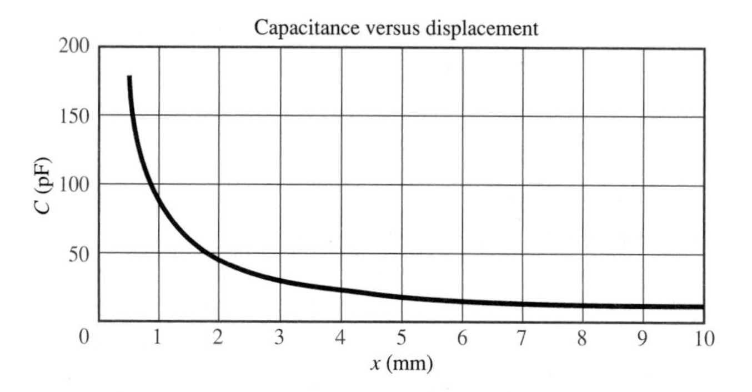

Figure 4.7 Response of a capacitive displacement transducer

This simple capacitive displacement transducer actually finds use in the popular *capacitive (or condenser) microphone,* in which the sound pressure waves act to displace one of the capacitor plates. The change in capacitance can then be converted into a change in voltage or current by means of a suitable circuit. An extension of this concept that permits measurement of differential pressures is shown in simplified form in Figure 4.8. In the figure, a three-terminal variable capacitor is shown to be made up of two fixed surfaces

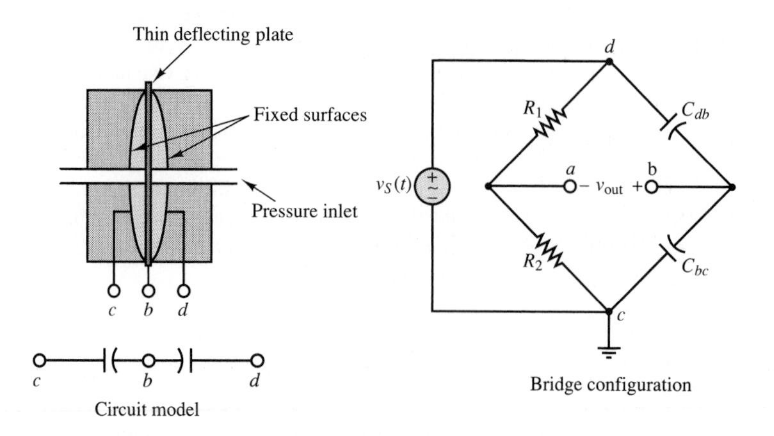

Figure 4.8 Capacitive pressure transducer, and related bridge circuit

(typically, spherical depressions ground into glass disks and coated with a conducting material) and of a deflecting plate (typically made of steel) sandwiched between the glass disks. Pressure inlet orifices are provided, so that the deflecting plate can come into contact with the fluid whose pressure it is measuring. When the pressure on both sides of the deflecting plate is the same, the capacitance between terminals b and d, C_{bd}, will be equal to that between terminals b and c, C_{bc}. If any pressure differential exists, the two capacitances will change, with an increase on the side where the deflecting plate has come closer to the fixed surface and a corresponding decrease on the other side.

This behavior is ideally suited for the application of a bridge circuit, similar to the Wheatstone bridge circuit illustrated in Example 2.12, and also shown in Figure 4.8. In the bridge circuit, the output voltage, v_{out}, is precisely balanced when the differential pressure across the transducer is zero, but it will deviate from zero whenever the two capacitances are not identical because of a pressure differential across the transducer. We shall analyze the bridge circuit later in Example 4.19.

The Ideal Inductor

The ideal inductor is an element that has the ability to store energy in a magnetic field. Inductors are typically made by winding a coil of wire around a **core,** which can be an insulator or a ferromagnetic material, as shown in Figure 4.9. When a current flows through the coil, a magnetic field is established, as you may recall from early physics experiments with electromagnets. In an ideal inductor, the resistance of the wire is zero, so that a constant current through the inductor will flow freely without causing a voltage drop. In other words, *the ideal inductor acts as a short circuit in the presence of DC currents.* If a time-varying voltage is established across the inductor, a corresponding current will result, according to the following relationship:

$$v_L(t) = L\frac{di_L}{dt} \tag{4.11}$$

where L is called the *inductance* of the coil and is measured in **henrys (H),** where

$$1 \text{ H} = 1 \text{ V-s/A} \tag{4.12}$$

Henrys are reasonable units for practical inductors; millihenrys (mH) and microhenrys (μH) are also used.

It is instructive to compare equation 4.11, which defines the behavior of an ideal inductor, with the expression relating capacitor current and voltage:

$$i_C(t) = C\frac{dv_C}{dt} \tag{4.13}$$

We note that the roles of voltage and current are reversed in the two elements, but that both are described by a differential equation of the same form. This *duality* between inductors and capacitors can be exploited to derive the same basic results for the inductor that we already have for the capacitor simply by replacing

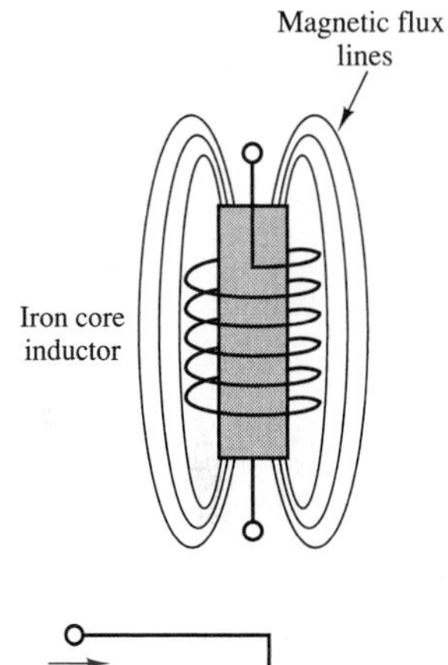

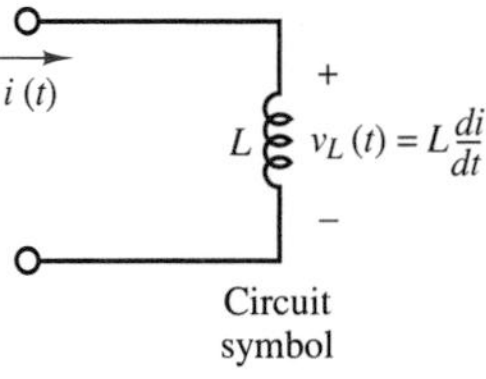

Figure 4.9 Iron-core inductor

the capacitance parameter, C, with the inductance, L, and voltage with current (and vice versa) in the equations we derived for the capacitor. Thus, the inductor current is found by integrating the voltage across the inductor:

$$i_L(t) = \frac{1}{L}\int_{-\infty}^{t} v_L\, dt \tag{4.14}$$

If the current flowing through the inductor at time $t = t_0$ is known to be I_0, with

$$I_0 = i_L(t = t_0) = \frac{1}{L}\int_{-\infty}^{t_0} v_L\, dt \tag{4.15}$$

then the inductor current can be found according to the equation

$$i_L(t) = \frac{1}{L}\int_{t_0}^{t} v_L\, dt + I_0 \qquad t \geq t_0 \tag{4.16}$$

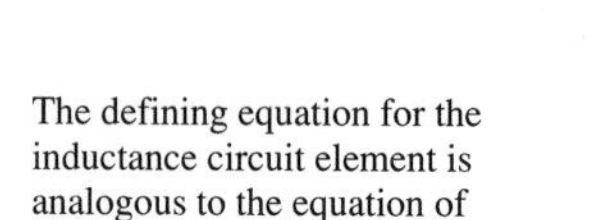

The defining equation for the inductance circuit element is analogous to the equation of motion of a spring acted upon by a force.

It is instructive to compare the behavior of the inductor with that of a linear spring subject to a force. Figure 4.10 illustrates the analogy between the circuit equation for the ideal inductor and the equation of motion of a linear spring with coefficient k, subject to an external force, f. Note the analogy between the force acting on the spring (equation 4.17) and the current through the inductor in equation 4.14:

$$f(t) = k\int u\, dt \tag{4.17}$$

This analogy will be developed further when the energy-storage properties of the ideal inductor are discussed. Series and parallel combinations of inductors behave like resistors, as illustrated in Figure 4.11, and stated as follows:

> Inductors in series add. Inductors in parallel combine according to the same rules used for resistors connected in parallel.

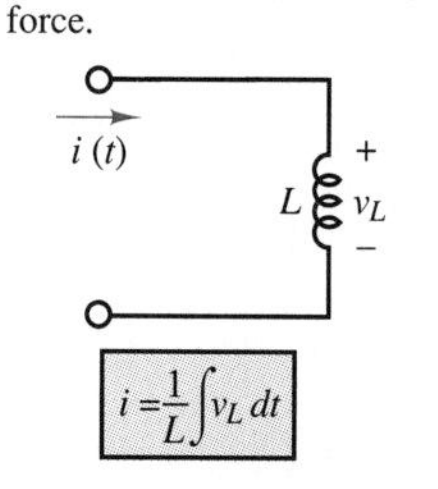

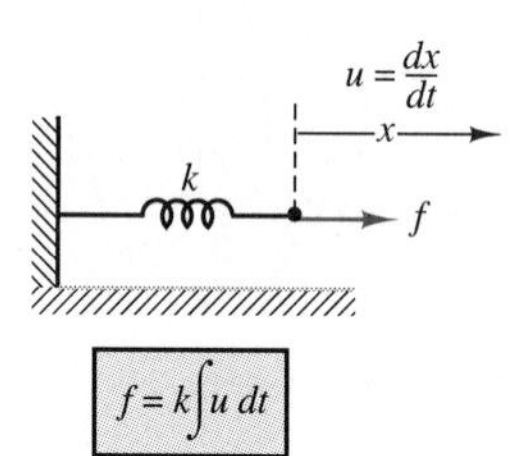

Figure 4.10 Defining equation for the ideal inductor and analogy with force-spring system

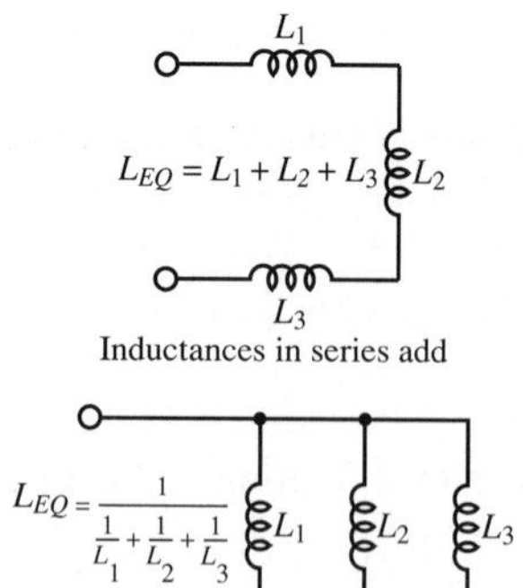

Figure 4.11 Combining inductors in a circuit

Example 4.5

An inductor with $L = 10$ mH is subject to the time-varying current $i(t)$ shown in Figure 4.12. Find the voltage across the inductor.

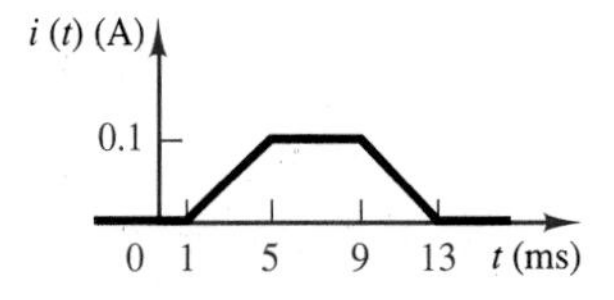

Solution:

The voltage may be found by differentiating the current in a piecewise fashion:

$$v_L(t) = \begin{cases} L\dfrac{di_L(t)}{dt} = 0 \text{ V} & t < 1 \text{ ms} \\ (0.01)\dfrac{0.1}{4 \times 10^{-3}} = 0.25 \text{ V} & 1 \le t \le 5 \text{ ms} \\ 0 & 5 < t \le 9 \text{ ms} \\ -(0.01)\dfrac{0.1}{4 \times 10^{-3}} = -0.25 \text{ V} & 9 \le t \le 13 \text{ ms} \\ 0 & t \ge 13 \text{ ms} \end{cases}$$

i (t) (A)
0.1
0 1 5 9 13 t (ms)

Figure 4.12

Graphically, we may represent the inductor voltage as shown in Figure 4.13.

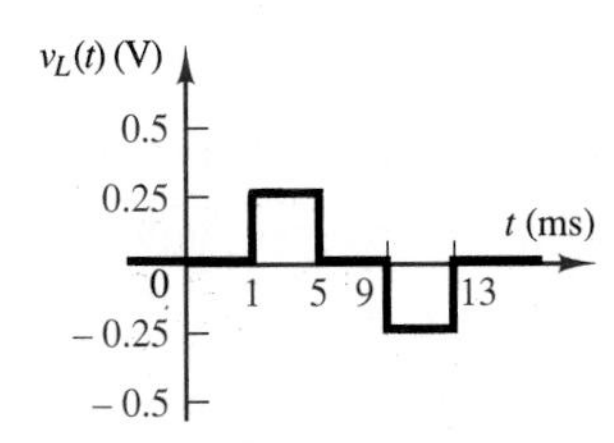

Figure 4.13

Example 4.6

Assume that the inductor of Example 4.5 is connected to a voltage source that provides a constant voltage equal to -10mV for a period of 1 s, starting at $t = 0$. If no current is flowing through the inductor prior to $t = 0$, sketch the inductor current as a function of time.

Solution:

Integrating the inductor voltage from $t = 0$ to $t = 1$ yields:

$$i_L(t) = \frac{1}{L}\int_0^t V\,dt + 0 \qquad t \ge 0$$

and

$$i_L(t) = \frac{V}{L}t \qquad 1 \ge t \ge 0$$

$$= -t$$

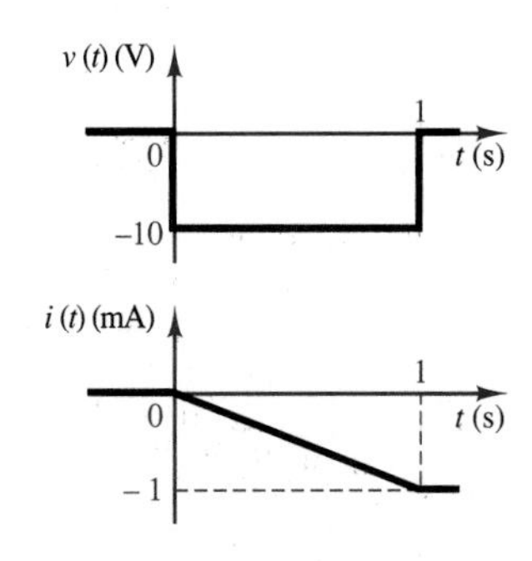

Figure 4.14

Thus, the inductor current will decrease linearly to a value of -1 A, as shown in Figure 4.14.

Energy Storage in Inductors

The magnetic energy stored in an ideal inductor may be found from a power calculation by following the same procedure employed for the ideal capacitor. The instantaneous power in the inductor is given by

$$\begin{aligned} P_L(t) &= i_L(t)v_L(t) \\ &= i_L(t)L\frac{di_L(t)}{dt} \\ &= \frac{d}{dt}\left[\frac{1}{2}Li_L^2(t)\right] \end{aligned} \tag{4.18}$$

Integrating the power, we obtain the total energy stored in the inductor, as shown in the following equation:

$$W_L(t) = \int P_L(t)\,dt = \int \frac{d}{dt}\left[\frac{1}{2}Li_L^2(t)\right]dt \tag{4.19}$$

$$W_L(t) = \tfrac{1}{2}Li_L^2(t) \quad \text{Energy stored in an inductor (J)}$$

Note, once again, the duality with the expression for the energy stored in a capacitor, in equation 4.9. Further, the expression for the energy stored in the inductor may be compared with the energy stored in the spring of Figure 4.10, given the following equation:

$$W_k = \tfrac{1}{2}kf^2 \tag{4.20}$$

Example 4.7

In this example we compute the energy stored in the inductor of Example 4.5. First, we compute the current for each part of the current waveform:

$$i_L(t) = \begin{cases} 0\ \text{A} & t < 1\ \text{ms} \\ \left(\dfrac{0.1}{0.004}\right)(t - 10^{-3}) & 1 \le t < 5\ \text{ms} \\ 0.1 & 5 \le t < 9\ \text{ms} \\ \left(-\dfrac{0.1}{0.004}\right)(t - 13 \times 10^{-3}) & 9 \le t < 13\ \text{ms} \\ 0 & t \ge 13\ \text{ms} \end{cases}$$

Then we compute the energy according to equation 4.19:

$$W_L(t) = \begin{cases} 0\ \text{J} & t < 1\ \text{ms} \\ 3.125(t - 10^{-3})^2 & 1 \le t < 5\ \text{ms} \\ 0.00005 & 5 \le t < 9\ \text{ms} \\ -3.125(t - 13 \times 10^{-3})^2 & 9 \le t < 13\ \text{ms} \\ 0 & t \ge 13\ \text{ms} \end{cases}$$

Note that energy is stored in the inductor only when a current is flowing through it. No energy is stored when $i_L(t) = 0$. It is also useful to observe that energy is stored in the inductor when a DC (constant) current flows through it, although there is no voltage induced across the inductor, since $v_L = L\frac{di_L}{dt} = 0$ for a constant current.

Analogy between Electrical and Mechanical Elements

To complete the analogy between electrical and mechanical systems developed in this section (see Figures 4.2 and 4.10), we again call attention to the role of the resistor as a power-dissipating element, and to its mechanical counterpart, the **damper.** Figure 4.15 illustrates the fundamental relationship between voltage and current in a resistor, on the one hand, and force and velocity in a damper, on the other. Note that the basic form of the equations is the same. The analogies between mechanical and electrical systems are summarized in Table 4.2.

The defining equation for the resistance circuit element is analogous to the equation of motion of a damper acted upon by a force.

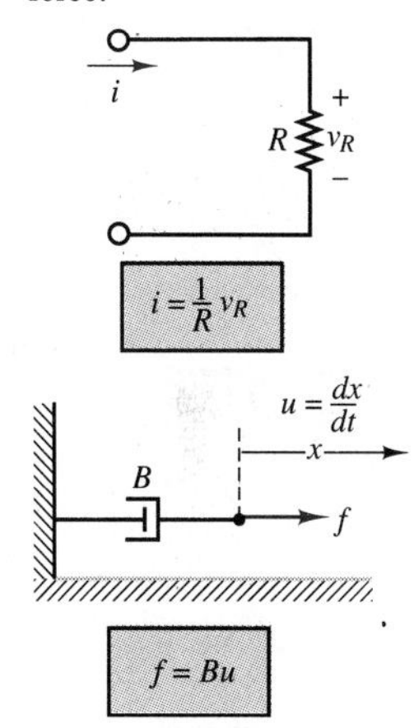

Figure 4.15 Analogy between electrical and mechanical elements

Table 4.2 **Analogy between electrical and mechanical variables**

Mechanical system	Electrical system
Force, f (N)	Current, i (A)
Velocity, u (m/s)	Voltage, v (V)
Damping, B (N-s/m)	Conductance, $\frac{1}{R}$ (S)
Compliance, $\frac{1}{k}$ (m/N)	Inductance, L (H)
Mass, M (kg)	Capacitance, C (F)

DRILL EXERCISES

4.1 The current waveform shown in Figure 4.16 flows through a 50-mH inductor. Plot the inductor voltage, $v_L(t)$.

4.2 The voltage waveform of Figure 4.17 appears across a 1,000-μF capacitor. Plot the capacitor current, $i_C(t)$.

4.3 Calculate the energy stored in the inductor (in joules) at $t = 3$ ms by the waveform of Drill Exercise 4.1. Assume $i(-\infty) = 0$.

4.4 Perform the calculation of Drill Exercise 4.3 for the capacitor if $v_C(-\infty) = 0$ V.

4.5 Compute and plot the inductor energy (in joules) and power (in watts) for the case of Drill Exercise 4.1.

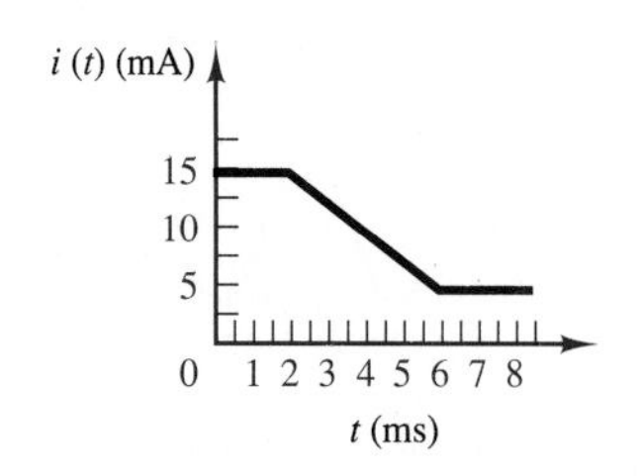

Figure 4.16

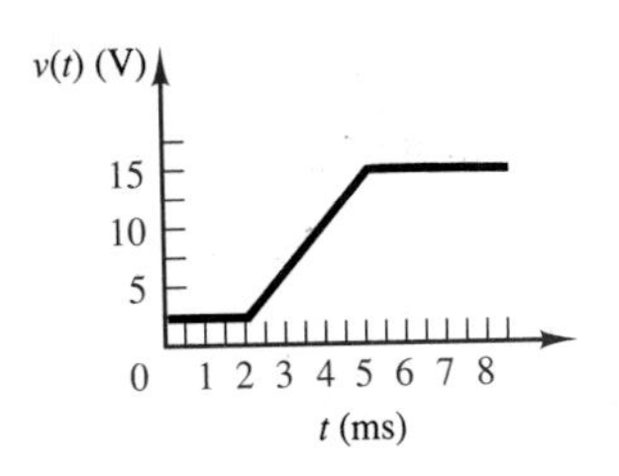

Figure 4.17

4.2 TIME-DEPENDENT SIGNAL SOURCES

In Chapter 2, the general concept of an ideal energy source was introduced. In the present chapter, it will be useful to specifically consider sources that generate time-varying voltages and currents and, in particular, sinusoidal sources. Figure 4.18 illustrates the convention that will be employed to denote time-dependent signal sources.

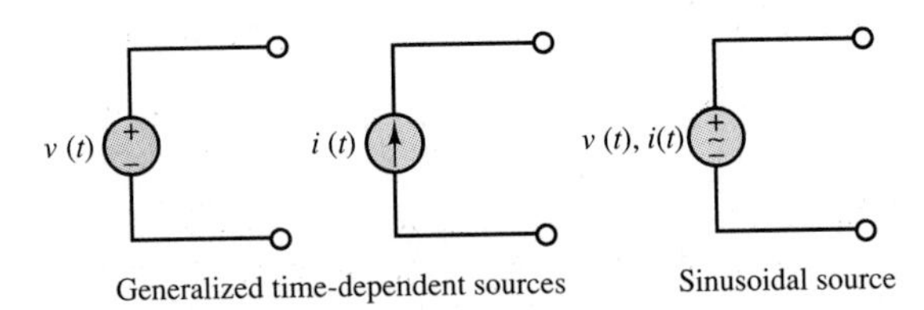

Figure 4.18 Time-dependent signal sources

One of the most important classes of time-dependent signals is that of **periodic signals.** These signals appear frequently in practical applications and are a useful approximation of many physical phenomena. A periodic signal $x(t)$ is a signal that satisfies the following equation:

$$x(t) = x(t + nT) \qquad n = 1, 2, 3, \ldots \tag{4.21}$$

where T is the **period** of $x(t)$. Figure 4.19 illustrates a number of the periodic waveforms that are typically encountered in the study of electrical circuits. Waveforms such as the sine, triangle, square, pulse, and sawtooth waves are provided in the form of voltages (or, less frequently, currents) by commercially available **signal** (or **waveform**) **generators.** Such instruments allow for selection of the waveform peak amplitude, and of its period.

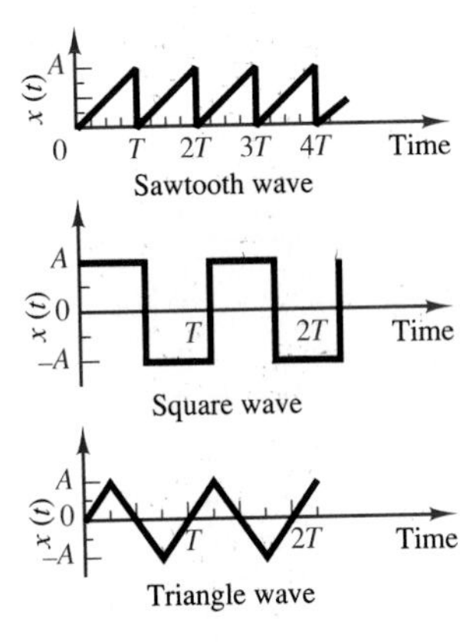

As stated in the introduction, sinusoidal waveforms constitute by far the most important class of time-dependent signals. Figure 4.20 depicts the relevant parameters of a sinusoidal waveform. A generalized sinusoid is defined as follows:

$$x(t) = A\cos(\omega t + \phi) \tag{4.22}$$

where A is the **amplitude,** ω the **radian frequency,** and ϕ the **phase.** Figure 4.20 summarizes the definitions of A, ω, and ϕ for the waveforms

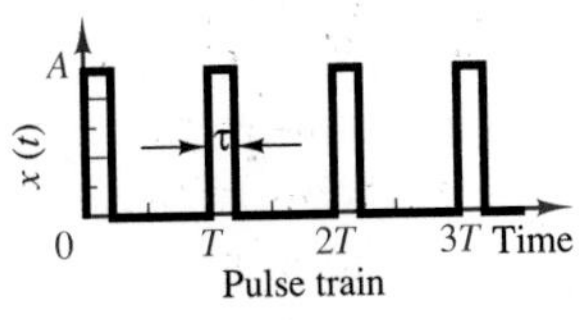

$$x_1(t) = A\cos(\omega t) \qquad \text{and} \qquad x_2(t) = A\cos(\omega t + \phi)$$

where

$$\begin{aligned} f &= \text{natural frequency} = \frac{1}{T}(\text{cycles/s, or Hz}) \\ \omega &= \text{radian frequency} = 2\pi f(\text{radians/s}) \\ \phi &= 2\pi\frac{\Delta T}{T}(\text{radians}) = 360\frac{\Delta T}{T}(\text{degrees}) \end{aligned} \tag{4.23}$$

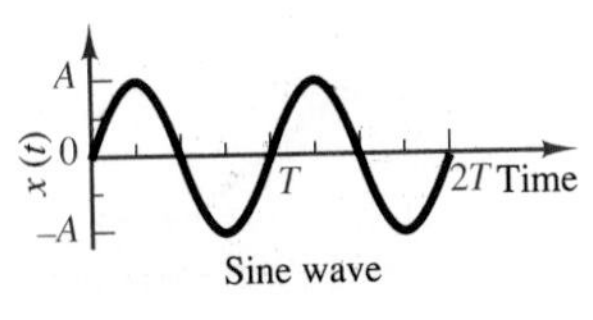

Figure 4.19 Periodic signal waveforms

The phase shift, ϕ, permits the representation of an arbitrary sinusoidal signal. Thus, the choice of the reference cosine function to represent sinusoidal signals—

arbitrary as it may appear at first—does not restrict the ability to represent all sinusoids. For example, one can represent a sine wave in terms of a cosine wave simply by introducing a phase shift of $\pi/2$ radians:

$$A\sin(\omega t) = A\cos\left(\omega t - \frac{\pi}{2}\right) \tag{4.24}$$

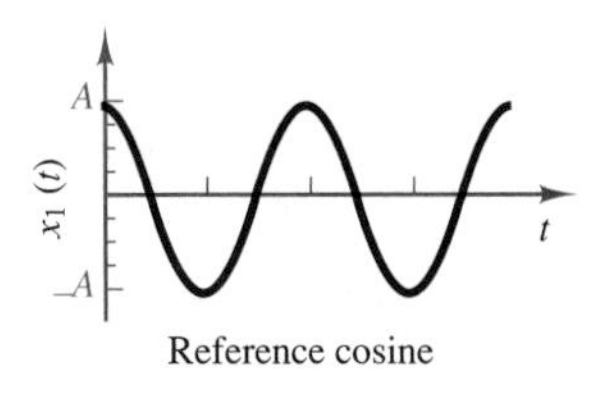

Reference cosine

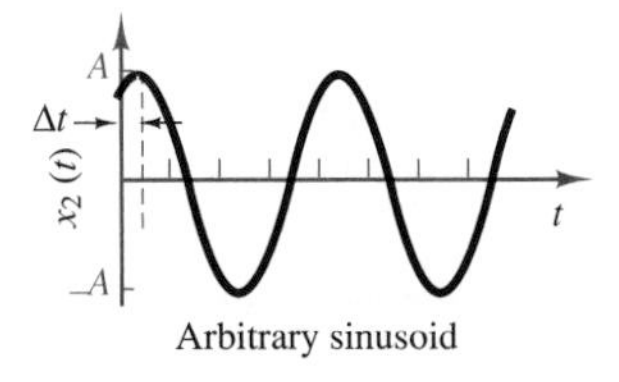

Arbitrary sinusoid

Figure 4.20 Sinusoidal waveforms

It is important to note that, although one usually employs the variable ω (in units of radians per second) to denote sinusoidal frequency, it is common to refer to natural frequency, f, in units of cycles per second, or **hertz (Hz).** The reader with some training in music theory knows that a sinusoid represents what in music is called a *pure tone;* an A-440, for example, is a tone at a frequency of 440 Hz. It is important to be aware of the factor of 2π that differentiates radian frequency (in units of rad/s) from natural frequency (in units of Hz). The distinction between the two units of frequency—which are otherwise completely equivalent—is whether one chooses to define frequency in terms of revolutions around a trigonometric circle (in which case the resulting units are rad/s), or to interpret frequency as a repetition rate (cycles/second), in which case the units are Hz. The relationship between the two is the following:

$$\boxed{\omega = 2\pi f} \tag{4.25}$$

Why Sinusoids?

You should by now have developed a healthy curiosity about why so much attention is being devoted to sinusoidal signals. Perhaps the simplest explanation is that the electric power used for industrial and household applications worldwide is generated and delivered in the form of either 50- or 60-Hz voltages and currents. Chapter 5 will provide more detail regarding the analysis of electric power circuits. The more ambitious reader may explore the box "Fourier Analysis" in Chapter 6 to obtain a more comprehensive explanation of the importance of sinusoidal signals. It should be remarked that the methods developed in this section and the subsequent sections apply to many engineering systems, not just to electrical circuits, and will be encountered again in the study of dynamic-system modeling and of control systems.

Average and RMS Values

Now that a number of different signal waveforms have been defined, it is appropriate to define suitable measurements for quantifying the strength of a time-varying electrical signal. The most common types of measurements are the **average** (or **DC**) **value** of a signal waveform—which corresponds to just measuring the mean voltage or current over a period of time—and the **root-mean-square** (or **rms**) **value,** which takes into account the fluctuations of the signal about its average value. Formally, the operation of computing the average value of a signal corresponds to integrating the signal waveform over some (presumably, suitably chosen) period of time. We define the time-averaged value of a signal $x(t)$ as

Figure 4.21 Averaging a signal waveform

$$\langle x(t) \rangle = \frac{1}{T}\int_0^T x(t)\,dt \tag{4.26}$$

where T is the period of integration. Figure 4.21 illustrates how this process does, in fact, correspond to computing the average amplitude of $x(t)$ over a period of T seconds.

Example 4.8

In this example, we illustrate the computation of the time average of a periodic signal by integration over one period. We seek to find the average value of the signal

$$x(t) = 10\cos(100t)$$

Solution:

Since the signal is periodic, with period

$$T = \frac{2\pi}{\omega} = \frac{2\pi}{100}$$

we need only integrate over one period of the waveform to obtain the average value:

$$\langle x(t) \rangle = \frac{100}{2\pi}\int_0^{2\pi/100} 10\cos(100t)\,dt = \frac{10}{2\pi}(\sin 2\pi - \sin 0) = 0$$

This result states an important conclusion: that the average value of the reference cosine is identically zero, independent of its amplitude and frequency.

The result of Example 4.8 can be generalized to state that

$$\langle A\cos(\omega t + \phi) \rangle = 0 \tag{4.27}$$

a result that might be perplexing at first: if any sinusoidal voltage or current has zero average value, is its average power equal to zero? Clearly, the answer must be no. Otherwise, it would be impossible to illuminate households and streets and power industrial machinery with 60-Hz sinusoidal current! There must be another way, then, of quantifying the strength of an AC signal.

Very conveniently, a useful measure of the voltage of an AC waveform is the root-mean-square, or rms, value of the signal, $x(t)$, defined as follows:

$$x_{\text{rms}} = \sqrt{\frac{1}{T}\int_0^T x^2(t)\,dt} \tag{4.28}$$

Note immediately that if $x(t)$ is a voltage, the resulting x_{rms} will also have units of volts. If you analyze equation 4.28, you can see that, in effect, the rms value consists of the square *r*oot of the average (or *m*ean) of the *s*quare of the signal. Thus, the notation *rms* indicates exactly the operations performed on $x(t)$ in order to obtain its rms value.

Example 4.9

In this example we illustrate the procedure one follows in computing the rms value of a signal, by computing the rms value of the sinusoidal current

$$i(t) = I\cos(\omega t)$$

Applying the definition of rms value, we compute

$$\begin{aligned} i_{\text{rms}} &= \sqrt{\frac{\omega}{2\pi}\int_0^{2\pi/\omega} I^2\cos^2\omega t\,dt} \\ &= \sqrt{\frac{\omega}{2\pi}\int_0^{2\pi/\omega} I^2\left(\frac{1}{2}+\frac{1}{2}\cos 2\omega t\right)dt} \\ &= \sqrt{\frac{I^2}{2}+\frac{\omega}{2\pi}\int_0^{2\pi/\omega}\frac{I^2}{2}\cos 2\omega t\,dt} \end{aligned}$$

At this point, we recognize that the integral under the square root sign is equal to zero, since we are averaging $\cos 2\omega t$ over two periods. Therefore, we conclude that the rms value of the sinusoidal current is

$$i_{\text{rms}} = \frac{I}{\sqrt{2}} = 0.707I$$

where I is the *peak value* of $i(t)$. This result is clearly independent of the frequency and phase of the waveform.

The preceding example illustrates how the rms value of a sinusoid is proportional to its peak amplitude. The factor of $0.707 = 1/\sqrt{2}$ is a useful number to remember, since it applies to any sinusoidal signal. It is not, however, generally applicable to signal waveforms other than sinusoids, as the Drill Exercises will illustrate.

DRILL EXERCISES

4.6 Express the voltage $v(t) = 155.6\sin(377t + 60°)$ in cosine form. You should note that the radian frequency $\omega = 377$ will recur very often, since $377 = 2\pi 60$; that is, 377 is the radian equivalent of the natural frequency of 60 cycles/second, which is the frequency of the electric power generated in North America.

4.7 Compute the average value of the sawtooth waveform shown in Figure 4.22.

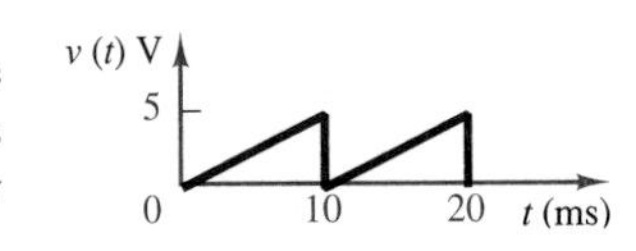

Figure 4.22

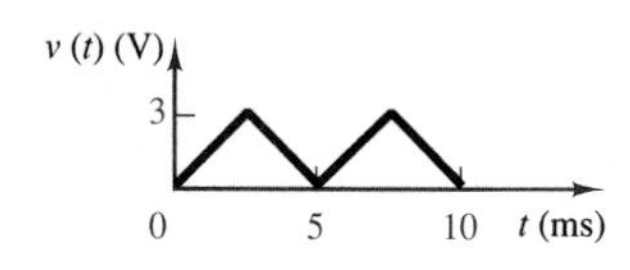

Figure 4.23

4.8 Compute the average value of the shifted triangle wave shown in Figure 4.23.

4.9 Find the rms value of the sawtooth wave of Drill Exercise 4.7.

4.10 Find the rms value of the half cosine wave shown in Figure 4.24.

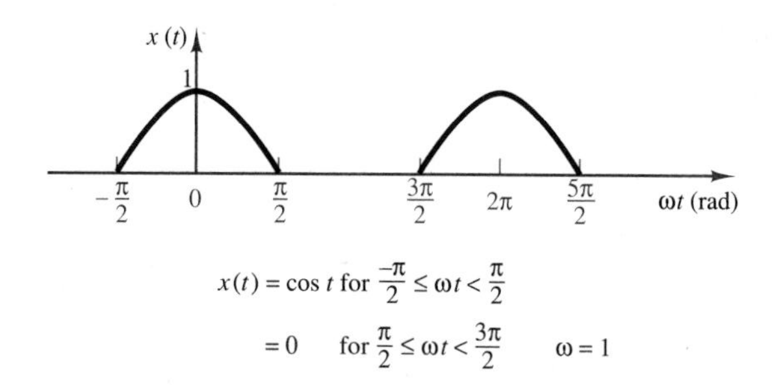

Figure 4.24

4.3 SOLUTION OF CIRCUITS CONTAINING DYNAMIC ELEMENTS

A circuit containing energy-storage elements is described by a differential equation. The differential equation describing the series *RC* circuit shown is

$$\frac{di_C}{dt} + \frac{1}{RC} i_C = \frac{dv_S}{dt}$$

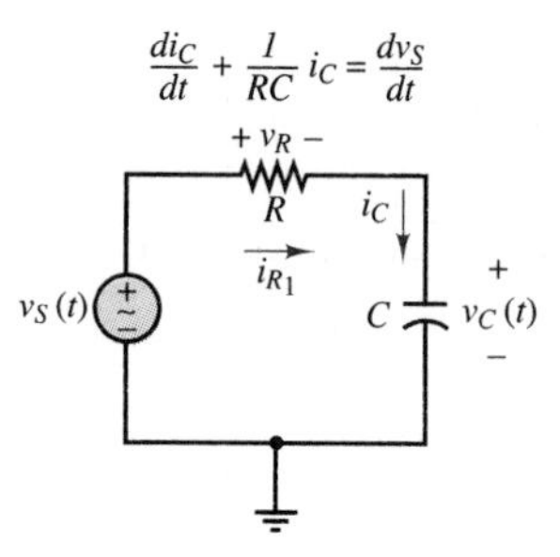

Figure 4.25 Circuit containing energy-storage element

The first two section of this chapter introduced energy-storage elements and time-dependent signal sources. The logical next task is to analyze the behavior of circuits containing such elements. The major difference between the analysis of the resistive circuits studied in Chapters 2 and 3 and the circuits we will explore in the remainder of this chapter is that now the equations that result from applying Kirchhoff's laws are differential equations, as opposed to the algebraic equations obtained in solving resistive circuits. Consider, for example, the circuit of Figure 4.25, which consists of the series connection of a voltage source, a resistor, and a capacitor. Applying KVL around the loop, we may obtain the following equation:

$$v_S(t) = v_R(t) + v_C(t) \tag{4.29}$$

Observing that $i_R = i_C$, equation 4.29 may be combined with the defining equation for the capacitor (equation 4.6) to obtain

$$v_S(t) = Ri_C(t) + \frac{1}{C}\int_{-\infty}^{t} i_C \, dt \tag{4.30}$$

Equation 4.30 is an integral equation, which may be converted to the more familiar form of a differential equation by differentiating both sides of the equation, and recalling that

$$\frac{d}{dt}\left(\int_{-\infty}^{t} i_C(t)\, dt\right) = i_C(t) \tag{4.31}$$

to obtain the following differential equation:

$$\frac{di_C}{dt} + \frac{1}{RC} i_C = \frac{1}{R}\frac{dv_S}{dt} \tag{4.32}$$

where the argument (t) has been dropped for ease of notation.

Observe that in equation 4.32, the independent variable is the series current flowing in the circuit, and that this is not the only equation that describes the series RC circuit. If, instead of applying KVL, for example, we had applied KCL at the node connecting the resistor to the capacitor, we would have obtained the following relationship:

$$i_R = \frac{v_S - v_C}{R} = i_C = C\frac{dv_C}{dt} \tag{4.33}$$

or

$$\frac{dv_C}{dt} + \frac{1}{RC}v_C = \frac{1}{RC}v_S \tag{4.34}$$

Note the similarity between equations 4.32 and 4.34. The left-hand side of both equations is identical, except for the independent variable, while the right-hand side takes a slightly different form. The solution of either equation is sufficient, however, to determine all voltages and currents in the circuit. The following example illustrates the derivation of the differential equation for another simple circuit containing an energy-storage element.

Example 4.10

In this example we illustrate the derivation of the differential equation governing the behavior of the circuit shown in Figure 4.26. Assume that $R_1 = 10\ \Omega, R_2 = 5\ \Omega$, and $L = 0.4$ H; $v_S(t)$ is an arbitrary function of time.

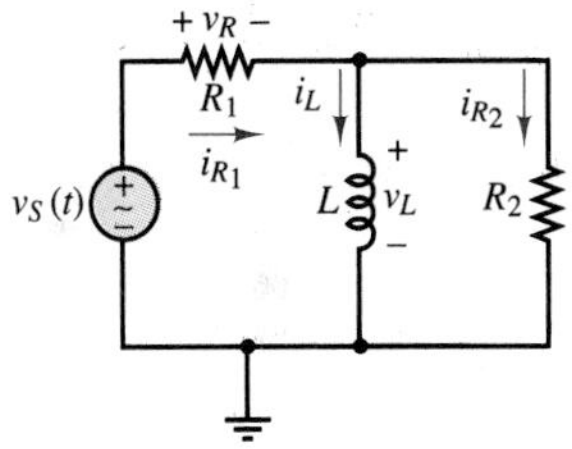

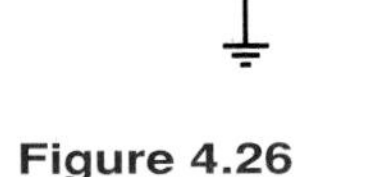

Figure 4.26

Solution:

One possible approach to deriving the differential equation for the circuit is to apply KCL to the node at the top right corner of the circuit, obtaining the equation

$$\frac{v_S - v_L}{R_1} = i_L + \frac{v_L}{R_2}$$

(since the voltage across R_2 is the same as the inductor voltage). Next, the i-v relationship for the inductor (equation 4.11) may be used to eliminate the variable v_L from the equation:

$$\frac{v_S}{R_1} - \frac{L}{R_1}\frac{di_L}{dt} = i_L + \frac{L}{R_2}\frac{di_L}{dt}$$

to obtain the final form of the equation:

$$\frac{di_L}{dt} + \frac{R_1 R_2}{L(R_1 + R_2)}i_L = \frac{R_2}{L(R_1 + R_2)}v_S$$

or

$$\frac{di_L}{dt} + 8.33 i_L = 0.833 v_S$$

We can generalize the results presented in the preceding pages by observing that any circuit containing a single energy-storage element can be described by a differential equation of the form

$$a_1 \frac{dy(t)}{dt} + a_0 y(t) = F(t) \tag{4.35}$$

where $y(t)$ represents the capacitor voltage in the circuit of Figure 4.25 and the inductor current in the circuit of Figure 4.26, and where the constants a_0 and a_1 consist of combinations of circuit element parameters. Equation 4.35 is a **first-order ordinary differential equation** with constant coefficients. The equation is said to be of first order because the highest derivative present is of first order; it is said to be ordinary because the derivative that appears in it is an ordinary derivative (in contrast to a *partial* derivative); and the coefficients of the differential equation are constant in that they depend only on the values of resistors, capacitors, or inductors in the circuit, and not, for example, on time, voltage, or current.

Consider now a circuit that contains two energy-storage elements, such as that shown in Figure 4.27. Application of KVL results in the following equation:

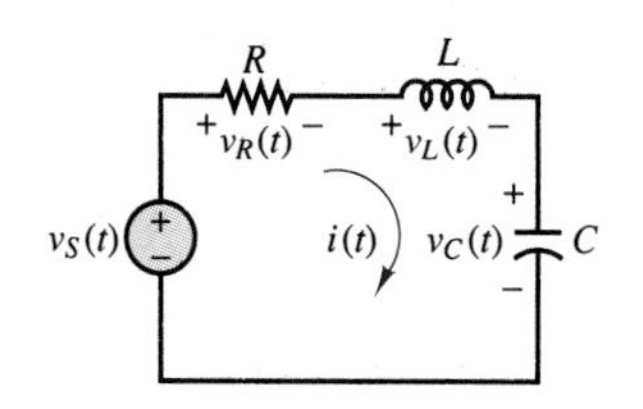

Figure 4.27 Second-order circuit

$$Ri(t) + L\frac{di(t)}{dt} + \frac{1}{C}\int_{-\infty}^{t} i(t)\,dt = v_S(t) \tag{4.36}$$

Equation 4.36 is called an integro-differential equation, because it contains both an integral and a derivative. This equation can be converted into a differential equation by differentiating both sides, to obtain:

$$R\frac{di(t)}{dt} + L\frac{d^2i(t)}{dt^2} + \frac{1}{C}i(t) = \frac{dv_S(t)}{dt} \tag{4.37}$$

or, equivalently, by observing that the current flowing in the series circuit is related to the capacitor voltage by $i(t) = C\frac{dv_C}{dt}$, and that equation 4.36 can be rewritten as:

$$RC\frac{dv_C}{dt} + LC\frac{d^2v_C(t)}{dt^2} + v_C(t) = v_S(t) \tag{4.38}$$

Note that, although different variables appear in the preceding differential equations, both equations 4.37 and 4.38 can be rearranged to appear in the same general form, as follows:

$$a_2\frac{d^2y(t)}{dt^2} + a_1\frac{dy(t)}{dt} + a_0 y(t) = F(t) \tag{4.39}$$

where the general variable $y(t)$ represents either the series current of the circuit of Figure 4.27 or the capacitor voltage. By analogy with equation 4.35, we call equation 4.39 a **second-order ordinary differential equation** with constant coefficients. As the number of energy-storage elements in a circuit increases, one can therefore expect that higher-order differential equations will result. For the purposes of this book, we shall concentrate on the solution of first- and second-order differential equations. Computer aids are often employed to solve differential equations of higher order; some of these software packages are specifically targeted at the solution of the equations that result from the analysis of electrical circuits (e.g., PSPICE™).

Example 4.11

In this example we demonstrate the derivation of the second-order differential equation for the circuit shown in Figure 4.28. Assume $R_1 = 10\ \text{k}\Omega$, $R_2 = 50\ \Omega$, $C = 0.1\ \mu\text{F}$, $L = 10\ \text{mH}$.

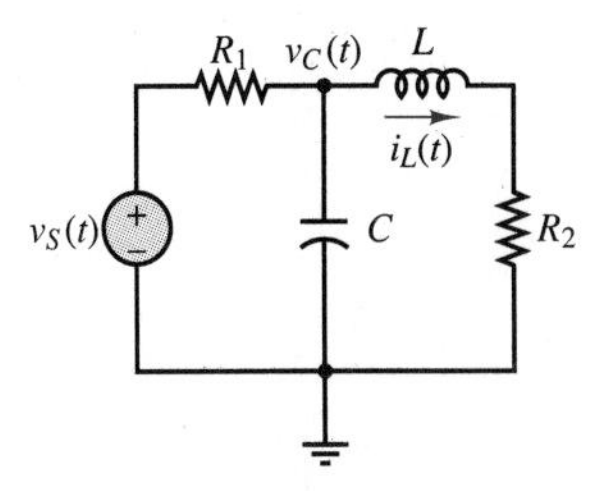

Figure 4.28 Second-order circuit of Example 4.11

Solution:

Let v_C denote the capacitor voltage. Then we can apply KCL at the node labeled v_C to obtain the following equation:

$$\frac{v_S(t)}{R_1} - \frac{v_C(t)}{R_1} = C\frac{dv_C(t)}{dt} + i_L(t)$$

Applying KVL to the right-hand-side mesh, we can obtain a second equation,

$$v_C(t) = L\frac{di_L(t)}{dt} + R_2 i_L(t)$$

which can be substituted into the first equation to obtain a single second-order differential equation:

$$\frac{v_S(t)}{R_1} - \frac{L}{R_1}\frac{di_L(t)}{dt} + \frac{R_2}{R_1}i_L(t) = LC\frac{d^2 i_L(t)}{dt^2} + R_2 C\frac{di_L(t)}{dt} + i_L(t)$$

This equation can be rearranged to appear in the more general form of equation 4.39, as follows:

$$R_1 LC\frac{d^2 i_L(t)}{dt^2} + (R_1 R_2 C + L)\frac{di_L(t)}{dt} + (R_1 + R_2)i_L(t) = v_S(t)$$

Note that it is also possible to derive a similar differential equation, with the capacitor voltage as the dependent variable; the choice of a particular variable in deriving a differential equation is dictated by the problem requirements. Note also that the variables chosen in either case are tied to the energy-storage equations: the energy stored in a capacitor is proportional to the square of its voltage, while the energy stored in an inductor is proportional to the square of its current. These two variables are the "natural" choice for writing differential equations. We shall return to this point in Chapter 6.

Solution of Differential Equations

The first- and second-order differential equations introduced in the preceding section can be systematically solved by determining the natural and forced responses of the circuit. The aim of this section is to introduce general methods for the solution of these differential equations. In the remainder of the chapter we shall focus on the very special case of sinusoidal forcing functions.

Natural Response of First-Order Circuits

The **natural** (or **free**) **response** of a circuit describes the behavior of the circuit when the external forcing functions are equal to zero. In this case, the response of the circuit is determined entirely by the energy stored in capacitors or inductors. Consider, for example, the case of a parallel *RC* circuit, shown in Figure 4.29.

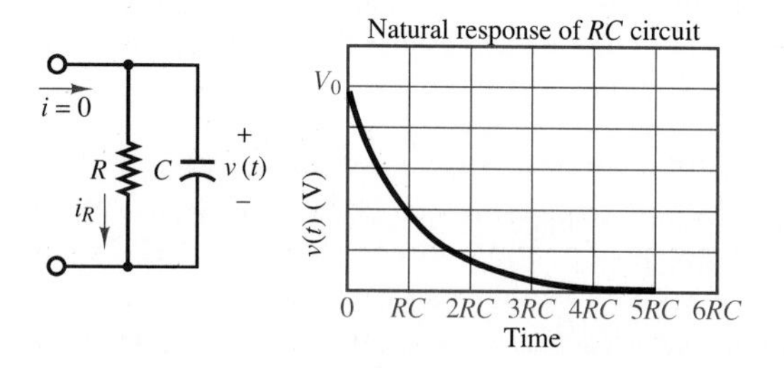

Figure 4.29 Unforced *RC* circuit and natural solution

If the capacitor has been charged to a voltage V_0 prior to $t = 0$, then a current will flow in the circuit for $t \geq 0$, as the energy stored in the capacitor is discharged through the resistor. This natural response is governed by the differential equation

$$C\frac{dv_N}{dt} + \frac{1}{R}v_N = 0 \tag{4.40}$$

where v_N is the natural response[2] component of the voltage $v(t)$. The evolution of the natural response in time is described by the solution to this equation, which is called a **homogeneous differential equation.** The natural response voltage can be determined by observing that, since

$$C\frac{dv_N}{dt} = -\frac{1}{R}v_N \tag{4.41}$$

that is, since the solution must be proportional to its own derivative, the solution of equation 4.40 must be of exponential form, since an exponential is the only function that is proportional to its derivative. We therefore postulate that

$$v_N(t) = Ke^{st} \tag{4.42}$$

where K and s are constants to be determined. If we substitute this solution into equation 4.41, we find that

$$sCKe^{st} = -\frac{1}{R}Ke^{st} \tag{4.43}$$

[2]Mathematicians call the natural response the **complementary solution.**

Therefore, equation 4.42 is a solution of equation 4.41 if $s = -\frac{1}{RC}$. To determine the other unknown constant, K, we use knowledge of the fact that at $t = 0$ the capacitor was charged to the voltage V_0. Using this **initial condition** at $t = 0$, we determine that

$$v_N(0) = Ke^0 = K = V_0 \tag{4.44}$$

Finally, we can write the natural solution,

$$v_N(t) = V_0 e^{-t/RC} \tag{4.45}$$

which is plotted in Figure 4.29.

Energy Storage in Capacitors and Inductors

Before delving into the complete solution of the differential equation describing the response of first-order circuits, it will be helpful to review some basic results pertaining to the response of energy-storage elements to DC sources. This knowledge will later greatly simplify the complete solution of the differential equation describing a *circuit*. Consider, first, a capacitor, which accumulates charge according to the relationship $Q = CV$. The charge accumulated in the capacitor leads to the storage of energy according to the following equation:

$$W_C = \frac{1}{2}Cv_C^2(t) \tag{4.46}$$

To understand the role of stored energy, consider, as an illustration, the simple circuit of Figure 4.30, where a capacitor is shown to have been connected to a battery, V_B, for a long time. The capacitor voltage is therefore equal to the battery voltage: $v_C(t) = V_B$. The charge stored in the capacitor (and the corresponding energy) can be directly determined using equation 4.46. Suppose, next, that at $t = 0$ the capacitor is disconnected from the battery and connected to a resistor, as shown by the action of the switches in Figure 4.30. The resulting circuit would be governed by the RC differential equation described earlier (see Figure 4.29), subject to the initial condition $v_C(t = 0) = V_B$. Thus, according to the results of the preceding section, the capacitor voltage would decay exponentially according to the following equation:

$$v_C(t) = V_B e^{-t/RC} \tag{4.47}$$

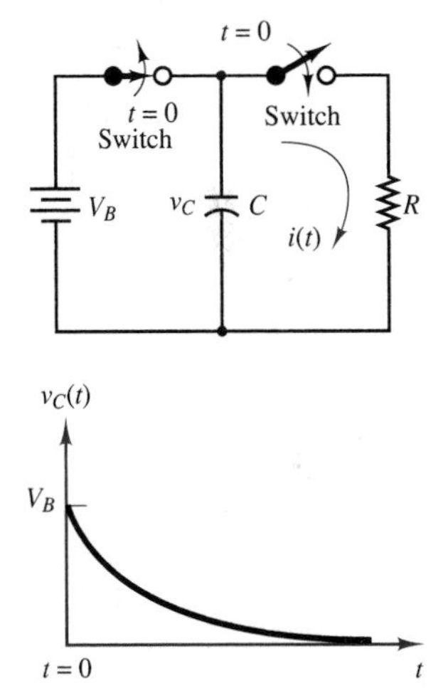

Figure 4.30 Decay through a resistor of energy stored in a capacitor

Physically, this exponential decay signifies that the energy stored in the capacitor at $t = 0$ is dissipated by the resistor at a rate determined by the time constant of the circuit, $\tau = RC$. Intuitively, the existence of a closed circuit path allows for the flow of a current, thus draining the capacitor of its charge. All of the energy initially stored in the capacitor is eventually dissipated by the resistor.

A very analogous reasoning process explains the behavior of an inductor. Recall that an inductor stores energy according to the expression

$$W_L = \frac{1}{2}Li_L^2(t) \tag{4.48}$$

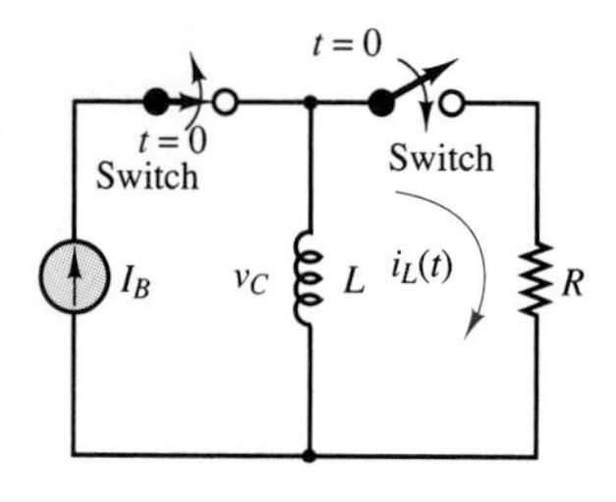

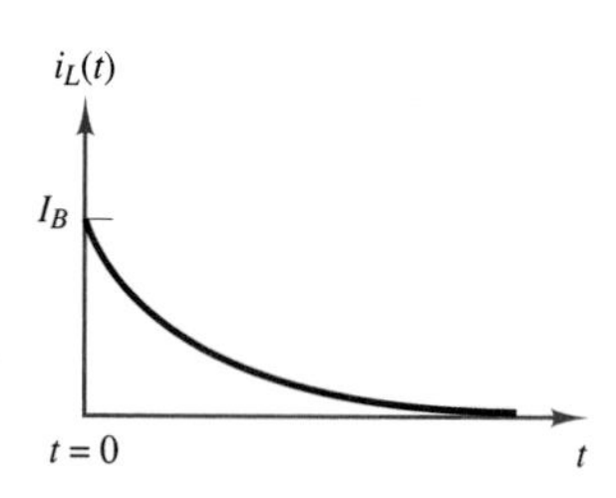

Figure 4.31 Decay through a resistor of energy stored in an inductor

Thus, in an inductor, energy storage is associated with the flow of a current (note the dual relationship between i_L and v_C). Consider the circuit of Figure 4.31, which is similar to that of Figure 4.30 except that the battery has been replaced with a current source and the capacitor with an inductor. For $t < 0$, the source current, I_B, flows through the inductor, and energy is thus stored; at $t = 0$, the inductor current is equal to I_B. At this point, the current source is disconnected by means of the left-hand switch and a resistor is simultaneously connected to the inductor, to form a closed circuit. The inductor current will now continue to flow through the resistor, which dissipates the energy stored in the inductor. By the reasoning in the preceding discussion, the inductor current will decay exponentially:

$$i_L(t) = I_B e^{-tR/L} \tag{4.49}$$

That is, the inductor current will decay exponentially from its initial condition, with a time constant $\tau = L/R$. Example 4.12 further illustrates the significance of the time constant in a first-order circuit.

Example 4.12

This example focuses on the concept of a time constant and aims to build an intuitive understanding of this very important electrical engineering concept. A certain radioactive sample has a half-life of 35 s. The half-life is defined as the time $(t_1 - t_0)$ it takes for the radioactivity to decay to one half of its initial value (its value at $t = t_0$). The radioactivity of this element is modeled by the inductor current, $i_L(t)$, in an RL circuit. We assume that the current $i_L(t)$ is directly proportional to the radioactivity of the element.

1. Find the time constant of radioactivity decay.
2. If $R = 0.1\ \Omega$ in the circuit model, what should L be?
3. If $i(0) = I_0$, find the current $i_L(t)$ at the following times: (a) $t = \tau$; (b) $t = 2\tau$; (c) $t = 3\tau$; (d) $t = 4\tau$; (e) $t = 5\tau$.

Solution:

Let the symbol γ denote the radioactivity of the sample.

1. The general form of the exponential radioactivity decay is $\gamma(t) = \gamma(0)e^{-t/\tau}$. Therefore, knowing the half-life of the radioactive sample, we can write

$$\frac{1}{2}\gamma(t_0) = \gamma(t_0)e^{-(t_1-t_0)/\tau}$$

and we can compute the time constant, τ, according to the equation

$$\frac{t_1 - t_0}{\tau} = \log_e\left(\frac{1}{2}\right)$$

Since $(t_1 - t_0)$ corresponds, by definition, to the half-life, which is 35 s, we can write

$$\frac{1}{\tau} = \frac{\log_e\left(\frac{1}{2}\right)}{-35}$$

or $\tau = t_{1/2} \log_e(2)$, where $t_{1/2}$ = half decay time. Thus,

$$\tau = 50.5 \text{ s}$$

2. In the circuit analogy, $\tau = L/R$, so that $L = \tau R = 5.05$ H.
3.

t	$i_L(t)$
τ	$0.368I_0$
2τ	$0.135I_0$
3τ	$0.0498I_0$
4τ	$0.0183I_0$
5τ	$0.006I_0$

Natural Response of Second-Order Circuits

The natural response of a second-order circuit may be obtained by observing that the homogeneous equation describing any second-order circuit is derived from equation 4.39, repeated below for convenience:

$$a_2 \frac{d^2 y_N}{dt^2} + a_1 \frac{d y_N(t)}{dt} + a_0 y_N(t) = 0 \qquad \textbf{(4.50)}$$

If we now assume an exponential solution of the form $v_N(t) = Ke^{st}$ and substitute into equation 4.50, we obtain the following expression:

$$(a_2 s^2 + a_1 s + a_0)Ke^{st} = 0 \qquad \textbf{(4.51)}$$

Equation 4.51 is called the **characteristic equation** of the second-order circuit; since it is quadratic, it has two roots:

$$s_{1,2} = -\frac{a_1 \pm \sqrt{a_1^2 - 4a_2 a_0}}{2a_2} \qquad \textbf{(4.52)}$$

Since both s_1 and s_2 are acceptable solutions of equation 4.51, a general solution for the natural response of a second-order circuit is given by the sum of the solutions corresponding to each of the roots:

$$y_N(t) = K_1 e^{s_1 t} + K_2 e^{s_2 t} \qquad \textbf{(4.53)}$$

To evaluate the constants K_1 and K_2, two initial conditions are required. This can be justified physically by considering that each energy-storage element in the second-order circuit stores energy and is therefore characterized by its own initial condition.

In dealing with first-order circuits, we determined that the natural response of these circuits must have a decaying exponential characteristic, because of the nature of the solution. What is the appearance of the natural response of a second-order circuit? Note that the roots of equation 4.52 can be complex numbers, depending on the relative size of the coefficients a_0, a_1, and a_2. Observe also that these coefficients are always positive numbers, since they consist of products or ratios of positive numbers (R, L, and C parameters). Thus, we can have three types of roots:

1. $a_1^2 > 4a_2a_0$: real and distinct roots
2. $a_1^2 = 4a_2a_0$: real and identical roots
3. $a_1^2 < 4a_2a_0$: complex conjugate roots

When the roots are real and distinct or real and identical, the natural response of the circuit is qualitatively similar to that of a first-order circuit, in the sense that capacitor voltages and inductor currents decay exponentially, since the natural response is the sum of two decaying exponential waveforms. In the case of complex conjugate roots, however, the natural response takes on a different appearance. Let $s_{1,2} = -\alpha \pm j\beta$, where α and β are positive numbers;[3] then we can write the natural response as follows:

$$y_N(t) = K_1 e^{-\alpha t} e^{j\beta t} + K_2 e^{-\alpha t} e^{-j\beta t}$$

This response, as will be shown in Section 4.4, consists of the product of a decaying exponential term, $e^{-\alpha t}$, with a *complex exponential* term. The latter will be shown to correspond to a sinusoidal term. Let us, for the sake of illustration, assume that $K_1 = K_2 = K$. Then, as will be explained in Section 4.4, we have

$$y_N(t) = Ke^{-\alpha t}(e^{j\beta t} + e^{-j\beta t}) = 2Ke^{-\alpha t}\cos(\beta t)$$

This natural response is plotted in Figure 4.32 for the case $K = 1$, $\alpha = 5$, and $\beta = 100$. A more detailed analysis of the response of second-order circuits will be shown in Chapter 6.

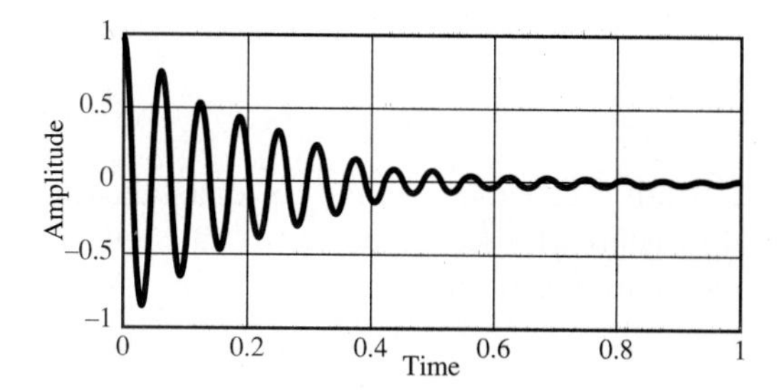

Figure 4.32 Natural response of second-order circuit with complex conjugate roots

[3]Note that in electrical engineering notation the symbol j, rather than i, is used to denote the square root of -1; this is required because the symbol i is used to denote current.

Forced and Complete Response

Once the natural response of the circuit has been obtained, the next step is to obtain the **forced response,**[4] which shall be generally denoted by $y_F(t)$. The forced response is the response of the circuit to external forcing functions, *without regard for the initial conditions,* and is the solution of the **inhomogeneous equation**

$$a_1 \frac{dy_F(t)}{dt} + a_0 y_F(t) = F(t) \tag{4.54}$$

for a first-order circuit and of the inhomogeneous equation

$$a_2 \frac{d^2 y_F}{dt^2} + a_1 \frac{dy_F(t)}{dt} + a_0 y_F(t) = F(t) \tag{4.55}$$

for a second-order circuit. To obtain the forced solution, one must know the form of the forcing function. A special case of great importance is that of sinusoidal forcing functions. We illustrate the determination of the forced response for this special case in the following subsection. The remainder of this chapter will be exclusively concerned with the analysis of circuits excited by sinusoidal sources. These circuits are commonly referred to as **AC circuits.**

Observe, now, that once the forced response has been obtained, we have knowledge of the circuit for the two separate conditions of no forcing function (natural response) and zero initial conditions (forced response). These two responses summarize the entire behavior of the circuit; that is, the complete response of the circuit is:

$$y(t) = y_N(t) + y_F(t) \tag{4.56}$$

We should also note that, since the natural response of first- and second-order circuits has been shown to decay to zero, the complete response of these circuits will eventually be equal to the forced response; i.e.,

$$y(t) \approx y_F(t)$$

When this condition has been reached, i.e., when the natural response has died out, the circuit is said to have reached **steady state.** The analysis of AC circuits at steady state is a very important part of electrical engineering and is presented in the remainder of this chapter.

Forced Response of Circuits Excited by Sinusoidal Sources

Consider again the circuit of Figure 4.25, where now the external source produces a sinusoidal voltage, described by the expression

$$v_S(t) = V \cos(\omega t) \tag{4.57}$$

---

[4] Mathematicians call the forced response the **particular solution.**

Substituting the expression $V\cos(\omega t)$ in place of the source voltage, $v_S(t)$, in the differential equation obtained earlier (equation 4.34), we obtain the following differential equation:

$$\frac{d}{dt}v_C + \frac{1}{RC}v_C = \frac{1}{RC}V\cos\omega t \tag{4.58}$$

Since the forcing function is a sinusoid, the solution may also be assumed to be of the same form. An expression for $v_C(t)$ is then the following:

$$v_C(t) = A\sin\omega t + B\cos\omega t \tag{4.59}$$

which is equivalent to

$$v_C(t) = C\cos(\omega t + \phi) \tag{4.60}$$

Substituting equation 4.59 in the differential equation for $v_C(t)$ and solving for the coefficients A and B yields the expression

$$A\omega\cos\omega t - B\omega\sin\omega t + \frac{1}{RC}(A\sin\omega t + B\cos\omega t) = \frac{1}{RC}V\cos\omega t \tag{4.61}$$

and if the coefficients of like terms are grouped, the following equation is obtained:

$$\left(\frac{A}{RC} - B\omega\right)\sin\omega t + \left(A\omega + \frac{B}{RC} - \frac{V}{RC}\right)\cos\omega t = 0 \tag{4.62}$$

The coefficients of $\sin\omega t$ and $\cos\omega t$ must both be identically zero in order for equation 4.62 to hold. Thus,

$$\frac{A}{RC} - B\omega = 0$$

and

$$A\omega + \frac{B}{RC} - \frac{V}{RC} = 0 \tag{4.63}$$

The unknown coefficients, A and B, may now be determined by solving equation 4.63, to obtain:

$$A = \frac{V\omega RC}{1 + \omega^2(RC)^2}$$

$$B = \frac{V}{1 + \omega^2(RC)^2} \tag{4.64}$$

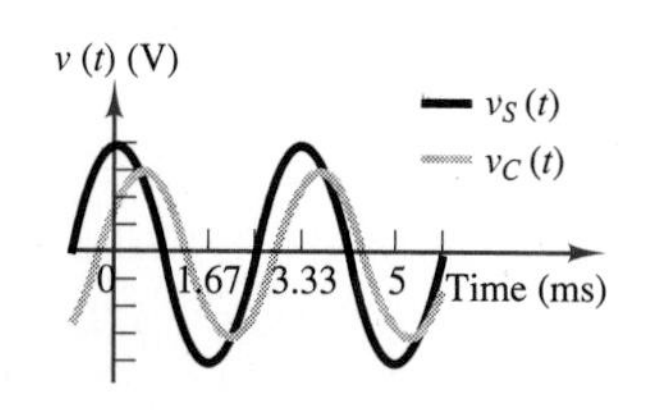

Figure 4.33 Waveforms for the AC circuit of Figure 4.25

Thus, the solution for $v_C(t)$ may be written as follows:

$$v_C(t) = \frac{V\omega RC}{1 + \omega^2(RC)^2}\sin\omega t + \frac{V}{1 + \omega^2(RC)^2}\cos\omega t \tag{4.65}$$

This response is plotted in Figure 4.33.

The solution method outlined in the previous paragraphs can become quite complicated for circuits containing a large number of elements; in particular, one may need to solve higher-order differential equations if more than one energy-storage element is present in the circuit. A simpler and preferred method for the solution of AC circuits will be presented in the next section. This brief section has provided a simple, but complete, illustration of the key elements of AC circuit analysis. These can be summarized in the following statement:

In a sinusoidally excited linear circuit, all branch voltages and currents are *sinusoids* at the *same frequency* as the excitation signal. The amplitudes of these voltages and currents are a *scaled* version of the excitation *amplitude*, and the voltages and currents may be *shifted in phase* with respect to the excitation signal.

These observations indicate that three parameters uniquely define a sinusoid: *frequency, amplitude,* and *phase*. But if this is the case, is it necessary to carry the "excess luggage," that is, the sinusoidal functions? Might it be possible to simply keep track of the three parameters just mentioned? Fortunately, the answers to these two questions are no and yes, respectively. The next section will describe the use of a notation that, with the aid of complex algebra, eliminates the need for the sinusoidal functions of time, and for the formulation and solution of differential equations, permitting the use of simpler algebraic methods.

DRILL EXERCISES

4.11 Show that the solution to either equation 4.32 or equation 4.34 is sufficient to compute all of the currents and voltages in the circuit of Figure 4.25.

4.12 Show that the equality

$$A \sin \omega t + B \cos \omega t = C \cos(\omega t + \phi)$$

holds if

$$A = -C \sin \phi$$

$$B = C \cos \phi$$

or, conversely, if

$$C = \sqrt{A^2 + B^2}$$

$$\phi = \tan^{-1}\left(\frac{-A}{B}\right)$$

4.13 Use the result of Drill Exercise 4.12 to compute C and ϕ as functions of V, ω, R, and C in equation 4.65.

4.14 Obtain a differential equation in v_C for the circuit of Example 4.11. [*Hint:* Substitute the equation obtained by KVL into the equation obtained by KCL.]

4.4 PHASORS AND IMPEDANCE

In this section, we introduce an efficient notation to make it possible to represent sinusoidal signals as *complex numbers,* and to eliminate the need for solving differential equations. The student who needs a brief review of complex algebra will find a reasonably complete treatment in Appendix A, including solved examples and drill exercises. For the remainder of the chapter, it will be assumed that you are familiar with both the rectangular and the polar forms of complex number coordinates, with the conversion between these two forms, and with the basic operations of addition, subtraction, multiplication, and division of complex numbers.

Leonhard Euler (1707–1783). *Photo courtesy of Deutsches Museum, Munich.*

Euler's Identity

Named after the Swiss mathematician Leonhard Euler (the last name is pronounced "Oiler"), Euler's identity forms the basis of phasor notation. Simply stated, the identity defines the **complex exponential** $e^{j\theta}$ as a point in the complex plane, which may be represented by real and imaginary components:

$$e^{j\theta} = \cos\theta + j\sin\theta \tag{4.66}$$

Figure 4.34 illustrates how the complex exponential may be visualized as a point (or vector, if referenced to the origin) in the complex plane. Note immediately that the magnitude of $e^{j\theta}$ is equal to 1:

$$|e^{j\theta}| = 1 \tag{4.67}$$

since

$$|\cos\theta + j\sin\theta| = \sqrt{\cos^2\theta + \sin^2\theta} = 1 \tag{4.68}$$

and note also that writing Euler's identity corresponds to equating the polar form of a complex number to its rectangular form. For example, consider a vector of length A making an angle θ with the real axis. The following equation illustrates the relationship between the rectangular and polar forms:

$$Ae^{j\theta} = A\cos\theta + jA\sin\theta \tag{4.69}$$

In effect, Euler's identity is simply a trigonometric relationship in the complex plane.

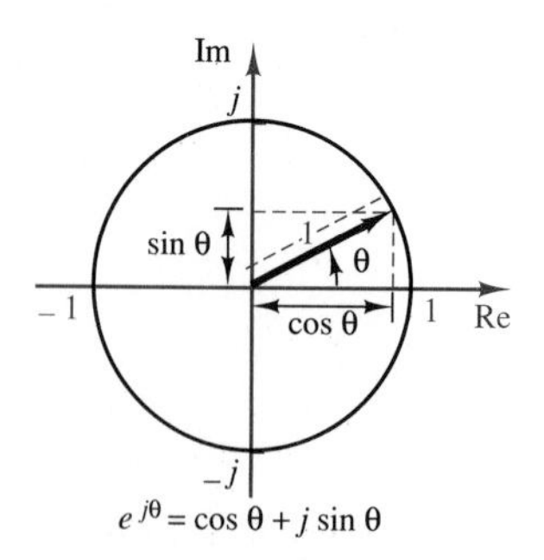

Figure 4.34 Euler's identity

Phasors

To see how complex numbers can be used to represent sinusoidal signals, rewrite the expression for a generalized sinusoid in light of Euler's equation:

$$A\cos(\omega t + \phi) = \text{Re}\,[Ae^{j(\omega t+\phi)}] \tag{4.70}$$

This equality is easily verified by expanding the right-hand side, as follows:

$$\begin{aligned}&\text{Re}\,[Ae^{j(\omega t+\phi)}]\\&\quad = \text{Re}\,[A\cos(\omega t + \phi) + jA\sin(\omega t + \phi)]\\&\quad = A\cos(\omega t + \phi)\end{aligned}$$

We see, then, that *it is possible to express a generalized sinusoid as the real part of a complex vector* whose **argument,** or **angle,** is given by $(\omega t + \phi)$ and whose length, or **magnitude,** is equal to the peak amplitude of the sinusoid. The **complex phasor** corresponding to the sinusoidal signal $A\cos(\omega t + \phi)$ is therefore defined to be the complex number $Ae^{j\phi}$:

$$Ae^{j\phi} = \text{complex phasor notation for } A\cos(\omega t + \phi) \tag{4.71}$$

It is important to explicitly point out that this is a *definition*. Phasor notation arises from equation 4.70; however, this expression is simplified (for convenience, as will be promptly shown) by removing the "real part of" operator (Re) and factoring out and deleting the term $e^{j\omega t}$. The next equation illustrates the simplification:

$$A\cos(\omega t + \phi) = \text{Re}\,[Ae^{j(\omega t+\phi)}] = \text{Re}\,[Ae^{j\phi}e^{j\omega t}] \tag{4.72}$$

The reason for this simplification is simply mathematical convenience, as will become apparent in the examples; you will have to remember that the $e^{j\omega t}$ term that was removed from the complex form of the sinusoid is really still present, indicating the specific frequency of the sinusoidal signal, ω. With these caveats, you should now be prepared to use the newly found phasor to analyze AC circuits. The following comments summarize the important points developed thus far in the section.

1. Any sinusoidal signal may be mathematically represented in one of two ways: a **time-domain form,**

   $$v(t) = A\cos(\omega t + \phi)$$

   and a **frequency-domain** (or **phasor**) **form,**

   $$\mathbf{V}(j\omega) = Ae^{j\phi}$$

   Note the "$j\omega$" in the notation $\mathbf{V}(j\omega)$, indicating the $e^{j\omega t}$ dependence of the phasor. In the remainder of this chapter, bold uppercase quantities will be employed to indicate phasor voltages or currents.
2. A phasor is a complex number, expressed in polar form, consisting of a *magnitude* equal to the peak amplitude of the sinusoidal signal and a *phase angle* equal to the phase shift of the sinusoidal signal *referenced to a cosine signal.*
3. When using phasor notation, it is important to make a note of the specific frequency, ω, of the sinusoidal signal, since this is not explicitly apparent in the phasor expression.

Example 4.13

Two sinusoidal sources of equal frequency, but of different phase, can be added using phasor notation. Let $v_1(t)$ and $v_2(t)$ be the source voltages defined by

$$v_1(t) = 15\cos\left(377t + \frac{\pi}{4}\right)$$

and

$$v_2(t) = 15\cos\left(377t + \frac{\pi}{12}\right)$$

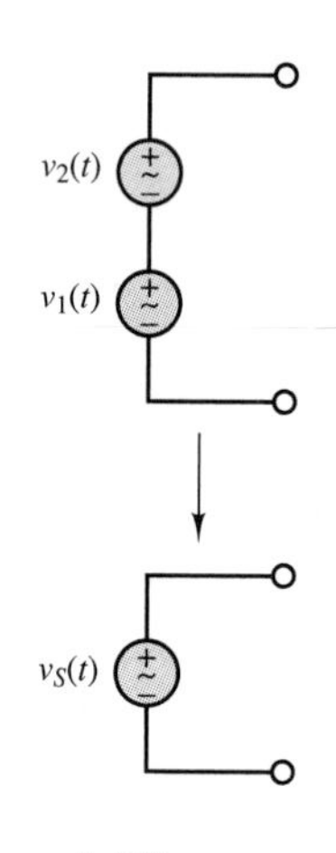

Figure 4.35

If the two sources are connected in series, as shown in Figure 4.35, what is the resultant voltage, $v_S(t)$?

Solution:

Write $v_1(t)$ and $v_2(t)$ in phasor form:

$$\mathbf{V}_1(j\omega) = 15e^{j45^\circ}$$
$$\mathbf{V}_2(j\omega) = 15e^{j15^\circ}$$

To compute the complex addition of $\mathbf{V}_1$ and $\mathbf{V}_2$, the phasors must first be converted to rectangular form:

$$\mathbf{V}_1 = 15e^{j45^\circ} = 10.61 + j10.61$$
$$\mathbf{V}_2 = 15e^{j15^\circ} = 14.49 + j3.88$$

so that

$$\mathbf{V}_S = \mathbf{V}_1 + \mathbf{V}_2 = 25.10 + j14.49 = 28.98e^{j30^\circ}$$

Having computed the phasor sum, we may convert $\mathbf{V}_S$ to time-domain form:

$$v_S(t) = 28.98\cos(377t + 30^\circ)$$

Note that the same result could have been obtained by adding the two sinusoids in the time domain, utilizing trigonometric identities. Expanding $v_1(t)$ and $v_2(t)$, we obtain

$$v_1(t) = 15\cos(\omega t + 45^\circ) = 15\cos(45^\circ)\cos(\omega t) - 15\sin(45^\circ)\sin(\omega t)$$
$$v_2(t) = 15\cos(\omega t + 15^\circ) = 15\cos(15^\circ)\cos(\omega t) - 15\sin(15^\circ)\sin(\omega t)$$

Then, combining like terms to get

$$v_1(t) + v_2(t) = 15(\cos 15^\circ + \cos 45^\circ)\cos\omega t - 15(\sin 15^\circ + \sin 45^\circ)\sin\omega t$$

we obtain

$$\begin{aligned} v_1(t) + v_2(t) &= 15(1.673\cos\omega t - 0.966\sin\omega t) \\ &= 15\sqrt{(1.673)^2 + (0.966)^2} \times \cos\left(\omega t + \tan^{-1}\left(\frac{0.966}{1.673}\right)\right) \\ &= 15(1.932\cos(\omega t + 30^\circ)) \\ &= 28.98\cos(\omega t + 30^\circ) \end{aligned}$$

which is, of course, the same answer we obtained using phasors. Which of the two solution methods required less labor?

It should be apparent by now that phasor notation can be a very efficient technique to solve AC circuit problems. The following sections will continue developing this new method to build your confidence in using it.

Superposition of AC Signals

Example 4.13 explored the combined effect of two sinusoidal sources of different phase and amplitude, but of the same frequency. It is important to realize that the simple answer obtained there does not apply to the superposition of two (or more) sinusoidal sources that *are not at the same frequency.* In this subsection, the case of two sinusoidal sources oscillating at different frequencies will be used to illustrate how phasor analysis can deal with this more general case.

The circuit shown in Figure 4.36 depicts a source excited by two current sources connected in parallel, where

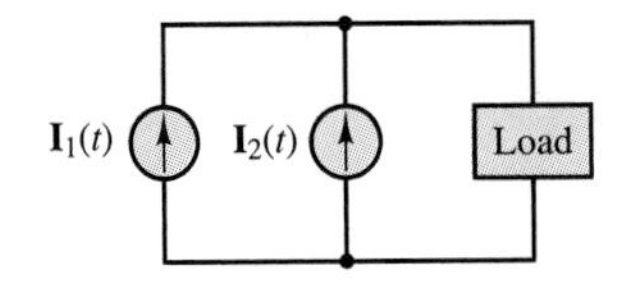

Figure 4.36 Superposition of AC currents

$$\begin{aligned} i_1(t) &= A_1 \cos(\omega_1 t) \\ i_2(t) &= A_2 \cos(\omega_2 t) \end{aligned} \tag{4.73}$$

The load current is equal to the sum of the two source currents; that is,

$$i_L(t) = i_1(t) + i_2(t) \tag{4.74}$$

or, in phasor form,

$$\mathbf{I}_L = \mathbf{I}_1 + \mathbf{I}_2 \tag{4.75}$$

At this point, you might be tempted to write $\mathbf{I}_1$ and $\mathbf{I}_2$ in a more explicit phasor form as

$$\begin{aligned} \mathbf{I}_1 &= A_1 e^{j0} \\ \mathbf{I}_2 &= A_2 e^{j0} \end{aligned} \tag{4.76}$$

and to add the two phasors using the familiar techniques of complex algebra. However, this approach *would be incorrect.* Whenever a sinusoidal signal is expressed in phasor notation, the term $e^{j\omega t}$ is implicitly present, where ω is the actual radian frequency of the signal. In our example, the two frequencies are not the same, as can be verified by writing the phasor currents in the form of equation 4.72:

$$\begin{aligned} \mathbf{I}_1 &= \text{Re}\,[A_1 e^{j0} e^{j\omega_1 t}] \\ \mathbf{I}_2 &= \text{Re}\,[A_2 e^{j0} e^{j\omega_2 t}] \end{aligned} \tag{4.77}$$

Since phasor notation does not *explicitly* include the $e^{j\omega t}$ factor, this can lead to serious errors if you are not careful! The two phasors of equation 4.76 cannot be added, but must be kept separate; thus, the only unambiguous expression for the load current in this case is equation 4.74. In order to complete the analysis of any circuit with multiple sinusoidal sources at different frequencies using phasors, it is necessary to solve the circuit separately for each signal and then add the individual answers obtained for the different excitation sources. Example 4.14 illustrates the response of a circuit with two separate AC excitations using AC superposition.

Example 4.14

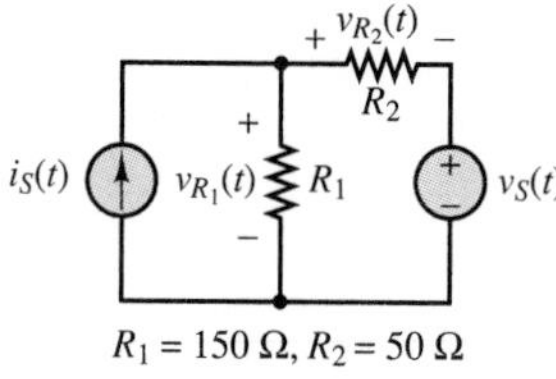

Figure 4.37

The circuit of Figure 4.37 is excited by two sources, both sinusoidal. The current source has a frequency of 100 Hz and can be expressed as

$$i_S(t) = 0.5\cos(2\pi 100t)$$

or

$$\mathbf{I}_S = 0.5\angle 0^\circ \qquad (\omega = 2\pi 100)$$

The voltage source has a frequency of 1,000 Hz and can be expressed as

$$v_S(t) = 20\cos(2\pi 1{,}000t)$$

or

$$\mathbf{V}_S = 20\angle 0^\circ \qquad (\omega = 2\pi 1{,}000)$$

Find the voltages $v_{R_2}(t)$ and $v_{R_1}(t)$.

Solution:

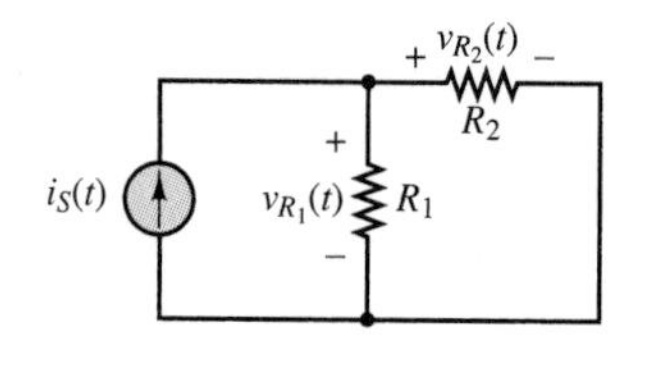

Figure 4.38

Since we have sources of different frequency, we are forced to find a separate solution for each source. First, we consider the effect of the current source. To do so, we replace the voltage source with a short circuit, as shown in Figure 4.38. It should be clear that we have reduced the circuit to a current divider, and that the voltages indicated in the figure are given by

$$\mathbf{V}_{R_1}(\mathbf{I}_S) = \mathbf{I}_{R_1} R_1$$

$$\mathbf{V}_{R_2}(\mathbf{I}_S) = \mathbf{I}_{R_2} R_2$$

$$\mathbf{V}_{R_1}(\mathbf{I}_S) = \mathbf{I}_S \frac{R_2}{R_2 + R_1} R_1 = 0.5\angle 0^\circ \left(\frac{50}{50 + 150}\right)(150)$$

$$= 18.75\angle 0^\circ \qquad (\omega = 2\pi 100)$$

$$\mathbf{V}_{R_2}(\mathbf{I}_S) = \mathbf{I}_S \frac{R_1}{R_2 + R_1} R_2 = 0.5\angle 0^\circ \left(\frac{150}{50 + 150}\right)(50)$$

$$= 18.75\angle 0^\circ \qquad (\omega = 2\pi 100)$$

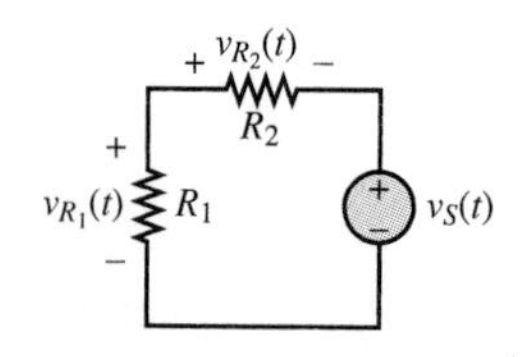

Figure 4.39

Next, we consider the effect of the voltage source. Replacing the current source with an open circuit, we obtain the voltage divider shown in Figure 4.39, from which we obtain the expressions

$$\mathbf{V}_{R_1}(\mathbf{V}_S) = \mathbf{V}_S \frac{R_1}{R_2 + R_1} = 20\angle 0^\circ \left(\frac{150}{50 + 150}\right)$$

$$= 15\angle 0^\circ \qquad (\omega = 2\pi 1{,}000)$$

$$\mathbf{V}_{R_2}(\mathbf{V}_S) = -\mathbf{V}_S \frac{R_2}{R_2 + R_1} = -20\angle 0^\circ \left(\frac{50}{50 + 150}\right)$$

$$= -5\angle 0^\circ = 5\angle 180^\circ \qquad (\omega = 2\pi 1{,}000)$$

Adding the contribution from each of the sources and converting to time-domain representation yields

$$\mathbf{V}_{R_1} = \mathbf{V}_{R_1}(\mathbf{I}_S) + \mathbf{V}_{R_1}(\mathbf{V}_S)$$

$$v_{R_1}(t) = 18.75\cos(2\pi 100t) + 15\cos(2\pi 1{,}000t)$$

and

$$\mathbf{V}_{R_2}(t) = \mathbf{V}_{R_2}(\mathbf{I}_S) + \mathbf{V}_{R_2}(\mathbf{V}_S)$$

$$v_{R_2}(t) = 18.75\cos(2\pi 100t) + 5\cos(2\pi 1{,}000t + 180^\circ)$$

It is important to observe that it is not possible to simplify these expressions any further, since the two sinusoidal components are at different frequencies.

Impedance

We now analyze the *i-v* relationship of the three ideal circuit elements in light of the new phasor notation. The result will be a new formulation in which resistors, capacitors, and inductors will be described in the same notation. A direct consequence of this result will be that the circuit theorems of Chapter 3 will be extended to AC circuits. In the context of AC circuits, any one of the three ideal circuit elements defined so far will be described by a parameter called **impedance**, which may be viewed as a *complex resistance*. The impedance concept is equivalent to stating that capacitors and inductors act as *frequency-dependent resistors*, that is, as resistors whose resistance is a function of the frequency of the sinusoidal excitation. Figure 4.40 depicts the same circuit represented in conventional form (top) and in phasor-impedance form (bottom); the latter representation explicitly shows phasor voltages and currents and treats the circuit element as a generalized "impedance." It will presently be shown that each of the three ideal circuit elements may be represented by one such impedance element.

Let the source voltage in the circuit of Figure 4.40 be defined by

$$v_S(t) = A\cos\omega t \qquad \text{or} \qquad \mathbf{V}_S(j\omega) = Ae^{j0^\circ} \tag{4.78}$$

without loss of generality. Then the current $i(t)$ is defined by the *i-v* relationship for each circuit element. Let us examine the frequency-dependent properties of the resistor, inductor, and capacitor, one at a time.

The Resistor

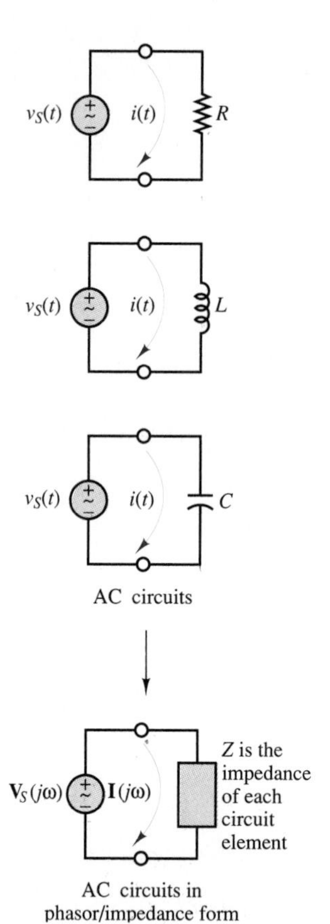

Figure 4.40 The impedance element

Ohm's law dictates the well-known relationship $v = iR$. In the case of sinusoidal sources, then, the current flowing through the resistor of Figure 4.40 may be expressed as

$$i(t) = \frac{v_S(t)}{R} = \frac{A}{R}\cos(\omega t) \tag{4.79}$$

Converting the voltage $v_S(t)$ and the current $i(t)$ to phasor notation, we obtain the following expressions:

$$\begin{aligned} \mathbf{V}_S(j\omega) &= Ae^{j0^\circ} \\ \mathbf{I}(j\omega) &= \frac{A}{R}e^{j0^\circ} \end{aligned} \tag{4.80}$$

Finally, the *impedance* of the resistor is defined as the ratio of the phasor voltage across the resistor to the phasor current flowing through it, and the symbol Z_R is used to denote it:

$$Z_R(j\omega) = \frac{\mathbf{V}_S(j\omega)}{\mathbf{I}(j\omega)} = R \quad \text{Impedance of a resistor} \tag{4.81}$$

Equation 4.81 corresponds to Ohm's law in phasor form, and the result should be intuitively appealing: Ohm's law applies to a resistor independent of the particular form of the voltages and currents (whether AC or DC, for instance). The ratio of phasor voltage to phasor current has a very simple form in the case of the resistor. In general, however, the impedance of an element is a complex function of frequency, as it must be, since it is the ratio of two phasor quantities, which are frequency-dependent. This property will become apparent when the impedances of the inductor and capacitor are defined.

The Inductor

Recall the defining relationships for the ideal inductor (equations 4.11 and 4.14), repeated here for convenience:

$$\begin{aligned} v_L(t) &= L\frac{di_L(t)}{dt} \\ i_L(t) &= \frac{1}{L}\int v_L\,dt \end{aligned} \tag{4.82}$$

Let $v_L(t) = v_S(t)$ and $i_L(t) = i(t)$ in the circuit of Figure 4.40. Then the following expression may be derived for the inductor current:

$$i_L(t) = i(t) = \frac{1}{L}\int v_S(t)\,dt$$

$$i_L(t) = \frac{1}{L}\int A\cos\omega t\,dt \qquad \textbf{(4.83)}$$

$$= \frac{A}{\omega L}\sin\omega t$$

Note how a dependence on the radian frequency of the source is clearly present in the expression for the inductor current. Further, the inductor current is shifted in phase (by 90°) with respect to the voltage. This fact can be seen by writing the inductor voltage and current in time-domain form:

$$v_S(t) = v_L(t) = A\cos\omega t$$

$$i(t) = i_L(t) = \frac{A}{\omega L}\cos(\omega t - 90°) \qquad \textbf{(4.84)}$$

It is evident that the current is not just a scaled version of the source voltage, as it was for the resistor. Its magnitude depends on the frequency, ω, and it is shifted (delayed) in phase by 90°. Using phasor notation, equation 4.84 becomes

$$\mathbf{V}_S(j\omega) = Ae^{j0°}$$

$$\mathbf{I}(j\omega) = \frac{A}{\omega L}e^{-j90°} \qquad \textbf{(4.85)}$$

Thus, the impedance of the inductor is defined as follows:

$$Z_L(j\omega) = \frac{\mathbf{V}_S(j\omega)}{\mathbf{I}(j\omega)} = \omega L e^{j90°} = j\omega L \qquad \text{Impedance of an inductor} \qquad \textbf{(4.86)}$$

Note that the inductor now appears to behave like a *complex frequency–dependent resistor,* and that the magnitude of this complex resistor, ωL, is proportional to the signal frequency, ω. Thus, an inductor will "impede" current flow in proportion to the sinusoidal frequency of the source signal. This means that at low signal frequencies, an inductor acts somewhat like a short circuit, while at high frequencies it tends to behave more as an open circuit. Another important point is that *the magnitude of the impedance of an inductor is always positive,* since both L and ω are positive numbers. You should verify that the units of this magnitude are also ohms.

The Capacitor

An analogous procedure may be followed to derive the equivalent result for a capacitor. Beginning with the defining relationships for the ideal capacitor,

$$i_C(t) = C\frac{dv_C(t)}{dt}$$
$$v_C(t) = \frac{1}{C}\int i_C(t)\,dt \tag{4.87}$$

with $i_C = i$ and $v_C = v_S$ in Figure 4.40, the capacitor current may be expressed as:

$$\begin{aligned} i_C(t) &= C\frac{dv_C(t)}{dt} \\ &= C\frac{d}{dt}(A\cos\omega t) \\ &= -C(A\omega \sin\omega t) \\ &= \omega C A \cos(\omega t + 90°) \end{aligned} \tag{4.88}$$

so that, in phasor form,

$$\begin{aligned} \mathbf{V}_S(j\omega) &= Ae^{j0°} \\ \mathbf{I}(j\omega) &= \omega C A e^{j90°} \end{aligned} \tag{4.89}$$

The impedance of the ideal capacitor, $Z_C(j\omega)$, is therefore defined as follows:

$$Z_C(j\omega) = \frac{\mathbf{V}_S(j\omega)}{\mathbf{I}(j\omega)} = \frac{1}{\omega C}e^{-j90°} = \frac{-j}{\omega C} = \frac{1}{j\omega C} \quad \text{Impedance of a capacitor} \tag{4.90}$$

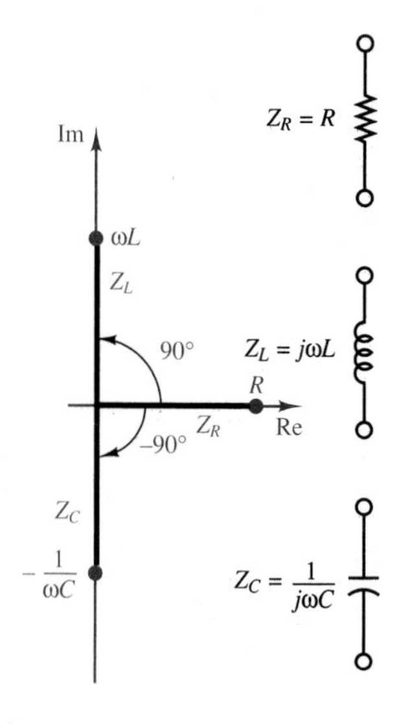

Figure 4.41 Impedances of R, L, and C in the complex plane

where we have used the fact that $1/j = e^{-j90°} = -j$. Thus, the impedance of a capacitor is also a frequency-dependent complex quantity, with the impedance of the capacitor varying as an inverse function of frequency; and so a capacitor acts like a short circuit at high frequencies, whereas it behaves more like an open circuit at low frequencies. Another important point is that *the impedance of a capacitor is always negative,* since both C and ω are positive numbers. You should verify that the units of impedance for a capacitor are ohms. Figure 4.41 depicts $Z_C(j\omega)$ in the complex plane, alongside $Z_R(j\omega)$ and $Z_L(j\omega)$.

The impedance parameter defined in this section is extremely useful in solving AC circuit analysis problems, because it will make it possible to take advantage of most of the network theorems developed for DC circuits by replacing resistances with complex-valued impedances. The examples that follow illustrate how branches containing series and parallel elements may be reduced to a single equivalent impedance, much in the same way resistive circuits were reduced to equivalent forms. It is important to emphasize that although the impedance of simple circuit elements is either purely real (for resistors) or purely imaginary (for capacitors and inductors), the general definition of impedance for an arbitrary circuit must allow for the possibility of having both a real and an imaginary part, since practical circuits are made up of more or less complex interconnections of different circuit elements. In its most general form, the impedance of a circuit element is defined as the sum of a real part and an imaginary part:

$$Z(j\omega) = R(j\omega) + jX(j\omega) \tag{4.91}$$

where R is called the **AC resistance** and X is called the **reactance**. The frequency dependence of R and X has been indicated explicitly, since it is possible for a circuit to have a frequency-dependent resistance. The examples illustrate how a complex impedance containing both real and imaginary parts arises in a circuit.

Example 4.15

Find the equivalent impedance of the branch shown in Figure 4.42 at the frequency $\omega = 377$ rad/s.

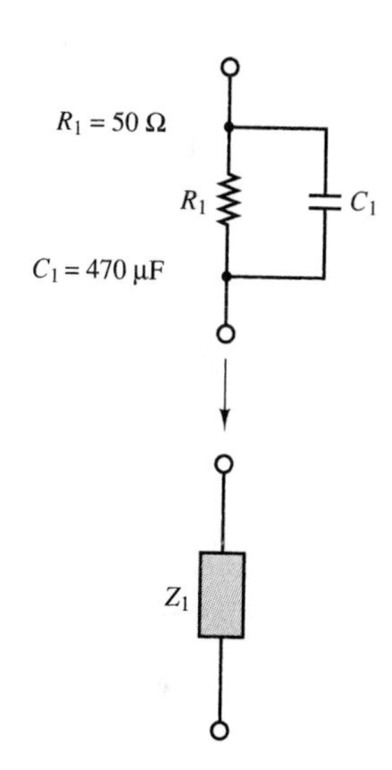

Figure 4.42

Solution:

To obtain the equivalent impedance, we combine the individual impedances of the two elements in parallel:

$$Z_1 = R_1 \parallel \frac{1}{j\omega C_1} = \frac{R_1 \frac{1}{j\omega C_1}}{R_1 + \frac{1}{j\omega C_1}} = \frac{R_1}{1 + j\omega C_1 R_1}$$

Thus,

$$Z_1(\omega = 377) = \frac{50}{1 + j8.86} = \frac{50(1 - j8.86)}{(1 + j8.86)(1 - j8.86)}$$

$$= \frac{50 - j443}{1 + 78.5} = 0.629 - j5.57$$

How would the answer change if the frequency doubled?

Note that the answer indicates that the impedance corresponding to the parallel RC circuit is equivalent to the series connection of a resistance ($0.629\ \Omega$) with a reactance ($-j5.57\ \Omega$). What circuit element corresponds to this reactance? Since the reactance is negative, the corresponding circuit element must be a capacitor; this is true because the impedance of a capacitor is always expressed by

$$Z_C = \frac{-j}{\omega C}$$

Thus, we have

$$\frac{1}{\omega C} = 5.57$$

and since $\omega = 377$, the capacitance corresponding to the given reactance is

$$C = \frac{1}{5.57\omega} = 476\ \mu\text{F}$$

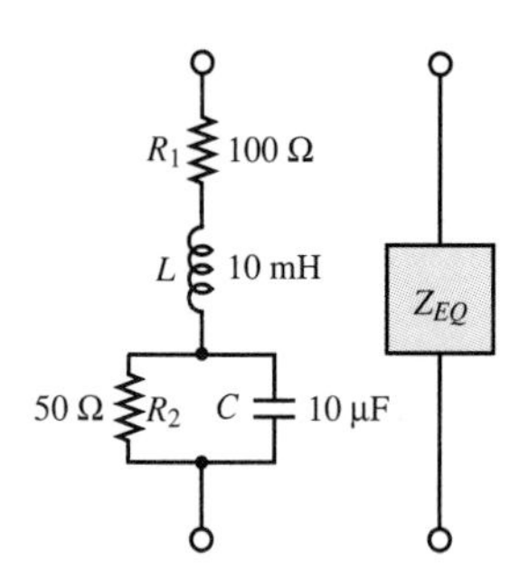

Figure 4.43

Example 4.16

Find the equivalent impedance of the circuit shown in Figure 4.43 at the frequency $\omega = 10,000$ rad/s.

Solution:

Let us first consider the parallel R_2-C combination, to obtain the equivalent parallel impedance $Z_\| = Z_{R_2} \| Z_C$:

$$Z_{R_2} \| Z_C = \frac{Z_{R_2} Z_C}{Z_{R_2} + Z_C} = Z_\|$$

$$Z_\| = \frac{50(-j10)}{50 - j10}$$

$$= \frac{500e^{-j90^\circ}}{51e^{-j11.31^\circ}} = 9.8e^{-j78.69^\circ}$$

$$= 1.92 - j9.61$$

Thus, $Z_\|$ may be considered the series combination of two elements, a 1.92-Ω resistor and a capacitor with reactance equal to

$$X_\| = \frac{1}{\omega C} = 9.61$$

which, at the frequency of 10,000 radians per second, corresponds to a capacitance

$$C = 10.41 \ \mu\text{F}$$

It is very instructive to stop for a second to consider the fact that the parallel combination of a resistor and a capacitor can be converted to a completely equivalent series combination *at a given frequency* and that the resulting equivalent capacitance and resistance differ from the original values. (In this particular case, the capacitance dominates, because of the high frequency. What would the equivalent resistance and capacitance be at a frequency of 10 rad/s?)

The resulting circuit is shown in Figure 4.44, with the associated complex impedance values. If we combine the real and imaginary parts of the series elements of the branch, we have the total equivalent impedance Z_{EQ}, given by

$$Z_{EQ} = (100 + 1.92) + j(100 - 9.61)$$

$$= 101.91 + j90.39$$

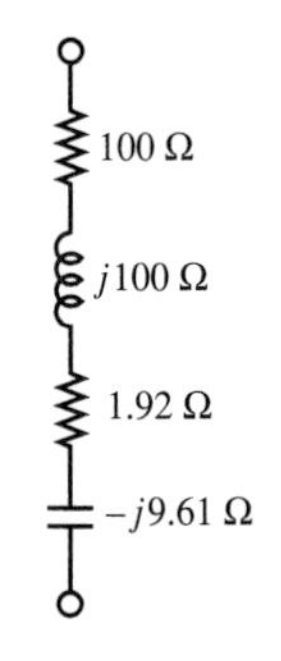

Figure 4.44

This impedance is the total equivalent impedance of the circuit branch. Note that this impedance is inductive, i.e., has a positive reactance (why?).

Example 4.17

The inductor, which we have so far modeled as an ideal circuit element, cannot always be idealized in practice, because of the finite resistance of the wire used in the coil. A more accurate model for an inductor is shown in Figure 4.45. The inductor model consists of an ideal inductor and two resistances that represent the resistance of the wire used in the windings of the inductor. R_1 and R_2 are usually considered to be equal, with each resistance equal to half the total winding resistance. Practical inductors are often constructed in the form of a *toroid* (doughnut), as shown in Figure 4.46.

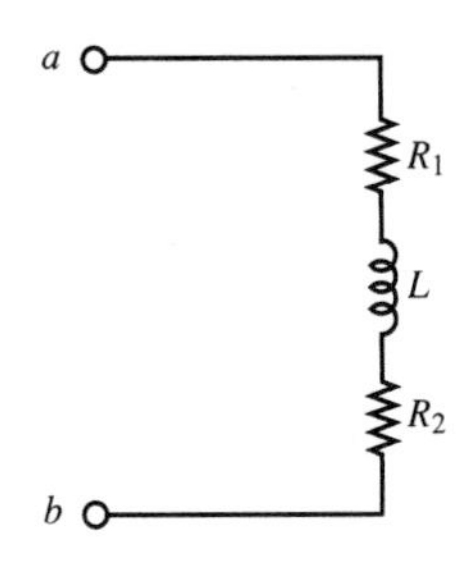

Figure 4.45

1. If the toroid is wound with 250 turns of 30 gauge wire, what is the total resistance of the windings, including lead length of 10 cm per lead? Find R_1 and R_2. (30 gauge wire has a resistance of 105 Ω/1,000 ft.)
2. If the inductance is 0.098 H and we assume that in order for the equivalent circuit to be "mostly inductive," the reactance of the inductor must be at least 10 times larger than the resistance of the lead wires, then over what frequency range will the equivalent circuit be considered "mostly inductive"?

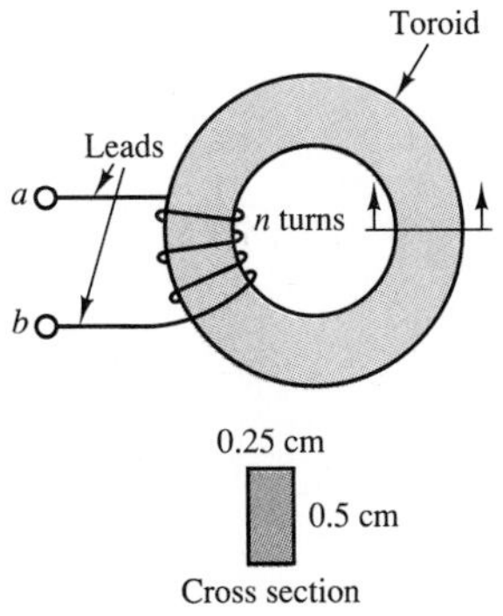

Figure 4.46 A practical inductor

Solution:

1. For the 250-turn coil, we can compute the length of wire used in the windings as l_w and lead length as l_c, with

$$l_w = 250 \times [2(0.25 \text{ cm}) + 2(0.5 \text{ cm})] = 375 \text{ cm}$$

$$l_c = 2(10) \text{ cm} = 20 \text{ cm}$$

$$\text{Total length} = 395 \text{ cm}$$

$$= 155.5 \text{ in}$$

$$= 12.96 \text{ ft}$$

Thus the total resistance is given by:

$$\text{Total resistance} = \frac{105\ \Omega}{1{,}000 \text{ ft}} \times 12.96 \text{ ft}$$

$$R_{\text{total}} = 1.361\ \Omega$$

$$R_1 = R_2 = \frac{R_{\text{total}}}{2} = 0.680\ \Omega$$

2. To determine the desired frequency range, we require that the reactance of the inductor be greater than or equal to 10 times the resistance of the lead wires, so that the resistive component of the impedance is negligible:

$$\omega L \geq 10(1.361)$$

$$\omega \geq \frac{10(1.361)}{0.098}$$

which in turn suggests that the desired frequency range is

$$\omega \geq 138.9 \text{ rad/s}$$

or

$$f \geq 22.1 \text{ Hz}$$

Admittance

In Chapter 3, it was suggested that the solution of certain circuit analysis problems was handled more easily in terms of conductances than resistances. In AC circuit analysis, an analogous quantity may be defined, the reciprocal of complex impedance. Just as the conductance, G, of a resistive element was defined as the inverse of the resistance, the **admittance** of a branch is defined as follows:

$$y = \frac{1}{Z} \text{ S} \tag{4.92}$$

Note immediately that whenever Z is purely real—that is, when $Z = R + j0$—the admittance Y is identical to the conductance G. In general, however, Y is the complex number

$$Y = G + jB \tag{4.93}$$

where G is called the **AC conductance** and B is called the **susceptance;** the latter plays a role analogous to that of reactance in the definition of impedance. Clearly, G and B are related to R and X. However, this relationship is not as simple as an inverse. Let $Z = R + jX$ be an arbitrary impedance. Then, the corresponding admittance is

$$Y = \frac{1}{Z} = \frac{1}{R + jX} \tag{4.94}$$

In order to express Y in the form $Y = G + jB$, we multiply numerator and denominator by $R - jX$:

$$\begin{aligned} Y &= \frac{1}{R + jX}\frac{R - jX}{R - jX} = \frac{R - jX}{R^2 + X^2} \\ &= \frac{R}{R^2 + X^2} - j\frac{X}{R^2 + X^2} \end{aligned} \tag{4.95}$$

and conclude that

$$\begin{aligned} G &= \frac{R}{R^2 + X^2} \\ B &= \frac{-X}{R^2 + X^2} \end{aligned} \tag{4.96}$$

Notice in particular that G *is not the reciprocal of* R in the general case!

The following example illustrates the determination of Y for some common circuits.

Example 4.18

This example illustrates how one can compute the admittance of a circuit. We consider the impedance seen at the input terminals of a one-port circuit and compute the admittance as the inverse of the impedance. The example is carried out for the two circuits of Figure 4.47.

Solution:

In Figure 4.47(a), the impedance Z_{ab} is

$$Z_{ab} = R + j\omega L$$

and therefore the admittance of the circuit is

$$Y_{ab} = \frac{1}{R + j\omega L} = \frac{R - j\omega L}{R^2 + (\omega L)^2}$$

In Figure 4.47(b), the impedance Z_{ab} is

$$Z_{ab} = R \parallel \frac{1}{j\omega C} = \frac{R}{1 + j\omega RC}$$

and so the admittance in the circuit is

$$Y_{ab} = \frac{1 + j\omega RC}{R} = \frac{1}{R} + j\omega C$$

If the source frequency were 1,000 Hz and the resistors had values of $R_a = 50\ \Omega$ and $R_b = 100\ \Omega$, with $L = 16$ mH and $C = 3\ \mu$F, we could compute the admittances in Figure 4.47(a) and (b) as follows:

$$Y_{ab\ (a)} = \frac{1}{50 + j100.5} = 8.91 \times 10^{-3}\angle -63.6° = 3.97 \times 10^{-3} - j7.98 \times 10^{-3}\ \text{S}$$

$$Y_{ab\ (b)} = \frac{1}{100} + j0.01885 = 0.0213\angle 62° = 1 \times 10^{-2} + j19 \times 10^{-3}\ \text{S}$$

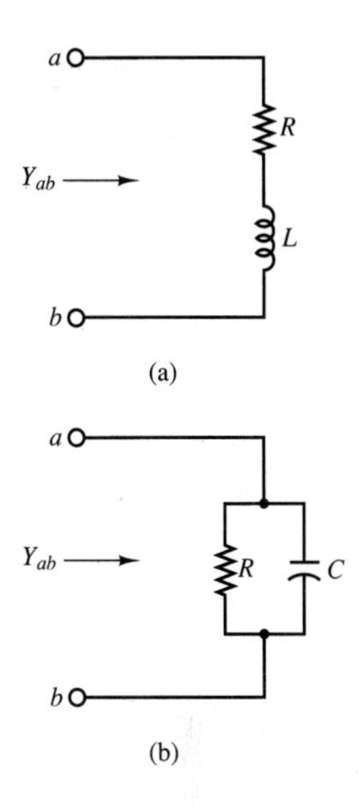

Figure 4.47

Example 4.19 Capacitive Displacement Transducer

In Example 4.4, the idea of a capacitive displacement transducer was introduced when we considered a parallel-plate capacitor composed of a fixed plate and a movable plate. The capacitance of this variable capacitor was shown to be a *nonlinear* function of the position of the movable plate, x (see Figure 4.7). In this example, we show that under certain conditions the impedance of the capacitor varies as a *linear* function of displacement—that is, the movable-plate capacitor can serve as a linear transducer.

Recall the expression derived in Example 4.4:

$$C = \frac{8.854 \times 10^{-3} A}{x}$$

where C is the capacitance in pF, A is the area of the plates in mm^2, and x is the (variable) distance in mm. If the capacitor is placed in an AC circuit, its impedance will be determined by the expression

$$Z_C = \frac{1}{j\omega C}$$

so that

$$Z_C = \frac{x}{j\omega 8.854 A}$$

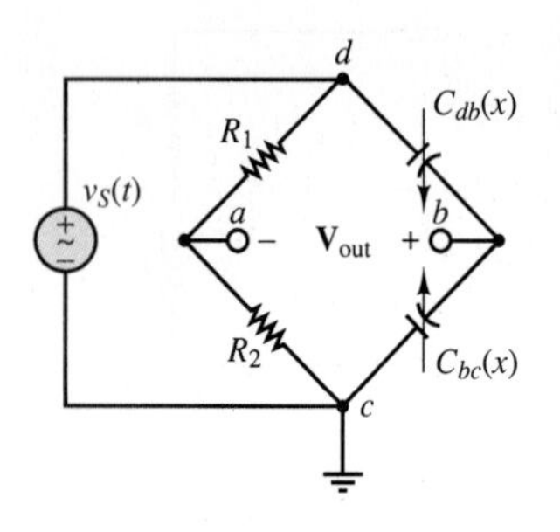

Figure 4.48 Bridge circuit for capacitive displacement transducer

Thus, at a fixed frequency ω, the impedance of the capacitor will vary linearly with displacement. This property may be exploited in the bridge circuit of Example 4.4, where a differential pressure transducer was shown as being made of two movable-plate capacitors, such that if the capacitance of one increased as a consequence of a pressure differential across the transducer, the capacitance of the other had to decrease by a corresponding amount (at least for small displacements). The circuit is shown again in Figure 4.48, where two resistors have been connected in the bridge along with the variable capacitors (denoted by $C(x)$). The bridge is excited by a sinusoidal source.

Using phasor notation, we can express the output voltage as follows:

$$\mathbf{V}_{out}(j\omega) = \mathbf{V}_S(j\omega)\left(\frac{Z_{C_{bc}(x)}}{Z_{C_{db}(x)} + Z_{C_{bc}(x)}} - \frac{R_2}{R_1 + R_2}\right)$$

If the nominal capacitance of each movable-plate capacitor with the diaphragm in the center position is given by

$$C = \frac{\varepsilon A}{d}$$

where d is the nominal (undisplaced) separation between the diaphragm and the fixed surfaces of the capacitors (in mm), the capacitors will see a change in capacitance given by

$$C_{db} = \frac{\varepsilon A}{d - x} \quad \text{and} \quad C_{bc} = \frac{\varepsilon A}{d + x}$$

when a pressure differential exists across the transducer, so that the impedances of the variable capacitors change according to the displacement:

$$Z_{C_{db}} = \frac{d - x}{j\omega 8.854A} \quad \text{and} \quad Z_{C_{bc}} = \frac{d + x}{j\omega 8.854A}$$

and we obtain the following expression for the phasor output voltage:

$$\begin{aligned}\mathbf{V}_{out}(j\omega) &= \mathbf{V}_S(j\omega)\left(\frac{\dfrac{d + x}{j\omega 8.854A}}{\dfrac{d - x}{j\omega 8.854A} + \dfrac{d + x}{j\omega 8.854A}} - \frac{R_2}{R_1 + R_2}\right)\\ &= \mathbf{V}_S(j\omega)\left(\frac{1}{2} + \frac{x}{2d} - \frac{R_2}{R_1 + R_2}\right)\\ &= \mathbf{V}_S(j\omega)\frac{x}{2d}\end{aligned}$$

if we choose $R_1 = R_2$. Thus, the output voltage will vary as a scaled version of the input voltage in proportion to the displacement. A typical $v_{out}(t)$ is displayed in Figure 4.49 for a 0.05-mm "triangular" diaphragm displacement, with $d = 0.5$ mm and $\mathbf{V}_S$ a 25-Hz sinusoid with 1 V amplitude.

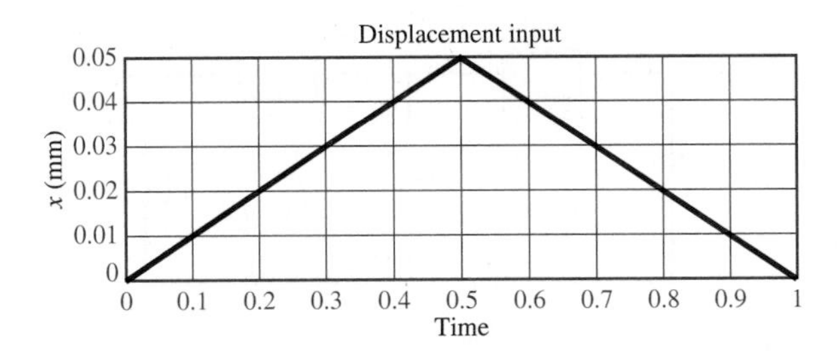

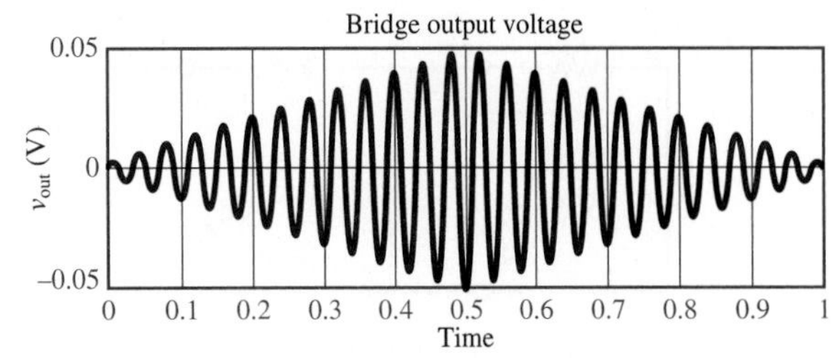

Figure 4.49 Displacement input and bridge output voltage for capacitive displacement transducer

DRILL EXERCISES

4.15 Add the sinusoidal voltages $v_1(t) = A\cos(\omega t + \phi)$ and $v_2(t) = B\cos(\omega t + \theta)$ using phasor notation, and then convert back to time-domain form, for:

a. $A = 1.5, \phi = 10°; B = 3.2, \theta = 25°$.
b. $A = 50, \phi = -60°; B = 24, \theta = 15°$.

4.16 Add the sinusoidal currents $i_1(t) = A\cos(\omega t + \phi)$ and $i_2(t) = B\cos(\omega t + \theta)$ for:

a. $A = 0.09, \phi = 72°; B = 0.12, \theta = 20°$.
b. $A = 0.82, \phi = -30°; B = 0.5, \theta = -36°$.

4.17 Compute the equivalent impedance of the circuit of Example 4.16 for $\omega = 1{,}000$ and 100,000 rad/s.

4.18 Compute the equivalent admittance of the circuit of Example 4.16.

4.19 Calculate the equivalent series capacitance of the parallel R_2-C circuit of Example 4.16 at the frequency $\omega = 10$ rad/s.

4.5 AC CIRCUIT ANALYSIS METHODS

This section will illustrate how the use of phasors and impedance facilitates the solution of AC circuits by making it possible to use the same solution methods developed in Chapter 3 for DC circuits. The AC circuit analysis problem of interest in this section consists of determining the unknown voltage (or currents) in a circuit containing linear passive circuit elements (R, L, C) and excited by a sinusoidal source. Figure 4.50 depicts one such circuit, represented in both conventional time-domain form and phasor-impedance form.

The first step in the analysis of an AC circuit is to note the frequency of the sinusoidal excitation. Next, all sources are converted to phasor form, and each circuit element to impedance form. This is illustrated in the phasor circuit of Figure 4.50. At this point, if the excitation frequency, ω, is known numerically, it will be possible to express each impedance in terms of a known amplitude and phase, and a numerical answer to the problem will be found. It does often

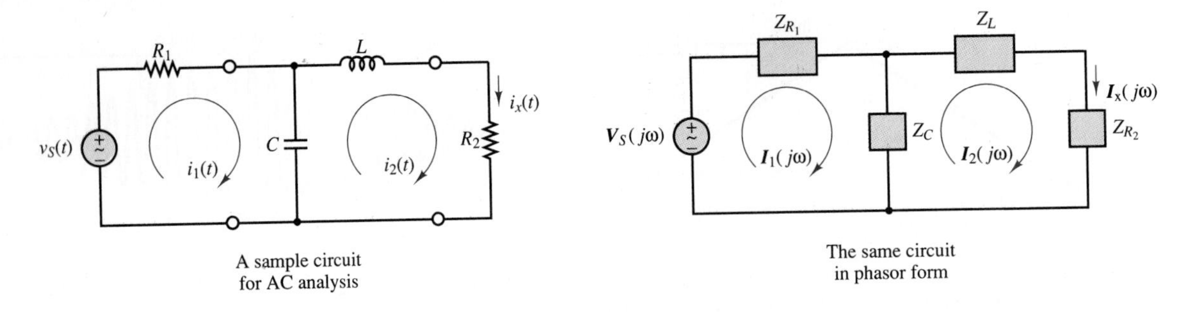

Figure 4.50 An AC circuit

happen, however, that one is interested in a more general circuit solution, valid for an arbitrary excitation frequency. In this latter case, the solution becomes a function of ω. This point will be developed further in Chapter 6, where the concept of sinusoidal frequency response is discussed.

With the problem formulated in phasor notation, the resulting solution will be in phasor form and will need to be converted to time-domain form. In effect, the use of phasor notation is but an intermediate step that greatly facilitates the computation of the final answer. In summary, here is the procedure that will be followed to solve an AC circuit analysis problem. Example 4.20 illustrates the various aspects of this method.

1. Identify the sinusoidal source(s) and note the excitation frequency.
2. Convert the source(s) to phasor form.
3. Represent each circuit element by its impedance.
4. Solve the resulting phasor circuit, using appropriate network analysis tools.
5. Convert the (phasor-form) answer to its time-domain equivalent, using equation 4.72.

Example 4.20

In this example, the method of circuit analysis described in this section is applied to solve for the unknown source current $i_S(t)$ in the circuit shown in Figure 4.51.

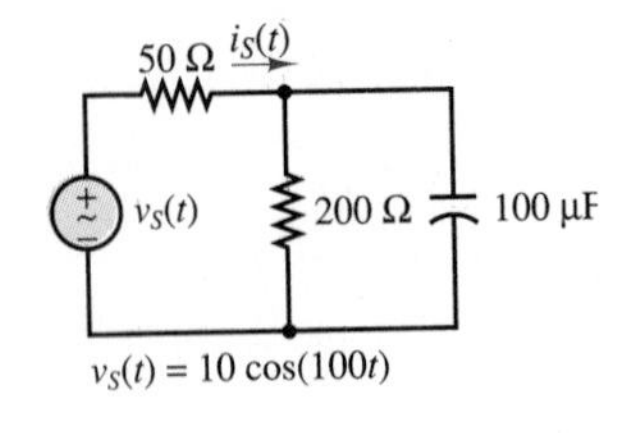

Figure 4.51

Solution:

From the time-domain form of the circuit, we deduce that

$$\omega = 100 \text{ rad/s}$$

$$\mathbf{V}_S(j\omega) = 10e^{j0} \text{ V}$$

$$Z_C(j\omega) = \frac{1}{j\omega C} = \frac{-j}{100 \times 100 \times 10^{-6}} = -j100$$

The corresponding phasor circuit is shown in Figure 4.52.

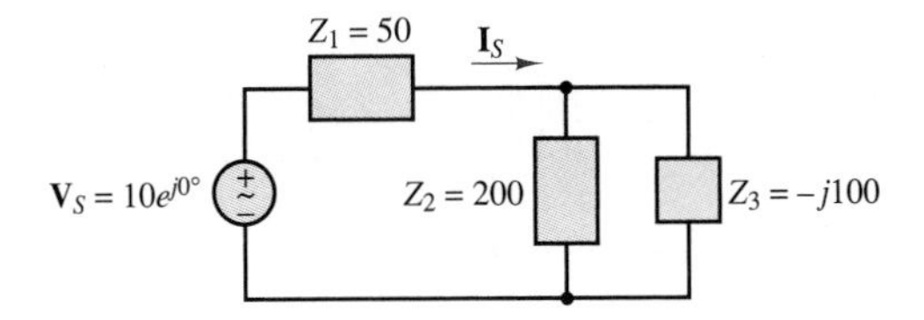

Figure 4.52

In order to solve the circuit by means of the techniques illustrated in the preceding sections, we recognize that the impedances Z_3 and Z_2 are in parallel; thus,

$$
\begin{aligned}
Z_2 \parallel Z_3 &= \frac{Z_2 Z_3}{Z_2 + Z_3} = \frac{-j2 \times 10^4}{200 - j100} \\
&= \frac{2 \times 10^4 e^{-j90°}}{223.6 e^{-j26.57°}} \\
&= 40 - j80
\end{aligned}
$$

The resulting equivalent circuit is shown in Figure 4.53, and the total series impedance seen by the source is given by the expression

$$Z_S = Z_1 + Z_2 \parallel Z_3 = 90 - j80$$

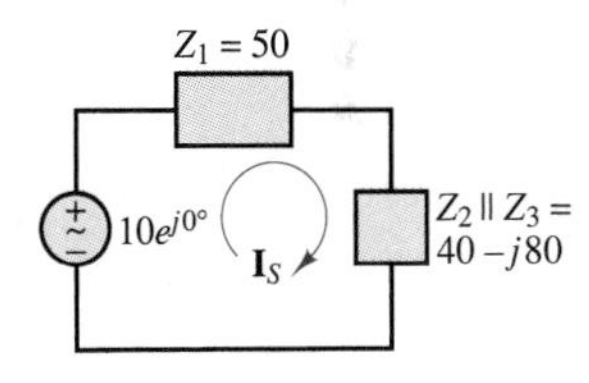

Figure 4.53

Finally, the phasor current $\mathbf{I}_S$ is given by the ratio of the source voltage to the equivalent impedance:

$$\mathbf{I}_S = \frac{10e^{j0°}}{90 - j80} = \frac{10e^{j0°}}{120e^{-j41.6°}} = 0.083e^{+j41.6°}$$

In order to complete the problem, we need to convert the phasor solution to time-domain form:

$$\mathbf{I}_S(\omega = 100) = 0.083e^{+j41.6°} \Longrightarrow i_S(t) = 0.083\cos(100t + 41.6°)$$

How would the answer differ if the excitation frequency were changed to $\omega = 200$?

Example 4.21

This example illustrates how to obtain the response of the circuit of Example 4.20 to a sinusoidal source of arbitrary amplitude A, phase ϕ, and frequency ω. Let the source excitation be $v_S(t) = A\cos(\omega t + \phi)$. The phasor form of the source will then be $\mathbf{V}_S(j\omega) = Ae^{j\phi}$.

Solution:

Using the same expression as in Example 4.20 to find the equivalent parallel impedance, we note that now the impedance of the capacitor is a function of frequency, since no value has been specified for ω:

$$Z_3 = \frac{1}{j\omega C} = \frac{10^4}{j\omega}$$

Using the results of the preceding example, we find the equivalent parallel impedance of Z_2 and Z_3 to be

$$Z_2 \parallel Z_3 = \frac{200 \times 10^4/j\omega}{200 + 10^4/j\omega} = \frac{2 \times 10^6}{10^4 + 200 \times j\omega}$$

The total series impedance may then be computed as follows:

$$Z_1 + Z_2 \parallel Z_3 = 50 + \frac{2 \times 10^6}{10^4 + 200 \times j\omega} = \frac{2.5 \times 10^6 + j\omega \times 10^4}{10^4 + 200 \times j\omega}$$

Thus, the phasor current $\mathbf{I}_S$ is given by

$$\mathbf{I}_S(j\omega) = \frac{\mathbf{V}_S(j\omega)}{Z_1 + Z_2 \parallel Z_3} = Ae^{j\phi}\frac{10^4 + 200 \times j\omega}{2.5 \times 10^6 + j\omega \times 10^4}$$

The numerical expression for $\mathbf{I}_S(j\omega)$ can now be obtained for any excitation amplitude A, frequency ω, and phase ϕ.

By now it should be apparent that the laws of network analysis introduced in Chapter 3 are also applicable to phasor voltages and currents. This fact suggests that it may be possible to extend the node and mesh analysis methods developed earlier to circuits containing phasor sources and impedances, although the resulting simultaneous complex equations are difficult to solve without the aid of a computer, even for relatively simple circuits. On the other hand, it is very useful to extend the concept of equivalent circuits to the AC case, and to define complex Thévenin (or Norton) equivalent impedances. The fundamental difference between resistive and AC equivalent circuits is that the AC Thévenin (or Norton) equivalent circuits will be frequency-dependent and complex-valued. In general, then, one may think of the resistive circuit analysis of Chapter 3 as a special case of AC analysis in which all impedances are real.

AC Equivalent Circuits

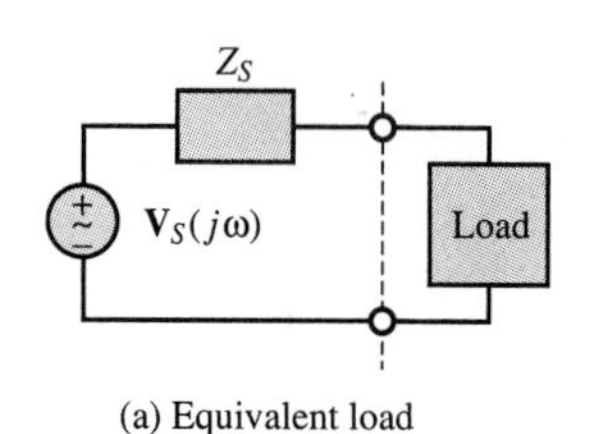

(a) Equivalent load

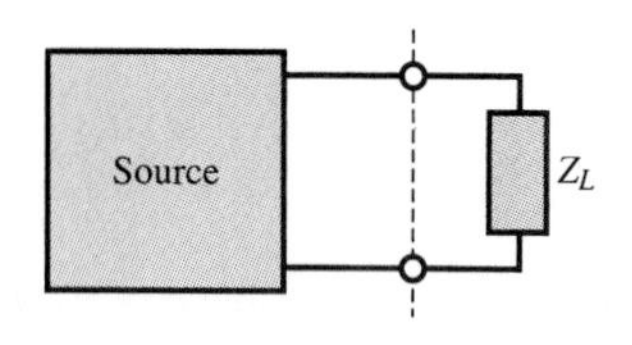

(b) Equivalent source

Figure 4.54 AC equivalent circuits

In Chapter 3, we demonstrated that it was convenient to compute equivalent circuits, especially in solving for load-related variables. Figure 4.54 depicts the two representations analogous to those developed in Chapter 3. Figure 4.54(a) shows an *equivalent load,* as viewed by the source, while Figure 4.54(b) shows an *equivalent source* circuit, from the perspective of the load.

In the case of linear resistive circuits, the equivalent load circuit can always be expressed by a single equivalent resistor, while the equivalent source circuit may take the form of a Norton or a Thévenin equivalent. This section extends these concepts to AC circuits and demonstrates that the notion of equivalent circuits applies to phasor sources and impedances as well. The techniques described in this section are all analogous to those used for resistive circuits, with resistances replaced by impedances, and arbitrary sources replaced by phasor sources. The principal difference between resistive and AC equivalent circuits will be that the latter are frequency-dependent. Figure 4.55 summarizes the fundamental principles used in computing an AC equivalent circuit. Note the definite analogy between impedance and resistance elements, and between conductance and admittance elements.

The computation of an equivalent impedance is carried out in the same way as that of equivalent resistance in the case of resistive circuits:

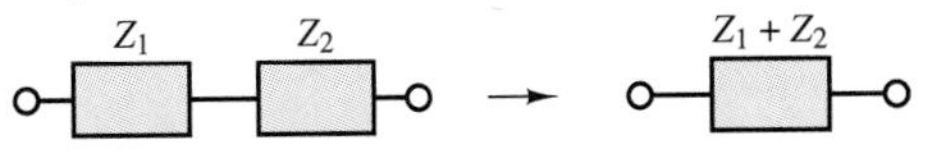

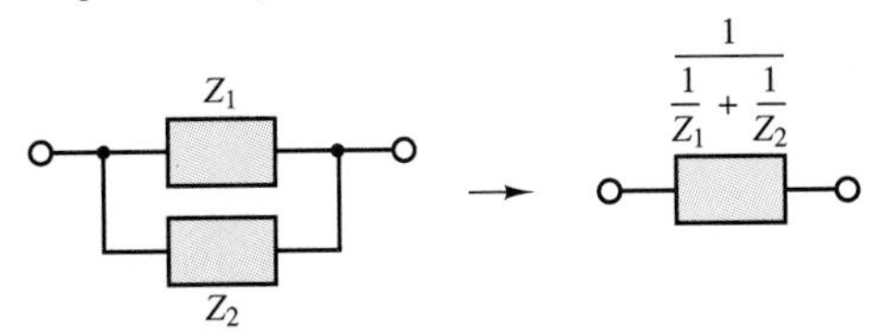

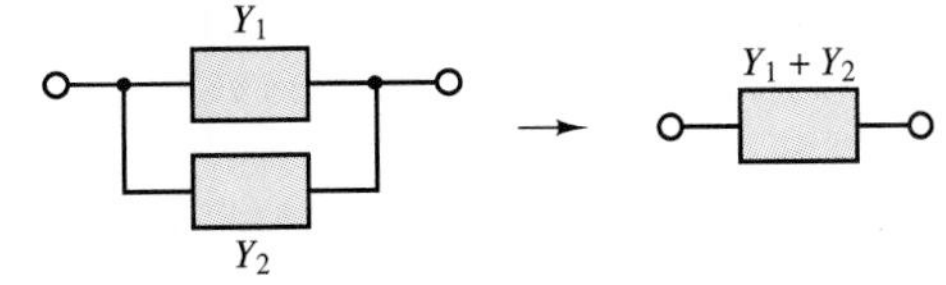

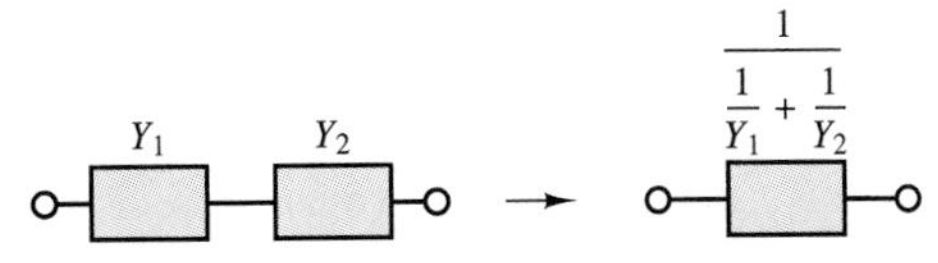

Figure 4.55 Rules for impedance and admittance reduction

1. Short-circuit all voltage sources, and open-circuit all current sources.
2. Compute the equivalent impedance between load terminals, with the load disconnected.

In order to compute the Thévenin or Norton equivalent form, we recognize that the Thévenin equivalent voltage source is the open-circuit voltage at the load terminals and the Norton equivalent current source is the short-circuit current (the current with the load replaced by a short circuit). Figure 4.56 illustrates

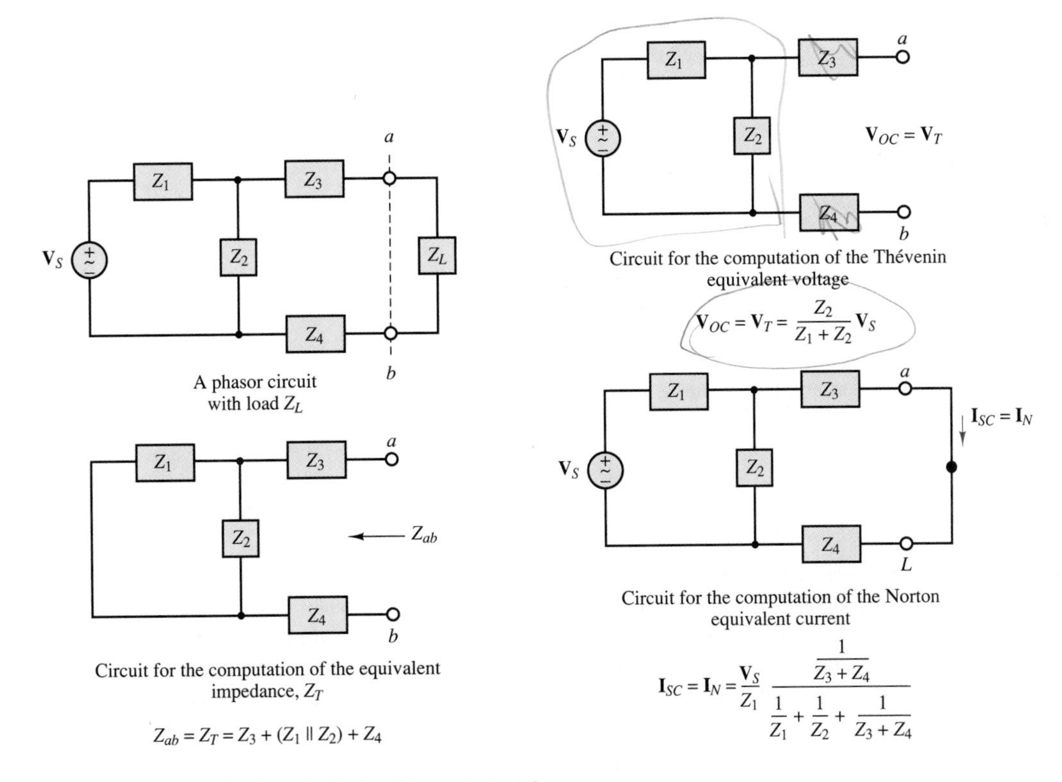

Figure 4.56 Reduction of AC circuit to equivalent form

these points by outlining the steps in the computation of an equivalent circuit. The remainder of the section will consist of examples aimed at clarifying some of the finer points in the calculation of such equivalent circuits. Note how the initial circuit reduction proceeds exactly as in the case of a resistive circuit; the details of the complex algebra required in the calculations are explored in the examples.

Example 4.22

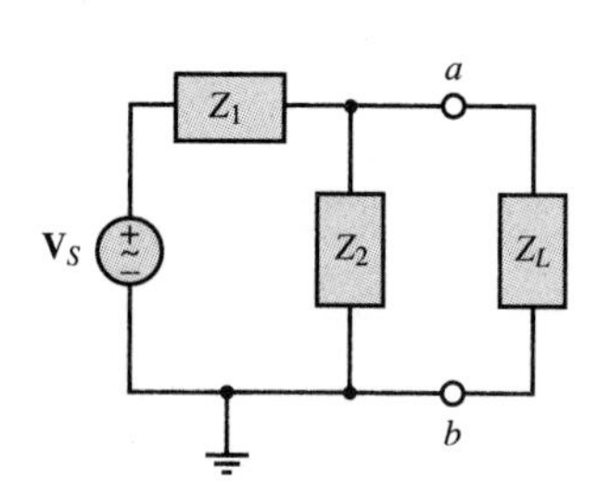

Figure 4.57

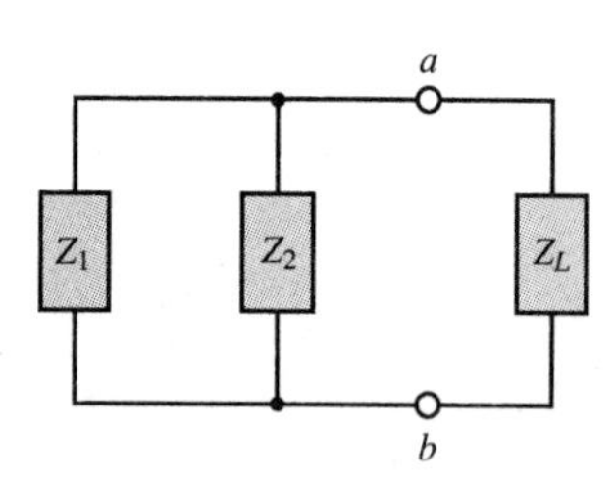

Figure 4.58

In this example, we illustrate the computation of the Thévenin equivalent of a simple AC circuit, shown in Figure 4.57.

Solution:

Note that, in solving problems of this nature, one may choose between two algebraic techniques, both leading to the same final answer. One uses the method of computing the ratio of two complex numbers in rectangular form; the other uses the polar form of the complex numbers. It should be remarked that, in general, the rectangular form is more convenient for adding or subtracting complex numbers, while the polar form is easier to use in computing products and ratios. However, these operations are frequently mixed together in the solution of AC circuit problems, and the preferred method depends on the user.

To compute the Thévenin impedance, Z_T, we replace each voltage source with a short circuit and each current source with an open circuit, obtaining the circuit of Figure 4.58. The Thévenin impedance is then computed much in the same way as Thévenin resistances were computed in Chapter 3. After removing the load impedance, Z_L, we see that the remaining impedances appear in parallel with respect to the load terminals:

$$Z_T = Z_1 \parallel Z_2$$

Since $Z_1 = 50\ \Omega$ and $Z_2 = j20\ \Omega$, we can write

$$Z_T = \frac{(5)(j20)}{5 + j20}$$

To obtain a physically meaningful expression for the impedance Z_T, we need to reduce this expression to one of the form $R_T + jX_T$; that is, we need to reduce the denominator of the expression to a real number. One method that will accomplish this consists of multiplying the numerator and the denominator by the complex conjugate of the denominator. The effect of this operation is to convert the denominator into a pure real quantity, reducing the ratio of two complex numbers to a single complex number divided by a real number:

$$Z_T = \frac{(5)(j20)}{(5 + j20)}\frac{(5 - j20)}{(5 - j20)} = \frac{2{,}000 + j500}{425} = 4.71 + j1.176$$

An alternative method of accomplishing the same objective consists of converting both numerator and denominator to polar form (i.e., magnitude and phase) and computing the ratio of the complex numbers in polar form. The latter is a simple task, since the ratio of two complex numbers in polar form is a complex number with magnitude equal to the ratio of the magnitudes and phase equal to the difference between the phases:

$$Z_T = \frac{(5)(j20)}{(5 + j20)} = \frac{100e^{j90°}}{20.615e^{j76°}} = 4.85e^{j14°} = 4.71 + j1.176$$

The two techniques are equivalent, although occasional savings in computation may be experienced by using one rather than the other.

Now, recall that the Thévenin voltage, V_T, was simply the open-circuit voltage between terminals a and b. If we were to replace the impedances Z_1 and Z_2 with resistors R_1 and R_2, then, using the methods of Chapter 3, we could compute the Thévenin voltage as

$$\mathbf{V}_T = \frac{R_2}{R_1 + R_2}\mathbf{V}_S$$

The Thévenin phasor voltage for a complex circuit is found in a similar fashion, by computing the (complex) voltage divider ratio

$$\mathbf{V}_T = \frac{Z_2}{Z_1 + Z_2}\mathbf{V}_S$$

or

$$\mathbf{V}_T = \frac{j20}{5 + j20}110e^{j0^\circ}$$

Note that the voltage source has a zero phase angle, so that the source is expressed by $\mathbf{V}_S = 110e^{j0^\circ}$.

The computation of $\mathbf{V}_T$ can also be carried out in two different ways:

$$\mathbf{V}_T = \frac{(j2{,}200)}{(5 + j20)}\frac{(5 - j20)}{(5 - j20)} = \frac{44{,}000 + j11{,}000}{425}$$

$$= 103.5 + j25.88 = 106.7e^{j14^\circ}$$

or

$$\mathbf{V}_T = \frac{j2{,}200}{5 + j20} = \frac{2{,}200e^{j90^\circ}}{20.616e^{j76^\circ}} = 106.7e^{j14^\circ}$$

The complete Thévenin equivalent circuit can finally be drawn as shown in Figure 4.59.

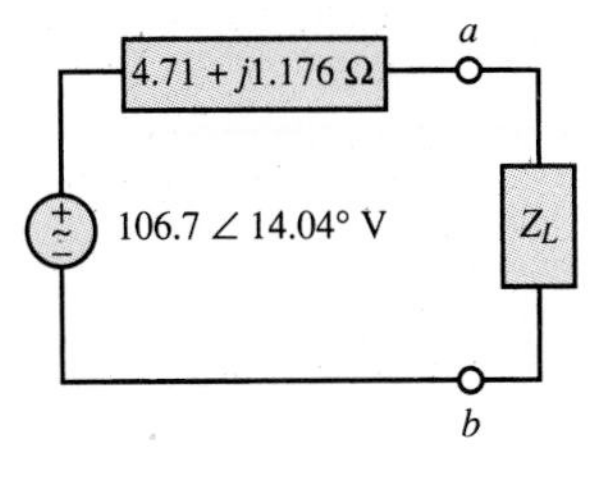

Figure 4.59

Example 4.23

This example illustrates the computation of a Norton equivalent. The objective is to find the Norton equivalent circuit seen by the load impedance, Z_L, for the circuit shown in Figure 4.60, given the following source and impedance values:

$$i_S(t) = 50\cos(\omega t + 45^\circ) \qquad Z_1 = 400\ \Omega \qquad Z_2 = j10\ \Omega \qquad Z_3 = -j20\ \Omega$$

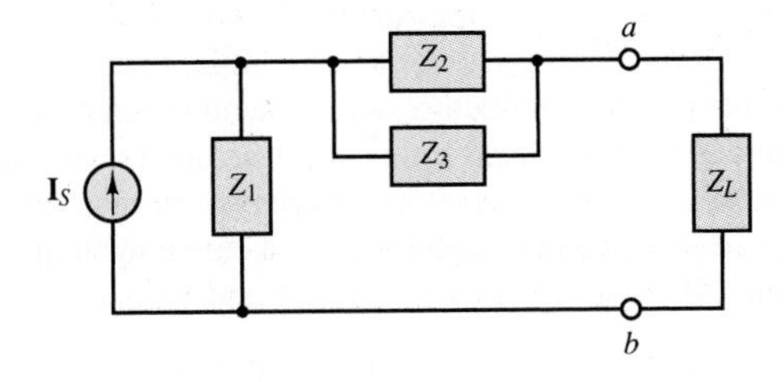

Figure 4.60

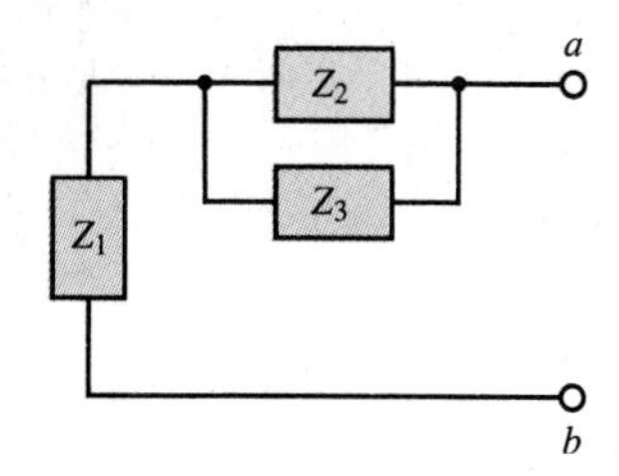

Figure 4.61

Solution:

Begin by determining the Thévenin (or Norton) impedance, Z_T. Replacing the current source with an open circuit yields the circuit of Figure 4.61, from which it should be apparent that the Thévenin impedance is given by

$$Z_T = Z_1 + Z_2 \parallel Z_3$$

or

$$Z_T = 400 + \frac{(j10)(-j20)}{(j10 - j20)} = 400 + \frac{200}{-j10} = 400 + j20$$

The Norton current source is simply the short-circuit current between the load terminals (replace Z_L with a short circuit). In the circuit of Figure 4.61, this current is determined by means of the current divider between impedance Z_1 and the parallel combination of impedances Z_3 and Z_2, with $\mathbf{I}_{SC}$ the current flowing through $Z_2 \parallel Z_3$:

$$\mathbf{I}_{SC} = \mathbf{I}_N = \frac{Z_1}{Z_1 + Z_2 \parallel Z_3}\mathbf{I}_S$$

$$= \frac{400}{400 + j20} 50e^{j45^\circ} = 49.94e^{j42.14^\circ}$$

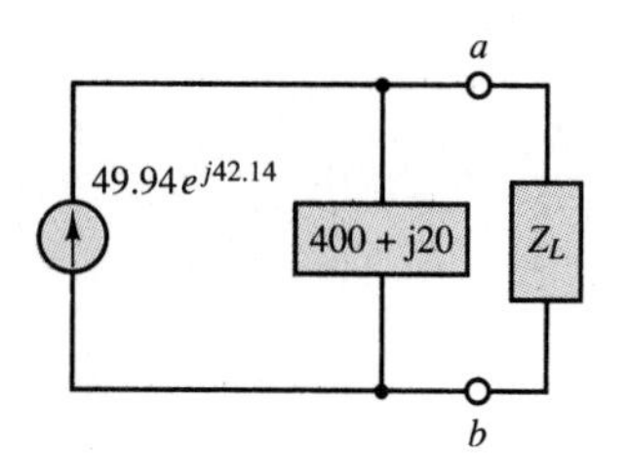

Figure 4.62

The final appearance of the Norton equivalent representation is shown in Figure 4.62.

DRILL EXERCISES

4.20 Compute the magnitude of the current $\mathbf{I}_S(j\omega)$ of Example 4.21 if $A = 1$ and $\phi = 0$, for $\omega = 10, 10^2, 10^3, 10^4$, and 10^5 rad/s. Can you explain these results intuitively? [*Hint:* Evaluate the impedance of the capacitor relative to that of the two resistors at each frequency.]

4.21 Find the voltage across the capacitor in Example 4.20.

4.22 Determine the Norton current in Example 4.22.

4.23 Determine the Thévenin voltage in Example 4.23.

CONCLUSION

In this chapter we have introduced concepts and tools useful in the analysis of AC circuits. The importance of AC circuit analysis cannot be overemphasized, for a number of reasons. First, circuits made up of resistors, inductors, and capacitors constitute reasonable models for more complex devices, such as transformers, electric motors, and electronic amplifiers. Second, sinusoidal signals are ever present in the analysis of many physical systems, not just circuits. The skills developed in Chapter 4 will be called upon in the remainder of the book. In particular, they form the basis of Chapters 5 and 6.

- In addition to elements that dissipate electric power, there are also electric energy-storage elements. The ideal inductor and capacitor are ideal elements that represent the energy-storage properties of electric circuits.

- Since the i-v relationship for the ideal capacitor and the ideal inductor consists of a differential equation, application of the fundamental circuit laws in the presence of such dynamic circuit elements leads to the formulation of differential equations.
- For the very special case of sinusoidal sources, the differential equations describing circuits containing dynamic elements can be converted into algebraic equations and solved using techniques similar to those employed in Chapter 3 for resistive circuits.
- Sinusoidal voltages and currents can be represented by means of complex phasors, which explicitly indicate the amplitude and phase of the sinusoidal signal and implicitly denote the sinusoidal frequency dependence.
- Circuit elements can be represented in terms of their impedance, which may be conceptualized as a frequency-dependent resistance. The rules of circuit analysis developed in Chapters 2 and 3 can then be employed to analyze AC circuits by using impedance elements as complex resistors. Thus, the only difference between the analysis of AC and resistive circuits lies in the use of complex algebra instead of real algebra.

EIT REVIEW: CAPACITANCE AND INDUCTANCE

Problems that pertain to the topics introduced in this chapter are typically oriented toward the calculation of voltages, current, and energy storage in capacitors and inductors. The following two problems are based on the waveform of Figure 4.63. A review of AC circuit problems can be found at the end of Chapter 5.

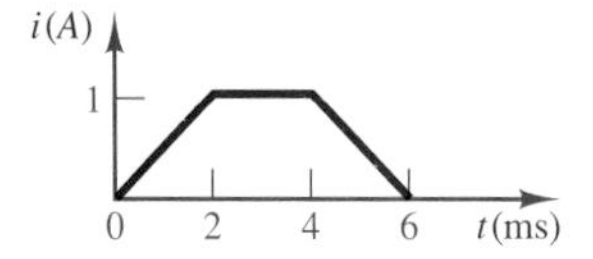

Figure 4.63

DRILL EXERCISES

4.24 A coil with inductance of 1 H and negligible resistance carries the current shown in Figure 4.63. The maximum energy stored in the inductor is:

a. 2 J b. 0.5 J c. 0.25 J d. 1 J e. 0.2 J

Solution:

The energy stored in an inductor is $W = \frac{1}{2}Li^2$ (see Section 4.1). Since the maximum current is 1 A, the maximum energy will be $W_{\text{max}} = \frac{1}{2}Li_{\text{max}}^2 = \frac{1}{2}$ J. Thus, b is the correct answer.

4.25 The maximum voltage that will appear across the coil is:

a. 5 V b. 100 V c. 250 V d. 500 V e. 5,000 V

Solution:

Since the voltage across an inductor is given by $v = L\frac{di}{dt}$, we need to find the maximum (positive) value of $\frac{di}{dt}$. This will occur anywhere between $t = 0$ and $t = 2$ ms. The corresponding slope is: $\frac{di}{dt}\big|_{\text{max}} = \frac{1}{2\times10^{-3}} = 500$. Therefore $v_{\text{max}} = 1 \times 500 = 500$ V, and the correct answer is d.

In addition to these sample problems, you may also find it useful to review the following examples in the text: Chapter 4, Examples 4.1, 4.2, 4.3, 4.5, 4.6, 4.7; Chapter 6, Examples 6.10, 6.11, 6.12, 6.13.

The problems marked "EIT" in the Homework Problems section of Chapter 4 will provide additional practice material. Answers to these problems will be found at the end of the book (Appendix B).

KEY TERMS

AC Circuits *145*
AC Conductance *160*
AC Resistance *157*
Admittance *160*
Amplitude *132*
Argument *149*
Average (or DC) Value *133*
Capacitance *120*
Characteristic Equation *143*
Complex Exponential *148*
Complex Phasor *149*
Damper *131*
Dielectric *120*
Farad (F) *121*
First-Order Ordinary Differential Equation *138*
Forced Response (Particular Solution) *145*
Frequency-Domain (Phasor) Form *149*
Henry (H) *127*
Hertz (Hz) *133*
Homogeneous Differential Equation *140*
Ideal Capacitor *120*
Ideal Inductor *120*
Impedance *153*
Inductance *120*
Inductive Core *127*
Inhomogeneous Equation *145*
Initial Condition *141*
Magnitude (of a Complex Vector) *149*
Natural (or Free) Response (Complementary Solution) *140*
Period *132*
Periodic Signal *132*
Permittivity *125*
Phase *132*
(Phase) Angle *149*
Radian Frequency *132*
Reactance *157*
Root-Mean-Square (or RMS) Value *133*
Second-Order Ordinary Differential Equation *138*
Signal (or Waveform) Generator *132*
Steady State *145*
Susceptance *160*
Time-Domain Form *149*

ANSWERS TO DRILL EXERCISES

4.1 Plot for Drill Exercise 4.1

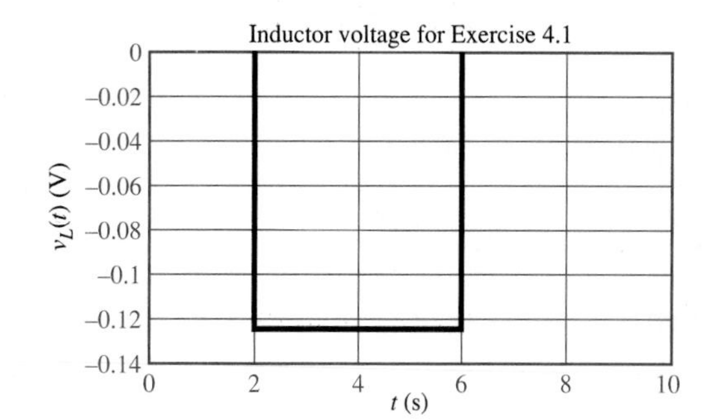

4.2 Plot for Drill Exercise 4.2

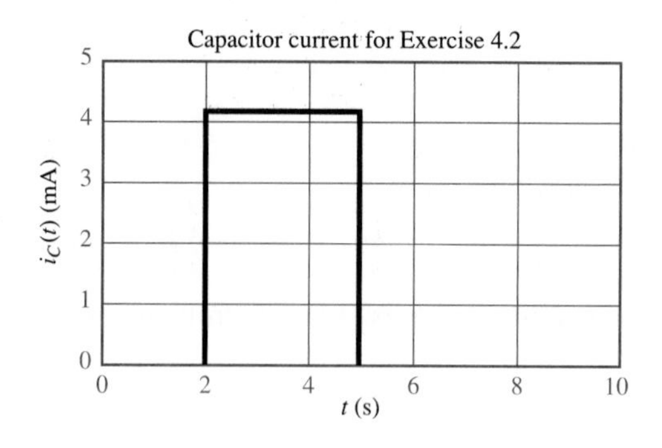

4.3 $w(t = 3 \text{ ms}) = 3.9\ \mu\text{J}$

4.4 $w(t = 3 \text{ ms}) = 22.22 \text{ mJ}$

4.5

$$w(t) = \begin{cases} 5.625 \times 10^{-6} \text{ J} & 0 \le t < 2 \text{ ms} \\ 0.156 \times 10^{-6} t^2 - 2.5 \times 10^{-6} t & \\ +10^{-5} & 2 \le t < 6 \text{ ms} \\ 0.625 \times 10^{-6} & t \ge 6 \text{ ms} \end{cases}$$

$$p(t) = \begin{cases} (20 \times 10^{-3} - 2.5t) \times (-0.125) \text{ W} & 2 \le t < 6 \text{ ms} \\ 0 & \text{otherwise} \end{cases}$$

4.6 $v(t) = 155.6 \cos(377t - 30°)$

4.7 $\langle v(t) \rangle = 2.5 \text{ V}$

4.8 $\langle v(t) \rangle = 1.5 \text{ V}$

4.9 2.89 V

4.10 0.5 V

4.13 $C = \dfrac{V}{\sqrt{1 + (\omega RC)^2}}$

$\phi = \tan^{-1}(-\omega RC)$

4.14 $R_1 LC \dfrac{d^2 v_C(t)}{dt^2} + (R_1 R_2 C + L)\dfrac{dv_C(t)}{dt} + (R_1 + R_2) v_C(t) = R_2 v_S(t) + L\dfrac{dv_S(t)}{dt}$

4.15 (a) $v_1 + v_2 = 4.67 \cos(\omega t + 20.2°)$; (b) $v_1 + v_2 = 60.8 \cos(\omega t - 37.6°)$

4.16 (a) $i_1 + i_2 = 0.19 \cos(\omega t + 42.0°)$; (b) $i_1 + i_2 = 1.32 \cos(\omega t - 32.3°)$

4.17 $Z(1{,}000) = 140 - j10$; $Z(100{,}000) = 100 + j999$

4.18 $Y_{EQ} = 5.492 \times 10^{-3} - j4.871 \times 10^{-3}$

4.19 $X_{\|} = 0.25$; $C = 0.4$ F

4.20 $|\mathbf{I}_S| = 0.0041$ A; 0.0083 A; 0.0194 A; 0.02 A; 0.02 A

4.21 $6.9e^{-j2.76°}$ V

4.22 $22e^{j0°}$ A

4.23 $20{,}000e^{j45°}$ V

HOMEWORK PROBLEMS

EIT 4.1 Find the rms value of the waveform shown in Figure P4.1.

4.2 Find the rms voltage of the waveform of Figure P4.2.

4.3 Find the rms value of the waveform shown in Figure P4.3.

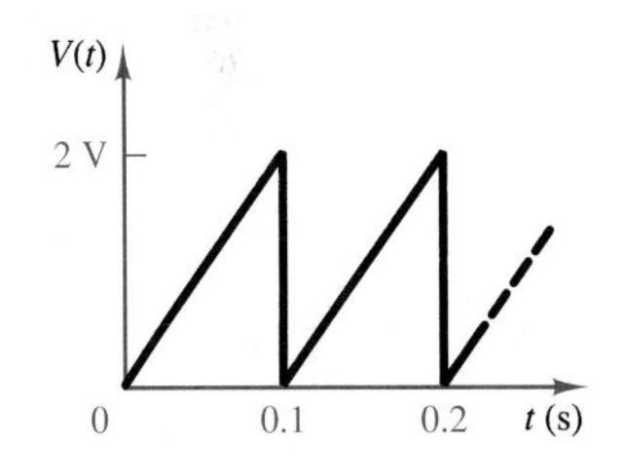

Figure P4.1

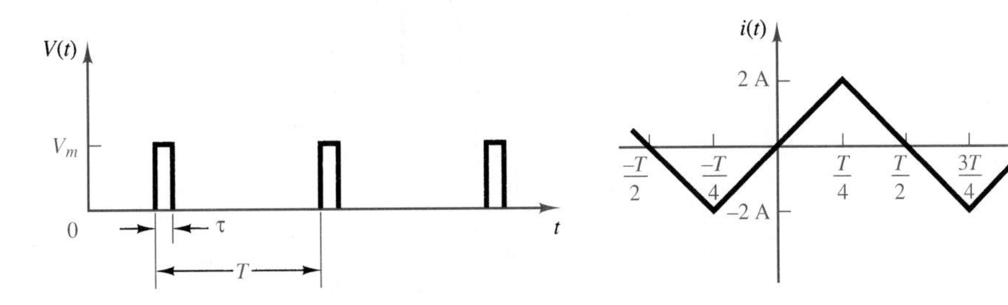

Figure P4.2

Figure P4.3

4.4 Write the differential equation for the circuit of Figure P4.4, and find the voltage across the capacitor for all time.

4.5 Write the differential equation for the circuit of Figure P4.5, and find the current through the inductor for all time.

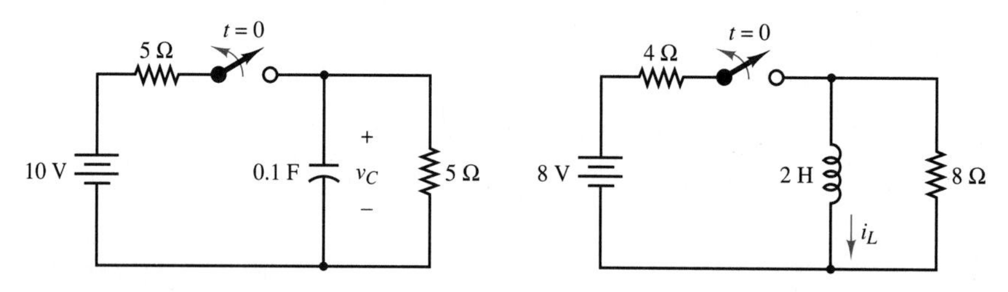

Figure P4.4

Figure P4.5

EIT 4.6 Find the rms value of the waveform of Figure P4.6.

4.7 Find the rms value of the waveform of Figure P4.7.

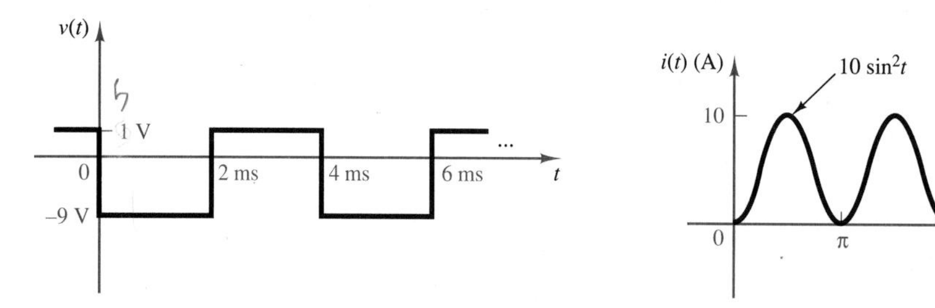

Figure P4.6

Figure P4.7

4.8 Find the rms value of $x(t)$ if $x(t)$ is a sinusoid that is offset by a DC value:

$$x(t) = 2\sin(\omega t) + 2.5$$

4.9 For the waveform of Figure P4.8:

a. Find the rms current.
b. If θ_1 is $\pi/2$, what is the rms current of this waveform?

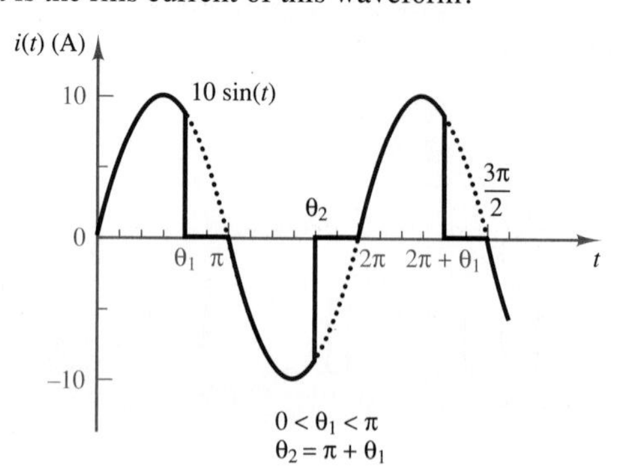

Figure P4.8

4.10 Write the differential equation for the circuit of Figure P4.9, in terms of the variables v and i.

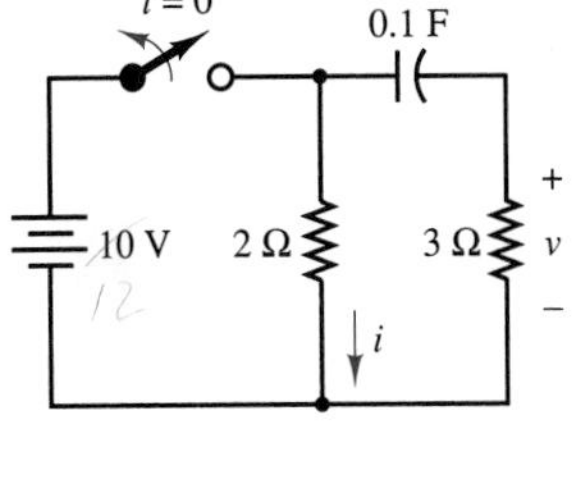

Figure P4.9

EIT 4.11 Find the average power dissipated by the resistor of the circuit of Figure P4.10 if

a. $v(t) = 10$ V.
b. $v(t) = 10\sqrt{2}\cos 200\pi t$ V.

Note that the DC voltage of part a is the same as the rms voltage of the source in part b.

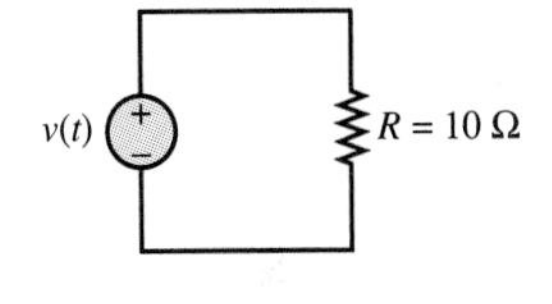

Figure P4.10

4.12 A capacitor carries a charge of 20 mC, and the voltage across it is 60 V. Find the capacitance and stored energy.

4.13 Use the defining law for a capacitor to find the current $i_C(t)$ corresponding to the voltage shown in Figure P4.11. Sketch your results.

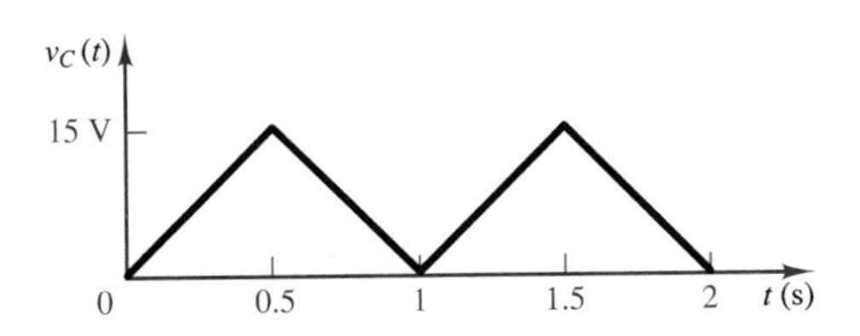

+ $v_C(t)$ − $i_C(t)$ C 0.01 F

Figure P4.11

4.14 Write the differential equation for the circuit of Figure P4.12, and find the voltage across the 10-Ω resistor and the current through the 5-Ω resistor for all time.

4.15 The current through a 0.5-H inductor is given by $i_L = 2\cos(377t + \pi/6)$. Write the expression for the voltage across the inductor.

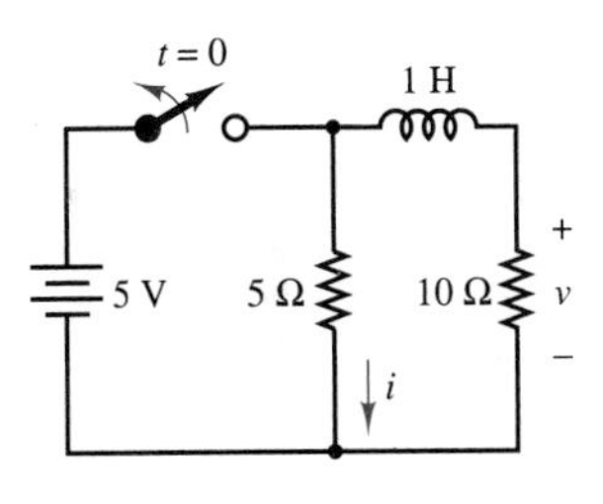

Figure P4.12

4.16 A spring-loaded scale mechanism is being used to measure the weight of an object placed on the platform shown in Figure P4.13. The position of the plunger is directly proportional to the force exerted on the platform by the weight of the object. The variable capacitor has a flat plate at its bottom that is electrically isolated from its case. The face of the variable capacitor's plunger is also flat and electrically isolated from the variable capacitor.

Find the capacitance of the parallel plates as a function of the distance x (in cm) shown in Figure P4.13 if the dielectric material is air.

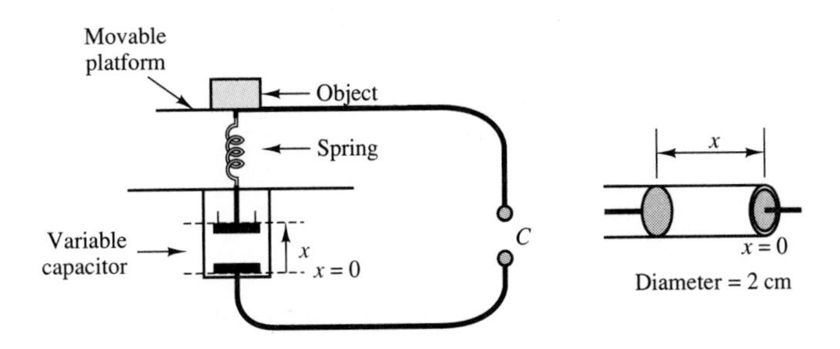

Figure P4.13

4.17 The voltage across a 100-μF capacitor takes the following values. Calculate the expression for the current through the capacitor in each case.

a. $v_C(t) = 40\cos(20t - \pi/2)$ V
b. $v_C(t) = 20\sin 100t$ V
c. $v_C(t) = -60\sin(80t + \pi/6)$ V
d. $v_C(t) = 30\cos(100t + \pi/4)$ V

4.18 The current through a 250-mH inductor takes the following values. Calculate the expression for the voltage across the inductor in each case.

a. $i_L(t) = 5\sin 25t$ A
b. $i_L(t) = -10\cos 50t$ A
c. $i_L(t) = 25\cos(100t + \pi/3)$ A
d. $i_L(t) = 20\sin(10t - \pi/12)$ A

EIT 4.19 Use the defining law for an inductor to find the current $i_L(t)$ as shown in Figure P4.14. Assume $i_L(0) = 0$ A. Sketch your results.

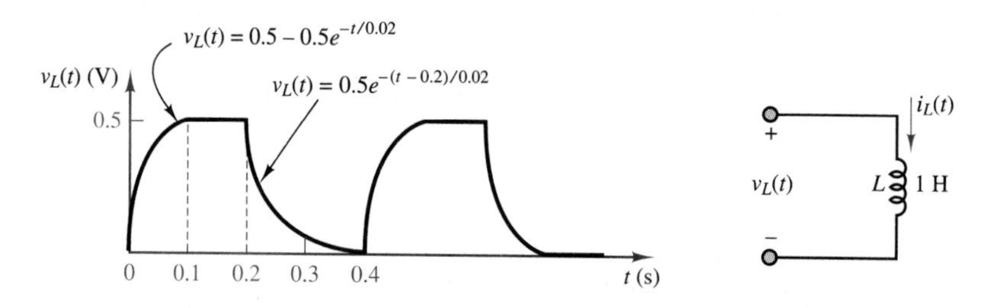

Figure P4.14

4.20 The voltage $v(t)$ shown in Figure P4.15 is applied to a 10-mH inductor. Find the current through the inductor. Assume $i_L(0) = 0$ A.

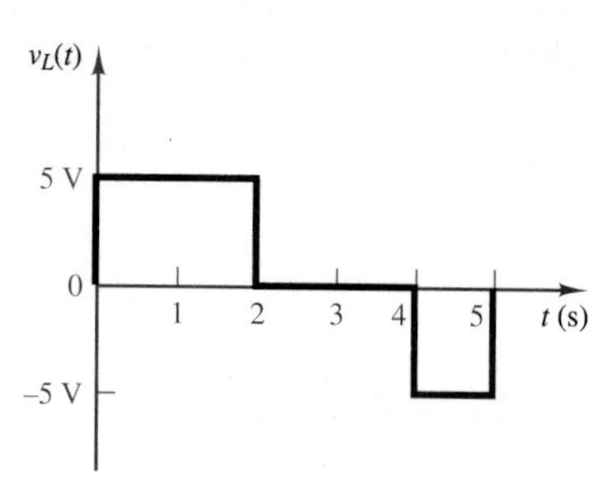

Figure P4.15

4.21 The current waveform shown in Figure P4.16 flows through a 2-H inductor. Plot the inductor voltage $v_L(t)$.

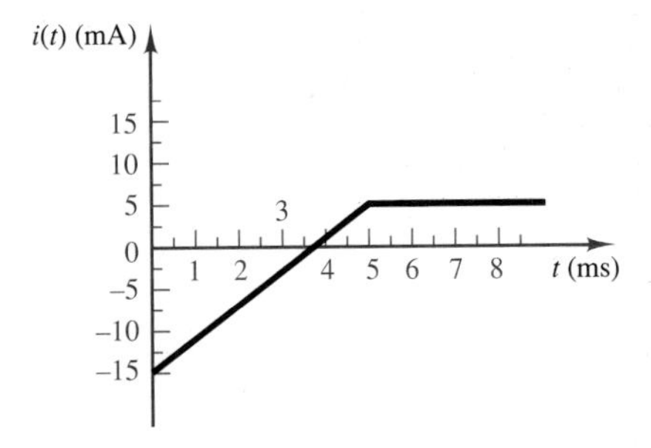

Figure P4.16

4.22 An exponential current $i(t) = 4e^{-t/10}$ A is applied to a 10-mH inductor at $t = 0$. Find the resulting voltage across the inductor, and draw the waveform.

EIT 4.23 The voltage waveform shown in Figure P4.17 appears across a 100-mH inductor and a 500-μF capacitor. Plot the capacitor and inductor currents, $i_C(t)$ and $i_L(t)$, assuming $i_L(0) = 0$ A.

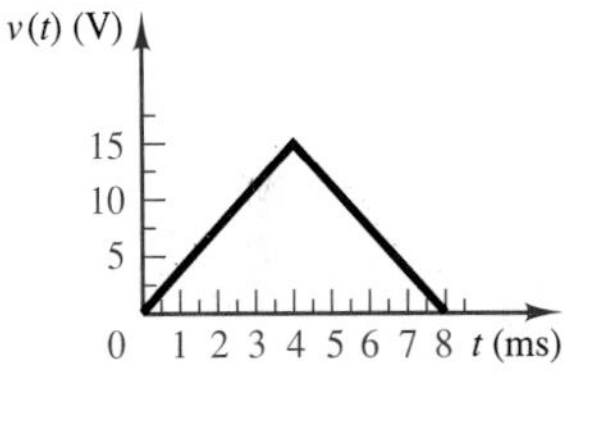

Figure P4.17

4.24 In the circuit shown in Figure P4.18, let

$$
\begin{aligned}
i(t) &= 0 && \text{for } -\infty < t < 0 \\
&= t && \text{for } 0 \le t < 1 \text{ s} \\
&= -(t-2) && \text{for } 1 \text{ s} \le t < 2 \text{ s} \\
&= 0 && \text{for } 2 \text{ s} \le t < \infty
\end{aligned}
$$

Find the energy stored in the inductor for all time.

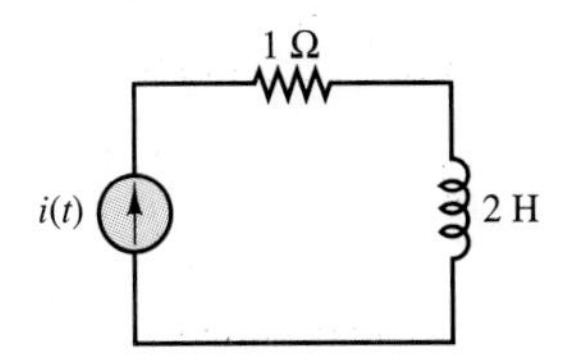

Figure P4.18

4.25 In the circuit shown in Figure P4.19, let

$$
\begin{aligned}
v(t) &= 0 && \text{for } -\infty < t < 0 \\
&= 2t && \text{for } 0 \le t < 1 \text{ s} \\
&= -(2t-4) && \text{for } 1 \text{ s} \le t < 2 \text{ s} \\
&= 0 && \text{for } 2 \text{ s} \le t < \infty
\end{aligned}
$$

Find the energy stored in the capacitor for all time.

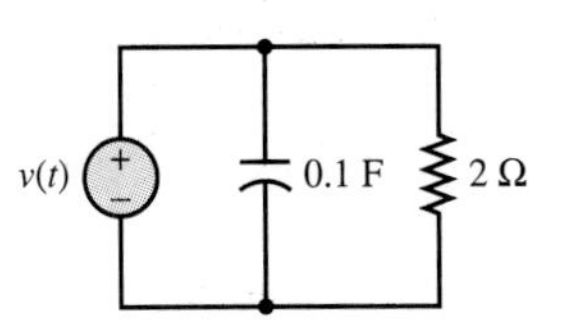

Figure P4.19

4.26 Find the energy stored in each capacitor and inductor, under steady-state conditions, in the circuit shown in Figure P4.20.

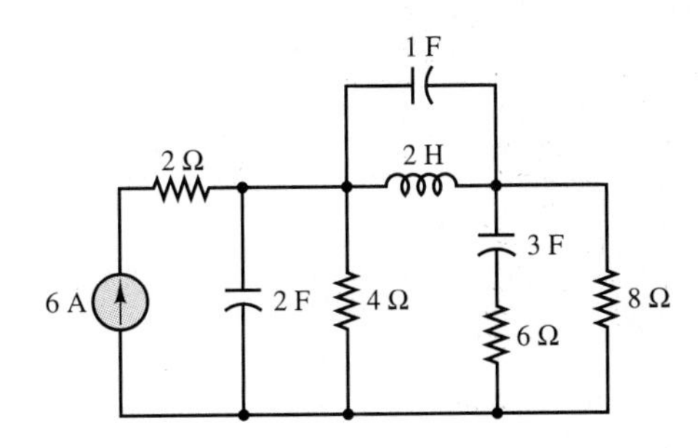

Figure P4.20

4.27 Find the energy stored in each capacitor and inductor, under steady-state conditions, in the circuit shown in Figure P4.21.

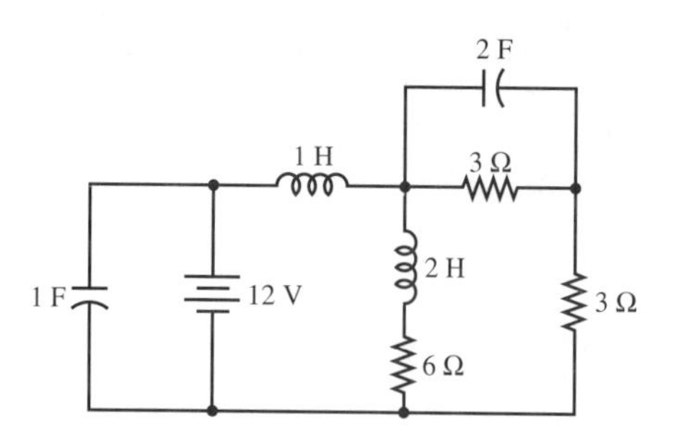

Figure P4.21

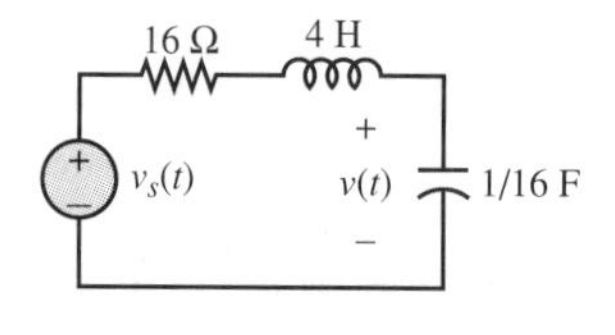

Figure P4.22

4.28 For the circuit shown in Figure P4.22, find $v(t)$ when the source is a unit step function occurring at time $t = 0$.

EIT 4.29 Write the differential equation for the circuit shown in Figure P4.23, using the capacitor voltage as the independent variable.

4.30 Find the output voltage, $v_{out}(t)$, for the circuit shown in Figure P4.24, by assuming that the solution has the form $v_{out}(t) = D\cos(\omega t + \phi)$.

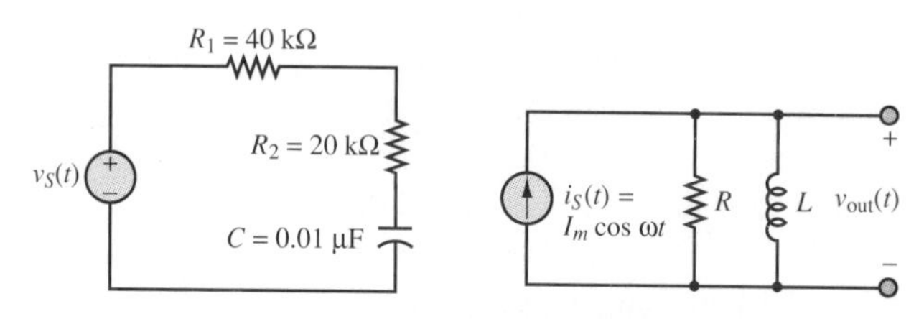

Figure P4.23 **Figure P4.24**

4.31 Refer to the circuit of Figure P4.19. If $R = 40\ \text{k}\Omega, C = 0.05\ \mu\text{F}, V_1 = 20$ V, and $\omega = 377$ rad/s, find $v_{out}(t)$ using phasor techniques.

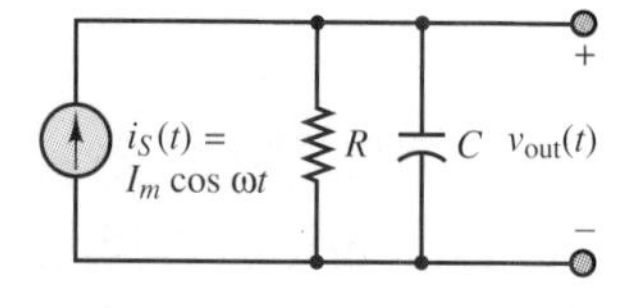

Figure P4.25

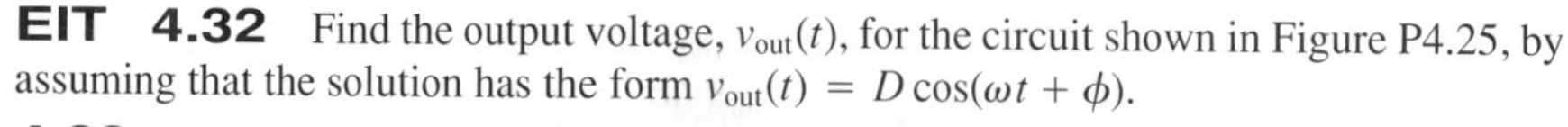

EIT 4.32 Find the output voltage, $v_{out}(t)$, for the circuit shown in Figure P4.25, by assuming that the solution has the form $v_{out}(t) = D\cos(\omega t + \phi)$.

4.33

a. Find the current flowing in the series *RLC* circuit shown in Figure P4.26, by writing the differential equation and assuming that the solution has the form $i(t) = A\sin\omega t + B\cos\omega t$.
b. What is the inductor voltage, $v_L(t)$?
c. What is the resistor voltage, $v_R(t)$?

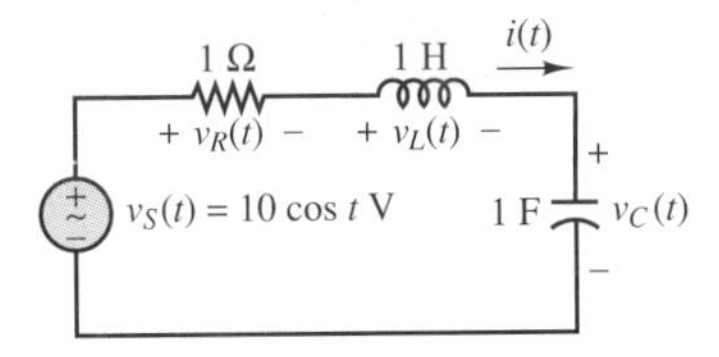

Figure P4.26

EIT 4.34

a. For the parallel *RLC* circuit shown in Figure P4.27, find the output voltage, $v_{out}(t)$, by writing the differential equation and assuming that $v_{out}(t)$ has the form $v_{out}(t) = A \sin \omega t + B \cos \omega t$.

b. What is the current $i_R(t)$?

c. What is the current $i_L(t)$?

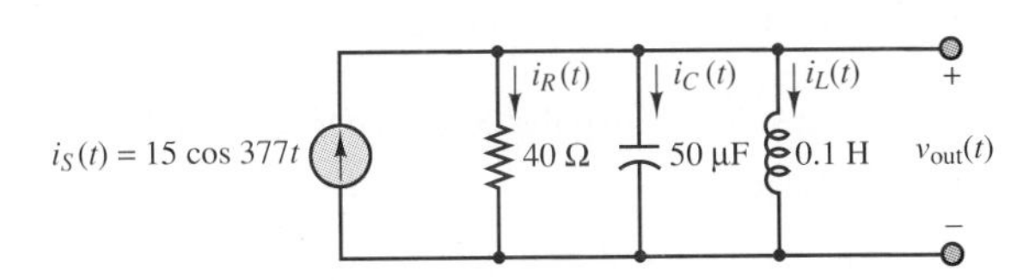

Figure P4.27

4.35 For the circuit shown in Figure P4.28, find $v(t)$ when the source is a unit step function occurring at time $t = 0$.

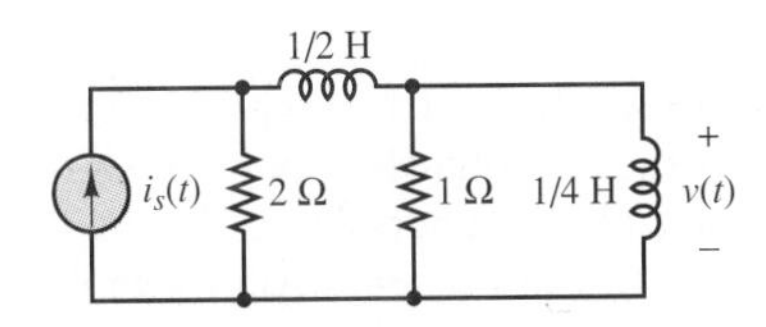

Figure P4.28

4.36 For the circuit shown in Figure P4.29, find $i(t)$.

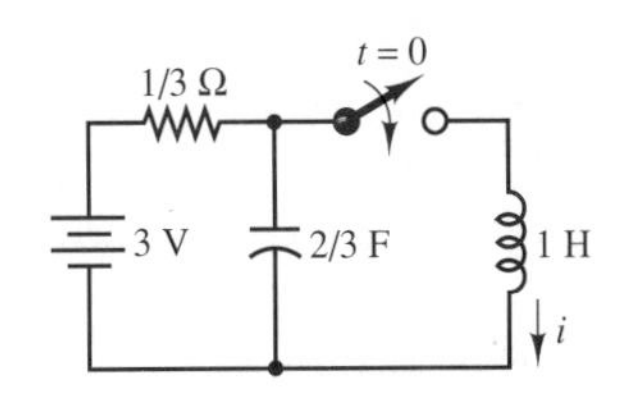

Figure P4.29

EIT 4.37 Express the following current in phasor notation:

$$i(t) = 0.5 \cos(2\pi 10t + 45°) \text{ A}$$

Sketch the waveform along the waveform $\cos(2\pi 10t)$ on the same axis as precisely as you can.

4.38 Find the phasor form of the following functions:

a. $v(t) = 155\cos(377t - 25°)$ V
b. $v(t) = 5\sin(1{,}000t - 40°)$ V
c. $i(t) = 10\cos(10t + 63°) + 15\cos(10t - 42°)$ A
d. $i(t) = 460\cos(500\pi t - 25°) - 220\sin(500\pi t + 15°)$ A

4.39 Convert the following complex numbers to polar form:

a. $4 + j4$
b. $-3 + j4$
c. $j + 2 - j4 - 3$

4.40 Convert the following to polar form by first multiplying rectangular coordinates and then converting to polar form:

a. $(50 + j10)(4 + j8)$
b. $(j2 - 2)(4 + j5)(2 + j7)$

EIT 4.41 Complete the following exercises in complex arithmetic.

a. Find the complex conjugate of $(4 + j4), (2 - j8), (-5 + j2)$.
b. Convert the following to polar form by multiplying the numerator and denominator by the complex conjugate of the denominator and then performing the conversion to polar coordinates: $\frac{1+j7}{4+j4}, \frac{j4}{2-j8}, \frac{1}{-5+j2}$.
c. Repeat part b but this time convert to polar coordinates before performing the division.

4.42 Given the two voltages

$$v_1(t) = 10\cos(\omega t + 30°) \quad \text{and} \quad v_2(t) = 20\cos(\omega t + 60°)$$

find $v(t) = v_1(t) + v_2(t)$ using

a. Trigonometric identities.
b. Phasors.

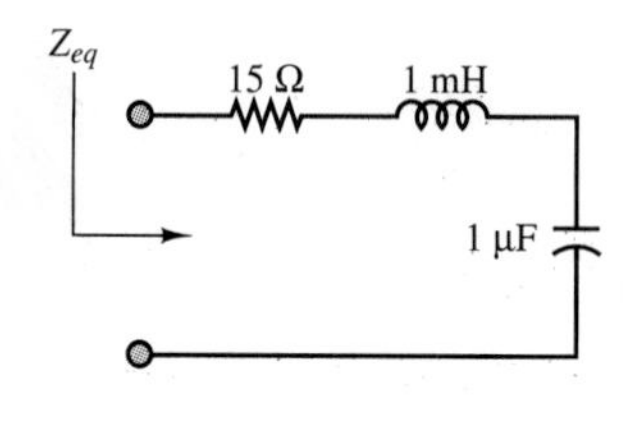

Figure P4.30

4.43 For the circuit shown in Figure P4.30, find the frequency that causes the equivalent impedance to appear to be purely resistive.

EIT 4.44

a. Find the equivalent impedance Z_L shown in Figure P4.31, as seen by the source, if the frequency is 377 rad/s.
b. If we wanted the source to see the load as completely resistive, what value of capacitance should we place between terminals a and b as shown in Figure P4.32? [*Hint:* Find an expression for the equivalent impedance Z_L, and then find C so that $\phi = 0$.]
c. What is the actual impedance that the source sees with the capacitor included in the circuit?

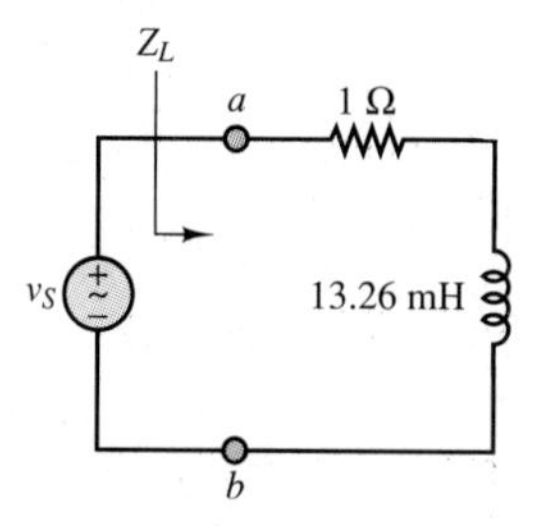

Figure P4.31

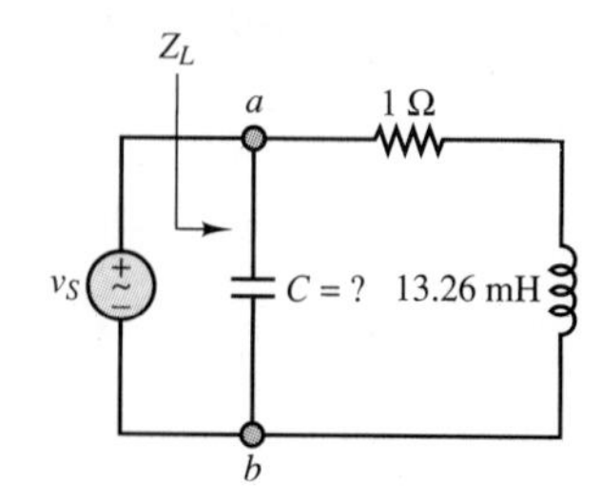

Figure P4.32

a. Assuming that the circuit is balanced, that is, that $v_{ab} = 0$, determine X_4 in terms of the circuit elements.
b. If $C_3 = 4.7\ \mu F, L_3 = 0.098$ H, $R_1 = 100\ \Omega, R_2 = 1\ \Omega, v_S(t) = 24 \sin(2{,}000t)$, and $v_{ab} = 0$, what is the reactance of the unknown circuit element? Is it a capacitor or an inductor? What is its value?
c. What frequency should be avoided by the source in this circuit, and why?

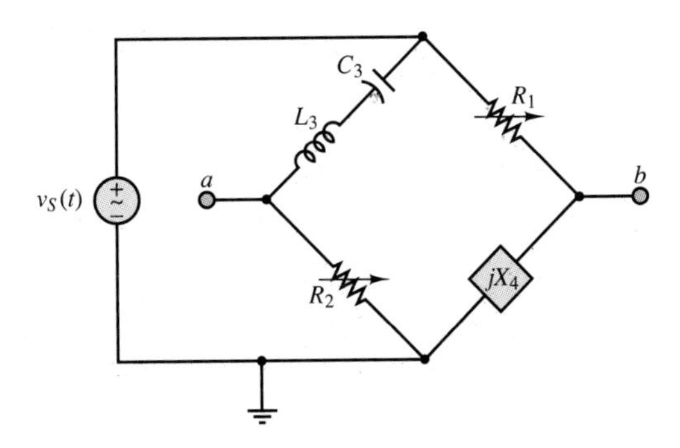

Figure P4.49

EIT 4.63 Compute the Thévenin impedance seen by resistor R_2 in Problem 4.59.

4.64 Compute the Thévenin voltage seen by the inductance, L, in Problem 4.61.

4.65 For the circuit shown in Figure P4.50, find the impedance Z, given $\omega = 4$ rad/s.

4.66 Find the Thévenin equivalent circuit as seen from terminals a-b for the circuit shown in Figure P4.51.

4.67 Find the admittance, Y, for the circuit shown in Figure P4.52, when $\omega = 5$ rad/s.

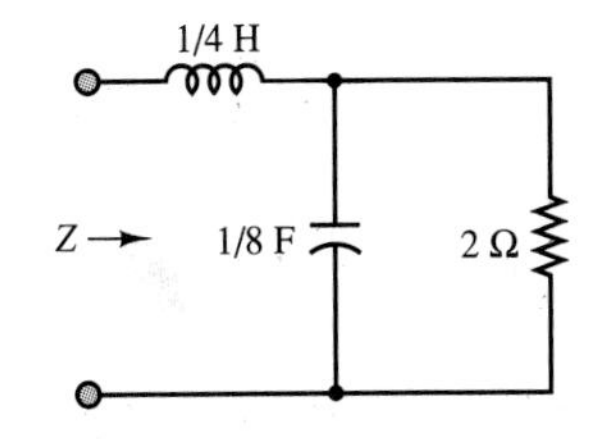

Figure P4.50

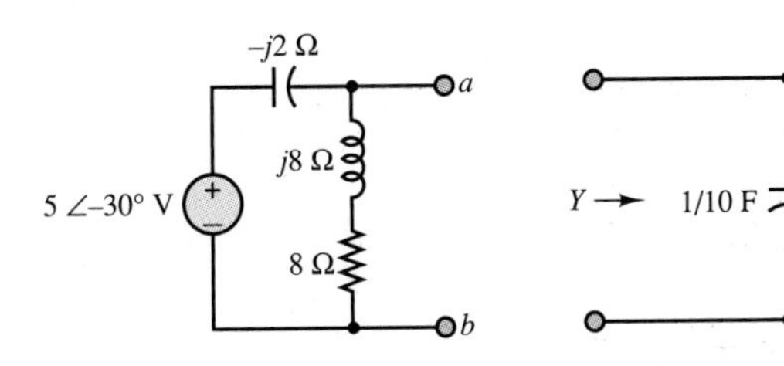

Figure P4.51

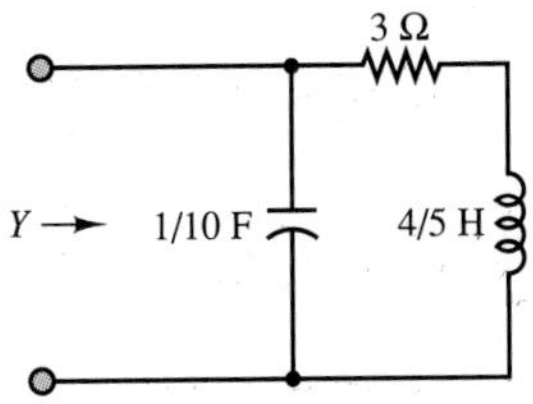

Figure P4.52

EIT 4.68 Compute the Thévenin voltage seen by resistor R_2 in Problem 4.59.

4.69 Find the Norton equivalent circuit seen by resistor R_2 in Problem 4.59.

and currents in the circuit (see Section 4.2). Use of the rms value eliminates the factor $\frac{1}{2}$ in power expressions and leads to considerable simplification. Thus, the following expressions will be used in this chapter:

$$V_{rms} = \frac{V}{\sqrt{2}} = \tilde{V} \tag{5.11}$$

$$I_{rms} = \frac{I}{\sqrt{2}} = \tilde{I} \tag{5.12}$$

$$\begin{aligned} P_{av} &= \frac{1}{2}\frac{V^2}{|Z|}\cos\theta = \frac{\tilde{V}^2}{|Z|}\cos\theta \\ &= \frac{1}{2}I^2|Z|\cos\theta = \tilde{I}^2|Z|\cos\theta = \tilde{V}\tilde{I}\cos\theta \end{aligned} \tag{5.13}$$

Figure 5.4 illustrates the so-called **impedance triangle,** which provides a convenient graphical interpretation of impedance as a vector in the complex plane. From the figure, it is simple to verify that

$$R = |Z|\cos\theta \tag{5.14}$$

$$X = |Z|\sin\theta \tag{5.15}$$

Finally, the amplitudes of phasor voltages and currents will be denoted throughout this chapter by means of the rms amplitude:

$$\mathbf{V} = V_{rms}e^{j\theta_V} = \tilde{V}e^{j\theta_V} = \tilde{V}\angle\theta_V \tag{5.16}$$

and

$$\mathbf{I} = I_{rms}e^{j\theta_I} = \tilde{I}e^{j\theta_I} = \tilde{I}\angle\theta_I \tag{5.17}$$

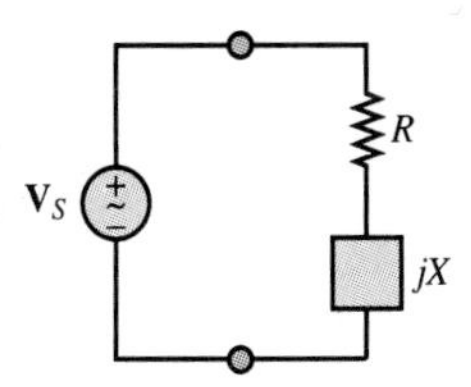

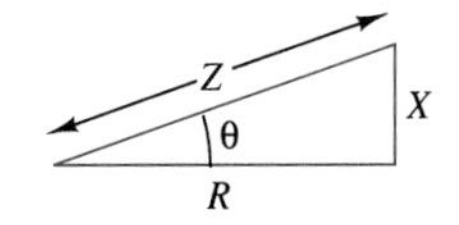

Figure 5.4 Impedance triangle

In other words, throughout the remainder of this chapter the symbols $\tilde{V}$ and $\tilde{I}$ will denote the rms value of a voltage or a current. Also note the use of the symbol $\angle$ to represent the complex exponential. Thus, the sinusoidal waveform corresponding to the phasor current $\mathbf{I} = \tilde{I}\angle\theta_I$ corresponds to the time-domain waveform

$$i(t) = \sqrt{2}\tilde{I}\cos(\omega t + \theta_I) \tag{5.18}$$

and the sinusoidal form of the phasor voltage $\mathbf{V} = \tilde{V}\angle\theta_V$ is

$$v(t) = \sqrt{2}\tilde{V}\cos(\omega t + \theta_V) \tag{5.19}$$

Example 5.1

This example illustrates the computation of average power for a typical load consisting of an inductor in series with a resistor. This type of load is actually quite common in practice, being representative of the circuit behavior of many electric motors. The circuit is shown in Figure 5.5.

The circuit can be analyzed by phasor notation to compute the current $i(t)$ that flows through the series RL load. The radian frequency of the source voltage is $\omega = 377$ rad/s; thus, the load impedance is:

$$\begin{aligned} Z &= 4 + j3 = \sqrt{16+9}e^{j\tan^{-1}(3/4)} \\ &= 5e^{j36.9^\circ} \end{aligned}$$

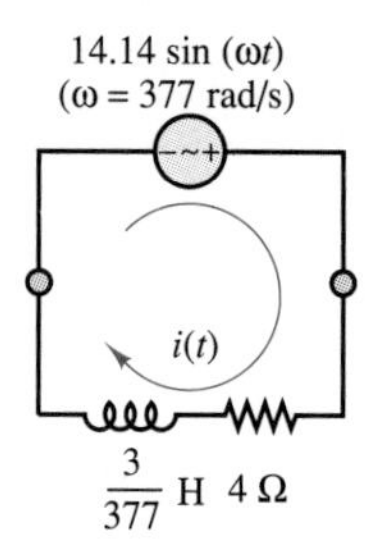

Figure 5.5

The phasor source voltage may be expressed in rms form as

$$\mathbf{V} = 10e^{-j90^\circ}$$

since

$$14.14 \sin \omega t = 10\sqrt{2}\cos(\omega t - 90^\circ)$$

Thus the phasor current must be

$$\mathbf{I} = \frac{\mathbf{V}}{Z} = \frac{10e^{-j90^\circ}}{5e^{j36.9^\circ}} = 2e^{-j126.9^\circ}$$

which corresponds to the time-domain current

$$i(t) = 2\sqrt{2}\cos(\omega t - 126.9^\circ)$$

The two functions

$$v(t) = 10\sqrt{2}\cos(\omega t - 90^\circ)$$

and

$$i(t) = 2\sqrt{2}\cos(\omega t - 126.9^\circ)$$

and the instantaneous power

$$p(t) = 2\sqrt{2}\cos(\omega t - 126.9^\circ)\, 10\sqrt{2}\cos(\omega t - 90^\circ)$$

are sketched in Figure 5.6.

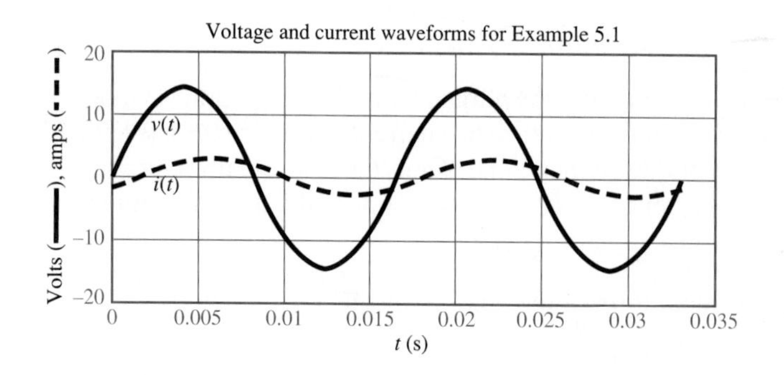

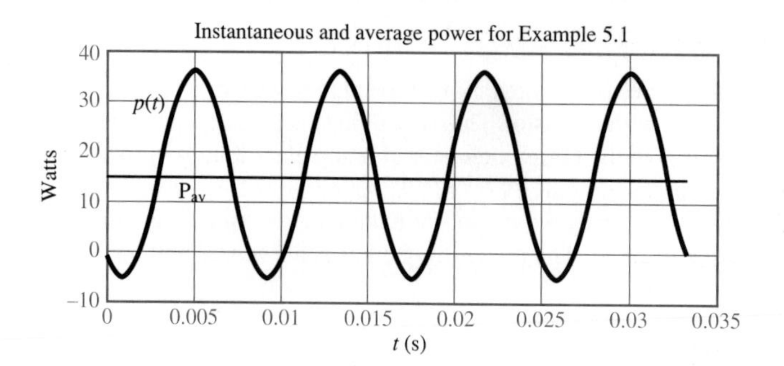

Figure 5.6

Example 5.2

This example illustrates the computation of average power for a parallel RC load. The circuit is shown in Figure 5.7. Note that in the circuit of Figure 5.7, the AC source is modeled as having an internal resistance; this resistance affects the magnitude and phase of the load voltage and causes a power loss.

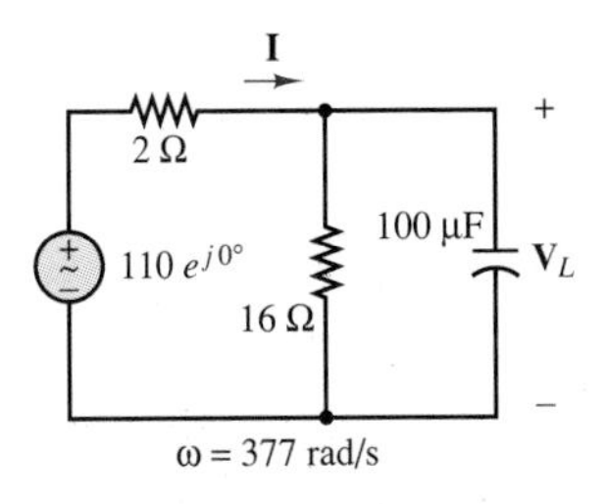

Figure 5.7

Solution:

The first step in the computation of average power consists of the calculation of the load impedance:

$$Z_L = R \parallel \frac{1}{j\omega C} = \frac{R}{1 + j\omega RC} = \frac{16}{1 + j0.6032} = 11.7315 - j7.0764 = 13.7e^{-j31.1°}$$

Thus, $|Z_L| = 13.7$ and $\theta = -31.1°$.

The second step consists of determining the load voltage. Application of the voltage divider rule yields the expression

$$\mathbf{V}_L = 110e^{j0°}\frac{Z_L}{2 + Z_L} = 110e^{j0°}\frac{11.7315 - j7.0764}{13.7315 - j7.0764}$$

$$= (110e^{j0°})(0.8869e^{-j3.83°}) = 97.6e^{-j3.83°}$$

Given the impedance and the load voltage, it is then possible to directly apply equation 5.9 to compute the average power:

$$P_{\text{av}} = \frac{\tilde{V}_L^2}{|Z_L|}\cos\theta = \frac{(97.6)^2}{13.6}\cos(-31.1°) = 594.7 \text{ W}$$

Alternatively, the same solution could be obtained by first determining the load current and using the expression

$$P_{\text{av}} = \tilde{I}^2 |Z| \cos\theta$$

Example 5.3

Find the power dissipated by the complex load impedance seen by the 60-Hz AC source in the circuit shown in Figure 5.8. Assume that the rms source voltage is $\mathbf{V} = 110e^{j0°}$.

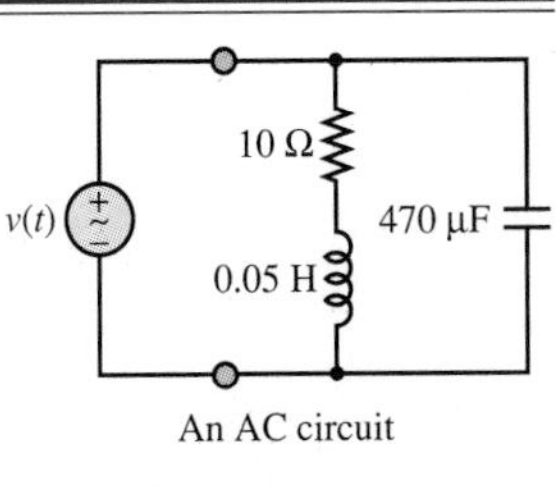

An AC circuit

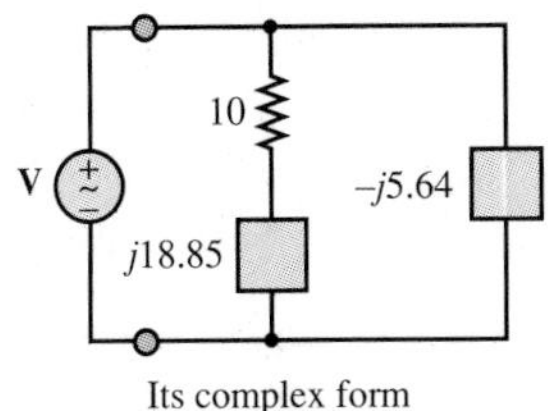

Its complex form

Figure 5.8

Solution:

The circuit is also shown in complex form in Figure 5.8. The equivalent impedance seen by the source is equal to the parallel combination of the two branches:

$$jX_C = \frac{1}{j\omega C} = -j\frac{1}{377 \times 470 \times 10^{-6}} = -j5.64$$

$$jX_L = j\omega L = j377 \times 0.05 = j18.85$$

$$Z_{EQ} = (10 + j18.85) \parallel (-j5.64)$$

$$= \frac{(10 + j18.85)(-j5.64)}{10 + j18.85 - j5.64}$$

$$= \frac{106.31 - j56.4}{10 + j13.21} = \frac{120.35\angle -27.95°}{16.568\angle 52.87°}$$

$$= 7.26\angle -80.8° = 1.16 - j7.17$$

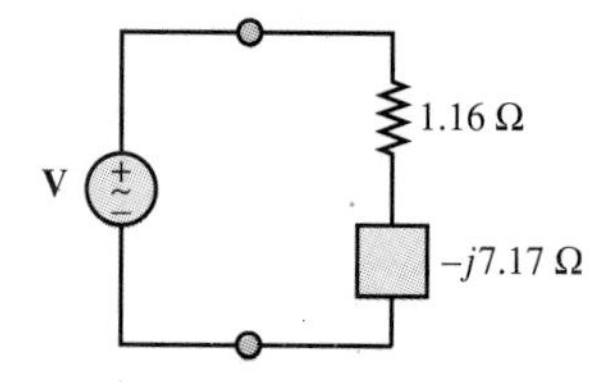

Figure 5.9

Thus, the equivalent impedance seen by the source consists of an AC resistance of 1.16 Ω and a (capacitive) reactance of −7.17 Ω, shown in Figure 5.9.

Finally, the average power dissipated by the load is

$$P_{av} = \frac{V^2}{|Z|}\cos(\theta) = \frac{110^2}{7.26}\cos(-80.8°) = 266.5 \text{ W}$$

Power Factor

The phase angle of the load impedance plays a very important role in the absorption of power by a load impedance. As illustrated in equation 5.13 and in the preceding examples, the average power dissipated by an AC load is dependent on the cosine of the angle of the impedance. To recognize the importance of this factor in AC power computations, the term $\cos(\theta)$ is referred to as the **power factor (pf)**. Note that the power factor is equal to 0 for a purely inductive or capacitive load and equal to 1 for a purely resistive load; in every other case,

$$0 < \text{pf} < 1 \qquad \textbf{(5.20)}$$

Two equivalent expressions for the power factor are given in the following:

$$\text{pf} = \cos(\theta) = \frac{P_{av}}{\tilde{V}\tilde{I}} \qquad \text{Power factor} \qquad \textbf{(5.21)}$$

where $\tilde{V}$ and $\tilde{I}$ are the rms values of the load voltage and current.

DRILL EXERCISES

5.1 Show that the equalities in equation 5.9 hold when phasor notation is used.

5.2 Consider the circuit shown in Figure 5.10. Find the load impedance of the circuit, and compute the average power dissipated by the load.

5.3 Use the expression $P_{av} = \tilde{I}^2|Z|\cos\theta$ to compute the average power dissipated by the load of Example 5.2.

5.4 Compute the power dissipated by the internal source resistance in Example 5.2.

5.5 Compute the power factor for an inductive load with $L = 100$ mH and $R = 0.4\ \Omega$. Assume $\omega = 377$ rad/s.

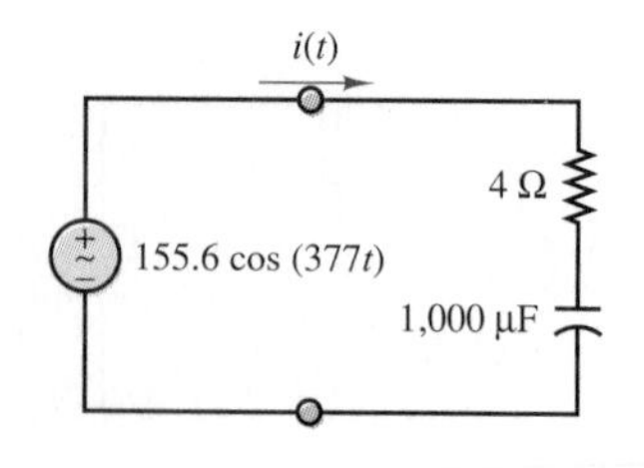

Figure 5.10

5.2 COMPLEX POWER

The expression for the instantaneous power given in equation 5.4 may be further expanded to provide further insight into AC power. Using trigonometric identities, we obtain the following expressions:

$$\begin{aligned} p(t) &= \frac{\tilde{V}^2}{|Z|}[\cos\theta + \cos\theta\cos(2\omega t) - \sin\theta\sin(2\omega t)] \\ &= \tilde{I}^2|Z|[\cos\theta + \cos\theta\cos(2\omega t) - \sin\theta\sin(2\omega t)] \\ &= \tilde{I}^2|Z|\cos\theta(1 + \cos(2\omega t)) - \tilde{I}^2|Z|\sin\theta\sin(2\omega t) \end{aligned} \qquad (5.22)$$

Recalling the geometric interpretation of the impedance Z of Figure 5.4, you may recognize that

$$|Z|\cos\theta = R$$

and **(5.23)**

$$|Z|\sin\theta = X$$

are the resistive and reactive components of the load impedance, respectively. On the basis of this fact, it becomes possible to write the instantaneous power as:

$$\begin{aligned} p(t) &= \tilde{I}^2R(1 + \cos(2\omega t)) - \tilde{I}^2X\sin(2\omega t) \\ &= \tilde{I}^2R + \tilde{I}^2R\cos(2\omega t) - \tilde{I}^2X\sin(2\omega t) \end{aligned} \qquad (5.24)$$

The physical interpretation of this expression for the instantaneous power should be intuitively appealing at this point. As equation 5.23 suggests, the instantaneous power dissipated by a complex load consists of the following three components:

1. An average component, which is constant; this is called the *average power* and is denoted by the symbol P_{av}:

   $$P_{av} = \tilde{I}^2R \qquad (5.25)$$

   where $R = \text{Re}(Z)$.
2. A time-varying (sinusoidal) component with zero average value that is contributed by the power fluctuations in the resistive component of the load and is denoted by $p_R(t)$:

   $$\begin{aligned} p_R(t) &= \tilde{I}^2R\cos 2\omega t \\ &= P_{av}\cos 2\omega t \end{aligned} \qquad (5.26)$$

3. A time-varying (sinusoidal) component with zero average value, due to the power fluctuation in the reactive component of the load and denoted by $p_X(t)$:

   $$\begin{aligned} p_X(t) &= -\tilde{I}^2X\sin(2\omega t) \\ &= Q\sin 2\omega t \end{aligned} \qquad (5.27)$$

where $X = \text{Im}(Z)$ and Q is called the **reactive power.** *Note that since reactive elements can only store energy and not dissipate it, there is no net average power absorbed by X.*

Since P_{av} corresponds to the power absorbed by the load resistance, it is also called the **real power,** measured in units of watts (W). On the other hand, Q takes the name of *reactive power,* since it is associated with the load reactance. Table 5.1 shows the general methods of calculating P and Q.

Table 5.1 **Real and reactive power**

Real power, P_{av}	Reactive power, Q
$\tilde{V}\tilde{I}\cos(\theta)$	$\tilde{V}\tilde{I}\sin(\theta)$
$\tilde{I}^2R$	$\tilde{I}^2X$

The units of Q are **volt-amperes reactive,** or **VAR.** Note that Q represents an exchange of energy between the source and the reactive part of the load; thus, no net power is gained or lost in the process, since the average reactive power is zero. In general, it is desirable to minimize the reactive power in a load. Example 5.5 will explain the reason for this statement.

The computation of AC power is greatly simplified by defining a fictitious but very useful quantity called the **complex power,** S:

$$S = \mathbf{VI}^* \quad \text{Complex power} \tag{5.28}$$

where the asterisk denotes the complex conjugate (see Appendix A). You may easily verify that this definition leads to the convenient expression

$$S = \tilde{V}\tilde{I}\cos\theta + j\tilde{V}\tilde{I}\sin\theta = \tilde{I}^2R + j\tilde{I}^2X = \tilde{I}^2Z$$

or (5.29)

$$S = P_{av} + jQ$$

The complex power S may be interpreted graphically as a vector in the complex plane, as shown in Figure 5.11.

The magnitude of S, $|S|$, is measured in units of **volt-amperes (VA)** and is called **apparent power,** because this is the quantity one would compute by measuring the rms load voltage and currents without regard for the phase angle of the load. Note that the right triangle of Figure 5.11 is similar to the right triangle of Figure 5.4, since θ is the load impedance angle. The complex power may also be expressed by the product of the square of the rms current through the load and the complex load impedance:

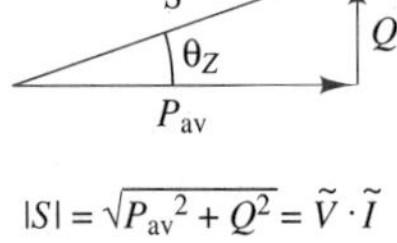

$$|S| = \sqrt{P_{av}^2 + Q^2} = \tilde{V}\cdot\tilde{I}$$

$$P_{av} = \tilde{V}\tilde{I}\cos\theta$$

$$Q = \tilde{V}\tilde{I}\sin\theta$$

Figure 5.11 The complex power triangle

$$S = \tilde{I}^2Z$$

or (5.30)

$$\tilde{I}^2R + j\tilde{I}^2X = \tilde{I}^2Z$$

or, equivalently, by the ratio of the rms voltage across the load to the complex conjugate of the load impedance:

$$S = \frac{\tilde{V}^2}{Z^*} \tag{5.31}$$

The power triangle and complex power greatly simplify load power calculations, as illustrated in the following examples.

Example 5.4

Calculate the complex power for the circuit shown in Figure 5.12, given the following voltage and current waveforms:

$$v(t) = 100\cos(\omega t + 15°) \qquad i(t) = 2\cos(\omega t - 15°)$$

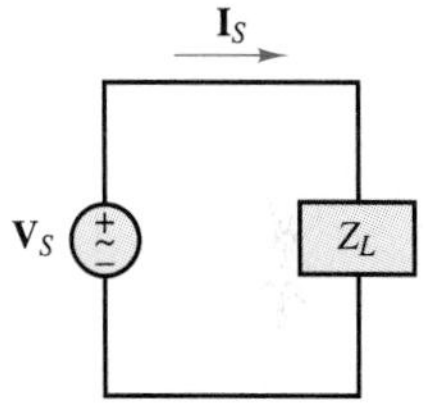

Figure 5.12

Solution:

First, we convert the voltage and current into rms phasor form:

$$\mathbf{V} = \frac{100}{\sqrt{2}}\angle 15° \qquad \mathbf{I} = \frac{2}{\sqrt{2}}\angle -15°$$

Then the real and reactive load power are computed using equation 5.13. The average power is found to be

$$P_{\text{av}} = \tilde{V}\tilde{I}\cos\theta \qquad (\theta = \theta_V - \theta_I)$$

$$= \frac{100}{\sqrt{2}} \times \frac{2}{\sqrt{2}} \times \cos(15° - (-15°)) = 86.6 \text{ W}$$

while the reactive power is

$$Q = \frac{100}{\sqrt{2}} \times \frac{2}{\sqrt{2}} \times \sin(15° - (-15°)) = 50 \text{ VAR}$$

Next, the same quantities are computed using equation 5.28:

$$S = \mathbf{V}\mathbf{I}^* = P_{\text{av}} + jQ$$

$$= \left(\frac{100}{\sqrt{2}}e^{j15°}\right)\left(\frac{2}{\sqrt{2}}e^{-j15°}\right)^* = 100e^{j30°} = 86.6 + j50$$

or

$$P_{\text{av}} = 86.6 \text{ W} \qquad Q = 50 \text{ VAR}$$

Note that the computation using complex power yields both real and reactive power in a single calculation.

Example 5.5

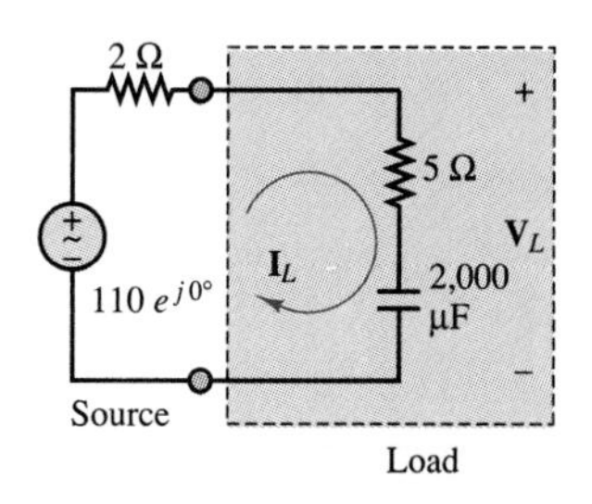

Figure 5.13

The objective of this example is to illustrate the effect of a reactive load element on the power dissipated by a load. We seek to find the real and reactive instantaneous power dissipated by the load in the circuit of Figure 5.13 (all voltages and currents are rms and have frequency 60 Hz).

Solution:

We start by expressing the load current as the ratio of the source voltage to the total series impedance of the circuit:

$$Z = R_S + R_L + X_L$$

$$= 2 + 5 + \frac{1}{j377 \times 2{,}000 \times 10^{-6}}$$

$$= 7.13e^{-j10.76°}$$

$$\mathbf{V}_S = 110e^{j0°}$$

$$\mathbf{I} = \frac{\mathbf{V}_S}{Z} = \frac{110e^{j0°}}{7.13e^{-j10.76°}}$$

$$= 15.44e^{j10.76°}$$

The power factor of the load is

$$\cos\theta = 0.982$$

The AC load resistance is equal to

$$R = 5\ \Omega$$

and the load reactance is

$$X = -1.33\ \Omega$$

while the rms load current is

$$\tilde{I} = 15.44\ \text{A}$$

Thus, we find the real power absorbed by the resistive component of the load:

$$P_{\text{av}} = \tilde{I}^2 R = 1{,}192\ \text{W}$$

and the reactive power:

$$Q = \tilde{I}^2 X = -316\ \text{VAR}$$

Although the reactive power does not contribute to any average power dissipation in the load, it may have an adverse effect on power consumption, because it increases the overall rms current flowing in the circuit. Recall from Example

5.2 that the presence of any source resistance (typically, the resistance of the line wires in AC power circuits) will cause a loss of power; the power loss due to this line resistance is unrecoverable and constitutes a net loss for the electric company, since the user never receives this power. The following example illustrates quantitatively the effect of such **line losses** in an AC circuit.

Example 5.6

This example illustrates how, by reducing the reactive power at the load, it is possible to improve the **power transfer** from a source to a load. Consider the AC source shown in Figure 5.14, modeled by an ideal source with internal resistance (including the wires connecting the source to the load) of 4 Ω. The load consists of a parallel RL combination, with $R = 10\ \Omega$ and $X = 6\ \Omega$.

1. Compute the load impedance seen by the source.
2. Find the real and reactive power of the load.
3. Determine the percent real power transfer from the source to the load with and without the inductor included in the circuit.

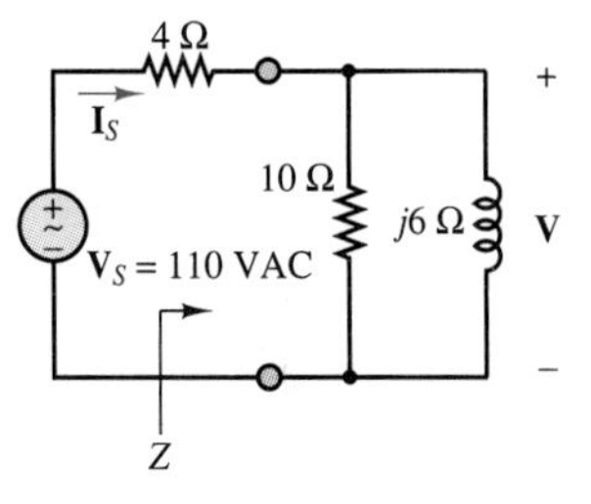

Figure 5.14

Solution:

1. The equivalent load impedance is given by

$$Z = 10 \parallel j6 = \frac{(j6)(10)}{j6 + 10} = \frac{60\angle 90^\circ}{11.66\angle 31^\circ}$$

$$= 5.145\angle 59^\circ\ \Omega$$

2. To find the real and reactive power, we compute the load voltage. By the voltage divider rule, we obtain:

$$\mathbf{V} = \frac{Z}{Z+4}\mathbf{V}_S = \frac{5.145\angle 59^\circ}{2.647 + j4.412 + 4}\mathbf{V}_S$$

$$= \frac{5.145\angle 59^\circ}{6.647 + j4.4118}\mathbf{V}_S = \frac{5.145\angle 59^\circ}{7.978\angle 33.57^\circ}\mathbf{V}_S$$

$$= 70.94\angle 25.46^\circ$$

The real power absorbed by the load is found to be

$$P = \frac{\tilde{V}^2}{R} = \frac{(70.94)^2}{10} = 503.2\ \text{W}$$

and the reactive power is

$$Q = \frac{\tilde{V}^2}{X} = \frac{(70.9)^2}{6} = 838.7\ \text{VAR}$$

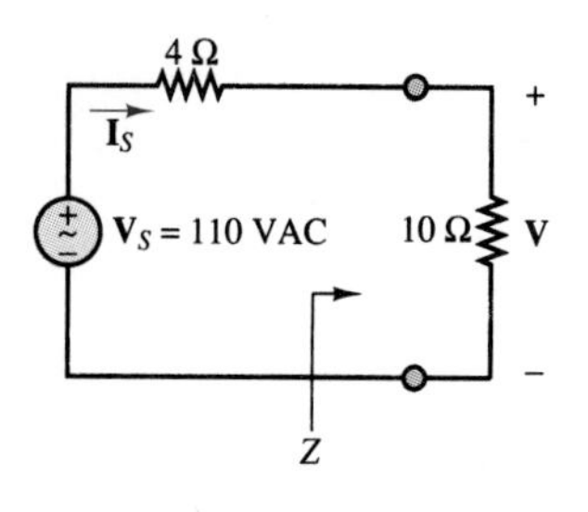

Figure 5.15

3. Without the inductor in the circuit, as shown in Figure 5.15, we would compute the following results:

$$P = \tilde{I}^2 10 = \left(\frac{\tilde{V}_S}{10+4}\right)^2 10 = \left(\frac{110}{14}\right)^2 (10) = 617.3 \text{ W}$$

and

$$Q = 0$$

The total power generated by the source is the product of the source current and the source voltage (since these are in phase when the inductor is not in the circuit):

$$P_S = \tilde{V}\tilde{I} = 110\left(\frac{110}{14}\right) \text{ W}$$

$$P_S = 864.3 \text{ W}$$

so that the percent power transfer in the absence of the inductive component of the load is

$$100\% \frac{P_L}{P_S} = 71.4\%$$

With the inductor in the circuit, we need to consider the phasor form of the source voltage and load current:

$$\mathbf{V}_S = 110\angle 0°$$

$$\mathbf{I}_S = \frac{\mathbf{V}_S}{Z+4} = 13.79\angle -33.57°$$

Hence,

$$P_S = 110 \times 13.79 \times \cos(-33.57°)$$
$$= 1{,}263.7 \text{ W}$$

Thus the percent power transfer is now

$$\% \text{ power transfer} = 100\% \frac{P_L}{P_S} = 100\% \frac{503.2}{1{,}263.7} = 39.82\%$$

Observe that, although the inductor does not dissipate any real power, its presence greatly reduces the power that can be absorbed by the load and increases the total power that must be generated by the source.

Power Factor, Revisited

The power factor, defined earlier as the cosine of the angle of the load impedance, plays a very important role in AC power. A power factor close to unity signifies an efficient transfer of energy from the AC source to the load, while a small power factor corresponds to inefficient use of energy, as illustrated in Example 5.6. It should be apparent that if a load requires a fixed amount of real power, P, the source will be providing the smallest amount of current when the power factor is the greatest, that is, when $\cos\theta = 1$. If the power factor is less than unity, some additional current will be drawn from the source, lowering the efficiency of power transfer from the source to the load. However, it will be shown shortly that it is possible to correct the power factor of a load by adding an appropriate reactive component to the load itself.

Since the reactive power, Q, is related to the reactive part of the load, its sign depends on whether the load reactance is inductive or capacitive. This leads to the following important statement:

If the load has an inductive reactance, then θ is positive and the current *lags* (or *follows*) the voltage. Thus, when θ and Q are positive, the corresponding power factor is termed *lagging*. Conversely, a capacitive load will have a negative Q, and hence a negative θ. This corresponds to a *leading* power factor, meaning that the load current *leads* the load voltage.

Table 5.2 illustrates the concept and summarizes all of the important points so far. In the table, the phasor voltage **V** has a zero phase angle and the current phasor is referenced to the phase of **V**.

Table 5.2 **Important facts related to complex power**

	Resistive load	**Capacitive load**	**Inductive load**
Ohm's law	$\mathbf{V}_L = Z_L\mathbf{I}_L$	$\mathbf{V}_L = Z_L\mathbf{I}_L$	$\mathbf{V}_L = Z_L\mathbf{I}_L$
Complex impedance	$Z_L = R_L$	$Z_L = R_L - jX_L$	$Z_L = R_L + jX_L$
Phase angle	$\theta_V - \theta_I = \theta = 0$	$\theta_V - \theta_I = \theta < 0$	$\theta_V - \theta_I = \theta > 0$
Complex plane sketch	Im, $\theta = 0$, **I** **V**, Re	Im, **I**, θ, **V**, Re	Im, **V**, θ, Re, **I**
Explanation	The current is in phase with the voltage.	The current "leads" the voltage.	The current "lags" the voltage.
Power factor	Unity	Leading, < 1	Lagging, < 1
Reactive power	0	Negative	Positive

The following examples illustrate the computation of complex power for a simple circuit.

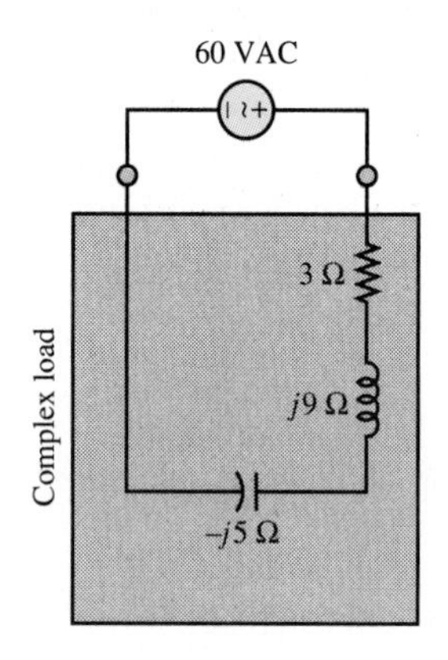

Figure 5.16

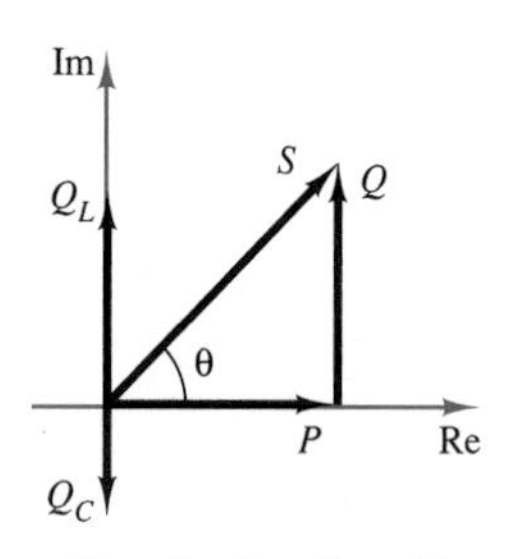

Note: $S = P_R + jQ_C + jQ_L$

Figure 5.17

Example 5.7

Find the reactive and real power dissipated by the circuit of Figure 5.16, and draw the associated power triangle.

Solution:

The phasor current **I** flowing in the series circuit is found as follows:

$$\mathbf{I} = \frac{60\angle 0^\circ}{Z_{EQ}} = \frac{60\angle 0^\circ}{3 + j9 - j5} = \frac{60\angle 0^\circ}{5\angle 53.13^\circ} = 12\angle -53.13^\circ \text{A}$$

and the complex power S is given by the expression

$$S = \tilde{I}^2 Z = (12^2)(3 + j4) = (144)(3 + j4) = 432 + j576$$
$$= P + jQ$$

If we observe that the total reactive power must be the sum of the reactive power in the inductor and in the capacitor, we can write

$$Q = Q_C + Q_L$$

and compute each quantity, as follows:

$$Q_C = (144)(-5) = -720 \text{ VAR}$$
$$Q_L = (144)(9) = 1{,}296 \text{ VAR}$$
$$Q = Q_L + Q_C = 1{,}296 - 720 = 576 \text{ VAR}$$

Next, we draw the power triangle (Figure 5.17), showing how the complex power, S, is a vector in the complex plane equal to the vector sum of the three components, corresponding to the three elements.

The distinction between leading and lagging power factors made in Table 5.2 is important, because it corresponds to opposite signs of the reactive power: Q is positive if the load is inductive ($\theta > 0$) and the power factor is lagging; Q is negative if the load is capacitive and the power factor is leading ($\theta < 0$). It is therefore possible to improve the power factor of a load according to a procedure called **power factor correction**—that is, by placing a suitable reactance in parallel with the load so that the reactive power component generated by the additional reactance is of opposite sign to the original load reactive power. Most often the need is to improve the power factor of an inductive load, because many common industrial loads consist of electric motors, which are predominantly inductive loads. This improvement may be accomplished by placing a capacitance in parallel with the load. The following example illustrates a typical power factor correction for an industrial load.

Example 5.8

For the circuit shown in Figure 5.18:

1. Calculate the complex power for the load.
2. Correct the power factor by adding a suitable reactance in parallel with the load.

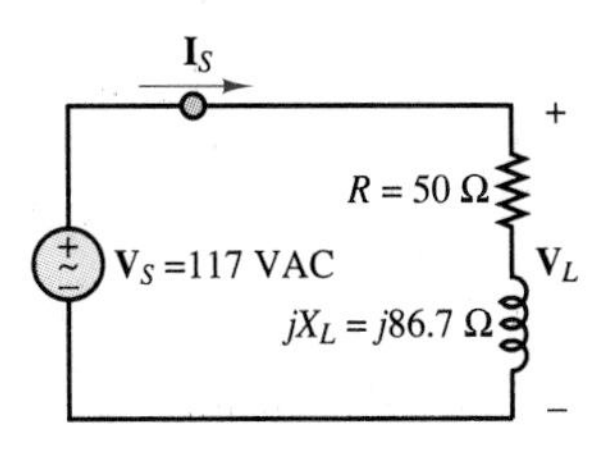

Figure 5.18

Solution:

1. The circuit of Figure 5.18 is an inductive load. The total impedance is

$$Z = R + jX_L = 50 + j86.7\ \Omega = 100\angle 60^\circ$$

The power factor is then

$$\text{pf} = \cos\theta = \cos 60^\circ = 0.5 \text{ (lagging)}$$

The current drawn from the source by the load is

$$\mathbf{I}_S = \frac{\mathbf{V}_S}{Z} = \frac{117\angle 0^\circ}{100\angle 60^\circ} = 1.17\angle -60^\circ$$

and the average power is found to be

$$P = \tilde{V}_S \tilde{I}_S \cos\theta = 117 \times 1.17 \cos 60^\circ = 68.4 \text{ W}$$

while the reactive power is

$$Q_L = \tilde{V}_S \tilde{I}_S \sin\theta = 117 \times 1.17 \sin 60^\circ = 119 \text{ VAR}$$

The power triangle is shown in Figure 5.19.

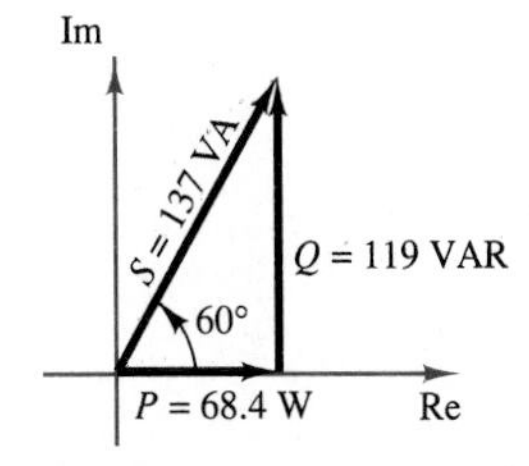

Figure 5.19

2. The unity power factor for the circuit can be obtained by simply reducing the power factor angle θ to 0°. This can be accomplished by adding a capacitor to the circuit that requires -119 VAR of reactive power. The capacitive power and the inductive power will then cancel each other in the power triangle, resulting in a unity power factor, as shown in Figure 5.20.

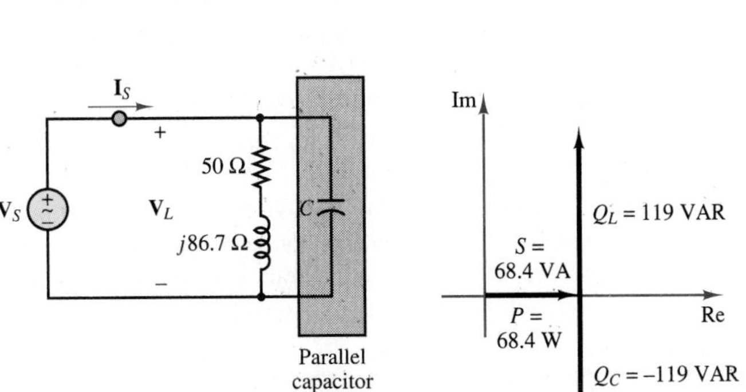

Figure 5.20 Power factor correction

The value of capacitive reactance, X_C, required to cancel the reactive power due to the inductance is found most easily by observing that the total reactive power in the circuit must be the sum of the reactive power due to the capacitance and that due to the inductance. Observing that the capacitor sees the same voltage as the RL load, because of the parallel connection, we can write

$$X_C = \frac{\tilde{V}_S^2}{Q_C} = \frac{117^2}{119} = 115\ \Omega$$

From the expression for the reactance, it is then possible to compute the value of the capacitor that will cancel the reactive power due to the inductor:

$$C = \frac{1}{\omega X_C} = \frac{1}{377 \times 115} = 23.1\ \mu\text{F}$$

The reactive component of power needed by the inductance is now balanced by the capacitance, and all the power delivered by the source is real power. The power factor is 1.

Example 5.9

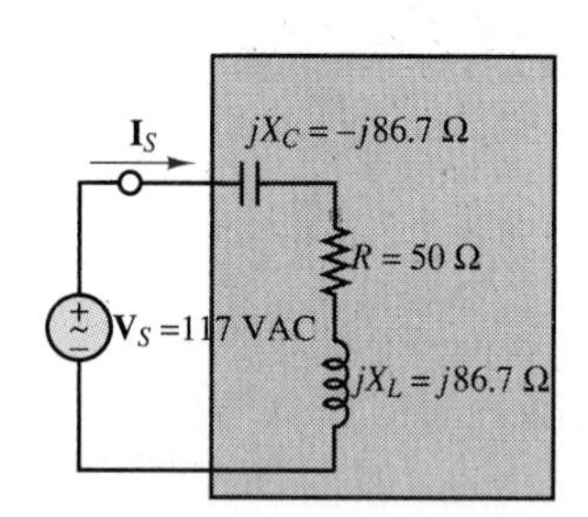

Figure 5.21

With reference to Example 5.8, consider the series capacitor connection shown in Figure 5.21, which is proposed as an alternative to the parallel connection to correct the power factor. Is this a feasible alternative to the solution of Example 5.8?

Solution:

The series load impedance, Z', may be computed as follows:

$$Z' = R + jX_L - jX_C = 50 + j86.7 - j86.7 = 50$$

while the series current in the circuit is

$$\mathbf{I}_S = \frac{\mathbf{V}_S}{Z'} = \frac{117\angle 0^\circ}{50} = 2.34\ \text{A}$$

Note the significant increase in the current drawn from the source! The addition of the series capacitor is not an acceptable method for power factor correction, because the load current has increased by a factor of 2, from 1.17 A to 2.34 A. It is important to observe that the voltage across the RL load has also increased to:

$$\mathbf{I}_S(R + j\omega L) = 2.34\angle 0^\circ \times 100\angle 60^\circ = 234\angle 60^\circ\ \text{V}$$

which is also approximately double the load voltage prior to the addition of the series capacitor. It should be apparent that the power factor correction of an inductive load must be accomplished by connecting a capacitance in parallel with the original load, or else the current drawn by the source will be increased excessively.

Example 5.10

The instrument used to measure power is called a *wattmeter*. The external part of a wattmeter consists of four connections and a metering mechanism that displays the amount of real power dissipated by a circuit. The external and internal appearance of a wattmeter are depicted in Figure 5.22. Inside the wattmeter are two coils: a current-sensing coil, and a voltage-sensing coil. In this example, we assume for simplicity that the impedance of the current-sensing coil, C_I, is zero and the impedance of the voltage-sensing coil, C_V, is infinite. In practice, this will not necessarily be true; some correction mechanism will be required to account for the impedance of the sensing coils.

A wattmeter should be connected as shown in Figure 5.23, to provide both current and voltage measurements. We see that the current-sensing coil is placed in series with the load and the voltage-sensing coil is placed in parallel with the load. In this manner, the wattmeter is seeing the current through and the voltage across the load. Remember that the power dissipated by a circuit element is related to these two quantities. The wattmeter, then, is constructed to provide a readout of the product of the rms values of the load current and the voltage, which is the real power absorbed by the load: $P = \text{Re}(S) = \text{Re}(\mathbf{VI}^*)$.

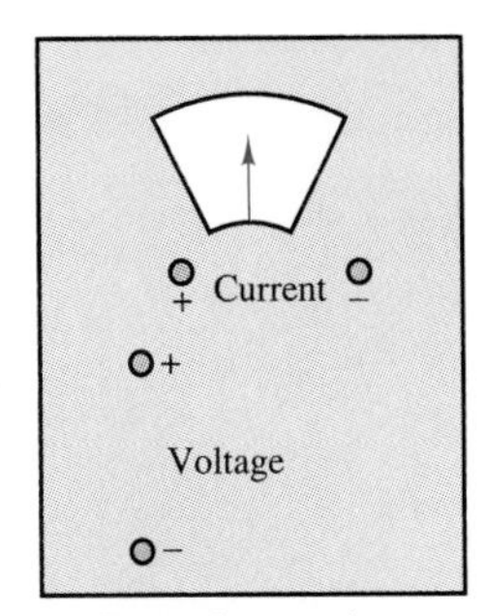

External connections

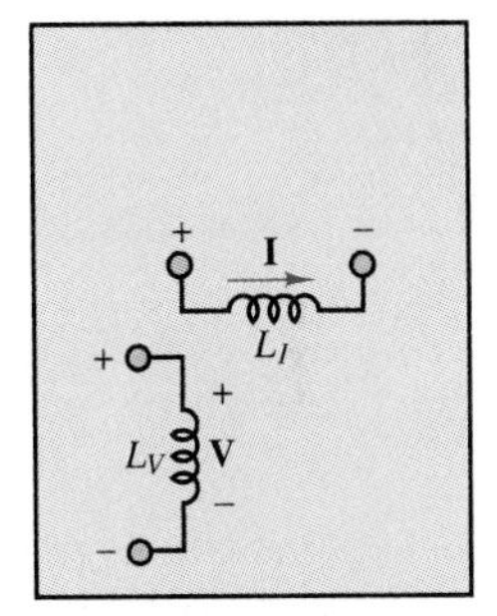

Wattmeter coils (inside)

Figure 5.22

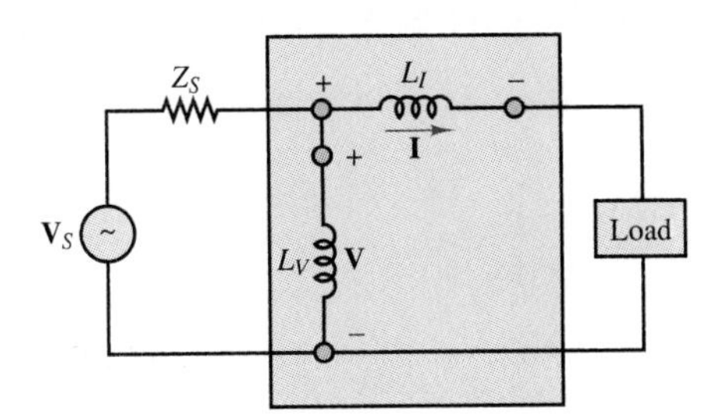

Figure 5.23

1. For the circuit shown in Figure 5.24, show the connections of the wattmeter, and find the power dissipated by the load.
2. Show the connections that will determine the power dissipated by R_2. What should the meter read?

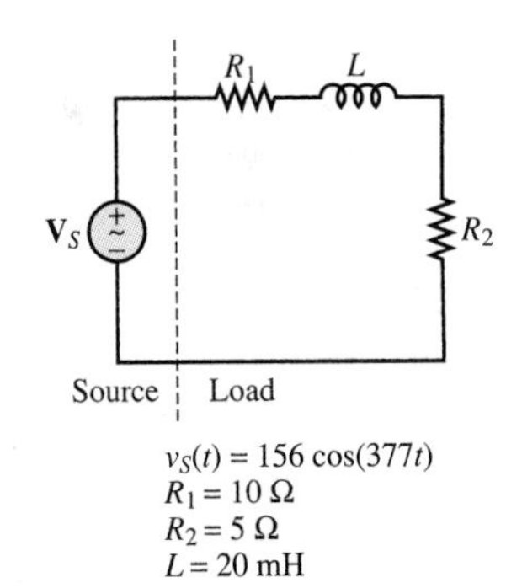

Figure 5.24

Solution:

1. To measure the power dissipated by the load, we must know the current through and the voltage across the entire load circuit. This means that the wattmeter must be connected as shown in Figure 5.25. The wattmeter should read:

$$P = \text{Re}\,(\mathbf{V}_S\mathbf{I}^*)$$

$$= \text{Re}\left\{\left(\frac{156}{\sqrt{2}}\angle 0^\circ\right)\left(\frac{\frac{156}{\sqrt{2}}\angle 0^\circ}{R_1 + R_2 + j\omega L}\right)^*\right\}$$

$$= \text{Re}\left\{110\angle 0^\circ\left(\frac{110\angle 0^\circ}{15 + j7.54}\right)^*\right\}$$

$$\text{Re}\left\{110\angle 0^\circ\left(\frac{110\angle 0^\circ}{16.79\angle 26.69^\circ}\right)^*\right\} = \text{Re}\left(\frac{110^2}{16.79\angle -26.69^\circ}\right)$$

$$= \text{Re}\,(720.67\angle 26.69^\circ)$$

$$= 643.88\ \text{W}$$

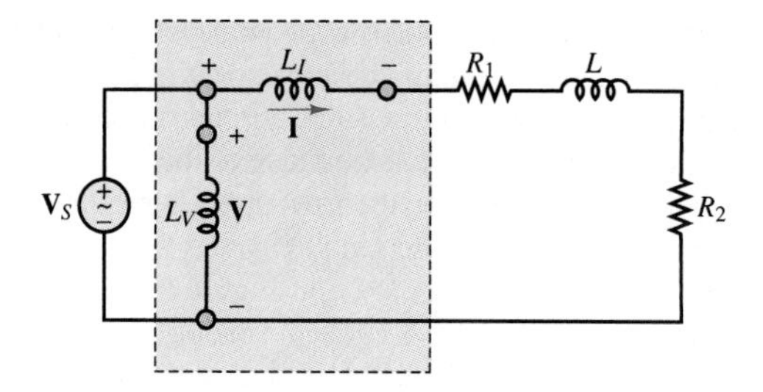

Figure 5.25

2. To measure the power dissipated by R_2 alone, we must measure the current through R_2 and the voltage across R_2 *alone*. The connection is shown in Figure 5.26. The meter will read

$$P = \tilde{I}^2 R_2$$

$$= \left(\frac{110}{(15^2 + 7.54^2)^{1/2}}\right)^2 \times 5$$

$$= \left(\frac{110^2}{15^2 + 7.54^2}\right) \times 5 = 215 \text{ W}$$

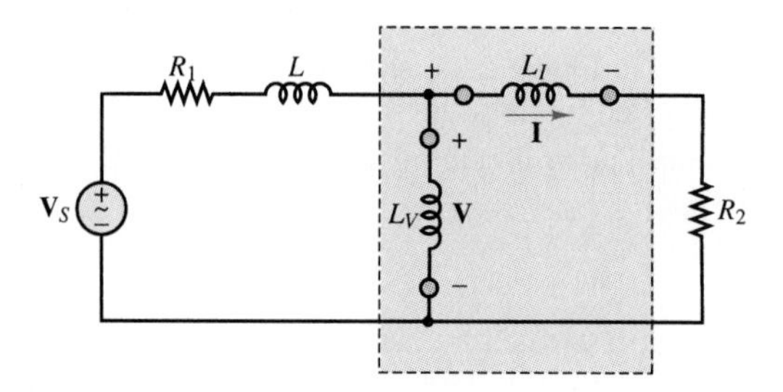

Figure 5.26

Example 5.11 How Hall-Effect Current Transducers Work*

In 1879, E. H. Hall noticed that if a conducting material is placed in a magnetic field perpendicular to a current flow, a voltage perpendicular to both the initial current flow and the magnetic field is developed. This voltage is called the *Hall voltage* and is directly proportional to both the strength of the magnetic field and the current. It results from the deflection of the moving charge carriers from their normal path by the magnetic field and its resulting transverse electric field.

---

*Courtesy Ohio Semitronics, Inc., Columbus, Ohio.

To illustrate the physics involved, consider a confined stream of free particles each having a charge e and an initial velocity u_x. A magnetic field in the Z direction will produce a deflection in the y direction. Therefore, a charge imbalance is created; this results in an electric field E_y. This electric field, the Hall field, will build up until the force it exerts on a charged particle counterbalances the force resulting from the magnetic field. Now subsequent particles of the same charge and velocity are no longer deflected. A steady state exists. Figure 5.27 depicts this effect.

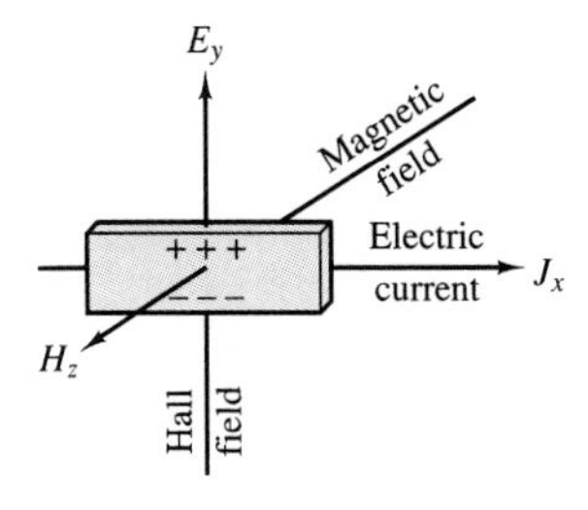

Figure 5.27 Hall effect

The Hall effect occurs in any conductor. In most conductors the Hall voltage is very small and is difficult to measure. Dr. Warren E. Bulman, working with others, developed semiconductor compounds in the early 1950s that made the Hall effect practical for measuring magnetic fields.

The choice of materials for the active Hall element of most Hall probes is indium arsenide (InAs). This semiconductor compound is manufactured from highly refined elemental arsenic and indium. From an ingot of the semiconductor compound, thin slices are taken. These slices are then diced by an ultrasonic cutter into small, rectangular chips. These chips of indium arsenide are then placed on a thin ceramic substrate and soldered to a conducting pattern on the substrate, as shown in Figure 5.28.

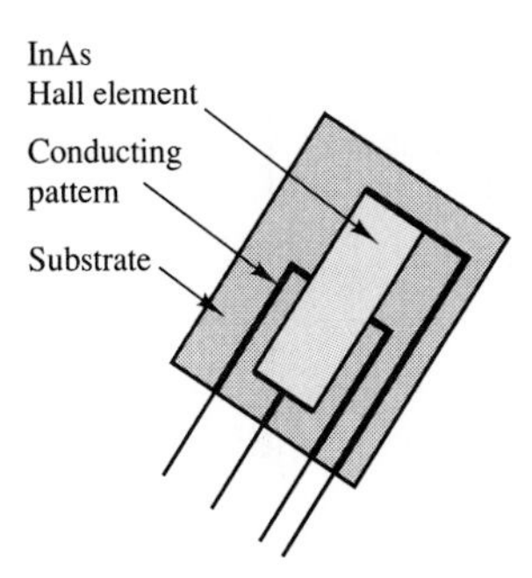

Figure 5.28 Hall-effect probe

Once made, the Hall probe is normally coated with epoxy to protect the semiconductor compound and other components.

To use the probe, an electric current is passed through the length of the InAs chip, as shown in Figure 5.29. Note that the contact areas for passing current through the Hall element are made larger than the ones for detecting the Hall voltage. Typically, current on the order of 10^{-1} A is passed through the Hall element. This is known as the control current. Care must be taken when using a Hall probe never to put the control current through the output. Because the solder contacts for the voltage sensing are very small, the control current can melt these solder joints. This may destroy or damage the Hall probe.

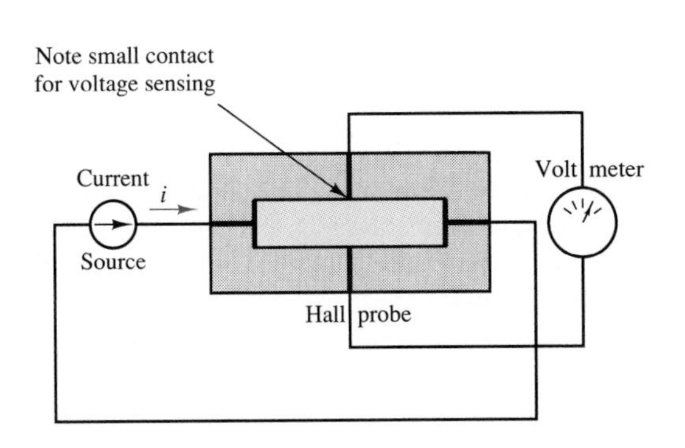

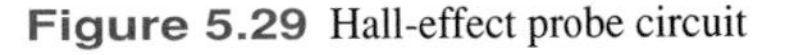
Figure 5.29 Hall-effect probe circuit

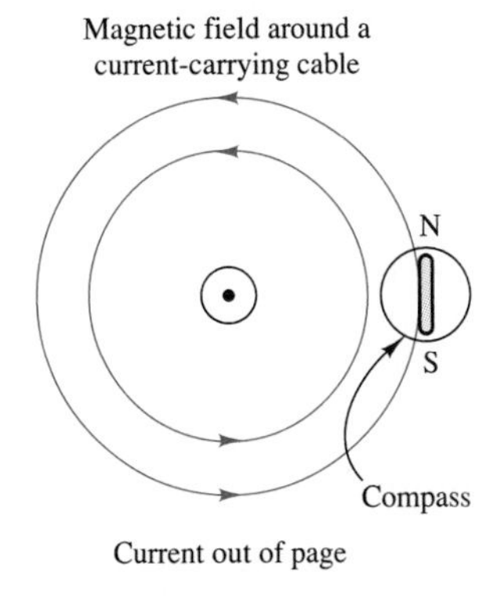

Figure 5.30

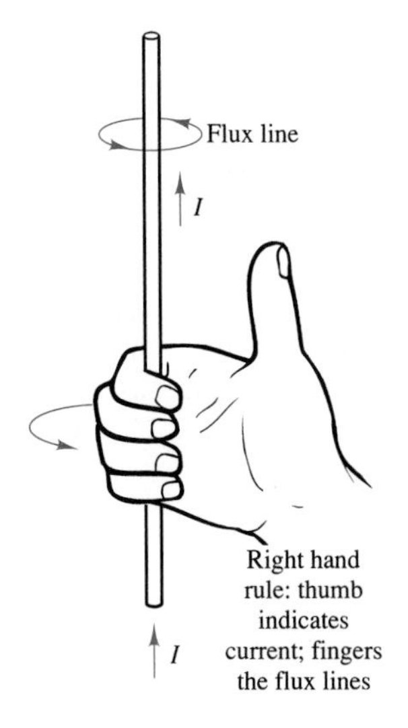

Figure 5.31 Right-hand rule

A wire carrying a current will have a closed magnetic field around it, as depicted in Figures 5.30 and 5.31. If a Hall probe is placed perpendicular to the magnetic flux lines around a current-carrying conductor, then the Hall probe will have a voltage output proportional to that magnetic field and the control current through the Hall probe. Since the magnetic field is directly proportional to the current, I, the output of the Hall probe is directly proportional to the current, I, and to the control current. We have a current transducer.

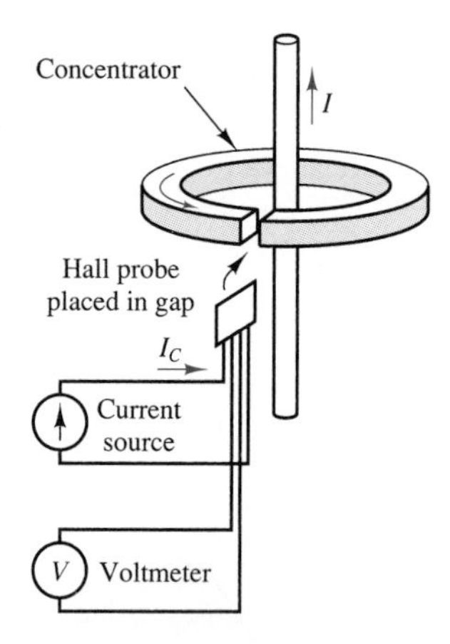

Figure 5.32

Unfortunately, this method will provide adequate output only if the current being measured, I, is of the order of 10^4 amperes. Also, the strength of the magnetic field is proportional to the inverse of the distance from the center of the conductor.

A practical current transducer (Figure 5.32) can be made by using a magnetic field concentrator with a Hall probe placed in a gap. Typically, a laminated iron core with very low magnetic retention is utilized. This arrangement makes a simple but very effective current transducer.

Unfortunately, a Hall probe is temperature-sensitive. Hence, the voltage output of the current transducer as described will be dependent upon the control current, I_c, the current through the window, I, and the temperature, T. To correct for this, a *thermistor*–resistor network is used to maintain temperature influence to a minimum for the operating range of the transducer. By careful selection of the thermistor and resistor used, temperature influence in the range of −40°C to +65°C can be mostly eliminated.

The measurement and correction of the power factor for the load are an extremely important aspect of any engineering application in industry that requires the use of substantial quantities of electric power. In particular, industrial plants, construction sites, heavy machinery, and other heavy users of electric power must be aware of the power factor their loads present to the electric utility company. As was already observed, a low power factor results in greater current draw from the electric utility, and in greater line losses. Thus, computations related to the power factor of complex loads are of great practical utility to any practicing engineer. To provide you with deeper insight into calculations related to power factor, a few more advanced examples are given in the remainder of the section.

Example 5.12

A capacitor is being used to correct the power factor to unity. The circuit is shown in Figure 5.33. The capacitor value is varied, and measurements of the total current are taken. Explain how it is possible to "zero in" on the capacitance value necessary to bring the power factor to unity just by monitoring the current $\mathbf{I}_S$.

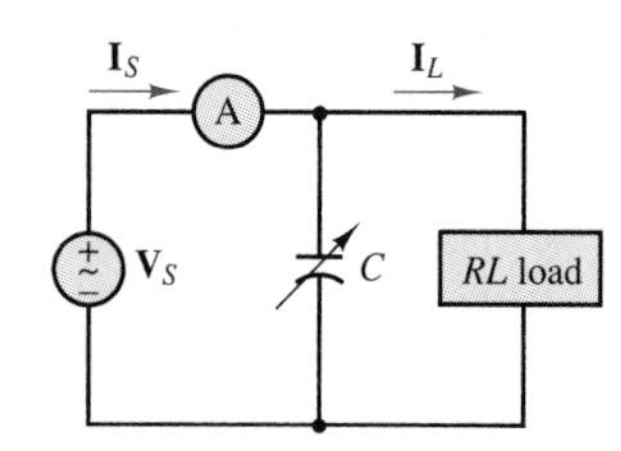

Figure 5.33

Solution:

The current through the load is

$$\mathbf{I}_L = \frac{\tilde{V}_S\angle 0^\circ}{R + j\omega L} = \frac{\tilde{V}_S}{R^2 + \omega^2 L^2}(R - j\omega L)$$

$$= \frac{\tilde{V}_S R}{R^2 + \omega^2 L^2} - j\frac{\tilde{V}_S \omega L}{R^2 + \omega^2 L^2}$$

The current through the capacitor is

$$\mathbf{I}_C = \frac{\tilde{V}_S\angle 0^\circ}{1/j\omega C} = j\tilde{V}_S\omega C$$

The source current to be measured is

$$\mathbf{I}_S = \mathbf{I}_L + \mathbf{I}_C = \frac{\tilde{V}_S R}{R^2 + \omega^2 L^2} + j\left(\tilde{V}_S \omega C - \frac{\tilde{V}_S \omega L}{R^2 + \omega^2 L^2}\right)$$

The magnitude of the source current is

$$\tilde{I}_S = \sqrt{\left(\frac{\tilde{V}_S R}{R^2 + \omega^2 L^2}\right)^2 + \left(\tilde{V}_S \omega C - \frac{\tilde{V}_S \omega L}{R^2 + \omega^2 L^2}\right)^2}$$

We know that when the load is a pure resistance, the current and voltage are in phase, the power factor is 1, and all the power delivered by the source is dissipated by the load as real power. This corresponds to equating the imaginary part of the expression for the source current to zero, or, equivalently, to the following expression:

$$\frac{\tilde{V}_S \omega L}{R^2 + \omega^2 L^2} = \tilde{V}_S \omega C$$

in the expression for $\tilde{I}_S$. Thus, the magnitude of the source current is actually a minimum when the power factor is unity! It is therefore possible to "tune" a load to a unity pf by observing the readout of the ammeter while changing the value of capacitor and selecting the capacitor value that corresponds to the lowest source current value.

Example 5.13

A 100-kW load with pf = 0.7 (lagging) is connected to a 480-V (rms) source. A capacitor is connected in parallel with the load to improve the power factor, as shown in Figure 5.34.

1. Determine the reactive power when no capacitor is connected to the load.
2. Determine the required capacitance to bring the power factor to unity.

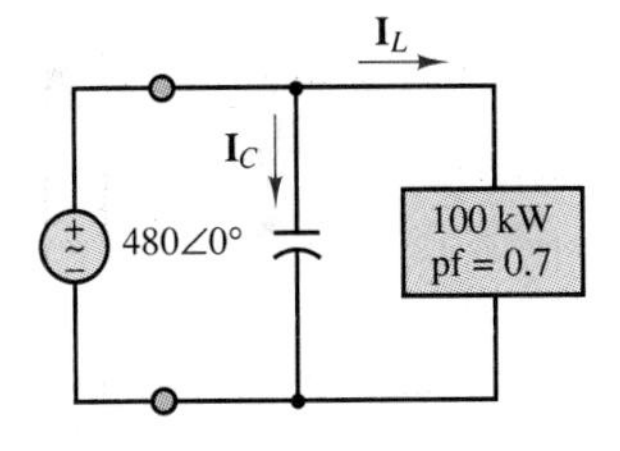

Figure 5.34

Solution:

1. The magnitude of the complex power, S, may be found by computing the ratio of the rated power to the power factor as shown in the power triangle.:

$$|S| = \frac{P}{\cos\theta} = \frac{P}{\text{pf}} = \frac{10^5}{0.7}$$
$$= 142{,}857 \text{ VA}$$

Since the power factor is lagging, the reactive power is positive, and we can compute the reactive power according to the relation

$$Q = \tilde{V}\tilde{I}\sin\theta = |S|\sin(\arccos(0.7))$$
$$= 102 \text{ kVAR}$$

2. There are two approaches to solving this part of the problem. One is to determine the value of a capacitor whose reactance will cancel that of the rest of the circuit. A second method, usually more tedious in calculations, is to find a capacitor whose value will make the load seem purely resistive. It is then simply a matter

of finding the capacitance that causes the impedance (or, equivalently, the admittance) of the load to be purely real; this corresponds to equating the imaginary part to zero.

Using the first method to solve this problem, we know that placing a capacitor in parallel with the load will not change the current in the load, nor will it change the voltage across the load. This makes it clear that the reactive power in the load will be 102 kVAR with or without the power factor correction capacitor. To cancel this reactive power, we require that

$$Q_C = -Q_L = -102 \text{ kVAR}$$

$$Q_C = -\frac{\tilde{V}^2}{X_C}$$

$$X_C = -\frac{1}{\omega C}$$

$$C = -\frac{Q_C}{\omega \tilde{V}^2} = \frac{102 \times 10^3}{377 \times 480^2} = 1{,}174\ \mu\text{F}$$

To determine the capacitance required to obtain a power factor of unity using the second method, we first compute the current flowing in the circuit without the capacitor:

$$\mathbf{I}_S^* = \frac{S}{\mathbf{V}} = \frac{100{,}000 + j102{,}000}{480\angle 0^\circ} = \frac{142{,}840\angle 45.6^\circ}{480\angle 0^\circ}$$

$$\mathbf{I}_S = 297.6\angle -45.6^\circ \text{A}$$

From knowledge of the source voltage and current, we can then infer the value of the load impedance:

$$Z_L = \frac{\mathbf{V}}{\mathbf{I}_S} = \frac{480\angle 0^\circ}{297.6\angle -45.6^\circ}$$

$$= 1.613\angle +45.6^\circ$$

$$= 1.13 + j1.15$$

With a capacitor in parallel with the load, the new load impedance is Z_L':

$$Z_L' = Z_L \parallel -jX_C$$

$$= \frac{-jX_C Z_L}{Z_L - jX_C}$$

$$= \frac{1.613 X_C \angle -44.43^\circ}{1.13 + j(1.15 - X_C)}$$

$$= \frac{(1.613 X_C \angle -44.43^\circ)[1.13 - j(1.15 - X_C)]}{1.13^2 + (1.15 - X_C)^2}$$

To bring the power factor to unity, we require the imaginary part of Z_L' to be zero. Neglecting the denominator in the expression for Z_L', the requirement $X_L' = 0$ is satisfied by the following expression:

$$-1.613X_C[(0.7)(1.13) + (1.15 - X_C)(0.714)] = 0$$

or

$$0.7(1.13) + 0.714(1.15 - X_C) = 0$$

Solving for the capacitive reactance, X_C, we obtain:

$$X_C = 2.26 = \frac{1}{\omega C}$$

$$C = \frac{1}{2.26(377)} = 1.174 \times 10^{-3} \text{ F}$$

$$= 1{,}174 \ \mu\text{F}$$

Thus, a 1,174-μF capacitor placed in parallel with the load will reduce the reactive power to zero. Following the power factor correction, the current supplied by the source is given by:

$$\mathbf{I}'_S = \frac{\mathbf{V}}{Z'_L}$$

where Z'_L can be calculated to be 5.77 Ω by substituting the value $X_C = 2.26$ in the earlier expression. Thus,

$$\mathbf{I}'_S = \frac{480\angle 0^\circ}{5.77} = 83.2 \text{ A}$$

Example 5.14

A second load, with $P = 50$ kW, pf = 0.95 (leading), is also connected to the source in Example 5.13, in parallel with the first load. The overall circuit is shown in Figure 5.35.

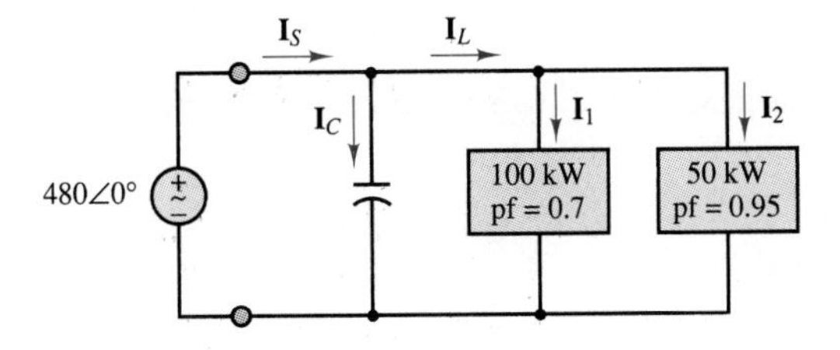

Figure 5.35

1. Determine the size of the capacitance needed to obtain an overall power factor of unity.
2. Draw the phasor diagram of the two load currents $\mathbf{I}_1$ and $\mathbf{I}_2$, the capacitor current $\mathbf{I}_C$, and the source current $\mathbf{I}_S$.

Solution:

1. The circuit of Example 5.13 is now modified as shown in Figure 5.35. We can find the currents $\mathbf{I}_1$ and $\mathbf{I}_2$ by the following calculations:

$$\mathbf{I} = \frac{P}{\text{pf}_1 \cdot \mathbf{V}} \angle \cos^{-1}(\text{pf})$$

$$\mathbf{I}_1 = \frac{10^5}{(0.7)480\angle 0^\circ} \angle -\cos^{-1}(0.7) = 297.6\angle -45.6^\circ \text{ A}$$

$$\mathbf{I}_2 = \frac{5 \times 10^4}{(0.95)480\angle 0^\circ} \angle \cos^{-1}(0.95) = 109.65\angle 18.2^\circ \text{ A}$$

so that the current drawn by the combined loads prior to connecting the power factor correction capacitor is

$$\mathbf{I}_S = \mathbf{I}_L = \mathbf{I}_2 + \mathbf{I}_1 = (104.17 + j34.23 + 208.3 - j212.5) \text{ A}$$
$$= 312.5 - j178.3 = 359.76\angle -29.7^\circ \text{ A}$$

To have unity power factor, we must cancel the reactive power of the loads. We already know the reactive power of the first load from the previous example:

$$Q_1 = 102{,}000 \text{ VAR}$$

For the second load,

$$|S_2| = \frac{P_2}{\text{pf}_2} = \frac{50 \text{ kW}}{0.95}$$
$$= 52{,}632 \text{ VA}$$

Since the power factor is leading, the reactive power is negative; thus,

$$Q_2 = |S_2| \sin(-\cos^{-1}(0.95))$$
$$= -16{,}434 \text{ VAR}$$

and the total reactive power of the combined loads is

$$Q_1 + Q_2 = 85{,}566 \text{ VAR}$$

To find the capacitance value required to cancel this reactive power, we apply the same method as in the previous example:

$$Q_C = -\frac{\tilde{V}^2}{X_C}$$

$$X_C = -\frac{1}{\omega C}$$

$$C = -\frac{Q_C}{\omega \tilde{V}^2} = \frac{85{,}566}{377 \times 480^2} = 985 \ \mu\text{F}$$

2. We need find only the capacitor current, since we know the two load currents already:

$$\mathbf{I}_C = \frac{\mathbf{V}_S}{Z_C} = \frac{480\angle 0^\circ}{2.69\angle -90^\circ} = 178.3\angle 90^\circ \text{ A}$$

Then the new source current, $\mathbf{I}'_S$, is the sum of the capacitor and load currents:

$$\begin{aligned}\mathbf{I}'_S &= \mathbf{I}_C + \mathbf{I}_L = \mathbf{I}_C + \mathbf{I}_1 + \mathbf{I}_2 \\ &= j178.3 + 312.5 - j178.3 \\ &= 312.5 \text{ A}\end{aligned}$$

Note that the current drawn from the source after power factor correction, $\mathbf{I}'_S$, is lower than the current required of the source prior to insertion of the capacitor, $\mathbf{I}_S$. The phasor diagram corresponding to these calculations is shown in Figure 5.36.

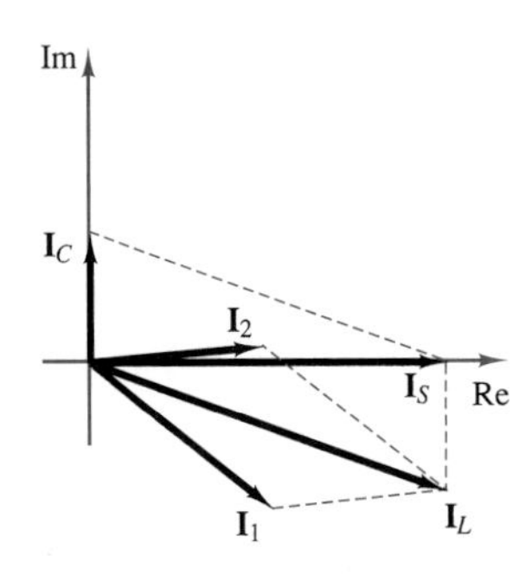

Figure 5.36

DRILL EXERCISES

5.6 Compute the power factor for the load of Example 5.6 with and without the inductance in the circuit.

5.7 Show that one can also express the instantaneous power for an arbitrary complex load $Z = |Z|\angle\theta$ as

$$p(t) = \tilde{I}^2|Z|\cos\theta + \tilde{I}^2|Z|\cos(2\omega t + \theta)$$

5.8 Determine the power factor for the load in the circuit of Figure 5.37, and state whether it is leading or lagging for the following conditions:

a. $v_S(t) = 540\cos(\omega t + 15°)$ V
$i(t) = 2\cos(\omega t + 47°)$ A

b. $v_S(t) = 155\cos(\omega t - 15°)$ V
$i(t) = 2\cos(\omega t - 22°)$ A

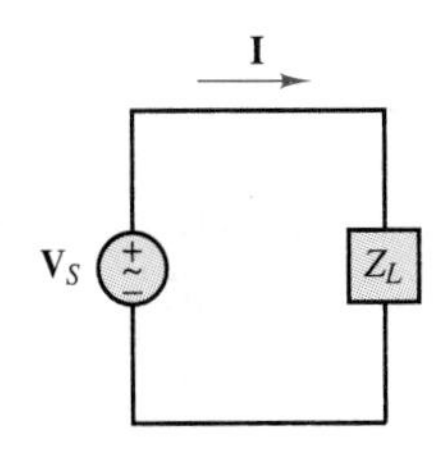

Figure 5.37

5.9 Determine whether the load is capacitive or inductive for the circuit of Figure 5.37 if

a. pf = 0.87 (leading)

b. pf = 0.42 (leading)

c. $v_S(t) = 42\cos(\omega t)$
$i(t) = 4.2\sin(\omega t)$

d. $v_S(t) = 10.4\cos(\omega t - 12°)$
$i(t) = 0.4\cos(\omega t - 12°)$

5.10 Prove that the power factor is indeed 1 after the addition of the parallel capacitor in Example 5.8.

5.11 Compute the magnitude of the current drawn from the source after the power factor correction in the circuit of Example 5.8.

5.3 TRANSFORMERS

AC circuits are very commonly connected to each other by means of **transformers.** A transformer is a device that couples two AC circuits magnetically rather than through any direct conductive connection and permits a "transformation" of the voltage and current between one circuit and the other (for example, by matching a high-voltage, low-current AC output to a circuit requiring a low-voltage, high-current source). Transformers play a major role in electric power engineering and are a necessary part of the electric power distribution network. The objective of this section is to introduce the ideal transformer and the concepts of impedance reflection and impedance matching. The physical operations of practical transformers, and more advanced models, will be discussed in Chapter 15.

The Ideal Transformer

The ideal transformer consists of two coils that are coupled to each other by some magnetic medium. There is no electrical connection between the coils. The coil on the input side is termed the **primary,** and that on the output side the **secondary.** The primary coil is wound so that it has n_1 turns, while the secondary has n_2 turns. We define the **turns ratio** N as

$$N = \frac{n_2}{n_1} \tag{5.32}$$

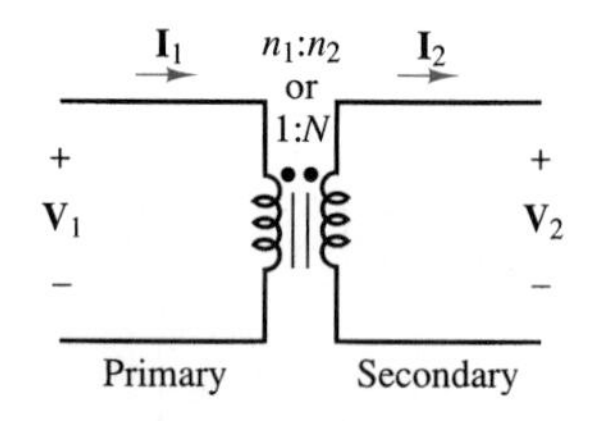

Figure 5.38 Ideal transformer

Figure 5.38 illustrates the convention by which voltages and currents are usually assigned at a transformer. The dots in Figure 5.38 are related to the polarity of the coil voltage: coil terminals marked with a dot have the same polarity.

Since an ideal inductor acts as a short circuit in the presence of DC currents, transformers do not perform any useful function when the primary voltage is DC. However, when a time-varying current flows in the primary winding, a corresponding time-varying voltage is generated in the secondary because of the magnetic coupling between the two coils. This behavior is due to Faraday's law, as will be explained in Chapter 15. The relationship between primary and secondary current in an ideal transformer is very simply stated as follows:

$$\begin{aligned} \mathbf{V}_2 &= N\mathbf{V}_1 \\ \mathbf{I}_2 &= \frac{\mathbf{I}_1}{N} \end{aligned} \tag{5.33}$$

An ideal transformer multiplies a sinusoidal input voltage by a factor of N and divides a sinusoidal input current by a factor of N.

If N is greater than 1, the output voltage is greater than the input voltage and the transformer is called a **step-up transformer.** If N is less than 1, then the transformer is called a **step-down transformer,** since $\mathbf{V}_2$ is now smaller than $\mathbf{V}_1$.

An ideal transformer can be used in either direction (i.e., either of its coils may be viewed as the input side or primary). Finally, a transformer with $N = 1$ is called an **isolation transformer** and may perform a very useful function if one needs to electrically isolate two circuits from each other; note that any DC currents at the primary will not appear at the secondary coil. An important property of ideal transformers is conservation of power; one can easily verify that an ideal transformer conserves power, since

$$S_1 = \mathbf{I}_1^*\mathbf{V}_1 = N\mathbf{I}_2^*\frac{\mathbf{V}_2}{N} = \mathbf{I}_2^*\mathbf{V}_2 = S_2 \tag{5.34}$$

That is, the power on the primary side equals that on the secondary.

In many practical circuits, the secondary is tapped at two different points, giving rise to two separate output circuits, as shown in Figure 5.39. The most common configuration is the **center-tapped transformer,** which splits the secondary voltage into two equal voltages of half the original amplitude. The most common occurrence of this type of transformer is found at the entry of a power line into a household, where the 240-VAC line is split into two 120-VAC lines (this may help explain why both 240-VAC and 120-VAC power are present in your house).

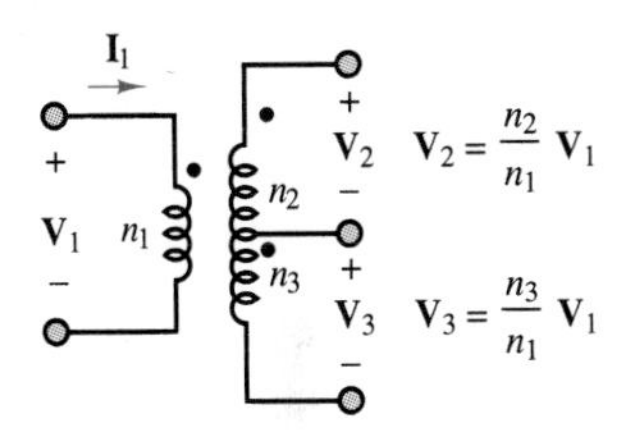

Figure 5.39 Center-tapped transformer

Impedance Reflection and Power Transfer

As stated in the preceding paragraphs, transformers are commonly used to couple one AC circuit to another. A very common and rather general situation is that depicted in Figure 5.40, where an AC source, represented by its Thévenin equivalent, is connected to an equivalent load impedance by means of a transformer.

It should be apparent that expressing the circuit in phasor form does not alter the basic properties of the ideal transformer, as illustrated in the following equation:

$$\mathbf{V}_1 = \frac{\mathbf{V}_2}{N} \qquad \mathbf{I}_1 = N\mathbf{I}_2$$
$$\mathbf{V}_2 = N\mathbf{V}_1 \qquad \mathbf{I}_2 = \frac{\mathbf{I}_1}{N} \tag{5.35}$$

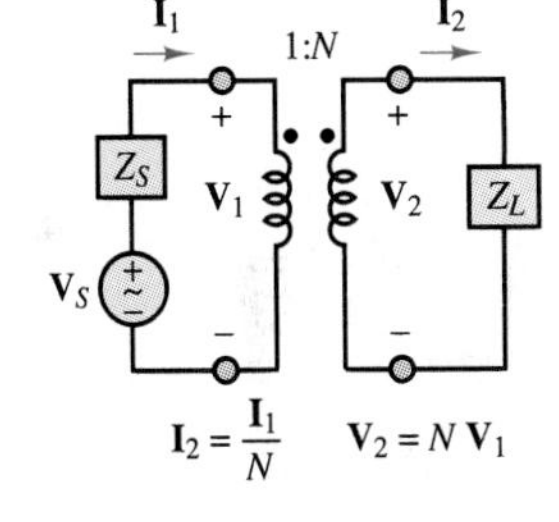

Figure 5.40 Operation of an ideal transformer

These expressions are very useful in determining the equivalent impedance seen by the source and by the load, on opposite sides of the transformer. At the primary connection, the equivalent impedance seen by the source must equal the ratio of $\mathbf{V}_1$ to $\mathbf{I}_1$:

$$Z' = \frac{\mathbf{V}_1}{\mathbf{I}_1} \tag{5.36}$$

which can be written as:

$$Z' = \frac{\mathbf{V}_2/N}{N\mathbf{I}_2} = \frac{1}{N^2}\frac{\mathbf{V}_2}{\mathbf{I}_2} \tag{5.37}$$

But the ratio $\mathbf{V}_2/\mathbf{I}_2$ is by definition the load impedance, Z_L. Thus,

$$Z' = \frac{1}{N^2} Z_L \tag{5.38}$$

That is, the AC source "sees" the load impedance reduced by a factor of $1/N^2$.

The load impedance also sees an equivalent source. The open-circuit voltage is given by the expression

$$\mathbf{V}_{OC} = N\mathbf{V}_1 = N\mathbf{V}_S \tag{5.39}$$

since there is no voltage drop across the source impedance in the circuit of Figure 5.40. The short-circuit current is given by the expression

$$\mathbf{I}_{SC} = \frac{\mathbf{V}_S}{Z_S}\frac{1}{N} \tag{5.40}$$

and the load sees a Thévenin impedance equal to

$$Z'' = \frac{\mathbf{V}_{OC}}{\mathbf{I}_{SC}} = \frac{N\mathbf{V}_S}{\frac{\mathbf{V}_S}{Z_S}\frac{1}{N}} = N^2 Z_S \tag{5.41}$$

Thus the load sees the source impedance multiplied by a factor of N^2. Figure 5.41 illustrates this **impedance reflection** across a transformer. It is very important to note that an ideal transformer changes the magnitude of the load impedance seen by the source by a factor of $1/N^2$. This property naturally leads to the discussion of power transfer, which we consider next.

Recall that in DC circuits, given a fixed equivalent source, maximum power is transferred to a resistive load when the latter is equal to the internal resistance of the source; achieving an analogous maximum power transfer condition in an

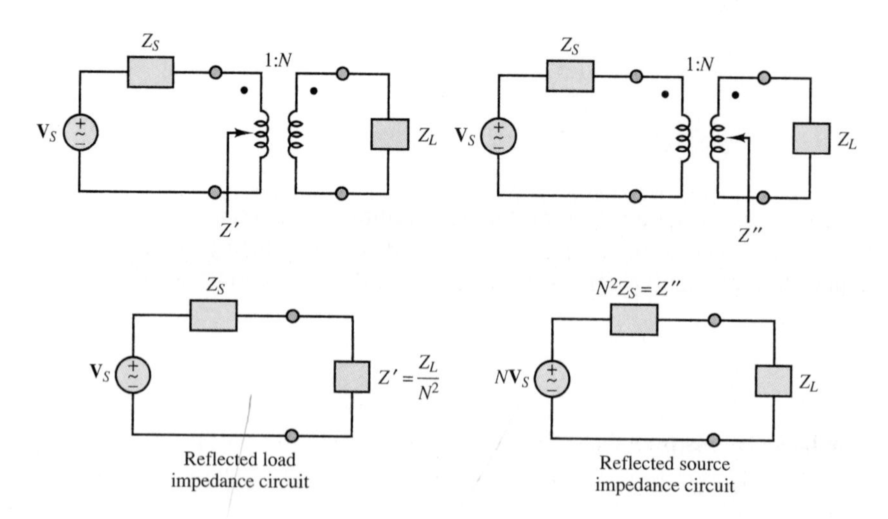

Figure 5.41 Impedance reflection across a transformer

AC circuit is referred to as **impedance matching.** Consider the general form of an AC circuit, shown in Figure 5.42, and assume that the source impedance, Z_S, is given by

$$Z_S = R_S + jX_S \tag{5.42}$$

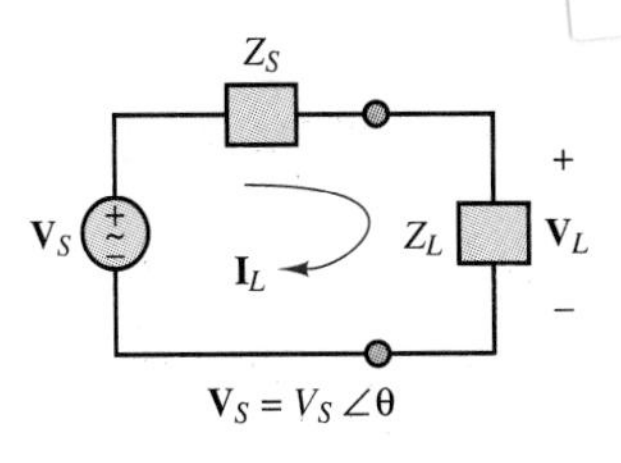

Figure 5.42 The maximum power transfer problem in AC circuits

The problem of interest is often that of selecting the load resistance and reactance that will maximize the real (average) power absorbed by the load. Note that the requirement is to maximize the real power absorbed by the load. Thus, the problem can be restated by expressing the real load power in terms of the impedance of the source and load. The real power absorbed by the load is

$$P_L = \tilde{V}_L \tilde{I}_L \cos\theta = \text{Re}\,(\mathbf{V}_L \mathbf{I}_L^*) \tag{5.43}$$

where

$$\mathbf{V}_L = \frac{Z_L}{Z_S + Z_L}\mathbf{V}_S \tag{5.44}$$

and

$$\mathbf{I}_L^* = \left(\frac{\mathbf{V}_S}{Z_S + Z_L}\right)^* = \frac{\mathbf{V}_S^*}{(Z_S + Z_L)^*} \tag{5.45}$$

Thus, the complex load power is given by

$$S_L = \mathbf{V}_L \mathbf{I}_L^* = \frac{Z_L \mathbf{V}_S}{Z_S + Z_L} \times \frac{\mathbf{V}_S^*}{(Z_S + Z_L)^*} = \frac{\mathbf{V}_S^2}{|Z_S + Z_L|^2} Z_L \tag{5.46}$$

and the average power by

$$\begin{aligned} P_L &= \text{Re}\,(\mathbf{V}_L \mathbf{I}_L^*) = \text{Re}\left(\frac{\mathbf{V}_S^2}{|Z_S + Z_L|^2}\right)\text{Re}\,(Z_L) \\ &= \frac{\tilde{V}_S^2}{(R_S + R_L)^2 + (X_S + X_L)^2}\,\text{Re}\,(Z_L) \\ &= \frac{\tilde{V}_S^2 R_L}{(R_S + R_L)^2 + (X_S + X_L)^2} \end{aligned} \tag{5.47}$$

The expression for P_L is maximized by selecting appropriate values of R_L and X_L; it can be shown that the average power is greatest when $R_L = R_S$ and $X_L = -X_S$, that is, when the load impedance is equal to the complex conjugate of the source impedance, as shown in the following equation:

$$Z_L = Z_S^* \qquad \text{i.e.,} \qquad R_L = R_S \qquad X_L = -X_S \tag{5.48}$$

When the load impedance is equal to the complex conjugate of the source impedance, the load and source impedances are matched and maximum power is transferred to the load.

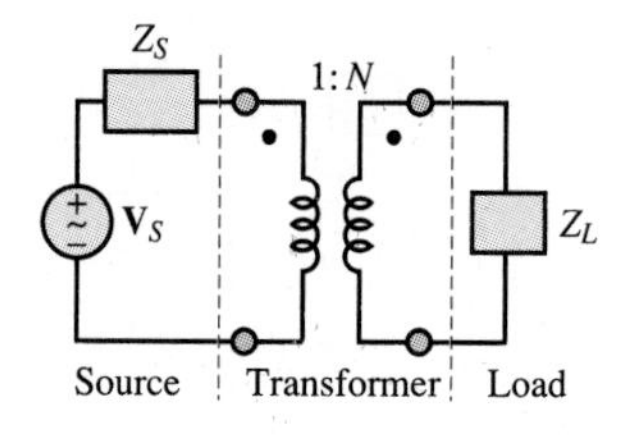

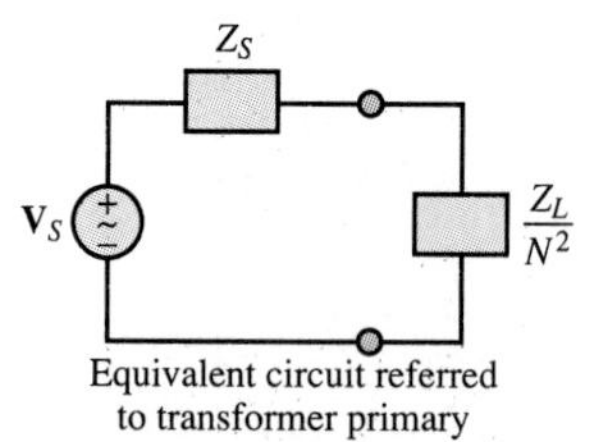

Figure 5.43 Maximum power transfer in an AC circuit with a transformer

In many cases, it may not be possible to select a matched load impedance, because of physical limitations in the selection of appropriate components. In these situations, it is possible to use the impedance reflection properties of a transformer to maximize the transfer of AC power to the load. The circuit of Figure 5.43 illustrates how the reflected load impedance, as seen by the source, is equal to Z_L/N^2, so that maximum power transfer occurs when

$$\begin{aligned} \frac{Z_L}{N^2} &= Z_S^* \\ R_L &= N^2 R_S \\ X_L &= -N^2 X_S \end{aligned} \tag{5.49}$$

Example 5.15

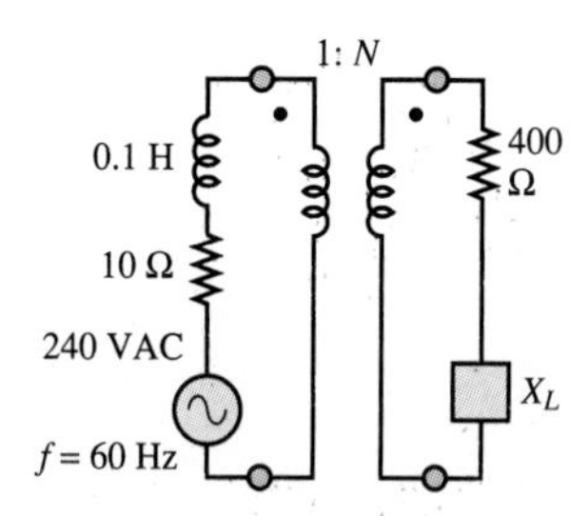

Figure 5.44

This example illustrates the transfer of maximum AC power with the aid of a transformer. A 240-VAC, 60-Hz source with the internal impedance shown in Figure 5.44 is connected to a load that has a resistive component of 400 Ω. Find the transformer turns ratio N and load reactance that allow for maximum power transfer.

Solution:

For maximum power transfer, we require that

$$\begin{aligned} R_L &= N^2 R_S \\ N^2 &= \frac{R_L}{R_S} = \frac{400}{10} = 40 \\ N &= \sqrt{40} = 6.325 \end{aligned}$$

and we must also have

$$X_L = -N^2 X_S$$

with

$$X_S = (\omega)(0.1 \text{ H}) = 37.7\ \Omega$$

Thus,

$$\begin{aligned} X_L &= -40 \times 37.7 \\ &= -1{,}508 \end{aligned}$$

This means that the load reactive element should be a capacitor whose value is

$$C_L = \frac{-1}{X_L \omega} = 1.76\ \mu\text{F}$$

DRILL EXERCISES

5.12 If the transformer shown in Figure 5.45 is ideal, find the turns ratio, N, that will ensure maximum power transfer to the load. Assume that $Z_S = 1{,}800\ \Omega$ and $Z_L = 8\ \Omega$.

5.13 If the circuit of Drill Exercise 5.12 has $Z_L = (2 + j10)\ \Omega$ and the turns ratio of the transformer is $N = 5.4$, what should Z_S be in order to have maximum power transfer?

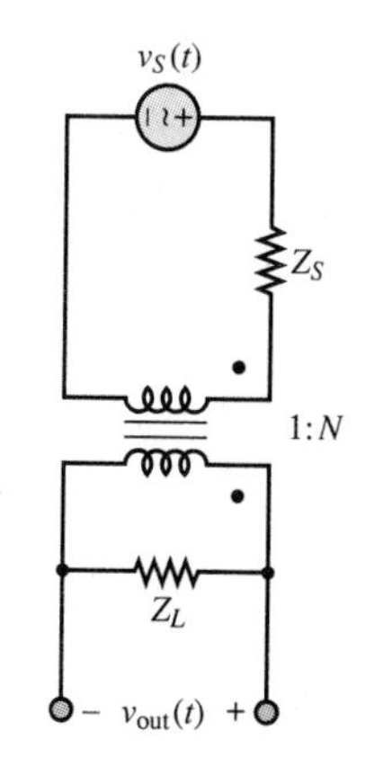

Figure 5.45

5.4 THREE-PHASE POWER

The material presented so far in this chapter has dealt exclusively with **single-phase AC power,** that is, with single sinusoidal sources. In fact, most of the AC power used today is generated and distributed as **three-phase power,** by means of an arrangement in which three sinusoidal voltages are generated out of phase with each other. The primary reason is efficiency: the weight of the conductors and other components in a three-phase system is much lower than in a single-phase system delivering the same amount of power. Further, while the power produced by a single-phase system has a pulsating nature (recall the results of Section 5.1), a three-phase system can deliver a steady, constant supply of power. For example, later in this section it will be shown that a three-phase generator producing three **balanced voltages**—that is, voltages of equal amplitude and frequency displaced in phase by 120°—has the property of delivering constant instantaneous power.

Another important advantage of three-phase power is that, as will be explained in Chapter 16, three-phase motors have a nonzero starting torque, unlike their single-phase counterpart. The change to three-phase AC power systems from the early DC system proposed by Edison was therefore due to a number of reasons: the efficiency resulting from transforming voltages up and down to minimize transmission losses over long distances; the ability to deliver constant power (an ability not shared by single- and two-phase AC systems); a more efficient use of conductors; and the ability to provide starting torque for industrial motors.

To begin the discussion of three-phase power, consider a three-phase source connected in the **wye** (or **Y**) **configuration,** as shown in Figure 5.46. Each of the

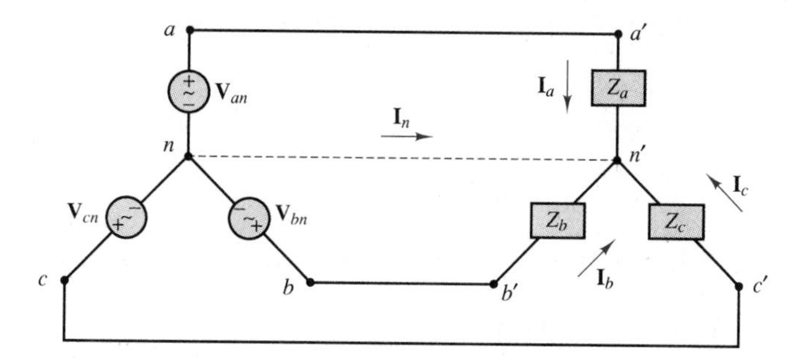

Figure 5.46 Balanced three-phase AC circuit

three voltages is 120° out of phase with the others, so that, using phasor notation, we may write:

$$\mathbf{V}_{an} = \tilde{V}_{an}\angle 0^\circ$$
$$\mathbf{V}_{bn} = \tilde{V}_{bn}\angle -120^\circ \quad (5.50)$$
$$\mathbf{V}_{cn} = \tilde{V}_{cn}\angle -240^\circ = \tilde{V}_{cn}\angle 120^\circ$$

where the quantities $\tilde{V}_{an}$, $\tilde{V}_{bn}$, and $\tilde{V}_{cn}$ are rms values and are equal to each other. To simplify the notation, it will be assumed from here on that

$$\tilde{V}_{an} = \tilde{V}_{bn} = \tilde{V}_{cn} = \tilde{V} \quad (5.51)$$

Chapter 16 will discuss how three-phase AC electric generators may be constructed to provide such balanced voltages. In the circuit of Figure 5.46, the resistive loads are also wye-connected and balanced (i.e., equal). The three AC sources are all connected together at a node called the *neutral node,* denoted by n. The voltages $\mathbf{V}_{an}$, $\mathbf{V}_{bn}$, and $\mathbf{V}_{cn}$ are called the **phase voltages** and form a balanced set in the sense that

$$\mathbf{V}_{an} + \mathbf{V}_{bn} + \mathbf{V}_{cn} = 0 \quad (5.52)$$

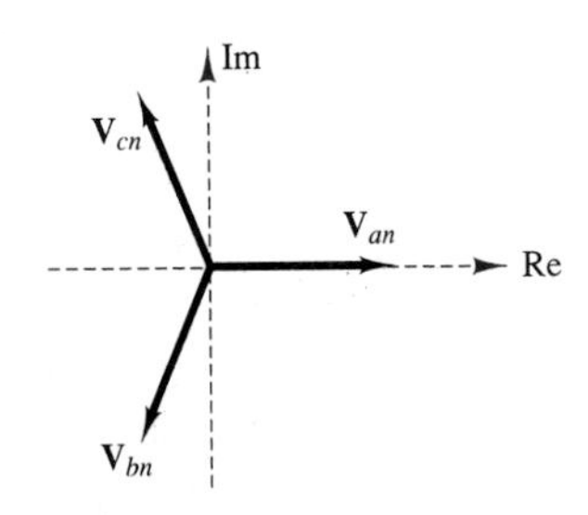

Figure 5.47 Positive, or *abc*, sequence for balanced three-phase voltages

This last statement is easily verified by sketching the phasor diagram. The sequence of phasor voltages shown in Figure 5.47 is usually referred to as the **positive** (or ***abc***) **sequence.**

Consider now the "lines" connecting each source to the load and observe that it is possible to also define **line voltages** (also called *line-to-line voltages*) by considering the voltages between the lines aa' and bb', aa' and cc', and bb' and cc'. Since the line voltage, say, between aa' and bb' is given by

$$\mathbf{V}_{ab} = \mathbf{V}_{an} + \mathbf{V}_{nb} = \mathbf{V}_{an} - \mathbf{V}_{bn} \quad (5.53)$$

the line voltages may be computed relative to the phase voltages as follows:

$$\mathbf{V}_{ab} = \tilde{V}\angle 0^\circ - \tilde{V}\angle -120^\circ = \sqrt{3}\tilde{V}\angle 30^\circ$$
$$\mathbf{V}_{bc} = \tilde{V}\angle -120^\circ - \tilde{V}\angle 120^\circ = \sqrt{3}\tilde{V}\angle -90^\circ \quad (5.54)$$
$$\mathbf{V}_{ca} = \tilde{V}\angle 120^\circ - \tilde{V}\angle 0^\circ = \sqrt{3}\tilde{V}\angle 150^\circ$$

It can be seen, then, that the magnitude of the line voltages is equal to $\sqrt{3}$ times the magnitude of the phase voltages. It is instructive, at least once, to point out that the circuit of Figure 5.46 can be redrawn to have the appearance of the circuit of Figure 5.48.

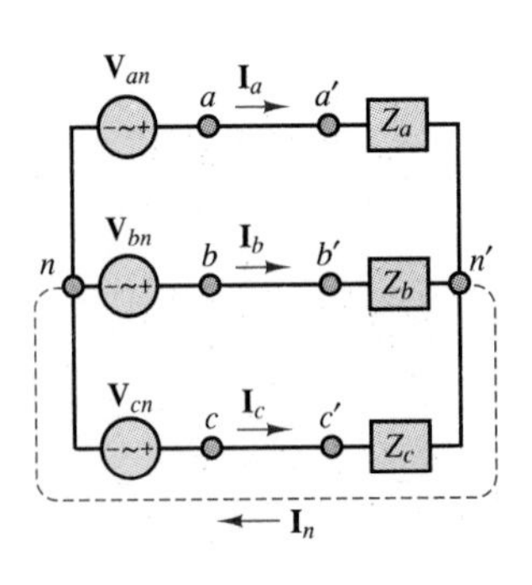

Figure 5.48 Balanced three-phase AC circuit (redrawn)

One of the important features of a balanced three-phase system is that it does not require a fourth wire (the neutral connection), since the current $\mathbf{I}_n$ is identically zero (for balanced load $Z_a = Z_b = Z_c = Z$). This can be shown by applying KCL at the neutral node n:

$$\mathbf{I}_n = (\mathbf{I}_a + \mathbf{I}_b + \mathbf{I}_c)$$
$$= \frac{1}{Z}(\mathbf{V}_{an} + \mathbf{V}_{bn} + \mathbf{V}_{cn}) \quad (5.55)$$
$$= 0$$

Another, more important characteristic of a balanced three-phase power system may be illustrated by simplifying the circuits of Figures 5.46 and 5.48 by replacing the balanced load impedances with three equal resistances, R. With this simplified configuration, one can show that the total power delivered to the balanced load by the three-phase generator is constant. This is an extremely important result, for a very practical reason: delivering power in a smooth fashion (as opposed to the pulsating nature of single-phase power) reduces the wear and stress on the generating equipment. Although we have not yet discussed the nature of the machines used to generate power, a useful analogy here is that of a single-cylinder engine versus a perfectly balanced V-8 engine. To show that the total power delivered by the three sources to a balanced resistive load is constant, consider the instantaneous power delivered by each source:

$$\begin{aligned} p_a(t) &= \frac{\tilde{V}^2}{R}(1 + \cos 2\omega t) \\ p_b(t) &= \frac{\tilde{V}^2}{R}[1 + \cos(2\omega t - 120°)] \\ p_c(t) &= \frac{\tilde{V}^2}{R}[1 + \cos(2\omega t + 120°)] \end{aligned} \qquad (5.56)$$

The total instantaneous load power is then given by the sum of the three contributions:

$$\begin{aligned} p(t) &= p_a(t) + p_b(t) + p_c(t) \\ &= \frac{3\tilde{V}^2}{R} + \frac{\tilde{V}^2}{R}[\cos 2\omega t + \cos(2\omega t - 120°) + \cos(2\omega t + 120°)] \qquad (5.57) \\ &= \frac{3\tilde{V}^2}{R} = \text{a constant!} \end{aligned}$$

You may wish to verify that the sum of the trigonometric terms inside the brackets is identically zero.

It is also possible to connect the three AC sources in a three-phase system in a so-called **delta** (or Δ) **connection,** although in practice this configuration is rarely used. Figure 5.49 depicts a set of three delta-connected generators.

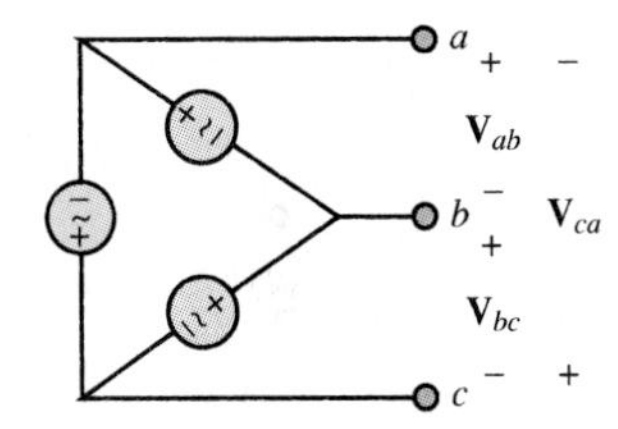

Figure 5.49 Delta-connected generators

Example 5.16

This example illustrates the effect of line impedance in a balanced wye-wye system, as well as a solution method that involves computations for only a single phase. Compute the power delivered to the load shown by the three-phase generator in the circuit shown in Figure 5.50, where the phase voltage is 480 V. Assume

$$Z_y = 2 + j4 = 4.47\angle 63.43°\ \Omega$$

$$\mathbf{V}_{an} = 480\angle 0° \qquad \mathbf{V}_{bn} = 480\angle -120° \qquad \mathbf{V}_{cn} = 480\angle 120°$$

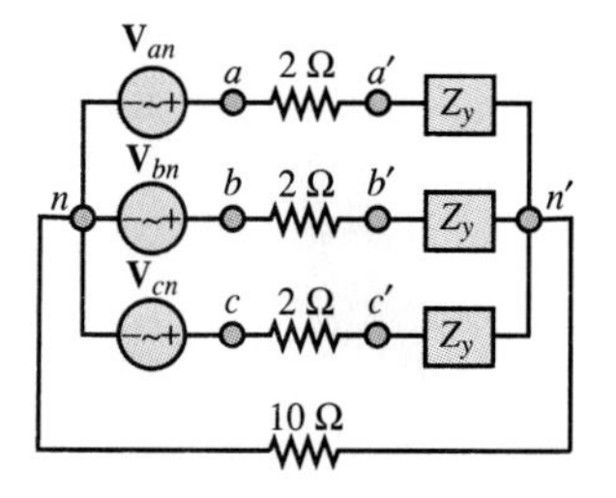

Figure 5.50

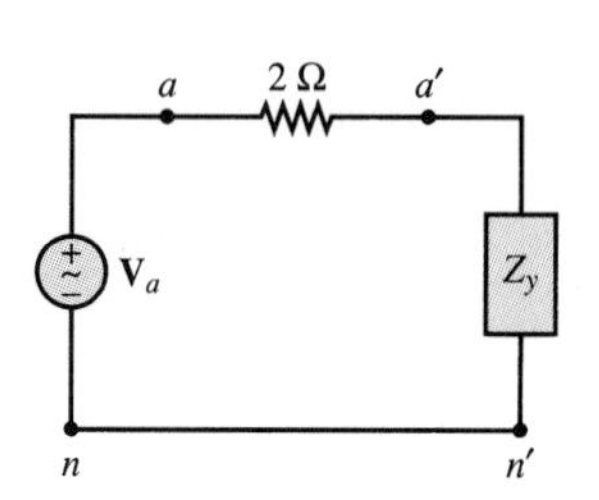

Figure 5.51 One phase of the three-phase circuit

Solution:

In approaching the solution of this problem, we note that, since the circuit is perfectly balanced, the current through the 10-Ω resistor is zero, and therefore the voltage across it is zero. Further, by virtue of the symmetry of the circuit, it is possible to solve the problem by considering just one phase, as shown in Figure 5.51.

Note than when we redraw the a phase of the circuit, we connect n' and n together. This is done to show that n and n' are at the same voltage for a balanced wye-connected load. The result is a simple series circuit representing one phase of the three-phase circuit.

The average power in phase a is given by

$$P_a = \tilde{I}^2 R_L$$

where the rms current $\tilde{I}$ is

$$|\mathbf{I}| = \left|\frac{\mathbf{V}_a}{Z_y + 2}\right| = \left|\frac{480\angle 0^\circ}{2 + 2 + j4}\right| = \left|\frac{480\angle 0^\circ}{5.66\angle 45^\circ}\right| = |84.85\angle -45^\circ|$$

so that

$$P_a = (84.85)^2 \times 2 = 14{,}400 \text{ W}$$

The calculations are similar for the b and c phases, so that

$$\begin{aligned} P_{\text{total}} &= P_a + P_b + P_c \\ &= 3P_a \\ &= 43.2 \text{ kW} \end{aligned}$$

Balanced Wye Loads

In the previous section we performed some power computations for a purely resistive balanced wye load. We shall now generalize those results for an arbitrary balanced complex load. Consider again the circuit of Figure 5.46, where now the balanced load consists of the three complex impedances

$$Z_a = Z_b = Z_c = Z_y = |Z_y|\angle\theta \tag{5.58}$$

From the diagram of Figure 5.46, it can be verified that each impedance sees the corresponding phase voltage across itself; thus, since the currents $\mathbf{I}_a$, $\mathbf{I}_b$, and $\mathbf{I}_c$ have the same rms value, $\tilde{I}$, the phase angles of the currents will differ by $\pm 120^\circ$. It is therefore possible to compute the power for each phase by considering the phase voltage (equal to the load voltage) for each impedance, and the associated line current. Let us denote the complex power for each phase by S:

$$S = \mathbf{V} \cdot \mathbf{I}^* \tag{5.59}$$

so that

$$\begin{aligned} S &= P + jQ \\ &= \tilde{V}\tilde{I}\cos\theta + j\tilde{V}\tilde{I}\sin\theta \end{aligned} \tag{5.60}$$

where $\tilde{V}$ and $\tilde{I}$ denote, once again, the rms values of each phase voltage and line current. Consequently, the total real power delivered to the balanced wye load is $3P$, and the total reactive power is $3Q$. Thus, the total complex power, S_T, is given by

$$\begin{aligned} S_T &= P_T + jQ_T = 3P + j3Q \\ &= \sqrt{(3P)^2 + (3Q)^2}\angle\theta \end{aligned} \tag{5.61}$$

and the apparent power is

$$\begin{aligned} |S_T| &= 3\sqrt{(VI)^2\cos^2\theta + (VI)^2\sin^2\theta} \\ &= 3VI \end{aligned}$$

and the total real and reactive power may be expressed in terms of the apparent power:

$$\begin{aligned} P_T &= |S_T|\cos\theta \\ Q_T &= |S_T|\sin\theta \end{aligned} \tag{5.62}$$

Balanced Delta Loads

In addition to a wye connection, it is also possible to connect a balanced load in the delta configuration. A wye-connected generator and a delta-connected load are shown in Figure 5.52.

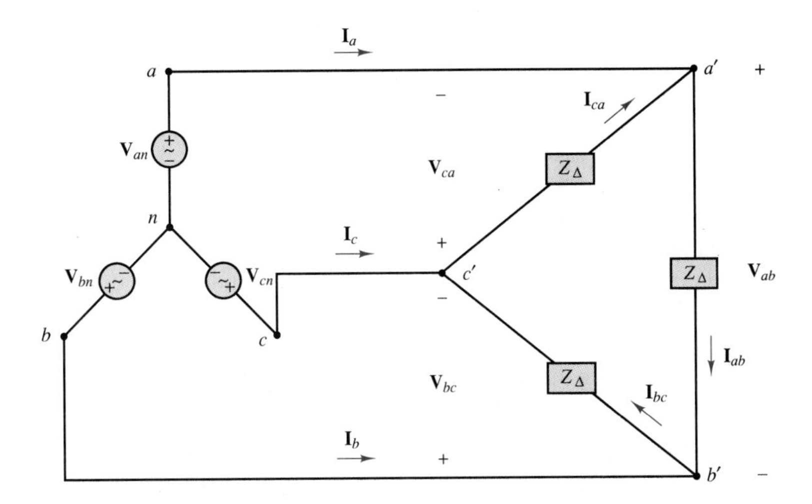

Figure 5.52 Balanced wye generators with balanced delta load

It should be noted immediately that now the corresponding line voltage (not phase voltage) appears across each impedance. For example, the voltage across $Z_{c'a'}$ is $\mathbf{V}_{ca}$. Thus, the three load currents are given by the following expressions:

$$
\begin{aligned}
\mathbf{I}_{ab} &= \frac{\mathbf{V}_{ab}}{Z_\Delta} = \frac{\sqrt{3}V\angle 30^\circ}{|Z_\Delta|\angle\theta} \\
\mathbf{I}_{bc} &= \frac{\mathbf{V}_{bc}}{Z_\Delta} = \frac{\sqrt{3}V\angle -90^\circ}{|Z_\Delta|\angle\theta} \\
\mathbf{I}_{ca} &= \frac{\mathbf{V}_{ca}}{Z_\Delta} = \frac{\sqrt{3}V\angle 150^\circ}{|Z_\Delta|\angle\theta}
\end{aligned} \tag{5.63}
$$

To understand the relationship between delta-connected and wye-connected loads, it is reasonable to ask the question, For what value of Z_Δ would a delta-connected load draw the same amount of current as a wye-connected load with impedance Z_y for a given source voltage? This is equivalent to asking what value of Z_Δ would make the line currents the same in both circuits (compare Figure 5.48 with Figure 5.52).

The line current drawn, say, in phase a by a wye-connected load is

$$
(\mathbf{I}_{an})_y = \frac{\mathbf{V}_{an}}{Z} = \frac{\tilde{V}}{|Z_y|}\angle -\theta \tag{5.64}
$$

while that drawn by the delta-connected load is

$$
\begin{aligned}
(\mathbf{I}_a)_\Delta &= \mathbf{I}_{ab} - \mathbf{I}_{ca} \\
&= \frac{\mathbf{V}_{ab}}{Z_\Delta} - \frac{\mathbf{V}_{ca}}{Z_\Delta} \\
&= \frac{1}{Z_\Delta}(\mathbf{V}_{an} - \mathbf{V}_{bn} - \mathbf{V}_{cn} + \mathbf{V}_{an}) \\
&= \frac{1}{Z_\Delta}(2\mathbf{V}_{an} - \mathbf{V}_{bn} - \mathbf{V}_{cn}) \\
&= \frac{3\mathbf{V}_{an}}{Z_\Delta} = \frac{3\mathbf{V}}{|Z_\Delta|}\angle -\theta
\end{aligned} \tag{5.65}
$$

One can readily verify that the two currents $(\mathbf{I}_a)_\Delta$ and $(\mathbf{I}_a)_y$ will be equal if the magnitude of the delta-connected impedance is 3 times larger than Z_y:

$$
Z_\Delta = 3Z_y \tag{5.66}
$$

This result also implies that a delta load will necessarily draw 3 times as much current (and therefore absorb 3 times as much power) as a wye load with the same branch impedance.

Example 5.17

The generator of Example 5.16 is now required to deliver power to an additional delta load, connected in parallel with the wye load of Example 5.16. If

$$Z_\Delta = 5 - j2$$

what is the total power absorbed by the two parallel loads?

Solution:

The circuit of Example 5.16 is sketched in Figure 5.53 with the additional delta load. Assume that

$$Z_y = 2 + j4\ \Omega$$

and

$$Z_\Delta = 5 - j2\ \Omega$$

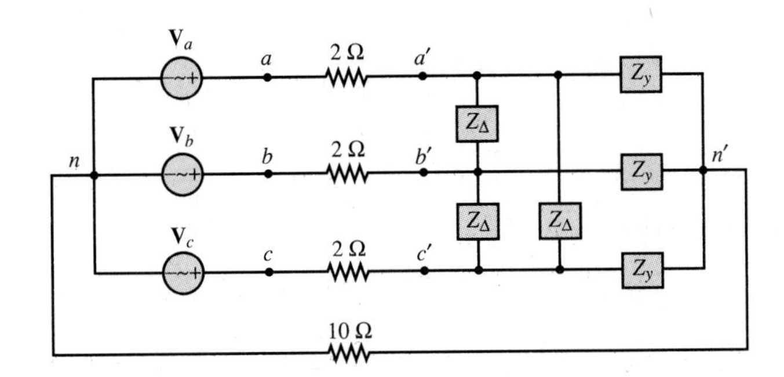

Figure 5.53 AC circuit with delta and wye loads

One method of solving this problem is to convert the balanced delta load to an equivalent wye load, using the method outlined in the preceding paragraphs. Figure 5.54 illustrates how this conversion can take place.

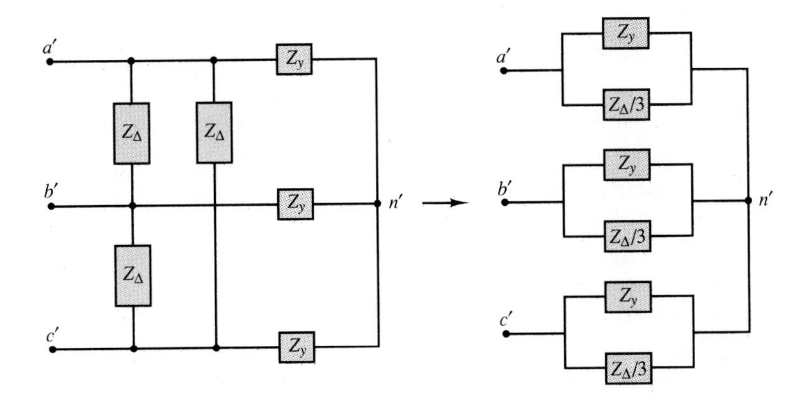

Figure 5.54 Conversion of delta load to equivalent wye load

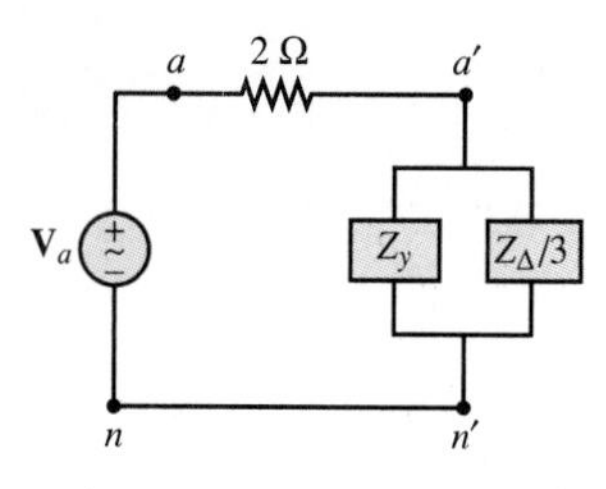

Figure 5.55 Per-phase circuit

Now we need to perform the calculations for only one phase, recognizing that the wye load appears in parallel with the delta-to-wye transformed load, as shown in Figure 5.55. If we compute the effective impedance at each phase, we have

$$Z_y \parallel \frac{Z_\Delta}{3} = (2 + j4) \parallel \left(\frac{5 - j2}{3}\right)$$

$$= \frac{(2 + j4)(\frac{5}{3} - j\frac{2}{3})}{(2 + \frac{5}{3}) + j(4 - \frac{2}{3})} = \frac{10 + 8 - j4 + j20}{11 + j10}$$

$$= \frac{18 + j16}{11 + j10} = \frac{24.08\angle 41.6^\circ}{14.87\angle 42.3^\circ} = 1.62\angle -0.64^\circ$$

and therefore a current

$$\mathbf{I}_a = \frac{480\angle 0^\circ}{2 + 1.62\angle -0.64^\circ} = \frac{480\angle 0^\circ}{2 + 1.62 - j0.018}$$

$$= \frac{480\angle 0^\circ}{3.62\angle -0.29^\circ} = 132.6\angle 0.29^\circ$$

yielding the single-phase power

$$P_a = \tilde{I}_a^2 R_L = (132.6)^2 \times 1.62 = 28{,}482 \text{ W}$$

The total power will then be 3 times P_a, or

$$P_{\text{total}} = 85.45 \text{ kW}$$

DRILL EXERCISES

5.14 Find the power lost in the lines in the circuit of Example 5.16.

5.15 Draw the phasor diagram and power triangle for a single phase and compute the power delivered to the balanced load of Example 5.16 if the lines have zero resistance and $Z_L = 1 + j3\ \Omega$.

5.16 Show that the voltage across each branch of the balanced wye load in Drill Exercise 5.15 is equal to the corresponding phase voltage (e.g., the voltage across Z_a is $\mathbf{V}_a$).

5.17 Prove that the sum of the instantaneous powers absorbed by the three branches in a balanced wye-connected load is constant and equal to $3\tilde{V}\tilde{I}\cos\theta$.

5.18 Derive an expression for the rms line current of a delta load in terms of the rms line current for a wye load with the same branch impedances (i.e., $Z_y = Z_\Delta$) and same source voltage. Assume $Z_S = 0$.

5.19 The equivalent wye load of Example 5.17 is connected in a delta configuration. Compute the line currents.

5.5 RESIDENTIAL WIRING; GROUNDING AND SAFETY

Common residential electric power service consists of a three-wire AC system supplied by the local power company. The three wires originate from a utility pole and consist of a neutral wire, which is connected to earth ground, and two "hot" wires. Each of the hot lines supplies 120 V rms to the residential circuits; the two lines are 180° out of phase, for reasons that will become apparent during the course of this discussion. The phasor line voltages, shown in Figure 5.56, are usually referred to by means of a subscript convention derived from the color of the insulation on the different wires: W for white (neutral), B for black (hot), and R for red (hot). This convention is adhered to uniformly.

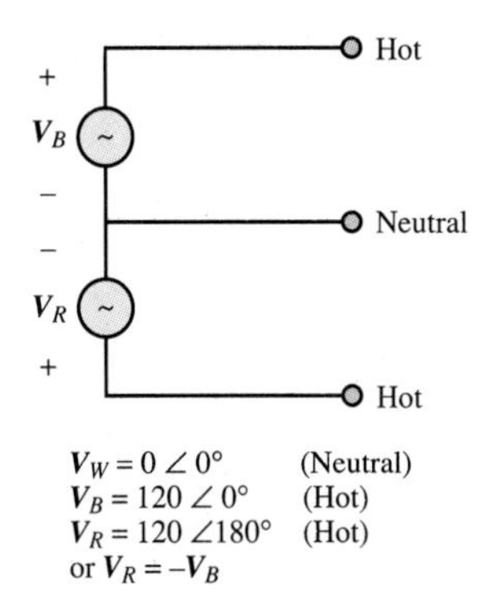

Figure 5.56 Line voltage convention for residential circuits

The voltages across the hot lines are given by:

$$\mathbf{V}_B - \mathbf{V}_R = \mathbf{V}_{BR} = \mathbf{V}_B - (-\mathbf{V}_B) = 2\mathbf{V}_B = 240\angle 0° \qquad (5.67)$$

Thus, the voltage between the hot wires is actually 240 V rms. Appliances such as electric stoves, air conditioners, and heaters are powered by the 240-V rms arrangement. On the other hand, lighting and all of the electric outlets in the house used for small appliances are powered by a single 120-V rms line.

The use of 240-V rms service for appliances that require a substantial amount of power to operate is dictated by power transfer considerations. Consider the two circuits shown in Figure 5.57. In delivering the necessary power to a load, a lower line loss will be incurred with the 240-V rms wiring, since the power loss in the lines (the **I^2R loss,** as it is commonly referred to) is directly related to the current required by the load. In an effort to minimize line losses, the size of the wires is increased for the lower-voltage case. This typically reduces the wire resistance by a factor of 2. In the top circuit, assuming $R_S/2 = 0.01\ \Omega$, the current required by the 10-kW load is approximately 83.3 A, while in the bottom circuit, with $R_S = 0.02\ \Omega$, it is approximately half as much (41.7 A). (You should be able to verify that the approximate I^2R losses are 69.4 W in the top circuit and 34.7 W in the bottom circuit.) Limiting the I^2R losses is important from the viewpoint of efficiency, besides reducing the amount of heat generated in the wiring for safety considerations. Figure 5.58 shows some typical wiring configurations for a home. Note that several circuits are wired and fused separately.

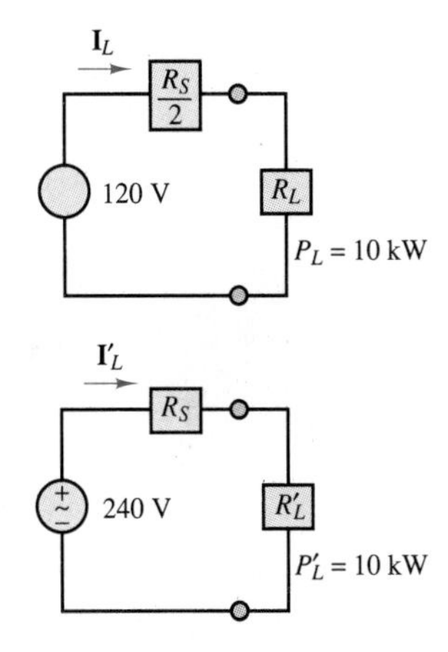

Figure 5.57 Line losses in 120-VAC and 240-VAC circuits

Today, most homes have three wire connections to their outlets. The outlets appear as sketched in Figure 5.59. Then why are both the ground and neutral connections needed in an outlet? The answer to this question is *safety:* the ground connection is used to connect the chassis of the appliance to earth ground. Without this provision, the appliance chassis could be at any potential with respect to ground, possibly even at the hot wire's potential if a segment of the hot wire were to lose some insulation and come in contact with the inside of the chassis! Poorly grounded appliances can thus be a significant hazard. Figure 5.60 illustrates schematically how, even though the chassis is intended to be insulated from the electric circuit, an unintended connection (represented by the dashed line) may

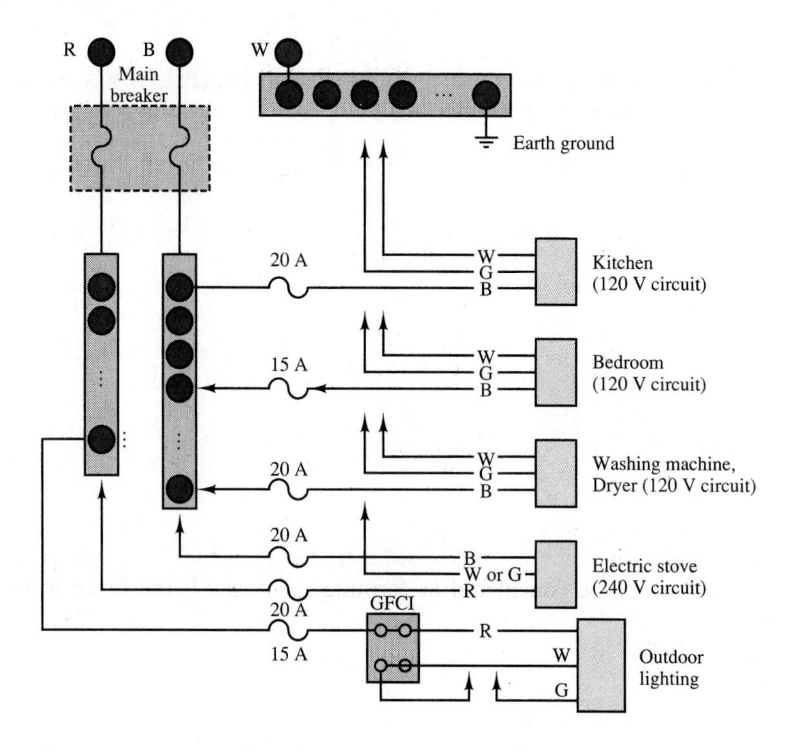

Figure 5.58 A typical residential wiring arrangement

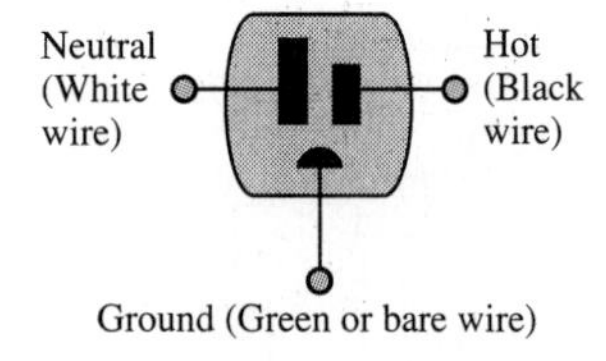

Figure 5.59 A three-wire outlet

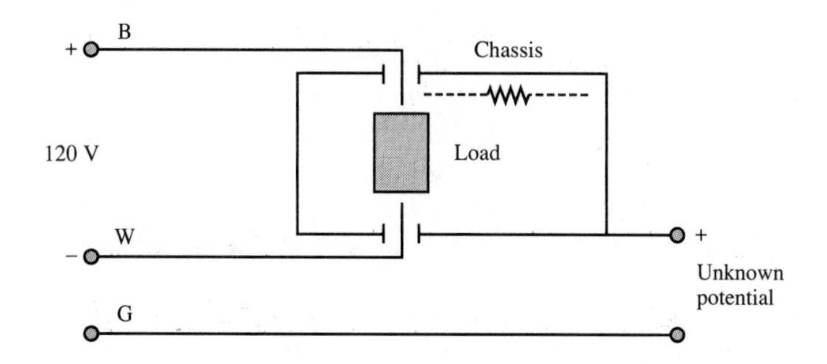

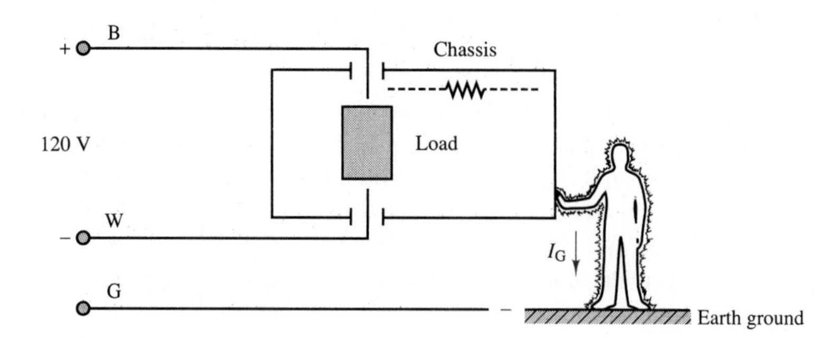

Figure 5.60 Unintended connection

occur, for example, because of corrosion or a loose mechanical connection. A path to ground might be provided by the body of a person touching the chassis with a hand. In the figure, such an undesired ground loop current is indicated by I_G. In this case, the ground current I_G would flow directly through the body to ground and could be harmful.

In some cases the danger posed by such undesired ground loops can be great, leading to death by electric shock. Figure 5.61 describes the effects of electric currents on an average male when the point of contact is dry skin. Particularly hazardous conditions are liable to occur whenever the natural resistance to current flow provided by the skin breaks down, as would happen in the presence of water. The **ground fault circuit interrupter,** labeled **GFCI** in Figure 5.58, is a special safety circuit used primarily with outdoor circuits and in bathrooms, where the risk of death by electric shock is greatest. Its application is best described by an example.

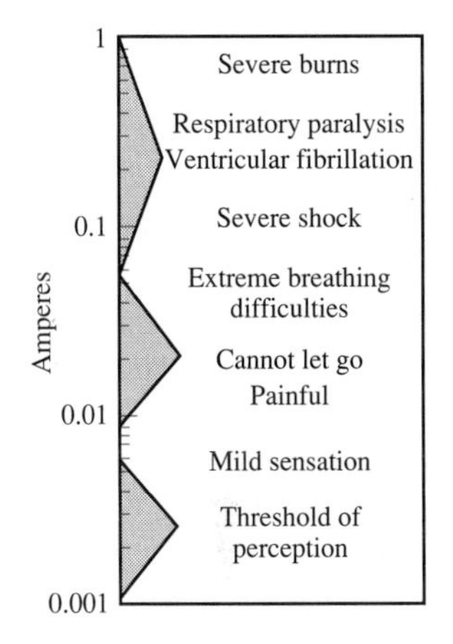

Figure 5.61 Physiological effects of electric currents

Consider the case of an outdoor pool surrounded by a metal fence, which uses an existing light pole for a post, as shown in Figure 5.62. The light pole and the metal fence can be considered as forming a chassis. If the fence were not properly grounded all the way around the pool and if the light fixture were poorly insulated from the pole, a path to ground could easily be created by an unaware swimmer reaching, say, for the metal gate. A GFCI provides protection from potentially lethal ground loops, such as this one, by sensing both the hot-wire (B) and the neutral (W) currents. If the difference between the hot-wire current, I_B, and the neutral current, I_W, is more than a few milliamperes, then the GFCI disconnects the circuit nearly instantaneously. Any significant difference between the hot and neutral (return-path) currents means that a second path to ground has been created

Figure 5.62 Outdoor pool

(by the unfortunate swimmer, in this example) and a potentially dangerous condition has arisen. Figure 5.63 illustrates the idea. GFCIs are typically resettable circuit breakers, so that one does not need to replace a fuse every time the GFCI circuit is enabled.

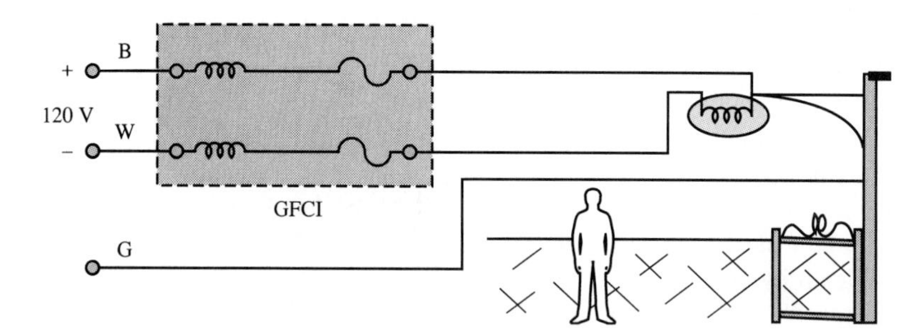

Figure 5.63 Use of a GFCI in a potentially hazardous setting

DRILL EXERCISE

5.20 Use the circuit of Figure 5.57 to show that the I^2R losses will be higher for a 120-V service appliance than a 240-V service appliance if both have the same power usage rating.

5.6 GENERATION AND DISTRIBUTION OF AC POWER

We now conclude the discussion of power systems with a brief description of the various elements of a power system. Electric power originates from a variety of sources; in Chapter 16, electric generators will be introduced as a means of producing electric power from a variety of energy-conversion processes. In general, electric power may be obtained from hydroelectric, thermoelectric, geothermal, wind, solar, and nuclear sources. The choice of a given source is typically dictated by the power requirement for the given application, and by economic and environmental factors. In this section, the structure of an AC power network, from the power-generating station to the residential circuits discussed in the previous section, is briefly outlined.

A typical generator will produce electric power at 18 kV, as shown in the diagram of Figure 5.64. To minimize losses along the conductors, the output of the generators is processed through a step-up transformer to achieve line voltages of hundreds of kilovolts (345 kV, in Figure 5.64). Without this transformation, the majority of the power generated would be lost in the **transmission lines** that carry the electric current from the power station.

The local electric company operates a power-generating plant that is capable of supplying several hundred megavolt-amperes (MVA) on a three-phase

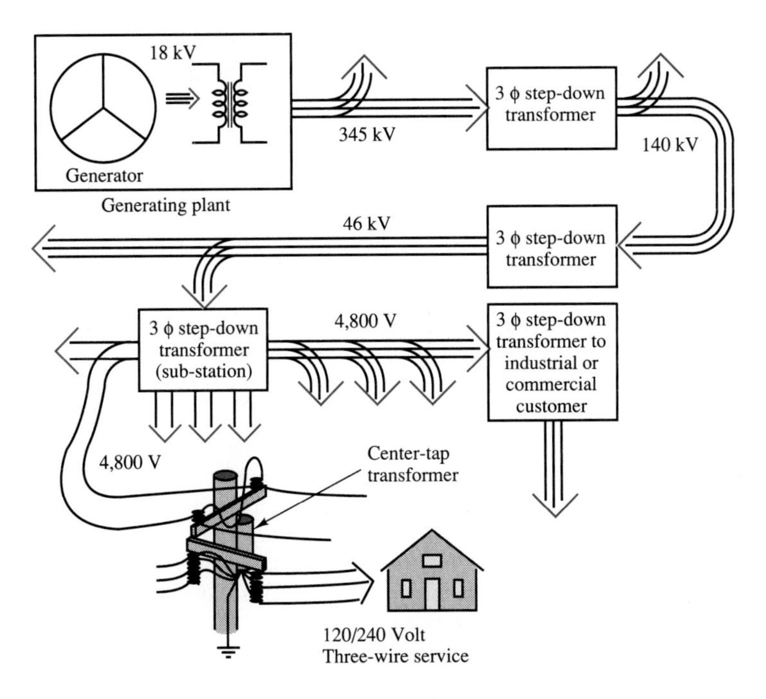

Figure 5.64 Structure of an AC power distribution network

basis. For this reason, the power company uses a three-phase step-up transformer at the generation plant to increase the line voltage to around 345 kV. One can immediately see that at the rated power of the generator (in MVA) there will be a significant reduction of current beyond the step-up transformer.

Beyond the generation plant, an electric power network distributes energy to several **substations.** This network is usually referred to as the **power grid.** At the substations, the voltage is stepped down to a lower level (10 to 150 kV, typically). Some very large loads (for example, an industrial plant) may be served directly from the power grid, although most loads are supplied by individual substations in the power grid. At the local substations (one of which you may have seen in your own neighborhood), the voltage is stepped down further by a three-phase step-down transformer to 4,800 V. These substations distribute the energy to residential and industrial customers. To further reduce the line voltage to levels that are safe for residential use, step-down transformers are mounted on utility poles. These drop the voltage to the 120/240-V three-wire single-phase residential service discussed in the previous section. Industrial and commercial customers receive 460- and/or 208-V three-phase service.

CONCLUSION

This chapter introduced the essential elements leading to the analysis of AC power systems. Single-phase AC power, ideal transformers, and three-phase power were discussed.

A brief review of residential circuit wiring and safety, and a description of an electric distribution network, were also given to underscore the importance of these concepts in electric power.

- The power dissipated by a load in an AC circuit consists of the sum of an average and a fluctuating component. In practice, the average power is the quantity of interest.
- AC power can best be analyzed with the aid of complex notation. Complex power is defined as the product of the phasor load voltage and the complex conjugate of the phasor load current. Complex power consists of the sum of a real component (the average, or real, power) and an imaginary component (reactive power). Real power corresponds to the electric power for which a user is billed by a utility company; reactive power corresponds to energy storage and cannot be directly used.
- Although reactive power is of no practical use, it does cause an undesirable increase in the current that must be generated by the electric company, resulting in additional line losses. Thus, it is customary to try to reduce reactive power. A measure of the presence of reactive power at a load is the power factor, equal to the cosine of the angle of the load impedance. By adding a suitable reactance to the load, it is possible to attain power factors close to ideal (unity). This procedure is called *power factor correction.*
- Electric power is most commonly generated in three-phase form, for reasons of efficiency. Three-phase power entails the generation of three 120° out-of-phase AC voltages of equal amplitude, so that the instantaneous power is actually constant. Three-phase sources and loads can be configured in either wye or delta configurations; of these, the wye form is more common. The calculation of currents, voltages, and power in three-phase circuits is greatly simplified if one uses per-phase calculations.

EIT REVIEW: AC CIRCUITS

One of the most important applications of electricity in the engineering profession is the analysis of AC electric power systems. This topic is a very important part of the EIT exam. The first part of this review section is devoted to the basic concepts of rms values, impedance, and phasors.

DRILL EXERCISES

5.21 A voltage sine wave of peak value 100 V is in phase with a current sine wave of peak value 4 A. When the phase angle is 60° later than a time at which the voltage and the current are both zero, the instantaneous power is most nearly

a. 250 W
b. 200 W
c. 400 W
d. 150 W
e. 100 W

Solution:

As discussed in Section 5.1, the instantaneous AC power $p(t)$ is

$$p(t) = \frac{VI}{2}\cos\theta + \frac{VI}{2}\cos(2\omega t + \theta_V + \theta_I)$$

In this problem, when the phase angle is 60° later than a "zero crossing," we have $\theta_V = \theta_I = 0$, $\theta = \theta_V - \theta_I = 0$, $2\omega t = 120°$. Thus, we can compute the power at this instant as

$$p = \frac{\left(\frac{100}{\sqrt{2}}\frac{4}{\sqrt{2}}\right)}{2} + \frac{\left(\frac{100}{\sqrt{2}}\frac{4}{\sqrt{2}}\right)}{2}\cos 120° = 250\ \text{W}$$

The correct answer is a.

5.22 A sinusoidal voltage whose amplitude is $20\sqrt{2}$ V is applied to a 5-Ω resistor. The root-mean-square value of the current is

a. 5.66 A
b. 4 A
c. 7.07 A
d. 8 A
e. 10 A

Solution:

From Section 4.2, we know that

$$V_{\text{rms}} = \frac{V}{\sqrt{2}} = \frac{20\sqrt{2}}{\sqrt{2}} = 20\ \text{V}$$

Thus, $I_{\text{rms}} = \frac{20}{5} = 4$ A. Therefore, b is the correct answer.

5.23 The magnitude of the steady-state root-mean-square voltage across the capacitor in the circuit of Figure 5.65 is

a. 30 V
b. 15 V
c. 10 V
d. 45 V
e. 60 V

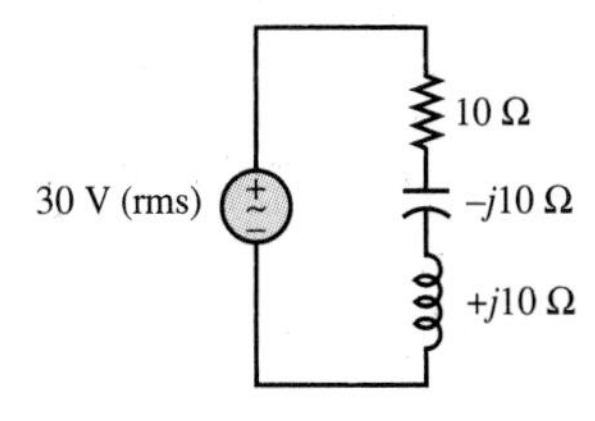

Figure 5.65

Solution:

This problem requires the use of impedances (Section 4.4). Using the voltage divider rule for impedances, we write the voltage across the capacitor as

$$\mathbf{V} = 30\angle 0° \times \frac{-j10}{10 - j10 + j10}$$
$$= 30\angle 0° \times (-j1) = 30\angle 0° \times 1\angle -90° = 30\angle -90°$$

Thus, the rms amplitude of the voltage across the capacitor is 30 V, and a is the correct answer. Note the importance of the phase angle in this kind of problem.

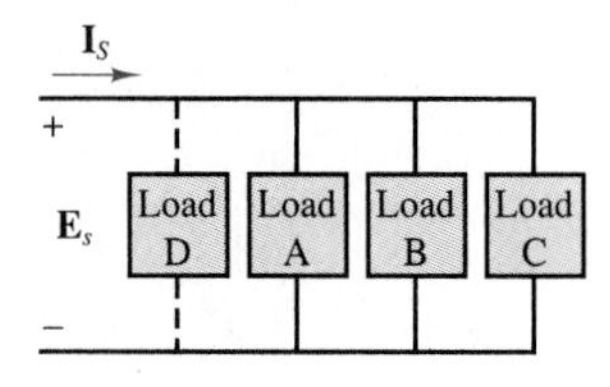

Figure 5.66

The next set of questions (Drill Exercises 5.24 to 5.28) pertain to single-phase AC power calculations and refer to the single-phase electrical network shown in Figure 5.66. In this figure, $\mathbf{E}_S = 480\angle 0°$ V; $\mathbf{I}_S = 100\angle -15°$ A; $\omega = 120\pi$ rad/s. Further, load A is a bank of single-phase induction machines. The bank has an efficiency (η) of 80 percent, a power factor of 0.70 lagging, and a load of 20 hp. Load B is a bank of overexcited single-phase synchronous machines. The machines draw 15 kVA and the load current leads the line voltage by 30 degrees. Load C is a lighting (resistive) load and absorbs 10 kW. Load D is a proposed single-phase capacitor that will correct the source power factor to unity. This material is covered in Sections 5.1 and 5.2.

DRILL EXERCISES

5.24 The root-mean-square magnitude of load A current, I_A, is most nearly

a. 44.4 A
b. 31.08 A
c. 60 A
d. 38.85 A
e. 55.5 A

Solution:

The output power P_O of the single-phase induction motor is: $P_O = 20 \times 746 = 14{,}920$ W. The input electric power P_{in} is:

$$P_{\text{in}} = \frac{P_O}{\eta} = \frac{14{,}920}{0.80} = 18{,}650 \text{ W}$$

P_{in} can be expressed as:

$$P_{\text{in}} = E_S I_A \cos\theta_A$$

Therefore, the rms magnitude of the current $\mathbf{I}_A$ is found as

$$I_A = \frac{P_{\text{in}}}{E_S \cos\theta_A} = \frac{18{,}650}{480 \times 0.70} = 55.5015 \approx 55.5 \text{ A}$$

Thus, the correct answer is e.

5.25 The phase angle of $\mathbf{I}_A$ with respect to the line voltage $\mathbf{E}_S$ is most nearly

a. 36.87°
b. 60°
c. 45.6°
d. 30°
e. 48°

Solution:

The phase angle between $\mathbf{I}_A$ and $\mathbf{E}_S$ is:

$$\theta = \cos^{-1} 0.70 = 45.57° \approx 45.6°$$

The correct answer is c.

5.26 The power absorbed by synchronous machines is most nearly

a. 20,000 W
b. 7,500 W
c. 13,000 W
d. 12,990 W
e. 15,000 W

Solution:

The apparent power, S, is known to be 15 kVA, and θ is 30°. From the power triangle, we have

$$P = S\cos\theta$$

Therefore, the power drawn by the bank of synchronous motors is:

$$P = 15{,}000 \times \cos 30° = 12{,}990.38 \approx 12.99 \text{ kW}$$

The answer is d.

5.27 The power factor of the system before load D is installed is most nearly

a. 0.70 lagging
b. 0.866 leading
c. 0.866 lagging
d. 0.966 leading
e. 0.966 lagging

Solution:

From the expression for the current $\mathbf{I}_S$, we have

$$\text{pf} = \cos\theta = \cos(0° - (-15°)) = \cos 15° = 0.966 \text{ lagging}$$

The correct answer is e.

5.28 The capacitance of the capacitor that will give a unity power factor of the system is most nearly

a. 219 μF
b. 187 μF
c. 132.7 μF
d. 240 μF
e. 132.7 pF

Solution:

The reactive power Q_A in load A is:

$$Q_A = P_A \times \tan\theta_A$$
$$\theta_A = \cos^{-1} 0.70 = 45.57°$$

Therefore,

$$Q_A = 18{,}650 \times \tan 45.57° = 19{,}024.82 \text{ VAR}$$

The total reactive power Q_B in load B is:

$$Q_B = S \times \sin\theta_B = 15{,}000 \times \sin(-30°) = -7{,}500 \text{ VAR}$$

The total reactive power Q is:

$$Q = Q_A + Q_B = 19{,}024.82 - 7{,}500 = 11{,}524.82 \text{ VAR}$$

To cancel this reactive power, we set

$$Q_C = -Q = -11{,}524.82 \text{ VAR}$$

and

$$Q_C = -\frac{E_S^2}{X_C} \text{ and } X_C = -\frac{1}{\omega C}$$

Therefore, the capacitance required to obtain a power factor of unity is:

$$C = -\frac{Q_C}{\omega E_S^2} = \frac{11{,}524.82}{120\pi \times 480^2} = 132.7 \ \mu\text{F}$$

The correct answer is c.

EIT REVIEW: THREE-PHASE CIRCUITS

Another aspect of circuit analysis that plays a prominent role in engineering practice is the use of three-phase AC circuits for power distribution and industrial applications. Three-phase power is therefore an important section of the EIT exam. This material is covered primarily in Section 5.4 of the text. The following sample problems are representative of the type of questions that might appear in the exam.

DRILL EXERCISES

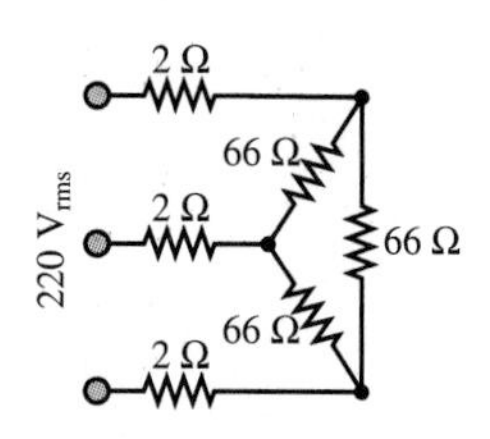

Figure 5.67

5.29 A 3-phase circuit is shown in Figure 5.67. Load resistors (66 Ω) are connected in delta and supplied by a 220-volt balanced three-phase source through three lines of 2-ohm resistance. The magnitude of the root-mean-square, line-to-line voltage across each 66-ohm resistor is most nearly

a. 198 V
b. 110 V
c. 201 V
d. 220 V
e. 120 V

Solution:

Since the load in this problem is Δ-connected, it must first be converted to an equivalent Y-form. The phase impedance of the Δ-connected load is 66 Ω, so the equivalent phase impedance of the corresponding Y-form is:

$$Z_Y = \frac{Z_\Delta}{3} = \frac{66}{3} = 22 \ \Omega$$

The phase voltage is:

$$\mathbf{V}_\phi = \frac{208}{\sqrt{3}}\angle 0^\circ \text{ V}$$

The per-phase equivalent circuit of this problem is shown in Figure 5.68. The load voltage is obtained by the voltage divider rule:

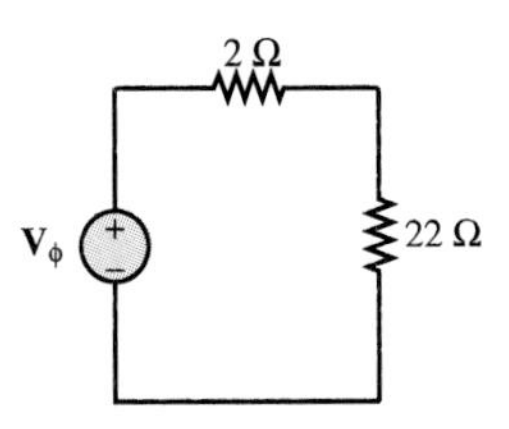

Figure 5.68

$$\mathbf{V}_L = \frac{22}{2+22} \times \mathbf{V}_\phi = \frac{208 \times 22}{\sqrt{3} \times 24} = \frac{201.67}{\sqrt{3}}\angle 0^\circ \text{ V}$$

The rms line-to-line voltage across the 66 Ω resistor therefore is:

$$\mathbf{V}_{66\Omega} = \sqrt{3}\mathbf{V}_L = 201.67\angle 0^\circ \text{ V}$$

The correct answer is c.

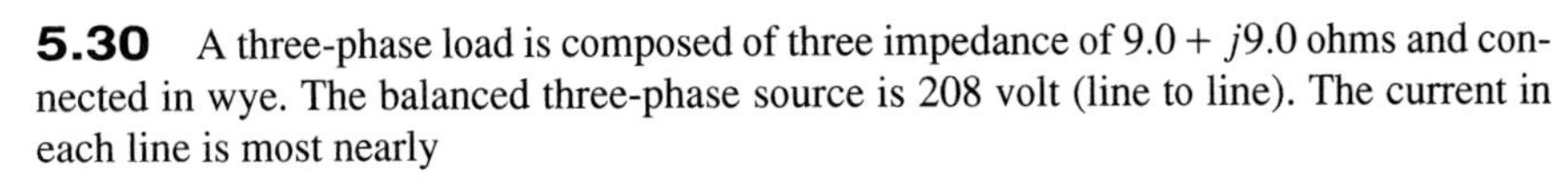

5.30 A three-phase load is composed of three impedance of 9.0 + j9.0 ohms and connected in wye. The balanced three-phase source is 208 volt (line to line). The current in each line is most nearly

a. 40 A
b. 16.3 A
c. 13.3 A
d. 9 A
e. 6 A

Solution:

The phase voltage is $\frac{208}{\sqrt{3}}$. Therefore, the magnitude of phase current I_{an} is

$$I_{an} = \frac{\left(\frac{208}{\sqrt{3}}\right)}{9+j9} = \frac{208}{\sqrt{3}}\frac{1}{12.73} = 9.43 \text{ A}$$

The line current is equal to the phase current in a Y-connected system. The answer is d.

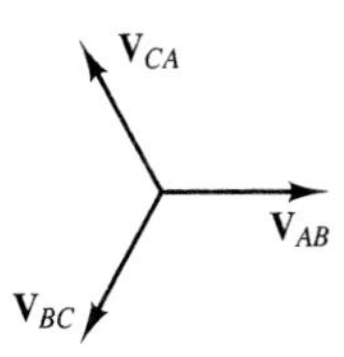

Figure 5.69

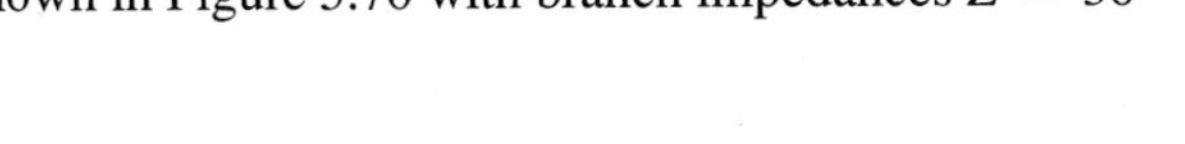

The next four questions refer to a three-phase system with line-to-line voltage of 220 V rms, with ABC phase sequence, and with phase reference V_{AB} shown in the phase diagram of Figure 5.69. The load is a balanced delta connection, shown in Figure 5.70 with branch impedances $Z = 30 - j40$ ohms, $j = \sqrt{-1}$.

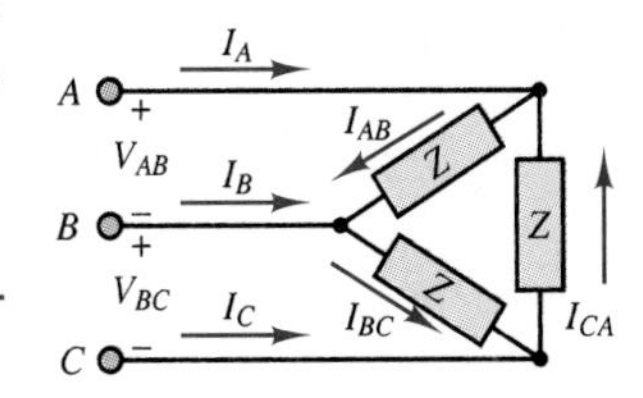

Figure 5.70

DRILL EXERCISES

5.31 The phase current is most nearly

a. $4.4\angle 53.13^\circ$ A
b. $2.4\angle 53.13^\circ$ A
c. $4.4\angle 0^\circ$ A
d. $4.4\angle -53.13^\circ$ A
e. $2.4\angle -53.13^\circ$ A

Solution:

The load current $\mathbf{I}_{AB}$ is given by

$$\mathbf{I}_{AB} = \frac{\mathbf{V}_{AB}}{Z} = \frac{220\angle 0^\circ}{30 - j40} = \frac{220\angle 0^\circ}{50\angle -53.13^\circ} = 4.4\angle 53.13^\circ \text{A}$$

The correct answer is a.

5.32 The line current I_A (in amperes) is most nearly

a. $4.4\angle -186.87^\circ$
b. $4.4\angle 23^\circ$
c. 7
d. $7.6\angle 23^\circ$
e. $7\angle -186.87^\circ$

Solution:

The current phasor diagram for this Δ-connected system is shown in Figure 5.71. From the relationship among three-phase currents, we have

$$\mathbf{I}_{CA} = \mathbf{I}_{AB}\angle -240^\circ = 4.4\angle(53.13^\circ - 240^\circ) = 4.4\angle -186.87^\circ \text{A}$$

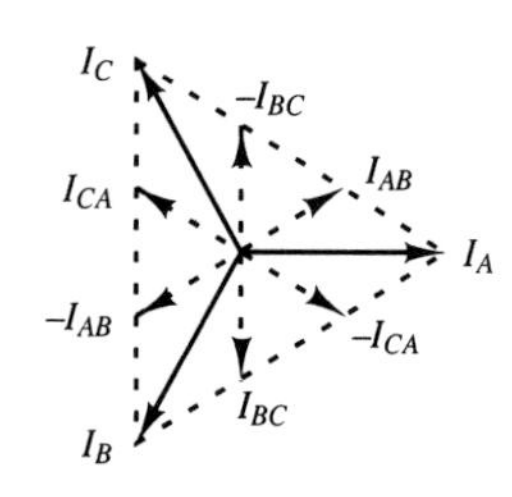

Figure 5.71

From the phasor diagram, we have

$$\begin{aligned}\mathbf{I}_A &= \mathbf{I}_{AB} - \mathbf{I}_{CA} = 4.4\angle 53.13^\circ - 4.4\angle -186.87^\circ \\ &= 2.64 + j3.52 - (-4.37 + j0.53) = 7.62\angle 23.1^\circ \text{A}\end{aligned}$$

Therefore, the correct answer is d.

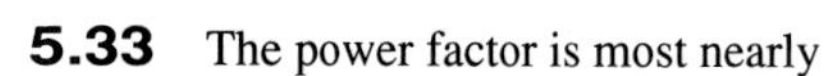

5.33 The power factor is most nearly

a. 1.0
b. 0.6 leading
c. 0.866 leading
d. 0
e. 0.8 lagging

Solution:

The impedance angle θ is:

$$\theta = \tan^{-1}\frac{40}{30} = 53.13^\circ$$

Therefore, the power factor is:

$$\text{pf} = \cos\theta = 0.6, \text{ leading}$$

We can also get the answer directly from the expression for the current $\mathbf{I}_{AB}$. The correct answer is b.

5.34 The total real power delivered from the source to the load is most nearly

a. 1496 W
b. 580 W
c. 1742 W
d. 2904 W
e. 850 W

Solution:

The total power P delivered to the balanced load is:

$$P = 3V_{AB}I_{AB}\cos\theta = 3 \times 220 \times 4.4 \times 0.6 = 1742.4 \approx 1742 \text{ W}$$

The answer is c.

In addition to these sample problems, you may also find it useful to review the following examples in the text: Chapter 4, Examples 4.8, 4.9, 4.13, 4.14, 4.15, 4.16, 4.18, 4.20, 4.22, 4.23; Chapter 5, Examples 5.1, 5.2, 5.3, 5.4, 5.5, 5.6, 5.7, 5.8, 5.9, 5.13, 5.15, and 5.16.

The problems marked "EIT" in the Homework Problems section of Chapters 4 and 5 will provide additional practice material. Answers to these problems will be found in Appendix B.

KEY TERMS

ANSWERS TO DRILL EXERCISES

5.2 $Z = 4.8e^{-j33.5^\circ}\ \Omega$; $P_{av} = 2{,}103.4$ W

5.3 See Example 5.2.

5.4 101.46 W

5.5 pf $= \cos 89.36^\circ = 0.0105$

5.6 0.514 lagging, 1

5.8 (a) 0.848, leading; (b) 0.9925, lagging

5.9 (a) capacitive; (b) capacitive; (c) inductive; (d) neither (resistive)

5.11 0.584 A

5.12 $N = 0.0667$

5.13 $Z_S = 0.0686 - j0.3429\ \Omega$

5.14 $P_{\text{loss}} = 43.2$ kW

5.15 $\mathbf{V}_a = 480\angle 0°$ V; $\mathbf{I}_a = 151.8\angle -71.6°$ A; $S_T = 69.12\text{ W} + j207.4 \times 10^3$ VA

5.18 $I_\Delta = 3I_y$

5.19 $\mathbf{I}_a = 189\angle 0°$ A; $\mathbf{I}_b = 189\angle -120°$ A; $\mathbf{I}_c = 189\angle 120°$ A

5.20 Losses for a 120-V circuit are approximately double the losses for a 240-V circuit of the same power rating.

HOMEWORK PROBLEMS

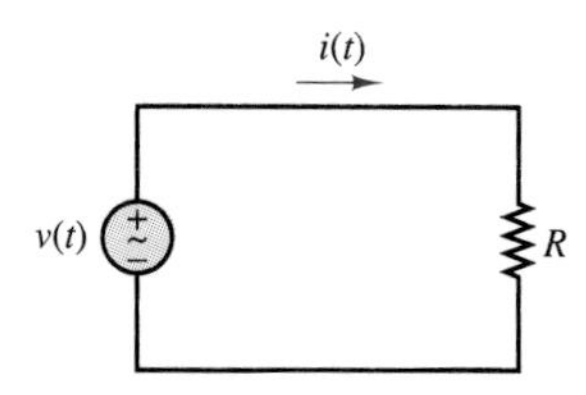

Figure P5.1

5.1 Find the peak instantaneous power, P, and the average power, P_{av}, if $v(t) = 80 \sin 377t$ and $R = 2$ kΩ for the circuit shown in Figure P5.1.

5.2 The heating element in a soldering iron has a resistance of 391 Ω. Find the average power dissipated in the soldering iron if it is connected to a voltage source of 117 V rms.

EIT 5.3 The heating element in an electric heater has a resistance of 10 Ω. Find the power dissipated in the heater when it is connected to a voltage source of 240 V rms.

5.4 Find the peak instantaneous power and the average power dissipated in a 200-Ω resistor connected across a voltage source $v(t) = 10 \sin 10^5 t$ V.

5.5 Find the average power supplied to a 5-Ω resistor by each of the periodic voltage waveforms shown in Figure P5.2.

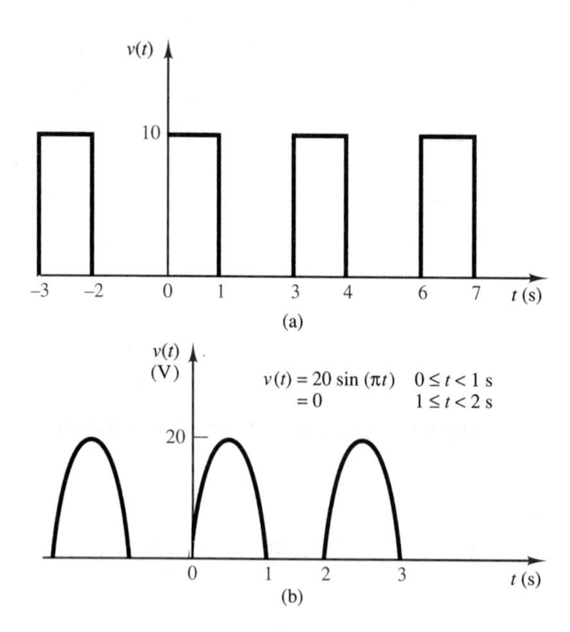

Figure P5.2

EIT 15.6 A current source $i(t)$ is connected to a 100-Ω resistor. Find the average power delivered to the resistor, given that $i(t)$ is:

a. $4\cos 100t$ A
b. $4\cos(100t - 50°)$ A
c. $4\cos 100t - 3\cos(100t - 50°)$ A
d. $4\cos 100t - 3$ A

5.7 Find the rms value of each of the following periodic currents:

a. $\cos 377t + \cos 377t$
b. $\cos 2t + \sin 2t$
c. $\cos 377t + 1$
d. $\cos 2t + \cos(2t + 135°)$
e. $\cos 2t + \cos 3t$

5.8 A current of 10 A rms flows when a single-phase circuit is placed across a 220-V rms source. The current lags the voltage by 60°. Find the power dissipated by the circuit and the power factor.

EIT 5.9 A single-phase circuit is placed across a 120-V rms, 60-Hz source, with an ammeter, a voltmeter, and a wattmeter connected. The instruments indicate 12 A, 120 V, and 800 W, respectively. Find

a. The power factor.
b. The phase angle.
c. The impedance.
d. The resistance.

5.10 The nameplate on a single-phase induction machine reads 2 horsepower (output), 110 V rms, 60 Hz, and 24 A rms. Find the power factor of the machine if the efficiency at rated output is 80 percent. [*Note:* 1 horsepower = 0.746 kW.]

5.11 For the following numerical values, determine the average power, P, the reactive power, Q, and the complex power, S, of the circuit shown in Figure P5.3. Note: phasor quantities are rms.

a. $v_S(t) = 650\cos(377t)$ V
   $i_L(t) = 20\cos(377t - 10°)$ A
b. $\mathbf{V}_S = 460\angle 0°$ V
   $\mathbf{I}_L = 14.14\angle -45°$ A
c. $\mathbf{V}_S = 100\angle 0°$ V
   $\mathbf{I}_L = 8.6\angle -86°$ A
d. $\mathbf{V}_S = 208\angle -30°$ V
   $\mathbf{I}_L = 2.3\angle -63°$ A

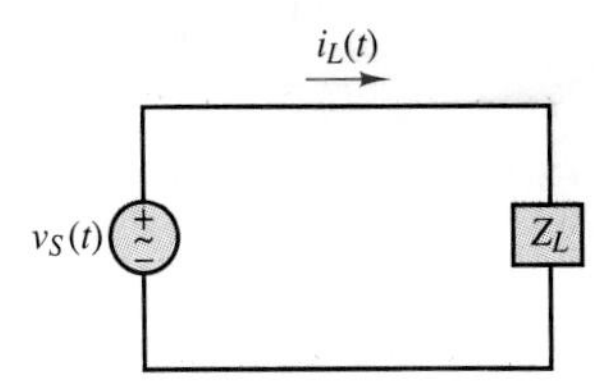

Figure P5.3

EIT 5.12 For the circuit of Figure P5.3, determine the power factor for the load and state whether it is leading or lagging for the following conditions:

a. $v_S(t) = 540\cos(\omega t + 15°)$ V
   $i_L(t) = 20\cos(\omega t + 47°)$ A
b. $v_S(t) = 155\cos(\omega t - 15°)$ V
   $i_L(t) = 20\cos(\omega t - 22°)$ A
c. $v_S(t) = 208\cos(\omega t)$ V
   $i_L(t) = 1.7\sin(\omega t + 175°)$ A
d. $Z_L = (48 + j16)$ Ω

EIT 5.13 For the circuit of Figure P5.3, determine whether the load is capacitive or inductive for the circuit shown if

a. pf = 0.87 (leading)
b. pf = 0.42 (leading)
c. $v_S(t) = 42\cos(\omega t)$
   $i_L(t) = 4.2\sin(\omega t)$
d. $v_S(t) = 10.4\cos(\omega t - 12°)$
   $i_L(t) = 0.4\cos(\omega t - 12°)$

5.14 Given the waveform of a voltage source shown in Figure P5.4, find:

a. the average and rms values of the voltage.
b. the average power supplied to a 10-Ω resistor connected across the voltage source.

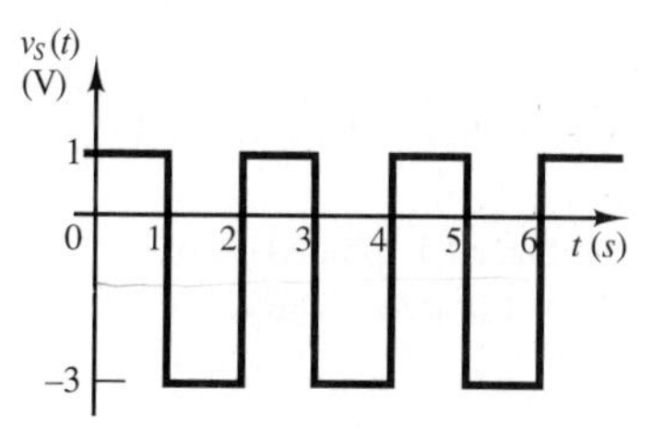

Figure P5.4

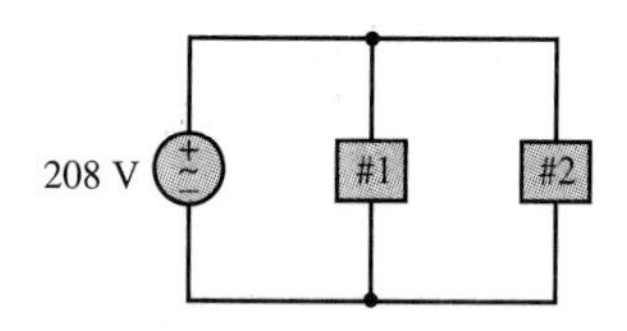

Figure P5.5

5.15 In the circuit shown in Figure P5.5, two loads are connected in parallel across a 208-V rms source. Load 1 absorbs 2 kW and 1.4 kVAR. Load 2 absorbs 4 kVA at 0.7 pf (leading). Find the impedance that is equivalent to the two parallel loads.

EIT 5.16 For the circuit of Figure P5.3, a source of strength $\mathbf{V}_S$ = 120 V rms delivers $\mathbf{I}_S = 1.94\angle 40°$ A rms to a load.

a. Determine the complex power delivered to the load.
b. Find the power dissipated by the load.
c. What is the equivalent impedance of the load?

5.17 The circuit shown in Figure P5.6(a) contains two resistors. The circuit shown in Figure P5.6(b) has an impedance of value $= -j4\ \Omega$ replacing R_1.

a. Find $\mathbf{V}_{out}$ for Figure P5.6(a).
b. What is the power dissipated by R_2 in this circuit?
c. Find $\mathbf{V}_{out}$ for Figure P5.6(b).
d. What is the power dissipated by R_2 in this circuit?
e. In which circuit does R_2 dissipate more power?

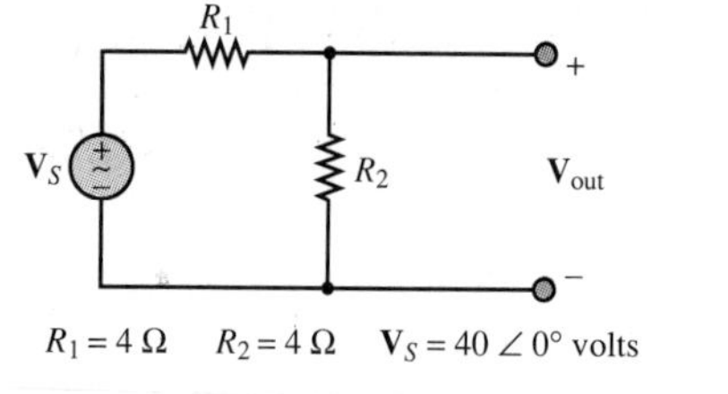

Figure P5.6(a)

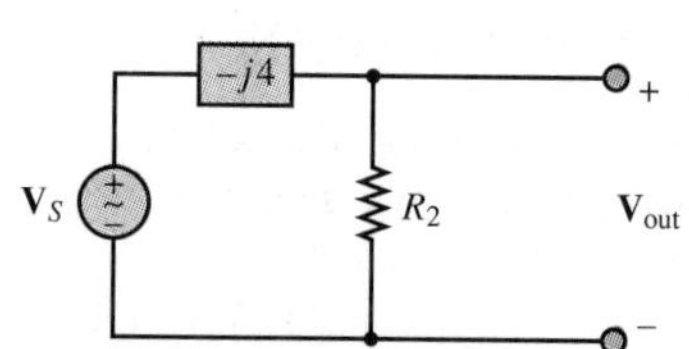

Figure P5.6(b)

5.18 Find the real and reactive power supplied by the source in the circuit shown in Figure P5.7.

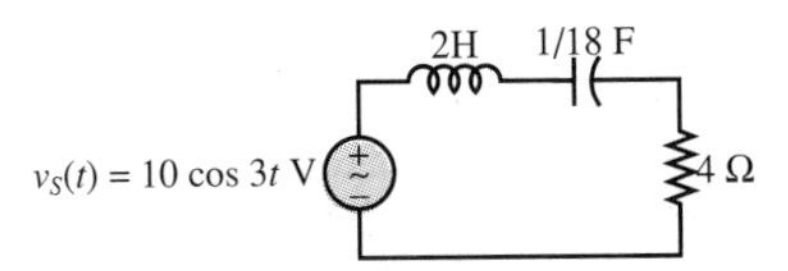

Figure P5.7

5.19 The circuit shown in Figure P5.8 is to be used on two different sources, each with the same amplitude but at different frequencies.

a. Find the real and reactive power if $v_S(t) = 120 \cos 377t$ (i.e., the frequency is 60 Hz).
b. Find the real and reactive power if $v_S(t) = 120 \cos 314t$ (i.e., the frequency is 50 Hz).

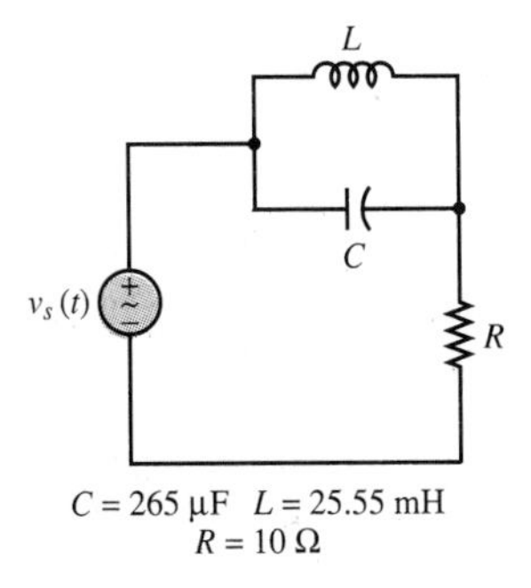

Figure P5.8

5.20 For the circuit shown in Figure P5.9, find the real and reactive power supplied by each source. The sources are $V_{s1} = 36\angle-60°$ V and $V_{s2} = 24\angle36.9°$ V.

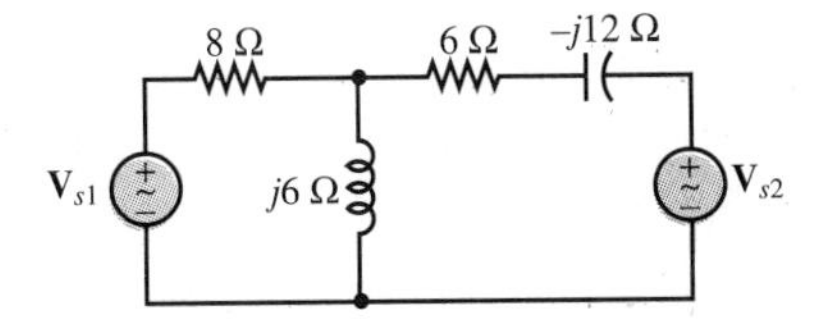

Figure P5.9

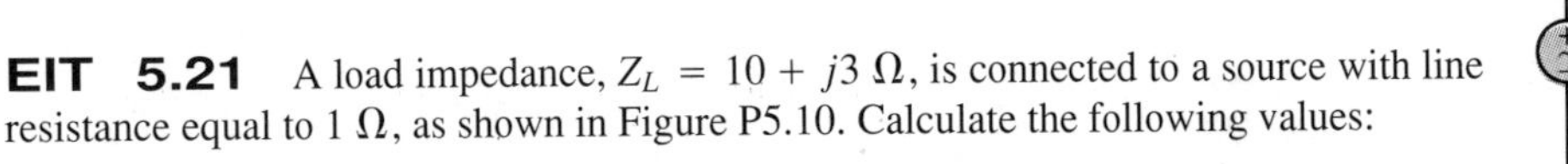

EIT 5.21 A load impedance, $Z_L = 10 + j3$ Ω, is connected to a source with line resistance equal to 1 Ω, as shown in Figure P5.10. Calculate the following values:

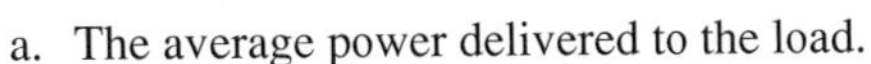

a. The average power delivered to the load.
b. The average power absorbed by the line.
c. The apparent power supplied by the generator.
d. The power factor of the load.
e. The power factor of line plus load.

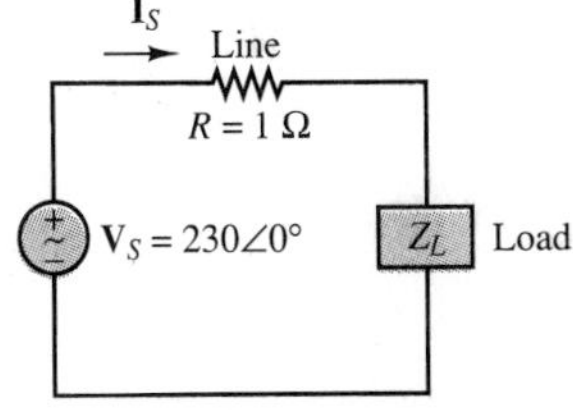

Figure P5.10

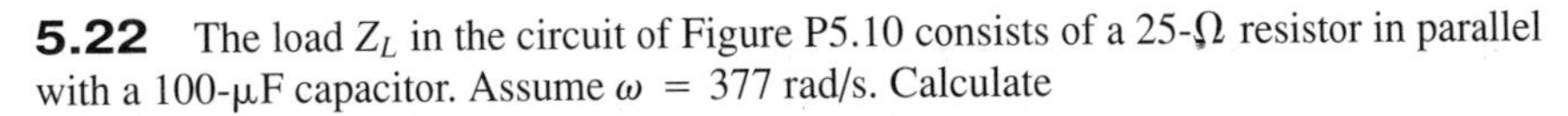

5.22 The load Z_L in the circuit of Figure P5.10 consists of a 25-Ω resistor in parallel with a 100-μF capacitor. Assume $\omega = 377$ rad/s. Calculate

a. The apparent power delivered to the load.
b. The apparent power supplied by the source.
c. The power factor of the load.

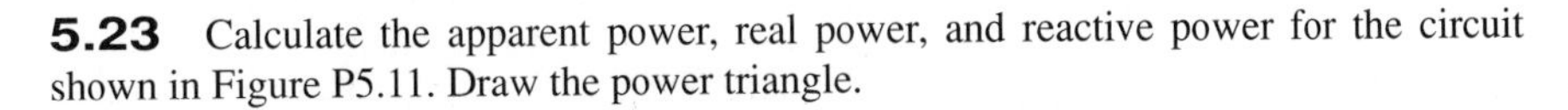

5.23 Calculate the apparent power, real power, and reactive power for the circuit shown in Figure P5.11. Draw the power triangle.

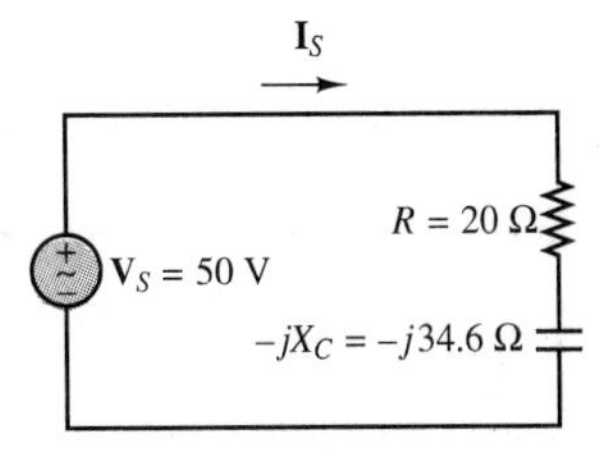

Figure P5.11

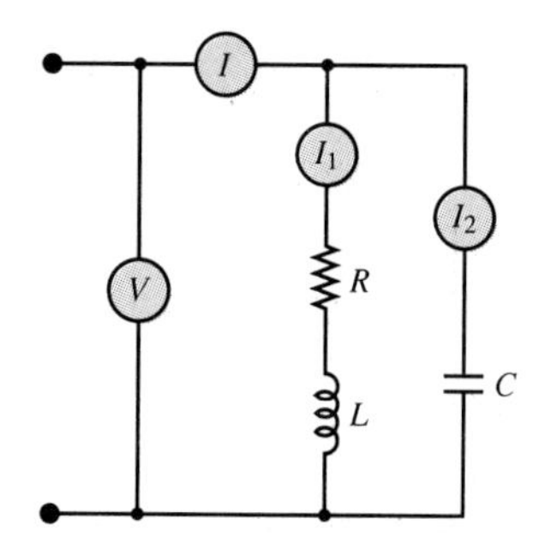

Figure P5.12

5.24 A single-phase motor draws 220 W at a power factor of 80 percent (lagging) when connected across a 200-V, 60-Hz source. A capacitor is connected in parallel with the load to give a unity power factor, as shown in Figure P5.12. Find the required capacitance.

EIT 5.25 Suppose that the electricity in your home has gone out and the power company will not be able to have you hooked up again for several days. The freezer in the basement contains several hundred dollars' worth of food that you cannot afford to let spoil. You have also been experiencing very hot, humid weather and would like to keep one room air-conditioned with a window air conditioner, as well as run the refrigerator in your kitchen. When the appliances are on, they draw the following currents (all values are rms):

Air conditioner:	9.6 A @ 120 V	pf = 0.90 (lagging)
Freezer:	4.2 A @ 120 V	pf = 0.87 (lagging)
Refrigerator:	3.5 A @ 120 V	pf = 0.80 (lagging)

In the worst-case scenario, how much power must an emergency generator supply?

5.26 The load on a single-phase three-wire system in a home is generally not balanced. For the system shown in Figure P5.13, let $V_{s1} = 115\angle 0°\ V_{\text{rms}}$ and $V_{s2} = 115\angle 0°\ V_{\text{rms}}$. Determine:

a. The total average power delivered to the connected loads: Z_{L1}, Z_{L2}, and Z_{L3}.
b. The total average power lost in the lines: Z_{g1}, Z_{g2}, and Z_n.
c. The average power supplied by each source.

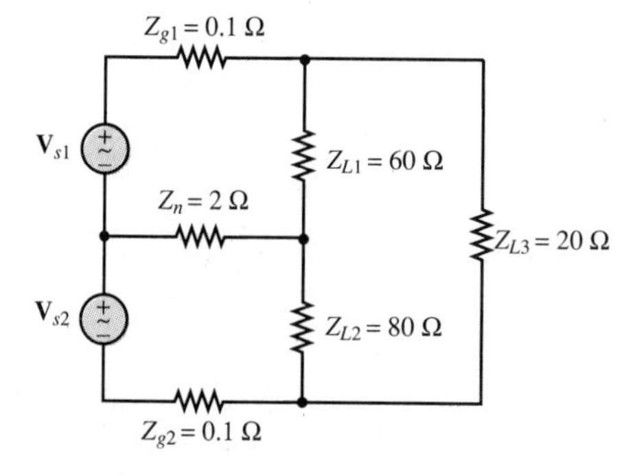

Figure P5.13

***5.27** Two motors are connected in parallel as shown in Figure P5.14. Both motors are rated for 6.8 A and 120 V. This means that each of these motors draws 6.8 A when the motor is supplying rated power to a mechanical system. The plots in Figure P5.15 show the power factor of the identical motors as a function of the mechanical load placed on the motor, and the current into each motor as a function of the same load.

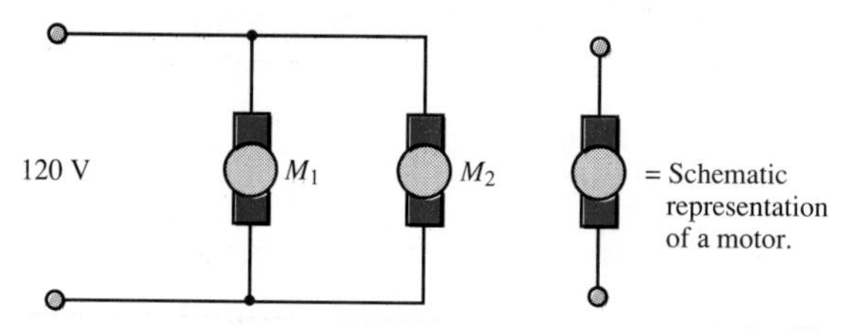

Figure P5.14

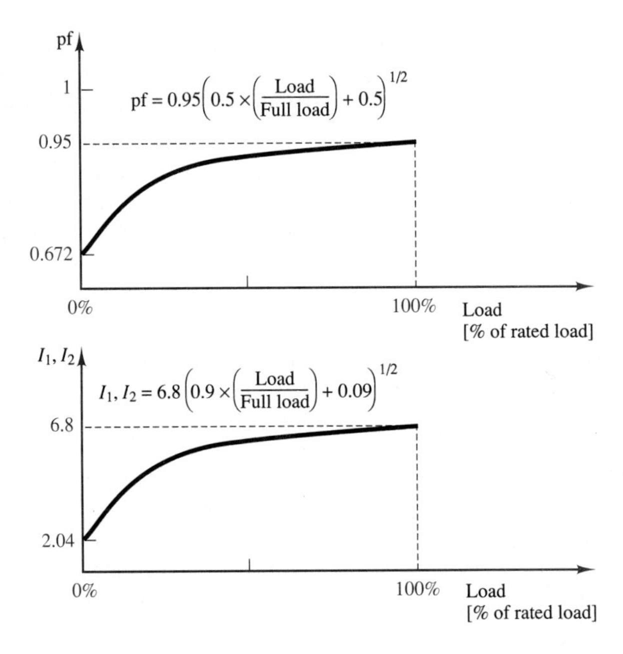

Figure P5.15

a. How much average power is lost if both motors are coasting (0 percent load)?
b. If the motors are 80 percent efficient at full load—that is, if 80 percent of the electrical power is converted to mechanical power—how much power in watts is delivered to the mechanical systems under full load?
c. If motor 1 is operating at 75 percent of full load capability and motor 2 is operating at 95 percent of full load, find the power factor for the system.

EIT 5.28 The motor inside a blender can be modeled as a resistance in series with an inductance, as shown in Figure P5.16.

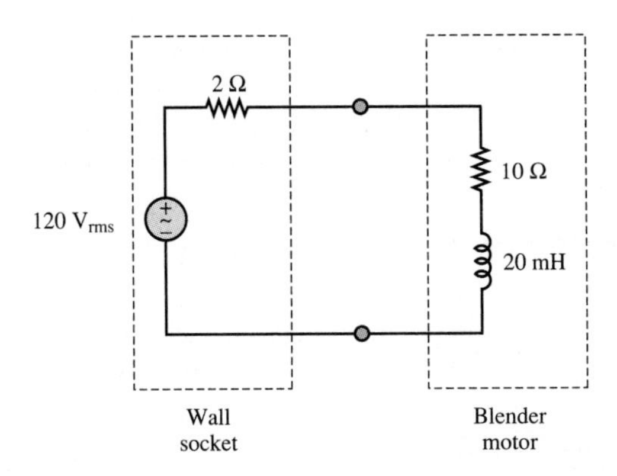

Figure P5.16

a. What is the average power, P_{av}, dissipated in the load?
b. What is the motor's power factor?
c. What value of capacitor when placed in parallel with the motor will change the power factor to 0.9 (lagging)?

5.29 A large consumer of electricity requires 10 kW of power at 230 V_{rms} at a pf angle of 60° lagging. The transmission line between the electric utility and the consumer has a resistance of 0.1 Ω. If the consumer can increase the pf from 0.5 to 0.9 lagging, determine the change in transmission line losses and load current.

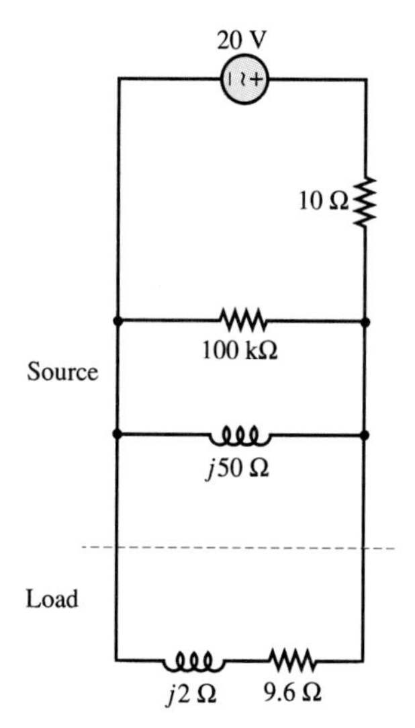

Figure P5.17

5.30 A 1000 W electric motor is connected to a source of 120 V_{rms}, 60 Hz, and the result is a lagging pf of 0.8. To correct the pf to 0.95 lagging, a capacitor is placed in parallel with the motor. Calculate the current drawn from the source with and without the capacitor connected. Determine the value of the capacitor required to make the correction.

5.31 For the circuit shown in Figure P5.17,

a. Find the Thévenin equivalent circuit for the source.
b. Find the power dissipated by the load resistor.
c. What value of load impedance would permit maximum power transfer?

5.32 A center-tap transformer has the schematic representation shown in Figure P5.18. The primary-side voltage is stepped down to a secondary-side voltage, $\mathbf{V}_{sec}$, by a ratio of $n:1$. On the secondary side, $\mathbf{V}_{sec1} = \mathbf{V}_{sec2} = \frac{1}{2}\mathbf{V}_{sec}$.

a. If $\mathbf{V}_{prim} = 120\angle 32°$ V and $n = 9$, find $\mathbf{V}_{sec}$, $\mathbf{V}_{sec1}$, and $\mathbf{V}_{sec2}$.
b. What must n be if $\mathbf{V}_{prim} = 208\angle 10°$ V and we desire $|\mathbf{V}_{sec2}|$ to be 8.7 V?

5.33 For the circuit shown in Figure P5.19, find:

a. The total resistance seen by the voltage source.
b. The voltage gain v_2/v_g.
c. The value to which the 16-Ω load resistance should be changed so it will absorb maximum power from the given source.

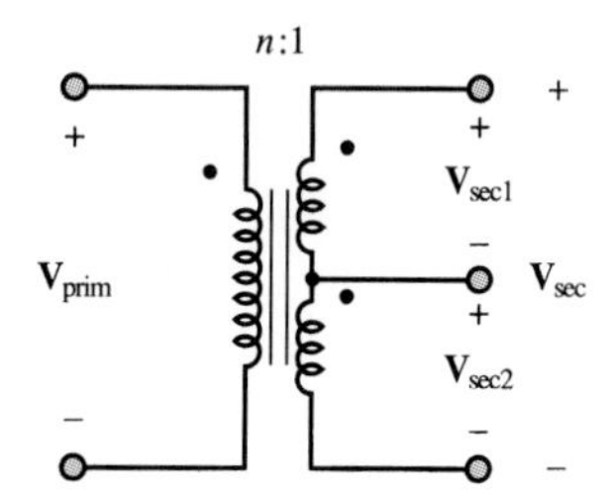

Figure P5.18

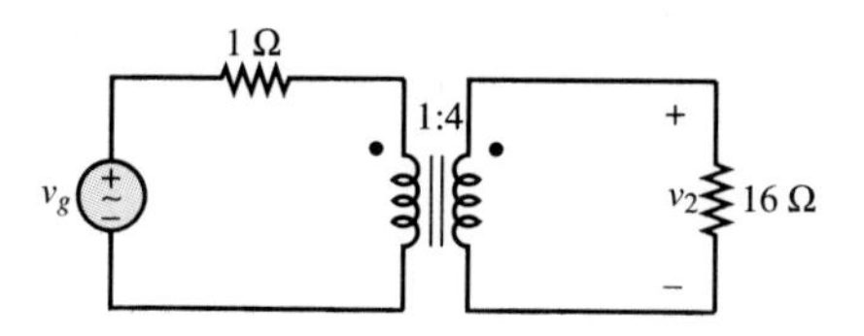

Figure P5.19

EIT 5.34 An ideal transformer is rated to deliver 400 kVA at 460 V to a customer as shown in Figure P5.20.

a. How much current can the transformer supply to the customer?
b. If the customer's load is purely resistive (i.e., if pf = 1), what is the maximum power that the customer can receive?
c. If the customer's power factor is 0.8 (lagging), what is the maximum usable power the customer can receive?

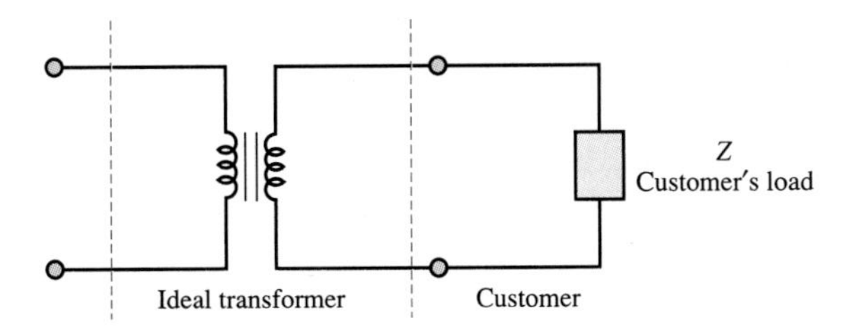

Figure P5.20

d. What is the maximum power if the pf is 0.7 (lagging)?
e. If the customer requires 300 kW to operate, what is the minimum power factor with the given size transformer?

5.35 For the ideal transformer shown in Figure P5.21, find $v_o(t)$ if $v_S(t)$ is 294 cos 377t.

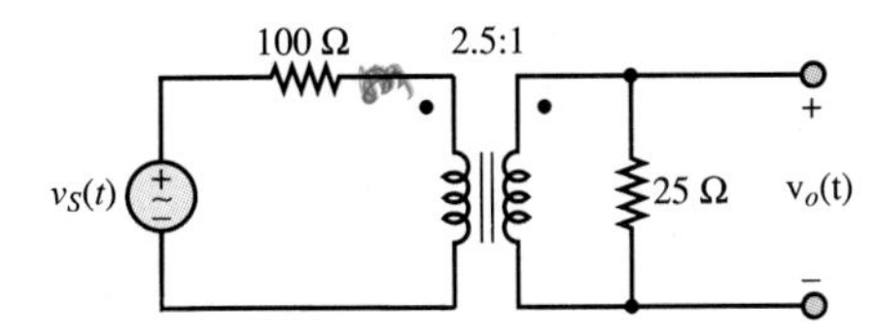

Figure P5.21

5.36 If the transformer shown in Figure P5.22 is ideal, find the turns ratio $N = \frac{1}{n}$ that will provide maximum power transfer to the load.

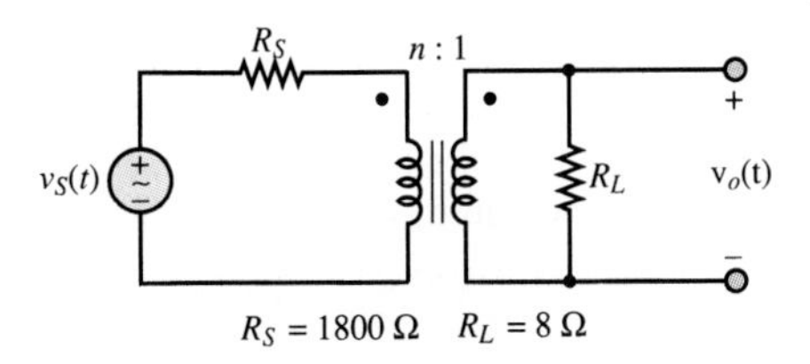

Figure P5.22

5.37 The wire that connects an antenna on your roof to the TV set in your den is a 300-Ω wire, as shown in Figure P5.23(a). This means that the impedance seen by the connections on your set is 300 Ω. Your TV, however, has a 75-Ω impedance connection, as shown in Figure P5.23(b). To achieve maximum power transfer from the antenna to the television set, you place an ideal transformer between the antenna and the TV as shown in Figure P5.23(c). What is the turns ratio, $N = \frac{1}{n}$, needed to obtain maximum power transfer?

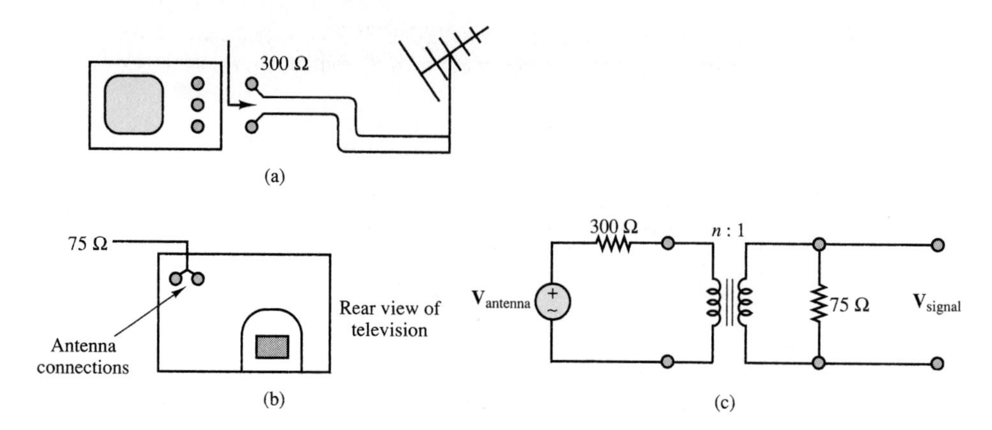

Figure P5.23

5.38 Assume the 8-Ω resistor is the load in the circuit shown in Figure P5.24. Assume a turns ratio of $1 : n$. What value of n will result in the load resistor absorbing maximum power from the source?

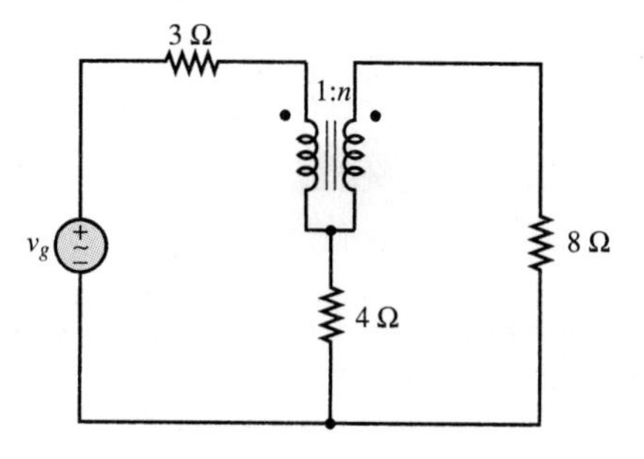

Figure P5.24

EIT 5.39 If a transformer is modeled as shown in Figure P5.25, find the actual voltage on the secondary side of the transformer.

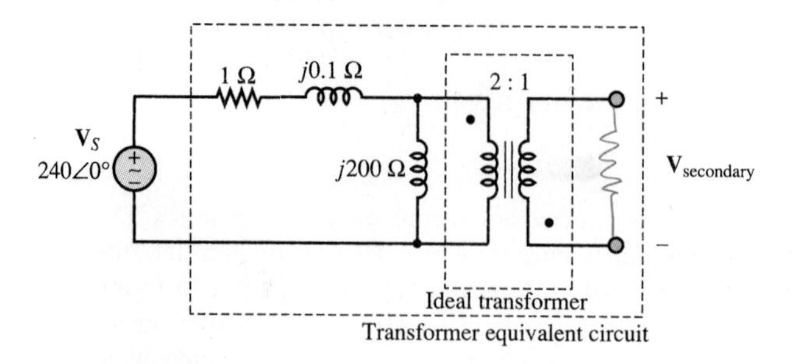

Figure P5.25

***5.40** If we knew that the transformer shown in Figure P5.26 was to deliver 50 A at 110 V rms with a certain resistive load, what rms phasor voltage source, $\mathbf{V}_S$, would provide this voltage and current?

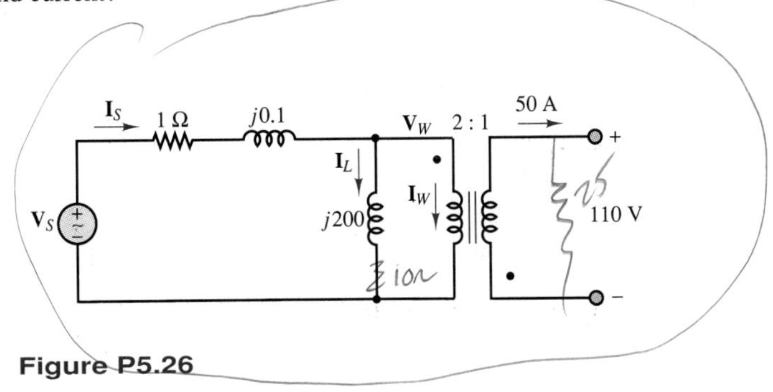

Figure P5.26

***5.41** A method for determining the equivalent circuit of a transformer consists of two tests: the open-circuit test and the short-circuit test. The open-circuit test, shown in Figure P5.27(a), is usually done by applying rated voltage to the primary side of the transformer while leaving the secondary side open. The current into the primary side is measured, as is the power dissipated.

The short-circuit test, shown in Figure P5.27(b), is performed by increasing the primary voltage until rated current is going into the transformer while the secondary side is short-circuited. The current into the transformer, the applied voltage, and the power dissipated are measured.

The equivalent circuit of a transformer is shown in Figure P5.27(c), where r_w and L_w represent the winding resistance and inductance, respectively, and r_c and L_c represent the losses in the core of the transformer and the inductance of the core. The ideal transformer is also included in the model.

With the open-circuit test, we may assume that $\mathbf{I}_P = \mathbf{I}_S = 0$. Then all of the current that is measured is directed through the parallel combination of r_c and L_c. We also assume that $|r_c \| j\omega L_c|$ is much greater than $r_w + j\omega L_w$. Using these assumptions and the open-circuit test data, we can find the resistance r_c and the inductance L_c.

In the short-circuit test, we assume that $\mathbf{V}_{\text{secondary}}$ is zero, so that the voltage on the primary side of the ideal transformer is also zero, causing no current flow through the $r_c - L_c$ parallel combination. Using this assumption with the short-circuit test data, we are able to find the resistance r_w and inductance L_w.

Using the following test data, find the equivalent circuit of the transformer:

Open-circuit test: $V = 241$ V

$I = 0.95$ A

$P = 32$ W

Short-circuit test: $V = 5$ V

$I = 5.25$ A

$P = 26$ W

Both tests were made at $\omega = 377$ rad/s.

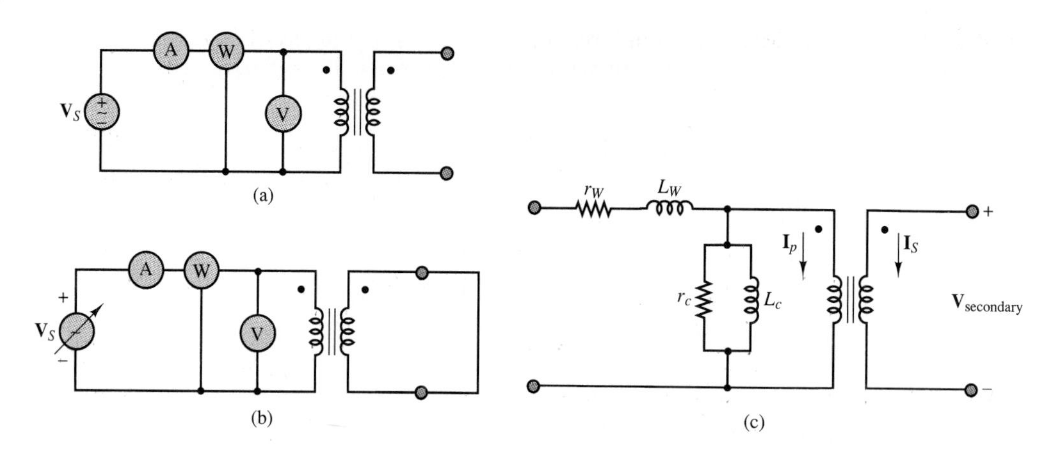

Figure P5.27

***5.42** Using the methods of Problem 5.41 and the following data, find the equivalent circuit of the transformer tested:

$$\text{Open-circuit test:} \quad V_P = 4{,}600\ \text{V}$$
$$I_{OC} = 0.7\ \text{A}$$
$$P = 200\ \text{W}$$
$$\text{Short-circuit test:} \quad P = 50\ \text{W}$$
$$V_P = 5.2\ \text{V}$$

The transformer is a 460-kVA transformer, and the tests are performed at 60 Hz.

5.43 The magnitude of the phase voltage of a balanced three-phase wye system is 100 V. Express each phase and line voltage in both polar and rectangular coordinates.

EIT 5.44 The phase currents in a four-wire wye-connected load are as follows:

$$\mathbf{I}_{an} = 10\angle 0°, \mathbf{I}_{bn} = 12\angle 150°, \mathbf{I}_{cn} = 8\angle 165°$$

Determine the current in the neutral wire.

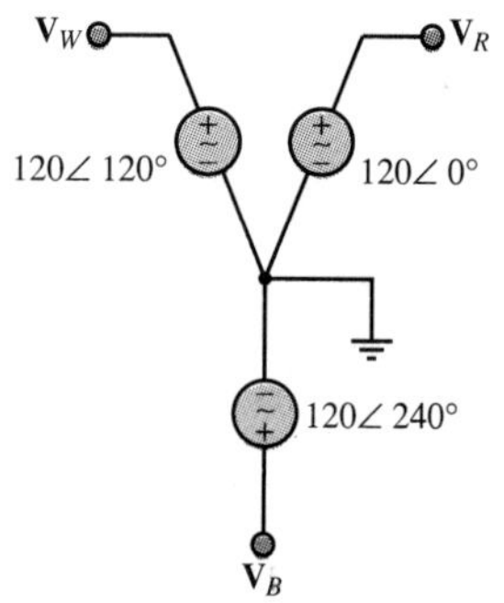

5.45 For the circuit shown in Figure P5.28, we see that each voltage source has a phase difference of 120° in relation to the others.

a. Find $\mathbf{V}_{RW}$, $\mathbf{V}_{WB}$, and $\mathbf{V}_{BR}$, where $\mathbf{V}_{RW} = \mathbf{V}_R - \mathbf{V}_W$, $\mathbf{V}_{WB} = \mathbf{V}_W - \mathbf{V}_B$, and $\mathbf{V}_{BR} = \mathbf{V}_B - \mathbf{V}_R$.
b. Repeat part a, using the calculations

$$\mathbf{V}_{RW} = \mathbf{V}_R \sqrt{3}\angle -30°$$
$$\mathbf{V}_{WB} = \mathbf{V}_W \sqrt{3}\angle -30°$$
$$\mathbf{V}_{BR} = \mathbf{V}_B \sqrt{3}\angle -30°$$

c. Compare the results of part a with the results of part b.

V_W V_R $120\angle 120°$ $120\angle 0°$ $120\angle 240°$ V_B

Figure P5.28

5.46 For the three-phase circuit shown in Figure P5.29, find the currents $\mathbf{I}_W$, $\mathbf{I}_B$, $\mathbf{I}_R$, and $\mathbf{I}_N$.

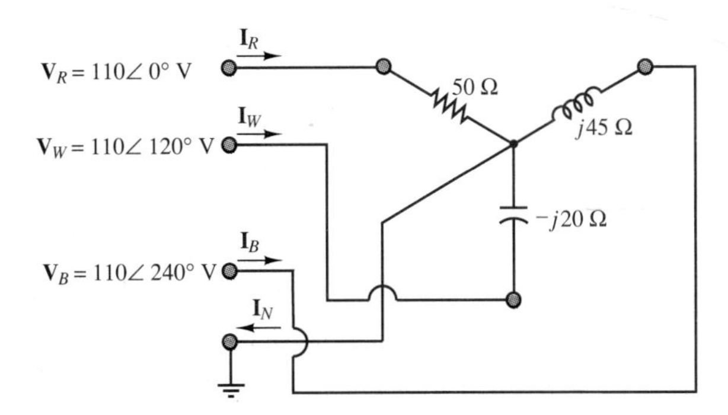

Figure P5.29

EIT 5.47 For the circuit shown in Figure P5.30, find the currents $\mathbf{I}_R$, $\mathbf{I}_W$, $\mathbf{I}_B$, and $\mathbf{I}_N$.

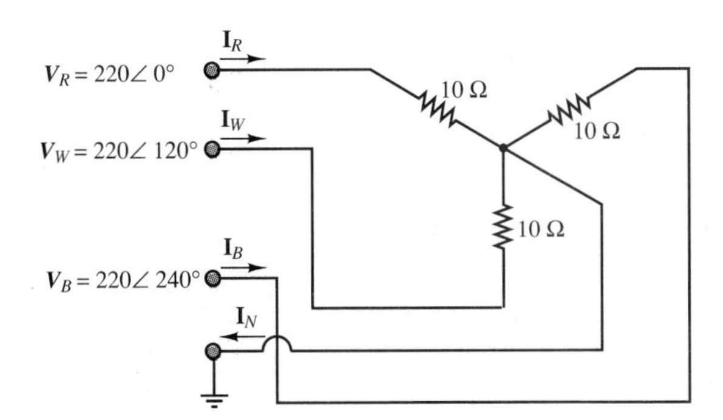

Figure P5.30

***5.48** If we model each winding of a three-phase motor like the circuit shown in Figure P5.31(a) and connect the windings as shown in Figure P5.31(b), we have the three-phase circuit shown in Figure P5.31(c). The motor can be constructed so that $R_1 = R_2 = R_3$ and $L_1 = L_2 = L_3$, as is the usual case. If we connect the motor as shown in Figure P5.31(c), find the currents $\mathbf{I}_R$, $\mathbf{I}_W$, $\mathbf{I}_B$, and $\mathbf{I}_N$, assuming that the resistances are 40 Ω each and each inductance is 5 mH. The frequency of each of the sources is 60 Hz.

5.49 With reference to the motor of Problem 5.48,

a. How much power (in watts) is delivered to the motor?
b. What is the motor's power factor?
c. Why is it common in industrial practice *not* to connect the ground lead to motors of this type?

EIT 5.50 A three-phase balanced load draws 40 A per terminal when connected to a 440-V three-phase system. The power factor of the load is 0.8, lagging. Determine the power drawn by the load.

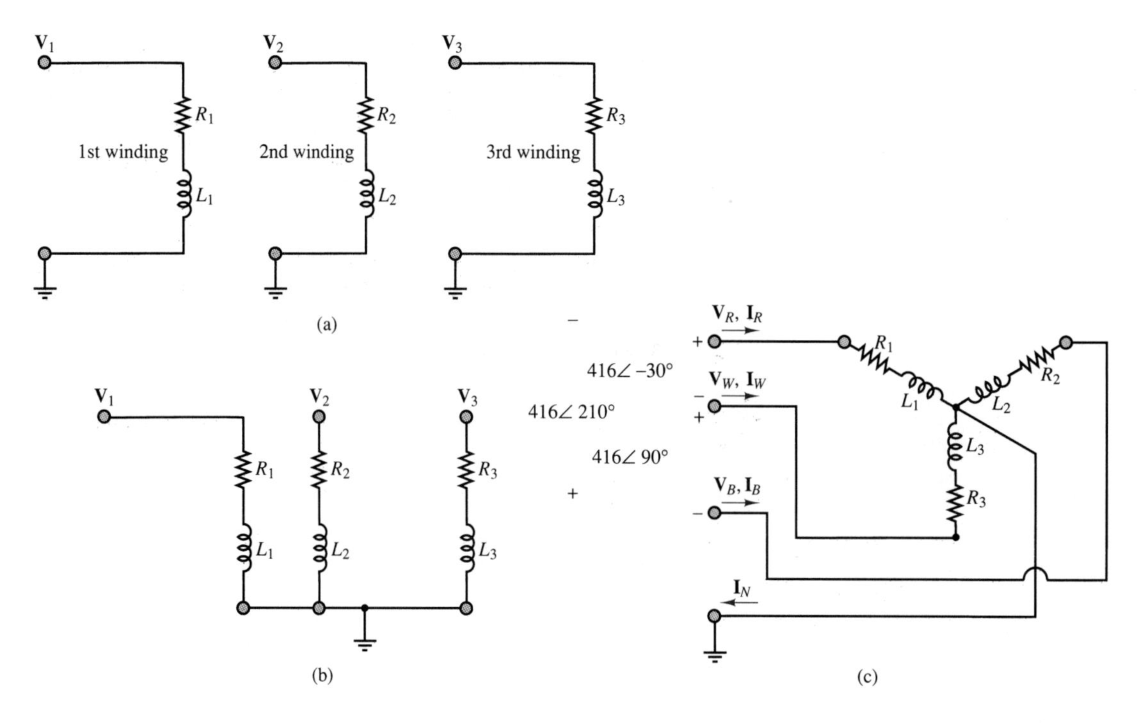

Figure P5.31

5.51 Find the apparent power and the real power delivered to the load in the Y-Δ circuit shown in Figure P5.32. What is the power factor? Assume rms values.

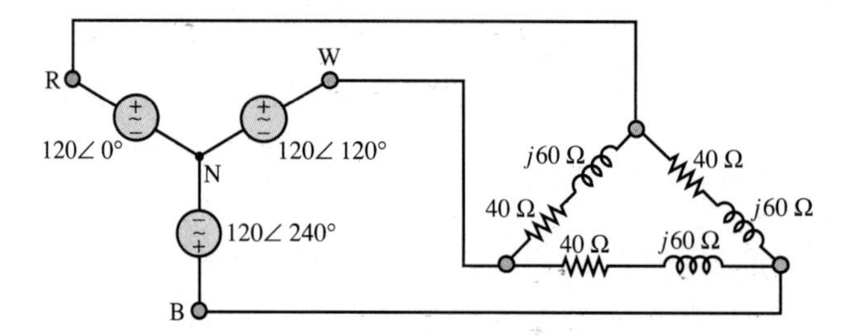

Figure P5.32

5.52 The electric power company is concerned with the loading of its transformers. Since it is responsible to a large number of customers, it must be certain that it can supply the demands of *all* customers. The power company's transformers will deliver rated kVA to the secondary load. However, if the demand were to increase to a point where greater than rated current were required, the secondary voltage would have to drop below rated value. Also, the current would increase, and with it the I^2R losses (due to winding resistance), possibly causing the transformer to overheat. Unreasonable current demand could be caused, for example, by excessively low power factors at the load.

The customer, on the other hand, is not greatly concerned with an inefficient power factor, provided that sufficient power reaches the load. To make the customer more aware

Table 5.3

Power factor	Penalty
0.850 and higher	None
0.8 to 0.849	1%
0.75 to 0.799	2%
0.7 to 0.749	3%

Courtesy of Detroit Edison.

of power factor considerations, the power company may install a penalty on the customer's bill. A typical penalty-power factor chart is shown in Table 5.3. Power factors below 0.7 are not permitted. A 25 percent penalty will be applied to any billing after two consecutive months in which the customer's power factor has remained below 0.7.

The Y-Y circuit shown in Figure P5.33 is representative of a three-phase motor load. Assume rms values.

a. Find the total power supplied to the motor.
b. Find the power converted to mechanical energy if the motor is 80 percent efficient.
c. Find the power factor.
d. Does the company risk facing a power factor penalty on its next bill if all the motors in the factory are similar to this one?

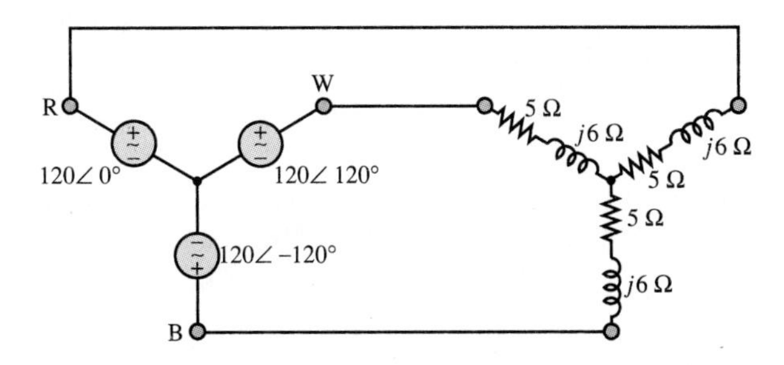

Figure P5.33

5.53 A balanced, three-phase Y-connected source with 230-V_{rms} line voltages has a balanced Y-connected load of $3 + j4\ \Omega$ per phase. For the case that the lines have zero impedance, find all three line currents and the total real power absorbed by the load.

EIT *5.54 For the wye circuit shown in Figure P5.34(a),

a. Find the total complex power.
b. Find the real power dissipated.

For the delta circuit shown in Figure P5.34(b),

c. Find the total complex power.
d. Find the real power dissipated.

[*Note:* A common method of starting a motor with reduced voltage is called the *star-delta,* or *wye-delta,* method, where the motor is first connected in a wye configuration and then switched to the delta after starting.]

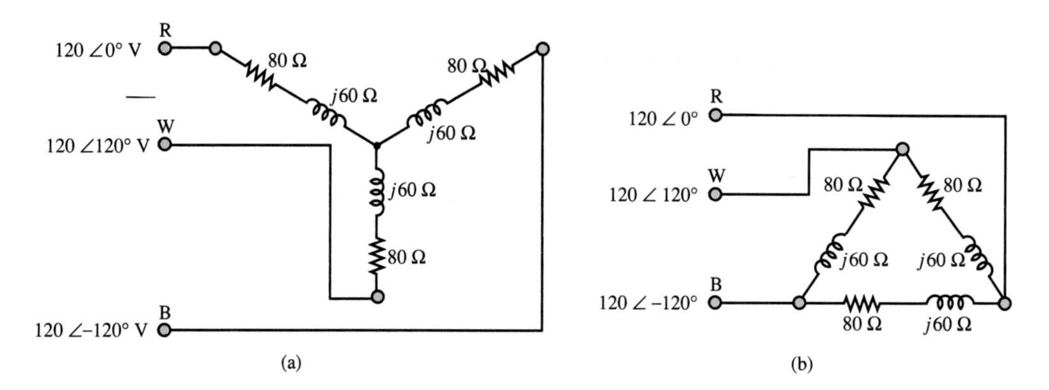

Figure P5.34

5.55 A balanced three-phase Y-connected source with 120-V_{rms} line voltages is loaded with a balanced Y-connected load of $12 - j5\ \Omega$ per phase and a balanced Δ-connected load of $3 + j4\ \Omega$ per phase. Find the total real power absorbed by the load and the overall load pf.

***5.56** The circuit shown in Figure P5.35 is a Y-Δ-Y connected three-phase circuit. The primaries of the transformers are wye-connected, the secondaries are delta-connected, and the load is wye-connected. Find the currents $\mathbf{I}_{RP}$, $\mathbf{I}_{WP}$, $\mathbf{I}_{BP}$, $\mathbf{I}_A$, $\mathbf{I}_B$, and $\mathbf{I}_C$.

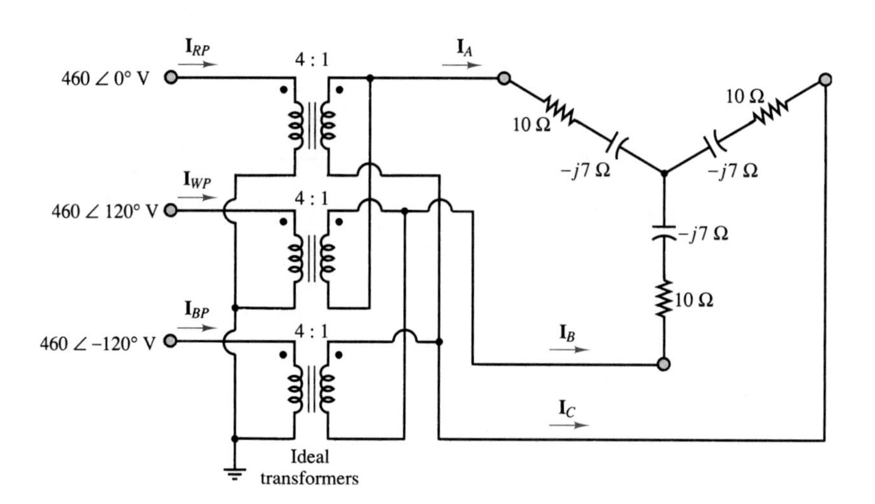

Figure P5.35

***5.57** A three-phase motor is modeled by the wye-connected circuit shown in Figure P5.36. At $t = t_1$, a line fuse is blown (modeled by the switch). Find the line currents $\mathbf{I}_R$, $\mathbf{I}_W$, and $\mathbf{I}_B$ and the power dissipated by the motor for the following conditions:

a. $t \ll t_1$
b. $t \gg t_1$

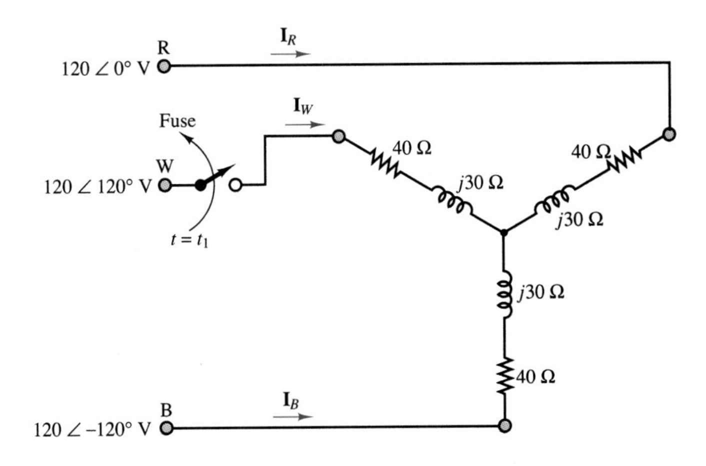

Figure P5.36

EIT 5.58 For the circuit shown in Figure P5.37, find the currents I_A, I_B, I_C, and I_N, and the real power dissipated by the load.

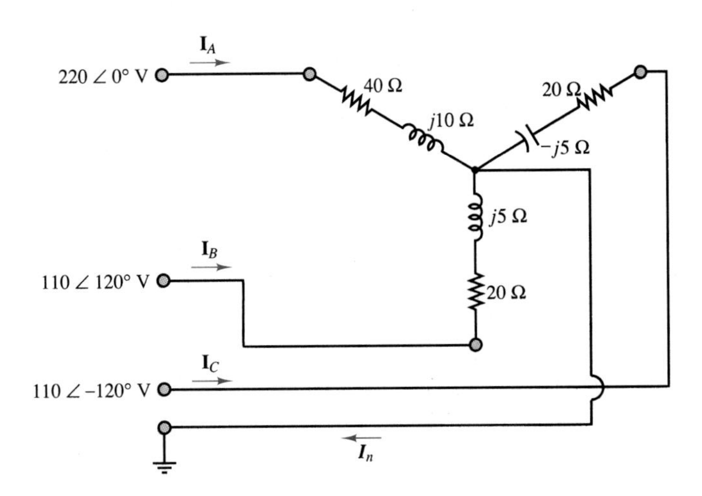

Figure P5.37

CHAPTER

6

Frequency Response and Transient Analysis

Chapter 4 introduced the notions of energy-storage elements and dynamic circuit equations and developed appropriate tools (complex algebra and phasors) for the solution of AC circuits. In Chapter 5, AC analysis techniques were applied to the study of electric power. The aim of the present chapter is twofold: first, to exploit AC circuit analysis methods to study the frequency response of electric circuits; and second, to continue the discussion of dynamic circuit equations for the purpose of analyzing the transient response of electrical circuits.

It is common, in engineering problems, to encounter phenomena that are frequency-dependent. For example, structures vibrate at a characteristic frequency when excited by wind forces (some high-rise buildings experience perceptible oscillation!). The propeller on a ship excites the shaft at a vibration frequency related to the engine's speed of rotation and to the number of blades on the propeller. An internal combustion engine is excited periodically by the combustion events in the individual cylinder, at a frequency determined by the firing of the cylinders. Wind blowing across a pipe excites a resonant vibration that is perceived as sound (wind instruments operate on this principle). Electrical circuits are no different from other dynamic systems in this respect, and a large body of knowledge has been developed for understanding the frequency response of electrical circuits, mostly based on the ideas behind phasors and impedance. These ideas, and the concept of filtering, will be explored in this chapter.

In addition to analyzing the frequency response of electrical circuits, it is often necessary to characterize the response of a circuit or system to abrupt changes in input, such as, for example, the change that occurs when power to a circuit is initially turned on. Sections 6.3 through 6.6 are devoted to the analysis of the transient response of simple first- and second-order linear circuits.

The ideas developed in this chapter will also be applied, by analogy, to the analysis of other physical systems (e.g., hydraulic, thermal, and mechanical systems), to illustrate the generality of the concepts. By the end of the chapter, you should be able to:

- Compute the frequency response function for an arbitrary circuit.
- Use knowledge of the frequency response to determine the output of a circuit.
- Compute expressions for the transient and steady-state response of first- and second-order circuits.
- Recognize the analogy between electrical circuits and other dynamic systems.

6.1 SINUSOIDAL FREQUENCY RESPONSE

The **sinusoidal frequency response** (or, simply, **frequency response**) of a circuit provides a measure of how the circuit responds to sinusoidal inputs of arbitrary frequency. In other words, given the input signal amplitude, phase, and frequency, knowledge of the frequency response of a circuit permits the computation of the output signal. The box "Fourier Analysis" provides further explanation of the importance of sinusoidal signals. Suppose, for example, that you wanted to determine how the load voltage or current varied in response to different excitation signal frequencies in the circuit of Figure 6.1. An analogy could be made, for example, with how a speaker (the load) responds to the audio signal generated by a CD player (the source) when an amplifier (the circuit) is placed between the two.[1] In the circuit of Figure 6.1, the signal source circuitry is represented by its Thévenin equivalent. Recall that the impedance Z_S presented by the source to the remainder of the circuit is a function of the frequency of the source signal (Section 4.4). For the purpose of illustration, the amplifier circuit is represented by the idealized connection of two impedances, Z_1 and Z_2, and the load is represented

[1] In reality, the circuitry in a hi-fi stereo system is far more complex than the circuits that will be discussed in this chapter and in the homework problems. However, from the standpoint of intuition and everyday experience, the audio analogy provides a useful example; it allows you to build a quick feeling for the idea of frequency response. Practically everyone has an intuitive idea of bass, mid range, and treble as coarsely defined frequency regions in the audio spectrum. The material presented in the next few sections should give you a more rigorous understanding of these concepts.

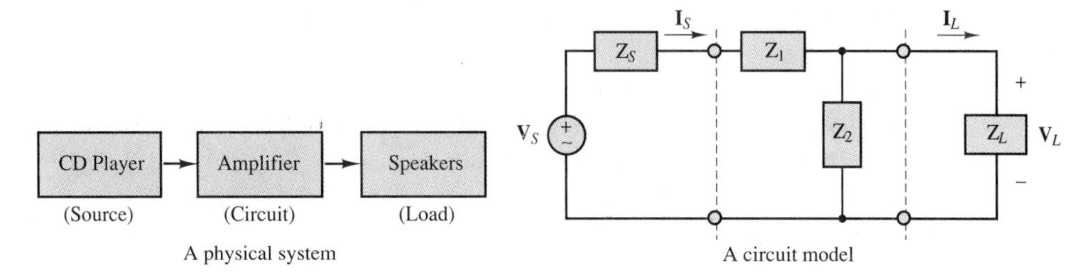

Figure 6.1 A circuit model

by an additional impedance, Z_L. What, then, is the frequency response of this circuit? The following is a fairly general definition:

> The frequency response of a circuit is a measure of the variation of a load-related voltage or current as a function of the amplitude, phase, and frequency of the excitation signal.

According to this definition, frequency response could be defined in a variety of ways. For example, we might be interested in determining how the load voltage varies as a function of the source voltage. Then, analysis of the circuit of Figure 6.1 might proceed as follows.

To express the frequency response of a circuit in terms of variation in output voltage as a function of source voltage, we use the general formula

$$H_V(j\omega) = \frac{\mathbf{V}_L(j\omega)}{\mathbf{V}_S(j\omega)} \tag{6.1}$$

One method that allows for representation of the load voltage as a function of the source voltage (this is, in effect, what the frequency response of a circuit implies) is to describe the source and attached circuit by means of the Thévenin equivalent circuit. (This is not the only useful technique; the node voltage or mesh current equations for the circuit could also be employed.) Figure 6.2 depicts the original circuit of Figure 6.1 with the load removed, ready for the computation of the Thévenin equivalent.

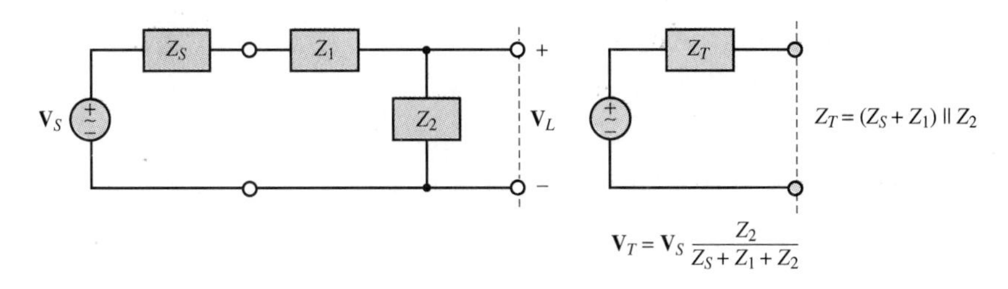

Figure 6.2 Thévenin equivalent source circuit

Fourier Analysis

In this brief introduction to **Fourier theory**, we shall explain in an intuitive manner how it is possible to represent many signals by means of the superposition of various sinusoidal signals of different amplitude, phase, and frequency. Any periodic finite-energy signals may be expressed by means of an infinite sum of sinusoids, as illustrated in the following paragraphs.

Consider a periodic waveform, $x(t)$. Its Fourier series representation is defined below by the infinite summation of sinusoids at the frequencies $n\omega_0$ (integer multiples of the **fundamental frequency**, ω_0), with amplitudes A_n and phases ϕ_n.

$$x(t) = x(t + T_0) \qquad T_0 = \text{period} \tag{6.2}$$

$$x(t) = \sum_{n=0}^{\infty} A_n \cos\left(\frac{2\pi n t}{T_0} + \phi_n\right) \tag{6.3}$$

One could also write the term $2\pi n/T_0$ as $n\omega_0$, where

$$\omega_0 = \frac{2\pi}{T_0} = 2\pi f_0 \tag{6.4}$$

is the fundamental (radian) frequency and the frequencies $2\omega_0, 3\omega_0, 4\omega_0$, and so on, are called its **harmonics**.

The notion that a signal may be represented by sinusoidal components is particularly useful, and not only in the study of electrical circuits—in the sense that we need only understand the response of a circuit to an arbitrary sinusoidal excitation in order to be able to infer the circuit's response to more complex signals. In fact, the frequently employed *sinusoidal frequency response* discussed in this chapter is a function that enables us to explain how a circuit would respond to a signal made up of a superposition of sinusoidal components at various frequencies. These sinusoidal components form the **spectrum** of the signal, that is, its frequency composition; the amplitude and phase of each of the sinusoids contribute to the overall "character" of the signal, in the same sense as the timbre of a musical instrument is made up of the different harmonics that are generated when a note is played (the timbre is what differentiates, for example, a viola from a cello or a violin). An example of the amplitude spectrum of a "square-wave" signal is shown in Figure 6.3. In order to further illustrate how the superposition of sinusoids can give rise to a signal that at first might appear substantially different from a sinusoid, the evolution of a sine wave into a square wave is displayed in

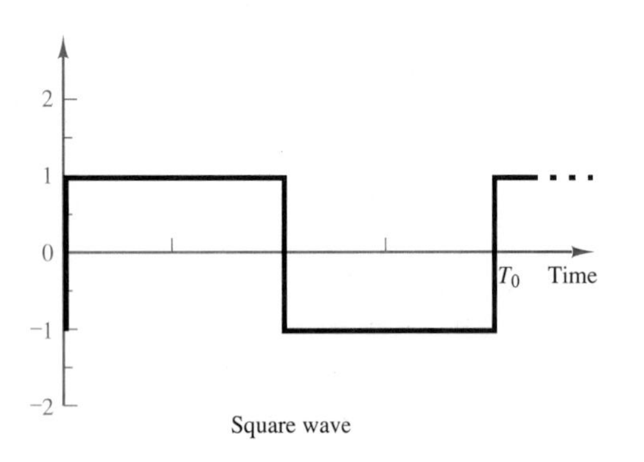

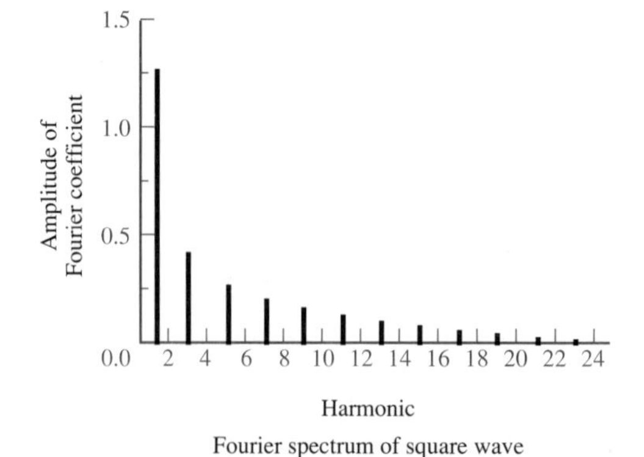

Figure 6.3 Amplitude spectrum of square wave

Figure 6.4, as more Fourier components are added. The first picture represents the fundamental component, that is, the sinusoid that has the same frequency as the square wave. Then one harmonic at a time is added, up to the fifth nonzero component (the ninth frequency component; see Figure 6.3), illustrating how, little by little, the rounded peaks of the sinusoid transform into the flat top of the square wave!

Although this book will not deal with the mathematical aspects of Fourier series, it is important to recognize that this analysis tool provides excellent motivation for the study of sinusoidal signals, and of the sinusoidal frequency response of electric circuits.

Jean Baptiste Joseph Fourier (1768–1830), French mathematician and physicist who formulated the Fourier series. *Photo courtesy of Deutsches Museum, Munich.*

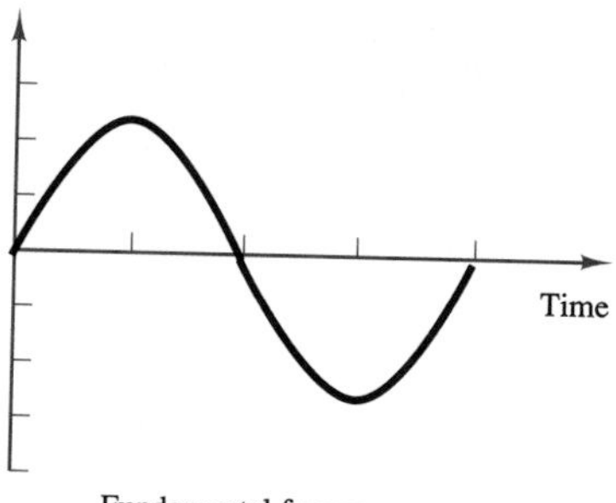

Fundamental frequency

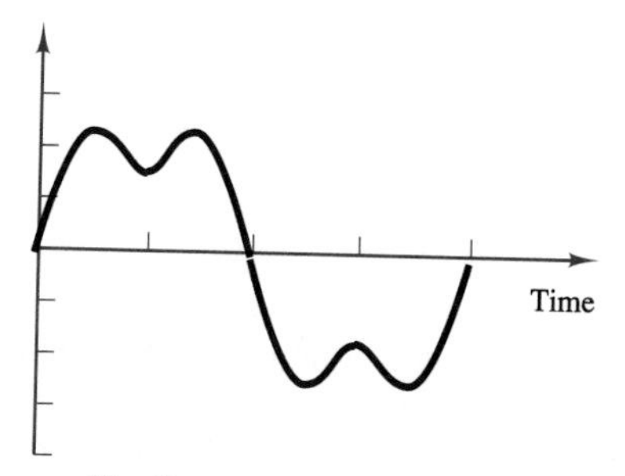

Two frequency components

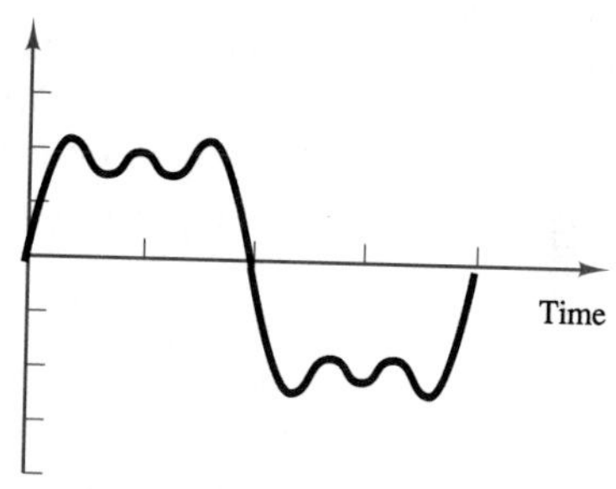

Three frequency components

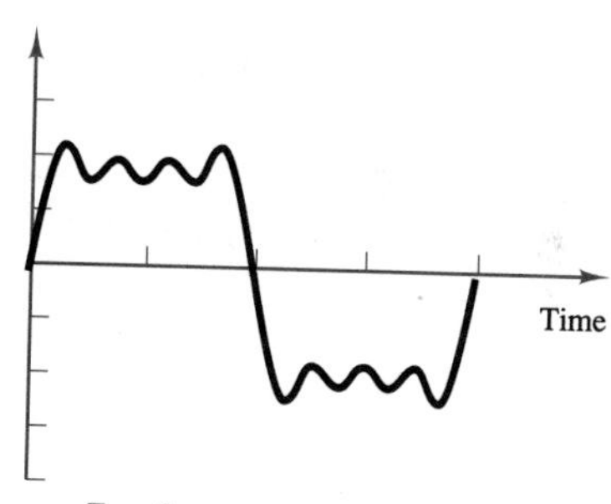

Four frequency components

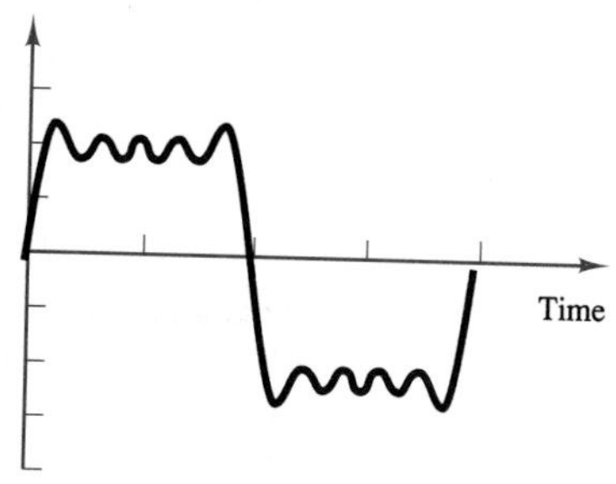

Five frequency components

Figure 6.4 Evolution of a square wave from its Fourier components

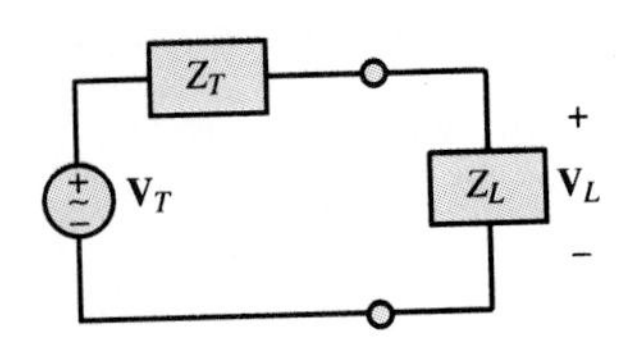

Figure 6.5 Complete equivalent circuit

Next, an expression for the load voltage, $\mathbf{V}_L$, may be found by connecting the load to the Thévenin equivalent source circuit and by computing the result of a simple voltage divider, as illustrated in Figure 6.5 and by the following equation:

$$\begin{aligned}\mathbf{V}_L &= \frac{Z_L}{Z_L + Z_T}\mathbf{V}_T \\ &= \frac{Z_L}{Z_L + \dfrac{(Z_S + Z_1)Z_2}{Z_S + Z_1 + Z_2}} \cdot \frac{Z_2}{Z_S + Z_1 + Z_2}\mathbf{V}_S \\ &= \frac{Z_L Z_2}{Z_L(Z_S + Z_1 + Z_2) + (Z_S + Z_1)Z_2}\mathbf{V}_S\end{aligned} \tag{6.5}$$

Thus, the frequency response of the circuit, as defined in equation 6.4, is given by the expression

$$\frac{\mathbf{V}_L}{\mathbf{V}_S}(j\omega) = H_V(j\omega) = \frac{Z_L Z_2}{Z_L(Z_S + Z_1 + Z_2) + (Z_S + Z_1)Z_2} \tag{6.6}$$

The expression for $H_V(j\omega)$ could be evaluated for any given $\mathbf{V}_S(j\omega)$ (i.e., for any given source signal amplitude, phase, and frequency) to determine what the resultant load voltage would be. Note that $H_V(j\omega)$ is a complex quantity (dimensionless, because it is the ratio of two voltages), and that it therefore follows that

$\mathbf{V}_L(j\omega)$ is a phase-shifted and amplitude-scaled version of $\mathbf{V}_S(j\omega)$:

$$\mathbf{V}_L(j\omega) = H_V(j\omega) \cdot \mathbf{V}_S(j\omega) \tag{6.7}$$

$$V_L e^{j\phi_L} = |H_V| e^{j\phi_H} \cdot V_S e^{j\phi_S} \tag{6.8}$$

or

$$V_L e^{j\phi_L} = |H_V| \mathbf{V}_S e^{j(\phi_H + \phi_S)} \tag{6.9}$$

where

$$V_L = |H_V| \cdot V_S$$

and

$$\phi_L = \phi_H + \phi_S \tag{6.10}$$

Thus, the effect of inserting a linear circuit between a source and a load is best understood by considering that, at any given frequency, ω, the load voltage is a sinusoid at the same frequency as the source voltage, with amplitude given by $V_L = |H_V| \cdot V_S$ and phase equal to $\phi_L = \phi_H + \phi_S$, where $|H_V|$ is the magnitude of the frequency response and ϕ_H its phase angle. Both $|H_V|$ and ϕ_H are functions of frequency.

Example 6.1

Compute the frequency response $H_V(j\omega)$ for the circuit of Figure 6.6.

Solution:

The impedances corresponding to the circuit elements of Figure 6.6 are $Z_1 = 1{,}000$, $Z_2 = \frac{1}{j10^{-5}\omega}$, $Z_L = 10{,}000$, and $Z_S = 0$. To represent the load voltage as a function of the source voltage, we need to describe the source and attached circuit by means of the Thévenin equivalent circuit. Figure 6.7 shows the equivalent circuit with the load removed, ready for the computation of the Thévenin circuit.

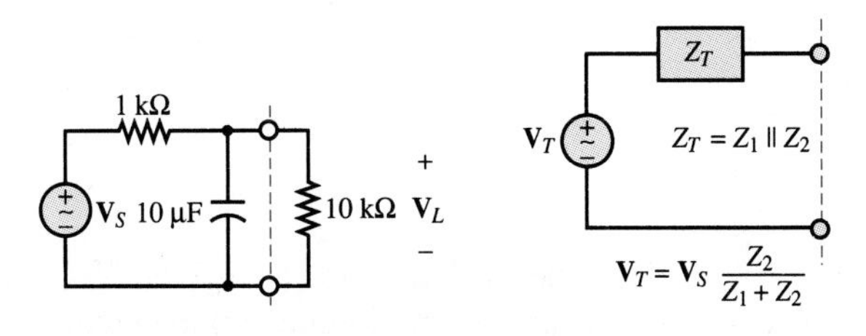

Figure 6.6

Figure 6.7

Next, we connect the load to the Thévenin equivalent and find an expression for $\mathbf{V}_L$ by computing the result of a voltage divider:

$$
\begin{aligned}
\mathbf{V}_L &= \frac{Z_L}{Z_L + Z_T}\mathbf{V}_T \\
&= \frac{Z_L}{Z_L + \dfrac{Z_1 Z_2}{Z_1 + Z_2}} \cdot \frac{Z_2}{Z_1 + Z_2}\mathbf{V}_S \\
&= \frac{Z_L Z_2}{Z_L(Z_1 + Z_2) + Z_1 Z_2}\mathbf{V}_S
\end{aligned}
$$

Thus, the frequency response of the circuit, as defined before, is given by the expression

$$
\frac{\mathbf{V}_L}{\mathbf{V}_S}(j\omega) = H_V(j\omega) = \frac{Z_L Z_2}{Z_L(Z_1 + Z_2) + Z_1 Z_2}
$$

Replacing Z_1, Z_2, and Z_L by their numerical values, we can obtain $H_V(j\omega)$ as

$$
H_V(j\omega) = \frac{100}{j\omega + 110}
$$

$H_V(j\omega)$ can also be represented in polar form as

$$
H_V(j\omega) = \frac{100}{\sqrt{\omega^2 + 110^2}\, e^{j\arctan(\omega/110)}}
$$

or

$$= \frac{100}{\sqrt{\omega^2 + 110^2}} e^{-j\arctan(\omega/110)} = \frac{100}{\sqrt{\omega^2 + 110^2}} \angle -\arctan(\omega/110)$$

The importance and usefulness of the frequency response concept lies in its ability to summarize the response of a circuit in a single function of frequency, $H(j\omega)$, which can predict the load voltage or current at any frequency, given the input. Note that the frequency response of a circuit can be defined in four different ways:

$$H_V(j\omega) = \frac{\mathbf{V}_L(j\omega)}{\mathbf{V}_S(j\omega)} \qquad H_I(j\omega) = \frac{\mathbf{I}_L(j\omega)}{\mathbf{I}_S(j\omega)}$$
$$H_Z(j\omega) = \frac{\mathbf{V}_L(j\omega)}{\mathbf{I}_S(j\omega)} \qquad H_Y(j\omega) = \frac{\mathbf{I}_L(j\omega)}{\mathbf{V}_S(j\omega)} \tag{6.11}$$

If $H_V(j\omega)$ and $H_I(j\omega)$ are known, one can directly derive the other two expressions:

$$H_Z(j\omega) = \frac{\mathbf{V}_L(j\omega)}{\mathbf{I}_S(j\omega)} = Z_L(j\omega)\frac{\mathbf{I}_L(j\omega)}{\mathbf{I}_S(j\omega)} = Z_L(j\omega)H_I(j\omega) \tag{6.12}$$

$$H_Y(j\omega) = \frac{\mathbf{I}_L(j\omega)}{\mathbf{V}_S(j\omega)} = \frac{1}{Z_L(j\omega)}\frac{\mathbf{V}_L(j\omega)}{\mathbf{V}_S(j\omega)} = \frac{1}{Z_L(j\omega)}H_V(j\omega) \tag{6.13}$$

With these definitions in hand, it is now possible to introduce one of the central concepts of electrical circuit analysis: **filters**. The concept of filtering an electrical signal will be discussed in the next section.

Example 6.2

Compute the frequency response $H_Z(j\omega)$ for the circuit of Figure 6.8.

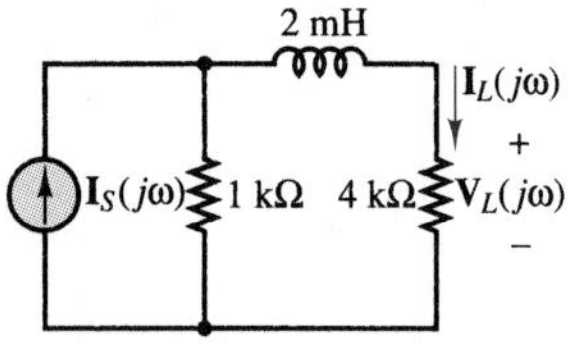

Figure 6.8

Solution:

The frequency response $H_Z(j\omega) = \mathbf{V}_L(j\omega)/\mathbf{I}_S(j\omega)$ may be found in a relatively simple manner by observing that (1) the load current is related to the source current by a current divider and (2) the load voltage is equal to the product of the 4-kΩ resistor and the load current. Formally,

$$\mathbf{I}_L(j\omega) = \frac{1{,}000}{1{,}000 + 4{,}000 + j\omega \times 2 \times 10^{-3}} \mathbf{I}_S(j\omega)$$

and

$$\mathbf{V}_L(j\omega) = 4{,}000\mathbf{I}_L(j\omega) = \frac{4 \times 10^6}{5 \times 10^3 + j\omega \times 2 \times 10^{-3}} \mathbf{I}_S(j\omega) = \frac{0.8 \times 10^3}{1 + j\omega \times 0.4 \times 10^{-6}} \mathbf{I}_S(j\omega)$$

Finally,

$$H_Z(j\omega) = \frac{\mathbf{V}_L(j\omega)}{\mathbf{I}_S(j\omega)} = \frac{1{,}000}{1.25 + j\omega \times 0.5 \times 10^{-6}}$$

6.2 FILTERS

There are a host of practical, everyday applications that involve filters of one kind or another. Just to mention two, filtration systems are used to eliminate impurities from drinking water, and sunglasses are used to filter out eye-damaging ultraviolet radiation and to reduce the intensity of sunlight reaching the eyes. An analogous concept applies to electrical circuits: it is possible to *attenuate* (i.e., reduce in amplitude) or altogether eliminate signals of unwanted frequencies, such as those that may be caused by electrical noise or other forms of interference. This section will be devoted to the analysis of electrical filters.

Low-Pass Filters

Figure 6.9 depicts a simple ***RC* filter** and denotes its input and output voltages by $\mathbf{V}_i$ and $\mathbf{V}_o$. The frequency response for the filter may be obtained by considering the function

RC low-pass filter. The circuit preserves lower frequencies while attenuating the frequencies above the cutoff frequency, $\omega_0 = 1/RC$. The voltages $\mathbf{V}_i$ and $\mathbf{V}_o$ are the filter input and output voltages, respectively.

$$H(j\omega) = \frac{\mathbf{V}_o}{\mathbf{V}_i}(j\omega) \quad \textbf{(6.14)}$$

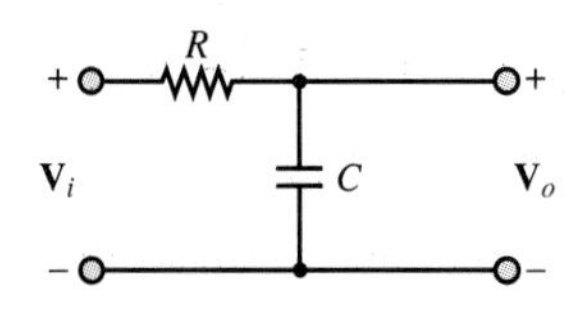

Figure 6.9 A simple *RC* filter

and noting that the output voltage may be expressed as a function of the input voltage by means of a voltage divider, as follows:

$$\mathbf{V}_o(j\omega) = \mathbf{V}_i(j\omega)\frac{1/j\omega C}{R + 1/j\omega C} = \mathbf{V}_i(j\omega)\frac{1}{1 + j\omega RC} \quad \textbf{(6.15)}$$

Thus, the frequency response of the *RC* filter is

$$\frac{\mathbf{V}_o}{\mathbf{V}_i}(j\omega) = \frac{1}{1 + j\omega CR} \quad \textbf{(6.16)}$$

An immediate observation upon studying this frequency response is that if the signal frequency, ω, is zero, the response of the filter is equal to 1. That is, the filter is passing all of the input. Why? To answer this question, we note that at $\omega = 0$, the impedance of the capacitor, $1/j\omega C$, becomes infinite. Thus, the capacitor acts as an open circuit, and the output voltage equals the input:

$$\mathbf{V}_o(j\omega = 0) = \mathbf{V}_i(j\omega = 0) \tag{6.17}$$

Since a signal at sinusoidal frequency equal to zero is a DC signal, this filter circuit does not in any way affect DC voltages and currents. As the signal frequency increases, the magnitude of the frequency response decreases, since the denominator increases with ω. More precisely, equations 6.18 to 6.21 describe the magnitude and phase of the frequency response of the RC filter:

$$\begin{aligned} H(j\omega) &= \frac{\mathbf{V}_o}{\mathbf{V}_i}(j\omega) = \frac{1}{1 + j\omega CR} \\ &= \frac{1}{\sqrt{1 + (\omega CR)^2}} \frac{e^{j0^\circ}}{e^{j \arctan(\omega CR/1)}} \\ &= \frac{1}{\sqrt{1 + (\omega CR)^2}} \cdot e^{-j \arctan(\omega CR)} \end{aligned} \tag{6.18}$$

or

$$H(j\omega) = |H(j\omega)| e^{j\phi_H(j\omega)} \tag{6.19}$$

with

$$|H(j\omega)| = \frac{1}{\sqrt{1 + (\omega CR)^2}} = \frac{1}{\sqrt{1 + (\omega/\omega_0)^2}} \tag{6.20}$$

and

$$\phi_H(j\omega) = -\arctan(\omega CR) = -\arctan\left(\frac{\omega}{\omega_0}\right) \tag{6.21}$$

with

$$\omega_0 = \frac{1}{RC} \tag{6.22}$$

The simplest way to envision the effect of the filter is to think of the phasor voltage $\mathbf{V}_i = V_i e^{j\phi_i}$ scaled by a factor of $|H|$ and shifted by a phase angle ϕ_H by the filter *at each frequency*, so that the resultant output is given by the phasor $V_o e^{j\phi_o}$, with

$$\begin{aligned} V_o &= |H| \cdot V_i \\ \phi_o &= \phi_H + \phi_i \end{aligned} \tag{6.23}$$

and where $|H|$ and ϕ_H are functions of frequency. The frequency ω_0 is called the **cutoff frequency** of the filter and, as will presently be shown, gives an indication of the filtering characteristics of the circuit.

It is customary to represent $H(j\omega)$ in two separate plots, representing $|H|$ and ϕ_H as functions of ω. These are shown in Figure 6.10 in normalized form—that is, with $|H|$ and ϕ_H plotted versus ω/ω_0, corresponding to a cutoff frequency $\omega_0 = 1$ rad/s. Note that, in the plot, the frequency axis has been scaled logarithmically. This is a common practice in electrical engineering, because it allows viewing a very broad range of frequencies on the same plot without excessively compressing the low-frequency end of the plot. The frequency response plots of Figure 6.10 are commonly employed to describe the frequency response of a circuit, since they can provide a clear idea at a glance of the effect of a filter on an excitation signal. For example, the RC filter of Figure 6.9 has the property of "passing" signals at low frequencies ($\omega \ll 1/RC$) and of filtering out signals at high frequencies ($\omega \gg 1/RC$). This type of filter is called a **low-pass filter**. The cutoff frequency $\omega = 1/RC$ has a special significance in that it represents—approximately—the point where the filter begins to filter out the higher-frequency signals. The value of $H(j\omega)$ at the cutoff frequency is $1/\sqrt{2} = 0.707$. Note how the cutoff frequency depends exclusively on the values of R and C. Therefore, one can adjust the filter response as desired simply by selecting appropriate values for C and R, and therefore choose the desired filtering characteristics.

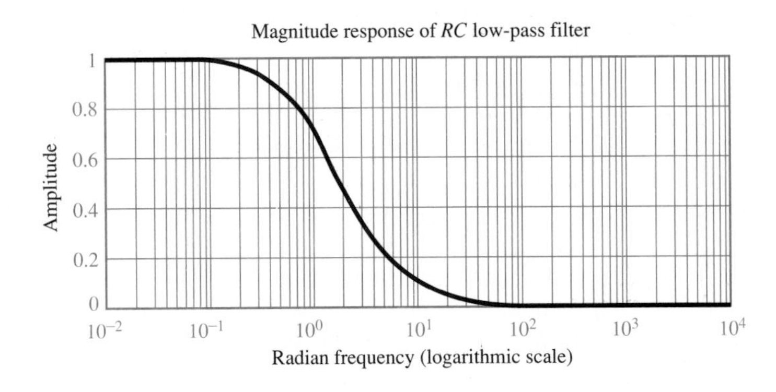

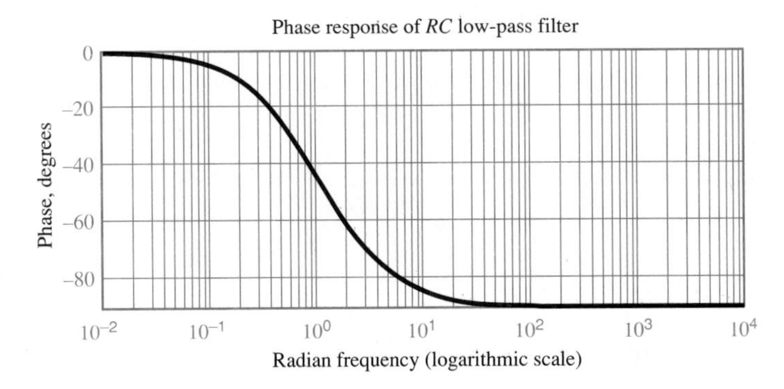

Figure 6.10 Magnitude and phase response plots for RC filter

Example 6.3

To develop a better understanding of frequency response, it is instructive to compute the effects of the RC filter circuit of Figure 6.9 on a sinusoidal input signal at frequencies above and below the cutoff frequency. Assume that $R = 1\ \text{k}\Omega$ and $C = 0.47\ \mu\text{F}$. Compute the response of the filter at the frequencies 60 Hz and 10,000 Hz. Also compute the output voltage if the input voltage (at each of these two frequencies) has a peak amplitude of 5 V and zero phase angle.

Solution:

Observe, first, that the cutoff frequency of the filter is

$$\omega_0 = \frac{1}{RC} = 2{,}127.7 \text{ rad/s}$$

or

$$f_0 = \frac{\omega_0}{2\pi} = 338.6 \text{ Hz}$$

Evaluating the amplitude at the first frequency, $\omega = 2\pi \times 60$, we obtain

$$\left|\frac{V_o}{V_i}(\omega = 2\pi \times 60)\right| = \frac{1}{\sqrt{1 + \left(\dfrac{\omega}{\omega_0}\right)^2}} = \frac{1}{\sqrt{1 + \left(\dfrac{377}{2{,}127.7}\right)^2}} = 0.9847$$

The phase angle may also be computed at the same frequency:

$$\phi(\omega = 2\pi \times 60) = -\arctan\left(\frac{377}{2{,}127.7}\right) = -10.05^\circ$$

This result corresponds to stating that if the input voltage to the filter were a sinusoid with zero phase and, say, a 5-V amplitude, the output voltage would consist of a sinusoid at the same frequency with amplitude 0.9847×5 volts and phase equal to -10.05°:

$$v_i(t) = 5\cos(377t)$$

$$v_o(t) = 4.9233\cos(377t - 10.05^\circ)$$

At the frequency $\omega = 20{,}000\pi$ rad/s (10,000 Hz), the magnitude and phase response are computed to be

$$\left|\frac{V_o}{V_i}(\omega = 2\pi \times 10{,}000)\right| = \frac{1}{\sqrt{1 + \left(\dfrac{20{,}000\pi}{2{,}127.7}\right)^2}} = 0.03384$$

and

$$\phi(\omega = 2\pi \times 10{,}000) = -\arctan\left(\frac{20{,}000\pi}{2{,}127.7}\right) = -88^\circ$$

so that the input and output voltages would be, respectively,

$$v_i(t) = 5\cos(20{,}000\pi t)$$
$$v_o(t) = 0.169\cos(20{,}000\pi t - 88°)$$

The magnitude and phase responses of the filter are plotted in Figure 6.11. It should be evident that only the low-frequency components of the signal are passed by the filter. This low-pass filter would only pass the *bass range* of the audio spectrum.

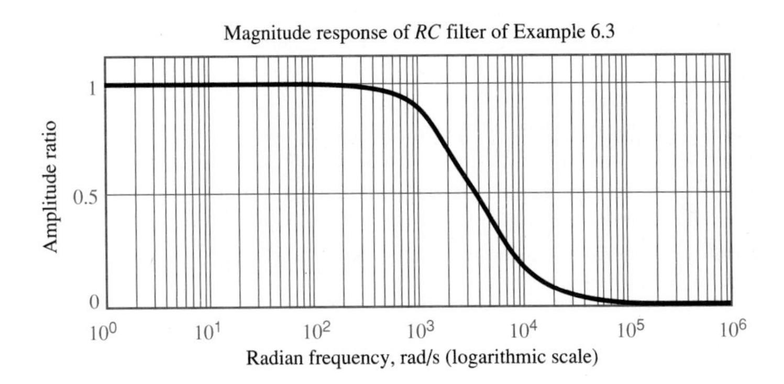

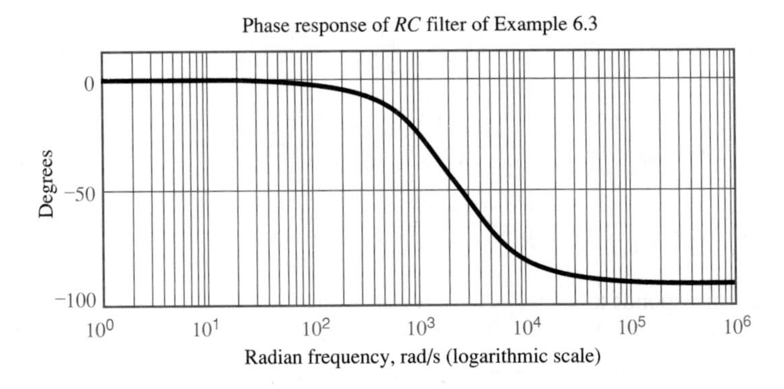

Figure 6.11 Response of *RC* filter of Example 6.3

Example 6.4

This example illustrates quantitatively the computation of the frequency response of the circuit of Figure 6.12. In this example, we assume that $R_S = 50\ \Omega$, $R_1 = 200\ \Omega$, $R_L = 500\ \Omega$, and $C = 10\ \mu$F.

Solution:

The circuit of Figure 6.12 represents a more realistic circuit than the simple *RC* filter discussed earlier in this section. It includes the internal resistance of the source (assumed to be 50 Ω) and a separate load resistance. Although the circuit of Figure 6.12 may seem

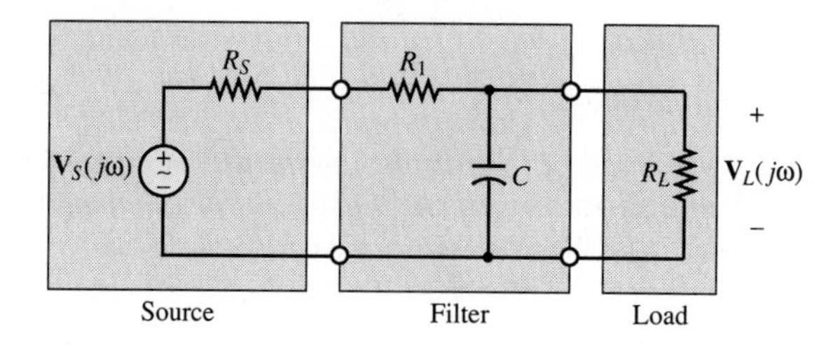

Figure 6.12 *RC* filter inserted in a circuit

more complex than that of Figure 6.9 at first, it will be shown that the two circuits may be reduced to the same form with judicious use of equivalent-circuit concepts. Observe, first, that the circuit may be represented directly in Thévenin equivalent form by combining resistors R_1 and R_S and representing the parallel combination of the capacitor and the load resistor as a single impedance, as shown in Figure 6.13. The equivalent circuit is described by the following circuit elements:

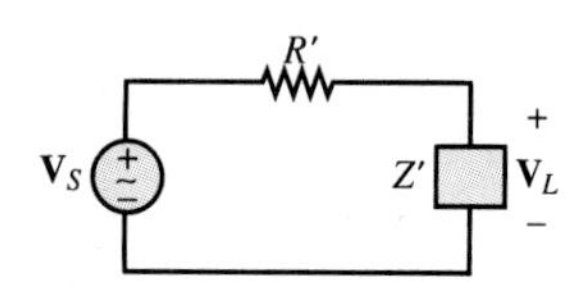

Figure 6.13 Equivalent-circuit representation of Figure 6.12

$$R' = R_S + R_1$$

$$Z' = R_L \parallel \frac{1}{j\omega C} = \frac{R_L}{1 + j\omega C R_L}$$

Now, the frequency response of the entire circuit—including the source, filter, and load—may be computed as follows:

$$\begin{aligned}
\frac{\mathbf{V}_L}{\mathbf{V}_S}(j\omega) &= \frac{Z'}{R' + Z'} = \frac{R_L/(1 + j\omega C R_L)}{R_S + R_1 + R_L/(1 + j\omega C R_L)} \\
&= \frac{R_L}{R_L + R_S + R_1 + j\omega C R_L(R_S + R_1)} \\
&= \frac{R_L/(R_L + R')}{1 + j\omega C(R_L \parallel R')}
\end{aligned}$$

Compare this last frequency response expression with equation 6.16:

$$H(j\omega) = \frac{R_L/(R_L + R')}{1 + j\omega C(R_L \parallel R')} = \frac{k}{1 + j\omega C R_{EQ}} = \frac{0.667}{1 + j(\omega/600)}$$

versus

$$\frac{\mathbf{V}_o}{\mathbf{V}_i}(j\omega) = \frac{1}{1 + j\omega C R} \tag{6.16}$$

Note the similarity between the two expressions: the denominator is virtually the same if we recognize that the parallel combination of R' and R_L (indicated as R_{EQ}) corresponds to the resistor R in equation 6.16 and the constant k replaces the 1 in the numerator. Further, the cutoff frequency takes the same form as in equation 6.22:

$$\omega_0 = \frac{1}{C(R_L \parallel R')} = \frac{1}{C R_{EQ}}$$

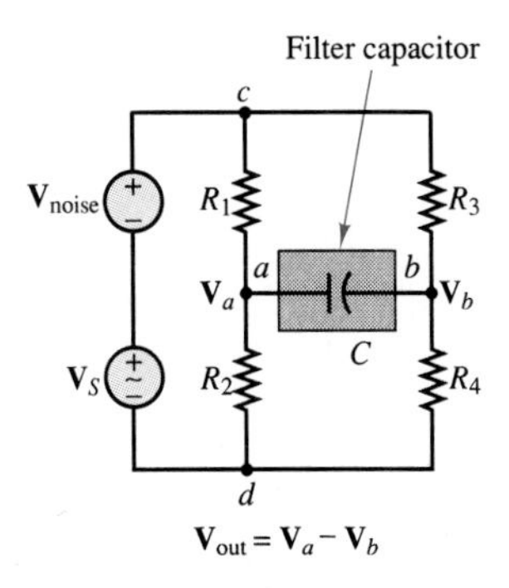

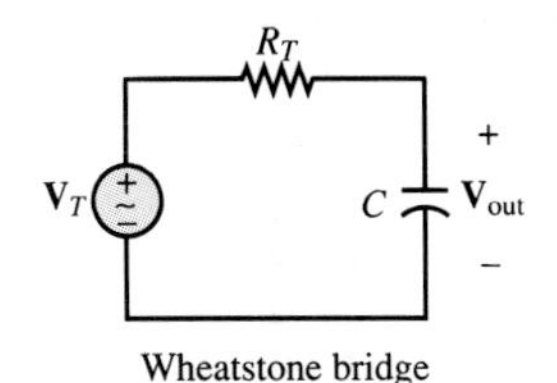

Figure 6.14 Wheatstone bridge with equivalent circuit and simple capacitive filter

Example 6.5 Wheatstone Bridge Filter

The Wheatstone bridge circuit of Examples 2.12 and 2.13 is used in a number of instrumentation applications, including the measurement of force (see Example 2.13, describing the strain gauge bridge). Figure 6.14 depicts the appearance of the bridge circuit. When undesired noise and interference are present in a measurement, it is often appropriate to use a low-pass filter to reduce the effect of the noise. The capacitor that is connected to the output terminals of the bridge in Figure 6.14 constitutes an effective and simple low-pass filter, in conjunction with the bridge resistance. Assume that the average resistance of each leg of the bridge is 350 Ω (a standard value for strain gauges) and that we desire to measure a sinusoidal force at a frequency of 30 Hz. From prior measurements, it has been determined that a filter with a cutoff frequency of 300 Hz is sufficient to reduce the effects of noise. Choose a capacitor that matches this filtering requirement.

Solution:

By evaluating the Thévenin equivalent circuit for the Wheatstone bridge, calculating the desired value for the filter capacitor becomes relatively simple, as illustrated at the bottom of Figure 6.14. The Thévenin resistance for the bridge circuit may be computed by short-circuiting the two voltage sources and removing the capacitor placed across the load terminals:

$$R_T = R_1 \| R_2 + R_3 \| R_4 = 350 \| 350 + 350 \| 350 = 350\ \Omega$$

Since the required cutoff frequency is 300 Hz, the capacitor value can be computed from the expression

$$\omega_0 = \frac{1}{R_T C} = 2\pi \times 300$$

or

$$C = \frac{1}{R_T \omega_0} = \frac{1}{350 \times 2\pi \times 300} = 1.51\ \mu\text{F}$$

The frequency response of the bridge circuit is of the same form as equation 6.16:

$$\frac{\mathbf{V}_{\text{out}}}{\mathbf{V}_T}(j\omega) = \frac{1}{1 + j\omega C R_T}$$

This response can be evaluated at the frequency of 30 Hz to verify that the attenuation and phase shift at the desired signal frequency are minimal:

$$\frac{\mathbf{V}_{\text{out}}}{\mathbf{V}_T}(j\omega = j2\pi \times 30) = \frac{1}{1 + j2\pi \times 30 \times 1.51 \times 10^{-6} \times 350}$$
$$= 0.9951\angle -5.7^\circ$$

Figure 6.15 depicts the appearance of a 30-Hz sinusoidal signal before and after the addition of the capacitor to the circuit.

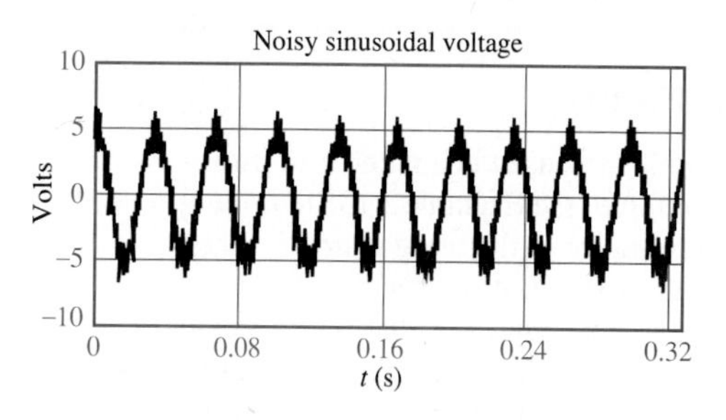

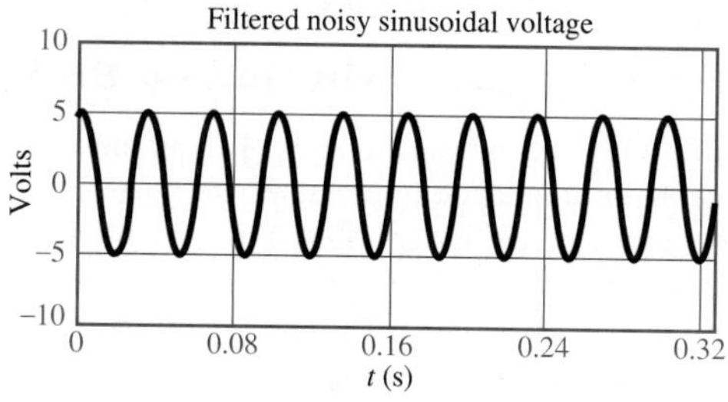

Figure 6.15 Unfiltered and filtered bridge output

Much more complex low-pass filters than the simple *RC* combinations shown so far can be designed by using appropriate combinations of various circuit elements. The synthesis of such advanced filter networks is beyond the scope of this book; however, we shall discuss the practical implementation of some commonly used filters in Chapters 11 and 14, in connection with the discussion of the operational amplifier. The next two sections extend the basic ideas introduced in the preceding pages to high- and band-pass filters—that is, to filters that emphasize the higher frequencies or a band of frequencies, respectively.

High-Pass Filters

Just as you can construct a simple filter that preserves low frequencies and attenuates higher frequencies, you can easily construct a **high-pass filter** that passes mainly those frequencies *above a certain cutoff frequency*. The analysis of a simple high-pass filter can be conducted by analogy with the preceding discussion of the low-pass filter. Consider the circuit shown in Figure 6.16. The frequency response for the high-pass filter, $H(j\omega) = \frac{\mathbf{V}_o}{\mathbf{V}_i}(j\omega)$, may be obtained by noting that

RC high-pass filter. The circuit preserves higher frequencies while attenuating the frequencies below the cutoff frequency, $\omega_0 = 1/RC$.

$$\mathbf{V}_o(j\omega) = \mathbf{V}_i(j\omega)\frac{R}{R + 1/j\omega C} = \mathbf{V}_i(j\omega)\frac{j\omega CR}{1 + j\omega CR} \tag{6.24}$$

Thus, the frequency response of the filter is:

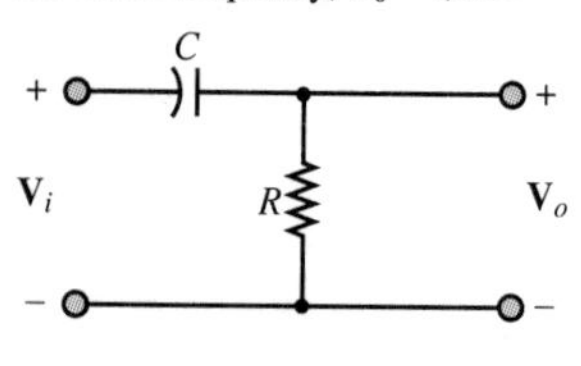

Figure 6.16 High-pass filter

$$\frac{\mathbf{V}_o}{\mathbf{V}_i}(j\omega) = \frac{j\omega CR}{1 + j\omega CR} \tag{6.25}$$

which can be expressed in magnitude-and-phase form by

$$H(j\omega) = \frac{\mathbf{V}_o}{\mathbf{V}_i}(j\omega) = \frac{j\omega CR}{1 + j\omega CR} = \frac{\omega CR e^{j90^\circ}}{\sqrt{1 + (\omega CR)^2} e^{j\arctan(\omega CR/1)}} \tag{6.26}$$

$$= \frac{\omega CR}{\sqrt{1 + (\omega CR)^2}} \cdot e^{j(90^\circ - \arctan(\omega CR))}$$

or

$$H(j\omega) = |H|e^{j\phi_H}$$

with

$$H(j\omega) = \frac{\omega CR}{\sqrt{1 + (\omega CR)^2}} \qquad \textbf{(6.27)}$$

$$\phi_H(j\omega) = 90° - \arctan(\omega CR)$$

You can verify by inspection that the amplitude response of the high-pass filter will be zero at $\omega = 0$ and will asymptotically approach 1 as ω approaches infinity, while the phase shift is 90° at $\omega = 0$ and tends to zero for increasing ω. Amplitude-and-phase response curves for the high-pass filter are shown in Figure 6.17. These plots have been normalized to have the filter cutoff frequency $\omega_0 = 1$ rad/s. Note that, once again, it is possible to define a cutoff frequency at $\omega_0 = 1/RC$ in the same way as was done for the low-pass filter.

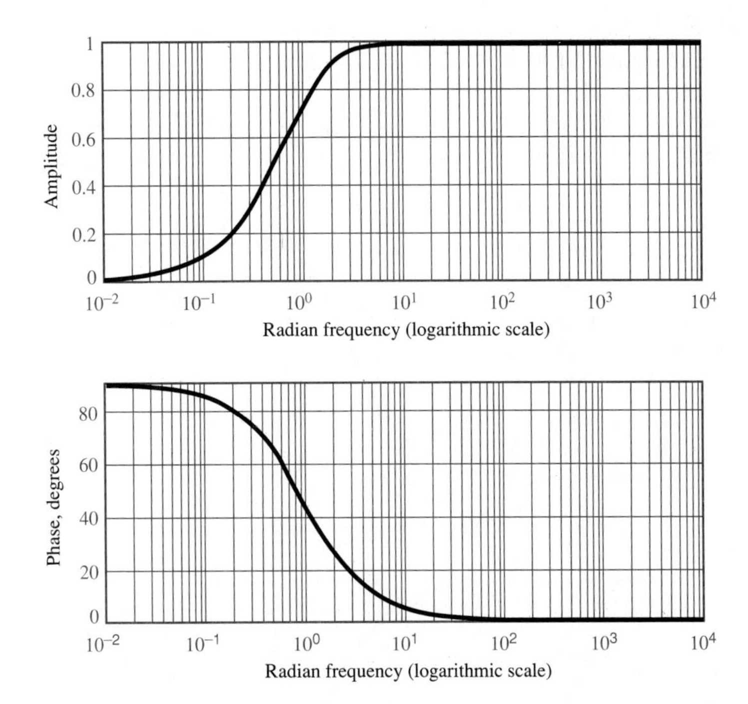

Figure 6.17 Frequency response of a high-pass filter

Example 6.6

This example will analyze the frequency response of a high-pass RC circuit. Assume that a 0.199-μF capacitor and a 200-Ω resistor are employed in constructing a high-pass filter similar to that of Figure 6.16. The cutoff frequency of the filter is

$$\omega_0 = \frac{1}{RC} = 2\pi \times 4{,}000 \text{ rad/s}$$

Thus, the frequency response of equation 6.25 becomes

$$\frac{\mathbf{V}_o}{\mathbf{V}_i}(j\omega) = \frac{j\omega/(2\pi \times 4{,}000)}{1 + j\omega/(2\pi \times 4{,}000)}$$

The magnitude and phase of this frequency response are given by

$$\left|\frac{\mathbf{V}_o}{\mathbf{V}_i}(j\omega)\right| = \frac{\omega/\omega_0}{\sqrt{1 + (\omega/\omega_0)^2}}$$

and

$$\phi(j\omega) = 90^\circ - \arctan\left(\frac{\omega}{\omega_0}\right)$$

and are plotted in Figure 6.18. We can compute the frequency response of the filter at the frequencies $\omega = 2\pi \times 100$ rad/s and $\omega = 2\pi \times 10{,}000$ rad/s to evaluate its performance:

$$\left|\frac{\mathbf{V}_o}{\mathbf{V}_i}(\omega = 2\pi \times 100)\right| = \frac{100/4{,}000}{\sqrt{1 + (100/4{,}000)^2}} = 0.025$$

$$\phi(\omega = 2\pi \times 100) = 90^\circ - \arctan\left(\frac{100}{4{,}000}\right) = 88.6^\circ$$

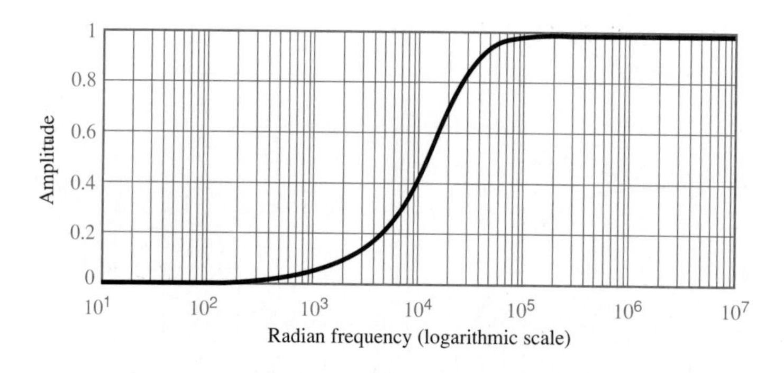

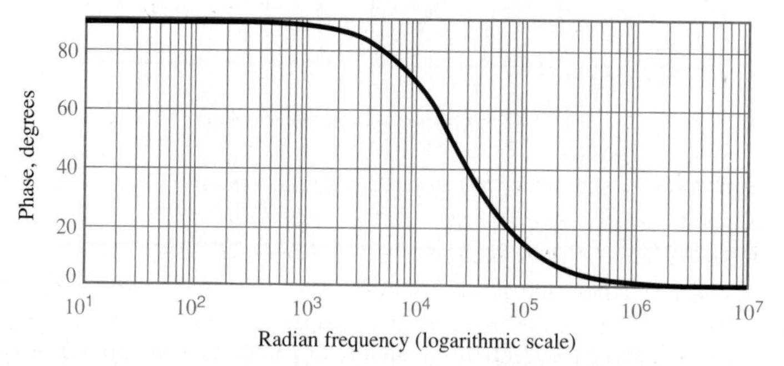

Figure 6.18 Response of high-pass filter of Example 6.6

and

$$\left|\frac{\mathbf{V}_o}{\mathbf{V}_i}(\omega = 2\pi \times 10{,}000)\right| = \frac{10{,}000/4{,}000}{\sqrt{1 + (10{,}000/4{,}000)^2}} = 0.928$$

$$\phi(\omega = 2\pi \times 10{,}000) = 90^\circ - \arctan\left(\frac{10{,}000}{4{,}000}\right) = 21.8^\circ$$

It should be apparent that, while at the higher frequency most of the amplitude of the input signal is preserved in the output, at the much lower frequency of 100 Hz less than 3 percent of the input signal amplitude passes through the filter. The high-pass filter analyzed in this example would "pass" only the *treble range* of the audio spectrum.

Band-Pass Filters

Building on the principles developed in the preceding sections, we can also construct a circuit that acts as a **band-pass filter**, passing mainly those frequencies *within a certain frequency range*. The analysis of a simple *second-order* band-pass filter (i.e., a filter with two energy-storage elements) can be conducted by analogy with the preceding discussions of the low-pass and high-pass filters. Consider the circuit shown in Figure 6.19, and the related frequency response function for the filter $H(j\omega) = \frac{\mathbf{V}_o}{\mathbf{V}_i}(j\omega)$. Noting that

RLC band-pass filter. The circuit preserves frequencies within a band.

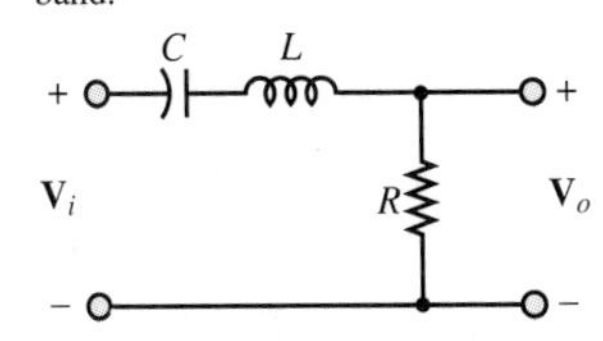

Figure 6.19 *RLC* band-pass filter

$$\begin{aligned}\mathbf{V}_o(j\omega) &= \mathbf{V}_i(j\omega) \cdot \frac{R}{R + 1/j\omega C + j\omega L} \\ &= \mathbf{V}_i(j\omega)\frac{j\omega CR}{1 + j\omega CR + (j\omega)^2 LC}\end{aligned} \tag{6.28}$$

we may write the frequency response of the filter as

$$\frac{\mathbf{V}_o}{\mathbf{V}_i}(j\omega) = \frac{j\omega CR}{1 + j\omega CR + (j\omega)^2 LC} \tag{6.29}$$

Equation 6.29 can often be factored into the following form:

$$\frac{\mathbf{V}_o}{\mathbf{V}_i}(j\omega) = \frac{jA\omega}{(j\omega/\omega_1 + 1)(j\omega/\omega_2 + 1)} \tag{6.30}$$

where ω_1 and ω_2 are the two frequencies that determine the **pass-band** (or **bandwidth**) of the filter—that is, the frequency range over which the filter "passes" the input signal—and A is a constant that results from the factoring. An immediate observation we can make is that if the signal frequency, ω, is zero, the response of the filter is equal to zero, since at $\omega = 0$ the impedance of the capacitor, $1/j\omega C$, becomes infinite. Thus, the capacitor acts as an open circuit, and the output voltage equals zero. Further, we note that the filter output in response to an input signal at sinusoidal frequency approaching infinity is again equal to zero. This result can be verified by considering that as ω approaches infinity, the impedance of the inductor becomes infinite, that is, an open circuit. Thus, the filter cannot pass signals at very high frequencies. In an intermediate band of frequencies, the band-pass filter circuit will provide a variable attenuation of the input signal, dependent on the frequency of the excitation. This may be verified by taking a closer look at equation 6.30:

$$
\begin{aligned}
H(j\omega) &= \frac{\mathbf{V}_o}{\mathbf{V}_i}(j\omega) = \frac{jA\omega}{(j\omega/\omega_1 + 1)(j\omega/\omega_2 + 1)} \\
&= \frac{A\omega e^{j90^\circ}}{\sqrt{1 + \left(\frac{\omega}{\omega_1}\right)^2}\sqrt{1 + \left(\frac{\omega}{\omega_2}\right)^2}\, e^{j\arctan(\omega/\omega_1)} e^{j\arctan(\omega/\omega_2)}} \\
&= \frac{A\omega}{\sqrt{\left[1 + \left(\frac{\omega}{\omega_1}\right)^2\right]\left[1 + \left(\frac{\omega}{\omega_2}\right)^2\right]}} \cdot e^{j[90^\circ - \arctan(\omega/\omega_1) - \arctan(\omega/\omega_2)]}
\end{aligned}
\tag{6.31}
$$

Equation 6.31 is of the form $H(j\omega) = |H|e^{j\phi_H}$, with

$$
|H(j\omega)| = \frac{A\omega}{\sqrt{\left[1 + \left(\frac{\omega}{\omega_1}\right)^2\right]\left[1 + \left(\frac{\omega}{\omega_2}\right)^2\right]}} \tag{6.32}
$$

and

$$
\phi_H(j\omega) = 90^\circ - \arctan\left(\frac{\omega}{\omega_1}\right) - \arctan\left(\frac{\omega}{\omega_2}\right)
$$

The magnitude and phase plots for the frequency response of the band-pass filter of Figure 6.19 are shown in Figure 6.20. These plots have been normalized to have the filter pass-band centered at the frequency $\omega = 1$ rad/s.

The frequency response plots of Figure 6.20 suggest that, in some sense, the band-pass filter acts as a combination of a high-pass and a low-pass filter. As illustrated in the previous cases, it should be evident that one can adjust the filter response as desired simply by selecting appropriate values for L, C, and R.

The expression for the frequency response of a second-order band-pass filter (equation 6.29) can also be rearranged to illustrate two important features of this circuit: the **quality factor**, Q, and the **resonant frequency**, ω_0. Let

$$
\omega_0 = \frac{1}{\sqrt{LC}} \qquad \text{and} \qquad Q = \omega_0 CR = \frac{R}{\omega_0 L} \tag{6.33}
$$

Then we can write $\omega CR = \omega_0 CR \frac{\omega}{\omega_0} = Q \frac{\omega}{\omega_0}$ and rearrange equation 6.29 as follows:

$$
\frac{\mathbf{V}_o}{\mathbf{V}_i}(j\omega) = \frac{jQ\frac{\omega}{\omega_0}}{\left(\frac{j\omega}{\omega_0}\right)^2 + jQ\frac{\omega}{\omega_0} + 1} \tag{6.34}
$$

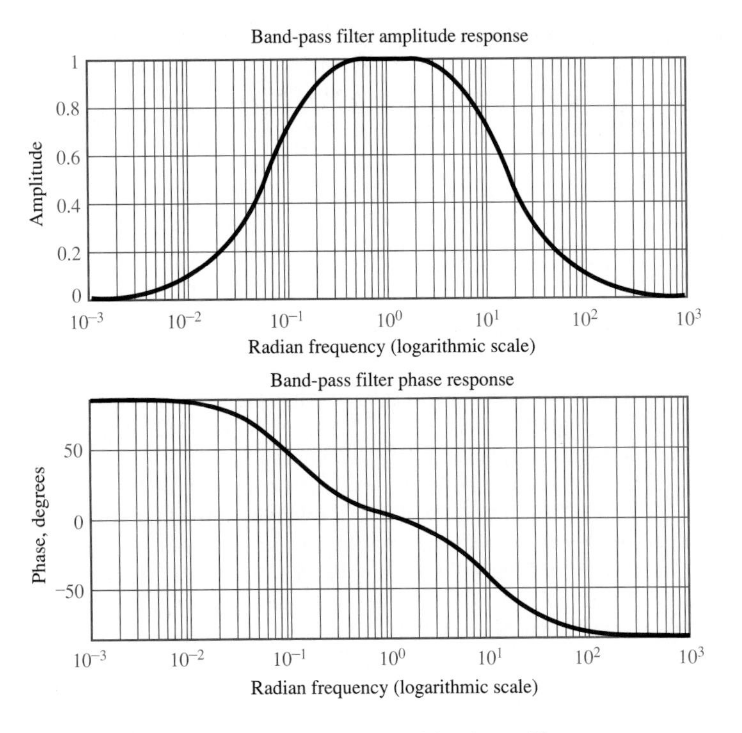

Figure 6.20 Frequency response of *RLC* band-pass filter

In equation 6.34, the resonant frequency, ω_0, corresponds to the center frequency of the filter, while Q, the quality factor, indicates the *sharpness* of the resonance, that is, how narrow or wide the shape of the pass-band of the filter is. The width of the pass-band is also referred to as the *bandwidth*, and it can easily be shown that the bandwidth of the filter is given by the expression

$$B = \frac{\omega_0}{Q} \tag{6.35}$$

Thus, a high-Q filter has a narrow bandwidth, while a low-Q filter has a large bandwidth and is therefore less selective. The quality factor of a filter provides an immediate indication of the nature of the filter. The following examples illustrate the significance of these parameters in the response of various *RLC* filters.

Example 6.7

This examples illustrates quantitatively the computation of the frequency response of a band-pass *RLC* circuit such as the one of Figure 6.19. We can write the frequency response of the band-pass filter as

$$\frac{\mathbf{V}_o}{\mathbf{V}_i}(j\omega) = \frac{j\omega CR}{1 + j\omega CR + (j\omega)^2 LC}$$

The general form of this response is

$$H(j\omega) = \frac{\mathbf{V}_o}{\mathbf{V}_i}(j\omega) = \frac{\omega CRe^{j90^\circ}}{\sqrt{(1-\omega^2LC)^2 + (\omega CR)^2}e^{j\arctan[\omega CR/(1-\omega^2LC)]}}$$

with

$$|H(j\omega)| = \frac{\omega CR}{\sqrt{(1-\omega^2LC)^2 + (\omega CR)^2}}$$

$$\phi(j\omega) = 90^\circ - \arctan\left(\frac{\omega CR}{1-\omega^2LC}\right)$$

It is very instructive to consider the effect of varying the series resistance on the width of the pass-band (or bandwidth) of the filter. Figures 6.21 and 6.22 depict the band-pass filter magnitude and phase response for the following two sets of component values: $R = 1\ \text{k}\Omega, L = 5$ mH, $C = 10\ \mu$F; and $R = 10\ \Omega$, $L = 5$ mH, $C = 10\ \mu$F. For the larger value of R (1 kΩ), the filter response allows a relatively broad band of frequencies to be passed; filters with a response similar to that shown in Figure 6.21 are thus usually referred to as **broad-band filters**. For the smaller R (10 Ω), only a very narrow band of frequencies is allowed through the filter. For this reason, filters such as the one shown in Figure 6.22 are referred to as **narrow-band filters**.

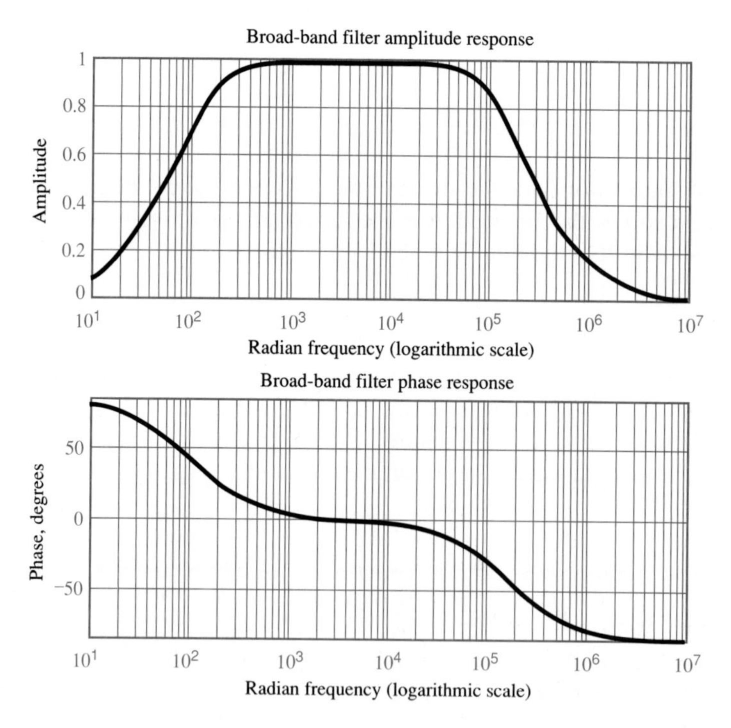

Figure 6.21 Frequency response of broad-band band-pass filter of Example 6.7

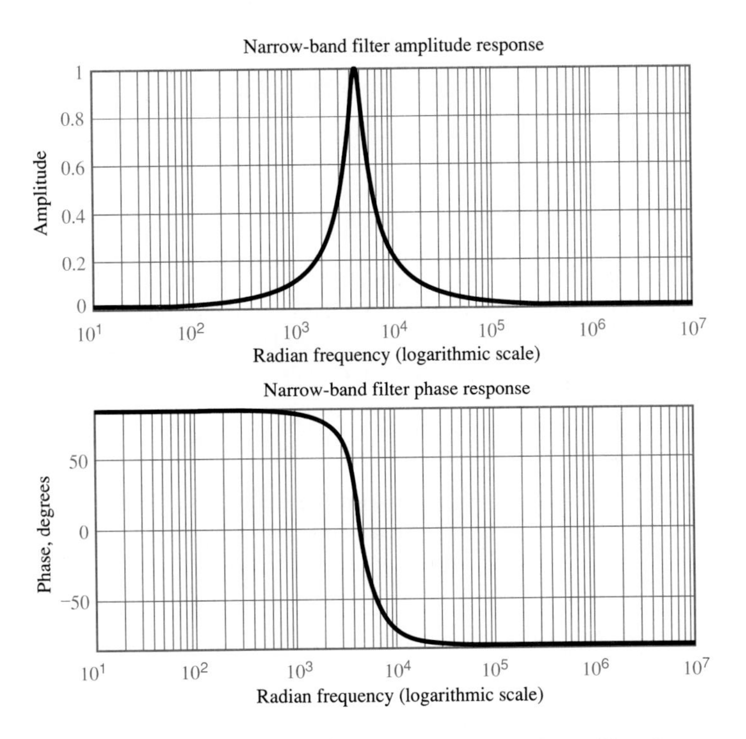

Figure 6.22 Frequency response of narrow-band band-pass filter of Example 6.7

Let us calculate the quality factor and resonant frequency of these two filters. Since L and C are the same for both circuits, the resonant frequency will also be the same: $\omega_0 = 1/\sqrt{LC} = 4.47 \times 10^3$ rad/s, or, approximately, 4,500 rad/s. On the other hand, the quality factor will differ significantly. In the narrow-band filter ($R = 1$ kΩ), $Q \approx 45$; therefore its bandwidth is $B = \omega_0/Q = 100$ rad/s. In the broad-band filter, $Q \approx 0.45$, and the bandwidth is 10,000 rad/s. The plots of Figures 6.21 and 6.22 confirm these calculations.

It should be apparent that, while at the higher and lower frequencies most of the amplitude of the input signal is filtered from the output, at the mid-band frequency (4,500 rad/s) most of the input signal amplitude passes through the filter. The first band-pass filter analyzed in this example would "pass" the *mid-band range* of the audio spectrum, while the second would pass only a very narrow band of frequencies around the **center frequency** of 4,500 rad/s. Such narrow-band filters find application in **tuning circuits**, such as those employed in conventional AM radios (although at frequencies much higher than that of the present example). In a tuning circuit, a narrow-band filter is used to tune in a frequency associated with the **carrier** of a radio station (for example, for a station found at a setting of "AM 820," the carrier wave transmitted by the radio station is at a frequency of 820 kHz). By using a variable capacitor, it is possible to tune in a range of carrier frequencies and therefore select the preferred station. Other circuits are then used to decode the actual speech or music signal modulated on the carrier wave; some of these will be discussed in Chapter 7.

Example 6.8 AC Line Interference Filter

One very useful application of narrow-band filters is in rejecting interference due to the AC line power. An undesired 60-Hz signal originating in the AC line power can cause serious interference in delicate instruments by disturbing the desired signals. In medical instruments such as the electrocardiograph, 60-Hz notch filters are often provided to reduce the effect of this interference upon cardiac measurements. Figure 6.23 depicts a circuit representing the situation just described by means of two series voltage sources: one representing the desired signal, the other representing the undesired 60-Hz interference. This example will illustrate the design of a 60-Hz notch filter capable of reducing the interference caused by the AC line power. Assume that $R_S = 50\ \Omega$ and $R_L = 150\ \Omega$ in the circuit of Figure 6.23.

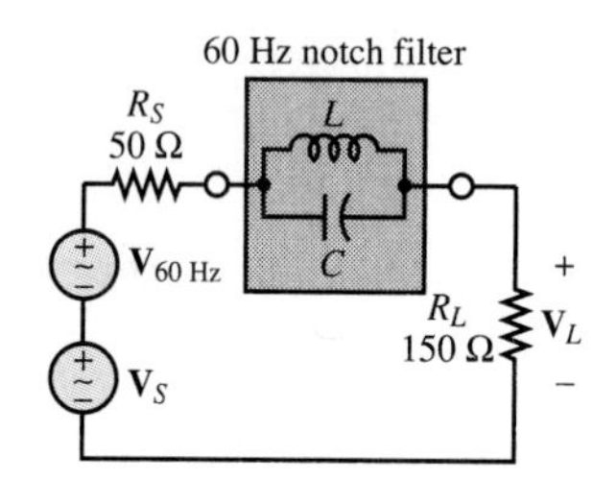

Figure 6.23 60-Hz notch filter

Solution:

The **notch filter** shown in Figure 6.23 utilizes the parallel combination of an inductor and a capacitor to attain the desired band-rejection effect. To understand how this takes place, consider the combined impedance of the parallel LC circuit:

$$Z_{\|} = Z_L \| Z_C = \frac{j\omega L / j\omega C}{j\omega L + 1/j\omega C} = \frac{j\omega L}{1 - \omega^2 LC}$$

and note that when $\omega^2 LC = 1$, or $\omega = 1/\sqrt{LC}$, the impedance of the circuit is infinite! The frequency

$$\omega_R = \frac{1}{\sqrt{LC}}$$

is the resonant frequency of the circuit. If the capacitor and the inductor were selected to have a resonant frequency of 60 Hz, then the series circuit would show an infinite impedance to 60-Hz currents while passing currents at other frequencies. Of course, we shall see that, because some finite resistance always exists in the circuit, some 60-Hz current will always pass through the notch filter. However, notch filters can be quite selective in rejecting one frequency component without excessively disturbing the desired signal components.

To analyze the complete circuit, we compute the frequency response of the circuit, $H_V(j\omega) = \frac{\mathbf{V}_o}{\mathbf{V}_i}(j\omega)$, where $\mathbf{V}_o = \mathbf{V}_L$ and $\mathbf{V}_i = \mathbf{V}_{60\ \text{Hz}} + \mathbf{V}_S$. $H_V(j\omega)$ can be determined by applying the voltage divider rule, to obtain

$$H_V(j\omega) = \frac{\mathbf{V}_o}{\mathbf{V}_i}(j\omega) = \frac{R_L}{R_L + R_S + Z_{\|}} = \frac{150}{200 + j\omega L/(1 - \omega^2 LC)}$$

To obtain the desired rejection of the 60-Hz signal, we need to select appropriate component values to satisfy the relationship

$$\omega_R = \frac{1}{\sqrt{LC}} = 2\pi \times 60$$

Selecting $L = 100$ mH leads to the following value for C:

$$C = \frac{1}{L\omega_R^2} = 70.36\ \mu\text{F}$$

The frequency response of the complete circuit is plotted in Figure 6.24. Note that the magnitude response takes a sharp dip around 60 Hz, and that the phase angle is equal to zero, except for a small region around 60 Hz.

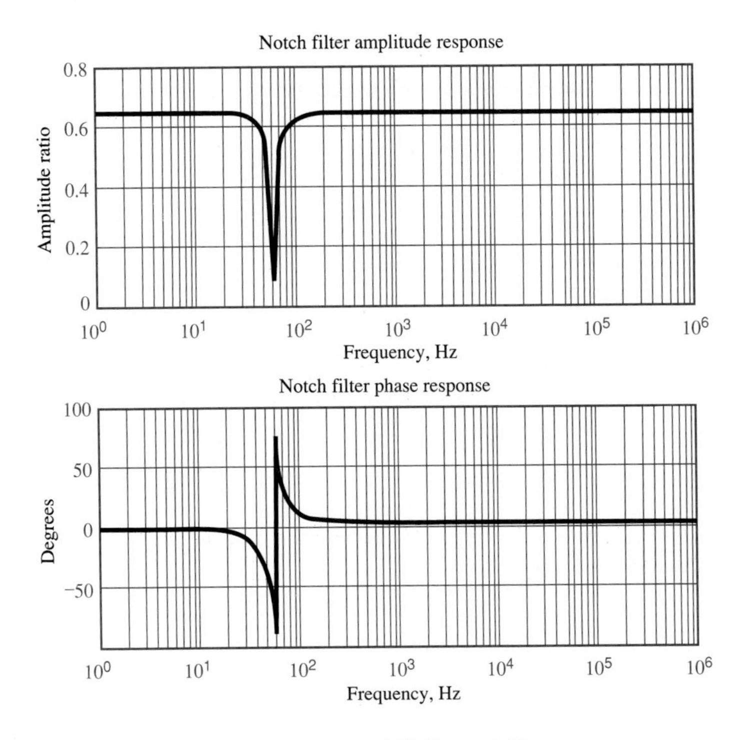

Figure 6.24 Frequency response of 60-Hz notch filter

Example 6.9 Displacement Transducer

This example illustrates the application of the frequency response idea to a practical displacement transducer. The frequency response of a *seismic displacement transducer* is analyzed, and it is shown that there is an analogy between the equations describing the mechanical transducer and those that describe a second-order electrical circuit.

The configuration of the transducer is shown in Figure 6.25. The transducer is housed in a case rigidly affixed to the surface of a body whose motion is to be measured. Thus, the case will experience the same displacement as the body, x_i. Inside the case, a small mass, M, rests on a spring characterized by stiffness K, placed in parallel with a damper, B. The wiper arm of a potentiometer is connected to the floating mass, M; the potentiometer is attached to the transducer case, so that the voltage V_o is proportional to the *relative displacement of the mass with respect to the case*, x_o.

The equation of motion for the mass-spring-damper system may be obtained by summing all the forces acting on the mass M:

$$Kx_o + B\frac{dx_o}{dt} = M\frac{d^2x_M}{dt^2} = M\left(\frac{d^2x_i}{dt^2} - \frac{d^2x_o}{dt^2}\right)$$

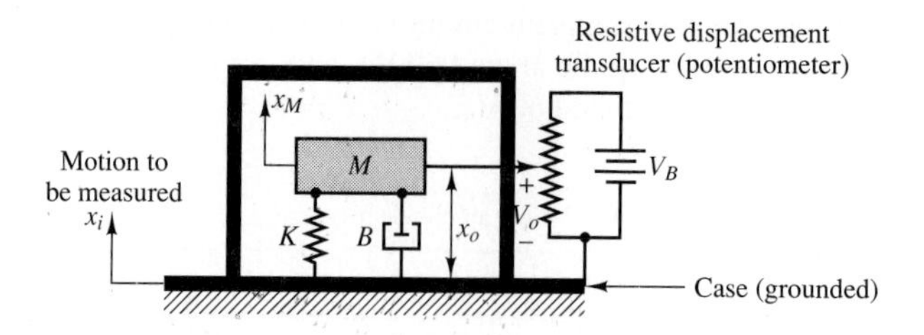

Figure 6.25 Seismic displacement transducer

where we have noted that the motion of the mass is equal to the difference between the motion of the case and the motion of the mass relative to the case itself; that is,

$$x_M = x_i - x_o$$

If we assume that the motion of the mass is sinusoidal, we may use phasor analysis to obtain the frequency response of the transducer by defining the phasor quantities

$$\mathbf{X}_i(j\omega) = |X_i|e^{j\phi_i} \qquad \text{and} \qquad \mathbf{X}_o(j\omega) = |X_o|e^{j\phi_o}$$

The assumption of a sinusoidal motion may be justified in light of the discussion of Fourier analysis in Section 6.1. If we then recall (from Chapter 4) that taking the derivative of a phasor corresponds to multiplying the phasor by $j\omega$, we can rewrite the second-order differential equation as follows:

$$M(j\omega)^2\mathbf{X}_o + B(j\omega)\mathbf{X}_o + K\mathbf{X}_o = M(j\omega)^2\mathbf{X}_i$$

$$(-\omega^2 M + j\omega B + K)\mathbf{X}_o = -\omega^2 M\mathbf{X}_i$$

and we can write an expression for the frequency response:

$$\frac{\mathbf{X}_o(j\omega)}{\mathbf{X}_i(j\omega)} = H(j\omega) = \frac{-\omega^2 M}{-\omega^2 M + j\omega B + K}$$

The frequency response of the transducer is plotted in Figure 6.26 for the component values $M = 0.005$ kg and $K = 1{,}000$ N/m and for three values of B:

$B = 10$ N/ms (dotted line)

$B = 2$ N/ms (dashed line)

and

$B = 1$ N/ms (solid line)

The transducer clearly displays a high-pass response, indicating that for a sufficiently high input signal frequency, the measured displacement (proportional to the voltage V_o) is equal to the input displacement, x_i, which is the desired quantity. Note how sensitive the frequency response of the transducer is to changes in damping: as B changes from 2 to 1, a sharp **resonant peak** appears around the frequency $\omega = 316$ rad/s (approximately 50 Hz). As B increases to a value of 10, the amplitude response curve shifts to the right. Thus, this transducer, with the preferred damping given by $B = 2$, would be capable of correctly measuring displacements at frequencies above a minimum value, about 1,000 rad/s (or 159 Hz). The choice of $B = 2$ as the preferred design may be explained by observing that, ideally, we would like to obtain a constant amplitude response at all frequencies. The magnitude response that most closely approximates the ideal case in Figure

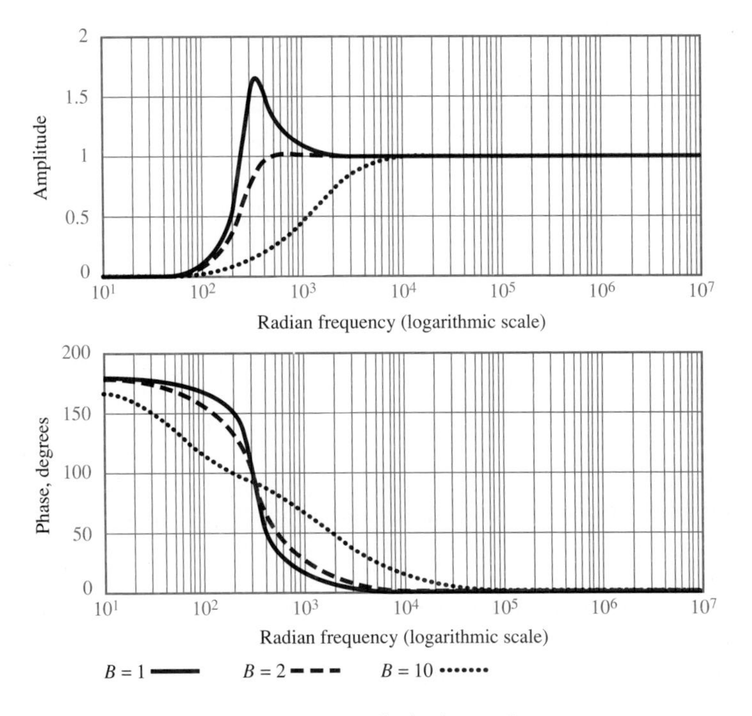

Figure 6.26 Frequency response of seismic transducer

6.26 corresponds to $B = 2$. This concept is commonly applied to a variety of vibration measurements.

We now illustrate how a second-order electrical circuit will exhibit the same type of response as the seismic transducer. Consider the circuit shown in Figure 6.27. The frequency response for the circuit may be obtained by using the principles developed in the preceding sections:

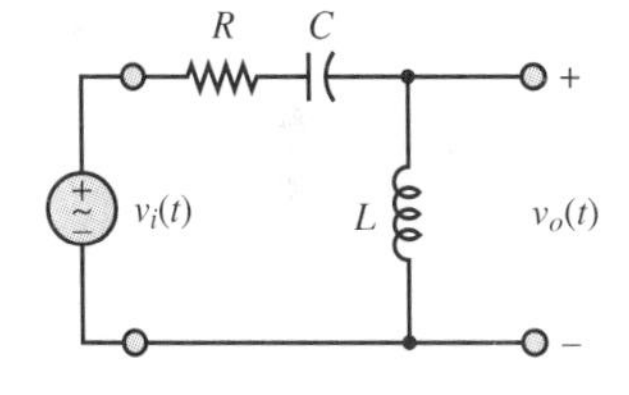

Figure 6.27 Electrical circuit analog of the seismic transducer

$$\frac{\mathbf{V}_o}{\mathbf{V}_i}(j\omega) = \frac{j\omega L}{R + 1/j\omega C + j\omega L} = \frac{(j\omega L)(j\omega C)}{j\omega CR + 1 + (j\omega L)(j\omega C)}$$

$$= \frac{-\omega^2 L}{-\omega^2 L + j\omega R + 1/C}$$

Comparing this expression with the frequency response of the seismic transducer,

$$\frac{\mathbf{X}_o(j\omega)}{\mathbf{X}_i(j\omega)} = H(j\omega) = \frac{-\omega^2 M}{-\omega^2 M + j\omega B + K}$$

we find that there is a definite resemblance between the two. In fact, it is possible to draw an analogy between input and output motions and input and output voltages. Note also that the mass, M, plays a role analogous to that of the inductance, L. The damper, B, acts in analogy with the resistor, R; and the spring, K, is analogous to the inverse of the capacitance, C. This analogy between the mechanical system and the electrical circuit derives simply from the fact that the equations describing the two systems have the same form. Engineers often use such analogies to construct electrical *models*, or *analogs*, of physical systems. For example, to study the behavior of a large mechanical system, it might be easier and less costly to start by modeling the mechanical system with an inexpensive electrical circuit and testing the model, rather than the full-scale mechanical system.

Decibel (dB) Plots

Frequency response plots are often displayed in the form of logarithmic plots, where the horizontal axis represents frequency on a logarithmic scale (to base 10) and the vertical axis represents the amplitude of the frequency response, in units of **decibels (dB)**. In a dB plot, the ratio $|\mathbf{V}_{out}/\mathbf{V}_{in}|$ is given in units of decibels (dB), where

$$\left|\frac{\mathbf{V}_{out}}{\mathbf{V}_{in}}\right|_{dB} = 20 \log_{10} \left|\frac{\mathbf{V}_{out}}{\mathbf{V}_{in}}\right| \tag{6.36}$$

and this is plotted as a function of frequency on a $\log_{10}$ scale. Note that the use of decibels implies that one is measuring a *ratio*. Decibel plots are usually displayed on semilogarithmic paper, with decibels on the linear axis and frequency on the logarithmic axis.

Let us examine the appearance of dB plots for typical low-pass and high-pass filter circuits. From Figure 6.28, we can see that both plots have a very simple appearance: either the low-frequency part of the plot (for a low-pass filter) or the

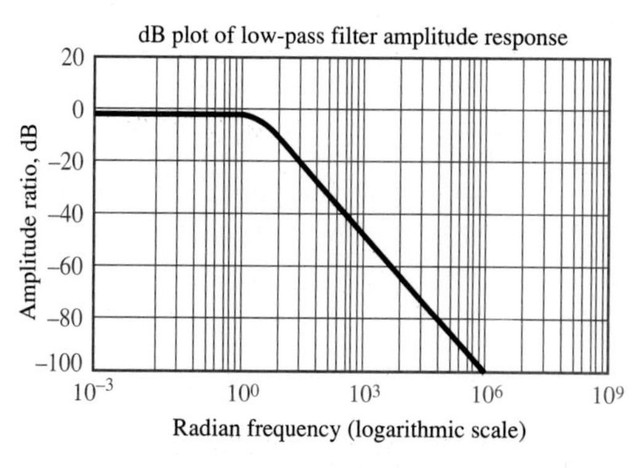

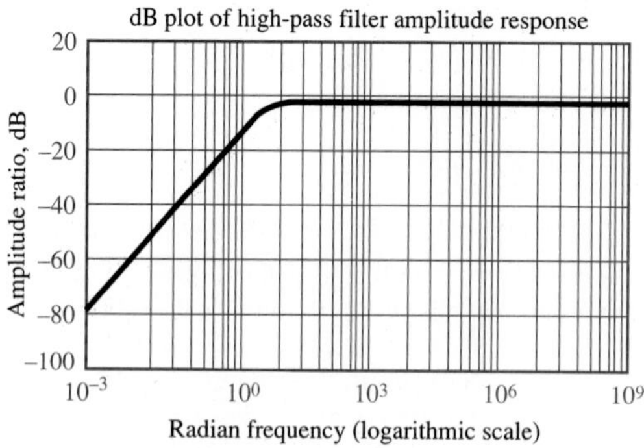

Figure 6.28 dB magnitude plots of low- and high-pass filters

high-frequency part (for the high-pass filter) is well approximated by a flat line, indicating that for some range of frequencies, the filter has a constant amplitude response, equal to 1. Further, the filter cutoff frequency, ω_0, appears quite clearly as the approximate frequency where the filter response starts to fall. The response of the circuit decreases (or increases) with a constant slope with respect to ω (on a logarithmic scale). For the high-pass and low-pass filters described earlier, this slope is equal to ± 20 dB/decade ($-$ for the low-pass filter, $+$ for the high-pass), where a **decade** is a range of frequencies f_1 to f_2 such that

$$\frac{f_2}{f_1} = 10 \tag{6.37}$$

What kind of decrease in gain is -20 dB/decade? The expression

$$|H(j\omega)|_{\text{dB}} = -20 \text{ dB} \tag{6.38}$$

means that

$$-20 = 20 \log_{10} |H(j\omega)|$$

or (6.39)

$$|H(j\omega)| = 0.1$$

That is, the gain decreases by a factor of 10 for every increase in frequency by a factor of 10. You see how natural these units are. Further, if ω_0 is known, a plot of $|H(j\omega)|_{\text{dB}}$ versus ω (on a logarithmic scale) may be readily sketched using the asymptotic approximations of two straight lines, one of slope zero and the other with slope equal to -20 dB/decade, and with intersection at ω_0. The homework problems and exercises provide a good number of practical examples of this technique.

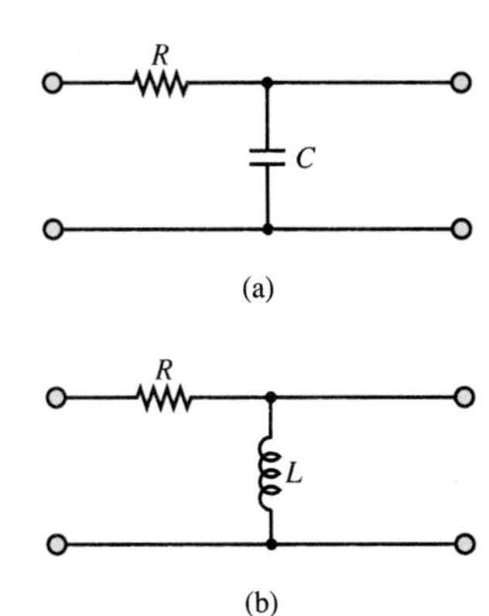

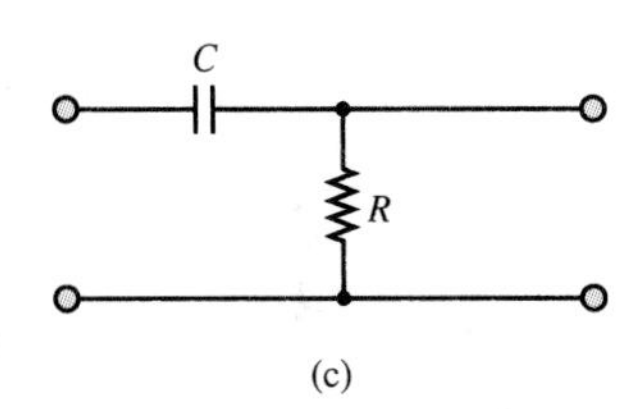

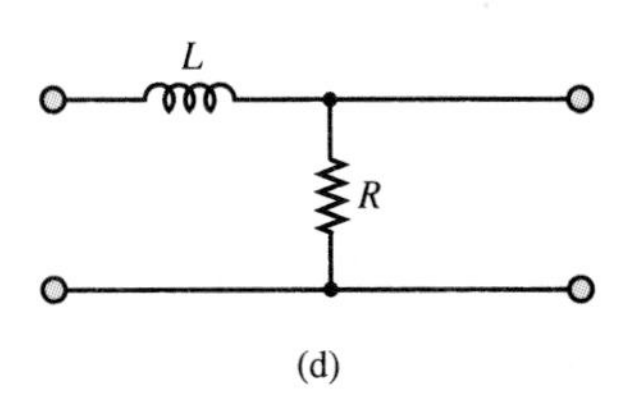

Figure 6.29

DRILL EXERCISES

6.1 Derive an expression for $H_I(j\omega) = \frac{\mathbf{I}_L}{\mathbf{I}_S}(j\omega)$ for the circuit of Figure 6.1.

6.2 Use the method of node voltages to derive $H_Y(j\omega)$ for the circuit of Figure 6.1.

6.3 Use the method of mesh currents to derive $H_V(j\omega)$ for the circuit of Figure 6.1.

6.4 Connect the filter of Example 6.3 to a 1-V sinusoidal source with internal resistance of 50 Ω to form a circuit similar to that of Figure 6.12. Determine the circuit cutoff frequency, ω_0.

6.5 Determine the cutoff frequency for each of the four "prototype" filters shown in Figure 6.29. Which are high-pass and which are low-pass?

6.6 Show that it is possible to obtain a high-pass filter response simply by substituting an inductor for the capacitor in the circuit of Figure 6.9. Derive the frequency response for the circuit.

6.7 Determine the cutoff frequency for the high-pass *RC* filter shown in Figure 6.30. [*Hint:* First find the frequency response in the form $j\omega a/(1 + j\omega b)$, where a and b are constants related to R_1, R_2, and C_1, and then solve numerically.] Sketch the amplitude and frequency responses.

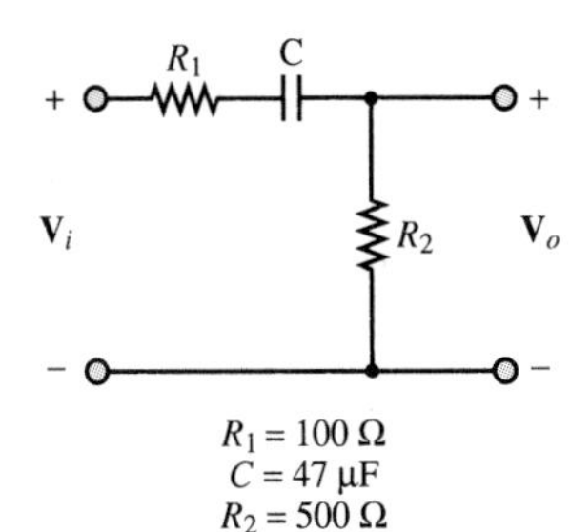

Figure 6.30

6.8 A simple *RC* low-pass filter is constructed using a 10-μF capacitor and a 2.2-kΩ resistor. Over what range of frequencies will the output of the filter be within 1 percent of the input signal amplitude (i.e., when will $V_L \geq 0.99V_S$)?

6.9 Compute the frequency at which the phase shift introduced by the circuit of Example 6.3 is equal to $-10°$.

6.10 Compute the frequency at which the output of the circuit of Example 6.3 is attenuated by 10 percent (i.e., $V_L = 0.9V_S$).

6.11 Compute the frequency at which the output of the circuit of Example 6.6 is attenuated by 10 percent (i.e., $V_L = 0.9V_S$).

6.12 Compute the frequency at which the phase shift introduced by the circuit of Example 6.6 is equal to 20°.

6.13 Compute the frequencies ω_1 and ω_2 for the band-pass filter of Example 6.7 (with $R = 1$ kΩ) for equating the magnitude of the band-pass filter frequency response to $1/\sqrt{2}$ (this will result in a quadratic equation in ω, which can be solved for the two frequencies).

6.3 TRANSIENT ANALYSIS

In analyzing the frequency response of AC circuits earlier in this chapter, we made the assumption that the particular form of the voltages and currents in the circuit was sinusoidal. For this very useful special case it was found that it was not necessary to describe the circuit by means of differential equations. Instead, by employing complex algebra and the phasor representation for sinusoids, the analysis of such circuits could be completed using the methods and techniques of Chapter 4. But how should one analyze circuits in which voltages and currents are not sinusoidal? In the case of periodic signals, it may be possible to resort to the idea of a Fourier series, which was briefly introduced in Section 6.1: a periodic signal can be expressed as the sum of a (possibly very large) number of sinusoids at various amplitudes, frequencies, and phases:

$$v(t) = \sum_{n=1}^{N} a_n \cos(\omega_n t + \phi_n) \tag{6.40}$$

In this case, sinusoidal analysis is still applicable, and the newly acquired understanding of the frequency response of a circuit may be used to obtain a complete solution by the superposition of the individual responses at each frequency.

There are many signals, however, for which the steady-state sinusoidal representation is not adequate. In particular, the Fourier method of analysis does not apply to **transient signals**, that is, voltages and currents that vary as a function of time as a consequence of a sudden change in the input. Figure 6.31 illustrates the appearance of the voltage across some hypothetical load when a DC and an AC source, respectively, are abruptly switched on at time $t = 2$ s. The waveforms in Figure 6.31 can be subdivided into three regions: a *steady-state region* for $0 \leq t \leq 0.2$ s; a *transient region*, for $0.2 \text{ s} \leq t \leq 2$ s (approximately); and a new steady-state region for $t > 2$ s, where the waveform reaches a new steady-

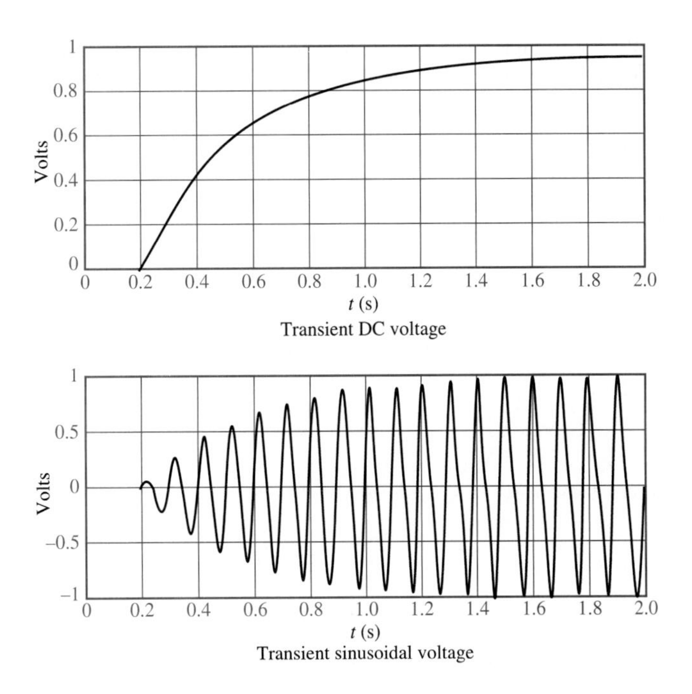

Figure 6.31 Examples of transient response

state DC or sinusoidal condition. The objective of **transient analysis** is to describe the behavior of a voltage or current during the transition that takes place between two different steady-state conditions.

You already know how to analyze circuits in a sinusoidal steady state by means of phasors. The material presented in the remainder of this chapter will provide the tools necessary to describe the *transient response* of circuits containing resistors, inductors, and capacitors. A general example of the type of circuit that will be discussed in this section is shown in Figure 6.32. The switch indicates that we turn the battery power on at time $t = 0$. Transient behavior may be expected whenever a source of electrical energy is switched on or off, whether it be AC or DC. A typical example of the transient response to a switched DC voltage would be what occurs when the ignition circuits in an automobile are turned on, so that a 12-V battery is suddenly connected to a large number of electrical circuits. The degree of complexity in transient analysis depends on the number of energy-storage elements in the circuit; the analysis can became quite involved for high-order circuits. In this chapter, we shall analyze only first- and second-order circuits—that is, circuits containing one or two energy-storage elements, respectively. In electrical engineering practice, we would typically resort to computer-aided analysis for higher-order circuits.

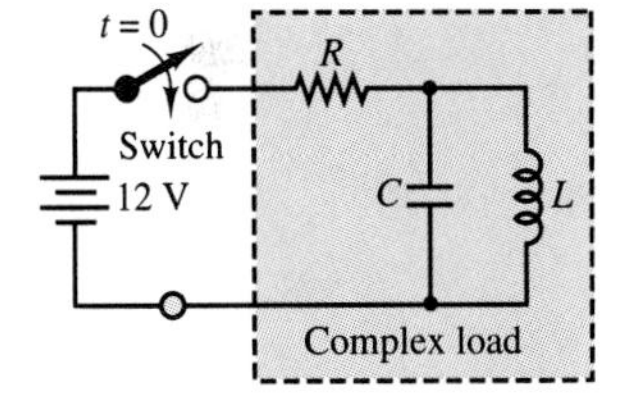

Figure 6.32 Circuit with switched DC excitation

A convenient starting point in approaching the transient response of electrical circuits is to consider the general model shown in Figure 6.33, where the circuits in the box consist of a combination of resistors connected to a *single energy-storage element*, either an inductor or a capacitor. Regardless of how many

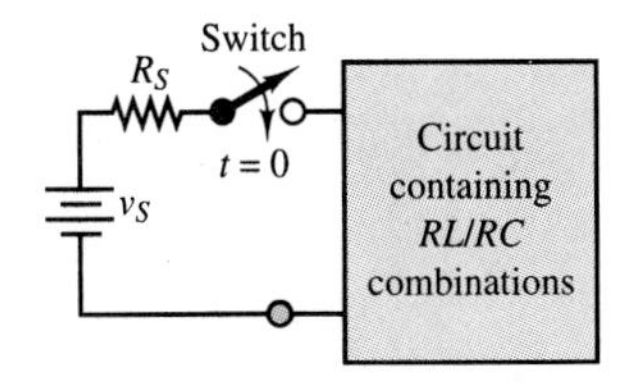

Figure 6.33 A general model of the transient analysis problem

resistors the circuit contains, it is a **first-order circuit**. In general, the response of a first-order circuit to a switched DC source will appear in one of the two forms shown in Figure 6.34, which represent, in order, a **decaying exponential** and a **rising exponential** waveform. In the next sections, we will systematically analyze these responses by recognizing that they are exponential in nature and can be computed very easily once we have the proper form of the differential equation describing the circuit.

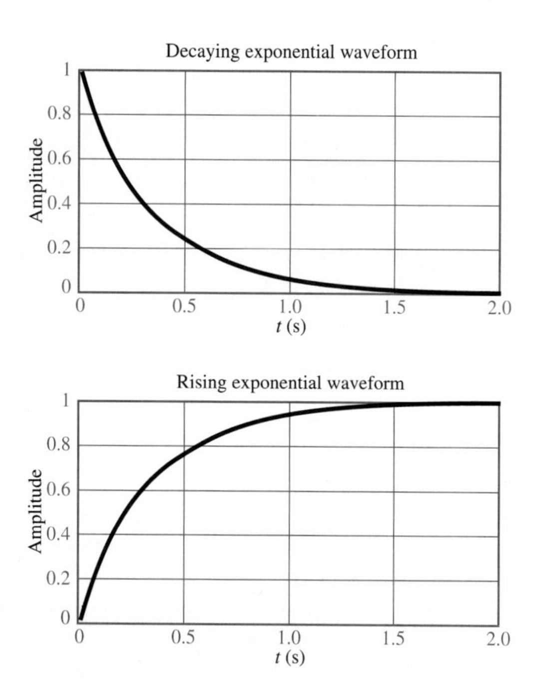

Figure 6.34 Decaying and rising exponential responses

6.4 TRANSIENT RESPONSE OF FIRST-ORDER CIRCUITS

Application of the basic circuit laws—Kirchhoff's laws, Ohm's law, and the defining relationships for the ideal inductor and capacitor—to circuits containing energy-storage elements naturally leads to the formulation of differential equations, as was explained in Section 4.3. In Section 4.4, it was found that whenever the voltages and currents in such a circuit are sinusoidal, it is possible to circumvent solving the resulting circuit differential equation by employing phasor notation. In general, however, *it is not possible to employ phasor analysis when the circuit currents and voltages are not steady-state sinusoids*. Thus, in the transient analysis of electrical circuits, it will be necessary to formulate a solution to the differential circuit equations. In the following sections, we shall revisit the solution of such equations for first- and second-order circuits. The reader is encouraged to review Section 4.3.

Consider the simple series RL circuit of Figure 6.35. Application of KVL yields the expression

$$v_S(t) = v_R(t) + v_L(t) \tag{6.41}$$

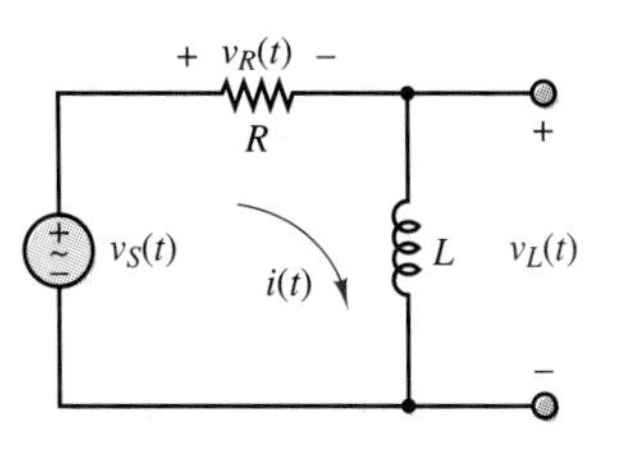

Figure 6.35 A simple RL circuit

Since the defining relationships for the ideal resistor and ideal inductor are $v_R(t) = Ri(t)$ and $v_L(t) = L\,di(t)/dt$, respectively, the following differential equation results from application of KVL:

$$\frac{di(t)}{dt} + \frac{R}{L}i(t) = \frac{1}{L}v_S(t) \tag{6.42}$$

This equation governs the relationship between the source voltage and the series current, and therefore it also indirectly explains the relationship between $v_S(t)$ and the resistor and inductor voltages. Similar equations can also be obtained for other simple RL and RC circuits (see Drill Exercises 6.14 to 6.16).

Natural Response of First-Order Circuits

Figure 6.36 compares the RL circuit of Figure 6.35 with the general form of the series RC circuit, showing the corresponding differential equation. From Figure 6.36, it is clear that equation 6.43 is in the general form of the equation for any first-order circuit:

$$\frac{dx(t)}{dt} + ax(t) = f(t) \tag{6.43}$$

where f is the **forcing function** and $x(t)$ represents either $v_C(t)$ or $i_L(t)$. The constant a is the inverse of the parameter τ, called the **time constant** of the system: $a = 1/\tau$.

To gain some insight into the solution of this equation, consider first the **natural solution**, or **natural response**, of the equation,[2] which is obtained by setting the forcing function equal to zero. This solution, in effect, describes the response of the circuit in the absence of a source and is therefore characteristic of all RL and RC circuits, regardless of the nature of the excitation. Thus, we are interested in the solution of the equation

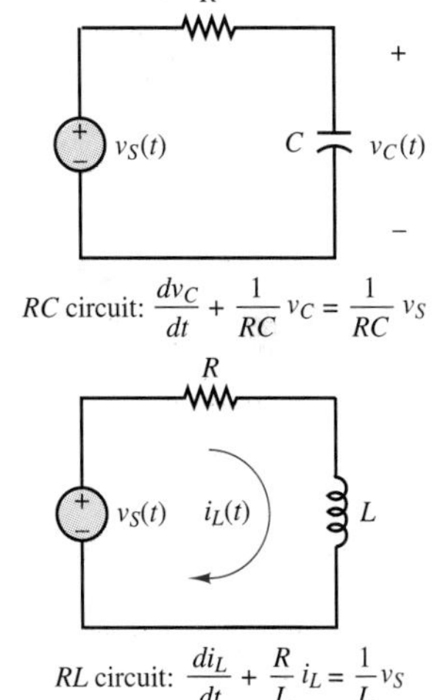

Figure 6.36 Differential equations of first-order circuits

$$\frac{dx_N(t)}{dt} + \frac{1}{\tau}x_N(t) = 0 \tag{6.44}$$

or

$$\frac{dx_N(t)}{dt} = -\frac{1}{\tau}x_N(t) \tag{6.45}$$

[2]Mathematicians usually refer to the unforced solution as the **homogeneous solution**.

where the subscript N has been chosen to denote the natural solution. One can easily verify by substitution that the general form of the solution of the homogeneous equation for a first-order circuit must be exponential in nature, that is, that

$$x_N(t) = Ke^{-at} = Ke^{-t/\tau} \tag{6.46}$$

To evaluate the constant K, we need to know the **initial condition**. The initial condition is related to the energy stored in the capacitor or inductor, as will be further explained shortly. Knowing the value of the capacitor voltage or inductor current at $t = 0$ allows for the computation of the constant K, as follows:

$$x_N(t = 0) = Ke^{-0} = K = x_0 \tag{6.47}$$

Thus, the natural solution, which depends on the initial condition of the circuit at $t = 0$, is given by the expression

$$x_N(t) = x_0 e^{-t/\tau} \tag{6.48}$$

where, once again, $x_N(t)$ represents either the capacitor voltage or the inductor current and x_0 is the initial condition (i.e., the value of the capacitor voltage or inductor current at $t = 0$).

Forced and Complete Response of First-Order Circuits

In the preceding section, the natural response of a first-order circuit was found by setting the forcing function equal to zero and considering the energy initially stored in the circuit as the driving force. The **forced response**, $x_F(t)$, of the **inhomogeneous equation**

$$\frac{dx_F(t)}{dt} + \frac{1}{\tau}x_F(t) = f(t) \tag{6.49}$$

is defined as the response to a particular forcing function $f(t)$, *without regard for the initial conditions*.[3] Thus, the forced response depends exclusively on the nature of the forcing function. The distinction between natural and forced response is particularly useful because it clarifies the nature of the transient response of a first-order circuit: the voltages and currents in the circuit are due to the superposition of two effects, the presence of *stored energy* (which can either decay or further accumulate) and the action of *external sources* (forcing functions). The natural response considers only the former, while the forced response describes the latter. The sum of these two responses forms the **complete response** of the circuit:

$$x(t) = x_N(t) + x_F(t) \tag{6.50}$$

[3]Mathematicians call this solution the **particular solution**.

The forced response depends, in general, on the form of the forcing function, $f(t)$. For the purpose of the present discussion, it will be assumed that $f(t)$ is a constant, applied at $t = 0$, that is, that

$$f(t) = F \qquad t \geq 0 \tag{6.51}$$

(Note that this is equivalent to turning a switch on or off.) In this case, the differential equation describing the circuit may be written as follows:

$$\frac{dx_F}{dt} = -\frac{x_F}{\tau} + F \qquad t \geq 0 \tag{6.52}$$

For the case of a DC forcing function, the form of the forced solution is also a constant. Substituting $x_F(t) = X_F =$ constant in the inhomogeneous differential equation, we obtain

$$0 = -\frac{x_F}{\tau} + F$$

or (6.53)

$$x_F = \tau F$$

Thus, the complete solution of the original differential equation subject to initial condition $x(t = 0) = x_0$ and to a DC forcing function F for $t \geq 0$ is

$$x(t) = x_N(t) + x_F(t)$$

or (6.54)

$$x(t) = Ke^{-t/\tau} + \tau F$$

where the constant K can be determined from the initial condition $x(t = 0) = x_0$:

$$x_0 = K + \tau F$$

(6.55)

$$K = x_0 - \tau F$$

Electrical engineers often classify this response as the sum of a **transient response** and a **steady-state response**, rather than a sum of a natural response and a forced response. The transient response is the response of the circuit following the switching action *before the exponential decay terms have died out;* that is, the transient response is the sum of the natural and forced responses during the transient readjustment period we have just described. The steady-state response is the response of the circuit *after all of the exponential terms have died out.* Equation 6.54 could therefore be rewritten as

$$x(t) = x_T(t) + x_{SS} \tag{6.56}$$

where

$$x_T(t) = (x_0 - \tau F)e^{-t/\tau}$$

and in the case of a DC excitation, F, (6.57)

$$x_{SS}(t) = \tau F = x_\infty$$

Note that the transient response is not equal to the natural response, but it includes part of the forced response. The representation in equations 6.57 is particularly convenient, because it allows for solution of the differential equation that results from describing the circuit *by inspection*. The key to solving first-order circuits subject to DC transients by inspection is in considering *two separate circuits:* the circuit prior to the switching action, to determine the initial condition, x_0; and the circuit following the switching action, to determine the time constant of the circuit, τ, and the steady-state (final) condition, x_∞. Having determined these three values, you can write the solution directly in the form of equation 6.56, and you can then evaluate it using the initial condition to determine the constant K.

To summarize, the transient behavior of a circuit can be characterized in three stages. Prior to the switching action, the circuit is in a steady-state condition (the initial condition, determined by x_0). For a period of time following the switching action, the circuit sees a transient readjustment, which is the sum of the effects of the natural response and of the forced response. Finally, after a suitably long time (which depends on the time constant of the system), the natural response decays to zero (i.e., the term $e^{-t/\tau} \to 0$ as $t \to \infty$) and the new steady-state condition of the circuit is equal to the forced response: as $t \to \infty$, $x(t) \to x_F(t)$. You may recall that this is exactly the sequence of events described in the introductory paragraphs of Section 6.3. Analysis of the circuit differential equation has formalized our understanding of the transient behavior of a circuit.

Continuity of Capacitor Voltages and Inductor Currents

As has already been stated, the primary variables employed in the analysis of circuits containing energy-storage elements are *capacitor voltages* and *inductor currents*. This choice stems from the fact that the energy-storage process in capacitors and inductors is closely related to these respective variables. The amount of charge stored in a capacitor is directly related to the voltage present across the capacitor, while the energy stored in an inductor is related to the current flowing through it. A fundamental property of inductor currents and capacitor voltages makes it easy to identify the initial condition and final value for the differential equation describing a circuit: *capacitor voltages and inductor currents cannot change instantaneously*. An instantaneous change in either of these variables would require an infinite amount of power. Since power equals energy per unit time, it follows that a truly instantaneous change in energy (i.e., a finite change in energy in zero time) would require infinite power.

Another approach to illustrating the same principle is as follows. Consider the defining equation for the capacitor:

$$i_C(t) = C\frac{dv_C(t)}{dt}$$

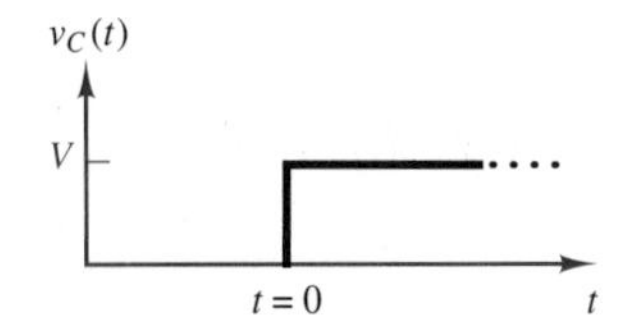

Figure 6.37 Abrupt change in capacitor voltage

and assume that the capacitor voltage, $v_C(t)$, can change instantaneously, say, from 0 to V volts, as shown in Figure 6.37. The value of dv_C/dt at $t = 0$ is simply the slope of the voltage, $v_C(t)$, at $t = 0$. Since the slope is infinite at that point, because of the instantaneous transition, it would require an infinite amount of current for the voltage across a capacitor to change instantaneously. But this is equivalent to requiring an infinite amount of power, since power is the product of voltage and current. A similar argument holds if we assume a "step" change in inductor current from, say, 0 to I amperes: an infinite voltage would be required to cause an instantaneous change in inductor current. This simple fact is extremely useful in determining the response of a circuit. Its immediate consequence is that *the value of an inductor current or a capacitor voltage just prior to the closing (or opening) of a switch is equal to the value just after the switch has been closed (or opened).* Formally,

$$v_C(0^+) = v_C(0^-) \quad \textbf{(6.58)}$$

$$i_L(0^+) = i_L(0^-) \quad \textbf{(6.59)}$$

where the notation 0^+ signifies "just after $t = 0$" and 0^- means "just before $t = 0$."

Example 6.10

The continuity of inductor currents can be used to determine the initial condition and the final value of the inductor current in a circuit. Consider the circuit of Figure 6.38, where the switch opens at $t = 0$, disconnecting the current source. Find the initial condition and final value of the inductor current. Assume that $I_S = 10$ mA.

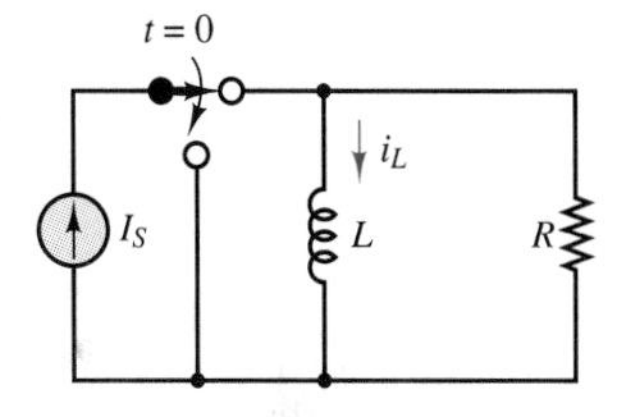

Figure 6.38

Solution:

At $t = 0^-$, the current source has been connected to the circuit for a very long time. Thus, the inductor acts as a short circuit, and all of the source current, I_S, flows through the inductor. Thus,

$$i_L(0^-) = I_S$$

Further, the voltage across the resistor must be zero, since

1. $v_L(t) = L\dfrac{di_L}{dt} = 0$, or
2. The current through R is zero.

At $t = 0$, the switch opens, and we can immediately state that

$$i_L(0^+) = i_L(0^-) = I_S$$

because of the principle of continuity of the inductor current. The new circuit, for $t \geq 0$, is shown in Figure 6.39, where the presence of a current through the inductor at $t = 0$ represents the initial condition for the circuit, indicating that a current was flowing through the inductor just before the switch was opened. Note that, since the inductor current cannot change instantaneously, its direction is as indicated in Figure 6.39.

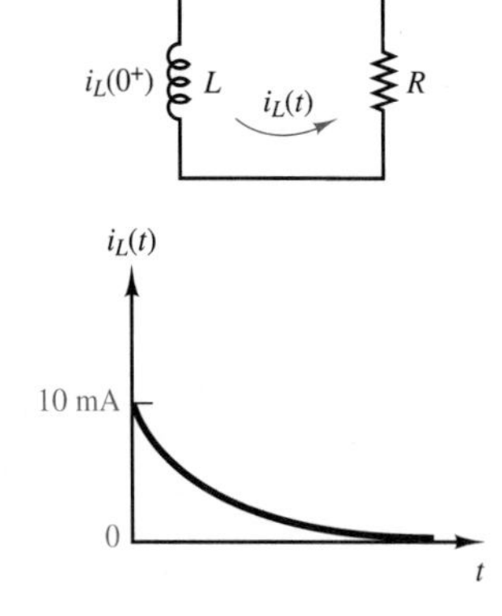

Figure 6.39

Complete Solution of First-Order Circuits

In this section, we illustrate the application of the principles put forth in the preceding sections by presenting a number of examples. The first example summarizes the complete solution of a simple *RC* circuit.

Example 6.11

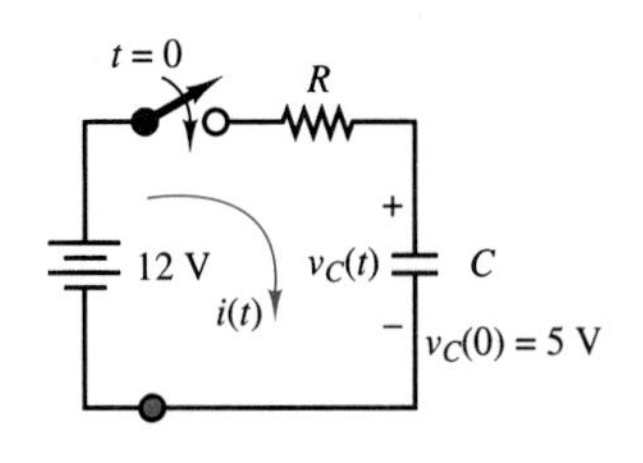

Figure 6.40

This example will illustrate the complete solution of a first-order circuit. In Figure 6.40, the capacitor is assumed to have been previously charged, so that $v_C(t = 0) = 5$ V. Find an expression for the capacitor voltage. Assume that $R = 1$ kΩ and $C = 470$ μF, and let the battery voltage be $V_B = 12$ V.

Solution:

To compute all of the voltages and currents in the circuit, we need to solve for the capacitor voltage; all other circuit variables of interest can be derived from $v_C(t)$. For $t \le 0$, the capacitor voltage is

$$v_C(t) = 5\text{ V} \qquad t \le 0$$

At $t = 0$, the switch is closed, resulting in the following differential equation:

$$\frac{dv_C}{dt} + \frac{1}{RC}v_C = \frac{1}{RC}V_B$$

where we recognize the following:

$$x = v_C \qquad \tau = RC \qquad F = \frac{1}{RC}V_B$$

The natural response of the circuit is of the form

$$x_N = v_{CN} = Ke^{-t/\tau} = Ke^{-t/RC}$$

while the forced response is

$$x_F = v_{CF} = \tau F = V_B$$

Thus, the complete response of the circuit is

$$x(t) = v_C(t) = Ke^{-t/RC} + V_B = Ke^{-t/RC} + 12$$

At this point, we must apply the initial condition, $v_C(0) = 5$, to determine the value of the constant K:

$$v_C(0) = K + V_B$$

or

$$K = v_C(0) - V_B = 5 - 12 = -7$$

We can then write the complete response as

$$v_C(t) = [v_C(0) - V_B]e^{-t/RC} + V_B$$
$$= -7e^{-t/0.47} + 12 = 12(1 - e^{-t/0.47}) + 5e^{-t/0.47}$$

The different components of the response are shown graphically in Figure 6.41.

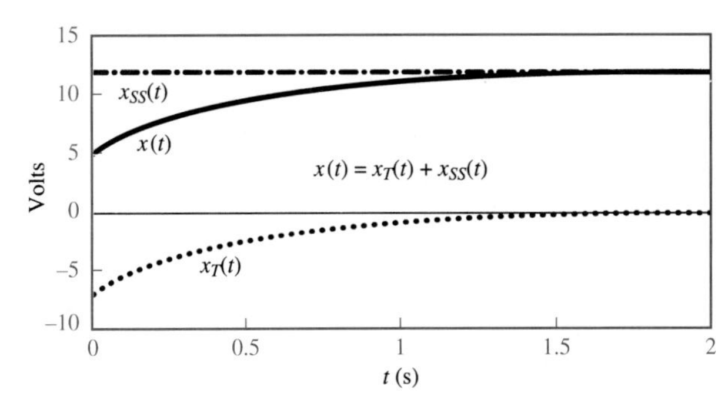

Figure 6.41

To summarize, the following steps have led to the solution:

1. Write the differential equation for the circuit for $t \geq 0$.
2. Determine the initial condition.
3. Compute the time constant of the circuit for $t \geq 0$.
4. Write the complete solution as the sum of natural and forced responses.
5. Apply the initial condition *to the complete solution* to determine the constant K.

Example 6.12

The transient solution for a first-order circuit can be easily determined by following the three simple steps outlined in the previous section. These steps consist of: (1) determining the initial condition; (2) determining the final value; and (3) writing the solution according to equations 6.56 and 6.57. The last step requires that we recognize the appropriate time constant. Assume that a capacitor has been charged to $v_C = 3$ V by means of a separate circuit (not shown). The capacitor is then connected to a resistive circuit at $t = 0$, as shown in Figure 6.42. Find $v_C(t)$ for $t \geq 0$.

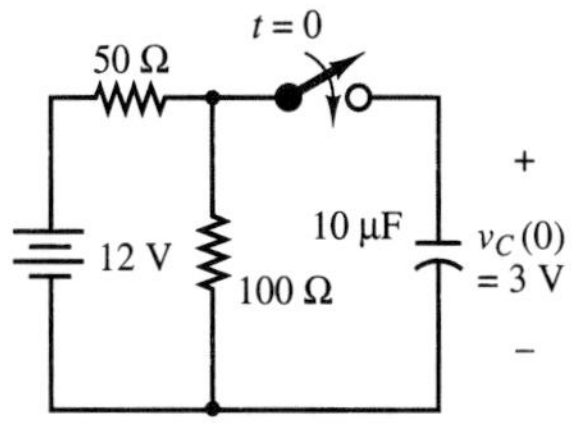

Figure 6.42

Solution:

We know immediately that $v_C(0^-) = v_C(0^+) = 3$ V. To determine the time constant of the circuit and the steady-state value of $v_C(t)$, it is helpful to compute the Thévenin equivalent for the resistive circuit for $t \geq 0$. This is shown in Figure 6.43, where:

$$V_T = \frac{100}{150} \times 12 \text{ V} = 8 \text{ V}$$

$$R_T = 50 \,\|\, 100 \ \Omega = 33.3 \ \Omega$$

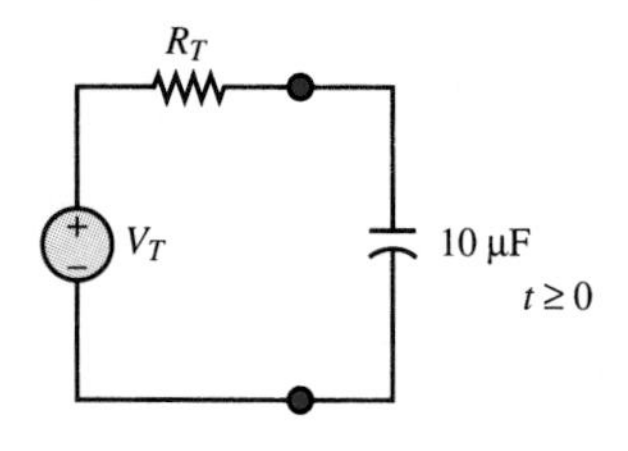

Figure 6.43

The differential equation for the RC circuit for $t \geq 0$ can now be written as

$$\frac{dv_C}{dt} + \frac{1}{R_T C} v_C = \frac{V_T}{R_T C}$$

Recognizing that the time constant is equal to the product of the Thévenin equivalent resistance and the capacitance, we find the time constant to be

$$\tau = 0.333 \times 10^{-3} \text{ s}$$

Further, we can also see that at steady state (i.e., as $t \to \infty$), the capacitor voltage will be equal to the Thévenin voltage, since the capacitor acts as an open circuit with respect to DC currents:

$$v_C(\infty) = V_{C\infty} = V_T = 8 \text{ V}$$

Then, by direct comparison with the general form of the solution (equation 6.57), we write

$$v_C(t) = (3 - 8)e^{-t/0.333\times10^{-3}} + 8 = 8(1 - e^{-t/0.333\times10^{-3}}) + 3e^{-t/0.333\times10^{-3}}$$
$$= 8 - 5e^{-t/0.333\times10^{-3}}$$

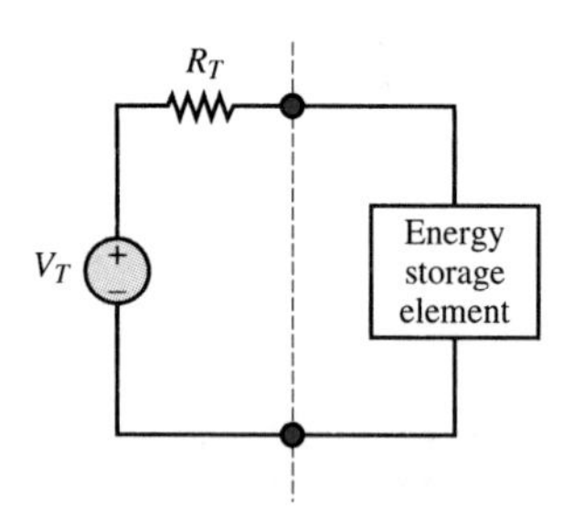

Figure 6.44 Equivalent-circuit representation of first-order circuits

The results of Example 6.12 indicate than even when faced with more involved circuits, we can obtain solutions of first-order circuits subject to DC transients by inspection,[4] provided that we carefully identify the Thévenin equivalent circuit seen by the energy-storage element. Figure 6.44 illustrates how every RL or RC circuit can effectively be reduced to a simple series connection of the energy-storage element to a Thévenin equivalent circuit (or reduced to a parallel connection to a Norton equivalent, by duality). Thus, once you have mastered the solution of the simple circuits presented so far, the extension of these ideas to more complex circuits simply requires identifying the Thévenin equivalent circuit seen by the energy-storage element *before and after* the switching action. It is important to emphasize that, in general, the two equivalent circuits seen by the energy-storage element are not the same, since the switching may disconnect or add several circuit elements. Therefore, it must be remarked, again, that the equivalent circuit prior to switching should be used to compute the initial condition, while the equivalent circuit after switching is useful in computing the steady-state value and the time constant of the circuit.

To illustrate the procedure, we next reduce an RC circuit to a form consisting of a Thévenin equivalent circuit with a capacitor appearing as the load. The objective is to find the capacitor voltage for $t \geq 0$ for the circuit of Figure 6.45. Before the switch closes—that is, at $t < 0$—the capacitor sees the series combination of V_2 and R_2 only; since this circuit is already in Thévenin form, we recognize that at steady state, $v_C(0^-) = V_2$. For $t \geq 0$, we resketch the circuit as

[4] We have examined only DC transients for conceptual clarity. The case of switched sinusoids is treated in a few advanced homework problems for this chapter.

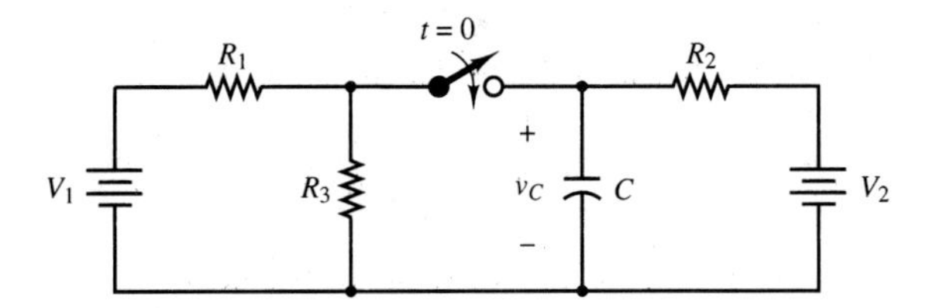

Figure 6.45 A more involved RC circuit

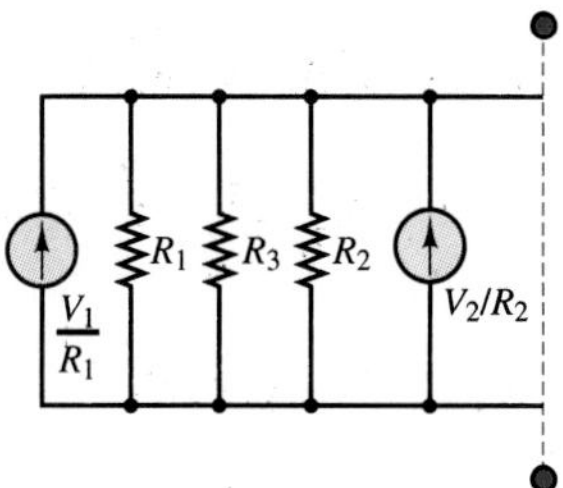

shown in Figure 6.46. Then we compute the Thévenin equivalent, using a source transformation technique (from Chapter 3), as shown in Figure 6.47, where

$$V_T = \left(\frac{V_1}{R_1} + \frac{V_2}{R_2}\right) \cdot R_T \tag{6.60}$$

and

$$R_T = R_1 \parallel R_2 \parallel R_3 \tag{6.61}$$

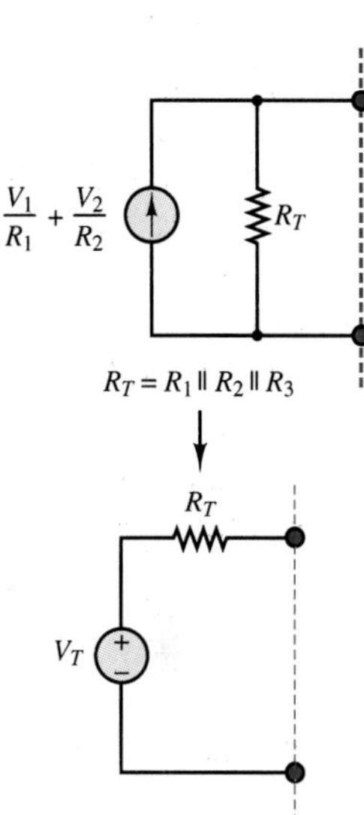

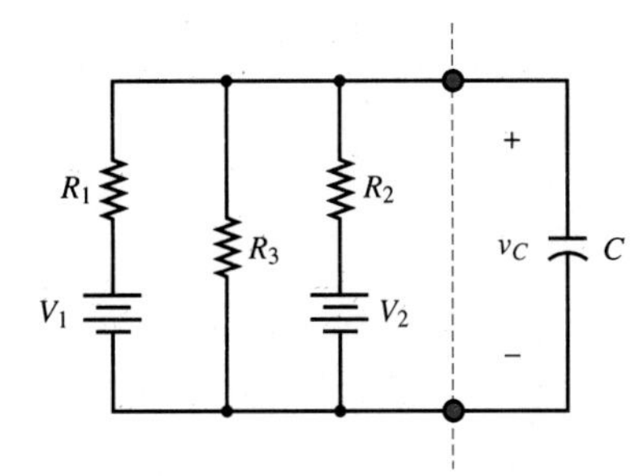

Figure 6.46 The circuit of Figure 6.45 for $t \geq 0$

Figure 6.47 Reduction of the circuit of Figure 6.46 to Thévenin equivalent form

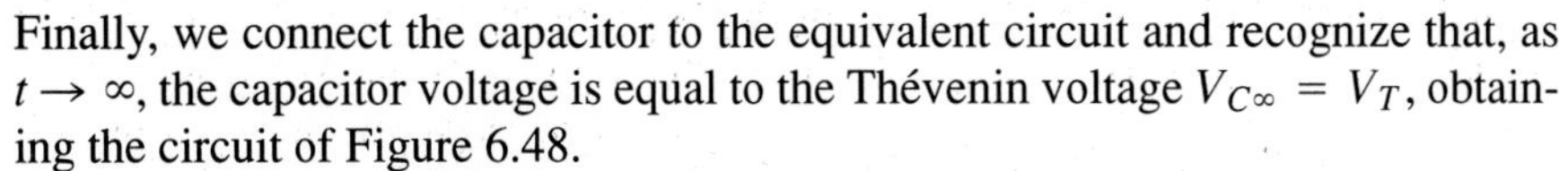

Finally, we connect the capacitor to the equivalent circuit and recognize that, as $t \to \infty$, the capacitor voltage is equal to the Thévenin voltage $V_{C\infty} = V_T$, obtaining the circuit of Figure 6.48.

To determine the time constant, we write the differential equation

$$\frac{dv_C}{dt} + \frac{1}{R_T C} v_C = V_T \tag{6.62}$$

which suggests (as should already be apparent) that $\tau = R_T C = (R_1 \parallel R_2 \parallel R_3)C$. The complete solution is given by the expression

$$v_C(t) = (V_2 - V_T)e^{-t/R_T C} + V_T \tag{6.63}$$

The procedure just shown may be followed for RL circuits as well as RC circuits to obtain a standard circuit that involves only a single equivalent source and its associated resistance.

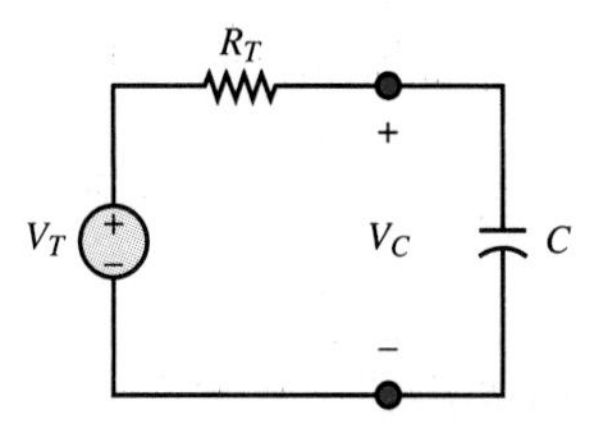

Figure 6.48 The circuit of Figure 6.45 in equivalent form for $t \geq 0$

Example 6.13

For a numerical example that shows how the methodology of Example 6.11 also applies to circuits containing inductors, determine the inductor current for $t > 0$ if the switch opens at $t = 0$ in the circuit of Figure 6.49. Assume that $L = 10$ mH.

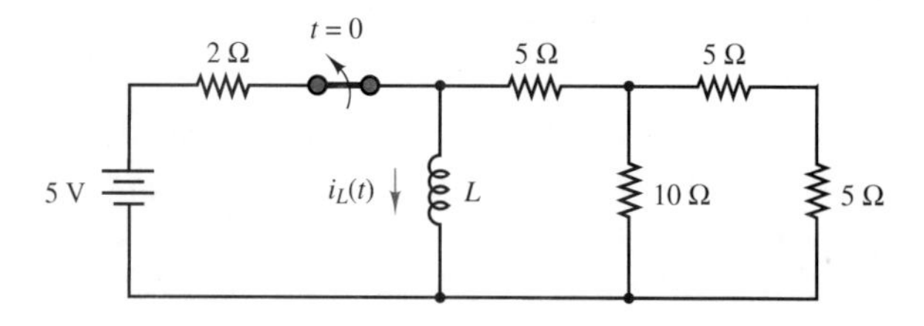

Figure 6.49

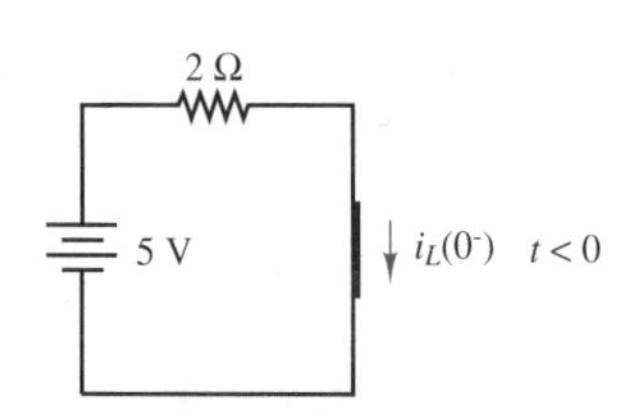

Figure 6.50

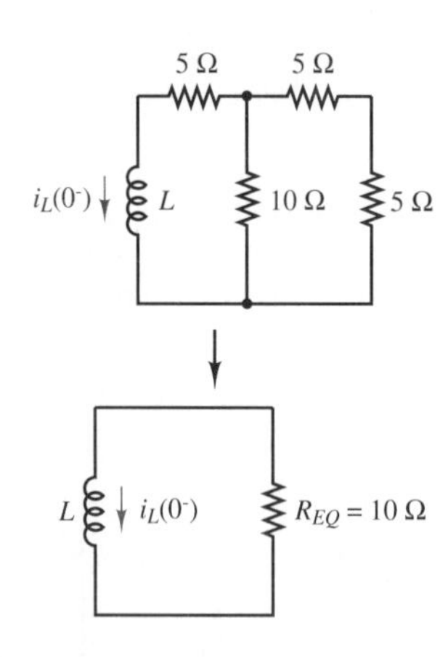

Figure 6.51

Solution:

Since the inductor acts as a short circuit for $t < 0$ (assuming that the switch has been closed for a long time), the equivalent circuit for $t < 0$ is the one shown in Figure 6.50, where the thick line indicates that the inductor has been replaced with a short circuit. To understand this simplification, recall that an inductor acts as a short circuit in the presence of steady-state DC currents and therefore no current can reach the circuit to the right of the inductor at $t < 0$.

On the basis of the circuit in Figure 6.50, we can compute the initial condition:

$$i_L(0^-) = \frac{5}{2} = 2.5 \text{ A}$$

At $t = 0$, the switch is opened, and we can sketch the equivalent circuit for $t \geq 0$, shown in Figure 6.51. Note that the direction of the inductor current cannot change, according to the continuity principle, and has been indicated accordingly. Because no source is present at $t \geq 0$, the inductor current must decay to zero; thus,

$$I_{L\infty} = 0$$

To write the differential equation, we apply KVL to the equivalent circuit for $t > 0$. The circuit is redrawn in Figure 6.52 to show the polarity of the voltages:

$$v_L(t) = -v_R(t)$$

(Convince yourself of the minus sign!) Since the resistor voltage is

$$v_R(t) = R_{EQ} i_L(t)$$

and since

$$v_L(t) = L\frac{di_L}{dt}$$

we obtain the equation

$$L\frac{di_L}{dt} + R_{EQ}i_L(t) = 0$$

or

$$\frac{di_L}{dt} + \frac{R_{EQ}}{L}i_L(t) = 0 \qquad t \ge 0$$

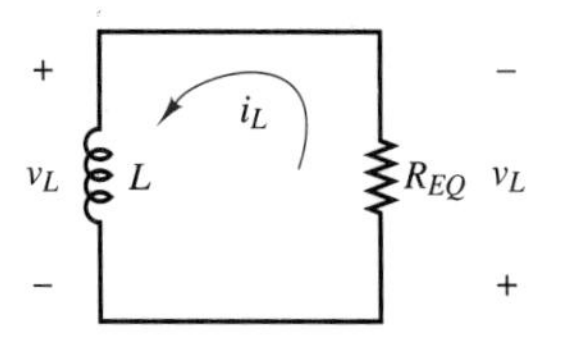

Figure 6.52

so that the time constant is found to be $\tau = L/R_{EQ} = 10^{-3}$ s. The complete solution is

$$\begin{aligned} i_L(t) &= (i_L(0) - i_{LSS})e^{-t/\tau} + i_{LSS} = i_{LSS}(1 - e^{-t/\tau}) + i_L(0)e^{-t/\tau} \\ &= 2.5e^{-1{,}000t} \end{aligned}$$

Example 6.14 Pulse Response

A problem of great practical importance is the transmission of voltage *pulses* along cables. Short voltage pulses are used to represent the two-level binary signals that are characteristic of digital computers; it is often necessary to transmit such voltage pulses over a long distance through **coaxial cables,** which are characterized by a finite resistance per unit length and by a certain capacitance per unit length, usually expressed in units of pF/m. A simplified model of a long coaxial cable is shown in Figure 6.53; it has the appearance of a low-pass filter. If a 100-m cable has a capacitance of 40 pF/m and a series resistance of 0.2 Ω/m, what will the output pulse look like after traveling the length of the cable? Assume the input pulse has a duration of 0.1 μs and has an amplitude of 5 V. The load resistance is 150 Ω.

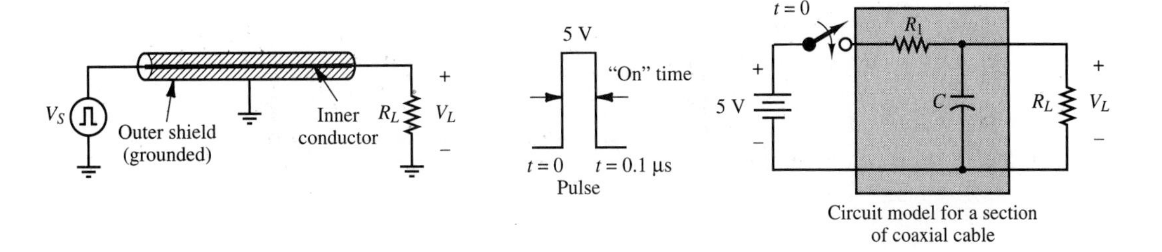

Figure 6.53 Pulse transmission in a coaxial cable

Solution:

The Thévenin equivalent circuit seen by the capacitor will vary, depending on whether the pulse is "on" or "off." In the former case, the equivalent resistance consists of the parallel combination of R_1 and R_L; in the latter case, since the switch is open, the capacitor is connected only to R_L. Thus, the effect of the pulse is to charge C through the parallel combination of R_1 and R_L during the "on" time ($0 \le t < 0.1$ μs); the capacitor will then discharge through R_L during the "off" time. This behavior is depicted by the circuit model of Figure 6.53, in which the pulse signal is represented by a 5-V battery in series with a switch.

The charging time constant of the coaxial cable equivalent circuit when the switch is closed is therefore given by

$$\tau_{\text{on}} = (R_1 \parallel R_L)C = 17.65 \times 4{,}000 \times 10^{-12} = 0.07\ \mu\text{s}$$

and the transient response of the cable during the "on" time is:

$$v_L(t) = 4.41(1 - e^{-t/\tau}) = 4.41(1 - e^{-1.42\times 10^7 t}) \qquad 0 \le t \le 0.1\ \mu\text{s}$$

where $V_T = \frac{R_L}{R_1+R_L} \times 5 = 4.41$ V. At $t = 0.1$ μs we calculate the load voltage to be:

$$v_L(0.1\ \mu\text{s}) = 4.41(1 - e^{-1.42}) = 3.35\ \text{V}$$

For $t \ge 0.1$ μs, the output voltage will naturally decay to zero, starting from the initial condition, $v_L(0.1\ \mu\text{s})$, with a time constant τ_{off} equal to:

$$\tau_{\text{off}} = R_L C = 150 \times 4{,}000 \times 10^{-12} = 0.6\ \mu\text{s}$$

The load voltage will therefore decay according to the following expression:

$$v_L(t) = 3.35\left(e^{-1.67\times 10^6(t-0.1\times 10^{-6})}\right) \qquad t > 0.1\ \mu\text{s}$$

The appearance of the response is shown in Figure 6.54. It should be apparent that as the cable becomes longer, both R_1 and C increase, and therefore the output voltage will respond more slowly to the input pulse; according to the simple model of a long coaxial cable given in this example, there is a limit to the maximum distance over which a voltage pulse can be transmitted by cable.

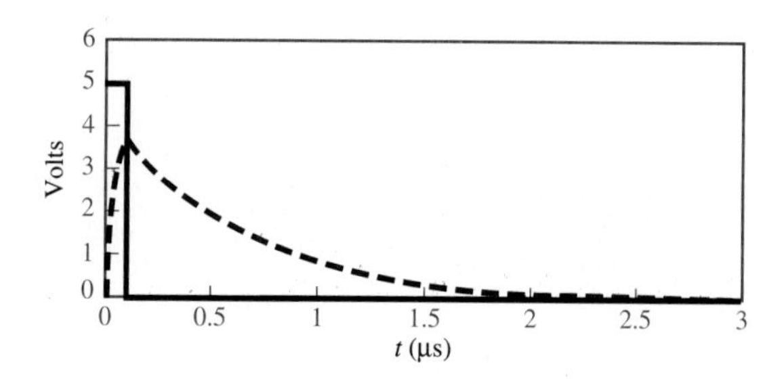

Figure 6.54 Pulse response of 100-m-long coaxial cable

The response of first-order circuits to switching transients is analogous to the behavior of other dynamical systems that are described by first-order differential equations. The following examples illustrate how the analysis methods developed so far in this chapter could also be employed to describe the dynamics of other systems. These examples further reinforce the notion that electrical circuit analysis methods are applicable to much more than just electrical circuits.

Example 6.15 Fluid Flow

Consider a tank filled with water, and with input flow rate equal to F_i. Let the output flow rate be equal to F_o. If we assume that the output flow rate is linearly related to the fluid level (or *head*), h, we can write $F_o = h/R$, where R is related to the diameter of the pipe. Figure 6.55 depicts the simple hydraulic system.

Figure 6.55 A hydraulic first-order system

In effect, the *fluid resistance* parameter R is analogous to the resistance parameter in an electrical circuit, since the diameter of the pipe limits the flow out of the tank much as a resistor limits the flow of an electrical current. In other words, an analogy could be constructed in which the pipe acts as a resistor and the fluid flow acts as current. Similarly, the head, h, corresponds to the electric potential in a circuit: it enables the fluid to flow, much as the potential difference between two points in a circuit enables a current flow.

But if fluid flow behaves in a manner analogous to current (which is defined as the flow of charge), we may think of the fluid as being analogous to charge. Then, in effect, the water tank acts as a capacitor; that is, it stores "charge" (fluid). We can already realize that there is a clear analogy between variables in the fluid flow problem and the circuit problem. This analogy is formally illustrated in Table 6.1.

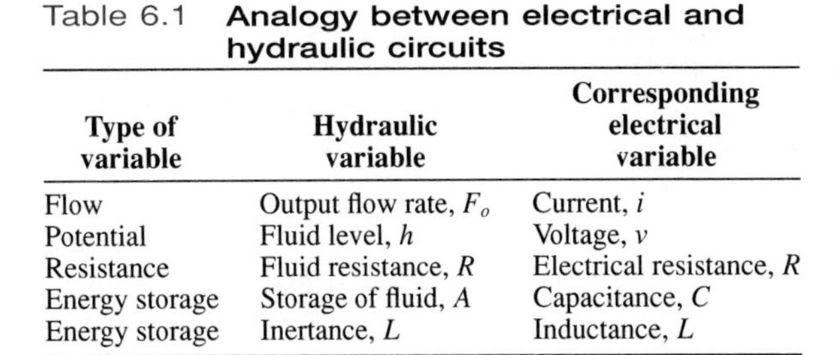

Table 6.1 **Analogy between electrical and hydraulic circuits**

Type of variable	Hydraulic variable	Corresponding electrical variable
Flow	Output flow rate, F_o	Current, i
Potential	Fluid level, h	Voltage, v
Resistance	Fluid resistance, R	Electrical resistance, R
Energy storage	Storage of fluid, A	Capacitance, C
Energy storage	Inertance, L	Inductance, L

To unravel the analogy between the fluid storage parameter, A, and the capacitance, C, let A be the cross-sectional area of the tank. Since the volume of the tank is equal to the product of the head and the cross-sectional area, we obtain the equation

$$\text{Volume} = Ah$$

which is analogous to the equation $Q = CV$ in a capacitor. If we write the mass balance equation, by equating the rate of change of fluid volume to the net difference between input and output flow, we have

$$\frac{d\text{Volume}}{dt} = \frac{d(Ah)}{dt} = F_i - F_o = F_i - \frac{h}{R}$$

or

$$A\frac{dh}{dt} + \frac{h}{R} = F_i$$

where F_i (the output flow rate) is the forcing function. Does this equation look familiar?

Example 6.16 Heat Flow

A simplified interpretation of a heat flow problem gives rise to another interesting analogy. Consider a house as a heat-storage element: the heat capacity of the house is analogous to capacitance, in the sense that the house retains part of the heat generated by the heating system. However, a temperature difference, ΔT, between the inside and outside of the house causes heat to flow outward from the house. The insulation present in the walls, the roof, and other areas limits the flow of heat. From this simple description, we may infer the analogies between electrical and thermal systems listed in Table 6.2.

Table 6.2 **Analogy between thermal and electrical systems**

Thermal	Electrical
Heat	Charge
Heat flow	Current
Temperature difference	Voltage
Ambient temperature	Ground
Heat capacity	Capacitance
Thermal resistance	Resistance
—	Inductance[a]

[a]No direct physical analogy.

One interpretation of the heat flow analogy is obtained by revisiting Newton's law of cooling (though, admittedly, in a greatly simplified form), which states that the heat flow, q, is proportional to the rate of change of temperature with respect to distance. We may approximate the derivative, to obtain a linear relationship between heat flow and change in temperature:

$$q = K\frac{dT}{dx} \approx K\frac{\Delta T}{\Delta x} = \frac{K}{\Delta x}\Delta T$$

Thus, $K/\Delta x$ plays the role of conductance in our thermal analogy, and Newton's law of cooling (although in a simplified form) turns out to be a thermal version of Ohm's law![5]

[5]In reality, heat flow is more properly described by partial differential equations, since it is a distributed phenomenon. However, the approximation used here is adequate to illustrate the analogy.

DRILL EXERCISES

6.14 Write the differential equation for the circuit shown in Figure 6.56.

6.15 Write the differential equation for the circuit shown in Figure 6.57.

6.16 Write the differential equation for the circuit shown in Figure 6.58.

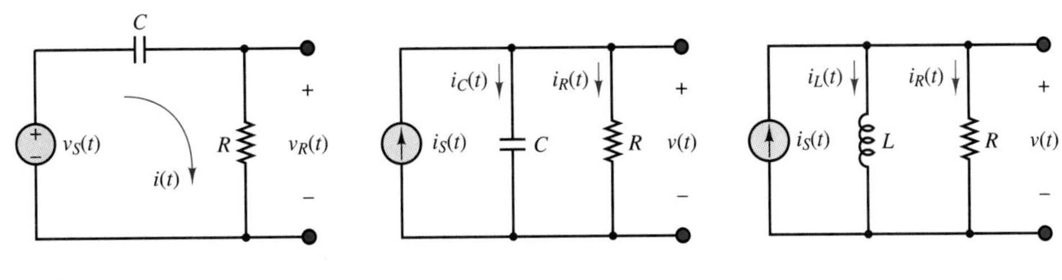

Figure 6.56 **Figure 6.57** **Figure 6.58**

6.17 It is instructive to repeat the analysis of Example 6.10 for a capacitive circuit. For the circuit shown in Figure 6.59, compute the quantities $v_C(0^-)$ and $i_R(0^+)$, and sketch the response of the circuit, that is, $v_C(t)$, if the switch opens at $t = 0$.

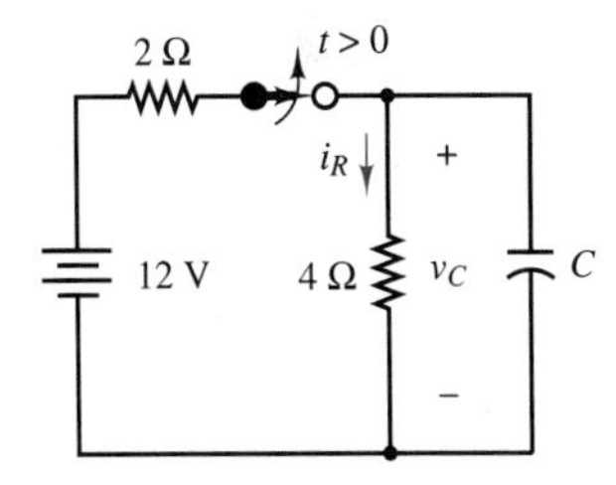

Figure 6.59

6.18 The circuit of Figure 6.60 has a switch that can be used to connect and disconnect a battery. The switch has been open for a very long time. At $t = 0$, the switch closes, and then at $t = 50$ ms, the switch opens again. Assume that $R_1 = R_2 = 1{,}000\ \Omega$, $R_3 = 500\ \Omega$, and $C = 25\ \mu\text{F}$.

a. Determine the capacitor voltage as a function of time.
b. Plot the capacitor voltage from $t = 0$ to $t = 100$ ms.

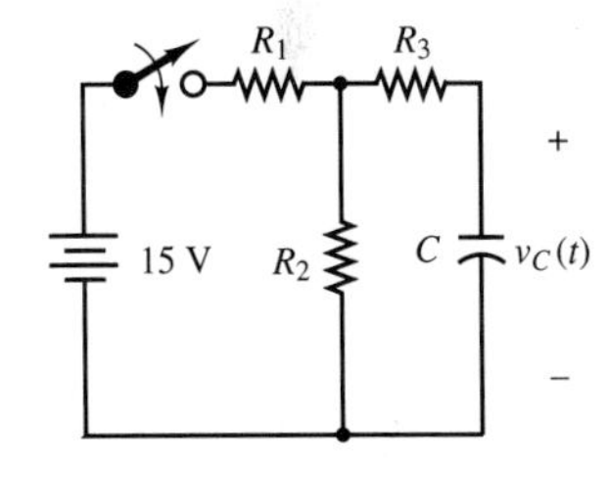

Figure 6.60

6.19 If the 10-mA current source is switched on at $t = 0$ in the circuit of Figure 6.61, how long will it take for the capacitor to charge to 90 percent of its final voltage?

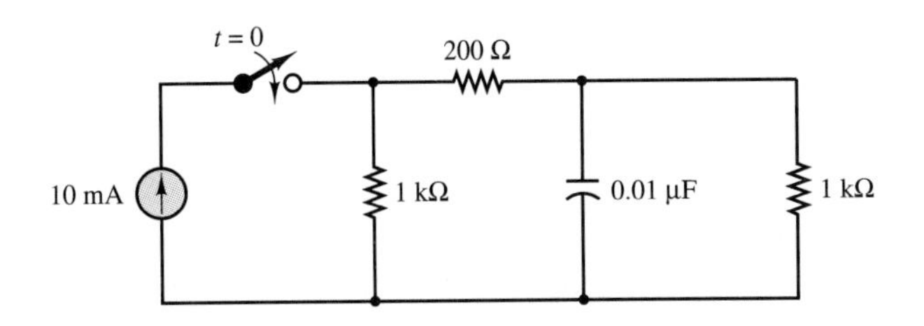

Figure 6.61

6.20 Find the time constant for the circuit shown in Figure 6.62.

6.21 Compute the maximum amplitude of the output pulse in Example 6.14 if the length of the coaxial cable is 300 m.

6.22 Repeat the calculations of Example 6.14 if the load resistance is 1,000 Ω. What is the effect of this change?

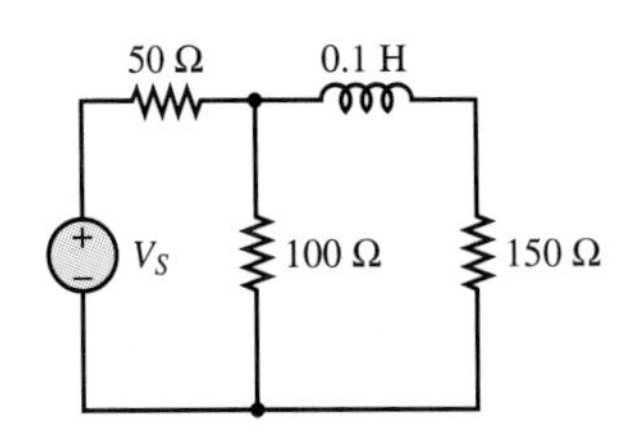

Figure 6.62

6.5 TRANSIENT RESPONSE OF SECOND-ORDER CIRCUITS

In many practical applications, understanding the behavior of first- and second-order systems is often all that is needed to describe the response of a physical system to external excitation. In this section, we discuss the solution of the second-order differential equations that characterize second-order circuits.

Deriving the Differential Equations for Second-Order Circuits

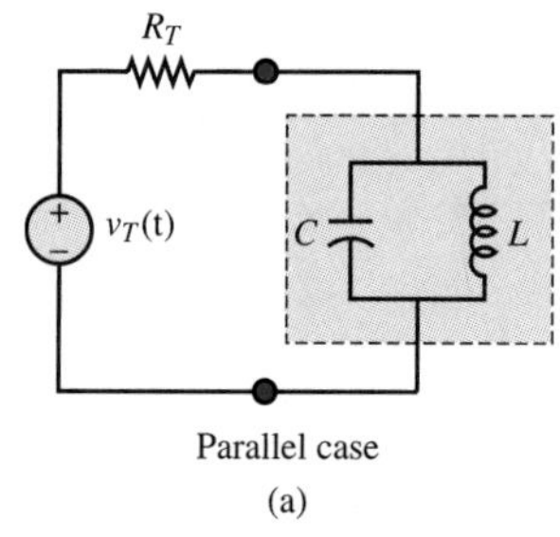

Parallel case
(a)

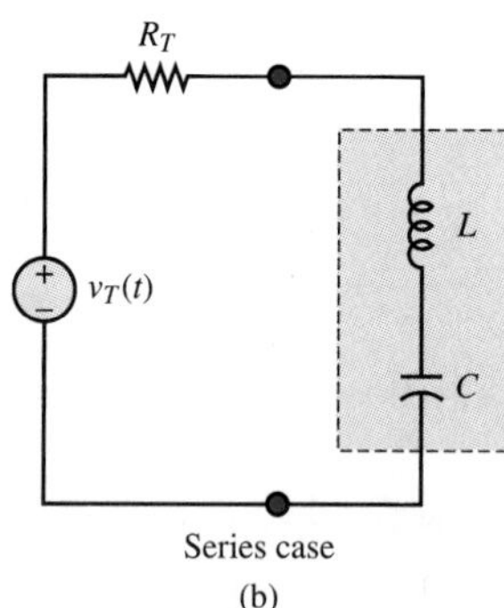

Series case
(b)

Figure 6.63 Second-order circuits

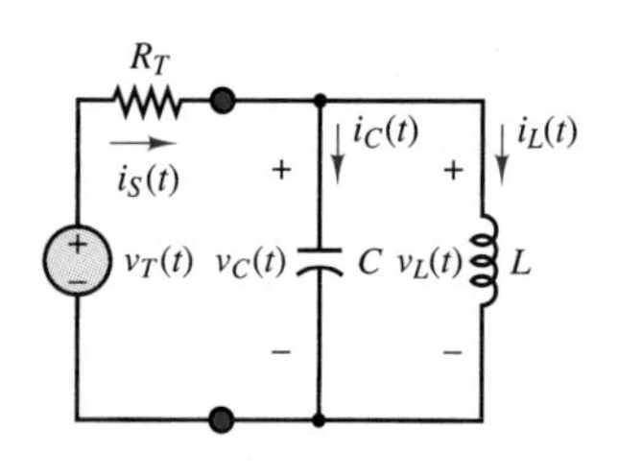

Figure 6.64 Parallel case

A simple way of introducing second-order circuits consists of replacing the box labeled "Circuit containing *RL*/*RC* combinations" in Figure 6.33 with a combination of two energy-storage elements, as shown in Figure 6.63. Note that two different cases are considered, depending on whether the energy-storage elements are connected in series or in parallel.

Consider the parallel case first, which has been redrawn in Figure 6.64 for clarity. Practice and experience will eventually suggest the best method for writing the circuit equations. At this point, the most sensible procedure consists of applying the basic circuit laws to the circuit of Figure 6.64. Start with KVL around the left-hand loop:

$$v_T(t) = R_T i_S(t) + v_C(t) \tag{6.64}$$

Then apply KCL to the top node, to obtain

$$i_S(t) = i_C(t) + i_L(t) \tag{6.65}$$

Further, KVL applied to the right-hand loop yields

$$v_C(t) = v_L(t) \tag{6.66}$$

It should be apparent that we have all the equations we need (in fact, more). Using the defining relationships for capacitor and inductor, we can express equation 6.65 as

$$i_S(t) = \frac{v_T(t) - v_C(t)}{R_T} = C\frac{dv_C}{dt} + i_L(t) \tag{6.67}$$

and equation 6.66 becomes

$$v_C(t) = L\frac{di_L}{dt} \tag{6.68}$$

Substituting equation 6.68 in equation 6.67, we can obtain a differential circuit equation in terms of the variable $i_L(t)$:

$$\frac{1}{R_T}v_T(t) - \frac{L}{R_T}\frac{di_L}{dt} = LC\frac{d^2i_L}{dt^2} + i_L(t) \tag{6.69}$$

or

$$\frac{d^2i_L}{dt^2} + \frac{1}{R_T C}\frac{di_L}{dt} + \frac{1}{LC}i_L = \frac{v_T(t)}{R_T LC} \tag{6.70}$$

The solution to this differential equation (which depends, as in the case of first-order circuits, on the initial conditions and on the forcing function) completely determines the behavior of the circuit. By now, two questions should have appeared in your mind:

1. Why is the differential equation expressed in terms of $i_L(t)$? (Why not $v_C(t)$?)
2. Why did we not use equation 6.64 in deriving equation 6.70?

In response to the first question, it is instructive to note that, knowing $i_L(t)$, we can certainly derive any one of the voltages and currents in the circuit. For example,

$$v_C(t) = v_L(t) = L\frac{di_L}{dt} \tag{6.71}$$

$$i_C(t) = C\frac{dv_C}{dt} = LC\frac{d^2 i_L}{dt^2} \tag{6.72}$$

To answer the second question, note that equation 6.70 is not the only form the differential circuit equation can take. By using equation 6.64 in conjunction with equation 6.65, one could obtain the following equation:

$$v_T(t) = R_T[i_C(t) + i_L(t)] + v_C(t) \tag{6.73}$$

Upon differentiating both sides of the equation and appropriately substituting from equation 6.67, the following second-order differential equation in v_C would be obtained:

$$\frac{d^2 v_C}{dt^2} + \frac{1}{R_T C}\frac{dv_C}{dt} + \frac{1}{LC}v_C = \frac{1}{R_T C}\frac{dv_T(t)}{dt} \tag{6.74}$$

Note that the left-hand side of the equation is identical to equation 6.70, except that v_C has been substituted for i_L. The right-hand side, however, differs substantially from equation 6.70, because the forcing function is the derivative of the equivalent voltage.

Since all of the desired circuit variables may be obtained either as a function of i_L or as a function of v_C, the choice of the preferred differential equation depends on the specific circuit application, and we conclude that there is no unique method to arrive at the final equation. As a case in point, consider the two circuits depicted in Figure 6.65. If the objective of the analysis were to determine the output voltage, v_{out}, then for the circuit in Figure 6.65(a), one would choose to write the differential equation in v_C, since $v_C = v_{\text{out}}$. In the case of Figure 6.65(b), however, the inductor current would be a better choice, since $v_{\text{out}} = R_T i_{\text{out}}$.

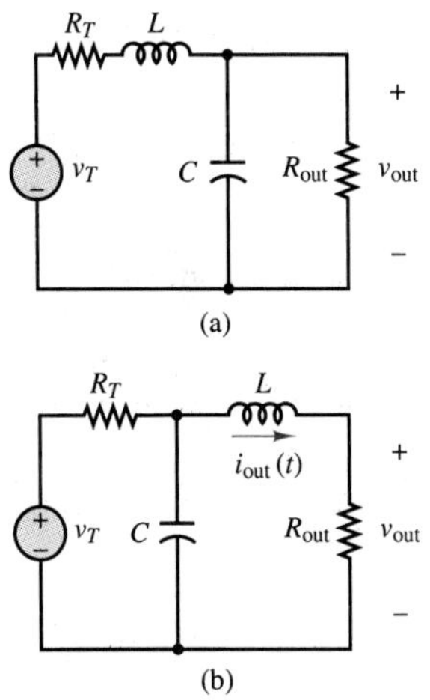

R_T L v_T C R_{out} v_{out} + – (a)

R_T L $i_{\text{out}}(t)$ v_T C R_{out} v_{out} + – (b)

Figure 6.65 Two second-order circuits

Natural Response of Second-Order Circuits

From the previous discussion, we can derive a general form for the governing equation of a second-order circuit:

$$a_2\frac{d^2x(t)}{dt^2} + a_1\frac{dx(t)}{dt} + a_0 x(t) = f(t) \tag{6.75}$$

It is now appropriate to derive a general form for the solution. The same classification used for first-order circuits is also valid for second-order circuits. Therefore, the complete solution of the second-order equation is the sum of the natural and forced responses:

$$x(t) = \underbrace{x_N(t)}_{\text{Natural response}} + \underbrace{x_F(t)}_{\text{Forced response}} \quad \textbf{(6.76)}$$

where the natural response is the solution of the homogeneous equation without regard for the forcing function (i.e., with $f(t) = 0$) and the forced response is the solution of the forced equation with no consideration of the effects of the initial conditions. Once the general form of the complete response is found, the unknown constants are evaluated subject to the initial conditions, and the solution can then be divided into transient and steady-state parts, with

$$x(t) = \underbrace{x_T(t)}_{\text{Transient part}} + \underbrace{x_{SS}(t)}_{\text{Steady-state part}} \quad \textbf{(6.77)}$$

The aim of this section is to determine the natural response, which satisfies the homogeneous equation:

$$\frac{d^2x_N(t)}{dt^2} + b\frac{dx_N(t)}{dt} + cx_N(t) = 0 \quad \textbf{(6.78)}$$

where $b = a_1/a_2$ and $c = a_0/a_2$. Just as in the case of first-order circuits, $x_N(t)$ takes on an exponential form:

$$x_N(t) = Ke^{st} \quad \textbf{(6.79)}$$

This is easily verifiable by direct substitution in the differential equation:

$$s^2Ke^{st} + bsKe^{st} + cKe^{st} = 0 \quad \textbf{(6.80)}$$

and since it is possible to divide both sides by Ke^{st}, the natural response of the differential equation is, in effect, determined by the solution of the quadratic equation

$$s^2 + bs + c = 0 \quad \textbf{(6.81)}$$

This polynomial in the variable s is called the **characteristic polynomial** of the differential equation. Thus, the natural response, $x_N(t)$, is of the form

$$x_N(t) = K_1e^{s_1t} + K_2e^{s_2t} \quad \textbf{(6.82)}$$

where the exponents s_1 and s_2 are found by applying the quadratic formula to the characteristic polynomial:

$$s_{1,2} = -\frac{b}{2} \pm \frac{1}{2}\sqrt{b^2 - 4c} \quad \textbf{(6.83)}$$

The exponential solution in terms of the exponents $s_{1,2}$ can take different forms depending on whether the roots of the quadratic equation are real or complex. As an example, consider the parallel circuit of Figure 6.64, and the governing differential equation, 6.70. The natural response for $i_L(t)$ in this case is the solution of the following equation:

$$\frac{d^2 i_L(t)}{dt^2} + \frac{1}{RC}\frac{di_L(t)}{dt} + \frac{1}{LC} i_L(t) = 0 \tag{6.84}$$

where $R = R_T$ in Figure 6.64. The solution of equation 6.84 is determined by solving the quadratic equation

$$s^2 + \frac{1}{RC}s + \frac{1}{LC} = 0 \tag{6.85}$$

The roots are

$$s_{1,2} = -\frac{1}{2RC} \pm \frac{1}{2}\sqrt{\left(\frac{1}{RC}\right)^2 - \frac{4}{LC}} \tag{6.86}$$

where

$$s_1 = -\frac{1}{2RC} + \frac{1}{2}\sqrt{\left(\frac{1}{RC}\right)^2 - \frac{4}{LC}} \tag{6.87a}$$

$$s_2 = -\frac{1}{2RC} - \frac{1}{2}\sqrt{\left(\frac{1}{RC}\right)^2 - \frac{4}{LC}} \tag{6.87b}$$

The key to interpreting this solution is to analyze the term under the square root sign; we can readily identify three cases:

- Case I:

$$\left(\frac{1}{RC}\right)^2 > \frac{4}{LC} \tag{6.88}$$

s_1 and s_2 are real and distinct roots: $s_1 = \alpha_1$ and $s_2 = \alpha_2$.

- Case II:

$$\left(\frac{1}{RC}\right)^2 = \frac{4}{LC} \tag{6.89}$$

s_1 and s_2 are real, repeated roots: $s_1 = s_2 = \alpha$.

- Case III:

$$\left(\frac{1}{RC}\right)^2 < \frac{4}{LC} \tag{6.90}$$

s_1, s_2 are complex conjugate roots: $s_1 = s_2^* = \alpha \pm j\beta$.

For each of these three cases, as we shall see, the solution of the differential equation takes a different form. The remainder of this section will explore the three different cases that can arise.

Example 6.17

Consider the circuit of Figure 6.66. Find the natural response for $i_L(t)$.

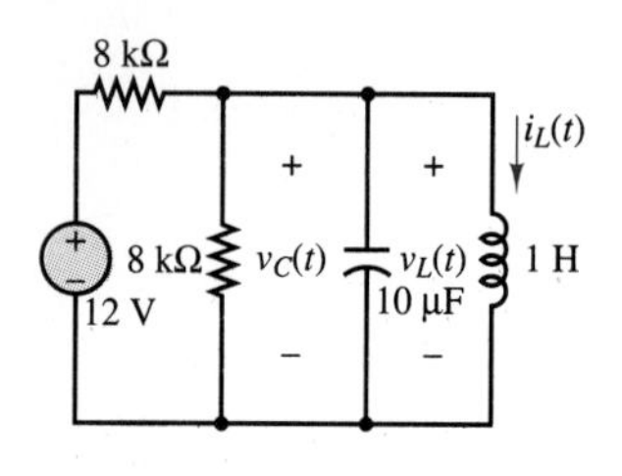

Figure 6.66

Solution:

Looking at the circuit, we can immediately see that

$$R_T = 8\ \text{k}\Omega \parallel 8\ \text{k}\Omega = 4\ \text{k}\Omega$$

The natural response for $i_L(t)$, in this case, is the solution to the homogeneous equation

$$\frac{d^2 i_L(t)}{dt^2} + \frac{1}{R_T C}\frac{di_L(t)}{dt} + \frac{1}{LC} i_L(t) = 0$$

This equation can be determined by solving the quadratic equation

$$s^2 + \frac{1}{R_T C} s + \frac{1}{LC} = 0$$

Thus, for the given values, we obtain the following quadratic equation:

$$s^2 + \frac{1}{0.04} s + \frac{1}{0.00001} = 0$$

which has roots

$$s_{1,2} = 12.5 \pm \frac{1}{2}\sqrt{\left(\frac{1}{0.04}\right)^2 - \frac{4}{10^{-5}}}$$

with

$$s_1 = 12.5 + j315.98$$
$$s_2 = 12.5 - j315.98$$

Finally, the natural response for $i_L(t)$ is of the form

$$i_L(t) = K_1 e^{s_1 t} + K_2 e^{s_2 t}$$

or

$$i_L(t) = K_1 e^{(12.5 + j315.98)t} + K_2 e^{(12.5 - j315.98)t}$$

Here, K_1 and K_2 are constants that can be found from given initial conditions.

Although the previous example dealt with a specific circuit, one can extend the result by stating that the natural response of any second-order system can be described by one of the following three expressions:

- Case I. Real, distinct roots: $s_1 = \alpha_1, s_2 = \alpha_2$.

$$x_N(t) = K_1 e^{\alpha_1 t} + K_2 e^{\alpha_2 t} \quad \textbf{(6.91)}$$

- Case II. Real, repeated roots: $s_1 = s_2 = \alpha$.

$$x_N(t) = K_1 e^{\alpha t} + K_2 t e^{\alpha t} \quad \textbf{(6.92)}$$

- Case III. Complex conjugate roots: $s_1 = \alpha + j\beta, s_2 = \alpha - j\beta$.

$$x_N(t) = K_1 e^{(\alpha + j\beta)t} + K_2 e^{(\alpha - j\beta)t} \quad \textbf{(6.93)}$$

The solution of the homogeneous second-order differential equation will now be discussed for each of the three cases.

Overdamped Solution

The case of real and distinct roots yields the so-called **overdamped solution,** which consists of a sum of real exponentials. An overdamped system naturally decays to zero in the absence of a forcing function, according to the expression

$$x_N(t) = K_1 e^{-\alpha_1 t} + K_2 e^{-\alpha_2 t} \quad \textbf{(6.94)}$$

Note that α_1 and α_2 are the reciprocals of two time constants:

$$\tau_1 = \frac{1}{\alpha_1} \qquad \tau_2 = \frac{1}{\alpha_2} \quad \textbf{(6.95)}$$

so that the behavior of an overdamped system may be portrayed, for example, as in Figure 6.67 ($K_1 = K_2 = 1, \alpha_1 = 5$, and $\alpha_2 = 2$ in the figure).

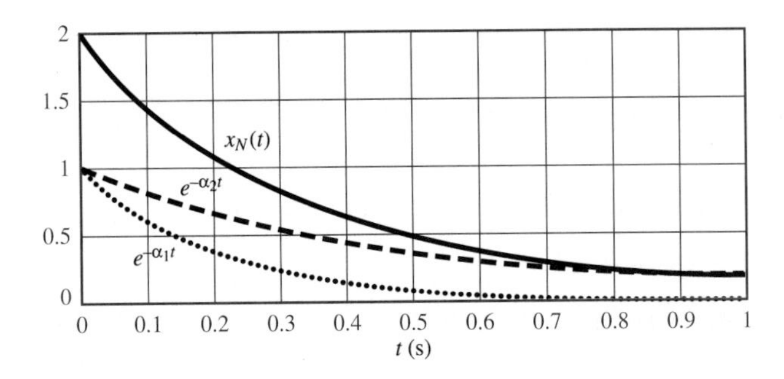

Figure 6.67 Response of overdamped second-order circuit

Critically Damped Solution

When the roots are real and repeated, the natural solution is said to be **critically damped,** and is of the form

$$x_N(t) = K_1 e^{-\alpha t} + K_2 t e^{-\alpha t} \quad \textbf{(6.96)}$$

The first term, $K_1 e^{-\alpha t}$, is the familiar exponential decay term. The term $K_2 t e^{-\alpha t}$, on the other hand, has a behavior that differs from a decaying exponential: for small t, the function t grows faster than $e^{-\alpha t}$ decays, so that the function initially increases, reaches a maximum at $t = 1/\alpha$, and finally decays to zero. Figure 6.68 depicts the critically damped solution for $K_1 = K_2 = 1, \alpha = 5$.

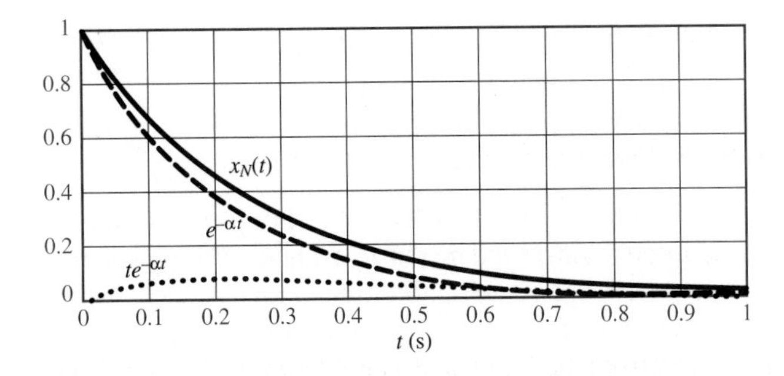

Figure 6.68 Response of critically damped second-order circuit

Underdamped Solution

A slightly more involved form of the natural response of a second-order circuit occurs when the roots of the characteristic polynomial form a complex conjugate pair, that is, $s_1 = s_2^*$. In this case, the solution is said to be **underdamped.** The solution for $x_N(t)$, then, is of the form

$$x_N(t) = K_1 e^{s_1 t} + K_2 e^{s_2 t} \quad \textbf{(6.97)}$$

or

$$x_N(t) = K_1 e^{\alpha t} e^{j\beta t} + K_2 e^{\alpha t} e^{-j\beta t} \quad \textbf{(6.98)}$$

where $s_1 = \alpha + j\beta$ and $s_2 = \alpha - j\beta$. What is the significance of the complex exponential in the case of underdamped natural response? Recall Euler's identity, which was introduced in Chapter 4:

$$e^{j\theta} = \cos\theta + j\sin\theta \quad \textbf{(6.99)}$$

If we assume for the moment that $K_1 = K_2 = K$, then the natural response takes the form

$$\begin{aligned} x_N(t) &= K e^{\alpha t}(e^{j\beta t} + e^{-j\beta t}) \\ &= K e^{\alpha t}(2\cos\beta t) \end{aligned} \quad \textbf{(6.100)}$$

Thus, in the case of complex roots, the natural response of a second-order circuit can have oscillatory behavior! The function $2Ke^{\alpha t}\cos\beta t$ is a **damped sinusoid;** it is depicted in Figure 6.69 for $\alpha = -5$, $\beta = 50$, and $K = 0.5$. Note that, in general, K_1 and K_2 will not be equal; nonetheless, the underdamped response will still display damped sinusoidal oscillations.

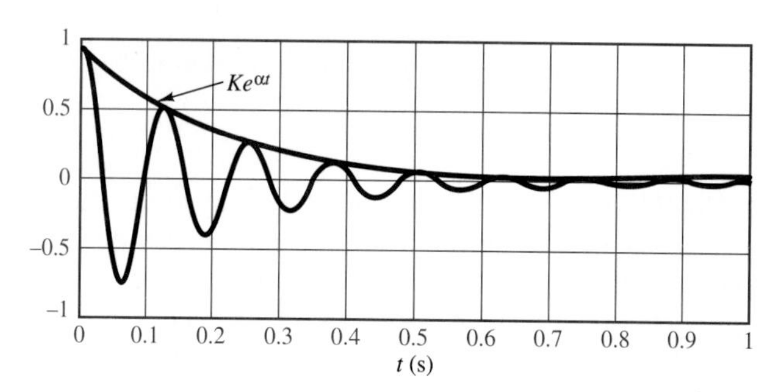

Figure 6.69 Response of underdamped second-order circuit

Forced and Complete Response of Second-Order Circuits

Once we obtain the natural response using the techniques described in the preceding section, we may find the forced response using the same method employed for first-order circuits. Once again, we shall limit our analysis to a switched DC forcing function, for the sake of simplicity (the form of the forced response when the forcing function is a switched sinusoid is explored in the homework problems). The form of the forced differential equation is

$$a\frac{d^2x(t)}{dt^2} + b\frac{dx(t)}{dt} + cx(t) = F \tag{6.101}$$

where F is a constant. Therefore we assume a solution of the form $x_F(t) = X_F =$ constant, and we substitute in the forced equation to find that

$$X_F = \frac{F}{c} \tag{6.102}$$

Finally, in order to compute the complete solution, we sum the natural and forced responses, to obtain

$$x(t) = x_N(t) + x_F(t) = K_1e^{s_1t} + K_2e^{s_2t} + \frac{F}{c} \tag{6.103}$$

For a second-order differential equation, we need two initial conditions to solve for the constants K_1 and K_2. These are the values of $x(t)$ at $t = 0$ and of the derivative of $x(t)$, dx/dt, at $t = 0$. To complete the solution, we therefore need to solve the two equations

$$x(t = 0) = x_0 = K_1 + K_2 + \frac{F}{c} \tag{6.104}$$

and

$$\frac{dx}{dt}(t = 0) = \dot{x}_0 = s_1K_1 + s_2K_2 \tag{6.105}$$

To summarize, we must follow these steps to obtain the complete solution of a second-order circuit excited by a switched DC source:

1. Write the differential equation for the circuit.
2. Find the roots of the characteristic polynomial.
3. Find the forced response.
4. Write the complete solution as the sum of natural and forced responses.
5. Apply the initial conditions *to the complete solution* to determine the constants K_1 and K_2.

Although these steps are straightforward, the successful application of this technique will require some practice, especially the determination of the initial conditions and the computation of the constants. There is no substitute for practice in gaining familiarity with these techniques! The following examples should be of help in illustrating the methods just described.

Example 6.18

Determine the complete response of the circuit shown in Figure 6.70. Assume that some energy is stored in the capacitor, so that $v_C(0^-) = 5$ V. No energy could be stored in the inductor for $t < 0$ since no current flows in the circuit until the switch is closed. The inductor current, $i(t)$, and the capacitor voltage, $v_C(t)$, are the outputs to be determined.

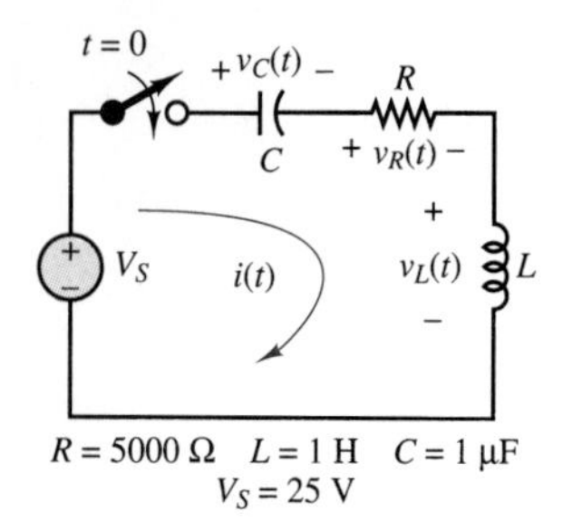

Figure 6.70

Solution:

We begin by applying KVL around the circuit for $t \geq 0$:

$$v_C(t) + v_R(t) + v_L(t) = V_S$$

or

$$\frac{1}{C}\int i\,dt + Ri + L\frac{di}{dt} = V_S$$

Differentiating once yields the second-order differential equation

$$\frac{d^2i}{dt^2} + \frac{R}{L}\frac{di}{dt} + \frac{1}{LC}i = \frac{1}{L}\frac{dV_S}{dt} = 0$$

from which we determine the characteristic polynomial

$$s^2 + \frac{R}{L}s + \frac{1}{LC} = 0$$

the solution to which is given by

$$s_{1,2} = -\frac{R}{2L} \pm \frac{\sqrt{\left(\frac{R}{L}\right)^2 - \frac{4}{LC}}}{2} = -2{,}500 \pm \frac{\sqrt{5{,}000^2 - 4 \times 10^6}}{2}$$

or

$$s_1 = -208.7, \qquad s_2 = -4{,}791.3$$

which are real, distinct roots. This indicates that we have an overdamped circuit with a natural response

$$i_N(t) = K_1 e^{-208.7t} + K_2 e^{-4{,}791.3t}$$

The forced response is found to be zero, since $F = dV_S/dt = 0$ for $t > 0$. This result should not be surprising, since the capacitor will act as an open circuit at steady state and therefore no current can flow through the inductor due to the forcing function alone (however, transient current will flow). Thus, the complete solution is

$$i(t) = K_1 e^{-208.7t} + K_2 e^{-4{,}791.3t}$$

To evaluate the coefficients K_1 and K_2, we consider the initial conditions, $i(0^+)$ and $\frac{di}{dt}(0^+)$. The current in the inductor cannot change instantaneously, so that

$$i(0^-) = i(0^+) = 0 \text{ A}$$

and

$$i(0^-) = K_1 e^0 + K_2 e^0 = 0$$

$$0 = K_1 + K_2$$

or

$$K_1 = -K_2$$

To find $\frac{di}{dt}(0^+)$, we observe that $\frac{di}{dt}(0^+) = \frac{1}{L}v_L(0^+)$, from the defining relationship for the inductor, $v_L = L\,di/dt$. Now, the voltage across the inductor can change instantaneously, unlike the inductor current, and so we cannot state that $v_L(0^-) = v_L(0^+)$. To find $v_L(0^+)$, however, we can apply KVL once again for $t = 0^+$:

$$V_S = v_C(0^+) + v_R(0^+) + v_L(0^+)$$

But

$$v_R(0^+) = i(0^+)R = 0$$

so that

$$v_L(0^+) = 25 - v_C(0^+) = 20$$

and we conclude that

$$\frac{di}{dt}(0^+) = \frac{v_L(0^+)}{L} = 20$$

Applying this second initial condition to the solution of the differential equation at $t = 0^+$, we obtain

$$\frac{di}{dt}(0^+) = K_1(-208.7) + K_2(-4{,}791.3) = 20$$

Solving the two equations in K_1 and K_2 thus obtained, we find that

$$-4{,}791.3K_2 + 208.7K_2 = 20$$

$$K_2 = -\frac{20}{4{,}582.6} = -0.0044$$

and

$$K_1 = -K_2 = 0.0044$$

Thus, the complete solution is

$$i(t) = 0.0044e^{-4{,}791.3t} + 0.0044e^{-208.7t}$$

Finally, to compute $v_C(t)$, we could integrate $i(t)$, but it is simpler to apply KVL once again, to obtain

$$\begin{aligned} v_C(t) &= V_S - v_R(t) - v_L(t) \\ &= 25 - Ri(t) - L\frac{di}{dt} \end{aligned}$$

where

$$L\frac{di}{dt} = -21.1e^{-4{,}791.3t} + 0.918e^{-208.7t}$$

and

$$Ri(t) = 22e^{-4{,}791.3t} - 22e^{-208.7t}$$

Thus,

$$\begin{aligned} v_C(t) &= 25 - 22e^{-4{,}791.3t} + 22e^{-208.7t} + 21.1e^{-4{,}791.3t} - 0.918e^{-208.7t} \\ &= 25 - 0.9e^{-4{,}791.3t} + 21.082e^{-208.7t} \end{aligned}$$

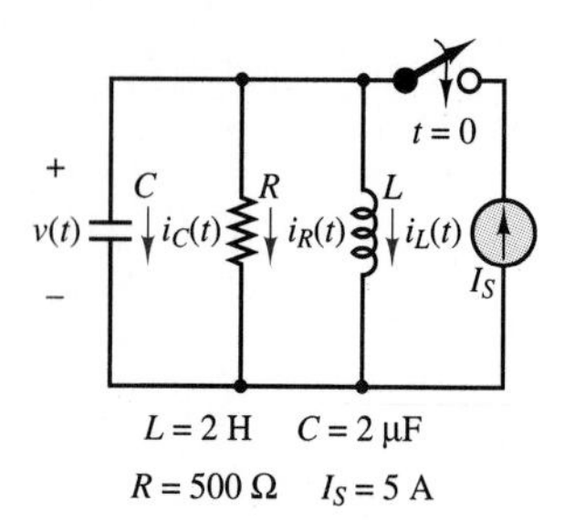

Figure 6.71

Example 6.19

This example illustrates the solution method for a critically damped circuit. Assuming that the initial inductor current and capacitor voltage are both zero, determine $v(t)$ for $t \geq 0$ in the circuit of Figure 6.71.

Solution:

Applying KCL for $t \geq 0$ results in the following equation:

$$i_L(t) + i_R(t) + i_C(t) = I_S$$

or

$$-\frac{1}{L}\int v(t)dt - \frac{v(t)}{R} - C\frac{dv(t)}{dt} = -I_S$$

Differentiating once gives us a second-order differential equation:

$$-C\frac{d^2v}{dt^2} - \frac{1}{R}\frac{dv}{dt} - \frac{1}{L}v = 0$$

or

$$\frac{d^2v}{dt^2} + \frac{1}{RC}\frac{dv}{dt} + \frac{1}{LC}v = 0$$

which results in the characteristic polynomial

$$s^2 + \frac{1}{RC}s + \frac{1}{LC} = 0$$

The solution to the quadratic equation is

$$s_{1,2} = -\frac{1}{2RC} \pm \frac{\sqrt{\left(\frac{1}{RC}\right)^2 - \frac{4}{LC}}}{2} = -500 \pm \frac{\sqrt{(10^3)^2 - 10^6}}{2} = -500 \pm 0$$

or

$$s_1 = s_2 = -500$$

Thus, we see that we have real, repeated roots. The general solution to the differential equation of this type was given by equation 6.96:

$$\begin{aligned} x(t) &= K_1e^{st} + K_2te^{st} \\ &= K_1e^{-500t} + K_2te^{-500t} \end{aligned}$$

The coefficients K_1 and K_2 are determined on the basis of the initial conditions. Since the voltage across the capacitor cannot change instantaneously, we know that

$$v(0^-) = v(0^+) = 0$$

and

$$v(0) = 0 = K_1e^0 + K_2(0)e^0$$

or

$$K_1 = 0$$

The capacitor current, however, can change instantaneously; thus, we cannot assume that $i_C(0^+) = i_C(0^-)$. By applying KCL, we can see that at $t = 0^+$,

$$i_C(0^+) + i_R(0^+) + i_L(0^+) = I_S$$

But

$$i_R(0^+) = \frac{v(0^+)}{R} = \frac{0}{R} = 0$$

and

$$i_L(0^-) = i_L(0^+) = 0$$

so that

$$i_C(0) = I_S$$

Since $i_C(t)$ is given by

$$i_C(t) = C\frac{dv_C}{dt}$$

we can write

$$i_C(t) = C[K_1(-500)e^{-500t} + K_2e^{-500t} + K_2(-500)te^{-500t}]$$
$$i_C(0) = C[K_1(-500)e^0 + K_2e^0 + K_2(-500)(0)e^0] = I_S$$
$$5 = C[K_1(-500) + K_2]$$

and since we have found K_1 to be zero, we can solve for K_2 as follows:

$$K_2 = \frac{5}{C} = 2.5 \times 10^6$$

Finally, the complete solution for $v(t)$ is

$$v(t) = 2.5 \times 10^6 te^{-500t}$$

Example 6.20

The circuit shown in Figure 6.72 is identical to the band-pass filter of Example 6.7. This examples illustrates the transient analysis of a second-order band-pass filter when subjected to a switched DC input. We shall consider the case where $R = 10\ \Omega$, $L = 5$ mH, and $C = 10\ \mu$F, which corresponds to an underdamped second-order circuit. Assume zero initial conditions for both energy-storage elements—that is, $v_C(0^-) = i_L(0^-) = 0$.

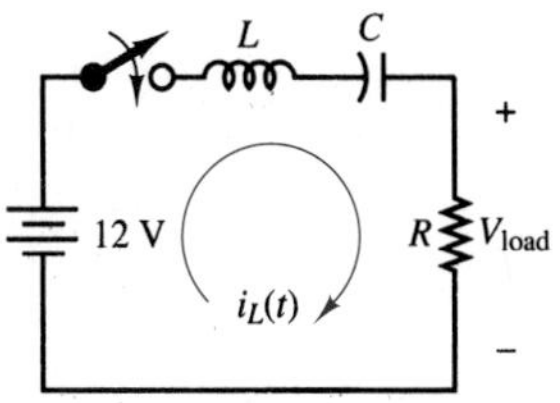

Figure 6.72

Solution:

To determine the load voltage, $v_{load}(t)$, at $t \geq 0$, we observe that the desired output is $v_L(t) = Ri_L(t)$. Thus, the preferred circuit variable in this case is the inductor current. Applying KVL for $t \geq 0$, we write

$$12 = L\frac{di_L}{dt} + v_C + Ri_L$$

Observing that

$$v_C = \frac{1}{C}\int i_L dt$$

we obtain the equation for $i_L(t)$:

$$12 = L\frac{di_L}{dt} + \frac{1}{C}\int i_L dt + Ri_L$$

which can be differentiated to obtain the second-order differential equation

$$\frac{d^2 i_L}{dt^2} + \frac{R}{L}\frac{di_L}{dt} + \frac{1}{LC}i_L = 0$$

Thus, the characteristic polynomial for the equation is given by

$$s^2 + \frac{R}{L}s + \frac{1}{LC} = 0$$

or

$$s^2 + 2{,}000s + 2 \times 10^7 = 0$$

The solution of the quadratic equation reveals that the characteristic polynomial is that of an underdamped circuit whose roots are

$$s_{1,2} = \alpha \pm j\beta = -1{,}000 \pm j4{,}359$$

Thus, the general form of the natural response is

$$i_L(t) = K_1 e^{(\alpha + j\beta)t} + K_2 e^{(\alpha - j\beta)t}$$

Since the steady-state response must be equal to zero (no current can flow through the series circuits as $t \to \infty$), the complete response is equal to the natural response. To find the constants K_1 and K_2, we must use the initial conditions for the inductor current, $i_L(0)$ and $\frac{di_L}{dt}(0)$. The first condition is easily identified:

$$i_L(0) = K_1 + K_2 = 0$$

or

$$K_1 = -K_2$$

To identify the second condition, we note that the inductor voltage is $v_L = L\,di_L/dt$. Thus, at $t = 0^+$, we can apply KVL to obtain

$$12 = L\frac{di_L}{dt}(0^+) + v_C(0^+) + Ri_L(0^+)$$

and since $v_C(0^-) = i_L(0^-) = 0$, we obtain $\frac{di_L}{dt}(0^+) = \frac{12}{L} = 2{,}400$. Now we can apply the second initial condition:

$$\frac{di_L}{dt}(0^+) = s_1 K_1 + s_2 K_2$$

$$2{,}400 = K_1[(-1{,}000 + j4{,}359) - (-1{,}000 - j4{,}359)]$$

or

$$K_1 = \frac{2{,}400}{j8{,}718} = -j0.2753$$

$$K_2 = -K_1 = j0.2753$$

The complete response is then given by

$$\begin{aligned} i_L(t) &= -j0.2753e^{(-1{,}000+j4{,}359)t} + j0.2753e^{(-1{,}000-j4{,}359)t} \\ &= 0.2753e^{-1{,}000t}(-je^{j4{,}359t} + je^{-j4{,}359t}) \\ &= (0.2753e^{-1{,}000t})2\sin(4{,}359t) \end{aligned}$$

The output voltage, $v_{\text{load}}(t) = Ri_L(t)$, is plotted in Figure 6.73.

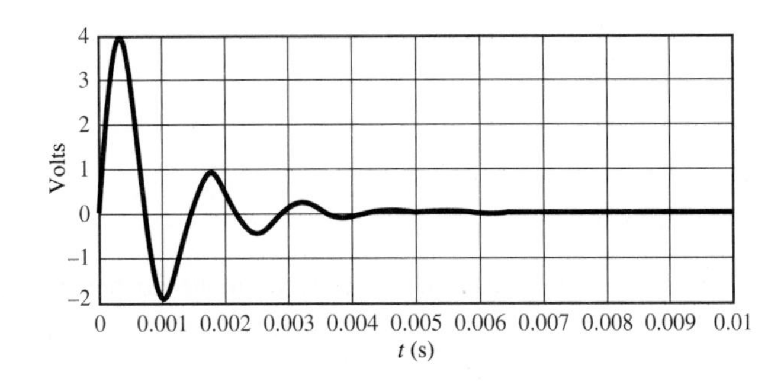

Figure 6.73 Transient response of band-pass filter

Example 6.21

Throughout the preceding examples, we have studied methods for computing the transient response of a second-order circuit. It is also possible, however, to determine the parameters of a circuit by studying its transient response to a known input. Suppose, for example, we wished to determine the (unknown) values of an inductor and a capacitor, by placing these elements in the circuit shown in Figure 6.74. The source voltage is $V_S = 10$ V, and at $t = 0$, the switch closes. The voltage across the known resistor ($R = 100\ \Omega$) is measured and recorded; the results are displayed in Figure 6.75. Is it possible to determine the values of C and L from the measured response?

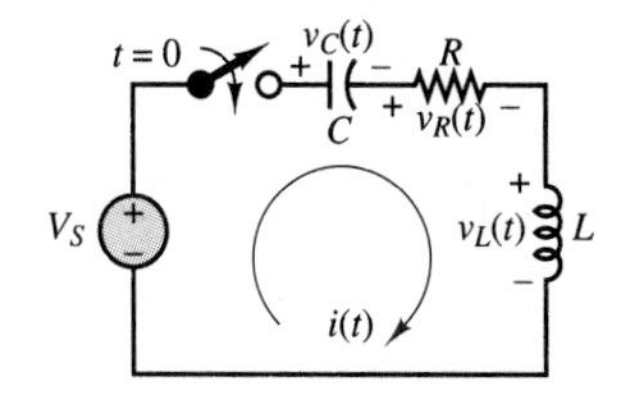

Figure 6.74 A second-order circuit with unknown L and C

Solution:

The differential equation for the series RLC circuit is:

$$\frac{d^2i}{dt^2} + \frac{R}{L}\frac{di}{dt} + \frac{1}{LC}i = 0$$

with the associated characteristic polynomial

$$s^2 + \frac{R}{L}s + \frac{1}{LC} = 0$$

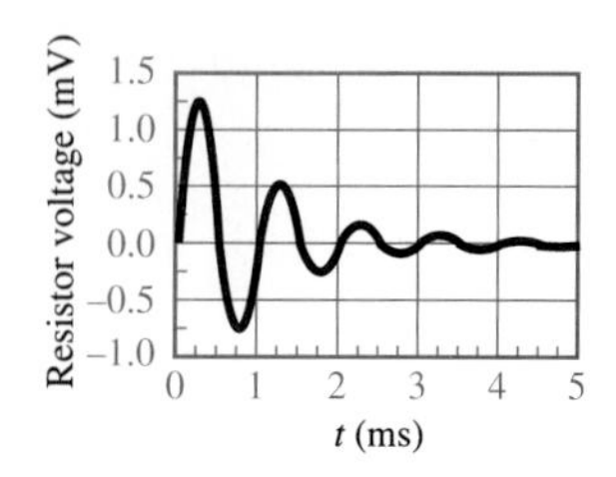

Figure 6.75 Response of second-order circuit with unknown parameters

The roots of this quadratic equation are

$$s_{1,2} = -\frac{R}{2L} \pm \frac{\sqrt{\left(\frac{R}{L}\right)^2 - \frac{4}{LC}}}{2}$$

From the appearance of the response measured across the 100-Ω resistor, it is clear that we are faced with an underdamped circuit. Thus, the response is described by an equation of the form

$$v(t) = K_1 e^{s_1 t} + K_2 e^{s_2 t}$$

where $s_1 = \alpha + j\beta$, $s_2 = \alpha - j\beta$, and

$$\alpha = -\frac{R}{2L}$$

$$\beta = \frac{\sqrt{(R/L)^2 - 4/LC}}{2}$$

This expression can also be written as

$$\begin{aligned} v(t) &= K_1 e^{\alpha t} \sin \beta t + K_2 e^{\alpha t} \cos \beta t \\ &= e^{\alpha t} A \cos(\beta t + \phi) \end{aligned}$$

where $e^{\alpha t}$ is the envelope of the response, β the frequency of the damped oscillations, and ϕ a phase angle resulting from combining the sine and cosine terms. From the measured resistor voltage, we see that the envelope can be characterized using the local maxima of the voltages at $t_0 = 0.2$ ms and $t_1 = 1.2$ ms:

$$v_R(t_0) = 1.24 \text{ V}$$

$$v_R(t_1) = 0.455 \text{ V}$$

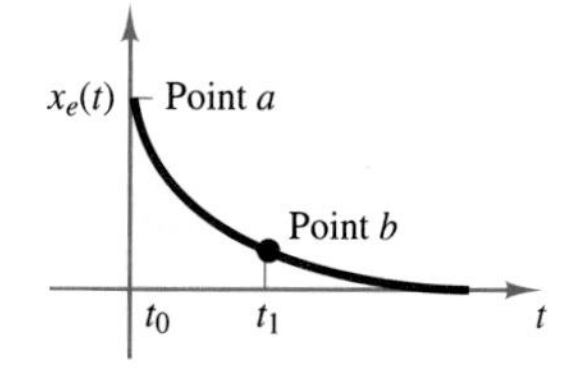

Figure 6.76

Since an envelope can be drawn as shown in Figure 6.76, we can write the equation of the envelope using two points:

$$v_{\text{en}}(t) = Ae^{\alpha(t-t_0)}$$

where A is the amplitude at the first local maximum. If we write the equation for point b, we have

$$v_{\text{en}}(t) = Ae^{\alpha(t_1-t_0)}$$

or

$$0.455 = 1.24e^{\alpha(1.2-0.2)\times 10^{-3}}$$

Now we solve for α:

$$10^{-3} \times \alpha = \log_e \left(\frac{0.455}{1.25}\right)$$

$$\begin{aligned} \alpha &= \frac{1}{0.001} \log_e \left(\frac{0.455}{1.24}\right) \\ &= -1{,}014 \end{aligned}$$

Since

$$\alpha = -\frac{R}{2L} = -1{,}014$$

and

$$R = 100\ \Omega$$

we obtain

$$L = \frac{R}{2(1{,}014)} = \frac{100}{2(1{,}014)}$$
$$= 0.05\ \text{H}$$

Now we consider the frequency of the damped oscillations, β. Since we have already found the local maxima, we will use these to determine the period of the sinusoid (we could also use the zero crossings, however). The period of the sinusoid is estimated to be

$$T = t_1 - t_0 = 1\ \text{ms}$$

and

$$f = \frac{1}{T} = 1{,}000\ \text{Hz}$$

Converting to rad/s, we have

$$\omega = 2\pi f$$
$$\omega = \beta = 6{,}283\ \text{rad/s}$$

Now, since

$$\beta = \frac{\sqrt{(R/L)^2 - 4/LC}}{2}$$

and we know R, L, and β, we can solve for C:

$$(2\beta)^2 = \left(\frac{R}{L}\right)^2 - \frac{4}{LC}$$

$$\frac{4}{LC} = \left(\frac{R}{L}\right)^2 - (2\beta)^2$$

$$C = \frac{4}{L}\left[\frac{1}{(R/L)^2 - (2\beta)^2}\right] = \frac{4}{0.05}\left[\frac{1}{(100/0.05)^2 - (2 \times 6{,}283)^2}\right]$$
$$= 0.494\ \mu\text{F}$$

Thus, the measured transient response of a circuit permits the determination of unknown circuit elements.

DRILL EXERCISES

6.23 Derive the differential equation for the series circuit of Figure 6.63(b). Show that one can write the equation either as

$$\frac{d^2 v_C}{dt^2} + \frac{R_T}{L}\frac{dv_C}{dt} + \frac{1}{LC}v_C = \frac{1}{LC}v_T(t)$$

or as

$$\frac{d^2 i_L}{dt^2} + \frac{R_T}{L}\frac{di_L}{dt} + \frac{1}{LC}i_L = \frac{1}{L}\frac{dv_T(t)}{dt}$$

6.24 Determine the roots of the characteristic equation of the series *RLC* circuit of Figure 6.63(b) with $R = 100\ \Omega$, $C = 10\ \mu\text{F}$, and $L = 1$ H.

6.25 For the series *RLC* circuit of Figure 6.63(b), with $L = 1$ H and $C = 10\ \mu\text{F}$, find the ranges of values of R for which the circuit response is overdamped and underdamped, respectively.

6.26 For the seismic transducer of Example 6.9, show that the three cases given by $B = 10$, $B = 2$, and $B = 1$ N/ms correspond to an overdamped, a critically damped, and an underdamped second-order system, respectively.

6.27 For what value of R is the band-pass filter of Figure 6.19 critically damped?

6.6 COMPLEX FREQUENCY AND THE LAPLACE TRANSFORM

The transient analysis methods illustrated in the preceding section for first- and second-order circuits can become rather cumbersome when applied to higher-order circuits. Moreover, solving the differential equations directly, as is done in Section 6.5, does not reveal the strong connection that exists between the transient response and the frequency response of a circuit. The aim of this section is to introduce an alternate solution method based on the notions of complex frequency and of the **Laplace transform**. The concepts presented in this section will demonstrate that the frequency response of linear circuits is but a special case of the general transient response of the circuit, when analyzed by means of Laplace methods. In addition, the use of the Laplace transform method allows the introduction of *systems* concepts, such as poles, zeros, and transfer functions, that cannot be otherwise recognized.

Complex Frequency

In Chapter 4, we considered circuits with sinusoidal excitations such as

$$v(t) = A\cos(\omega t + \phi) \qquad \textbf{(6.106)}$$

which we also wrote in the equivalent phasor form

$$\mathbf{V}(j\omega) = Ae^{j\phi} = A\angle\phi \qquad (6.107)$$

The two expressions just given are related by

$$v(t) = \text{Re}(\mathbf{V}e^{j\omega t}) \qquad (6.108)$$

As was shown in Chapter 4, phasor notation is extremely useful in solving AC steady-state circuits, in which the voltages and currents are *steady-state sinusoids* of the form of equation 6.106. We now consider a different class of waveforms, useful in the transient analysis of circuits, namely, *damped sinusoids*. The most general form of a damped sinusoid is

$$v(t) = Ae^{\sigma t}\cos(\omega t + \phi) \qquad (6.109)$$

As one can see, a damped sinusoid is a sinusoid multiplied by a real exponential, $e^{\sigma t}$. The constant σ is real and is usually zero or negative in most practical circuits. Figures 6.77(a) and (b) depict the case of a damped sinusoid with negative σ and with positive σ, respectively. Note that the case $\sigma = 0$ corresponds exactly to a sinusoidal waveform. The definition of phasor voltages and currents given in Chapter 4 can easily be extended to account for the case of damped sinusoidal waveforms by defining a new variable, s, called the *complex frequency*:

$$s = \sigma + j\omega \qquad (6.110)$$

You may wish to compare this expression with the term $\alpha \pm j\beta$ in equation 6.93. Note that the special case $\sigma = 0$ corresponds to $s = j\omega$, that is, the familiar steady-state sinusoidal (phasor) case. We shall now refer to the complex variable $\mathbf{V}(s)$ as the **complex frequency domain** representation of $v(t)$. It should be observed that from the viewpoint of circuit analysis, the use of the Laplace transform is, in all, analogous to phasor analysis; that is, substituting the variable s wherever $j\omega$ was used is the only step required to describe a circuit using the new notation.

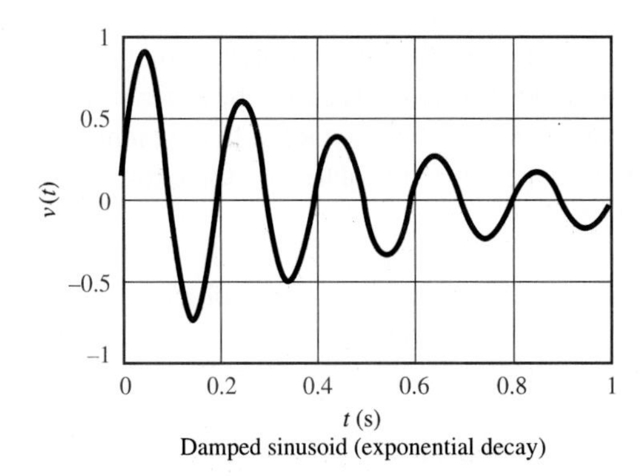

Figure 6.77(a) Damped sinusoid for negative σ

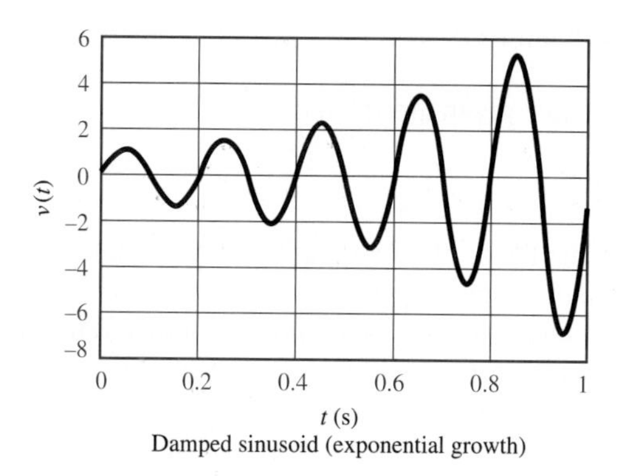

Figure 6.77(b) Damped sinusoid for positive σ

DRILL EXERCISES

6.28 Find the complex frequencies that are associated with

a. $5e^{-4t}$
b. $\cos 2\omega t$
c. $\sin(\omega t + 2\theta)$
d. $4e^{-2t}\sin(3t - 50°)$
e. $e^{-3t}(2 + \cos 4t)$

6.29 Find s and $\mathbf{V}(s)$ if $v(t)$ is given by

a. $5e^{-2t}$
b. $5e^{-2t}\cos(4t + 10°)$
c. $4\cos(2t - 20°)$

6.30 Find $v(t)$ if

a. $s = -2, \mathbf{V} = 2\angle 0°$
b. $s = j2, \mathbf{V} = 12\angle -30°$
c. $s = -4 + j3, \mathbf{V} = 6\angle 10°$

All the concepts and rules used in AC network analysis (see Chapter 4), such as impedance, admittance, KVL, KCL, and Thévenin's and Norton's theorems, carry over to the damped sinusoid case exactly. In the complex frequency domain, the current $\mathbf{I}(s)$ and voltage $\mathbf{V}(s)$ are related by the expression

$$\mathbf{V}(s) = Z(s)\mathbf{I}(s) \tag{6.111}$$

where $Z(s)$ is the familiar impedance, with s replacing $j\omega$. We may obtain $Z(s)$ from $Z(j\omega)$ by simply replacing $j\omega$ by s. For a resistance, R, the impedance is

$$Z_R(s) = R \tag{6.112}$$

For an inductance, L, the impedance is

$$Z_L(s) = sL \tag{6.113}$$

For a capacitance, C, it is

$$Z_C(s) = \frac{1}{sC} \tag{6.114}$$

Impedances in series or parallel are combined in exactly the same way as in the AC steady-state case, since we only replace $j\omega$ by s.

Example 6.22

Consider a series RL circuit with $R = 4\ \Omega$ and $L = 2$ H, excited by a voltage source $v(t) = 10e^{-2t}\cos 5t$ V. Find the impedance seen by the source, and determine a time-domain expression for the series current flowing in the circuit.

Solution:

The impedance seen by the source terminal is given by

$$Z(s) = 4 + 2s$$

Thus, since the input voltage phasor is

$$\mathbf{V}(s) = 10\angle 0^\circ \text{ V}$$

we have

$$\mathbf{I}(s) = \frac{\mathbf{V}(s)}{Z(s)} = \frac{10\angle 0^\circ}{2s+4} = \frac{10\angle 0^\circ}{2(-2+j5)+4} = 1\angle{-90^\circ}\text{A}$$

Therefore, the time-domain expression for the current is

$$i(t) = e^{-2t}\cos(5t - 90^\circ) \text{ A}$$

Just as frequency response functions $H(j\omega)$ were defined in this chapter, it is possible to define a **transfer function**, $H(s)$. This can be a ratio of a voltage to a current, a ratio of a voltage to a voltage, a ratio of a current to a current, or a ratio of a current to a voltage. The transfer function $H(s)$ is a function of network elements and their interconnections. Using the transfer function and knowing the input (voltage or current) to a circuit, we can find an expression for the output either in the complex frequency domain or in the time domain. As an example, suppose $\mathbf{V}_i(s)$ and $\mathbf{V}_o(s)$ are the input and output voltages to a circuit, respectively, in complex frequency notation. Then

$$H(s) = \frac{\mathbf{V}_o(s)}{\mathbf{V}_i(s)} \tag{6.115}$$

from which we can obtain the output in the complex frequency domain by computing

$$\mathbf{V}_o(s) = H(s)\mathbf{V}_i(s) \tag{6.116}$$

If $\mathbf{V}_i(s)$ is a known damped sinusoid, we can then proceed to determine $v_o(t)$ by means of the method illustrated earlier in this section.

DRILL EXERCISES

6.31 Given the transfer function $H(s) = \frac{3(s+2)}{s^2+2s+3}$ and the input $\mathbf{V}_i(s) = 4\angle 0°$, find the forced response $v_o(t)$ if

a. $s = -1$
b. $s = -1 + j1$
c. $s = -2 + j1$

6.32 Given the transfer function $H(s) = \frac{2(s+4)}{s^2+4s+5}$ and the input $\mathbf{V}_i(s) = 6\angle 30°$, find the forced response $v_o(t)$ if

a. $s = -4 + j1$
b. $s = -2 + j2$

The Laplace Transform

The Laplace transform, named after the French mathematician and astronomer Pierre Simon de Laplace, is defined by

$$\mathcal{L}[f(t)] = F(s) = \int_0^{\infty} f(t)e^{-st}dt \tag{6.117}$$

The function $F(s)$ is the Laplace transform of $f(t)$ and is a function of the complex frequency, $s = \sigma + j\omega$, considered earlier in this section. Note that the function $f(t)$ is defined only for $t \geq 0$. This definition of the Laplace transform applies to what is known as the **one-sided** or **unilateral Laplace transform,** since $f(t)$ is evaluated only for positive t. In order to conveniently express arbitrary functions only for positive time, we introduce a special function called the **unit step function,** $u(t)$, defined by the expression

$$u(t) = \begin{cases} 0 & t < 0 \\ 1 & t > 0 \end{cases} \tag{6.118}$$

Example 6.23

Find the Laplace transform of $f(t) = e^{-at}u(t)$.

Solution:

The Laplace transform of $f(t)$ can be obtained by applying equation 6.117:

$$F(s) = \int_0^{\infty} e^{-at}e^{-st}\,dt = \int_0^{\infty} e^{-(s+a)t}\,dt = \frac{1}{s+a}e^{-(s+a)t}\Big|_0^{\infty} = \frac{1}{s+a}$$

Thus,

$$\mathcal{L}[e^{-at}] = \frac{1}{s+a}$$

Example 6.24

Find the Laplace transform of $f(t) = \cos(\omega t)u(t)$.

Solution:

Using Euler's formula, we can expand $f(t)$ into: $f(t) = \frac{1}{2}(e^{j\omega t} + e^{-j\omega t})u(t)$. We can now apply equation 6.117 to find the Laplace transform of $f(t)$.

$$F(s) = \int_0^\infty \frac{1}{2}(e^{j\omega t} + e^{-j\omega t})e^{-st}\, dt = \frac{1}{2}\left(\int_0^\infty e^{-(s+j\omega)t}\, dt + \int_0^\infty e^{-(s-j\omega)t}\, dt\right)$$

Evaluating the integrals, we find the Laplace transform of $f(t)$ to be

$$\mathcal{L}[\cos\omega t] = \frac{s}{s^2+\omega^2}$$

DRILL EXERCISES

6.33 Find the Laplace transform of the following functions:

a. $u(t)$
b. $\sin(\omega t)u(t)$
c. $tu(t)$

6.34 Find the Laplace transform of the following functions:

a. $e^{-at}\sin\omega t\, u(t)$
b. $e^{-at}\cos\omega t\, u(t)$

From what has been said so far about the Laplace transform, it is obvious that we may compile a lengthy table of functions and their Laplace transforms by repeated application of equation 6.117 for various functions of time, $f(t)$. Then, we could obtain a wide variety of inverse transforms by matching entries in the table. Table 6.3 lists some of the more common **Laplace transform pairs**. The computation of the **inverse Laplace transform** is in general rather complex if one wishes to consider arbitrary functions of s. In many practical cases, however, it is possible to use combinations of known transform pairs to obtain the desired result.

Table 6.3 **Laplace transform pairs**

$f(t)$	$F(s)$
$\delta(t)$ (unit impulse)	1
$u(t)$ (unit step)	$\frac{1}{s}$
$e^{-at}u(t)$	$\frac{1}{s+a}$
$\sin \omega t\, u(t)$	$\frac{\omega}{s^2+\omega^2}$
$\cos \omega t\, u(t)$	$\frac{s}{s^2+\omega^2}$
$e^{-at} \sin \omega t\, u(t)$	$\frac{\omega}{(s+a)^2+\omega^2}$
$e^{-at} \cos \omega t\, u(t)$	$\frac{s+a}{(s+a)^2+\omega^2}$
$tu(t)$	$\frac{1}{s^2}$

Example 6.25

Find the inverse of the transform

$$F(s) = \frac{2}{s+3} + \frac{4}{s^2+4} - \frac{4}{s}$$

Solution:

We need to find the inverse transform of $F_1(s)$, $F_2(s)$, and $F_3(s)$ where

$$F_1(s) = \frac{2}{s+3},\ F_2(s) = \frac{4}{s^2+4},\ \text{and}\ F_3(s) = \frac{4}{s}$$

Looking at Table 6.3, we can see that

$$f_1(t) = 2\mathcal{L}^{-1}\left(\frac{1}{s+3}\right) = 2e^{-3t}u(t)$$

$$f_2(t) = F_2(s) = 2\mathcal{L}^{-1}\left(\frac{2}{s^2+2^2}\right) = 2\sin 2t\ u(t)$$

$$f_3(t) = 4\mathcal{L}^{-1}\left(\frac{1}{s}\right) = 4u(t)$$

Thus,

$$f(t) = f_1(t) + f_2(t) + f_3(t) = (2e^{-3t} + 2\sin 2t - 4)u(t)$$

Example 6.26

Find the inverse of the transform

$$F(s) = \frac{2s + 5}{s^2 + 5s + 6}$$

Solution:

As you can see, there is no direct entry in Table 6.3 that we can use to obtain $f(t)$. However, we can obtain a partial fraction expansion of $F(s)$ and apply the table to each term in the expansion. Obtaining a partial fraction expansion is the opposite operation of obtaining a common denominator. We illustrate this by obtaining the partial fraction expansion of $F(s)$:

$$F(s) = \frac{2s + 5}{s^2 + 5s + 6} = \frac{A}{s + 2} + \frac{B}{s + 3}$$

The next step is to determine the constants A and B. There are many ways of determining A and B. However, the simplest means of determining A and B is to note that

$$(s + 2)F(s) = \frac{2s + 5}{s + 3} = A + \frac{B(s + 2)}{s + 3}$$

and

$$(s + 3)F(s) = \frac{2s + 5}{s + 2} = \frac{A(s + 3)}{s + 2} + B$$

Therefore,

$$A = (s + 2)F(s) = \left.\frac{2s + 5}{s + 3}\right|_{s=-2} = 1$$

$$B = (s + 3)F(s) = \left.\frac{2s + 5}{s + 2}\right|_{s=-3} = 1$$

Thus,

$$F(s) = \frac{2s + 5}{s^2 + 5s + 6} = \frac{1}{s + 2} + \frac{1}{s + 3}$$

We can now use Table 6.3 to find the inverse transform of $F(s)$. Using Table 6.3:

$$f(t) = (e^{-2t} + e^{-3t})u(t)$$

DRILL EXERCISE

6.35 Find the inverse Laplace transform of each of the following functions:

a. $F(s) = \dfrac{1}{s^2 + 5s + 6}$

b. $F(s) = \dfrac{s - 1}{s(s + 2)}$

c. $F(s) = \dfrac{3s}{(s^2 + 1)(s^2 + 4)}$

d. $F(s) = \dfrac{1}{(s + 2)(s + 1)^2}$

Transfer Functions, Poles, and Zeros

It should be clear that the Laplace transform can be quite a convenient tool for analyzing the transient response of a circuit. The Laplace variable, s, is an extension of the steady-state frequency response variable $j\omega$ already encountered in this chapter. Thus, it is possible to describe the input-output behavior of a circuit using Laplace transform ideas in the same way in which we used frequency response ideas earlier. Now, we can define voltages and currents in the complex frequency domain as $\mathbf{V}(s)$ and $\mathbf{I}(s)$, and denote impedances by the notation $Z(s)$, where s replaces the familiar $j\omega$. We define an extension of the frequency response of a circuit, called the transfer function, as the ratio of any input variable to any output variable, i.e.:

$$H_1(s) = \frac{\mathbf{V}_o(s)}{\mathbf{V}_i(s)} \quad \text{or} \quad H_2(s) = \frac{\mathbf{I}_o(s)}{\mathbf{V}_i(s)}, \quad \text{etc.} \tag{6.119}$$

As an example, consider the circuit of Figure 6.78. We can analyze it using a method analogous to phasor analysis by defining impedances

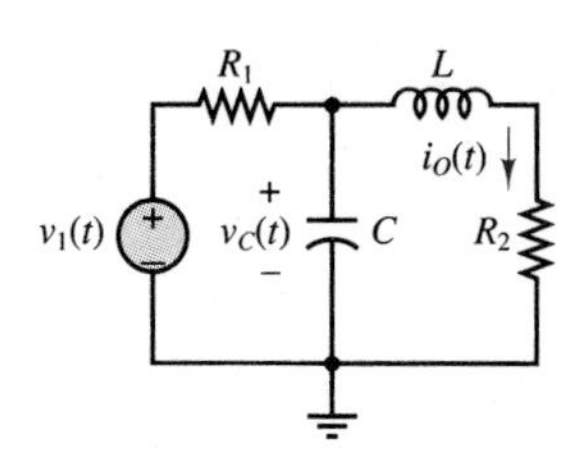

$$Z_1 = R_1, Z_C = \frac{1}{sC}, Z_L = sL, \text{ and } Z_2 = R_2 \tag{6.120}$$

Then, we can use mesh analysis methods to determine that

$$\mathbf{I}_o(s) = \mathbf{V}_i(s)\frac{Z_C}{(Z_1 + Z_C)(Z_C + Z_L + Z_2) + Z_C^2} \tag{6.121}$$

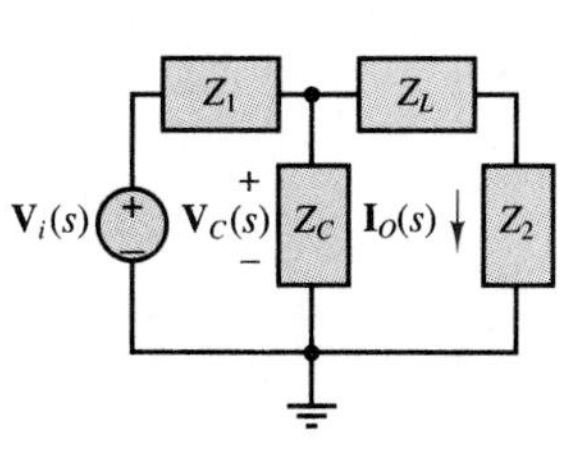

Figure 6.78 A circuit and its Laplace transform domain equivalent

or, upon simplifying and substituting the relationships of equation 6.120,

$$H_2(s) = \frac{\mathbf{I}_o(s)}{\mathbf{V}_i(s)} = \frac{1}{R_1LCs^2 + (R_1R_2C + LC)s + R_1 + R_2} \tag{6.122}$$

If we were interested in the relationship between the input voltages and, say, the capacitor voltage, we could similarly calculate

$$H_1(s) = \frac{\mathbf{V}_C(s)}{\mathbf{V}_i(s)} = \frac{sL + R_2}{R_1LCs^2 + (R_1R_2C + LC)s + R_1 + R_2} \tag{6.123}$$

Note that a transfer function consists of a *ratio of polynomials;* this ratio can also be expressed in factored form, leading to the discovery of additional important properties of the circuit. Let us, for the sake of simplicity, choose numerical values for the components of the circuit of Figure 6.78. For example, let $R_1 = 0.5\ \Omega$, $C = \frac{1}{4}$ F, $L = 0.5$ H, and $R_2 = 2\ \Omega$. Then we can substitute these values into equation 6.123 to obtain

$$H_1(s) = \frac{0.5s + 2}{0.0625s^2 + 0.375s + 2.5} = 8\left(\frac{s + 4}{s^2 + 6s + 40}\right) \tag{6.124}$$

Equation 6.124 can be factored into products of first-order terms as follows:

$$H_1(s) = 8\left[\frac{s + 4}{s - 3.0000 + j5.5678)(s - 3.0000 - j5.5678)}\right] \tag{6.125}$$

where it is apparent that the response of the circuit has very special characteristics for three values of s: $s = -4$; $s = -3.0000 + j5.5678$; and $s = -3.0000 - j5.5678$. In the first case, at the complex frequency $s = -4$, the numerator of the transfer function becomes zero, and the response of the circuit is zero, regardless of how large the input voltage is. We call this particular value of s a **zero** of the transfer function. In the latter two cases, for $s = -3.0000 \pm j5.5678$, the response of the circuit becomes infinite, and we refer to these values of s as **poles** of the transfer function.

It is customary to represent the response of electric circuits in terms of poles and zeros, since knowledge of the location of these poles and zeros is equivalent to knowing the transfer function and provides complete information regarding the response of the circuit. Further, if the poles and zeros of the transfer function of a circuit are plotted in the complex plane, it is possible to visualize the response of the circuit very effectively. Figure 6.79 depicts the pole–zero plot of the circuit of Figure 6.78; in plots of this type it is customary to denote zeros by a small circle, and poles by an "x."

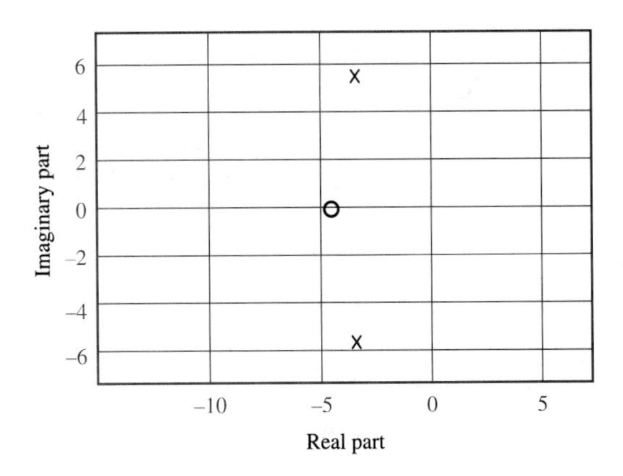

Figure 6.79 Zero–pole plot for the circuit of Figure 6.78

The poles of a transfer function have a special significance, in that they are equal to the roots of the natural response of the system. They are also called the **natural frequencies** of the circuit. Example 6.27 illustrates this point.

Example 6.27

Consider the circuit of Example 6.18. In that example, we found the differential equation describing the circuit to be:

$$\frac{d^2i}{dt^2} + \frac{R}{L}\frac{di}{dt} + \frac{1}{LC}i = 0$$

The characteristic equation corresponding to this differential equation is

$$s^2 + \frac{R}{L}s + \frac{1}{LC} = 0$$

If we compute the transfer function for this circuit, say, $\frac{\mathbf{V}_L(s)}{\mathbf{V}_S(s)}$, we find that

$$\frac{\mathbf{V}_L(s)}{\mathbf{V}_S(s)} = \frac{sL}{\frac{1}{sC} + R + sL} = \frac{s^2}{s^2 + \frac{R}{L}s + \frac{1}{LC}}$$

Note how the denominator of the transfer function, which determines the poles, is identical to the characteristic equation of the circuit. Thus, the poles of the transfer function are the same as the roots that determine the natural response of the circuit:

$$s_{1,2} = \frac{R}{2L} \pm \frac{\sqrt{\left(\frac{R}{L}\right)^2 - \frac{-4}{LC}}}{2}$$

You can verify that, for the circuit in question, the roots correspond to an overdamped system (real and distinct roots) and the natural response is given by the expression $K_1e^{-s_1t} + K_2e^{-s_2t}$. Thus, knowledge of the poles of a transfer function is completely equivalent to knowledge of the natural response of the corresponding circuit.

CONCLUSION

- In many practical applications it is important to analyze the *frequency response* of a circuit, that is, the response of the circuit to sinusoidal signals of different frequencies. This can be accomplished quite effectively by means of the phasor analysis methods developed in Chapter 4, where the radian frequency, ω, is now a variable. The frequency response of a circuit is then defined as the ratio of an output phasor quantity (voltage or current) to an input phasor quantity (voltage or current), as a function of frequency.

- One of the primary applications of frequency analysis is in the study of electrical *filters,* that is, circuits that can selectively attenuate signals in certain frequency regions. Filters can be designed, using standard resistors, inductors, and capacitors, to have one of four types of characteristics: low-pass, high-pass, band-pass, and band-reject. Such filters find widespread application in many practical engineering applications that involve signal conditioning. Filters will be studied in more depth in Chapters 11 and 14.
- Although the analysis of electrical circuits by means of phasors—that is, steady-state sinusoidal voltages and currents—is quite useful in many applications, there are situations where these methods are not appropriate. In particular, when a circuit (or another system) is subjected to an abrupt change in input voltage or current, different analysis methods must be employed to determine the *transient response* of the circuit. In this chapter, we have studied the analysis methods that are required to determine the transient response of first- and second-order circuits (that is, circuits containing one or two energy-storage elements, respectively). One method involves identifying the differential equation that describes the circuit during the transient period and recognizing important parameters, such as the time constant of a first-order circuit and the damping ratio and natural frequency of a second-order circuit. A second method exploits the idea of complex frequency and the Laplace transform.

EIT REVIEW: TRANSIENTS

The analysis of transient circuits is discussed in the latter parts of Chapter 6 (Sections 6.3, 6.4, 6.5, 6.6). One of the fundamental topics in transient analysis is the solution of first-order circuits. The following problem is a representative sample.

DRILL EXERCISE

6.36 The expression for the current in the 2-Ω resistor in Figure 6.80 for time greater than zero is:

a. $-3e^{-0.5t} + 3$ A
b. $3e^{-0.5t} + 3$ A
c. $-3e^{0.5t} + 3$ A
d. $-6e^{-0.5t} + 6$ A
e. $6e^{-0.5t} + 6$ A

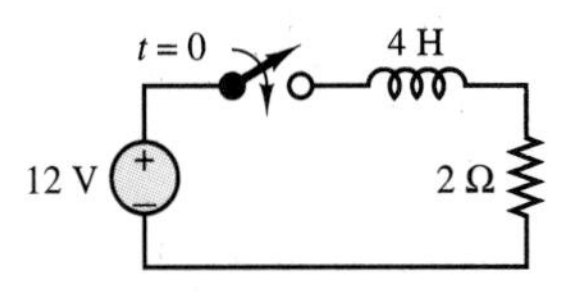

Figure 6.80

Solution:

Applying KVL to the circuit when the switch is closed ($t \geq 0$), we have

$$4\frac{di}{dt} + 2i = 12$$

Solving the differential equation:

$$i(t) = Ke^{-0.5t} + 6 \text{ A} \qquad t \geq 0$$

$$i(0^-) = i(0^+) = 0$$

Therefore,

$$i(t) = -6e^{-0.5t} + 6 \text{ A} \qquad t \geq 0$$

Therefore, the correct answer is d.

Additional solved review problems related to transient analysis may be found in Examples 6.10, 6.11, 6.12, 6.17, 6.18, 6.19, and 6.20. The problems marked "EIT" in the Homework Problems section of Chapter 6 will provide additional practice material. Answers to these problems may be found in Appendix B.

KEY TERMS

ANSWERS TO DRILL EXERCISES

6.1 $H_I(j\omega) = \dfrac{Z_2}{Z_L + Z_2}$

6.4 $\omega_0 = 2{,}026.3$ rad/s

6.5 (a) $\omega_0 = \dfrac{1}{RC}$ (low); (b) $\omega_0 = \dfrac{R}{L}$ (high);
(c) $\omega_0 = \dfrac{1}{RC}$ (high); (d) $\omega_0 = \dfrac{R}{L}$ (low)

6.6 $H(j\omega) = \dfrac{\omega L/R}{\sqrt{1 + (\omega L/R)^2}}$

$\phi(j\omega) = 90° + \arctan \dfrac{-\omega L}{R}$

6.7 $\omega_0 = 35.46$ rad/s

6.8 $0 \le \omega \le 6.48$ rad/s

6.9 $\omega = 375.17$ rad/s

6.10 $\omega = 1{,}030.49$ rad/s

6.11 $\omega = 51{,}878$ rad/s

6.12 $\omega = 69{,}032$ rad/s

6.13 $\omega_1 = 99.95$ rad/s; $\omega_2 = 200.1$ krad/s

6.14 $\dfrac{dv_C}{dt} + \dfrac{1}{RC}v_C = \dfrac{1}{RC}v_S$

6.15 $\dfrac{dv}{dt} + \dfrac{1}{RC}v = \dfrac{1}{C}i_S$

6.16 $\dfrac{di_L}{dt} + \dfrac{R}{L}i_L = \dfrac{R}{L}i_S$

6.17 $v_C(0^-) = 8$ V and $i_R(0^+) = 2$ A

6.18 $v_C = 7.5 - 7.5e^{-t/0.025}$ V, $0 \le t < 0.05$ s;
$v_C = 6.485e^{-(t-0.05)/0.0375}$ V, $t \ge 0.05$ s

6.19 $t_{90\%} = 12.5\ \mu$s

6.20 545 μs

6.21 0.631 V

6.22 The output pulse has a higher peak.

6.24 $-50 \pm j312.25$

6.25 Overdamped: $R > 632.46\ \Omega$; underdamped: $R < 632.46\ \Omega$

6.27 $2\sqrt{L/C}$

6.28 a. -4; b. $\pm j2\omega$; c. $\pm j\omega$; d. $-2 \pm j3$; e. -3 and $-3 \pm j4$

6.29 a. -2, $5\angle 0°$; b. $-2 + j4$, $5\angle 10°$; c. $j2$, $4\angle -20°$

6.30 a. $2e^{-2t}$; b. $12\cos(2t - 30°)$; c. $6e^{-4t}\cos(3t + 10°)$

6.31 a. $6e^{-t}$; b. $12\sqrt{2}e^{-t}\cos(t + 45°)$; c. $6e^{-2t}\cos(t + 135°)$

6.32 a. $3e^{-4t}\cos(t + 165°)$; b. $8\sqrt{2}e^{-2t}\cos(2t - 105°)$

6.33 a. $\dfrac{1}{s}$; b. $\dfrac{\omega}{s^2 + \omega^2}$; c. $\dfrac{1}{s^2}$

6.34 a. $\dfrac{\omega}{(s + a)^2 + \omega^2}$; b. $\dfrac{(s + a)}{(s + a)^2 + \omega^2}$

6.35 a. $f(t) = (e^{-2t} - e^{-3t})(u(t)$;
b. $f(t) = \left(\frac{3}{2}e^{-2t} - \frac{1}{2}\right)u(t)$; c. $f(t) = (\cos t - \cos 2t)u(t)$; d. $f(t) = (e^{-2t} + te^{-t} - e^{-t})u(t)$

HOMEWORK PROBLEMS

6.1

a. Determine the frequency response $\mathbf{V}_{out}(j\omega)/\mathbf{V}_{in}(j\omega)$ for the circuit of Figure P6.1.
b. Plot the magnitude and phase of the circuit for frequencies between 1 and 100 rad/s on graph paper, with a linear scale for frequency.
c. Repeat part b, using semilog paper. (Place the frequency on the logarithmic axis.)
d. Plot the magnitude response on semilog paper with magnitude in dB.

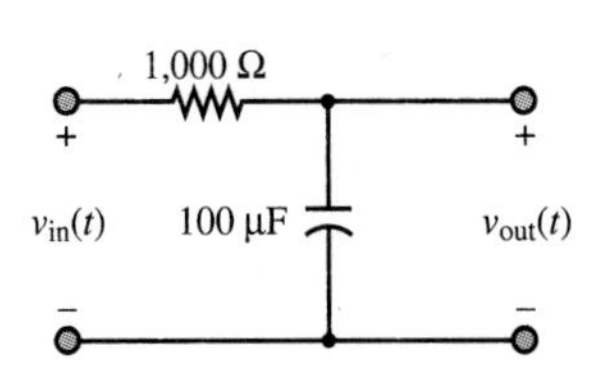

Figure P6.1

6.2 Consider the circuit shown in Figure P6.2.

a. Sketch the amplitude response of $Y = I/V_S$.
b. Sketch the amplitude response of V_1/V_S.
c. Sketch the amplitude response of V_2/V_S.

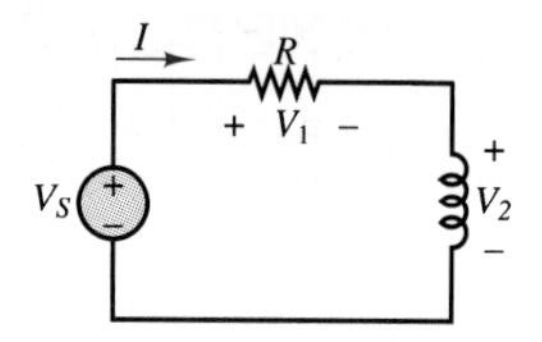

Figure P6.2

EIT 6.3 Repeat Problem 6.1 for the circuit of Figure P6.3.

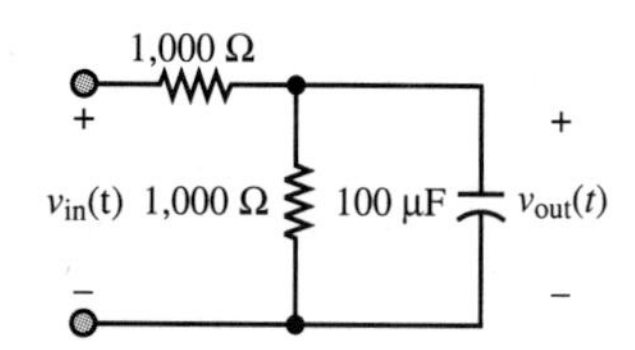

Figure P6.3

6.4 Repeat Problem 6.1 for the circuit of Figure P6.4.

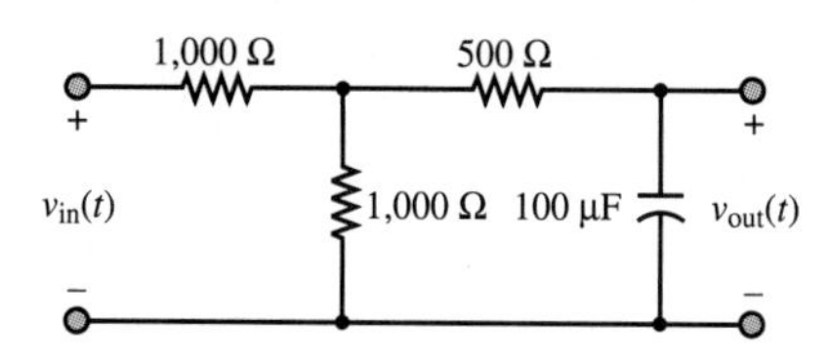

Figure P6.4

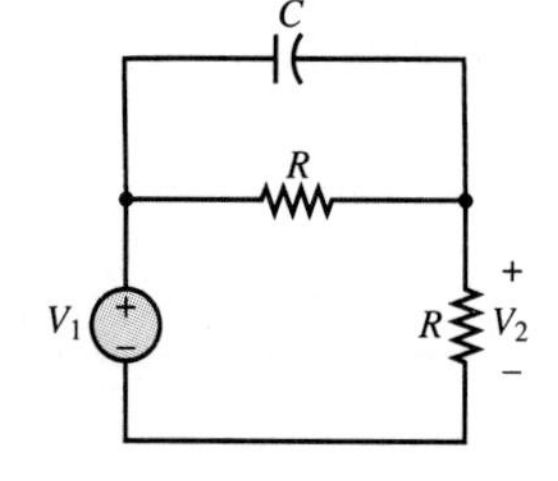

Figure P6.5

6.5 For the circuit shown in Figure P6.5, sketch the amplitude response of V_2/V_1 and indicate the half-power frequency.

6.6 Repeat problem 6.5 for the circuit of Figure P6.6.

EIT 6.7

a. Determine the frequency response, $\mathbf{V}_{out}(j\omega)/\mathbf{V}_{in}(j\omega)$, for the circuit of Figure P6.7.
b. Plot the magnitude and phase of the circuit for frequencies between 100 krad/s and 10 Mrad/s on graph paper, with a linear scale for frequency.
c. Repeat part b, using semilog paper. (Place frequency on the logarithmic axis.)
d. Plot the magnitude response on semilog paper with magnitude in dB.

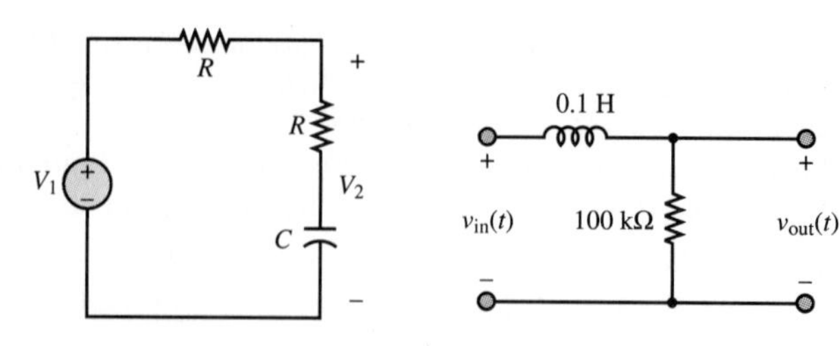

Figure P6.6 **Figure P6.7**

6.8 Repeat Problem 6.7 for the circuit of Figure P6.8.

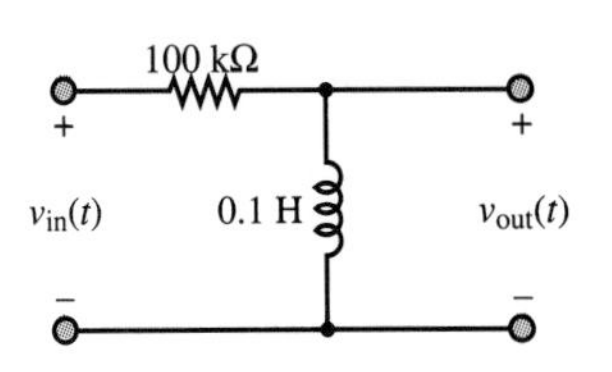

Figure P6.8

6.9 Assume in a certain frequency range that the ratio of output amplitude to input amplitude is proportional to $1/\omega^2$. What is the slope of the Bode plot in this frequency range, expressed in dB per decade?

6.10 Determine the frequency response $\mathbf{I}_L(j\omega)/\mathbf{V}_S(j\omega)$ for the circuit of Figure P6.9.

6.11 Assume that the output amplitude of a circuit depends on frequency according to:

$$V = \frac{A\omega}{\sqrt{B + C\omega^2}}$$

Find:

a. The break frequency.
b. The slope of the Bode plot (in dB per decade) above the break frequency.
c. The slope of the Bode plot below the break frequency.
d. The high-frequency limit of V.

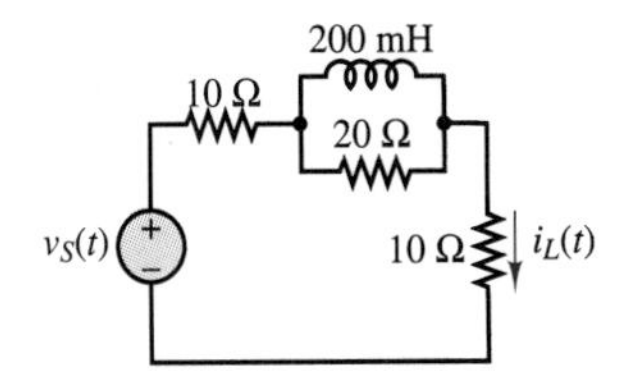

Figure P6.9

6.12 The function of a loudspeaker *crossover network* is to channel frequencies higher than a given crossover frequency, f_c, into the high-frequency speaker (tweeter) and frequencies below f_c into the low-frequency speaker (woofer). Figure P6.10 shows an approximate equivalent circuit where the amplifier is represented as a voltage source with zero internal resistance and each speaker acts as an 8 Ω resistance. If the crossover frequency is chosen to be 1200 Hz, evaluate C and L. Hint: The break frequency would be a reasonable value to set as the crossover frequency.

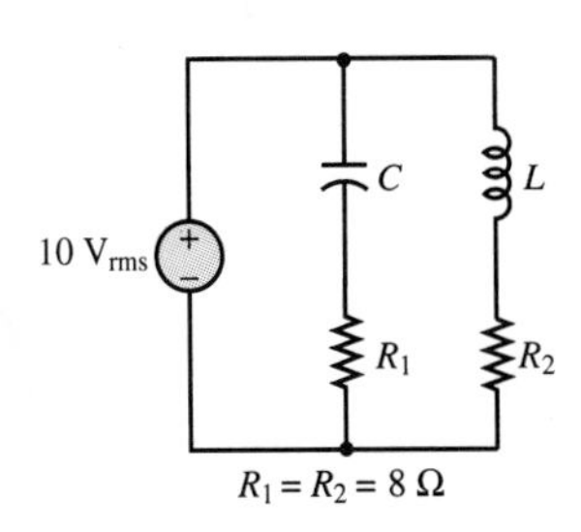

Figure P6.10

EIT 6.13

a. Determine the frequency response, $\mathbf{V}_{out}(j\omega)/\mathbf{V}_{in}(j\omega)$, for the circuit of Figure 6.19 (in the text), where $R_1 = 1{,}000\ \Omega$, $C = 1\ \mu F$, and $L = 1$ H.
b. What is the center frequency of this band-pass filter circuit?
c. Plot the magnitude frequency response of this circuit on semilog paper in units of dB.

6.14 Consider the circuit shown in Figure P6.11. Determine the resonance frequency and the bandwidth for the circuit.

6.15 Repeat Problem 6.13 for $R = 500\ \Omega$, $C = 20\ \mu F$, and $L = 1$ H.

6.16 Repeat Problem 6.14 for the circuit of Figure P6.12.

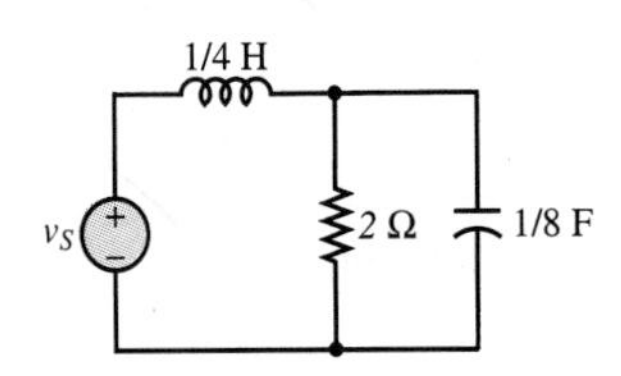

Figure P6.11

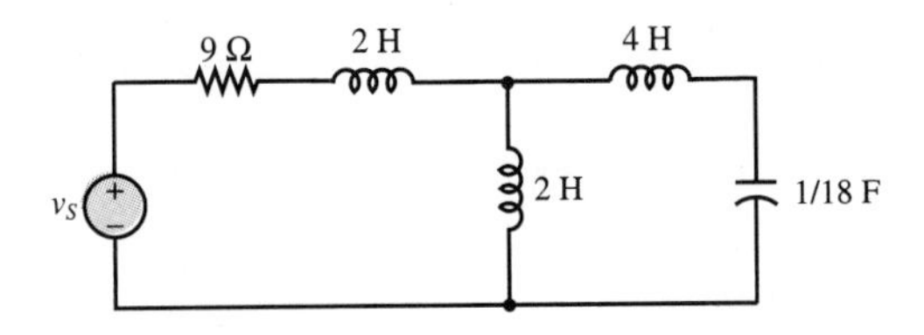

Figure P6.12

EIT 6.17

a. What is the equivalent impedance, Z_{ab}, of the filter of Figure P6.13?
b. At what frequency does the magnitude of the impedance go to infinity?

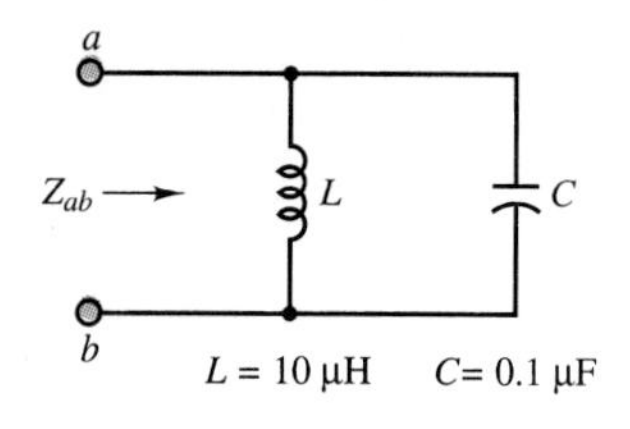

Figure P6.13

6.18 Repeat Problem 6.17 for $C = 10\ \mu\text{F}$ and $L = 0.1$ mH.

6.19 What is the frequency response, $\mathbf{V}_{\text{out}}(\omega)/\mathbf{V}_S(\omega)$, for the circuit of Figure P6.14? Sketch the frequency response of the circuit (magnitude and phase) if $R_S = R_L = 5{,}000\ \Omega$, $L = 10\ \mu\text{H}$, and $C = 0.1\ \mu\text{F}$.

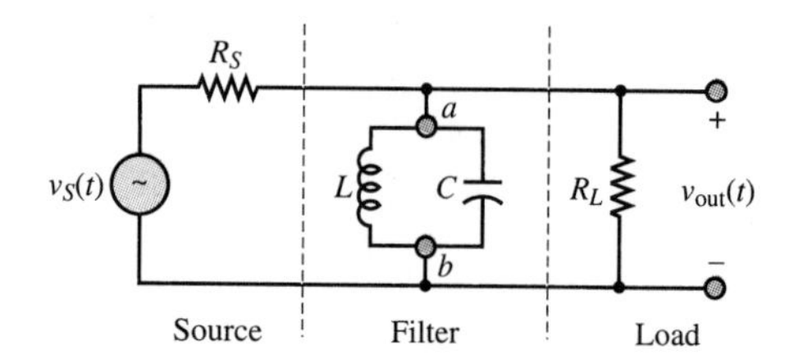

Figure P6.14

6.20 A 60 Hz notch filter was discussed in Example 6.8. If the inductor has a 0.2 Ω series resistance, and the capacitor has a 10 MΩ parallel resistance,

a. What is the impedance of the *nonideal* notch filter at 60 Hz?
b. How much of the 60 Hz interference signal will appear at $\mathbf{V}_L$?

***6.21** Many stereo speakers are two-way speaker systems; that is, they have a woofer for low-frequency sounds and a tweeter for high-frequency sounds. To get the proper separation of frequencies going to the woofer and to the tweeter, crossover circuitry is used. A crossover circuit is effectively a band-pass, high-pass, or low-pass filter. The system model is shown in Figure P6.15.

a. If $L = 2$ mH, $C = 125\ \mu\text{F}$, and $R_S = 4\ \Omega$, find the load impedance as a function of frequency. At what frequency is maximum power transfer obtained?
b. Plot the magnitude and phase responses of the currents through the woofer and tweeter as a function of frequency.

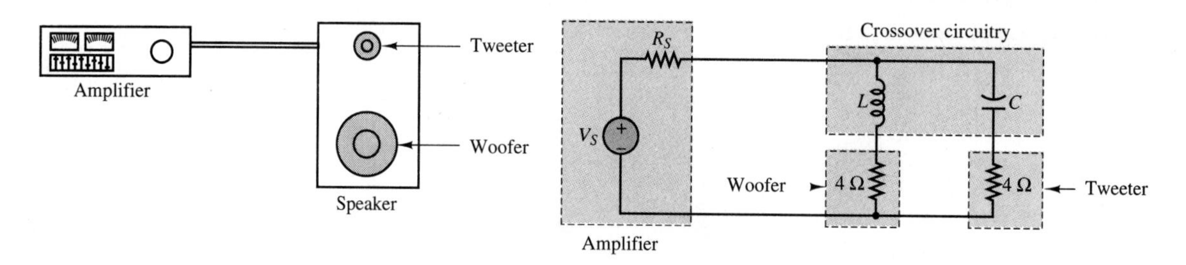

Figure P6.15

EIT 6.22 The same *LC* filter section of Problem 6.19 is used in the circuit of Figure P6.16.

a. Compute the frequency response of this circuit, $\mathbf{V}_{\text{out}}(j\omega)/\mathbf{V}_S(j\omega)$.
b. Plot the frequency response of the circuit.

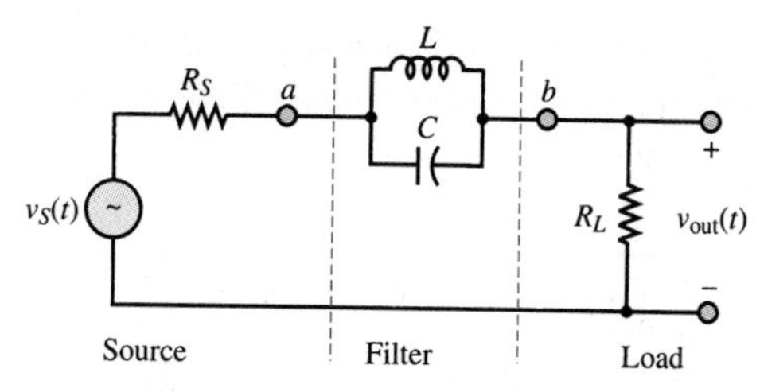

$R_S = R_L = 500\ \Omega$; $L = 10$ mH ; $C = 0.1\ \mu$F

Figure P6.16

6.23 It is very common to see interference caused by the power lines, at a frequency of 60 Hz. This problem outlines the design of a notch filter, shown in Figure P6.17, to reject a band of frequencies around 60 Hz.

a. Write the impedance function for the filter of Figure P6.17 (the resistor r_L represents the internal resistance of a practical inductor).
b. For what value of C will the center frequency of the filter equal 60 Hz if $L = 100$ mH and $r_L = 5\ \Omega$?
c. Would the "sharpness," or selectivity, of the filter increase or decrease if r_L were to increase?
d. Assume that the filter is used to eliminate the 60-Hz noise from a signal generator with output frequency of 1 kHz. Evaluate the frequency response $\frac{\mathbf{V}_L}{\mathbf{V}_{in}}(j\omega)$ at both frequencies if:

$$v_g(t) = \sin(2\pi 1{,}000t)\ \text{V} \qquad r_g = 50\ \Omega$$
$$v_n(t) = 3\sin(2\pi 60t) \qquad R_L = 300\ \Omega$$

and if L and C are as in part b.

e. Plot the magnitude frequency response $\left|\frac{\mathbf{V}_L(j\omega)}{\mathbf{V}_{in}}\right|$ in dB versus $\log(j\omega)$, and indicate the value of $\left|\frac{\mathbf{V}_L}{\mathbf{V}_{in}}\right|_{dB}$ at the frequencies 60 Hz and 1,000 Hz on your plot.

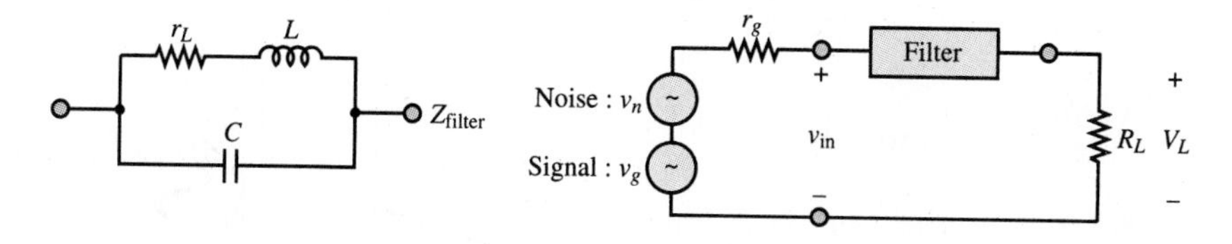

Figure P6.17

***6.24** The circuit of Figure P6.18 is representative of an amplifier–speaker connection. The crossover circuit (filter) is a low-pass filter that is connected to a woofer. The filter's topography is known as a π network.

a. Find the frequency response $\mathbf{V}_o(j\omega)/\mathbf{V}_S(j\omega)$.

b. If $C_1 = C_2 = C$, $R_S = R_L = 600\ \Omega$, and $1/\sqrt{LC} = R/L = 1/RC = 2{,}000\pi$, plot $|\mathbf{V}_o(j\omega)/\mathbf{V}_S(j\omega)|$ in dB versus frequency (logarithmic scale) in the range 100 Hz $\leq f \leq$ 10,000 Hz.

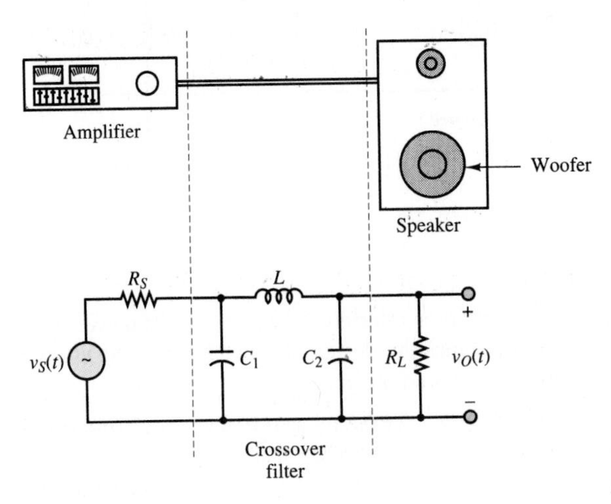

Figure P6.18

***6.25** The π filter of the circuit of Figure P6.19 is a high-pass filter that may be used to pass signals to the tweeter portion of a speaker.

a. Find the frequency response $\mathbf{V}_o(j\omega)/\mathbf{V}_S(j\omega)$.
b. If $L_1 = L_2 = L$, $R_S = R_L = 600\ \Omega$, and $1/\sqrt{LC} = R/L = 1/RC = 2{,}000\pi$, plot $|\mathbf{V}_o(j\omega)/\mathbf{V}_S(j\omega)|$ in dB versus frequency (logarithmic scale) in the range 100 Hz $\leq f \leq$ 10,000 Hz.

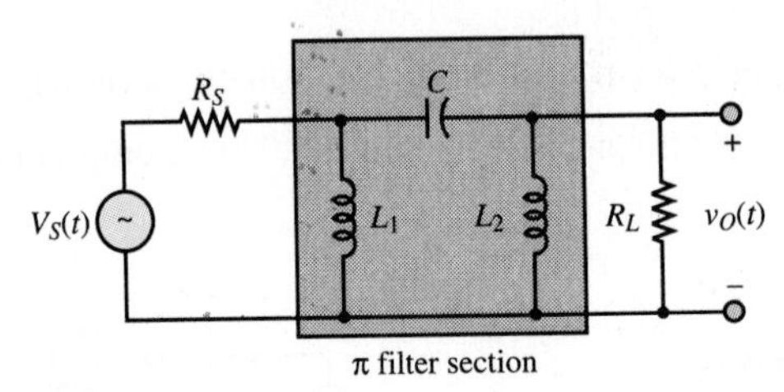

Figure P6.19

***6.26** The circuit of Figure P6.20 is representative of an amplifier–speaker connection (see the left side of Figure P6.20). The crossover circuit (filter) is a high-pass filter that is connected to a tweeter. The filter's topography is known as a T network.

a. Find the frequency response $\mathbf{V}_o(j\omega)/\mathbf{V}_S(j\omega)$.
b. If $C_1 = C_2 = C$, $R_L = R_S = 600\ \Omega$, and $1/\sqrt{LC} = R/L = 1/RC = 2{,}000\pi$, plot $|\mathbf{V}_o(j\omega)/\mathbf{V}_S(j\omega)|$ in dB versus frequency (logarithmic scale) in the range 100 Hz $\leq f \leq$ 10,000 Hz.

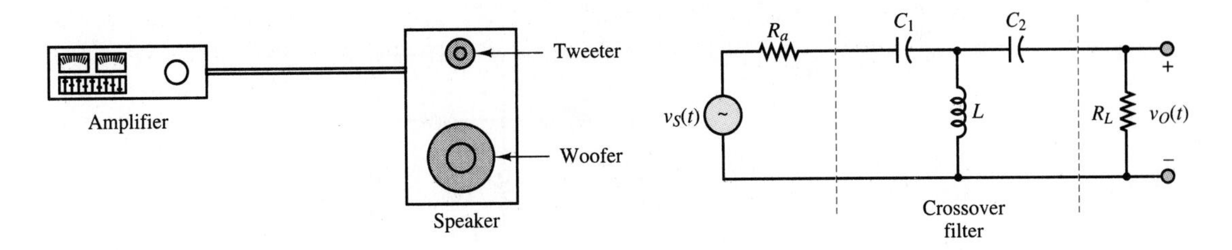

Figure P6.20

***6.27** The T filter of the circuit of Figure P6.21 is a low-pass filter that may be used to pass signals to the woofer portion of a speaker.

a. Find the frequency response $\mathbf{V}_o(j\omega)/\mathbf{V}_S(j\omega)$.
b. If $L_1 = L_2 = L$, $R_S = R_L = 600\ \Omega$, and $1/\sqrt{LC} = R/L = 1/RC = 2{,}000\pi$, plot $|\mathbf{V}_o(j\omega)/\mathbf{V}_S(j\omega)|$ in dB versus frequency (logarithmic scale) in the range 100 Hz $\leq f \leq$ 10,000 Hz.

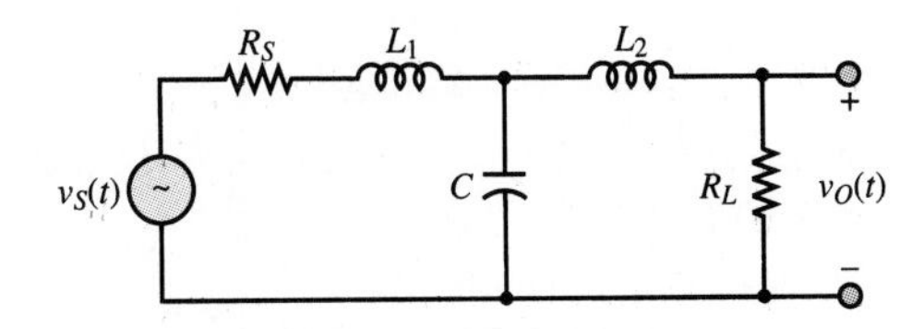

Figure P6.21

6.28 Assume the circuit of Figure P6.22 initially stores no energy. Determine an expression for the capacitor voltage for $t \geq 0$.

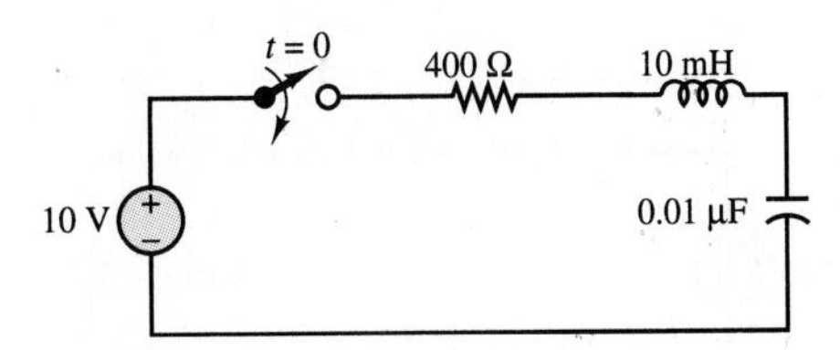

Figure P6.22

6.29 Assume the switch in the circuit of Figure P6.23 has been closed for a very long time. It is suddenly opened at $t = 0$. Determine an expression for the inductor current for $t \geq 0$.

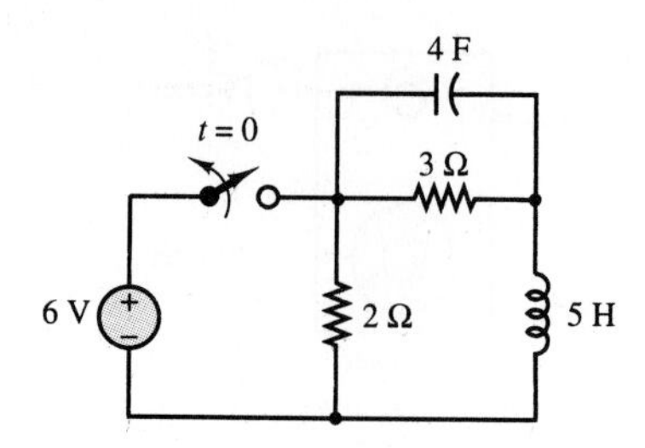

Figure P6.23

6.30 Assume the switch in the circuit of Figure P6.24 has been closed for a very long time. It is suddenly opened at $t = 0$, and then reclosed at $t = 5$ s. Determine an expression for the inductor current for $t \geq 0$.

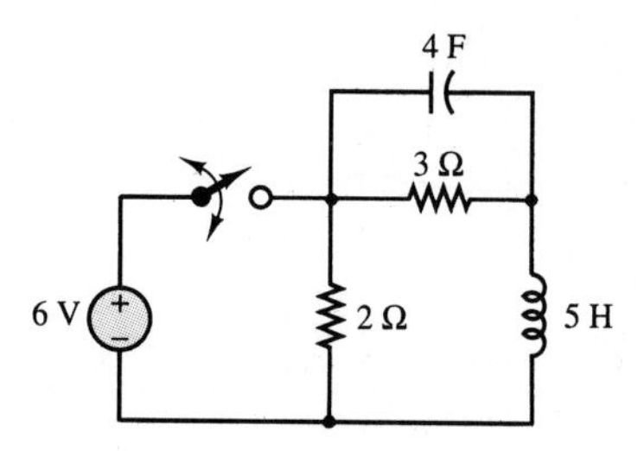

Figure P6.24

6.31 Assume the circuit of Figure P6.25 initially stores no energy. The switch is closed at $t = 0$, and then reopened at $t = 50\ \mu$s. Determine an expression for the capacitor voltage for $t \geq 0$.

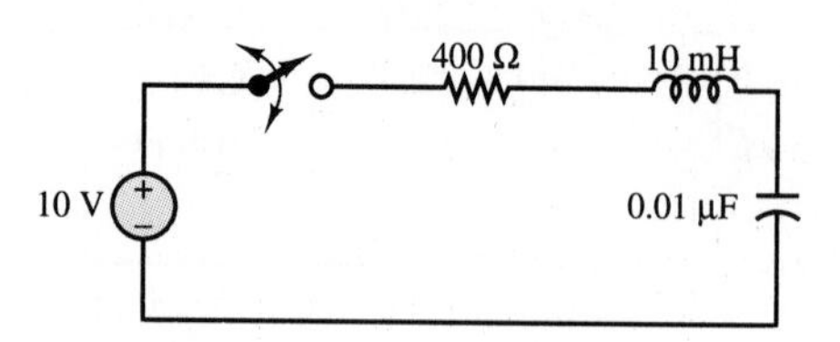

Figure P6.25

6.32 Assume the circuit of Figure P6.26 initially stores no energy. Switch S_1 is open, and S_2 is closed. Switch S_1 is closed at $t = 0$, and switch S_2 is opened at $t = 5$ s. Determine an expression for the capacitor voltage for $t \geq 0$.

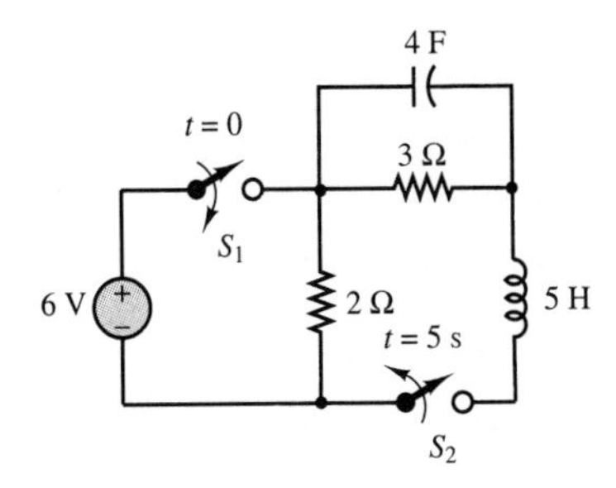

Figure P6.26

EIT 6.33 For the circuit shown in Figure P6.27, assume that switch S_1 is always open and that switch S_2 closes at $t = 0$.

a. Find the capacitor voltage, $v_C(t)$, at $t = 0^+$.
b. Find the time constant, τ, for $t \geq 0$.
c. Find an expression for $v_C(t)$ and sketch the function.
d. Find $v_C(t)$ for each of the following values of t: $0, \tau, 2\tau, 5\tau, 10\tau$.

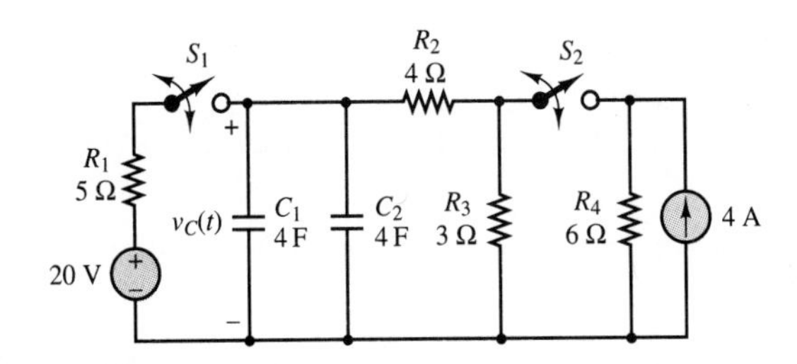

Figure P6.27

6.34 For the circuit shown in Figure P6.27, assume that switch S_1 is always open; switch S_2 has been closed for a long time, and opens at $t = 0$.

a. Find the capacitor voltage, $v_C(t)$, at $t = 0^+$.
b. Find the time constant, τ, for $t \geq 0$.
c. Find an expression for $v_C(t)$ and sketch the function.
d. Find $v_C(t)$ for each of the following values of t: $0, \tau, 2\tau, 5\tau, 10\tau$.

EIT 6.35 For the circuit of Figure P6.27, assume that switch S_2 is always open, and that switch S_1 has been closed for a long time and opens at $t = 0$. At $t = t_1 = 3\tau$, switch S_1 closes again.

a. Find the capacitor voltage, $v_C(t)$, at $t = 0^+$.
b. Find an expression for $v_C(t)$ for $t > 0$ and sketch the function.

6.36 Assume both switches S_1 and S_2 in Figure P6.27 close at $t = 0$.

a. Find the capacitor voltage, $v_C(t)$, at $t = 0^+$.
b. Find the time constant, τ, for $t \geq 0$.
c. Find an expression for $v_C(t)$ and sketch the function.
d. Find $v_C(t)$ for each of the following values of t: $0, \tau, 2\tau, 5\tau, 10\tau$.

6.37 Assume both switches S_1 and S_2 in Figure P6.27 have been closed for a long time and switch S_2 opens at $t = 0^+$.

a. Find the capacitor voltage, $v_C(t)$, at $t = 0^+$.
b. Find an expression for $v_C(t)$ and sketch the function.
c. Find $v_C(t)$ for each of the following values of t: $0, \tau, 2\tau, 5\tau, 10\tau$.

EIT 6.38 For the circuit of Figure P6.28, determine the time constants τ and τ' before and after the switch opens, respectively. $R_S = 4\ \text{k}\Omega$, $R_1 = 2\ \text{k}\Omega$, $R_2 = R_3 = 6\ \text{k}\Omega$, and $C = 1\ \mu\text{F}$.

6.39 We use an analogy between electrical circuits and thermal conduction to analyze the behavior of a pot heating on an electric stove. We can model the heating element as shown in the circuit of Figure P6.29. Find the "heat capacity" of the burner, C_S, if the burner reaches 90 percent of the desired temperature in 10 seconds.

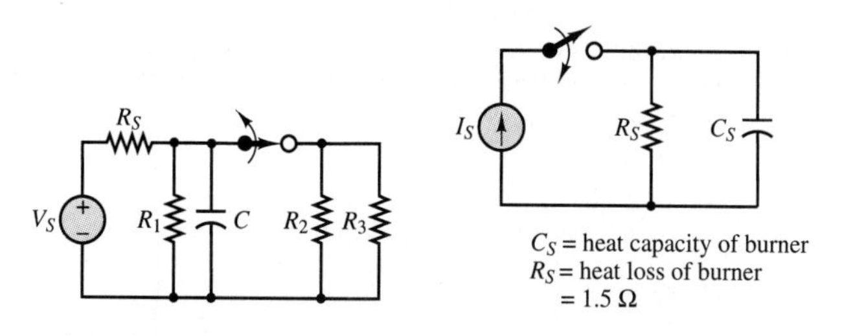

Figure P6.28 **Figure P6.29**

6.40 With a pot placed on the burner of Problem 6.39, we can model the system as shown in the circuit of Figure P6.30. Some loss between the burner and pot is modeled by the series resistance R_L. The pot is modeled by a heat-storage element, C_p, and a loss element, R_p.

a. Find the final temperature of the water in the pot—that is, find $v(t)$ as $t \to \infty$—if $I_S = 75$ A, $C_p = 80$ F, $R_L = 0.8\ \Omega$, $R_p = 2.5\ \Omega$, and the burner is the same as in the previous problem.
b. How long will it take for the water to reach 80 percent of its final temperature? [*Hint:* Neglect C_S, since $C_S \ll C_p$.]

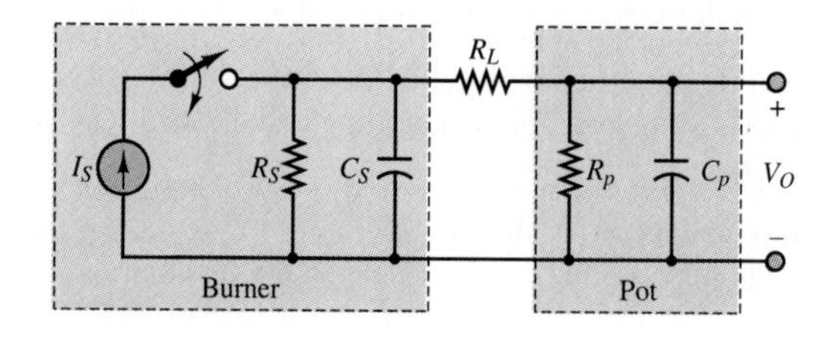

Figure P6.30

6.41 Consider a parallel *RLC* network, with $R = 10\ \Omega$, $L = 1$ H, and $C = 1$ F. The circuit is driven by a time-varying source, $i_s(t) = 10 \cos t$ A, which is turned on at $t = 0$, and has initial conditions $v_c(0) = 2$ V and $i_L(0) = 5$ A. Determine the complete solution for $v_c(t)$, and indicate clearly the transient and steady-state parts of the solution.

6.42 The model shown in Figure P6.31 represents a clock circuit driving a gate in a digital circuit. Determine an expression for the gate input voltage for $0 < t < 2\ \mu$s. Sketch the function over the same range.

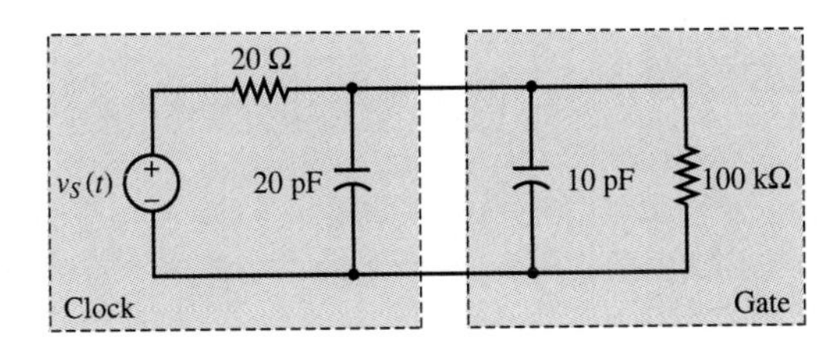

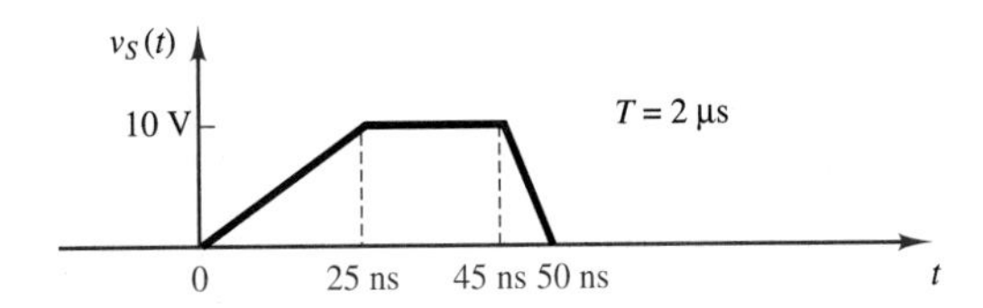

Figure P6.31

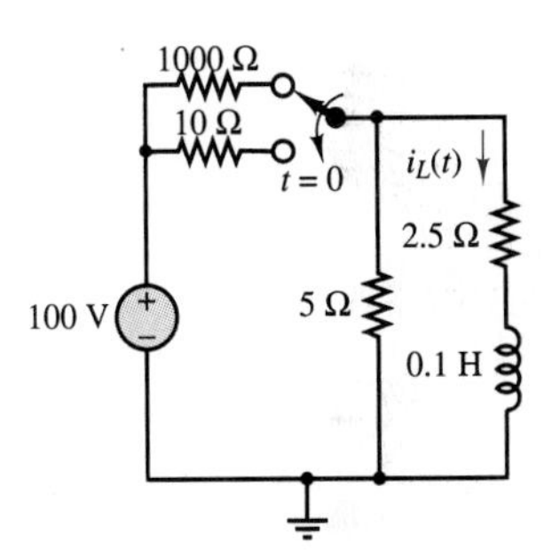

Figure P6.32

6.43 For the circuit of Figure P6.32, find the initial current through the inductor, the final current through the inductor, and the expression for $i_L(t)$ for $t \geq 0$.

EIT 6.44 At $t = 0$, the switch in the circuit of Figure P6.33 opens. At $t = 10$ s, the switch closes.

a. What is the time constant for $0 < t < 10$ s?
b. What is the time constant for $t > 10$ s?

6.45 The circuit of Figure P6.34 includes a model of a voltage-controlled switch. When the voltage across the capacitor reaches 7 V, the switch is closed. When the capacitor voltage reaches 0.5 V, the switch opens. Assume that the capacitor voltage is initially $V_C = 0.5$ V and that the switch has just opened.

a. Sketch the capacitor voltage versus time, showing explicitly the periods when the switch is open and when the switch is closed.
b. What is the period of the voltage waveform across the 10-Ω resistor?

EIT 6.46 The circuit of Figure P6.35 models the charging circuit of an electronic flash. When you turn on the flash, you must wait until a light turns on to tell you that the capacitor is completely charged.

a. If the light that tells you that the flash is ready turns on when $V_C = 0.99(7.5\text{ V})$, how long will you have to wait to take your picture?

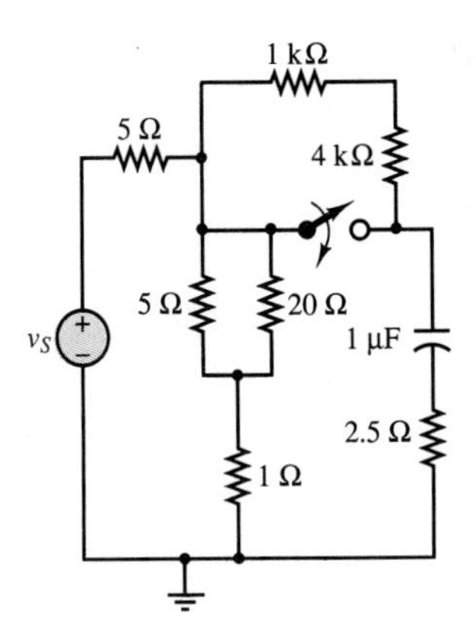

Figure P6.33

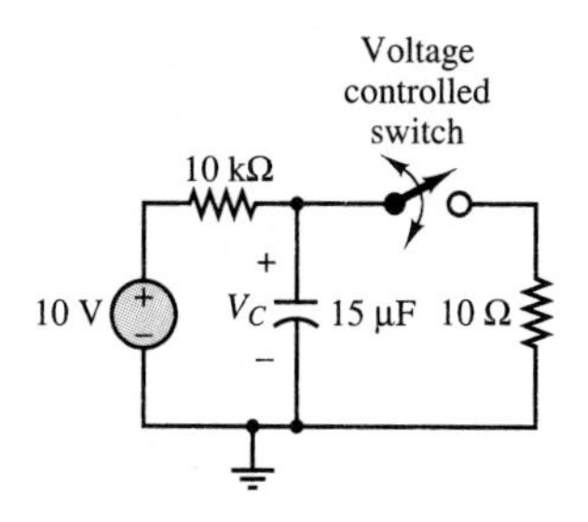

Figure P6.34

b. If the shutter button stays closed for $\frac{1}{30}$ s, how much energy is delivered to the flashbulb (R_2)? Assume that the capacitor has completely charged.
c. If you don't wait till the flash is fully charged and you take the picture 3 seconds after the flash is turned on, how much energy is delivered to R_2?

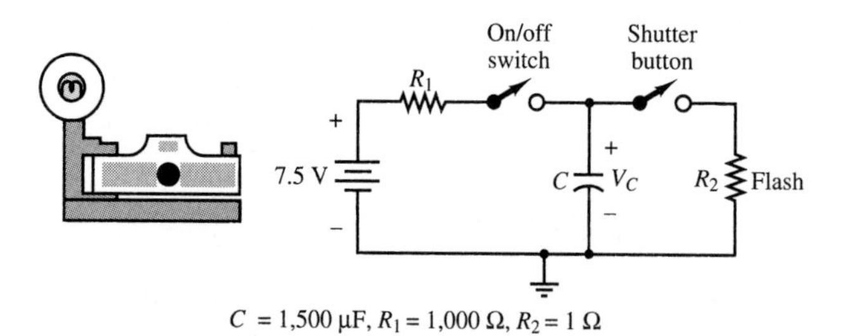

Figure P6.35

6.47 Assume that the circuit shown in Figure P6.36 is underdamped and that the circuit initially has no energy stored. It has been observed that, after the switch is closed at $t = 0$, the capacitor voltage reaches an initial peak value of 70 V when $t = 5\pi/3$ μs, a second peak value of 53.2 V when $t = 5\pi$ μs, and eventually approaches a steady-state value of 50 V. If $C = 1.6$ nF, what are the values of R and L?

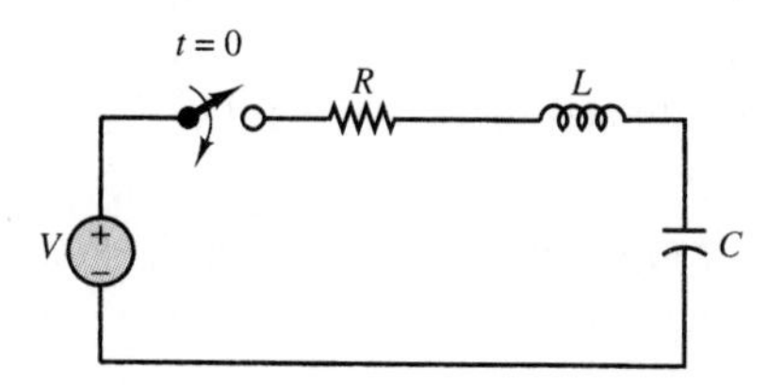

Figure P6.36

6.48 Given the information provided in Problem 6.47, explain how to modify the circuit so that the first two peaks occur at 5π μs and 15π μs. Assume that C cannot be changed.

6.49 At $t = 0$, the switch in the circuit of Figure P6.37 closes. Assume that $i_L(0) = 0$ A. For $t \geq 0$,

a. Find $i_L(t)$.
b. Find $v_{L_1}(t)$.

EIT 6.50 Repeat Drill Exercise 6.23 for the circuit of Figure 6.63(a) with $R_T = 1$ kΩ, $C = 10$ μF, $L = 1$ H, and $v_T = 15$ V.

6.51 Repeat Drill Exercise 6.23 for the circuit of Figure 6.63(b) with $R_T = 10$ kΩ, $C = 100$ μF, $L = 10$ mH, and $v_T = 12$ V.

6.52 Find i for $t > 0$ in the circuit of Figure P6.38 if $i(0) = 4$ A and $v(0) = 6$ V.

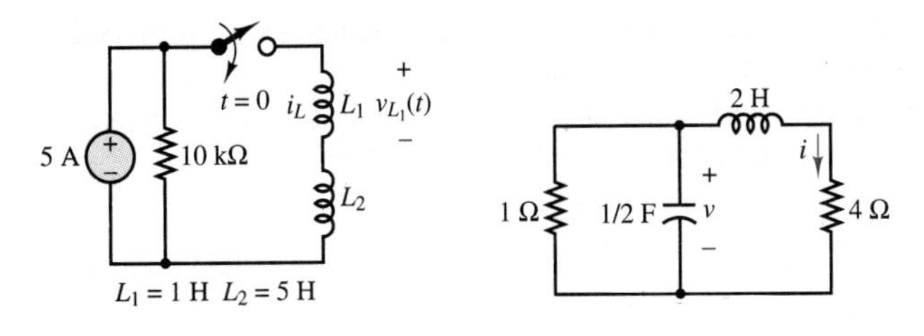

Figure P6.37

Figure P6.38

EIT 6.53 Find v for $t > 0$ in the circuit of Figure P6.39 if the circuit is in steady state at $t = 0^-$.

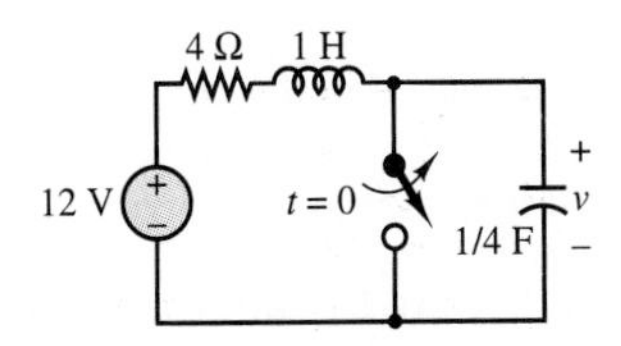

Figure P6.39

6.54 Find i for $t > 0$ in the circuit of Figure P6.40 if the circuit is in steady state at $t = 0^-$.

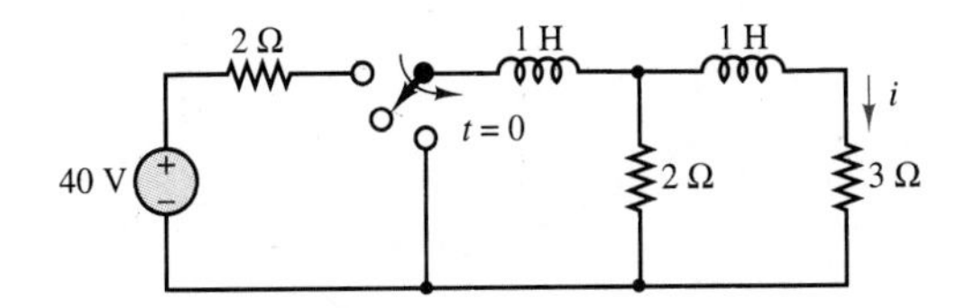

Figure P6.40

6.55 Find i for $t > 0$ in the circuit of Figure P6.41 if the circuit is in steady state at $t = 0^-$.

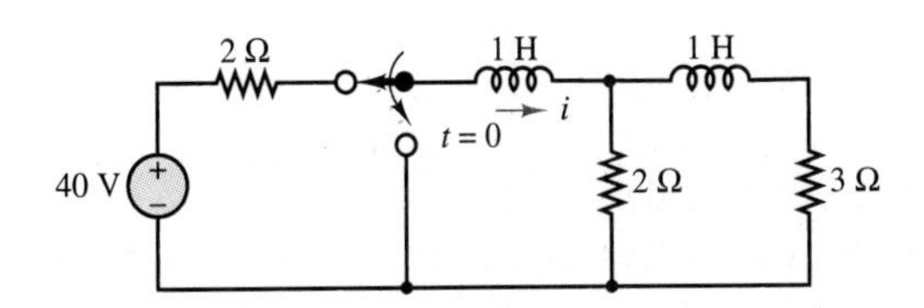

Figure P6.41

EIT 6.56 Find v for $t > 0$ in the circuit of Figure P6.42 if the circuit is in steady state at $t = 0^-$.

6.57 The circuit of Figure P6.43 is in steady state at $t = 0^-$. Find v for $t > 0$ if L is (a) 2.4 H, (b) 3 H, and (c) 4 H.

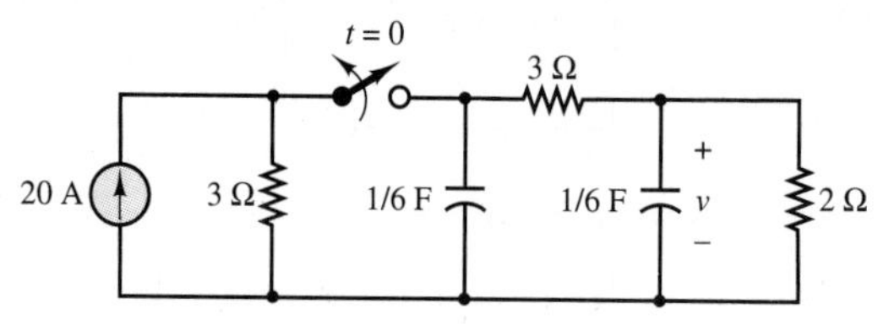

Figure P6.42

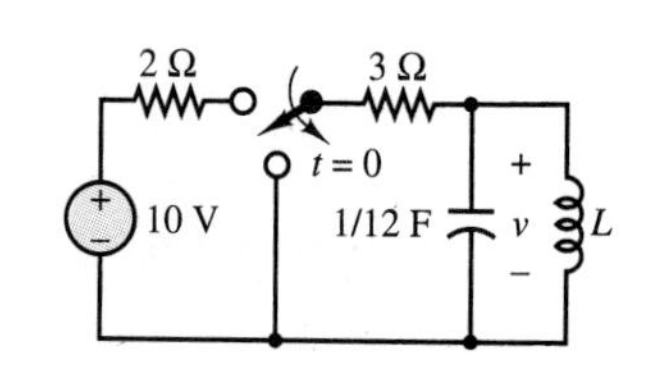

Figure P6.43

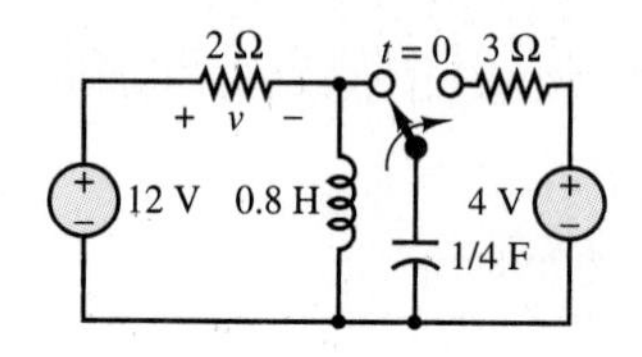

Figure P6.44

6.58 Find v for $t > 0$ in the circuit of Figure P6.44 if the circuit is in steady state at $t = 0^-$.

6.59 An unknown element is placed in the circuit of Figure P6.45(a). At $t = 0$, the switch is closed. If $i(t)$ has the waveforms shown in Figure P6.45(b), state what the element is, and give its value for each of the cases in the figure.

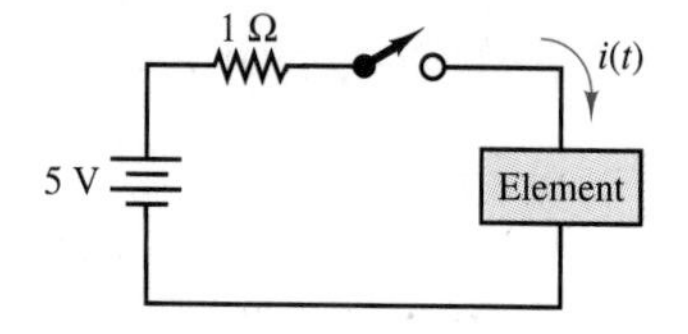

Figure P6.45(a)

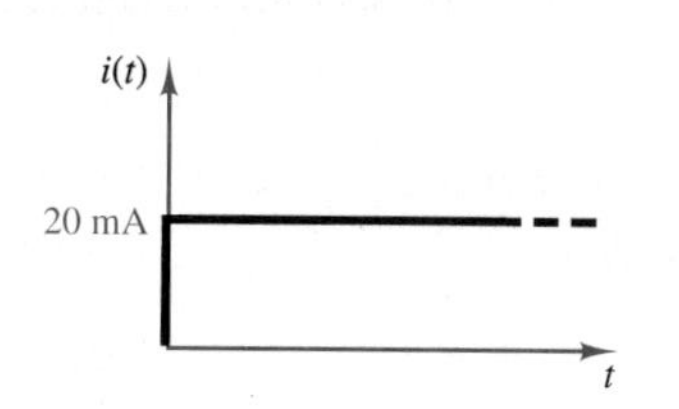

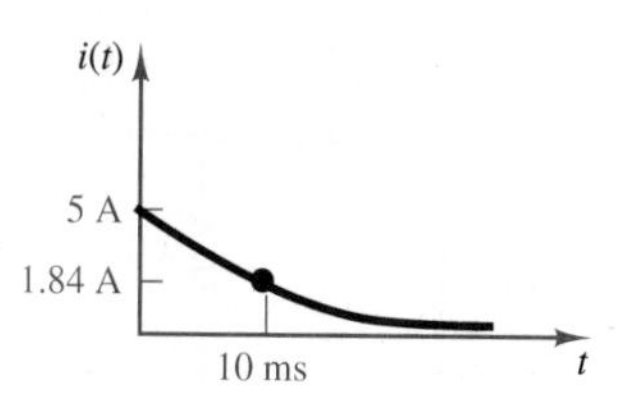

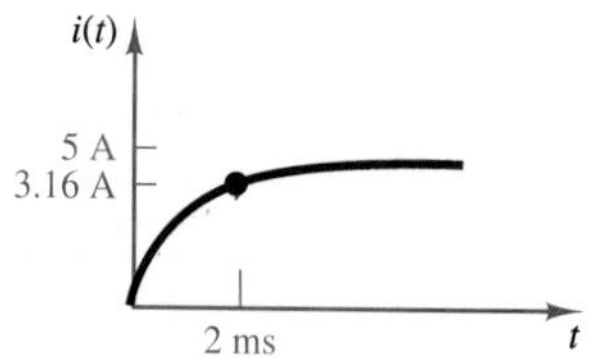

Figure P6.45(b)

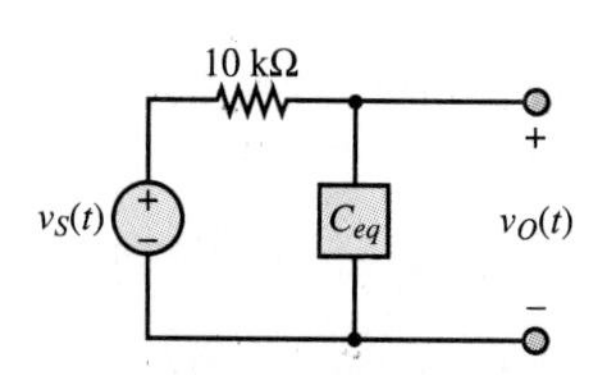

Figure P6.46

EIT 6.60 A simple RC circuit must be constructed with a time constant $\tau = RC_{eq} = 37.5$ ms. The circuit should have the form shown in Figure P6.46. Assume that capacitors are available in the following sizes: 10 μF, 0.5 μF, and 5 μF. Show the values in the RC circuit that will provide this time constant.

ELECTRONICS

CHAPTER 7

Semiconductors and Diodes

This chapter introduces semiconductor-based electronic devices, and in so doing, it provides a transition between the fundamentals of electrical circuit analysis and the study of electronic circuits. Although the theme of this chapter may seem somewhat different from the circuit analysis of the first six chapters, it is important to recognize that the analysis of electrical circuits is still at the core of the material. For example, we will explain the operation of diodes using load-line analysis (explored in Chapter 3), in conjunction with linear circuit models containing resistors, capacitors, and voltage and current sources. In fact, the primary emphasis in this chapter and Chapters 8 through 10 will be the use of linear circuit models for understanding and analyzing the behavior of more complex nonlinear electronic devices; we will show how it is possible to construct models of devices having nonlinear *i-v* characteristics by means of linear circuits. The alternative to the modeling approach would be to conduct an in-depth study of the physics of each class of device: diodes, bipolar transistors, field-effect devices, and other types of semiconductors. Such an approach is neither practical nor fruitful from the viewpoint of this book, since it would entail lengthy explanations and require a significant background in semiconductor physics. Thus, the approach here will be first to provide a qualitative understanding of the physics of each family of devices, and then to quickly turn to describing the devices in terms of their *i-v* characteristics and simple circuit models, to illustrate their analysis and applications.

The chapter starts with a discussion of semiconductors and of the *pn* junction and the semiconductor diode. The second part of this chapter is devoted to a study of diode circuit models, and numerous practical applications. By the end of Chapter 7, you should have accomplished the following objectives:

- A qualitative understanding of electrical conduction in semiconductor materials.
- The ability to explain the *i-v* characteristic of a semiconductor diode (or of a *pn* junction).
- The ability to use the ideal, offset, and piecewise linear diode models in simple circuits.
- The ability to analyze diode rectifier, peak limiter, peak detector, and regulator circuits and the behavior of LEDs and photocells.

7.1 ELECTRICAL CONDUCTION IN SEMICONDUCTOR DEVICES

This section briefly introduces the mechanism of conduction in a class of materials called **semiconductors.** Semiconductors are typically materials consisting of elements from group IV of the periodic table and having electrical properties falling somewhere between those of conducting and of insulating materials. As an example, consider the conductivity of three common materials. Copper, a good conductor, has a conductivity of 0.59×10^6 S/cm; glass, a common insulator, may range between 10^{-16} and 10^{-13} S/cm; while silicon, a semiconductor, has a conductivity that varies from 10^{-8} to 10^{-1} S/cm. You see, then, that the name *semiconductor* is an appropriate one.

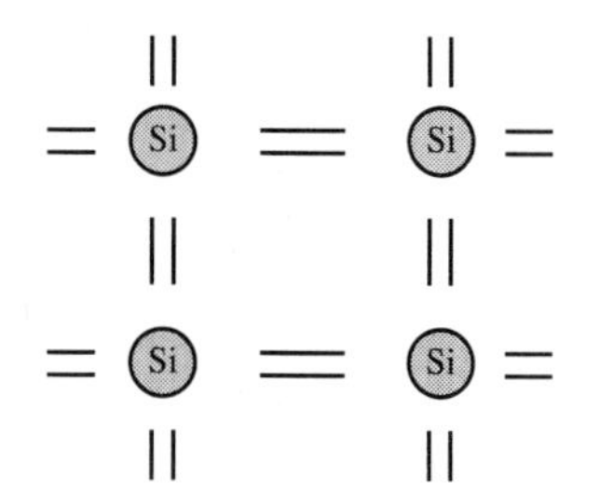

Figure 7.1 Lattice structure of silicon, with four valence electrons

A conducting material is characterized by a large number of conduction-band electrons, which have a very weak bond with the basic structure of the material. Thus, an electric field easily imparts energy to the outer electrons in a conductor and enables the flow of electric current. In a semiconductor, on the other hand, one needs to consider the lattice structure of the material, which in this case is characterized by **covalent bonding.** Figure 7.1 depicts the lattice arrangement for silicon (Si), one of the more common semiconductors. At sufficiently high temperatures, thermal energy causes the atoms in the lattice to vibrate; when sufficient kinetic energy is present, some of the valence electrons break their bonds with the lattice structure and become available as conduction electrons. These **free electrons** enable current flow in the semiconductor. It should be noted that in a conductor valence electrons have a very loose bond with the nucleus and are therefore available for conduction to a much greater extent than valence electrons in a semiconductor. One important aspect of this type of conduction is that the number of charge carriers depends on the amount of thermal energy present in the structure. Thus, many semiconductor properties are a function of temperature.

The free valence electrons are not the only mechanism of conduction in a semiconductor, however. Whenever a free electron leaves the lattice structure, it creates a corresponding positive charge within the lattice. Figure 7.2 depicts the situation in which a covalent bond is missing because of the departure of a free electron from the structure. The vacancy caused by the departure of a free electron is called a **hole.** Note that whenever a hole is present, we have, in effect, a positive charge. The positive charges also contribute to the conduction process, in the sense that if a valence-band electron "jumps" to fill a neighboring hole, thereby neutralizing a positive charge, it correspondingly creates a new hole at a different location. Thus, the effect is equivalent to that of a positive charge moving to the right, in the sketch of Figure 7.2. This phenomenon becomes relevant when an external electric field is applied to the material. It is important to point out here that the **mobility**—that is, the ease with which charge carriers move across the lattice—differs greatly for the two types of carriers. Free electrons can move far more easily around the lattice than holes. To appreciate this, consider the fact that a free electron has already broken the covalent bond, whereas for a hole to travel through the structure, an electron must overcome the covalent bond each time the hole jumps to a new position.

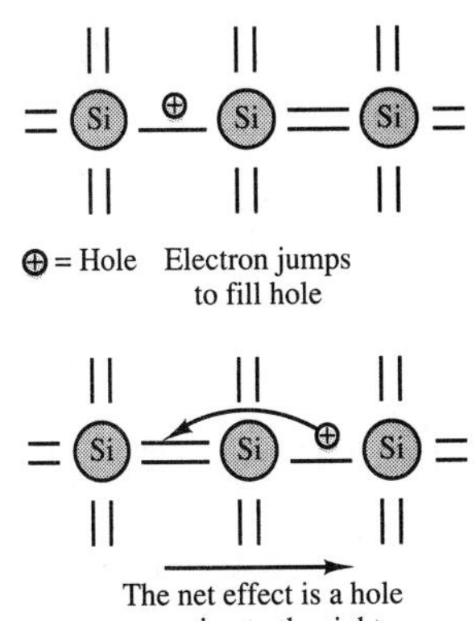

Figure 7.2 Free electrons and "holes" in the lattice structure

According to this relatively simplified view of semiconductor materials, we can envision a semiconductor as having two types of charge carriers—holes and free electrons—which travel in opposite directions when the semiconductor is subjected to an external electric field, giving rise to a net flow of current in the direction of the electric field. Figure 7.3 illustrates the concept.

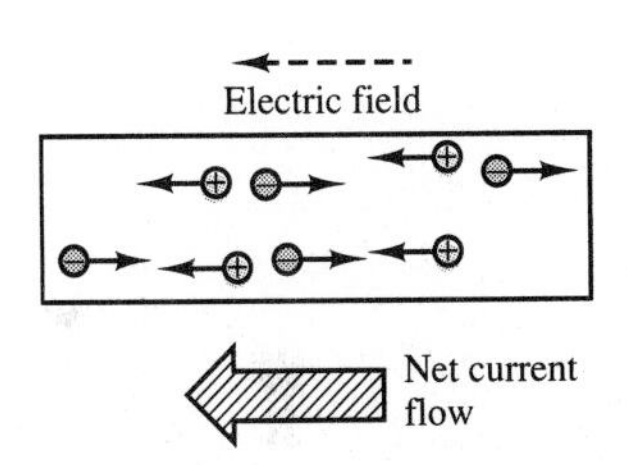

Figure 7.3 Current flow in a semiconductor

An additional phenomenon, called **recombination,** reduces the number of charge carriers in a semiconductor. Occasionally, a free electron traveling in the immediate neighborhood of a hole will recombine with the hole, to form a covalent bond. Whenever this phenomenon takes place, two charge carriers are lost. However, in spite of recombination, the net balance is such that a number of free electrons always exist at a given temperature. These electrons are therefore available for conduction. The number of free electrons available for a given material is called the **intrinsic concentration,** n_i. For example, at room temperature, silicon has

$$n_i = 1.5 \times 10^{16} \text{ electrons/m}^3 \tag{7.1}$$

Note that there must be an equivalent number of holes present as well.

Semiconductor technology rarely employs pure, or intrinsic, semiconductors. To control the number of charge carriers in a semiconductor, the process of **doping** is usually employed. Doping consists of adding impurities to the crystalline structure of the semiconductor. The amount of these impurities is controlled, and the impurities can be of one of two types. If the dopant is an element from the fifth column of the periodic table (e.g., arsenic), the end result is that wherever an impurity is present, an additional free electron is available for conduction. Figure 7.4 illustrates the concept. The elements providing the impurities are called **donors** in the case of group V elements, since they "donate" an additional free electron to the lattice structure. An equivalent situation arises when

An additional free electron is created when Si is "doped" with a group V element.

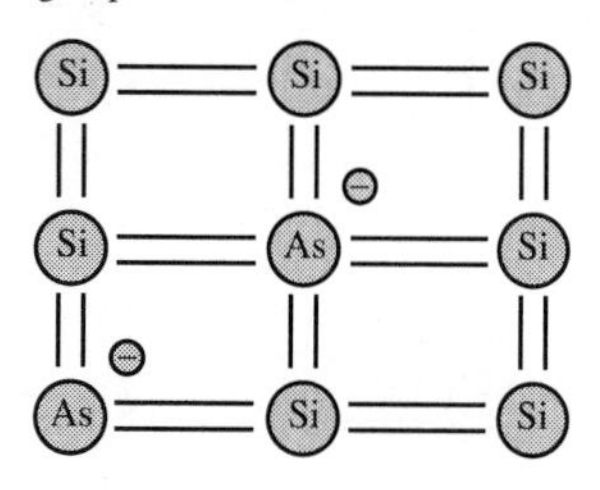

Figure 7.4 Doped semiconductor

group III elements (e.g., indium) are used to dope silicon. In this case, however, an additional hole is created by the doping element, which is called an **acceptor,** since it accepts a free electron from the structure and generates a hole in doing so.

Semiconductors doped with donor elements conduct current predominantly by means of free electrons and are therefore called ***n*-type semiconductors.** When an acceptor element is used as the dopant, holes constitute the most common carrier, and the resulting semiconductor is said to be a ***p*-type semiconductor.** Doping usually takes place at such levels that the concentration of carriers due to the dopant is significantly greater than the intrinsic concentration of the original semiconductor. If n is the total number of free electrons and p that of holes, then in an n-type doped semiconductor, we have

$$n \gg n_i \tag{7.2}$$

and

$$p \ll p_i \tag{7.3}$$

Thus, free electrons are the **majority carriers** in an n-type material, while holes are the **minority carriers.** In a p-type material, the majority and minority carriers are reversed.

Doping is a standard practice for a number of reasons. Among these are the ability to control the concentration of charge carriers, and the increase in the conductivity of the material that results from doping.

7.2 THE *pn* JUNCTION AND THE SEMICONDUCTOR DIODE

A simple section of semiconductor material does not in and of itself possess properties that make it useful for the construction of electronic circuits. However, when a section of p-type material and a section of n-type material are brought in contact to form a ***pn* junction,** a number of interesting properties arise. The pn junction forms the basis of the **semiconductor diode**, a widely used circuit element.

Figure 7.5 depicts an idealized pn junction, where on the p side, we see a dominance of positive charge carriers, or holes, and on the n side, the free electrons

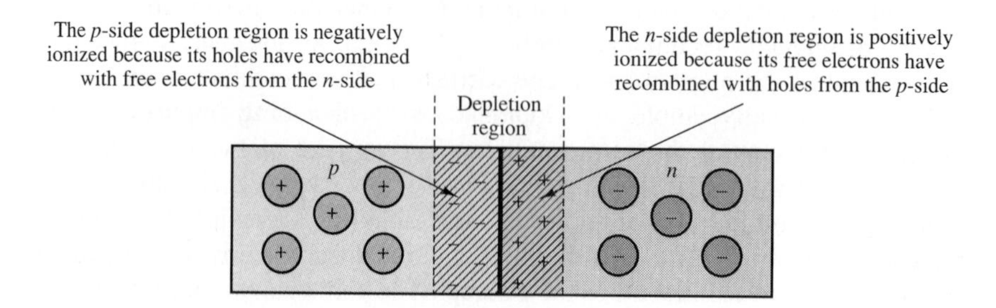

Figure 7.5 A *pn* junction

dominate. Now, in the neighborhood of the junction, in a small section called the **depletion region**, the mobile charge carriers (holes and free electrons) come into contact with each other and recombine, thus leaving virtually no charge carriers at the junction. What is left in the depletion region, in the absence of the charge carriers, is the lattice structure of the n-type material on the right, and of the p-type material on the left. But the n-type material, deprived of the free electrons, which have recombined with holes in the neighborhood of the junction, is now positively ionized. Similarly, the p-type material at the junction is negatively ionized, because holes have been lost to recombination. The net effect is that, while most of the material (p- or n-type) is charge-neutral because the lattice structure and the charge carriers neutralize each other (on average), the depletion region sees a separation of charge, giving rise to an electric field pointing from the n side to the p side. The charge separation therefore causes a **contact potential** to exist at the junction. This potential is typically on the order of a few tenths of a volt and depends on the material (about 0.6 to 0.7 V for silicon). The contact potential is also called the *offset voltage*, V_γ.

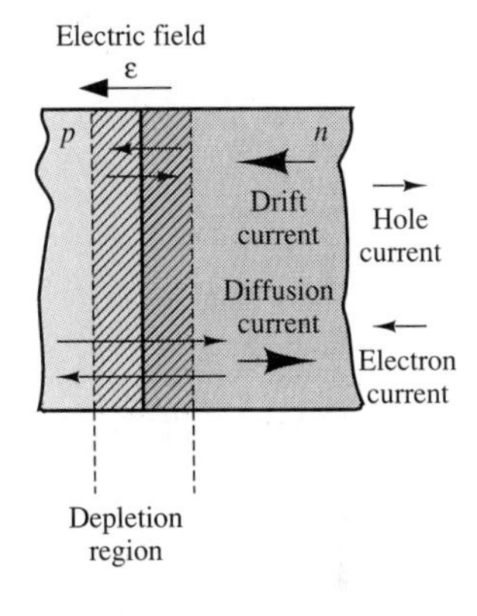

Figure 7.6 Drift and diffusion currents in a *pn* junction

In effect, then, if one were to connect the two terminals of the *pn* junction to each other, to form a closed circuit, two currents would be present. First, a small current, called **reverse saturation current,** I_0, exists because of the presence of the contact potential and the associated electric field. In addition, it also happens that holes and free electrons with sufficient thermal energy can cross the junction. This current across the junction flows opposite to the reverse saturation current and is called **diffusion current,** I_d. Of course, if a hole from the p side enters the n side, it is quite likely that it will quickly recombine with one of the n-type carriers on the n side. One way to explain diffusion current is to visualize the diffusion of a gas in a room: gas molecules naturally tend to diffuse from a region of higher concentration to one of lower concentration. Similarly, the p-type material—for example—has a much greater concentration of holes than the n-type material. Thus, some holes will tend to diffuse into the n-type material across the junction, although only those that have sufficient (thermal) energy to do so will succeed. Figure 7.6 illustrates this process.

The phenomena of drift and diffusion help explain how a *pn* junction behaves when it is connected to an external energy source. Consider the diagrams of Figure 7.7, where a battery has been connected to a *pn* junction in the **reverse-biased** direction (Figure 7.7(a)), and in the **forward-biased** direction (Figure 7.7(b)). We assume that some suitable form of contact between the battery wires and the semiconductor material can be established (this is called an **ohmic contact**). The effect of a reverse bias is to increase the contact potential at the junction. Now, the majority carriers trying to diffuse across the junction need to overcome a greater barrier (a larger potential) and a wider depletion region. Thus, the diffusion current becomes negligible. The only current that flows under reverse bias is the very small reverse saturation current, so that the diode current, i_D (defined in the figure), is

$$i_D = -I_0 \quad (7.4)$$

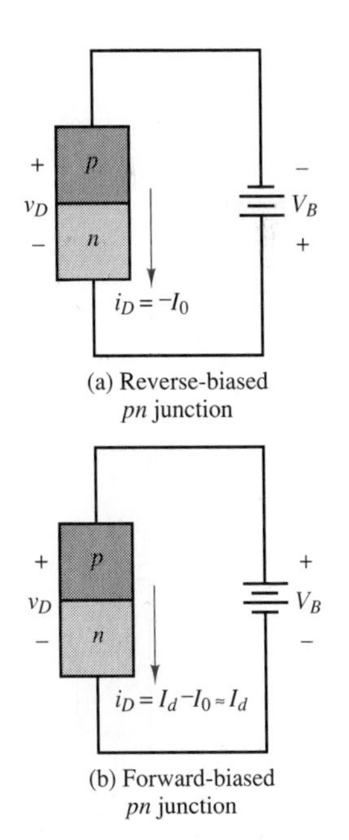

Figure 7.7 Forward- and reverse-biased *pn* junctions

When the *pn* junction is forward-biased, the contact potential across the junction is lowered (note that V_B acts in opposition to the contact potential). Now, the diffusion of majority carriers is aided by the external voltage source; in fact, the diffusion current increases as a function of the applied voltage, according to the expression

$$I_d = I_0 e^{qv_D/kT} \tag{7.5}$$

where v_D is the voltage across the *pn* junction, $k = 1.381 \times 10^{-23}$ J/K is Boltzmann's constant, q the charge of one electron, and T the temperature of the material in kelvins (K). The quantity kT/q is constant at a given temperature and is approximately equal to 25 mV at room temperature. The net diode current under forward bias is given by the expression

$$i_D = I_d - I_0 = I_0(e^{qv_D/kT} - 1) \quad \text{Diode equation} \tag{7.6}$$

which is known as the **diode equation.** Figure 7.8 depicts the diode *i-v* characteristic described by the diode equation for a fairly typical silicon diode for positive diode voltages. Since the reverse saturation current, I_0, is typically very small (10^{-9} to 10^{-15} A), the expression

$$i_D = I_0 e^{qv_D/kT} \tag{7.7}$$

is a good approximation if the diode voltage, v_D, is greater than a few tenths of a volt.

The arrow in the circuit symbol for the diode indicates the direction of current flow when the diode is forward-biased.

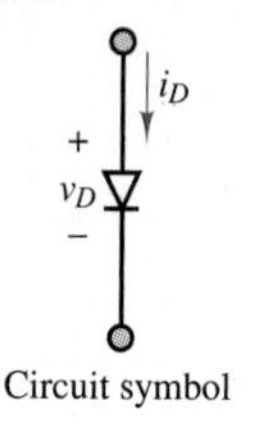

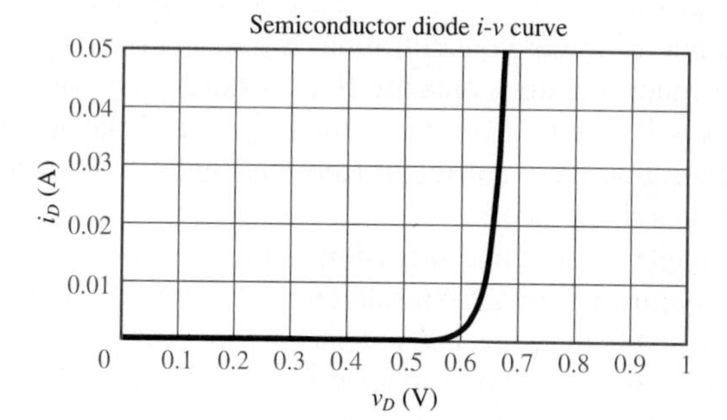

Figure 7.8 Semiconductor diode *i-v* characteristic

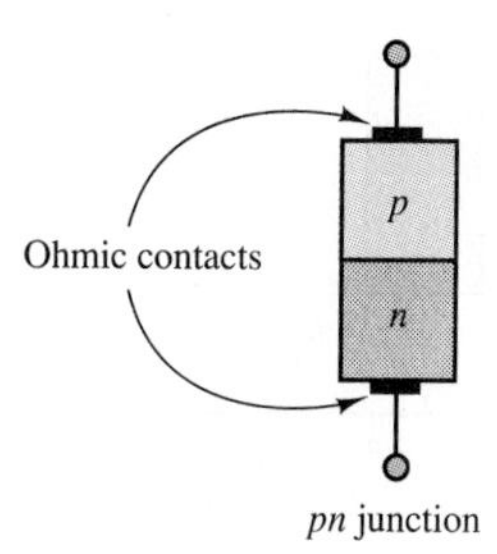

Figure 7.9 Semiconductor diode circuit symbol

The ability of the *pn* junction to essentially conduct current in only one direction—that is, to conduct only when the junction is forward-biased—makes it valuable in circuit applications. A device having a single *pn* junction and ohmic contacts at its terminals, as described in the preceding paragraphs, is called a *semiconductor diode*, or simply *diode*. As will be shown later in this chapter, it finds use in many practical circuits. The circuit symbol for the diode is shown in Figure 7.9, alongside with a sketch of the *pn* junction.

Figure 7.10 summarizes the behavior of the semiconductor diode in terms of its *i-v* characteristic; it will become apparent later that this *i-v* characteristic plays an important role in constructing circuit models for the diode. Note that a third region appears in the diode *i-v* curve that has not been discussed yet. The **reverse-breakdown** region to the far left of the curve represents the behavior of the diode when a sufficiently high reverse bias is applied. Under such a large reverse bias (greater in magnitude than the voltage V_Z, a quantity that will be explained shortly), the diode conducts current again, this time *in the reverse direction.* To explain the mechanism of reverse conduction, one needs to visualize the phenomenon of *avalanche breakdown.* When a very large negative bias is applied to the *pn* junction, sufficient energy is imparted to charge carriers that reverse current can flow, well beyond the normal reverse saturation current. In addition, because of the large electric field, electrons are energized to such levels that if they collide with other charge carriers at a lower energy level, some of their energy is transferred to the carriers with lower energy, and these can now contribute to the reverse conduction process, as well. This process is called *impact ionization.* Now, these new carriers may also have enough energy to energize other low-energy electrons by impact ionization, so that once a sufficiently high reverse bias is provided, this process of conduction takes place very much like an avalanche: a single electron can ionize several others.

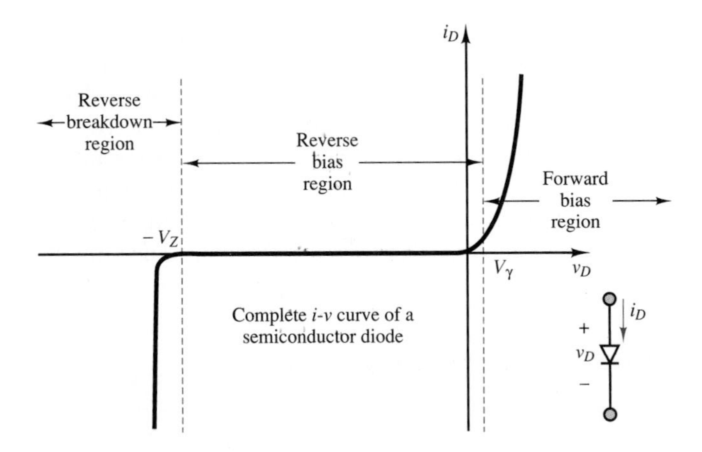

Figure 7.10 *i-v* characteristic of semiconductor diode

The phenomenon of **Zener breakdown** is related to avalanche breakdown. It is usually achieved by means of heavily doped regions in the neighborhood of the metal-semiconductor junction (the ohmic contact). The high density of charge carriers provides the means for a substantial reverse breakdown current to be sustained, at a nearly constant reverse bias, the **Zener voltage,** V_Z. This phenomenon is very useful in applications where one would like to hold some load voltage constant—for example, in **voltage regulators,** which are discussed in a later section.

To summarize the behavior of the semiconductor diode, it is useful to refer to the sketch of Figure 7.10, observing that when the voltage across the diode, v_D, is greater than the offset voltage, V_γ, the diode is said to be forward-biased and acts nearly as a short circuit, readily conducting current. When v_D is between V_γ and the Zener breakdown voltage, $-V_Z$, the diode acts very much like an open circuit, conducting a small reverse current, I_0, of the order of only nanoamperes (nA). Finally, if the voltage v_D is more negative than the Zener voltage, $-V_Z$, the diode conducts again, this time in the reverse direction.

7.3 CIRCUIT MODELS FOR THE SEMICONDUCTOR DIODE

From the viewpoint of a *user* of electronic circuits (as opposed to a *designer*), it is often sufficient to characterize a device in terms of its *i-v* characteristic, using either load-line analysis or appropriate circuit models to determine the operating currents and voltages. This section will show how it is possible to use the *i-v* characteristics of the semiconductor diode to construct simple yet useful *circuit models*. Depending on the desired level of detail, it is possible to construct *large-signal models* of the diode, which describe the gross behavior of the device in the presence of relatively large voltages and currents; or *small-signal models*, which are capable of describing the behavior of the diode in finer detail and, in particular, the response of the diode to small changes in the average diode voltage and current. From the user's standpoint, these circuit models greatly simplify the analysis of diode circuits and make it possible to effectively analyze relatively "difficult" circuits simply by using the familiar circuit analysis tools of Chapter 3. The first two major divisions of this section will describe different diode models and the assumptions under which they are obtained, to provide the knowledge you will need to select and use the appropriate model for a given application.

Large-Signal Diode Models

Ideal Diode Model

Our first large-signal model treats the diode as a simple on-off device (much like a check valve in hydraulic circuits). Figure 7.11 illustrates how, on a large scale, the *i-v* characteristic of a typical diode may be approximated by an open circuit when $v_D < 0$ and by a short circuit when $v_D \geq 0$ (recall the *i-v* curves of the ideal short and open circuits presented in Chapter 2). The analysis of a circuit containing a diode may be greatly simplified by using the short-circuit–open-circuit model. From here on, this diode model will be known as the **ideal diode model.** In spite of its simplicity, the ideal diode model (indicated by the symbol shown in Figure 7.11) can be very useful in analyzing diode circuits. Consider the circuit shown in

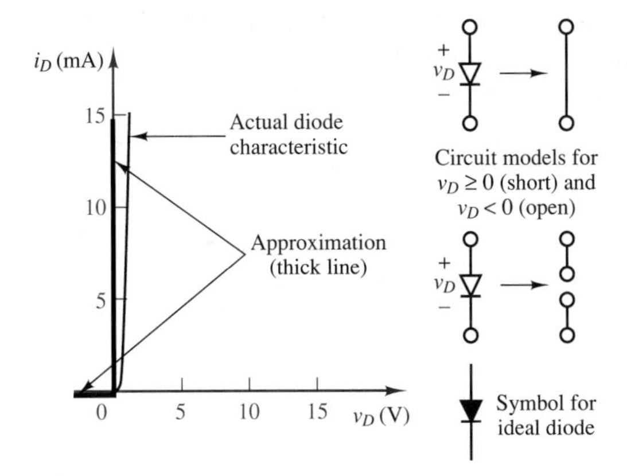

Figure 7.11 Large-signal on-off diode model

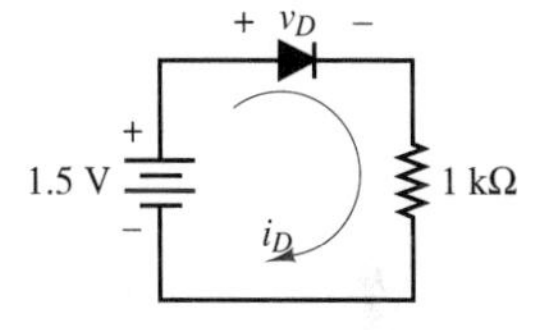

Figure 7.12 Circuit containing ideal diode

Figure 7.12, which contains a 1.5-V battery, an ideal diode, and a 1-kΩ resistor. A technique will now be developed to determine whether the diode is conducting or not, with the aid of the ideal diode model.

Assume first that the diode is conducting (or, equivalently, that $v_D \geq 0$). This enables us to substitute a short circuit in place of the diode, as shown in Figure 7.13. Since the diode is now represented by a short circuit, $v_D = 0$; this is consistent with the initial assumption (i.e., diode "on"), since the diode is assumed to conduct for $v_D \geq 0$ and since $v_D = 0$ does not contradict the assumption. The series current in the circuit (and through the diode) is $i_D = 1.5/1{,}000 = 1.5$ mA. To summarize, the assumption that the diode is on in the circuit of Figure 7.13 allows us to assume a positive (clockwise) current in the circuit. Since the direction of the current and the diode voltage are consistent with the assumption that the diode is on ($v_D \geq 0, i_D > 0$), it must be concluded that the diode is conducting.

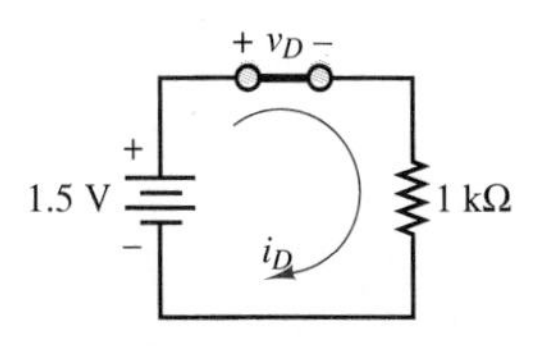

Figure 7.13 Circuit of Figure 7.12, assuming that the ideal diode conducts

Suppose, now, that the diode had been assumed to be off. In this case, the diode would be represented by an open circuit, as shown in Figure 7.14. Applying KVL to the circuit of Figure 7.14 reveals that the voltage v_D must equal the battery voltage, or $v_D = 1.5$ V, since the diode is assumed to be an open circuit and no current flows through the circuit:

$$1.5 = v_D + 1{,}000 i_D = v_D \tag{7.8}$$

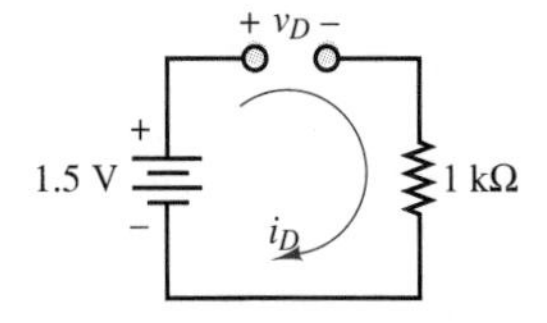

Figure 7.14 Circuit of Figure 7.12, assuming that the ideal diode does not conduct

But the result $v_D = 1.5$ V is contrary to the initial assumption (i.e., $v_D < 0$). Thus, assuming that the diode is an open circuit leads to an inconsistent answer. Clearly, the assumption must be incorrect, and therefore the diode must be conducting.

This method may seem somewhat trivial, because of the simplicity of the circuit used in the illustration, but it can be very useful in more involved circuits, where it is not quite so obvious whether a diode is seeing a positive or a negative bias. The method is particularly effective in these cases, since one can make an educated guess whether the diode is on or off and solve the resulting circuit to verify the correctness of the initial assumption. Some solved examples are perhaps the best way to illustrate the concept.

Example 7.1

The method just illustrated in this section will now be applied to a slightly more complicated circuit, shown in Figure 7.15. Determine whether the ideal diode in the circuit is conducting.

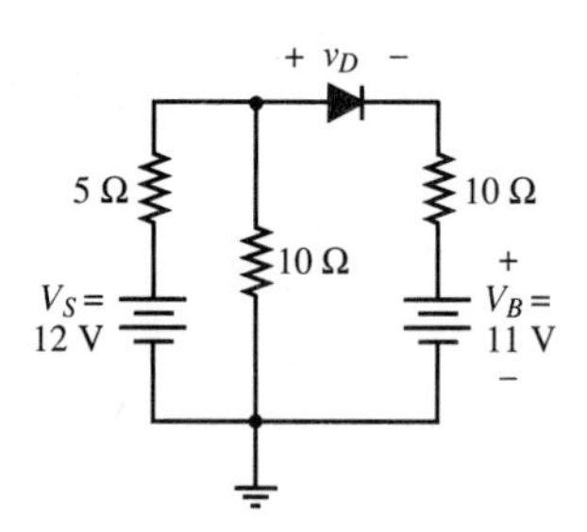

Figure 7.15

Solution:

To determine the condition of the ideal diode, let us initially assume that it does not conduct, and let us replace the diode with an open circuit, as shown in Figure 7.16. Since no current flows in the mesh on the right-hand side of the circuit of Figure 7.16, we may compute the voltage across the (shunt) 10-Ω resistor to be 8 V, by the voltage divider rule. Then, applying KVL around the right-hand mesh, we obtain:

$$8 = v_D + 11$$

or

$$v_D = -3$$

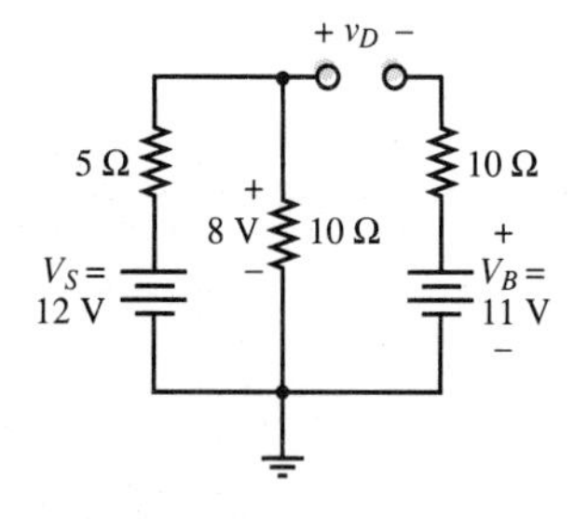

Figure 7.16

This result is consistent with the initial assumption (diode not conducting, or $v_D < 0$). Therefore, the diode does not conduct.

To verify the result just obtained, we may reverse the original assumption and presume that the diode is conducting. Figure 7.17 depicts the appearance of the circuit with the diode replaced by a short circuit. If this assumption were correct, we should be able to verify that the mesh current i_2 (equal to the diode current) is positive. With a short circuit in place of the diode, this is a simple two-mesh problem:

$$(10 + 5)i_1 + (-10)i_2 = 12$$

$$(-10)i_1 + (10 + 10)i_2 = -11$$

which can be solved for i_2 by using Cramer's rule (see Appendix A):

$$i_2 = \frac{\det\begin{bmatrix} 15 & 12 \\ -10 & -11 \end{bmatrix}}{\det\begin{bmatrix} 15 & -10 \\ -10 & 20 \end{bmatrix}} = -0.225 \text{ A}$$

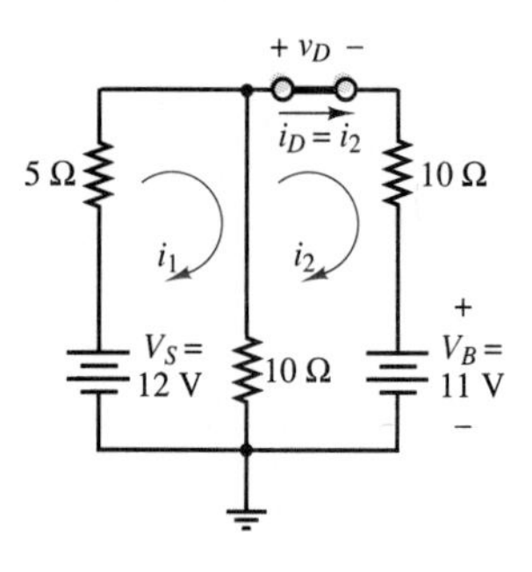

Figure 7.17

or by substituting one equation in the other, as preferred. In any case, the resulting current is $i_2 = -0.225$ A. This result corresponds to stating that a reverse current is flowing through the diode, which is not possible. Thus, the assumption that the diode is on was incorrect, and the diode cannot conduct.

Example 7.2

Determine whether the ideal diode in the circuit of Figure 7.18 is conducting.

Solution:

Assume that the diode is off, and replace it with an open circuit, as shown in Figure 7.19. Then compute the current flowing in the single loop that remains:

$$i = \frac{(12 - 11)}{(5 + 4)} = \frac{1}{9}\text{ A}$$

Thus, the potential at the "node" between resistors is $12 - \frac{5}{9} = 11 + \frac{4}{9}V$. This corresponds to a strong reverse bias for the diode, since the negative terminal of the diode is well above ground. Thus, the diode cannot conduct, and the assumption is correct.

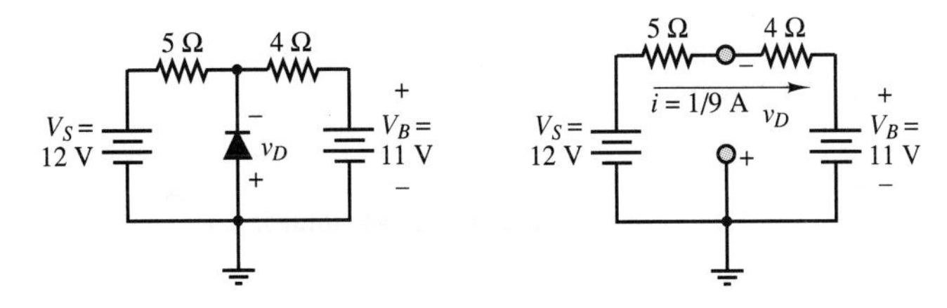

Figure 7.18 **Figure 7.19**

One of the important applications of the semiconductor diode is **rectification** of AC signals, that is, the ability to convert an AC signal with zero average (DC) value to a signal with a nonzero DC value. The application of the semiconductor diode as a rectifier is very useful in obtaining DC voltage supplies from the readily available AC line voltage. Here, we illustrate the basic principle of rectification, using an ideal diode—for simplicity, and also because the large-signal model is appropriate when the diode is used in applications involving large AC voltage and current levels.

Consider the circuit of Figure 7.20, where an AC source, $v_i = 155.56 \cdot \sin \omega t$, is connected to a load by means of a series ideal diode. From the analysis of Example 7.1, it should be apparent that the diode will conduct only during the positive half-cycle of the sinusoidal voltage—that is, that the condition $v_D \geq 0$ will be satisfied only when the AC source voltage is positive—and that it will act as an open circuit during the negative half-cycle of the sinusoid ($v_D < 0$). Thus, the appearance of the load voltage will be as shown in Figure 7.21, with the negative portion of the sinusoidal waveform cut off. The rectified waveform clearly has a nonzero DC (average) voltage, whereas the average input waveform voltage was zero. When the diode is conducting, or $v_D \geq 0$, the unknowns v_L and i_D can be found by using the following equations:

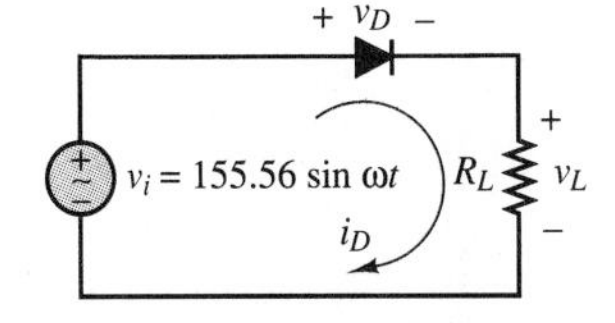

Figure 7.20

$$i_D = \frac{v_i}{R_L} \qquad \text{when} \qquad v_i > 0 \tag{7.9}$$

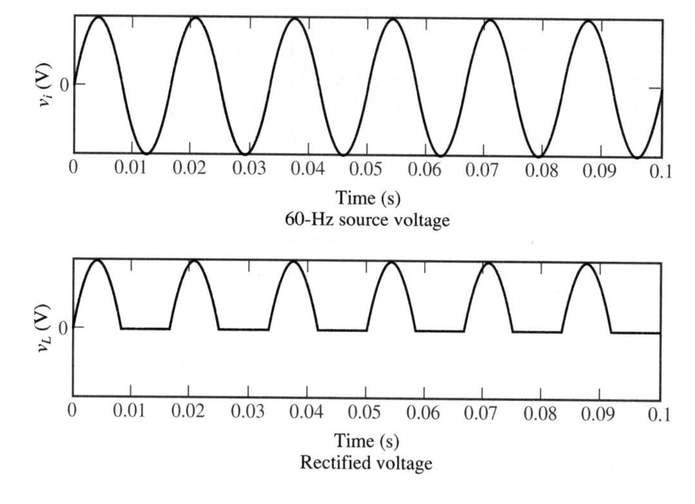

Figure 7.21 Ideal diode rectifier input and output voltages

and

$$v_L = i_D R_L \tag{7.10}$$

The load voltage, v_L, and the input voltage, v_i, are sketched in Figure 7.21. From equation 7.10, it is obvious that the current waveform has the same shape as the load voltage. The average value of the load voltage is obtained by integrating the load voltage over one period and dividing by the period:

$$v_{\text{load,DC}} = \frac{\omega}{2\pi}\int_0^{\frac{\pi}{\omega}} 155.56 \sin \omega t \, dt = \frac{155.56}{\pi} = 49.52 \text{ V} \tag{7.11}$$

The circuit of Figure 7.20 is called a **half-wave rectifier,** since it preserves only half of the waveform. This is not usually a very efficient way of rectifying an AC signal, since half the energy in the AC signal is not recovered. It will be shown in a later section that it is possible to recover also the negative half of the AC waveform by means of a *full-wave rectifier.*

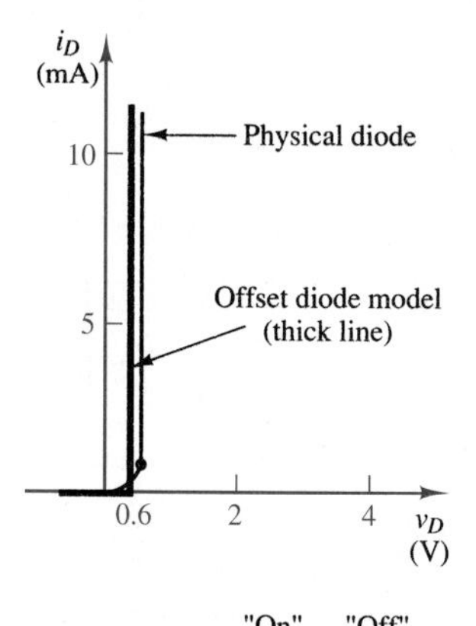

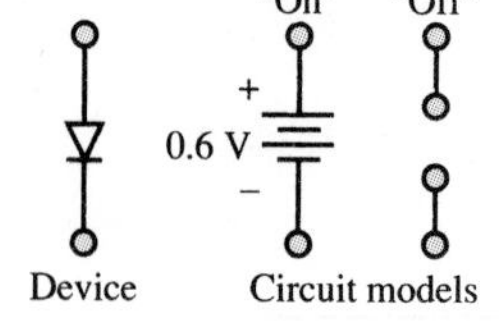

Figure 7.22

Offset Diode Model

While the ideal diode model is useful in approximating the large-scale characteristics of a physical diode, it does not account for the presence of an offset voltage, which is an unavoidable component in semiconductor diodes (recall the discussion of the contact potential in Section 7.2). The **offset diode model** consists of an ideal diode in series with a battery of strength equal to the offset voltage (we shall use the value $V_\gamma = 0.6$ V for silicon diodes, unless otherwise indicated). The effect of the battery is to shift the characteristic of the ideal diode to the right on the voltage axis, as shown in Figure 7.22. This model is a better approximation of the large-signal behavior of a semiconductor diode than the ideal diode model.

According to the offset diode model, the diode of Figure 7.22 acts as an open circuit for $v_D < 0.6$ V, and it behaves like a 0.6-V battery for $v_D \geq 0.6$ V. The equations describing the offset diode model are as follows:

$$\begin{aligned} v_D &\geq 0.6 \text{ V} \qquad \text{Diode} \rightarrow \text{0.6-V battery} \\ v_D &< 0.6 \text{ V} \qquad \text{Diode} \rightarrow \text{Open circuit} \end{aligned} \tag{7.12}$$

The diode offset model may be represented by an ideal diode in series with a 0.6-V ideal battery, as shown in Figure 7.23. Use of the offset diode model is best described by means of examples.

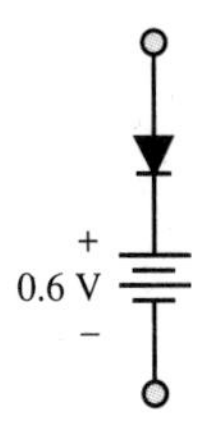

Figure 7.23 Offset diode as an extension of ideal diode model

Example 7.3

In this example, we illustrate the effect of the offset voltage on a diode half-wave rectifier using the offset diode model. Consider the circuit of Figure 7.24, where a sinusoidal source of peak amplitude 3 V is exciting a series-connected diode-resistor circuit. Below the circuit we show its equivalent representation when the diode is replaced by the equivalent circuit of the offset diode model. Note that, in the equivalent circuit, we need only determine whether the voltage, v_D, across the ideal diode, is positive or negative to compute a value for i. In other words, once the circuit model has replaced the diode in the circuit, the problem resembles an ideal diode circuit problem, except for the presence of an additional voltage source, representing the diode offset voltage.

Assume that the diode is initially off. Then the circuit will appear as shown in Figure 7.25(a). In this case, since no current flows around the circuit, the voltage drop across the diode must be

$$v_D = v_S - 0.6 \text{ V}$$

and for our answer to be consistent with the assumption that the diode is off, the voltage v_D must be negative:

$$v_D < 0 \qquad \text{or} \qquad v_S < 0.6 \text{ V}$$

Thus, whenever the source voltage is less than the offset voltage (0.6 V), the diode does not conduct, and no current flows in the circuit. If, on the other hand, the source voltage is greater than 0.6 V, the diode conducts, and the circuit of Figure 7.25(b) applies. In this case, the current flowing in the circuit is

$$i = \frac{v_S - 0.6}{R}$$

The load voltage, v_R, must then be given by

$$v_R = iR = v_S - 0.6$$

The quantities v_S and v_R are sketched in Figure 7.26.

A glance at Figure 7.26 reveals that, in addition to rectifying the sinusoidal waveform, the offset diode model predicts a downward shift of the load waveform, caused by the offset voltage. This shift is particularly visible in this case, because the excitation signal amplitude is of the same order of magnitude as the offset voltage. If the source voltage were, say, of the order of hundreds of volts, the 0.6-V shift would hardly be noticeable.

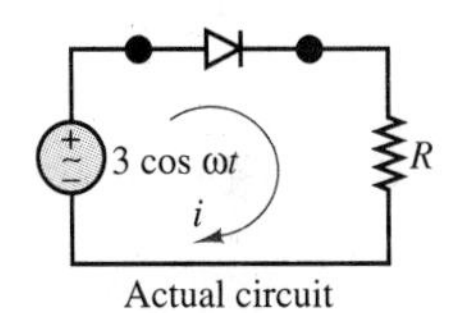

Actual circuit

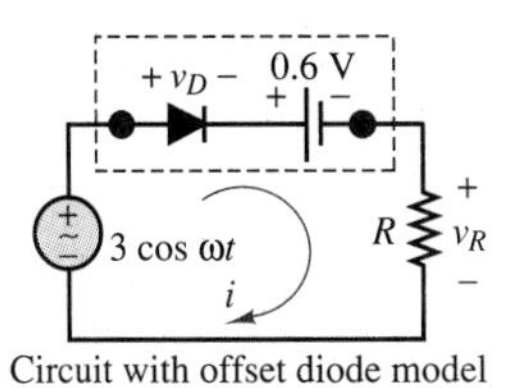

Circuit with offset diode model

Figure 7.24

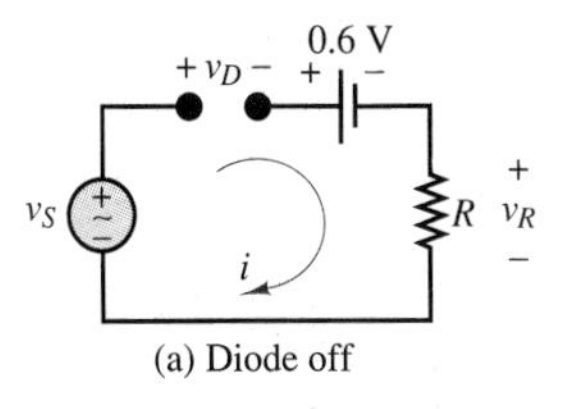

(a) Diode off

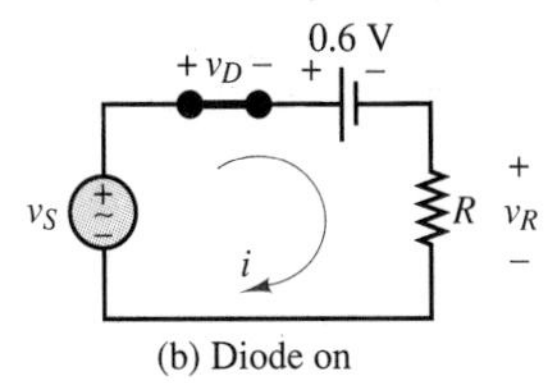

(b) Diode on

Figure 7.25

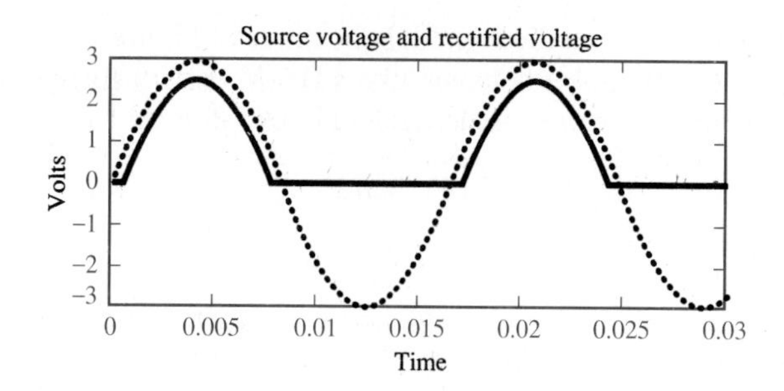

Figure 7.26 Source voltage (dotted curve) and rectified voltage (solid curve) for the circuit of Figure 7.24

Example 7.4

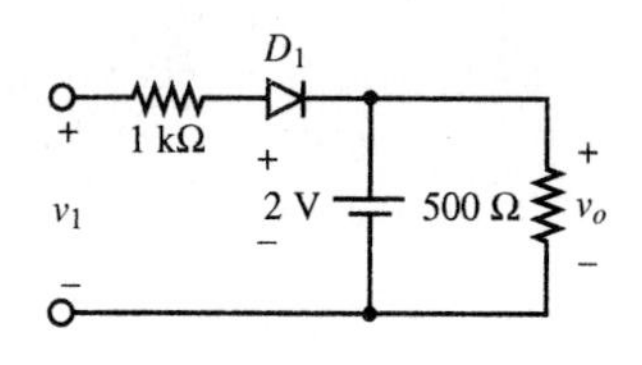

Figure 7.27

Use the offset diode model to determine the value of v_1 for which the diode D_1 will first conduct in the circuit of Figure 7.27.

Solution:

The circuit of Figure 7.27 is shown in Figure 7.28 with the diode replaced by the circuit model. It is safe to assume that for v_1 zero or negative, the diode is off. Thus, a reasonable approach to determine when the diode will turn on is to assume that the diode is off and to write the corresponding circuit equations. Then, it will be possible to solve for the values of v_1 that will cause v_{D_1} to become greater than zero. Note that assuming the diode to be initially off simplifies the problem, since no current flows in the circuit. Applying KVL to each of the loops with this assumption, we obtain

$$v_1 = v_{D_1} + 0.6 + 2$$

and

$$v_0 = 2$$

According to the first equation, then,

$$v_{D_1} = v_1 - 2.6$$

and the condition for the diode to conduct is

$$v_1 > 2.6 \text{ V}$$

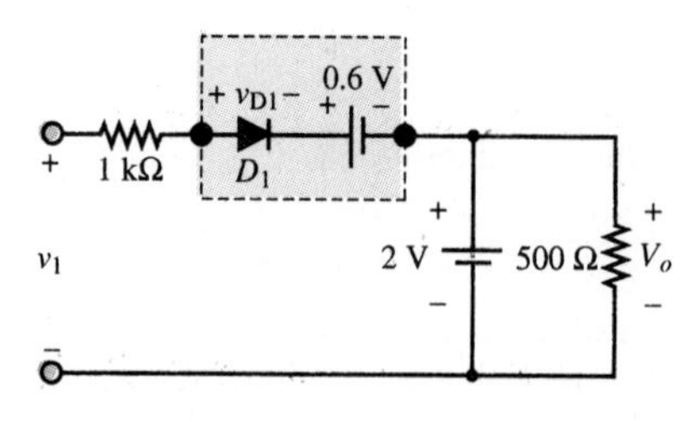

Figure 7.28

Small-Signal Diode Models

As one examines the diode *i-v* characteristic more closely, it becomes apparent that the short-circuit approximation is not adequate to represent the *small-signal behavior* of the diode. The term *small-signal behavior* usually signifies the response of the diode to small time-varying signals that may be superimposed on the average diode current and voltage. Figure 7.8 was a close-up view of a silicon diode *i-v* curve. From this figure, it should be apparent that the short-circuit approximation is not very accurate when a diode's behavior is viewed on an expanded scale. To a first-order approximation, however, the *i-v* characteristic resembles that of a resistor (i.e., is linear) for voltages greater than the offset voltage. Thus, it may be reasonable to model the diode as a resistor (instead of a short circuit) *once it is conducting*, to account for the slope of its *i-v* curve. In the following discussion, the method of load-line analysis (which was introduced in Chapter 3) will be exploited to determine the **small-signal resistance** of a diode.

Consider the circuit of Figure 7.29, which represents the Thévenin equivalent circuit of an arbitrary linear resistive circuit connected to a diode. The equations describing the operation of the circuit are

$$v_T = i_D R_T + v_D \tag{7.13}$$

which arises from application of KVL; and the diode equation,

$$i_D = I_0(e^{qv_D/kT} - 1) \tag{7.14}$$

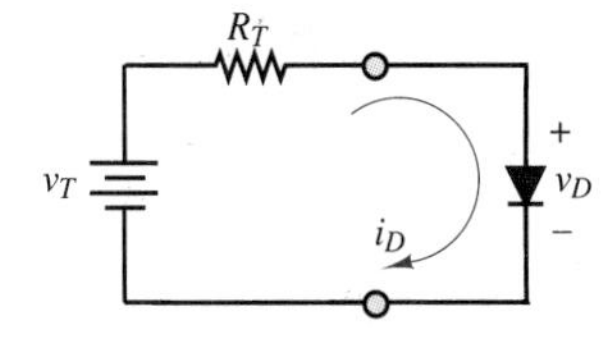

Figure 7.29 Diode circuit for illustration of load-line analysis

Although we thus have two equations in two unknowns, these cannot be solved analytically, since one of the equations contains v_D in exponential form. As discussed in Chapter 3, two methods exist for the solution of *transcendental equations* of this type: graphical and numerical. In the present case, only the graphical solution shall be considered. The graphical solution is best understood if we associate a curve in the i_D-v_D plane with each of the two preceding equations. The diode equation gives rise to the familiar curve of Figure 7.8. The *load-line equation*, obtained by KVL, is the equation of a line with slope $-1/R$ and ordinate intercept given by V_T/R_T.

$$i_D = -\frac{1}{R_T}v_D + \frac{1}{R_T}V_T \qquad \text{Load line equation} \tag{7.15}$$

The superposition of these two curves gives rise to the plot of Figure 7.30, where the solution to the two equations is graphically found to be the pair of values I_Q, V_Q). The intersection of the two curves is called the **quiescent (operating) point,** or ***Q* point**. The voltage $v_D = V_Q$ and the current $i_D = I_Q$ are the actual diode voltage and current when the diode is connected as in the circuit of Figure 7.29. Note that this method is also useful for circuits containing a larger number of elements, provided that we can represent these circuits by their Thévenin equivalents, with the diode appearing as the load.

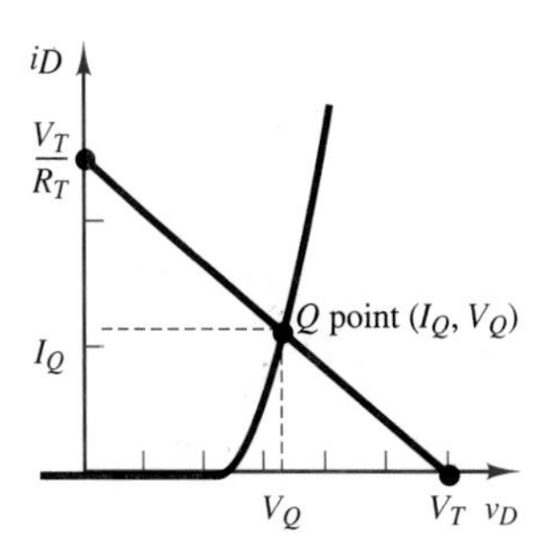

Figure 7.30 Graphical solution of equations 7.13 and 7.14

Example 7.5

In this example, load-line analysis is employed to find the diode current and voltage in order to compute the total power output of the battery in the circuit of Figure 7.31. The diode used is an MV122, which has the small-signal i-v characteristic shown in Figure 7.32.

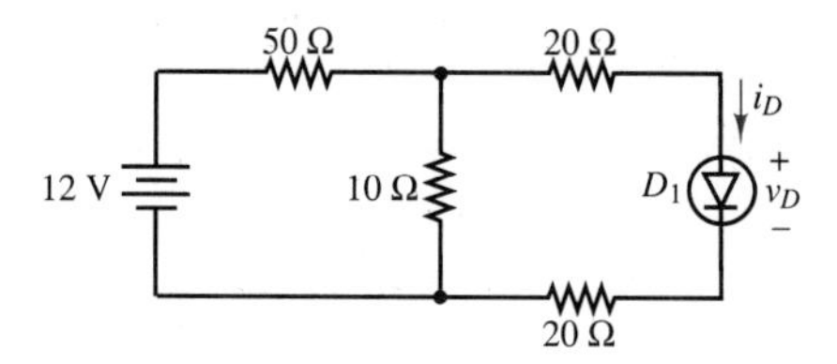

Figure 7.31

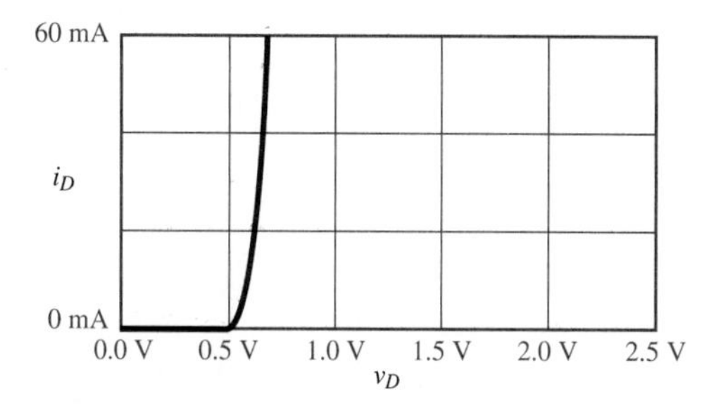

Figure 7.32 MV122 diode i-v curve

Solution:

First, the Thévenin equivalent circuit as seen by the diode is computed:

$$R_T = 20 + 20 + (10 \parallel 50) = 48.33\ \Omega$$

and

$$V_T = \frac{10}{60}(12) = 2\ \text{V}$$

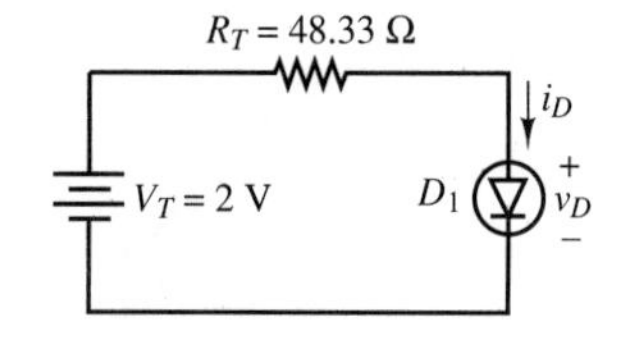

Figure 7.33

The Thévenin equivalent circuit is now shown in Figure 7.33.

Next, the load line is plotted, superimposed on the i-v characteristic curve; the combined curves are shown in Figure 7.34. The line intercepts the v_D axis at 2.0 V and the

i_D axis at 41 mA. We see from the sketch that the load line intersects the diode curve at approximately 0.67 V and 27.5 mA. The voltage across the 10-Ω resistor is

$$V_{10\,\Omega} = 40I_Q + V_Q = 1.77\text{ V}$$

Thus, the current through the 10-Ω resistor is 0.177 A. The total current out of the source is therefore computed to be

$$I_B = 0.177\text{ A} + 0.0275\text{ A} = 0.2043\text{ A}$$

and therefore the total power supplied to the circuit by the battery is

$$P_B = 12 \times 0.2043 = 2.452\text{ W}$$

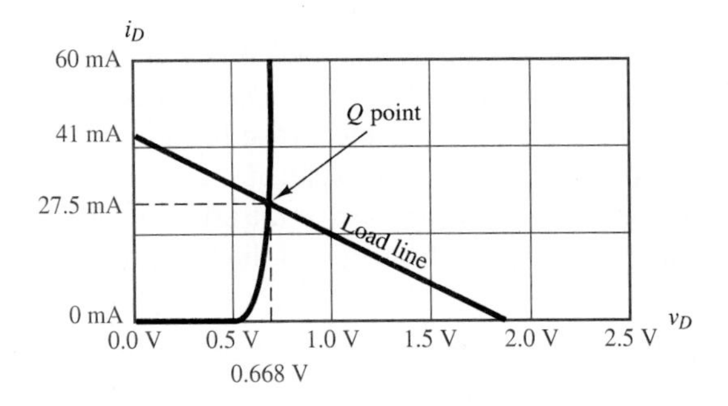

Figure 7.34 Superposition of load line and diode *i-v* curve

Piecewise Linear Diode Model

The graphical solution of diode circuits can be somewhat tedious, and its accuracy is limited by the resolution of the graph; it does, however, provide insight into the **piecewise linear diode model.** In the piecewise linear model, the diode is treated as an open circuit in its off state, and as a linear resistor in series with V_γ in the on state. Figure 7.35 illustrates the graphical appearance of this model. Note that the straight line that approximates the "on" part of the diode characteristic is tangent to the Q point. Thus, in the neighborhood of the Q point, the diode does act as a linear small-signal resistance, with slope given by $\frac{1}{r_D}$, where

$$\frac{1}{r_D} = \left.\frac{\partial i_D}{\partial v_D}\right|_{(I_Q, V_Q)} \tag{7.16}$$

That is, it acts as a linear resistance whose *i-v* characteristic is the tangent to the diode curve at the operating point. The tangent is extended to meet the voltage

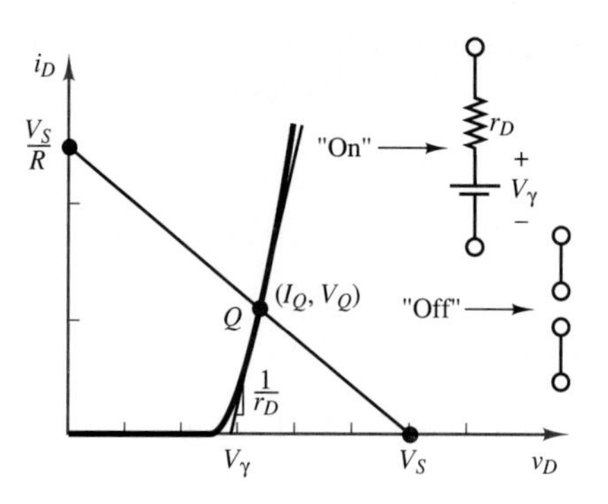

Figure 7.35 Piecewise linear diode model

axis, thus defining the intersection as the diode offset voltage. Thus, rather than represent the diode by a short circuit in its forward-biased state, we treat it as a linear resistor, with resistance r_D. The piecewise linear model offers the convenience of a linear representation once the state of the diode is established, and of a more accurate model than either the ideal or the offset diode model. This model is very useful in illustrating the performance of diodes in real-world applications.

Example 7.6

In this example, we compute the incremental (small-signal) resistance of a diode at a given operating point. Assume that a certain diode has $I_0 = 10^{-14}$ A and $kT/q = \alpha = 0.025$ V at room temperature. Use the approximate exponential diode equation to compute the incremental resistance of the diode at the operating point defined by $I_Q = 50$ mA.

Solution:

The approximate exponential diode equation is given by

$$i_D = I_0 e^{v_D/\alpha}$$

while the incremental resistance can be found from the expression

$$\frac{1}{r_D} = \left.\frac{\partial i_D}{\partial v_D}\right|_{(I_Q,V_Q)}$$

Differentiating the expression for i_D with respect to v_D, we can obtain

$$\frac{1}{r_D} = \frac{I_0}{\alpha} e^{v_D/\alpha}$$

To determine the operating voltage of the diode, we may solve the diode equation with $i_D = 50$ mA:

$$\log_e\left(\frac{0.05}{10^{-14}}\right) = \frac{v_D}{0.025}$$

which can be evaluated to obtain

$$v_D = 0.731 \text{ V}$$

so that

$$\frac{1}{r_D} = \frac{10^{-14}}{0.025} e^{0.731/0.025} = 2.00$$

or

$$r_D = 0.5\ \Omega$$

Example 7.7

In this example, we use the piecewise linear model of Figure 7.35 to analyze a slightly more realistic rectifier circuit, shown in Figure 7.36. Determine the rectified load voltage, v_L, if the source voltage is $v_S(t) = 10\cos(\omega t)$.

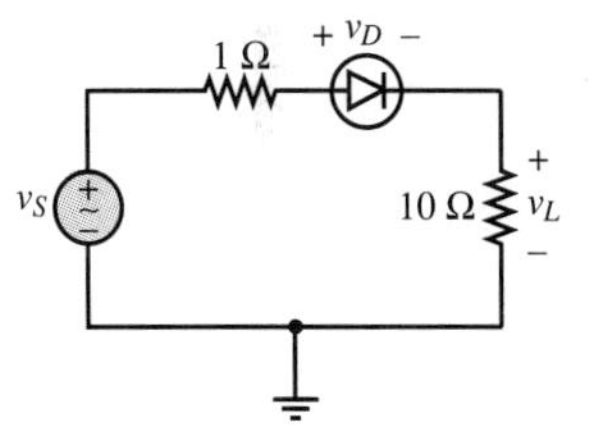

Figure 7.36

Solution:

Figure 7.37 depicts the rectifier circuit with the diode replaced by its piecewise linear model. Note that the circuit of Figure 7.37 contains an ideal diode with linear resistors

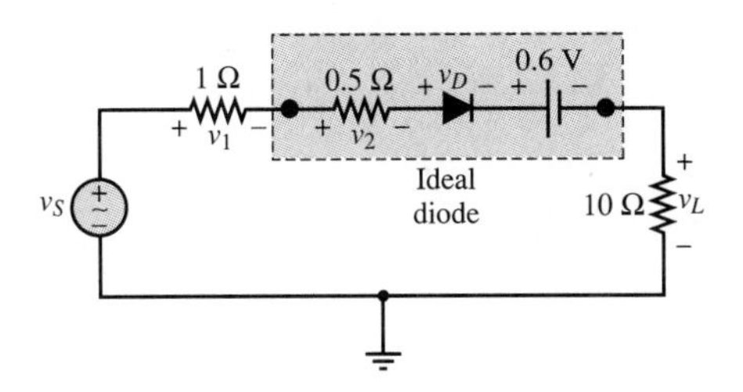

Figure 7.37

and voltage sources, and we can therefore determine the state of the diode using the same method explored earlier for the ideal and offset diode models. In particular, if KVL is applied to the circuit, the following expression is obtained:

$$v_S = v_1 + v_2 + v_D + 0.6 + v_L$$

or

$$v_D = v_S - v_1 - v_2 - 0.6 - v_L$$

To determine the state of the diode, we reason as follows. Since, for negative voltages (i.e., during the negative half-cycle of the sine wave), the diode is off, it will be sufficient to determine the point at which the diode first turns on. Assume, therefore, that the diode is off (corresponding to the negative half-cycle of the source voltage); then, no current flows in the series circuit, and the voltages v_1, v_2, and v_L are all zero. Thus, when the diode is not conducting, the following expression holds:

$$v_D = v_S - 0.6$$

and the condition for the ideal diode to conduct is

$$v_D \geq 0 \qquad \text{or} \qquad v_S \geq 0.6 \text{ V}$$

Thus, the diode will conduct when $v_S \geq 0.6$ V. Once the diode conducts, the expression for the load voltage can be easily obtained by the voltage divider rule, considering that the ideal diode behaves like a short circuit. On the other hand, the load voltage is zero for $v_S < 0.6$ V, since the diode does not conduct and no current reaches the load. The complete expression for the load voltage is therefore given by:

$$v_L = \frac{10}{10 + 1 + 0.5}(v_S - 0.6) = 8.7\cos(\omega t) - 0.52 \qquad v_S \geq 0.6 \text{ V}$$

$$v_L = 0 \qquad v_S < 0.6 \text{ V}$$

The source and load voltages are plotted in Figure 7.38. Note that, because of the combined effect of the offset voltage and the voltage divider, the peak voltage of the rectified waveform is only 8.18 V.

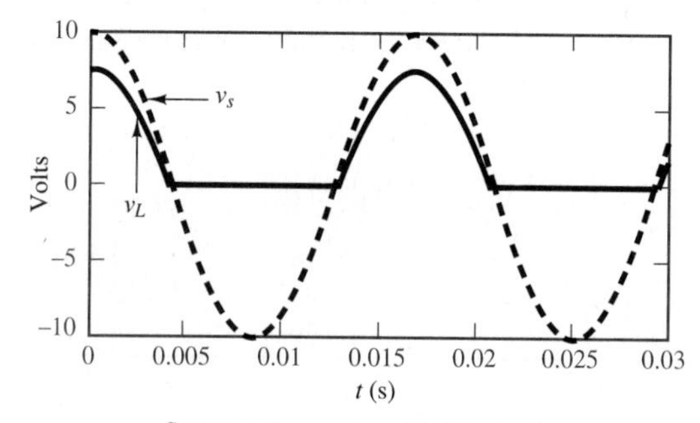

Figure 7.38 Source voltage and rectified load voltage

DRILL EXERCISES

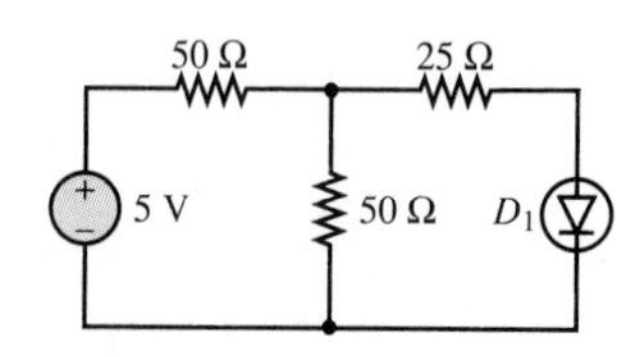

Figure 7.39

7.1 Repeat Example 7.2, assuming that the diode is conducting, and show that this assumption leads to an inconsistent result.

7.2 Compute the DC value of the rectified waveform for the circuit of Figure 7.20 for $v_i = 52\cos\omega t$ V.

7.3 Use load-line analysis to determine the operating point (Q point) of the diode in the circuit of Figure 7.39. The diode has the characteristic curve shown in Figure 7.32.

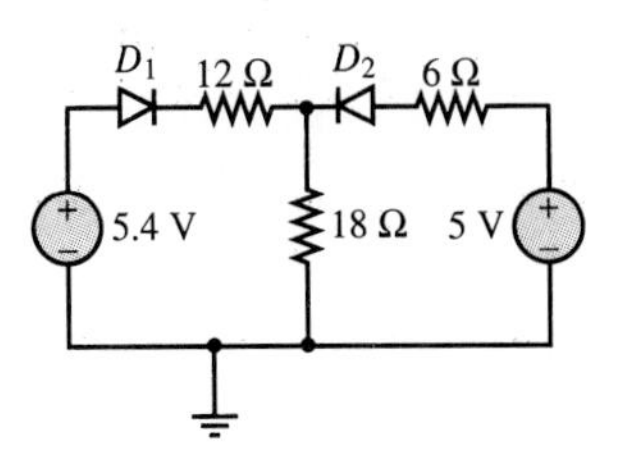

Figure 7.40

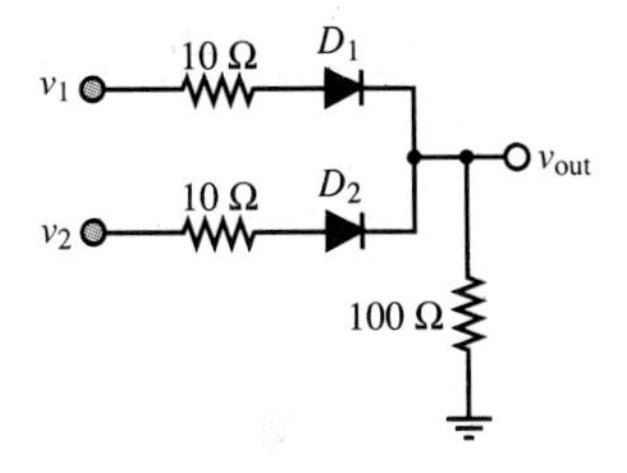

Figure 7.41

7.4 Compute the incremental resistance of the diode of Example 7.6 if the current through the diode is 250 mA.

7.5 Consider a half-wave rectifier similar to that of Figure 7.20, with $v_i(t) = 18\cos(t)$, and a 4-Ω load resistor. Sketch the output waveform if the piecewise linear diode model is used to describe the operation of the diode, with $V_\gamma = 0.6$ V and $r_D = 1\ \Omega$. What is the peak value of the rectifier waveform?

7.6 Determine which of the diodes in the circuit of Figure 7.40 conducts. Each diode has an offset voltage of 0.6 V.

7.7 Determine which of the diodes in Figure 7.41 conduct for the following voltages (in V): (a) $v_1 = 0, v_2 = 0$; (b) $v_1 = 5, v_2 = 5$; (c) $v_1 = 0, v_2 = 5$; (d) $v_1 = 5, v_2 = 0$. Treat the diodes as ideal.

7.4 PRACTICAL DIODE CIRCUITS

This section illustrates some of the applications of diodes to practical engineering circuits. The nonlinear behavior of diodes, especially the rectification property, makes these devices valuable in a number of applications. In this section, more advanced rectifier circuits (the **full-wave rectifier** and the **bridge rectifier**) will be explored, as well as **limiter** and *peak detector* circuits. These circuits will be analyzed by making use of the circuit models developed in the preceding sections; as stated earlier, these models are more than adequate to develop an understanding of the operation of diode circuits.

In addition to the operation of diodes as rectifiers and limiters, there is another useful class of applications that takes advantage of the reverse-breakdown characteristic of the semiconductor diode discussed in the opening section. The phenomenon of Zener breakdown is exploited in a class of devices called **Zener diodes,** which enjoy the property of a sharp reverse-bias breakdown with relatively constant breakdown voltage. These devices are used as voltage regulators, that is, to provide a nearly constant output (DC) voltage from a voltage source whose output might ordinarily fluctuate substantially (for example, a rectified sinusoid).

The Full-Wave Rectifier

The half-wave rectifier discussed earlier is one simple method of converting AC energy to DC energy. The need for converting one form of electrical energy into the other arises frequently in practice. The most readily available form of electric power is AC (the standard 110- or 220-V rms AC line power), but one frequently needs a DC power supply, for applications ranging from the control of certain types of electric motors to the operation of electronic circuits such as those that will be discussed in Chapters 8 through 14.

The half-wave rectifier, however, is not a very efficient AC-DC conversion circuit, because it fails to utilize half the energy available in the AC waveform, by not conducting current during the negative half-cycle of the AC waveform.

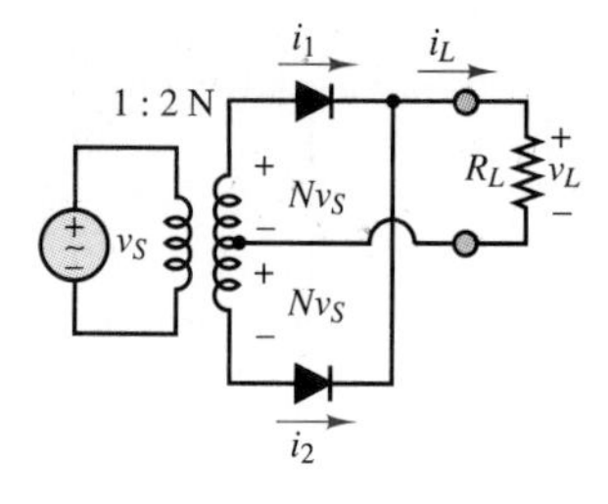

Figure 7.42 Full-wave rectifier

The full-wave rectifier shown in Figure 7.42 offers a substantial improvement in efficiency over the half-wave rectifier. The first section of the full-wave rectifier circuit includes an AC source and a center-tapped transformer (see Chapter 5) with 1:2*N* turns ratio. The purpose of the transformer is to obtain the desired voltage amplitude prior to rectification. Thus, if the peak amplitude of the AC source voltage is v_S, the amplitude of the voltage across each half of the output side of the transformer will be Nv_S; this scheme permits scaling the source voltage up or down (depending on whether *N* is greater or less than 1), according to the specific requirements of the application. In addition to scaling the source voltage, the transformer also isolates the rectifier circuit from the AC source voltage, since there is no direct electrical connection between the input and output of a transformer.

In the analysis of the full-wave rectifier, the diodes will be treated as ideal, since in most cases the source voltage is the AC line voltage (110 V rms, 60 Hz) and therefore the offset voltage is negligible in comparison. The key to the operation of the full-wave rectifier is to note that during the positive half-cycle of v_S, the top diode is forward-biased while the bottom diode is reverse-biased; therefore, the load current during the positive half-cycle is

$$i_L = i_1 = \frac{Nv_S}{R_L} \qquad v_S \geq 0 \tag{7.17}$$

while during the negative half-cycle, the bottom diode conducts and the top diode is off, and the load current is given by

$$i_L = i_2 = \frac{-Nv_S}{R_L} \qquad v_S < 0 \tag{7.18}$$

Note that the direction of i_L is always positive, because of the manner of connecting the diodes (when the top diode is off, i_2 is forced to flow from + to − across R_L).

The source voltage, the load voltage, and the currents i_1 and i_2 are shown in Figure 7.43 for a load resistance $R_L = 1\ \Omega$ and $N = 1$. The full-wave rectifier results in a twofold improvement in efficiency over the half-wave rectifier introduced earlier.

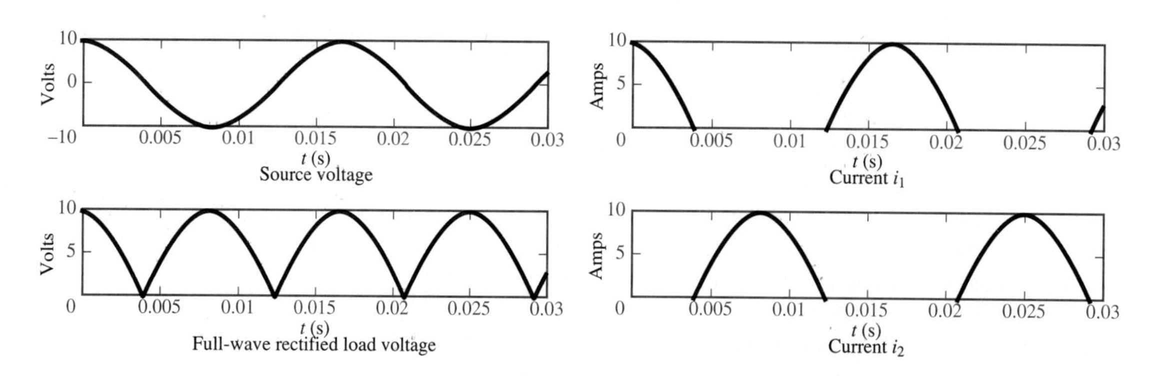

Figure 7.43 Full-wave rectifier current and voltage waveforms ($R_L = 1\ \Omega$)

The Bridge Rectifier

Another rectifier circuit commonly available "off the shelf" as a single *integrated circuit package*[1] is the *bridge rectifier*, which employs four diodes in a bridge configuration, similar to the Wheatstone bridge already explored in Chapter 2. Figure 7.44 depicts the bridge rectifier, along with the associated integrated circuit (IC) package.

The analysis of the bridge rectifier is simple to understand by visualizing the operation of the rectifier for the two half-cycles of the AC waveform separately. The key is that, as illustrated in Figure 7.45, diodes D_1 and D_3 conduct during the positive half-cycle, while diodes D_2 and D_4 conduct during the negative half-cycle. Because of the structure of the bridge, the flow of current through the load resistor is in the same direction (from c to d) during both halves of the cycle; hence, the full-wave rectification of the waveform. The original and rectified waveforms are shown in Figure 7.46(a) for the case of ideal diodes and a 30-V peak AC source. Figure 7.46(b) depicts the rectified waveform if we assume

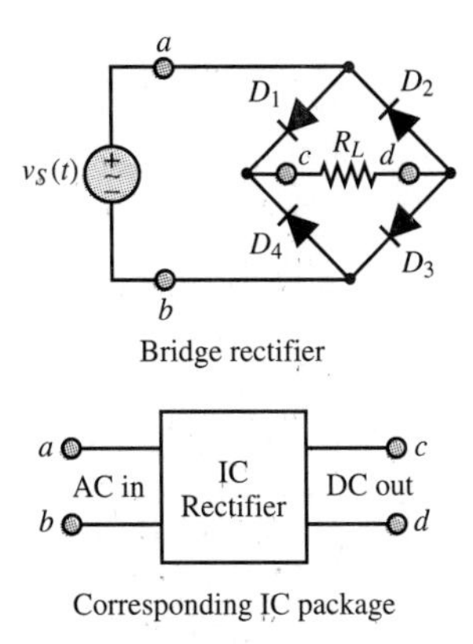

Figure 7.44 Full-wave bridge rectifier

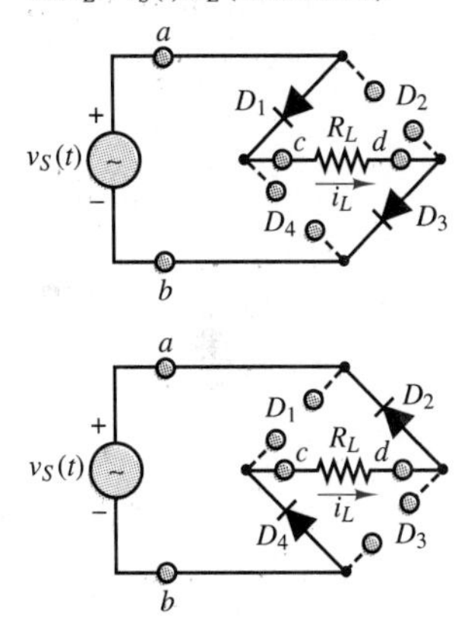

Figure 7.45 Diodes conduction paths for diode bridge: positive half-cycle (top); negative half-cycle (bottom)

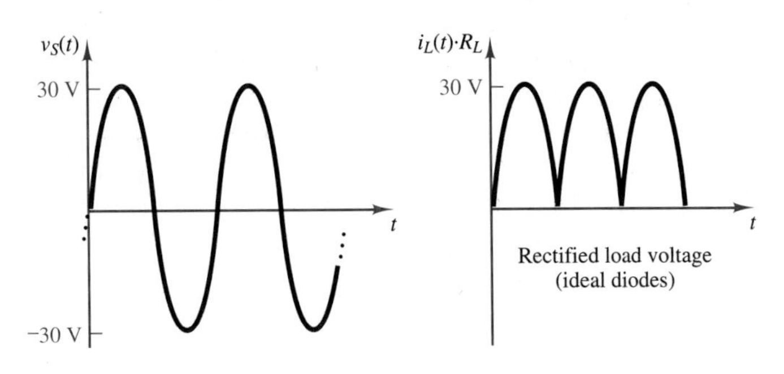

Figure 7.46(a) Bridge rectifier output with ideal diodes

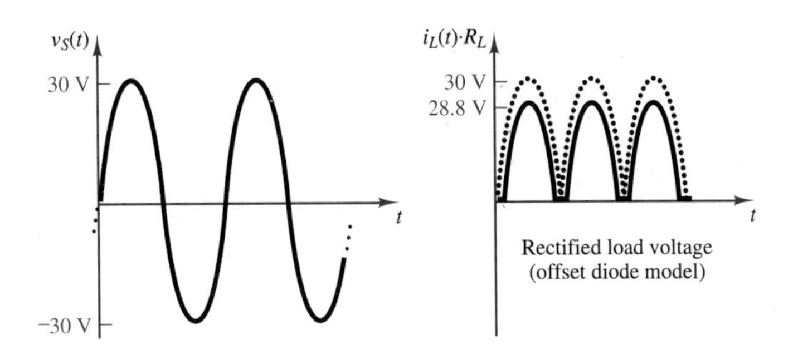

Figure 7.46(b) Bridge rectifier output accounting for diode offset voltage

[1] An integrated circuit is a collection of electronic devices interconnected on a single silicon chip.

diodes with a 0.6-V offset voltage. Note that the waveform of Figure 7.46(b) is not a pure rectified sinusoid any longer: the effect of the offset voltage is to shift the waveform downward by twice the offset voltage. This is most easily understood by considering that the load seen by the source during either half-cycle consists of two diodes in series with the load resistor.

Although the conventional and bridge full-wave rectifier circuits effectively convert AC signals that have zero average, or DC, value to a signal with a nonzero average voltage, either rectifier's output is still an oscillating waveform. Rather than provide a smooth, constant voltage, the full-wave rectifier generates a sequence of sinusoidal pulses at a frequency double that of the original AC signal. The **ripple**—that is, the fluctuation about the mean voltage that is characteristic of these rectifier circuits—is undesirable if one desires a true DC supply. A simple yet effective means of eliminating most of the ripple (i.e., AC component) associated with the output of a rectifier is to take advantage of the energy-storage properties of capacitors to filter out the ripple component of the load voltage. A low-pass filter that preserves the DC component of the rectified voltage while filtering out components at frequencies at or above twice the AC signal frequency would be an appropriate choice to remove the ripple component from the rectified voltage. In most practical applications of rectifier circuits, the signal waveform to be rectified is the 60-Hz, 110-V rms line voltage. The ripple frequency is, therefore, $f_{\text{ripple}} = 120$ Hz, or $\omega_{\text{ripple}} = 2\pi \cdot 120$ rad/s. A low-pass filter is required for which

$$\omega_0 \ll \omega_{\text{ripple}} \tag{7.19}$$

For example, the filter could be characterized by

$$\omega_0 = 2\pi \cdot 2 \text{ rad/s}$$

A simple low-pass filter circuit similar to those studied in Chapter 6 that accomplishes this task is shown in Figure 7.47.

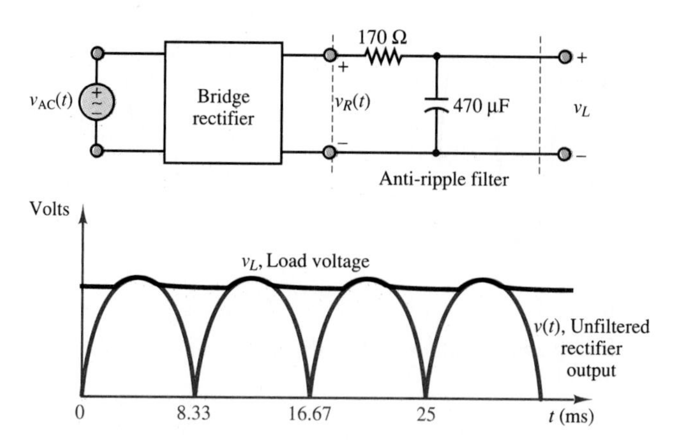

Figure 7.47 Bridge rectifier with filter circuit

DC Power Supplies and Voltage Regulation

The principal application of rectifier circuits is in the conversion of AC to DC power. A circuit that accomplishes this conversion is usually called a **DC power supply.** In power supply applications, transformers are employed to obtain an AC voltage that is reasonably close to the desired DC supply voltage. DC power supplies are very useful in practice: many familiar electrical and electronic appliances (e.g., radios, personal computers, TVs) require DC power to operate. For most applications, it is desirable that the DC supply be as steady and ripple-free as possible. To ensure that the DC voltage generated by a DC supply is constant, the DC supply is made up of voltage regulators, that is, devices that can hold a DC load voltage relatively constant in spite of possible fluctuations in the DC supply. This section will describe the fundamentals of voltage regulators.

A typical DC power supply is made up of the components shown in Figure 7.48. In the figure, a transformer is shown connecting the AC source to the rectifier circuit to permit scaling of the AC voltage to the desired level. For example, one might wish to step the 110-V rms line voltage down to 24 V rms by means of a transformer prior to rectification and filtering, to eventually obtain a 12-VDC regulated supply (*regulated* here means that the output voltage is a DC voltage that is constant and independent of load and supply variations). Following the step-down transformer are a bridge rectifier, a filter capacitor, a voltage regulator, and, finally, the load.

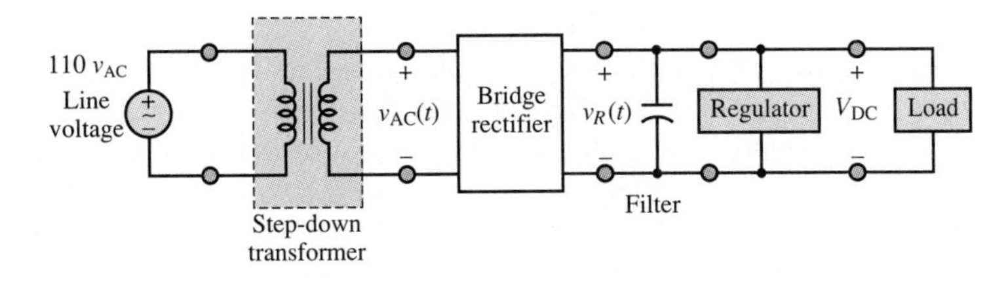

Figure 7.48 DC power supply

The most common device employed in voltage regulation schemes is the Zener diode. Zener diodes function on the basis of the reverse portion of the *i-v* characteristic of the diode discussed in Section 7.2. Figure 7.10 in Section 7.2 illustrates the general characteristic of a diode, with forward offset voltage V_γ and **reverse Zener voltage** V_Z. Note how steep the *i-v* characteristic is at the Zener breakdown voltage, indicating that in the Zener breakdown region the diode can hold a very nearly constant voltage for a large range of currents. This property makes it possible to use the Zener diode as a voltage regulator.

The operation of the Zener diode may be analyzed by considering three modes of operation:

1. For $v_D \geq V_\gamma$, the device acts as a conventional forward-biased diode (Figure 7.49).

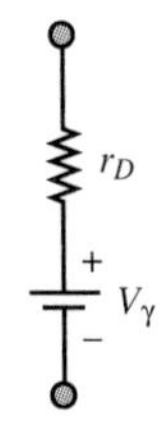

Figure 7.49 Zener diode model for forward bias

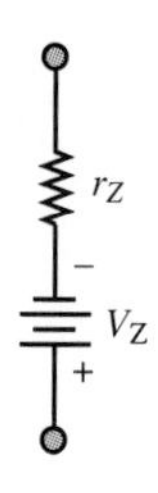

Figure 7.50 Zener diode model for reverse bias

2. For $V_Z < v_D < V_\gamma$, the diode is reverse-biased but Zener breakdown has not taken place yet. Thus, it acts as an open circuit.
3. For $v_D \le V_Z$, Zener breakdown occurs and the device holds a nearly constant voltage, $-V_Z$ (Figure 7.50).

The combined effect of forward and reverse bias may be lumped into a single model with the aid of ideal diodes, as shown in Figure 7.51.

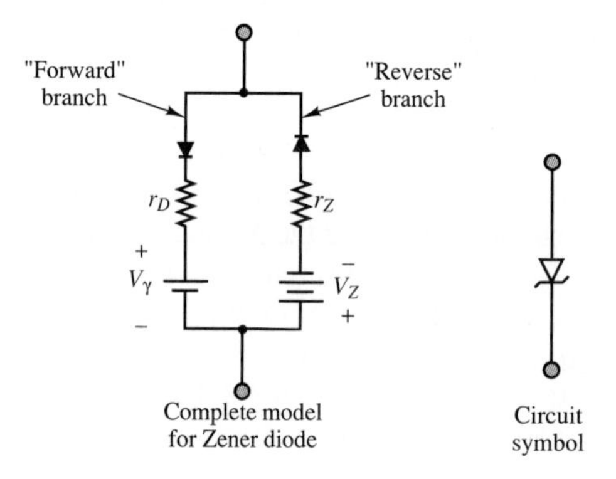

Figure 7.51 Complete model for Zener diode

To illustrate the operation of a Zener diode as a voltage regulator, consider the circuit of Figure 7.52, where the unregulated DC source, V_S, is regulated to the value of the Zener voltage V_Z. Note how the diode must be connected "upside down" to obtain a positive regulated voltage. Note also that if v_S is greater than V_Z, it follows that the Zener diode is in its reverse-breakdown mode. Thus, one need not worry whether the diode is conducting or not in simple voltage regulator problems, provided that the unregulated supply voltage is guaranteed to stay above V_Z (a problem arises, however, if the unregulated supply can drop below the Zener voltage). Assuming that the resistance r_Z is negligible with respect to R_S and R_L, we replace the Zener diode with the simplified circuit model of Figure 7.53, consisting of a battery of strength V_Z (the effects of the nonzero Zener resistance are explored in the examples and homework problems).

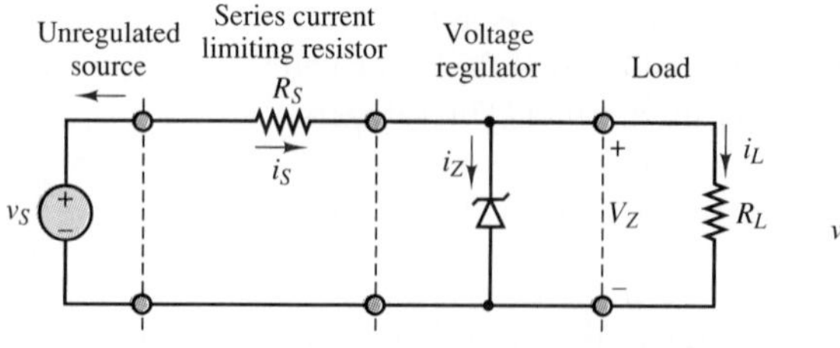

Figure 7.52 A Zener diode voltage regulator

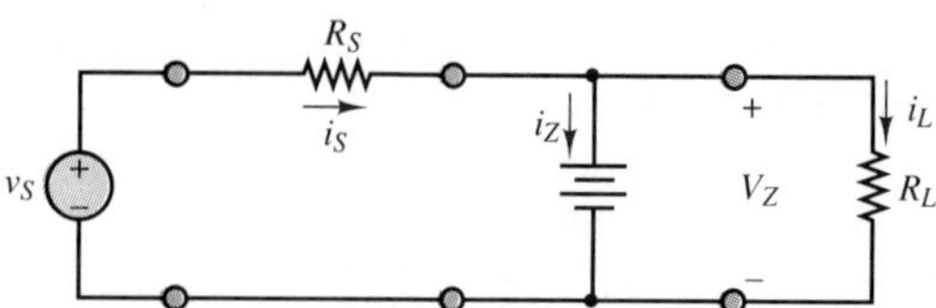

Figure 7.53 Simplified circuit for Zener regulator

Three simple observations are sufficient to explain the operation of this voltage regulator:

1. The load voltage must equal V_Z, as long as the Zener diode is in the reverse-breakdown mode. Then,

$$i_L = \frac{V_Z}{R_L} \tag{7.20}$$

2. The load current (which should be constant if the load voltage is to be regulated to sustain V_Z) is the difference between the unregulated supply current, i_S, and the diode current, i_Z:

$$i_L = i_S - i_Z \tag{7.21}$$

   This second point explains intuitively how a Zener diode operates: any current in excess of that required to keep the load at the constant voltage V_Z is "dumped" to ground through the diode. Thus, the Zener diode acts as a sink to the undesired source current.

3. The source current is given by

$$i_S = \frac{v_S - V_Z}{R_S} \tag{7.22}$$

In the ideal case, the operation of a Zener voltage regulator can be explained very simply on the basis of this model. The examples and exercises will illustrate the effects of the practical limitations that arise in the design of a practical voltage regulator; the general principles will be discussed in the following paragraphs.

The Zener diode is usually rated in terms of its maximum allowable power dissipation. The power dissipated by the diode, P_Z, may be computed from

$$P_Z = i_Z V_Z \tag{7.23}$$

Thus, one needs to worry about the possibility that i_Z will become too large. This may occur either if the supply current is very large (perhaps because of an unexpected upward fluctuation of the unregulated supply), or if the load is suddenly removed and all of the supply current sinks through the diode. The latter case, of an open-circuit load, is an important design consideration.

Another significant limitation occurs when the load resistance is small, thus requiring large amounts of current from the unregulated supply. In this case, the Zener diode is hardly taxed at all in terms of power dissipation, but the unregulated supply may not be able to provide the current required to sustain the load voltage. In this case, regulation fails to take place. Thus, in practice, the range of load resistances for which load voltage regulation may be attained is constrained to a finite interval:

$$R_{L\,\min} \le R_L \le R_{L\,\max} \tag{7.24}$$

where $R_{L\,\max}$ is typically limited by the Zener diode power dissipation and $R_{L\,\min}$ by the maximum supply current. The following examples illustrate these concepts.

Example 7.8

In this example, we illustrate the selection of the power rating for a Zener diode. Consider the design of a regulator similar to that of Figure 7.52. Find the safe minimum power rating that will ensure that the Zener diode will not be damaged under worst-case conditions. Assume that $R_S = 50\ \Omega, R_L = 250\ \Omega$, the unregulated supply is 24 V, and the desired regulated voltage is $V_Z = 12$ V.

Solution:

Under normal operation, with a 250-Ω load, we may compute the source current to be

$$i_S = \frac{v_S - V_Z}{R_S} = \frac{12}{50} = 0.24 \text{ A}$$

and the load current to be

$$i_L = \frac{V_Z}{R_L} = \frac{12}{250} = 0.048 \text{ A}$$

Thus, the Zener current would be

$$i_Z = i_S - i_L = 0.192 \text{ A}$$

and the nominal Zener power dissipation could be computed to be

$$P_Z = i_Z V_Z = 0.192 \times 12 = 2.304 \text{ W}$$

However, if the load were accidentally disconnected from the regulator, all of the load current would be diverted to flow through the Zener diode. Thus, the worst-case Zener current is actually equal to i_S:

$$i_{Z\max} = i_S = \frac{v_S - V_Z}{R_S} = 0.24 \text{ A}$$

and therefore the maximum Zener diode power dissipation is

$$P_{Z\max} = i_{Z\max} V_Z = 2.88 \text{ W}$$

A safe design would include a Zener diode with a power rating at least equal to $P_{Z\max}$.

Example 7.9

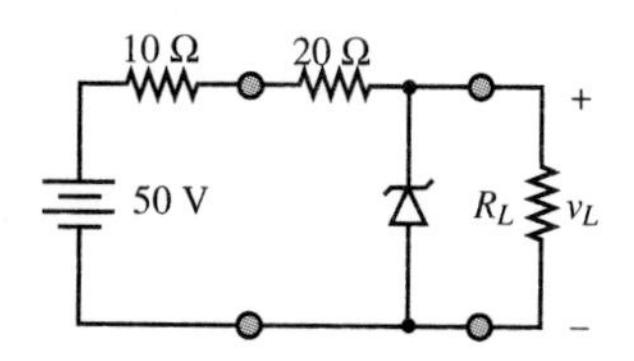

Figure 7.54

This example illustrates the calculation of the range of allowable load resistances for a Zener regulator design. For the Zener regulator shown in Figure 7.54, we want to maintain the load voltage at 14 V. Find the range of load resistances for which regulation can be obtained if the Zener diode is rated at 14 V, 5 W.

Solution:

The minimum load resistance for which a regulated load voltage of 14 V may be attained is found by requiring that the load voltage be 14 V and applying KVL subject to this constraint:

$$50\left(\frac{R_{L\min}}{R_{L\min}+30}\right) = 14$$

$$R_{L\min} = \frac{14}{50}(R_{L\min}+30)$$

$$= 11.7\ \Omega$$

The maximum current through the Zener diode that does not exceed the diode power rating may be computed by considering the 5-W power rating:

$$i_{Z\max} = \frac{5}{14} = 0.357\ \text{A}$$

The current through the 20-Ω resistor will be

$$\frac{50-14}{30} = 1.2\ \text{A}$$

so that the maximum load resistance for which regulation occurs is

$$R_{L\max} = \frac{14}{1.2-0.357}$$

$$= 16.6\ \Omega$$

Finally, the range of allowable load resistances is:

$$16.6\ \Omega \geq R_L \geq 11.7\ \Omega$$

Example 7.10

This example illustrates the effect of nonzero Zener resistance in a voltage regulator. We assume that, because of imperfect filtering, some ripple is present in the unregulated voltage source signal. This is usually the case in practice. The goal is to determine how much of the undesired ripple signal will appear at the output of the regulator (ideally, it should be completely removed by the regulator). Assume that 100 mV of ripple is present in addition to the unregulated 14-V (nominal) source. The circuit is sketched in Figure 7.55, and the source voltage has the appearance of the sketch of Figure 7.56. How much of the ripple will reach the load resistor? What is the actual DC load voltage? The Zener diode has $V_Z = 8$ V and a Zener resistance $r_Z = 10\ \Omega$. $R_S = 50\ \Omega$ and $R_L = 150\ \Omega$.

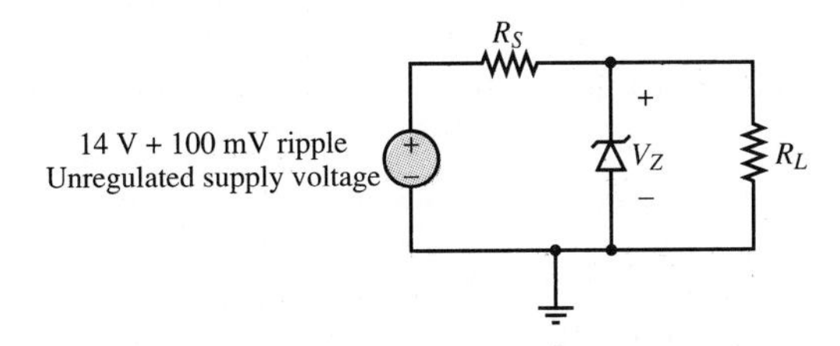

Figure 7.55

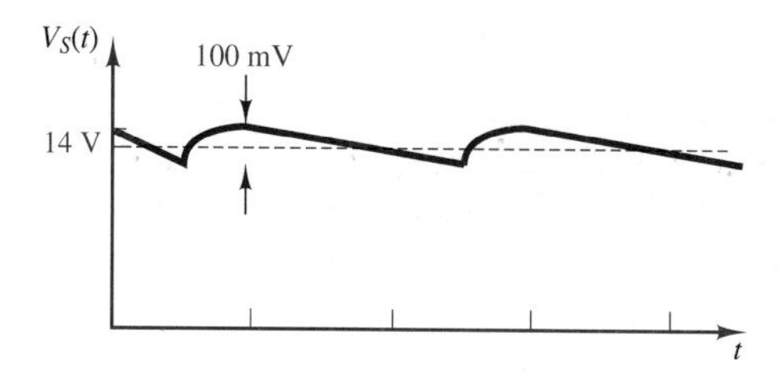

Figure 7.56

Solution:

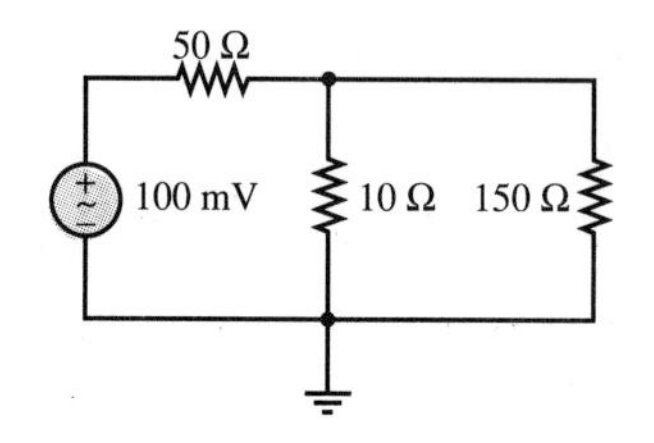

AC equivalent circuit

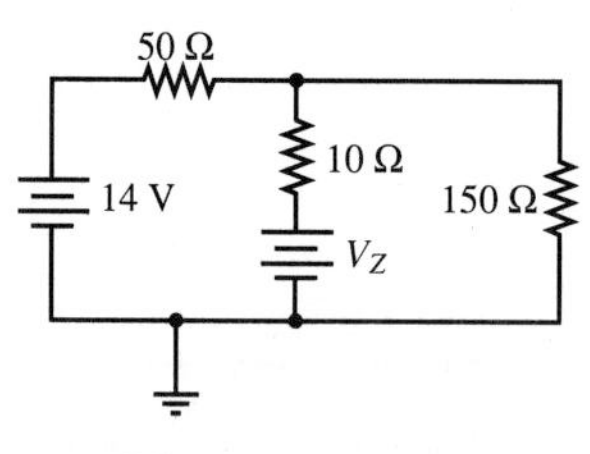

DC equivalent circuit

Figure 7.57

Since the unregulated supply voltage is significantly larger than the Zener voltage, we are assured that the diode is operating in the Zener mode. To determine the load voltage, we can analyze the AC and DC behavior of the circuit separately, using the principle of superposition. Figure 7.57 depicts the AC and DC equivalent circuits, respectively.

The DC equivalent circuit shown in Figure 7.57 reveals that the actual DC load voltage is given by the sum of the contribution due to the DC portion of the unregulated supply and the contribution of the Zener diode:

$$V_L = 14\left(\frac{10 \parallel 150}{(10 \parallel 150) + 50}\right) + 8\left(\frac{50 \parallel 150}{(50 \parallel 150) + 10}\right)$$
$$= 2.21 + 6.32 = 8.53 \text{ V}$$

On the other hand, the AC component of the load voltage may be computed from the AC circuit to be

$$v_L = 0.1\left(\frac{10 \parallel 150}{(10 \parallel 150) + 50}\right) = 0.016 \text{ V}$$

Thus, the load resistor will see approximately 16 mV of ripple, or about one sixth of what is present in the unregulated supply.

Note that accounting for the Zener resistance complicates the calculations to some extent, because the resistance introduces a coupling between the unregulated supply and the load, whereas in the case of the ideal Zener diode, such coupling does not exist. If the Zener resistance is significantly smaller than both R_S and R_L, these effects are less pronounced, and the behavior of the Zener regulator is much closer to the ideal case (see Drill Exercise 7.10).

DRILL EXERCISES

7.8 Show that the DC voltage output of the full-wave rectifier of Figure 7.42 is $2Nv_{S\,\max}/\pi$.

7.9 Compute the peak voltage output of the bridge rectifier of Figure 7.44, assuming diodes with 0.6-V offset voltage and a 110-V rms AC supply.

7.10 Compute the actual DC load voltage and the percentage of the ripple reaching the load (relative to the initial 100-mV ripple) for the circuit of Example 7.10 if $r_Z = 1\ \Omega$.

Signal-Processing Applications

Among the numerous applications of diodes, there are a number of interesting signal-conditioning or signal-processing applications that are made possible by the nonlinear nature of the device. We explore three such applications here: the **diode limiter,** or **clipper;** the **diode clamp;** and the **peak detector.** Other applications are left for the homework problems.

The Diode Clipper (Limiter)

The *diode clipper* is a relatively simple diode circuit that is often employed to protect loads against excessive voltages. The objective of the clipper circuit is to keep the load voltage within a range—say, $-V_{\max} \le v_L(t) \le V_{\max}$—so that the maximum allowable load voltage (or power) is never exceeded. The circuit of Figure 7.58 accomplishes this goal.

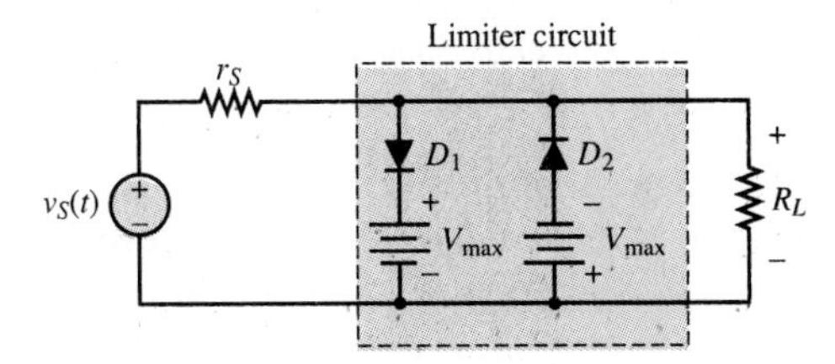

Figure 7.58 Two-sided diode clipper

The circuit of Figure 7.58 is most easily analyzed by first considering just the branch containing D_1. This corresponds to clipping only the positive peak voltages; the analysis of the negative voltage limiter is left as a drill exercise. The circuit containing the D_1 branch is sketched in Figure 7.59; note that we have exchanged the location of the D_1 branch and that of the load branch for convenience. Further, the circuit is reduced to Thévenin equivalent form. Having reduced the circuit to a simpler form, we can now analyze its operation for two distinct cases: the ideal diode and the piecewise linear diode.

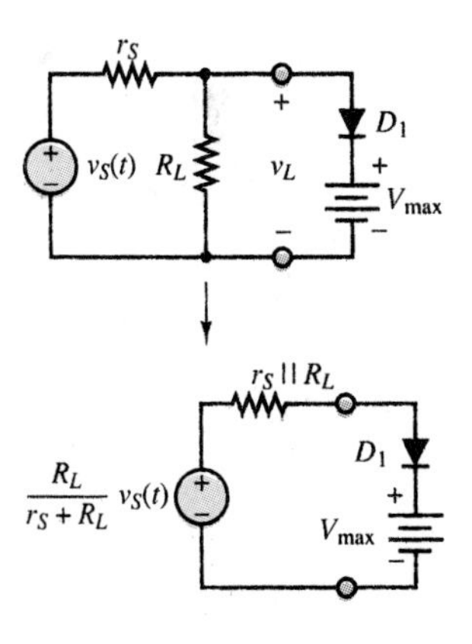

Figure 7.59 Circuit model for the diode clipper

1. Ideal diode model For the ideal diode case, we see immediately that D_1 conducts if

$$\frac{R_L}{r_S + R_L} v_S(t) \ge V_{\max} \tag{7.25}$$

and that if this condition occurs, then (D_1 being a short circuit) the load voltage, v_L, becomes equal to $V_{\max}$. The equivalent circuit for the "on" condition is shown in Figure 7.60.

If, on the other hand, the source voltage is such that

$$\frac{R_L}{r_S + R_L} v_S(t) < V_{\max} \tag{7.26}$$

then D_1 is an open circuit and the load voltage is simply

$$v_L(t) = \frac{R_L}{r_S + R_L} v_S(t) \tag{7.27}$$

The equivalent circuit for this case is depicted in Figure 7.61.

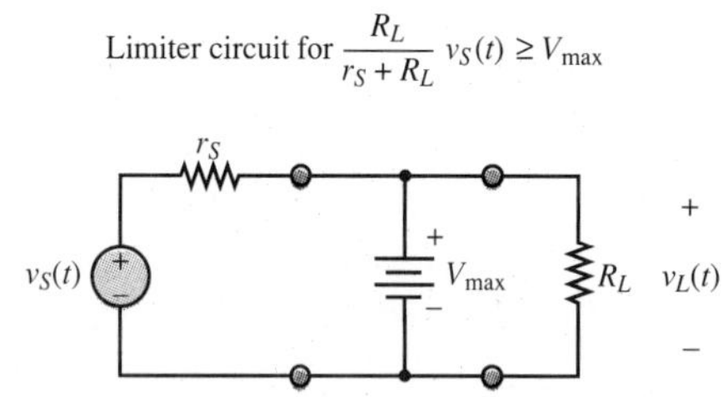

Limiter circuit for $\frac{R_L}{r_S + R_L} v_S(t) \geq V_{max}$

r_S $v_S(t)$ + − V_{max} R_L + $v_L(t)$ −

Figure 7.60 Equivalent circuit for the one-sided limiter (diode on)

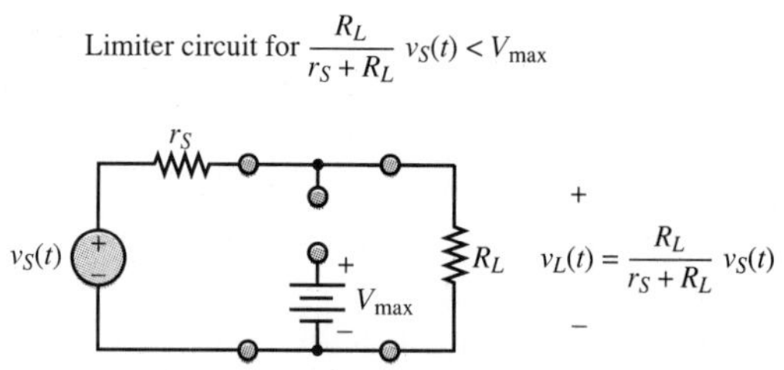

Limiter circuit for $\frac{R_L}{r_S + R_L} v_S(t) < V_{max}$

r_S $v_S(t)$ + − V_{max} R_L + $v_L(t) = \frac{R_L}{r_S + R_L} v_S(t)$ −

Figure 7.61 Equivalent circuit for the one-sided limiter (diode off)

The analysis for the negative branch of the circuit of Figure 7.58 can be conducted by analogy with the preceding derivation, resulting in the waveform for the two-sided clipper shown in Figure 7.62. Note how the load voltage is drastically "clipped" by the limiter in the waveform of Figure 7.62. In reality, such hard clipping does not occur, because the actual diode characteristic does not have the sharp on-off breakpoint the ideal diode model implies. One can develop a reasonable representation of the operation of a physical diode limiter by using the piecewise linear model.

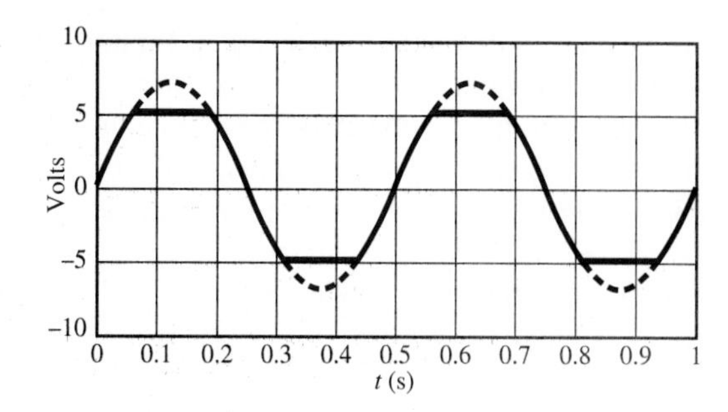

Figure 7.62 Two-sided (ideal diode) clipper input and output voltages

2. Piecewise linear diode model To avoid unnecessary complexity in the analysis, assume that V_{max} is much greater than the diode offset voltage, and therefore assume that $V_\gamma \approx 0$. We do, however, consider the finite diode resistance r_D. The circuit of Figure 7.59 still applies, and thus the determination of the diode on-off state is still based on whether $\frac{R_L}{r_S+R_L}v_S(t)$ is greater or less than V_{max}. When D_1 is open, the load voltage is still given by

$$v_L(t) = \frac{R_L}{r_S + R_L}v_S(t) \tag{7.28}$$

When D_1 is conducting, however, the corresponding circuit is as shown in Figure 7.63.

The effect of finite diode resistance on the limiter circuit.

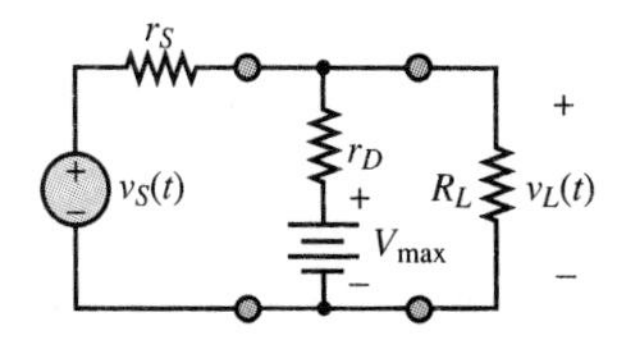

Figure 7.63 Circuit model for the diode clipper (piecewise linear diode model)

The primary effect the diode resistance has on the load waveform is that some of the source voltage will reach the load even when the diode is conducting. This is most easily verified by applying superposition; it can be readily shown that the load voltage is now composed of two parts, one due to the voltage V_{max}, and one proportional to $v_S(t)$:

$$v_L(t) = \frac{R_L \parallel r_S}{r_D + (R_L \parallel r_S)}V_{max} + \frac{r_D \parallel R_L}{r_S + (r_D \parallel R_L)}v_S(t) \tag{7.29}$$

It may easily be verified that as $r_D \to 0$, the expression for $v_L(t)$ is the same as for the ideal diode case. The effect of the diode resistance on the limiter circuit is depicted in Figure 7.64. Note how the clipping has a softer, more rounded appearance.

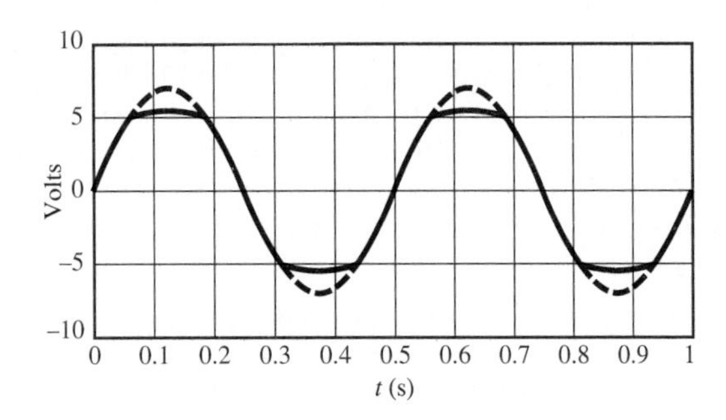

Figure 7.64 Voltages for the diode clipper (piecewise linear diode model)

The Diode Peak Detector

Another common application of semiconductor diodes, the *peak detector*, is very similar in appearance to the half-wave rectifier with capacitive filtering described in an earlier section. One of its more classic applications is in the demodulation of amplitude-modulated (AM) signals. We study this circuit in the following example.

Example 7.11 Peak Detector Circuit

In Chapter 4, a capacitive displacement transducer was introduced. It took the form of a parallel-plate capacitor composed of a fixed plate and a movable plate. The capacitance of this variable capacitor was shown to be a function of displacement; that is, it was shown that a movable-plate capacitor can serve as a linear transducer. Recall the expression derived in Example 4.19:

$$C = \frac{8.854 \times 10^{-3} A}{x}$$

where C is the capacitance in pF, A is the area of the plates in mm^2, and x is the (variable) distance in mm. If the capacitor is placed in an AC circuit, its impedance will be determined by the expression

$$Z_C = \frac{1}{j\omega C}$$

so that

$$Z_C = \frac{x}{j\omega 8.854 \times 10^{-3} A}$$

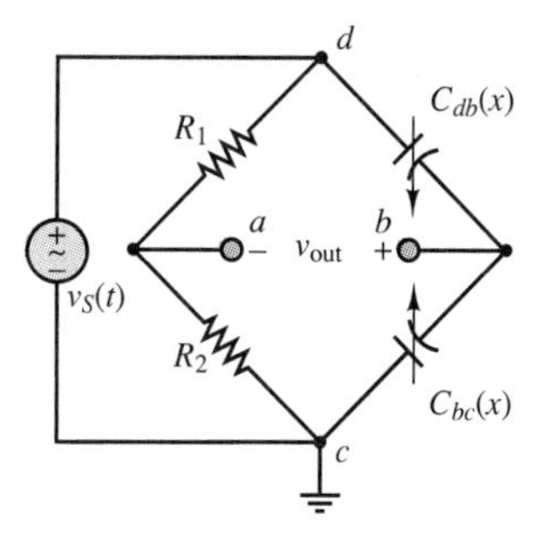

Figure 7.65 Bridge circuit for displacement transducer

Thus, at a fixed frequency ω, the impedance of the capacitor will vary linearly with displacement. This property may be exploited in the bridge circuit of Figure 7.65, where a differential-pressure transducer is shown made of two movable-plate capacitors. If the capacitance of one of these capacitors increases as a consequence of a pressure differential across the transducer, the capacitance of the other must decrease by a corresponding amount, at least for small displacements (you may wish to refer to Example 4.4 for a picture of this transducer). The bridge is excited by a sinusoidal source.

Using phasor notation, Example 4.19 showed that the output voltage of the bridge circuit is given by

$$\mathbf{V}_{out}(j\omega) = \mathbf{V}_S(j\omega)\frac{x}{2d}$$

provided that $R_1 = R_2$. Thus, the output voltage will vary as a scaled version of the input voltage in proportion to the displacement. A typical $v_{out}(t)$ is displayed in Figure 7.66 for a 0.05-mm "triangle" diaphragm displacement, with $d = 0.5$ mm and $\mathbf{V}_S$ a 50-Hz sinusoid with 1-V amplitude. Clearly, although the output voltage is a function of the displacement, x, it is not in a convenient form, since the displacement is proportional to the amplitude of the sinusoidal peaks.

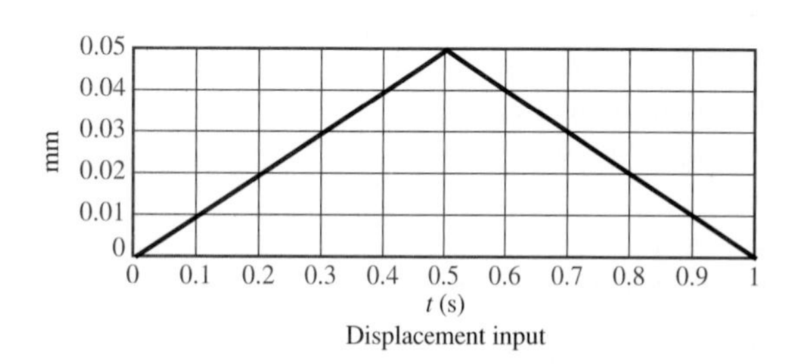

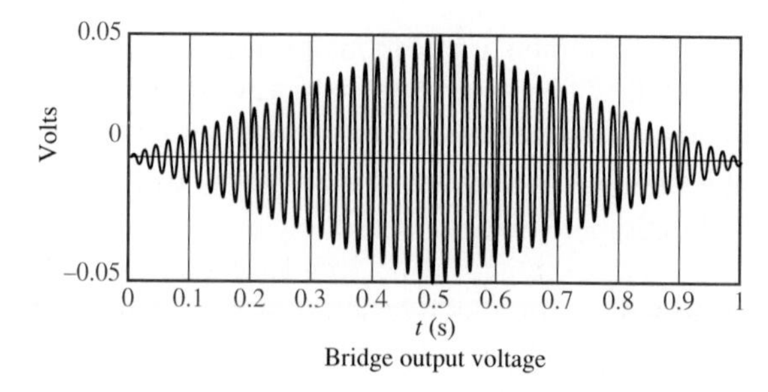

Figure 7.66 Displacement and bridge output voltage waveforms

The diode peak detector is a circuit capable of tracking the sinusoidal peaks without exhibiting the oscillations of the bridge output voltage. The peak detector operates by rectifying and filtering the bridge output in a manner similar to that of the circuit of Figure 7.47. The ideal peak detector circuit is shown in Figure 7.67, and the response of a practical peak detector is shown in Figure 7.68. Its operation is based on the rectification property of the diode, coupled with the filtering effect of the shunt capacitor, which acts as a low-pass filter.

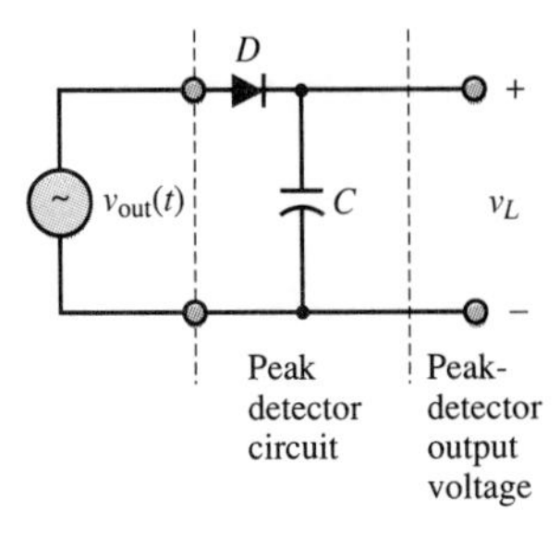

Figure 7.67 Peak detector circuit

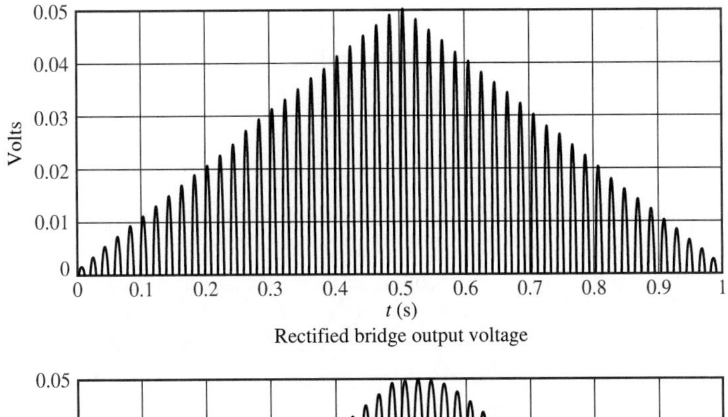

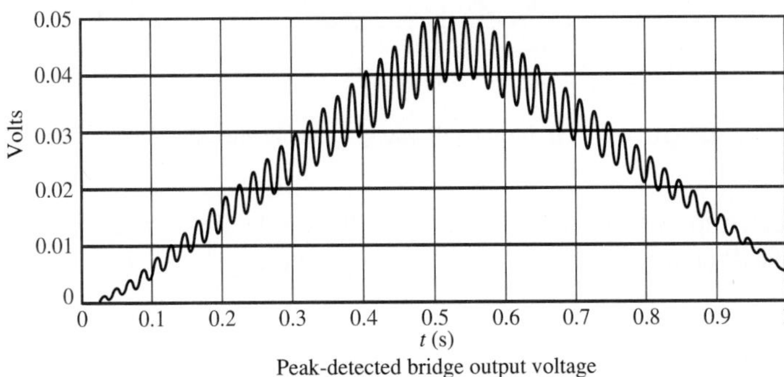

Figure 7.68 Rectified and peak-detected bridge output voltage waveforms

The Diode Clamp

Another circuit that finds common application is the *diode clamp*, which permits "clamping" a waveform to a fixed DC value. Figure 7.69 depicts two different types of clamp circuits.

The operation of the simple clamp circuit is based on the notion that the diode will conduct current only in the forward direction, and that therefore the capacitor will charge during the positive half-cycle of $v_S(t)$ but will not discharge during the negative half-cycle. Thus, the capacitor will eventually charge up to the peak voltage of $v_S(t)$, V_{peak}. The DC voltage across the capacitor has the effect

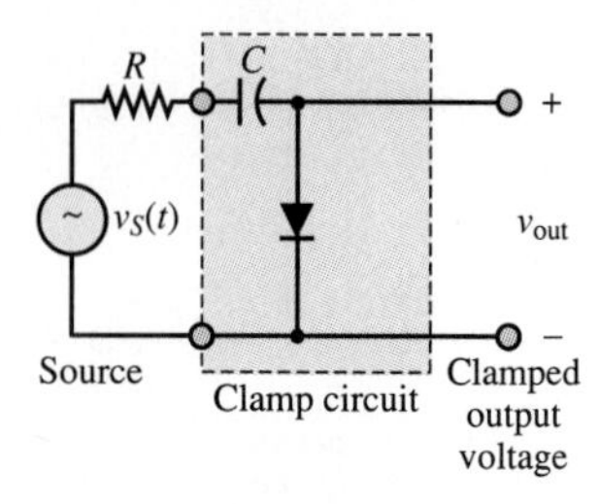

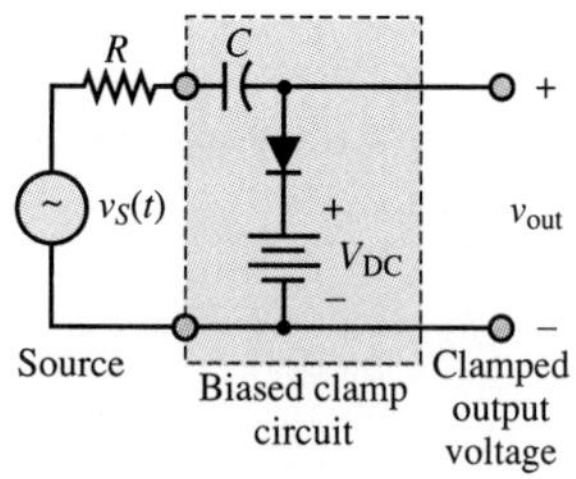

Figure 7.69 Diode clamp circuits

of shifting the source waveform down by V_{peak}, so that, after the initial transient period, the output voltage is

$$v_{out}(t) = v_S(t) - V_{peak} \tag{7.30}$$

and the positive peaks of $v_S(t)$ are now clamped at 0 V. For equation 7.30 to be accurate, it is important that the RC time constant be greater than the period, T, of $v_S(t)$:

$$RC \gg T \tag{7.31}$$

Figure 7.70 depicts the behavior of the diode clamp for a sinusoidal input waveform, where the dashed line is the source voltage and the solid line represents the clamped voltage.

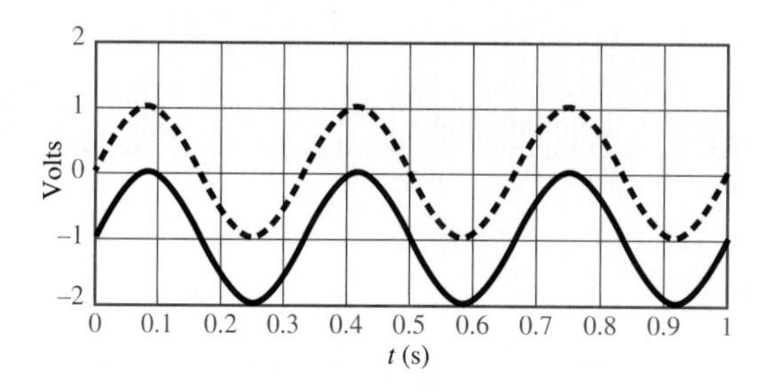

Figure 7.70 Ideal diode clamp input and output voltages

The clamp circuit can also work with the diode in the reverse direction; the capacitor will charge to $-V_{peak}$ with the output voltage given by

$$v_{out}(t) = v_S(t) + V_{peak} \tag{7.32}$$

Now the output voltage has its negative peaks clamped to zero, since the entire waveform is shifted upward by V_{peak} volts. Note that in either case, the diode clamp has the effect of introducing a DC component in a waveform that does not originally have one. It is also possible to shift the input waveform by a voltage different from V_{peak} by connecting a battery, V_{DC}, in series with the diode, provided that

$$V_{DC} < V_{peak} \tag{7.33}$$

The resulting circuit is called a *biased diode clamp*; it is discussed in Example 7.12.

Example 7.12 Biased Diode Clamp

In this example, we design a biased diode clamp with the aim of clamping the signal

$$v_S(t) = 5 \sin(\omega t)$$

to a 3-VDC level (i.e., shifting the average value of the signal from 0 V to 3 V). Consider the biased clamp circuit of Figure 7.69. Once the capacitor has charged to $V_{peak} - V_{DC}$, the output voltage will be given by

$$v_{out} = v_S(t) - V_{peak} + V_{DC}$$

But since V_{DC} must be smaller than V_{peak} (otherwise the diode would never conduct!), this circuit would never permit raising the DC level of v_{out}. To resolve this problem, we must invert both the diode and the battery, as shown in the circuit of Figure 7.71. Now the output voltage is given by

$$v_{out} = v_S(t) + V_{peak} - V_{DC}$$

and, to have a DC level of 3 V, we choose $V_{DC} = 2$ V. The resulting waveforms are shown in Figure 7.72.

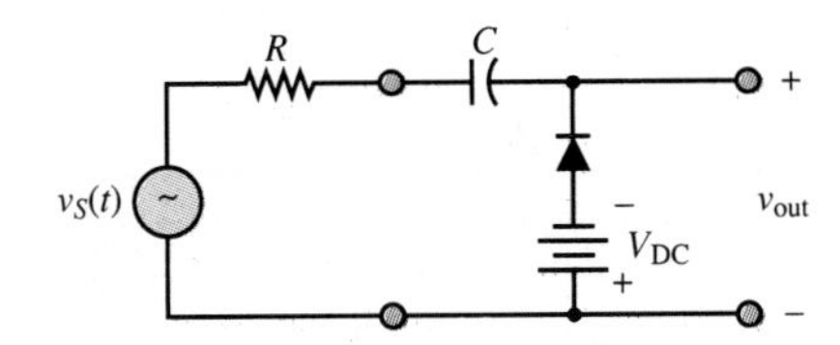

Figure 7.71

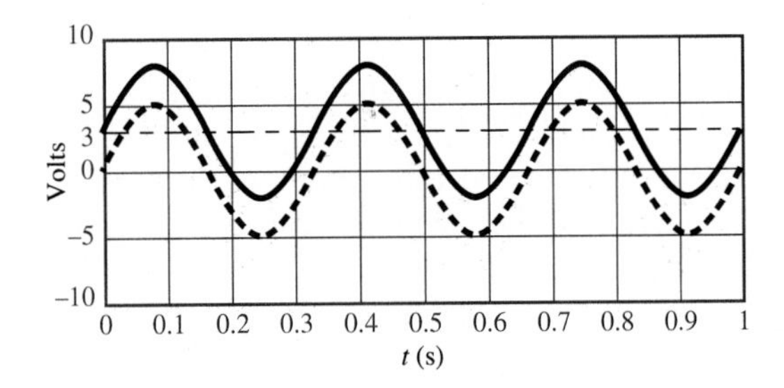

Figure 7.72

Photodiodes

Another property of semiconductor materials that finds common application in measurement systems is their response to light energy. In appropriately fabricated diodes, called **photodiodes,** when light reaches the depletion region of a *pn* junction, photons cause hole-electron pairs to be generated by a process called *photo-ionization*. This effect can be achieved by using a surface material that is transparent to light. As a consequence, the reverse saturation current depends on the light intensity (i.e., on the number of incident photons), in addition to the other factors mentioned earlier, in Section 7.2. In a photodiode, the reverse current is given by $-(I_0 + I_p)$, where I_p is the additional current generated by photo-ionization. The result is depicted in the family of curves of Figure 7.73, where the diode characteristic is shifted downward by an amount related to the additional current generated by photo-ionization. Figure 7.73 depicts the appearance of the i-v characteristic of a photodiode for various values of I_p, where the i-v curve is shifted to lower values for progressively larger values of I_p. The circuit symbol is depicted in Figure 7.74.

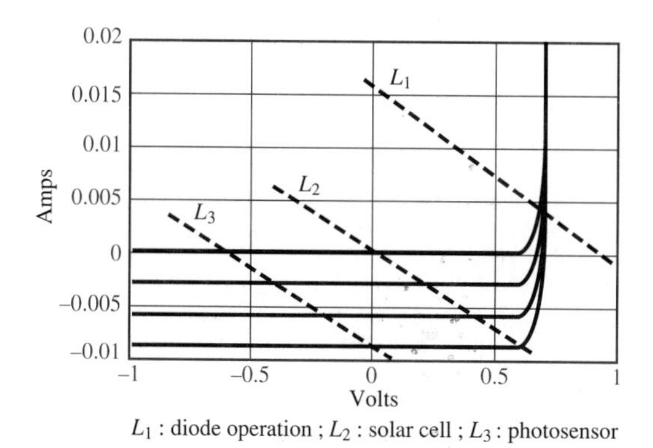

Figure 7.73 Photodiode i-v curves

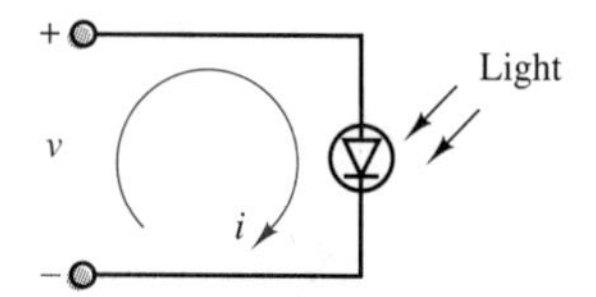

Figure 7.74 Photodiode circuit symbol

Also displayed in Figure 7.73 are three load lines, which depict the three modes of operation of a photodiode. Curve L_1 represents normal diode operation, under forward bias. Note that the operating point of the device is in the positive i, positive v (first) quadrant of the i-v plane; thus, the diode dissipates positive power in this mode, and is therefore a passive device, as we already know. On the other hand, load line L_2 represents operation of the photodiode as a **solar cell;** in this mode, the operating point is in the negative i, positive v, or fourth, quadrant, and therefore the power dissipated by the diode is *negative*. In other words, the photodiode is generating power by converting light energy to electrical energy. Note further that the load line intersects the voltage axis at zero, meaning that no supply voltage is required to bias the photodiode in the solar-cell mode. Finally, load line L_3 represents the operation of the diode as a light sensor: when the diode is reverse-biased, the current flowing through the diode is determined by the light intensity; thus, the diode current changes in response to changes in the incident light intensity.

The operation of the photodiode can also be reversed by forward-biasing the diode and causing a significant level of recombination to take place in the depletion region. Some of the energy released is converted to light energy by emission of photons. Thus, a diode operating in this mode emits light when forward-biased. Photodiodes used in this way are called **light-emitting diodes (LEDs);** they exhibit a forward (offset) voltage of 1 to 2 volts. The circuit symbol for the LED is shown in Figure 7.75.

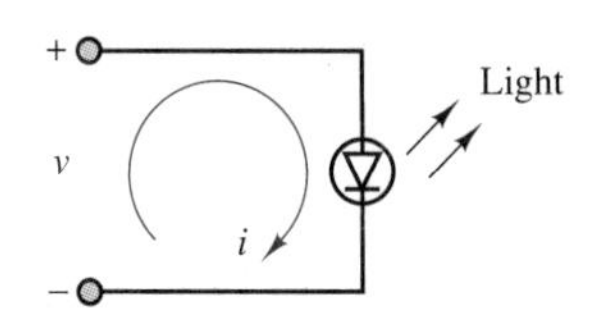

Figure 7.75 Light-emitting diode (LED) circuit symbol

Gallium arsenide (GaAs) is one of the more popular substrates for creating LEDs; gallium phosphide (GaP) and the alloy $GaAs_{1-x}P_x$ are also quite common. Table 7.1 lists combinations of materials and dopants used for common LEDs and the colors they emit. The dopants are used to create the necessary *pn* junction.

Table 7.1 **LED materials and wavelengths**

Material	Dopant	Wavelength	Color
GaAs	Zn	900 nm	Infrared
GaAs	Si	910–1,020 nm	Infrared
GaP	N	570 nm	Green
GaP	N	590 nm	Yellow
GaP	Zn, O	700 nm	Red
$GaAs_{0.6}P_{0.4}$		650 nm	Red
$GaAs_{0.35}P_{0.65}$	N	632 nm	Orange
$GaAs_{0.15}P_{0.85}$	N	589 nm	Yellow

The construction of a typical LED is shown in Figure 7.76, along with the schematic representation for an LED. A shallow *pn* junction is created with electrical contacts made to both *p* and *n* regions. As much of the upper surface of the *p* material is uncovered as possible, so that light can leave the device unimpeded. It is important to note that, actually, only a relatively small fraction of the emitted light leaves the device; the majority stays inside the semiconductor. A photon that stays inside the device will eventually collide with an electron in the valence band, and the collision will force the electron into the conduction band, emitting an electron-hole pair and absorbing the photon. To minimize the probability that a photon will be absorbed before it has an opportunity to leave the LED, the depth of the *p*-doped region is left very thin. Also, it is advantageous to have most of the recombinations that emit photons occur as close to the surface of the diode as possible. This is made possible by various doping schemes, but even so, of all of the carriers going through the diode, only a small fraction emit photons that are able to leave the semiconductor.

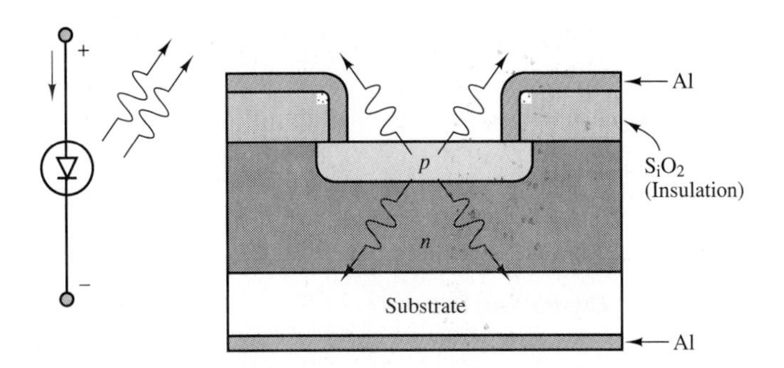

Figure 7.76 Light-emitting diode (LED)

A simple LED drive circuit is shown in Figure 7.77. From the standpoint of circuit analysis, LED characteristics are very similar to those of the silicon diode, except that the offset voltage is usually quite a bit larger. Typical values of V_γ can be in the range of 1.2 to 2 volts, and operating currents can range from 20 mA to 100 mA. Manufacturers usually specify an LED's characteristics by giving the rated operating-point current and voltage.

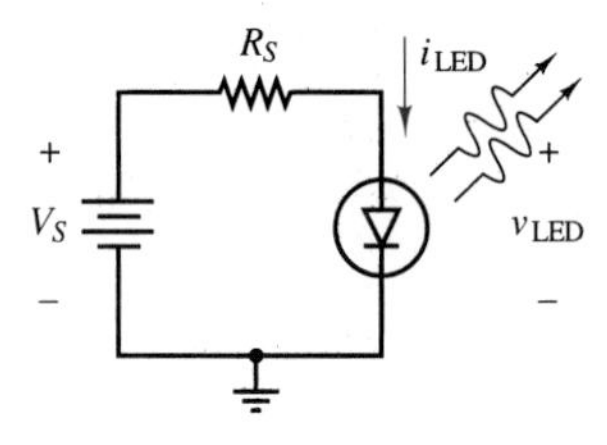

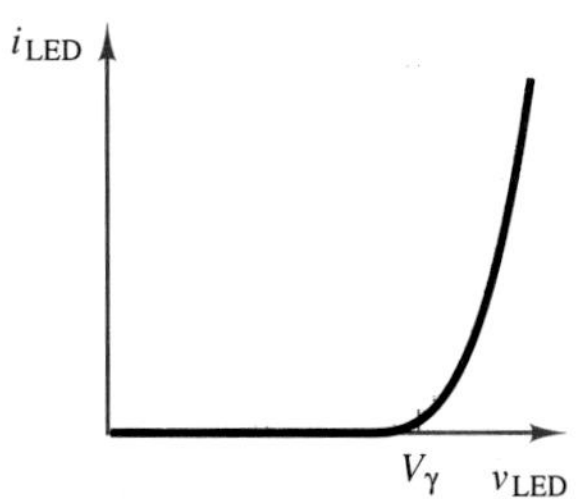

Figure 7.77 LED drive circuit and i-v characteristic

Example 7.13

The manufacturer of a certain LED suggests an operating point of 1.7 V and 40 mA.

1. What is the total power consumption of the LED when operating?
2. If the diode is to be used in the circuit of Figure 7.77 and $V_S = 5$ V, find the resistance R_S.
3. How much power must the voltage source be able to provide?

Solution:

1. The power consumption of the LED can be determined directly from the specifications:

$$P_{LED} = V_{LED} I_{LED} = (1.7)(40 \times 10^{-3}) = 68 \text{ mW}$$

2. Writing KVL around the circuit of Figure 7.77, we obtain the equation

$$V_S = I_{LED}(R_S) + V_{LED}$$

or

$$R_S = \frac{V_S - V_{LED}}{I_{LED}} = \frac{(5 - 1.7)}{40 \times 10^{-3}}$$

from which the required value of R_S may be computed:

$$R_S = 82.5 \ \Omega$$

3. To satisfy the power requirements of the circuit, the source must be able to provide 40 mA of current to both the resistor and the diode. Thus, the required power is

$$P_S = V_S I_{LED} = 5(40 \times 10^{-3})$$
$$= 200 \text{ mW}$$

Example 7.14 Opto-Isolators

One of the common applications of photodiodes and LEDs is the **opto-coupler,** or **opto-isolator.** This device, which is usually enclosed in a sealed package, uses the light-to-current and current-to-light conversion property of photodiodes and LEDs to provide signal connection between two circuits without any need for electrical connections. Figure 7.78 depicts the circuit symbol for the opto-isolator.

Because diodes are nonlinear devices, the opto-isolator is not used in transmitting analog signals: the signals would be distorted because of the nonlinear diode *i-v* characteristic. However, opto-isolators find a very important application when on-off signals need to be transmitted from high-power machinery to delicate computer control circuitry. The optical interface ensures that potentially damaging large currents cannot reach delicate instrumentation and computer circuits.

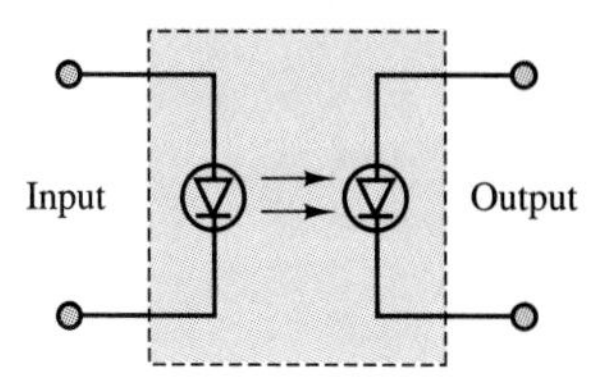

Figure 7.78 Opto-isolator

Example 7.15 Solar-Cell Array

As explained earlier in this section, a solar cell is a device that operates on the same principle as that of an LED. Instead of applying a forward bias to the terminals of an LED and generating a light in response to the diode current, the solar cell receives energy from light and converts this energy into an electrical current. The *i-v* characteristic of a solar (or photovoltaic) cell is shown in Figure 7.79.

Note that the curve has the same shape as the diode curve but, instead of passing through the origin, is shifted down by the amount I_p. I_p is the strength of the current generated by the incident light on the cell. We can express the current and voltage relationship of the solar cell using the diode equation:

$$i_D = I_0(e^{qv_D/kT} - 1) - I_p$$

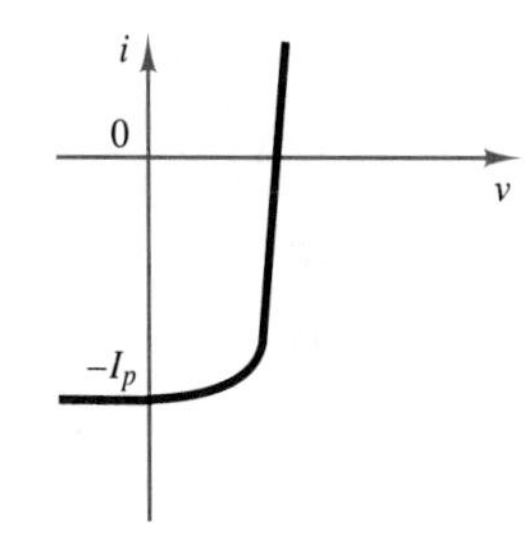

Figure 7.79 *i-v* curve of a solar cell

To increase the power output to a level sufficient for practical applications, solar cells can be connected in series (to increase output voltage) and in parallel (to increase output current). Such arrangements are commonly called **solar arrays.** The solar cell to be used in this example has the following characteristics when illuminated: $I_p = 20$ mA, $I_0 = 50$ pA, $R_S = 0.7\ \Omega$. The array has the minimum output requirements

$$V_{\text{out}} = 10 \text{ V} \qquad \text{when} \qquad I = 0.5 \text{ A}$$

Design an array that will meet these requirements, using the circuit model for a solar cell shown in Figure 7.80.

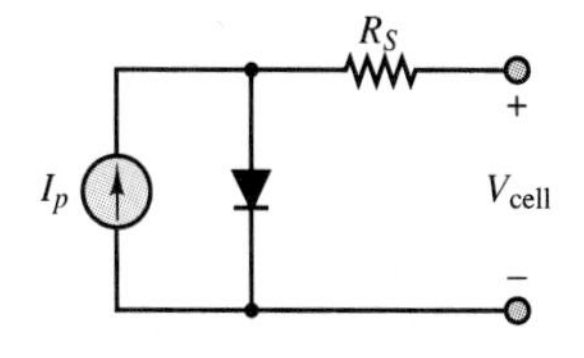

Figure 7.80 Circuit model for solar cell

Solution:

The array must be able to generate 0.5 A. Each photovoltaic cell can, at a maximum, generate 20 mA. This would seem to indicate that we must arrange the cells in parallel to increase the total current output. To find the minimum number, n_{min}, of branches needed to output the required current, we use the following expression:

$$n_{\text{min}} \times 20 \text{ mA} = 0.5 \text{ A}$$

$$n_{\text{min}} = \frac{0.5}{0.02} = 25$$

Since a total output voltage of 10 V is required, each parallel branch must consist of a series connection of m photovoltaic cells. The required number of series cells in each branch, m_{min}, is found by determining the open-circuit voltage of a cell from the diode equation:

$$I = I_0(e^{qv_D/kT} - 1) - I_p$$

Since the current is zero under open-circuit conditions,

$$I_p = I_0(e^{qv_{OC}/kT} - 1)$$

and

$$v_{OC} = \left[\log_e\left(\frac{I_p}{I_0} + 1\right)\right]\frac{kT}{q}$$
$$= \left[\log_e\left(\frac{20\text{ mA}}{50\text{ pA}} + 1\right)\right]0.025$$

Thus, we can compute the open-circuit voltage to be

$$v_{OC} = 0.495\text{ V}$$

and since

$$m_{\min} v_{OC} = 10\text{ V}$$

we conclude that, at least nominally,

$$m_{\min} = \frac{10\text{ V}}{0.495\text{ V}} = 20.195 \approx 21$$

That is, 21 solar cells will be needed in each branch.

The next step in the design process is to verify whether 25 branches of 21 cells each are sufficient to satisfy the original requirements of the problem. The appearance of the array is depicted in Figure 7.81. It should be pointed out that, because the open-circuit voltage and short-circuit current were used in the initial calculations, it is likely that the preliminary calculation is somewhat optimistic. Figure 7.82 depicts the *i-v* curve of one solar cell, clearly showing that the nominal values of short-circuit current (20 mA)

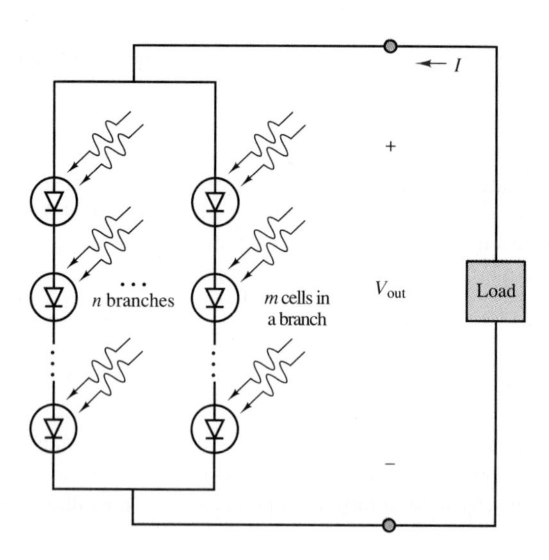

Figure 7.81 Solar array

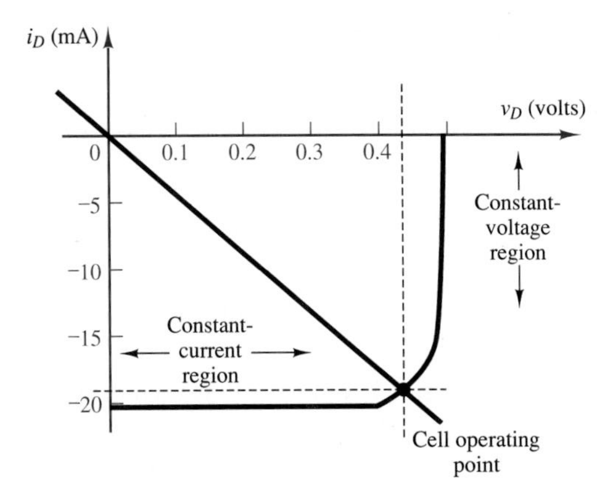

Figure 7.82 *i-v* curve of a solar cell

and open-circuit voltage (0.495 V) cannot be achieved simultaneously, but that a compromise must be made, as represented by the operating point shown in the picture. To clarify this point further, consider that the cell output voltage can be at a maximum 0.4 V if the cell current is 20 mA. To determine an acceptable compromise, let us assume that one more branch is added to the array, so that now each branch must supply $\frac{1}{26}$ of the rated current, or

$$I_{\text{branch}} = \frac{0.5 \text{ A}}{n_{\min} + 1} = \frac{0.5 \text{ A}}{26} = 19.23 \text{ mA}$$

The voltage supplied by a cell at this current level is then given by the expression

$$I_{\text{cell}} = I_0(e^{q(V_{\text{cell}} - IR_S)/kT} - 1) - I_p$$

or

$$V_{\text{cell}} = IR_S + \frac{kT}{q}\left[\log_e\left(\frac{I_p + I_{\text{cell}}}{I_0} + 1\right)\right]$$

where I is referenced as entering the cell (I is a negative value, since the cell is supplying the power to the load).

Thus,

$$V_{\text{cell}} = (-19.23 \text{ mA})0.7 + 0.025 \log_e\left(\frac{20 \text{ mA} - 19.23 \text{ mA}}{50 \text{ pA}} + 1\right) = 0.400 \text{ V}$$

and the number of series cells required to obtain the desired 10-V output rating is given by

$$m = \frac{10}{0.400} = 25$$

Thus, the array will consist of 26 branches, each containing 25 cells in series, or a total of 650 cells. Note that this is not the only solution to the problem. For example, it may be desirable to operate in either the constant-current or the constant-voltage region of the cell *i-v* curve, since in these regions the design is less sensitive to load variations.

DRILL EXERCISES

7.11 Repeat the analysis of the diode clipper of Figure 7.58 for the branch containing D_2.

7.12 For the one-sided diode clipper of Figure 7.59, find the percentage of the source voltage that reaches the load if $R_L = 150\ \Omega, r_S = 50\ \Omega$, and $r_D = 5\ \Omega$. Assume that the diode is conducting, and use the circuit model of Figure 7.63.

7.13 How would the diode clipper output waveform change if we used the offset diode model instead of the piecewise linear model in the analysis? [*Hint:* Compare Figures 7.26 and 7.38.]

7.14 Determine the value of load resistance that is compatible with the solar array design in Example 7.15. [*Hint:* Consider the characteristic curve in Figure 7.82.]

7.15 Compute the power output of each solar cell in Example 7.15. Would the cell generate more power if operated at the operating point shown, in the constant-voltage region (say, for $I = 12$ mA), or in the constant-current region (say, for $V = 0.3$ V)?

CONCLUSION

- Semiconductor materials have conductive properties that fall between those of conductors and insulators. These properties make such materials useful in the construction of many electronic devices that exhibit nonlinear *i-v* characteristics. Of these devices, the diode is one of the most commonly employed.
- The semiconductor diode acts like a one-way current valve, permitting the flow of current only when biased in the forward direction. Although the behavior of the diode is described by an exponential equation, it is possible to approximate the operation of the diode by means of simple circuit models. The simplest model treats the diode as either a short circuit or an open circuit (the on-off, or ideal, model). The ideal model can be extended to include an offset voltage (usually 0.2 to 0.7 V), which represents the contact potential at the diode junction. A slightly more realistic model, the piecewise linear diode model, accounts for the effects of the diode forward resistance. With the aid of these circuit models it becomes possible to analyze diode circuits using the DC and AC circuit analysis techniques developed in earlier chapters.
- One of the most important properties of the semiconductor diode is that of rectification, which permits the conversion of AC voltages and currents to DC voltages and currents. Diode rectifiers can be of the half-wave type, or they can be full-wave. Full-wave rectifiers can be constructed in a conventional two-diode configuration, or in a bridge configuration. Diode rectifiers are an essential part of DC power supplies and are usually employed in conjunction with filter capacitors to obtain a relatively smooth DC voltage waveform. In addition to rectification and smoothing, it is also necessary to regulate the output of

a DC power supply; Zener diodes accomplish this task by holding a constant voltage when reverse-biased above the Zener voltage.

- In addition to power supply applications, diodes find use in many signal-processing and signal-conditioning circuits. Of these, the diode limiter, peak detector, and clamp have been explored in the chapter. Further, since semiconductor material properties are also affected by light intensity, certain types of diodes, known as *photodiodes*, find application as light detectors, solar cells, or light-emitting diodes.

EIT REVIEW: DIODE APPLICATIONS

Diodes are used in many practical circuits in electric power systems and in signal-processing applications. Chapter 7 has been devoted entirely to diodes; Section 7.3, in particular, discusses circuit models that may be useful in the analysis of diode circuits, while Section 7.4 presents several applications. Drill Exercises 7.16 and 7.17 refer to the circuit of Figure 7.83.

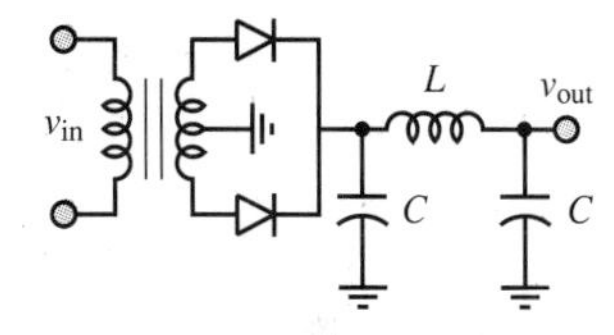

Figure 7.83

DRILL EXERCISES

7.16 The circuit of Figure 7.83 is a

a. Peak detector.
b. Half-wave rectifier.
c. Bridge rectifier.
d. Voltage doubler.
e. Full-wave rectifier.

Solution:

The correct answer is e.

7.17 The inductor L and the capacitor C serve the function of

a. Converting the AC input to DC output.
b. Increasing the peak value of the output voltage.
c. Protecting the diodes.
d. A high-pass filter.
e. Reducing the ripple component of the output voltage.

Solution:

The correct answer is e.

In addition to the preceding two sample problems, you may wish to review the following examples in Chapter 7: 7.1, 7.2, 7.3, 7.7, 7.9, and 7.12. The problems marked "EIT" in the Homework Problems section of Chapter 7 will provide additional practice material. Answers to these problems will be found in Appendix B.

KEY TERMS

Acceptor *352*
Bridge Rectifier *369*
Contact Potential *353*
Covalent Bonding *350*
DC Power Supply *373*
Depletion Region *353*
Diffusion Current *353*
Diode Clamp *379*
Diode Equation *354*
Diode Limiter (Clipper) *379*
Donor *351*
Doping *351*
Forward Biasing *353*
Free Electron *350*
Full-Wave Rectifier *369*
Half-Wave Rectifier *360*
Hole *351*
Ideal Diode Model *356*
Intrinsic Concentration *351*
Light-Emitting Diode (LED) *386*
Limiter *369*
Majority Carrier *352*
Minority Carrier *352*
Mobility *351*
n-Type Semiconductor *352*
Offset Diode Model *360*
Ohmic Contact *353*
Opto-Coupler (Opto-Isolator) *388*
p-type Semiconductor *352*
Peak Detector *379*
Photodiode *385*
Piecewise Linear Diode Model *365*
pn Junction *352*
Quiescent (Operating) Point, or *Q* Point *363*
Recombination *351*
Rectification *359*
Reverse Biasing *353*
Reverse Breakdown *355*
Reverse Saturation Current *353*
Reverse Zener Voltage *373*
Ripple *372*
Semiconductor *350*
Semiconductor Diode *352*
Small-Signal Resistance *363*
Solar Array *389*
Solar Cell *386*
Voltage Regulator *356*
Zener Breakdown *355*
Zener Diode *369*
Zener Voltage *355*

ANSWERS TO DRILL EXERCISES

7.2 16.55 V
7.3 $V_Q = 0.65$ V; $I_Q = 37$ mA
7.4 0.1 Ω
7.5 13.92 V
7.6 Both diodes conduct.
7.7 (a) Neither conducts. (b) Both conduct. (c) Only D_2 conducts. (d) Only D_1 conducts.
7.9 154.36 V
7.10 8.06 V; 2%
7.12 8.8%
7.14 20.8 Ω
7.15 $P_{cell} = 7.69$ mW; at the operating point shown.

DEVICE DATA SHEETS

1N461A through 1N661 General Purpose Diodes

Device No.	Package No.	V_{RRM} (V) Min	I_R (nA) Max	@ V_R (V)	V_F (V) Min	V_F (V) Max	@ I_F (mA)	C (pF) Max	t_{rr} (ns) Max	Test Cond.
1N461A	DO-35	30	500	25		1.0	100			
1N462A	DO-35	70	500	60		1.0	100			
1N463A	DO-35	200	500	175		1.0	100			
1N659	DO-35	60	5000	50		1.0	6.0		300	(Note 1)
1N660	DO-35	120	5000	100		1.0	6.0		300	(Note 1)
1N661	DO-35	240	10000	200		1.0	6.0		300	(Note 1)

Note 1: $V_R = 35$ V, $I_F = 30$ mA, $R_L = 2.0$ kΩ, $C_L = 10$ pF, recovery to 400 kΩ.

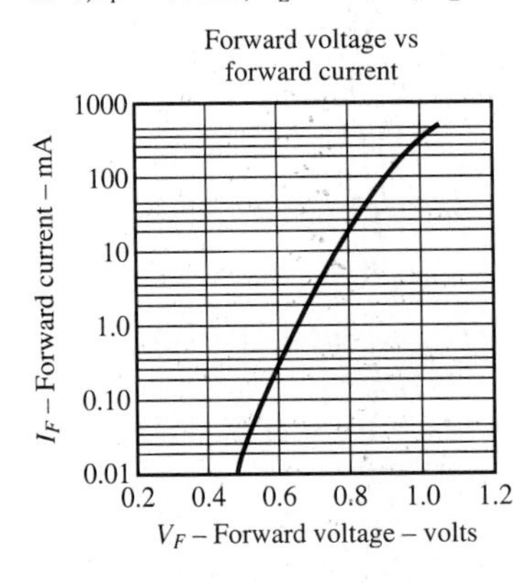

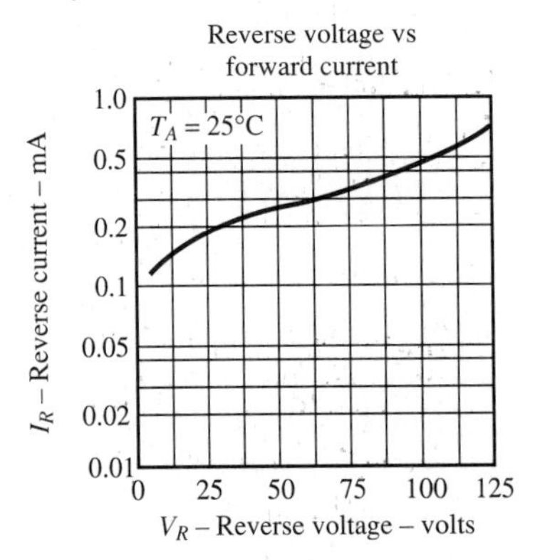

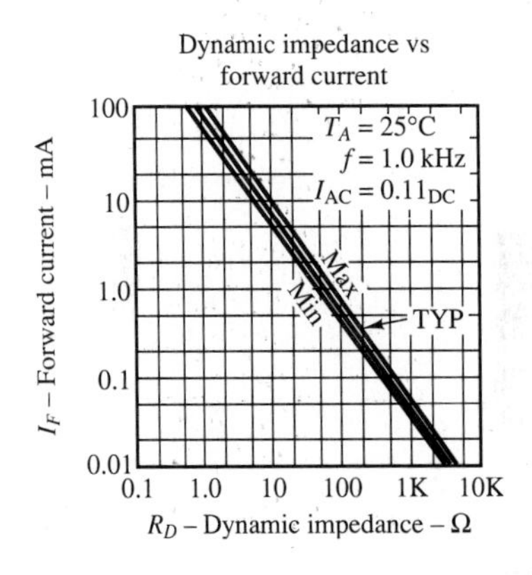

1N4370A through 1N759A Zener Diodes

Electrical Characteristics ($T_A = 25°C$, $V_F = 1.5$ V Max at 200 mA for all types)						
					Maximum Reverse Leakage Current	
Type Number (Note 1)	Nominal Zener Voltage V_Z @ I_{ZT} (Note 2) Volts	Test Current I_{ZT} mA	Maximum Zener Impedance Z_{ZT} @ I_{ZT} (Note 3) Ohms	Maximum DC Zener Current I_{ZM} (Note 4) mA	$T_A = 25°C$ I_R @ $V_R = 1$ V (μA)	$T_A = 150°C$ I_R @ $V_R = 1$ V (μA)
1N4370A	2.4	20	30	150	100	200
1N4371A	2.7	20	30	135	75	150
1N4372A	3	20	29	120	50	100
1N746A	3.3	20	28	110	10	30
1N747A	3.6	20	24	100	10	30
1N748A	3.9	20	23	95	10	30
1N749A	4.3	20	22	85	2	30
1N750A	4.7	20	19	75	2	30
1N751A	5.1	20	17	70	1	20
1N752A	5.6	20	11	65	1	20
1N753A	6.2	20	7	60	0.1	20
1N754A	6.8	20	5	55	0.1	20
1N755A	7.5	20	6	50	0.1	20
1N756A	8.2	20	8	45	0.1	20
1N757A	9.1	20	10	40	0.1	20
1N758A	10	20	17	35	0.1	20
1N759A	12	20	30	30	0.1	20

HOMEWORK PROBLEMS

7.1 Consider the circuit of Figure P7.1. Determine whether the diode is conducting or not. Assume that the diode is an ideal diode.

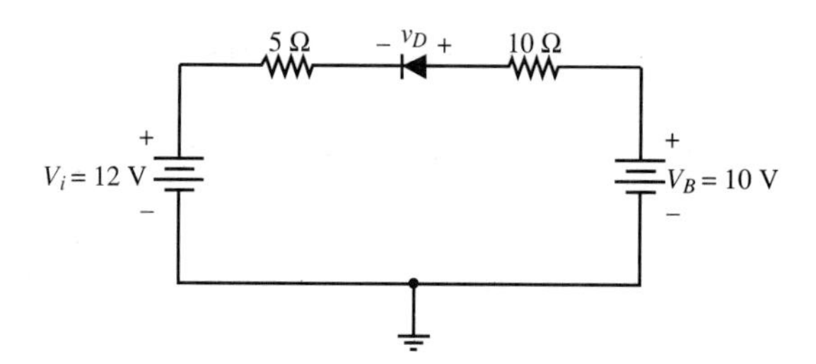

Figure P7.1

7.2 Find voltage v_L in the circuit of Figure P7.2, where D is an ideal diode. Use values of $v_S <$ and > 0.

7.3 In the circuit of Figure P7.2, $v_S = 6$ V and $R_1 = R_S = R_L = 500\ \Omega$. Determine i_D and v_D graphically, using the diode characteristic of the 1N461A.

7.4 Assume that the diode in Figure P7.3 requires a minimum current of 1 mA to be above the knee of its *i*-*v* characteristic.

a. What should be the value of R to establish 5 mA in the circuit?
b. With the value of R determined in part a, what is the minimum value to which the voltage E could be reduced and still maintain diode current above the knee? Use $V_\gamma = 0.7$ V.

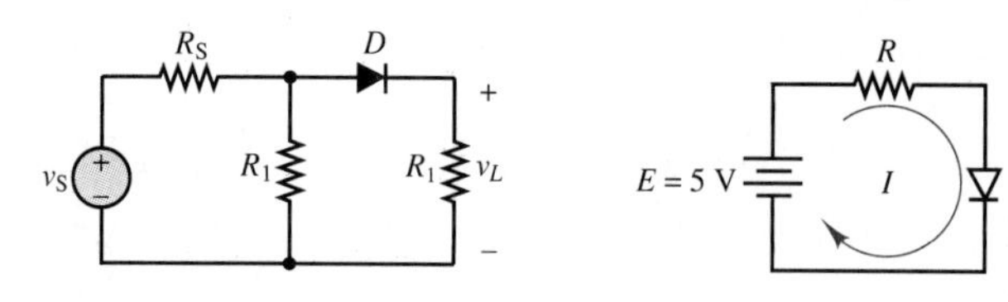

Figure P7.2 **Figure P7.3**

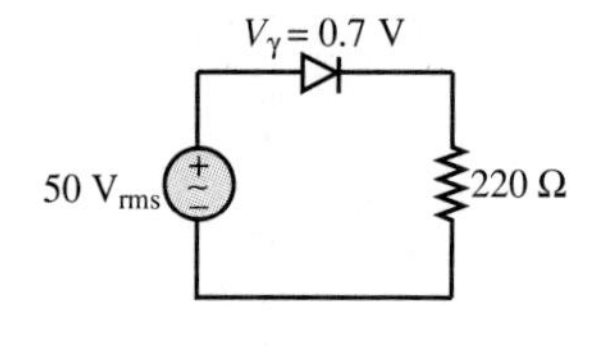

Figure P7.4

7.5 In Figure P7.4, a sinusoidal source of 50 V rms drives the circuit. Use the offset diode model for a silicon diode.

a. What is the maximum forward current?
b. What is the peak inverse voltage across the diode?

7.6 Determine which diodes are forward biased and which are reverse biased in each of the configurations shown in Figure P7.5.

EIT 7.7 For the circuit of Figure P7.6, find the range of V_{in} for which D_1 is forward-biased. Assume an ideal diode.

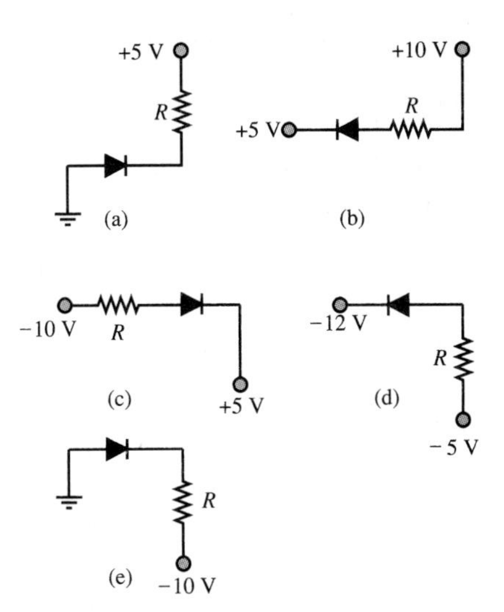

Figure P7.5

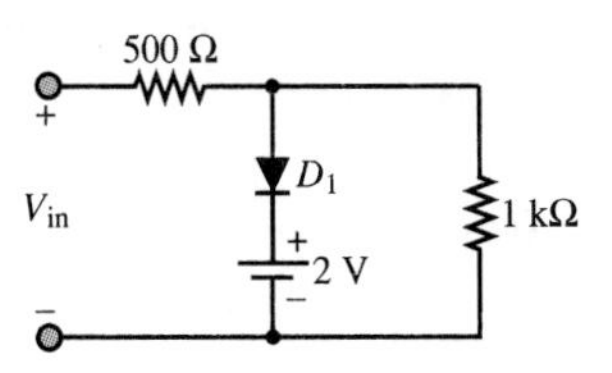

Figure P7.6

7.8 In the circuit of Figure P7.7, find the range of V_{in} for which D_1 is forward-biased. Assume ideal diodes.

7.9 Determine which diodes are forward biased and which are reverse-biased in the configurations shown in Figure P7.8. Assuming a 0.7 V drop across each forward-biased diode, determine the output voltage.

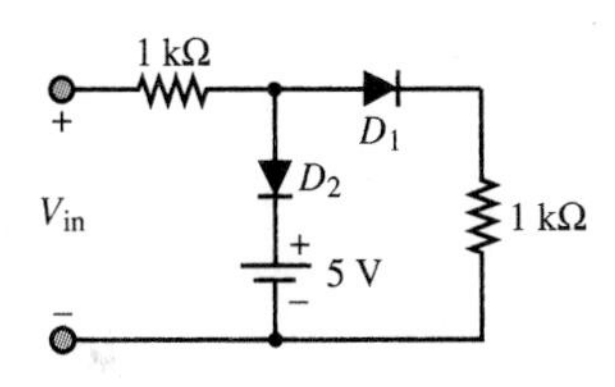

Figure P7.7

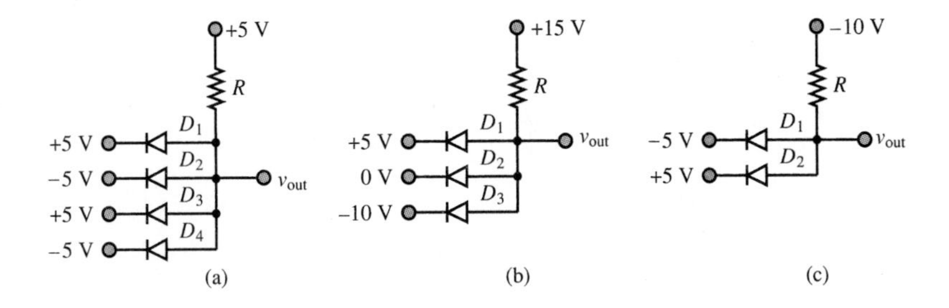

Figure P7.8

7.10 For the circuit of Figure P7.9, find the range of V_{in} for which D_1 is forward-biased. Assume an ideal diode.

EIT 7.11 Find V_o when $V_{in} = 8$ V in the circuit of Figure P7.9.

7.12 For the circuit of Figure P7.10, find the output voltage, V_{out}, if the input voltages V_1 and V_2 are as follows:

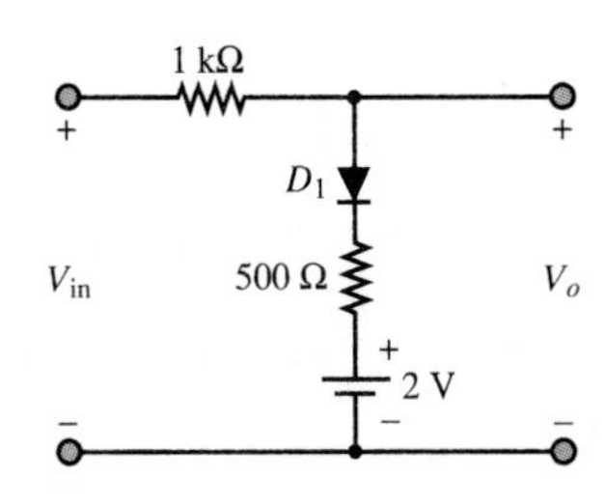

Figure P7.9

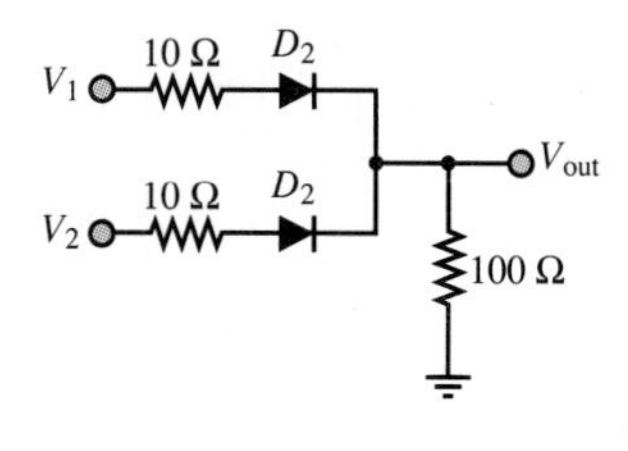

Figure P7.10

a. $V_1 = 0$ V, $V_2 = 0$ V
b. $V_1 = 5$ V, $V_2 = 5$ V
c. $V_1 = 5$ V, $V_2 = 0$ V

Assume ideal diodes.

***7.13** For the circuit of Figure P7.11, $V_{in} = V_0 \sin 2\pi 60t$, and for each diode, $V_\gamma = 0.7$ V when the diode is forward-biased. It is required that $V_o = 5$ V with less than 1 mV of ripple.

a. What is the minimum value R_L can have?
b. What is the value of V_{in} to achieve this specification?
c. If D_1 is destroyed—that is, becomes an open circuit—what happens to V_o?

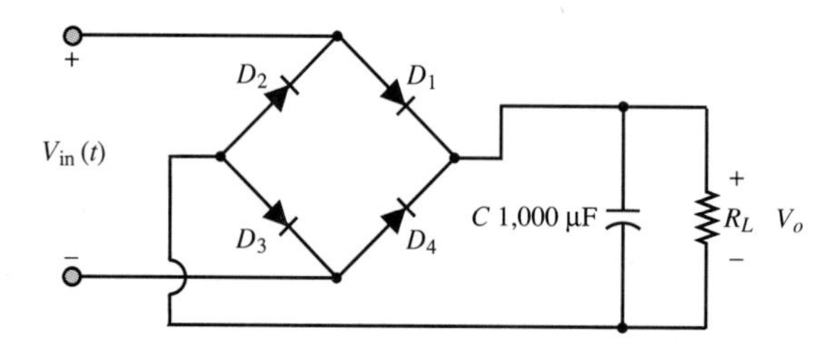

Figure P7.11

EIT 7.14 Find the operating point of the diode for the circuit of Figure P7.12 if the resistance R is

a. 1,000 Ω
b. 5,000 Ω

Find the power dissipated by the diode in each case.

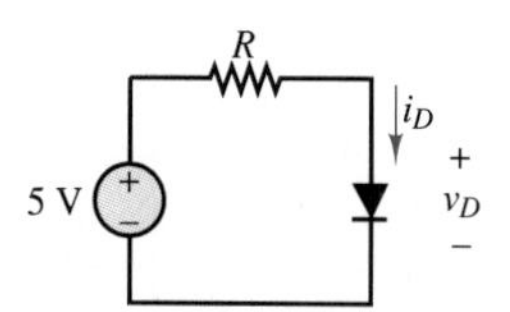

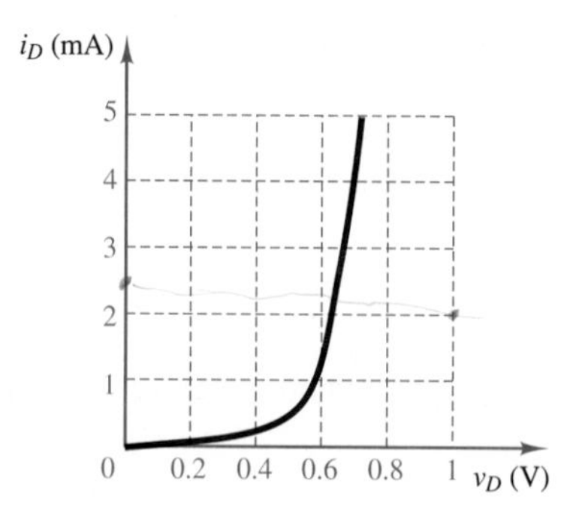

Figure P7.12

7.15 A Zener diode's ideal *i-v* characteristic is shown in Figure P7.13(a). Given a Zener voltage, V_Z of 7.7 V, find the output voltage, V_{out}, for the circuit of Figure P7.13(b) if V_S is

a. 12 V
b. 20 V

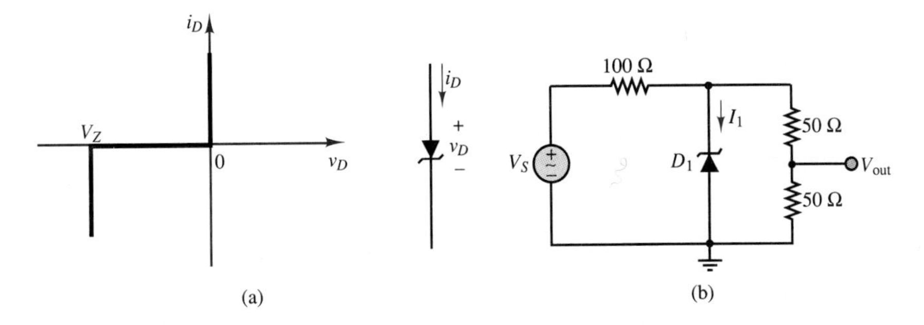

Figure P7.13

7.16 Find the minimum value of R_L in the circuit shown in Figure P7.14 for which the output voltage remains at just 5.6 V.

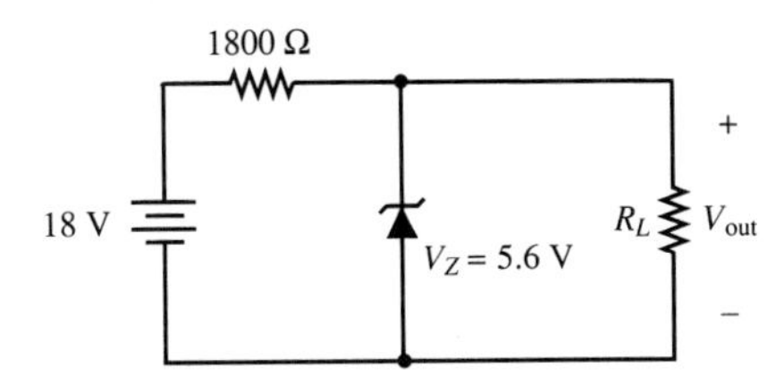

Figure P7.14

7.17 One of the more interesting applications of a diode, based on the diode equation, is an electronic thermometer. The concept is based on the empirical observation that if the current through a diode is nearly constant, the offset voltage is nearly a linear function of temperature, as shown in Figure P7.15(a).

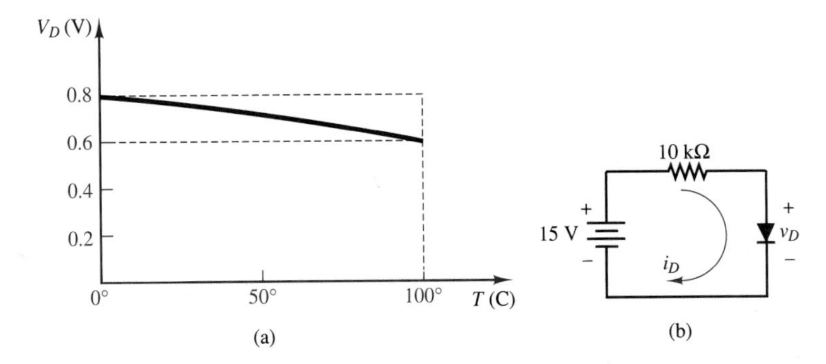

Figure P7.15

a. Show that i_D in the circuit of Figure P7.15(b) is nearly constant in the face of variations in the diode voltage, v_D. This can be done by computing the percent change in i_D for a given percent change in v_D. Assume that v_D changes by 10 percent, from 0.6 to 0.66 V.

b. On the basis of the graph of Figure P7.15(a), write an equation for $v_D(T^\circ)$ of the form

$$v_D = \alpha T^\circ + \beta$$

EIT 7.18 For the circuit of Figure P7.16, sketch $i_D(t)$ for the following conditions:

a. Use the ideal diode model.
b. Use the ideal diode model with offset ($V_\gamma = 0.6$ V).
c. Use the piecewise linear approximation with

$$r_D = 1\ \text{k}\Omega$$
$$V_\gamma = 0.6\ \text{V}$$

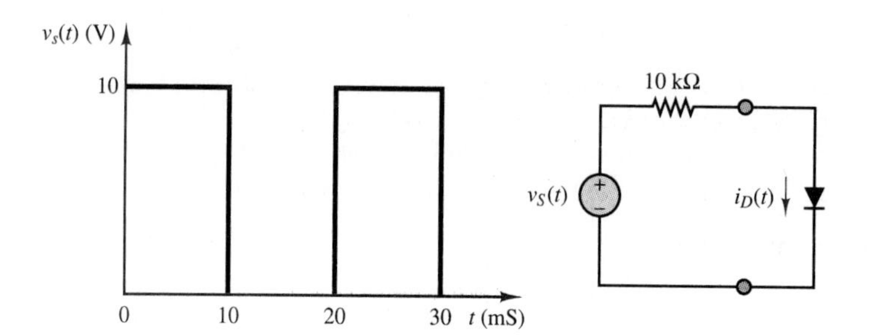

Figure P7.16

7.19 For the Zener regulator shown in the circuit of Figure P7.17, we desire to hold the load voltage to 14 V. Find the range of load resistances for which regulation can be obtained if the Zener diode is rated at 14 V, 5 W.

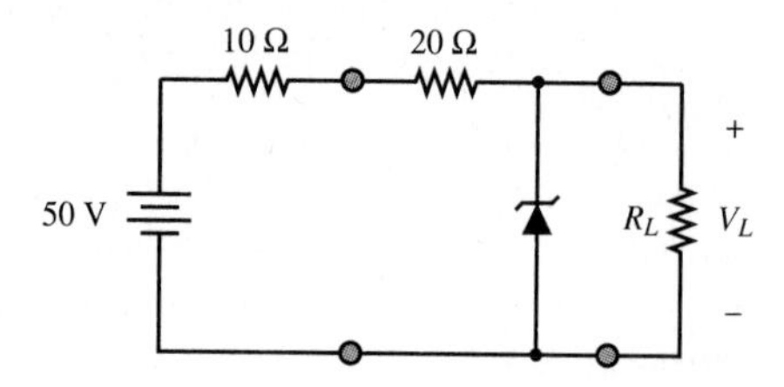

Figure P7.17

EIT 7.20 You have been asked to design a full-wave bridge rectifier for a power supply. A step-down transformer has already been chosen. It will supply 12 V rms to your rectifier. The full-wave rectifier is shown in the circuit of Figure P7.18.

a. If the diodes have an offset voltage of 0.6 V, sketch the input source voltage, $v_S(t)$, and the output voltage, $v_L(t)$, and state which diodes are on and which are off in the appropriate cycles of $v_S(t)$. The frequency of the source is 60 Hz.

b. If $R_L = 1{,}000\ \Omega$ and a capacitor, placed across R_L to provide some filtering, has a value of 8 μF, sketch the output voltage, $v_L(t)$.
c. Repeat part b, with the capacitance equal to 100 μF.

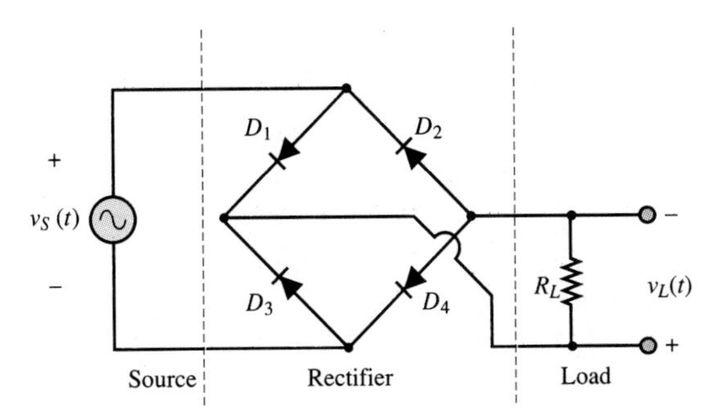

Figure P7.18

7.21 For the peak detector shown in Figure P7.19, find the charging and discharging time constants, given the diode *i-v* characteristic shown.

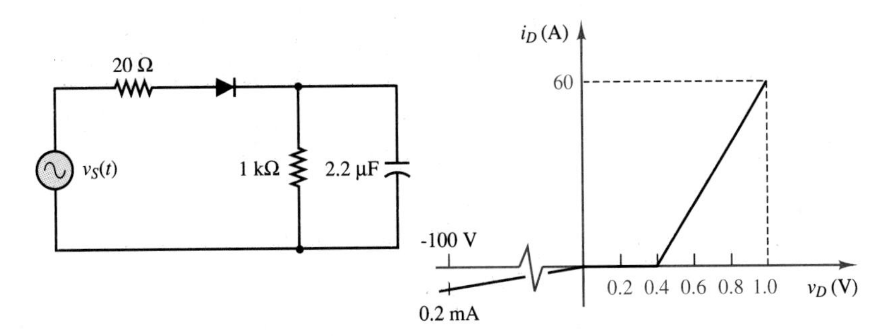

Figure P7.19

7.22 A piecewise linear model of a diode is shown in Figure P7.20. If $r_{D\text{ off}} = 100\ \text{k}\Omega$, $r_{D\text{ on}} = 100\ \Omega$, and $V_\gamma = 0.6$ V, sketch the *i-v* characteristics for $-100\ \text{V} \le v_D \le 1\ \text{V}$.

7.23 The diode shown in Figure P7.21 has a piecewise linear characteristic that passes through the points (−10 V, −5 μA), (0, 0), (0.5 V, 5 mA), and (1 V, 50 mA). Determine the piecewise linear model, and, using that model, solve for *i* and *v*.

7.24 Sketch the output waveform for the circuit of Figure P7.22. Assume ideal diode characteristics, $v_S(t) = 10 \sin(2{,}000\pi t)$.

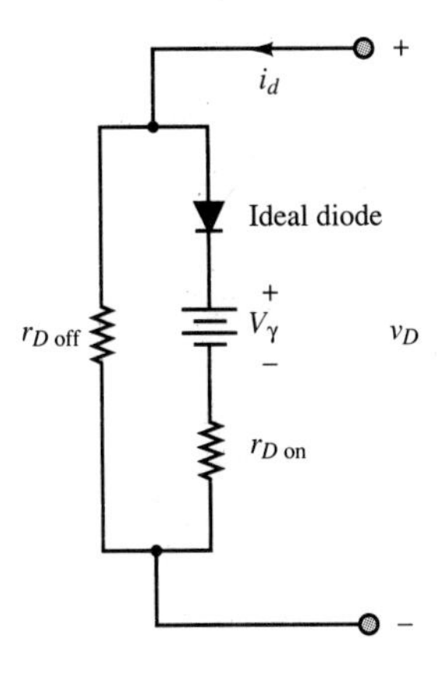

Figure P7.20

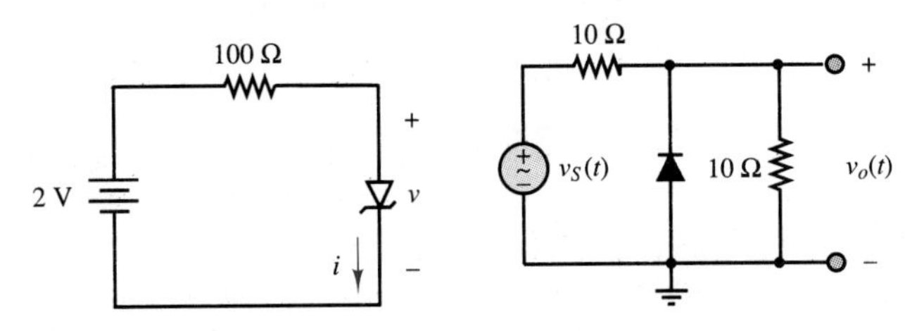

Figure P7.21 **Figure P7.22**

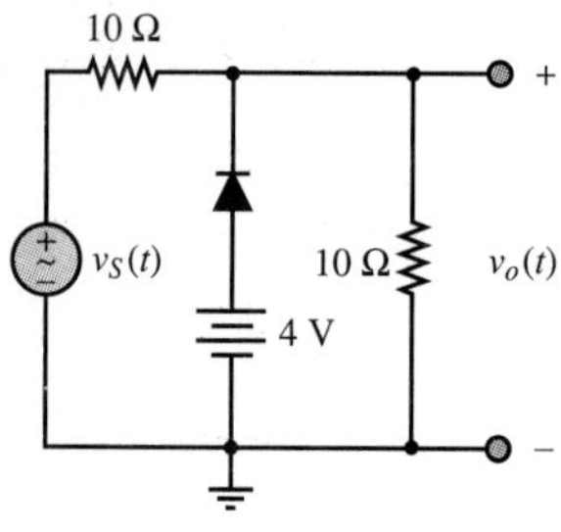
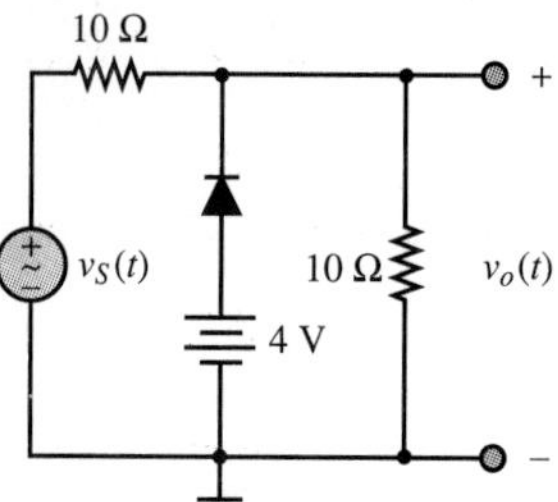

Figure P7.23

7.25 Sketch the output waveform for the circuit of Figure P7.23. Assume ideal diode characteristics, $v_S(t) = 10\sin(2{,}000\pi t)$.

EIT 7.26 Develop a series diode clipper circuit which will separate a signal as shown in Figure P7.24.

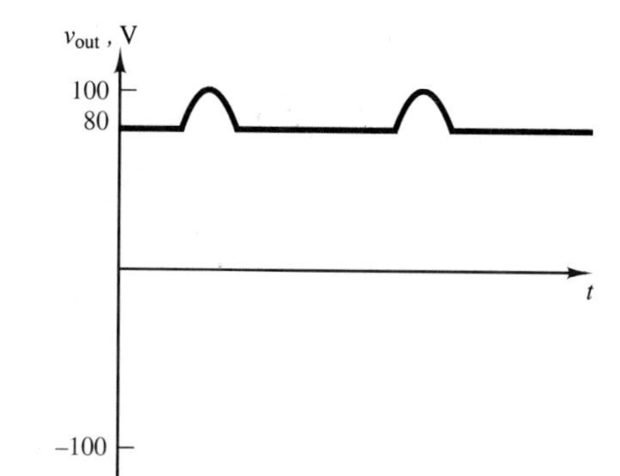

Figure P7.24

Figure P7.25

7.27 In the circuit of Figure P7.25, if the diodes D_1 and D_3 have an "on" resistance equal to 20 Ω and an offset voltage of 0.6 V, and if R_L is 3.3 kΩ, find v_{out} for the cases listed.

a. $v_1 = 0$ V, $v_2 = 0$ V
b. $v_1 = 5$ V, $v_2 = 5$ V
c. $v_1 = 0$ V, $v_2 = 5$ V
d. $v_1 = 5$ V, $v_2 = 0$ V

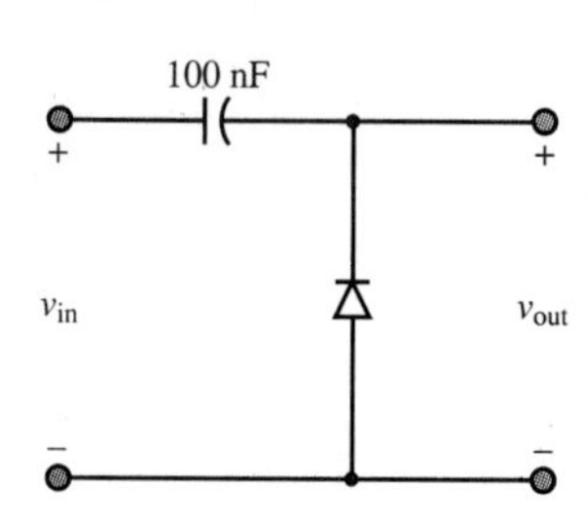

Figure P7.26

7.28 Find the average value of the output voltage for the circuit of Figure P7.26 if the input voltage is sinusoidal with an amplitude of 5 V. Let $V_\gamma = 0.7$ V.

7.29 Find the output waveform for the circuit of Figure P7.27. The diode is ideal.

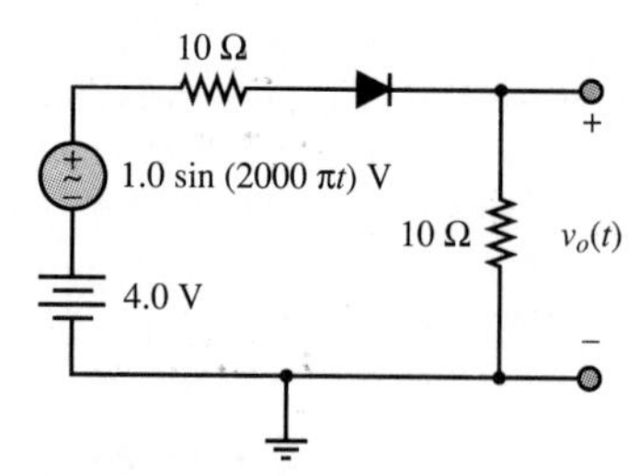

Figure P7.27

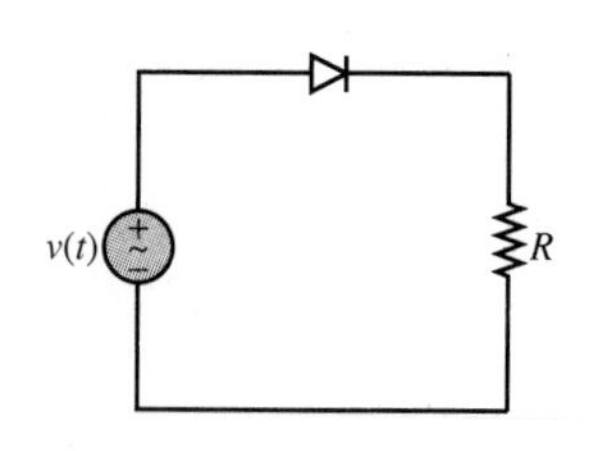

Figure P7.28

7.30 In the rectifier circuit shown in Figure P7.28, $v(t) = A\sin(2\pi 100)t$ V. Assume a forward voltage drop of 0.7 V across the diode when it is conducting. If conduction must begin during each positive half-cycle at an angle no greater than 5°, what is the minimum peak value, A, that the AC source must produce?

EIT 7.31 A half-wave rectifier is to provide an average voltage of 50 V at its output.

a. Draw a schematic diagram of the circuit.
b. Sketch the output voltage waveshape.
c. Determine the peak value of the output voltage.
d. Sketch the input voltage waveshape.
e. What is the rms voltage at the input?

7.32 We can estimate the dynamic resistance in the forward direction of a silicon diode by differentiating the equation

$$i_D = I_0(e^{v_D/V_T} - 1) \approx I_0 e^{v_D/V_T}$$

$$\frac{di_D}{dv_D} = \frac{1}{V_T} I_0 e^{v_D/V_T} \approx \frac{i_D}{V_T}$$

where $V_T = \frac{kT}{q}$, and inverting the result and evaluating r_D at the operating point:

$$r_D \approx \left.\frac{dv_D}{di_D}\right|_{Q\text{ point}} \approx \frac{V_T}{I_{DQ_{\text{point}}}} \approx \frac{25.6\text{ mV}}{I_{DQ_{\text{point}}}}$$

Use this method to determine the dynamic resistance of the diode in the circuit of Figure P7.29 if the diode has the small-signal *i*-*v* characteristic shown in Figure 7.32.

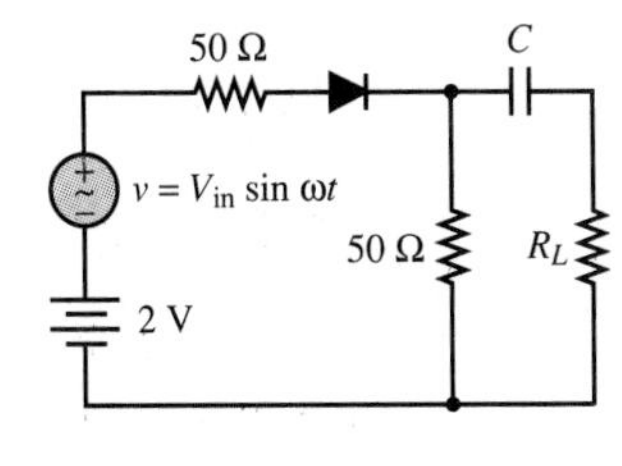

Figure P7.29

***7.33** Assume that the operating point of the diode in the circuit of Figure P7.30 is $I_Q = 7.0$ mA, $V_Q = 0.8$ V.

a. Estimate the small-signal diode resistance, r_D.
b. Estimate the peak values of the AC component of the diode current, using graphical analysis and circuit analysis methods.

Assume $C_C = \infty$ (open circuit for DC, short circuit for AC).

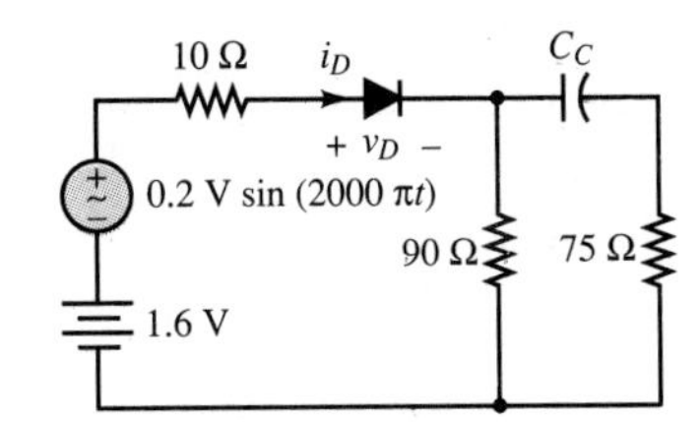

Figure P7.30

7.34 Assuming that the diodes are ideal, determine and sketch the *i*-*v* characteristics for the circuit of Figure P7.31. Consider the range $10 \geq v \geq 0$.

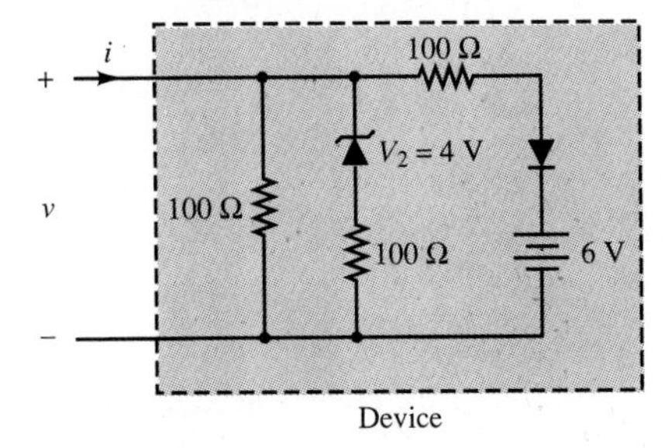

Figure P7.31

7.35 For the circuit of Figure P7.32, sketch the waveform of $v_L(t)$ if $v_S(t) = 15 \sin 2{,}000\pi t$, the Zener diode has $V_Z = 9$ V, $V_1 = 12$ V, $R_S = 50\ \Omega$, and $R_L = 150\ \Omega$. Consider the diodes to be ideal.

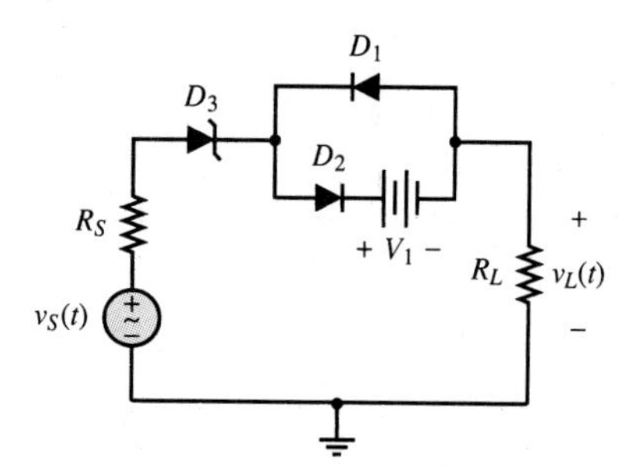

Figure P7.32

EIT 7.36 For the voltage limiter circuit of Figure P7.33, plot v_L versus v_S for $-20 < v_S < 20$ V. Assume that

$R_S = 10\ \Omega$ $\quad V_1 = 15$ V $\quad R_2 = 5\ \Omega$

$R_1 = 10\ \Omega$ $\quad V_2 = -12$ V $\quad R_L = 50\ \Omega$

Treat all diodes as ideal diodes.

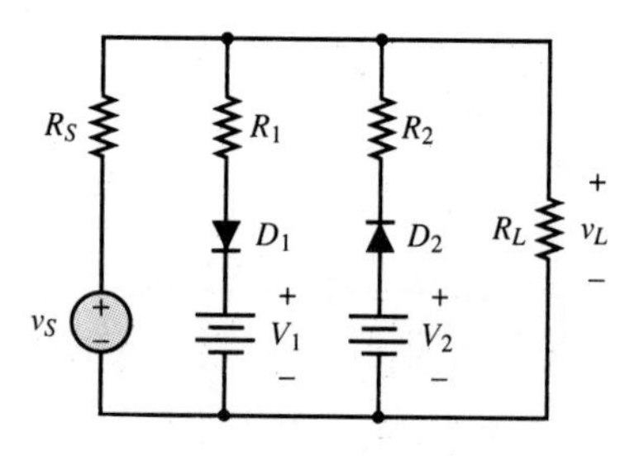

Figure P7.33

7.37 Given the input voltage waveform and the circuit shown in Figure P7.34, sketch the output voltage.

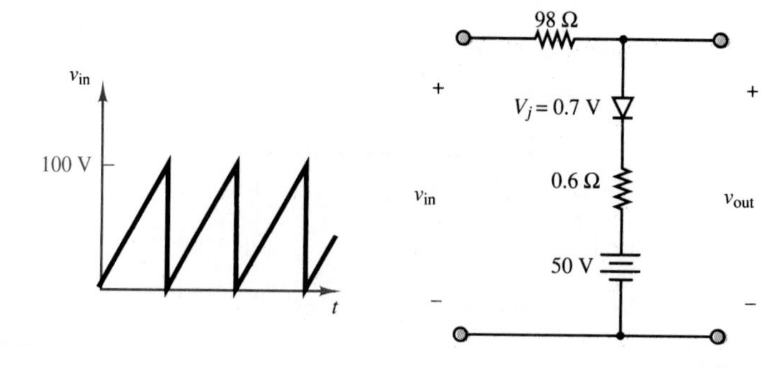

Figure P7.34

7.38 The circuit of Figure P7.35(a) demonstrates the use of a diode as a path for current in an inductive circuit. At $t = 0$, the switch closes, causing the inductor to become a short circuit as $t \to \infty$. $R = 100\ \Omega$, $L = 100$ mH, and $V_S = 15$ V.

a. Assume that $v_L(t) = 0$ and the switch, S_1, has been closed for a long period of time. At $t = t_1$, the switch opens. Determine the voltage $v_L(t)$ for $t > t_1$.
b. Using the assumptions of part a, now consider the same circuit with a diode placed as in the circuit of Figure P7.35(a). Use the diode model of Figure P7.35(b), with $r_{\text{on}} = r_D = 5\ \Omega$. At $t = t_1$, the switch opens. Determine the voltage $v_L(t)$ for $t \geq t_1$.

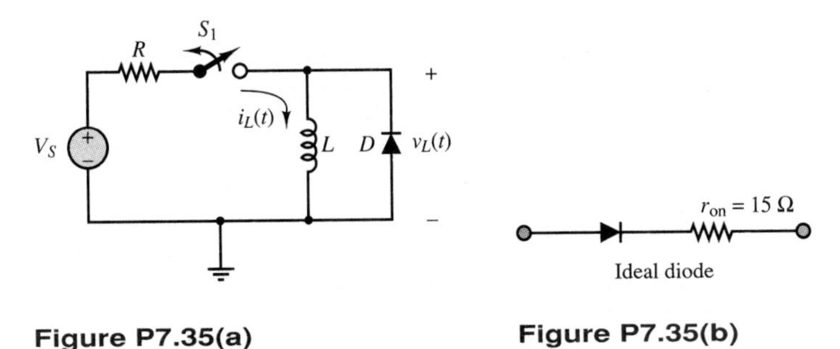

Figure P7.35(a) **Figure P7.35(b)**

EIT 7.39 We are using a voltage source to charge an automotive battery as shown in the circuit of Figure P7.36(a). At $t = t_1$, the protective circuitry of the source causes switch S_1 to close, and the source voltage goes to zero. Find the currents, I_S, I_B, and I_{SW}, for the following conditions:

a. $t = t_1^-$
b. $t = t_1^+$
c. What will happen to the battery after the switch closes?

Now we are going to charge the battery, using the circuit of Figure P7.36(b). Repeat parts a and b if the diode has an offset voltage of 0.6 V.

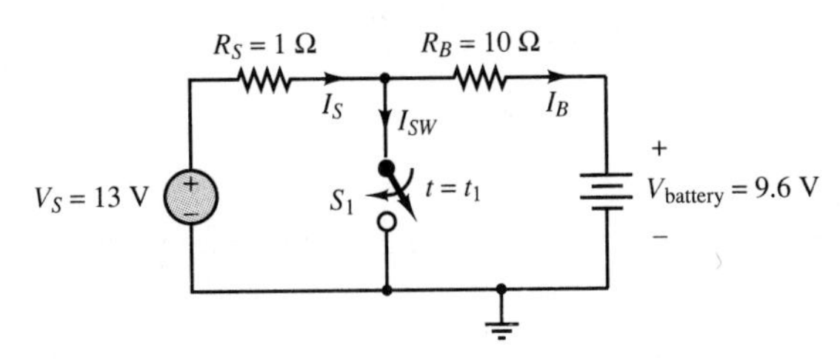

Figure P7.36(a)

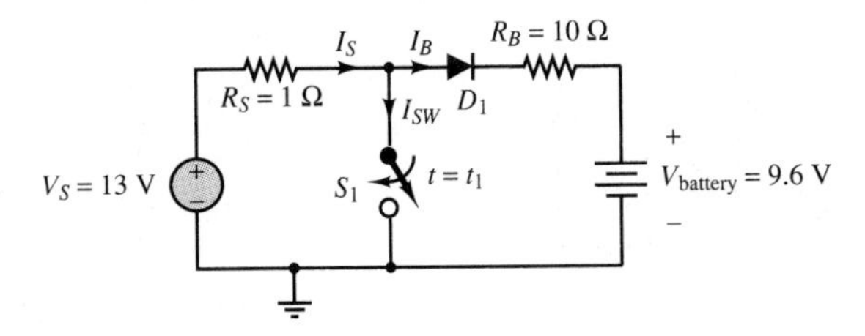

Figure P7.36(b)

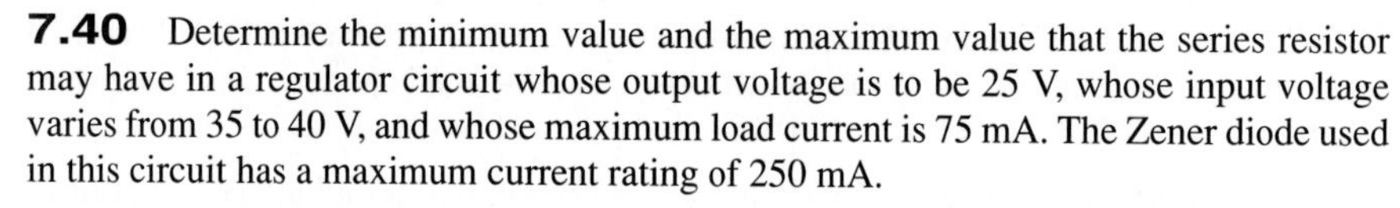

7.40 Determine the minimum value and the maximum value that the series resistor may have in a regulator circuit whose output voltage is to be 25 V, whose input voltage varies from 35 to 40 V, and whose maximum load current is 75 mA. The Zener diode used in this circuit has a maximum current rating of 250 mA.

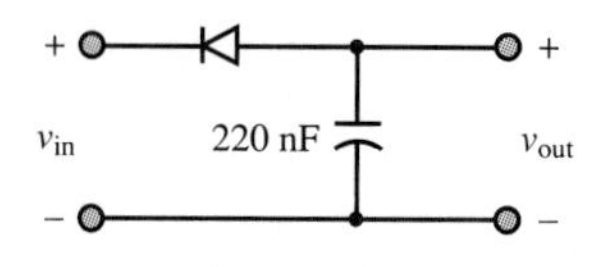

Figure P7.37

7.41 Find the output voltage of the peak detector shown in Figure P7.37. Use sinusoidal input voltages with amplitude 6, 1.5, and 0.4 V and zero average value. Let $V_\gamma = 0.7$ V.

7.42 Repeat the design of the solar array of Example 7.15 under the assumption that the solar array will be operated in the constant-current region.

7.43 Repeat the design of the solar array of Example 7.15 under the assumption that the solar array will be operated in the constant-voltage region.

Transistor Fundamentals

Chapter 8 continues the discussion of electronic devices that began in Chapter 7 with the semiconductor diode. This chapter will describe the operating characteristics of the two major families of electronic devices: bipolar and field-effect transistors. The first half of Chapter 8 is devoted to a brief, qualitative discussion of the physics and operation of the bipolar junction transistor (BJT), which naturally follows the discussion of the *pn* junction in Chapter 7. The *i-v* characteristics of bipolar transistors and their operating states are presented. Large-signal circuit models for the BJT are then introduced, in order to illustrate how one can analyze transistor circuits using basic circuit analysis methods. A few practical examples are discussed to illustrate the use of the circuit models. The second half of the chapter focuses on field-effect transistors; the basic operation and *i-v* characteristics of enhancement- and depletion-mode MOS transistors and of junction field-effect transistors are presented. Universal curves for each of these devices and large-signal circuit models are also discussed. By the end of Chapter 8, you should be able to:

- Describe the basic operation of bipolar junction transistors.
- Interpret BJT characteristic curves and extract large-signal model parameters from these curves.

- Identify the operating state of a BJT from measured data and determine its operating point.
- Analyze simple large-signal BJT amplifiers.
- Describe the basic operation of enhancement- and depletion-mode metal-oxide-semiconductor field-effect transistors (MOSFETs) and of junction field-effect transistors (JFETs).
- Interpret the universal curves for these devices and extract linear (small-signal) models for simple amplifiers from device curves and data sheets.
- Identify the operating state of a field-effect transistor from measured data and determine its operating point.
- Analyze simple large-signal FET amplifiers.

8.1 TRANSISTORS AS AMPLIFIERS AND SWITCHES

A transistor is a three-terminal semiconductor device that can perform two functions that are fundamental to the design of electronic circuits: **amplification** and **switching.** Put simply, amplification consists of magnifying a signal by transferring energy to it from an external source; whereas a transistor switch is a device for controlling a relatively large current between or voltage across two terminals by means of a small control current or voltage applied at a third terminal. In this chapter, we provide an introduction to the two major families of transistors: *bipolar junction transistors,* or *BJTs;* and *field-effect transistors,* or *FETs.*

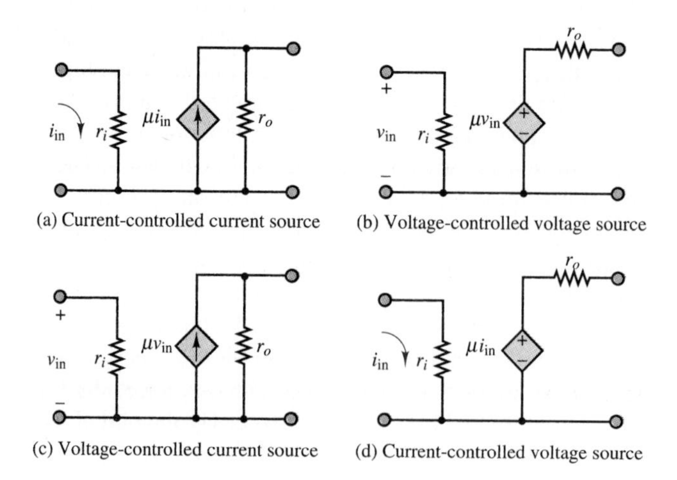

Figure 8.1 Controlled-source models of linear amplifier transistor operation

The operation of the transistor as a linear amplifier can be explained qualitatively by the sketch of Figure 8.1, in which the four possible modes of operation of a transistor are illustrated by means of circuit models employing controlled sources (you may wish to review the section on controlled sources in Chapter 2). In Figure 8.1, controlled voltage and current sources are shown to generate an output proportional to an input current or voltage; the proportionality constant, μ, is called the internal *gain* of the transistor. As will be shown, the BJT acts essentially as a current-controlled device, while the FET behaves as a voltage-controlled device.

Transistors can also act in a nonlinear mode, as voltage- or current-controlled switches. When a transistor operates as a switch, a small voltage or current is used to control the flow of current between two of the transistor terminals in an on-off fashion. Figure 8.2 depicts the idealized operation of the transistor as a switch, suggesting that the switch is closed (on) whenever a control voltage or current is greater than zero and is open (off) otherwise. It will later become apparent that the conditions for the switch to be on or off need not necessarily be those depicted in Figure 8.2. The operation of transistors as switches will be discussed in more detail in Chapter 9.

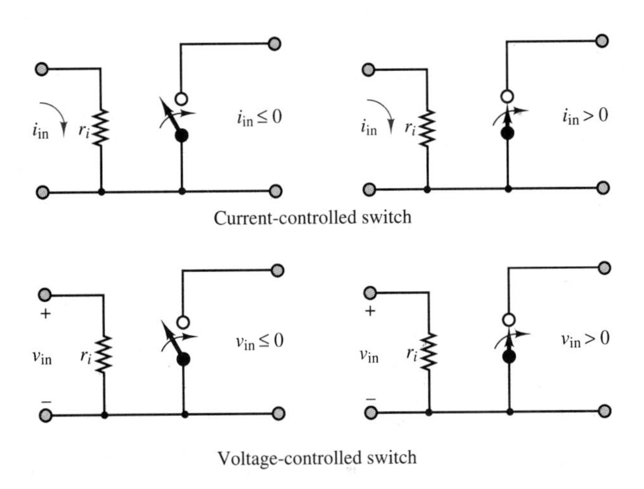

Figure 8.2 Models of ideal transistor switches

Example 8.1

Consider the simplified model of a transistor amplifier shown in Figure 8.3, where a signal source, $v_S(t)$, with internal resistance R_S, and a load resistance, R_L, have been connected to one of the four controlled-source models of Figure 8.1. The resistances r_i and r_o and the gain, μ, are internal properties of the transistor, as will be shown in the next section. Determine the actual gain of this amplifier, $A_V = v_L/v_S$, on the basis of the circuit model of Figure 8.3.

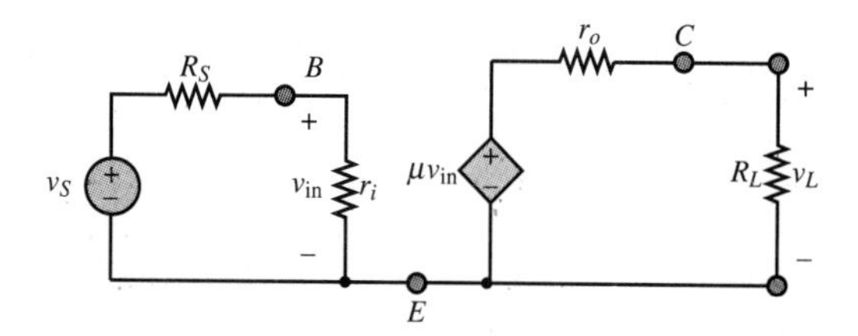

Figure 8.3

Solution:

Since the output of the controlled voltage source is determined by the voltage v_{in}, we need to determine this quantity first:

$$v_{in} = v_S \frac{r_i}{r_i + R_S}$$

Thus, the output of the controlled current source will be

$$\mu v_{in} = \mu v_S \frac{r_i}{r_i + R_S}$$

and the output voltage, v_L, can be found by computing the voltage divider result:

$$v_L = \mu v_S \frac{r_i}{r_i + R_S} \frac{R_L}{r_o + R_L}$$

Finally, the amplifier voltage gain is found to be

$$A_V = \frac{v_L}{v_S} = \mu \frac{r_i}{r_i + R_S} \cdot \frac{R_L}{r_o + R_L}$$

Note that this gain, A_V, is necessarily less than the internal transistor voltage gain, μ! You can now show that if the conditions

$$r_i \gg R_S$$
$$r_o \ll R_L$$

apply, then the gain of the amplifier becomes approximately equal to the internal gain of the transistor. An important observation, then, is that, given a transistor with known internal parameters, r_i, r_o, and μ, the gain of the amplifier will depend on the resistance of the source and on the load resistance.

DRILL EXERCISES

8.1 Repeat the analysis of Example 8.1 for the current-controlled voltage source model of Figure 8.1. What is the amplifier voltage gain? Under what conditions would the gain A be equal to $\frac{\mu}{R_S}$?

8.2 Repeat the analysis of Example 8.1 for the current-controlled current source model of Figure 8.1. What is the amplifier voltage gain?

8.3 Repeat the analysis of Example 8.1 for the voltage-controlled current source model of Figure 8.1. What is the amplifier voltage gain?

8.2 THE BIPOLAR JUNCTION TRANSISTOR (BJT)

The *pn* junction studied in Chapter 7 forms the basis of a large number of semiconductor devices. The semiconductor diode, a two-terminal device, is the most direct application of the *pn* junction. In this section, we introduce the **bipolar junction transistor (BJT).** As we did in analyzing the diode, we will introduce the physics of transistor devices as intuitively as possible, resorting to an analysis of their *i-v* characteristics to discover important properties and applications.

A BJT is formed by joining three sections of semiconductor material, each with a different doping concentration. The three sections can be either a thin n region sandwiched between p^+ and p layers, or a p region between n and n^+ layers, where the superscript "plus" indicates more heavily doped material. The resulting BJTs are called *pnp* and *npn* transistors, respectively; we shall discuss only the latter in this chapter. Figure 8.4 illustrates the approximate construction, symbols, and nomenclature for the two types of BJTs.

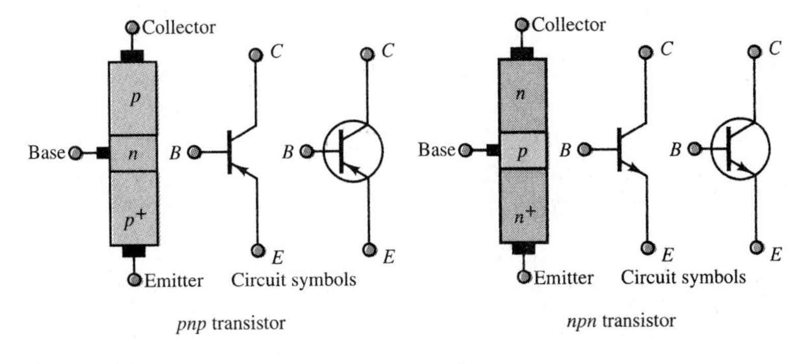

Figure 8.4 Bipolar junction transistors

The operation of the *npn* BJT may be explained by considering the transistor as consisting of two back-to-back *pn* junctions. The **base-emitter (*BE*)**

The *BE* junction acts very much as an ordinary diode when the collector is open. In this case, $I_B = I_E$.

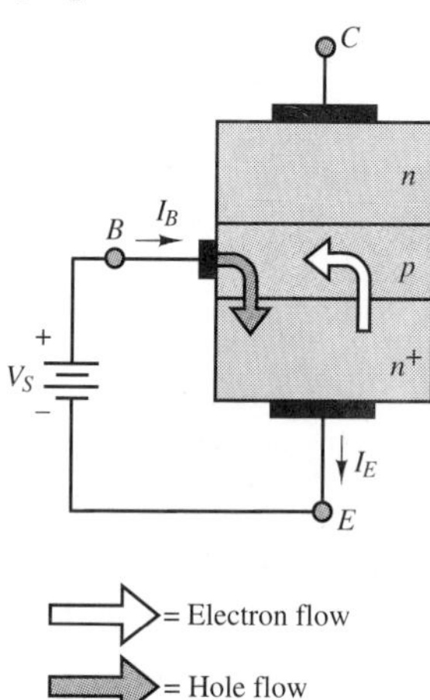

Figure 8.5 Current flow in an *npn* BJT

When the *BC* junction is reverse-biased, the electrons from the emitter region are swept across the base into the collector.

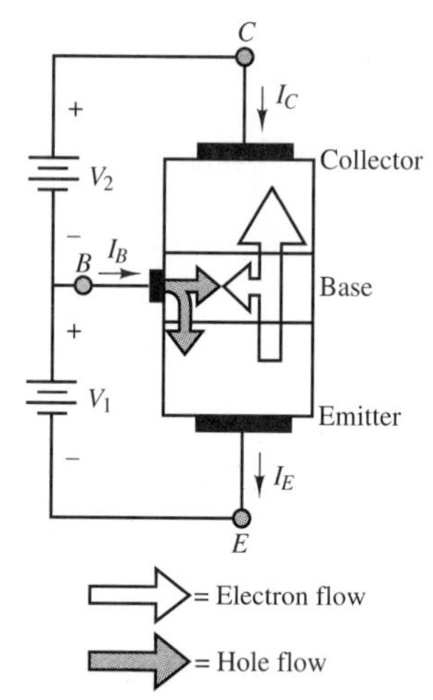

Figure 8.6 Flow of emitter electrons into the collector in an *npn* BJT

junction acts very much like a diode when it is forward-biased; thus, one can picture the corresponding flow of hole and electron currents from base to emitter when the collector is open and the *BE* junction is forward-biased, as depicted in Figure 8.5. Note that the electron current has been shown larger than the hole current, because of the heavier doping of the *n* side of the junction. Some of the electron-hole pairs in the base will recombine; the remaining charge carriers will give rise to a net flow of current from base to emitter. It is also important to observe that the base is much narrower than the emitter section of the transistor.

Imagine, now, connecting the **base-collector (*BC*) junction** so that it is reverse-biased. In this case, an interesting phenomenon takes place: the electrons "emitted" by the emitter with the *BE* junction forward-biased reach the very narrow base region, and after a few are lost to recombination in the base, most of these electrons are "collected" by the collector. Figure 8.6 illustrates how the reverse bias across the *BC* junction is in such a direction as to sweep the electrons from the emitter into the collector. This phenomenon can take place because the base region is kept particularly narrow. Since the base is narrow, there is a high probability that the electrons will have gathered enough momentum from the electric field to cross the reverse-biased collector-base junction and make it into the collector. The result is that there is a net flow of current from collector to emitter (opposite in direction to the flow of electrons), in addition to the hole current from base to emitter. The electron current flowing into the collector through the base is substantially larger than that which flows into the base from the external circuit. One can see from Figure 8.6 that if KCL is to be satisfied, we must have

$$I_E = I_B + I_C \tag{8.1}$$

The most important property of the bipolar transistor is that the small base current controls the amount of the much larger collector current

$$I_C = \beta I_B \tag{8.2}$$

where β is a current amplification factor dependent on the physical properties of the transistor. Typical values of β range from 20 to 200. The operation of a *pnp* transistor is completely analogous to that of the *npn* device, with the roles of the charge carriers (and therefore the signs of the currents) reversed. The symbol for a *pnp* transistor was shown in Figure 8.4.

The detailed operation of bipolar transistors can be explained by resorting to a detailed physical analysis of the *npn* or *pnp* structure of these devices. The reader interested in such a discussion of transistors is referred to any one of a number of excellent books on semiconductor electronics. The aim of this book, on the other hand, is to provide an introduction to the basic principles of transistor operation by means of simple linear circuit models based on the device *i-v* characteristic. Although it is certainly useful for the non–electrical engineer to understand the basic principles of operation of electronic devices, it is unlikely that most readers will engage in the design of high-performance electronic circuits or will need a detailed understanding of the operation of each device. The present chapter will therefore serve as a compendium of the basic ideas enabling an engineer to read and understand electronic circuit diagrams, and to specify the requirements of electronic instrumentation systems. The reader interested in furthering his or her knowledge of electronic circuits will find the fundamentals provided by this chapter very useful when undertaking the study of a more advanced textbook. The focus of this section will be on the analysis of the *i-v* characteristic of the *npn* BJT, based on the circuit notation defined in Figure 8.7. The device *i-v* characteristics will be presented qualitatively, without deriving the underlying equations, and will be utilized in constructing circuit models for the device.

The operation of the BJT is defined in terms of two currents and two voltages: i_B, i_C, v_{CE}, and v_{BE}.

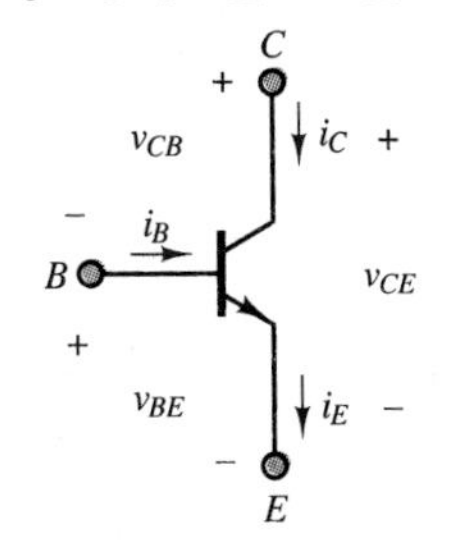

Figure 8.7 Definition of BJT voltages and currents

The number of independent variables required to uniquely define the operation of the transistor may be determined by applying KVL and KCL to the circuit of Figure 8.7. It should be apparent that two voltages and two currents are sufficient to specify the operation of the device. Note that, since the BJT is a three-terminal device, it will not be sufficient to deal with a single *i-v* characteristic; it will soon become apparent that two such characteristics are required to explain the operation of this device. One of these characteristics relates the base current, i_B, to the base-emitter voltage, v_{BE}; the other relates the collector current, i_C, to the collector-emitter voltage, v_{CE}. As will be shown, the latter characteristic actually consists of a *family* of curves. To determine these *i-v* characteristics, consider the *i-v* curves of Figures 8.8 and 8.9, using the circuit notation of Figure 8.7. In Figure 8.8, the collector is open and the *BE* junction is shown to be very similar to a diode. The ideal current source, I_{BB}, injects a base current, which causes the junction to be forward-biased. By varying I_{BB}, one can obtain the open-collector *BE* junction *i-v* curve shown in the figure.

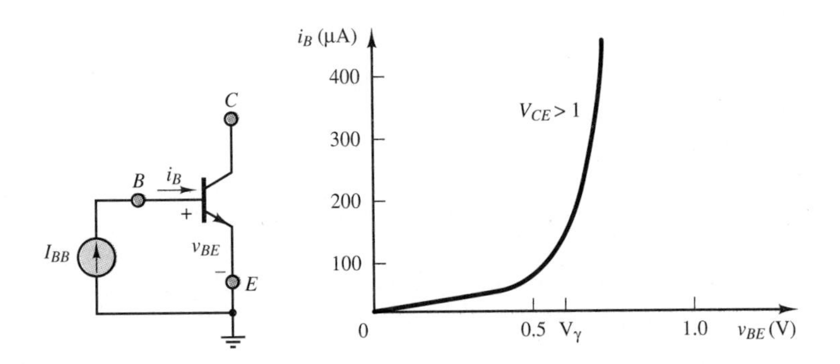

Figure 8.8 Determining the *BE* junction open-collector *i-v* characteristic

If a voltage source were now to be connected to the collector circuit, the voltage v_{CE} and, therefore, the collector current, i_C, could be varied, in addition to the base current, i_B. The resulting circuit is depicted in Figure 8.9(a). By varying both the base current and the collector-emitter voltage, one could then generate a plot of the device **collector characteristic.** This is also shown in Figure 8.9(b). Note that this figure depicts not just a single i_C-v_{CE} curve, but an entire family, since for each value of the base current, i_B, an i_C-v_{CE} curve can be generated. Four regions are identified in the collector characteristic:

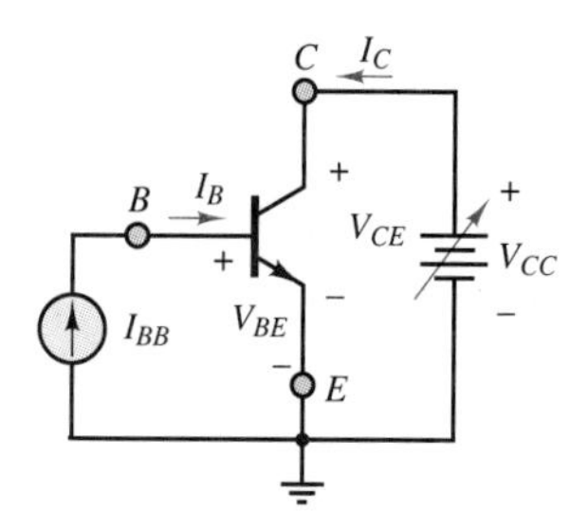

Figure 8.9(a) Ideal test circuit to determine the *i-v* characteristic of a BJT

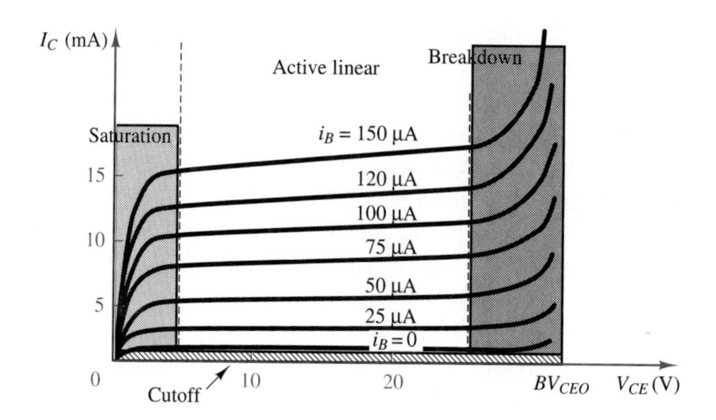

Figure 8.9(b) The collector-emitter output characteristics of a BJT

1. The **cutoff region,** where both junctions are reverse-biased, the base current is very small, and essentially no collector current flows.
2. The **active linear region,** in which the transistor can act as a linear amplifier, where the *BE* junction is forward-biased and the *CB* junction is reverse-biased.
3. The **saturation region,** in which both junctions are forward-biased.
4. The **breakdown region,** which determines the physical limit of operation of the device.

From the curves of Figure 8.9(b), we note that as v_{CE} is increased, the collector current increases rapidly, until it reaches a nearly constant value; this condition holds until the collector junction breakdown voltage, BV_{CEO}, is reached (for the purposes of this book, we shall not concern ourselves with the phenomenon of breakdown, except in noting that there are maximum allowable voltages and currents in a transistor). If we were to repeat the same measurement for a set of different values of i_B, the corresponding value of i_C would change accordingly; hence, the family of collector characteristic curves.

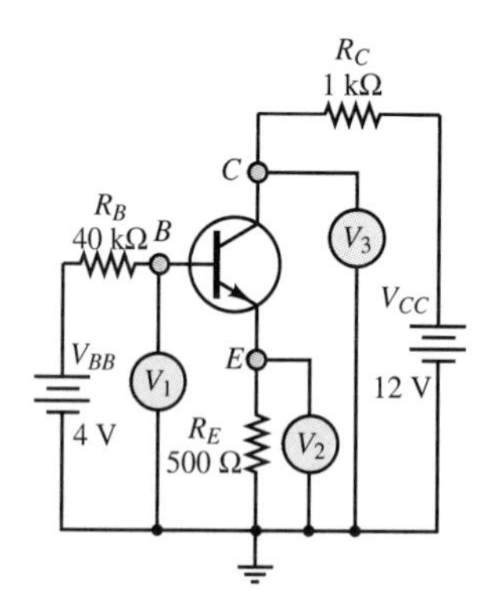

Figure 8.10 Determination of the operating state of a BJT

Determining the Operating State of a BJT

Before we discuss common circuit models for the BJT, it will be useful to consider the problem of determining the operating state of the transistor. A few simple voltage measurements permit a quick determination of the state of a transistor placed in a circuit. Consider, for example, the BJT described by the curves of Figure 8.9 if it is placed in the circuit of Figure 8.10. In this figure, voltmeters are used to measure the value of the collector, emitter, and base voltages. Can these simple measurements identify the operating state of the transistor? Assume that the measurements reveal the following conditions:

$$V_1 = 2 \text{ V}$$
$$V_2 = 1.3 \text{ V}$$
$$V_3 = 8 \text{ V}$$

What can be said about the operating state of the transistor?

The first observation is that knowing V_B and V_E permits determination of V_{BE}: $V_{BE} = 0.7$ V. Thus, we know that the BE junction is forward-biased. Another quick calculation permits determination of the relationship between base and collector current: the base current is equal to

$$I_B = \frac{4 - V_B}{40{,}000} = 50 \ \mu\text{A}$$

while the collector current is

$$I_C = \frac{12 - V_C}{1{,}000} = 4 \text{ mA}$$

Thus, the current amplification (or gain) factor for the transistor is

$$\frac{I_C}{I_B} = 80$$

Such a value for the current gain suggests that the transistor is in the linear active region, because substantial current amplification is taking place (typical values of current gain range from 20 to 200). Finally, the collector-to-emitter voltage, V_{CE}, is found to be: $V_{CE} = V_C - V_E = 8 - 1.3 = 6.7$ V.

At this point, you should be able to locate the operating point of the transistor on the curves of Figures 8.8 and 8.9. The currents I_B and I_C and the voltage V_{CE} uniquely determine the state of the transistor in the I_C-V_{CE} and I_B-V_{BE} characteristic curves. What would happen if the transistor were not in the linear active region? The following examples answer this question and provide further insight into the operation of the bipolar transistor.

Example 8.2

The objective of this example is to determine the operating state of a BJT. The circuit is identical to the one in Figure 8.10, except that now the base voltage source, V_{BB}, is short-circuited.

Solution:

Since the base voltage source is short-circuited, the base will be at zero volts, the emitter-base junction will not conduct, and the emitter current will be nearly zero.

From equation 8.1, it may also be concluded that the collector current must be zero. A glance at Figure 8.9(b) reveals that the transistor is in the cutoff state. The readings of the three voltmeters will be zero for V_B and V_E and $+12$ V for V_C, because there is no voltage drop across the 1-kΩ resistor.

In general, if the base supply voltage is not sufficient to forward-bias the base-emitter junction, the transistor will be in the cutoff region.

Example 8.3

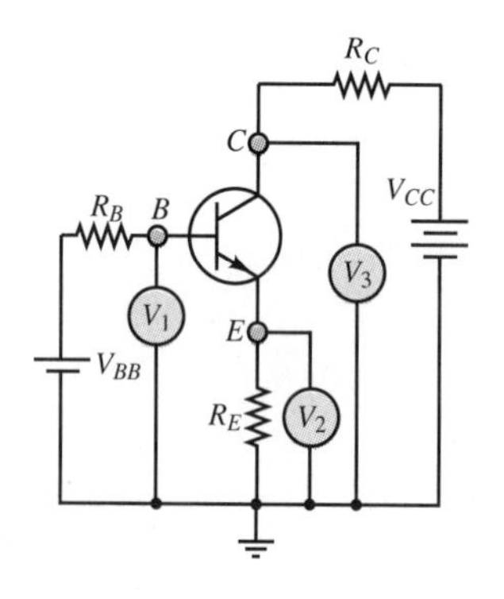

Figure 8.11

For the circuit in Figure 8.11, the voltmeter readings are $V_1 = 2.7$ V, $V_2 = 2$ V, and $V_3 = 2.3$ V. Check the operating state of the transistor.

Solution:

One way to check the state of a transistor is to calculate V_{BC} and V_{BE} to see whether the BE and CB junctions are forward- or reverse-biased. Saturation is characterized by operation of both junctions with forward bias and with very small voltage drops, whereas active operation is characterized by a forward-biased BE junction and a reverse-biased BC junction. Thus, for an npn transistor in saturation, $V_{BE} > 0$ and $V_{BC} > 0$.

The circuit of Figure 8.11 has

$$V_{BE} = V_B - V_E = 2.7 - 2.0 = 0.7 \text{ V}$$

and

$$V_{BC} = V_B - V_C = 2.7 - 2.3 = 0.4 \text{ V}$$

Thus, the transistor is operating in the saturation region. The value $V_{CE} = V_C - V_E = 2.3 - 2 = 0.3$ V is very small. This is usually a good indication that a BJT is operating in saturation. You should now try to locate the operating point of this transistor in Figure 8.9(b), observing that $I_C = \frac{V_{CC} - V_3}{R_C} = \frac{12 - 2.3}{1{,}000} = 9.7$ mA.

Selecting an Operating Point for a BJT

By appropriate choice of I_{BB}, R_C and V_{CC}, the desired Q point may be selected.

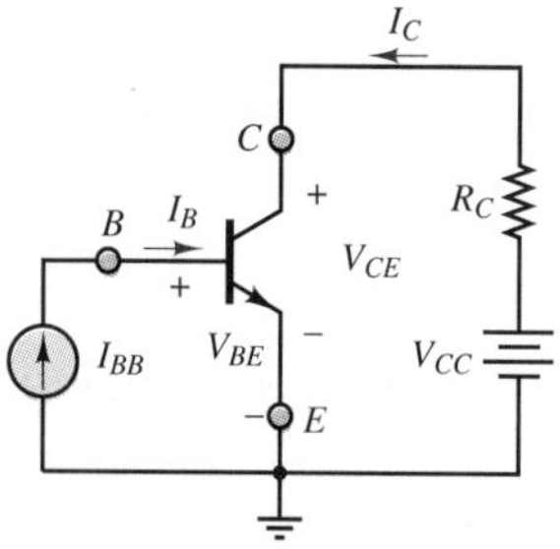

Figure 8.12 A simplified bias circuit for a BJT amplifier

The family of curves shown for the collector i-v characteristic in Figure 8.9(b) reflects the dependence of the collector current on the base current. For each value of the base current, i_B, there exists a corresponding i_C-v_{CE} curve. Thus, by appropriately selecting the base current and collector current (or collector-emitter voltage), we can determine the operating point, or ***Q* point,** of the transistor. The Q point of a device is defined in terms of the **quiescent** (or **idle**) **currents** and **voltages** that are present at the terminals of the device when DC supplies are connected to it. The circuit of Figure 8.12 illustrates an ideal **DC bias circuit,** used to set the Q point of the BJT in the approximate center of the collector characteristic. The circuit shown in Figure 8.12 is not a practical DC bias circuit for a BJT amplifier, but it is very useful for the purpose of introducing the relevant concepts. A practical bias circuit will be discussed later in this section.

Applying KVL around the base-emitter and collector circuits, we obtain the following equations:

$$I_B = I_{BB} \quad \textbf{(8.3)}$$

and

$$V_{CE} = V_{CC} - I_C R_C \quad \textbf{(8.4)}$$

which can be rewritten as

$$I_C = \frac{V_{CC}}{R_C} - \frac{V_{CE}}{R_C} \tag{8.5}$$

Note the similarity of equation 8.5 to the load-line curves of Chapters 3 and 7. Equation 8.5 represents a line that intersects the I_C axis at $I_C = V_{CC}/R_C$ and the V_{CE} axis at $V_{CE} = V_{CC}$. The slope of the load line is $-1/R_C$. Since the base current, I_B, is equal to the source current, I_{BB}, the operating point may be determined by noting that the load line intersects the entire collector family of curves. The intersection point at the curve that corresponds to the base current $I_B = I_{BB}$ constitutes the operating, or Q, point. The load line corresponding to the circuit of Figure 8.12 is shown in Figure 8.13, superimposed on the collector curves for the 2N3904 transistor (data sheets for the 2N3904 transistor are included at the end of Chapter 9). In Figure 8.13, $V_{CC} = 15$ V, $\frac{V_{CC}}{R_C} = 40$ mA, and $I_{BB} = 150$ μA; thus, the Q point is determined by the intersection of the load line with the I_C-V_{CE} curve corresponding to a base current of 150 μA.

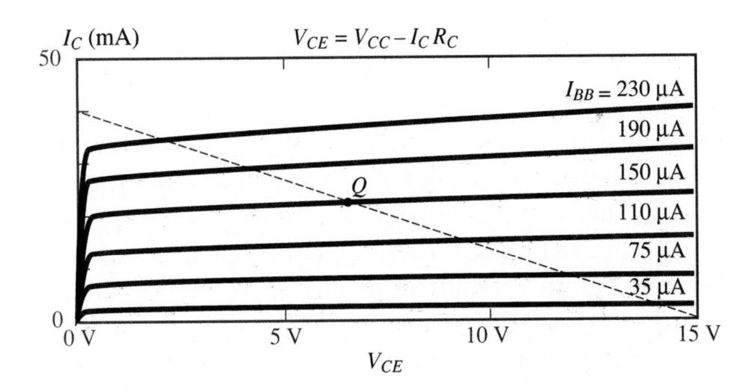

Figure 8.13 Load-line analysis of a simplified BJT amplifier

Once an operating point is established and DC currents I_{CQ} and I_{BQ} are flowing into the collector and base, respectively, the BJT can serve as a linear amplifier, as will be explained in Section 8.3. Example 8.4 serves as an illustration of the DC biasing procedures just described.

Example 8.4

The objective of this example is to illustrate numerically the DC biasing procedure for a BJT. For practical reasons, instead of an ideal current source, the base circuit shown in Figure 8.14 consists of a battery in series with a base resistance, R_B.

Assuming a base resistance $R_B = 62.7$ kΩ, a collector resistance $R_C = 375$ Ω, a collector DC voltage source $V_{CC} = 15$ V, and a base voltage supply $V_{BB} = 10$ V, we can sketch the load line shown in Figure 8.13. The equation for the load line is

$$V_{CE} = V_{CC} - R_C I_C$$

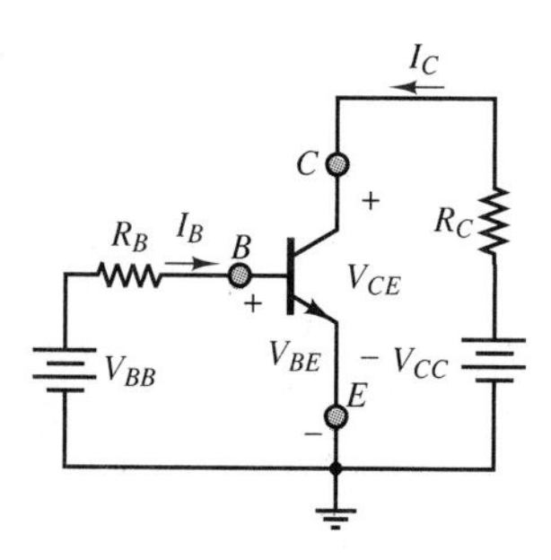

Figure 8.14

or

$$V_{CE} = 15 - 375 I_C$$

The value of the base current will then determine the Q point according to the approximate expression

$$I_B = \frac{V_{BB} - V_{BE}}{R_B} = \frac{10 - 0.6}{62{,}700} = 150\ \mu\text{A}$$

in which we have assumed the BE junction is forward-biased and is therefore behaving like a diode with offset voltage equal to 0.6 V. This corresponds to a base current of 150 μA. The intersection of the load line with the collector i-v curve for a base current of 150 μA corresponds, approximately, to $V_{CE} = 7$ V and $I_C = 22$ mA. These values define the Q point of the transistor as

$$I_{BQ} = 150\ \mu\text{A};\ I_{CQ} = 22\ \text{mA};\ V_{CEQ} = 7\ \text{V}$$

where the additional subscript Q is used to indicate the Q point. It is important to note that the quiescent DC base and collector currents flowing through the transistor at all times are necessary for amplification to occur. Thus, a transistor dissipates power continuously when subject to a DC bias, whether it is actually amplifying an input signal or not.

How can a transistor amplify a signal, then, given the V_{BE}-I_B and V_{CE}-I_C curves discussed in this section? The small-signal amplifier properties of the transistor are best illustrated by analyzing the effect of a small sinusoidal current superimposed on the DC current flowing into the base. The circuit of Figure 8.15 illustrates the idea, by including a small-signal AC source, of strength ΔV_B, in series with the base circuit. The effect of this AC source is to cause sinusoidal oscillations ΔI_B about the Q point, that is, around I_{BQ}. A study of the collector characteristic indicates that for a sinusoidal oscillation in I_B, a corresponding, but larger, oscillation will take place in the collector current. Figure 8.16 illustrates the concept. Note that the base current oscillates between 110 and 190 μA, causing the collector current to correspondingly fluctuate between 15.3 and 28.6 mA.

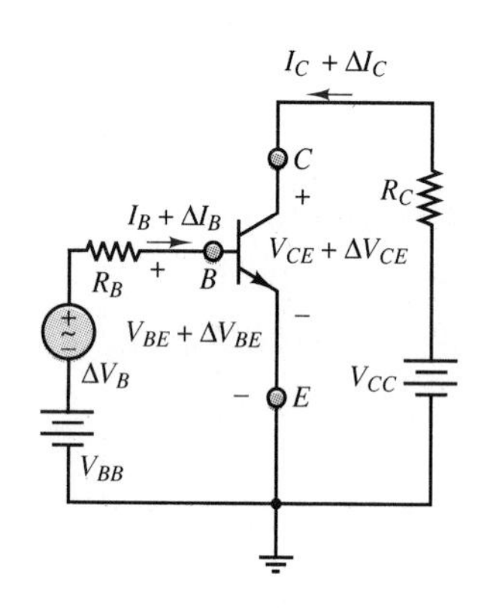

Figure 8.15 Circuit illustrating the amplification effect in a BJT

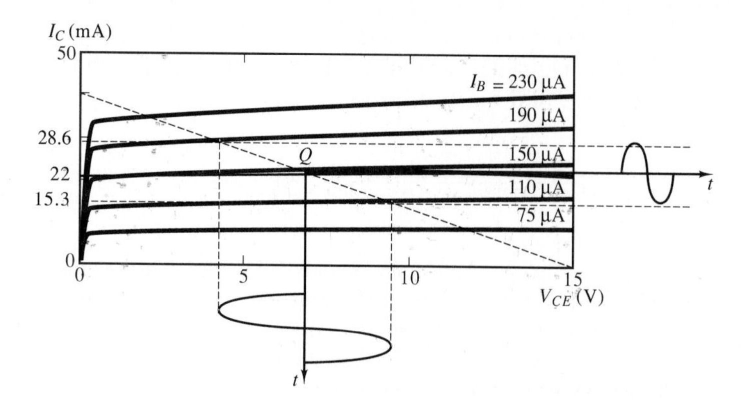

Figure 8.16 Amplification of sinusoidal oscillations in a BJT

The notation that will be used to differentiate between DC and AC (or fluctuating) components of transistor voltages and currents is as follows: DC (or quiescent) currents and voltages will be denoted by uppercase symbols; for example: I_B, I_C, V_{BE}, V_{CE}. AC components will be preceded by a "Δ": $\Delta I_B(t)$, $\Delta I_C(t)$, $\Delta V_{BE}(t)$, $\Delta V_{CE}(t)$. The complete expression for one of these quantities will therefore include both a DC term and a time-varying, or AC, term. For example, the collector current may be expressed by $i_C(t) = I_C + \Delta I_C(t)$.

The *i-v* characteristic of Figure 8.16 illustrates how this increase in collector current follows the same sinusoidal pattern of the base current but is greatly amplified. Thus, the BJT acts as a *current amplifier,* in the sense that any oscillations in the base current appear amplified in the collector current. Since the voltage across the collector resistance, R_C, is proportional to the collector current, one can see how the collector voltage is also affected by the amplification process. Example 8.5 illustrates numerically the effective amplification of the small AC signal that takes place in the circuit of Figure 8.15.

Example 8.5

The present example shows how amplification of small signals takes place in a BJT circuit. Let us assume that a DC operating point has been established—for example, as was illustrated in Example 8.4. In the circuit of Figure 8.17, the AC voltage source represents the signal to be amplified (e.g., the output of a transducer). The AC source induces an oscillation in the base current (think of this in terms of the principle of superposition). Correspondingly, a fluctuation in the collector current takes place, proportional to the base current oscillation. The current amplification factor may be deduced from the collector characteristic curve. Thus, the transistor acts as a current amplifier. In this particular configuration, voltage amplification takes place as well.

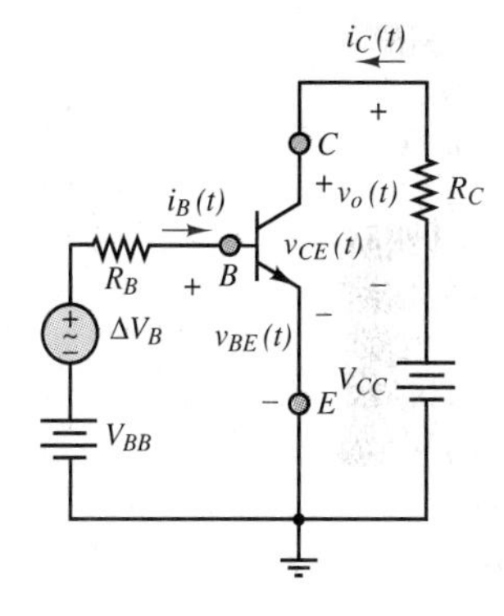

Figure 8.17

In the circuit of Figure 8.17, assume that $V_{CC} = 15$ V, $V_{BB} = 2.1$ V, $R_B = 10$ kΩ, and $R_C = 375$ Ω.

1. Determine the operating point of the transistor.
2. Determine the current gain $\beta = \frac{I_C}{I_B}$ at the quiescent point of the transistor.
3. Find the AC voltage gain of the circuit if the voltage gain A is defined as

$$A = \frac{\Delta V_o}{\Delta V_B}$$

Assume that each current and voltage consists of a DC and an AC component; for example,

$$i_C(t) = I_C + \Delta I_C(t)$$

where I_C is the DC component and $\Delta I_C(t) = \Delta I_C \cos(\omega t)$.

Solution:

1. The current into the base may be expressed as

$$i_B(t) = I_B + \Delta I_B(t)$$

where $\Delta I_B(t)$ is a small, time-dependent change in the base current given by

$$\Delta I_B(t) = \frac{\Delta V_B(t)}{10 \text{ k}\Omega}$$

and where it is assumed that the base-emitter voltage stays relatively constant and the BE junction forward resistance is negligible, relative to the 10-kΩ resistor. Both assumptions are quite reasonable (at least initially) on the basis of the circuit model for the diode presented in Chapter 7, since the BE junction acts, in effect, as a diode. Thus, we are in effect assuming that the BE junction acts as an offset diode and that, therefore, $\Delta V_{BE} \approx 0$.

Consider, now, the DC base circuit. Assuming that the BE junction voltage is 0.6 V when the junction is forward-biased, and neglecting the AC source (in effect, this is equivalent to applying the principle of superposition to determine the voltages and currents due exclusively to the DC source), we may simplify the base circuit as shown in Figure 8.18. Application of KVL around the base circuit reveals that the DC base current is

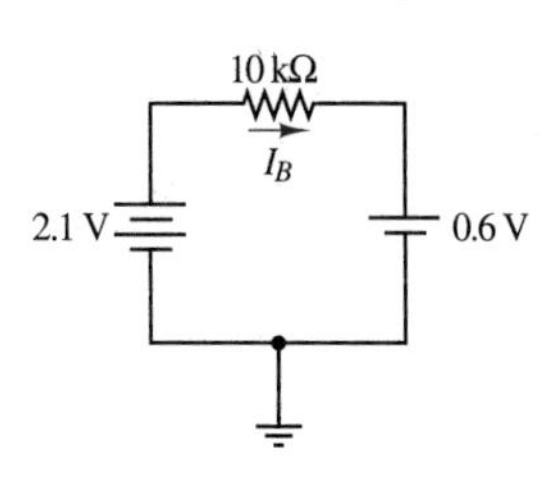

Figure 8.18

$$I_B = \frac{2.1 - 0.6}{10{,}000} = 150 \ \mu\text{A}$$

The operating point can then be found by plotting the load line on the characteristic curves shown in Figure 8.19. The Q point is found to be

$$I_{BQ} = 150 \ \mu\text{A}; V_{CEQ} = 7.2 \text{ V}; I_{CQ} = 22 \text{ mA}$$

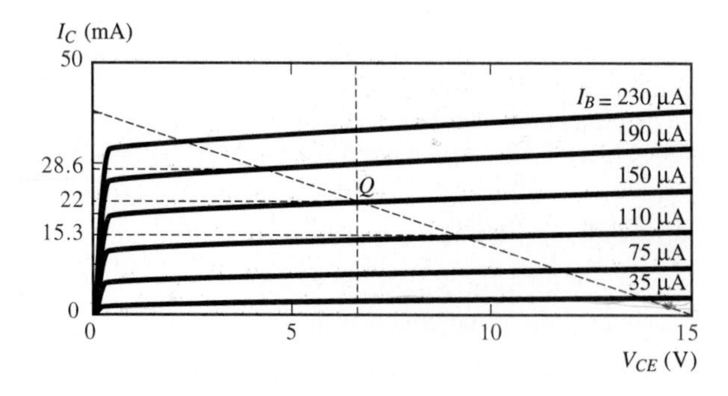

Figure 8.19 Operating point on the characteristic curve

2. The current gain, β, is found by considering the effect of an oscillation in the base current on the collector current. If we follow the load line between, say, base currents of 110 and 190 μA, we note that at the intercepts with these two curves, the corresponding collector currents are 15.3 mA and 28.6 mA, respectively. Thus, for a change in base current of 80 μA, we see a corresponding change in collector current of 13.3 mA. The ratio between these two numbers is the current gain of the transistor: $\beta = 166.25$.

3. The total collector current is the sum of the DC current due to the DC bias circuit and the AC current caused by the amplified oscillations in the base current:

$$i_C(t) = 0.022 + 166.25\,\Delta I_B(t)$$

Thus, the voltage across the collector resistor, which may be viewed as the output voltage in this example, also has corresponding DC and AC components:

$$\begin{aligned} v_o(t) &= V_o + \Delta V_o(t) \\ &= -i_C(t)R_C \\ &= -I_{CQ}R_C - 166.25\,\Delta I_B(t)\,R_C \end{aligned}$$

If we consider just the AC component of the output voltage, $\Delta V_o = \beta \Delta I_B R_C$, we see that this is also an amplified version of the AC base voltage, ΔV_B. The AC voltage gain is therefore given by

$$A_V = \frac{\Delta V_o}{\Delta V_B} = \frac{-\Delta I_C R_C}{\Delta I_B R_B} = \frac{-\beta\,\Delta I_B R_C}{\Delta I_B R_B} = -\beta \frac{R_C}{R_B} = -6.23$$

Note how the voltage gain is very much dependent on the base and collector resistors in the circuit.

In discussing the DC biasing procedure for the BJT, we pointed out that the simple circuit of Figure 8.12 would not be a practical one to use in an application circuit. In fact, the more realistic circuit of Example 8.4 is also not a practical biasing circuit. The reasons for this statement are that two different supplies are required (V_{CC} and V_{BB})—a requirement that is not very practical—and that the resulting DC bias (operating) point is not very stable. This latter point may be made clearer by pointing out that the location of the operating point could vary significantly if, say, the current gain of the transistor, β, were to vary from device to device. A circuit that provides great improvement on both counts is shown in Figure 8.20. Observe, first, that the voltage supply, V_{CC}, appears across the pair of resistors R_1 and R_2, and that therefore the base terminal for the transistor will see the Thévenin equivalent circuit composed of the equivalent voltage source,

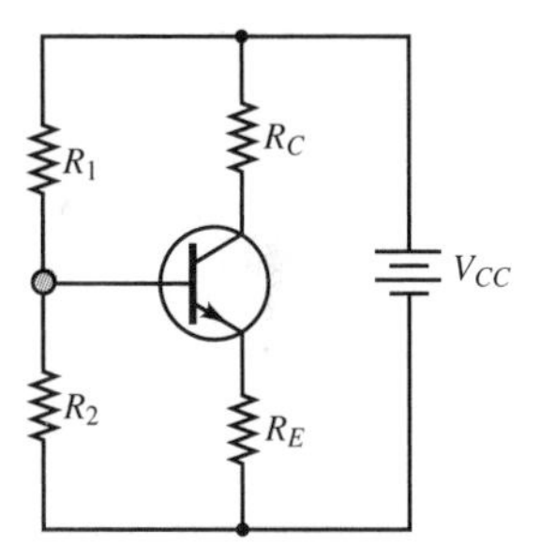

Figure 8.20 Practical BJT self-bias DC circuit

$$V_{BB} = \frac{R_2}{R_1 + R_2} V_{CC} \tag{8.6}$$

and of the equivalent resistance,

$$R_B = R_1 \,\|\, R_2 \tag{8.7}$$

Figure 8.21(b) shows a redrawn DC bias circuit that makes this observation more evident. The circuit to the left of the dashed line in Figure 8.21(a) is represented in Figure 8.21(b) by the equivalent circuit composed of V_{BB} and R_B.

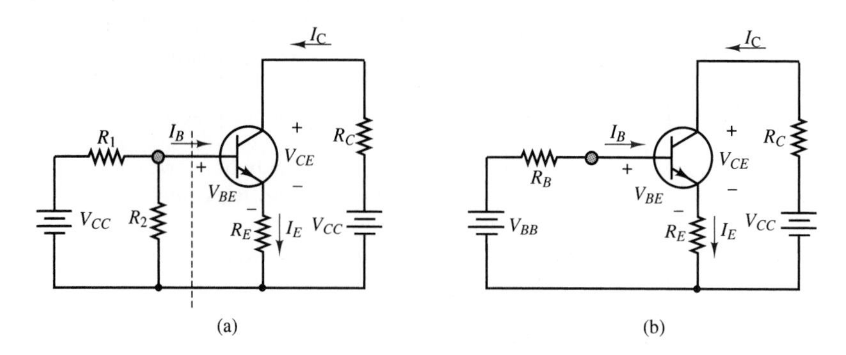

Figure 8.21 DC self-bias circuit represented in equivalent-circuit form

Recalling that the BE junction acts much as a diode, the following equations describe the DC operating point of the self-bias circuit. Around the base-emitter circuit,

$$V_{BB} = I_B R_B + V_{BE} + I_E R_E = [R_B + (\beta + 1)R_E]I_B + V_{BE} \quad \textbf{(8.8)}$$

where V_{BE} is the BE junction voltage (diode forward voltage) and $I_E = (\beta+1)I_B$. Around the collector circuit, on the other hand, the following equation applies:

$$V_{CC} = I_C R_C + V_{CE} + I_E R_E = I_C\left(R_C + \frac{\beta + 1}{\beta}R_E\right) + V_{CE} \quad \textbf{(8.9)}$$

since $I_E = I_B + I_C = (\frac{1}{\beta} + 1)I_C$. These two equations may be solved to obtain an expression for the base current,

$$I_B = \frac{V_{BB} - V_{BE}}{R_B + (\beta + 1)R_E} \quad \textbf{(8.10)}$$

from which the collector current can be determined as $I_C = \beta I_B$; and an expression for the collector-emitter voltage:

$$V_{CE} = V_{CC} - I_C\left(R_C + \frac{\beta + 1}{\beta}R_E\right) \quad \textbf{(8.11)}$$

This last equation is the load-line equation for the bias circuit. Note that the effective load resistance seen by the DC collector circuit is no longer just R_C, but is now given by

$$R_C + \frac{\beta + 1}{\beta}R_E \approx R_C + R_E$$

The following example provides a numerical illustration of the analysis of a DC self-bias circuit for a BJT.

Example 8.6

Find the DC operating point of the circuit shown in Figure 8.20, given $R_1 = 100\ \text{k}\Omega$, $R_2 = 50\ \text{k}\Omega$, $R_C = 5\ \text{k}\Omega$, $R_E = 3\ \text{k}\Omega$, $V_{CC} = 15$ V, $\beta = 100$, and $V_{BE} = 0.7$ V.

Solution:

The first step is to determine the equivalent base supply voltage, V_{BB}, and the equivalent base resistance, R_B. From the expression

$$V_{BB} = \frac{R_2}{R_1 + R_2} V_{CC}$$

we obtain

$$V_{BB} = \frac{50{,}000}{100{,}000 + 50{,}000}(15\text{ V}) = 5\text{ V}$$

and the equivalent base resistance can be computed to be

$$R_B = R_1 \parallel R_2 = \frac{100{,}000 \times 50{,}000}{100{,}000 + 50{,}000} = 33.3\ \text{k}\Omega$$

Now, it is possible to compute the base current:

$$I_B = \frac{V_{BB} - V_{BE}}{R_B + (\beta + 1)R_E} = \frac{5 - 0.7}{33{,}333 + (100 + 1) \times 3{,}000}\text{ A}$$
$$= 12.8\ \mu\text{A}$$

As a check, we can recompute the base voltage supply:

$$V_{BB} = V_{BE} + [R_B + (\beta + 1)R_E]I_B$$
$$= 0.7 + [33{,}333 + (100 + 1) \times 3{,}000] \times 12.8 \times 10^{-6}$$
$$= 5.0\text{ V}$$

Knowing the current amplification factor, β, we can easily find the collector current:

$$I_C = \beta I_B = 1.28\text{ mA}$$

Finally, the collector-to-emitter voltage can be computed from the expression

$$V_{CE} = V_{CC} - I_C\left(R_C + \frac{\beta + 1}{\beta}R_E\right)$$
$$= 15 - 1.28 \times 10^{-3} \times \left(5{,}000 + \frac{101}{100} \times 3{,}000\right)$$
$$= 4.72\text{ V}$$

Thus, the Q point of the transistor of Figure 8.20, given the component values of this example, is

$$I_{BQ} = 12.8\ \mu\text{A} \qquad I_{CQ} = 1.28\text{ mA} \qquad V_{CEQ} = 4.72\text{ V}$$

The material presented in this section has illustrated the basic principles that underlie the operation of a BJT and the determination of its Q point. In the next section and later, in Chapter 9, we shall develop some simple circuit models for the BJT that will enable us to analyze the transistor amplifier in the linear active region using the familiar tools of linear circuit analysis.

DRILL EXERCISES

8.4 Describe the operation of a *pnp* transistor in the active region, by analogy with that of the *npn* transistor.

8.5 For the circuit given in Figure 8.11, the readings are $V_1 = 3$ V, $V_2 = 2.4$ V, and $V_3 = 2.7$ V. Determine the operating state of the transistor.

8.6 For the circuit given in Figure 8.21, find the value of V_{BB} that yields a collector current $I_C = 6.3$ mA. What is the corresponding collector-emitter voltage (assume that $V_{BE} = 0.6$ V and that the transistor is in the active region)? Assume that $R_B = 50\ \text{k}\Omega$, $R_E = 200\ \Omega$, $R_C = 1\ \text{k}\Omega$, $\beta = 100$, and $V_{CC} = 14$ V.

8.7 What percent change in collector current would result if β were changed to 150 in Example 8.6? Why does the collector current increase by less than 50 percent?

8.3 BJT LARGE-SIGNAL MODEL

The *i-v* characteristics and the simple circuits of the previous sections indicate that the BJT acts very much as a current-controlled current source: a small amount of current injected into the base can cause a much larger current to flow into the collector. This conceptual model, although somewhat idealized, is useful in describing a **large-signal model** for the BJT, that is, a model that describes the behavior of the BJT in the presence of relatively large base and collector currents, close to the limit of operation of the device. A more careful analysis of the collector curves in Chapter 9 will reveal that it is also possible to generate a small-signal model, a model that describes the operation of the transistor as a linear amplifier of small AC signals. These models are certainly not a complete description of the properties of the BJT, nor do they accurately depict all of the effects that characterize the operation of such devices (for example, temperature effects, saturation, and cutoff); however, they are adequate for the intended objectives of this book, in that they provide a good, qualitative feel for the important features of transistor amplifiers.

Large-Signal Model of the *npn* BJT

The large-signal model for the BJT recognizes three basic operating modes of the transistor. When the BE junction is reverse-biased, no base current (and therefore no forward collector current) flows, and the transistor acts virtually as an open circuit; the transistor is said to be in the *cutoff region*. In practice, there is always a leakage current flowing through the collector, even when $V_{BE} = 0$ and $I_B = 0$. This leakage current is denoted by I_{CEO}. When the BE junction becomes forward-biased, the transistor is said to be in the *active region,* and the base current is amplified by a factor of β at the collector:

$$I_C = \beta I_B \tag{8.12}$$

Since the collector current is controlled by the base current, the controlled-source symbol is used to represent the collector current. Finally, when the base current becomes sufficiently large, the collector-emitter voltage, V_{CE}, reaches its saturation limit, and the collector current is no longer proportional to the base current; this is called the *saturation region.* The three conditions are described in Figure 8.22 in terms of simple circuit models. The corresponding collector curves are shown in Figure 8.23.

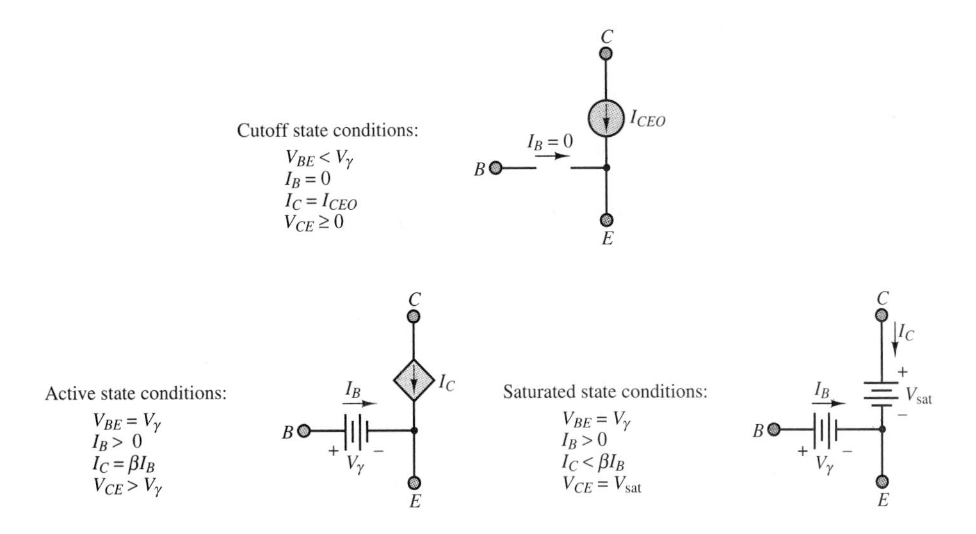

Figure 8.22 *npn* BJT large-signal model

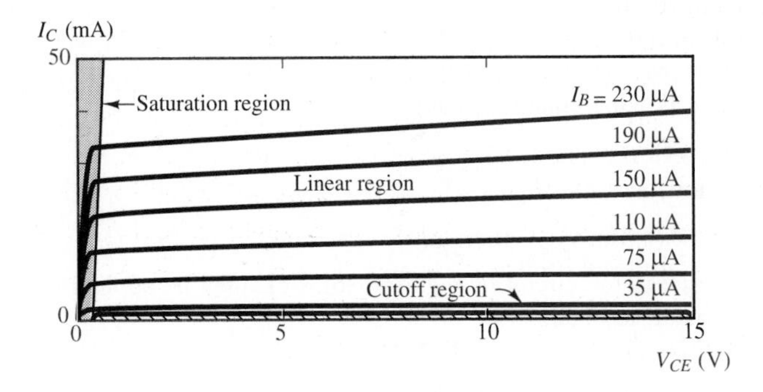

Figure 8.23 BJT collector characteristic

Example 8.7 illustrates the application of this large-signal model in a practical circuit and illustrates how to determine which of the three states is applicable, using relatively simple analysis.

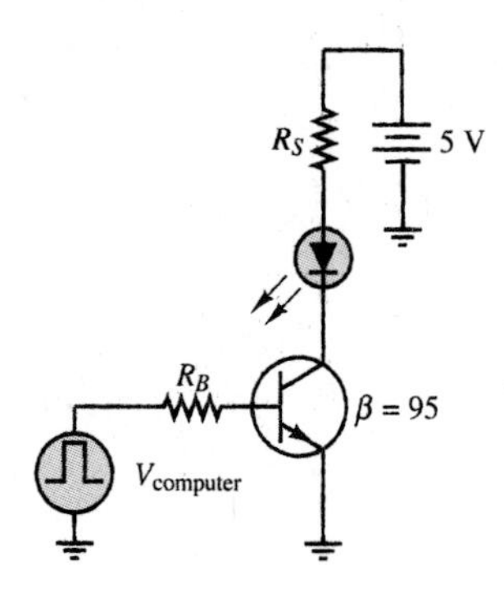

Figure 8.24 LED driver circuit

Example 8.7 LED Driver

The circuit shown in Figure 8.24 is being used to "drive" (i.e., provide power for) a light-emitting diode (LED) from a desktop computer. The signal available from the computer consists of a 5-V low-current output. The reason for using a BJT (rather than driving the LED directly with the signal available from the computer) is that the LED requires a substantial amount of current (at least 15 mA for the device used in this example) and the computer output is limited to 5 mA. Thus, some current amplification is required. The LED has an offset voltage of 1.4 V and a maximum power dissipation rating of 280 mW. The circuit has been designed so that the transistor will be either in cutoff, when the LED is to be turned off, or in saturation, when the LED is to be turned on.

Assume that the base-emitter junction of the transistor has an offset voltage of 0.7 V and that $R_B = 1\ \text{k}\Omega$ and $R_S = 42.5\ \Omega$.

1. When the computer is supplying 5 V to the circuit, is the transistor in cutoff, in saturation, or in the linear region of operation?
2. Determine the current in the LED when "turned on," and, thus, state whether the LED will exceed its power rating.

Solution:

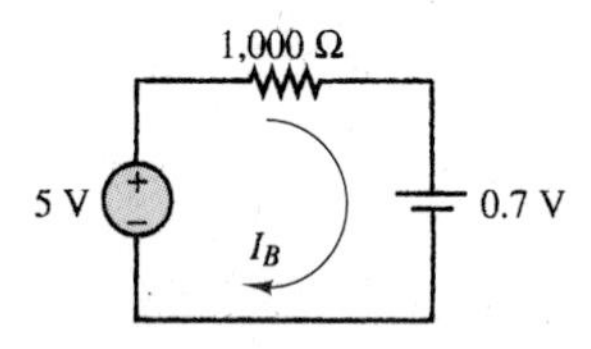

Figure 8.25 *BE* circuit for LED driver

1. The base-emitter circuit is considered first, to determine whether the *BE* junction is forward-biased. The equivalent circuit is shown in Figure 8.25.

   Writing KVL around the base circuit, we obtain

   $$(5 - 0.7) = I_B(1{,}000)$$

   or

   $$I_B = 4.3\text{ mA}$$

   Since this current is greater than zero (i.e., since positive current flows into the base), the transistor is not in cutoff.

   Next, we need to determine whether the transistor is in the linear active or in the saturation region. One method that can be employed to determine whether the device is in the active region is to assume that it is and to solve for the circuit that results from the assumed condition. If the resulting equations are consistent, then the assumption is justified. Assuming that the transistor is in its active region, the following equations apply:

   $$I_C = \beta I_B$$

   or

   $$I_C = 95(4.3\text{ mA}) = 408.5\text{ mA}$$

   With reference to the circuit of Figure 8.26, KVL may be applied around the collector circuit, to obtain

   $$5 = 1.4 + V_{CE} + I_C R_S$$

   or

   $$(5 - 1.4) - 408.5(42.5) = V_{CE}$$

   or

   $$V_{CE} = -13.76\text{ V}$$

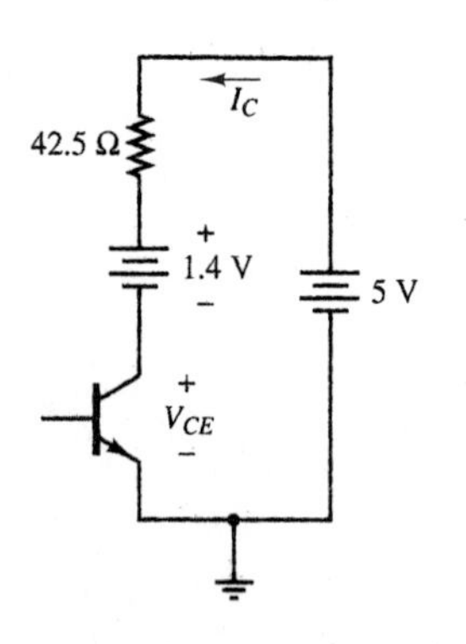

Figure 8.26 Equivalent collector circuit of LED driver, assuming that the BJT is in the linear active region

This result is clearly not possible, since the supply voltage is only 5 V! It must therefore be concluded that the transistor is not in the linear region and must be in saturation. To test this hypothesis, we can substitute the nominal saturation voltage for the BJT in place of V_{CE} (a typical saturation voltage would be $V_{CE\ \text{sat}} = 0.2$ V) and verify that in this case we obtain a consistent solution:

$$I_C = \frac{(5 - 1.4 - 0.2)}{42.5} = 80 \text{ mA}$$

This is a reasonable result, stating that the collector current in the saturation region for the given circuit is of the order of 80 mA.

2. In part 1, it was determined that the transistor was in saturation. Using the circuit model for saturation, we may draw the equivalent circuit of Figure 8.27. Since $I_C = 80$ mA, the power dissipated by the LED may be calculated as follows:

$$P_{\text{LED}} = I_C V_{\text{LED}} = 80 \text{ mA} \times 1.4 \text{ V} = 112 \text{ mW}$$

Thus, the power limitation of the LED will not be exceeded.

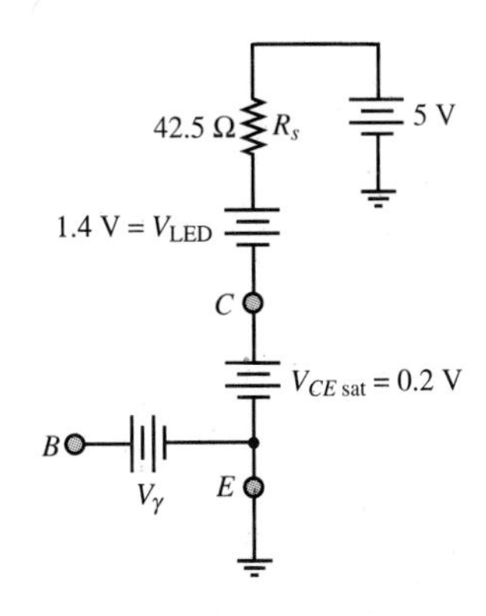

Figure 8.27 LED driver equivalent collector circuit, assuming that the BJT is in the saturation region

Example 8.8

This example demonstrates the use of the large-signal model for the BJT in a switching circuit. The input voltage, V_{in}, switches between two values, representing the circuit on and off conditions. In the example, we analyze the behavior of the transistor circuit of Figure 8.28 for different values of V_{in}, assuming a nominal transistor current gain $\beta = 60$. Find the output voltage, V_{out}, for: (1) $V_{\text{in}} = 0.2$ V; (2) $V_{\text{in}} = 0.6$ V; and (3) $V_{\text{in}} = 0.8$ V. The collector characteristic of transistor T is shown in Figure 8.9(b), and the base characteristic is shown in Figure 8.8.

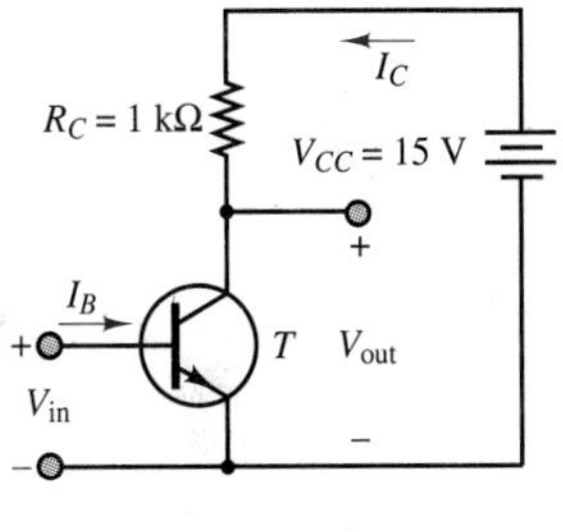

Figure 8.28

Solution:

1. For $V_{\text{in}} = V_{BE} = 0.2$ V, the base-emitter junction is not forward-biased. Thus,

$$I_B \approx 0 \qquad I_C \approx 0 \qquad V_{\text{out}} = 15 \text{ V}$$

The transistor is in the cutoff state.

2. For $V_{\text{in}} = V_{BE} = 0.6$ V, the BE junction is forward-biased. From Figure 8.8, we find that $I_B \approx 125$ μA and consequently compute:

$$I_C = \beta I_B = 60 \times 0.125 = 7.5 \text{ mA}$$

Thus, the output voltage is

$$V_{\text{out}} = V_{CE} = 15 - 1{,}000 \times 0.0075 = 7.5 \text{ V}$$

3. For $V_{\text{in}} = V_{BE} = 0.8$ V, $I_B = 400$ μA. If the basic relation $I_C = \beta I_B$ were to hold here, we should have $I_C = 24$ mA and $V_{CE} = V_{CC} - I_C R_C = 15 - 0.024 \times 1{,}000 = -9$ V. This is an impossible situation. For an npn transistor, V_{CE} must be positive. Thus, the linear active region model does not apply, and the transistor is in saturation. Using the saturation model of the BJT, we draw the

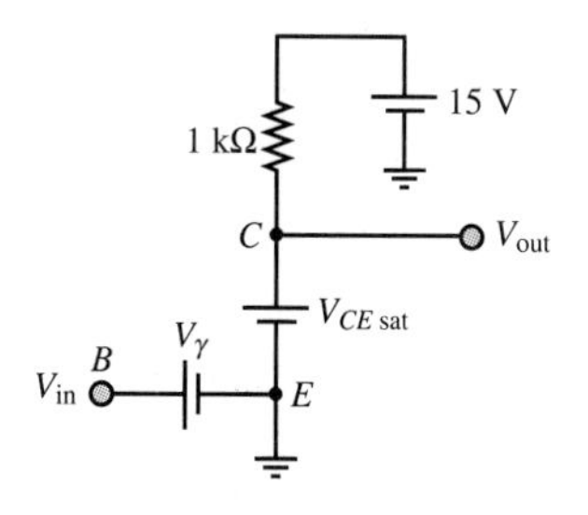

Figure 8.29 Transistor equivalent circuit using saturation model

equivalent circuit shown in Figure 8.29 and conclude that, for a nominal saturation voltage of 0.2 V, the output voltage is

$$V_{\text{out}} = 0.2 \text{ V}$$

The collector current is given by the expression

$$I_C = \frac{V_{CC} - V_{CE\ \text{sat}}}{R_C} = \frac{15 - 0.2}{1 \times 10^3} = 14.8 \text{ mA}$$

Note also that the effective β in this particular saturation condition is $I_C/I_B = 14.8/0.4 = 37$, which is smaller than the nominal value of 60, as is always the case for a transistor in saturation.

It is clear that a change as small as 0.6 V in the input voltage, from 0.2 V to 0.8 V, results in a corresponding change in output voltage from 15 V to 0.2 V, that is, from the cutoff state to the saturation state. This property is very useful in switching circuits. This type of circuit finds common application in digital logic devices, to be discussed in Chapter 12.

The large-signal model of the BJT presented in this section treats the *BE* junction as an offset diode and assumes that the BJT in the linear active region acts as an ideal controlled current source. In reality, the *BE* junction is better modeled by considering the forward resistance of the *pn* junction; further, the BJT does not act quite like an ideal current-controlled current source. These phenomena will be partially taken into account in the small-signal model introduced in Chapter 9.

DRILL EXERCISES

8.8 Repeat the analysis of Example 8.7 if $R_S = 400\ \Omega$. Which region is the transistor in? What is the collector current?

8.9 What is the power dissipated by the LED of Example 8.7 if $R_S = 30\ \Omega$?

Enhancement MOS

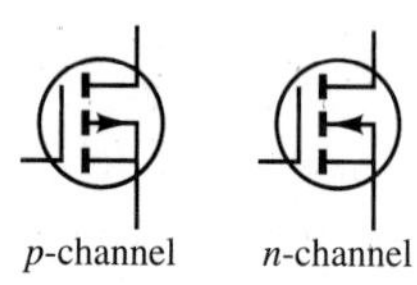

p-channel *n*-channel

Depletion MOS

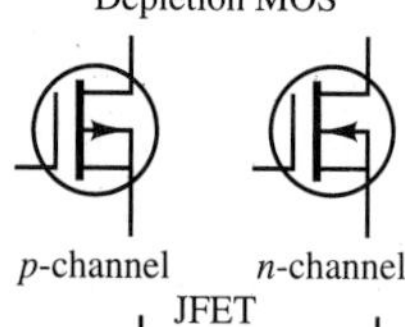

p-channel *n*-channel

JFET

p-channel *n*-channel

Figure 8.30 Classification of field-effect transistors

8.4 FIELD-EFFECT TRANSISTORS

The second transistor family discussed in this chapter operates on the basis of a principle that is quite different from that of the *pn* junction devices. The concept that forms the basis of the operation of the **field-effect transistor,** or **FET,** is that the width of a conducting channel in a semiconductor may be varied by the external application of an electric field. Thus, FETs behave as *voltage-controlled resistors*. This family of electronic devices can be subdivided into three groups, all of which will be introduced in the remainder of this chapter. Figure 8.30 depicts the classification of field-effect transistors, as well as the more commonly used symbols for these devices. These devices can be grouped into three major categories. The first two categories are both types of **metal-oxide-semiconductor field-effect transistors,** or **MOSFETs: enhancement-mode MOSFETs** and **depletion-mode MOSFETs.** The third category consists of **junction field-effect transistors,** or **JFETs.** In addition, each of these devices can be fabricated either as an ***n*-channel** device or as a ***p*-channel** device, where the n or p designation indicates the nature of the doping in the semiconductor channel. All these transistors behave in a very similar fashion, and we shall predominantly discuss enhancement MOSFETs in this chapter, although some discussion of depletion devices and JFETs will also be included.

Operation of the *n*-Channel Enhancement-Mode MOSFET

The construction of the MOSFET we will be examining is shown in Figure 8.31, along with its circuit symbol. The device consists of a metal **gate,** insulated from the bulk of the p-type semiconductor material by an oxide layer (thus the terminology *metal-oxide-semiconductor*). Two n^+ regions on either side of the gate are called **source** and **drain,** respectively. An electric field can be applied between the gate and the *bulk* of the semiconductor by connecting the gate to a voltage source. The effect of the electric field, the intensity of which depends on the strength of the gate voltage, is to push positive charge carriers away from the surface of the bulk material. As a consequence of this effect, the p-type material has a lower concentration of positive charge carriers near its surface, and it behaves progressively more like intrinsic semiconductor material and then, finally, like n-type material as the electric field strength increases. Thus, in a narrow layer between n^+ regions, the p-type material is *inverted* and n-type charge carriers become available for conduction. This layer is called the *channel*. The thickness of the channel is, to a certain extent, controlled by the intensity of the field: the stronger the field, the farther into the bulk material the channel extends. This behavior is illustrated in Figure 8.32, where the n channel thus formed is shown for different field intensities. Since the channel thus formed is n-type, the device of Figure 8.31 takes the name *n-channel MOSFET*. It is also called an *enhancement-mode* device, because the applied electric field "enhances" the conduction in the channel by attracting n-type charge carriers. There are also *depletion-mode* devices, in which the application of an electric field "depletes" the channel of charge carriers, reducing the effective channel width. It is useful to think of enhancement-mode devices as being normally off: current cannot flow from drain to source unless a gate voltage is applied. On the other hand, depletion-mode devices are normally on; that is, the gate voltage is used to reduce the conduction of current from drain to source.

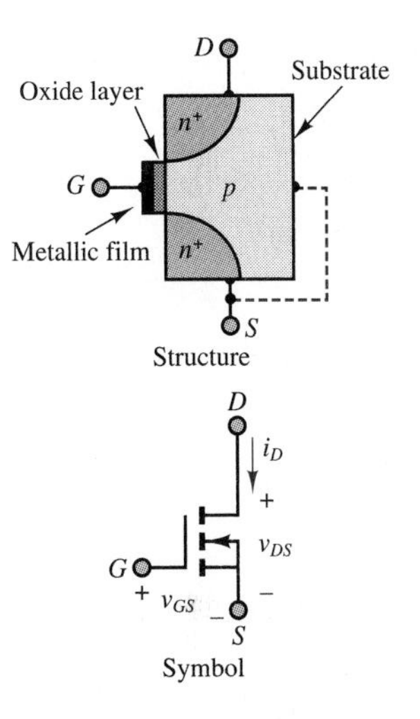

Figure 8.31 n-channel enhancement MOSFET

An analogous discussion could be carried out for *p-channel MOSFETs*. In a p-channel device, conduction in the channel occurs through positive charge carriers. The correspondence between n-channel and p-channel devices is akin to that between *npn* and *pnp* bipolar devices.

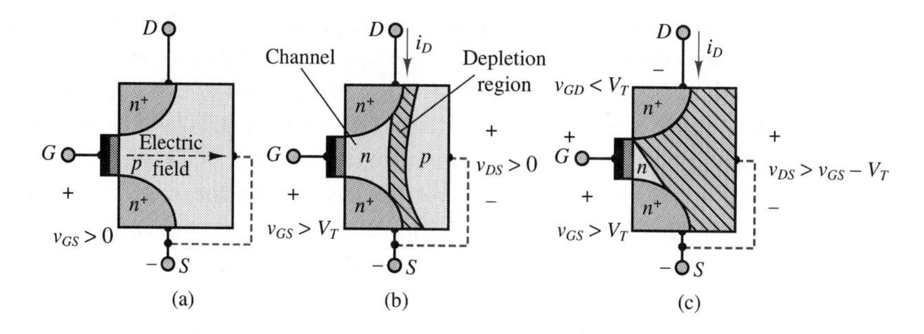

Figure 8.32 (a) Electric field in an FET; (b) n-channel formation; (c) pinched-down channel

Let us now return to the n-channel enhancement MOSFET and to the diagrams of Figure 8.32 to better understand the operation of this device. Figure 8.32(a) shows the direction of the electric field resulting from the application of a positive drain-to-source voltage. When this voltage, v_{GS}, exceeds a **threshold voltage** denoted by V_T, an n-type **channel** begins to form, as shown in Figure 8.32(b). Between the n-type channel and the p-type substrate, a depletion region forms, indicated by the shaded region in Figure 8.32(b) and (c). If the gate-to-source voltage is above V_T and the drain-to-source voltage, v_{DS}, is positive, then electrons from the n^+ source region are injected into the channel and eventually reach the drain (note the analogy between the source and the emitter region in the BJT, and the drain and the collector region in the BJT). Unlike what occurs in bipolar devices, conduction is due exclusively to majority carriers (electrons, in this case), and thus FETs are *unipolar* devices. Since electrons are negative charge carriers, the drain current is in the direction indicated in the figure. Now the p-type substrate forms a pn junction with respect to the n-type channel. The existence of such a junction is denoted by the arrow in the MOSFET symbol of Figure 8.31.

As the gate voltage increases above V_T, so does the strength of the electric field; and therefore the channel width increases; thus, the applied field *enhances* conduction. Note that enhancement-mode devices are normally *off*, and that they can conduct only when an external field is applied. Conversely, a depletion-mode device will conduct in the absence of an external field.

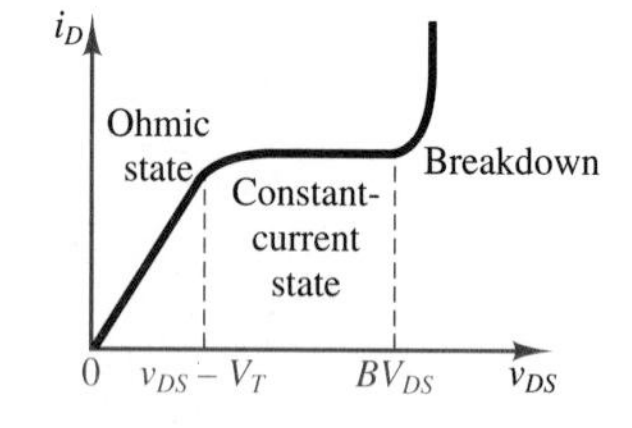

Figure 8.33 MOSFET i-v curve

We are now ready to describe qualitatively the i-v characteristic of this enhancement-mode MOSFET. For small values of drain-to-source voltage, v_{DS}, and for constant v_{GS}, the channel has essentially constant width (as shown in Figure 8.32(b)) and therefore acts as a constant resistance. As the gate voltage is changed, this resistance can be varied over a certain range. This mode of operation, for small drain voltages, is called the **ohmic state,** and, as depicted in Figure 8.33, it corresponds to a linear i-v curve for fixed v_{GS}, as would be expected of a resistor. Thus, in the ohmic state the MOSFET acts as a **voltage-controlled resistor.**

As the drain voltage is increased, the gate-to-drain voltage, v_{GD}, decreases, reducing the electric field strength at the drain end of the device. Consequently, the channel is narrower toward the drain than it is at the source end, as illustrated in Figure 8.32(c). When $v_{GD} = v_{GS} - v_{DS} < V_T$ and $v_{DS} > v_{GS} - V_T$, the channel is *pinched down,* and the electron flow into the drain is physically limited, so that the drain current becomes essentially constant. This phenomenon is clearly reflected in the curve of Figure 8.33, where it is shown that for drain-to-source voltages above $v_{DS} > v_{GS} - V_T$, the drain current becomes constant, independent of v_{DS}. This mode of operation is the **constant-current state.** If v_{DS} exceeds a given **drain breakdown voltage,** BV_{DS} (usually between 20 and 50 V), avalanche breakdown occurs and the drain current increases substantially. Operation in this **breakdown state** can lead to permanent damage because of the excessive heat caused by the large drain current. Finally, if v_{GS} exceeds the **gate breakdown voltage** (around 50 V), permanent damage to the oxide layer can occur. It is important to know that it is possible to generate *static* voltages of magnitude sufficient to exceed this breakdown voltage just by handling the device; thus, some attention must usually be paid to static voltage buildup in handling MOS circuits.

p-Channel MOSFETs and CMOS Devices

As the designation indicates, a *p*-channel MOSFET is characterized by *p*-type doping; the construction and symbol are shown in Figure 8.34. Note the opposite direction of the arrow to indicate that the *pn* junction formed by the channel and substrate is now in the opposite direction. The direction of drain current is opposite; therefore v_{DS} and v_{GS} are now negative. A more convenient reference is obtained if voltages are defined in the direction opposite to that for the *n*-channel device: if one defines $v_{SD} = -v_{DS}$ and $v_{SG} = -v_{GS}$, then these voltages will be positive for the drain current direction indicated in Figure 8.34. The carriers are holes in this device, since the channel, when formed, is *p*-type. Aside from the nature of the charge carriers and the direction of current and polarity of the voltages, the *p*-channel and *n*-channel transistors behave in conceptually the same way; however, since holes are in general less mobile (recall the discussion of carrier *mobility* in Section 7.1), *p*-channel MOSFETs are not used very much by themselves. They do find widespread application in **complementary metal-oxide-semiconductor (CMOS) devices.**

CMOS devices take advantage of the complementary symmetry of *p*- and *n*-channel transistors built on the same integrated circuit. Example 8.12 later in this chapter will illustrate the operation of a common CMOS device.

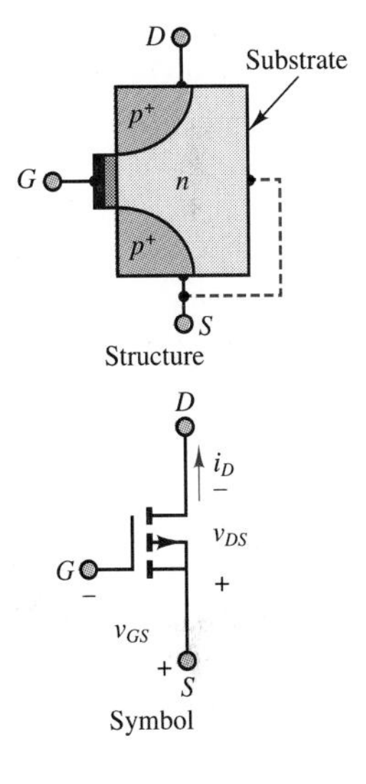

Figure 8.34 *p*-channel enhancement MOSFET

8.5 MOSFET CHARACTERISTIC CURVES AND LARGE-SIGNAL MODEL

The information contained in the curve of Figure 8.33 is usually presented as a family of curves, where the third parameter, in addition to the drain current and drain-source voltage, is the gate-to-source voltage, v_{GS}. For example, one could use the test circuit of Figure 8.35 to obtain a set of *i-v* curves such as that

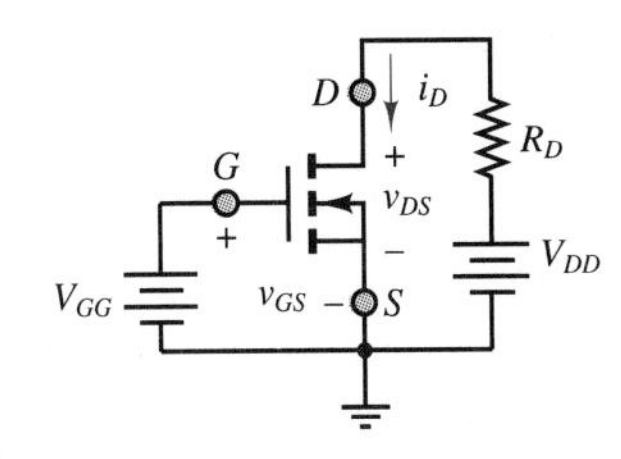

Figure 8.35 Test circuit to obtain MOSFET drain characteristic curves

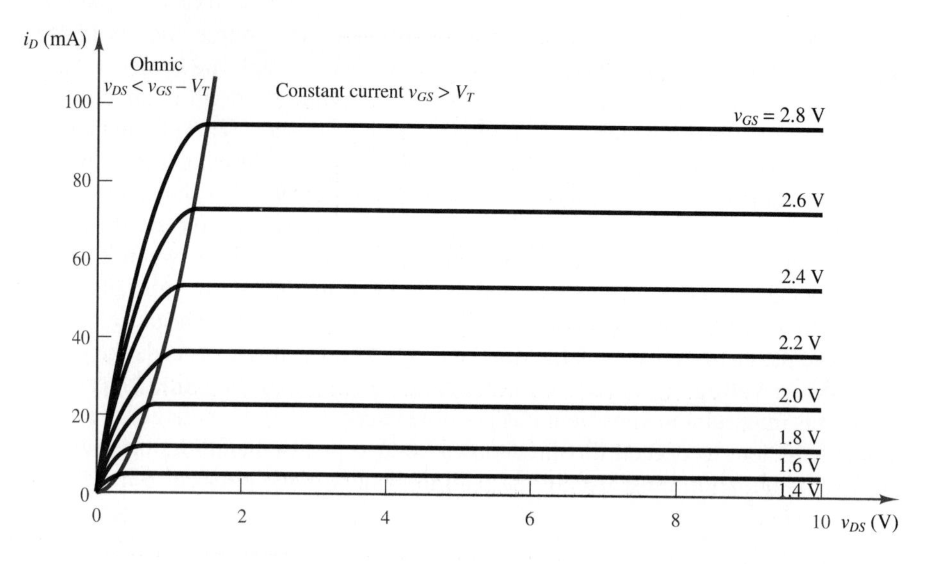

Figure 8.36 *n*-channel MOSFET drain characteristic

of Figure 8.36, where each *i*-*v* curve corresponds to one value of v_{GS}. Note the similarity between these drain curves and the BJT collector characteristic curves. There are, however, several differences: (1) the controlling parameter in the case of the MOSFET is a voltage, whereas for the BJT it was the base current; (2) the spacing between the curves is nonlinear in v_{GS}, whereas it was very nearly linear in I_B in the case of the BJT collector curves; and (3) the drain characteristic is sufficient by itself to determine the operating point of the MOSFET, whereas for the BJT we also required knowledge of the base *i*-*v* curve. These points illustrate that there are fundamental differences and similarities between bipolar and field-effect transistors; we shall describe some of these in greater detail as we proceed with the present discussion. Example 8.9 illustrates the use of the drain characteristic curves in the determination of the Q point of a MOSFET circuit.

Example 8.9

To illustrate the determination of the Q point for a MOSFET, we employ the circuit shown in Figure 8.37. Applying KVL around the drain circuit yields:

$$10 = v_{DS} + 100 i_D$$

The resulting load-line curve is sketched in Figure 8.37, superimposed on the drain characteristic. The Q point that corresponds to a gate voltage of 2.4 V can then be determined to be approximately given by $i_{DQ} = 50$ mA, $v_{DSQ} = 4.75$ V.

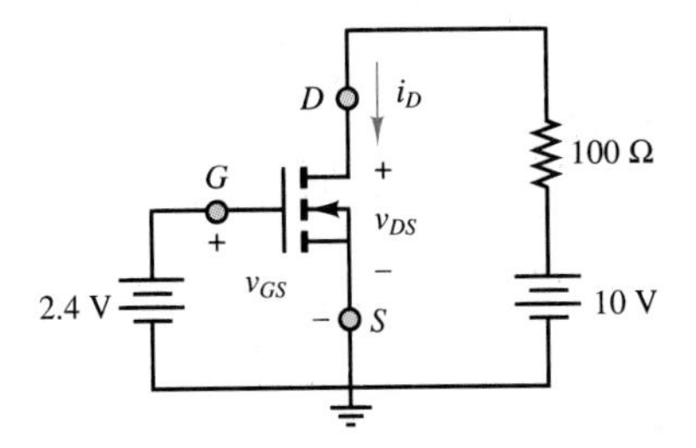

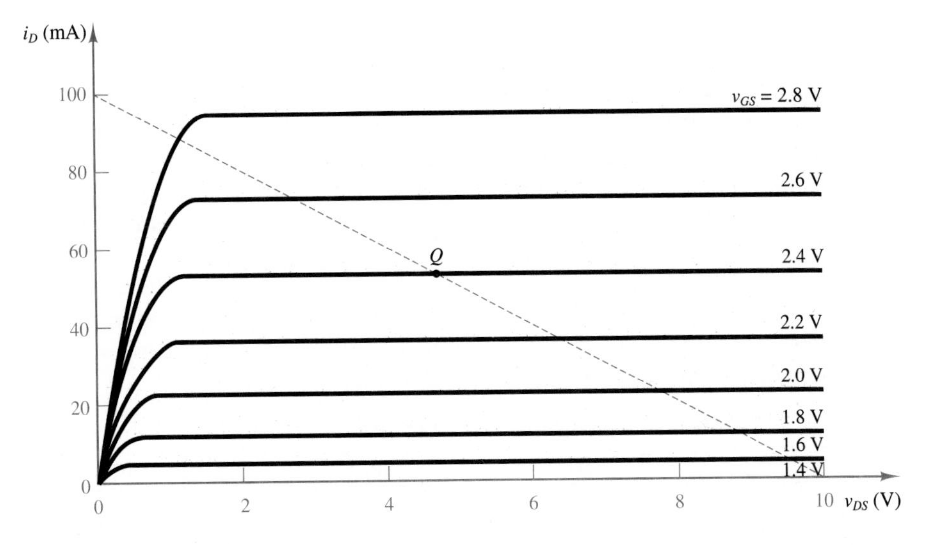

Figure 8.37 *n*-channel enhancement MOSFET circuit and drain characteristic for Example 8.9

In analyzing the curves of Figure 8.36, we see that for values of $v_{GS} \leq 1.4$ V the drain current is zero; thus, the threshold voltage for this MOSFET is $V_T = 1.4$ V. Note that the gate-source voltage is then increased in equal increments of 0.2 V, but that the spacing between the resulting i_D-v_D curves is not even, instead increasing progressively more as v_{GS} increases. This is what we mean when we say that the MOSFET characteristic has a nonlinear dependence on v_{GS}. As v_{DS} is increased (for each value of v_{GS}) we first see the linear-resistor behavior of the MOSFET in the ohmic state; then, after passing through a transition region (knee), the drain curves flatten out, as the MOSFET moves to the constant-current state described earlier. If the drain voltage were increased further, the MOSFET would then enter the breakdown state.These operating states of the *n*-channel enhancement MOSFET correspond to various regions of operation of the device:

1. The **cutoff region,** where the drain current is near zero and the drain voltage is large: $i_D \ll V_{DD}/R_D$ and $v_{DS} \approx V_{DD}$. This is the normally off state of the device.

2. The **saturation region,** where the drain current is small and the drain voltage is small: $i_D \approx V_{DD}/R_D$ and $v_{DS} \ll V_{DD}$. This corresponds to the ohmic state of the MOSFET.
3. The **active region,** where the drain voltage and current values are between those of the preceding two cases and the drain current is controlled by the gate voltage. This corresponds to the constant-current state of the MOSFET.
4. The **breakdown region,** which corresponds to the breakdown state of the MOSFET.

Note that the same terminology used for bipolar transistors was used to define the operating *regions* of a MOSFET, although the operating *states* of the latter device are quite different, physically, from those of a BJT. It is important to observe that, although the devices have fundamental physical differences, there is a substantial similarity in their operation, as described by the characteristic curves. That is why we have chosen to describe the different regions of operation by the same names in both cases.

The MOSFET drain characteristics can also be described quite effectively by a concise set of equations that lead to the so-called **universal curves.** These equations are valid for any enhancement-mode MOSFET that is operated below breakdown. The equations are summarized in Table 8.1.

Table 8.1 ***n*-channel enhancement MOSFET equations**

State	Conditions	Equation	Remarks
Constant current (active region)	$v_{DS} > v_{GS} - V_T$, $v_{GS} > V_T$	$i_D = \frac{I_{DSS}}{V_T^2}(v_{GS} - V_T)^2 = I_{DSS}\left(\frac{v_{GS}}{V_T} - 1\right)^2$	Nonlinear voltage-controlled current source operation.
Ohmic (saturation region)	$\lvert v_{DS}\rvert < \frac{1}{4}(v_{GS} - V_T)$, $v_{GS} > V_T$	$i_D \approx \frac{v_{DS}}{R_{DS}}$, $R_{DS} = \frac{V_T^2}{2I_{DSS}(v_{GS} - V_T)}$	Voltage-controlled resistor operation. R_{DS} is the equivalent drain-to-source resistance.

A few additional notes should be made with regard to the equations of Table 8.1. The parameter I_{DSS} represents the value of i_D when $v_{GS} = 2V_T$. The active-region equation can also be written as $i_D = k(v_{GS} - V_T)^2$, where $k = \frac{I_{DSS}}{V_T^2}$. Note also that operation as a voltage-controlled resistor can take place for both positive and negative drain current values (hence the absolute value sign around v_{DS}); this fact makes it possible for the MOSFET in the voltage-controlled resistor mode to conduct current in either direction.

The equations of Table 8.1 can be used to construct universal curves for the *n*-channel enhancement-mode MOSFET, shown in Figure 8.38. The dashed curve in Figure 8.38(a) denotes the boundary of the active region, corresponding to $v_{DS} = v_{GS} - V_T$ and $i_D = I_{DSS}(\frac{v_{DS}}{V_T})^2$. Another useful curve, shown in Figure 8.38(b), is the **transfer characteristic** of the MOSFET in the active region; recall

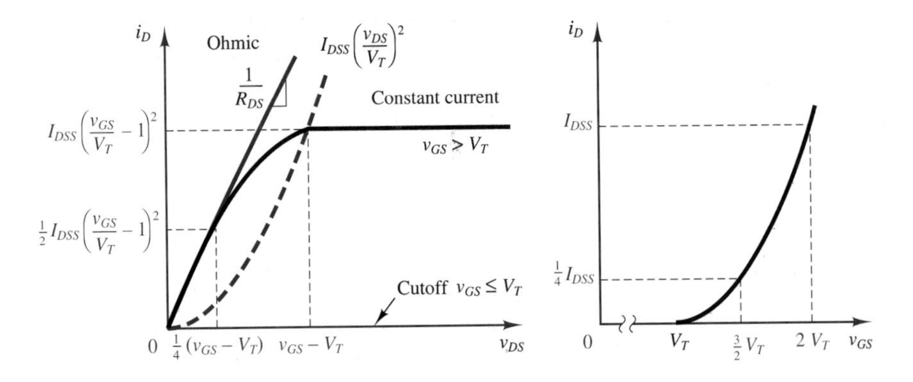

Figure 8.38 (a) Universal curves for n-channel enhancement MOSFET; (b) transfer characteristic in active region

that in this region the transistor operates as a voltage-controlled current source, as shown in the figure. Observe also that the relationship between gate voltage and drain current is nonlinear, unlike the case for the BJT, which had a linear relationship between base and collector currents in the active region.

To summarize this brief discussion of MOSFET characteristic curves and the large-signal model, we wish to emphasize that the equations of Table 8.1, which are reflected in the curves of Figure 8.38, permit large-signal analysis of MOSFET circuits with relative ease, and that only two parameters need to be known: I_{DSS} and V_T. The values of these parameters vary depending on the physical characteristics of the device; the threshold voltage can be in the range of 1 to 6 volts, while I_{DSS} can range from tens of microamperes to tens of amperes for power devices. It is important, however, to understand that the values of these two parameters can vary significantly (by as much as 100%) from transistor to transistor.

Example 8.10

The aim of this example is to repeat the calculations of Example 8.9, in which a graphical approach was used, by means of the universal equations of Table 8.1. The curves of Figure 8.38 permit the estimation of the two MOSFET parameters described in the preceding section. Since the drain current is zero for gate voltages less than or equal to 1.4 V, we know that the threshold voltage is $V_T = 1.4$ V. The drain current corresponding to $v_{GS} = 2V_T$ is approximately 95 mA; therefore $I_{DSS} = 95$ mA. From these parameters, and knowing that the gate voltage is 2.4 V, we can calculate the Q point to be:

$$i_{DQ} = I_{DSS}\left(\frac{v_{GS}}{V_T} - 1\right)^2 = 95 \times \left(\frac{2.4}{1.4} - 1\right)^2 = 48.5 \text{ mA}$$

$$v_{DSQ} = 10 - R_D i_{DQ} = 5.15 \text{ V}$$

These results differ only slightly from the (approximate) graphical analysis of Example 8.9. It is clear that the MOSFET is in the active region.

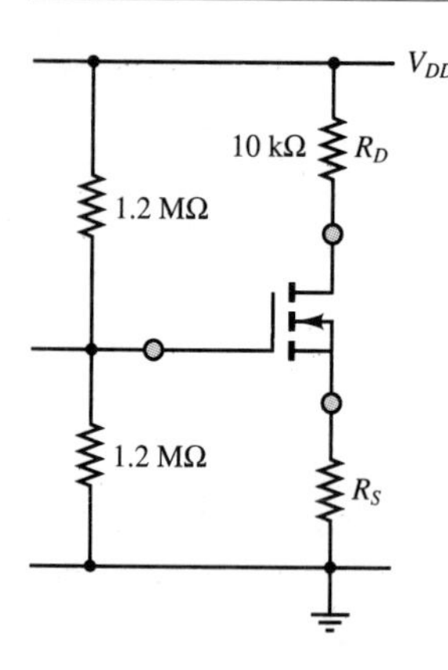

Figure 8.39 Self-bias circuit for Example 8.11

Example 8.11 Self-Bias Circuit for a MOSFET

The circuit of Figure 8.39 depicts a practical self-bias circuit for a MOSFET. Let $V_T = 4$ V, $I_{DSS} = 7.2$ mA, and the supply voltage $V_{DD} = 30$ V. What is the proper choice of R_S such that the operating point of the transistor is $v_{DSQ} = 8$ V? What are i_{DQ} and v_{GSQ}?

Solution:

Let all resistances be in kΩ and all currents in mA. The equation for the gate circuit is:

$$V_{GG} = v_{GSQ} + R_S i_{DQ} \tag{a}$$

where $V_{GG} = \frac{V_{DD}}{2} = 15$ V. We also know from the MOSFET equations that

$$i_{DQ} = I_{DSS}\left(\frac{v_{GS}}{V_T} - 1\right)^2 = 7.2\left(\frac{v_{GSQ}}{4} - 1\right)^2 = \frac{7.2}{16}v_{GSQ}^2 - \frac{14.4}{4}v_{GSQ} + 7.2 \tag{b}$$

Finally, the drain circuit equation is:

$$v_{DSQ} = 8 = 30 - 10i_{DQ} - R_S i_{DQ} \tag{c}$$

The three unknowns we are seeking to solve for are i_{DQ}, v_{GSQ}, and R_S; since we have three distinct equations, a solution exists. Substituting $R_S i_{DQ} = 15 - v_{GSQ}$ from equation a into equation c, we obtain:

$$8 = 30 - i_{DQ}(10) - 15 + v_{GSQ}$$

or

$$i_{DQ} = 0.7 + 0.1v_{GSQ} \tag{d}$$

Equation d can be substituted into equation b to obtain a quadratic equation in v_{GSQ}. It is typical of FET problems to require the solution of quadratic equations, because of the nature of the defining relationships. In our case, the equation for v_{GSQ} is:

$$0.45v_{GSQ}^2 - 3.7v_{GSQ} + 6.5 = 0 \tag{e}$$

leading to the two roots $v_{GSQ} = 5.68$ V and $v_{GSQ} = 2.54$ V. Since it is impossible to decide which of these two solutions is the correct one, we shall try both. Substituting the first one into equation d, we obtain $i_{DQ} = 1.268$ mA; the second root yields $i_{DQ} = 0.954$ mA. Each of these roots can be substituted into the drain equation c, to obtain two values of R_S:

$$8 = 30 - (10 + R_S)i_{DQ}$$

leading to $R_S = 7.35$ kΩ and $R_S = 13$ kΩ. Both of these values are physically possible; therefore we conclude that the desired drain-source voltage of 8 V can be attained with either of these two solutions for R_S.

Example 8.12 MOSFET Bidirectional Gate

The variable-resistor feature of MOSFETs in the ohmic state finds application in the **analog transmission gate.** The circuit shown in Figure 8.40 depicts a circuit constructed using CMOS technology. The circuit operates on the basis of a control voltage, v, that can be either "low" (say, 0 V), or "high" ($v > V_T$), where V_T is the threshold voltage for the n-channel MOSFET and $-V_T$ is the threshold voltage for the p-channel MOSFET.

The circuit operates in one of two modes. When the gate of Q_1 is connected to the high voltage and the gate of Q_2 is connected to the low voltage, the path between v_{in} and v_{out} is a relatively small resistance, and the transmission gate conducts. When the gate of Q_1 is connected to the low voltage and the gate of Q_2 is connected to the high voltage, the transmission gate acts like a very large resistance and is an open circuit for all practical purposes. A more precise analysis follows.

Let $v = V > V_T$ and $\overline{v} = 0$. Assume that the input voltage, v_{in}, is in the range $0 \le v_{\text{in}} \le V$. To determine the state of the transmission gate, we shall consider only the extreme cases $v_{\text{in}} = 0$ and $v_{\text{in}} = V$. When $v_{\text{in}} = 0$, $v_{GS1} = v - v_{\text{in}} = V - 0 = V > V_T$. Since V is above the threshold voltage, MOSFET Q_1 conducts (in the ohmic state). Further, $v_{GS2} = \overline{v} - v_{\text{in}} = 0 > -V_T$. Since the gate-source voltage is not more negative than the threshold voltage, Q_2 is in cutoff and does not conduct. Since one of the two possible paths between v_{in} and v_{out} is conducting, the transmission gate is on. Now consider the other extreme, where $v_{\text{in}} = V$. By reversing the previous argument, we can see that Q_1 is now off, since $v_{GS1} = 0 < V_T$. However, now Q_2 is in the ohmic state, because $v_{GS2} = \overline{v} - v_{\text{in}} = 0 - V < -V_T$. In this case, then, it is Q_2 that provides a conducting path between the input and the output of the transmission gate, and the transmission gate is also on. We have therefore concluded that when $v = V$ and $\overline{v} = 0$, the transmission gate conducts and provides a near zero-resistance (typically tens of ohms) connection between the input and the output of the transmission gate, for values of the input ranging from 0 to V.

Let us now reverse the control voltages and set $v = 0$ and $\overline{v} = V > V_T$. It is very straightforward to show that in this case, regardless of the value of v_{in}, both Q_1 and Q_2 are always off; therefore, the transmission gate is essentially an open circuit.

The analog transmission gate finds common application in *analog multiplexers* and *sample-and-hold* circuits, to be discussed in Chapter 14.

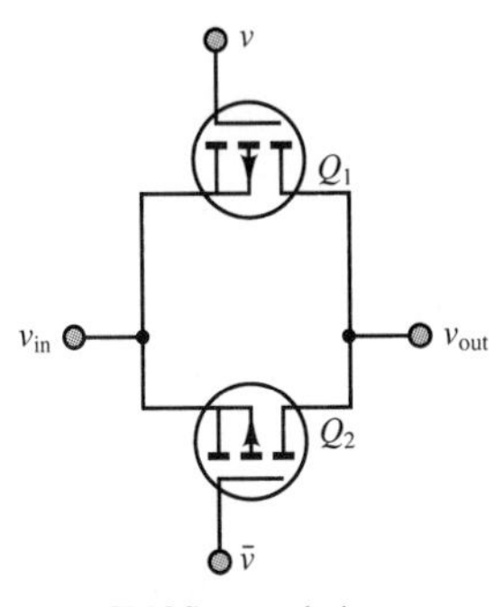

(a) CMOS transmission gate

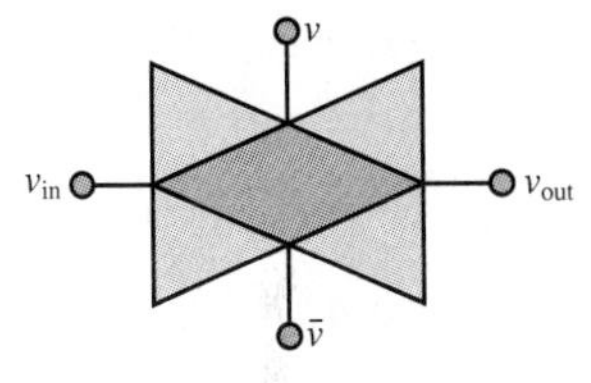

(b) CMOS transmission gate circuit symbol

Figure 8.40 Analog transmission gate

DRILL EXERCISES

8.10 Determine the operating state of the MOSFET of Example 8.10 when $v_{GS} = 3.5$ V.

8.11 Determine the appropriate value of R_S if we wish to move the operating point of the MOSFET of Example 8.11 to $v_{DSQ} = 12$ V. Also find the values of v_{GSQ} and i_{DQ}. Are these values unique?

8.12 Show that the CMOS bidirectional gate of Example 8.12 is off for all values of v_{in} between 0 and V whenever $v = 0$ and $\overline{v} = V > V_T$.

8.13 Find the lowest value of R_D for the MOSFET of Example 8.10 that will place the MOSFET in the ohmic region.

8.6 DEPLETION MOSFETs AND JFETs

The two remaining types of field-effect transistors are normally "on" devices and have very similar characteristics; in both depletion MOSFETs and JFETs, the external field reduces conduction by depleting an existing channel.

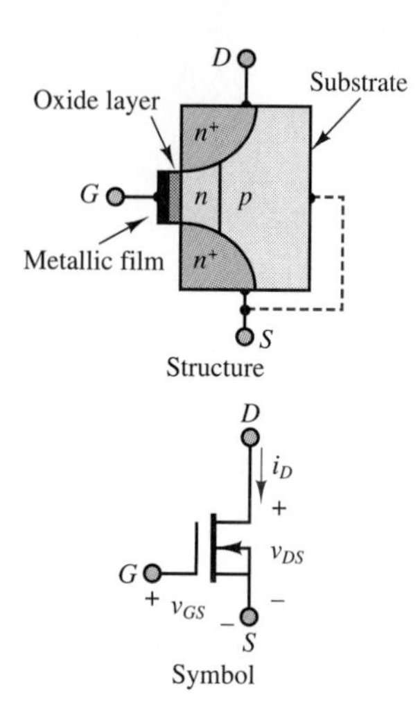

Figure 8.41 Depletion MOSFET

Depletion MOSFETs

The structure of an n-channel depletion MOSFET is shown in Figure 8.41. Note that the only difference with respect to the enhancement type is the addition of a lightly doped n-type region. This region in effect constitutes a channel; thus, a field is not required to initiate conduction between drain and source. If a positive gate voltage is applied, electrons will move from the substrate to the channel, further enhancing conduction; however, a negative gate voltage has the opposite effect, pushing electrons into the substrate and drawing holes into the channel. Negative gate voltages therefore have the effect of depleting the channel of charge carriers and of forming two depletion regions, shown in Figure 8.42(a). Thus, a negative gate voltage has the effect of reducing the drain current; when the gate voltage reaches the **pinch-off voltage,** $-V_P$, the channel is "pinched off," and no drain current flows. Note that the quantity V_P is positive; thus, $-V_P$ is a negative voltage. This condition is shown in Figure 8.42(b). On the other hand, the channel is only partially blocked (or *pinched down*) if $v_{GS} > -V_P$ but $v_{DS} > v_{GS} + V_P$, as shown in Figure 8.42(c). This latter condition corresponds to the constant-current state of the MOSFET.

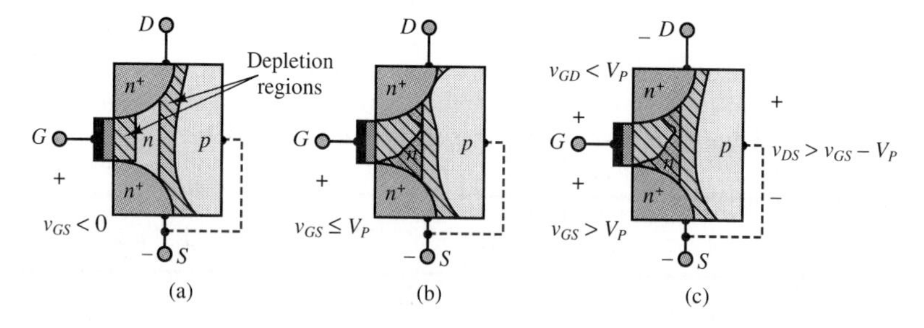

Figure 8.42 Channel formation in depletion MOSFET

Figure 8.43 depicts the drain characteristic for an n-channel depletion MOSFET. In analogy with the analysis of the enhancement MOSFET performed earlier, we see that the pinch-off voltage is $V_P = 4$ V, since no drain current flows for gate voltages lower than -4 V. For gate voltages between -4 and 0 V, the MOSFET operates in the depletion mode; when $v_{GS} > 0$, it operates in the enhancement mode. Regardless of whether in the depletion or the enhancement mode, for small drain voltages the MOSFET is in the ohmic state, and acts as a linear resistor, until the channel is pinched down. For all practical purposes the drain characteristic of a depletion MOSFET is very similar to that of an enhancement MOSFET, with two principal differences:

1. The depletion MOSFET allows negative as well as positive gate voltages.
2. The depletion MOSFET can be in the active region when $v_{GS} = 0$, whereas the enhancement MOSFET requires gate voltages above V_T.

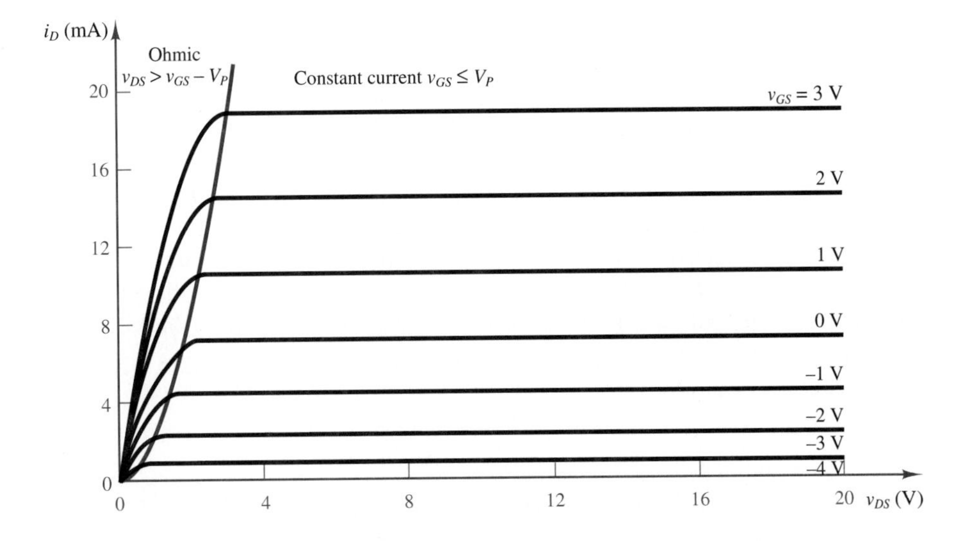

Figure 8.43 n-channel depletion MOSFET drain characteristic

JFETs

The last class of field-effect transistors discussed in this chapter is the junction field-effect transistor, or JFET. These devices are similar in construction to MOS devices, except for the fact that instead of a metal-oxide interface at the gate, a pn junction is used for the gate contact, as shown in Figure 8.44(a). Note that the gate is not insulated, as was the case for MOSFETs; therefore, a gate current, i_G, is also indicated in the figure. In a JFET, the application of a reverse bias to the gate junction ($v_{GS} < 0$) causes the creation of a depletion region, shown as the dashed area in Figure 8.44(c); the depletion region reduces the width of the channel and, consequently, the drain current. Depending on the values of v_{DS} and v_{GS}, the channel may be pinched down, or pinched off altogether. Thus, the

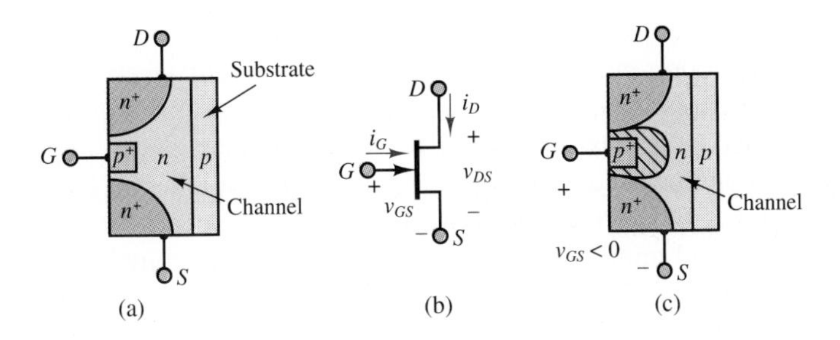

Figure 8.44 n-channel JFET

JFET behaves in a manner very similar to a depletion MOSFET. The principal differences between depletion MOSFETs and JFETs are the following:

1. Since the gate junction is a *pn* junction, there will exist a reverse saturation current for $v_{GS} < 0$. This current is usually denoted by $i_G = -I_{GSS}$.
2. Positive gate voltages above the junction forward bias potential (≈ 0.6 V) will cause a large current to flow into the gate. This is not compatible with the intended operation of the JFET, and thus *positive gate voltages cannot be applied,* and therefore the JFET cannot operate in the enhancement mode, but only in the depletion mode, unlike depletion MOSFETs.
3. JFETs will generally carry larger drain currents than MOSFETs, given the same geometry.

8.7 UNIVERSAL CURVES FOR DEPLETION MOSFETs AND JFETs

Given the remarkable similarity between JFETs and depletion MOSFETs, the universal curves for these devices and the corresponding equations are virtually identical. Table 8.2 summarizes the relevant equations. Note that now the

Table 8.2 ***n*-channel depletion MOSFET and JFET equations**

State	Conditions	Equation	Remarks
Constant current (active region)	$v_{DS} > v_{GS} + V_P$, $v_{GS} > -V_P$	$i_D = \frac{I_{DSS}}{V_P^2}(v_{GS} + V_P)^2 = I_{DSS}\left(\frac{v_{GS}}{V_P} + 1\right)^2$	Nonlinear voltage-controlled current source operation.
Ohmic (saturation region)	$\lvert v_{DS}\rvert < \frac{1}{4}(v_{GS} + V_P)$, $v_{GS} > -V_P$	$i_D \approx \frac{v_{DS}}{R_{DS}}$ $R_{DS} = \frac{V_P^2}{2I_{DSS}(v_{GS} + V_P)}$	Voltage-controlled resistor operation. R_{DS} is the equivalent drain-to-source resistance.

parameter I_{DSS} represents the value of the drain current when $v_{GS} = 0$. Figure 8.45 summarizes these equations in the form of universal curves. Note that the transfer curve of Figure 8.45(b) displays two regions: the solid portion corresponds to depletion-mode operation ($v_{GS} < 0$) and applies to both depletion MOSFETs and JFETs, while the dashed section corresponds to enhancement-mode operation and applies only to depletion MOSFETs. Finally, if v_{GS} is replaced by v_{SG}, and v_{DS} by v_{SD}, the universal curves are also valid for *p*-channel devices.

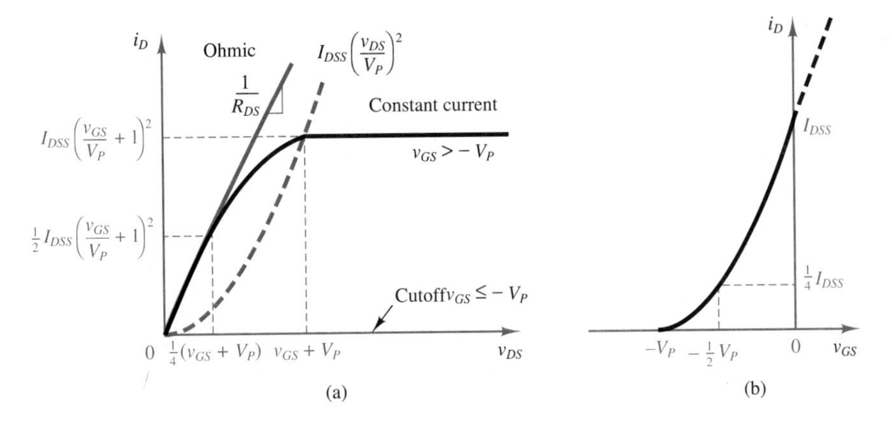

Figure 8.45 (a) Universal curves for *n*-channel depletion MOSFET and JFET; (b) transfer characteristic in active region

---

Example 8.13 Operating State of JFETs

We wish to determine the operating region for each of the circuits shown in Figure 8.46. Assume $I_{DSS} = 10$ mA and $V_P = 4$ V.

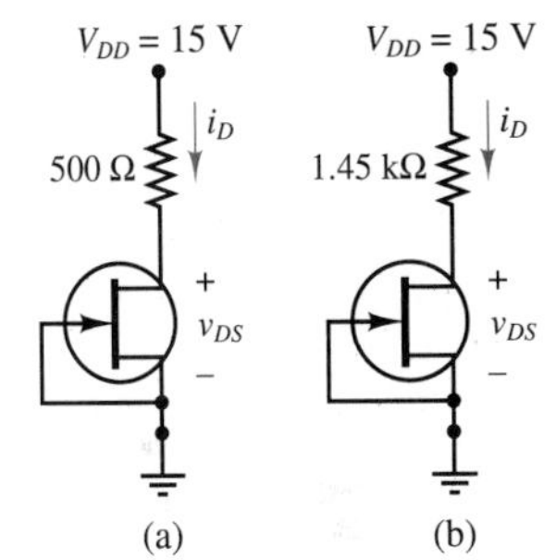

Figure 8.46

Solution:

Since $v_{GS} = 0$, we know that $i_D = I_{DSS} = 10$ mA. Therefore, the drain circuit equation for the circuit of Figure 8.46(a) is:

$$15 = 500 i_D + v_{DS} = 5 + v_{DS}$$

and $v_{DS} = 10$ V. Now, the condition required for operation in the active region can be found in Table 8.2 to be $v_{DS} > v_{GS} + V_P$; since $10 > 0 + 4$, we conclude that the JFET is in the active region.

Repeating the analysis for the circuit of Figure 8.46(b), we find that

$$15 = 1{,}450 i_D + v_{DS} = 14 + v_{DS}$$

and thus, $v_{DS} = 0.5$ V. In this case, since $v_{DS} < v_{GS} + V_P = 4$, we suspect that the JFET might be operating in the ohmic state; to confirm this hypothesis, we check to see if $|v_{DS}| < \frac{1}{4}(v_{GS} + V_P)$:

$$0.5 < \tfrac{1}{4}(v_{GS} + V_P) = 1$$

Therefore, the JFET is in the ohmic region.

Example 8.14 JFET Bias Circuit

We wish to bias the circuit shown in Figure 8.47 so that it will have a drain-source voltage of approximately 10 V and a drain current of 4 mA. The JFET has a pinch-off voltage of 3 V, and $I_{DSS} = 6$ mA. The breakdown voltage is 30 V.

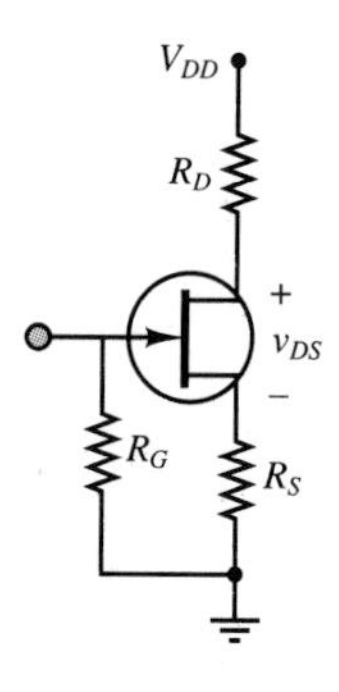

Figure 8.47

Solution:

Since

$$i_D = I_{DSS}\left(\frac{v_{GS}}{V_P} + 1\right)^2$$

and we desire $i_D = 4$ mA, we can calculate the desired v_{GS}:

$$0.66v_{GS}^2 + 4v_{GS} + 2 = 0$$

resulting in the roots $v_{GS} = -5.45$ V and $v_{GS} = -0.55$ V. Since the pinch-off voltage is 3 V, this JFET cannot conduct for gate-source voltages less than -3 V; therefore, the only acceptable solution is $v_{GS} = -0.55$ V.

The resistance R_G has the purpose of tying the gate close to ground; since the gate leakage current is very small (of the order of nanoamperes), R_G can be chosen to be around 1 MΩ. This will ensure that the gate is a few millivolts above ground (or at ground, for practical purposes). R_S must be chosen so that the proper v_{GS} is obtained. Knowing that the gate is effectively tied to ground ($v_G \approx 0$) and that the desired current is 4 mA, we can calculate the required source voltage to be $v_S = i_D R_S = 0.55$ V. Thus, $v_{GS} = -0.55$ V. Solving for R_S, we find $R_S = 0.55/0.004 = 137.5\ \Omega$.

Finally, a supply voltage must be selected. Since the breakdown voltage is 30 V, 24 V is a reasonable choice for V_{DD}. Now we can use the drain circuit equation to determine R_D such that $v_{DS} = 10$ V, as required:

$$V_{DD} = v_{DS} + i_D(R_D + R_S)$$

$$24 = 10 + 0.04R_D + 0.55$$

or

$$R_D = \frac{13.45}{0.04} = 3{,}362.5\ \Omega$$

Thus, using standard resistor values, we choose $R_S = 150\ \Omega$ and $R_G = 3.3$ kΩ for the final design.

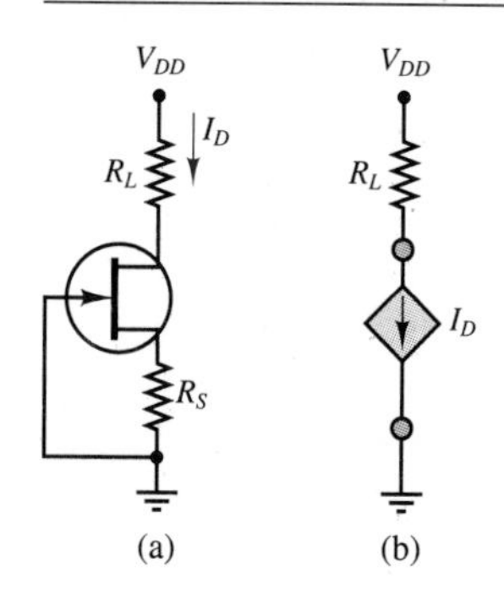

Figure 8.48

Example 8.15 JFET Current Source

JFETs are capable of providing fairly constant current if biased in the active region. Since a JFET can be biased in the active region for gate voltages equal to zero, it is possible to construct very simple current sources using JFETs. Figure 8.48 depicts such a current source and its equivalent circuit. In this circuit, the value of the source current is equal to I_{DSS}, since $v_{GS} = 0$. Although this circuit approximates an ideal current source quite well, the source current is subject to significant uncertainty because the parameter I_{DSS} can vary by as much as a factor of 5. In order to adjust the operating point of the JFET, to provide the desired source current, a source resistor is often employed.

Typical ranges of current values that may be obtained by JFET current sources are between a few tenths of a milliampere and a few milliamperes.

DRILL EXERCISES

8.14 What is R_{DS} for the circuit of Figure 8.46(b)?

8.15 What is the drain current for the circuit of Figure 8.46(b)?

8.16 Determine the actual operating point of the JFET of Example 8.14, given the choice of standard resistor values.

8.17 Repeat the design of Example 8.14 (i.e., calculate R_S and R_D) if the required drain current is halved.

CONCLUSION

- Transistors are three-terminal electronic semiconductor devices that can serve as linear amplifiers or switches.
- The bipolar junction transistor (BJT) acts as a current-controlled current source, amplifying a small base current by a factor ranging from 20 to 200. The operation of the BJT can be explained in terms of the device base-emitter and collector *i-v* characteristics. Large-signal linear circuit models for the BJT can be obtained by treating the transistor as a controlled current source.
- Field-effect transistors (FETs) can be grouped into three major families: enhancement MOSFETs, depletion MOSFETs, and JFETs. All FETs behave like voltage-controlled current sources. FET *i-v* characteristics are intrinsically nonlinear, characterized by a quadratic dependence of the drain current on gate voltage. The nonlinear equations that describe FET drain characteristics can be summarized in a set of universal curves for each family.

KEY TERMS

Active Linear Region of a BJT *414*
Active Region of a MOSFET *434*
Amplification *408*
Analog Transmission Gate *436*
Base-Collector (*BC*) Junction *412*
Base-Emitter (*BE*) Junction *411*
Bipolar Junction Transistor (BJT) *411*
Breakdown Region of a BJT *414*
Breakdown Region of a MOSFET *434*
Breakdown State of a MOSFET *431*
Channel *430*
Collector Characteristic *413*
Complementary Metal-Oxide-Semiconductor (CMOS) Device *431*
Constant-Current State of a MOSFET *431*
Cutoff Region of a BJT *414*
Cutoff Region of a MOSFET *433*
DC Bias Circuit *416*
Depletion-Mode MOSFET *429*
Drain *429*
Drain Breakdown Voltage *431*
Enhancement-Mode MOSFET *429*
Field-Effect Transistor (FET) *428*
Gate *429*
Gate Breakdown Voltage *431*
Junction Field-Effect Transistor (JFET) *429*
Large-Signal Model of a BJT *424*
Metal-Oxide-Semiconductor Field-Effect Transistor (MOSFET) *429*

ANSWERS TO DRILL EXERCISES

8.1 $A = \mu \frac{1}{(r_i+R_S)} \frac{R_L}{(r_o+R_L)}; r_i \to 0, r_o \to 0$

8.2 $A = \frac{1}{(r_i+R_S)} \frac{r_o R_L}{(r_o+R_L)} \mu$

8.3 $A = \mu \frac{r_i}{(r_i+R_S)} \frac{r_o R_L}{(r_o+R_L)}$

8.5 Saturation

8.6 $V_{BB} = 5$ V; $V_{CE} = 6.44$ V

8.7 3.74%; because R_E provides a negative feedback action that will keep I_C and I_E almost constant.

8.8 Saturation; 8.5 mA

8.9 159 mW

8.10 The MOSFET is in the ohmic state.

8.11 Choosing the smaller value of v_{GS}, $R_S = 20.7$ kΩ, $v_{GS} = 2.86$ V, $i_D = 0.586$ mA. The answer is not unique: selecting the larger gate voltage, we find $R_S = 11.5$ kΩ.

8.13 Approximately 200 Ω

8.14 200 Ω

8.15 Approximately 2.5 mA

8.16 $i_D = 3.89$ mA; $v_{DS} = 10.57$ V; $v_{GS} = -0.58$ V

8.17 $R_S = 634$ Ω; $R_D = 6.366$ kΩ

DEVICE DATA SHEETS

BJT Low-Level Amplifiers

Type No.	V_{CBO} (V) Min	V_{CEO} (V) Min	V_{EBO} (V) Min	I_{CBO} (nA) Max	@ V_{CB} (V)	h_{FE} Min	h_{FE} Max	@ I_C (mA)	& V_{CE} (V)	$V_{CE(SAT)}$ (V) Max	& $V_{BE(SAT)}$ (V) Min	$V_{BE(SAT)}$ (V) Max	@ I_C (mA)	C_{ob} (pF) Max
2N4410	120	80	5	10	100	60 60	400	10 1	1 1	0.2		0.8	1	12
2N4966	50	40	6	25	25	40 50	200	0.01 10	5 5	0.4			10	6
2N4967	50	40	6	25	25	100 120	600	0.01 10	5 5	0.4			10	6
2N4968	30	25	6	50	25	40 50	200	0.01 10	5 5	0.4			10	6
2N5088	35	30		50	20	300 350 300	 900	10 1 100 μA	5 5 5	0.5			10	4
2N5089	30	25		50	15	400 450 400	 1200	10 1 100 μA	5 5 5	0.5			10	4

BJT General Purpose Amplifiers and Switches

Type No.	V_{CBO} (V) Min	V_{CEO} (V) Min	V_{EBO} (V) Min	I_{CBO} (nA) Max	@ V_{CB} (V)	h_{FE} Min	h_{FE} Max	@ I_C (mA)	& V_{CE} (V)	$V_{CE(SAT)}$ (V) Max	& $V_{BE(SAT)}$ (V) Min	$V_{BE(SAT)}$ (V) Max	@ I_C (mA)	C_{ob} (pF) Max
2N3903	60	40	6			15		100	1	0.2	0.6	0.85	10	4
						30		50	1					
						50	150	10	1	0.3		0.95	50	
						35		1	1					
						20		100 μA	1					
2N3904	60	40	6		30	30		100	1	0.2	0.65	0.85	10	4
						60		50	1					
						100	300	10	1	0.3		0.95	50	
						70		1	1					
						40		100 μA	1					
2N3946	60	40	6			20		50	1	0.2	0.6	0.9	10	4
						50	150	10	1					
						45		1	1	0.3		1.0	50	
						30		100 μA	1					

Field Effect Transistors Maximum Ratings at T_A = 25°C (Observe MOS Handling)

ECG Type	Description and Application	Trans-conductance gfs Type μmhos	Gate to Source Cutoff Voltage V_{GS} (off) Max V	Zero-Gate Voltage Drain Current I_{DSS} mA Min–Max	Gate To Source Breakdown Voltage BV_{GSS} Min V	Input Cap C_{iss} Max pf	Transfer Cap C_{rss} Max pf	Device Diss. P_D Max mW
ECG220	MOSFET, N-Ch, VHF Amp/ Mix, NF 5dB Max at 200 MHz	7,500	8	5–25	20	7	0.35	330
ECG221	Dual Gate MOSFET, N-Ch, VHF Amp/Mix, NF 5dB Max at 200 MHz	15,000	6	18 typ	20	5.5	0.03	400
ECG222	Dual Gate MOSFET, N-Ch, VHF Amp/Mix, NF 6dB Max at 200 MHz Gate Protected	12,000	4	5–35	20	6 typ	0.03	330
ECG452	JFET, N-Ch, UHF/VHF Amp, NF 4dB at 400 MHz	5,500	6	5–15	30	4	0.8	300
ECG461	Matched Dual JFET, N-Ch, DC Amp/Sampler/Chopper	3,500	4.5	0.5–8	50	6	2	400
ECG462	MOSFET, N-Ch, AF Amp, NF 3.8dB at 1.0 kHz	2,250	7	2–6	20	8	0.8	300

HOMEWORK PROBLEMS

8.1 Determine the region of operation for the following transistors:

a. *npn*, $V_{BE} = 0.8$ V, $V_{CE} = 0.4$ V
b. *npn*, $V_{CB} = 1.4$ V, $V_{CE} = 2.1$ V
c. *pnp*, $V_{CB} = 0.9$ V, $V_{CE} = 0.4$ V
d. *npn*, $V_{BE} = -1.2$ V, $V_{CB} = 0.6$ V

8.2 For the circuit in Figure 8.11 in the text, $R_C = 5\ \text{k}\Omega$, $R_E = 3\ \text{k}\Omega$, $R_B = 10\ \text{k}\Omega$, $V_{BB} = 4$ V, and $V_{CC} = 12$ V. Assume $V_{BE} = 0.7$ V. A single measurement indicates the reading of V_2 is 2.0 V. Find V_1, V_3, I_E, I_B, I_C, and β.

8.3 For the circuit in Figure 8.20 in the text, $V_{CC} = 20$ V, $R_C = 5\ \text{k}\Omega$, and $R_E = 1\ \text{k}\Omega$. Determine the region of operation of the transistor if:

a. $I_C = 1$ mA, $I_B = 20\ \mu$A, $V_{BE} = 0.7$ V
b. $I_C = 3.2$ mA, $I_B = 0.3$ mA, $V_{BE} = 0.8$ V
c. $I_C = 3$ mA, $I_B = 1.5$ mA, $V_{BE} = 0.85$ V

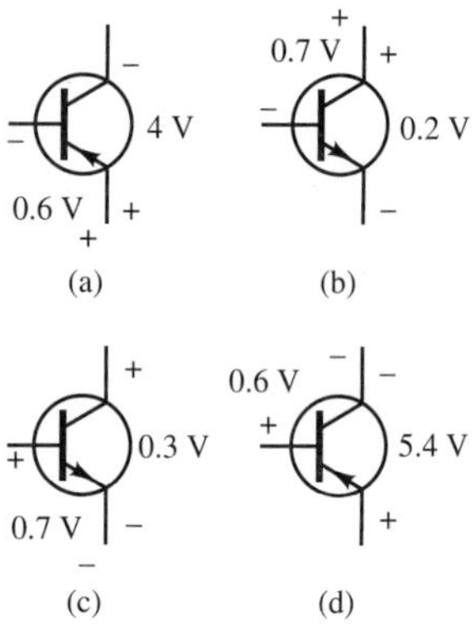

8.4 For each transistor shown in Figure P8.1, determine whether the *BC* and *BE* junctions are forward- or reverse-biased, and also determine the operating state.

Figure P8.1

8.5 Given the circuit of Figure P8.2, determine the operating point of the transistor. Assume the BJT is a silicon device with $\beta = 100$. What is the state of the transistor?

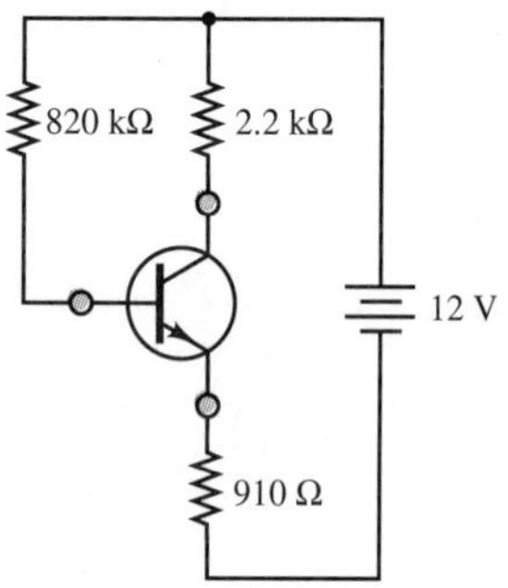

Figure P8.2

8.6 The magnitudes of a *pnp* transistor's emitter and base currents are 6 mA and 0.1 mA, respectively. The magnitudes of the voltages across the emitter-base and collector-base junctions are 0.65 and 7.3 V. Find

a. V_{CE}.
b. I_C.
c. The total power dissipated in the transistor, defined here as $P = V_{CE}I_C + V_{BE}I_B$.

8.7 Given the circuit of Figure P8.3, determine the emitter current and the collector-base voltage. Assume the BJT has $V_\gamma = 0.6$V.

8.8 Given the circuit of Figure P8.4, determine the operating point of the transistor. Assume a 0.6-V offset voltage and $\beta = 150$. What is the state of the transistor?

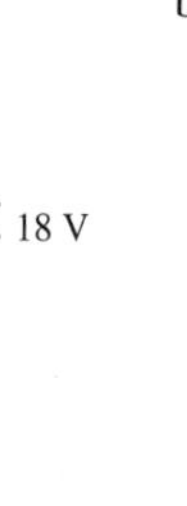

Figure P8.4

8.9 Given the circuit of Figure P8.5, determine the emitter current and the collector-base voltage. Assume the BJT has a 0.6-V offset voltage at the *BE* junction.

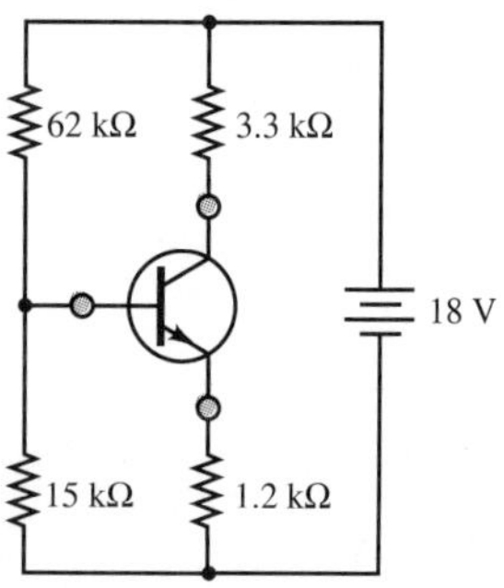

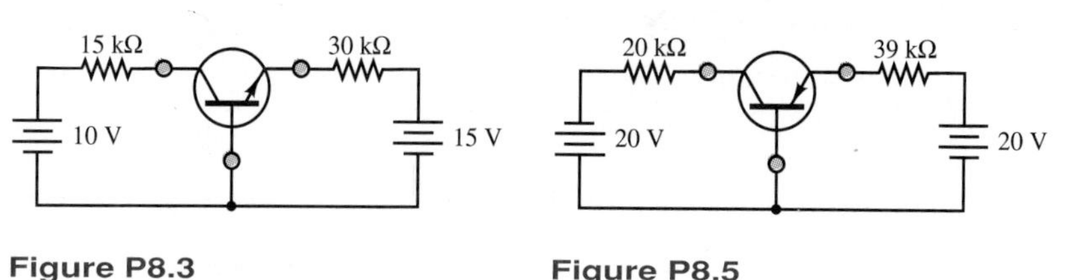

Figure P8.3 **Figure P8.5**

8.10 If the emitter resistor in Problem 8.9 (Figure P8.5) is changed to 22 kΩ, how does the operating point of the BJT change?

8.11 For the circuit given in Figure P8.6, verify that the transistor operates in the saturation region by computing the ratio of collector current to base current. (Hint: With reference to Figure 8.22, $V_\gamma = 0.6$ V, $V_{\text{sat}} = 0.2$ V.)

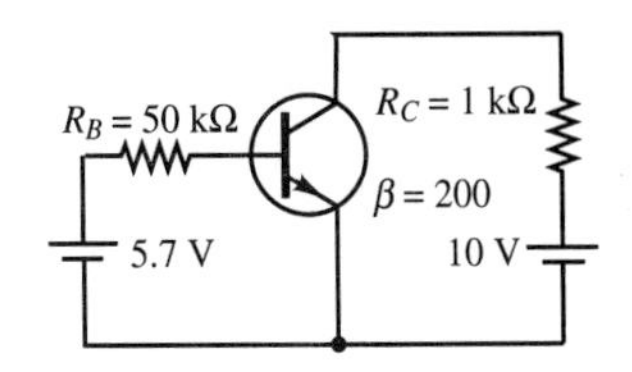

Figure P8.6

8.12 For the circuit shown in Figure P8.7, determine the base voltage V_{BB} required to saturate the transistor. Assume that $V_{CE\ \text{sat}} = 0.1$ V, $V_{BE\ \text{sat}} = 0.6$ V, and $\beta = 50$.

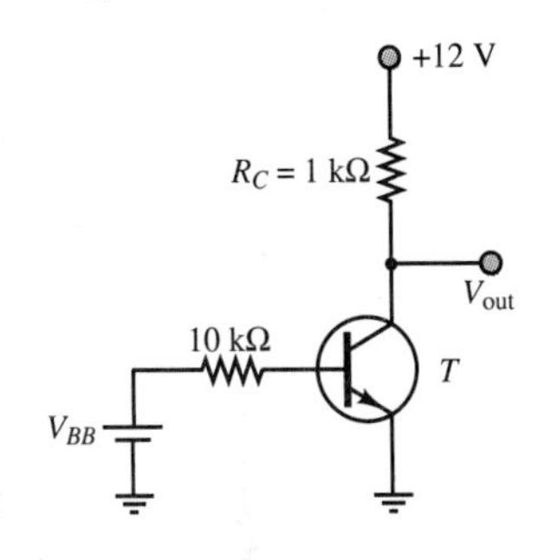

Figure P8.7

8.13 For the circuit shown in Figure 8.20 in the text, calculate the quiescent values of V_{BE}, I_B, V_{CE}, I_C for $V_{CC} = 15$ V, $R_1 = 10$ kΩ, $R_2 = 47$ kΩ, $R_C = 2$ kΩ, $R_E = 4$ kΩ, and $\beta = 75$.

8.14 An *npn* transistor is operated in the active region with the collector current 60 times the base current and with junction voltages of $V_{BE} = 0.6$ V and $V_{CB} = 7.2$ V. If $|I_E| = 4$ mA, find (a) I_B and (b) V_{CE}.

8.15 The collector characteristics for a certain transistor are shown in Figure P8.8.

a. Find the ratio I_C/I_B for $V_{CE} = 10$ V and $I_B = 100\,\mu$A, 200 μA, and 600 μA.
b. The maximum allowable collector power dissipation is 0.5 W for $I_B = 500\,\mu$A. Find V_{CE}.

[*Hint:* A reasonable approximation for the power dissipated at the collector is the product of the collector voltage and current:

$$P = I_C V_{CE}$$

where P is the permissible power dissipation,
I_C is the quiescent collector current,
V_{CE} is the operating point collector-emitter voltage.]

8.16 Given the circuit of Figure P8.9, assume both transistors are silicon-based with $\beta = 100$. Determine:

a. I_{C1}, V_{C1}, V_{CE1}
b. I_{C2}, V_{C2}, V_{CE2}

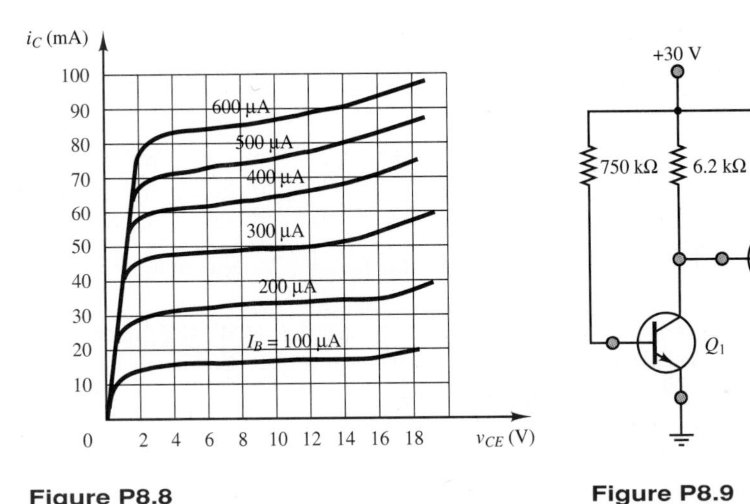

Figure P8.8 **Figure P8.9**

8.17 Use the collector characteristics of the 2N3904 *npn* transistor shown on p. 506 to determine the operating point (I_{CQ}, V_{CEQ}) of the transistor in Figure P8.10. What is the value of β at this point?

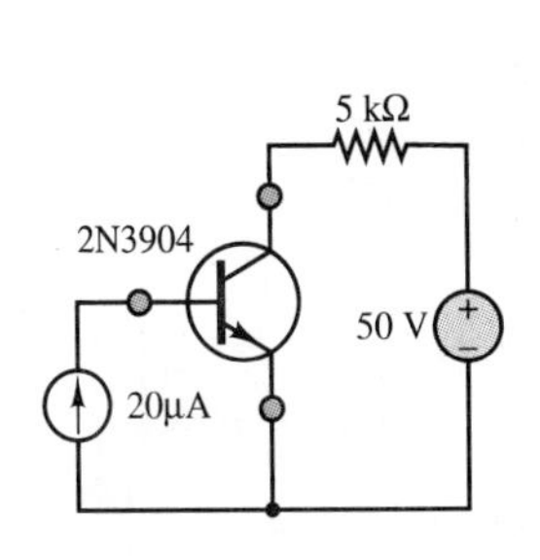

Figure P8.10

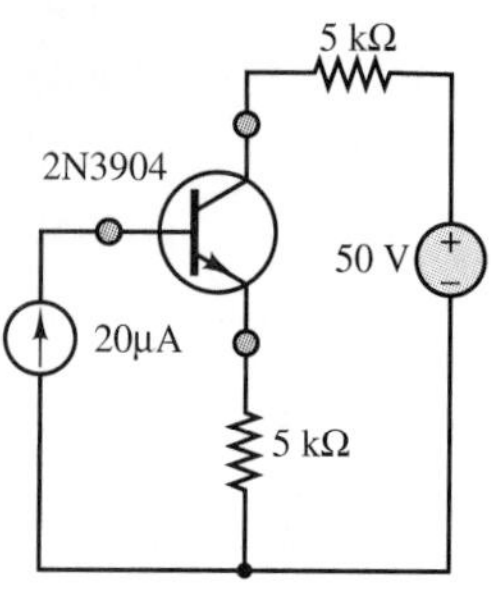
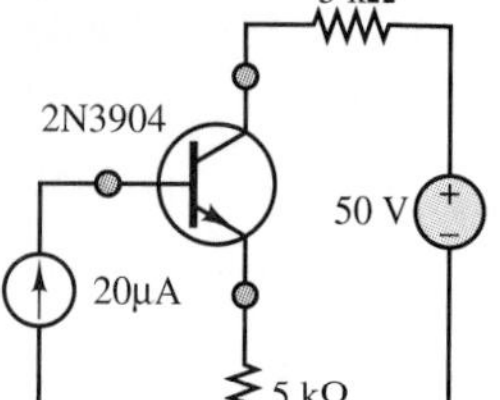

Figure P8.11

8.18 Use the collector characteristics of the 2N3904 *npn* transistor shown on p. 506 to determine the operating point (I_{CQ}, V_{CEQ}) of the transistor in Figure P8.11. What is the value of β at this point? How does your solution differ from that of Problem 8.17?

8.19 For the circuit in Figure 8.14 in the text, $R_C = 1\ \text{k}\Omega$, $V_{BB} = 5$ V, $\beta_{\text{min}} = 50$, and $V_{CC} = 10$ V. Find the range of R_B so that the transistor is in the saturation state.

8.20 For the circuit in Figure 8.14 in the text, $V_{CC} = 5$ V, $R_C = 1\ \text{k}\Omega$, $R_B = 10\ \text{k}\Omega$, and $\beta_{\text{min}} = 50$. Find the range of values of V_{BB} so that the transistor is in saturation.

8.21 Use the collector characteristics of the 2N3904 *npn* transistor shown on p. 506 to determine the operating point (I_{CQ}, V_{CEQ}) of the transistor in Figure P8.12. What is the value of β at this point?

8.22 For the circuit in Figure 8.12 in the text, $I_{BB} = 20\ \mu\text{A}$, $R_C = 2\ \text{k}\Omega$, $V_{CC} = 10$ V, and $\beta = 100$. Find I_C, I_E, V_{CE}, and V_{CB}.

8.23 For the circuit shown in Figure P8.13, find the maximum achievable voltage gain. Assume that $\beta = 100$.

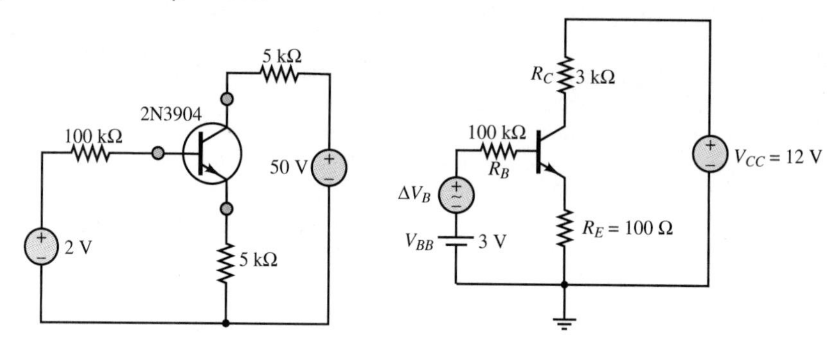

Figure P8.12 Figure P8.13

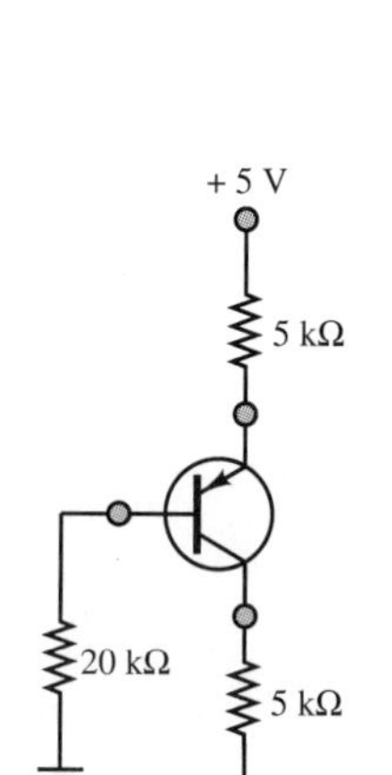

Figure P8.14

8.24 It has been found that V_E in the circuit of Figure P8.14 is 1 V. If the transistor has $V_\gamma = 0.6$ V, determine:

a. V_B
b. I_B
c. I_E
d. I_C
e. β
f. α

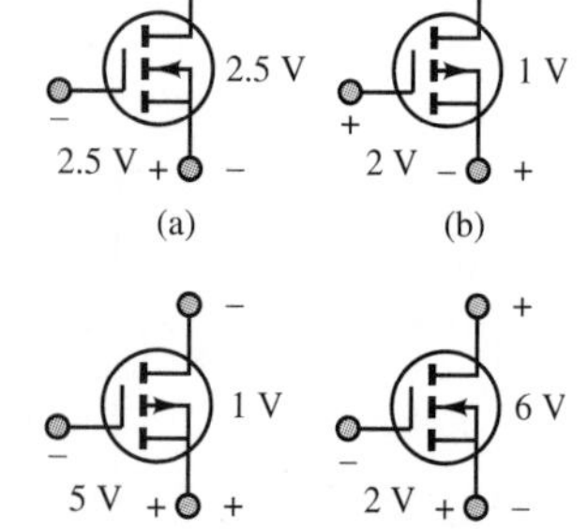

Figure P8.15

8.25 The transistors shown in Figure P8.15 have $|V_T| = 3$ V. Determine the operating state.

8.26 The three terminals of an *n*-channel enhancement-mode MOSFET are at potentials of 4 V, 5 V, and 10 V with respect to ground. Draw the circuit symbol, with the appropriate voltages at each terminal, if the device is operating

a. In the ohmic state.
b. In the active region.

8.27 An enhancement-type NMOS transistor with $V_T = 2$ V has its source grounded and a 3-VDC source connected to the gate. Determine the operating state if

a. $v_D = 0.5$ V
b. $v_D = 1$ V
c. $v_D = 5$ V

8.28 The characteristic curves of an enhancement-type MOSFET are shown in Figure 8.36 in the text. Assume that $k = 0.1$ mA/V^2 and $V_T = 1.5$ V and that the transistor is operated at $v_{GS} = 3.5$ V. Find the drain current obtained at

a. $v_{DS} = 2$ V
b. $v_{DS} = 10$ V

8.29 An n-channel enhancement MOSFET is found to have a measured drain current of 4 mA at $v_{GS} = v_{DS} = 10$ V and of 1 mA at $v_{GS} = v_{DS} = 6$ V. Find the parameters I_{DSS} and V_T.

8.30 In the circuit shown in Figure P8.16, the p-channel transistor has $V_T = 2$ V and $k = 10$ mA/V^2. Find R and v_D for $i_D = 0.4$ mA.

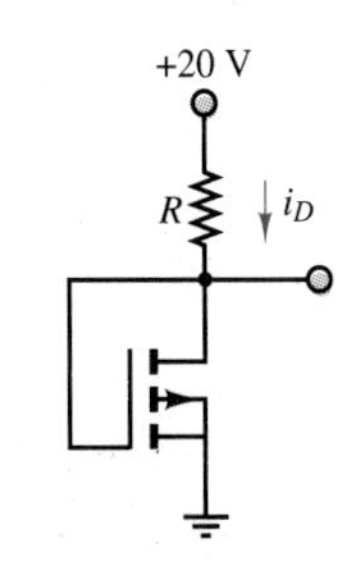

Figure P8.16

8.31 An enhancement-type NMOS transistor has $V_T = 2$ V and $i_D = 1$ mA when $v_{GS} = v_{DS} = 3$ V. Find the value of i_D for $v_{GS} = 4$ V.

8.32 For $v_{DS} = 3$ V, $V_T = -4$ V, and $k = 0.625$ mA/V^2, find the change in i_D corresponding to a change in v_{GS} from 2.1 V to -1.5 V.

8.33 In the circuit shown in Figure P8.17, a drain voltage of 0.1 V is established. Find the current i_D for $V_T = 1$ V and $k = 0.5$ mA/V^2.

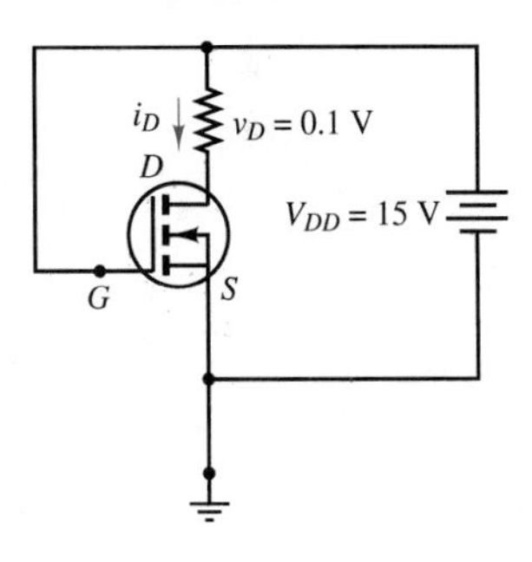

Figure P8.17

8.34 An n-channel enhancement-mode MOSFET is operated in the ohmic region, with $v_{DS} = 0.4$ V and $V_T = 3.2$ V. The effective resistance of the channel is given by $R_{DS} = 500/(V_{GS} - 3.2)\ \Omega$. Find i_D when $v_{GS} = 5$ V, $R_{DS} = 500\ \Omega$, and $v_{GD} = 4$ V.

8.35 In the circuit shown in Figure P8.18, the MOSFET operates in the active region, for $i_D = 0.5$ mA and $v_D = 3$ V. This enhancement-type PMOS has $V_T = -1$ V, $k = 0.5$ mA/V^2. Find

a. R_D.
b. The largest allowable value of R_D for the MOSFET to remain in the saturation region.

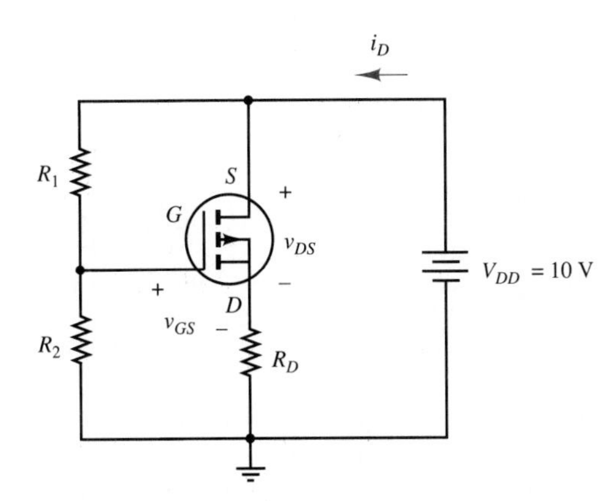

Figure P8.18

8.36 An n-channel JFET has $V_p = -2.8$ V. It is operating in the ohmic region, with $v_{GS} = -1$ V, $v_{DS} = 0.05$ V, and $i_D = 0.3$ mA. Find i_D if

a. $v_{GS} = -1$ V, $v_{DS} = 0.08$ V
b. $v_{GS} = 0$ V, $v_{DS} = 0.1$ V
c. $v_{GS} = -3.2$ V, $v_{DS} = 0.06$ V

8.37 An enhancement-type MOSFET has the parameters $k = 0.5\ \text{mA/V}^2$ and $V_T = 1.5$ V and the transistor is operated at $v_{GS} = 3.5$ V. Find the drain current obtained at

a. $v_{DS} = 3$ V
b. $v_{DS} = 10$ V

8.38 An *n*-channel enhancement MOSFET has $i_D = 6$ mA at $v_{GS} = v_{DS} = 12$ V and $i_D = 1.5$ mA at $v_{GS} = v_{DS} = 6$ V. Find parameters k and V_T.

8.39 An enhancement-type NMOS transistor with $V_T = 2.5$ V has its source grounded and a 4-VDC source connected to the gate. Find the operating region of the device if

a. $v_D = 0.5$ V
b. $v_D = 1.5$ V

8.40 An enhancement-type NMOS transistor has $V_T = 4$ V, $i_D = 1$ mA when $v_{GS} = v_{DS} = 6$ V. Neglecting the dependence of i_D on v_{DS} in saturation, find the value of i_D for $v_{GS} = 5$ V.

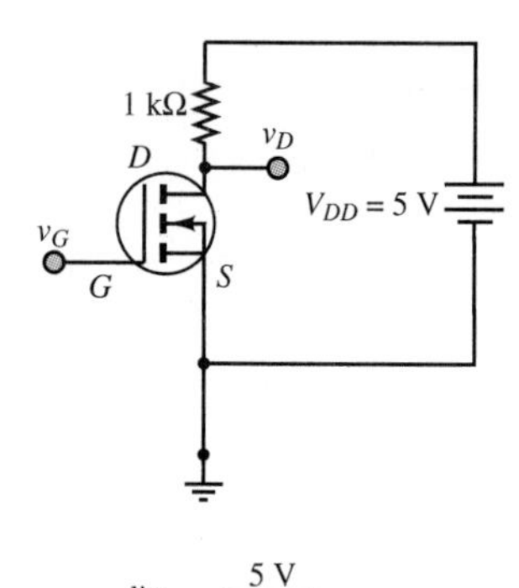

Figure P8.19

8.41 The NMOS transistor shown in Figure P8.19 has $V_T = 1.5$ V, $k = 0.4\ \text{mA/V}^2$. Now if v_G is a pulse with 0 V to 5 V, find the voltage levels of the pulse signal at the drain output.

8.42 A JFET having $V_p = -2$ V and $I_{DSS} = 8$ mA is operating at $v_{GS} = -1$ V and a very small v_{DS}. Find

a. r_{DS}
b. v_{GS} at which r_{DS} is half of its value in (a)

8.43 Analyze the circuit shown in Figure P8.20. Find drain current and drain voltage. The depletion MOSFET has $V_T = -1$ V, $k = 0.5\ \text{mA/V}^2$.

8.44 The transistor shown in Figure P8.21 has $V_T = 2$ V, $k = 0.2\ \text{mA/V}^2$. Find the DC voltage v_{out}.

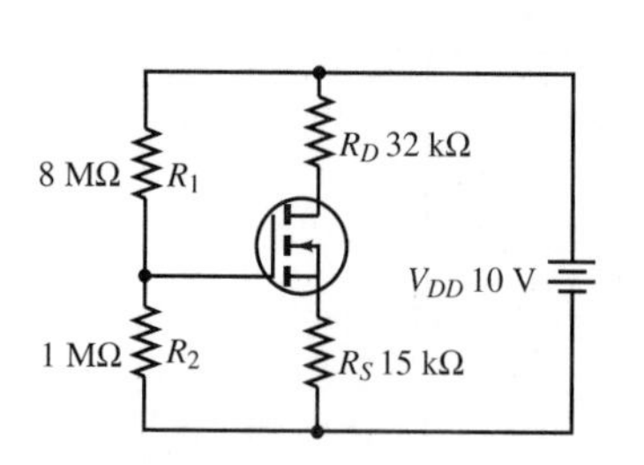

Figure P8.20

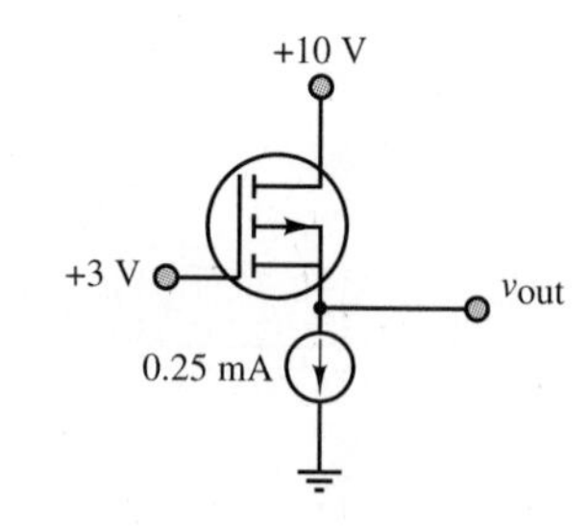

Figure P8.21

CHAPTER

9

Transistor Amplifiers and Switches

The aim of this chapter is to describe the application of transistors as amplifiers and switches. Small-signal transistor amplifiers can be analyzed by means of linear small-signal models that make it possible, through the use of linear circuit analysis techniques, to determine an amplifier's principal characteristics, such as input and output impedance and current and voltage gain. Small-signal models of transistor amplifier circuits also permit analysis of multistage amplifiers.

The chapter begins with the analysis of the BJT h parameters; these are linear approximations that are valid in the neighborhood of an operating point and are directly derived from the base and collector characteristics. Subsequently, the common-emitter BJT amplifier is discussed in some detail, and the common-base and emitter-follower amplifiers are briefly introduced. Next, a similar analysis is conducted for MOSFET amplifiers. The material on amplifiers closes with a general discussion of multistage transistor amplifiers and of amplifier frequency response.

In addition to serving as the essential component of electronic amplifiers, transistors find common application in switching circuits and logic gates. The last section of the chapter describes BJT and MOSFET inverters and gates and introduces the two major families of logic devices, TTL and CMOS.

Upon completing this chapter, you should be able to:

- Use small signal models of bipolar and field-effect transistors to construct small signal amplifier models, from which voltage and current gain and input and output resistance can be computed.
- Qualitatively evaluate the frequency response characteristics of a transistor amplifier and understand the major mechanisms limiting an amplifier's frequency response.
- Understand the major requirements in the design of multistage amplifiers.
- Understand the switching characteristics of BJTs and MOSFETs and be able to analyze the fundamental behavior of TTL and CMOS logic gates.

9.1 SMALL-SIGNAL MODELS OF THE BJT

Small-signal models for the BJT take advantage of the relative linearity of the base and collector curves in the vicinity of an operating point. These linear circuit models work very effectively provided that the transistor voltages and currents remain within some region around the operating point. This condition is usually satisfied in small-signal amplifiers used to magnify low-level signals (e.g., outputs of transducers). For the purpose of our discussion, it is reasonable to use the **hybrid-parameter (*h*-parameter) small-signal model** of the BJT, to be discussed presently.

Note that the small-signal model assumes that the DC bias point of the transistor has been established. As was done in Chapter 8, the following convention will be used: each voltage and current is assumed to be the superposition of a DC component (the quiescent voltage or current) and a small-signal AC component. The former is denoted by an uppercase letter, and the latter by an uppercase letter preceded by the symbol Δ. Thus,

$$
\begin{aligned}
i_B &= I_{BQ} + \Delta I_B \\
i_C &= I_{CQ} + \Delta I_C \\
v_{CE} &= V_{CEQ} + \Delta V_{CE}
\end{aligned}
$$

Figure 9.1 depicts the appearance of the collector current $i_C(t)$ when $I_{CQ} = 15 \times 10^{-3}$ A and $\Delta I_C(t) = 0.1 \times 10^{-3} \sin \omega t$ A.

Imagine the collector curves of Figure 8.19 magnified about the Q point. Figure 9.2 graphically depicts the interpretation of each of the h parameters relative to the operating point of the BJT.

The parameter h_{ie} is approximately equal to the ratio $\frac{\Delta V_{BE}}{\Delta I_B}$ in the neighborhood of the Q point; Figure 9.2(a) illustrates how this parameter is equal to the reciprocal of the slope of the I_B-V_{BE} curve at the operating point. Physically, this parameter represents the forward resistance of the BE junction.

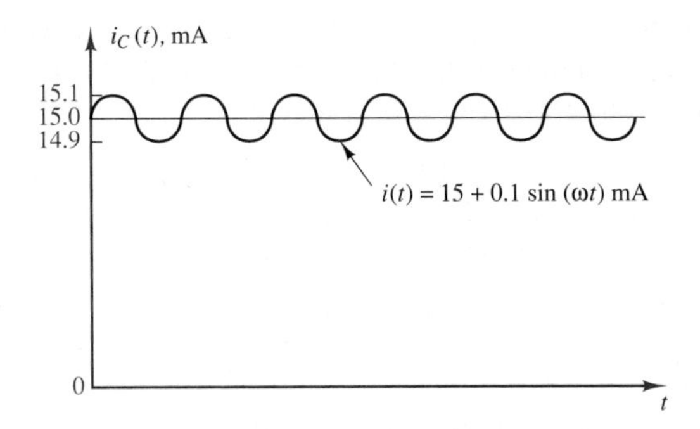

Figure 9.1 Superposition of AC and DC signals

The parameter h_{re} is representative of the fact that the I_B-V_{BE} curve is slightly dependent on the actual value of the collector-emitter voltage, V_{CE}. However, this effect is virtually negligible in any applications of interest to us. Thus, we shall assume that $h_{re} \approx 0$. Figure 9.2(b) depicts the shift in the I_B-V_{BE} curves represented by h_{re}. A typical value of h_{re} for $V_{CE} \geq 1$ V is around 10^{-2}.

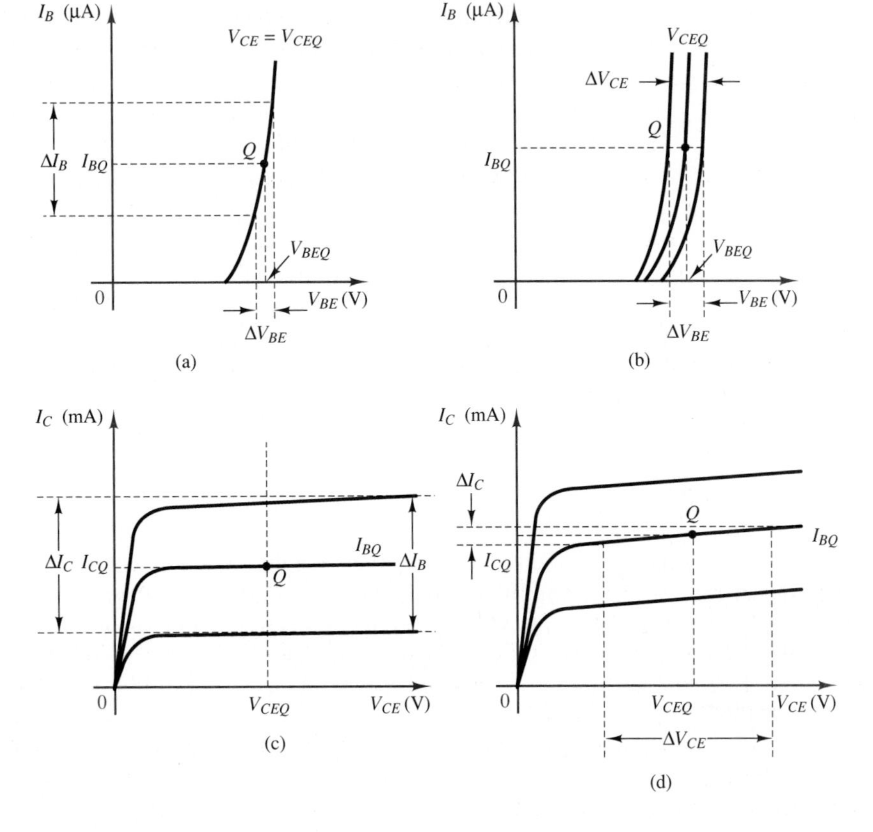

Figure 9.2

The parameter h_{fe} is approximated in Figure 9.2(c) by the current ratio $\frac{\Delta I_C}{\Delta I_B}$. This parameter represents the current gain of the transistor and is approximately equivalent to the parameter β introduced earlier. For the purpose of our discussion, β and h_{fe} will be interchangeable, although they are not exactly identical.

The parameter h_{oe} may be calculated as $h_{oe} = \frac{\Delta I_C}{\Delta V_{CE}}$ from the collector characteristic curves, as shown in Figure 9.2(d). This parameter is a physical indication of the fact that the I_C-V_{CE} curves in the linear active region are not exactly flat; h_{oe} represents the upward slope of these curves and therefore has units of conductance (S). Typical values of h_{oe} are around 10^{-5} S. We shall often assume that this effect is negligible.

To be more precise, the h parameters are defined by the following set of equations:

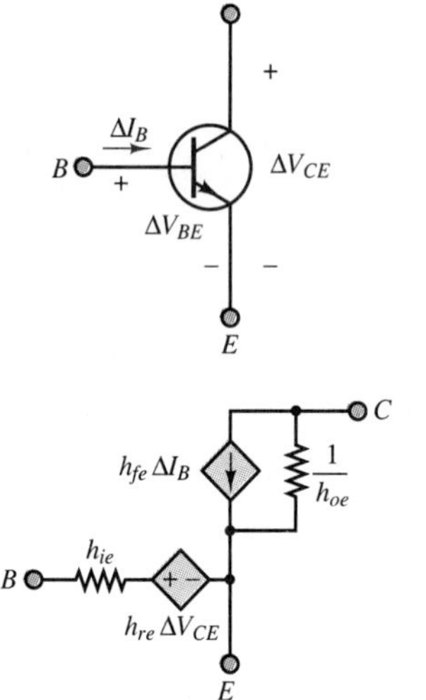

Figure 9.3 h-parameter small-signal model for BJT

$$h_{ie} = \left.\frac{\partial v_{BE}}{\partial i_B}\right|_{I_{BQ}} \quad (\Omega) \tag{9.1}$$

$$h_{oe} = \left.\frac{\partial i_C}{\partial v_{CE}}\right|_{V_{CEQ}} \quad (\text{S}) \tag{9.2}$$

$$h_{fe} = \left.\frac{\partial i_C}{\partial i_B}\right|_{I_{BQ}} \quad \left(\frac{\text{A}}{\text{A}}\right) \tag{9.3}$$

$$h_{re} = \left.\frac{\partial v_{BE}}{\partial v_{CE}}\right|_{V_{CEQ}} \quad \left(\frac{\text{V}}{\text{V}}\right) \tag{9.4}$$

The circuit of Figure 9.3 illustrates the small-signal model side by side with the BJT circuit symbol. Representative parameters for a small-signal transistor are listed in Table 9.1.

Table 9.1 ***h* parameters for the 2N2222A BJT**

Parameter	Minimum	Maximum
h_{ie}(kΩ)	2	4
$h_{re}(\times 10^{-4})$		8
h_{fe}	50	300
h_{oe}(μS)	5	35

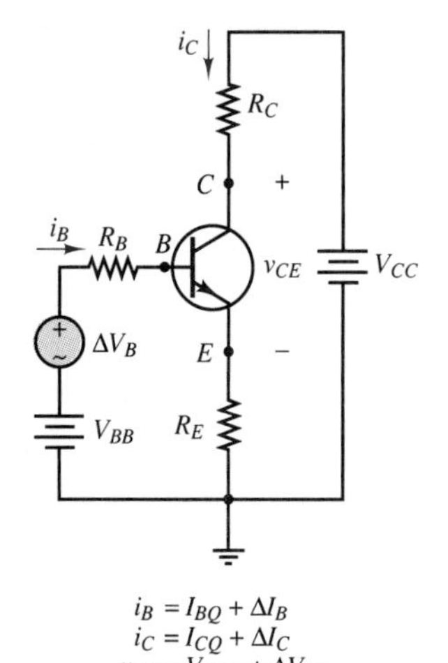

Figure 9.4 BJT amplifier

To illustrate the application of the h-parameter small-signal model of Figure 9.3, consider the transistor amplifier circuit shown in Figure 9.4. This circuit is most readily analyzed if DC and AC equivalent circuits are treated separately. To obtain the DC circuit, the AC source is replaced by a short circuit. The resulting DC circuit is shown in Figure 9.5. The DC circuit may be employed to carry out a Q-point analysis similar to that of Examples 8.4 and 8.6. Since our objective at present is to illustrate the AC circuit model for the transistor amplifier, we shall assume that the DC analysis (i.e., selection of the appropriate Q point) has already been carried out, and that a suitable operating point has been established.

Replacing the DC voltage sources with short circuits, we obtain the AC equivalent circuit of Figure 9.6. The transistor may now be replaced by its h-parameter small-signal model, also shown in Figure 9.6. We may simplify the model by observing that h_{oe}^{-1} is a very large resistance and that if the load resistance R_L (in parallel with h_{oe}^{-1}) is small (i.e., if $R_L h_{oe} \leq 0.1$), the resistor h_{oe}^{-1} in the model may be ignored. The linear AC equivalent circuit makes it possible to take advantage of the circuit analysis techniques developed in Chapters 2 and 3 to analyze the operation of the amplifier. For example, application of KVL around the base circuit loop yields the following equation:

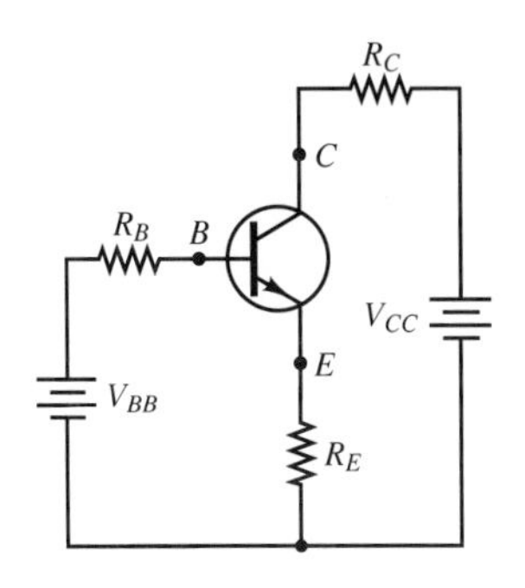

Figure 9.5 DC equivalent circuit for the BJT amplifier of Figure 9.4

$$\Delta V_B = \Delta I_B R_B + \Delta I_B h_{ie} + (h_{fe} + 1)\Delta I_B R_E \tag{9.5}$$

which can be solved to obtain

$$\frac{\Delta I_B}{\Delta V_B} = \frac{1}{R_B + h_{ie} + (h_{fe} + 1)R_E} \tag{9.6}$$

and

$$\begin{aligned} \Delta V_C &= -\Delta I_C R_C \\ &= -h_{fe} \Delta I_B R_C = \frac{-h_{fe} R_C}{R_B + h_{ie} + (h_{fe} + 1)R_E} \Delta V_B \end{aligned} \tag{9.7}$$

Then, the **AC/open-loop voltage gain** of the amplifier is given by the expression

$$\mu = \frac{\Delta V_C}{\Delta V_B} = \frac{-h_{fe} R_C}{R_B + h_{ie} + (h_{fe} + 1)R_E} \tag{9.8}$$

The small-signal model for the BJT will be further explored in the next section.

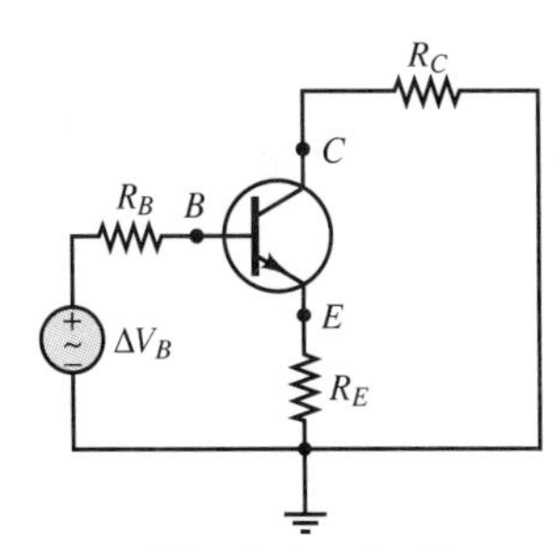

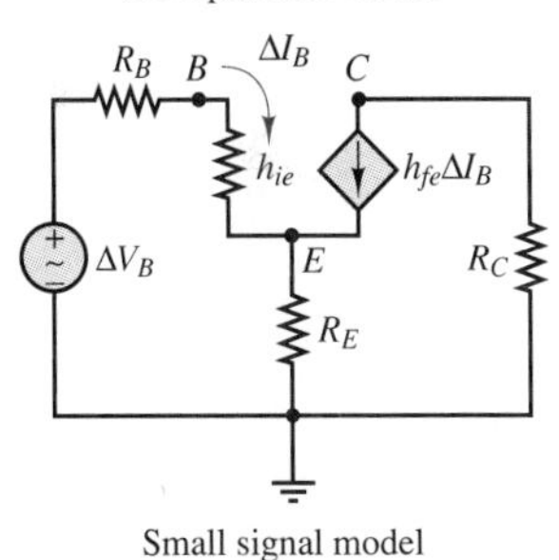

Figure 9.6 AC equivalent circuit and small-signal model for the amplifier of Figure 9.4

Transconductance

In addition to the h parameters described above, another useful small-signal transistor parameter is the **transfer conductance,** or **transconductance**. The transconductance of a bipolar transistor is defined as the local slope of the collector current base-emitter voltage curve:

$$g_m = \frac{\partial i_C}{\partial v_{BE}} \tag{9.9}$$

and it can be expressed in terms of the h parameters if we observe that we can write

$$g_m = \frac{\partial i_C}{\partial v_{BE}} = \frac{\partial i_C}{\partial i_B}\frac{\partial i_B}{\partial v_{BE}} = \frac{h_{fe}}{h_{ie}} \tag{9.10}$$

It can be shown that the expression for the transconductance can be approximated by

$$g_m \cong \frac{I_{CQ}}{\frac{kT}{q}} = \frac{I_{CQ}}{0.026} = 39 I_{CQ} \frac{\text{mA}}{\text{V}} \qquad \textbf{(9.11)}$$

at room temperature, where k is Boltzmann's constant, T is the temperature in degrees Kelvin, and q the electron charge (see Chapter 7 for a review of the *pn* junction equation). Now, the transconductance is an important measure of the voltage amplification of a BJT amplifier, because it relates small oscillations in the base-emitter junction voltage to the corresponding oscillations in the collector current. We shall see in the next sections that this parameter can be related to the voltage gain of the transistor. Note that the approximation of equation 9.11 suggests that the transconductance parameter is dependent on the operating point.

Example 9.1

This example illustrates the computation of the AC equivalent circuit model for a typical *npn* transistor, the 2N5088. This transistor is designed for low-level, low-noise amplifier applications. Using the information $h_{fe} = 350$, $h_{ie} = 1.4\ \text{k}\Omega$, and $h_{oe} = 150\ \mu\text{S}$ and the AC equivalent circuit of Figure 9.3, determine the nominal AC/open-loop voltage gain of the transistor amplifier of Figure 9.4. Assume that $V_{CC} = 12$ V, $V_{BB} = 6$ V, $R_B = 100\ \text{k}\Omega$, $R_C = 500\ \Omega$, and $R_E = 100\ \Omega$.

Solution:

Before the AC model can be constructed, the operating point of the transistor must be determined. Using the procedure of Example 8.4, we establish the DC operating point as follows:

$$V_{CE} = V_{CC} - R_C I_C - R_E(I_C + I_B) \approx 12 - 600 I_C$$

With $h_{fe} = \beta = 350$, we obtain

$$I_B = \frac{V_{BB} - 0.6}{R_B + (1+\beta)R_E} = 40\ \mu\text{A}$$

$$I_C = \beta I_B = 14\ \text{mA}$$

and can therefore solve for V_{CE}:

$$V_{CE} = 12 - 600 I_C = 3.6\ \text{V}$$

Thus, the Q point for the circuit is

$$I_{BQ} = 40\ \mu\text{A} \quad I_{CQ} = 14\ \text{mA} \quad V_{CEQ} = 3.6\ \text{V}$$

Finally, from equation 9.7, we have

$$\mu = \frac{-h_{fe} R_C}{R_B + h_{ie} + (h_{fe} + 1)R_E} = \frac{-175}{136.5} = -1.28$$

DRILL EXERCISE

9.1 Determine the AC/open-loop voltage gain of a 2N2222A transistor, using the results of Example 9.1 and Table 9.1. Use maximum values of the h parameters, and assume that $I_{CQ} = 50$ mA, $R_C = 1$ kΩ, $R_B = 100$ kΩ, and $R_E = 100$ Ω.

9.2 BJT SMALL-SIGNAL AMPLIFIERS

The h-parameter model developed in the previous section is very useful in the small-signal analysis of various configurations of BJT amplifiers. In this section, a set of techniques will be developed to enable you to first establish the Q point of a transistor amplifier, then construct the small-signal model, and finally use the small-signal model for analyzing the small-signal behavior of the amplifier. A major portion of this section will be devoted to the analysis of the **common-emitter amplifier,** using this circuit as a case study to illustrate the analysis methods. At the end of the section, we will look briefly at two other common amplifier circuits: the **voltage follower,** or **common-collector amplifier;** and the **common-base amplifier.** A table will summarize the input and output resistance and gains of all three types of amplifiers. The discussion of the common-emitter amplifier will also provide an occasion to introduce, albeit qualitatively, the important issue of transistor amplifier frequency response. A detailed treatment of this last topic is beyond the intended scope of this book.

A complete common-emitter amplifier circuit is shown in Figure 9.7. The circuit may appear to be significantly different from the simple examples studied in the previous sections; however, it will soon become apparent that all the machinery necessary to understand the operation of a complete transistor amplifier is

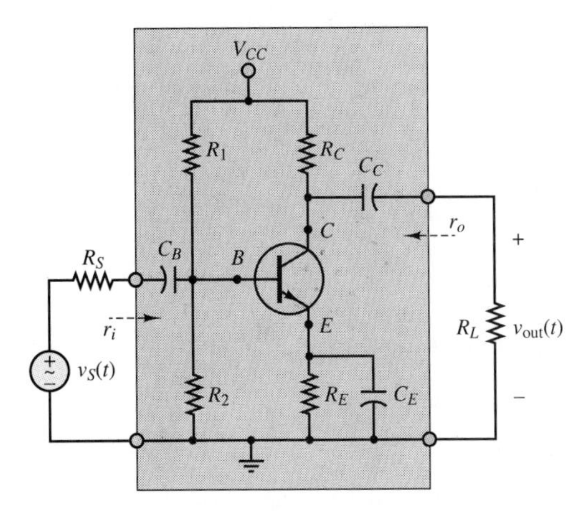

Figure 9.7 The BJT common-emitter amplifier

already available. The task at hand consists simply of piecing together the results obtained in the preceding sections.

The analysis of this circuit will be aimed at developing a simple small-signal linear AC equivalent-circuit model for the amplifier, to permit defining the open-circuit voltage and short-circuit current gains of the amplifier and its input and output resistances. To create this model, we shall develop a simple **two-port**[1] **equivalent circuit** for the amplifier; this equivalent circuit can then be used in connection with equivalent circuits for the load and source circuits to determine the actual gains of the amplifier as a function of the load and source impedances.

Figure 9.8 depicts the appearance of this simplified representation for a transistor amplifier, where r_i and r_o represent the input and output resistance of the amplifier, respectively, and μ is the open-loop voltage gain of the amplifier. Throughout this section, it will be shown how such a model can be obtained for a variety of amplifier configurations. You may wish to compare this model with that shown in Figure 8.3 in Section 8.1. Note that the model of Figure 9.8 makes use of the simple Thévenin equivalent-circuit model developed in Chapter 3 in representing the input to the amplifier as a single equivalent resistance, r_i. Similarly, the circuit seen by the load consists of a Thévenin equivalent circuit. In the remainder of this section, it will be shown how the values of r_i, r_o, and μ may be computed, given a transistor amplifier design. It is also useful to define the overall voltage and current gains for the amplifier model of Figure 9.8 as follows. The amplifier voltage gain is defined as

$$A_V = \frac{v_{\text{out}}}{v_S} = \frac{R_L}{R_L + r_o}\mu \cdot \frac{r_i}{R_S + r_i} \tag{9.12}$$

while the current gain is

$$A_I = \frac{i_{\text{out}}}{i_{\text{in}}} = \frac{v_{\text{out}}/R_L}{v_s/(R_S + r_i)} = \frac{R_S + r_i}{R_L} \cdot A_V \tag{9.13}$$

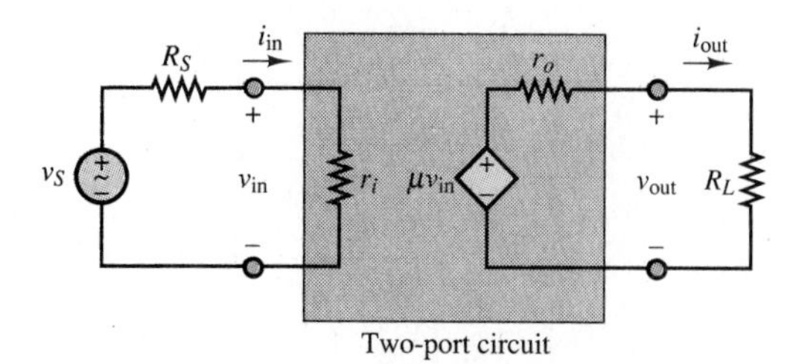

Figure 9.8 Equivalent-circuit model of voltage amplifier

[1]A two-port circuit is a circuit that has an input and an output port, in contrast with the one-port circuits studied in Chapter 3, which had only a single port connecting the source to a load. The amplifier configuration shown in Figure 9.7 is representative of a general two-port circuit.

The first observation that can be made regarding the circuit of Figure 9.7 is that the AC input signal, $v_S(t)$, has been "coupled" to the remainder of the circuit through a capacitor, C_C (called a **coupling capacitor**). Similarly, the load resistance, R_L, has been connected to the circuit by means of an identical coupling capacitor. The reason for the use of coupling capacitors is that they provide separate paths for DC and AC currents in the circuit. In particular, *the quiescent DC currents do not reach the source or the load.* This is especially useful, since the aim of the circuit is to amplify the AC input signal only, and it would be undesirable to have DC currents flowing through the load. In fact, the presence of DC currents would cause unnecessary and undesired power consumption at the load. The operation of the coupling capacitors is best explained by observing that a capacitor acts as an open circuit to DC currents, while—if the capacitance is sufficiently large—it will act as a short circuit at the frequency of the input signal. Thus, in general, one wishes to make C_C as large as possible, within reason. The **emitter bypass capacitor,** C_E, serves a similar purpose, by "bypassing" the emitter resistance R_E *insofar as AC currents are concerned,* since the capacitor acts as a short circuit at the signal frequency. On the other hand, C_E is an open circuit to DC currents, and therefore the quiescent current will flow through the emitter resistor, R_E. Thus, the emitter resistor can be chosen to select a given Q point, but it will not appear in the calculations of the AC gains. This dual role served by coupling and bypass capacitors in transistor amplifiers is of fundamental importance in their practical operation.

Figure 9.9 depicts the path taken by AC and DC currents in the circuit of Figure 9.7. Example 9.2 further explains the use of coupling capacitors.

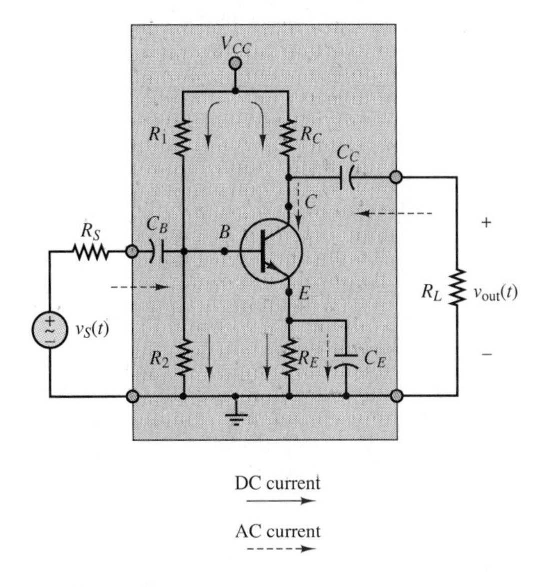

Figure 9.9 Effect of coupling capacitors on DC and AC current paths

Example 9.2

In this example, the numerical computation of the appropriate value for a coupling capacitor is illustrated for a signal $v_S(t)$ in the audio frequency range:

$$40\pi \le \omega_S \le 40{,}000\pi \text{ rad/s}$$

If we select a 1,000-μF coupling capacitor, its series impedance, Z_C, has magnitude

$$|Z_C| = \frac{1}{\omega C} = \frac{1}{40\pi \times 0.001} = 7.96\ \Omega$$

at

$$\omega = 40\pi \text{ rad/s}$$

and magnitude 1,000 times smaller at the high end of the audio spectrum (i.e., $\omega = 40{,}000\pi$). On the other hand, the impedance of the coupling capacitor at the frequency $\omega = 0$ (DC) is infinite. Thus, the coupling capacitor offers negligible resistance to the flow of current from the source over the frequency range of interest, while completely isolating the AC source from the DC supply.

The amplifier of Figure 9.7 employs a single DC supply, V_{CC}, as does the DC self-bias circuit of Example 8.6. The resistors R_1 and R_2, in effect, act as a voltage divider that provides a suitable bias for the *BE* junction. This effect is most readily understood if separate DC and AC equivalent circuits for the common-emitter amplifier are portrayed as in Figure 9.10.

To properly interpret the DC and AC equivalent circuits of Figure 9.10, a few comments are in order. Consider, first, the DC circuit. As far as DC currents are concerned, the two coupling capacitors and the emitter bypass capacitor are open circuits. Further, note that the supply voltage, V_{CC}, appears across two branches, the first consisting of the emitter and collector resistors and the *CE* "junction," and of the base resistors. This DC equivalent circuit is used in determining the Q point of the amplifier—that is, the quantities I_{BQ}, V_{CEQ}, and I_{CQ}. These are the DC quiescent currents and voltages that define the operating point

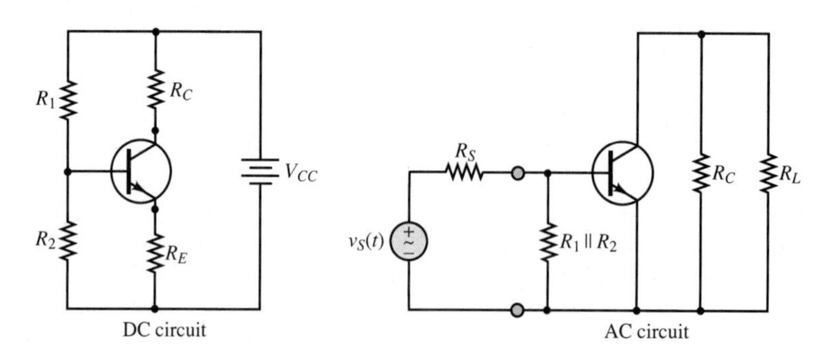

Figure 9.10 DC and AC circuits for the common-emitter amplifier

of the amplifier. In drawing the AC equivalent circuit, on the other hand, each of the capacitors has been replaced by a short circuit, as has the DC supply. The effect of the latter substitution (which applies only to AC signals) is to create a direct path to ground for the resistors R_1 and R_C. Thus, R_1 appears in parallel with R_2. Note, also, that in the AC equivalent circuit, the collector resistance R_C appears in parallel with the load. This will have an important effect on the overall gain of the amplifier.

DC Analysis of the Common-Emitter Amplifier

It is useful in analyzing the DC circuit of Figure 9.10 to redraw it in a slightly different form, recognizing that the Thévenin equivalent circuit seen by the base consists of the equivalent voltage

$$V_{BB} = \frac{R_2}{R_1 + R_2} V_{CC} \tag{9.14}$$

and of the equivalent resistance

$$R_B = R_1 \parallel R_2 \tag{9.15}$$

The resulting circuit is sketched in Figure 9.11. Application of KVL around the base and collector-emitter circuits yields the following equations, the solution of which determines the Q point of the transistor amplifier:

$$V_{CEQ} = V_{CC} - I_{CQ}R_C - \frac{\beta + 1}{\beta} I_{CQ}R_E \tag{9.16}$$

$$V_{BEQ} = V_{BB} - I_{BQ}R_B - \frac{\beta + 1}{\beta} I_{CQ}R_E \tag{9.17}$$

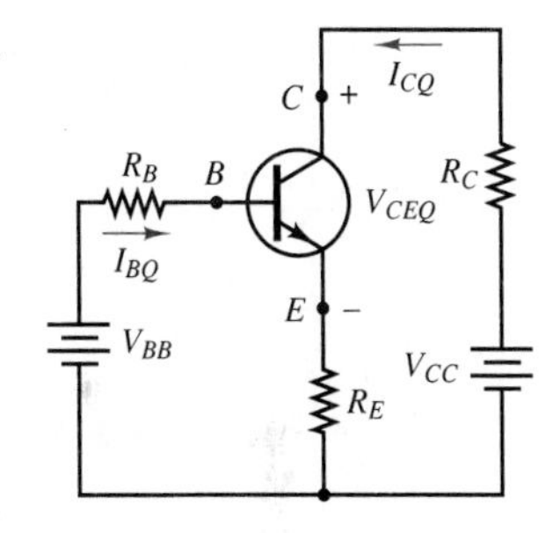

Figure 9.11 DC bias circuit for the common-emitter amplifier

In these equations, the quiescent emitter current, I_{EQ}, has been expressed in terms of the collector current, I_{CQ}, according to the relation

$$I_{EQ} = \frac{\beta + 1}{\beta} I_{CQ} \tag{9.18}$$

The next two examples illustrate a number of practical issues in the choice of DC bias point, and in the determination of other important features of the common-emitter transistor amplifier.

Example 9.3

One of the important choices in the design of a transistor amplifier is the location of the operating point. Even when the cutoff and saturation limits of the transistor are taken into account, there is no unique design for a given transistor amplifier. In this example, we analyze two similar designs, leading to two rather different common-emitter transistor amplifiers. The two amplifier circuits are shown in Figure 9.12. In analyzing the designs, we will assume that $V_\gamma = 0.7$ V and $V_{CE\text{ sat}} = 0.2$ V. Which is the preferred circuit, and why?

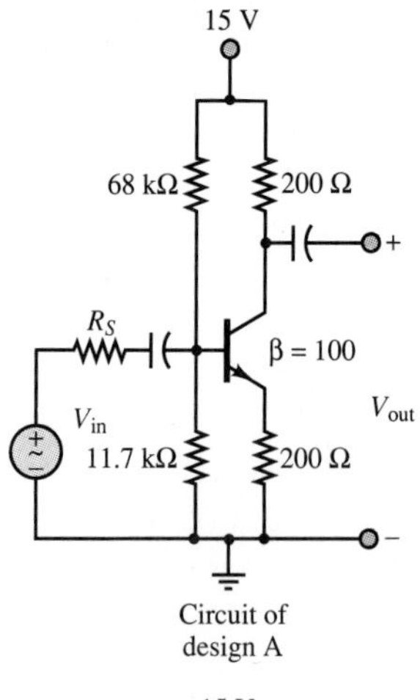

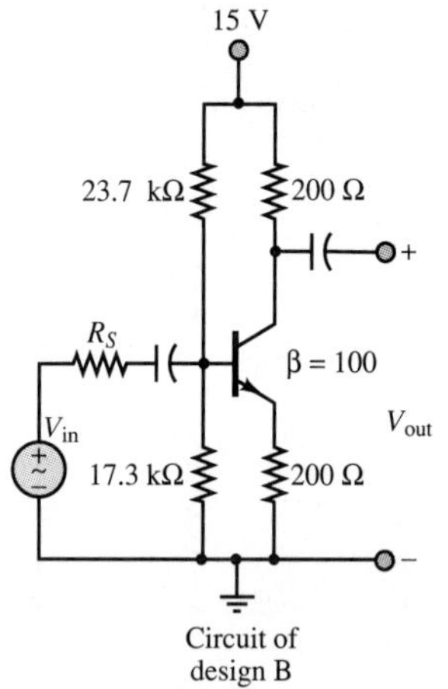

Figure 9.12 Two similar common-emitter amplifiers

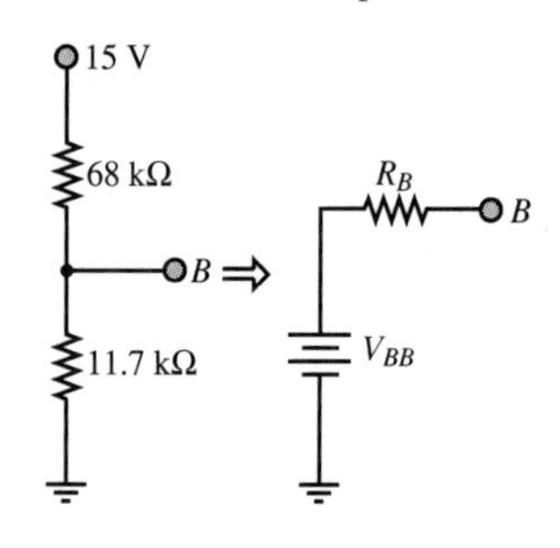

Figure 9.13 Equivalent base supply circuit of design A

Solution:

1. *Design A:* The equivalent circuit representing the base circuit of design A is shown in Figure 9.13, with

$$V_{BB} = \frac{11.7}{11.7 + 68}(15) = 2.2 \text{ V}$$

and

$$R_B = 68 \text{ k}\Omega \parallel 11.7 \text{ k}\Omega$$
$$\approx 10 \text{ k}\Omega$$

Next, we consider the base-emitter circuit, depicted in Figure 9.14, to find that

$$V_{BB} - V_{BE} = I_B R_B + I_E R_E = I_B R_B + I_B(\beta + 1)R_E$$

or

$$2.2 - 0.7 = I_B[10{,}000 + 101(200)]$$

Solving for the base current yields

$$I_B = 50 \ \mu\text{A}$$

and therefore

$$I_C = \beta I_B = 100(50 \ \mu\text{A}) = 5 \text{ mA}$$

Now, writing KVL around the collector-emitter loop, shown in Figure 9.15, we can solve for V_{CE}:

$$\begin{aligned} V_{CE} &= 15 - I_C R_C - I_E R_E \\ &= 15 - I_C R_C - I_C\left(\frac{\beta + 1}{\beta}\right)R_E \\ &= 15 - 0.005\left[200 + \frac{101}{100}(200)\right] \end{aligned}$$

so that

$$V_{CE} = 13 \text{ V}$$

Thus, we conclude that the Q point for design A is given by

$$I_{BQ} = 50 \ \mu\text{A} \qquad V_{CEQ} = 13 \text{ V} \qquad I_{CQ} = 5 \text{ mA}$$

2. *Design B:* Repeating the analysis for design B and using the same form for the equivalent circuits, we can describe the base equivalent circuit by

$$V_{BB} = 6.33 \text{ V} \qquad R_B = 10 \text{ k}\Omega$$

Application of KVL around the base-emitter circuit reveals that

$$6.33 - 0.7 = I_B(10{,}000) + I_B(200)(101)$$

and therefore

$$I_B = 186.4\ \mu\text{A}$$

and

$$I_C = \beta I_B = 18.6\ \text{mA}$$

Finally, writing KVL around the collector-emitter loop yields

$$V_{CE} = 15 - I_C\left[200 + \frac{101}{100}(200)\right] = 7.5\ \text{V}$$

Thus, the Q point for circuit B is

$$I_{BQ} = 186\ \mu\text{A} \qquad I_{CQ} = 18.6\ \text{mA} \qquad V_{CEQ} = 7.5\ \text{V}$$

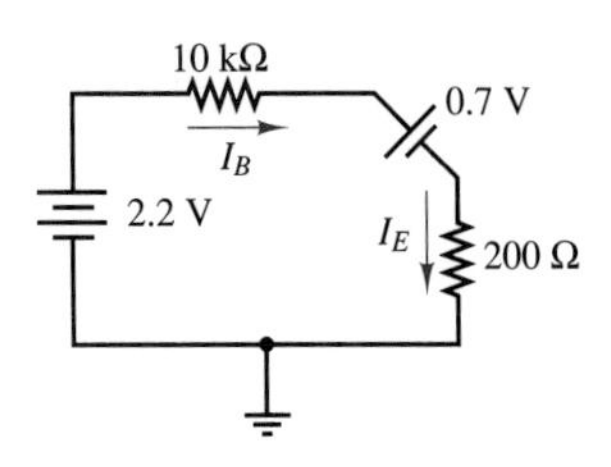

Figure 9.14 Equivalent base-emitter circuit of design A

Having completed the preliminary analysis, we plot the Q points on the load line for the circuit, to compare the results. The load-line equation for the collector circuit is given by the expression (obtained by applying KVL around the collector loop)

$$V_{CE} = V_{CC} - I_C(R_C + R_E)$$

and can most easily be sketched by finding the I_C and V_{CE} axis intercepts:

If $V_{CE} = 0$ V, then

$$I_C = \frac{15\ \text{V}}{R_C + R_E} = 37.5\ \text{mA}$$

If $I_C = 0$, then

$$V_{CE} = 15\ \text{V}$$

The results are displayed in Figure 9.16.

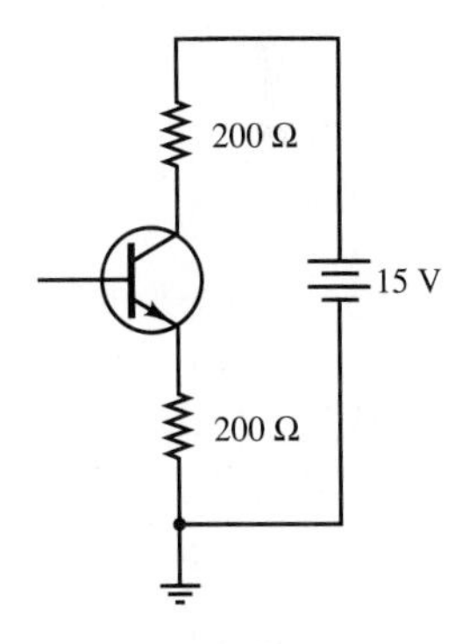

Figure 9.15 Collector-emitter circuit of design A

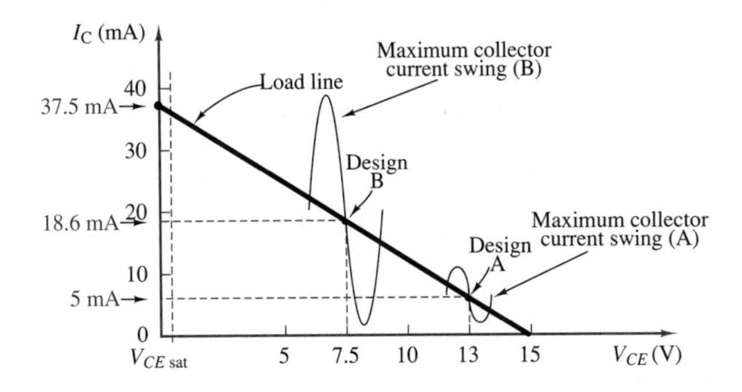

Figure 9.16 Operating points for designs A and B

We know that the AC input voltage, v_{in}, will cause the base current—and therefore the collector current—to oscillate around the operating point along the load line. Clearly, design A is much too close to the cutoff region and allows the collector current to swing undistorted only over a range of less than 10 mA before entering the cutoff region. On the other hand, design B permits a much greater swing for the collector current, thus offering a much larger dynamic range for the output voltage. Thus, the clear choice is the circuit of design B, because of the large symmetric swing in I_C and V_{CE}.

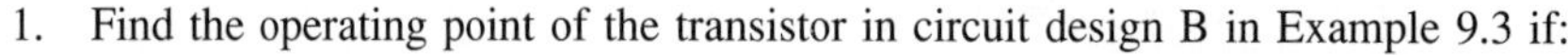

Example 9.4

One of the practical limitations in the design of BJT amplifiers arises from the variability of the current gain, β, even among transistors of the same family. A typical specification for a family of bipolar transistors—for example, the 2N3904—includes a β ranging from 75 to 150! It is therefore important, in designing an amplifier, to be able to compensate for such variability. It is especially important to obtain a stable operating point, independent of the variation in current gain from device to device. A common practice used to stabilize the operating point with respect to a range of β for a given transistor consists of designing the equivalent base supply circuit so that

$$R_B = \frac{\beta_{\min} R_E}{10}$$

where R_B is the equivalent base-circuit resistance and R_E is the emitter resistor. This choice reduces the variation of the collector current with respect to β, as can be shown using equation 8.10.

1. Find the operating point of the transistor in circuit design B in Example 9.3 if:
   a. $\beta = \beta_{\min} = 75$
   b. $\beta = \beta_{\max} = 150$
2. Use the rule of thumb stated in this example to set the desired operating point at $I_C = 18.6$ mA, and thus find R_B, V_{BB}, and R_1 and R_2.

The DC circuit for this example is shown in Figure 9.17.

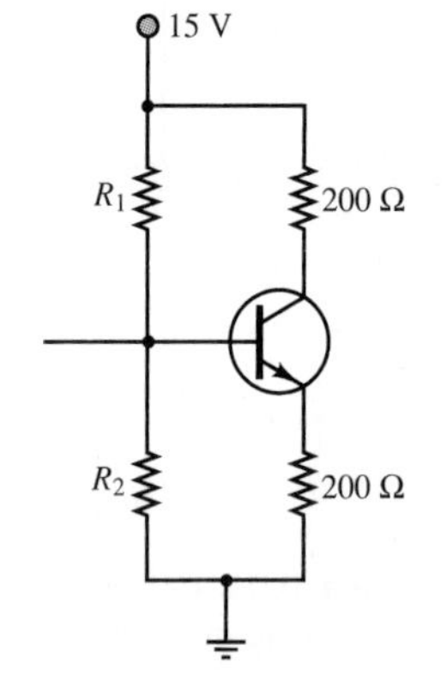

Figure 9.17 DC circuit for Example 9.4

Solution:

1. a. If $\beta = \beta_{\min} = 75$, we can derive the equivalent base circuit by repeating the analysis of Example 9.3. Then, by KVL, we find the base current from the expression

$$6.33 - 0.7 = I_B(10{,}000) + I_B(\beta_{\min} + 1)200$$

or

$$I_B = \frac{6.33 - 0.7}{10{,}000 + 76(200)} = 233\ \mu\text{A}$$

The corresponding collector current is then

$$I_C = \beta_{\min} I_B = 16.725\text{ mA}$$

Applying KVL around the collector-emitter loop, we determine the collector-emitter voltage,

$$V_{CE} = 15 - I_C(200) - I_C\left(\frac{\beta_{\min} + 1}{\beta_{\min}}\right)(200) = 8.3 \text{ V}$$

and the operating point:

$$I_{BQ} = 223\ \mu\text{A} \qquad I_{CQ} = 16.7 \text{ mA} \qquad V_{CEQ} = 8.3 \text{ V}$$

b. Now let us repeat the analysis for $\beta = \beta_{\max}$. Using KVL around the base circuit, we obtain

$$6.33 - 0.7 = I_B[10{,}000 + (\beta_{\max} + 1)200]$$

leading to

$$I_B = \frac{6.33 - 0.7}{10{,}000 + 151 \times 200} = 140\ \mu\text{A}$$

and therefore

$$I_C = \beta_{\max} I_B = 21 \text{ mA}$$

Repeating the analysis for the collector circuit, we find that

$$V_{CE} = 15 - I_C\left[R_C + \left(\frac{\beta_{\max} + 1}{\beta_{\max}}\right)R_E\right]$$

$$= 15 - (21 \text{ mA})\left[200 + \frac{151}{150}(200)\right] = 6.57 \text{ V}$$

so that the operating point is

$$I_{BQ} = 140\ \mu\text{A} \qquad I_{CQ} = 21 \text{ mA} \qquad V_{CEQ} = 6.57 \text{ V}$$

The two Q points are shown in Figure 9.18, as they appear on the load line.

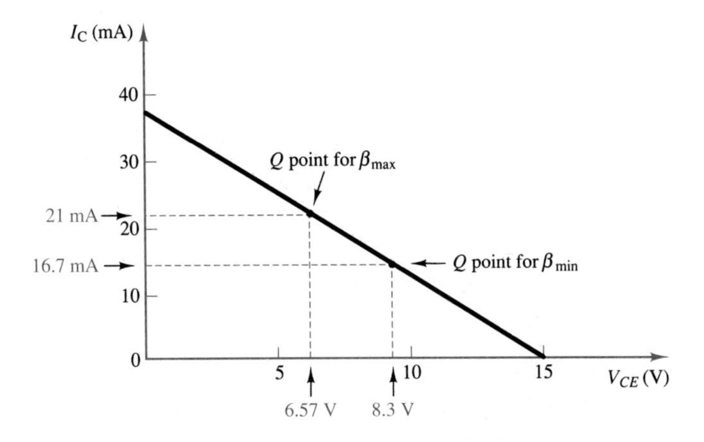

Figure 9.18 Variability in Q point due to change in β

The difference in operating point could then be as large as

$$\delta V_{CEQ} = 8.3 - 6.57 = 1.73 \text{ V}$$

and

$$\delta I_{CQ} = 21 - 16.7 = 4.3 \text{ mA}$$

for two identical amplifiers designed to operate at the same Q point.

2. Using the design rule of thumb, we find R_B to be

$$R_B = \frac{\beta_{\min} R_E}{10} = \frac{75(200)}{10} = 1{,}500 \ \Omega$$

By writing KVL around the base-emitter loop and using the relationship

$$I_E = \frac{\beta_{\min} + 1}{\beta_{\min}} I_C$$

with I_C = desired I_C = 18.6 mA, we arrive at the following equation:

$$\begin{aligned} V_{BB} &= I_B R_B + 0.7 + I_E R_E \\ &= \frac{18.6 \text{ mA}}{\beta_{\min}}(1{,}500) + 0.7 + \frac{\beta_{\min} + 1}{\beta_{\min}}(18.6 \text{ mA})(200) \\ &= 4.84 \text{ V} \end{aligned}$$

Now, since

$$V_{BB} = \left(\frac{R_2}{R_1 + R_2}\right) 15 \text{ V}$$

and

$$R_B = \frac{R_2 R_1}{R_2 + R_1} = 1{,}500 \ \Omega$$

two equations in two unknowns may be obtained to solve for R_1 and R_2:

$$4.84 = \left(\frac{R_2}{R_1 + R_2}\right) 15$$

$$1{,}500 = \frac{R_1 R_2}{R_1 + R_2}$$

We obtain

$$R_1 = 4.65 \text{ k}\Omega \qquad R_2 = 2.21 \text{ k}\Omega$$

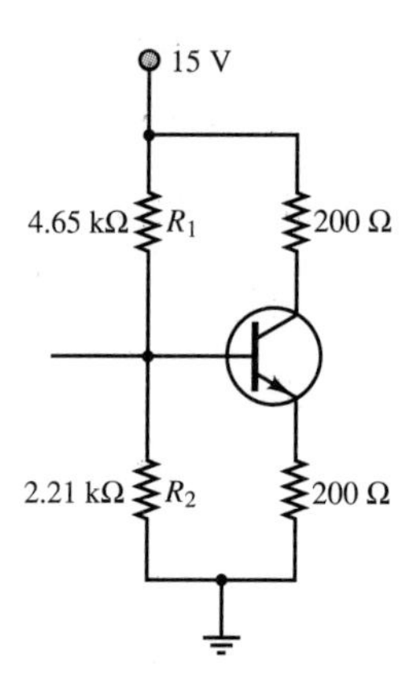

Figure 9.19 DC circuit resulting from application of rule of thumb

Thus, the final DC circuit will be as shown in Figure 9.19.

This design for the bias point will result in a far less sensitive Q point in the face of variations in the transistor current gain. The numerical answer is provided in Drill Exercise 9.3.

DRILL EXERCISES

9.2 Find the operating point of the circuit of design B in Example 9.3 if (a) $\beta = 90$ and (b) $\beta = 120$.

9.3 Verify that the operating point in design 2 of Example 9.4 is more stable than that of design 1.

9.4 Repeat all parts of Example 9.4 if $\beta = \beta_{\min} = 60$, $\beta = \beta_{\max} = 200$, and $I_C = 10$ mA.

AC Analysis of the Common-Emitter Amplifier

To analyze the AC circuit of the common-emitter amplifier, we substitute the hybrid-parameter small-signal model of Figure 9.3 in the AC circuit of Figure 9.10, obtaining the linear AC equivalent circuit of Figure 9.20. Note how the emitter resistance is bypassed by the emitter bypass capacitor at AC frequencies; this, in turn, implies that the base-to-emitter (BE) junction appears in parallel with the equivalent resistance $R_B = R_1 \parallel R_2$. Similarly, the collector-to-emitter (CE) junction is replaced by the parallel combination of the controlled current source, $h_{fe}\,\Delta I_B$, with the resistance $1/h_{oe}$. Once again, since the DC supply provides a direct path to ground for the AC currents, the collector resistance appears in parallel with the load. It is important to understand why we have defined input and output voltages, v_{in} and v_{out}, rather than use the complete circuit containing the source, v_S, its internal resistance, R_S, and the load resistance, R_L. The AC circuit model is most useful if an equivalent input resistance and the equivalent output resistance and amplifier gain are defined as quantities independent of the signal source and load properties. This approach permits the computation of the parameters r_i, r_o, and μ shown in Figure 9.8, and therefore provides a circuit model that may be called upon for any source-and-load configuration. The equivalent circuit shown in Figure 9.20 allows viewing the transistor amplifier as a single equivalent circuit either from the source or from the load end, as will presently be illustrated.

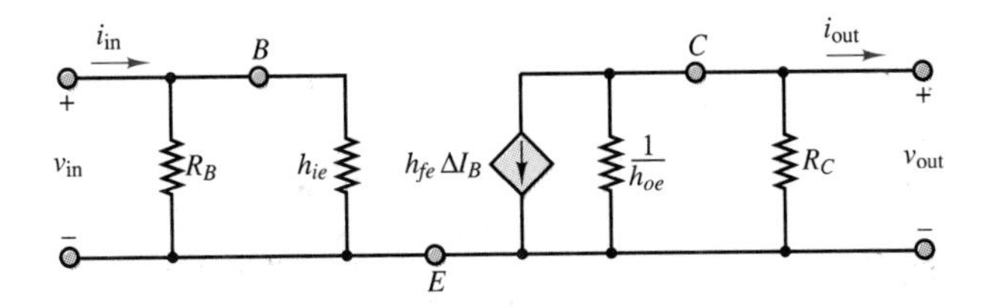

Figure 9.20 AC equivalent-circuit model for the common-emitter amplifier

The analysis of this circuit can easily be carried out by resorting to the DC circuit analysis tools of Chapters 2 and 3. First, compute the AC input current,

$$i_{\text{in}} = \frac{v_{\text{in}}}{h_{ie} \parallel R_B} \tag{9.19}$$

and the AC base current,

$$\Delta I_B = \frac{v_{\text{in}}}{h_{ie}} \tag{9.20}$$

noting, further, that the input resistance of the circuit consists of the parallel combination of h_{ie} and R_B:

$$r_i = h_{ie} \parallel R_B \tag{9.21}$$

Next, observe that the AC base current is amplified by a factor of h_{fe} and that it flows through the parallel combination of the collector resistance, R_C, and the *CE* junction resistance, $1/h_{oe}$. The latter term is usually negligible with respect to load and collector resistances, since the slope of the collector *i-v* curves is very shallow; in the remainder of this chapter it will be assumed that h_{oe} is negligible and that the parallel resistance $1/h_{oe}$ is infinite. Thus, the AC short-circuit output current i_{out} is given by the expression

$$i_{\text{out}} = -h_{fe}\,\Delta I_B = -h_{fe}\frac{v_{\text{in}}}{h_{ie}} = -g_m v_{\text{in}} \tag{9.22}$$

and the AC open-circuit output voltage is given by the expression

$$v_{\text{out}} = -h_{fe}\,\Delta I_B\,R_C = -h_{fe}\frac{v_{\text{in}}}{h_{ie}}R_C = -g_m v_{\text{in}} R_C \tag{9.23}$$

Note the negative sign, due to the direction of the controlled current source. This sign reversal is a typical characteristic of the common-emitter amplifier. Knowing the open-circuit output voltage and the short-circuit output current, we can find the output resistance of the amplifier as the ratio of these two quantities:

$$r_o = \frac{v_{\text{out}}}{i_{\text{out}}} = R_C \tag{9.24}$$

Next, if we define the **AC open-circuit voltage gain** of the amplifier, μ, by the expression

$$\mu = \frac{v_{\text{out}}}{v_{\text{in}}} \tag{9.25}$$

we find that

$$\mu = -h_{fe}\frac{R_C}{h_{ie}} = -g_m R_C \tag{9.26}$$

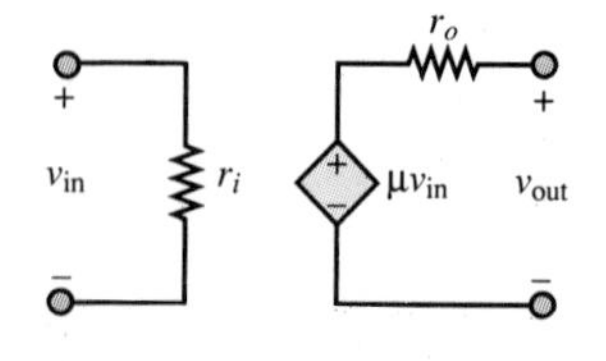

Figure 9.21 Simplified equivalent circuit for the common-emitter amplifier

At this point, it is possible to take advantage of the equivalent-circuit representation of Figure 9.8, with the expressions for μ, r_o, and r_i just given. Figure 9.21 illustrates the equivalent circuit for the common-emitter amplifier. This circuit replaces the transistor amplifier and enables us to calculate the actual voltage and current gain of the amplifier for any given load-and-source pair. Referring to the circuit of Figure 9.8 and to the associated definition of the voltage and current gains, A_V and A_I, we can compute these gains to be

$$A_V = \frac{v_{\text{out}}}{v_{\text{in}}} = \frac{R_L}{R_L + r_o}\mu = -\frac{R_L}{R_L + r_o}h_{fe}\frac{R_C}{h_{ie}} = -g_m R_L \parallel R_C \tag{9.27}$$

and

$$A_I = \frac{i_{\text{out}}}{i_{\text{in}}} = \frac{r_i}{R_L}A_V \tag{9.28}$$

Example 9.5 provides numerical values for a typical small-signal common-emitter amplifier.

Example 9.5

A common-emitter amplifier is shown in Figure 9.22. The transistor has the following h parameters: $h_{ie} = 1{,}400\ \Omega$, $h_{fe} = 100$, and $h_{oe} = 125\ \mu\text{S}$. Resistors R_1 and R_2 determine the DC operating point. The source has an internal resistance of 500 Ω. Find the actual voltage and current gains of this amplifier.

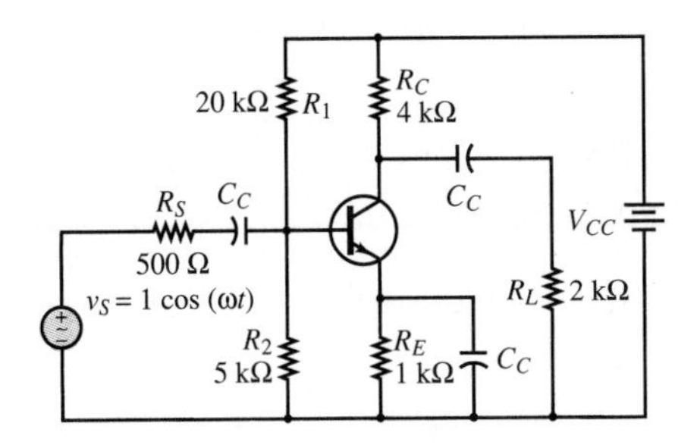

Figure 9.22 A common-emitter amplifier

Solution:

Replacing the BJT with its equivalent-circuit model results in the amplifier AC equivalent circuit shown in Figure 9.23, where

$$R_B = \frac{R_1 R_2}{R_1 + R_2} = \frac{20{,}000 \times 5{,}000}{20{,}000 + 5{,}000} = 4\ \text{k}\Omega$$

The input resistance r_i is determined from inspection of the circuit in Figure 9.23 to be

$$r_i = R_B \parallel h_{ie} = \frac{4{,}000 \times 1{,}400}{4{,}000 + 1{,}400} = \frac{56{,}000}{54} = 1{,}037\ \Omega$$

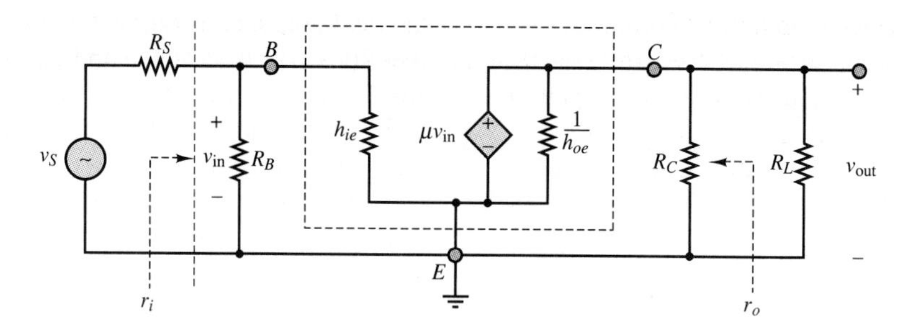

Figure 9.23

To determine the output resistance of the amplifier, we need to set $v_S = 0$. This results in $v_{in} = 0$ and the controlled-source input $\mu v_{in} = 0$. Thus,

$$r_o = R_C \parallel \frac{1}{h_{oe}} = \frac{4{,}000 \times 8{,}000}{4{,}000 + 8{,}000} = 2.67 \text{ k}\Omega$$

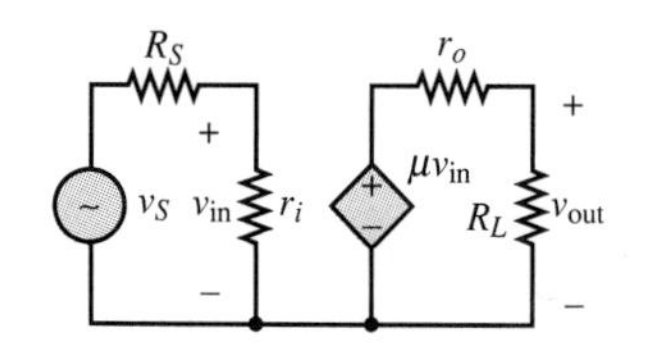

Figure 9.24

The complete AC model of the amplifier with source and load circuit is shown in Figure 9.24.

The open-circuit voltage gain is $\mu = -285.7$, from equation 9.26. The actual voltage gain, A_V, can be found by using equation 9.27:

$$A_V = \frac{v_{out}}{v_{in}} = \frac{-R_L}{R_L + r_o}\mu = -\frac{2 \times 10^3}{4.67 \times 10^3} \times 100 \times \frac{4 \times 10^3}{1.4 \times 10^3} = -122.4$$

The current gain, A_I, can be found by using equation 9.28:

$$A_I = \frac{r_i}{R_L} A_V = \frac{1{,}037}{2{,}000} \times (-122.4) = -63.46$$

DRILL EXERCISES

9.5 Compute the actual voltage gain, v_{out}/v_S, of the amplifier of Example 9.5 for the following source-load pairs:

a. $R_S = 50\ \Omega,\ R_L = 150\ \Omega$
b. $R_S = 50\ \Omega,\ R_L = 1{,}500\ \Omega$
c. $R_S = 500\ \Omega,\ R_L = 150\ \Omega$

What conclusions can you draw from these results?

9.6 Calculate the current gains for the amplifier parameters of Drill Exercise 9.5.

9.7 Repeat Example 9.5 for $h_{ie} = 2\text{ k}\Omega$ and $h_{fe} = 60$.

Other BJT Amplifier Circuits

The common-emitter amplifier is a commonly employed configuration but is by no means the only type of BJT amplifier. Other amplifier configurations are also used, depending on the specific application requirements. Each type of amplifier is classified in terms of properties such as input and output resistance, and voltage and current gain. Rather than duplicate the detailed analysis just conducted for the common-emitter amplifier, we summarize the properties of the three more common BJT amplifier circuits in Table 9.2, which depicts the amplifier circuits, the corresponding small-signal models, and the expressions for the input and output resistance and open-circuit voltage gain. The methodology employed to derive these results is completely analogous to that surveyed in the previous section: the

Table 9.2 **BJT amplifier configurations**

	Common emitter	Common collector	Common base
r_i	$h_{ie} \parallel R_B$	$R_B \parallel [h_{ie} + (h_{fe} + 1)R_E]$	$\frac{h_{ie}}{h_{fe} + 1} + R_E$
r_o	$R_C \parallel \frac{1}{h_{oe}}$	$\frac{\left(R_E \parallel \frac{1}{h_{oe}}\right)(h_{ie} + R_E)}{\left(R_E \parallel \frac{1}{h_{oe}}\right)(h_{fe} + 1) + h_{ie} + R_B}$	$R_C \parallel \frac{1 + h_{fe}}{h_{oe}}$
$\lvert A_V \rvert$	$\frac{h_{fe}}{h_{ie}}\left(R_C \parallel \frac{1}{h_{oe}}\right)$	$\frac{(h_{fe} + 1)\left(R_E \parallel \frac{1}{h_{oe}}\right)}{(h_{fe} + 1)\left(R_E \parallel \frac{1}{h_{oe}}\right) + h_{ie}}$	$\frac{h_{fe}R_C(R_E \parallel h_{ie})}{R_E h_{ie}}$
$\lvert A_I \rvert$	$h_{fe}\frac{R_B}{R_B + h_{ie}}$	$\frac{(1 + h_{fe})R_B}{R_B + h_{ie}}$	$\frac{h_{fe}}{h_{fe} + 1}$
Amplifier circuit	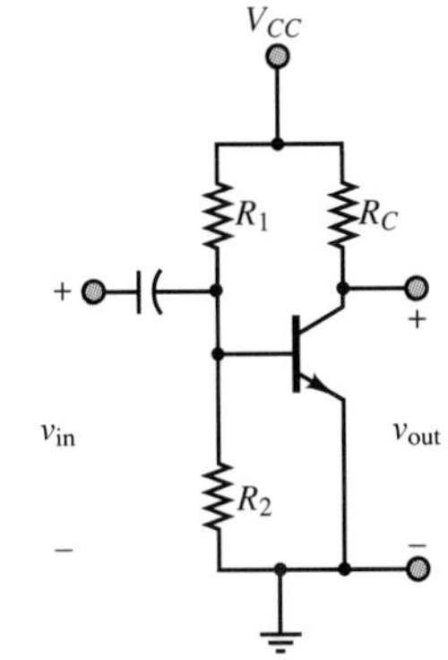	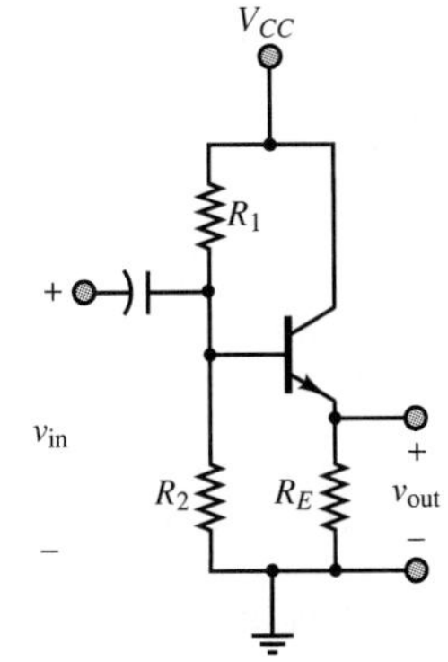	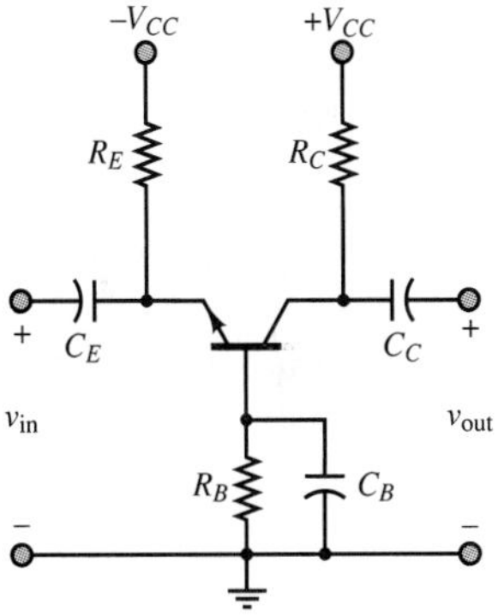
AC equivalent circuit	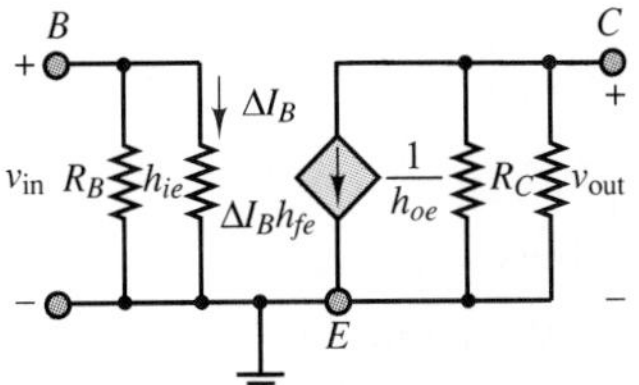	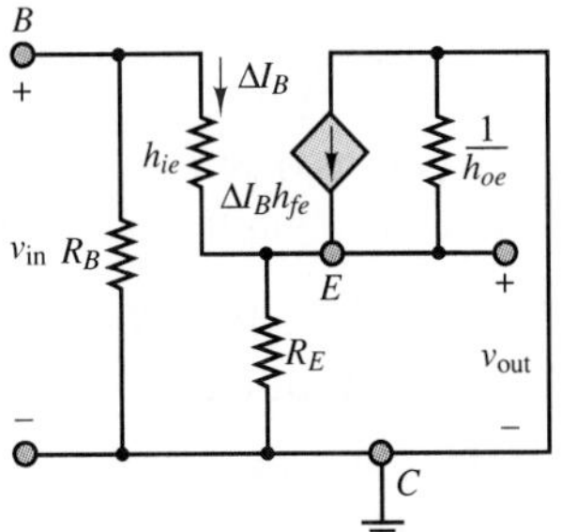	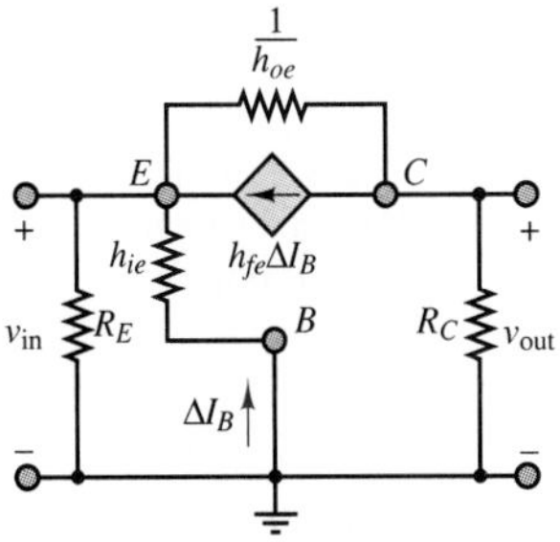
Properties	Input impedance: medium Output impedance: medium Voltage gain: high Current gain: high Phase shift: 180°	Input impedance: high Output impedance: low Voltage gain: low Current gain: medium Phase shift: 0°	Input impedance: low Output impedance: high Voltage gain: high Current gain: low Phase shift: 0°

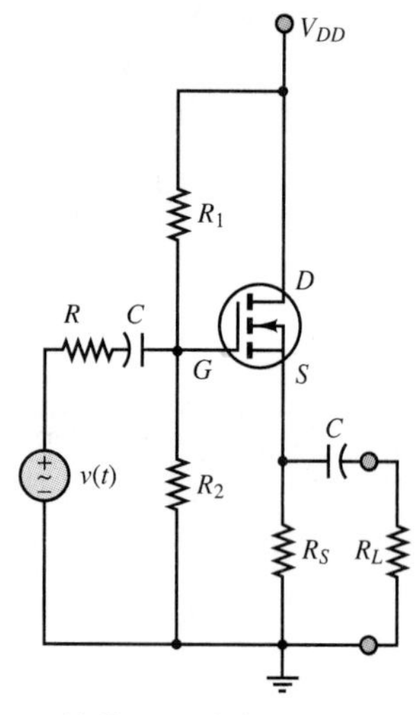

(a) Common-drain amplifier

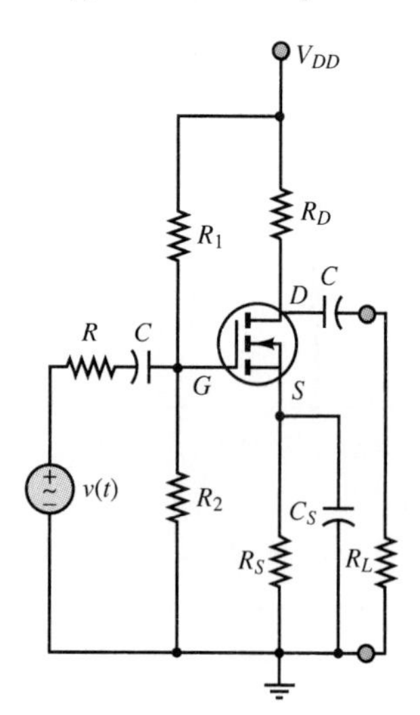

(b) Common-source amplifier

Figure 9.25 Typical common-drain and common-source MOSFET amplifiers

AC and DC circuits are identified first, then the transistor is replaced with its h-parameter model in the AC circuit, and standard circuit analysis techniques are used in deriving the desired expressions. We leave the derivation of the results of Table 9.2 to the student in a series of homework problems and recommend these exercises as an excellent means of practicing the newly acquired knowledge.

9.3 FET SMALL-SIGNAL AMPLIFIERS

The discussion of FETs as amplifiers is analogous to that of BJT amplifiers. In particular, the **common-source amplifier** circuit is equivalent in structure to the common-emitter amplifier circuit studied earlier, and the **common-drain amplifier (source follower)** is analogous to the common-collector amplifier (emitter follower). In this section, we discuss the general features of FET amplifiers; to simplify the discussion, we have selected the n-channel enhancement MOSFET to represent the FET family. Although some of the details differ depending on the specific device, the discussion that follows applies in general to all FET amplifiers. A summary of FET symbols was given in Figure 8.30 in Chapter 8; reviewing it will help you recognize a specific device in a circuit diagram.

Figure 9.25 depicts typical common-drain and common-source amplifiers, including coupling and bypass capacitors. One of the great features of FETs, and especially MOSFETs, is the high input impedance that can be achieved because the gate is effectively insulated from the substrate material. We shall illustrate this property in analyzing the source-follower circuit. Before proceeding with the analysis of FET amplifiers, though, we shall discuss how one can construct a small-signal model analogous to the one that was obtained for the BJT.

In the case of MOSFETs, we can make use of the analytic relation between drain current and gate-source voltage,

$$i_D = k(v_{GS} - V_T)^2 \qquad k = I_{DSS}/V_T^2 \tag{9.29}$$

to establish the Q point for the transistor, which is defined by the quiescent voltages, V_{GSQ} and V_{DSQ}, and by the quiescent current, I_{DQ}. The quadratic relationship allows us to determine the drain current, I_{DQ}, that will flow, given that $v_{GS} = V_{GSQ}$; the parameters k and V_T are a property of any given device. Thus,

$$I_{DQ} = k(V_{GSQ} - V_T)^2 \tag{9.30}$$

and it is now possible to determine the quiescent drain-to-source voltage from the load-line equation for the drain circuit:

$$V_{DD} = V_{DSQ} + R_D I_{DQ} + R_S I_{DQ} \tag{9.31}$$

In a MOSFET, it is possible to approximate the small-signal behavior of the device as a linear relationship by using the transconductance parameter g_m, where g_m is the slope of the i_D-v_{GS} curve at the Q point, as shown in Figure 9.26. Formally, we can write

$$g_m = \left.\frac{\partial i_D}{\partial v_{GS}}\right|_{I_{DQ},V_{GSQ}} \tag{9.32}$$

Since an analytical expression for the drain current is known (equation 9.26), we can actually determine g_m as given by

$$g_m = \frac{\partial}{\partial v_{GS}}[k(v_{GS} - V_T)^2] \quad (9.33)$$

at the operating point (I_{DQ}, V_{CSQ}). Thus, evaluating the expression for the transconductance parameter, we obtain the expression

$$\begin{aligned} g_m &= 2k(v_{GS} - V_T)\big|_{I_{DQ},V_{GSQ}} \\ &= 2\sqrt{kI_{DQ}} = 2\frac{\sqrt{I_{DSS}I_{DQ}}}{V_T} \end{aligned} \quad (9.34)$$

where g_m is a function of the quiescent drain current, as expected, since the tangent to the i_D-v_{GS} curve of Figure 9.26 has a slope that is dependent on the value of I_{DQ}.

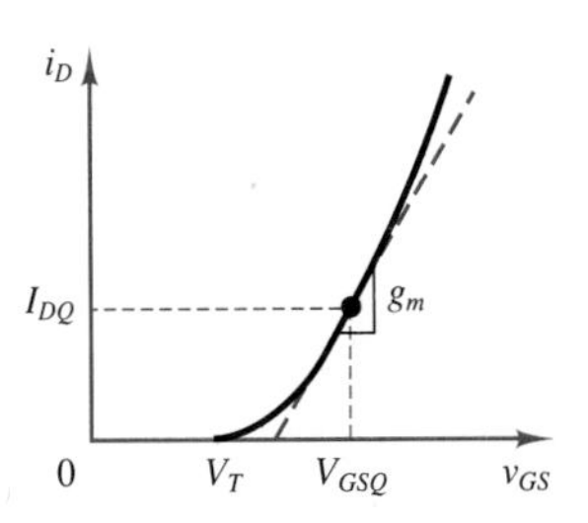

Figure 9.26 MOSFET transconductance parameter

Example 9.6

Let the Q point of a MOSFET be defined by $I_{DQ} = 55$ mA, $V_{DSQ} = 3.5$ V, $V_{GSQ} = 2.4$ V. Assume the values $V_T = 1.4$ V and $k = 0.125$ mA/V^2 for the MOSFET used in this example. Then, it follows that the transconductance parameter is given by

$$\begin{aligned} g_m &= 2\sqrt{kI_{DQ}} = 2\sqrt{0.125 \times 10^{-3} \times 55 \times 10^{-3}} \\ &= 5.24 \text{ mA/V} \end{aligned}$$

The significance of this value of the transconductance g_m is that for every volt of incremental oscillation around the gate-to-source voltage, there will result a corresponding drain current fluctuation of 5.24 mA; for example, if $\Delta V_{GS} = \cos(\omega t)$, then $\Delta I_D = 0.00524 \cos(\omega t)$. Thus, the MOSFET acts as a *voltage-controlled current source*. We shall see shortly how this behavior results in amplification.

Example 9.7

Determine all of the relevant transistor voltages and currents for the amplifier circuit shown in Figure 9.27. Assume that $V_T = 1$ V and $k = 0.5$ mA/V^2.

Solution:

Since the gate current is zero, the voltage between the gate and ground is

$$v_G = (10)\frac{10 \times 10^6}{10 \times 10^6 + 10 \times 10^6} = 5 \text{ V}$$

Let us assume for the moment that this gate voltage is such that the MOSFET is operating in the active region; we will proceed according to this assumption, and verify it once the

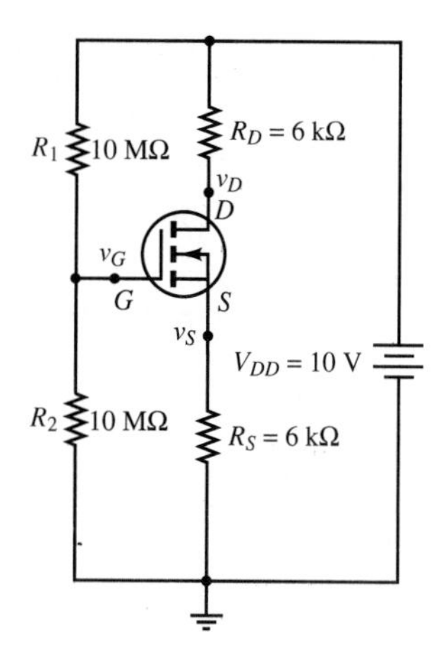

Figure 9.27

analysis is complete. The voltage between the source and ground is $v_S = 6 \times 10^3 i_D$, so that the gate-source voltage is given by

$$v_{GS} = 5 - 6i_D$$

where i_D is in mA. The drain current may be obtained from equation 9.29:

$$i_D = k(v_{GS} - V_T)^2 = 0.5(5 - 6i_D - 1)^2$$

Solving this quadratic equation, we obtain two solutions:

$$i_{D1} = 0.89 \text{ mA} \qquad \text{and} \qquad i_{D2} = 0.5 \text{ mA}$$

Which of the two solutions is the correct one? One way of determining the correct solution is to solve for the gate-source voltage for each case. For $i_{D1} = 0.89$ mA, the gate-source voltage is $v_{GS} = 5 - 6i_D = -0.34$ V. For $i_{D2} = 0.5$ mA, $v_{GS} = 2$ V. Since v_{GS} needs to be greater than V_T for the MOSFET to be in the active region, we must select

$$i_D = 0.5 \text{ mA}$$

to be consistent with the initial assumption. Correspondingly, the drain voltage is found to be

$$v_D = 10 - 6 \times 10^3 \times 0.5 \times 10^{-3} = 7 \text{ V}$$

and therefore $v_{DS} = v_D - v_S = 7 - 3 = 4$ V. Now it is possible to verify the initial assumption. Recall (equation 9.29) that the MOSFET is in the saturation region if $v_{DS} > v_{GS} - V_T$. From the calculations just carried out and the given value of V_T, $v_{DS} = 4 \text{ V} > v_{GS} - V_T = 1$ V, and therefore the MOSFET is indeed in saturation. Thus, the initial assumption was correct.

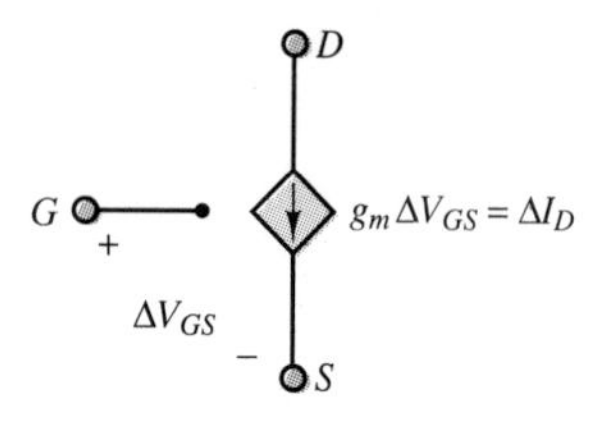

MOSFET small signal model

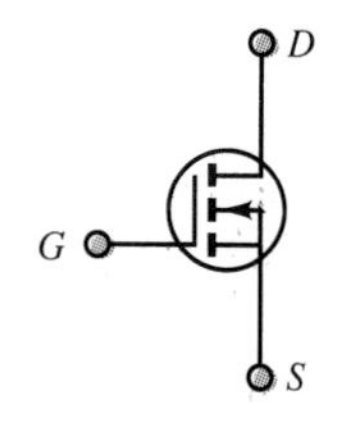

MOSFET circuit symbol

Figure 9.28 MOSFET small-signal model

The transconductance parameter allows us to define a very simple model for the MOSFET during small-signal operation: we replace the input circuit (gate) by an open circuit, since no current can flow into the insulated gate, and model the drain-to-source circuit by a controlled current source, $g_m \Delta V_{GS}$. This small-signal model is depicted in Figure 9.28.

It is important to appreciate the fact that the transconductance, g_m, is dependent on the quiescent value of the drain current, and therefore any MOSFET amplifier design is going to be very strongly dependent on the operating point.

The MOSFET Common-Source Amplifier

It is useful at this stage to compare the performance of the common-source amplifier of Figure 9.25(b) with that of the BJT common-emitter amplifier. The DC equivalent circuit is shown in Figure 9.29; note the remarkable similarity with the BJT common-emitter amplifier DC circuit.

The equations for the DC equivalent circuit are obtained most easily by reducing the gate circuit to a Thévenin equivalent with $V_{GG} = \frac{R_2}{R_1+R_2} V_{DD}$ and $R_G = R_1 \parallel R_2$. Then the gate circuit is described by the equation

$$V_{GG} = V_{GSQ} + I_{DQ} R_S \tag{9.35}$$

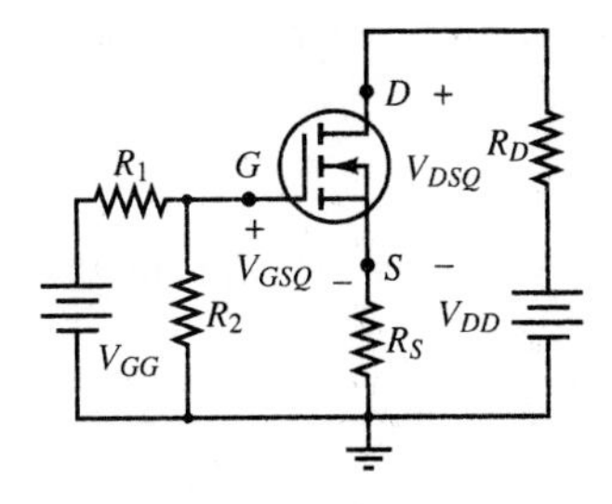

Figure 9.29 DC circuit for the common-source amplifier

and the drain circuit by

$$V_{DD} = V_{DSQ} + I_{DQ}(R_D + R_S) \tag{9.36}$$

Note that, given V_{DD}, any value for V_{GG} may be achieved by appropriate selection of R_1 and R_2. Example 9.8 illustrates the computation of the Q point for a MOSFET amplifier.

Example 9.8

In this example we use the MOSFET of Example 8.10 (with $V_T = 1.4$ V and $I_{DSS} = 95$ mA) to design a common-source amplifier like that of Figure 9.25(b) with the following Q point: $I_{DQ} = 55$ mA, $V_{DSQ} = 3.5$ V, $V_{GSQ} = 2.4$ V. A single voltage supply, $V_{DD} = 10$ V, is available.

Solution:

The gate equation resulting from KVL requires that

$$V_{GG} = V_{GSQ} + I_{DQ}R_S$$

or

$$V_{GG} = 2.4 + 0.055R_S$$

while the drain circuit imposes a requirement that

$$V_{DD} = V_{DSQ} + I_{DQ}(R_D + R_S)$$

or

$$10 = 3.5 + 0.055(R_D + R_S)$$

that is,

$$R_D + R_S = 118\ \Omega$$

If we select $R_D = 100\ \Omega$ and $R_S = 18\ \Omega$ for this operating point, we can solve for V_{GG}:

$$V_{GG} = 3.4\text{ V}$$

In order to choose resistances R_1 and R_2 such that $V_{GG} = 3.4$ V, we can use the equation

$$V_{GG} = \frac{R_2}{R_1 + R_2} V_{DD}$$

and arbitrarily choose $R_G = R_1 \parallel R_2 = 100$ kΩ to obtain $R_1 = 294$ kΩ and $R_2 = 151$ kΩ.

Substituting the small-signal model in the common-source amplifier circuit of Figure 9.25(b), we obtain the small-signal AC equivalent circuit of Figure 9.30, where we have assumed the coupling and bypass capacitors to be short circuits

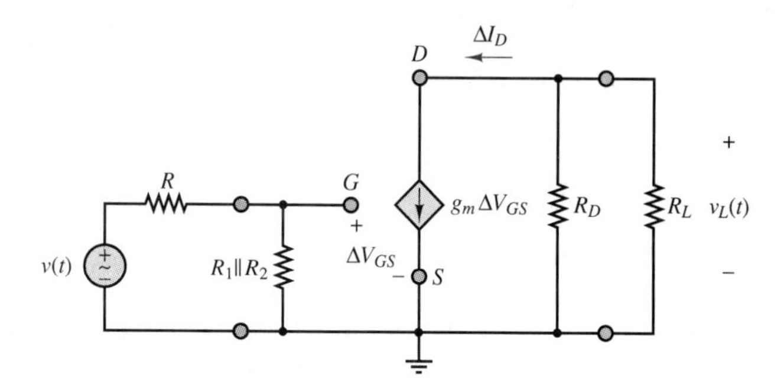

Figure 9.30 AC circuit for the common-source MOSFET amplifier

at the frequency of the input signal $v(t)$. The circuit is analyzed as follows. The load voltage, $v_L(t)$, is given by the expression

$$v_L(t) = -\Delta I_D \cdot (R_D \parallel R_L) \qquad \textbf{(9.37)}$$

where

$$\Delta I_D = g_m \Delta V_{GS} \qquad \textbf{(9.38)}$$

Thus, we need to determine ΔV_{GS} to write expressions for the voltage and current gains of the amplifier. Since the gate circuit is equivalent to an open circuit, we have:

$$\Delta V_{GS}(t) = \frac{(R_1 \parallel R_2)}{R + (R_1 \parallel R_2)} v(t) \qquad \textbf{(9.39)}$$

and

$$\Delta V_{GS}(t) \approx v(t) \qquad \textbf{(9.40)}$$

because we have purposely selected R_1 and R_2 to be quite large in the design of the DC bias circuit and therefore $R_1 \parallel R_2 \gg R$. Thus, we can express the load (output) voltage in terms of the input voltage $v(t)$ as

$$v_L(t) = -g_m \cdot (R_D \parallel R_L) \cdot v(t) \tag{9.41}$$

This expression corresponds to a voltage gain of

$$A_V = \frac{v_L(t)}{v(t)} = -g_m(R_D \parallel R_L) \tag{9.42}$$

Note that we have an upper bound on the voltage gain for this amplifier, $A_{V_{\max}}$, that is independent of the value of the load resistance:

$$|A_{V_{\max}}| \leq g_m R_D \tag{9.43}$$

This upper bound is achieved, of course, only when the load is an open circuit. This open-circuit voltage gain plays the same role in the analysis of the MOSFET amplifier as the parameter μ we defined in equation 9.26 for BJT amplifiers:

$$\mu = -g_m R_D \tag{9.44}$$

If we compute the value of g_m for the MOSFET amplifier of Example 9.8 ($g_m = 0.1033$ A/V), we find that the open-loop voltage gain of this amplifier is $\mu = -g_m R_D = -10.33$. The open-circuit voltage gain of a MOSFET, however, is highly dependent on the quiescent drain current, as illustrated in equation 9.34, and could be quite small if the MOSFET is biased for the small DC drain current. In comparison, the open-circuit voltage gain of a BJT is much less dependent on the operating point. This seeming drawback of MOSFET amplifiers is amply compensated by their high-input-resistance property. In the MOSFET common-source amplifier, the current drawn by the input is given by the expression

$$i_{\text{in}} = \frac{v(t)}{R + R_G} \approx \frac{v(t)}{R_G} \tag{9.45}$$

while the output current is given by

$$i_L(t) = \frac{v_L(t)}{R_L} = \frac{A_V \cdot v(t)}{R_L} \tag{9.46}$$

Thus, the current gain is given by the following expression:

$$\begin{aligned} A_I &= \frac{i_L}{i_{\text{in}}} = \frac{A_V \cdot R_G}{R_L} = -g_m \frac{R_D R_L}{R_D + R_L} \frac{R_G}{R_L} \\ &= -g_m \frac{R_D R_G}{R_D + R_L} \end{aligned} \tag{9.47}$$

Note that we have purposely made R_G large (100 kΩ), to limit the current required of the signal source, $v(t)$. Thus, the current gain for a common-source amplifier can be significant. For a 100-Ω load, we find that the effective voltage gain for the common-source amplifier is

$$A_V = -g_m R_D \parallel R_L \approx -5$$

and the corresponding current gain is

$$A_I = -g_m \frac{R_D R_G}{R_D + R_L} = -5{,}000$$

for the design values quoted in Example 9.8. Thus, the net **power gain** of the transistor, A_P, is quite significant.

$$A_P = A_V \cdot A_I = 25{,}000$$

This brief discussion of the common-source amplifier has pointed to some important features of MOSFETs. We first noted that the transconductance of a MOSFET is highly variable. However, MOSFETs make up for this drawback by their inherently high input impedance, which requires very little current of a signal source and thus affords substantial current gains. Finally, the output resistance of this amplifier can be made quite small by design. Example 9.9 provides an illustration of the amplification characteristics of a power MOSFET.

Example 9.9

The n-channel enhancement-mode small-signal MOSFET BS170 is designed for high-voltage, high-speed power switching applications such as line drivers, relay drivers, microprocessor to high-voltage interface, and high-voltage display drivers. The typical value of the transconductance parameter for this type of device is $g_m = 200$ mS for $V_{DS} = 10$ V and $I_D = 250$ mA, and the minimum gate threshold voltage is 0.8 V at $V_{DS} = V_{GS}$ and $I_D = 1.0$ mA. Assume that the desired operating point is at $V_{DSQ} = 10$ V, $I_{DQ} = 250$ mA, $V_{GSQ} = 9$ V, and $V_{DD} = 25$ V. Complete the design of the common-source amplifier shown in Figure 9.25(b) for this particular transistor, and calculate the gains A_V and A_I.

Solution:

The equation for the gate circuit is

$$V_{GG} = V_{GSQ} + I_{DQ} R_S = 9 + 0.25 R_S$$

while for the drain circuit, we obtain

$$\begin{aligned} V_{DD} &= V_{DSQ} + I_{DQ}(R_D + R_S) \\ &= 10 + 0.25(R_D + R_S) = 25 \text{ V} \end{aligned}$$

Thus,

$$R_D + R_S = 60 \ \Omega$$

If we select $R_S = 2\ \Omega$ and $R_D = 58\ \Omega$, a gate supply voltage may be selected as follows:

$$V_{GG} = 9 + 0.25 \times 2$$
$$= 25\left(\frac{R_2}{R_1 + R_2}\right) = 9.5\ \text{V}$$

If we choose $R_G = R_1 \| R_2 = 100\ \text{k}\Omega$ and solve the preceding equation, we get

$$R_1 \approx 263\ \text{k}\Omega \qquad R_2 \approx 161\ \text{k}\Omega$$

The voltage gain for an 80-Ω load may now be computed to be

$$A_V = -g_m(R_D \| R_L) = -200 \times 10^{-3} \times 33.6 = -6.7$$

and the current gain is

$$A_I = -g_m \frac{R_D R_G}{R_D + R_L} = -8{,}400$$

You can see that this power MOSFET can afford substantial power amplification.

The MOSFET Source Follower

FET source followers are commonly used as input stages to many common instruments, because of their very high input impedance. A commonly used source follower is shown in Figure 9.31. The desirable feature of this amplifier configuration is that, as we shall soon verify, it provides a high input impedance and a low output impedance. These features are both very useful and compensate for the fact that the voltage gain of the device is at most unity.

As usual, the Q point of the amplifier is found by writing KVL around the drain circuit:

$$V_{DSQ} = V_{DD} - I_{DQ}R_S \tag{9.48}$$

and by assuming that no current flows into the gate. The voltage V_{GSQ}, which controls the gate-to-source bias, is then given by the expression

$$V_{GSQ} = V_{GG} - I_{DQ}R_S \tag{9.49}$$

so that the desired Q point may be established by selecting appropriate values of R_1, R_2, and R_S (Example 9.10 will illustrate a typical design). The AC equivalent circuit is shown in Figure 9.32.

Since it is possible to select R_1 and R_2 arbitrarily, we see that the input resistance, R_G, of the source follower can be made quite large. Another observation is that, for this type of amplifier, the voltage gain is always less than 1. This fact may be verified by considering that

$$\Delta V_{GS} = v_{\text{in}}(t) - g_m\,\Delta V_{GS}\,R_S \tag{9.50}$$

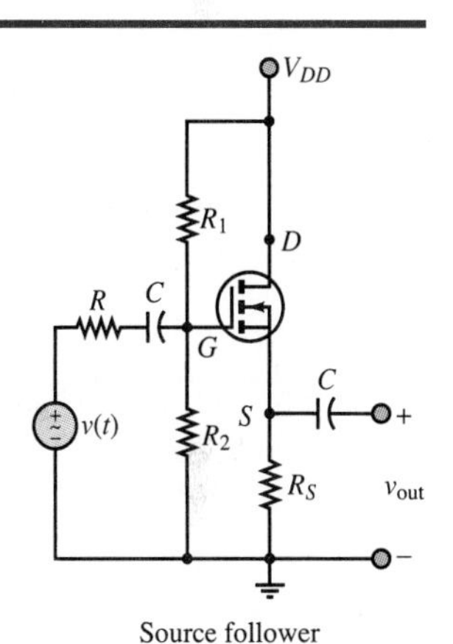

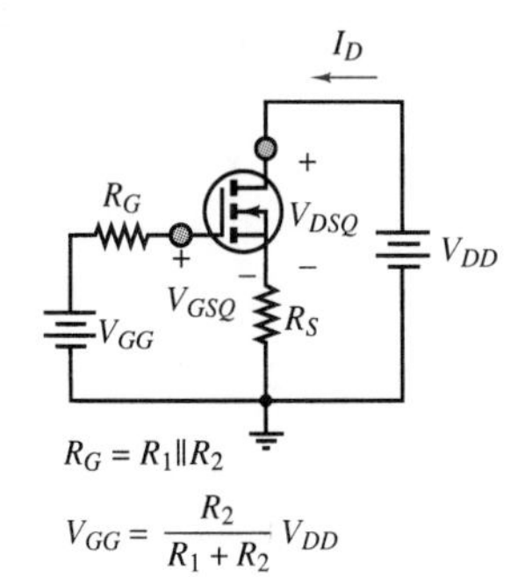

Figure 9.31 MOSFET source follower and DC circuit

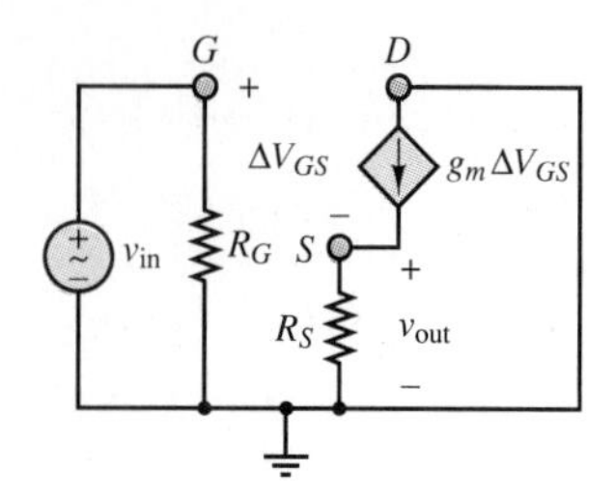

Figure 9.32 AC equivalent circuit for the MOSFET source follower

which implies

$$\frac{\Delta V_{GS}}{v_{\text{in}}(t)} = \frac{1}{1 + g_m R_S} \tag{9.51}$$

Since the AC output voltage is related to ΔV_{GS} by

$$v_{\text{out}} = g_m \, \Delta V_{GS} \, R_S \tag{9.52}$$

it follows that

$$\frac{v_{\text{out}}}{\Delta V_{GS}} = g_m R_S \tag{9.53}$$

Thus, the open-circuit voltage gain, μ, may be obtained as follows:

$$\begin{aligned} \frac{v_{\text{out}}}{v_{\text{in}}} &= \frac{v_{\text{out}}}{\Delta V_{GS}} = (g_m R_S)\left(\frac{1}{1 + g_m R_S}\right) \\ &= \frac{g_m R_S}{g_m R_S + 1} < 1 \end{aligned} \tag{9.54}$$

We can see that the open-circuit voltage gain will always be less than 1. The output resistance can be found by inspection from the circuit of Figure 9.32 to be equal to R_S.

In summary, the source-follower amplifier has an input resistance, $r_i = R_G = R_1 \parallel R_2$, which can be made arbitrarily large; a voltage gain less than unity; and an output resistance, $r_o = R_S$, which can be made small by appropriate design.

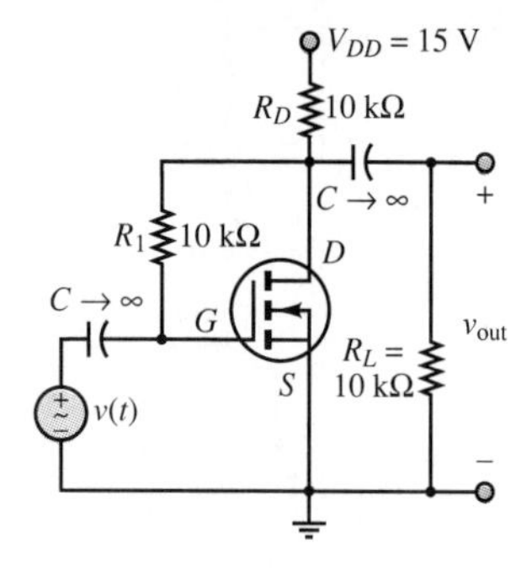

Figure 9.33

Example 9.10

An enhancement MOSFET source follower is shown in Figure 9.33. Given $V_T = 1.5$ V, $k = 0.125$ mA/V^2, find the small-signal voltage gain and the input resistance.

Solution:

The DC operating point for the amplifier is found by using the equation relating the drain current to the gate-source voltage:

$$\begin{aligned} I_{DQ} &= k(V_{GSQ} - V_T)^2 \\ &= 0.125 \times 10^{-3}(V_{GSQ} - 1.5)^2 \end{aligned}$$

Because the DC gate current is zero and no DC voltage will drop across R_G, it may be concluded that the gate-source voltage is equal to the drain voltage:

$$V_{GSQ} = V_{DQ}$$

Thus, the drain current is given by

$$I_{DQ} = 0.125 \times 10^{-3}(V_{DQ} - 1.5)^2$$

and the drain-source voltage is found from the load-line equation to be

$$V_{DSQ} = V_{DD} - R_D I_{DQ} = 15 - 10 \times 10^3 I_{DQ}$$

Solving the two preceding equations, we have

$$I_{DQ} = 1.06 \text{ mA} \qquad V_{DQ} = 4.4 \text{ V}$$

The transconductance parameter may also be found, once the operating point of the MOSFET is known:

$$g_m = 2k(V_{GSQ} - V_T) = 2 \times 0.125(4.4 - 1.5) = 0.725 \text{ mA/V}$$

Finally, the small-signal equivalent circuit is shown in Figure 9.34.

From the equivalent circuit, we may derive the following expression for the output voltage:

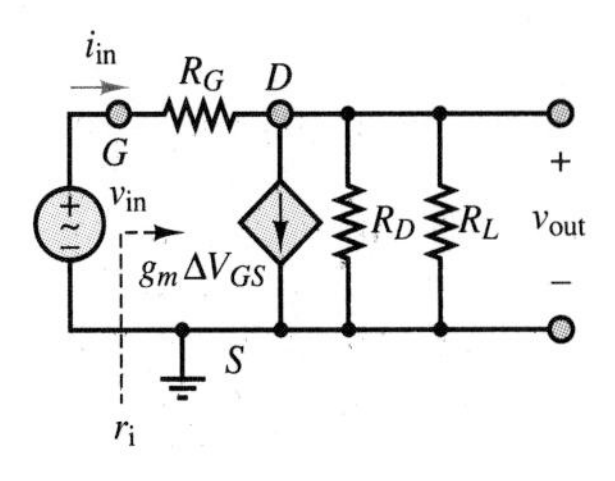

Figure 9.34

$$v_{out} = \Delta V_{DS} = -g_m \Delta V_{GS} (R_D \parallel R_L)$$

Since $\Delta V_{GS} = v_{in}(t)$, the voltage gain is

$$\frac{v_{out}}{v_{in}} = -g_m(R_D \parallel R_L)$$

$$= -0.725 \times 10^{-3} \times (10 \times 10^3 \parallel 10 \times 10^3) = -3.625$$

The input current, i_{in}, is found to be

$$i_{in} = \frac{v_{in} - v_{out}}{R_G} = \frac{v_{in}}{R_G}\left(1 - \frac{v_{out}}{v}\right)$$

$$= \frac{v_{in}}{R_G}[1 - (-3.625)] = \frac{4.625 v_{in}}{R_G}$$

and from the expressions for the input current and input voltage, we can now compute the input resistance of the amplifier:

$$r_i = \frac{v_{in}}{i_{in}} = \frac{R_G}{4.625} = 2.16 \text{ M}\Omega$$

You see that it is possible to design source followers with remarkably high input resistance.

DRILL EXERCISES

9.8 Compute the quiescent drain current and the small-signal transconductance parameter for a MOSFET with $V_T = 1.4$ V, $k = 2$ mA/V^2, for the following values of V_{GS}: 1.8 V, 2.0 V, 2.2 V, 2.4 V, 2.6 V, 2.8 V, 3.0 V.

9.9 Find the open-circuit AC voltage gain, μ, for the amplifier design of Example 9.8. What is the effective voltage gain of the amplifier if the load resistance is 1 kΩ?

9.10 Repeat Drill Exercise 9.9 for the values of g_m found in Drill Exercise 9.8.

9.4 TRANSISTOR AMPLIFIERS

The design of a practical transistor amplifier is a more complex process than what has been described in the preceding sections. A number of issues must be addressed in order to produce a useful amplifier design; such a detailed discussion is beyond the scope of this book, primarily because the intended audience of this text is not likely to be involved in the design of advanced amplifier circuits. However, it is important to briefly mention two topics: frequency response limitations of transistor amplifiers; and the need for **multistage amplifiers** consisting of several transistor amplifier stages—that is, several individual amplifiers like the ones described in the preceding sections. Other issues that will not be addressed in greater detail in this book include: amplifier input and output impedance; impedance matching between amplifier stages, and between source, amplifier and load; *direct-coupled* and *transformer-coupled amplifiers; differential amplifiers;* and *feedback*. However, some of these issues will be indirectly addressed in Chapter 11, devoted to *operational amplifiers,* where it will be shown that for many instrumentation and signal-conditioning needs, a non–electrical engineer can be satisfied with the use of *integrated circuit amplifiers,* consisting of multistage transistor amplifiers completely assembled into an integrated circuit "chip."

Frequency Response of Small-Signal Amplifiers

When the idea of a coupling capacitor was introduced earlier in this chapter, the observation was made that the input and output coupling capacitors (see Figure 9.7, for example) acted very nearly like a short circuit at AC frequencies. In fact, the presence of a series capacitor in the AC small-signal equivalent input circuit has very much the effect of a high-pass filter. Consider the small-signal equivalent circuit of Figure 9.35, which corresponds to a BJT common-emitter amplifier. In this circuit, a single coupling capacitor has been placed at the input, for simplicity.

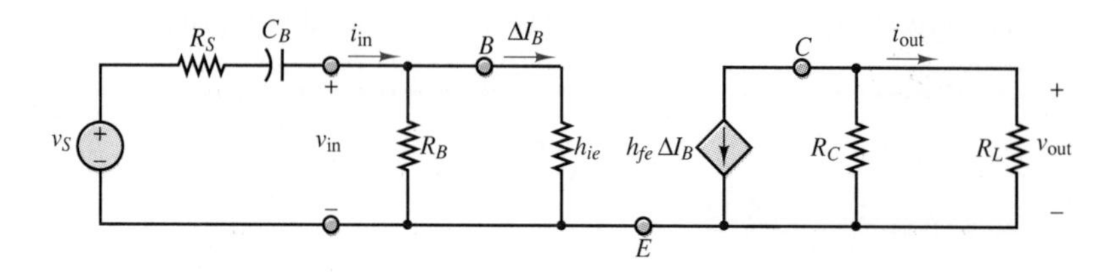

Figure 9.35 Small-signal equivalent circuit

Let us compute the frequency response of the small-signal equivalent circuit of Figure 9.34. If we assume that $R_B \gg R_S$ and $R_B \gg h_{ie}$, then we can approximate the base current by

$$\Delta i_B = \frac{v_{in}}{h_{ie}} \tag{9.55}$$

and

$$v_{in} = \frac{h_{ie}}{h_{ie} + R_S + \dfrac{1}{j\omega C_B}} v_S \tag{9.56}$$

so that the output voltage is given by

$$v_{\text{out}} = -R_C \parallel R_L h_{fe} \Delta i_B = R_C \parallel R_L \frac{v_{\text{in}}}{h_{ie}} \tag{9.57}$$

or

$$\frac{v_{\text{out}}}{v_S} = R_C \parallel R_L \frac{h_{fe}}{h_{ie} + R_S + \dfrac{1}{j\omega C_B}} \tag{9.58}$$

This expression for v_{out} is clearly a function of frequency, approaching a constant as ω tends to infinity. Note that, in effect, the cutoff frequency is determined at the input circuit by the effective RC combination:

$$\omega_B = \frac{1}{(R_S + h_{ie})C_B} \tag{9.59}$$

Clearly, if we wished to design an *audio amplifier* (that is, an amplifier that can provide an undistorted frequency response over the range of frequencies audible to the human ear), we would select a coupling capacitor such that the lower cutoff frequency in the circuit of Figure 9.35 would be greater than $2\pi \times 20$ rad/s. A similar effect can be attributed to the output coupling capacitor, C_C, in Figure 9.7, and to the emitter bypass capacitor, C_E, resulting in the cutoff frequencies

$$\omega_C = \frac{1}{(R_C + R_L)C_C} \tag{9.60}$$

and

$$\omega_E = \frac{1}{r_o C_E} \tag{9.61}$$

where r_o is the output impedance of the amplifier (see Table 9.2). The highest of the three frequencies given in equations 9.59 through 9.61 is often ω_E, and this is therefore the lower cutoff frequency of the amplifier. Figure 9.36 illustrates the general effect of a nonideal coupling capacitor on the frequency response of the common-emitter amplifier.

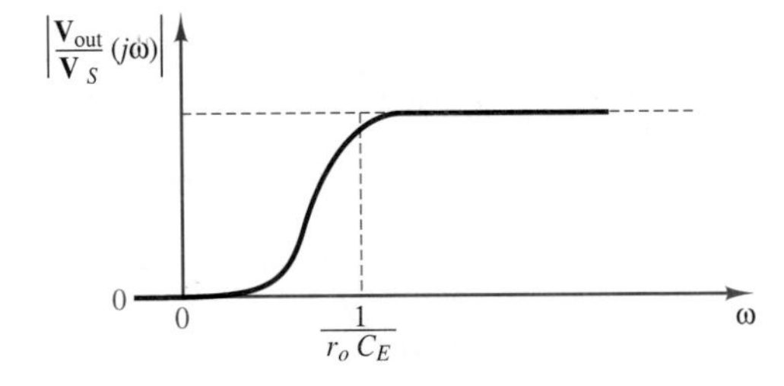

Figure 9.36 Approximate low-end frequency response of a common-emitter amplifier, assuming finite coupling capacitors

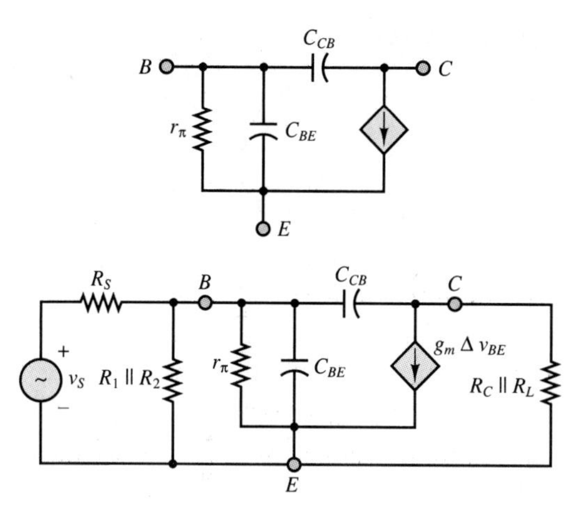

Figure 9.37 High-frequency BJT model and equivalent circuit

The frequency response of a bipolar transistor amplifier is generally limited at the high-frequency end by the so-called **parasitic capacitance** present between pairs of terminals at the transistor (don't forget that any two conductors separated by a dielectric—air, for example—form a capacitor). Figure 9.37 depicts a high-frequency small-signal model (called the **hybrid-pi model**) of the BJT, which includes the effect of two capacitances, C_{BE} and C_{CB}. These two capacitances can be safely ignored (by treating them as open circuits) at low and mid frequencies. It can be shown that at high frequency both capacitances contribute to a low-pass response with cutoff frequency given by:

$$\omega_T = \frac{1}{R_T C_T} \tag{9.62}$$

where $R_T = R_S \| R_1 \| R_2 \| r_\pi$ and C_T is an equivalent capacitance related to C_{CB} and C_{BE}. The calculation of C_T requires making use of Miller's theorem, which is well beyond the scope of this book. The effect of these capacitances on the frequency response of the amplifier is depicted in Figure 9.38. Figure 9.39 summa-

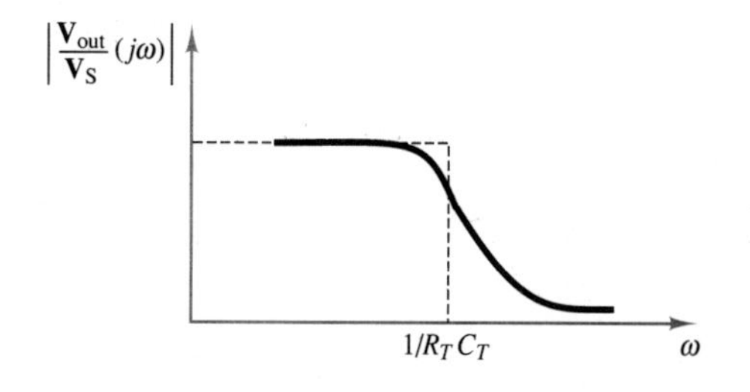

Figure 9.38 Low-pass filter effect of parasitic capacitance in a common-emitter amplifier

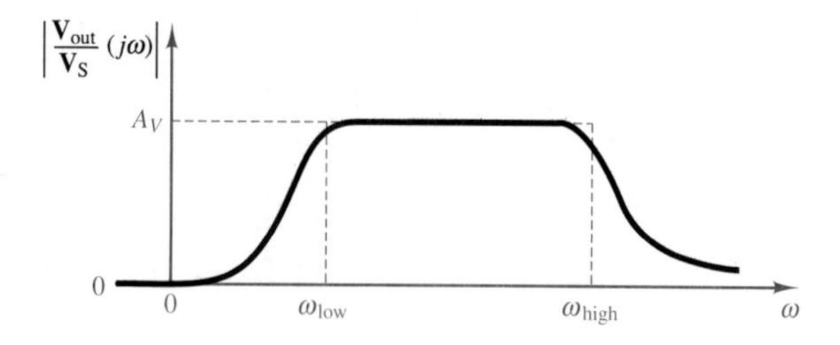

Figure 9.39 Frequency response of a common-emitter amplifier

rizes the discussion by illustrating that a practical BJT common-emitter amplifier stage will be characterized by a band-pass frequency response with low cutoff frequency ω_{low} and high cutoff frequency ω_{high}.

A completely analogous discussion could be made for FET amplifiers, where the effect of the input and output coupling capacitors and of the source bypass capacitor is to reduce the amplifier frequency response at the low end, creating a high-pass effect. Similarly, a high-frequency small-signal model valid for all FETs will include the effect of parasitic capacitances between gate, drain, and source terminals. This model is shown in Figure 9.40.

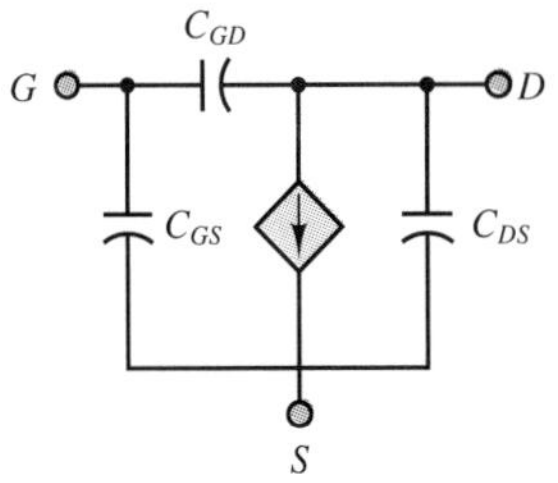

Figure 9.40 High-frequency FET model

Multistage Amplifiers

The design of a practical amplifier involves a variety of issues, as mentioned earlier in this chapter. To resolve the various design trade-offs and to obtain acceptable performance characteristics, it is usually necessary (except in the simplest applications) to design an amplifier in various stages. In general, three stages are needed to address three important issues:

1. Choosing an appropriate input impedance for the amplifier, so as not to load the small-signal source. The input impedance should, in general, be large.
2. Providing suitable gain.
3. Matching the output impedance of the amplifier to the load. This usually requires choosing a low output impedance.

Each of these tasks can be accomplished in a different manner and with more than one amplifier stage, depending on the intended application. Although the task of designing a multistage amplifier is very advanced, and beyond the scope of this book, the minimum necessary tools to understand such a design process have already been introduced. Each amplifier stage can, through the use of small-signal models, be represented in the form of a two-port circuit and characterized by a gain and an input and an output impedance; the overall response of the amplifier can then be obtained by cascading the individual two-port blocks, as shown in Figure 9.41.

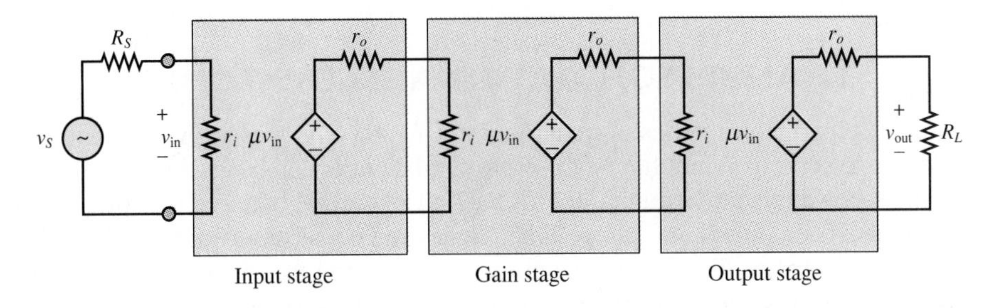

Figure 9.41 Block diagram of a multistage amplifier

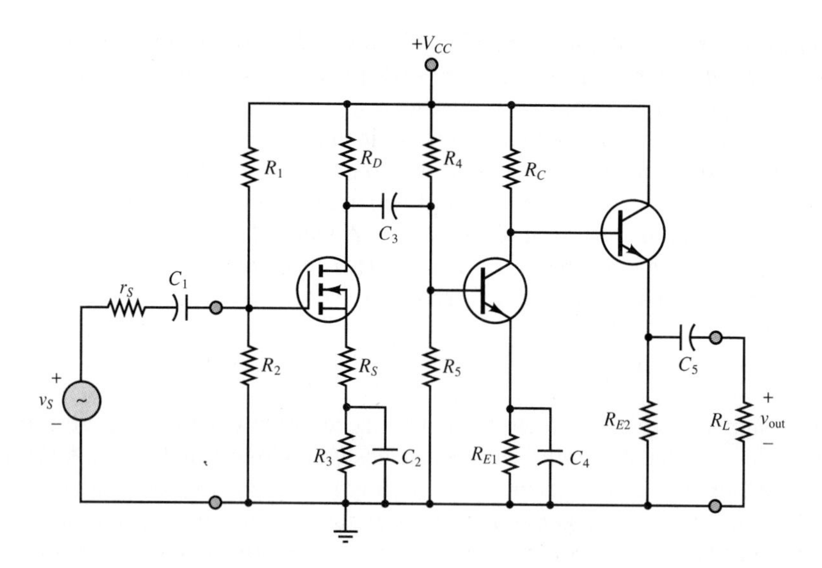

Figure 9.42 Three-stage amplifier

A three-stage amplifier is shown in Figure 9.42. The input stage consists of a MOSFET amplifier; the choice of a MOSFET for the input stage is quite natural, because of the high input impedance of this device. Note that the first stage is AC-coupled; thus, the amplifier will have a band-pass characteristic, as discussed in the preceding section. The choice of a MOSFET as the input stage of a linear amplifier is acceptable in spite of the nonlinearity of the MOSFET transfer characteristic, because the role of the first stage is to amplify a very low-amplitude signal, and therefore the MOSFET is required to amplify in only a relatively small operating region. Thus, the linear transconductance approximation will be valid, and relatively little distortion should be expected. The second stage consists of a BJT common-emitter stage, which is also AC-coupled to the first stage and provides most of the gain. The output stage is a BJT emitter follower, which produces no additional gain but has a relatively low output impedance, needed to match the load. Note that this last stage is DC-coupled to the preceding stage but is AC-coupled to the load. The amplifier of Figure 9.42 will be further explored in the homework problems.

9.5 TRANSISTOR GATES AND SWITCHES

In describing the properties of transistors in Chapter 8, it was suggested that, in addition to serving as amplifiers, three-terminal devices can be used as electronic switches in which one terminal controls the flow of current between the other two. It had also been hinted in Chapter 7 that diodes can act as on-off devices as well. In this section, we discuss the operation of diodes and transistors as electronic switches, illustrating the use of these electronic devices as the switching circuits that are at the heart of **analog** and **digital gates.** Transistor switching circuits form the basis of digital logic circuits, which will be discussed in more detail in the next chapter. The objective of this section is to discuss the internal operation of these circuits and to provide the reader interested in the internal workings of digital circuits with an adequate understanding of the basic principles.

An **electronic gate** is a device that, on the basis of one or more input signals, produces one of two or more prescribed outputs; as will be seen shortly, one can construct both digital and analog gates. A word of explanation is required, first, regarding the meaning of the words *analog* and *digital*. An analog voltage or current—or, more generally, an analog signal—is one that varies in a continuous fashion over time, in *analogy* (hence the expression *analog*) with a physical quantity. An example of an analog signal is a sensor voltage corresponding to ambient temperature on any given day, which may fluctuate between, say, 30° and 50°F. A digital signal, on the other hand, is a signal that can take only a finite number of values; in particular, a commonly encountered class of digital signals consists of **binary signals,** which can take only one of two values (for example, 1 and 0). A typical example of a binary signal would be the control signal for the furnace in a home heating system controlled by a conventional thermostat, where one can think of this signal as being "on" (or 1) if the temperature of the house has dropped below the thermostat setting (desired value), or "off" (or 0) if the house temperature is greater than or equal to the set temperature (say, 68°F). Figure 9.43 illustrates the appearance of the analog and digital signals in this furnace example.

The discussion of digital signals will be continued and expanded in Chapters 12, 13, and 14. Digital circuits are an especially important topic, because a large part of today's industrial and consumer electronics is realized in digital form.

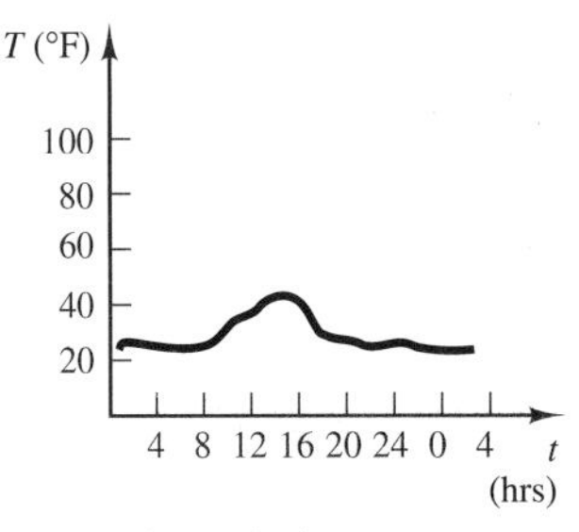

Atmospheric temperature over a 24-hour period

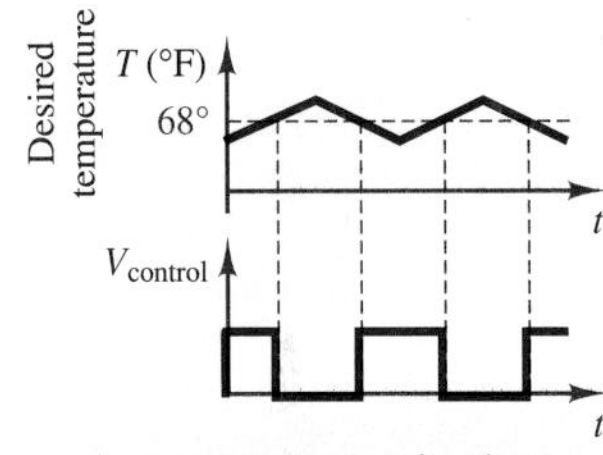

Average temperature in a house and related digital control voltage furnace

Figure 9.43 Illustration of analog and digital signals

Analog Gates

A common form of analog gate—probably the most important, in practice—employs an FET and takes advantage of the fact that current can flow in either direction in an FET biased in the ohmic region. Recall that the drain characteristic of the MOSFET discussed in Chapter 8 consists of three regions: ohmic, active, and breakdown. A MOSFET amplifier is operated in the active region, where the drain current is nearly constant for any given value of v_{GS}. On the other hand, a MOSFET biased in the ohmic state acts very much as a linear resistor. For example, for an n-channel enhancement MOSFET, the conditions for the transistor to be in the ohmic region are:

$$v_{GS} > V_T \qquad \text{and} \qquad |v_{DS}| \le \frac{1}{4}(v_{GS} - V_T) \tag{9.63}$$

As long as the FET is biased within these conditions, it acts simply as a linear resistor, and it can conduct current in either direction (provided that v_{DS} does not exceed the limits stated in equation 9.63). In particular, the resistance of the channel in the ohmic region is found to be

$$R_{DS} = \frac{V_T^2}{2I_{DSS}(v_{GS} - V_T)} \tag{9.64}$$

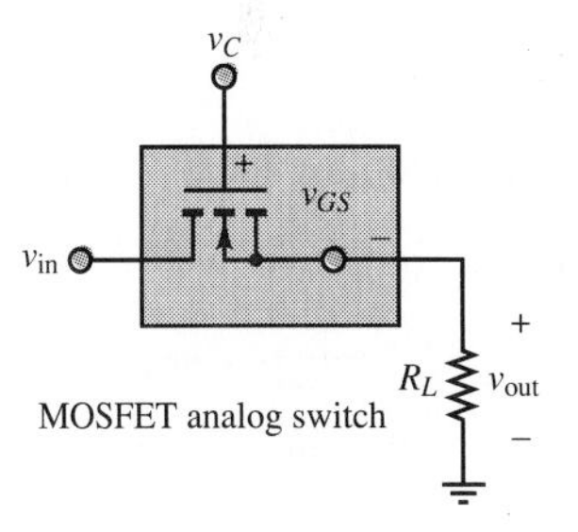

MOSFET analog switch

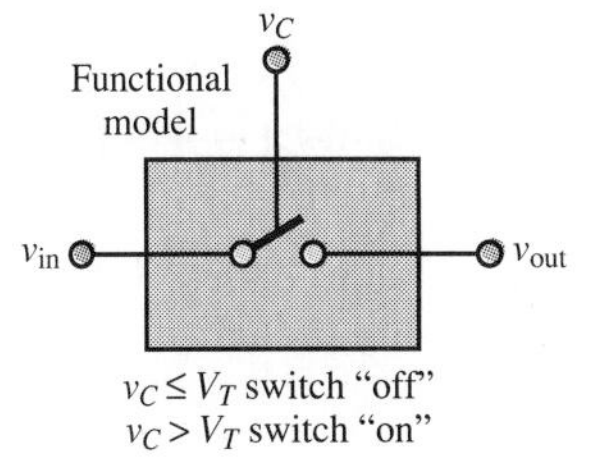

Figure 9.44 MOSFET analog switch

so that the drain current is equal to

$$i_D \approx \frac{v_{DS}}{r_{DS}} \quad \text{for} \quad |v_{DS}| \leq \frac{1}{4}(v_{GS} - V_T) \quad \text{and} \quad v_{GS} > V_T \tag{9.65}$$

The most important feature of the MOSFET operating in the ohmic region, then, is that it acts as a voltage-controlled resistor, with the gate-source voltage, v_{GS}, controlling the channel resistance, R_{DS}. The use of the MOSFET as a switch in the ohmic region, then, consists of providing a gate-source voltage that can either hold the MOSFET in the cutoff region ($v_{GS} \leq V_T$) or bring it into the ohmic region. In this fashion, v_{GS} acts as a control voltage for the transistor.

Consider the circuit shown in Figure 9.44, where we presume that v_C can be varied externally and that v_{in} is some analog signal source that we may wish to connect to the load R_L at some appropriate time. The operation of the switch is as follows. When $v_C \leq V_T$, the FET is in the cutoff region and acts as an open circuit. When $v_C > V_T$ (with a value of v_{GS} such that the MOSFET is in the ohmic region), the transistor acts as a linear resistance, R_{DS}. If $R_{DS} \ll R_L$, then $v_{out} \approx v_{in}$. By using a pair of MOSFETs, it is possible to improve the dynamic range of signals one can transmit through this analog gate.

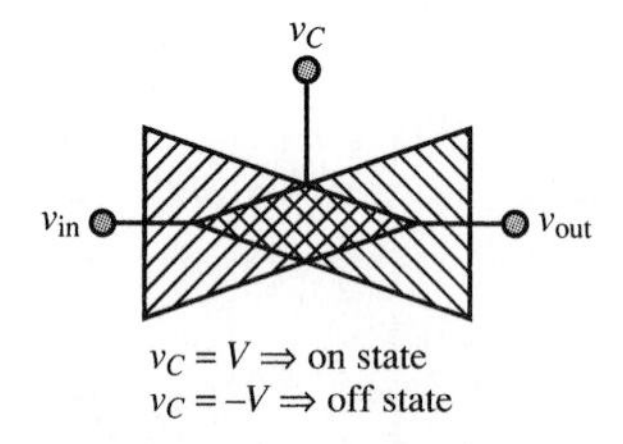

Figure 9.45 Symbol for bilateral FET analog gate

MOSFET analog switches are usually produced in integrated circuit (IC) form and denoted by the symbol shown in Figure 9.45.

Digital Gates

In this section, we explore the operation of diodes, BJTs, and FETs as digital gates. It will soon become apparent that, in general, digital circuits are simpler to analyze than analog circuits, because one is simply interested in whether a given device is conducting or not.

Diode Gates

You will recall that a diode conducts current when it is forward-biased and otherwise acts very much as an open circuit. Thus, the diode can serve as a switch if properly employed. The circuit of Figure 9.46 is called an **OR gate;** it operates as follows. Let voltage levels greater than, say, 2 V correspond to a "logic 1" and voltages less than 2 V represent a "logic 0." Suppose, then, that the input voltages v_A and v_B can be equal to either 0 V or 5 V. If $v_A = 5$ V, diode D_A will conduct; if $v_A = 0$ V, D_A will act as an open circuit. The same argument holds for D_B. It should be apparent, then, that the voltage across the resistor R will be 0 V, or logic 0, if both v_A and v_B are 0. If either v_A or v_B is equal to 5 V, though, the corresponding diode will conduct, and—assuming an offset model for the diode with $V_\gamma = 0.6$ V—we find that $v_{out} = 4.4$ V, or logic 1. Similar analysis yields an equivalent result if both v_A and v_B are equal to 5 V.

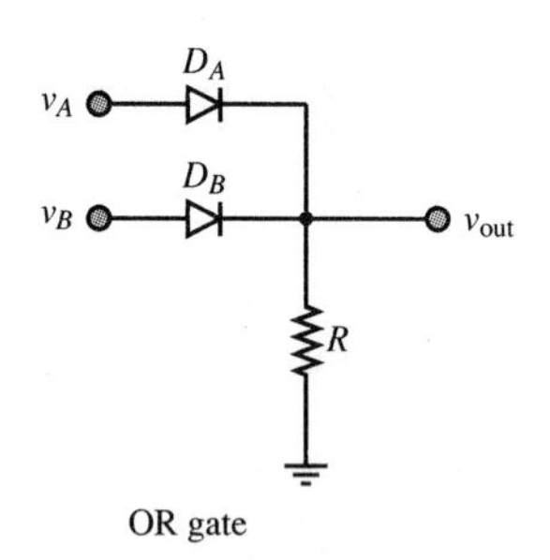

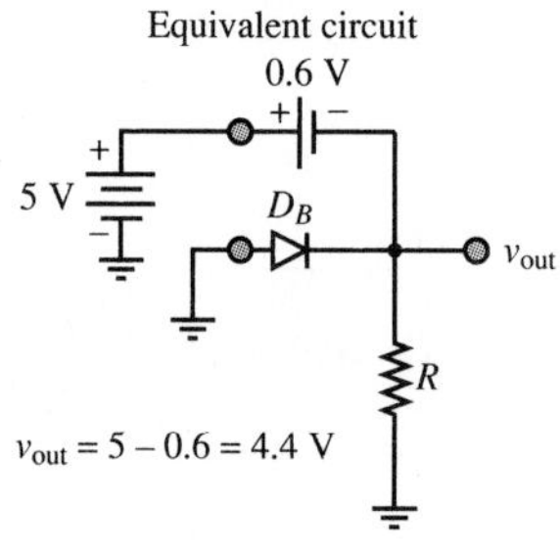

Figure 9.46 Diode OR gate

This type of gate is called an OR gate, because v_{out} is equal to logic 1 (or "high") if either v_A *or* v_B is on, while it is logic 0 (or "low") if neither v_A nor v_B is on. Other functions can also be implemented; however, the discussion of diode gates will be limited to this simple introduction, because diode gate circuits, such as the one of Figure 9.46, are rarely if ever employed in practice. Most modern digital circuits employ transistors to implement switching and gate functions.

BJT Gates

In discussing large-signal models for the BJT, we observed that the *i*-*v* characteristic of this family of devices includes a *cutoff* region, where virtually no current flows through the transistor. On the other hand, when a sufficient amount of current is injected into the base of the transistor, a bipolar transistor will reach *saturation*, and a substantial amount of collector current will flow. This behavior is quite well suited to the design of electronic gates and switches and can be visualized by superimposing a load line on the collector characteristic, as shown in Figure 9.47.

The operation of the simple **BJT switch** is illustrated in Figure 9.47, by means of load-line analysis. Writing the load-line equation at the collector circuit, we have

$$v_{CE} = V_{CC} - i_C R_C \tag{9.66}$$

and

$$v_{out} = v_{CE} \tag{9.67}$$

Thus, when the input voltage, v_{in}, is low (say, 0 V, for example) the transistor is in the cutoff region and little or no current flows, and

$$v_{out} = v_{CE} = V_{CC} \tag{9.68}$$

so that the output is "logic high."

When v_{in} is large enough to drive the transistor into the saturation region, a substantial amount of collector current will flow and the collector-emitter voltage will be reduced to the small saturation value, $V_{CE\,sat}$, which is typically a fraction of a volt. This corresponds to the point labeled *B* on the load line. For the input voltage v_{in} to drive the BJT of Figure 9.47 into saturation, a base current of approximately 50 μA will be required. Suppose, then, that the voltage v_{in} could take the values 0 V or 5 V. Then, if $v_{in} = 0$ V, v_{out} will be nearly equal to V_{CC}, or, again, 5 V. If, on the other hand, $v_{in} = 5$ V and R_B is, say, equal to 85 kΩ (so that the base current required for saturation flows into the base: $i_B \approx (5\text{ V} - 0.7\text{ V})/R_B = 50.6\ \mu\text{A}$), we have the BJT in saturation, and $v_{out} = V_{CE\,sat} \approx 0.2$ V.

Thus, you see that whenever v_{in} corresponds to a logic high (or logic 1), v_{out} takes a value close to 0 V, or logic low (or 0); conversely, v_{in} = "0" (logic "low") leads to v_{out} = "1." The values of 5 V and 0 V for the two logic levels 1 and 0 are quite common in practice and are the standard values used in a family of logic circuits denoted by the acronym **TTL,** which stands for **transistor-transistor logic.** One of the more common TTL blocks is the **inverter** shown in Figure 9.47, so called because it "inverts" the input by providing a low output for a high input, and vice versa. This type of inverting, or "negative," logic behavior is quite typical of BJT gates (and of transistor gates in general).

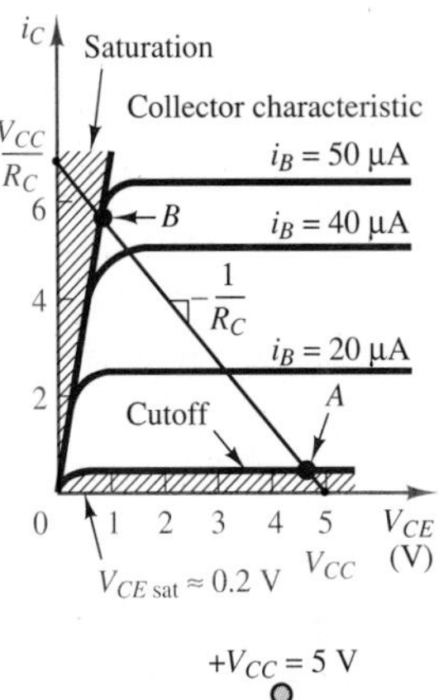

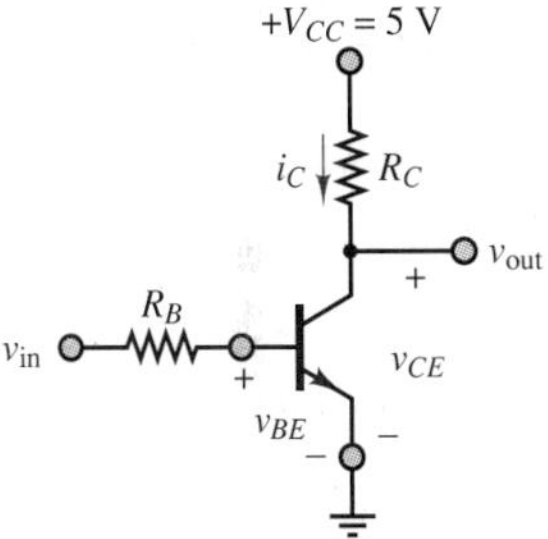

Elementary BJT inverter

Figure 9.47 BJT switching characteristic

In the following paragraphs, we introduce some elementary BJT logic gates, similar to the diode gates described previously; the theory and application of digital logic circuits is discussed in Chapter 12. Example 9.11 illustrates the operation of a **NAND gate,** that is, a logic gate that acts as an inverted AND gate (thus the prefix N in NAND, which stands for NOT).

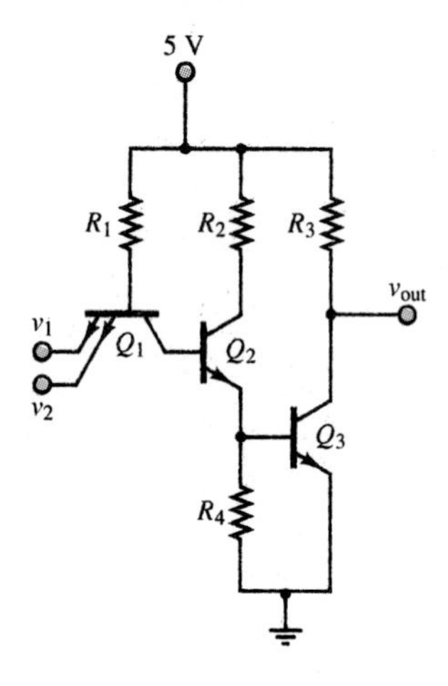

Figure 9.48 TTL NAND gate

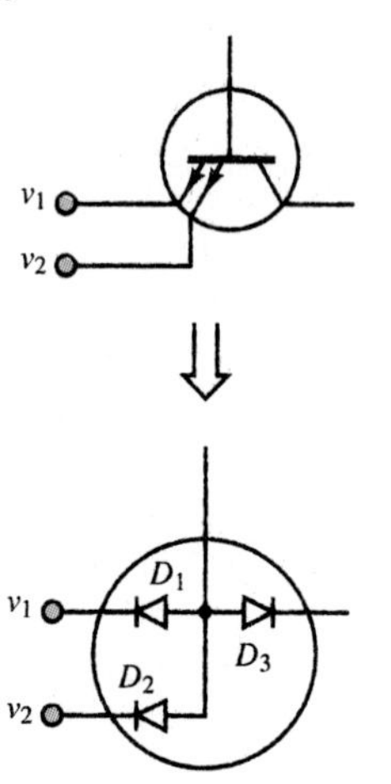

Figure 9.49

Example 9.11 TTL NAND Gate

The transistors in the circuit of Figure 9.48 are connected in such a fashion as to achieve logic gate operation. Given values of v_1 and v_2, complete the following table:

		State of		
v_1	v_2	Q_2	Q_3	v_{out}
0.2 V	0.2 V			
0.2 V	5.0 V			
5.0 V	0.2 V			
5.0 V	5.0 V			

Assume that transistors Q_1, Q_2, and Q_3 have a current gain $\beta = 50$, $R_1 = 5.7\ \text{k}\Omega$, $R_2 = 2\ \text{k}\Omega$, $R_3 = 2.4\ \text{k}\Omega$, $R_4 = 1.8\ \text{k}\Omega$, $V_\gamma = 0.6$ V, and $V_{CE\ \text{sat}} = 0.2$ V. Assume, further, that Q_1 can be modeled with the base-emitter and base-collector junctions as diodes, which may be considered ideal except for the offset voltage, assumed to be 0.6 V. Figure 9.49 depicts this model for transistor Q_1.

Solution:

Consider, first, the case $v_1 = v_2 = 0.2$ V. We will break the circuit into two parts to determine the voltages at different points. First, we consider the diode equivalent circuit of Q_1, shown in Figure 9.50; it should be apparent that both D_1 and D_2 will conduct. This means that

$$v_a = V_\gamma + 0.2 = 0.6 + 0.2 = 0.8\ \text{V}$$

Consider, next, the circuit including Q_2, which is shown in Figure 9.51. Writing KVL, as indicated in the figure, and assuming that Q_2 is turned on, we obtain

$$v_a - V_\gamma - V_{BE} = v_{E2}$$

or

$$v_{E2} = 0.8 - 0.6 - 0.6 = -0.4\ \text{V}$$

This result is inconsistent, since the collector voltage cannot be negative. Therefore, we conclude that Q_2 is actually in cutoff. With Q_2 in cutoff, the current through R_4 will be zero, causing v_{E2} to be zero as well.

Now consider the final stage of the circuit, shown in Figure 9.52. With $v_{E2} = 0$ V, Q_3 is clearly in cutoff, so that no current flows through R_3. Then

$$v_o = 5 - v_{R3} = 5\ \text{V}$$

Now consider the case where $v_1 = 0.2$ V and $v_2 = 5$ V. Again, consider the diode equivalent circuit of Q_1, shown in Figure 9.53. Once again, we can see that D_1 will

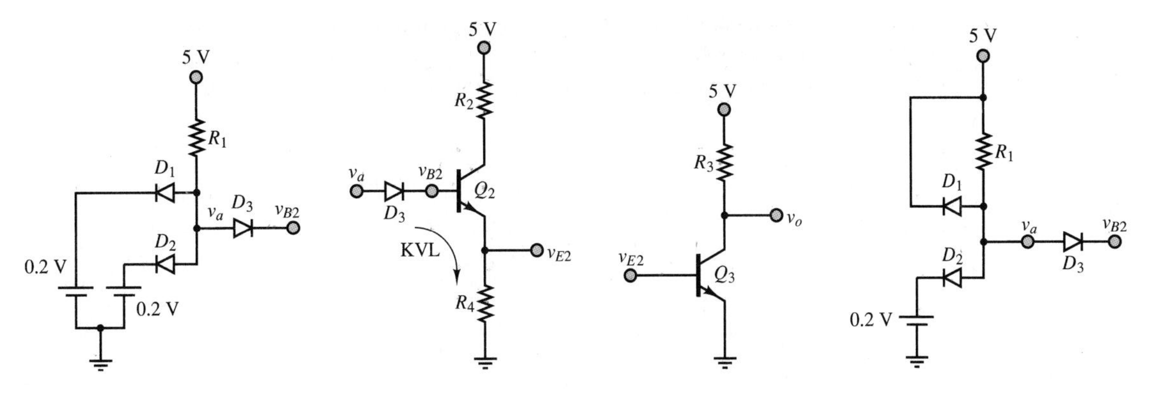

Figure 9.50 Figure 9.51 Figure 9.52 Figure 9.53

conduct, being sufficiently forward-biased, while D_2 does not conduct, since it is reverse-biased. Thus,

$$v_a = v_1 + V_\gamma = 0.2 + 0.6 = 0.8 \text{ V}$$

which is exactly the same result obtained when we analyzed the case where $v_1 = v_2 = 0.2$ V. Thus, we conclude that the transistors Q_2 and Q_3 will be in the cutoff state, causing v_{out} to be 5 V. The same result is obtained for the case where the input voltages are $v_1 = 5$ V and $v_2 = 0.2$ V.

The final condition of interest is where $v_1 = v_2 = 5$ V, shown in Figure 9.54. Clearly, the diodes D_1 and D_2 are reverse-biased, and thus do not conduct. Writing KVL around the circuit indicated in Figure 9.54 and assuming that Q_2 and Q_3 are turned on, we obtain

$$5 - I_{B2}R_1 - V_\gamma - v_{BE2} - v_{BE3} = 0$$

or

$$5 - I_{B2}5.7 \times 10^3 - 0.6 - 0.6 - 0.6 = 0$$

or

$$I_{B2} = \frac{3.2}{5.7 \times 10^3} = 561 \ \mu\text{A} = I_{B1}$$

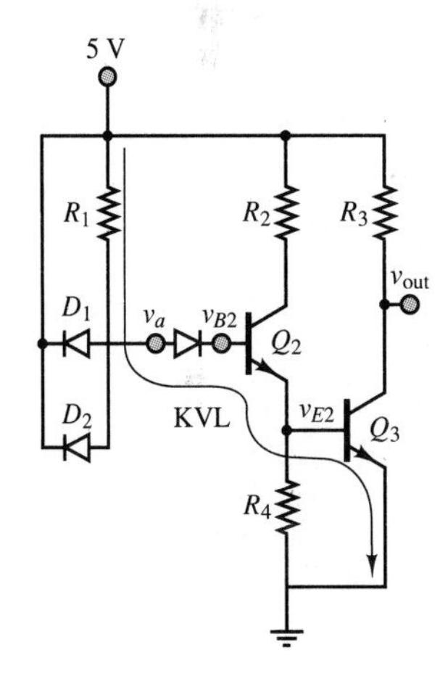

Figure 9.54

Assume now that Q_3 is in saturation. Then, $v_{\text{out}} = 0.2$ V, implying that Q_2 might possibly be in saturation. Assuming that this is the case, the following expression holds:

$$\begin{aligned} v_{CE2} &= 5 - I_{C2}R_2 - V_{BE3} = 0.2 \text{ V} \\ &= 5 - I_{C2} \times 2 \times 10^3 - 0.6 = 0.2 \end{aligned}$$

leading to

$$I_{C2} = 2.1 \text{ mA}$$

Now, the validity of the assumptions regarding the state of Q_2 and Q_3 may be verified by checking the value of β. In fact,

$$\beta = \frac{I_{C2}}{I_{B2}} = \frac{2.1 \times 10^{-3}}{561 \times 10^{-6}} = 3.74$$

and we conclude that Q_2 is indeed in saturation, since $3.74 \ll 50$. Thus, the results obtained above hold.

Having completed the analysis, we may finally fill out the chart:

		State of		
v_1	v_2	Q_2	Q_3	v_{out}
0.2 V	0.2 V	Cutoff	Cutoff	5.0 V
0.2 V	5.0 V	Cutoff	Cutoff	5.0 V
5.0 V	0.2 V	Cutoff	Cutoff	5.0 V
5.0 V	5.0 V	Saturation	Saturation	0.2 V

You should have no difficulty in verifying that the output voltage is related to the inputs in the fashion of an inverted AND (or a NAND) gate.

The analysis method employed in Example 9.11 can be used to analyze any TTL gate. With a little practice, the calculations of this example will become familiar. The drill exercises and homework problems will reinforce the concepts developed in this section.

MOSFET Logic Gates

Having discussed the BJT as a switching element, we might quite naturally suspect that FETs may similarly serve as logic gates. In fact, in some respects, FETs are better suited to be employed as logic gates than BJTs. The *n*-channel enhancement MOSFET, discussed in Chapter 8, serves as an excellent illustration: because of its physical construction, it is normally off (that is, it is off until a sufficient gate voltage is provided), and therefore it does not require much current from the input signal source. Further, MOS devices offer the additional advantage of easy fabrication into integrated circuit form, making production economical in large volume. On the other hand, MOS devices cannot provide as much current as BJTs, and their switching speeds are not quite as fast—although these last statements may not hold true for long, because great improvements are taking place in MOS technology. Overall, it is certainly true that in recent years it has become increasingly common to design logic circuits based on MOS technology. In particular, a successful family of logic gates called *CMOS* (for *complementary metal-oxide-semiconductor*) takes advantage of both *p*- and *n*-channel enhancement-mode MOSFETs to exploit the best features of both types of transistors. CMOS logic gates (and many other types of digital circuits constructed using the same technology) consume very little supply power, and have become the mainstay in pocket calculators, wristwatches, portable computers, and many other consumer electronics products. Without delving into the details of CMOS technology (a brief introduction is provided in Chapter 8), we shall briefly illustrate the properties of MOSFET logic gates and of CMOS gates, in the remainder of this section.

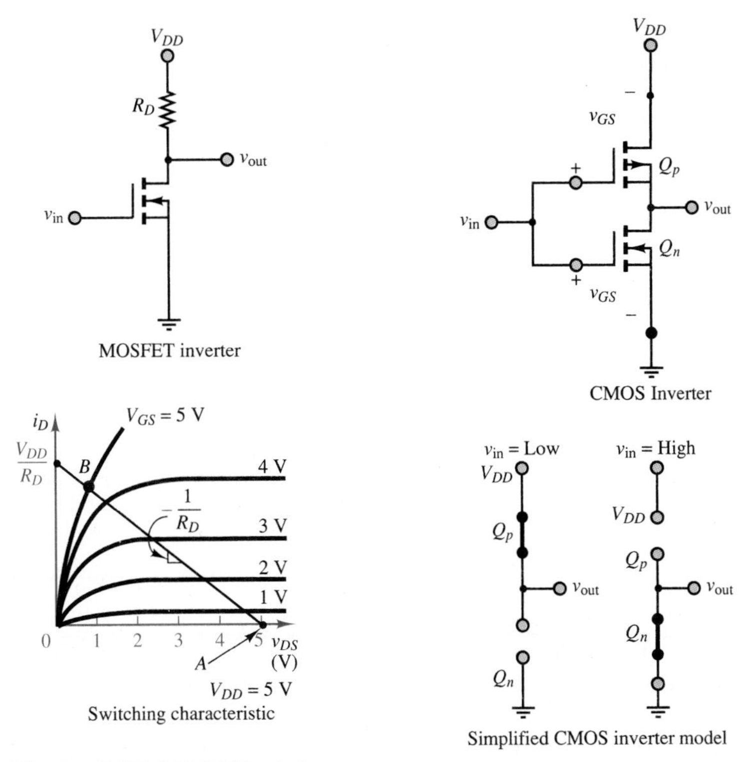

Figure 9.55 MOSFET switching characteristic

Figure 9.56

Figure 9.55 depicts a MOSFET switch with its drain *i*-*v* characteristic. Note the general similarity with the switching characteristic of the BJT shown in the previous section. When the input voltage, v_{in}, is zero, the MOSFET conducts virtually no current, and the output voltage, v_{out}, is equal to V_{DD}. When v_{in} is equal to 5 V, the MOSFET Q point moves from point A to point B along the load line, with $v_{DS} = 0.5$ V. Thus, the circuit acts as an inverter. Much as in the case of the BJT, the inverter forms the basis of all MOS logic gates.

An elementary CMOS inverter is shown in Figure 9.56. Note first the simplicity of this configuration, which simply employs two enhancement-mode MOSFETs: *p*-channel at the top, denoted by the symbol Q_p, and *n*-channel at the bottom, denoted by Q_n. Recall from Chapter 8 that when v_{in} is low, transistor Q_n is off. However, transistor Q_p sees a gate-to-source voltage $v_{GS} = v_{in} - V_{DD} = -V_{DD}$; in a *p*-channel device, this condition is the counterpart of having $v_{GS} = V_{DD}$ for an *n*-channel MOSFET. Thus, when Q_n is off, Q_p is on and acts very much as a small resistance. In summary, when v_{in} is low, the output is $v_{out} \approx V_{DD}$. When v_{in} is high, the situation is reversed: Q_n is now on and acts nearly as a short circuit, while Q_p is open (since $v_{GS} = 0$ for Q_p). Thus, $v_{out} \approx 0$. The complementary MOS operation is depicted in Figure 9.56 in simplified form by showing each transistor as either a short or an open circuit, depending on its state. This simplified analysis is sufficient for the purpose of a qualitative analysis. The following examples illustrate methods for analyzing MOS switches and gates.

Example 9.12

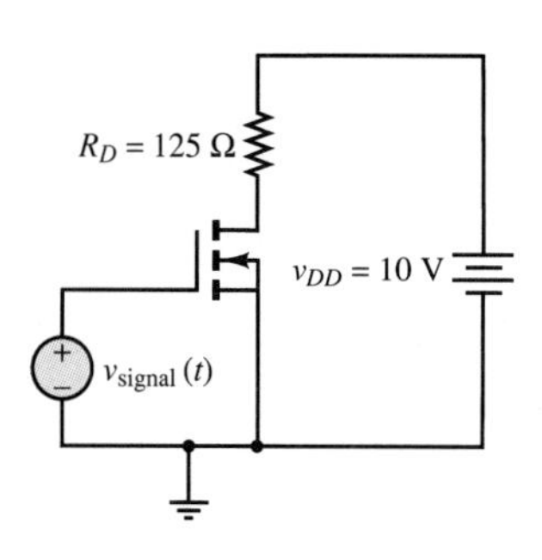

Figure 9.57

An n-channel MOSFET is used as a switch as shown in the circuit of Figure 9.57. The switching source signal is a 2.6-V signal applied to the gate at time t_1. The signal is zero for $t < t_1$.

1. Determine the operating point of the circuit for $t < t_1$.
2. Determine the operating point of the circuit for $t > t_1$.

Solution:

To determine the operating point, we draw the load line on the device characteristic curves, which are shown in Figure 9.58. The load line intercepts the v_{DS} axis at $v_{DS} = V_{DD} =$ 10 V and the i_D axis at $v_{DS}/R_D = 10/125$, or 80 mA.

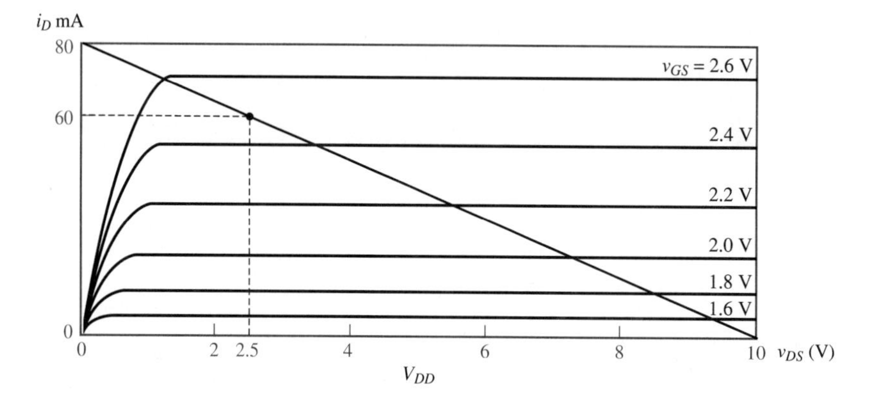

Figure 9.58

1. Next, to determine the operating point for $t < t_1$, we see that the gate voltage is 0 V and therefore the MOSFET is off. The operating point is then given by

$$v_{DS} = 10 \text{ V} \qquad i_D = 0 \text{ A}$$

2. For $t > t_1$, the gate voltage is 2.6 V, causing the transistor's operating point to move to a new position on the characteristic curves, determined by v_{GS}. The new operating point for the transistor is approximately

$$v_{DS} = 1.1 \text{ V} \qquad i_D = 66.5 \text{ mA}$$

Thus, the switching action at the gate causes the current through the MOSFET to switch from zero to approximately 60 mA. Consequently, the voltage across the drain resistor, $R_D = 125\ \Omega$, is equal to 7.5 V, and the voltage across the MOSFET must be 2.5 V, by KVL.

The transistors in this circuit show the substrate for each transistor connected to its respective gate. In a true CMOS IC, the substrates for the p-channel transistors are connected to 5 V and the substrates of the n-channel transistors are connected to ground.

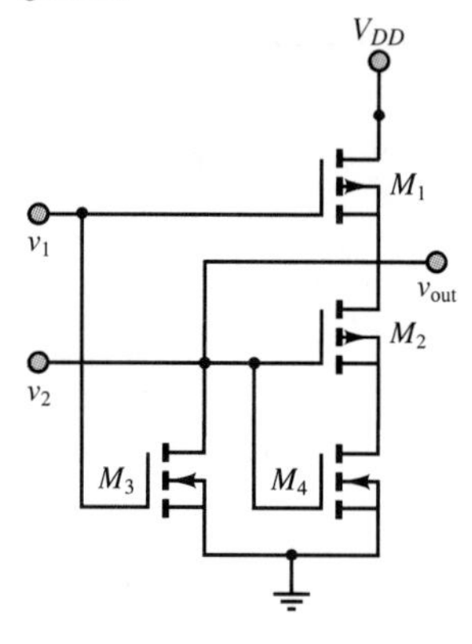

Figure 9.59

Example 9.13

The circuit shown in Figure 9.59 forms a CMOS gate. It is made of p-channel and n-channel MOSFETs. For analysis considerations, it is assumed that the transistors may be either on or off, and that $V_T = 1.7$ V for the MOSFETs. Find the output voltage for each of the following cases:

1. $v_1 = 0, v_2 = 0$
2. $v_1 = 5$ V, $v_2 = 0$ V
3. $v_1 = 0$ V, $v_2 = 5$ V
4. $v_1 = 5$ V, $v_2 = 5$ V

Solution:

To analyze each of these cases, we will represent the transistors that are on as linear resistors and those that are off as open circuits.

1. $$v_1 = 0 \text{ V}, \ v_2 = 0 \text{ V}$$

   Figure 9.60 shows that no current will flow through the resistors representing M_1 and M_2, since M_3 and M_4 are both open circuits. This means that there will be no voltage drop across M_1 and M_2, and so

   $$v_{\text{out}} = 5 \text{ V}$$

2. $$v_1 = 5 \text{ V}, \ v_2 = 0 \text{ V}$$

   M_1 and M_4 will be turned off, since $v_{GS1,4} < V_T$, and M_2 and M_3 will be turned on, leading to the representation shown in Figure 9.61. Once again, no current will flow, since M_1 is an open circuit, causing $v_{\text{out}} = 0$ V.

3. $$v_1 = 0 \text{ V}, \ v_2 = 5 \text{ V}$$

   Once again, we refer to Figure 9.59 and see that when we use the reasoning employed in part 2, M_1 and M_4 will be turned on and M_2 and M_3 will be in cutoff, as depicted in Figure 9.62.

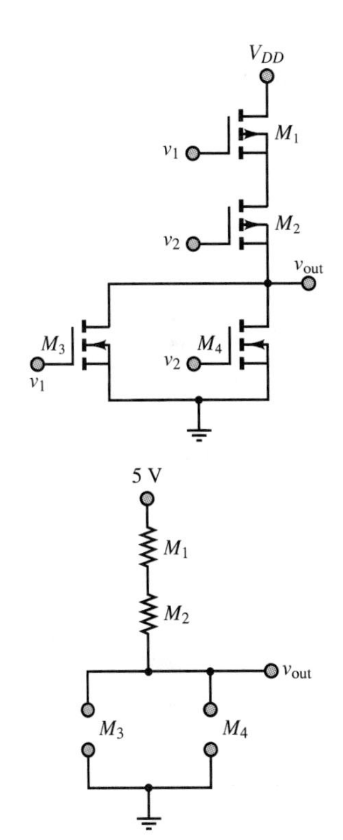

With both v_1 and v_2 at zero volts, M_3 and M_4 will be turned off (in cutoff), since v_{GS} is less than V_T (0 V < 1.7 V). M_1 and M_2 will be turned on, since the gate-to-source voltages will be greater than V_T.

Figure 9.60

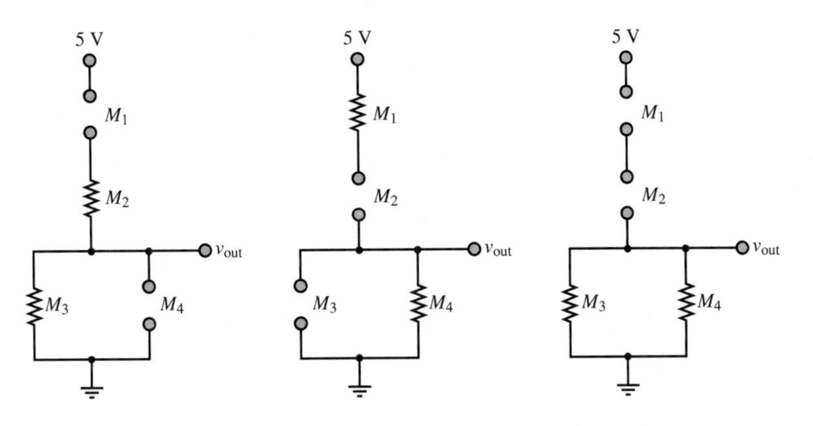

Figure 9.61 **Figure 9.62** **Figure 9.63**

4. $v_1 = v_2 = 5$ V

This time, we see from Figure 9.63 that M_3 and M_4 will both be turned on and M_1 and M_2 will be in cutoff. Clearly, $v_{out} = 0$ V.

A table may now be assembled, showing v_1, v_2, and v_{out}:

v_1	v_2	v_{out}
0	0	5
5	0	0
0	5	0
5	5	0

The table suggests that the circuit discussed in this example is a NOR gate—that is, an inverted OR gate.

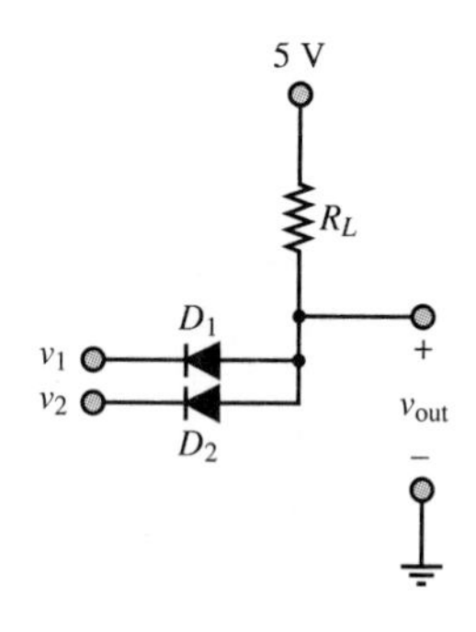

Figure 9.64 Diode AND gate

DRILL EXERCISES

9.11 Show that both v_1 and v_2 must be high for the AND gate circuit shown in Figure 9.64 to give a logic high output.

9.12 Show that the circuit in Figure 9.65 acts as an AND gate, and construct a truth table as in Example 9.11.

9.13 What value of R_D would ensure a drain-to-source voltage, v_{DS}, of 5 V in the circuit of Example 9.12?

9.14 Analyze the CMOS gate of Figure 9.66 and find the output voltages for the following conditions: (a) $v_1 = 0$, $v_2 = 0$; (b) $v_1 = 5$ V, $v_2 = 0$; (c) $v_1 = 0$, $v_2 = 5$ V; (d) $v_1 = 5$ V, $v_2 = 5$ V. Identify the logic function accomplished by the circuit.

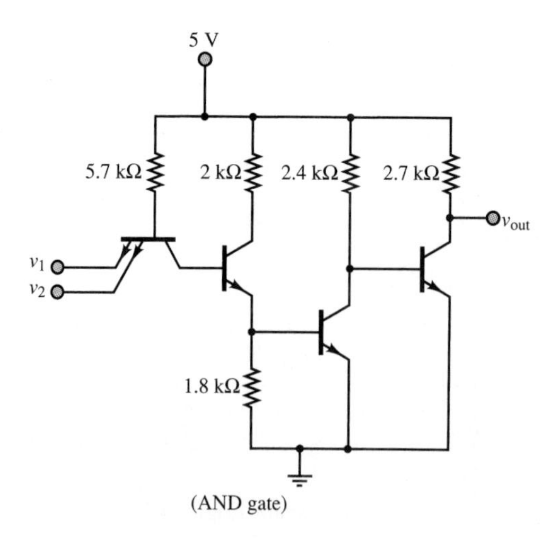

Figure 9.65 TTL AND gate

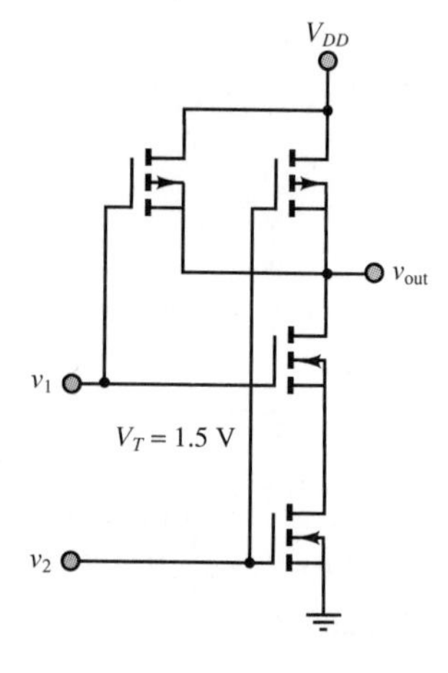

Figure 9.66 CMOS NAND gate

9.13 A DC relay coil can be modeled as an inductor with series resistance due to the windings, as shown in Figure P9.7(a). The relay in Figure P9.7(b) is being driven by a transistor circuit. When V_R is greater than 7.2 VDC, the relay switch will close if it is in the open state. If the switch is closed, it will open when V_R is less than 2.4 VDC. The diode "on" resistance is 10 Ω, and its off resistance is ∞. The relay is rated to dissipate 0.5 W when $V_R = 10$ V. If $R_E = 40\ \Omega$, $R_B = 450\ \Omega$, $R_S = 20\ \Omega$, $L = 100$ mH, $R_{w1} = R_{w2} = 100\ \Omega$, $V_{CE\ \text{sat}} = 0.3$ V, $V_\gamma = 0.75$ V, and $V_S(t)$ is a square wave that ranges from 0 volts to 4.8 volts and has a maximum current of 20 mA.

a. Determine the maximum power dissipated by this circuit.
b. Determine the time it takes for the relay to switch from the closed to the open position if V_S has been at 4.8 V for a very long time.

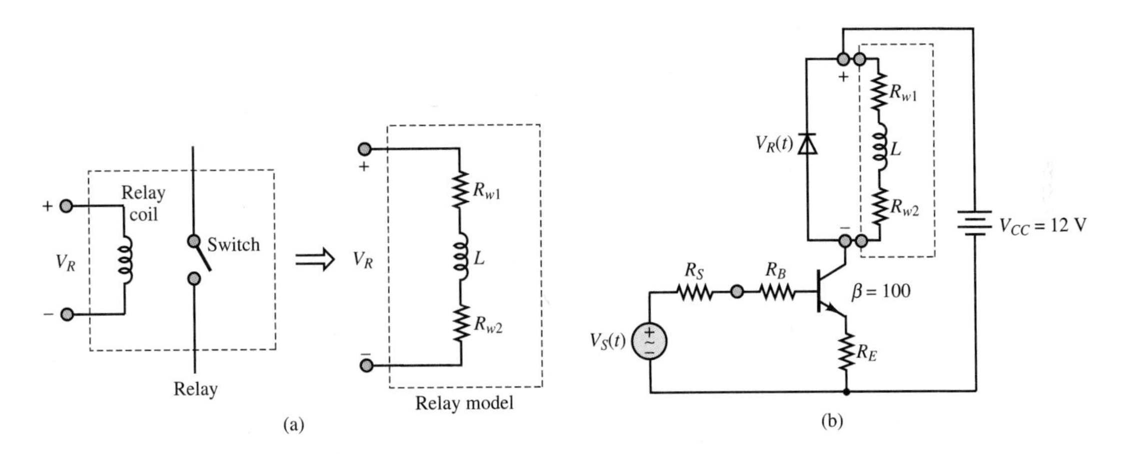

Figure P9.7

9.14 The circuit shown in Figure P9.8 is used to switch a relay that turns a light off and on under the control of a computer. The relay dissipates 0.5 W at 5 VDC. It switches on at 3 VDC and off at 1.0 VDC. What is the maximum frequency with which the light can be switched? The inductance of the relay is 5 mH, and the transistor saturates at 0.2 V, $V_\gamma = 0.8$ V.

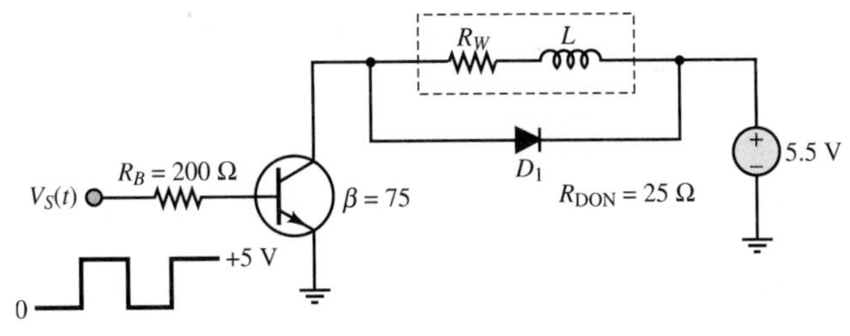

Figure P9.8

9.15 A Darlington pair of transistors is connected as shown in Figure P9.9. The transistor parameters for small-signal operation are Q_1: $h_{ie} = 1.5\ \text{k}\Omega$, $h_{re} = 4 \times 10^{-4}$, $h_{oe} = 110\ \mu\text{A/V}$, and $h_{fe} = 130$; Q_2: $h_{ie} = 200\ \Omega$, $h_{re} = 10^{-3}$, $h_{oe} = 500\ \mu\text{A/V}$, and $h_{fe} = 70$. Calculate:

a. The overall current gain.
b. The input impedance.

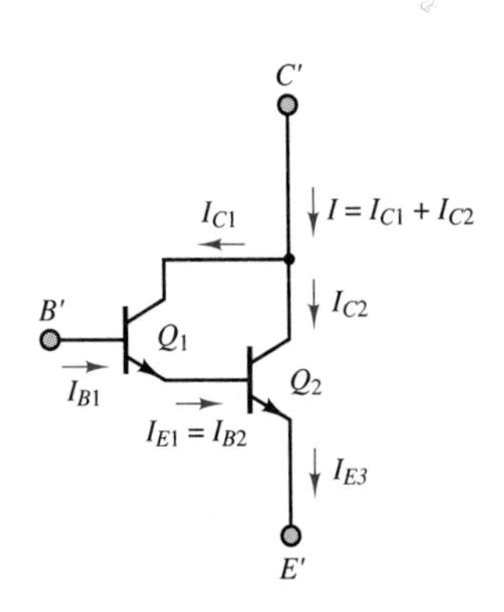

Figure P9.9

9.16 Given the common-emitter amplifier shown in Figure P9.10, where the transistor has the following h parameters:

	Maximum	Minimum
h_{ie}	15 kΩ	1 kΩ
h_{fe}	500	40
h_{re}	8×10^{-4}	0.1×10^{-4}
h_{oe}	30 μS	1×10^{-6} μS

Determine maximum and minimum values for:

a. The open-circuit voltage gain A_V.
b. The open-circuit current gain A_I.

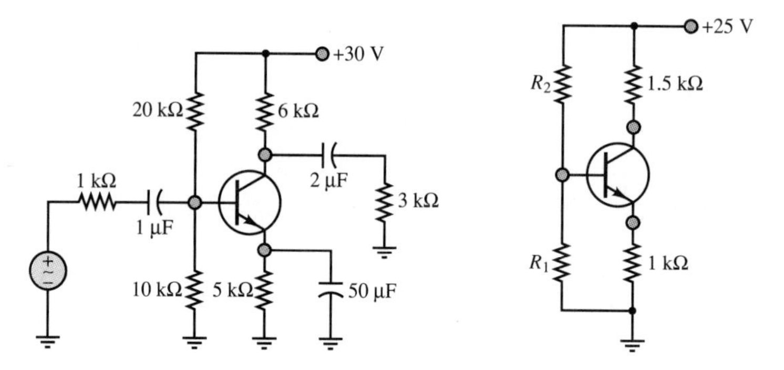

Figure P9.10 Figure P9.11

9.17 The transistor shown in Figure P9.11 has $V_x = 0.6$V. Determine values for R_1 and R_2 such that

a. The quiescent collector-emitter voltage, V_{CEQ}, is 5V.
b. The quiescent collector current, I_{CQ}, will vary no more than 10% as β varies from 20 to 50.

9.18 Consider again the amplifier of Figure P9.11. Determine values of R_1 and R_2 which will permit maximum symmetrical swing in the collector current. Assume $\delta = 100$.

9.19 A depletion MOSFET amplifier is shown in Figure P9.12. Determine the quiescent operating point if $V_{DD} = 30$ V and $R_D = 500\ \Omega$.

9.20 Consider again the amplifier of Figure P9.12. Determine the quiescent operating point if $V_{DD} = 15$ V and $R_D = 330\ \Omega$.

9.21 Again consider the amplifier of Figure P9.12. Determine the quiescent operating point if $V_{DD} = 50$ V and $R_D = 1.5$ kΩ.

9.22 Consider the amplifier of Figure P9.12. Let $R_D = 1$ kΩ. If the quiescent operating point is $I_{DQ} = -25$ mA, $V_{DSQ} = -12.5$ V, determine the value of V_{DD}.

9.23 Consider the amplifier of Figure P9.12. Let $R_D = 2$ kΩ. If the quiescent operating point is $I_{DQ} = -25$ mA, $V_{DSQ} = -12.5$ V, determine the value of V_{DD}.

9.24 Consider again the amplifier of Figure P9.12 with component values as specified in Problem 9.19. Let $v_{GS} = 1 \sin \omega t$ V. Sketch the output voltage and current, and estimate the voltage amplification.

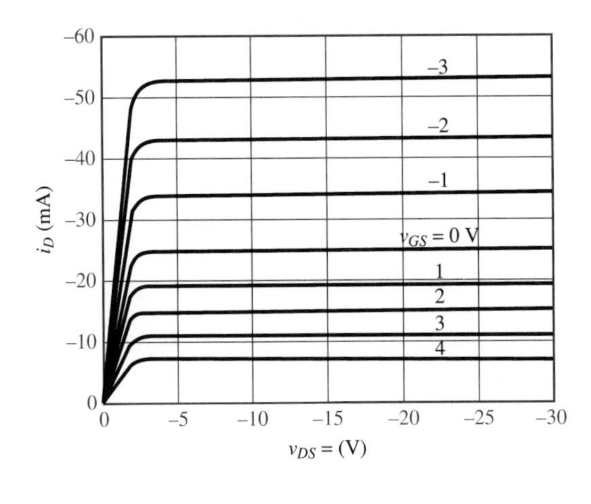

Figure P9.12(a)

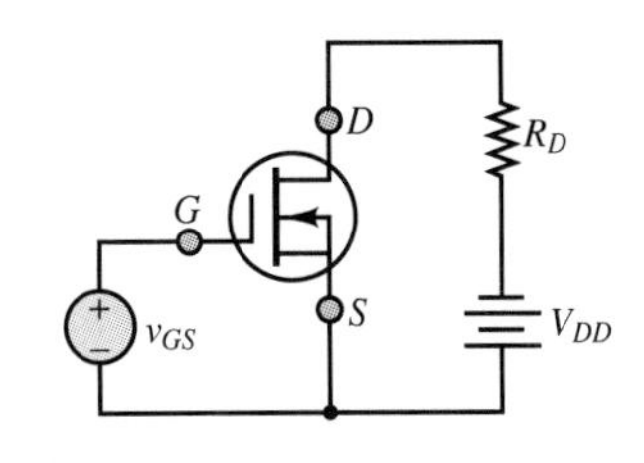

Figure P9.12(b)

9.25 Consider again the amplifier of Figure P9.12 with component values as specified in Problem 9.19. Let $v_{GS} = 3 \sin \omega t$ V. Sketch the output voltage and current, and estimate the voltage amplification.

9.26 Consider again the amplifier of Figure P9.12 with component values as specified in Problem 9.20. Let $v_{GS} = 1 \sin \omega t$ V. Sketch the output voltage and current, and estimate the voltage amplification.

9.27 Consider again the amplifier of Figure P9.12 with component values as specified in Problem 9.20. Let $v_{GS} = 3 \sin \omega t$ V. Sketch the output voltage and current, and estimate the voltage amplification.

9.28 Consider again the amplifier of Figure P9.12 with component values as specified in Problem 9.19. Estimate μ and g_m at the quiescent operating point.

9.29 Show that the circuit of Figure P9.13 functions as an OR gate if the output is taken at v_{o1}.

9.30 Show that the circuit of Figure P9.13 functions as a NOR gate if the output is taken at v_{o2}.

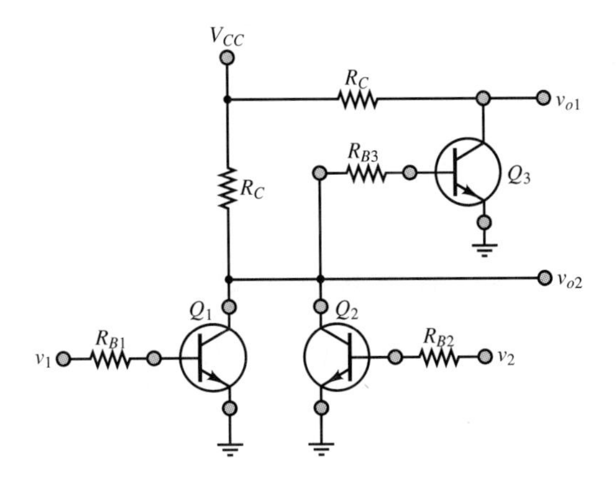

Figure P9.13

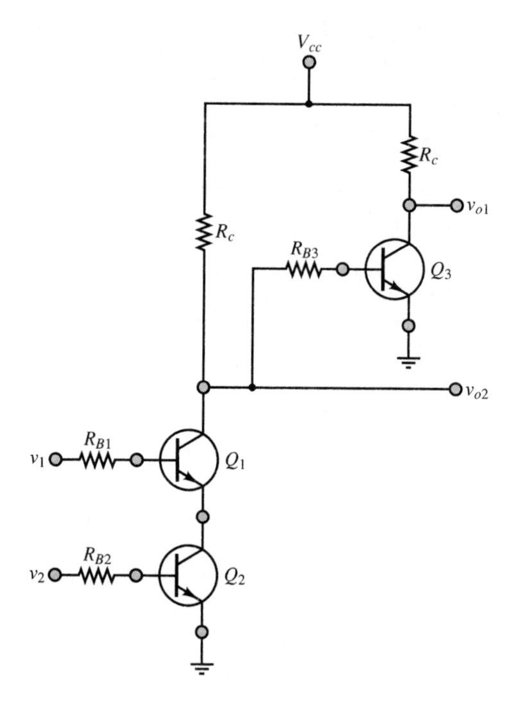

Figure P9.14

9.31 Show that the circuit of Figure P9.14 functions as an AND gate if the output is taken at v_{o1}.

9.32 Show that the circuit of Figure P9.14 functions as a NAND gate if the output is taken at v_{o2}.

9.33 In Figure P9.15, the minimum value of v_{in} for a high input is 2.0 V. Assume that transistor Q_1 has a β of at least 10. Find the range for resistor R_B that can guarantee that the transistor Q_1 is on.

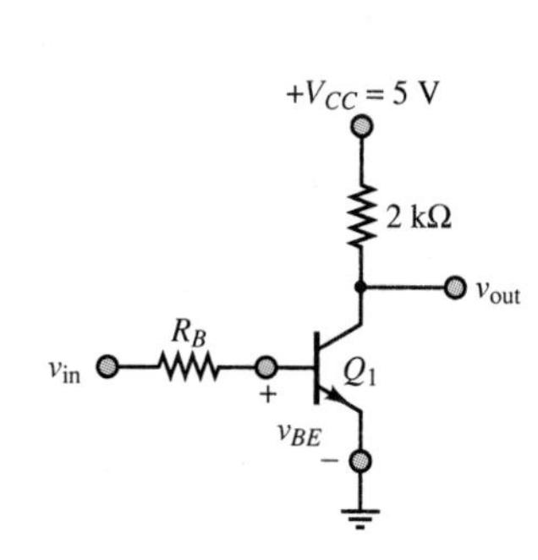

Figure P9.15

9.34 Figure P9.16 shows a circuit with two transistor inverters connected in series, where $R_{1C} = R_{2C} = 10\ \text{k}\Omega$ and $R_{1B} = R_{2B} = 27\ \text{k}\Omega$.

a. Find v_B, v_{out}, and the state of transistor Q_1 when v_{in} is low.
b. Find v_B, v_{out}, and the state of transistor Q_1 when v_{in} is high.

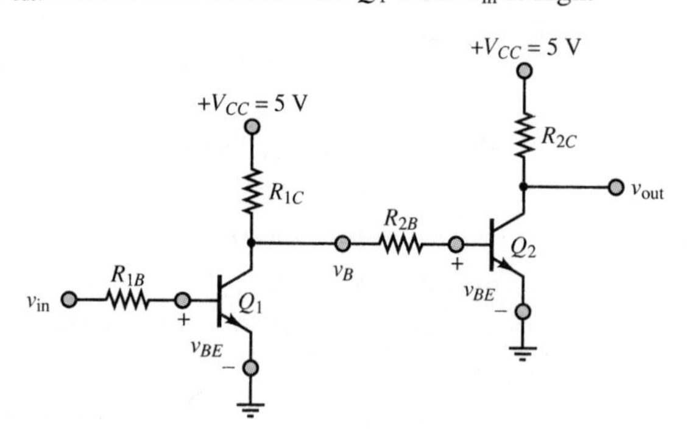

Figure P9.16

9.35 For the inverter of Figure P9.17, $R_B = 5\ \text{k}\Omega$ and $R_{C1} = R_{C2} = 2\ \text{k}\Omega$. Find the minimum values of β_1 and β_2 to ensure that Q_1 and Q_2 saturate when v_{in} is high.

9.36 For the inverter of Figure P9.17, $R_B = 4\ \text{k}\Omega$, $R_{C1} = 2.5\ \text{k}\Omega$, and $\beta_1 = \beta_2 = 4$. Show that Q_1 saturates when v_{in} is high. Find a condition for R_{C2} to ensure that Q_2 also saturates.

9.37 The basic circuit of a TTL gate is shown in the circuit of Figure P9.18. Determine the logic function performed by this circuit.

9.38 Figure P9.19 is a circuit diagram for a three-input TTL NAND gate. Assuming that all the input voltages are high, find v_{B1}, v_{B2}, v_{B3}, v_{C2}, and v_{out}. Also, indicate the state of each transistor.

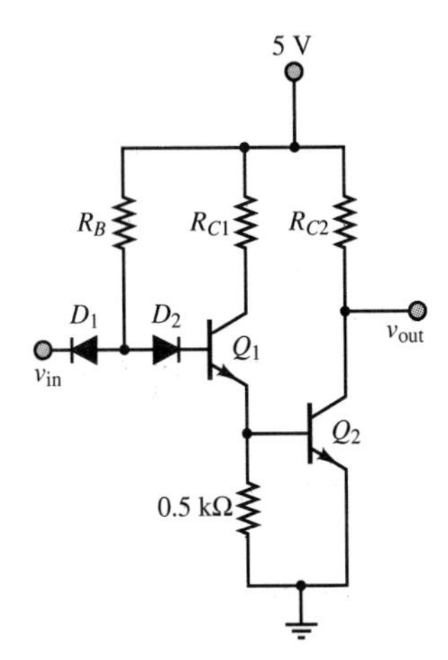

Figure P9.17

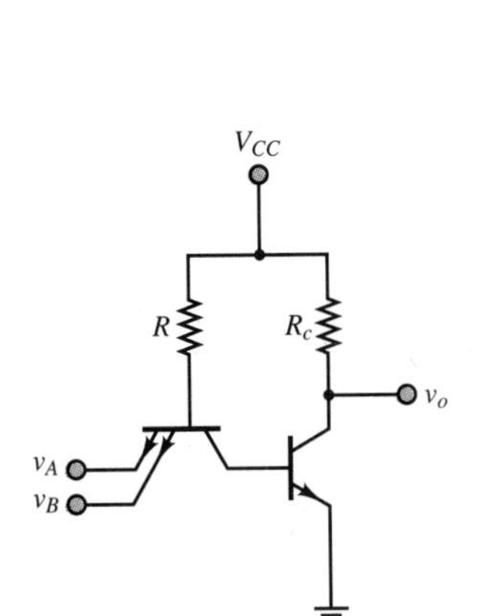

Figure P9.18

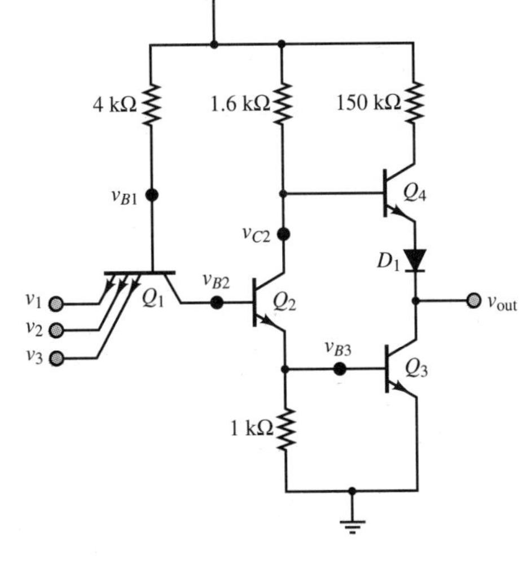

Figure P9.19

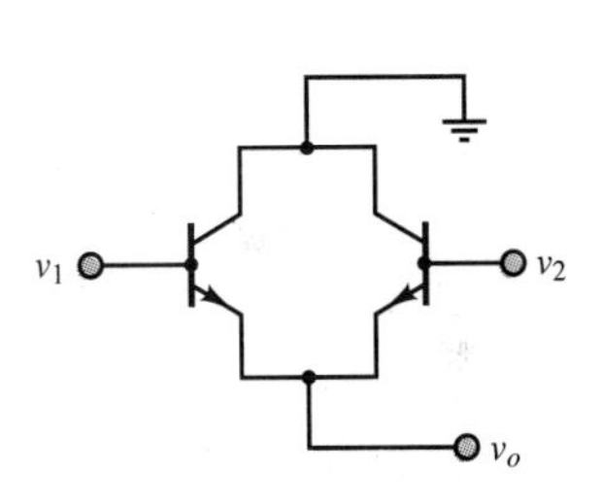

Figure P9.20

9.39 Show that when two or more emitter-follower outputs are connected to a common load, as shown in the circuit of Figure P9.20, the OR operation results; that is, $v_o = v_1$ OR v_2.

9.40 For the CMOS NAND gate of Drill Exercise 9.14 identify the state of each transistor for $v_1 = v_2 = 5$ V.

9.41 Repeat Problem 9.40 for $v_1 = 5$ V and $v_2 = 0$ V.

9.42 Draw the schematic diagram of a two-input CMOS OR gate.

9.43 Draw the schematic diagram of a two-input CMOS AND gate.

9.44 Draw the schematic diagram of a two-input TTL OR gate.

9.45 Draw the schematic diagram of a two-input TTL AND gate.

9.46 Show that the circuit of Figure P9.21 functions as a logic inverter.

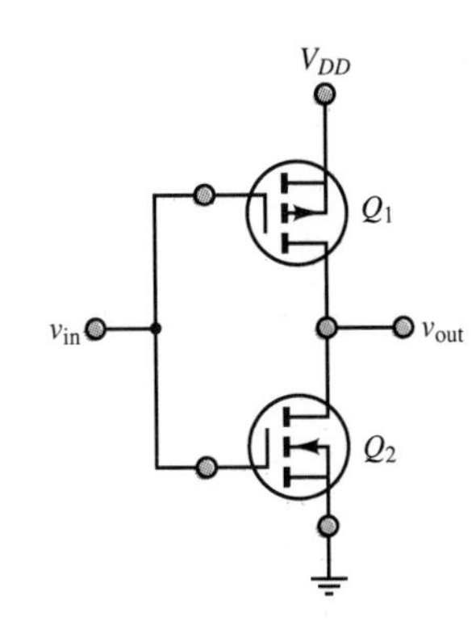

Figure P9.21

9.47 Show that the circuit of Figure P9.22 functions as a NOR gate.

9.48 Show that the circuit of Figure P9.23 functions as a NAND gate.

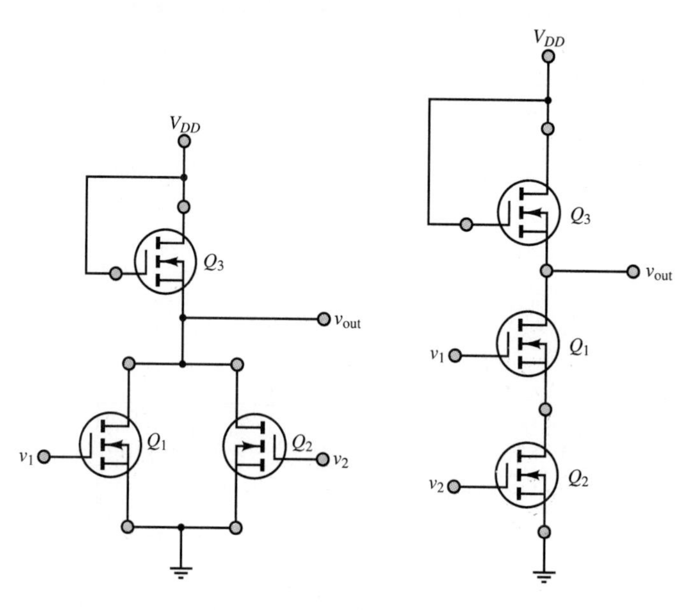

Figure P9.22 **Figure P9.23**

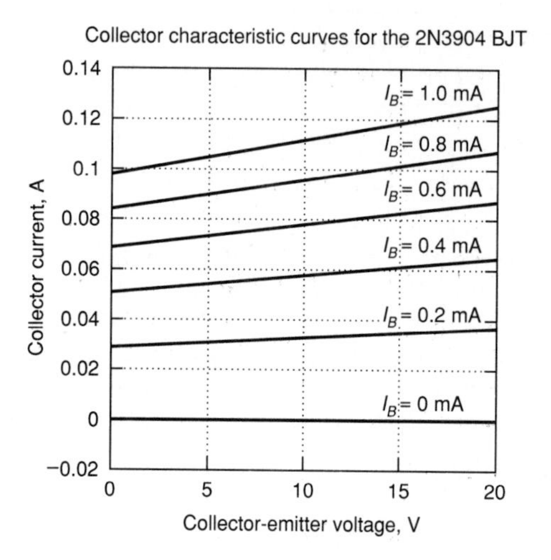

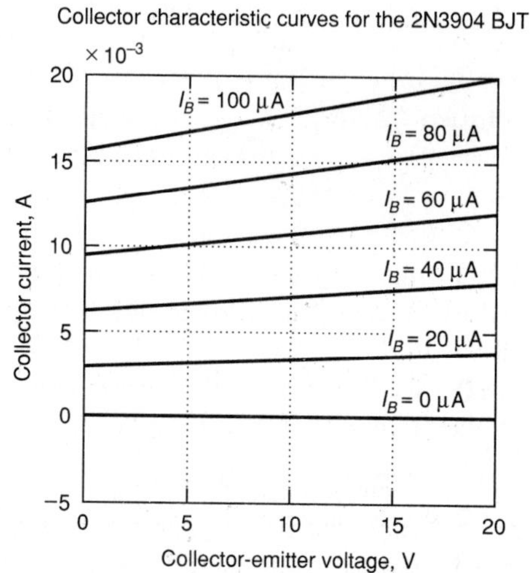

CHAPTER

10

Power Electronics

The objective of this chapter is to present a survey of power electronic devices and systems. Power electronic devices form the "muscle" of many electromechanical systems. For example, one finds such devices in many appliances, in industrial machinery, and virtually wherever an electric motor is found, since one of the foremost applications of power electronic devices is to supply and control the currents and voltages required to power electric machines, such as those introduced in Part III of this book.

Power electronic devices are specially designed diodes and transistors that have the ability to carry large currents and sustain large voltages; thus, the basis for this chapter is the material on diodes and transistors introduced in Chapters 7 through 9. A detailed understanding of diode and transistor small-signal models is not necessary for acquiring an essential knowledge of power semiconductor devices.

This chapter will describe the basic properties of each type of power electronic device, and it will illustrate the application of a selected few, especially in electric motor power supplies. After completing the chapter, you should be able to recognize the symbols for the major power semiconductor devices and understand their principles of operation. You should also understand the operation of the principal electronic power supplies for DC and AC motors.

Upon completing this chapter, you should be able to:

- Provide a classification of power electronic devices and circuits.
- Understand the operation of voltage regulators, transistor power amplifiers, and power switches.
- Analyze rectifier and controlled rectifier circuits.
- Understand the basic principles behind DC and AC electric motor drives.

10.1 CLASSIFICATION OF POWER ELECTRONIC DEVICES

Power semiconductors can be broadly subdivided into five groups: (1) power diodes, (2) thyristors, (3) power bipolar junction transistors (BJTs), (4) insulated-gate bipolar transistors (IGBTs), and (5) static induction transistors (SITs). Figure 10.1 depicts the symbols for the most common power electronic devices.

Power diodes are functionally identical to the diodes introduced in Chapter 7, except for their ability to carry much larger currents. You will recall that a diode conducts in the forward-biased mode when the anode voltage (V_A) is higher than the cathode voltage (V_K). Three types of power diodes exist: general-purpose, high-speed (fast-recovery), and Schottky. Typical ranges of voltage and current are 3,000 V and 3,500 A for general-purpose diodes and 3,000 V and 1,000 A for high-speed devices. The latter have switching times as low as a fraction of a microsecond. Schottky diodes can switch much faster (in the nanosecond range) but are limited to around 100 V and 300 A. The forward voltage drop of power diodes is not much higher than that of low-power diodes, being between 0.5 and 1.2 V. Since power diodes are used with rather large voltages, the forward bias voltage is usually considered negligible relative to other voltages in the circuit, and the switching characteristics of power diodes may be considered near ideal. The principal consideration in choosing power diodes is their power rating.

Thyristors function like power diodes with an additional gate terminal that controls the time when the device begins conducting; a thyristor starts to conduct when a small gate current is injected into the gate terminal, provided that the anode voltage is greater than the cathode voltage (or $V_{AK} > 0$ V). The forward voltage drop of a thyristor is of the order of 0.5 to 2 V. Once conduction is initiated, the gate current has no further control. To stop conduction, the device must be reverse-biased; that is, one must ensure that $V_{AK} \leq 0$ V. Thyristors can be rated at up to 6,000 V and 3,500 A. The **turn-off time** is an important characteristic of thyristors; it represents the time required for the device current to return to zero after external switching of V_{AK}. The fastest turn-off times available are in the range of 10 μs; however, such turn-off times are achieved only in devices with slightly lower power ratings (1,200 V, 1,000 A). Thyristors can be subclassified into the following groups: force-commutated and line-commutated thyristors, gate turn-off thyristors (GTOs), reverse-conducting thyristors (RCTs), static induction thyristors (SITHs), gate-assisted turn-off thyristors (GATTs), light-activated silicon controlled rectifiers (LASCRs), and MOS controlled thyristors (MCTs). It is beyond the scope of this chapter to go into a detailed description of each of these types of devices; their operation is typically a slight modification of the basic

Device	Device symbol
Diode	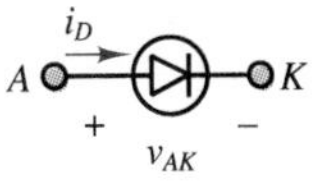 i_D A K + − v_{AK}
Thyristor	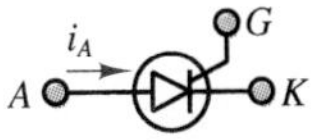 i_A G A K
Gate Turn-Off Thyristor (GTO)	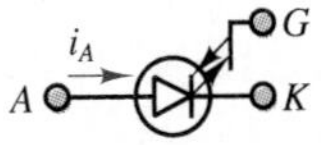 i_A G A K
Triac	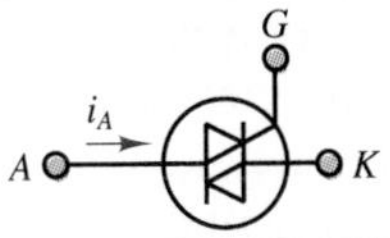 G i_A A K
npn BJT	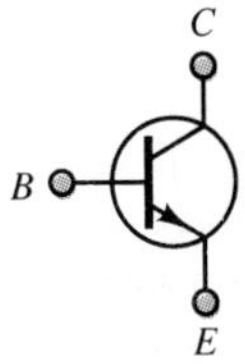 C B E
IGBT	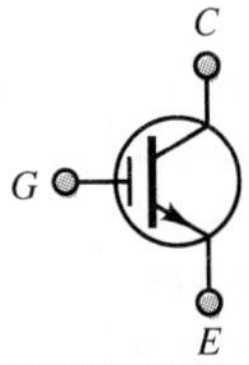 C G E
n-channel MOSFET	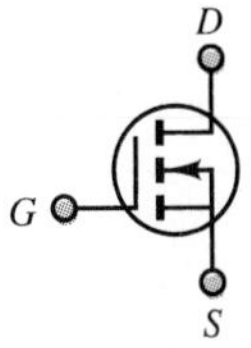 D G S

Figure 10.1 Classification of power electronic devices

operation of the thyristor. The reader who wishes to gain greater insight into this topic may refer to one of a number of excellent books specifically devoted to the subject of power electronics.

Two types of thyristor-based device deserve some more attention. The **triac,** as can be seen in Figure 10.1, consists of a pair of thyristors connected back to back, with a single gate; this allows for current control in either direction. Thus, a triac may be thought of as a bidirectional thyristor. The gate turn-off thyristor (GTO), on the other hand, can be turned on by applying a short positive pulse to the gate, like a thyristor, and can also be turned off by application of a short negative pulse. Thus, GTOs are very convenient in that they do not require separate commutation circuits to be turned on and off.

Power BJTs can reach ratings up to 1,200 V and 400 A, and they operate in much the same way as a conventional BJT. Power BJTs are used in power converter applications at frequencies up to around 10 kHz. **Power MOSFETs** can operate at somewhat higher frequencies (a few to several tens of kHz), but are limited in power (typically up to 1,000 V, 50 A). **Insulated-gate bipolar transistors (IGBTs)** are voltage-controlled (because of their insulated gate, reminiscent of insulated-gate FETs) power transistors that offer superior speed with respect to BJTs but are not quite as fast as power MOSFETs.

10.2 CLASSIFICATION OF POWER ELECTRONIC CIRCUITS

The devices that will be discussed in the present chapter find application in a variety of **power electronic circuits.** This section will briefly summarize the principal types of power electronic circuits and will qualitatively describe their operation. The following sections will describe the devices and their operation in these circuits in more detail.

One possible classification of power electronic circuits is given in Table 10.1. Many of the types of circuits are similar to circuits that were introduced in earlier chapters. Voltage regulators were introduced in Chapter 7 (see Fig. 7.52); this chapter will present a more detailed discussion of practical regulators. Power electronic switches function exactly like the transistor switches described in Chapter 9 (see Figures 9.47 and 9.55); their function is to act as voltage- or current-controlled switches to turn AC or DC supplies on and off. Transistor power amplifiers are the high-power version of the BJT and MOSFET amplifiers studied in Chapters 8 and 9; it is important to consider power limitations and signal distortion more carefully in power amplifiers than in the small-signal amplifiers described in Chapter 9.

Table 10.1 **Power electronic circuits**

Circuit type	Essential features
Voltage regulators	Regulate a DC supply to a fixed voltage output
Power amplifiers	Large-signal amplification of voltages and currents
Switches	Electronic switches (for example, transistor switches)
Diode rectifier	Converts fixed AC voltage (single- or multiphase) to fixed DC voltage
AC-DC converter (controlled rectifier)	Converts fixed AC voltage (single- or multiphase) to variable DC voltage
AC-AC converter (AC voltage controller)	Converts fixed AC voltage to variable AC voltage (single- or multiphase)
DC-DC converter (chopper)	Converts fixed DC voltage to variable DC voltage
DC-AC converter (inverter)	Converts fixed DC voltage to variable AC voltage (single- or multiphase)

Diode rectifiers were discussed in Chapter 7 in their single-phase form (see Figures 7.20, 7.42, and 7.44); similar rectifiers can also be designed to operate with three-phase sources. The operation of a single-phase full-wave rectifier was summarized in Figure 7.43. AC-DC converters are also rectifiers, but they take advantage of the controlled properties of thyristors. The thyristor gate current can be timed to "fire" conduction at variable times, resulting in a variable DC output, as illustrated in Figure 10.2, which shows the circuit and behavior of a single-phase AC-DC converter. This type of converter is very commonly used as a supply for DC electric motors. In Figure 10.2 α is the firing angle of thyristor T_1, where the device starts to conduct.

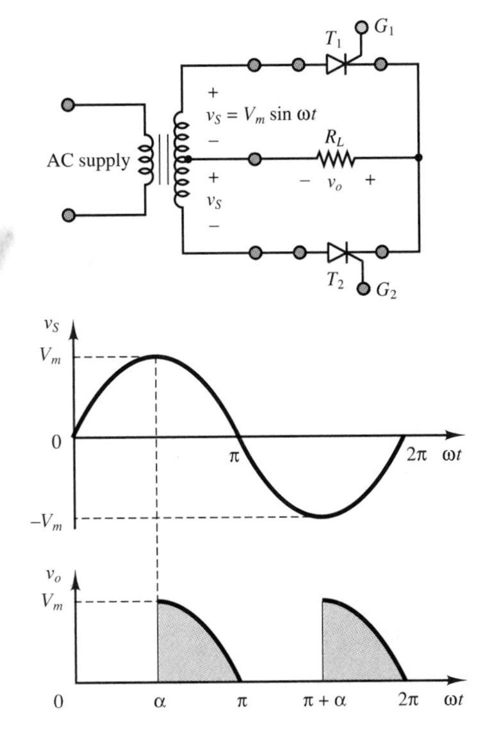

Figure 10.2 AC-DC converter circuit and waveform

AC-AC converters are used to obtain a variable AC voltage from a fixed AC source. Figure 10.3 shows a triac-based AC-AC converter, which takes advantage of the bidirectional capability of triacs to control the rms value of an alternating voltage. Note in particular that the resulting AC waveform is no longer a pure sinusoid even though its fundamental period (frequency) is unchanged. A DC-DC converter, also known as a *chopper,* or *switching regulator,* permits conversion of a fixed DC source to a variable DC supply. Figure 10.4 shows how such an effect may be obtained by controlling the base-emitter voltage of a bipolar transistor, enabling conduction at the desired time. This results in the conversion of the DC input voltage to a variable–duty-cycle output voltage, whose average value can be controlled by selecting the "on" time of the transistor. DC-DC converters find application as variable voltage supplies for DC electric motors used in electric vehicles.

Finally, DC-AC supplies, or inverters, are used to convert a fixed DC supply to a variable AC supply; they find application in AC motor control. The operation of these circuits is rather complex; it is illustrated conceptually in the waveforms

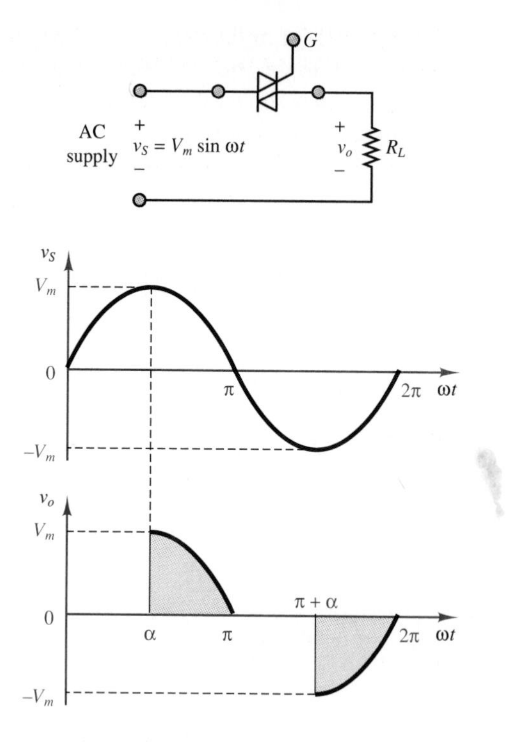

Figure 10.3 AC-AC converter circuit and waveform

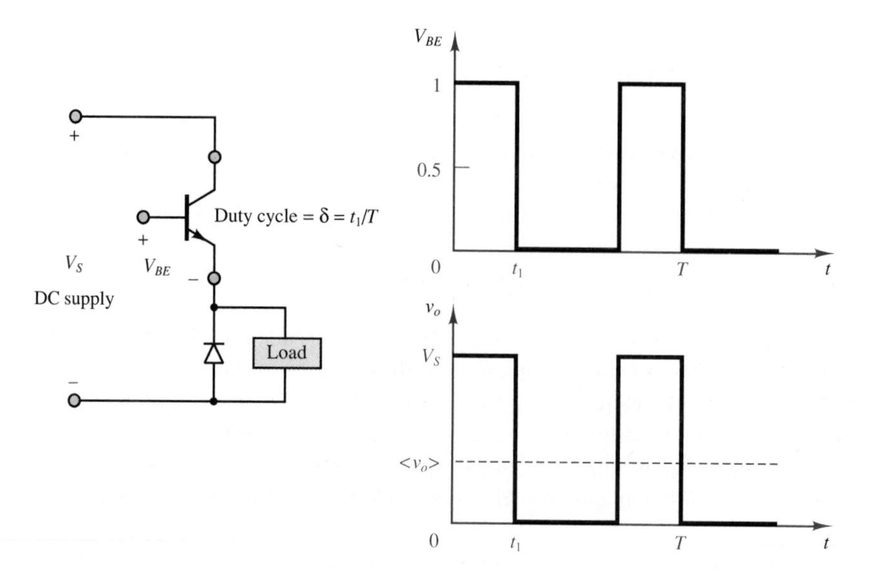

Figure 10.4 DC-DC converter circuit and waveform

power amplifiers, although in recent years semiconductor technology has made power MOSFETs competitive with the performance of bipolar devices.

You may recall the notion of breakdown from the introductory discussion of BJTs. In practice, a bipolar transistor is limited in its operation by three factors: the maximum collector current, the maximum collector-emitter voltage, and the maximum power dissipation, which is the product of I_C and V_{CE}. Figure 10.13 illustrates graphically the power limitation of a BJT by showing the region where the maximum capabilities of the transistor are exceeded. (1) Exceeding the maximum allowable current $I_{C\ max}$ on a continuous basis will result in melting the wires that bond the device to the package terminals. (2) Maximum power dissipation is the locus of points for which $V_{CE}I_C = P_{max}$ at a case temperature of 25°C. The average power dissipation should not exceed P_{max}. (3) The instantaneous value of v_{CE} should not exceed $V_{CE\ max}$; otherwise, avalanche breakdown of the collector-base junction may occur. It is important to note that the linear operation of the transistor as an amplifier is also limited by the saturation and cutoff limits.

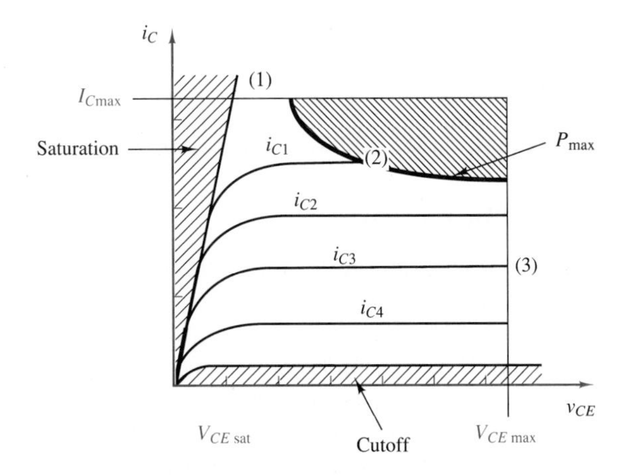

Figure 10.13 Limitations of a BJT amplifier

Thus, in effect, the operation of a BJT as a linear amplifier is rather severely limited by these factors. Consider, first, the effect of driving an amplifier beyond the limits of the linear active region, into saturation or cutoff. The result will be signal distortion. For example, a sinusoid amplified by a transistor amplifier that is forced into saturation, either by a large input or by an excessive gain, will be compressed around the peaks, because of the decreasing device gain in the extreme regions. Thus, to satisfy these limitations—and to fully take advantage of the relatively distortion-free linear active region of operation for a BJT—the Q point should be placed in the center of the device characteristic to obtain the **maximum symmetrical swing.** This point has already been discussed in Chapter 9 (see Example 9.3 and Figure 9.16, in particular).

The maximum power dissipation of the device, of course, presents a more drastic limitation on the performance of the amplifier, in that the transistor can be irreparably damaged if its power rating is exceeded. Values of the maximum allowable collector current, $I_{C\ \max}$, of the maximum allowable transistor power dissipation, $P_{\max}$, and of other relevant power BJT parameters are given in Table 10.2 for a few typical devices. Because of their large geometry and high operating currents, power transistors have typical parameters quite different from those of small-signal transistors.

From Table 10.2, we can find some of these differences:

1. β is low. It can be as low as 5; the typical value is 20 to 80.
2. $I_{C\ \max}$ is typically in the ampere range; it can be as high as 100 A.
3. V_{CEO} is usually 40 to 100 V, but it can reach 500 V.

Table 10.2 **Typical parameters for representative power BJTs**

	2N4921	2N6306	2N5683
Type	*npn*	*npn*	*pnp*
Maximum I_C (continuous)	3 A	16 A	50 A
V_{CEO}	40 V	250 V	60 V
Power rating	30 W	125 W	300 W
β	10 @ I_C = 1 A	25 @ I_C = 3 A	20 @ I_C = 25 A
$V_{CE\ \text{sat}}$	0.6 V	0.8 V	1–5 V
$V_{BE\ \text{on}}$	1.3 V	1.3 V	2 V

BJT Switching Characteristics

In addition to their application in power amplifiers, power BJTs can also serve as controlled power switches, taking advantage of the switching characteristic described in Chapter 9 (see Figure 9.47). In addition to the properties already discussed, it is important to understand the phenomena that limit the switching speed of bipolar devices. The parasitic capacitances C_{CB} and C_{BE} that exist at the CB and BE junctions have the effect of imposing a charging time constant; since the transistor is also characterized by an internal resistance, you see that it is impossible for the transistor to switch from the cutoff to the saturation region instantaneously, because the inherent RC circuits physically present inside the transistor must first be charged. Figure 10.14 illustrates the behavior of the base and collector currents in response to a step change in base voltage. If a step voltage up to amplitude V_1 is applied to the base of the transistor and a base current begins to flow, the collector current will not begin to flow until after a delay, because the base capacitance needs to charge up before the BE junction voltage reaches V_γ; this **delay time,** t_d, is an important parameter. After the BE junction finally becomes forward-biased, the collector current will rise to the final value in a finite time, called **rise time,** t_r. An analogous process (though the physics are different) takes place when the base voltage is reversed to drive the BJT into cutoff. Now the excess charge that had been accumulated in the base must be discharged before the BE junction can be reverse-biased. This discharge takes place over a **storage time,** t_s. To accelerate this process, the base voltage is usually driven to negative values ($-V_2$), so that the negative base current can accelerate the discharge of

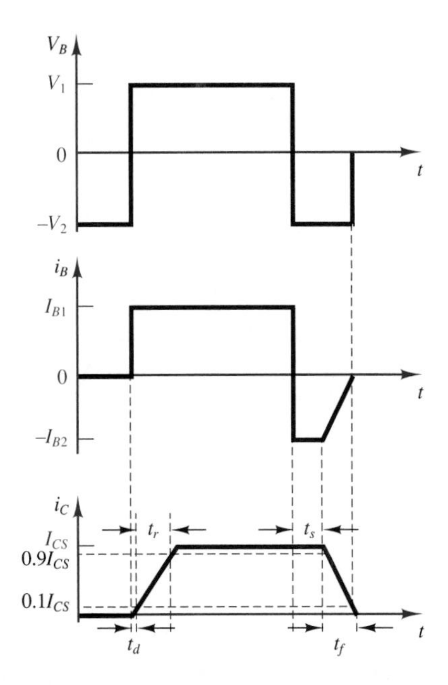

Figure 10.14 BJT switching waveforms

the charge stored in the base. Finally, the reverse-biased *BE* junction capacitance must now be charged to the negative base voltage value before the switching transient is complete; this process takes place during the fall time, t_f. In the figure, I_{CS} represents the collector saturation current. Thus, the turn-on time of the BJT is given by:

$$t_{\text{on}} = t_d + t_r \tag{10.3}$$

and the turn-off time by

$$t_{\text{off}} = t_s + t_f \tag{10.4}$$

Power MOSFETs

MOSFETs can also be used as power switches, like BJTs. The preferred mode of operation of a power MOSFET when operated as a switch is in the ohmic region, where substantial drain current can flow for relatively low drain voltages (see Table 8.1 and Figure 9.55). Thus, a MOSFET switch is driven from cutoff to the ohmic state by the gate voltage. In an enhancement MOSFET, positive gate voltages are required to turn the transistor on; in depletion MOSFETs, either positive or negative voltages can be used.

To understand the switching behavior of MOSFETs, recall once again the parasitic capacitances that exist between pairs of terminals: C_{GS}, C_{GD}, and C_{DS}. As a consequence of these capacitances, the transistor experiences a **turn-on delay,** $t_{d(\text{on})}$, corresponding to the time required to charge the equivalent input

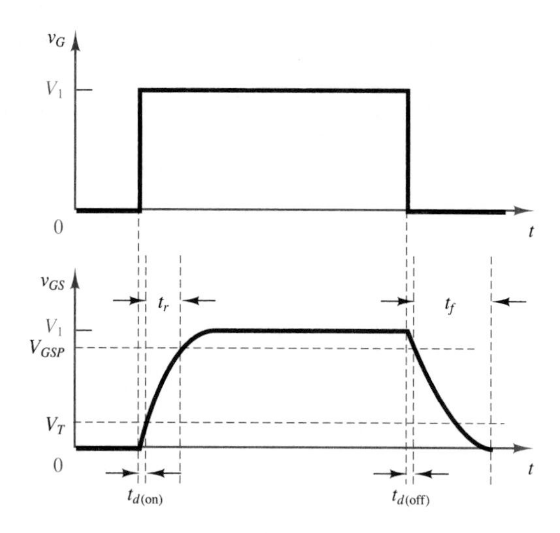

Figure 10.15 MOSFET switching waveforms

capacitance to the threshold voltage, V_T. As shown in Figure 10.15, the rise time, t_r, is defined as the time it takes to charge the gate from the threshold voltage to the gate voltage required to have the MOSFET in the ohmic state, V_{GSP}. The **turn-off delay time,** $t_{d(\text{off})}$, is the time required for the input capacitance to discharge, so that the gate voltage can drop and v_{DS} can begin to rise. As v_{GS} continues to decrease, we define the **fall time,** t_f, which is the time required for v_{GS} to drop below the threshold voltage and turn the transistor off.

Insulated-Gate Bipolar Transistors (IGBTs)

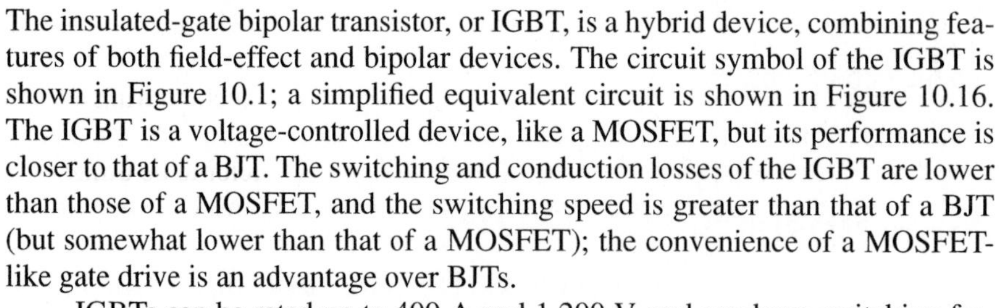
The insulated-gate bipolar transistor, or IGBT, is a hybrid device, combining features of both field-effect and bipolar devices. The circuit symbol of the IGBT is shown in Figure 10.1; a simplified equivalent circuit is shown in Figure 10.16. The IGBT is a voltage-controlled device, like a MOSFET, but its performance is closer to that of a BJT. The switching and conduction losses of the IGBT are lower than those of a MOSFET, and the switching speed is greater than that of a BJT (but somewhat lower than that of a MOSFET); the convenience of a MOSFET-like gate drive is an advantage over BJTs.

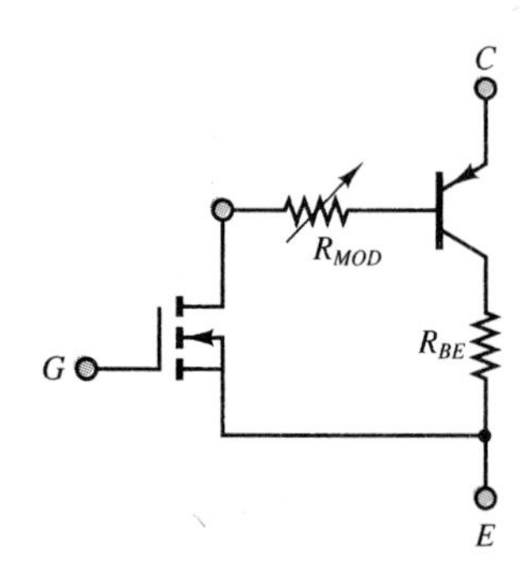

Figure 10.16 IGBT simplified equivalent circuit

IGBTs can be rated up to 400 A and 1,200 V, and can have switching frequencies as high as 20 kHz. These devices have in recent years found increasing application in medium-power applications, such as AC and DC motor drives.

10.5 RECTIFIERS AND CONTROLLED RECTIFIERS (AC-DC CONVERTERS)

As explained in Chapter 7, one of the most immediate applications of the semiconductor diode is rectification of AC voltages and currents, to convert AC waveforms to DC. Rectification can be achieved both with conventional diodes and with controlled diodes, such as thyristors. A simple diode rectifier can provide only a fixed DC voltage level; however, variable DC supplies can be easily obtained with the aid of thyristors. The aim of this section is to illustrate the basic features of diode rectifiers, and to introduce thyristor-based controlled rectifiers.

The basic diode half-wave rectifier and also full-wave and bridge rectifiers were discussed in Sections 7.3 and 7.4. In addition to the considerations noted in Chapter 7, one often has to take into account the nature of the load seen by such DC supplies.

In practice, loads are not always resistive, as will be seen in Chapters 15 through 17, where circuit models for electromechanical actuators and electric motors are introduced. A very common occurrence consists of a DC voltage supply providing current to a *DC motor*. For the purpose of the present discussion, it will suffice to state that a DC motor presents an inductive impedance to the voltage supply and requires a constant current from the supply to operate at a constant speed. The circuit of Figure 10.17 illustrates, as an example, a simple half-wave rectifier connected to an RL load.

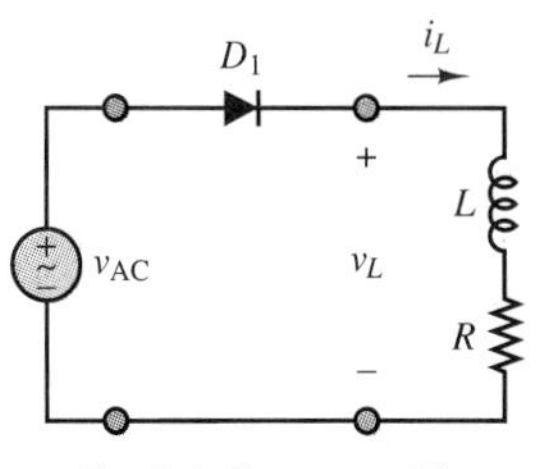

Simple half-wave rectifier

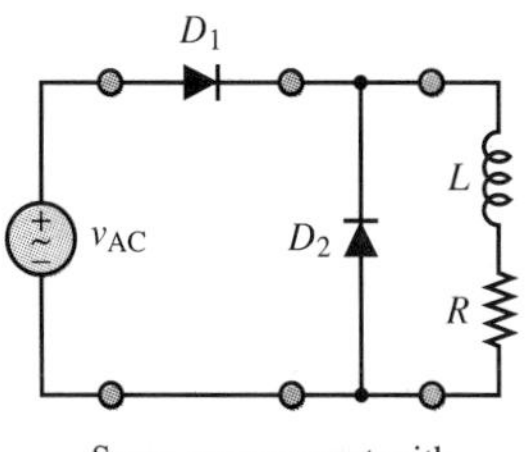

Same arrangement with free-wheeling diode

Figure 10.17 Rectifier connected to an inductive load

The circuit on top in Figure 10.17, assuming an ideal diode, would present a serious problem during the negative half-cycle of the source voltage, since the requirement for continuity of current in the inductor (recall the discussion on continuity of inductor currents and capacitor voltages in Chapter 6) would be violated with D_1 off. Whenever the current flow through the inductor is interrupted (during the negative half-cycles of v_{AC}), the inductor attempts to build a **flyback voltage** proportional to di_L/dt. Since the rectifier does not provide any current during the negative half-cycle of the source voltage, the instantaneous inductor voltage could be very large and could lead to serious damage to either the motor or the rectifier.

The circuit shown on the bottom in Figure 10.17 contains a so-called **free-wheeling diode,** D_2. The role of D_2 is to provide continuity of current when D_1 is in the off state. D_2 is off during the positive half-cycle but turns on when D_1 ceases to conduct, because of the flyback voltage, Ldi_L/dt. Rather than build up a large voltage, the inductor now has a path for current to flow, through D_2, when D_1 is off. Thus, the energy stored by the inductor during the positive half-cycle of v_{AC} is utilized to preserve a continuous current through the inductor during the off period. Figure 10.18 depicts the load current for the circuit including the diode. Note that D_2 allows the energy-storage properties of the inductor to be utilized to smooth the pulselike supply current and to produce a nearly constant load current.

Analyzing the circuit on the bottom of Figure 10.17, with

$$v_{AC}(t) = A\sin(\omega t) \tag{10.5}$$

(and assuming that both D_1 and D_2 are ideal), we conclude that the DC component of the load voltage, V_L, must appear across the load resistor, R (no steady-state DC voltage can appear across the inductor, since $v_L = Ldi_L/dt$). Thus, the approximate DC current flowing through the load is given by

$$I_L = \frac{A}{\pi R} \tag{10.6}$$

since the average output voltage of a half-wave rectifier is A/π V for an AC source of peak amplitude A (see Chapter 7). The AC component of the load current (or "ripple" current) is not as simple to compute, since it is due to the AC component

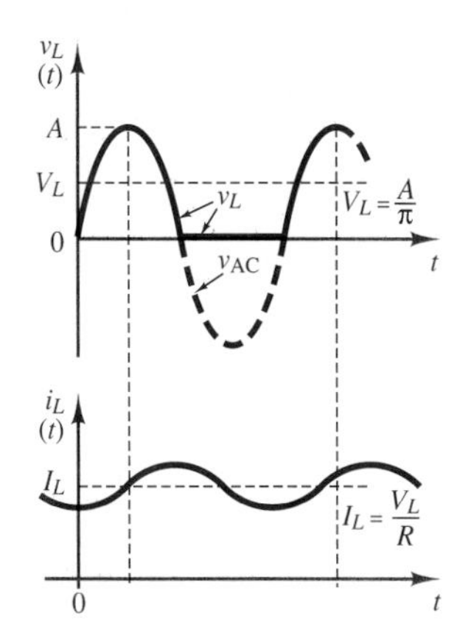

Figure 10.18 Operation of a free-wheeling diode

of v_L, which is not a pure sinusoid. The exact analysis would require the use of a Fourier series expansion. For the purposes of this discussion, it is not unreasonable to assume that most of the energy is at a frequency equal to that of the AC source:

$$i_L(t) \approx I_L + I_{AC} \cos(\omega t + \theta) \tag{10.7}$$

where I_L is the average load current, I_{AC} is the peak value of the ripple current, and θ is its phase. An acceptable approximation from which the amplitude of I_{AC} may be computed is

$$v_L(t) \approx \frac{A}{2\pi} + \frac{A}{2\pi} \sin \omega t \tag{10.8}$$

Figure 10.19 graphically illustrates the extent of this approximation.

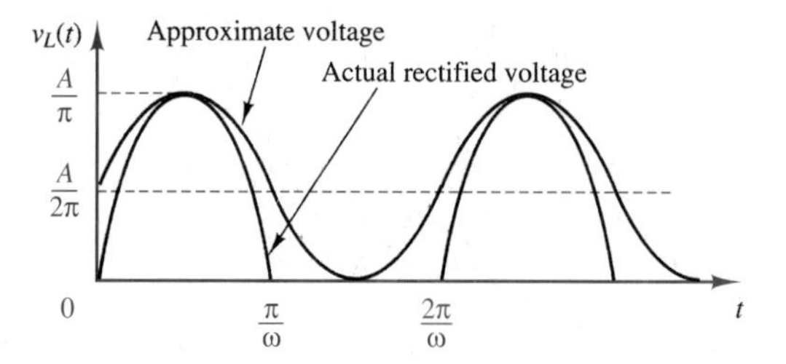

Figure 10.19 Approximation of ripple voltage for a half-wave rectifier

A common alternative to the half-wave rectifier is the full-wave rectifier, which was discussed in Chapter 7.

Three-Phase Rectifiers

It is important to realize that the same type of circuit that can be used for single-phase rectifiers can also be employed to design multiphase rectifiers. Recall the analysis of three-phase AC power systems in Section 5.4. In many high-power applications, three-phase voltages need to be rectified to give rise to a single DC supply; such rectification can be achieved by means of an extension of the bridge rectifier. Consider the balanced three-phase circuit shown in Figure 10.20. The three-phase wye-connected source is connected to a resistive load by means of a three-phase transformer, with a delta-connected primary and a wye-connected secondary. The circuit could also operate without the transformer. The three secondary currents, i_a, i_b, and i_c, flow through pairs of diodes D_1 to D_6 in a manner very similar to the single-phase bridge rectifier described in Figure 7.45. The diodes will conduct in pairs depending on the relative line voltages, according to the following sequence: D_1-D_2, D_2-D_3, D_3-D_4, D_4-D_5, D_5-D_6, and D_6-D_1. Recall from the analysis of Chapter 5 that the line-to-line voltage is $\sqrt{3}$ times the phase voltage in a three-phase wye-connected source. The instantaneous source voltages and the related diode conduction periods, as well as the load voltage, are shown in Figure 10.21.

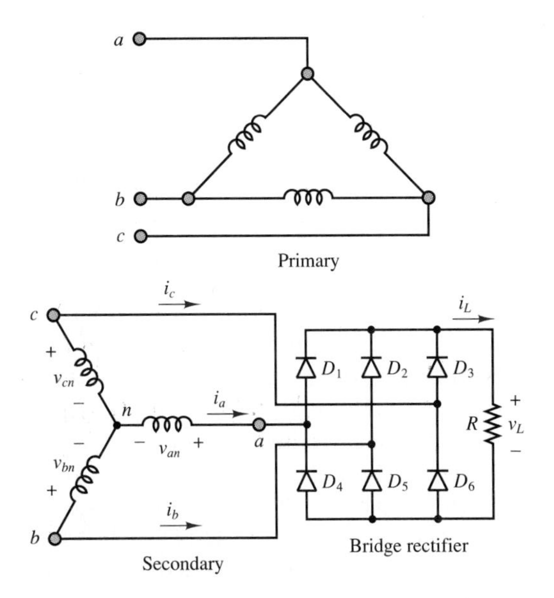

Figure 10.20 Three-phase diode bridge rectifier

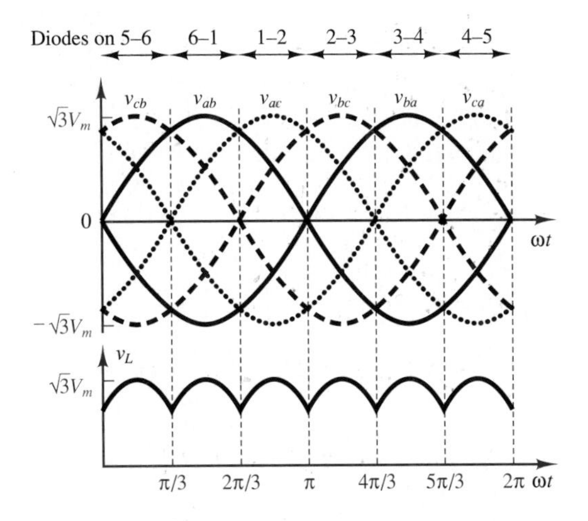

Figure 10.21 Waveforms and conduction times of three-phase bridge rectifier

It can be shown that the average output voltage is given by the expression:

$$V_L = \frac{2}{2\pi/6}\int_0^{\pi/6} \sqrt{3}V_m \cos\omega t \, d(\omega t) = \frac{3\sqrt{3}}{\pi}V_m = 1.654V_m \qquad \textbf{(10.9)}$$

where V_m is the peak phase voltage. The rms output voltage can be calculated to be

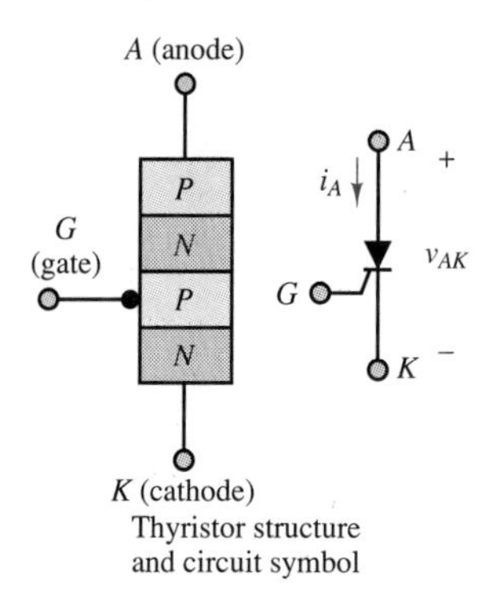

Figure 10.22 Thyristor structure and circuit symbol

$$V_{\text{rms}} = \sqrt{\frac{2}{2\pi/6}\int_0^{\pi/6} \sqrt{3}V_m^2 \cos^2 \omega t \, d(\omega t)}$$

$$= \left(\frac{3}{2} + \frac{9\sqrt{3}}{4\pi}\right)V_m = 1.6554V_m \qquad \textbf{(10.10)}$$

Thyristors and Controlled Rectifiers

In a number of applications, it is useful to be able to externally control the amount of current flowing from an AC source to the load. A family of power semiconductor devices called **controlled rectifiers** allows for control of the rectifier state by means of a third input, called the **gate.** Figure 10.22 depicts the appearance of a **thyristor,** or **silicon controlled rectifier (SCR),** illustrating how the physical structure of this device consists of four layers, alternating p-type and n-type material. Note that the circuit symbol for the thyristor suggests that this device acts as a diode, with provision for an additional external control signal.

The operation of the thyristor can be explained in an intuitive fashion as follows. When the voltage v_{AK} is negative (i.e., providing reverse bias), the thyristor acts just like a conventional pn junction in the off state. When v_{AK} is forward-biased *and* a small amount of current is injected into the gate, the thyristor conducts forward current. The thyristor then continues to conduct (even in the absence of gate current), provided that v_{AK} remains positive. Figure 10.23 depicts the i-v curve for the thyristor. Note that the thyristor has two stable states, determined by the bias v_{AK} and by the gate current. In summary, the thyristor acts as a diode with a control gate that determines the time when conduction begins.

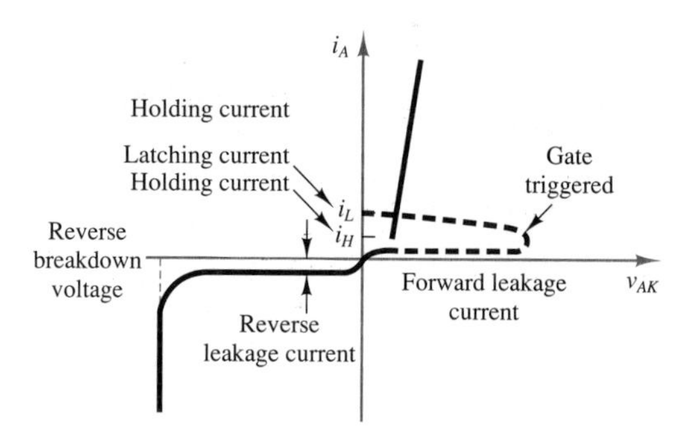

Figure 10.23 Thyristor i-v characteristic

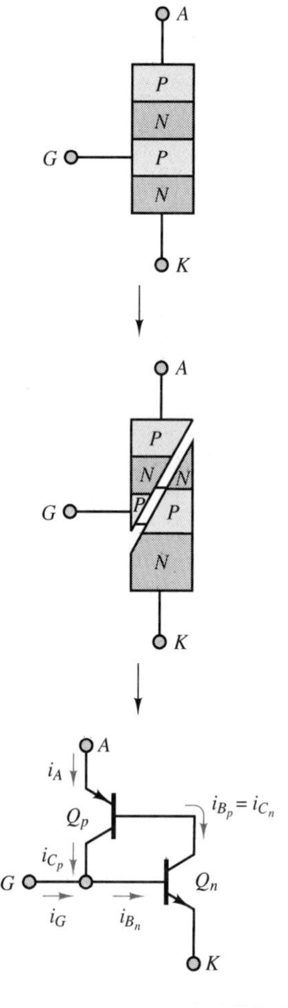

Figure 10.24 Thyristor two-transistor model

A somewhat more accurate description of thyristor operation may be provided if we realize that the four-layer $pnpn$ device can be modeled as a pnp transistor connected to an npn transistor. Figure 10.24 clearly shows that, physically, this is a realistic representation. Note that the anode current, i_A, is equal to the emitter current of the pnp transistor (labeled Q_p) and the base current of Q_p is equal to the collector current of the npn transistor, Q_n. Likewise, the base current of Q_n is the sum of the gate current and the collector current of Q_p. The behavior of this transistor model is explained as follows. Suppose, initially, i_G and i_{B_n} are both zero. Then it follows that Q_n is in cutoff, and therefore $i_{C_n} = 0$. But if $i_{C_n} = 0$, then the base current going into Q_p is also zero and Q_p is also in cutoff, and $i_{C_p} = 0$, consistent with our initial assumption. Thus, this is a stable state, in the sense that unless an external condition perturbs the thyristor, it will remain off.

Now, suppose a small pulse of current is injected at the gate. Then $i_{B_n} > 0$, and Q_n starts to conduct, provided, of course, that $v_{AK} > 0$. At this point, i_{C_n}, and therefore i_{B_p}, must be greater than zero, so that Q_p conducts. It is important to note that once the gate current has turned Q_n on, Q_p also conducts, so that $i_{C_P} > 0$. Thus, even though i_G may cease, once this "on" state is reached, $i_{C_p} = i_{B_n}$ continues to drive Q_n so that the on state is also self-sustaining. The only condition that will cause the thyristor to revert to the off state is the condition in which v_{AK} becomes negative; in this case, both transistors return to the cutoff state.

In a typical controlled rectifier application, the device is used as a half-wave rectifier that conducts only after a trigger pulse is applied to the gate. Without concerning ourselves with how the trigger pulse is generated, we can analyze the general waveforms for the circuit of Figure 10.25 as follows. Let the voltage v_{trigger} be applied to the gate of the thyristor at $t = \tau$. The voltage v_{trigger} can be a short pulse, provided by a suitable trigger-timing circuit (Chapter 12 will discuss timing and switching circuits). At $t = \tau$, the thyristor begins to conduct, and it continues to do so until the AC source enters its negative cycle. Figure 10.26 depicts the relevant waveforms.

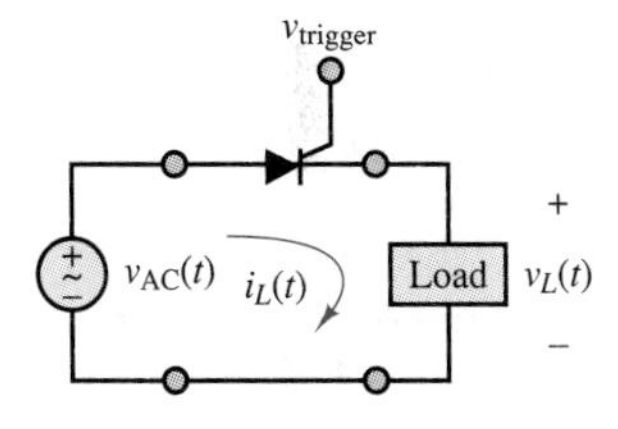

Figure 10.25 Controlled rectifier circuit

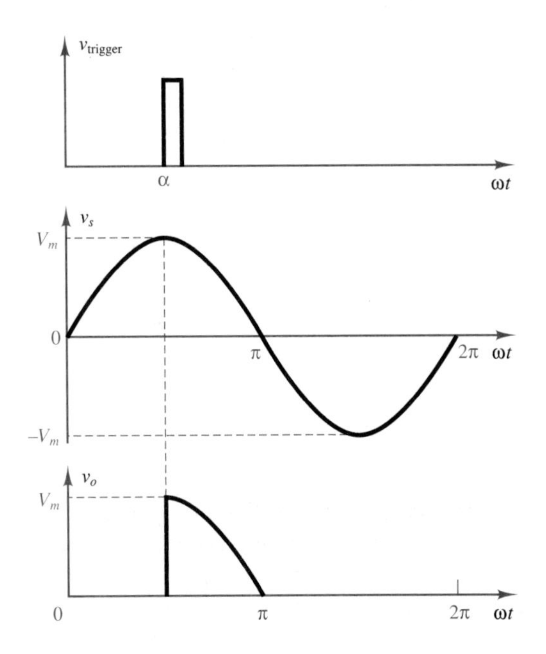

Figure 10.26 Half-wave controlled rectifier waveforms

Note how the DC load voltage is controlled by the firing time τ, according to the following expression:

$$\langle v_L \rangle = V_L = \frac{1}{T}\int_{\tau}^{T/2} v_{AC}(t)\,dt \qquad \textbf{(10.11)}$$

where T is the period of $v_{AC}(t)$. Now, if we let

$$v_{AC}(t) = A \sin \omega t \qquad \textbf{(10.12)}$$

we can express the average (DC) value of the load voltage

$$V_L = \frac{1}{T}\int_{\tau}^{T/2} A \sin \omega t\,dt = (1 + \cos \omega t)\frac{A}{2\pi} \qquad \textbf{(10.13)}$$

in terms of the **firing angle,** α, defined as

$$\alpha = \omega\tau \qquad \textbf{(10.14)}$$

By evaluating the integral of equation 10.13, we can see that the (DC) load voltage amplitude depends on the firing angle, α:

$$V_L = (1 + \cos \alpha)\frac{A}{2\pi} \qquad \textbf{(10.15)}$$

Example 10.2

Thyristors are often used in variable DC power supplies. The circuit of Figure 10.27 is used to control power to a resistive load, such as a light bulb. The control circuitry will provide a pulse of current $i_G(t)$ to the thyristor at $\omega t = \alpha + n360$ (where n is an integer 0,1,2,3,...) based upon the position of the control knob. At the instant that $i_G(t)$ becomes nonzero, the thyristor behaves like a forward-biased diode; that is, if $v_{AK}(t)$ is positive, then the thyristor will allow current to pass in the forward direction. The gate current is pulsed with the timing shown in Figure 10.28.

1. Find the rms load voltage as a function of the firing angle, α, if $v_{AK\text{on}} = 0$ V.
2. Find the power supplied to the light bulb if its internal resistance is 240 Ω and $\alpha = 0$.

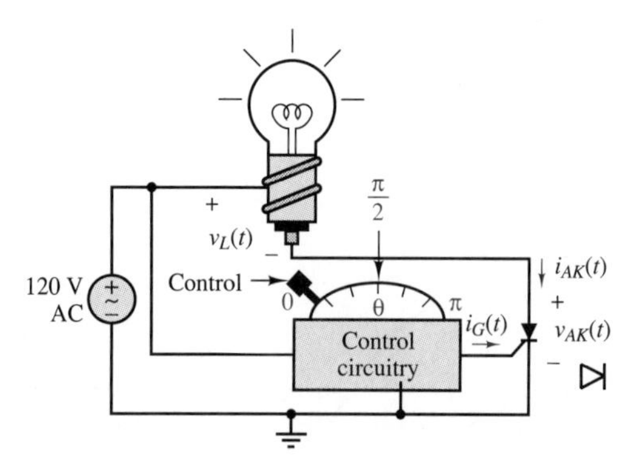

Figure 10.27

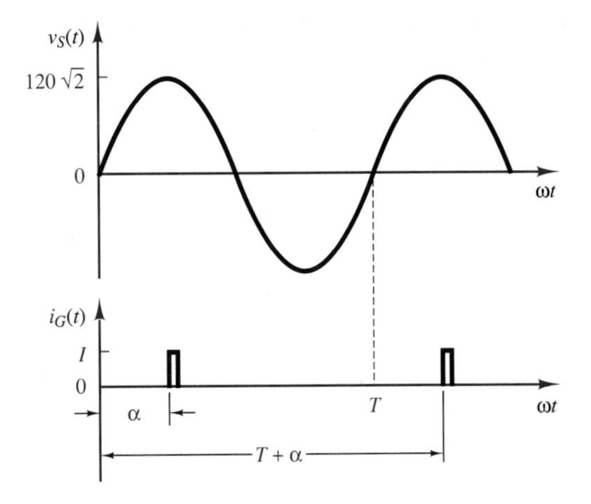

Figure 10.28

3. Repeat part 2 for $\alpha = \pi$.
4. Repeat part 2 for $\alpha = \pi/2$.

Solution:

1. The load voltage will appear as shown in Figure 10.29, and the rms value of the load voltage may be computed as follows:

$$V_{L\text{ rms}} = \sqrt{\frac{(120\sqrt{2})^2}{2\pi}\int_{\alpha}^{\pi}\sin^2\omega t\, d(\omega t)}$$

$$= \sqrt{\frac{(120\sqrt{2})^2}{2\pi}\int_{\pi}^{\alpha}\left(\frac{1}{2}-\frac{1}{2}\cos 2\omega t\right)d(\omega t)}$$

$$= \sqrt{\frac{(120\sqrt{2})^2}{4} - \frac{(120\sqrt{2})^2}{2\pi}\left(\frac{\alpha}{2}\right) + \frac{(120\sqrt{2})^2}{4\times 2\pi}\sin 2\alpha}$$

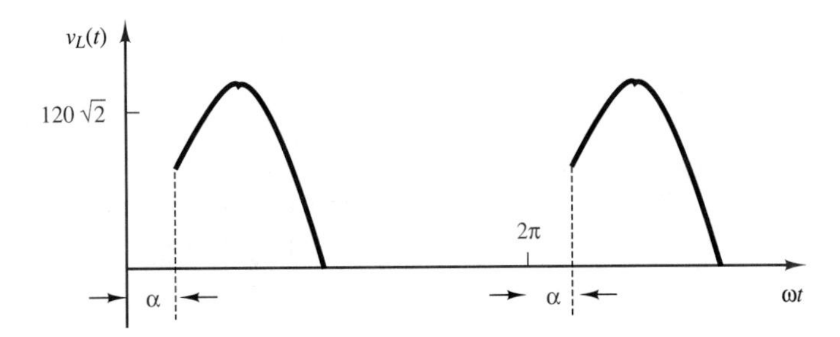

Figure 10.29

After some simplification, the load voltage may be expressed as a function of the firing angle, as follows:

$$V_L(\alpha) = \frac{120\sqrt{2}}{2}\sqrt{1 - \frac{\alpha}{\pi} + \sin 2\alpha}$$

The load power may now be computed for the three cases:

2.
$$P = \frac{V_{\text{rms}}^2}{R} = \frac{(120\sqrt{2}/2)^2}{240}$$
$$= 30 \text{ W}$$

3.
$$V_{\text{rms}} = 0$$
$$P = 0$$

4.
$$P = \frac{\left(\frac{120\sqrt{2}}{2}\sqrt{1 - \frac{1}{2}}\right)^2}{240}$$
$$= 15 \text{ W}$$

Note that, since the thyristor is off whenever power is not reaching the load, no power is wasted. This would not be the case if one simply used a resistive voltage divider to reduce the load voltage.

Example 10.3

An interesting application that involves the use of various power devices is an automotive battery charger. The charging circuit shown in Figure 10.30 is connected to a standard 120-V single-phase supply. Diodes D_1 and D_2 form a full-wave rectifier (see Chapter 7); resistors R_1 and R_2 and thyristor T_2 form a (variable) voltage divider.

Assume that T_2 is not in the conducting state, and that the anode voltage of D_3 is such that D_3 conducts. Then, T_1 will be fired near the beginning of the positive half-cycle of the AC voltage, and its period of conduction will be fairly long, thus providing a substantial charging current for the battery (R_4 and R_p are large enough that most of the current flowing through T_1 will go to the battery). The potentiometer R_p is set so that

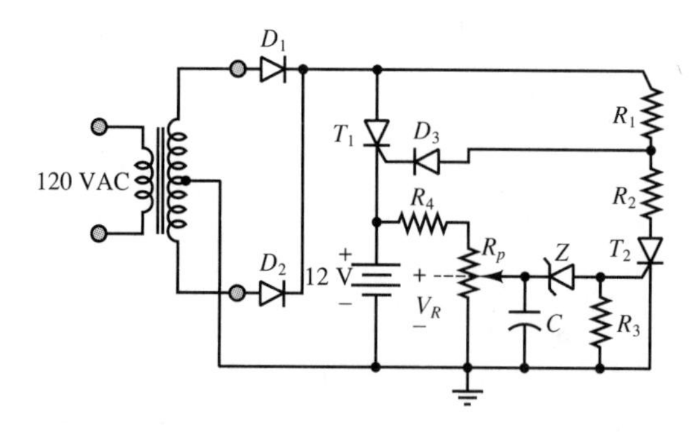

Figure 10.30 Automotive battery charger

when the battery voltage is low, the voltage V_R is not sufficient to turn on the Zener diode, Z. Thus, Z is effectively an open circuit, and T_2 remains off (recall that we had initially assumed T_2 to be off—the analysis confirms the validity of the assumption). As the battery charges to a higher voltage, the Zener diode will eventually conduct; when Z conducts, a gate current is injected into T_2, which is then turned on. When T_2 conducts, the voltage across the R_2-T_2 series connection becomes significantly lower, because the thyristor is now nearly a short circuit. Resistors R_1 and R_2 are selected so that when T_2 conducts, D_3 becomes reverse-biased. Once this condition occurs, T_1 is turned off, and charging stops. You see that this circuit will prevent battery overcharging.

Example 10.4

Assume that the thyristor in the circuit of Figure 10.31 is ideal (i.e., zero voltage drop upon conduction, and negligible turn-on gate current). Determine an appropriate value of R such that the average current through the thyristor will be 1 A.

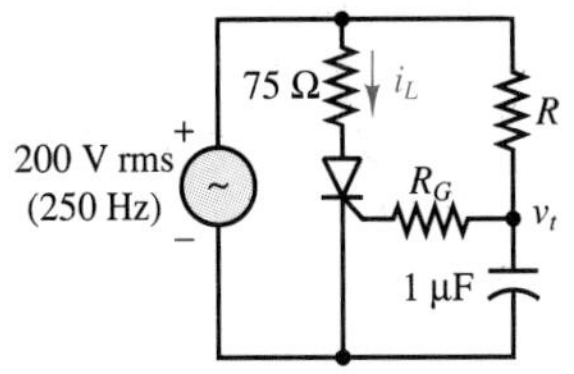

Figure 10.31

Solution:

The waveforms shown in Figure 10.32 illustrate the relative timing of the source voltage (v_S), thyristor current (i_L), and triggering voltage (v_t). The expression for the source voltage is:

$$v_S(t) = \sqrt{2}(200)\sin[2\pi(250)t] \text{ V}$$

and the load current (through the 75-Ω resistor) is:

$$i_L(t) = \frac{\sqrt{2}(200)}{75}\sin[2\pi(250)t] \text{ A} \qquad \alpha \le \omega t \le \pi$$

$$= 0 \qquad \pi \le \omega t \le 2\pi$$

and the triggering voltage is:

$$v_t(t) = V_t \sin[2\pi(250)t - \alpha] \text{ V}$$

Note that the trigger voltage is required to go positive at the desired firing angle, α, so that a positive current can be injected into the gate at this time. Thus, the requirement that $I_L = \langle i_L \rangle = 1$ A translates into finding α such that

$$I_L = \frac{1}{2\pi}\int_\alpha^\pi \sqrt{2}\frac{(200)}{75}\sin(\omega t)\, d(\omega t) = 1 \text{ A}$$

or

$$\frac{200 \times \sqrt{2}}{2\pi \times 75}(1 + \cos\alpha) = 1$$

which can be solved to find $\alpha = 48.23°$.

To determine R, we observe that the AC source voltage appears across the RC circuit; thus, v_t can be computed from an impedance voltage divider using phasor techniques:

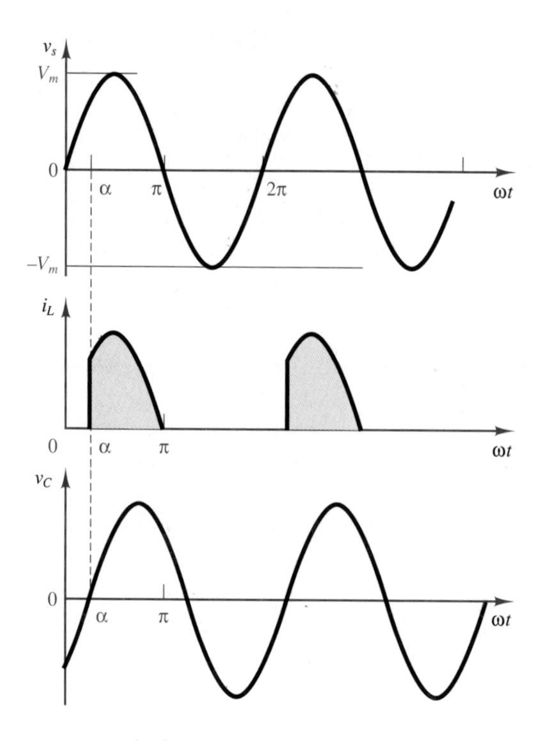

Figure 10.32

$$\mathbf{V}_t(j\omega) = V_S \frac{\frac{1}{j\omega C}}{R + \frac{1}{j\omega C}} = \frac{V_S}{\sqrt{1 + \omega^2 R^2 C^2}} \angle -\arctan(\omega RC)$$

and since the phase of $\mathbf{V}_t(j\omega) = -\arctan(\omega RC)$, we have

$$-\arctan(\omega RC) = \alpha = 48.23°$$

which can be solved to obtain $R = 713\ \Omega$.

DRILL EXERCISES

10.2. Using the approximation given in equation 10.8, find the DC and AC load currents for the circuit of Figure 10.17 if $R = 10\ \Omega, L = 0.3$ H, $A = 170$ V, and $\omega = 377$ rad/s.

10.3 Calculate the load voltage in Figure 10.26 for $A = 100, \alpha = \pi/3$.

10.4 For the circuit in Example 10.2, the input AC voltage is 240 V. Find the rms value of the load voltage and the power at the firing angle $\alpha = \pi/2$.

10.6 ELECTRIC MOTOR DRIVES

The advent of high-power semiconductor devices has made it possible to design effective and relatively low-cost electronic supplies that take full advantage of the capabilities of the devices introduced in this chapter. Electronic power supplies for DC and AC motors have become one of the major fields of application of power electronic devices. The last section of this chapter is devoted to an introduction to two families of power supplies, or **electric drives: choppers,** or **DC-DC converters;** and **inverters,** or **DC-AC converters.** These circuits find widespread application in the control of AC and DC motors in a variety of applications and power ranges.

Before we delve into the discussion of the electronic supplies, it will be helpful to introduce the concept of quadrants of operation of a drive. Depending on the direction of current flow, and on the polarity of the voltage, an electronic drive can operate in one of four possible modes, as indicated in Figure 10.33.

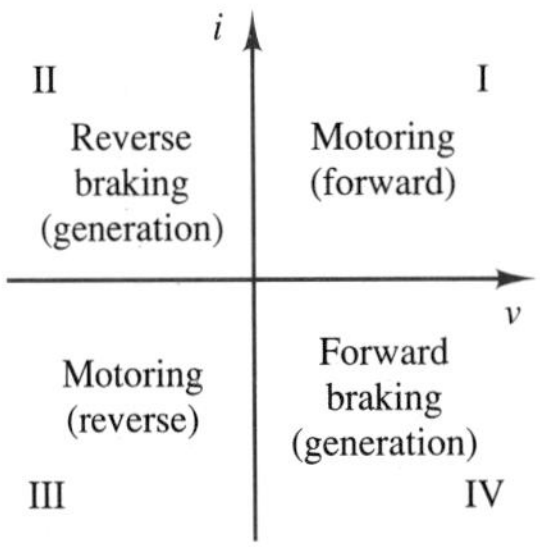

i
II
Reverse braking (generation)
I
Motoring (forward)
v
Motoring (reverse)
III
Forward braking (generation)
IV

Figure 10.33 The four quadrants of an electric drive

Choppers (DC-DC Converters)

As the name suggests, a DC-DC converter is capable of converting a fixed DC supply to a variable DC supply. This feature is particularly useful in the control of the speed of a DC motor (described in greater detail in Chapter 16). In a DC motor, shown schematically in Figure 10.34, the developed torque, T_m, is proportional to the current supplied to the motor **armature,** I_a, while the **electromotive force (emf),** E_a, which is the voltage developed across the armature, is proportional to the speed of rotation of the motor, ω_m. A DC motor is an electromechanical energy-conversion system; that is, it converts electrical to mechanical energy (or vice versa if it is used as a generator); if we recall that the product of torque and speed is equal to power in the mechanical domain, and that current times voltage is equal to power in the electrical domain, we conclude that in the ideal case of complete energy conversion, we have

$$E_a \times I_a = T_m \times \omega_m \qquad \textbf{(10.16)}$$

Naturally, such ideal energy conversion cannot take place; however we can see that there is a correspondence between the four electrical quadrants of Figure 10.33 and the mechanical power output of the motor: namely, if the voltage and current are both positive or both negative, the electrical power will be positive, and so will the mechanical power. This corresponds to the **forward** (i, v both positive) and **reverse** (i, v both negative) **motoring** operation. Forward motoring corresponds to quadrant I, and reverse motoring to quadrant III in Figure 10.33. If the voltage and current are of opposite polarity (quadrants II and IV), electrical energy is flowing back to the electric drive; in mechanical terms this corresponds to a braking condition. Operation in the fourth quadrant can lead to **regenerative braking,** so called because power is regenerated by making current flow back to the source. This mode could be useful, for example, to recharge a battery supply, because the braking energy can be regenerated by returning it to the electric supply.

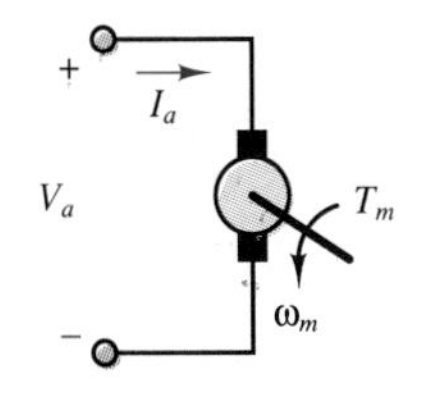

Figure 10.34 DC motor

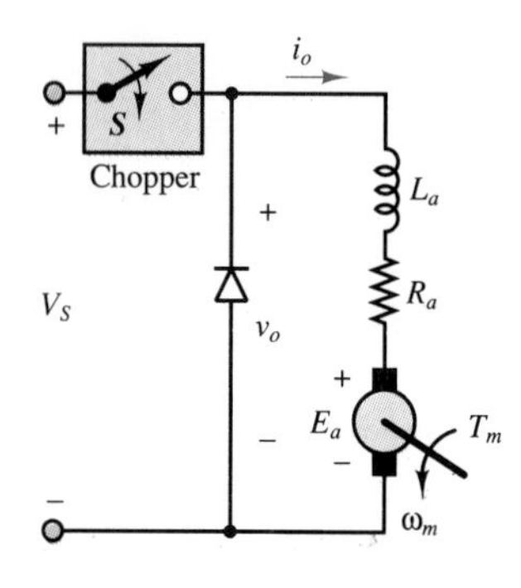

Figure 10.35 Step-down chopper (buck converter)

A simple circuit that can accomplish the task of providing a variable DC supply from a fixed DC source is the **step-down chopper (buck converter),** shown in Figure 10.35. The circuit consists of a "chopper" switch, denoted by the symbol S, and a free-wheeling diode, such as the one described in Section 10.5. The switch can be any of the power switches described in this chapter, for example, a power BJT or MOSFET, or a thyristor; see, for example, the BJT switch of Figure 10.4. The circuit to the right of the diode is a model of a DC motor, including the inductance and resistance of the armature windings, and the effect of the back emf E_a. When the switch is turned on (say, at $t = 0$), the supply V_S is connected to the load and $v_o = V_S$. The load current, i_o, is determined by the motor parameters. When the switch is turned off, the load current continues to flow through the free-wheeling diode, but the output voltage is now $v_o = 0$. At time T, the switch is turned on again, and the cycle repeats.

Figure 10.36 depicts the v_o and i_o waveforms. The average value of the output voltage, $\langle v_o \rangle$, is given by the expression

$$\langle v_o \rangle = \frac{t_1}{T} V_S = \delta V_S \tag{10.17}$$

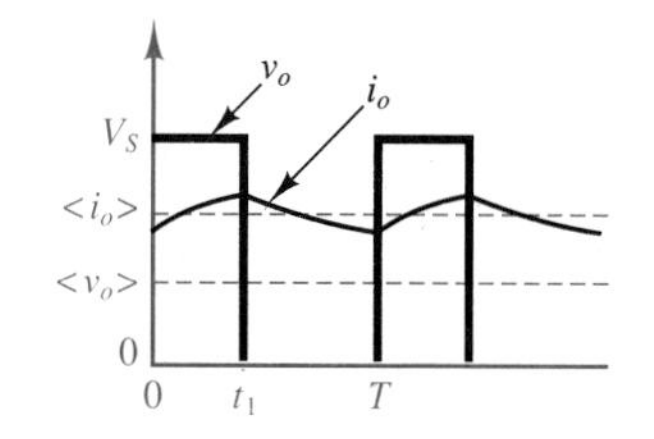

Figure 10.36 Step-down chopper waveforms

where δ is the **duty cycle** of the chopper. The step-down chopper has a useful range

$$0 \leq \langle v_o \rangle \leq V_S \tag{10.18}$$

It is also possible to increase the range of a DC-DC converter to above the supply voltage by making use of the energy-storage properties of an inductor; the resulting circuit is shown in Figure 10.37. When the chopper switch, S, is on, the supply current flows through the inductor and the closed switch, storing energy in the inductor; the output voltage, v_o, is zero, since the switch is a short circuit. When the switch is open, the supply current will flow through the load via the diode; but the inductor voltage is negative during the transient following the opening of the switch and therefore adds to the source voltage: the energy stored in the inductor while the switch was closed is now released and transferred to the load. This stored energy makes it possible for the output voltage to be higher than the supply voltage for a finite period of time.

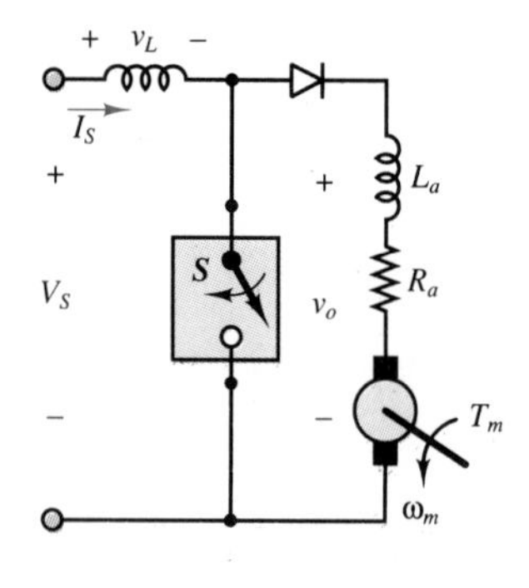

Figure 10.37 Step-up chopper (boost converter)

Let t_1 once again be the time during which the chopper conducts; neglecting for the moment the ripple in the supply current, the energy stored in the inductor during this period is:

$$W_i = V_S I_S t_1 \tag{10.19}$$

When the chopper is off, the energy released to the load is

$$W_i = (\langle v_o \rangle - V_S) I_S (T - t_1) \tag{10.20}$$

If the system is lossless, the two energy expressions must be equal:

$$V_S I_S t_1 = (\langle v_o \rangle - V_S) I_S (T - t_1) \tag{10.21}$$

and we can determine the average output voltage to be

$$\langle v_o \rangle = V_S \frac{t_1 + T - t_1}{T - t_1} = V_S \frac{T}{T - t_1} = V_S \frac{1}{1 - \delta} \qquad \textbf{(10.22)}$$

Since the duty cycle, δ, is always less than 1, the theoretical range of the supply is

$$V_S \leq \langle v_o \rangle < \infty \qquad \textbf{(10.23)}$$

The waveforms for the boost converter are shown in Figure 10.38.

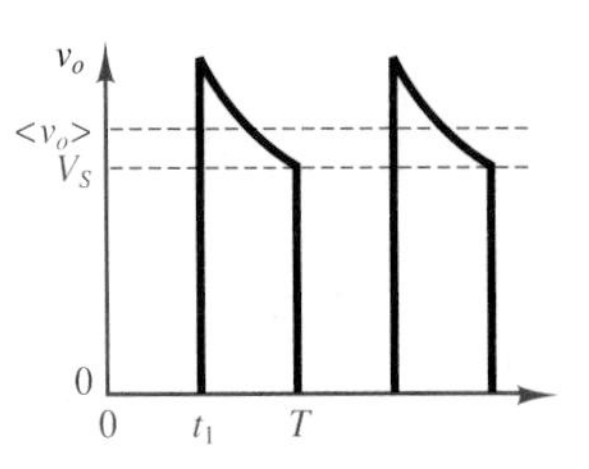

Figure 10.38 Step-up chopper output voltage waveform

A step-up chopper can also be used to provide regenerative braking: if the "supply" voltage is the motor armature voltage and the output voltage is the fixed DC supply (battery) voltage, then power can be made to flow from the motor to the DC supply (i.e., recharging the battery). This configuration is shown in Figure 10.39.

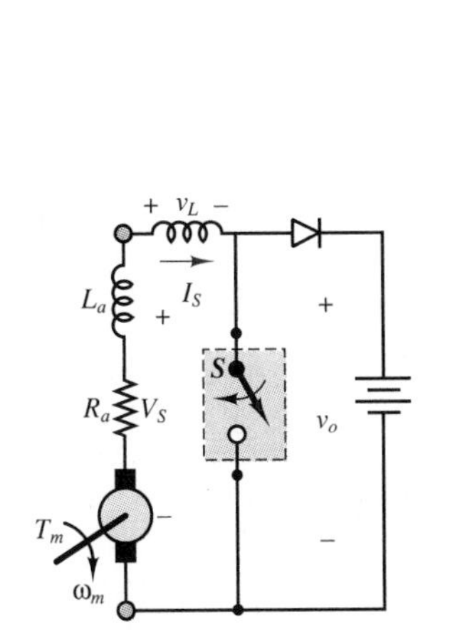

Figure 10.39 Step-up chopper used for regenerative braking

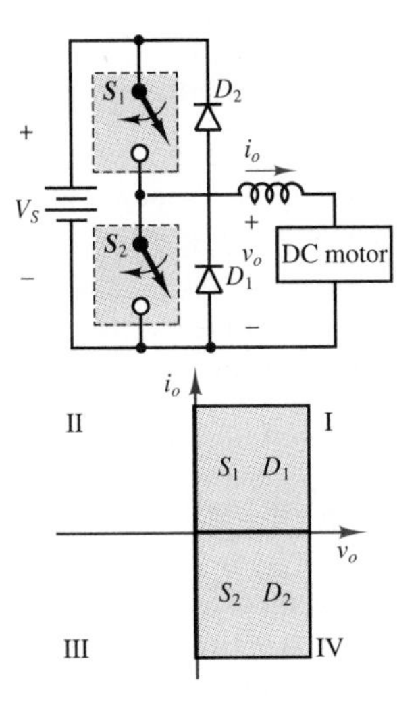

Figure 10.40 Two-quadrant chopper

Finally, the operation of the step-down and step-up choppers can be combined into a **two-quadrant chopper,** shown in Figure 10.40. The circuit shown schematically in Figure 10.40 can provide both regenerative braking and motoring operation in a DC motor. When switch S_2 is open, and switch S_1 serves as a chopper, the circuit operates as a step-down chopper, precisely as was described earlier in this section (convince yourself of this by redrawing the circuit with S_2 and D_2 replaced by open circuits). Thus, the drive and motor operate in the first quadrant (motoring operation). The output voltage, v_o, will switch between V_S and zero, as shown in Figure 10.36, and the load current will flow in the direction indicated by the arrow in Figure 10.40; diode D_1 free-wheels whenever S_1 is open. Since both output voltage and current are positive, the system operates in the first quadrant.

When switch S_1 is open and switch S_2 serves as a chopper, the circuit resembles a step-up chopper. The source is the motor emf, E_a, and the load is the battery; this is the situation depicted in Figure 10.39. The current will now be negative, since the sum of the motor emf and the voltage across the inductor (corresponding to the energy stored during the "on" cycle of S_2) is greater than the battery voltage. Thus, the drive operates in the fourth quadrant.

Example 10.5

The two-quadrant chopper of Figure 10.40 is used to control the speed of a DC motor and for regenerative braking. The supply is 120 VDC. The motor emf is given by the expression $E_a = 0.1n$, where n is the motor speed in rev/min. Assume that the chopping frequency is 300 Hz, that the motor armature resistance is $R_a = 0.2\ \Omega$, and that the armature inductance, L_a, is sufficiently large that the load current is ripple-free.

1. In the motoring mode, $n = 500$ rev/min and $i_o = 90$ A. Determine the turn-on time of the chopper, the power absorbed by the motor, the power absorbed by the armature, and the power delivered by the source.
2. In the regenerative braking mode, $n = 380$ rev/min and $i_o = -90$ A. Determine the turn-on time of the chopper, the power delivered by the motor, the power absorbed by the armature, and the power absorbed by the source.

Solution:

1. According to Figure 10.35,

$$v_o = E_a + i_o R_a = 0.1 \times 500 + 90 \times 0.2 = 68 \text{ V}$$

and from equation 10.17 we find

$$68 = \frac{t_1}{T} \times 120 \qquad \text{or} \qquad \delta = 0.5667$$

Since the switching frequency is 300 Hz, $T = 3.33$ ms, and $t_1 = 1.89$ ms.
The power absorbed by the motor is:

$$P_m = E_a i_o = (0.1 \times 500) \times 90 = 4{,}500 \text{ W}$$

The power lost in the armature resistance is:

$$P_R = R_a (i_o)^2 = (90)^2 \times 0.2 = 1{,}620 \text{ W}$$

and the power delivered by the source is:

$$P_S = \delta V_S i_o = 0.5667 \times 120 \times 90 = 6{,}120 \text{ W}$$

Note that $P_S = P_m + P_R$, as should be the case.

2. Since the current is flowing in the reverse direction,

$$v_o = E_a + i_o R_a = 0.1 \times 380 + (-90) \times 0.2 = 20 \text{ V}$$

According to equation 10.22, and observing that v_0 is now the source,

$$120 = 20\frac{1}{1 - \frac{t_1}{T}}$$

Since the switching frequency is 300 Hz, $T = 3.33$ ms, and $t_1 = 2.8$ ms. Note that now the duty cycle, or δ, is $1 - t_1/T = 1 - \frac{5}{6} = 0.1667$ (See Figure 10.38).

The power absorbed by the motor is:

$$P_m = E_a i_o = (0.1 \times 380) \times (-90) = -3{,}420 \text{ W}$$

Since this is a negative value, the motor is actually regenerating power. The power lost in the armature resistance is:

$$P_R = R_a(i_o)^2 = (-90)^2 \times 0.2 = 1{,}620 \text{ W}$$

and the power delivered by the source is:

$$P_S = \delta V_S i_o = 0.1667 \times 120 \times (-90) = -1{,}800 \text{ W}$$

Since this is a negative number, the source is actually absorbing power. The power regenerated by the motor equals the sum of the power lost to the armature resistance and the power absorbed by the source, as should be the case for operation in quadrant IV.

Inverters (DC-AC Converters)

As will be explained in Chapter 16, variable-speed drives for AC motors require a multiphase variable-frequency, variable-voltage supply. Such drives are called *DC-AC converters,* or *inverters.* Inverter circuits can be quite complex, so the objective of this section is to present a brief introduction to the subject, with the aim of illustrating the basic principles. A **voltage source inverter (VSI)** converts the output of a fixed DC supply (e.g., a battery) to a variable-frequency AC supply. Figure 10.41 depicts a **half-bridge VSI;** once again, the switches can be either bipolar or MOS transistors, or thyristors. The operation of this circuit is as follows. When switch S_1 is turned on, the output voltage is in the positive half-cycle, and $v_o = V_S/2$. To generate the negative half-cycle, switch S_2 is turned on, and $v_o = -V_S/2$. The switching sequence of S_1 and S_2 is shown in Figure 10.42. It is important that each switch be turned off before the other is turned on; otherwise, the DC supply would be short-circuited. Since the load is always going to be inductive in the case of a motor drive, it is important to observe that the load current, i_o, will lag the voltage waveform, as shown in Figure 10.42. As shown in this figure, there will be some portions of the cycle in which the voltage is positive but the current is negative. The function of diodes D_1 and D_2 is precisely to conduct the load current whenever it is of direction opposite to the polarity of the voltage. Without these diodes, there would be no load current in this case. Figure 10.42 also shows which element is conducting in each portion of the cycle.

A full-bridge version of the VSI can also be designed as shown in Figure 10.43; the associated output voltage waveform is shown in Figure 10.44. The

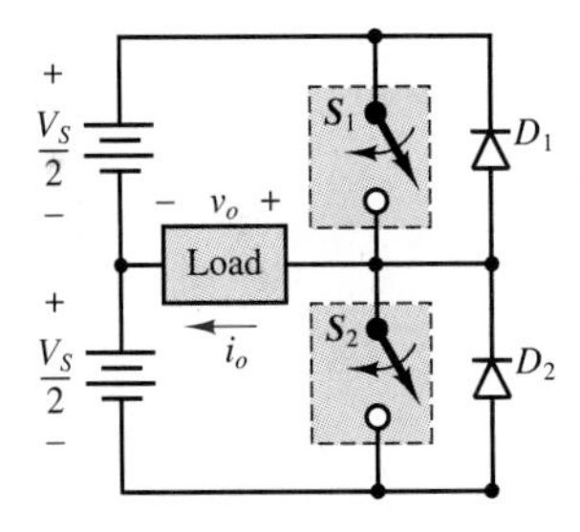

Figure 10.41 Half-bridge voltage source inverter

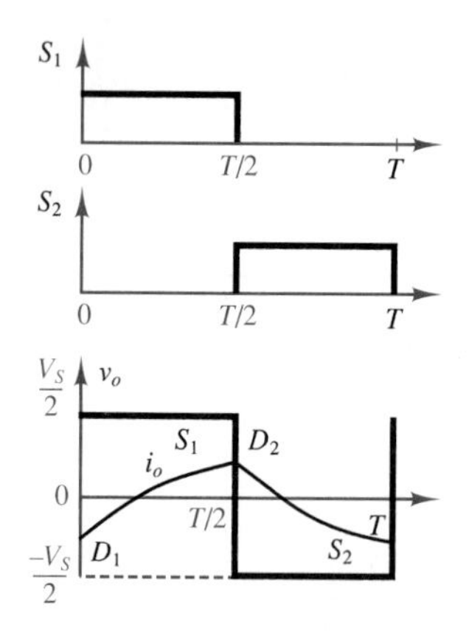

Figure 10.42 Half-bridge voltage source inverter waveforms

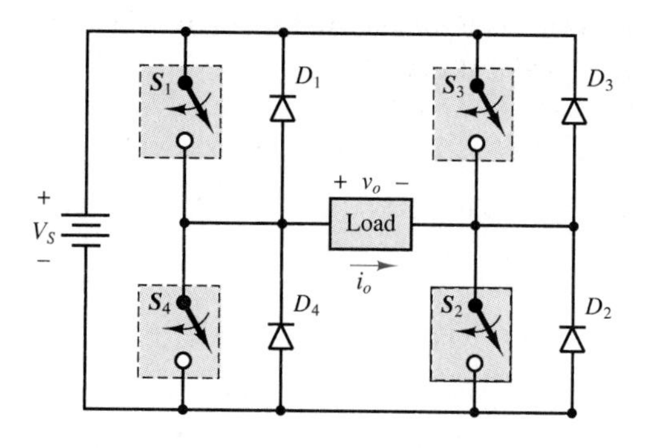

Figure 10.43 Full-bridge voltage source inverter

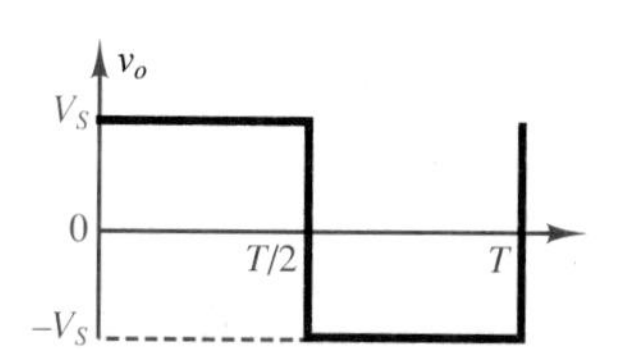

Figure 10.44 Half-bridge voltage source inverter output voltage waveform

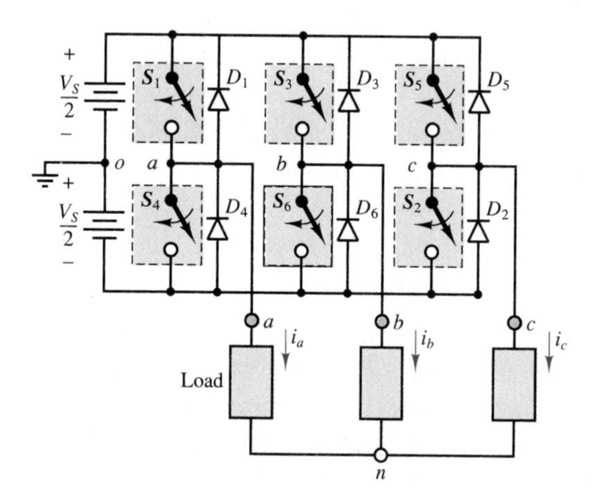

Figure 10.45 Three-phase voltage source inverter

operation of this circuit is analogous to that of the half-bridge VSI; switches S_1 and S_2 are fired during the first half-cycle, and switches S_3 and S_4 during the second half. Note that the full-bridge configuration allows the output voltage to swing from V_S to $-V_S$. The diodes provide a path for the load current whenever the load voltage and current are of opposite polarity.

A three-phase version of the VSI is shown in Figure 10.45. Once again, the operation is analogous to that of the VSI circuits just presented. The related waveforms are shown in Figure 10.46. The top three waveforms depict the **pole voltages,** which are referenced to the DC supply neutral point, o. The pole voltages are obtained by firing the switches S_1 through S_6 at appropriate times. For example, if S_1 is fired at $\omega t = 0$, then pole a is connected to the positive side of the DC supply, and $v_{ao} = V_S/2$; if S_4 is subsequently turned on at $\omega t = \pi$, then pole a is connected to the negative side of the DC supply, and $v_{ao} = -V_S/2$.

The other pairs of switches are then fired in an analogous sequence, shifted by 120 electrical degrees with respect to each other, to obtain the waveforms shown in the top three graphs of Figure 10.46. The **line voltages** are obtained from the pole voltages using the following relations:

$$
\begin{aligned}
v_{ab} &= v_{ao} - v_{bo} \\
v_{bc} &= v_{co} - v_{co} \\
v_{ca} &= v_{co} - v_{ao}
\end{aligned}
\tag{10.24}
$$

and are shown in the second set of three diagrams in Figure 10.46. These are also phase-shifted by 120°. Now, we can also express the pole voltages in terms of the **load phase voltages,** v_{an}, v_{bn}, and v_{cn}:

$$
\begin{aligned}
v_{ao} &= v_{an} - v_{no} \\
v_{bo} &= v_{bn} - v_{no} \\
v_{co} &= v_{cn} - v_{no}
\end{aligned}
\tag{10.25}
$$

and since we must have $v_{an} + v_{bn} + v_{cn} = 0$ for balanced operation (see Chapter 5), we can derive the following relationship for the DC **supply neutral** (o) to **load neutral** (n) voltage:

$$
v_{no} = \frac{v_{ao} + v_{bo} + v_{co}}{3} \tag{10.26}
$$

This voltage is also shown to be a square wave switching three times as fast as the inverter output voltage. Finally, to obtain the phase voltages, we make use of the relations

$$
\begin{aligned}
v_{an} &= v_{ao} - v_{no} = \tfrac{2}{3}v_{ao} - \tfrac{1}{3}(v_{bo} + v_{co}) \\
v_{bn} &= v_{bo} - v_{no} = \tfrac{2}{3}v_{bo} - \tfrac{1}{3}(v_{ao} + v_{co}) \\
v_{cn} &= v_{co} - v_{no} = \tfrac{2}{3}v_{bo} - \tfrac{1}{3}(v_{ao} + v_{bo})
\end{aligned}
\tag{10.27}
$$

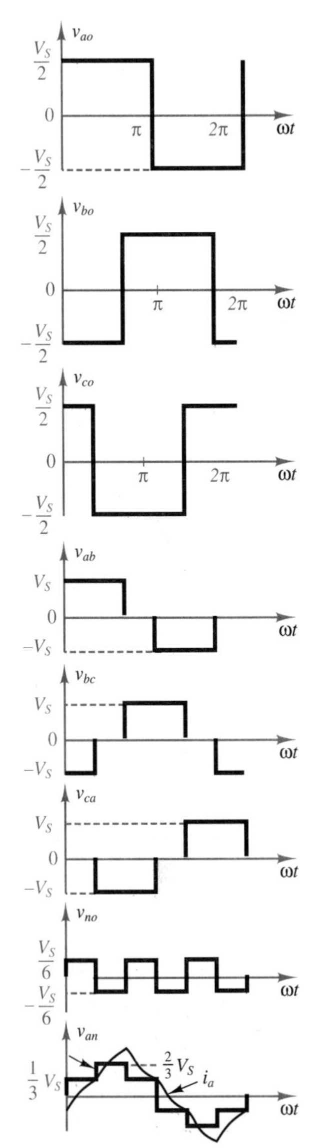

Figure 10.46 Three-phase voltage source inverter waveforms

Only one phase voltage, v_{an}, is shown in the picture; however, it is straightforward to construct the other two phase voltages using equation 10.27. Note that the load phase voltage waveform shown in Figure 10.46 is a coarse stepwise approximation of a sinusoidal waveform; the corresponding load current, i_a, is a filtered version of the load voltage, since the load is inductive in nature, and is therefore somewhat smoothed with respect to the voltage waveform. The discontinuous nature of these waveforms creates a significant higher harmonic spectrum (see the box "Fourier Analysis" in Chapter 6), at frequencies that are integer multiples of the inverter output frequency; this is an unavoidable property of all inverters that employ switching circuits, but the problem can be reduced by using more complex switching schemes. Another major shortcoming of this AC supply is that if the DC supply is fixed, the amplitude of the inverter output is fixed.

The VSI circuit described in the foregoing paragraphs can provide a variable-frequency supply provided that the commutation frequency of the electronic switches can be varied. Thus, in general, it is necessary to also provide the capability for timing circuits that can provide variable switching rates; this is often accomplished with a microprocessor (discussed in Chapter 13).

The limitations of the VSI of Figure 10.45 can be overcome with the use of more advanced switching schemes, such as *pulse-width modulation (PWM)* and *sinusoidal PWM*. The complexity of these schemes is beyond the scope of this book, and the interested reader is invited to explore a more advanced power electronics text to learn about advanced inverter circuits. We shall simply mention that it is possible to significantly reduce the harmonic content of the inverter waveforms and to provide variable-frequency, variable-amplitude, three-phase supplies for AC motors by means of power switching circuits under microprocessor control. These advances are finding a growing field of application in the electric vehicle arena. This subject will be approached again in Chapter 16.

CONCLUSION

- Power electronic devices can handle up to a few thousand volts and up to several hundred amperes and have a host of industrial applications. Various families of power electronic circuits and their application were discussed in this chapter.
- Voltage regulators are used in DC power supplies to provide a stable DC voltage output. The principal element of a voltage regulator is the Zener diode.
- Transistors find application both as power amplifiers and as switches; BJTs, MOSFETs, and IGBTs are all commonly employed, especially for switching functions. Each of these devices offers specific advantages, such as greater current capability, or faster response. Device technology is rapidly improving, especially among power MOSFETs.
- Power diodes and various types of thyristors find widespread application in rectifiers and controlled rectifiers, both for single- and three-phase circuits. Rectifiers are a necessary element of DC power supplies; controlled rectifiers also find application as DC motor drives and in many other variable-voltage applications.
- Electric motor drives based on power electronic devices allow for the implementation of sophisticated motor controls. DC motor drives include controlled regulators and choppers (DC-DC converters), while AC motor drives consist of inverter circuits (DC-AC converters). Both of the latter circuits make extensive use of high-power switching elements, such as MOSFETs, thyristors, BJTs, and IGBTs.

KEY TERMS

Armature *531*
Chopper (DC-DC Converter) *531*
Controlled Rectifier *524*
Delay Time *518*
Duty Cycle *532*
Electric Drive *531*
Electromotive Force (emf) *531*
Fall Time *519*

ANSWERS TO DRILL EXERCISES

10.1 $P = 1.3$ W

10.2 $I_L = 5.4$ A; $I_{AC} = 0.75$ A; $\alpha = 84.95°$

10.3 $V_L = 23.87$ V

10.4 120 V, 60 W

HOMEWORK PROBLEMS

10.1 Repeat Example 10.1 for a 7-V Zener diode.

10.2 For the current regulator circuit shown in Figure P10.1, find the expression for R_S.

10.3 For the shunt-type voltage regulator shown in Figure P10.2, find the expression for the output voltage, V_{out}.

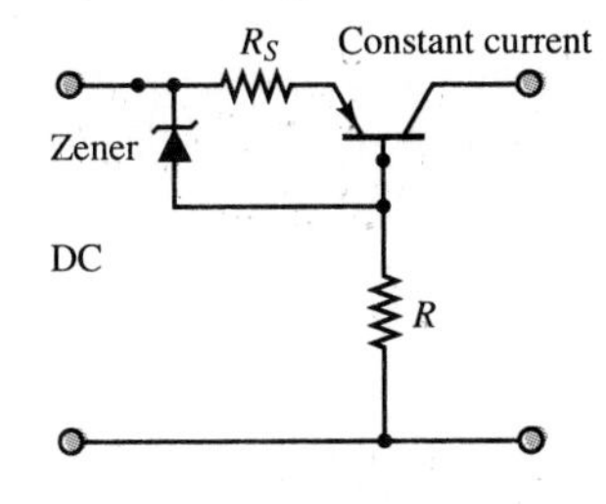

Figure P10.1

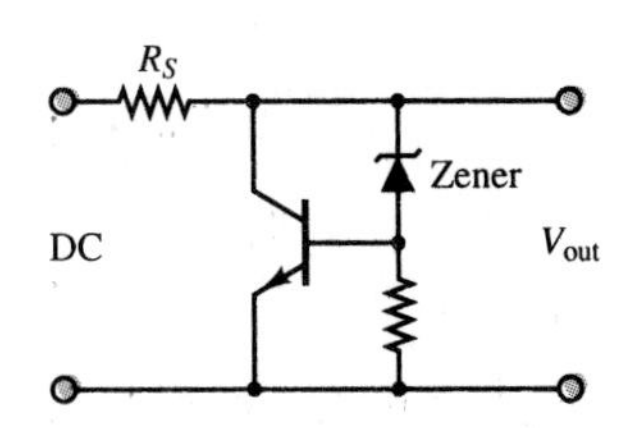

Figure P10.2

10.4 For the circuit shown in Figure 10.17, if the LR load is replaced by a capacitor, draw the output waveform and label the values.

10.5 Draw $v_L(t)$ and label the values for the circuit in Figure 10.17 if the diode forward resistance is 50 Ω, the forward bias voltage is 0.7 V, and the load consists of a resistor $R = 10\ \Omega$ and an inductor $L = 2$ H.

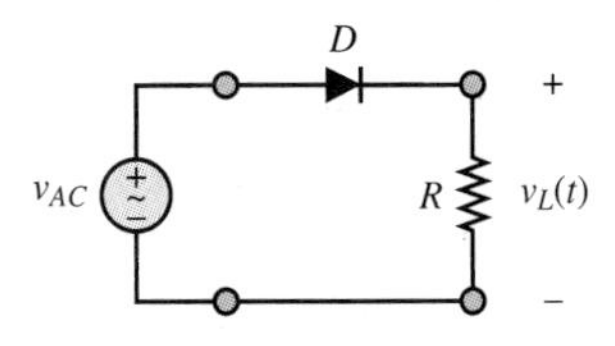

Figure P10.3

10.6 For the circuit shown in Figure P10.3, v_{AC} is a sinusoid with 10 V peak amplitude, $R = 2$ kΩ, and the forward-conducting voltage of D is 0.7 V.

a. Sketch the waveform of $v_L(t)$.
b. Find the average value of $v_L(t)$.

10.7 A vehicle battery charge circuit is shown in Figure P10.4. Describe the circuit, and draw the output waveform (L_1 and L_2 represent the inductances of the windings of the alternator).

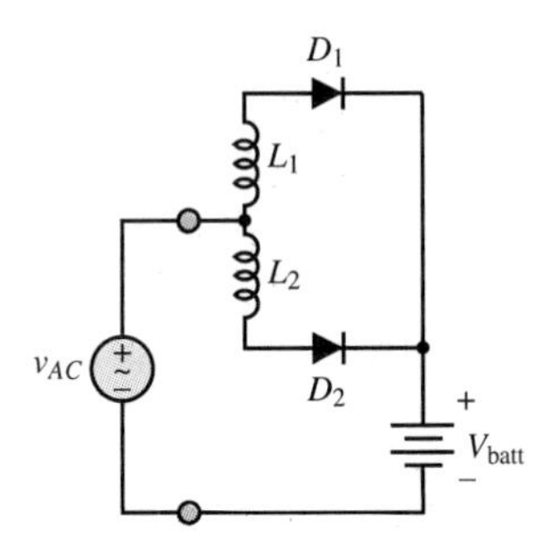

Figure P10.4

10.8 Repeat Example 10.2 for $\alpha = \pi/3$ and $\pi/6$.

10.9 The circuit shown in Figure P10.5 is a speed control system for a DC motor. Assume that the thyristors are fired at $\alpha = 60°$ and that the motor current is 20 A and is ripple free. Assume $v_{AC} = 80\ V_{rms}$.

a. Sketch the output voltage waveform, v_o.
b. Compute the power absorbed by the motor.
c. Determine the volt-amperes generated by the supply.

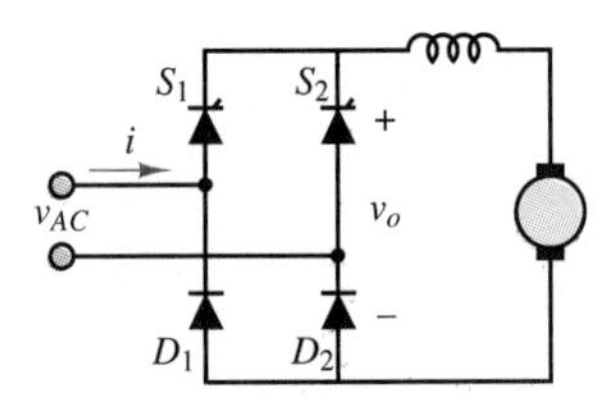

Figure P10.5

10.10 A full wave, single-phase controlled rectifier is used to control the speed of a DC motor. The circuit is similar to that of Figure 10.2, except for replacing the resistive load with a DC motor. The motor operates at 110 V and absorbs 4 kW of power. The AC supply is 120 V, 60 Hz. Assume that the motor inductance is very large (i.e., the motor current is ripple free), and that the motor constant is 0.055 V/rev/min. If the motor runs at 1,000 rev/min at rated current:

a. Determine the firing angle of the converter.
b. Determine the rms value of the supply current.

10.11 For the light dimmer circuit of Example 10.2, determine the load power at firing angles $\alpha = 0°, 30°, 60°, 90°, 120°, 150°, 180°$, and plot the load power as a function of α.

10.12 The chopper of Figure 10.35 is used to control the speed of a DC motor. Let the supply voltage be 120 V and the armature resistance of the motor be 0.15 Ω. The motor back emf constant is 0.05 V/rev/min and the chopper frequency is 250 Hz. Assume that the motor current is free of ripple and equal to 125 A at 120 rev/min.

a. Determine the duty cycle of the chopper, δ, and the chopper on time, t_1.
b. Determine the power absorbed by the motor.
c. Determine the power generated by the supply.

10.13 The circuit of Figure 10.39 is used to provide regenerative braking in a traction motor. The motor constant is 0.3 V/rev/min, $R_a = 0.2\ \Omega$, and the supply voltage is 600 V. If the motor speed is 800 rev/min and the motor current is 300 A:

a. Determine the duty cycle, δ, of the chopper.
b. Determine the power fed back to the supply (battery).

CHAPTER 11

Operational Amplifiers

Chapter 11 introduces a building block of modern electronic instrumentation. As will be shown, electronic amplifiers are not limited to the amplification of electrical signals; they form the basis of many other instrumentation circuits—for example, active filters. In this chapter we will analyze the properties of the ideal amplifier and then carefully explore the features of a general-purpose amplifier circuit known as the *operational amplifier*. Understanding the gain and frequency response properties of the operational amplifier is essential for the user of electronic instrumentation. Fortunately, the availability of operational amplifiers in integrated circuit form has made the task of analyzing such circuits quite simple. The models presented in this chapter are based on concepts that have already been explored at length in earlier chapters, namely, Thévenin and Norton equivalent circuits and frequency response.

Mastery of operational amplifier fundamentals is essential in any practical application of electronics. This chapter is aimed at developing your understanding of the fundamental properties of practical operational amplifiers. A number of useful applications are introduced in the examples and homework problems. Upon completion of the chapter, you should be able to:

- Analyze and design simple signal-conditioning circuits based on op-amps.
- Analyze and design simple active filters.
- Understand the operation of analog computers.
- Assess and understand the practical limitations of operational amplifiers.

11.1 AMPLIFIERS

One of the most important functions in electronic instrumentation is that of amplification. The need to amplify low-level electrical signals arises frequently in a number of applications. Perhaps the most familiar use of amplifiers arises in converting the low-voltage signal from a cassette tape player, a turntable, or a compact disk player to a level suitable for driving a pair of speakers. Figure 11.1 depicts a typical arrangement. Amplifiers have a number of applications of interest to the non–electrical engineer, such as the amplification of low-power signals from transducers (e.g., bioelectrodes, strain gauges, thermistors, and accelerometers) and other, less obvious functions that will be reviewed in this chapter—for example, filtering and impedance isolation. We turn first to the general features and characteristics of amplifiers, before delving into the analysis of the operational amplifier.

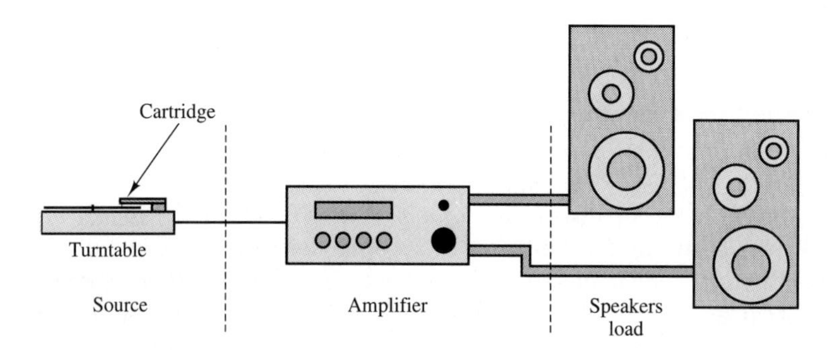

Figure 11.1

Ideal Amplifier Characteristics

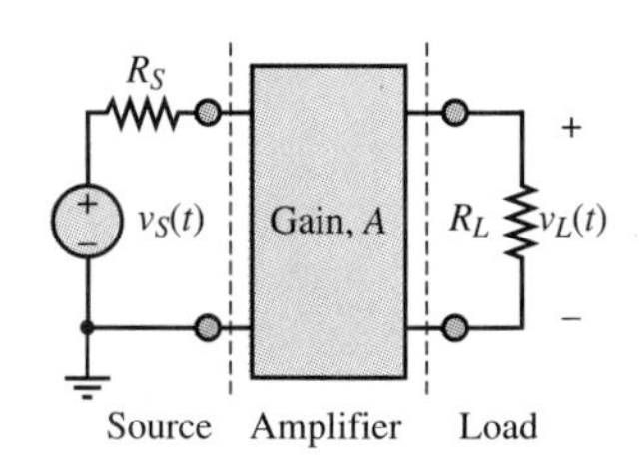

Figure 11.2 A voltage amplifier

The simplest model for an amplifier is depicted in Figure 11.2, where a signal, $v_S(t)$, is shown being amplified by a constant factor A, called the *gain* of the amplifier. Ideally, the load voltage should be given by the expression

$$v_L(t) = Av_S(t) \tag{11.1}$$

Note that the source has been modeled as a Thévenin equivalent, and the load as an equivalent resistance. Thévenin's theorem guarantees that this picture can be representative of more complex circuits. Hence, the equivalent source circuit is the circuit the amplifier "sees" from its input port; and R_L, the load, is the equivalent resistance seen from the output port of the amplifier.

What would happen if the roles were reversed? That is, what does the source see when it "looks" into the input port of the amplifier, and what does the load see when it "looks" into the output port of the amplifier? While it is not clear at this point how one might characterize the internal circuitry of an amplifier (which is rather complex), it can be presumed that the amplifier will act as an equivalent load with respect to the source, and as an equivalent source with respect to the load. After all, this is a direct application of Thévenin's theorem. Figure 11.3 provides a pictorial representation of this simplified characterization of an amplifier. The "black box" of Figure 11.2 is now represented as an equivalent circuit with the following behavior. The input circuit has equivalent resistance R_{in}, so that the input voltage, v_{in}, is given by

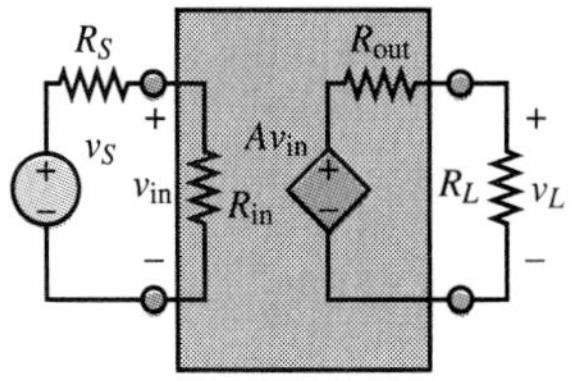

Figure 11.3 Simple voltage amplifier model

$$v_{\text{in}} = \frac{R_{\text{in}}}{R_S + R_{\text{in}}} v_S \tag{11.2}$$

The equivalent input voltage seen by the amplifier is then amplified by a constant factor, A. This is represented by the controlled voltage source Av_{in}. The controlled source appears in series with an internal resistor, R_{out}, denoting the internal (output) resistance of the amplifier. Thus, the voltage presented to the load is

$$v_L = Av_{\text{in}} \frac{R_L}{R_{\text{out}} + R_L} \tag{11.3}$$

or, substituting the equation for v_{in},

$$v_L = \left(A \frac{R_{\text{in}}}{R_S + R_{\text{in}}} \frac{R_L}{R_{\text{out}} + R_L} \right) v_S \tag{11.4}$$

In other words, the load voltage is an amplified version of the source voltage.

Unfortunately, the amplification factor is now dependent on both the source and load impedances, and on the input and output resistance of the amplifier. Thus, a given amplifier would perform differently with different loads or sources. What are the desirable characteristics for a voltage amplifier that would make its performance relatively independent of source and load impedances? Consider, once again, the expression for v_{in}. If the input resistance of the amplifier, R_{in}, were very large, the source voltage, v_S, and the input voltage, v_{in}, would be approximately equal:

$$v_{\text{in}} \approx v_S \tag{11.5}$$

since

$$\lim_{R_{\text{in}} \to \infty} \left(\frac{R_{\text{in}}}{R_{\text{in}} + R_S} \right) = 1 \tag{11.6}$$

By an analogous argument, it can also be seen that the desired output resistance for the amplifier, R_{out}, should be very small, since for an amplifier with $R_{\text{out}} = 0$, the load voltage would be

$$v_L = Av_{\text{in}} \tag{11.7}$$

Combining these two results, we can see that as R_{in} approaches infinity and R_{out} approaches zero, the ideal amplifier magnifies the source voltage by a factor of A:

$$v_L = Av_S \tag{11.8}$$

just as was indicated in the "black box" amplifier of Figure 11.2.

Thus, two desirable characteristics for a general-purpose voltage amplifier are a very large input impedance and a very small output impedance. In the next sections it will be shown how operational amplifiers provide these desired characteristics.

11.2 THE OPERATIONAL AMPLIFIER

An **operational amplifier** is an **integrated circuit,** that is, a large collection of individual electrical and electronic circuits integrated on a single silicon wafer. An operational amplifier—or op-amp—can perform a great number of operations, such as addition, filtering, or integration, which are all based on the properties of ideal amplifiers and of ideal circuit elements. The introduction of the operational amplifier in integrated circuit form marked the beginning of a new era in modern electronics. Since the introduction of the first IC op-amp, the trend in electronic instrumentation has been to move away from the discrete (individual-component) design of electronic circuits, toward the use of integrated circuits for a large number of applications. This statement is particularly true for applications of the type the non–electrical engineer is likely to encounter: op-amps are found in most measurement and instrumentation applications, serving as extremely versatile building blocks for any application that requires the processing of electrical signals.

In the following pages, simple circuit models of the op-amp will be introduced. The simplicity of the models will permit the use of the op-amp as a circuit element, or building block, without the need to describe its internal workings in detail. Integrated circuit technology has today reached such an advanced stage of development that it can be safely stated that for the purpose of many instrumentation applications, the op-amp can be treated as an ideal device. Following the introductory material presented in this chapter, more advanced instrumentation applications will be explored in Chapter 14.

The Open-Loop Model

The ideal operational amplifier behaves very much as an ideal **difference amplifier,** that is, a device that amplifies the difference between two input voltages. Operational amplifiers are characterized by near-infinite input resistance and very small output resistance. As shown in Figure 11.4, the output of the op-amp is an amplified version of the difference between the voltages present at the two inputs:[1]

$$v_{\text{out}} = A_{V(OL)}(v^+ - v^-) \tag{11.9}$$

---

[1]The amplifier of Figure 11.4 is a *voltage amplifier;* another type of operational amplifier, called a *current* or *transconductance amplifier*, is described in the homework problems.

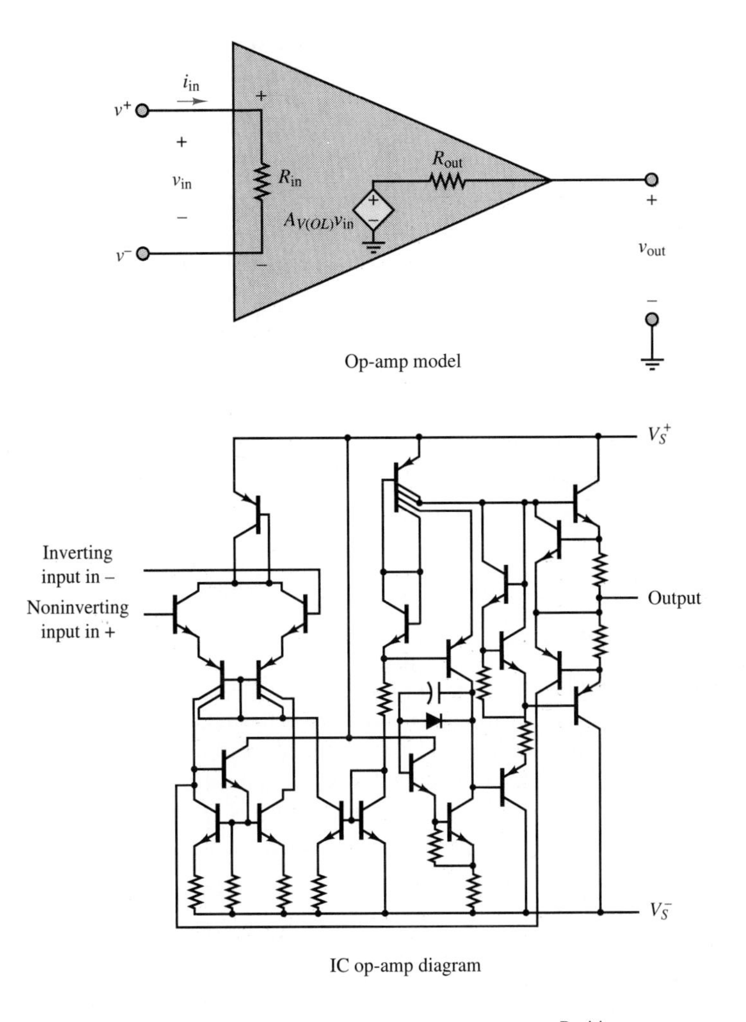

Offset null 1
Inverting input 2
Noninverting input 3
V_S^- 4
8 No connection
7 V_S^+
6 Output
5 Offset null

Integrated circuit operational amplifier (IC op-amp)

Positive power supply
V_S^+
Inverting input
Output
Noninverting input
V_S^-
Negative power supply

Simplified circuit symbol

Figure 11.4 Operational amplifier model symbols, and circuit diagram

The input denoted by a positive sign is called the **noninverting input** (or terminal), while that represented with a negative sign is termed the **inverting input** (or terminal). The amplification factor, or gain, $A_{V(OL)}$, is called the **open-loop voltage gain** and is quite large by design, typically of the order of 10^5 to 10^7; it will soon become apparent why a large open-loop gain is a desirable characteristic. Together with the high input resistance and low output resistance, the effect of a large amplifier open-loop voltage gain, $A_{V(OL)}$, is such that op-amp circuits can be designed to perform very nearly as ideal voltage or current amplifiers. In effect, to analyze the performance of an op-amp circuit, only one assumption will be needed: that the current flowing into the input circuit of the amplifier is zero, or

$$i_{in} = 0 \tag{11.10}$$

This assumption is justified by the large input resistance and large open-loop gain of the operational amplifier. The model just introduced will be used to analyze three amplifier circuits in the next part of this section.

The Operational Amplifier in the Closed-Loop Mode

The Inverting Amplifier

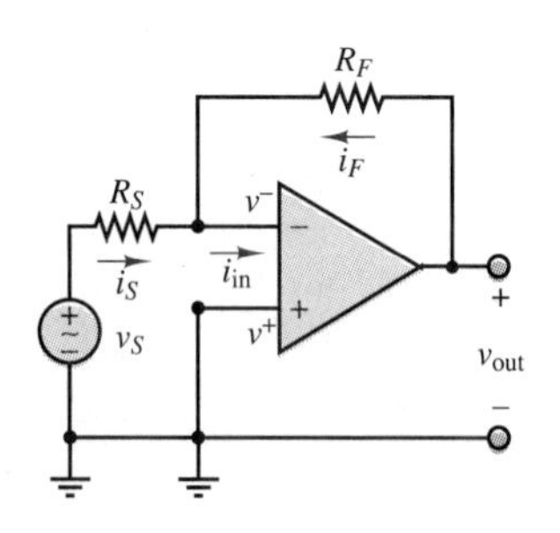

Figure 11.5 Inverting amplifier

One of the more popular circuit configurations of the op-amp, because of its simplicity, is the so-called inverting amplifier, shown in Figure 11.5. The input signal to be amplified is connected to the inverting terminal, while the noninverting terminal is grounded. It will now be shown how it is possible to choose an (almost) arbitrary gain for this amplifier by selecting the ratio of two resistors. The analysis is begun by noting that at the inverting input node, KCL requires that

$$i_S + i_F = i_{in} \tag{11.11}$$

The current i_F, which flows back to the inverting terminal from the output, is appropriately termed **feedback current,** because it represents an input to the amplifier that is "fed back" from the output. Applying Ohm's law, we may determine each of the three currents shown in Figure 11.5:

$$i_S = \frac{v_S - v^-}{R_S} \qquad i_F = \frac{v_{out} - v^-}{R_F} \qquad i_{in} = 0 \tag{11.12}$$

(the last by assumption, as stated earlier). The voltage at the noninverting input, v^+, is easily identified as zero, since it is directly connected to ground: $v^+ = 0$. Now, the **open-loop model for the op-amp** requires that

$$v_{out} = A_{V(OL)}(v^+ - v^-) = -A_{V(OL)}v^- \tag{11.13}$$

or

$$v^- = -\frac{v_{out}}{A_{V(OL)}} \tag{11.14}$$

Having solved for the voltage present at the inverting input, v^-, in terms of v_{out}, we may now compute an expression for the amplifier gain, v_{out}/v_S. This quantity is called the **closed-loop gain,** because the presence of a feedback connection between the output and the input constitutes a closed loop.[2] Combining equations 11.11 and 11.12, we can write

$$i_S = -i_F \tag{11.15}$$

and

$$\frac{v_S}{R_S} + \frac{v_{out}}{A_{V(OL)}R_S} = -\frac{v_{out}}{R_F} - \frac{v_{out}}{A_{V(OL)}R_F} \tag{11.16}$$

leading to the expression

$$\frac{v_S}{R_S} = -\frac{v_{out}}{R_F} - \frac{v_{out}}{A_{V(OL)}R_F} - \frac{v_{out}}{A_{V(OL)}R_S} \tag{11.17}$$

or

$$v_S = -v_{out}\left(\frac{1}{R_F/R_S} + \frac{1}{A_{V(OL)}R_F/R_S} + \frac{1}{A_{V(OL)}}\right) \tag{11.18}$$

If the open-loop gain of the amplifier, $A_{V(OL)}$, is sufficiently large, the terms $\frac{1}{A_{V(OL)}R_F/R_S}$ and $1/A_{V(OL)}$ are essentially negligible, compared with $1/(R_F/R_S)$. As stated earlier, typical values of $A_{V(OL)}$ range from 10^5 to 10^7, and thus it is reasonable to conclude that, to a close approximation, the following expression describes the closed-loop gain of the inverting amplifier:

$$v_{out} = -\frac{R_F}{R_S}v_S \quad \text{Inverting amplifier closed-loop gain} \tag{11.19}$$

That is, the closed-loop gain of an inverting amplifier may be selected simply by the appropriate choice of two externally connected resistors. The price for this extremely simple result is an inversion of the output with respect to the input—that is, a negative sign.

Next, we show that by making an additional assumption it is possible to simplify the analysis considerably. Consider that, as was shown for the inverting amplifier, the inverting terminal voltage is given by

$$v^- = -\frac{v_{out}}{A_{V(OL)}} \tag{11.20}$$

Clearly, as $A_{V(OL)}$ approaches infinity, the inverting-terminal voltage is going to be very small (practically, of the order of microvolts). It may then be assumed that *in the inverting amplifier*, v^- is virtually zero:

$$v^- \approx 0 \tag{11.21}$$

This assumption prompts an interesting observation (which may not yet appear obvious at this point):

[2]This terminology is borrowed from the field of automatic controls, of which the theory of closed-loop feedback systems forms the foundation.

The effect of the feedback connection from output to inverting input is to force the voltage at the inverting input to be equal to that at the noninverting input.

This is equivalent to stating that for an op-amp *with negative feedback*,

$$v^- \approx v^+ \tag{11.22}$$

Why Feedback?

Why is such emphasis placed on the notion of an amplifier with a very large open-loop gain and with negative feedback? Why not just design an amplifier with a reasonable gain, say, ×10, or ×100, and just use it as such, without using feedback connections? In these paragraphs, we hope to answer these and other questions, introducing the concept of **negative feedback** in an intuitive fashion.

The fundamental reason for designing an amplifier with a very large open-loop gain is the flexibility it provides in the design of amplifiers with an (almost) arbitrary gain; it has already been shown that the gain of the inverting amplifier is determined by the choice of two external resistors—undoubtedly a convenient feature! It is important to appreciate that negative feedback is the mechanism that enables us to enjoy such flexibility in the design of linear amplifiers.

To understand the role of feedback in the operational amplifier, consider the internal structure of the op-amp shown in Figure 11.4. The large open-loop gain causes any difference in voltage at the input terminals to appear greatly amplified at the output. When a negative feedback connection is provided, as shown, for example, in the inverting amplifier of Figure 11.5, the output voltage, v_{out}, causes a current, i_F, to flow through the feedback resistance so that KCL is satisfied at the inverting node. Assume, for a moment, that the differential voltage

$$\Delta v = v^+ - v^-$$

is identically zero. Then, the output voltage will continue to be such that KCL is satisfied at the inverting node, that is, such that the current i_F is equal to the current i_S.

Suppose, now, that a small imbalance in voltage, Δv, is present at the input to the op-amp. Then the output voltage will be increased by an amount $A_{V(OL)}\Delta v$. Thus, an incremental current approximately equal to $A_{V(OL)}\Delta v/R_F$ will flow from output to input via the feedback resistor. The effect of this incremental current is to reduce the voltage difference Δv to zero, so as to restore the original balance in the circuit. One way of viewing negative feedback, then, is to consider it a self-balancing mechanism, which allows the amplifier to preserve zero potential difference between its input terminals.

A practical example that illustrates a common application of negative feedback is the thermostat. This simple temperature control system operates by comparing the desired ambient temperature and the temperature measured by a thermometer and turns a heat source on and off to maintain the difference between actual and desired temperature as close to zero as possible. An analogy may be made with the inverting amplifier if we consider that, in this case, negative feedback is used to keep the inverting-terminal voltage as close as possible to the noninverting-terminal voltage. The latter voltage is analogous to the desired ambient temperature in your home, while the former plays a role akin to that of the actual ambient temperature. The open-loop gain of the amplifier forces the two voltages to be close to each other, much the way the furnace raises the heat in the house to match the desired ambient temperature.

The analysis of the operational amplifier can now be greatly simplified if the following two assumptions are made:

1. $i_{in} = 0$
2. $v^- = v^+$ **(11.23)**

This technique will be tested in the next subsection by analyzing a noninverting amplifier configuration.

Example 11.1

Consider the inverting amplifier circuit of Figure 11.5. If $v_S(t) = 0.015 \cos(50t)$, $R_F = 10\ \text{k}\Omega$, and $R_S = 1\ \text{k}\Omega$, what is the output voltage, $v_{out}(t)$? Sketch both input and output voltage.

Solution:

Let $|v_{out}|$ and $|v_S|$ denote the peak output and source voltages, respectively. Then,

$$|v_{out}| = -\left(\frac{R_F}{R_S}\right)|v_S| = -10 \times 0.015 = -150\ \text{mV}$$

and

$$v_{out}(t) = -0.15 \cos(50t)$$

The input and output voltage waveforms are shown in Figure 11.6.

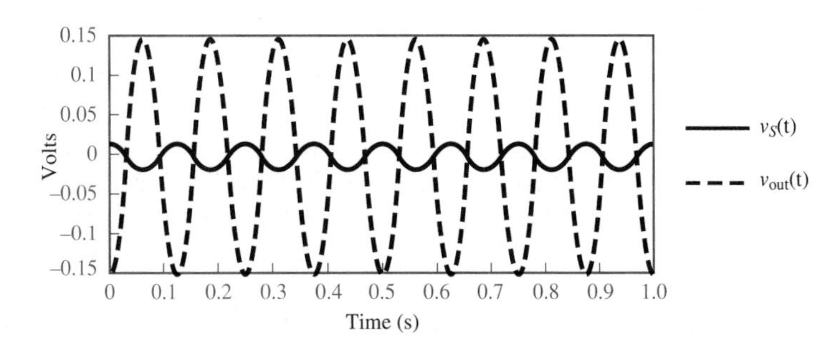

Figure 11.6

A useful op-amp circuit that is based on the inverting amplifier is the **op-amp summer,** or **summing amplifier.** This circuit, shown in Figure 11.7, is used to add signal sources. The primary advantage of using the op-amp as a summer is that the summation occurs independently of load and source impedances, so that sources with different internal impedances will not interact with each other.

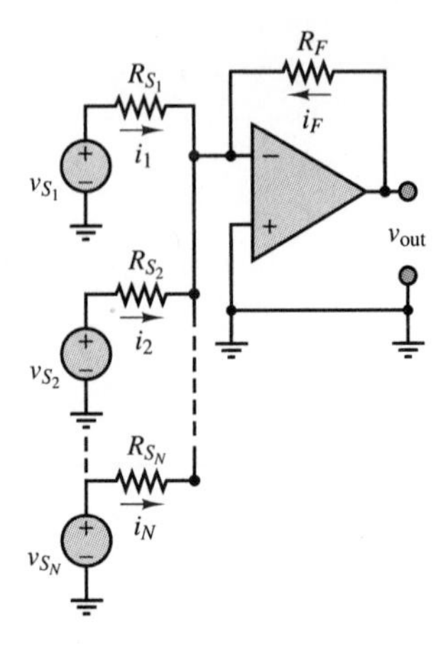

Figure 11.7 Summing amplifier

The operation of the summing amplifier is best understood by application of KCL at the inverting node: the sum of the N source currents and the feedback current must equal zero, so that

$$i_1 + i_2 + \cdots + i_N = -i_F \tag{11.24}$$

But each of the source currents is given by the following expression:

$$i_n = \frac{v_{S_n}}{R_{S_n}} \qquad n = 1, 2, \ldots, N \tag{11.25}$$

while the feedback current is

$$i_F = \frac{v_{out}}{R_F} \tag{11.26}$$

Combining equations 11.25 and 11.26, and using equation 11.15, we obtain the following result:

$$\sum_{n=1}^{N} \frac{v_{S_n}}{R_{S_n}} = -\frac{v_{out}}{R_F} \tag{11.27}$$

or

$$v_{out} = -\sum_{n=1}^{N} \frac{R_F}{R_{S_n}} v_{S_n} \tag{11.28}$$

That is, the output consists of the weighted sum of N input signal sources, with the weighting factor for each source equal to the ratio of the feedback resistance to the source resistance.

The Noninverting Amplifier

To avoid the negative gain (i.e., phase inversion) introduced by the inverting amplifier, a **noninverting amplifier** configuration is often employed. A typical noninverting amplifier is shown in Figure 11.8; note that the input signal is applied to the noninverting terminal this time.

The noninverting amplifier can be analyzed in much the same way as the inverting amplifier. Writing KCL at the inverting node yields

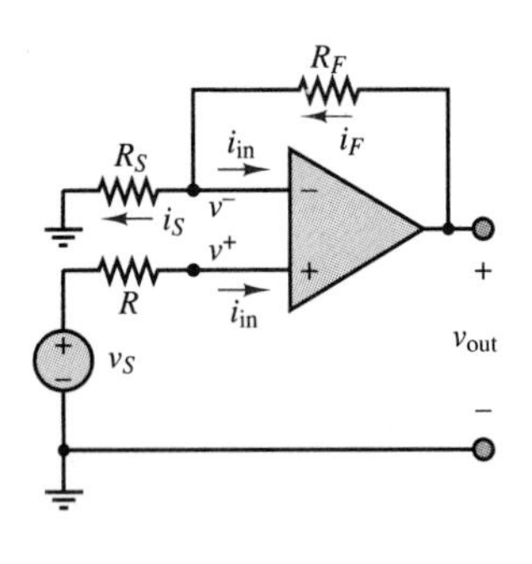

Figure 11.8 Noninverting amplifier

$$i_F = i_S + i_{in} \approx i_S \tag{11.29}$$

where

$$i_F = \frac{v_{out} - v^-}{R_F} \tag{11.30}$$

$$i_S = \frac{v^-}{R_S} \tag{11.31}$$

Now, since $i_{in} = 0$, the voltage drop across the source resistance, R, is equal to zero. Thus,

$$v^+ = v_s \quad \textbf{(11.32)}$$

and, using equation 11.22,

$$v^- = v^+ = v_S \quad \textbf{(11.33)}$$

Substituting this result in equations 11.29 and 11.30, we can easily show that

$$i_F = i_S \quad \textbf{(11.34)}$$

or

$$\frac{v_{out} - v_S}{R_F} = \frac{v_S}{R_S} \quad \textbf{(11.35)}$$

It is easy to manipulate equation 11.35 to obtain the result

$$\frac{v_{out}}{v_S} = 1 + \frac{R_F}{R_S} \quad \text{Noninverting amplifier closed-loop gain} \quad \textbf{(11.36)}$$

which is the closed-loop gain expression for a noninverting amplifier. Note that the gain of this type of amplifier is always positive and greater than (or equal to) 1.

The same result could have been obtained without making the assumption $v^+ = v^-$, at the expense of some additional work. The procedure one would follow in this latter case is analogous to the derivation carried out earlier for the inverting amplifier, and is left as an exercise.

In summary, in the preceding pages it has been shown that by constructing a nonideal amplifier with very large gain and near-infinite input resistance, it is possible to design amplifiers that have near-ideal performance and provide a variable range of gains, easily controlled by the selection of external resistors. The mechanism that allows this is negative feedback. From here on, unless otherwise noted, it will be reasonable and sufficient in analyzing new op-amp configurations to utilize the two assumptions

1. $i_{in} = 0$
2. $v^- = v^+$

Approximations used for ideal op-amps with negative feedback **(11.37)**

Example 11.2

The circuit of Figure 11.9 is called a **voltage-follower** circuit. This terminology is due to the fact that the output voltage is always equal to the input voltage, that is, the output

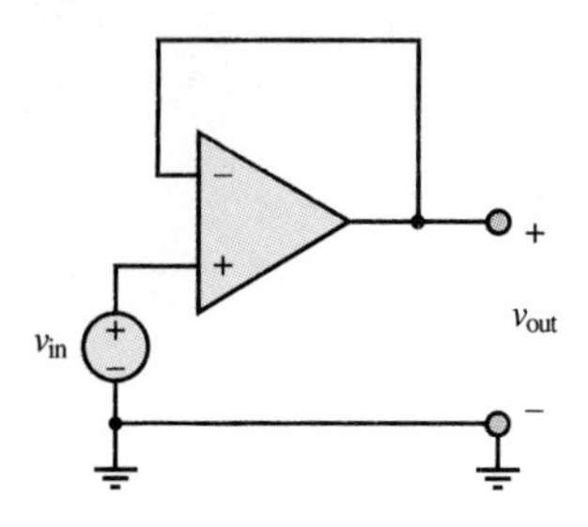

Figure 11.9 Voltage follower

"follows" the input. Using the two assumptions just stated for op-amp analyses in this text, show that the output voltage does in fact equal the input voltage.

Solution:

We have assumed that the current $i_{in} = 0$ and that $v^+ = v^-$. Therefore, with $v^- = v^+ = v_{in}$, and since v_{out} is short-circuited to v^-,

$$v_{out} = v_{in}$$

This amplifier circuit is commonly used to isolate circuits from each other. An application of this circuit is illustrated in the homework problems.

The Differential Amplifier

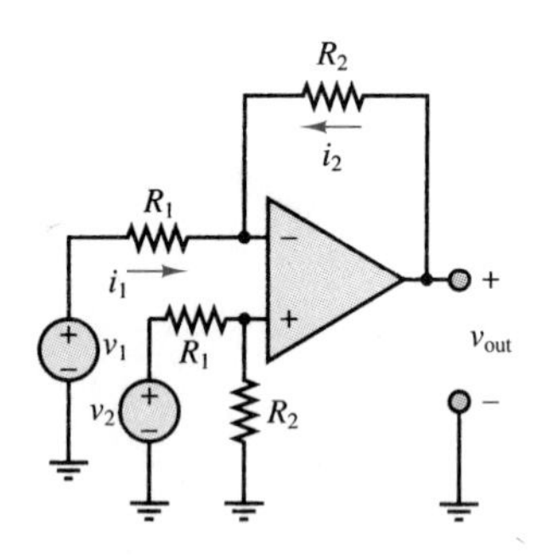

Figure 11.10 Differential amplifier

The third closed-loop model examined in this chapter is a combination of the inverting and noninverting amplifiers; it finds frequent use in situations where the difference between two signals needs to be amplified. The basic **differential amplifier** circuit is shown in Figure 11.10, where the two sources, v_1 and v_2, may be independent of each other, or may originate from the same process, as they do in Example 11.3.

The analysis of the differential amplifier may be approached by various methods; the one we select to use at this stage consists of:

1. Computing the noninverting- and inverting-terminal voltages, v^+ and v^-.
2. Equating the inverting and noninverting input voltages, $v^- = v^+$.
3. Applying KCL at the inverting node, where $i_2 = -i_1$.

Since it has been assumed that no current flows into the amplifier, the noninverting-terminal voltage is given by the following voltage divider:

$$v^+ = \frac{R_2}{R_1 + R_2} v_2 \qquad \textbf{(11.38)}$$

If the inverting-terminal voltage is assumed equal to v^+, then the currents i_1 and i_2 are found to be

$$i_1 = \frac{v_1 - v^+}{R_1} \qquad \textbf{(11.39)}$$

and

$$i_2 = \frac{v_{out} - v^+}{R_2} \qquad \textbf{(11.40)}$$

and since

$$i_2 = -i_1 \qquad \textbf{(11.41)}$$

the following expression for the output voltage is obtained:

$$\begin{aligned} v_{\text{out}} &= R_2\left[\frac{-v_1}{R_1} + \frac{1}{R_1 + R_2}v_2 + \frac{R_2}{R_1(R_1 + R_2)}v_2\right] \\ &= \frac{R_2}{R_1}(v_2 - v_1) \end{aligned} \quad \textbf{(11.42)}$$

Thus, the differential amplifier magnifies the difference between the two input signals by the closed-loop gain R_2/R_1.

In practice, it is often necessary to amplify the difference between two signals that are both corrupted by noise or some other form of interference. In such cases, the differential amplifier provides an invaluable tool in amplifying the desired signal while rejecting the noise. Example 11.3 provides a realistic look at a very common application of the differential amplifier.

Example 11.3 An EKG Amplifier

This example illustrates the principle behind a two-lead electrocardiogram (EKG) measurement. The desired cardiac waveform is given by the difference between the potentials measured by two electrodes suitably placed on the patient's chest, as shown in Figure 11.11. A healthy, noise-free EKG waveform, $v_1 - v_2$, is shown in Figure 11.12.

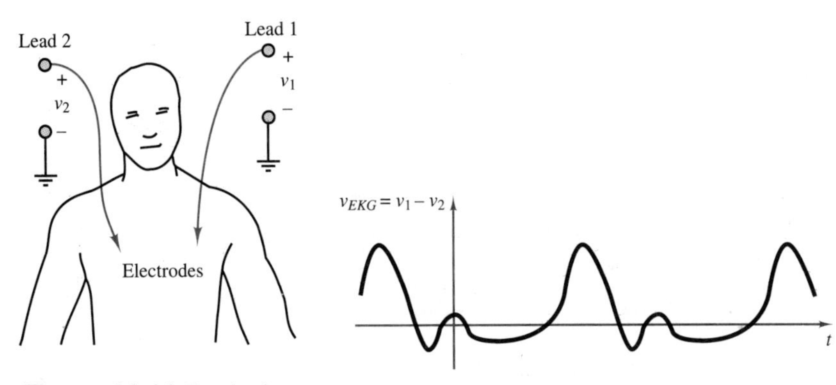

Figure 11.11 Two-lead electrocardiogram

Figure 11.12 EKG waveform

Unfortunately, the presence of electrical equipment powered by the 60-Hz, 110-VAC line current causes undesired interference at the electrode leads: the lead wires act as antennas and pick up some of the 60-Hz signal in addition to the desired EKG voltage. In effect, instead of recording the desired EKG signals, v_1 and v_2, the two electrodes provide the following inputs to the EKG amplifier, shown in Figure 11.13:

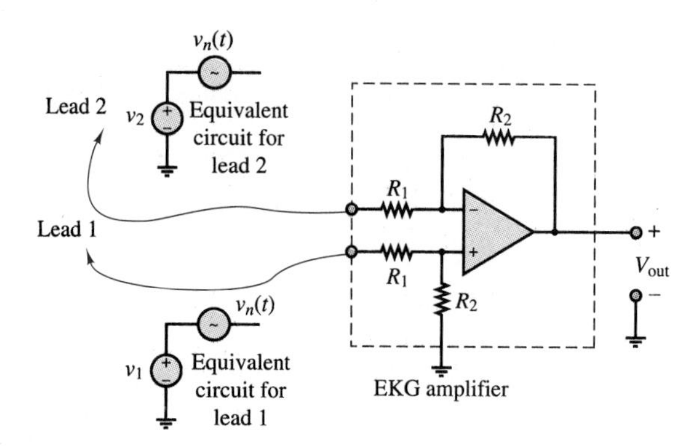

Figure 11.13 EKG amplifier

Lead 1:

$$v_1(t) + v_n(t) = v_1(t) + V_n \cos(377t + \phi_n)$$

Lead 2:

$$v_2(t) + v_n(t) = v_2(t) + V_n \cos(377t + \phi_n)$$

The interference signal, $V_n \cos(377t + \phi_n)$, is approximately the same at both leads, because the electrodes are chosen to be identical (e.g., they have the same lead lengths) and are in close proximity to each other. Further, the nature of the interference signal is such that it is common to both leads, since it is a property of the environment the EKG instrument is embedded in. On the basis of the analysis presented earlier, then,

$$v_{out} = \frac{R_2}{R_1}\left[(v_1 + v_n(t)) - (v_2 + v_n(t))\right]$$

or

$$v_{out} = \frac{R_2}{R_1}(v_1 - v_2)$$

Thus, the differential amplifier nullifies the effect of the 60-Hz interference, while amplifying the desired EKG waveform.

The preceding example introduced the concept of so-called **common-mode** and **differential-mode signals.** In the EKG example, the desired differential-mode EKG signal was amplified by the op-amp while the common-mode disturbance was canceled. Thus, the differential amplifier provides the ability to reject common-mode signal components (such as noise or undesired DC offsets) while amplifying the differential-mode components. This is a very desirable feature in instrumentation systems. In practice, rejection of the common-mode signal is not always complete: some of the common-mode signal component will always appear in the output. This fact gives rise to a figure of merit called the *common-mode rejection ratio*, which is discussed in Section 11.6.

Often, to provide impedance isolation between bridge transducers and the differential amplifier stage, the signals v_1 and v_2 are amplified separately. This technique gives rise to the so-called **instrumentation amplifier (IA),** shown in Figure 11.14. Example 11.4 illustrates the calculation of the closed-loop gain for a typical instrumentation amplifier.

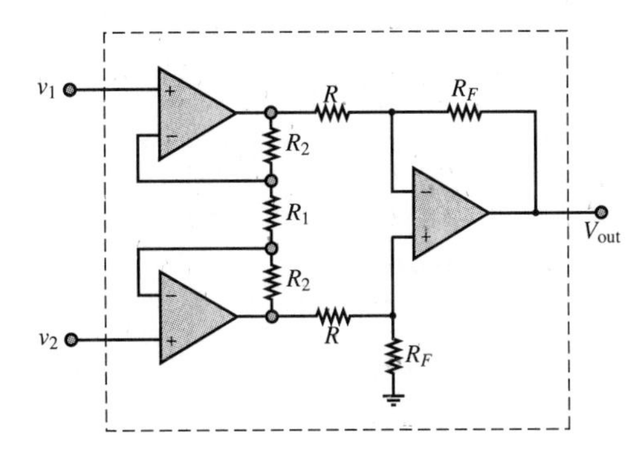

Figure 11.14 Instrumentation amplifier

Example 11.4 Instrumentation Amplifier

In this example, we compute the closed-loop gain of the instrumentation amplifier of Figure 11.14.

Solution:

To carry out the desired analysis as easily as possible, it is helpful to observe that resistor R_1 is shared by the two input amplifiers. This corresponds to having each amplifier connected to a resistor equal to $R_1/2$, as shown in Figure 11.15(a). Because of the symmetry of the circuit, one can view the shared resistor as being connected to ground in the center, so that the circuit takes the form of a noninverting amplifier, with closed-loop gain given by

$$A = 1 + \frac{2R_2}{R_1} \qquad (11.43)$$

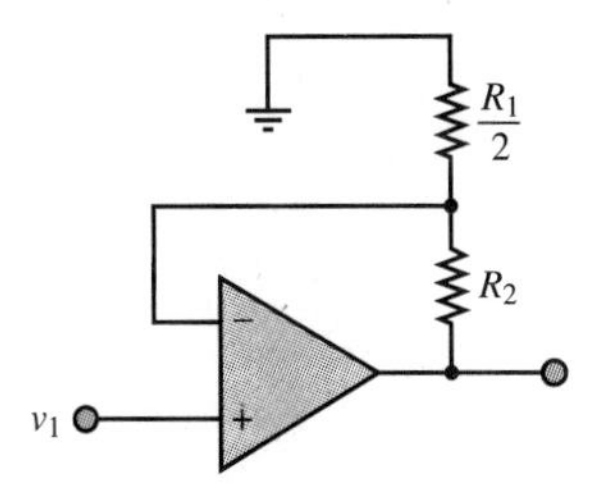

Figure 11.15(a)

Thus, each of the input voltages is amplified by this gain, and the overall gain of the instrumentation amplifier can then be computed by considering that the voltage difference $(Av_1 - Av_2)$ is then amplified by the differential amplifier stage, with gain R_F/R, as shown in Figure 11.15(b):

$$v_{out} = \frac{R_F}{R}(Av_1 - Av_2) = \frac{R_F}{R}\left(1 + \frac{2R_2}{R_1}\right)(v_1 - v_2) \qquad (11.44)$$

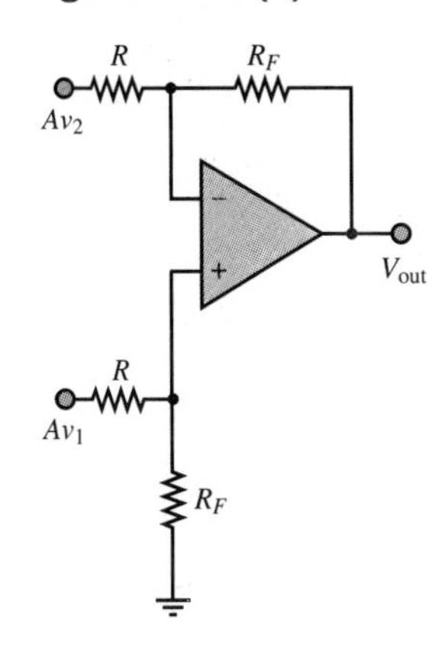

Figure 11.15(b)

Because the instrumentation amplifier has widespread application—and in order to ensure the best possible match between resistors—the entire circuit of Figure 11.14 is often packaged as a single integrated circuit. The advantage of this configuration is that the resistors R_1 and R_2 can be matched much more precisely in an integrated circuit than would be possible using discrete components. A typical, commercially available integrated circuit package is the AD625. Data sheets for this device are provided at the end of Chapter 14.

Another simple op-amp circuit that finds widespread application in electronic instrumentation is the **level shifter.** The following two examples discuss its operation and its application in a transducer calibration circuit.

Example 11.5 Level Shifter

It is often desirable to add or subtract a DC voltage to or from a signal. A simple op-amp circuit called the *level shifter* accomplishes this function very elegantly. As an illustration, consider a sensor with output given by the voltage

$$v_{\text{sensor}}(t) = 1.8 + 0.1 \cos(\omega t)$$

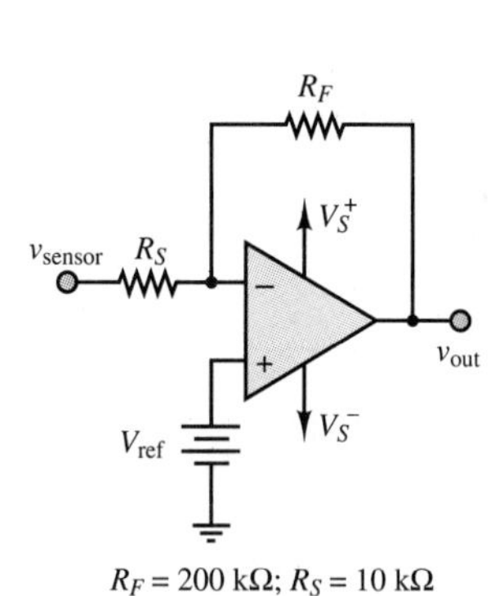

Figure 11.16 Level shifter

To remove the 1.8-VDC offset, a level shifter circuit such as the one shown in Figure 11.16 may be employed. We can see that this circuit is somewhat similar to the differential amplifier in that it has two sources. If we consider the two input sources separately, we can see by superposition that the sensor output voltage is amplified by an inverting amplifier with gain $-R_F/R_S$, while the reference battery sees a noninverting amplifier with gain $(1 + R_F/R_S)$. Thus, the complete output is given by the expression

$$v_{\text{out}} = -\frac{R_F}{R_S} v_{\text{sensor}} + \left(1 + \frac{R_F}{R_S}\right) V_{\text{ref}}$$

and substituting the expression for v_{sensor}, we obtain

$$v_{\text{out}} = -\frac{R_F}{R_S}(1.8 + 0.1 \cos \omega t) + \left(1 + \frac{R_F}{R_S}\right) V_{\text{ref}}$$

We can rewrite v_{out} as

$$v_{\text{out}} = -\frac{R_F}{R_S}(0.1) \cos \omega t - \frac{R_F}{R_S}(1.8) + \left(1 + \frac{R_F}{R_S}\right) V_{\text{ref}}$$

Since the intent is to remove the DC offset, we require that

$$-\frac{R_F}{R_S}(1.8) + \left(1 + \frac{R_F}{R_S}\right) V_{\text{ref}} = 0$$

or

$$\begin{aligned} V_{\text{ref}} &= (1.8) \frac{R_F/R_S}{1 + R_F/R_S} \\ &= (1.8) \frac{20}{1 + 20} \\ &= 1.714 \text{ V} \end{aligned}$$

It would be rather inconvenient, however, to require a variable external voltage supply or a precision battery to be included in what is ideally to be a simple (and possibly adjustable) level-shifting circuit. To solve this problem, we introduce a common technique for obtaining a variable voltage supply, using no more than a potentiometer and the ordinary voltage supplies that would be used for the op-amp package. Figure 11.17 shows the potentiometer connection between the op-amp voltage supplies. It is a simple exercise to show that the voltage V_{ref}, referenced to ground, is given by the following expression:

$$V_{\text{ref}} = \left(\frac{R + \Delta R}{2R + R_P}\right)V_S^+ + \left(\frac{R + R_P - \Delta R}{2R + R_P}\right)V_S^-$$

or, if the voltage supplies are symmetrical (i.e., if $V_S^+ = -V_S^-$), then

$$V_{\text{ref}} = V_S^+\left(\frac{2\Delta R - R_P}{2R + R_P}\right)$$

One can clearly see, then, that the potentiometer arrangement is capable of providing the desired reference voltage and may also be adjusted, if necessary. Thus, any undesired DC offset can be removed simply by adjusting a potentiometer setting.

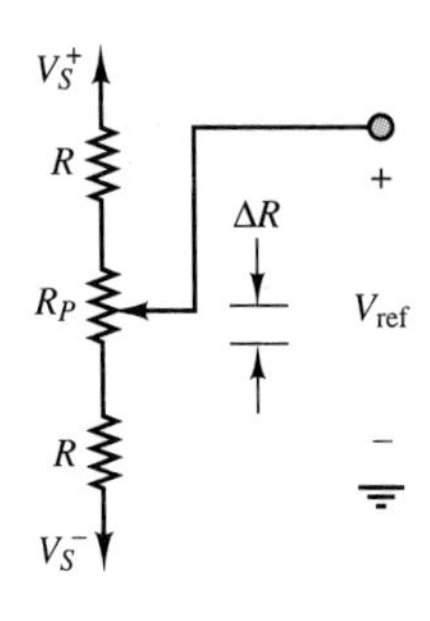

Figure 11.17

Example 11.6 Transducer Calibration Circuit

In many practical instances, the output of a transducer is related to the physical variable we wish to measure in a form that requires some signal conditioning before it is completely useful. The most desirable form is one in which the electrical output of the transducer (for example, voltage) is related to the physical variable by a constant factor. Such a relationship is depicted in Figure 11.18(a), where k is the calibration constant relating voltage to temperature. Note that k is a positive number, and that the *calibration curve* passes through the (0, 0) point. On the other hand, the transducer characteristic of Figure 11.18(b) is best described by the following equation:

$$v_{\text{sensor}} = -\beta T + V_0$$

It is possible to modify the transducer calibration curve of Figure 11.18(b) to the more desirable one of Figure 11.18(a) by means of the simple circuit displayed in Figure 11.19. This circuit provides the desired calibration constant k by a simple gain adjustment, while the zero (or bias) offset is adjusted by means of a potentiometer connected to the voltage supplies. The detailed operation of the circuit is described in the following paragraphs.

As noted before, the nonideal characteristic can be described by the following equation:

$$v_{\text{sensor}} = -\beta T + V_0$$

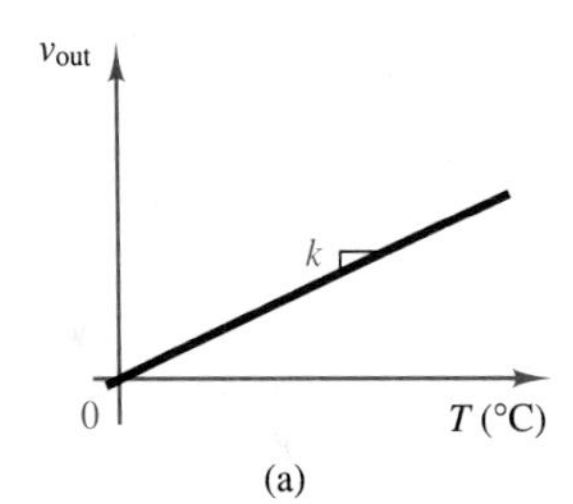

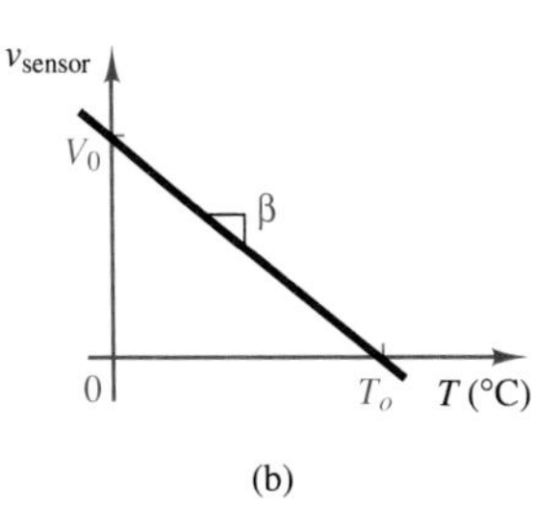

Figure 11.18 Transducer calibration curves

Then, the output of the op-amp circuit of Figure 11.19 may be determined by using the principle of superposition:

$$v_{\text{out}} = -\frac{R_F}{F_S}v_{\text{sensor}} + \left(1 + \frac{R_F}{R_S}\right)V_{\text{ref}}$$

$$= -\frac{R_F}{R_S}(-\beta T + V_0) + \left(1 + \frac{R_F}{R_S}\right)V_{\text{ref}}$$

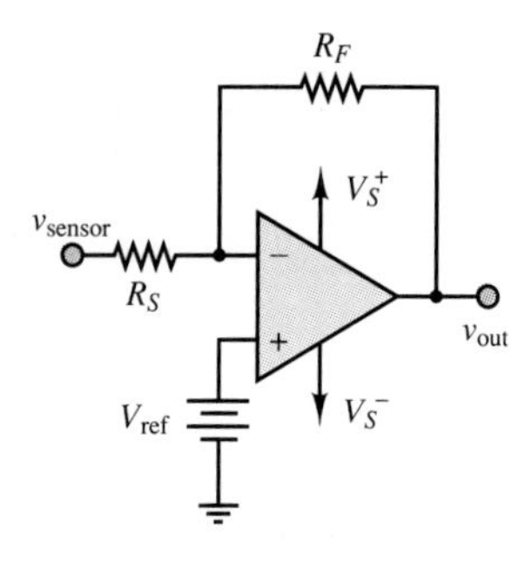

Figure 11.19 Transducer calibration circuit

After substituting the expression for the transducer voltage and after some manipulation, we see that by suitable choice of resistors, and of the reference voltage source, we can compensate for the nonideal transducer characteristic. We want the following expression to hold:

$$v_{\text{out}} = \frac{R_F}{R_S}\beta T + \left(1 + \frac{R_F}{R_S}\right)V_{\text{ref}} - \frac{R_F}{R_S}V_0 = kT$$

If we choose

$$\frac{R_F}{R_S}\beta = k$$

and

$$V_{\text{ref}} = \frac{R_F/R_S}{1 + R_F/R_S}V_0$$

then $v_{\text{out}} = kT$.

Note that

$$V_{\text{ref}} \approx V_0 \qquad \text{if} \qquad \frac{R_F}{R_S} \gg 1$$

and we can directly convert the characteristic of Figure 11.18(b) to that of Figure 11.18(a). Clearly, the effect of selecting the gain resistors is to change the magnitude of the slope of the calibration curve. The fact that the sign of the slope changes is purely a consequence of the inverting configuration of the amplifier. The reference voltage source simply shifts the DC level of the characteristic, so that the curve passes through the origin.

DRILL EXERCISES

11.1 Consider an op-amp connected in the inverting configuration with a nominal closed-loop gain of $-R_F/R_S = -1{,}000$ (this would be the gain if the op-amp had an infinite open-loop gain). Determine the value of the closed-loop gain that includes the open-loop gain as a parameter, and compute the closed-loop gain for the following values of $A_{V(OL)}$: 10^7, 10^6, 10^5, and 10^4. How large do you think the open-loop gain should be for this op-amp, to achieve the desired closed-loop gain? [*Hint:* Do not assume that $A_{V(OL)}$ is negligible in Equation 11.18.]

11.2 Repeat Drill Exercise 11.1 for $-R_F/R_S = -100$. What is the smallest value of $A_{V(OL)}$ you would recommend in this case?

11.3 Derive the result given for the differential amplifier by utilizing the principle of superposition. (Think of the differential amplifier as the combination of an inverting amplifier with input $= v_2$, plus a noninverting amplifier with input $= v_1$.)

11.4 For Example 11.5, find ΔR if the supply voltages are symmetrical at ± 15 V and a 10-kΩ potentiometer is tied to two 10-kΩ resistors.

11.5 For the circuit of Example 11.5, find the range of values of V_{ref} if the supply voltages are symmetrical at 15 V and a 1-kΩ potentiometer is tied to two 10-kΩ resistors.

11.6 Find the numerical values of R_F/R_S and V_{ref} if the temperature sensor of Example 11.6 has $\beta = 0.235$ and $V_0 = 0.7$ V and the desired relationship is $v_{\text{out}} = 10T$.

11.3 ACTIVE FILTERS

The range of useful applications of an operational amplifier is greatly expanded if energy-storage elements are introduced into the design; the frequency-dependent properties of these elements, studied in Chapter 4, will prove useful in the design of various types of op-amp circuits. In particular, it will be shown that it is possible to shape the frequency response of an operational amplifier by appropriate use of complex impedances in the input and feedback circuits. The class of filters one can obtain by means of op-amp designs is called **active filters,** because op-amps can provide amplification (gain) in addition to the filtering effects already studied in Chapter 6 for passive circuits (i.e., circuits comprising exclusively resistors, capacitors, and inductors).

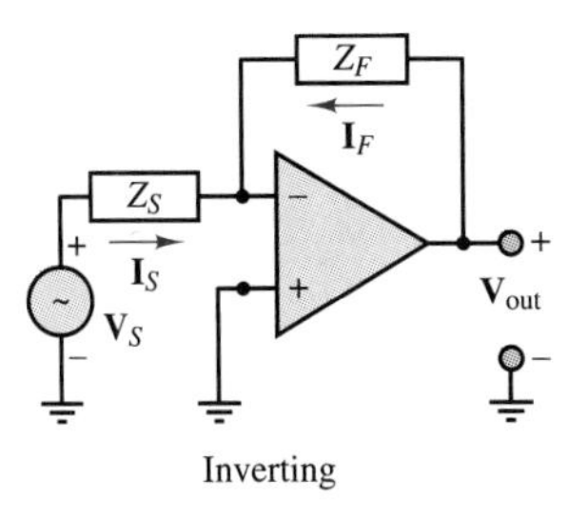

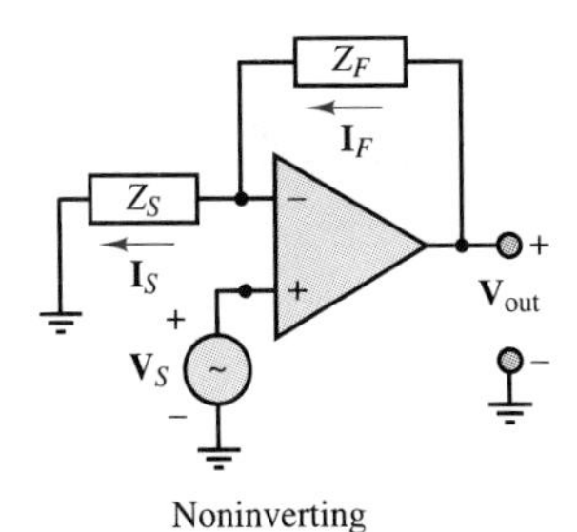

Figure 11.20 Op-amp circuits employing complex impedances

The easiest way to see how the frequency response of an op-amp can be shaped (almost) arbitrarily is to replace the resistors R_F and R_S in Figures 11.5 and 11.8 with impedances Z_F and Z_S, as shown in Figure 11.20. It is a straightforward matter to show that in the case of the inverting amplifier, the expression for the closed loop gain is given by

$$\frac{\mathbf{V}_{\text{out}}}{\mathbf{V}_S}(j\omega) = -\frac{Z_F}{Z_S} \tag{11.45}$$

whereas for the noninverting case, the gain is

$$\frac{\mathbf{V}_{\text{out}}}{\mathbf{V}_S}(j\omega) = 1 + \frac{Z_F}{Z_S} \tag{11.46}$$

where Z_F and Z_S can be arbitrarily complex impedance functions and where $\mathbf{V}_S$, $\mathbf{V}_{\text{out}}$, $\mathbf{I}_F$, and $\mathbf{I}_S$ are all phasors. Thus, it is possible to shape the frequency response of an ideal op-amp filter simply by selecting suitable ratios of feedback impedance to source impedance. By connecting a circuit similar to the low-pass filters studied in Chapter 6 in the feedback loop of an op-amp, the same filtering effect can be achieved and, in addition, the signal can be amplified.

The simplest op-amp low-pass filter is shown in Figure 11.21. Its analysis is quite simple if we take advantage of the fact that the closed-loop gain, as a function of frequency, is given by

$$A_{LP}(j\omega) = -\frac{Z_F}{Z_S} \tag{11.47}$$

where

$$Z_F = R_F \,\|\, \frac{1}{j\omega C_F} = \frac{R_F}{1 + j\omega C_F R_F} \tag{11.48}$$

and

$$Z_S = R_S \tag{11.49}$$

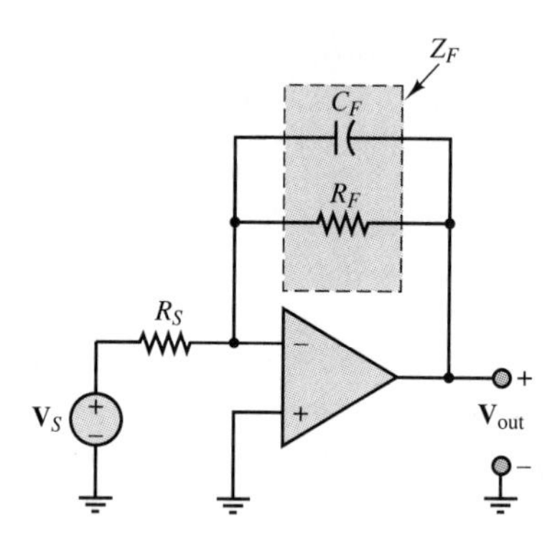

Figure 11.21 Active low-pass filter

Note the similarity between Z_F and the low-pass characteristic of the passive RC circuit! The closed-loop gain $A_{LP}(j\omega)$ is then computed to be

$$A_{LP}(j\omega) = -\frac{Z_F}{Z_S} = -\frac{R_F/R_S}{1 + j\omega C_F R_F} \tag{11.50}$$

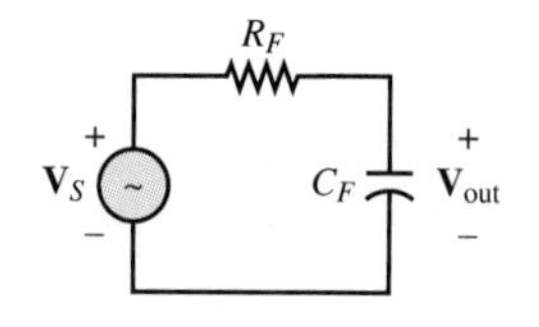

Figure 11.22 Passive low-pass filter

This expression can be factored into two terms. The first is an amplification factor analogous to the amplification that would be obtained with a simple inverting amplifier (i.e., the same circuit as that of Figure 11.21 with the capacitor removed); the second is a low-pass filter, with a cutoff frequency dictated by the parallel combination of R_F and C_F in the feedback loop. The filtering effect is completely analogous to that which would be attained by the passive circuit shown in Figure 11.22. However, the op-amp filter also provides amplification by a factor of R_F/R_S.

It should be apparent that the response of this op-amp filter is just an amplified version of that of the passive filter. Figure 11.23 depicts the amplitude response of the active low-pass filter (in the figure, $R_F/R_S = 10$ and $1/R_F C_F = 1$) in two different graphs; the first plots the amplitude ratio $\mathbf{V}_{out}/\mathbf{V}_S$ versus radian frequency, ω, on a logarithmic scale, while the second plots the amplitude ratio $20\log_{10}(\mathbf{V}_{out}/\mathbf{V}_S)$ (in units of dB), also versus ω on a logarithmic scale. You will recall from Chapter 6 that dB frequency response plots are encountered very frequently. Note that in the dB plot, the slope of the filter response for frequencies significantly higher than the cutoff frequency,

$$\omega_0 = \frac{1}{R_F C_F} \tag{11.51}$$

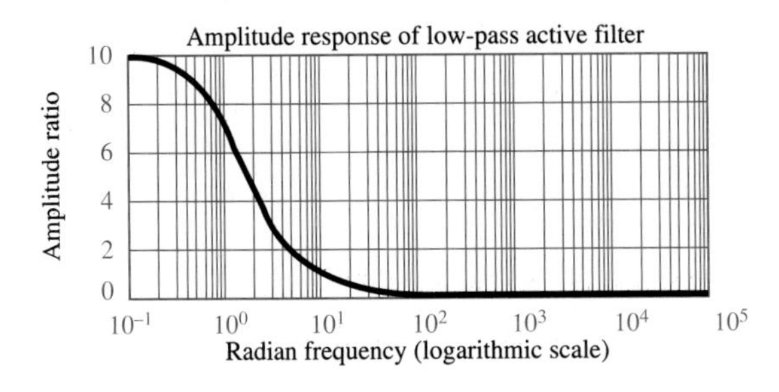

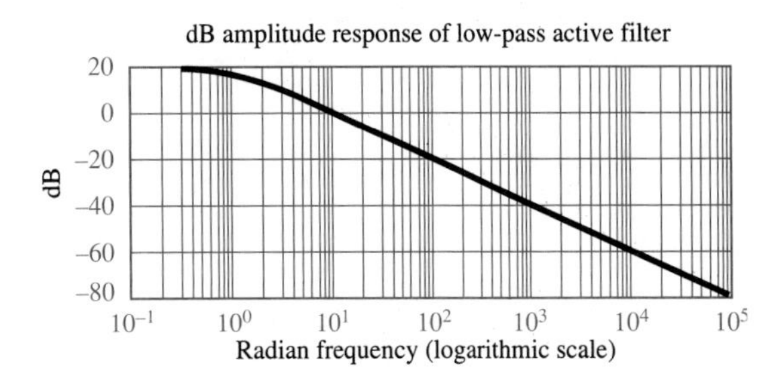

Figure 11.23 Normalized response of active low-pass filter

is -20 dB/decade, while the slope for frequencies significantly lower than this cutoff frequency is equal to zero. The value of the response at the cutoff frequency is found to be, in units of dB,

$$|A_{LP}(j\omega_0)|_{\text{dB}} = 20\log_{10}\frac{R_F}{R_S} - 20\log_{10}\sqrt{2} \tag{11.52}$$

where

$$-20\log_{10}\sqrt{2} = -3\text{ dB} \tag{11.53}$$

Thus, ω_0 is also called the **3-dB frequency.**

Among the advantages of such low-pass active filters is the ease with which the gain and the bandwidth can be adjusted by controlling the ratios R_F/R_S and $1/R_F C_F$, respectively.

It should be apparent at this point that it is possible to construct other types of filters by suitably connecting resistors and energy-storage elements to an op-amp. For example, a high-pass active filter can easily be obtained by using the circuit shown in Figure 11.24. Observe that the impedance of the input circuit is

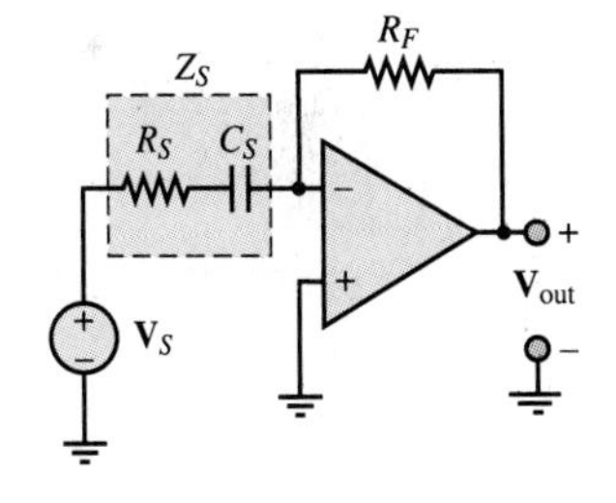

Figure 11.24 Active high-pass filter

$$Z_S = R_S + \frac{1}{j\omega C_S} \tag{11.54}$$

and that of the feedback circuit is

$$Z_F = R_F \tag{11.55}$$

Then, the following gain function for the op-amp circuit can be derived:

$$\begin{aligned} A_{HP}(j\omega) &= -\frac{Z_F}{Z_S} = -\frac{R_F}{R_S + 1/j\omega C_S} \\ &= -\frac{j\omega C_S R_F}{1 + j\omega R_S C_S} \end{aligned} \tag{11.56}$$

As ω approaches zero, so does the response of the filter, whereas as ω approaches infinity, according to the gain expression of equation 11.56, the gain of the amplifier approaches a constant:

$$\lim_{\omega \to \infty} A_{HP}(j\omega) = -\frac{R_F}{R_S} \quad \textbf{(11.57)}$$

That is, above a certain frequency range, the circuit acts as a linear amplifier. This is exactly the behavior one would expect of a high-pass filter. The high-pass response is depicted in Figure 11.25, in both linear and dB plots (in the figure, $R_F/R_S = 10$, $1/R_SC = 1$). Note that in the dB plot, the slope of the filter response for frequencies significantly lower than $\omega = 1/R_SC_S = 1$ is +20 dB/decade, while the slope for frequencies significantly higher than this cutoff (or 3-dB) frequency is equal to zero.

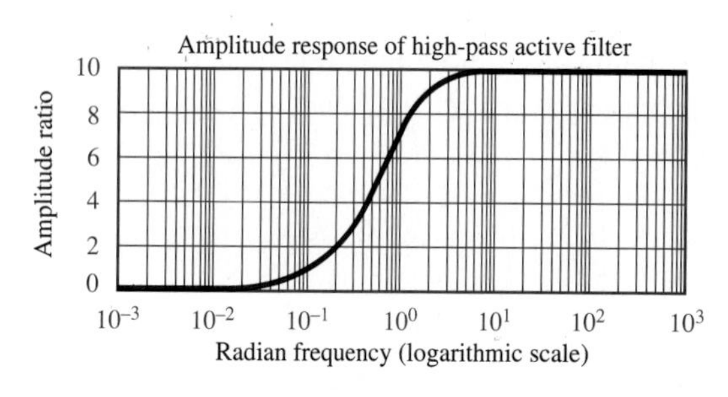

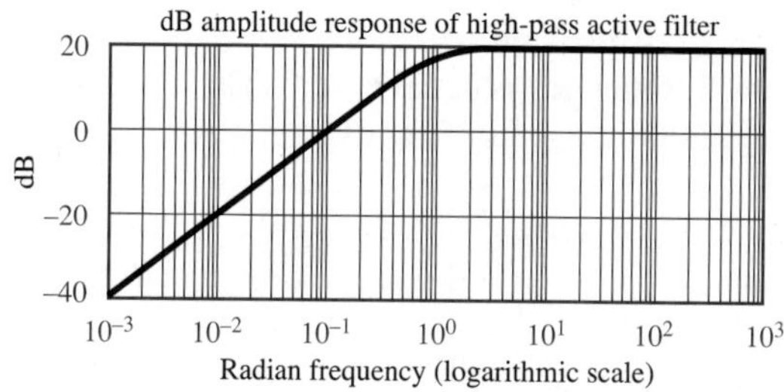

Figure 11.25 Normalized response of active high-pass filter

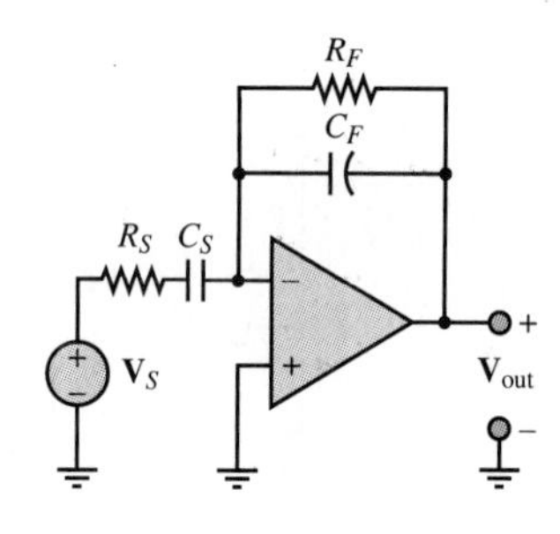

Figure 11.26 Active band-pass filter

As a final example of active filters, let us look at a simple active band-pass filter configuration. This type of response may be realized simply by combining the high- and low-pass filters we examined earlier. The circuit is shown in Figure 11.26.

The analysis of the band-pass circuit follows the same structure used in previous examples. First, we evaluate the feedback and input impedances:

$$Z_F = R_F \parallel \frac{1}{j\omega C_F} = \frac{R_F}{1 + j\omega C_F R_F} \quad \textbf{(11.58)}$$

$$Z_S = R_S + \frac{1}{j\omega C_S} = \frac{1 + j\omega C_S R_S}{j\omega C_S} \quad \textbf{(11.59)}$$

Next, we compute the closed-loop frequency response of the op-amp, as follows:

$$A_{BP}(j\omega) = -\frac{Z_F}{Z_S} = -\frac{j\omega C_S R_F}{(1 + j\omega C_F R_F)(1 + j\omega C_S R_S)} \quad \textbf{(11.60)}$$

The form of the op-amp response we just obtained should not appear as a surprise. It is very similar (although not identical) to the product of the low-pass and high-pass responses:

$$A_{LP}(j\omega) = -\frac{R_F/R_S}{1 + j\omega C_F R_F} \quad \textbf{(11.61)}$$

$$A_{HP}(j\omega) = -\frac{j\omega C_S R_F}{1 + j\omega R_S C_S} \tag{11.62}$$

In particular, the denominator of $A_{BP}(j\omega)$ is exactly the product of the denominators of $A_{LP}(j\omega)$ and $A_{HP}(j\omega)$. It is particularly enlightening to rewrite $A_{LP}(j\omega)$ in a slightly different form, after making the observation that each RC product corresponds to some "critical" frequency:

$$\omega_1 = \frac{1}{R_F C_S} \qquad \omega_{LP} = \frac{1}{R_F C_F} \qquad \omega_{HP} = \frac{1}{R_S C_S} \tag{11.63}$$

It is easy to verify that for the case where

$$\omega_{HP} > \omega_{LP} \tag{11.64}$$

the response of the op-amp filter may be represented as shown in Figure 11.27 in both linear and dB plots (in the figure, $\omega_1 = 1$, $\omega_{HP} = 1{,}000$, $\omega_{LP} = 10$). The dB plot is very revealing, for it shows that, in effect, the band-pass response is the graphical superposition of the low-pass and high-pass responses shown earlier. The two 3-dB (or cutoff) frequencies are the same as in $A_{LP}(j\omega)$, $1/R_F C_F$; and in $A_{HP}(j\omega)$, $1/R_S C_S$. The third frequency, $\omega_1 = 1/R_F C_S$, represents the point where the response of the filter crosses the 0-dB axis (rising slope). Since 0 dB corresponds to a gain of 1, this frequency is called the **unity gain frequency.**

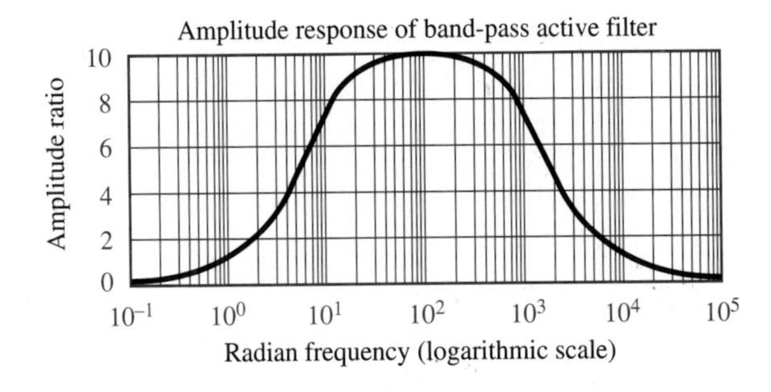

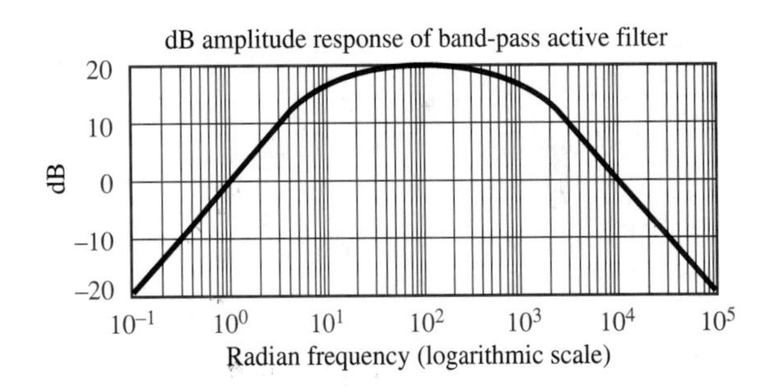

Figure 11.27 Normalized response of active band-pass filter

The ideas developed thus far can be employed to construct more complex functions of frequency. In fact, most active filters one encounters in practical applications are based on circuits involving more than one or two energy-storage elements. By constructing suitable functions for Z_F and Z_S, it is possible to realize filters with greater frequency selectivity (i.e., sharpness of cutoff), as well as flatter band-pass or band-rejection functions (that is, filters that either allow or reject signals in a limited band of frequencies). A few simple applications are investigated in the homework problems. One remark that should be made in passing, though, pertains to the exclusive use of capacitors in the circuits analyzed thus far. One of the advantages of op-amp filters is that it is not necessary to use both capacitors and inductors to obtain a band-pass response. Suitable connections of capacitors can accomplish that task in an op-amp. This seemingly minor fact is of great importance in practice, because inductors are expensive to mass-produce to close tolerances and exact specifications and are often bulkier than capacitors with equivalent energy-storage capabilities. On the other hand, capacitors are easy to manufacture in a wide variety of tolerances and values, and in relatively compact packages, including in integrated circuit form.

Example 11.7 illustrates how it is possible to construct active filters with greater frequency selectivity by adding energy-storage elements to the design.

Example 11.7

The op-amp filter shown in Figure 11.28 provides more effective low-pass filtering than is obtained by using a single energy-storage element. In particular, if $R_2C = R_1/L = 1/\omega_0$, one can obtain a relatively simple expression for $A(j\omega)$. Find $A(j\omega)$ for this special case.

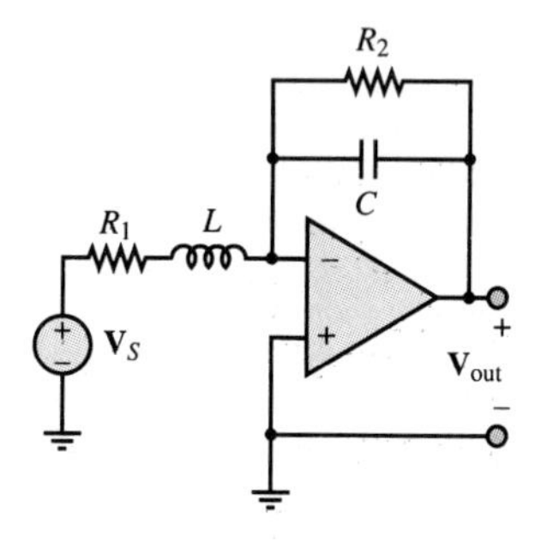

Figure 11.28

Solution:

Recognizing that the gain of this active filter is given by

$$A(j\omega) = \frac{\mathbf{V}_{out}}{\mathbf{V}_S} = -\frac{Z_F}{Z_S}$$

we can write

$$Z_F = R_2 \parallel \frac{1}{j\omega C} = \frac{R_2}{1 + j\omega C R_2} = \frac{R_2}{1 + j\omega/\omega_0}$$

and

$$Z_S = R_1 + j\omega L = R_1\left(1 + j\omega\frac{L}{R_1}\right) = R_1\left(1 + \frac{j\omega}{\omega_0}\right)$$

so that the closed-loop gain is given by the expression

$$A(j\omega) = \frac{R_2/R_1}{(1 + j\omega/\omega_0)^2}$$

Note the similarity between this filter response and the response of the simple low-pass filter presented earlier in this section. The only difference is that the denominator is squared in this example. The linear and dB responses of this filter are compared in Figure 11.29 with those of the simple low-pass filter of Figure 11.21. Note that in the dB plot, the slope of the filter presented in this example is −40 dB/decade, that is, double that of the filter of Figure 11.21. Although the use of an inductor in the filter circuit is not practical, this example serves the purpose of illustrating the advantage of using multiple storage elements in a filter.

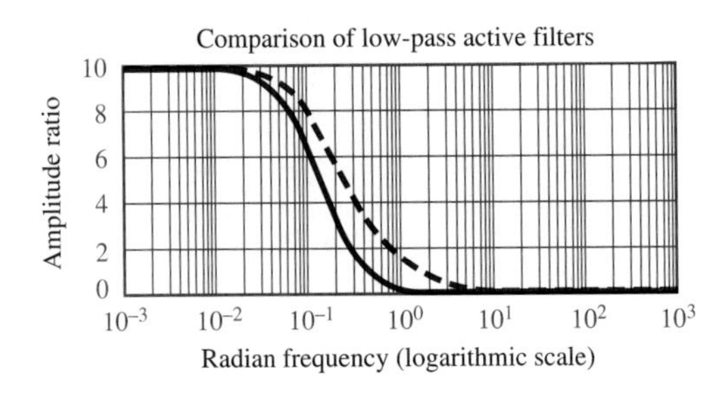

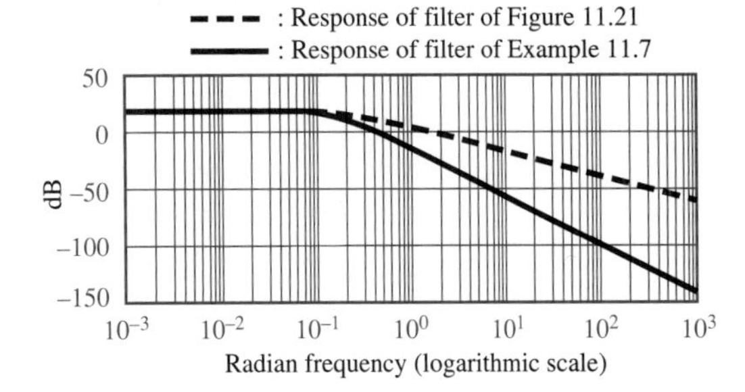

Figure 11.29

DRILL EXERCISES

11.7 Design a low-pass filter with closed-loop gain of 100 and cutoff (3-dB) frequency equal to 800 Hz. Assume that only 0.01-μF capacitors are available. Find R_F and R_S.

11.8 Repeat the design of Drill Exercise 11.7 for a high-pass filter with cutoff frequency of 2,000 Hz. This time, however, assume that only standard values of resistors are available (see Table 2.1 for a table of standard values). Select the nearest component values, and calculate the percent error in gain and cutoff frequency with respect to the desired values.

11.9 Find the frequency corresponding to attenuation of 1 dB (with respect to the maximum value of the amplitude response) for the filter of Drill Exercise 11.7.

11.10 What is the dB gain for the filter of Example 11.7 at the cutoff frequency, ω_0? Find the 3-dB frequency for this filter in terms of the cutoff frequency, ω_0, and note that the two are not the same.

11.4 INTEGRATOR AND DIFFERENTIATOR CIRCUITS

In the preceding sections, we examined the frequency response of op-amp circuits for sinusoidal inputs. However, certain op-amp circuits containing energy-storage elements reveal some of their more general properties if we analyze their response to inputs that are time-varying but not necessarily sinusoidal. Among such circuits are the commonly used integrator and differentiator; the analysis of these circuits is presented in the following paragraphs.

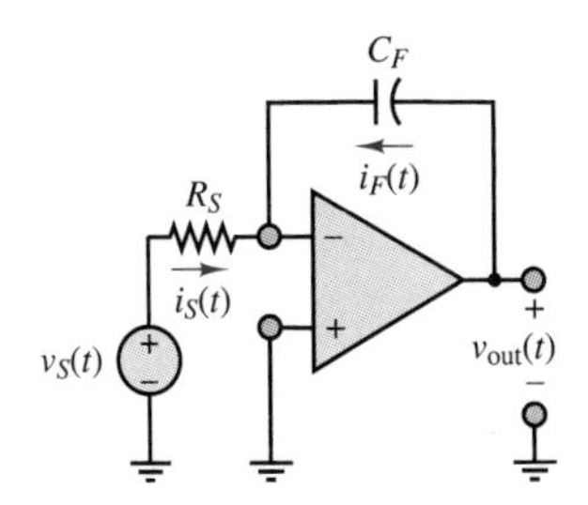

Figure 11.30 Op-amp integrator

The Ideal Integrator

Consider the circuit of Figure 11.30, where $v_S(t)$ is an arbitrary function of time (e.g., a pulse train, a triangular wave, or a square wave). The op-amp circuit shown provides an output that is proportional to the integral of $v_S(t)$. The analysis of the integrator circuit is, as always, based on the observation that

$$i_S(t) = -i_F(t) \tag{11.65}$$

where

$$i_S(t) = \frac{v_S(t)}{R_S} \tag{11.66}$$

It is also known that

$$i_F(t) = C_F \frac{dv_{\text{out}}(t)}{dt} \tag{11.67}$$

from the fundamental definition of the capacitor. The source voltage can then be expressed as a function of the derivative of the output voltage:

$$\frac{1}{R_S C_F} v_S(t) = -\frac{dv_{\text{out}}(t)}{dt} \tag{11.68}$$

By integrating both sides of equation 11.68, we obtain the following result:

$$v_{\text{out}}(t) = -\frac{1}{R_S C_F} \int_{-\infty}^{t} v_S(t')\, dt' \tag{11.69}$$

This equation states that the output voltage is the integral of the input voltage.

There are numerous applications of the op-amp integrator, most notably the **analog computer,** which will be discussed in Section 11.5. The following example illustrates the operation of the op-amp integrator.

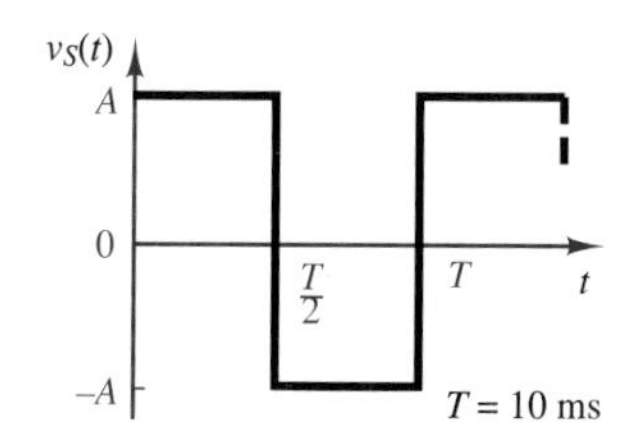

Figure 11.31

Example 11.8

Let $v_S(t)$ be a square wave of amplitude A and period $T = 10$ ms, as shown in Figure 11.31. Find $v_{\text{out}}(t)$ for an integrator with $C_F = 1\ \mu$F, $R_S = 10$ kΩ.

Solution:

The output of the ideal integrator is given by the expression

$$v_{\text{out}}(t) = -\frac{1}{R_S C_F} \int_{-\infty}^{t} v_S(t')\, dt' = -100 \int_{-\infty}^{t} v_S(t')\, dt'$$

To integrate a square wave, we note that each period consists, in effect, of two segments, which are constant at $+A$ and $-A$, respectively. We further assume that the square wave starts at $t = 0$, and that no energy is stored in the capacitor for $t \leq 0$. The technique we shall follow consists of integrating the waveform piecewise, taking into consideration the initial condition. For example,

$$\int_{-\infty}^{t} v_S(t')\, dt' = v_S(0) + \int_{0}^{t} v_S(t')\, dt'$$

Consider the first half-cycle of the square wave, starting at $t = 0$:

$$v_{out}(t) = v_{out}(0) - 100 \int_0^{T/2} A\, dt$$

$$= 0 - 100At \qquad 0 < t \le \frac{T}{2}$$

Now, to integrate over the second half-cycle, we must first evaluate $v_{out}(T/2)$ to obtain the initial condition for the next segment. Thus,

$$v_{out}(t) = v_{out}\left(\frac{T}{2}\right) - 100 \int_{T/2}^{t} -A\, dt$$

$$= -100A\frac{T}{2}\left[+100A\left(t - \frac{T}{2}\right)\right] \qquad \frac{T}{2} < t \le T$$

Since the function is periodic, this waveform will repeat with period T, as shown in Figure 11.32.

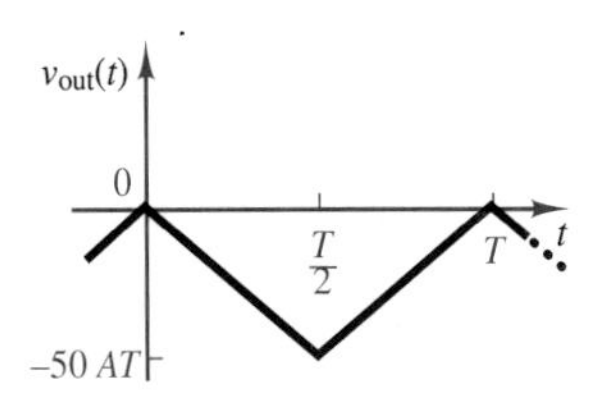

Figure 11.32

Example 11.9 Charge Amplifiers

One of the most common families of transducers for the measurement of force, pressure, and acceleration is that of **piezoelectric transducers.** These transducers contain a piezoelectric crystal, a crystal that generates an electric charge in response to deformation. Thus, if a force is applied to the crystal (leading to a displacement), a charge is generated within the crystal. If the external force generates a displacement x_i, then the transducer will generate a charge q according to the expression

$$q = K_p x_i$$

Figure 11.33 depicts the basic structure of the piezoelectric transducer, and a simple circuit model. The model consists of a current source in parallel with a capacitor, where the current source represents the rate of change of the charge generated in response to an external force and the capacitance is a consequence of the structure of the transducer, which consists of a piezoelectric crystal (e.g., quartz or Rochelle salt) sandwiched between conducting electrodes (in effect, this is a parallel-plate capacitor).

Although it is possible, in principle, to employ a conventional voltage amplifier to amplify the transducer output voltage, v_o, given by

$$v_o = \frac{1}{C}\int i\, dt = \frac{1}{C}\int \frac{dq}{dt} dt = \frac{q}{C} = \frac{K_p x_i}{C}$$

it is often advantageous to use a **charge amplifier.** The charge amplifier is essentially an integrator circuit, as shown in Figure 11.34, characterized by an extremely high input impedance.[3] The high impedance is essential; otherwise, the charge generated by the transducer would leak to ground through the input resistance of the amplifier.

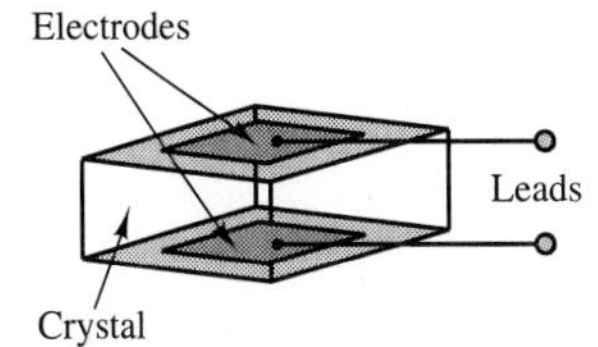

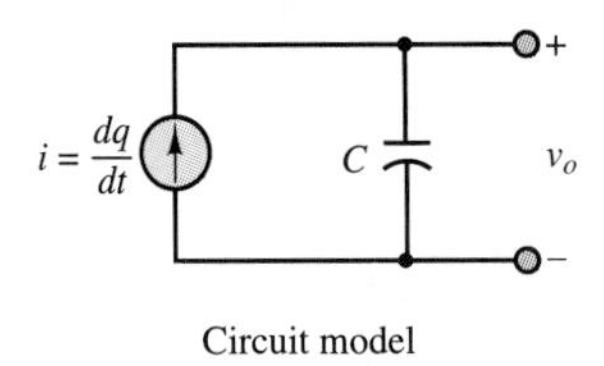

Figure 11.33 Piezoelectric transducer

[3]Special op-amps are employed to achieve extremely high input impedance, through FET input circuits.

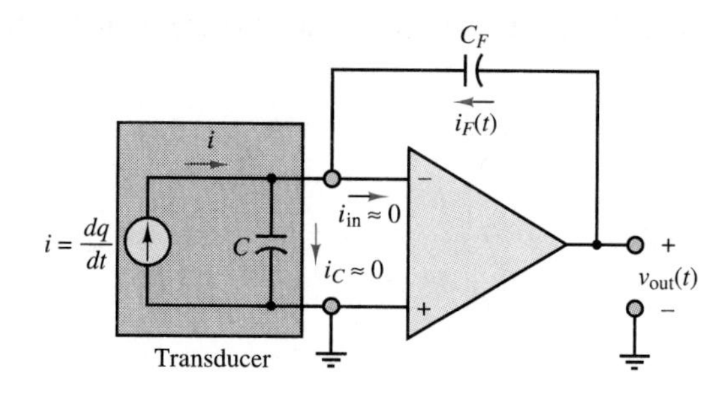

Figure 11.34 Charge amplifier

Because of the high input impedance, the input current into the amplifier is negligible; further, because of the high open-loop gain of the amplifier, the inverting-terminal voltage is essentially at ground potential. Thus, *the voltage across the transducer is effectively zero*. As a consequence, to satisfy KCL, the feedback current, $i_F(t)$, must be equal and opposite to the transducer current, i:

$$i_F(t) = -i$$

and since

$$v_{out}(t) = \frac{1}{C_F}\int i_F(t)\,dt$$

it follows that the output voltage is proportional to the charge generated by the transducer, and therefore to the displacement:

$$v_{out}(t) = \frac{1}{C_F}\int -i\,dt = \frac{1}{C_F}\int -\frac{dq}{dt}dt = -\frac{q}{C_F} = -\frac{K_p x_i}{C_F}$$

Since the displacement is caused by an external force or pressure, this sensing principle is widely adopted in the measurement of force and pressure.

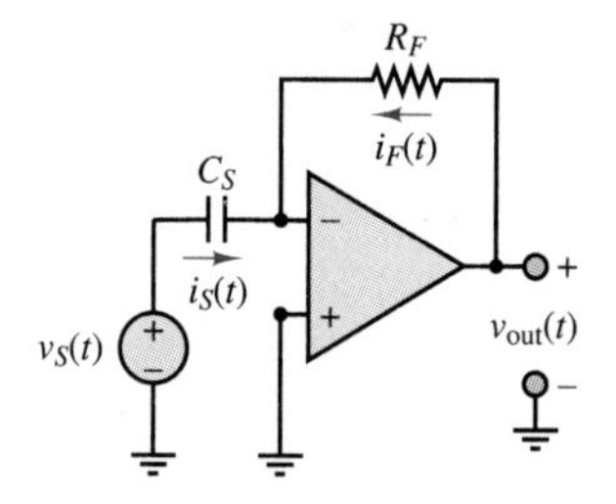

Figure 11.35 Op-amp differentiator

The Ideal Differentiator

Using an argument similar to that employed for the integrator, we can derive a result for the ideal differentiator circuit of Figure 11.35. The relationship between input and output is obtained by observing that

$$i_S(t) = C_S\frac{dv_S(t)}{dt} \tag{11.70}$$

and

$$i_F(t) = \frac{v_{out}(t)}{R_F} \tag{11.71}$$

so that the output of the differentiator circuit is proportional to the derivative of the input:

$$v_{out}(t) = -R_F C_S \frac{dv_S(t)}{dt} \tag{11.72}$$

Although mathematically attractive, the differentiation property of this op-amp circuit is seldom used in practice, because differentiation tends to amplify any noise that may be present in a signal.

DRILL EXERCISES

11.11 Plot the frequency response of the ideal integrator in dB plot. Determine the slope of the curve in dB/decade. You may assume $R_S C_F = 10$.

11.12 Plot the frequency response of the ideal differentiator in a dB plot. Determine the slope of the curve in dB/decade. You may assume $R_F C_S = 100$.

11.13 Verify that if the triangular wave of Example 11.8 is the input to the ideal differentiator, the resulting output is the original square wave.

11.5 ANALOG COMPUTERS

Prior to the advent of digital computers, the solution of differential equations and the simulation of complex dynamic systems were conducted exclusively by means of analog computers. Analog computers still find application in engineering practice in the simulation of dynamic systems. The analog computer is a device that is based on three op-amp circuits introduced earlier in this chapter: the amplifier, the summer, and the integrator. These three building blocks permit the construction of circuits that can be used to solve differential equations and to simulate dynamic systems. Figure 11.36 depicts the three symbols that are typically employed to represent the principal functions of an analog computer.

The simplest way to discuss the operation of the analog computer is to present an example. Consider the simple second-order mechanical system, shown in Figure 11.37, that represents, albeit in a greatly simplified fashion, one corner of an automobile suspension system. The mass, *M*, represents the mass of one quarter of the vehicle; the damper, *B*, represents the shock absorber; and the spring, *K*, represents the suspension spring (or strut). The differential equation of the system may be derived as follows:

$$M\frac{d^2x_M}{dt^2} + B\left(\frac{dx_M}{dt} - \frac{dx_R}{dt}\right) + K(x_M - x_R) = 0 \tag{11.73}$$

Rearranging terms, we obtain the following equation, in which the terms related to the road displacement and velocity—x_R and dx_R/dt, respectively—are the forcing functions:

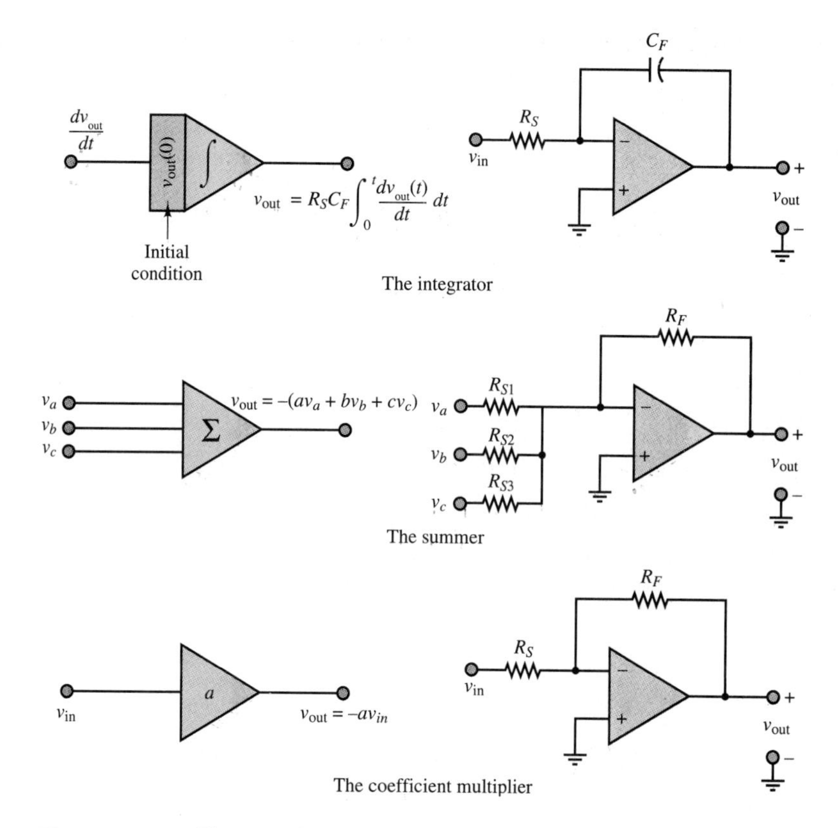

Figure 11.36 Elements of the analog computer

$$M\frac{d^2x_M}{dt^2} + B\frac{dx_M}{dt} + Kx_M = B\frac{dx_R}{dt} + Kx_R \tag{11.74}$$

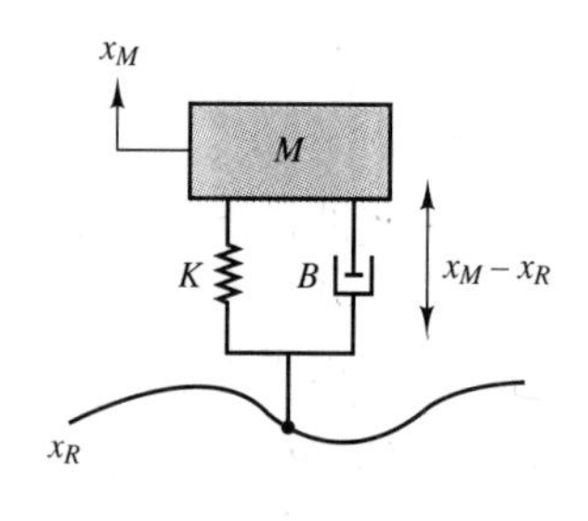

Figure 11.37 Model of automobile suspension

Assume that the car is traveling over a "washboard" surface on an unpaved road, such that the road profile is approximately described by the expression

$$x_R(t) = X\sin(\omega t) \tag{11.75}$$

It follows, then, that the vertical velocity input to the suspension is given by the expression

$$\frac{dx_R}{dt} = \omega X\cos(\omega t) \tag{11.76}$$

and we can write the equation for the suspension system in the form

$$M\frac{d^2x_M}{dt^2} + B\frac{dx_M}{dt} + Kx_M = B\omega X\cos(\omega t) + KX\sin(\omega t) \tag{11.77}$$

It would be desirable to solve the equation for the displacement, x_M, which represents the motion of the vehicle mass in response to the road excitation. The solution can be used as an aid in designing the suspension system that best absorbs the road vibration, providing a comfortable ride for the passengers. Equation 11.77 may be rearranged to obtain

$$\frac{d^2x_M}{dt^2} = -\frac{B}{M}\frac{dx_M}{dt} - \frac{K}{M}x_M + \frac{B}{M}\omega X\cos(\omega t) + \frac{K}{M}X\sin(\omega t) \qquad \textbf{(11.78)}$$

This equation is now in a form appropriate for solution by repeated integration, since we have isolated the highest derivative term; thus, it will be sufficient to integrate the right-hand side twice to obtain the solution for the displacement of the vehicle mass, x_M.

Figure 11.38 depicts the three basic operations that need to be performed to integrate the differential equation describing the motion of the mass, M. Note that in each of the three blocks—the summer and the two integrators—the inversion due to the inverting amplifier configuration used for the integrator is already accounted for. Finally, the basic summing and integrating blocks together with three coefficient multipliers (inverting amplifiers) are connected in the configuration that corresponds to the preceding differential equation (equation 11.78). You can easily verify that the analog computer circuit of Figure 11.39 does indeed solve the differential equation in x_M by repeated integration.

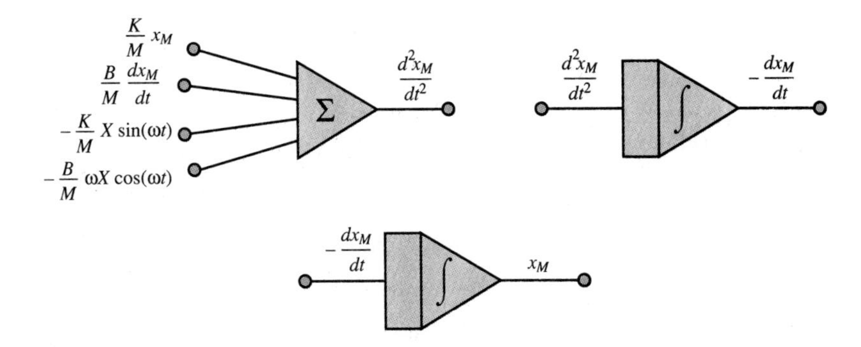

Figure 11.38 Solution by repeated integration

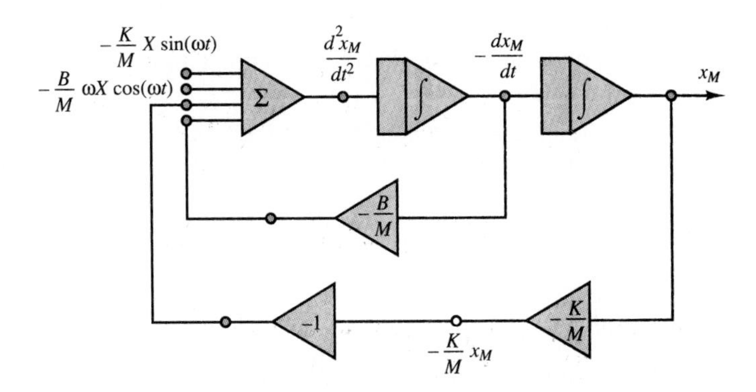

Figure 11.39 Analog computer simulation of suspension system

Scaling in Analog Computers

One of the important issues in analog computing is that of scaling. Since the analog computer implements an electrical analog of a physical system, there is no guarantee that the voltages and currents in the analog computer circuits will be of the same order of magnitude as the physical variables (e.g., velocity, displacement, temperature, or flow) they simulate. Further, it is not necessary that the computer simulate the physical system with the same time scale; it may be desirable in practice to speed up or slow down the simulation. Thus, the interest in time scaling and magnitude scaling.

Table 11.1 **Actual and simulated variables in analog computers**

Physical system	Analog simulation
Physical variable, x	Voltage, v
Time variable, t	Simulated time, τ

Let Table 11.1 represent the physical and simulation variables. Considering time scaling first, let t denote real time and τ, the computer time variable. Then the time derivative of a physical variable can be expressed as

$$\frac{dx}{dt} = \frac{dx}{d\tau}\frac{d\tau}{dt} = \alpha\frac{dx}{d\tau} \tag{11.79}$$

where α is the scaling factor between real time and computer time:

$$\tau = \alpha t \tag{11.80}$$

For higher-order derivatives, the following relationship will hold:

$$\frac{d^n x}{dt^n} = \alpha^n \frac{d^n x}{d\tau^n} \tag{11.81}$$

While time scaling is likely to be prompted by a desire to speed up or slow down a computation, magnitude scaling is motivated by several different factors:

1. The relationship between physical variables and computer voltages (e.g., calibration constants).
2. Overloading of the op-amp circuits (we shall see in Section 11.6 that one of the fundamental limitations of the operational amplifier is its voltage range).
3. Loss of accuracy if voltages are too small (errors are usually expressed as a percentage of the full-scale range).

Thus, if the relationship between a physical variable and the computer voltage is $v = \beta x$, where β is a magnitude scaling factor, the derivative terms will be affected according to the relation

$$\frac{dx}{dt} = \frac{1}{\beta}\frac{dv}{dt} \tag{11.82}$$

Note that different scaling factors may be introduced at each point in the analog computer simulation, and so there is no general rule with regard to magnitude scaling. For example, if $v = \beta_0 x$, it is entirely possible to have $\frac{dx}{dt} = \frac{1}{\beta_1}\frac{dv}{dt}$.

Example 11.10

In this example we consider the implementation of the analog computer simulation of the suspension system described in Section 11.5. Let $M = 400$ kg, $B = 20{,}000$ N/m-s, and $K = 16 \times 10^5$ N/m. We wish to find the appropriate values of resistors and capacitors for each of the op-amp circuits.

Solution:

It is generally convenient to express resistances in MΩ and capacitances in μF; since the output of an integrator is scaled by the factor $1/RC$, this means that an ideal integrator has $RC = 1$, for $R = 1$ MΩ and $C = 1$ μF. Thus, we shall use these values for each integrator. The resulting circuit for each integrator will therefore be as shown in Figure 11.40(a). Similarly, the summer will employ 1-MΩ resistors everywhere, as shown in the figure. On the other hand, the two coefficient multipliers will require resistor ratios of $K/M = 4{,}000$ and $B/M = 50$. Possible solutions are shown in Figure 11.40(c) and (d). Finally, the unity gain inverting amplifier can be realized with two 1-MΩ resistors.

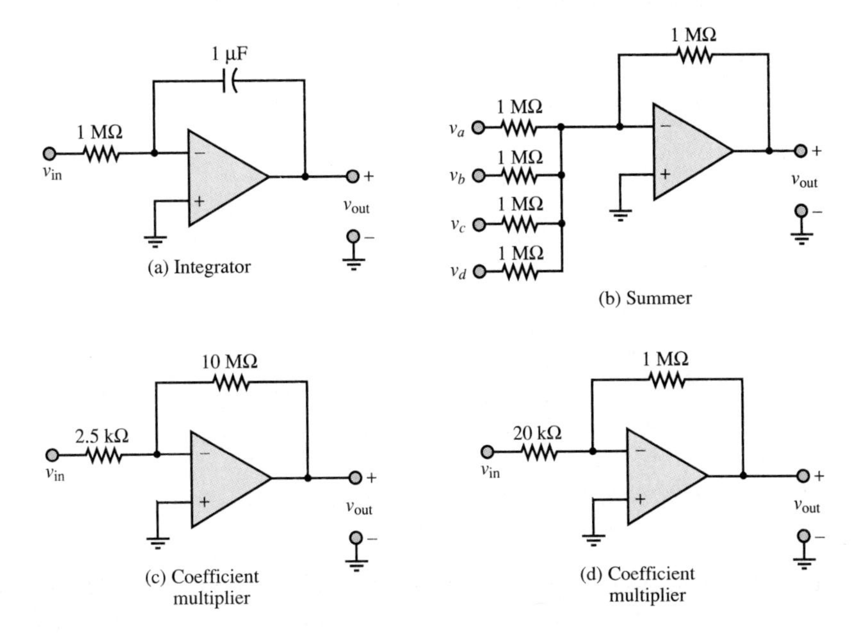

Figure 11.40 Analog computer simulation of suspension system

Note that it is not strictly necessary to use five op-amps in the simulation of the suspension system. For example, the summer and the first integrator could be combined into a single op-amp circuit, and one of the two coefficient multipliers could be eliminated. This idea is explored further in the homework problems.

Example 11.11

In this example, we derive the differential equation of a system from the analog computer simulation circuit of Figure 11.41.

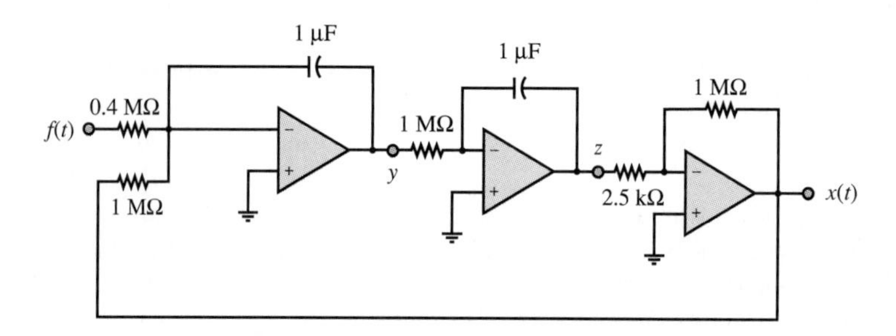

Figure 11.41 Analog computer simulation of unknown system

Solution:

We note first that the summing, scaling, and integrating functions have all been combined in the leftmost op-amp. Let the output of the first (leftmost) op-amp be denoted by y, that of the next by z, and the final output by x. Then, the following relationships hold:

$$x = -400z$$

and

$$y = -\frac{dz}{dt} = \frac{1}{400}\frac{dx}{dt}$$

Further, noting that

$$y = -\int [x + 2.5f(t)]\, dt$$

we can write

$$\frac{dy}{dt} = -x - 2.5f(t)$$

and finally obtain the equation

$$\frac{1}{400}\frac{d^2x}{dt^2} + x = -2.5f(t)$$

DRILL EXERCISES

11.14 Modify the gains of the coefficient multipliers in Example 11.10 if we wish to slow down the simulation by a factor of 10 (i.e., if $\alpha = 0.1$).

11.15 For the simulation of Example 11.10, what will the largest magnitude of the voltage analog of x_M be for a road displacement $x_R = 0.01 \sin(100t)$? [*Hint:* Use phasor techniques to compute the frequency response $x_M(\omega)/F(\omega)$, where $f(t) = B\, dx_R/dt + Kx_R$, and evaluate the output voltage by multiplying the input by the magnitude of the frequency response at $\omega = 100$.]

11.16 Change the parameters of the analog computer simulation of Example 11.11 to simulate the differential equation $d^2x/dt^2 + 2x = -10f(t)$.

11.6 PHYSICAL LIMITATIONS OF OP-AMPS

Thus far, the operational amplifier has been treated as an ideal device, characterized by infinite input resistance, zero output resistance, and infinite open-loop voltage gain. Although this model is adequate to represent the behavior of the op-amp in a large number of applications, it is important to realize that practical operational amplifiers are not ideal devices but exhibit a number of limitations that should be considered in the design of instrumentation. In particular, in dealing with relatively large voltages and currents, and in the presence of high-frequency signals, it is important to be aware of the nonideal properties of the op-amp. In the present section, we examine the principal limitations of the operational amplifier.

Voltage Supply Limits

As indicated in Figure 11.4, operational amplifiers (and all amplifiers, in general) are powered by external DC voltage supplies, V_S^+ and V_S^-, which are usually symmetrical and of the order of ± 10 V to ± 20 V. Some op-amps are especially designed to operate from a single voltage supply, but for the sake of simplicity we shall from here on consider only symmetrical supplies. The effect of limiting supply voltages is that amplifiers are capable of amplifying signals *only within the range of their supply voltages;* it would be physically impossible for an amplifier to generate a voltage greater than V_S^+ or less than V_S^-. This limitation may be stated as follows:

$$V_S^- < v_{\text{out}} < V_S^+ \tag{11.83}$$

For most op-amps, the limit is actually approximately 1.5 V less than the supply voltages. How does this practically affect the performance of an amplifier circuit? An example will best illustrate the idea.

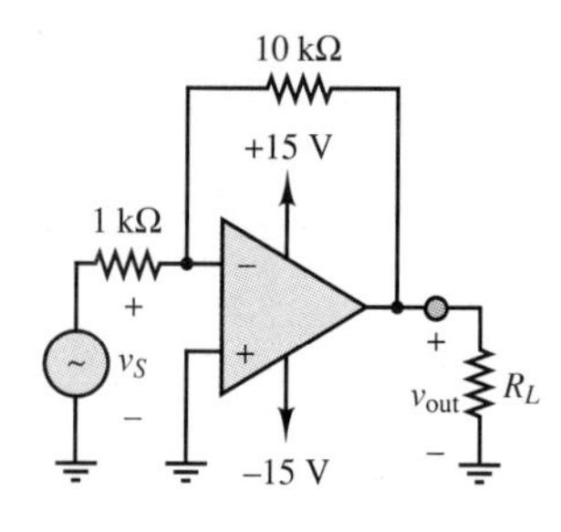

Figure 11.42

Example 11.12

This example deals with the voltage supply limitations of a practical op-amp. Consider the inverting amplifier shown in Figure 11.42. If the input to the amplifier is $v_S(t) = 2\sin(2{,}000t)$, calculate and sketch the output.

Solution:

For an ideal op-amp, the output would be

$$v_{out}(t) = -20\sin 2{,}000t \text{ V}$$

However, the voltage supplies are ± 15 V, and therefore $v_{out}(t)$ is limited to "swing" strictly between $+15$ V and -15 V. Figure 11.43 depicts the ideal and actual responses of the amplifier. For a practical op-amp, output saturation would most likely be reached somewhat before the voltage supply limits were reached, that is, around ± 13.5 V.

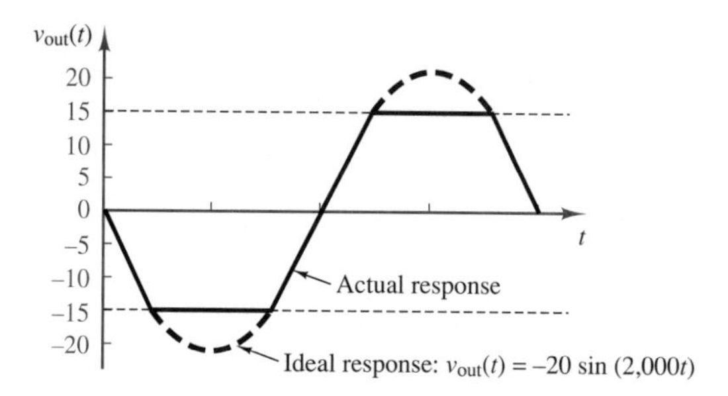

Figure 11.43 Op-amp output with voltage supply limit

Note how the voltage supply limit actually causes the peaks of the sine wave to be clipped in an abrupt fashion. This type of hard nonlinearity changes the characteristics of the signal quite radically, and could lead to significant errors if not taken into account. Just to give an intuitive idea of how such clipping can affect a signal, have you ever wondered why rock guitar has a characteristic sound that is very different from the sound of classical or jazz guitar? The reason is that the "rock sound" is obtained by overamplifying the signal, attempting to exceed the voltage supply limits, and causing clipping similar in quality to the distortion introduced by voltage supply limits in an op-amp. This clipping broadens the spectral content of each tone and causes the sound to be distorted.

One of the circuits most directly affected by supply voltage limitations is the op-amp integrator. The following example illustrates how saturation of an integrator circuit can lead to severe signal distortion.

Example 11.13

Consider the op-amp integrator of Figure 11.30, with $C_F = 20\ \mu\text{F}$ and $R_S = 10\ \text{k}\Omega$. Let the input to the integrator be a sinusoid with a small DC offset:

$$v_S(t) = 0.5 + 3\cos(10t)$$

Compute and sketch the output of the integrator, considering the voltage supply limitation of the op-amp.

Solution:

A noninverting ideal integrator would produce the following output:

$$v_{\text{out}}(t) = \frac{1}{R_S C_F}\int [0.5 + 3\cos(10t)]\,dt = 2.5t + 1.5\sin(10t)$$

Note that the presence of the DC offset in the input generates a term that increases linearly with time in the output and will eventually lead to saturation. Figure 11.44 depicts the op-amp input and output, the latter subject to ± 15-V supply limits. Note that it is possible to eliminate this problem by modifying the design of the op-amp integrator to include a feedback resistor in parallel with the capacitor. This approach is discussed in the homework problems.

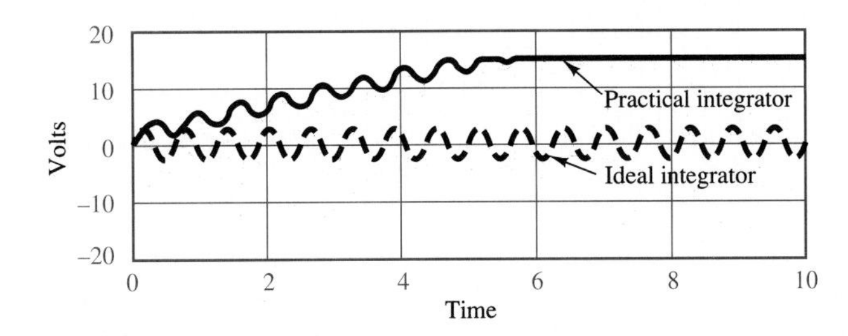

Figure 11.44 Output of practical integrator with input DC offset

Frequency Response Limits

Another property of all amplifiers that may pose severe limitations to the op-amp is their finite bandwidth. We have so far assumed, in our ideal op-amp model, that the open-loop gain is a very large constant. In reality, $A_{V(OL)}$ is a function of frequency and is characterized by a low-pass response. For a typical op-amp,

$$A_{V(OL)}(j\omega) = \frac{A_0}{1 + j\omega/\omega_0} \tag{11.84}$$

The cutoff frequency of the op-amp open-loop gain, ω_0, represents approximately the point where the amplifier response starts to drop off as a function of frequency, and is analogous to the cutoff frequencies of the RC and RL circuits of Chapter 6. Figure 11.45 depicts $A_{V(OL)}(j\omega)$ in both linear and dB plots for the fairly typical values $A_0 = 10^6$ and $\omega_0 = 10\pi$. It should be apparent from this figure that the assumption of a very large open-loop gain becomes less and less accurate for increasing frequency. Recall the initial derivation of the closed-loop gain for the inverting amplifier: in obtaining the final result, $\mathbf{V}_{out}/\mathbf{V}_S = -R_F/R_S$, it was assumed that $A_{V(OL)} \to \infty$. This assumption is clearly inadequate at the higher frequencies.

The finite bandwidth of the practical op-amp results in a fixed **gain-bandwidth product** for any given amplifier. The effect of a constant gain-bandwidth product is that as the closed-loop gain of the amplifier is increased, its 3-dB bandwidth is proportionally reduced, until, in the limit, if the amplifier

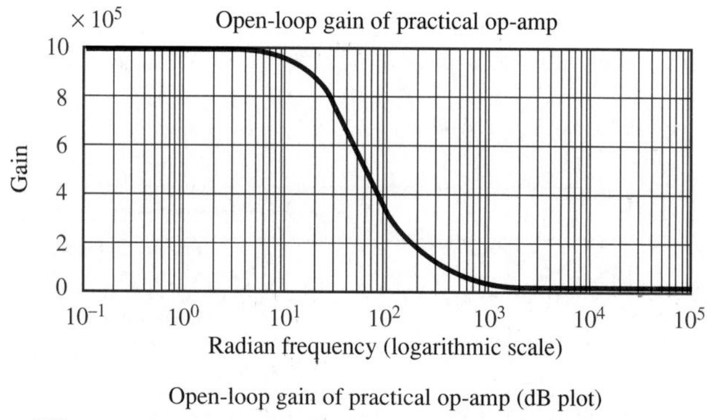

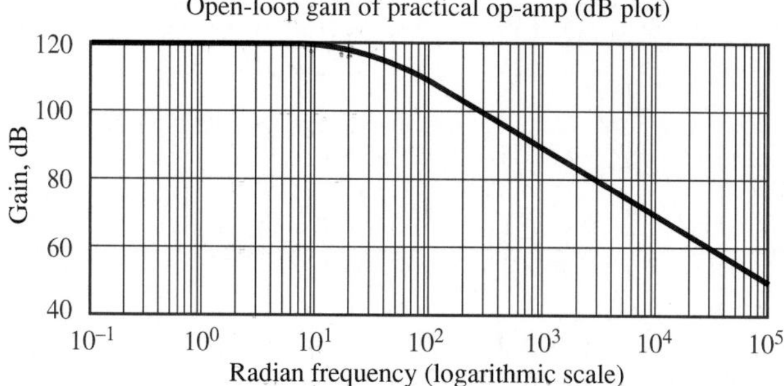

Figure 11.45 Open-loop gain of practical op-amp

were used in the open-loop mode, its gain would be equal to A_0 and its 3-dB bandwidth would be equal to ω_0. The constant gain-bandwidth product is therefore equal to the product of the open-loop gain and the open-loop bandwidth of the amplifier: $A_0 \times \omega_0 = K$. When the amplifier is connected in a closed-loop configuration (e.g., as an inverting amplifier), its gain is typically much less than the open-loop gain and the 3-dB bandwidth of the amplifier is proportionally increased. To explain this further, Figure 11.46 depicts the case in which two different linear amplifiers (achieved through any two different negative feedback configurations) have been designed for the same op-amp. The first has closed-loop gain A_1, and the second has closed-loop gain A_2. The bold line in the figure indicates the open-loop frequency response, with gain A_0 and cutoff frequency ω_0. As the gain decreases from the open-loop gain, A_0, to A_1, we see that the cutoff frequency increases from ω_0 to ω_1. If we further reduce the gain to A_2, we can expect the bandwidth to increase to ω_2. Thus, *the product of gain and bandwidth in any given op-amp is constant*. That is,

$$A_0 \times \omega_0 = A_1 \times \omega_1 = A_2 \times \omega_2 = K \tag{11.85}$$

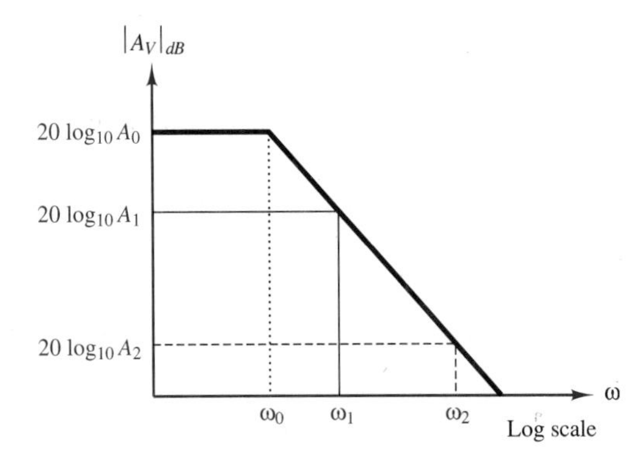

Figure 11.46

Example 11.14

The gain-bandwidth product is usually given in the manufacturer's specifications for a given op-amp. Suppose $A_0 = 10^6$ and $\omega_0 = 2\pi \times 5$ rad/s for a commercially available op-amp. If we wished to design an audio preamplifier with a bandwidth of 20 kHz, what is the maximum gain we should expect of the amplifier?

Solution:

The gain-bandwidth product for the amplifier is given by

$$A_0 \times \omega_0 = 2\pi \times 5 \times 10^6 = \pi \times 10^7 = K$$

Since the desired bandwidth is $\omega_1 = 2\pi \times 20{,}000$, the maximum allowable gain will be

$$A_1 = \frac{K}{\omega_1} = \frac{\pi \times 10^7}{2\pi \times 20{,}000} = 250$$

If we desire a gain larger than 250, we will have to purchase an op-amp with a greater gain-bandwidth product, or use two amplifiers in cascade, as shown in the next example.

Example 11.15 Cascade Amplifier

If a single amplifier is insufficient to obtain the desired amount of gain, because of its gain-bandwidth product limitation, one can resort to connecting two amplifiers in cascade, as shown in Figure 11.47. This example illustrates how a cascade amplifier can serve to increase the gain attainable for a given bandwidth. Let $K = 4\pi \times 10^6$ for both of the amplifiers in Figure 11.47. What is the 3-dB bandwidth of the cascade amplifier if $R_F/R_S = 100$ for each amplifier?

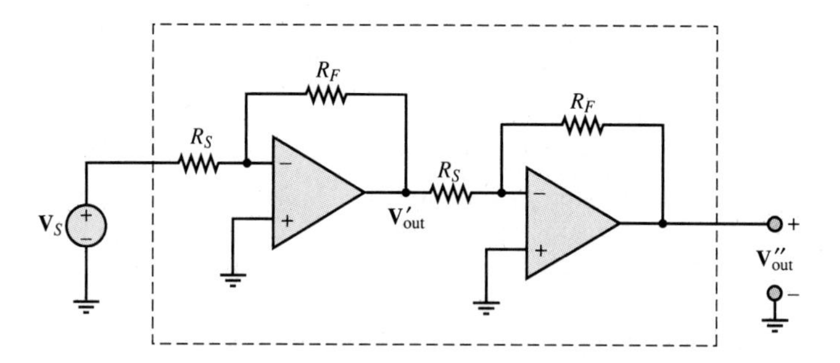

Figure 11.47 Cascade amplifier

Solution:

Let A_1 and ω_1 denote the gain and the 3-dB bandwidth of the first amplifier, and A_2 and ω_2 those of the second amplifier. Then the first amplifier has a 3-dB bandwidth given by

$$\omega_1 = \frac{A_0\omega_0}{A_1} = \frac{4\pi \times 10^6}{100} = 4\pi \times 10^4$$

Assuming an identical device, the second stage has the same approximate bandwidth:

$$\omega_2 = \frac{A_0\omega_0}{A_2} = \frac{4\pi \times 10^6}{100} = 4\pi \times 10^4$$

Thus, the *approximate* bandwidth of the cascade amplifier is $4\pi \times 10^4$. However, the gain of the cascade amplifier is equal to the product of the gains: $A_1 \times A_2 = 10{,}000$. Had we attempted to obtain the same amplification with a single op-amp by using resistors R'_F and R'_S such that

$$\frac{R'_F}{R'_S} = 10{,}000 = A'$$

we could have achieved a maximum bandwidth of only

$$\omega' = \frac{A_0\omega_0}{A'} = \frac{4\pi \times 10^6}{10{,}000} = 4\pi \times 10^2$$

In practice, the actual 3-dB bandwidth of the cascade amplifier is less than that of the single amplifiers, because the gain of each amplifier starts dropping off at frequencies somewhat lower than the nominal cutoff frequency. This problem is explored further in Drill Exercise 11.17.

Input Offset Voltage

Another limitation of practical op-amps results because even in the absence of any external inputs, it is possible that an **offset voltage** will be present at the input of an op-amp. This voltage is usually denoted by $\pm V_{os}$ and it is caused by mismatches in the internal circuitry of the op-amp. The offset voltage appears as a differential input voltage between the inverting and noninverting input terminals. The presence of an additional input voltage will cause a DC bias error in the amplifier output, as illustrated in Example 11.16. Typical and maximum values of V_{os} are quoted in manufacturers' data sheets. The worst-case effects due to the presence of offset voltages can therefore be predicted for any given application.

Example 11.16

To illustrate the phenomenon of DC offset, consider the case of a noninverting amplifier with a DC offset voltage $V_{os} = 1.5$ mV and a nominal (design) gain of $(1 + R_2/R_1) = 100$. The effect of the offset voltage is to cause a DC output even in the absence of any input. For the circuit shown in Figure 11.48, this DC error output voltage is

$$V_{\text{out},os} = \left(1 + \frac{R_2}{R_1}\right)V_{os} = 150 \text{ mV}$$

Thus, we can expect any AC output signal to be DC-shifted by 150 mV.

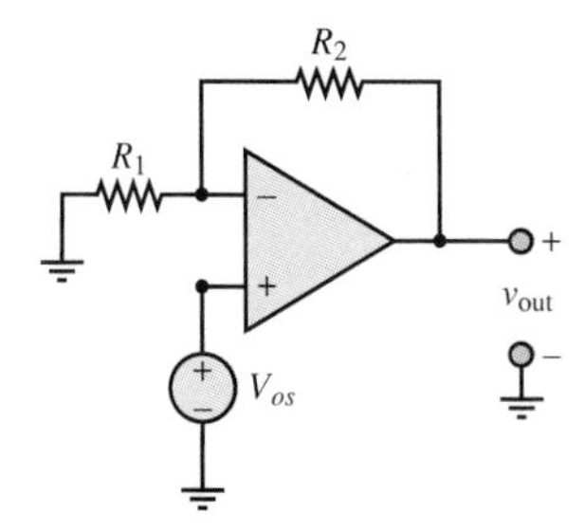

Figure 11.48 Op-amp input offset voltage

Input Bias Currents

Another nonideal characteristic of op-amps results from the presence of small input bias currents at the inverting and noninverting terminals. Once again, these

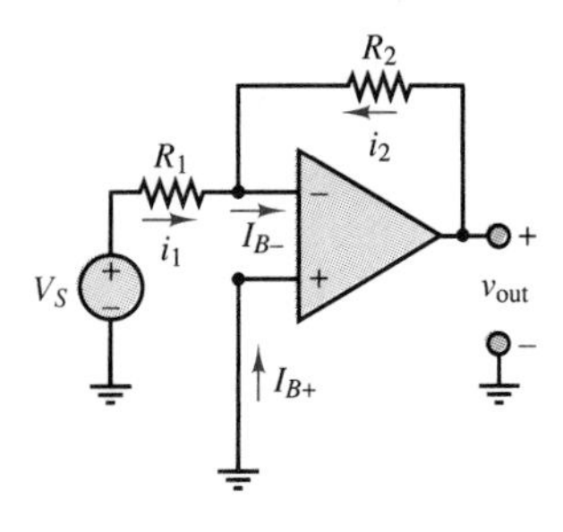

Figure 11.49

are due to the internal construction of the input stage of an operational amplifier. Figure 11.49 illustrates the presence of nonzero input bias currents (I_B) going into an op-amp.

Typical values of I_B depend on the semiconductor technology employed in the construction of the op-amp. Op-amps with bipolar transistor input stages may see input bias currents as large as 1 μA, while for FET input devices, the input bias currents are less than 1 nA. Since these currents depend on the internal design of the op-amp, they are not necessarily equal. One often designates the **input offset current** I_{os} as

$$I_{os} = I_{B+} - I_{B-} \tag{11.86}$$

The latter parameter is sometimes more convenient from the standpoint of analysis. The following example illustrates the effect of the nonzero input bias current on a practical amplifier design.

Example 11.17

The circuit shown in Figure 11.50 illustrates the effects of the input offset current I_{os}. Find the output voltage caused by the input offset currents in the absence of external inputs to the amplifier.

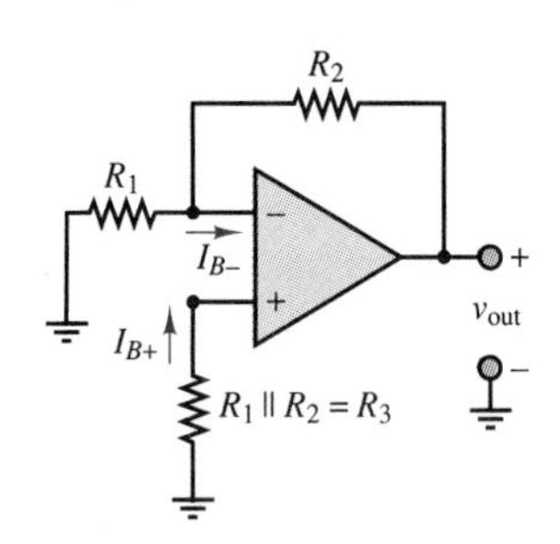

Figure 11.50

Solution:

Let the noninverting terminal resistance be $R_3 = R_1 \| R_2$. Then

$$v^+ = R_3 I_{B+}$$

and we can compute v_{out} by superposition:

$$\begin{aligned} v_{out} &= -\left(1 + \frac{R_2}{R_1}\right)v^+ + R_2 I_{B-} \\ &= -\left(1 + \frac{R_2}{R_1}\right)\left(\frac{R_1 R_2}{R_1 + R_2}\right)I_{B+} + R_2 I_{B-} \\ &= -R_2 I_{os} \end{aligned}$$

For a 100-kΩ feedback resistance and (a very worst case!) $I_{os} = 1$ μA, the DC error voltage at the op-amp output could be as large as

$$v_{out}(I_{os}) = 10^5 \times 10^{-6} = 100 \text{ mV}$$

Output Offset Adjustment

Both the offset voltage and the input offset current contribute to an output offset voltage $V_{out,os}$. Some op-amps provide a means for minimizing $V_{out,os}$. For example, the μA741 op-amp provides a connection for this procedure. Figure 11.51 shows a typical pin configuration for an op-amp in an eight-pin dual-in-line package (DIP) and the circuit used for nulling the output offset voltage. The variable resistor is adjusted until v_{out} reaches a minimum (ideally, 0 volts). Nulling the output voltage in this manner removes the effect of both input offset voltage and current on the output.

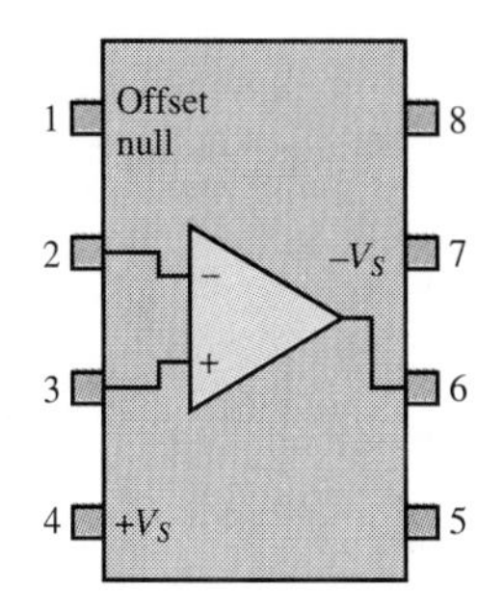

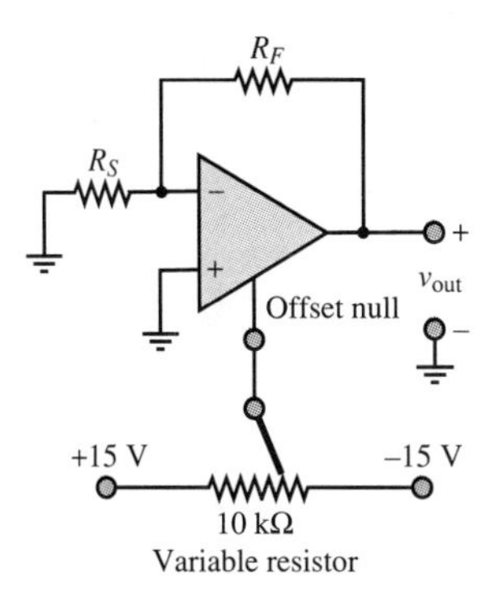

Figure 11.51 Output offset voltage adjustment

Slew Rate Limit

Another important restriction in the performance of a practical op-amp is associated with rapid changes in voltage. The op-amp can produce only a finite rate of change at its output. This limit rate is called the **slew rate.** Consider an ideal step input, where at $t = 0$ the input voltage is switched from zero to V volts. Then we would expect the output to switch from 0 to $A \cdot V$ volts, where A is the amplifier gain. However, $v_{out}(t)$ can change at only a finite rate; thus,

$$\left|\frac{dv_{out}(t)}{dt}\right|_{max} = S_0 = \text{Slew rate} \tag{11.87}$$

Figure 11.52 shows the response of an op-amp to an ideal step change in input voltage. Here, S_0, the slope of $v_{out}(t)$, represents the slew rate.

The slew rate limitation can affect sinusoidal signals, as well as signals that display abrupt changes, as does the step voltage of Figure 11.52. This may not be obvious until we examine the sinusoidal response more closely. It should be apparent that the maximum rate of change for a sinusoid occurs at the zero crossing, as shown by Figure 11.53. To evaluate the slope of the waveform at the zero crossing, let

$$v(t) = A \sin \omega t \tag{11.88}$$

so that

$$\frac{dv(t)}{dt} = \omega A \cos \omega t \tag{11.89}$$

The maximum slope of the sinusoidal signal will therefore occur at $\omega t = 0, \pi, 2\pi, \ldots,$ so that

$$\left|\frac{dv(t)}{dt}\right|_{max} = \omega \times A = S_0 \tag{11.90}$$

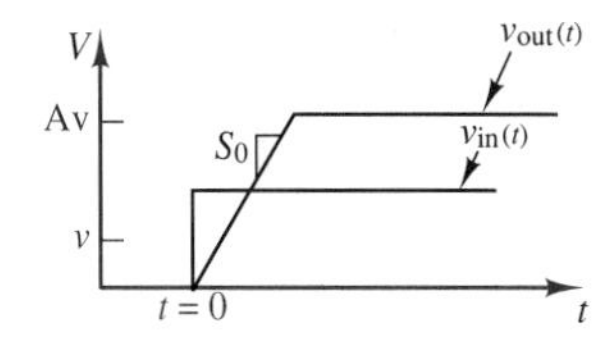

Figure 11.52 Slew rate limit in op-amps

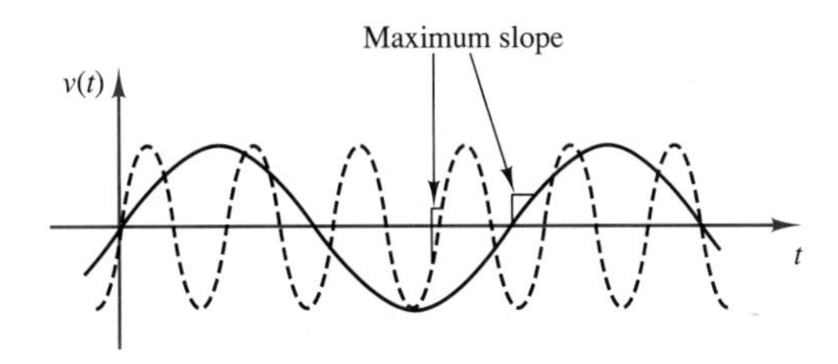

Figure 11.53 The maximum slope of a sinusoidal signal varies with the signal frequency

Thus, the maximum slope of a sinusoid is proportional to both the signal frequency and the amplitude. The curve shown by a dashed line in Figure 11.53 should indicate that as ω increases, so does the slope of $v(t)$ at the zero crossings. What is the direct consequence of this result, then? Example 11.18 gives an illustration of the effects of this slew rate limit.

Example 11.18

The manufacturer of an op-amp states that the slew rate of its device is

$$S_0 = 1 \text{ V}/\mu\text{s}$$

Would distortion be introduced in the output if we sought to amplify a 100,000-rad/s sinusoid of 1-V peak amplitude by a factor of 10, using an inverting amplifier configuration?

Solution:

Since

$$v_S(t) = 1 \sin(2\pi \times 10^5 t)$$

we can assume that the ideal output voltage would be

$$v_{\text{out}}(t) = -10 \sin(2\pi \times 10^5 t)$$

The maximum slope of $v_{\text{out}}(t)$ is

$$\left|\frac{dv(t)}{dt}\right|_{\text{max}} = A\omega = 10 \times 2\pi \times 10^5 = 6.28 \times 10^6 = 6.28 \text{ V}/\mu\text{s}$$

This slope far exceeds the slew rate of the op-amp, $S_0 = 1 \text{ V}/\mu\text{s}$. Figure 11.54 depicts the desired output (dashed line) and the actual output generated by the op-amp. Note how the distortion introduced by the slew rate limitation is so severe that the op-amp output resembles a triangular wave!

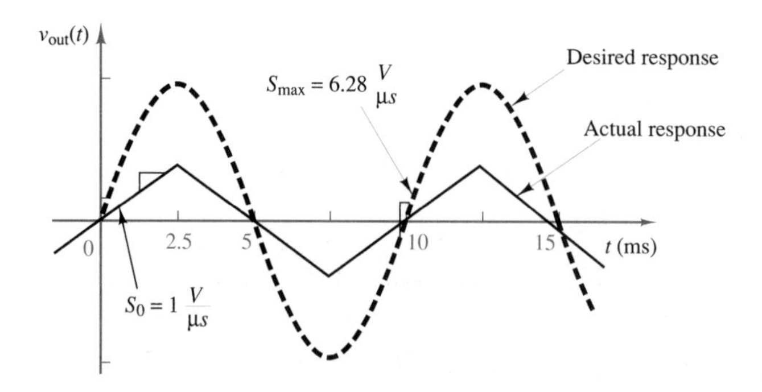

Figure 11.54 Distortion introduced by slew rate limit

Short-Circuit Output Current

Recall the model for the op-amp introduced in Section 11.2, which represented the internal circuits of the op-amp in terms of an equivalent input resistance, R_{in}, and a controlled voltage source, $A_V v_{in}$. In practice, the internal source is not ideal, because it cannot provide an infinite amount of current (either to the load, or to the feedback connection, or both). The immediate consequence of this nonideal op-amp characteristic is that the maximum output current of the amplifier is limited by the so-called short-circuit output current, I_{SC}:

$$|I_{out}| < I_{SC} \qquad \textbf{(11.91)}$$

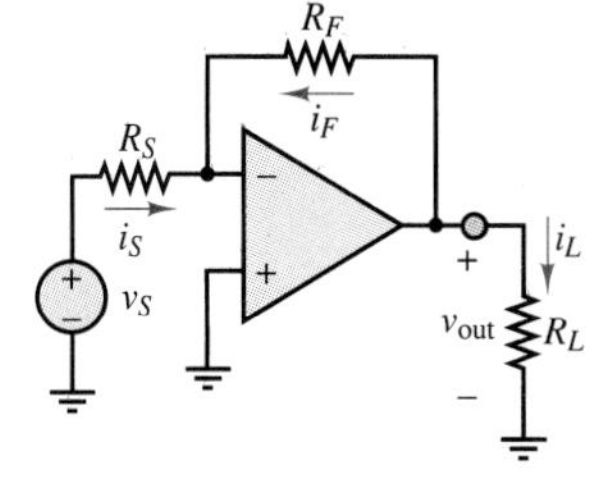

Figure 11.55

To further explain this point, consider that the op-amp needs to provide current to the feedback path (in order to "zero" the voltage differential at the input) and to whatever load resistance, R_L, may be connected to the output. Figure 11.55 illustrates this idea for the case of an inverting amplifier, where I_{SC} is the load current that would be provided to a short-circuit load ($R_L = 0$).

Example 11.19

This example illustrates the effects of the short-circuit current limit. For an inverting amplifier, assume that $v_S = 0.05 \sin \omega t$ and $R_F/R_S = 100$. Then, we would expect to see the output voltage

$$v_{out}(t) = -5 \sin \omega t$$

If the manufacturer's data sheet rates the short-circuit output current at $I_{SC} = 50$ mA, what is the smallest load resistance, R_L, for which the op-amp will generate an undistorted output $v_{out}(t) = -5 \sin \omega t$ V?

Solution:

Consider the peak values, for simplicity:

$$v_{out,\ peak} = 5 \text{ V}$$

$$I_{SC,\ peak} = 50 \text{ mA}$$

$$R_{L,\ min} = \frac{v_{out,\ peak}}{I_{SC,\ peak}} = \frac{5 \text{ V}}{50 \text{ mA}} = 100\ \Omega$$

For any load resistance smaller than 100 Ω, the output of the op-amp will be distorted. For example, if $R_L = 75\ \Omega$, we can calculate that the peak output voltage (which should ideally be 5 V) is only

$$v_{out,\ peak} = I_{SC} \times R_L = 50 \text{ mA} \times 75\ \Omega = 3.75 \text{ V}$$

The effect of this distortion is to *compress* the extreme values of the waveform. Unlike the voltage supply limit, this distortion is not a hard clipping, but rather a soft rounding of the peak. Figure 11.56 illustrates this point.

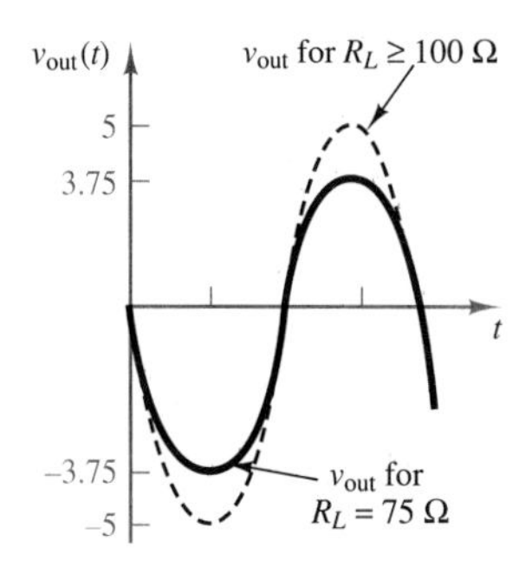

Figure 11.56 Distortion introduced by short-circuit current limit

Example 11.20

Consider an inverting amplifier with $R_S/R_F = 10$. If the input voltage is $v_S(t) = 1\cos\omega t$ V, find the lowest value of R_L such that the short-circuit current limit is not exceeded, if $I_{SC} = 20$ mA.

Solution:

The nominal output voltage of the amplifier is

$$v_{\text{out}} = -10\cos\omega t$$

Thus, the maximum amplitude of the output voltage is

$$|v_{\text{out}}|_{\text{max}} = 10 \text{ V}$$

and the minimum load resistance that will not cause the output short-circuit current limit to be exceeded may be found as follows:

$$(R_L)_{\text{min}} = \frac{|v_{\text{out}}|_{\text{max}}}{I_{SC}} = \frac{10}{0.02}\ \Omega$$

$$\text{Minimum } R_L = 500\ \Omega$$

Common-Mode Rejection Ratio (CMRR)

Example 11.3 introduced the notion of differential-mode and common-mode signals. If we define A_{dm} as the **differential-mode gain** and A_{cm} as the **common-mode gain** of the op-amp, the output of an op-amp can then be expressed as follows:

$$v_{\text{out}} = A_{dm}(v_2 - v_1) + A_{cm}\left(\frac{v_2 + v_1}{2}\right) \qquad \textbf{(11.92)}$$

Under ideal conditions, A_{cm} should be exactly zero, since the differential amplifier should completely reject common-mode signals. The departure from this ideal condition is a figure of merit for a differential amplifier and is measured by defining a quantity called the **common-mode rejection ratio (CMRR).** The CMRR is defined as the ratio of the differential-mode gain to the common-mode gain and should ideally be infinite:

$$\text{CMRR} = \frac{A_{dm}}{A_{cm}}$$

The CMRR is often expressed in units of decibels (dB). The common-mode rejection ratio idea is explored further in the problems at the end of the chapter.

DRILL EXERCISES

11.17 In Example 11.15, we implicitly assumed that the gain of each amplifier was constant for frequencies up to the cutoff frequency. This is, in practice, not true, since the individual op-amp closed-loop gain starts dropping below the DC gain value according to the equation

$$A(j\omega) = \frac{A_1}{1 + j\omega/\omega_1}$$

Thus, the calculations carried out in the example are only approximate. Find an expression for the closed-loop gain of the cascade amplifier. [*Hint:* The combined gain is equal to the product of the individual closed-loop gains.] What is the actual gain in dB at the cutoff frequency, ω_0, for the cascade amplifier?

11.18 What is the 3-dB bandwidth of the cascade amplifier of Example 11.15? [*Hint:* The gain of the cascade amplifier is the product of the individual op-amp frequency responses. Compute the magnitude of this product and set the magnitude of the product of the individual frequency responses equal to $(1/\sqrt{2}) \times 10{,}000$, and then solve for ω.]

Manufacturers generally supply values for the parameters discussed in this section in their device data specifications. Typical data sheets for a common op-amp are given at the end of the chapter.

CONCLUSION

This chapter has described the fundamental properties and limitations of the operational amplifier.

- Ideal amplifiers represent fundamental building blocks of electronic instrumentation. With the concept of the ideal amplifiers in mind, one can design practical amplifiers, filters, integrators, and other useful signal-processing circuits.
- The operational amplifier closely approximates the characteristics of an ideal amplifier. The analysis of op-amp circuits may be carried out very easily, if it is assumed that the op-amp's input resistance and open-loop gain are very large. The inverting, noninverting, and differential amplifier configurations permit the design of useful electronic amplifiers simply by selecting a few external resistors.
- If energy-storage elements are used in the construction of op-amp circuits, it is possible to accomplish the functions of filtering, integration, and differentiation.
- The properties of summing amplifiers and integrators make it possible to build analog computers, which serve as an aid in the solution of differential equations, and in the simulation of dynamic systems.
- When op-amps are employed in more advanced applications, it is important to know that there are limitations on their performance that are not predicted by the simple op-amp model introduced at the beginning of the chapter. These include voltage supply limits, frequency response limits, offset voltages and currents, slew rate limits, and finite common-mode rejection ratio. In general, it is not difficult to compensate for these limitations in the design of op-amp circuits.

EIT REVIEW: OPERATIONAL AMPLIFIERS

The content of Chapter 11, Sections 1 through 5, matches quite closely the requirements of the EIT exam in this area; Section 11.6 does not apply because EIT exam questions are targeted at ideal op-amps. For a review of this material we recommend revisiting Examples 11.1 through 11.9, and the specially marked homework problems, the solutions of which are supplied in Appendix B.

KEY TERMS

ANSWERS TO DRILL EXERCISES

11.1 $A_{V(CL)} = 999.9, 999.0, 990.1, 909.1$
$A_{V(OL)\ \min} = 10^6$ for 0.1% error in $A_{V(CL)}$.

11.2 $A_{V(CL)} = 99.99, 99.99, 99.90, 99.00$
$A_{V(OL)\ \min} = 10^5$ for 0.1% error in $A_{V(CL)}$.

11.4 $\Delta R = 6{,}714\ \Omega$

11.5 V_{ref} can have values between ± 0.714 V.

11.6 $R_F/R_S = 42.55$; $V_{\text{ref}} = 0.684$ V

11.7 $R_F = 19.9\ \text{k}\Omega$, $R_S = 199\ \Omega$

11.8 $R_F = 820\ \text{k}\Omega$, $R_S = 8.2\ \text{k}\Omega$; error: gain $=$ 0%, $\omega_{3\ \text{dB}} = 2.95\%$

11.9 407 Hz

11.10 -6 dB; $\omega_{3\ \text{dB}} = 0.642\omega_0$

11.11 -20 dB/decade

11.12 $+20$ dB/decade

11.14 $B/M = 5$; $K/M = 40$

11.15 $x_{M\ \max} = 0.0082$ m

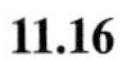

11.16

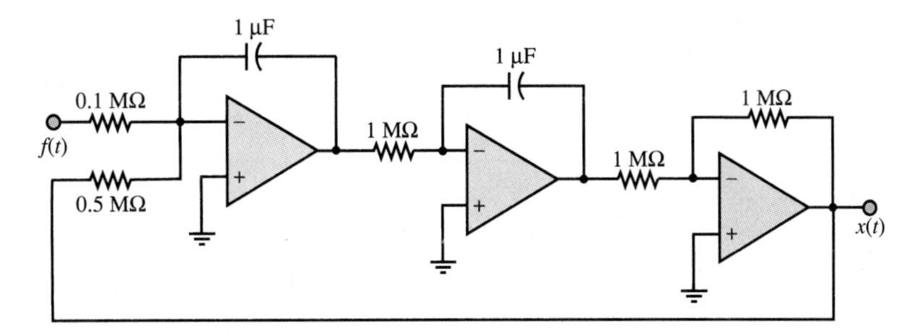

11.17 74 dB

11.18 $\omega_{3\text{ dB}} = 2\pi \times 12{,}800$ rad/s

DEVICE DATA SHEETS

LM741C Operational Amplifier

Absolute maximum ratings

	LM741C
Supply Voltage	±18 V
Power Dissipation (Note 1)	500 mW
Differential Input Voltage	±30 V
Input Voltage (Note 2)	±15 V
Output Short Circuit Duration	Continuous
Operating Temperature Range	0°C to +70°C
Storage Temperature Range	−65°C to +150°C
Junction Temperature	100°C

Electrical characteristics

Parameter	Conditions	LM741C			Units
		Min	Typ	Max	
Input offset voltage	$T_A = 25°C$ $R_S \leq 10\text{ k}\Omega$		2.0	6.0	mV
	$T_{A\text{ min}} \leq T_A \leq T_{A\text{ max}}$ $R_S \leq 10\text{ k}\Omega$			7.5	mV
Average input offset voltage drift					μV/°C
Input offset voltage adjustment range	$T_A = 25°C, V_S = \pm 20V$		±15		mV
Input offset current	$T_A = 25°C$		20	200	nA
	$T_{A\text{ min}} \leq T_A \leq T_{A\text{ max}}$			300	nA
Average input offset current drift					nA/°C
Input bias current	$T_A = 25°C$		80	500	nA
	$T_{A\text{ min}} \leq T_A \leq T_{A\text{ max}}$			0.8	μA

LM741C Operational Amplifier (continued)

Parameter	Conditions	LM741C Min	LM741C Typ	LM741C Max	Units
Input resistance	$T_A = 25°C, V_S = \pm 20$ V	0.3	2.0		MΩ
Input voltage range	$T_A = 25°C$	±12	±13		V
Large-signal voltage gain	$T_A = 25°C, R_L \geq 2$ kΩ $V_S = \pm 20$ V, $V_o = \pm 15$ V $V_S = \pm 15$ V, $V_o = \pm 10$ V	20	200		V/mV V/mV
	$T_{A\,\min} \leq T_A \leq T_{A\,\max}$, $R_L \geq 2$ kΩ, $V_S = \pm 15$ V, $V_o = \pm 10$ V	15			V/mV
Output voltage swing	$V_S = \pm 15$ V $R_L \geq 10$ kΩ $R_L \geq 2$ kΩ	±12 ±10	±14 ±13		V V
Output short-circuit current	$T_A = 25°C$		25		mA
Common-mode rejection ratio	$T_{A\,\min} \leq T_A \leq T_{A\,\max}$ $R_S \leq 10$ kΩ, $V_{cm} = \pm 12$ V	70	90		dB
Supply voltage rejection ratio	$T_{A\,\min} \leq T_A \leq T_{A\,\max}$, $V_S = \pm 20$ V to $V_S = \pm 5$ V $R_S \leq 10$ Ω	77	96		dB
Transient response rise time overshoot	$T_A = 25°C$, unity gain		0.3 5		μs %
Bandwidth	$T_A = 25°C$				MHz
Slew rate	$T_A = 25°C$, unity gain		0.5		V/μs
Supply current	$T_A = 25°C$		1.7	2.8	mA

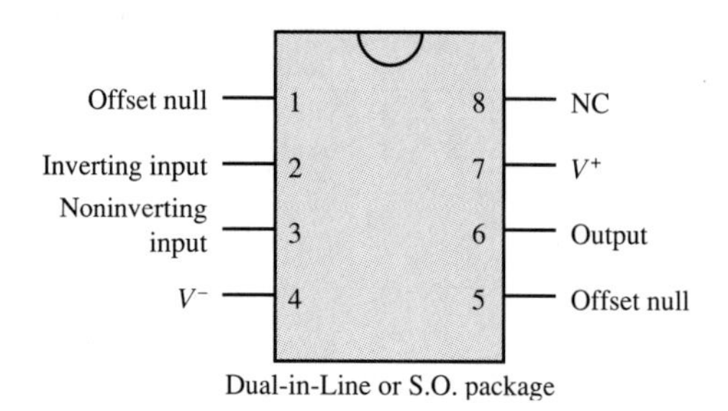

Dual-in-Line or S.O. package

HOMEWORK PROBLEMS

11.1 Find v_1 in Figure P11.1(a) and (b). Note how the voltage follower holds v_1 in Figure P11.1(b) to $v_g/2$, while the 3-kΩ resistor "loads" the output in Figure P11.1(a).

11.2 Determine an expression for the overall gain $A_V = v_o/v_i$ for the circuit of Figure P11.2. Assume the op-amp is ideal.

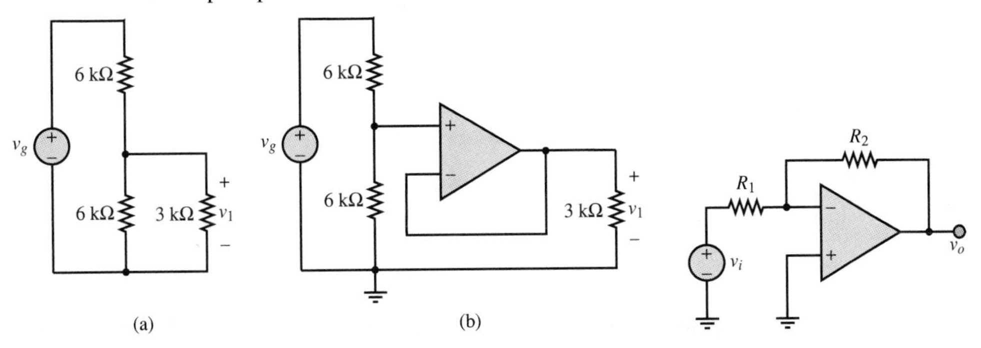

Figure P11.1

Figure P11.2

EIT 11.3 In the circuit of Figure P11.3, find the voltage v.

11.4 Determine an expression for the overall gain $A_V = v_o/v_i$ for the circuit of Figure P11.4. Assume the op-amp is ideal.

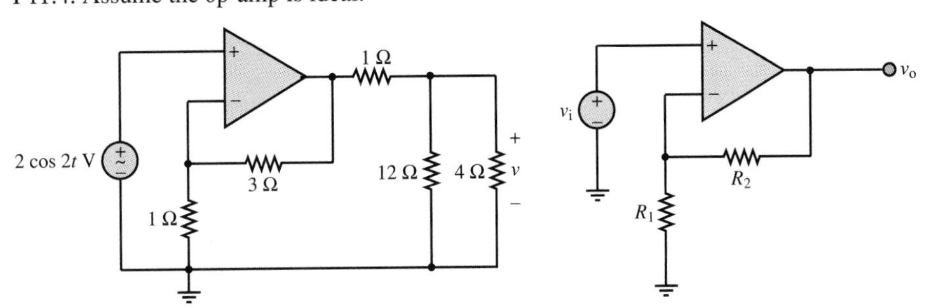

Figure P11.3

Figure P11.4

11.5 In the circuit of Figure P11.5, find the current i.

EIT 11.6 In the circuit of Figure P11.6, find the voltage v_o.

EIT 11.7 In the circuit of Figure P11.7, find the current i.

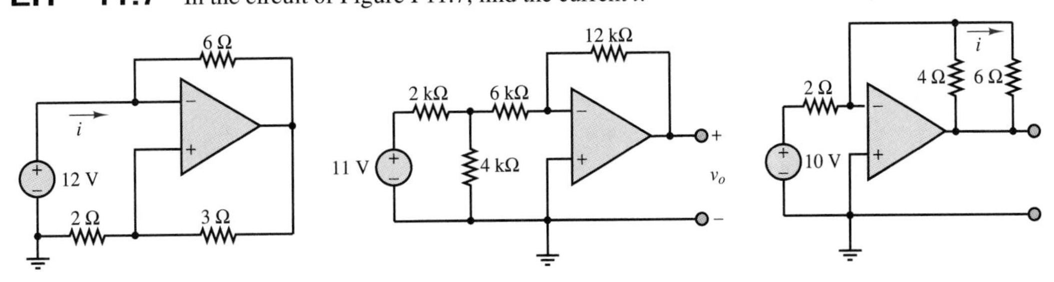

Figure P11.5

Figure P11.6

Figure P11.7

11.8 Determine an expression for the output voltage v_o in terms of the input voltages v_1 and v_2 in the circuit of Figure P11.8. Assume the op-amp is ideal.

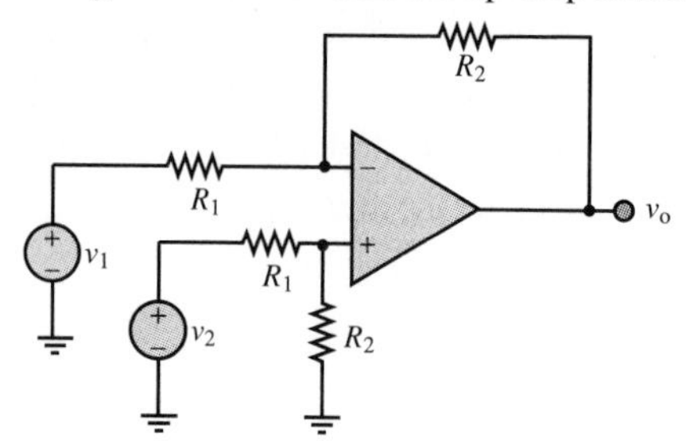

Figure P11.8

11.9 In the circuit of Figure P11.9, find the current i.

11.10 Show that the circuit of Figure P11.10 is a noninverting summer.

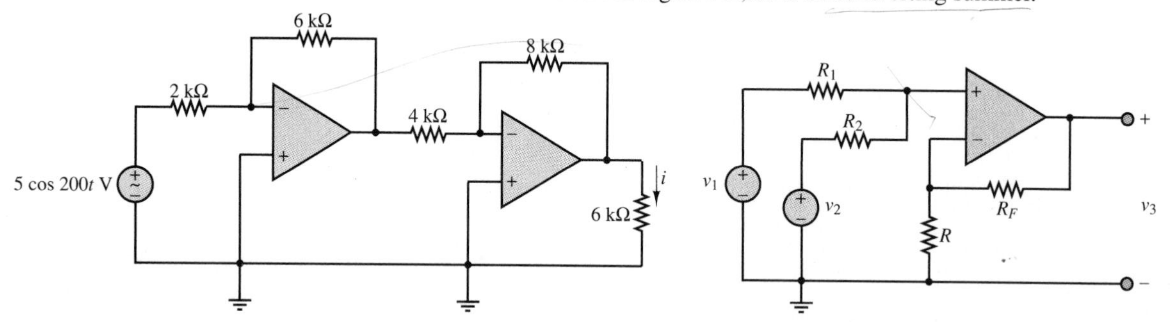

Figure P11.9

Figure P11.10

11.11 Determine an expression for the overall gain $A_V = v_o/v_i$ for the circuit of Figure P11.11. Find the conductance $G = i_i/v_i$ seen by the voltage source, v_i. Assume the op-amps are ideal.

11.12 Determine an expression for the overall gain $A_V = v_o/v_i$ for the circuit of Figure P11.12. Find the conductance $G = i_i/v_i$ seen by the voltage source, v_i. Assume the op-amp is ideal.

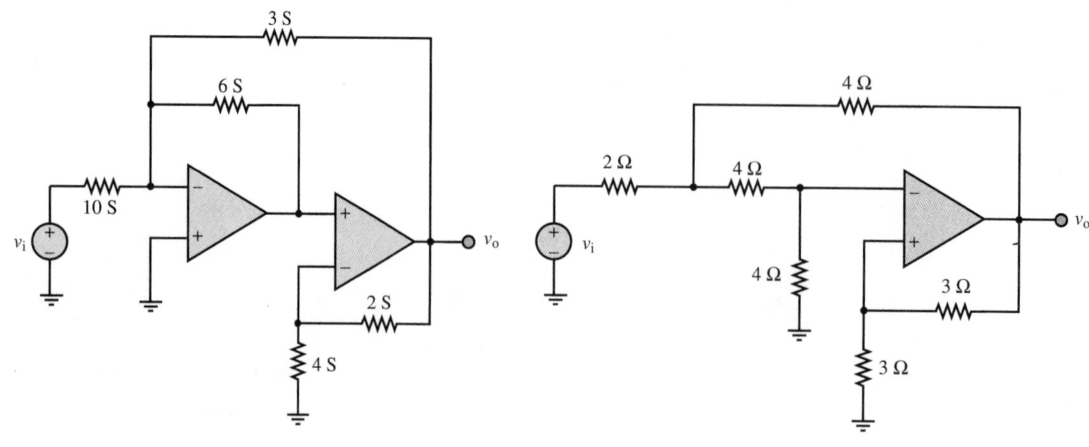

Figure P11.11

Figure P11.12

11.13 Find the Thévenin equivalent of the op-amp circuit to the left of terminals *a-b* in Figure P11.13. Assume the op-amp is ideal.

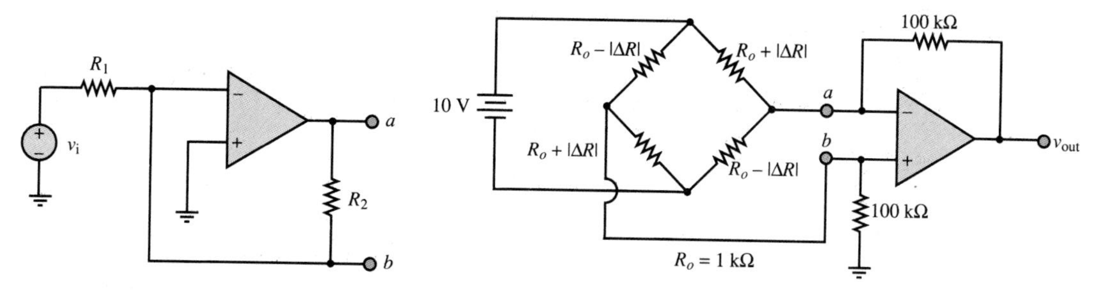

Figure P11.13 **Figure P11.14**

EIT 11.14 Differential amplifiers are often used in conjunction with the Wheatstone bridge. Consider the bridge shown in Figure P11.14, where each resistor is a temperature-sensing element and the change in resistance is directly proportional to a change in temperature—that is, $\Delta R = \alpha(\pm\Delta T)$, where the sign is determined by the positive or negative temperature coefficient of the resistive element.

a. Find the Thévenin equivalent that the amplifier sees at point *a* and at point *b*. Assume that $|\Delta R|^2 \ll R_o$.

b. If $|\Delta R| = K\Delta T$, with *K* a numerical constant, find an expression for $v_{out}(\Delta T)$, that is, v_{out} as a function of change in temperature.

11.15 In the circuit of Figure P11.15, it is critical that the gain remain within 2 percent of its nominal value, 16. Find the resistor, R_S, that will accomplish the nominal gain requirement, and state what the maximum and minimum values of R_S can be. Will a *standard* 5 percent tolerance resistor be adequate to satisfy this requirement? (See Chapter 2 for resistor standard values.)

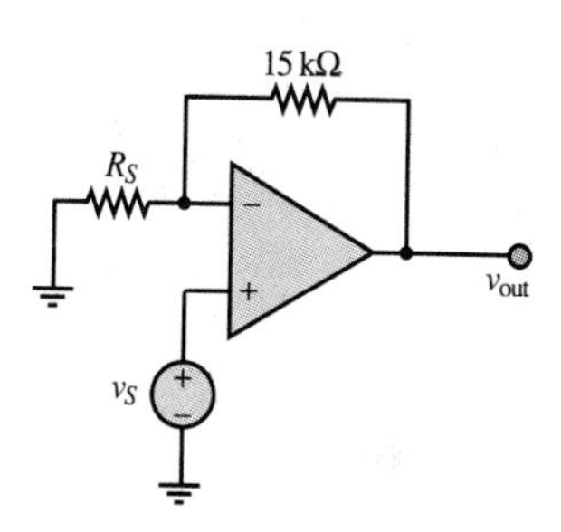

Figure P11.15

11.16 An inverting amplifier uses two 10 percent tolerance resistors: $R_F = 33$ kΩ, and $R_S = 1.2$ kΩ.

a. What is the nominal gain of the amplifier?
b. What is the maximum value of $|A_V|$?
c. What is the minimum value of $|A_V|$?

EIT 11.17 Find the voltage gain, $A_V = v_{out}/v_S$, for the op-amp circuit of Figure P11.16. $R_{S1} = 10$ kΩ, $R_{F1} = 47$ kΩ, $R_{F2} = 10$ kΩ, and $R_{S2} = 10$ kΩ.

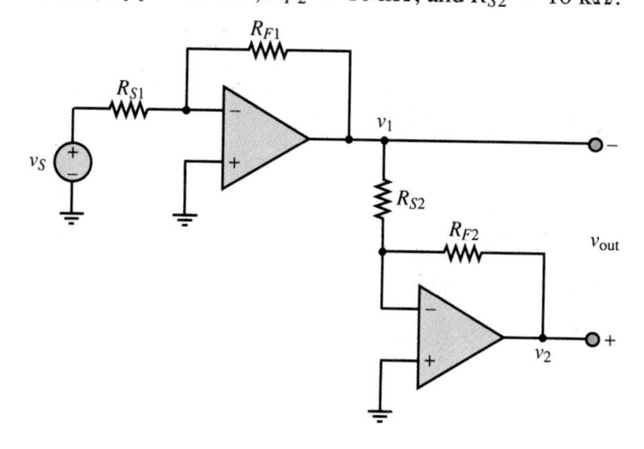

Figure P11.16

11.18 The circuit of Figure P11.17 includes a combination of resistors and op-amps. If $v_S(t) = 0.01 \cos \omega t$, find the output voltage.

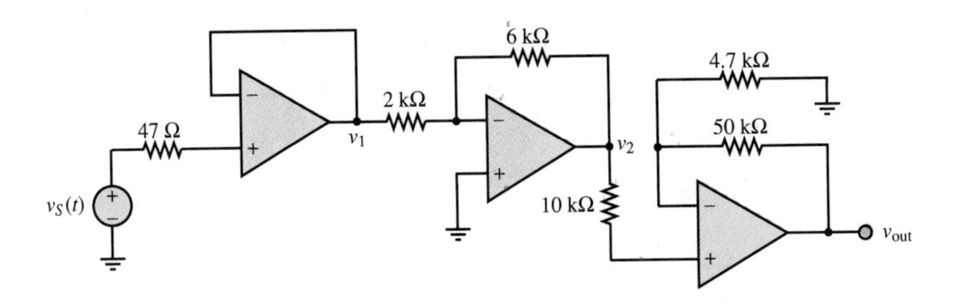

Figure P11.17

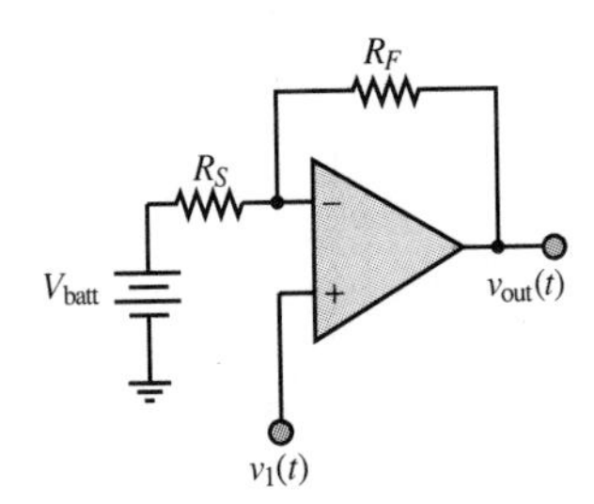

Figure P11.18

11.19 The circuit of Figure P11.18 will remove the DC portion of the input voltage, $v_1(t)$, while amplifying the AC portion. Let $v_1(t) = 10 + 10^{-3} \sin \omega t$ V, $R_F = 10$ kΩ, and $V_{batt} = 20$ V.

a. Find R_S such that no DC voltage appears at the output.
b. What is $v_{out}(t)$, using R_S from part a?

11.20 Figure P11.19 shows a simple practical amplifier that uses the 741 op-amp. Pin numbers are as indicated. Assume the input resistance is $R = 2$ MΩ, the open-loop gain $K = 200{,}000$ and output resistance $R_o = 75$ Ω. Find the gain $A_V = v_o/v_i$ approximately.

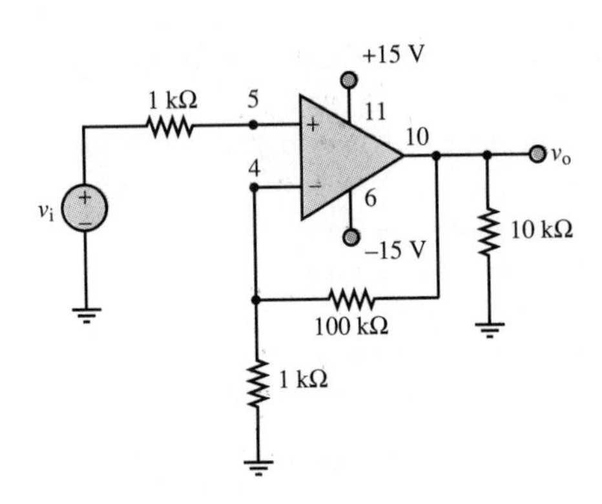

Figure P11.19

11.21 Use an inverting summing amplifier to obtain the following weighted sum of four different signal sources:

$$v_{out} = -\left(\frac{1}{4} \sin \omega_1 t + 5 \sin \omega_2 t + 2 \sin \omega_3 t + 16 \sin \omega_4 t\right)$$

Assume that $R_F = 10$ kΩ, and determine the required source resistors.

11.22 Figure P11.20 shows the s-domain representation for a low-pass filter.

a. Find the voltage gain $A_V(s) = V_o(s)/V_I(s)$.
b. Plot the voltage gain in dB vs. ω.

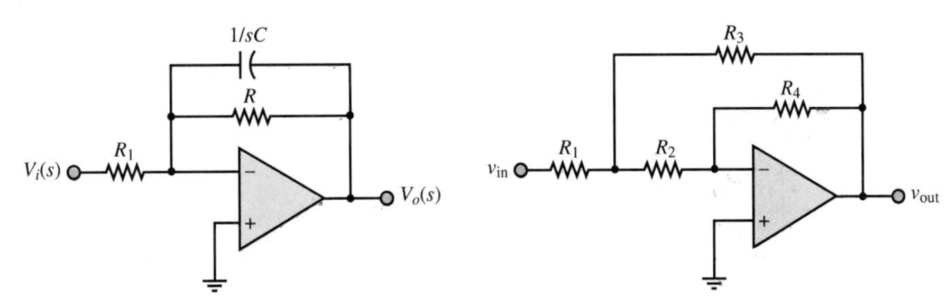

Figure P11.20 **Figure P11.21**

EIT 11.23 Find the gain of the amplifier circuit in Figure P11.21. Assume that $R_1 = R_2 = R_3 = 3$ kΩ and $R_4 = 1$ kΩ.

11.24 The unity-follower amplifier of Figure P11.22 has a voltage gain of 1 and the output is in phase with the input. It also has an extremely high input impedance leading to its use as a *buffer.* Assume a practical op-amp having an open-loop gain of 10^6 and input resistance of 1 MΩ.

a. Show that $v_o \approx v_i$.
b. Find an expression for the amplifier input impedance.

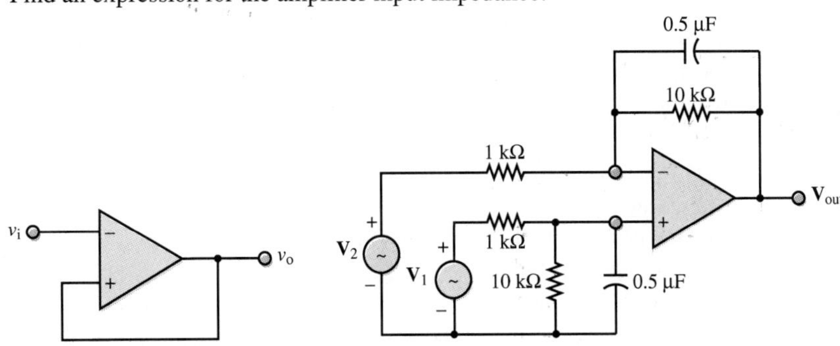

Figure P11.22 **Figure P11.23**

11.25 Consider the circuit of Figure P11.23.

a. If $v_1 - v_2 = 1 \cos(1{,}000t)$ V, find the peak amplitude of v_{out}.
b. Find the phase shift of v_{out}.

[*Hint:* Use phasor analysis.]

EIT 11.26 The inverting amplifier shown in Figure P11.24 can be used as a high-pass filter.

a. Derive frequency response of the circuit.
b. If $R_1 = R_2 = 10$ kΩ and $C = 1$ μF, compute attenuation in dB at $\omega = 1{,}000$ rad/s.
c. Compute gain and phase at $\omega = 1{,}000$ rad/s.
d. Find range of frequencies over which the attenuation is less than 2 dB.

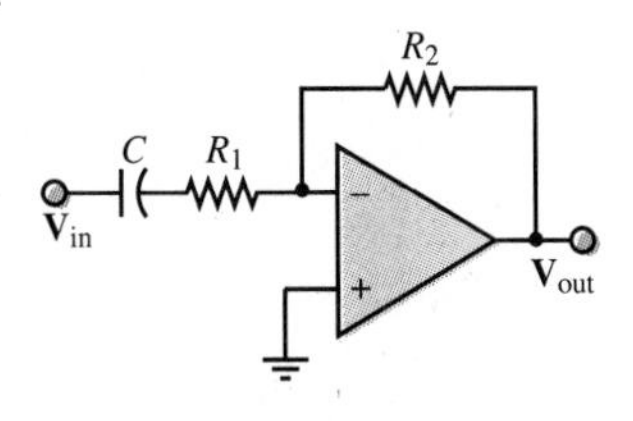

Figure P11.24

11.27 Repeat Problem 11.26 for $R_1 = 20$ kΩ, $R_2 = 10$ kΩ, and $C = 0.5$ μF.

11.28 The circuit shown in Figure P11.25 is an integrator. The capacitor is initially uncharged, and the source voltage is

$$v_{in}(t) = 10 \times 10^{-3} + \sin(2{,}000\pi t) \text{ V}$$

a. At $t = 0$, the switch, S_1, is closed. How long does it take before clipping occurs at the output if $R_s = 10$ kΩ and $C_F = 0.008$ μF?
b. At what times does the integration of the DC input cause the op-amp to saturate fully?

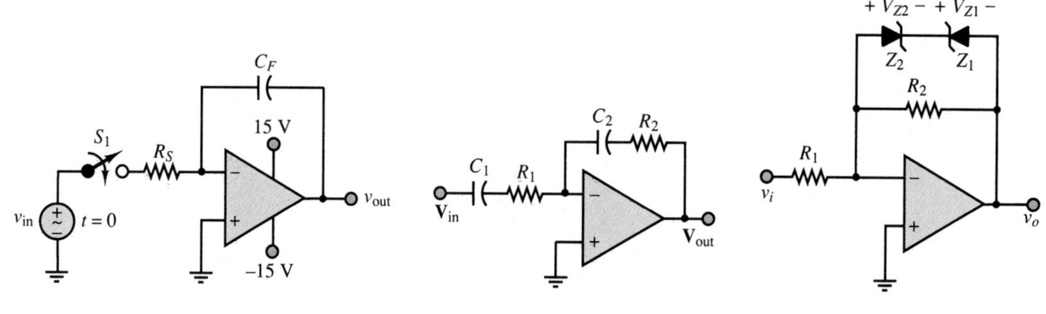

Figure P11.25 **Figure P11.26** **Figure P11.27**

11.29 A practical integrator is shown in Figure 11.21 in the text. Note that the resistor in parallel with the feedback capacitor provides a path for the capacitor to discharge the DC voltage. Usually, the time constant $R_F C_F$ is chosen to be long enough not to interfere with the integration.

a. If $R_S = 10\ \text{k}\Omega$, $R_F = 2\ \text{M}\Omega$, $C_F = 0.008\ \mu\text{F}$, and $v_S(t) = 10\ \text{V} + \sin(2{,}000\pi t)$ V, find $v_{out}(t)$ using phasor analysis.

b. Repeat part a if $R_F = 200\ \text{k}\Omega$, and if $R_F = 20\ \text{k}\Omega$.

c. Compare the time constants $R_F C_F$ with the period of the waveform for parts a and b. What can you say about the time constant and the ability of the circuit to integrate?

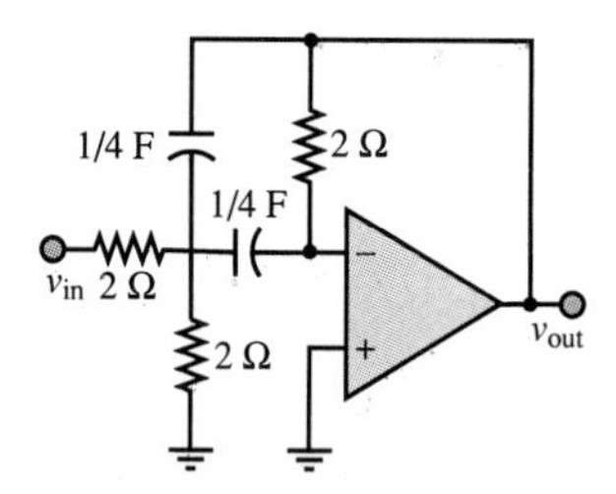

Figure P11.28

EIT 11.30 For the circuit of Figure P11.26,

a. Derive the frequency response of the circuit.

b. If $R_1 = 10\ \text{k}\Omega$, $R_2 = 20\ \text{k}\Omega$, $C = 1\ \mu\text{F}$, and $C_2 = 2\ \mu\text{F}$, compute the attenuation in dB at $\omega = 1{,}500$ rad/s.

c. Compute gain and phase shift at $\omega = 2{,}000$ rad/s.

11.31 In some signal-processing applications, a *clamping* circuit is used to hold the output at a certain level even when the input continues to increase. One such circuit is shown in Figure P11.27. Assume the Zener diodes and op-amp are ideal. Determine the relationship between v_o and v_i and sketch it.

11.32 Compute the frequency response of the circuit shown in Figure P11.28.

11.33 The circuit of Figure 11.26 in the text is a practical differentiator. Assuming an ideal op-amp with $v_S(t) = 10 \times 10^{-3} \sin(2{,}000\pi t)$ V, $C_S = 100\ \mu\text{F}$, $C_F = 0.008\ \mu\text{F}$, $R_F = 2\ \text{M}\Omega$, and $R_S = 10\ \text{k}\Omega$,

a. Determine the frequency response, $\frac{V_o}{V_S}(\omega)$.

b. Use superposition to find the actual output voltage (remember that DC = 0 Hz).

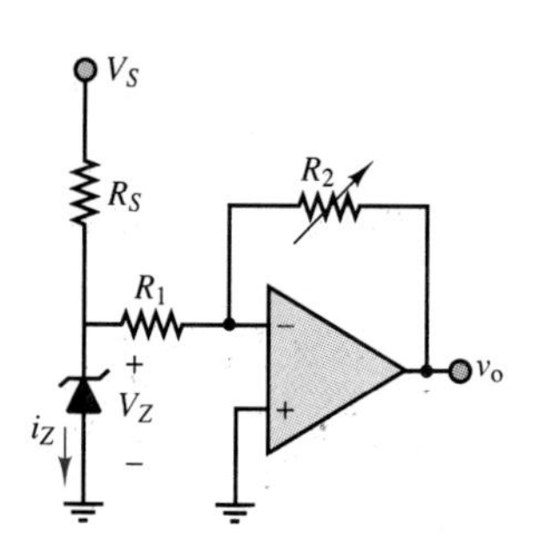

Figure P11.29

11.34 The circuit of Figure P11.29 serves as a voltage regulator whose output can be varied. Assume an ideal op-amp and that the Zener diode will hold its terminal voltage provided $i_Z \geq 0.1 I_Z$.

a. Find an expression for v_o in terms of V_Z.

b. If R_S, R_1, V_Z, and I_Z are known, specify the range of V_S over which the circuit could regulate.

EIT 11.35 The inverting amplifier shown in Figure P11.30 can be used as a low-pass filter.

a. Derive frequency response of the circuit.

b. If $R_1 = R_2 = 100\ \text{k}\Omega$ and $C = 0.1\ \mu\text{F}$, compute attenuation in dB at $\omega = 1{,}000$ rad/s.

c. Compute gain and phase at $\omega = 2{,}500$ rad/s.

d. Find range of frequencies over which the attenuation is less than 1 dB.

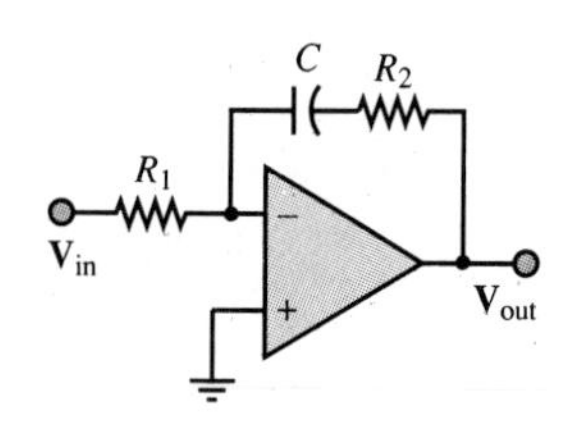

Figure P11.30

11.36 An op-amp voltmeter circuit as in Figure P11.31 is required to measure a maximum input of $E = 20$ mV. The op-amp input current is $I_B = 0.2\ \mu$A, and the meter circuit has $I_m = 100\ \mu$A full-scale deflection and $r_m = 10$ kΩ. Determine suitable values for R_3 and R_4.

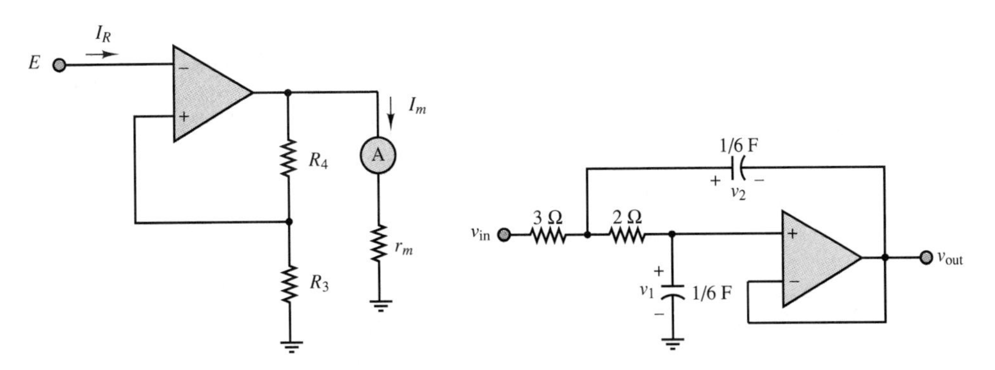

Figure P11.31 **Figure P11.32**

11.37 Find an expression for the gain of the circuit of Figure P11.32.

11.38 Reduce the number of op-amps in the analog computer simulation of Figure 11.39 by combining the summer and the first integrator into a single op-amp circuit, and by eliminating one of the two coefficient multipliers.

EIT 11.39 Derive the differential equation corresponding to the analog computer simulation circuit of Figure P11.33.

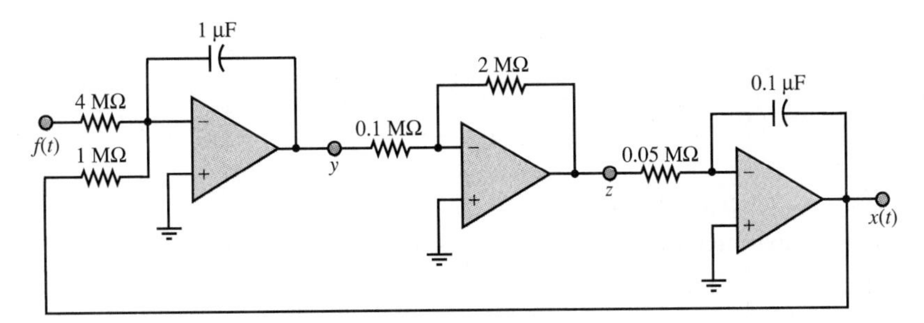

Figure P11.33

11.40 Construct the analog computer simulation corresponding to the following differential equation:

$$\frac{d^2x}{dt^2} + 100\frac{dx}{dt} + 10x = -5f(t)$$

11.41 The ideal charge amplifier discussed in Example 11.9 will saturate in the presence of any DC offsets, as discussed in Section 11.6. The circuit of Figure P11.34 represents a practical charge amplifier, in which the user is provided with a choice of three time constants—$\tau_{\text{long}} = R_L C_F$, $\tau_{\text{medium}} = R_M C_F$, $\tau_{\text{short}} = R_S C_F$—which can be selected by means of a switch. Assume that $R_L = 10$ MΩ, $R_M = 1$ MΩ, $R_S = 0.1$ MΩ, and $C_F = 0.1\ \mu$F. Analyze the frequency response of the practical charge amplifier for each case, and determine the lowest input signal frequency that can be amplified without excessive distortion for each case. Can this circuit amplify a DC signal?

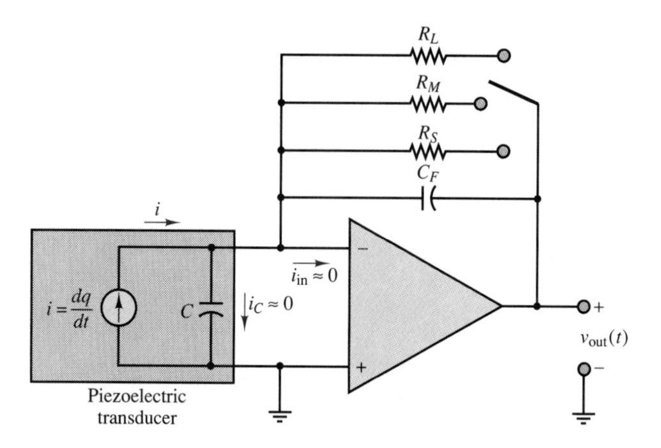

Figure P11.34

11.42 Consider a differential amplifier. We would desire the common-mode output to be less than 1 percent of the differential-mode output. Find the minimum dB common-mode rejection ratio (CMRR) to fulfill this requirement if the differential mode gain $A_{dm} = 1{,}000$. Let

$$v_1 = \sin(2{,}000\pi t) + 0.1 \sin(120\pi t) \text{ V}$$

$$v_2 = \sin(2{,}000\pi t + 180°) + 0.1 \sin(120\pi t) \text{ V}$$

$$v_o = A_{dm}(v_1 - v_2) + A_{cm}\left(\frac{v_1 + v_2}{2}\right)$$

11.43 Consider a standard inverting amplifier.

a. Find a relationship showing that the closed-loop gain, $A_{V(CL)}$, is a function of R_F, R_S, and $A_{V(OL)}$; that is, find the closed-loop gain as a function of the resistors R_F and R_S and the open-loop gain.
b. Plot $|A_{V(CL)}|$ versus $A_{V(OL)}$ on a range of $1 \le A_{V(OL)} \le 10^6$ if $R_F/R_S = 9$. Use a logarithmic scale for the open-loop gain axis.
c. What conclusion can you draw from these results?

11.44 Use equation 11.92 to show that if $A_{dm} \gg A_{cm}$ and $A_{dm} \gg 1 + R_2/R_1$ in the circuit of Figure 11.10 in the text, then

$$v_{out} = \frac{R_2}{R_1}\left[(v_2 - v_1) + \frac{1}{\text{CMRR}}\frac{v_1 + v_2}{2}\right]$$

If $R_1 = R_2$, $v_2 = 5$ V, $v_1 = 3$ V, and CMRR = 40 dB, find v_{out}.

EIT **11.45** For the circuit of Figure P11.35,

a. Derive frequency response of the circuit.
b. If $R_1 = R_2 = 10$ kΩ and $C = 0.1$ μF, compute the attenuation in dB at $\omega = 1{,}500$ rad/s.
c. Compute the amplitude gain and phase shift at $\omega = 2{,}000$ rad/s.
d. Find range of frequencies over which the attenuation is less than 5 percent.
e. What is the attenuation of part d in dB?

[*Hint:* Find the transfer function in the form of $\frac{\mathbf{V}_{out}}{(\mathbf{V}_1 - \mathbf{V}_2)}(j\omega)$.]

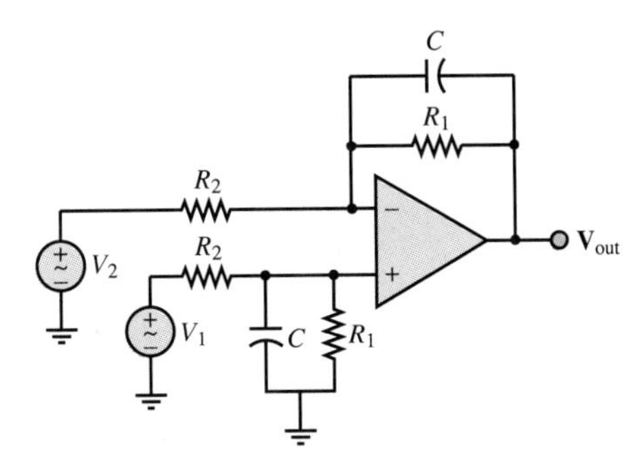

Figure P11.35

11.46 Sketch the amplitude response of V_2/V_1 for the circuit of Figure P11.36. Indicate the half-power frequency. Assume the op-amp is ideal.

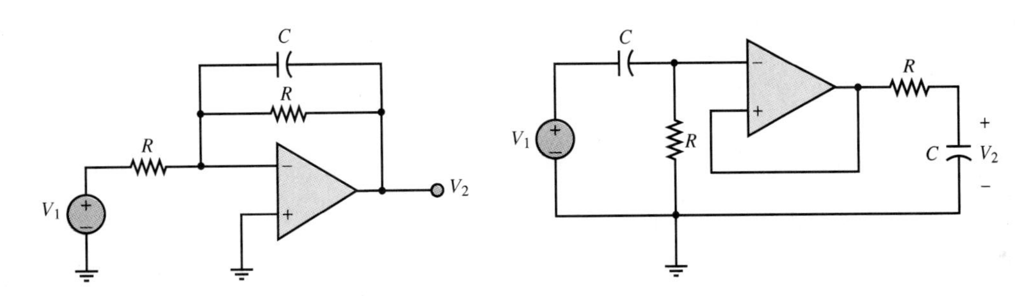

Figure P11.36 **Figure P11.37**

11.47 For the circuit of Figure P11.37, sketch the amplitude response of V_2/V_1, indicating the half-power frequencies. Assume the op-amp is ideal.

EIT 11.48

a. Given the voltage follower of Example 11.2, find an expression for the input impedance of the circuit in terms of R_{in}, $A_{V(OL)}$, and R_{out}.
b. Using the parameters of the data sheet for the μA741 op-amp, determine the input resistance of the voltage follower.

11.49 Square wave testing can be used with operational amplifiers to estimate the *slew rate,* which is defined as the maximum rate at which the output can change (in V/μs). Input and output waveforms for a noninverting op-amp circuit are shown in Figure P11.38. As indicated, the rise time, t_R, of the output waveform is defined as the time it takes for that waveform to increase from 10 percent to 90 percent of its final value; i.e.,

$$t_R \triangleq t_B - t_A = -\tau(\ln 0.1 - \ln 0.9) = 2.2\tau$$

where τ is the circuit time constant. Estimate the slew rate for the op-amp.

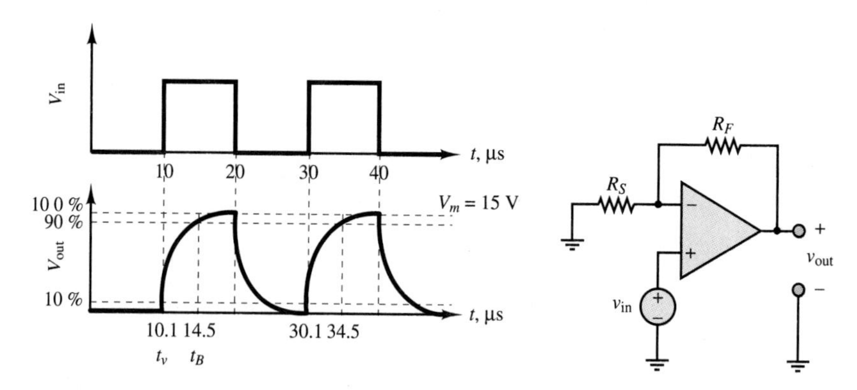

Figure P11.38 **Figure P11.39**

11.50 Consider an inverting amplifier with open-loop gain 10^5. With reference to equation 11.18,

a. If $R_S = 10\ \text{k}\Omega$ and $R_F = 1\ \text{M}\Omega$, find the voltage gain $A_{V(CL)}$.
b. Repeat part a if $R_S = 10\ \text{k}\Omega$ and $R_F = 10\ \text{M}\Omega$.
c. Repeat part a if $R_S = 10\ \text{k}\Omega$ and $R_F = 100\ \text{M}\Omega$.
d. Using the resistor values of part c, find $A_{V(CL)}$ if $A_{V(OL)} \to \infty$.

11.51

a. If the op-amp of Figure P11.39 has an open-loop gain of 45×10^5, find the closed-loop gain for $R_F = R_S = 7.5\ \text{k}\Omega$. Hint: Derive an equation similar to 11.18 that is valid for a non-inverting amplifier.
b. Repeat part a if $R_F = 5(R_S) = 37{,}500\ \Omega$.

EIT 11.52

a. Determine the closed-loop gain of the circuit shown in Figure P11.40 as a function of the resistors and of $A_{V(OL)}$, if $A_{V(OL)} = 10^6$.
b. If $R_2 = R_3 = 642.8\ \Omega$, $R_1 = R_4 = 6{,}224\ \Omega$, and $v_{in}(t) = 4.68 \cos 27.6t$, find $v_{out}(t)$.

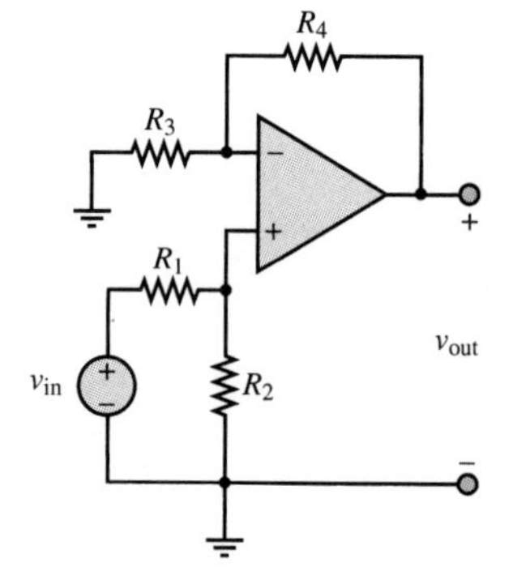

Figure P11.40

11.53 Given the unity-gain bandwidth for an ideal op-amp equal to 5.0 MHz, find the voltage gain at a frequency of $f = 500$ kHz.

CHAPTER

12

Digital Logic Circuits

Digital computers have taken a prominent place in engineering and science over the last two decades, performing a number of essential functions such as numerical computations and data acquisition. It is not necessary to further stress the importance of these electronic systems in this book, since you have certainly already had some encounters with digital computers and programming languages. The objective of the chapter is to discuss the essential features of digital logic circuits, which are at the heart of digital computers, by presenting an introduction to *combinational logic circuits*.

The chapter starts with a discussion of the binary number system, and continues with an introduction to Boolean algebra. The self-contained treatment of Boolean algebra will enable you to design simple logic functions using the techniques of combinational logic, and several practical examples are provided to demonstrate that even simple combinations of logic gates can serve to implement useful circuits in engineering practice. In a later section, we introduce a number of logic modules which can be described using simple logic gates but which provide more advanced functions. Among these, we discuss read-only memories, multiplexers, and decoders. Throughout the chapter, simple examples are given to demonstrate the usefulness of digital logic circuits in various engineering applications.

Chapter 12 provides the background needed to address the study of digital systems, which will be undertaken in Chapter 13. Upon completion of the chapter, you should be able to:

- Perform operations using the binary number system.
- Design simple combinational logic circuits using logic gates.
- Use Karnaugh maps to realize logical expressions.
- Interpret data sheets for multiplexers, decoders, and memory ICs.

12.1 ANALOG AND DIGITAL SIGNALS

One of the fundamental distinctions in the study of electronic circuits (and in the analysis of any signals derived from physical measurements) is that between analog and digital signals. As discussed in the preceding chapter, an **analog signal** is an electrical signal whose value varies in analogy with a physical quantity (e.g., temperature, force, or acceleration). For example, a voltage proportional to a measured variable pressure or to a vibration naturally varies in an analog fashion. Figure 12.1 depicts an arbitrary analog function of time, $f(t)$. We note immediately that for each value of time, t, $f(t)$ can take one value among any of the values in a given range. For example, in the case of the output voltage of an op-amp, we expect the signal to take any value between $+V_{\text{sat}}$ and $-V_{\text{sat}}$, where V_{sat} is the supply-imposed saturation voltage.

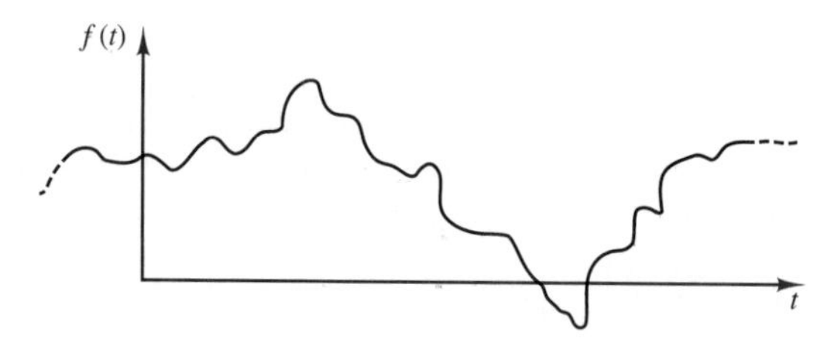

Figure 12.1 Analog signal

A **digital signal,** on the other hand, can take only a *finite number of values.* This is an extremely important distinction, as will be shown shortly. An example of a digital signal is a signal that allows display of a temperature measurement on a digital readout. Let us hypothesize that the digital readout is three digits long and can display numbers from 0 to 100, and let us assume that the temperature sensor is correctly calibrated to measure temperatures from 0 to 100°F. Further, the output of the sensor ranges from 0 to 5 volts, where 0 V corresponds to 0°F and 5 V to 100°F. Therefore, the calibration constant of the sensor is $k_T = \frac{100° - 0°}{5 - 0} = 20°/\text{V}$. Clearly, the output of the sensor is an analog signal; however, the display can show only a finite number of readouts (101, to be precise). Because the display itself can only take a value out of a discrete set of states—the integers from 0 to 100—we call it a digital display, indicating that the variable displayed is expressed in digital form.

Now, each temperature on the display corresponds to a *range of voltages:* each digit on the display represents one hundredth of the 5-volt range of the sensor, or 0.05 V $=$ 50 mV. Thus, the display will read 0 if the sensor voltage is between 0 and 49 mV, 1 if it is between 50 and 99 mV, and so on. Figure 12.2 depicts the staircase function relationship between the analog voltage and the digital readout. This **quantization** of the sensor output voltage is in effect an approximation. If one wished to know the temperature with greater precision, a greater number of display digits could be employed.

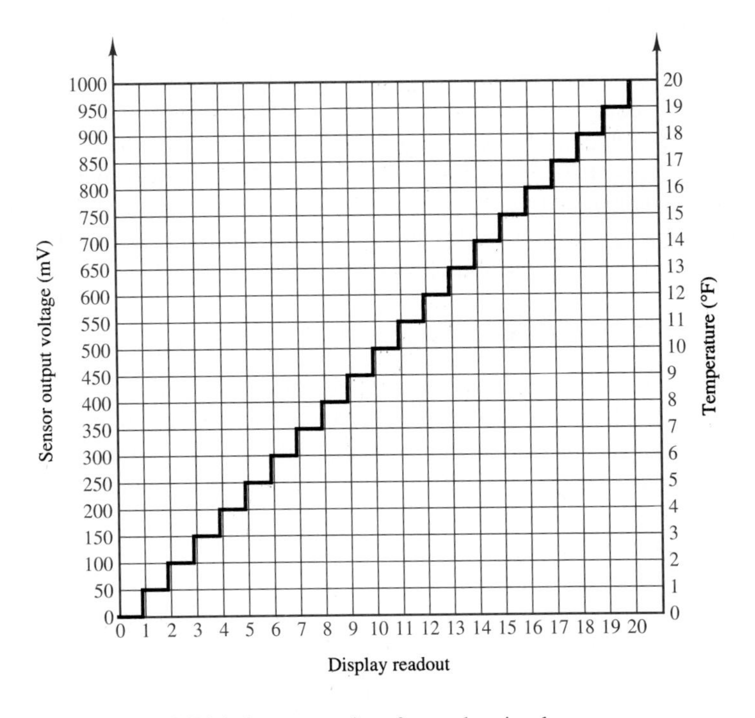

Figure 12.2 Digital representation of an analog signal

The most common digital signals are binary signals. A **binary signal** is a signal that can take only one of two discrete values and is therefore characterized by transitions between two states. Figure 12.3 displays a typical binary signal. In binary arithmetic (which we discuss in the next section), the two discrete values f_1 and f_0 are represented by the numbers 1 and 0. In binary voltage waveforms, these values are represented by two voltage levels. For example, in the TTL convention (see Chapter 9), these values are (nominally) 5 V and 0 V, respectively;

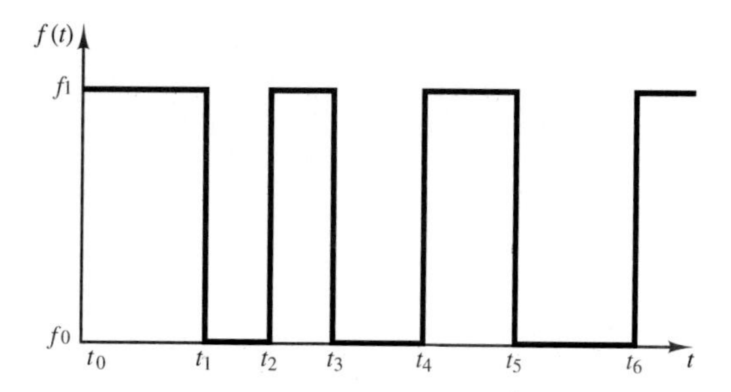

Figure 12.3 A binary signal

in CMOS circuits, these values can vary substantially. Other conventions are also used, including reversing the assignment—for example, by letting a 0-V level represent a logic 1 and a 5-V level represent a logic 0. Note that in a binary waveform, knowledge of the transition between one state and another (e.g., from f_0 to f_1 at $t = t_2$) is equivalent to knowledge of the state. Thus, digital logic circuits can operate by detecting transitions between voltage levels. The transitions are often called **edges** and can be positive (f_0 to f_1) or negative (f_1 to f_0). Virtually all of the signals handled by a computer are binary. From here on, whenever we speak of digital signals, you may assume that the text is referring to signals of the binary type, unless otherwise indicated.

12.2 THE BINARY NUMBER SYSTEM

The binary number system is a natural choice for representing the behavior of circuits that operate in one of two states (on or off, 1 or 0, or the like). The diode and transistor gates and switches studied in Chapter 9 fall in this category. Table 12.1 shows the correspondence between decimal and binary number systems for decimal numbers up to 16.

Binary numbers are based on powers of 2, whereas the decimal system is based on powers of 10. For example, the number 372 in the decimal system can be expressed as

$$372 = (3 \times 10^2) + (7 \times 10^1) + (2 \times 10^0)$$

while the binary number 10110 corresponds to the following combination of powers of 2:

$$10110 = (1 \times 2^4) + (0 \times 2^3) + (1 \times 2^2) + (1 \times 2^1) + (0 \times 2^0)$$

It is relatively simple to see the correspondence between the two number systems if we add the terms on the right-hand side of the previous expression. Let n_2 represent the number n **base 2** (i.e., in the binary system) and n_{10} the same number **base 10.** Then, our notation will be as follows:

$$10110_2 = 16 + 0 + 4 + 2 + 0 = 22_{10}$$

Table 12.1 **Conversion from decimal to binary**

Decimal number, n_{10}	Binary number, n_2
0	0
1	1
2	10
3	11
4	100
5	101
6	110
7	111
8	1000
9	1001
10	1010
11	1011
12	1100
13	1101
14	1110
15	1111
16	10000

Note that a fractional number can also be similarly represented. For example, the number 3.25 in the decimal system may be represented as

$$3.25_{10} = 3 \times 10^0 + 2 \times 10^{-1} + 5 \times 10^{-2}$$

while in the binary system the number 10.011 corresponds to

$$10.011_2 = 1 \times 2^1 + 0 \times 2^0 + 0 \times 2^{-1} + 1 \times 2^{-2} + 1 \times 2^{-3}$$

$$= 2 + 0 + 0 + \tfrac{1}{4} + \tfrac{1}{8} = 2.375_{10}$$

Table 12.1 shows that it takes four binary digits, also called **bits,** to represent the decimal numbers up to 15. Usually, the rightmost bit is called the **least significant bit,** or **LSB,** and the leftmost bit is called the **most significant bit,** or **MSB.** Since binary numbers clearly require a larger number of digits than decimal numbers, the digits are usually grouped in sets of four, eight, or sixteen. Four bits are usually termed a **nibble,** eight bits are called a **byte,** and sixteen bits (or two bytes) form a **word.**

Addition and Subtraction

The operations of addition and subtraction are based on the simple rules shown in Table 12.2. Note that, just as is done in the decimal system, a carry is generated whenever the sum of two digits exceeds the largest single-digit number in the given number system, which is 1 in the binary system. The carry is treated exactly

Table 12.2 **Rules for addition**

$0 + 0 = 0$
$0 + 1 = 1$
$1 + 0 = 1$
$1 + 1 = 0$ (with a carry of 1)

Decimal	Binary	Decimal	Binary	Decimal	Binary
5	101	15	1111	3.25	11.01
+6	+110	+20	+10100	+5.75	+101.11
11	1011	35	100011	9.00	1001.00

(Note that in this example, $3.25 = 3\frac{1}{4}$ and $5.75 = 5\frac{3}{4}$.)

Figure 12.4 Examples of binary addition

Table 12.3 **Rules for subtraction**

$0 - 0 = 0$
$1 - 0 = 1$
$1 - 1 = 0$
$0 - 1 = 1$ (with a borrow of 1)

as in the decimal system. A few examples of binary addition are shown in Figure 12.4, with their decimal counterparts.

The procedure for subtracting binary numbers is based on the rules of Table 12.3. A few examples of binary subtraction are given in Figure 12.5, with their decimal counterparts.

Decimal	Binary	Decimal	Binary	Decimal	Binary
9	1001	16	10000	6.25	110.01
−5	−101	−3	−11	−4.50	−100.10
4	0100	13	01101	1.75	001.11

Figure 12.5 Examples of binary subtraction

Table 12.4 **Rules for multiplication**

$0 \times 0 = 0$
$0 \times 1 = 0$
$1 \times 0 = 0$
$1 \times 1 = 1$

Table 12.5 **Rules for division**

$0 \div 1 = 0$
$1 \div 1 = 1$

Multiplication and Division

Whereas in the decimal system the multiplication table consists of $10^2 = 100$ entries, in the binary system we only have $2^2 = 4$ entries. Table 12.4 represents the complete multiplication table for the binary number system.

Division in the binary system is also based on rules analogous to those of the decimal system, with the two basic laws given in Table 12.5. Once again, we need be concerned with only two cases, and just as in the decimal system, division by zero is not contemplated.

Remainder
$49 \div 2 = 24 + 1$
$24 \div 2 = 12 + 0$
$12 \div 2 = 6 + 0$
$6 \div 2 = 3 + 0$
$3 \div 2 = 1 + 1$
$1 \div 2 = 0 + 1$
$49_{10} = 110001_2$

Figure 12.6 Example of conversion from decimal to binary

Conversion from Decimal to Binary

The conversion of a decimal number to its binary equivalent is performed by successive division of the decimal number by 2, checking for the remainder each time. Figure 12.6 illustrates this idea with an example. The result obtained in Figure 12.6 may be easily verified by performing the opposite conversion, from binary to decimal:

$$110001 = 2^5 + 2^4 + 2^0 = 32 + 16 + 1 = 49$$

The same technique can be used for converting decimal fractional numbers to their binary form, provided that the whole number is separated from the fractional part and each is converted to binary form (separately), with the results added at the end. Figure 12.7 outlines this procedure by converting the number 37.53 to binary form. The procedure is outlined in two steps. First, the integer part is converted; then, to convert the fractional part, one simple technique consists of multiplying the decimal fraction by 2 in successive stages. If the result exceeds 1, a 1 is needed to the right of the binary fraction being formed (100101 . . ., in our example). Otherwise, a 0 is added. This procedure is continued until no fractional terms are left. In this case, the decimal part is 0.53_{10}, and Figure 12.7 illustrates the succession of calculations. Stopping the procedure outlined in Figure 12.7 after 11 digits results in the following approximation:

$$37.53_{10} = 100101.10000111101$$

Greater precision could be attained by continuing to add binary digits, at the expense of added complexity. Note that an infinite number of binary digits may be required to represent a decimal number *exactly.*

Remainder
$37 \div 2 = 18 + 1$
$18 \div 2 = 9 + 0$
$9 \div 2 = 4 + 1$
$4 \div 2 = 2 + 0$
$2 \div 2 = 1 + 0$
$1 \div 2 = 0 + 1$
$37_{10} = 100101_2$

$2 \times 0.53 = 1.06 \rightarrow 1$
$2 \times 0.06 = 0.12 \rightarrow 0$
$2 \times 0.12 = 0.24 \rightarrow 0$
$2 \times 0.24 = 0.48 \rightarrow 0$
$2 \times 0.48 = 0.96 \rightarrow 0$
$2 \times 0.96 = 1.92 \rightarrow 1$
$2 \times 0.92 = 1.84 \rightarrow 1$
$2 \times 0.84 = 1.68 \rightarrow 1$
$2 \times 0.68 = 1.36 \rightarrow 1$
$2 \times 0.36 = 0.72 \rightarrow 0$
$2 \times 0.72 = 1.44 \rightarrow 1$
$0.53_{10} = 0.10000111101$

Figure 12.7 Conversion from decimal to binary

Complements and Negative Numbers

To simplify the operation of subtraction in digital computers, **complements** are used almost exclusively. In practice, this corresponds to replacing the operation $X - Y$ with the operation $X + (-Y)$. This procedure results in considerable simplification, since the computer hardware need include only adding circuitry. Two types of complements are used with binary numbers: the **one's complement** and the **two's complement.**

The one's complement of an n-bit binary number is obtained by subtracting the number itself from $(2^n - 1)$. Two examples are as follows:

$$\begin{aligned} a &= 0101 \\ \text{One's complement of } a &= (2^4 - 1) - a \\ &= (1111) - (0101) \\ &= 1010 \end{aligned}$$

$$\begin{aligned} b &= 101101 \\ \text{One's complement of } b &= (2^6 - 1) - b \\ &= (111111) - (101101) \\ &= 010010 \end{aligned}$$

The two's complement of an n-bit binary number is obtained by subtracting the number itself from 2^n. Two's complements of the same numbers a and b used in the preceding illustration are computed as follows:

$$
\begin{aligned}
a &= 0101 \\
\text{Two's complement of } a &= 2^4 - a \\
&= (10000) - (0101) \\
&= 1011
\end{aligned}
$$

$$
\begin{aligned}
b &= 101101 \\
\text{Two's complement of } b &= 2^6 - b \\
&= (1000000) - (101101) \\
&= 010011
\end{aligned}
$$

A simple rule that may be used to obtain the two's complement directly from a binary number is the following: starting at the least significant (rightmost) bit, copy each bit *until the first 1 has been copied,* and then replace each successive 1 by a 0 and each 0 by a 1. You may wish to try this rule on the two previous examples to verify that it is much easier to use than the subtraction from 2^n.

Different conventions exist in the binary system to represent whether a number is negative or positive. One convention, called the **sign-magnitude convention,** makes use of a *sign bit,* usually positioned at the beginning of the number, for which a value of 1 represents a minus sign and a value of 0, a plus sign. Thus, an eight-bit binary number would consist of a sign bit followed by seven *magnitude bits,* as shown in Figure 12.8(a). In a digital system that uses eight-bit signed integer words, we could represent integer numbers (decimal) in the range

$$-(2^7 - 1) \leq N \leq +(2^7 - 1)$$

or

$$-127 \leq N \leq +127$$

A second convention uses the one's complement notation. In this convention, a sign bit is also used to indicate whether the number is positive (sign bit = 0) or negative (sign bit = 1). However, the magnitude of the binary number is represented by the true magnitude if the number is positive, and by its *one's complement* if the number is negative. Figure 12.8(b) illustrates the convention. For example, the number $(91)_{10}$ would be represented by the seven-bit binary number $(1011011)_2$ with a leading 0 (the sign bit): $(\mathbf{0}1011011)_2$. On the other hand, the number $(-91)_{10}$ would be represented by the seven-bit one's complement binary number $(0100100)_2$ with a leading 1 (the sign bit): $(\mathbf{1}0100100)_2$.

Most digital computers use the two's complement convention in performing integer arithmetic operations. The two's complement convention represents positive numbers by a sign bit of 0, followed by the true binary magnitude; negative numbers are represented by a sign bit of 1, *followed by the two's complement of the binary number,* as shown in Figure 12.8(c). The advantage of the two's complement convention is that the algebraic sum of two's complement binary numbers is carried out very simply by adding the two numbers *including the sign bit.* Example 12.1 illustrates two's complement addition.

Sign bit b_7	b_6	b_5	b_4	b_3	b_2	b_1	b_0
	← Actual magnitude of binary number →						

Figure 12.8(a) Eight-bit sign-magnitude binary number

Sign bit b_7	b_6	b_5	b_4	b_3	b_2	b_1	b_0
	← Actual magnitude of binary number (if $b_7 = 0$) → ← One's complement of binary number (if $b_7 = 1$) →						

Figure 12.8(b) Eight-bit one's complement binary number

Sign bit b_7	b_6	b_5	b_4	b_3	b_2	b_1	b_0
	← Actual magnitude of binary number (if $b_7 = 0$) → ← Two's complement of binary number (if $b_7 = 1$) →						

Figure 12.8(c) Eight-bit two's complement binary number

Example 12.1 Two's Complement Operations

Perform the following subtractions in the binary system using two's complements:

1. 1011100 − 1110010
2. 10101111 − 01110011

Solution:

1. As discussed in the preceding section, the operation $X - Y$ where $X = 1011100$ and $Y = 1110010$ can be replaced with the operation $X + (-Y)$. The next step is to find the two's complement of Y and add the result to X. Therefore,

$$1011100 - 1110010 = 1011100 + (2^7 - 1110010) = 1011100 + 0001110$$
$$= 1101010$$

2. For this part, $X = 10101111$ and $Y = 01110011$. Using the same procedure as in part 1, $10101111 - 01110011 = 10101111 + (2^8 - 01110011) = 10101111 + 10001101 = 00111100$.

The Hexadecimal System

It should be apparent by now that representing numbers in base 2 and base 10 systems is purely a matter of convenience, given a specific application. Another

Table 12.6
Hexadecimal code

0	0000
1	0001
2	0010
3	0011
4	0100
5	0101
6	0110
7	0111
8	1000
9	1001
A	1010
B	1011
C	1100
D	1101
E	1110
F	1111

Table 12.7
BCD code

0	0000
1	0001
2	0010
3	0011
4	0100
5	0101
6	0110
7	0111
8	1000
9	1001

Table 12.8
Three-bit Gray code

Binary	Gray
000	000
001	001
010	011
011	010
100	110
101	111
110	101
111	100

base frequently used is the **hexadecimal system,** a direct derivation of the binary number system. In the hexadecimal (or hex) code, the bits in a binary number are subdivided into groups of four. Since there are 16 possible combinations for a four-bit number, the natural digits in the decimal system (0 through 9) are insufficient to represent a hex digit. To solve this problem, the first six letters of the alphabet are used, as shown in Table 12.6. Thus, in hex code, an eight-bit word corresponds to just two digits; for example:

$$1010\,0111_2 = A7_{16}$$

$$0010\,1001_2 = 29_{16}$$

Binary Codes

In this subsection, we describe two common binary codes that are often used for practical reasons. The first is a method of representing decimal numbers in digital logic circuits that is referred to as **binary-coded decimal,** or **BCD, representation.** In effect, the simplest BCD representation is just a sequence of four-bit binary numbers that stops after the first 10 entries, as shown in Table 12.7. There are also other BCD codes, all reflecting the same principle: that each decimal digit is represented by a fixed-length binary word. One should realize that although this method is attractive because of its direct correspondence with the decimal system, it is not efficient. Consider, for example, the decimal number 68. Its binary representation by direct conversion is the seven-bit number 1000100. On the other hand, the corresponding BCD representation would require eight bits:

$$68_{10} = 01101000_{BCD}$$

Another code that finds many applications is the **Gray code.** This is simply a reshuffling of the binary code with the property that any two consecutive numbers differ only by one bit. Table 12.8 illustrates the three-bit Gray code. The Gray code can be very useful in practical applications, because in counting up or down according to this code, the binary representation of a number changes only one bit at a time. The next example illustrates an application of the Gray code to a practical engineering problem.

Example 12.2 Position Encoders

Position encoders are devices that output a digital signal proportional to their (linear or angular) position. These devices are very useful in measuring instantaneous position in *motion control* applications. Motion control is a technique that is used when it is necessary to accurately control the motion of a moving object; examples are found in robotics, machine tools, and servomechanisms. For example, in positioning the arm of a robot to pick up an object, it is very important to know its exact position at all times. Since one is usually interested in both rotational and translational motion, two types of encoders are discussed in this example: *linear* and *angular* position encoders.

An optical position encoder consists of an *encoder pad,* which is either a strip (for translational motion) or a disk (for rotational motion) with alternating black and white areas. These areas are arranged to reproduce some binary code, as shown in Figure 12.9, where both the conventional binary and Gray codes are depicted for a four-bit linear encoder pad. A fixed array of photodiodes (see Chapter 7) senses the reflected light from each of the cells across a row of the encoder path; depending on the amount of light reflected, each photodiode circuit will output a voltage corresponding to a binary 1 or 0. Thus, a different four-bit word is generated for each row of the encoder.

Suppose the encoder pad is 100 mm in length. Then its resolution can be computed as follows. The pad will be divided into $2^4 = 16$ segments, and each segment corresponds to an increment of $\frac{100}{16}$ mm $= 6.25$ mm. If greater resolution were necessary, more bits could be employed: an eight-bit pad of the same length would attain a resolution of $\frac{100}{256}$ mm $=$ 0.39 mm.

A similar construction can be employed for the five-bit angular encoder of Figure 12.10. In this case, the angular resolution can be expressed in degrees of rotation, where $2^5 = 32$ sections correspond to 360°. Thus, the resolution would be $360°/32 = 11.25°$. Once again, greater angular resolution could be obtained by employing a larger number of bits.

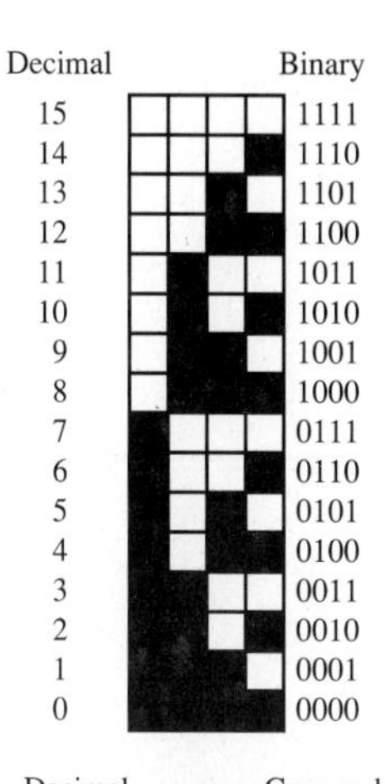

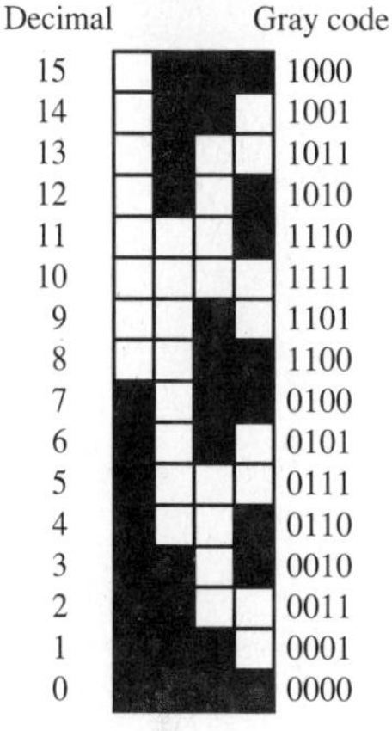

Figure 12.9 Binary and Gray code patterns for linear position encoders

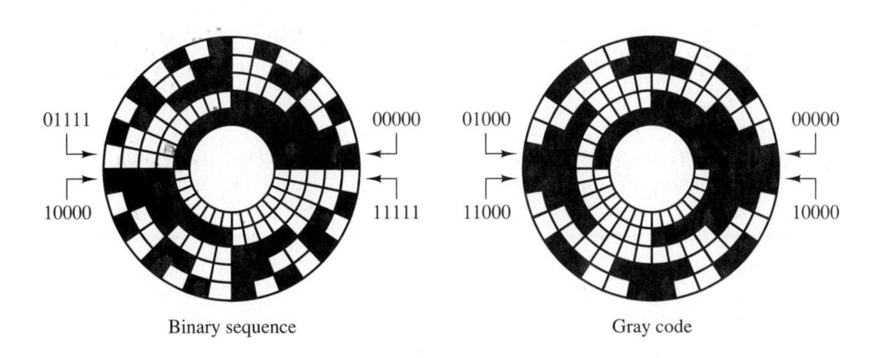

Figure 12.10 Binary and Gray code patterns for angular position encoders

Example 12.3

Convert the following binary numbers to hexadecimal numbers:

1. 100111
2. 1011101
3. 11001101

Solution:

The simplest method of conversion from a binary-base number to a hexadecimal-base number consists of grouping the binary number in subgroups of four starting from the right; that is, (1) 0010, 0111; (2) 0101, 1101; (3) 1100, 1101. Then, replacing each subgroup by its equivalent number in hexadecimal base:

1. 27_{16}
2. $5D_{16}$
3. CD_{16}

[*Note:* If you want to convert a hex number to a binary number, you need only replace each digit of the hex number by its equivalent four-bit binary number.]

Example 12.4

Convert the following binary numbers to hexadecimal numbers:

1. 101101111001
2. 100110110
3. 1101011011

Solution:

Replace each subgroup by its equivalent number in hexadecimal base:

1. $B79_{16}$
2. 136_{16}
3. $35B_{16}$

The hexadecimal base is widely used in digital computing because of its efficient representation of long strings of binary numbers.

DRILL EXERCISES

12.1 Convert the following decimal numbers to binary form:

a. 39
b. 59
c. 512
d. 0.4475
e. $\frac{25}{32}$
f. 0.796875
g. 256.75
h. 129.5625
i. 4,096.90625

12.2 Convert the following binary numbers to decimal:

a. 1101
b. 11011
c. 10111
d. 0.1011

e. 0.001101
f. 0.001101101
g. 111011.1011
h. 1011011.001101
i. 10110.0101011101

12.3 Perform the following additions and subtractions. Express the answer in decimal form for problems (a)–(d) and in binary form for problems (e)–(h).

a. $1001.1_2 + 1011.01_2$
b. $100101_2 + 100101_2$
c. $0.1011_2 + 0.1101_2$
d. $1011.01_2 + 1001.11_2$
e. $64_{10} - 32_{10}$
f. $127_{10} - 63_{10}$
g. $93.5_{10} - 42.75_{10}$
h. $(84\frac{9}{32})_{10} - (48\frac{5}{16})_{10}$

12.4 How many possible numbers can be represented in a 12-bit word?

12.5 If we use an eight-bit word with a sign bit (seven magnitude bits plus one sign bit) to represent voltages -5 V and $+5$ V, what is the smallest increment of voltage that can be represented?

12.6 Convert the following numbers from hex to binary or from binary to hex:

a. F83
b. 3C9
c. A6
d. 110101110_2
e. 10111001_2
f. 11011101101_2

12.7 Find the two's complement of the following binary numbers:

a. 11101001
b. 10010111
c. 1011110

12.8 Convert the following numbers from hex to binary, and find their two's complements:

a. F43
b. 2B9
c. A6

12.3 BOOLEAN ALGEBRA

The mathematics associated with the binary number system (and with the more general field of logic) is called *Boolean,* in honor of the English mathematician George Boole, who published a treatise in 1854 entitled *An Investigation of the Laws of Thought, on Which Are Founded the Mathematical Theories of Logic and Probabilities.* The development of a *logical algebra,* as Boole called it, is one of the results of his investigations. The variables in a Boolean, or logic, expression can take only one of two values, usually represented by the numbers 0 and 1. These variables are sometimes referred to as true (1) and false (0). This convention is normally referred to as **positive logic.** There is also a **negative logic** convention in which the roles of logic 1 and logic 0 are reversed. In this book we shall employ only positive logic.

Analysis of **logic functions,** that is, functions of logical (Boolean) variables, can be carried out in terms of truth tables. A truth table is a listing of all the possible values each of the Boolean variables can take, and of the corresponding value of the desired function. In the following paragraphs we shall define the basic logic functions upon which Boolean algebra is founded, and we shall describe each in terms of a set of rules and a truth table; in addition, we shall also introduce **logic gates**. Logic gates are physical devices (see Chapter 9) that can be used to implement logic functions.

AND and OR Gates

Table 12.9 **Rules for logical addition (OR)**

$0 + 0 = 0$
$0 + 1 = 1$
$1 + 0 = 1$
$1 + 1 = 1$

The basis of **Boolean algebra** lies in the operations of **logical addition,** or the **OR** operation; and **logical multiplication,** or the **AND** operation. Both of these find a correspondence in simple logic gates, as we shall presently illustrate. Logical addition, although represented by the symbol +, differs from conventional algebraic addition, as shown in the last rule listed in Table 12.9. Note that this rule also differs from the last rule of binary addition studied in the previous section. Logical addition can be represented by the logic gate called an **OR gate,** whose symbol and whose inputs and outputs are shown in Figure 12.11. The OR gate represents the following logical statement:

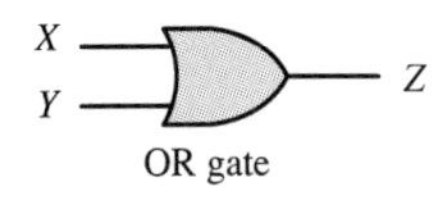

OR gate

X	Y	Z
0	0	0
0	1	1
1	0	1
1	1	1

Truth table

Figure 12.11 Logical addition and the OR gate

If either X or Y is true (1), then Z is true (1). **(12.1)**

This rule is embodied in the electronic gates discussed in Chapter 9, in which a logic 1 corresponds, say, to a 5-V signal and a logic 0 to a 0-V signal.

Logical multiplication is denoted by the center dot ($\cdot$) and is defined by the rules of Table 12.10. Figure 12.12 depicts the **AND gate**, which corresponds to this operation. The AND gate corresponds to the following logical statement:

If both X and Y are true (1), then Z is true (1). **(12.2)**

Table 12.10 **Rules for logical multiplication (AND)**

$0 \cdot 0 = 0$
$0 \cdot 1 = 0$
$1 \cdot 0 = 0$
$1 \cdot 1 = 1$

One can easily envision logic gates (AND and OR) with an arbitrary number of inputs; three- and four-input gates are not uncommon.

The rules that define a logic function are often represented in tabular form by means of a **truth table.** Truth tables for the AND and OR gates are shown in Figures 12.11 and 12.12. A truth table is nothing more than a tabular summary of all of the possible outputs of a logic gate, given all the possible input values. If the number of inputs is 3, the number of possible combinations grows from 4 to 8, but the basic idea is unchanged. Truth tables are very useful in defining logic functions. A typical logic design problem might specify requirements such as "the output Z shall be logic 1 only when the condition ($X = 1$ AND $Y = 1$) OR ($W = 1$) occurs, and shall be logic 0 otherwise." The truth table for this particular logic function is shown in Figure 12.13 as an illustration. The design consists, then, of determining the combination of logic gates that exactly implements the required logic function. Truth tables can greatly simplify this procedure.

The AND and OR gates form the basis of all logic design in conjunction with the **NOT gate.** The NOT gate is essentially an inverter (which can be constructed using bipolar or field-effect transistors, as discussed in Chapter 9), and it provides the complement of the logic variable connected to its input. The complement of a logic variable X is denoted by $\overline{X}$. The NOT gate has only one input, as shown in Figure 12.14.

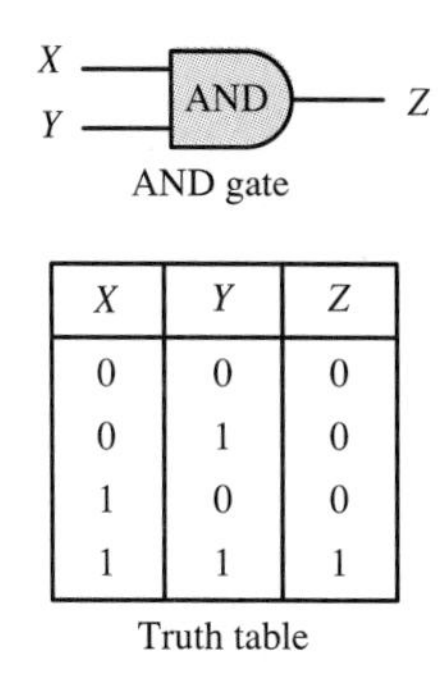

X	Y	Z
0	0	0
0	1	0
1	0	0
1	1	1

Truth table

Figure 12.12 Logical multiplication and the AND gate

Logic gate realization of the statement "the output Z shall be logic 1 only when the condition ($X = 1$ AND $Y = 1$) OR ($W = 1$) occurs, and shall be logic 0 otherwise."

X	Y	W	Z
0	0	0	0
0	0	1	1
0	1	0	0
0	1	1	1
1	0	0	0
1	0	1	1
1	1	0	1
1	1	1	1

Truth table

X Y AND W OR Z

Solution using logic gates

Figure 12.13 Example of logic function implementation with logic gates

X NOT $\overline{X}$

NOT gate

X	$\overline{X}$
1	0
0	1

Truth table for NOT gate

Figure 12.14 Complements and the NOT gate

To illustrate the use of the NOT gate, or inverter, we return to the design example of Figure 12.13, where we required that the output of a logic circuit be $Z = 1$ only if $X = 0$ AND $Y = 1$ OR if $W = 1$. We recognize that except for the requirement $X = 0$, this problem would be identical if we stated it as follows: "The output Z shall be logic 1 only when the condition ($\overline{X} = 1$ AND $Y = 1$) OR ($W = 1$) occurs, and shall be logic 0 otherwise." If we use an inverter to convert X to $\overline{X}$, we see that the required condition becomes ($\overline{X} = 1$ AND $Y = 1$) OR ($W = 1$). The formal solution to this elementary design exercise is illustrated in Figure 12.15.

In the course of the discussion of logic gates, extensive use will be made of truth tables to evaluate logic expressions. A set of basic rules will facilitate this

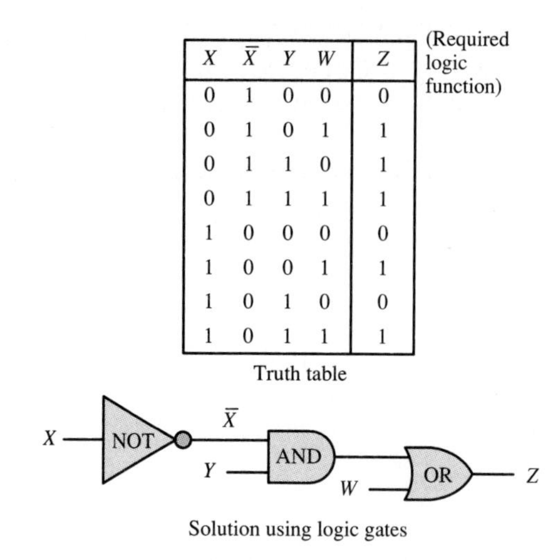

X	$\overline{X}$	Y	W	Z
0	1	0	0	0
0	1	0	1	1
0	1	1	0	1
0	1	1	1	1
1	0	0	0	0
1	0	0	1	1
1	0	1	0	0
1	0	1	1	1

(Required logic function)

Truth table

Solution using logic gates

Figure 12.15 Solution of a logic problem using logic gates

task. Table 12.11 lists some of the rules of Boolean algebra; each of these can be proven by using a truth table, as will be shown in examples and exercises. An example proof for rule 16 is given in Figure 12.16 in the form of a truth table. This technique can be employed to prove any of the laws of Table 12.11. From the simple truth table in Figure 12.16, which was obtained step by step, we can clearly see that indeed $X \cdot (X + Y) = X$. This methodology for proving the validity of logical equations is called **proof by perfect induction.** The 19 rules of Table 12.11 can be used to simplify logic expressions.

Table 12.11 **Rules of Boolean algebra**

1. $0 + X = X$	
2. $1 + X = 1$	
3. $X + X = X$	
4. $X + \overline{X} = 1$	
5. $0 \cdot X = 0$	
6. $1 \cdot X = X$	
7. $X \cdot X = X$	
8. $X \cdot \overline{X} = 0$	
9. $\overline{\overline{X}} = X$	
10. $X + Y = Y + X$	Commutative law
11. $X \cdot Y = Y \cdot X$	
12. $X + (Y + Z) = (X + Y) + Z$	Associative law
13. $X \cdot (Y \cdot Z) = (X \cdot Y) \cdot Z$	
14. $X \cdot (Y + Z) = X \cdot Y + X \cdot Z$	Distributive law
15. $X + X \cdot Z = X$	Absorption law
16. $X \cdot (X + Y) = X$	
17. $(X + Y) \cdot (X + Z) = X + Y \cdot Z$	
18. $X + \overline{X} \cdot Y = X + Y$	
19. $X \cdot Y + Y \cdot Z + \overline{X} \cdot Z = X \cdot Y + \overline{X} \cdot Z$	

X	Y	$(X+Y)$	$X \cdot (X+Y)$
0	0	0	0
0	1	1	0
1	0	1	1
1	1	1	1

Figure 12.16 Proof of rule 16 by perfect induction

To complete the introductory material on Boolean algebra, a few paragraphs need to be devoted to two very important theorems, called **De Morgan's theorems.** These are stated here in the form of logic functions:

$$\overline{(X+Y)} = \overline{X} \cdot \overline{Y} \qquad \textbf{(12.3)}$$

$$\overline{(X \cdot Y)} = \overline{X} + \overline{Y} \qquad \textbf{(12.4)}$$

These two laws state a very important property of logic functions:

> Any logic function can be implemented using only OR and NOT gates, or using only AND and NOT gates.

De Morgan's laws can easily be visualized in terms of logic gates, as shown in Figure 12.17. The associated truth tables are proof of these theorems.

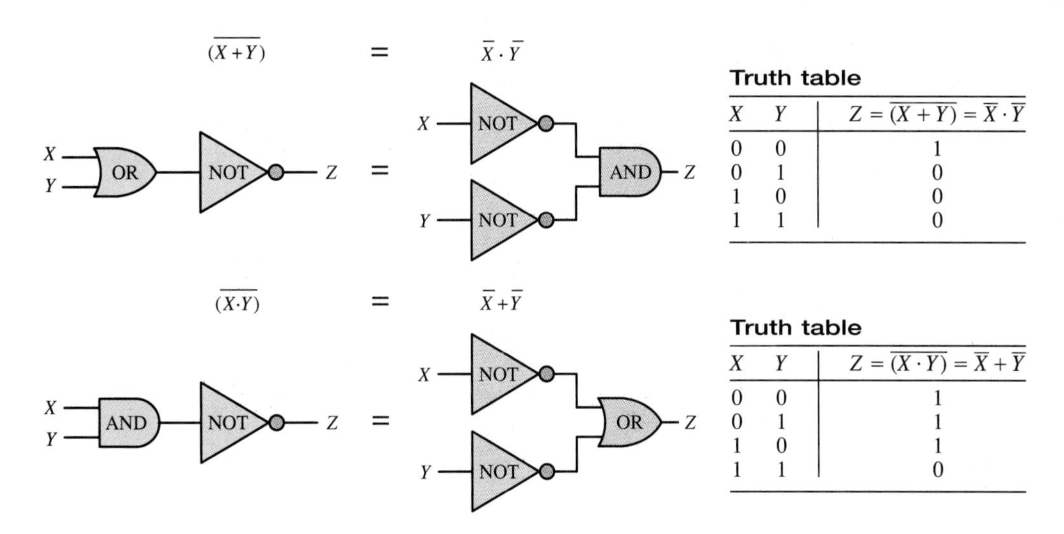

X	Y	$Z = \overline{(X+Y)} = \overline{X} \cdot \overline{Y}$
0	0	1
0	1	0
1	0	0
1	1	0

Truth table

X	Y	$Z = \overline{(X \cdot Y)} = \overline{X} + \overline{Y}$
0	0	1
0	1	1
1	0	1
1	1	0

Figure 12.17 De Morgan's laws

The importance of De Morgan's laws is in the statement of the **duality** that exists between AND and OR operations: any function can be realized by just one of the two basic operations, plus the complement operation. This gives rise to two families of logic functions: **sums of products** and **products of sums,** as shown in Figure 12.18. Any logical expression can be reduced to either one of these two forms. Although the two forms are equivalent, it may well be true that one of the two has a simpler implementation (fewer gates). Example 12.6 illustrates this point.

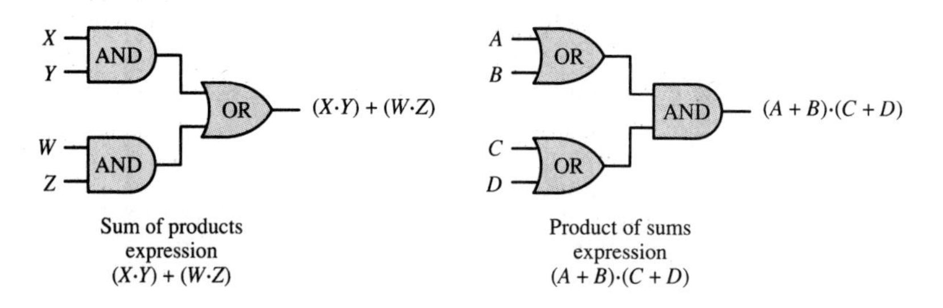

Figure 12.18 Sum-of-products and product-of-sums logic functions

---

Example 12.5

Using the rules of Boolean algebra, simplify the following logic function:

$$f(A,B,C,D) = \overline{A}\cdot\overline{B}\cdot D + \overline{A}\cdot B\cdot D + B\cdot C\cdot D + A\cdot C\cdot D$$

Solution:

$f = \overline{A}\cdot\overline{B}\cdot D + \overline{A}\cdot B\cdot D + B\cdot C\cdot D + A\cdot C\cdot D$	
$f = \overline{A}\cdot D\cdot(B + \overline{B}) + B\cdot C\cdot D + A\cdot C\cdot D$	Rule 14
$f = \overline{A}\cdot D + B\cdot C\cdot D + A\cdot C\cdot D$	Rule 4
$f = (\overline{A} + A\cdot C)\cdot D + B\cdot C\cdot D$	Rule 14
$f = (\overline{A} + C)\cdot D + B\cdot C\cdot D$	Rules 14 and 18
$f = \overline{A}\cdot D + C\cdot D + B\cdot C\cdot D$	Rule 14
$f = \overline{A}\cdot D + C\cdot D\cdot(1 + B)$	Rule 14
$f = \overline{A}\cdot D + C\cdot D$	Rules 2 and 6

Example 12.6

This example aims to illustrate the significance of De Morgan's laws and of the duality of the sum-of-products and product-of-sums forms. Suppose that a fail-safe autopilot system in a commercial aircraft requires that, prior to initiating a takeoff or landing maneuver, the following check must be passed: two of three possible pilots must be available. The three possibilities are the pilot, the co-pilot, and the autopilot. Imagine further that there exist switches in the pilot and co-pilot seats that are turned on by the weight of the crew, and that a self-check circuit exists to verify the proper operation of the autopilot system. Let the variable X denote the pilot state (1 if the pilot is sitting at the controls), Y denote the same condition for the co-pilot, and Z denote the state of the autopilot, where $Z = 1$ indicates that the autopilot is functioning. Then, since we wish two of these conditions to be active before the maneuver can be initiated, the logic function corresponding to "system ready" is:

$$f = X \cdot Y + X \cdot Z + Y \cdot Z$$

This can also be verified by the truth table shown below.

Pilot	**Co-pilot**	**Autopilot**	**System ready**
0	0	0	0
0	0	1	0
0	1	0	0
0	1	1	1
1	0	0	0
1	0	1	1
1	1	0	1
1	1	1	1

The function f defined above is based on the notion of a *positive check;* that is, it indicates when the system is ready. Let us now apply De Morgan's laws to the function f, which is in sum-of-products form:

$$\overline{f} = g = \overline{X \cdot Y + X \cdot Z + Y \cdot Z} = (\overline{X} + \overline{Y}) \cdot (\overline{X} + \overline{Z}) \cdot (\overline{Y} + \overline{Z})$$

The function g, in product-of-sums form, conveys exactly the same information as the function f, but it performs a negative check; in other words, g verifies the *system not ready condition.* You see then that whether one chooses to implement the function in one form or another is simply a matter of choice; the two forms give exactly the same information.

Example 12.7

Realize the function described by the truth table shown.

A	B	C	y
0	0	0	0
0	0	1	1
0	1	0	0
0	1	1	1
1	0	0	1
1	0	1	1
1	1	0	1
1	1	1	1

Solution:

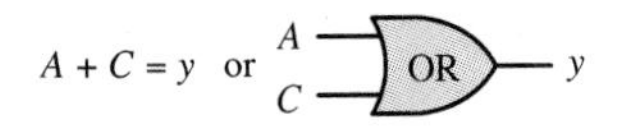

Figure 12.19

We may find a very simple solution to this problem by noting that the only occurrence of 0s in the output is when A and C are both 0, and that therefore the output y is 1 whenever either A or C or both are 1. But this is, in effect, the description of a two-input OR gate! Thus, the variable B is not necessary to realize the desired function. The solution is shown in Figure 12.19. Please note that in this problem, the function to be realized was simple, and the solution could be obtained by inspection. In general, however, formal techniques will be required to obtain the solution for more complex functions. You may wish to verify that in this example the logic function y is

$$y = \overline{A} \cdot \overline{B} \cdot C + \overline{A} \cdot B \cdot C + A \cdot \overline{B} \cdot \overline{C} + A \cdot \overline{B} \cdot C + A \cdot B \cdot \overline{C} + A \cdot B \cdot C$$

and that the expression can be simplified to $y = A + C$ by making use of the rules of Table 12.11. In the next sections, we shall develop formal techniques for the analysis and realization of more complex logic functions.

Example 12.8

Use De Morgan's theorem to realize the function $y = A + (B \cdot C)$ as a product-of-sums expression, and implement it using AND, OR, and NOT gates.

Solution:

Knowing that $\overline{\overline{y}} = y$, we can apply the first of De Morgan's laws to the complement of the function y to obtain the expression

$$\overline{y} = \overline{A + (B \cdot C)} = \overline{A} \cdot \overline{(B \cdot C)} = \overline{A} \cdot (\overline{B} + \overline{C})$$

Thus,

$$\overline{\overline{y}} = y = \overline{\overline{A} \cdot (\overline{B} + \overline{C})}$$

Using logic gates, we can then implement the function as shown in Figure 12.20.

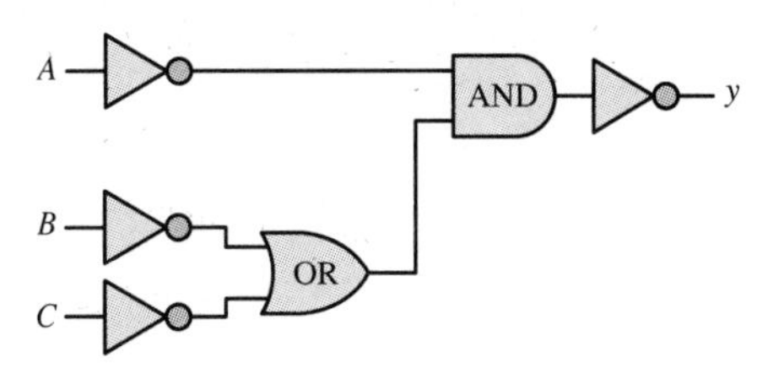

Figure 12.20

NAND and NOR Gates

In addition to the AND and OR gates we have just analyzed, the complementary forms of these gates, called NAND and NOR, are very commonly used in practice. In fact, NAND and NOR gates form the basis of most practical logic circuits. Figure 12.21 depicts these two gates, and illustrates how they can be easily interpreted in terms of AND, OR, and NOT gates by virtue of De Morgan's laws. You can readily verify that the logic function implemented by the NAND and NOR gates corresponds, respectively, to AND and OR gates followed by an inverter. It is very important to note that, by De Morgan's laws, the NAND gate performs a *logical addition* on the *complements* of the inputs, while the NOR gate performs a *logical multiplication* on the *complements* of the inputs. Functionally, then, any logic function could be implemented with either NOR or NAND gates only.

In the next section we shall learn how to systematically approach the design of logic functions. First, we provide a few examples to illustrate logic design with NAND and NOR gates.

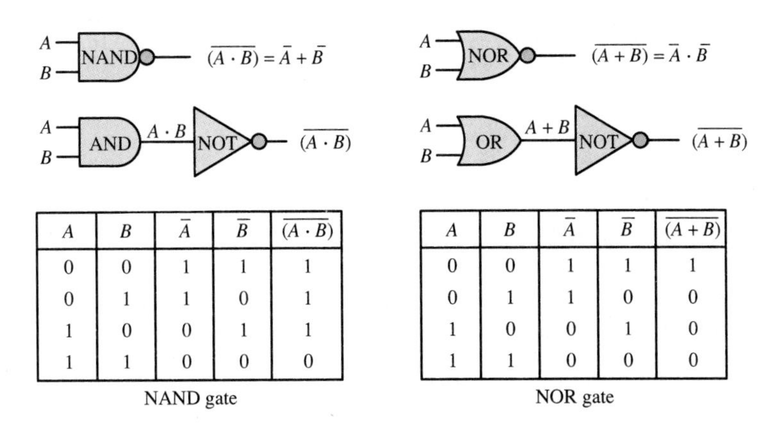

A	B	$\bar{A}$	$\bar{B}$	$\overline{(A \cdot B)}$
0	0	1	1	1
0	1	1	0	1
1	0	0	1	1
1	1	0	0	0

NAND gate

A	B	$\bar{A}$	$\bar{B}$	$\overline{(A + B)}$
0	0	1	1	1
0	1	1	0	0
1	0	0	1	0
1	1	0	0	0

NOR gate

Figure 12.21 Equivalence of NAND and NOR gates with AND and OR gates

---

Example 12.9

Show that an AND gate can be obtained as a combination of NAND gates.

Solution:

Let us show this for two arbitrary inputs, A and B. The desired truth table is as follows:

A	$B(=A)$	$A \cdot B$	$\overline{(A \cdot B)}$
0	0	0	1
1	1	1	0

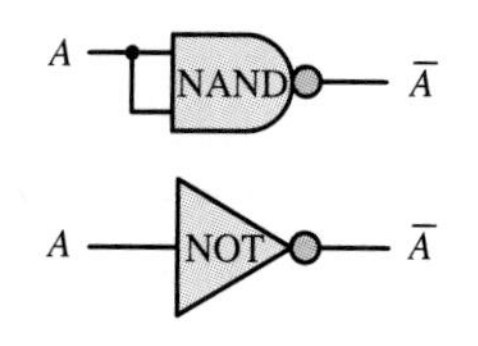

Figure 12.22 NAND gate as an inverter

A	B	NAND $\overline{A \cdot B}$	AND $A \cdot B$
0	0	1	0
0	1	1	0
1	0	1	0
1	1	0	1

Looking at the truth table, we can see that we need only invert the output of the NAND gate. This is easily accomplished if we realize that a NAND gate can serve as an inverter when all of its inputs are tied together, as illustrated in Figure 12.22. The desired logic circuit, which is functionally equivalent to an AND gate, is then shown in Figure 12.23.

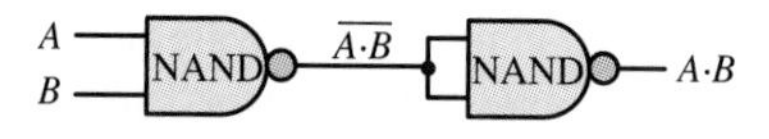

Figure 12.23

Example 12.10

Show that one can obtain an AND gate using a combination of NOR gates.

Solution:

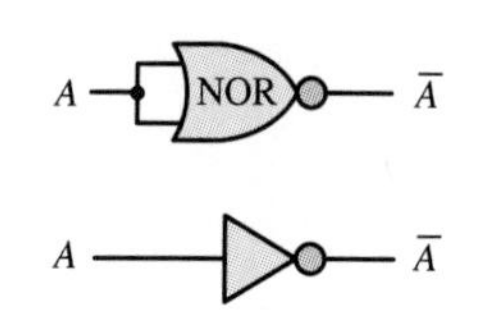

A	$B(=A)$	$(A + B)$	$\overline{(A + B)}$
0	0	0	1
1	1	1	0

Figure 12.24 NOR gate as an inverter

The simplest way of solving this problem is by using De Morgan's theorems. We want to build an AND gate using only NOR gates. The output of an AND gate, f, for two arbitrary inputs A and B is $f = A \cdot B$. Knowing that $\overline{\overline{f}} = f$, we use De Morgan's law in the form $f = \overline{\overline{A} + \overline{B}}$ to obtain a function that can be implemented using a NOR gate with inputs $\overline{A}$ and $\overline{B}$. Also, we observe that a NOR gate can act as an inverter when its inputs are tied together. Figure 12.24 illustrates the behavior of a NOR gate as an inverter.

Next, we recall the statement made earlier in this section suggesting that a NOR gate performs a logical multiplication on the complements of its inputs. Upon a moment's reflection, we see that the logic circuit of Figure 12.25 provides the answer to our question.

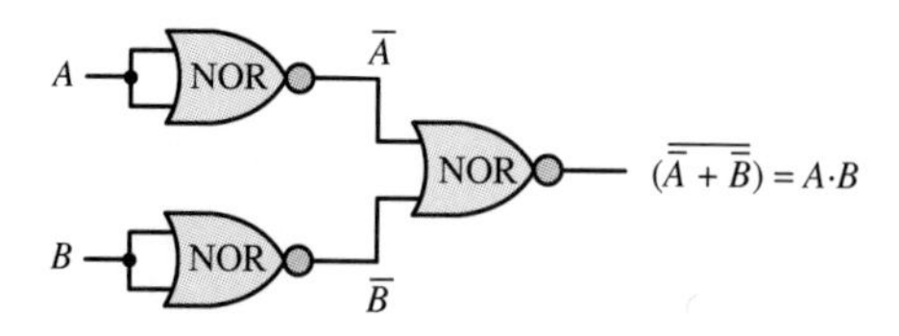

Figure 12.25

Example 12.11

Realize the following function using only NAND and NOR gates:

$$y = \overline{(A \cdot B) + C}$$

Solution:

Since the term in parentheses appears as a complement product, it can be obtained by means of a NAND gate. Further, once the function $\overline{(A \cdot B)}$ has been realized, we can see that y is the complemented sum of two terms—that is, it can be obtained directly with a NOR gate. The resulting logic circuit is shown in Figure 12.26.

Can you find another solution to this problem that employs only two gates?

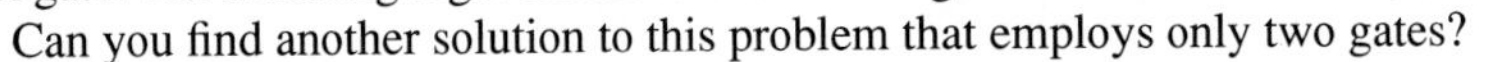

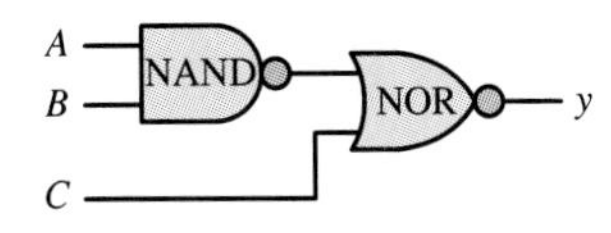

Figure 12.26

Example 12.12

Realize the logic function described by the following truth table, using only NAND and NOR gates.

A	B	C	y
0	0	0	1
0	0	1	0
0	1	0	1
0	1	1	0
1	0	0	0
1	0	1	0
1	1	0	0
1	1	1	0

Solution:

Note that the output has 1s occurring only when A and C are both 0, and that this corresponds to the function $\overline{A} \cdot \overline{C}$. Using De Morgan's theorems, we obtain:

$$y = \overline{A + C}$$

This function corresponds to a single NOR gate, as shown in Figure 12.27.

Figure 12.27

Example 12.13

Use De Morgan's theorems to implement the function $y = \overline{(A \cdot B) + C}$ using at least one NOR gate.

Solution:

To simplify the function, De Morgan's laws are applied to obtain

$$y = \overline{\overline{(A \cdot B)} + C} = \overline{\overline{A \cdot B}} \cdot \overline{C}$$
$$= \overline{\overline{A} + \overline{B}} \cdot \overline{C}$$

resulting in the circuit shown in Figure 12.28.

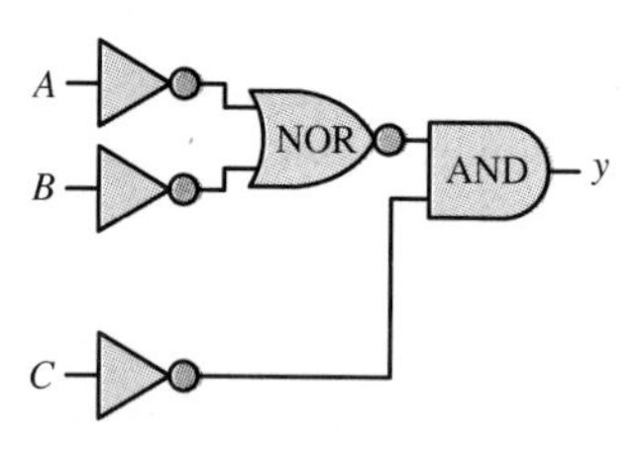

Figure 12.28

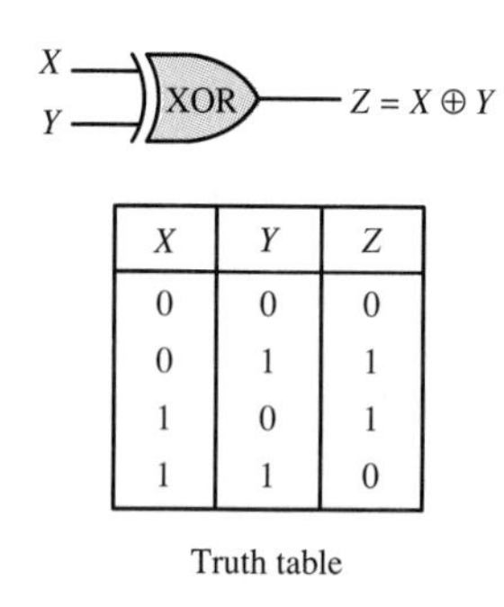

X	Y	Z
0	0	0
0	1	1
1	0	1
1	1	0

Truth table

Figure 12.29 XOR gate

The XOR (Exclusive OR) Gate

It is rather common practice for a manufacturer of integrated circuits to provide common combinations of logic circuits in a single integrated circuit package. We shall review many of these common **logic modules** in Section 12.5. An example of this idea is provided by the **exclusive OR (XOR) gate,** which provides a logic function similar, but not identical, to the OR gate we have already studied. The XOR gate acts as an OR gate, except when its inputs are all logic 1s; in this case, the output is a logic 0 (thus the term *exclusive*). Figure 12.29 shows the logic circuit symbol adopted for this gate, and the corresponding truth table. The logic function implemented by the XOR gate is the following: "either X or Y, but not both." This description can be extended to an arbitrary number of inputs.

The symbol adopted for the exclusive OR operation is ⊕, and so we shall write

$$Z = X \oplus Y$$

to denote this logic operation. The XOR gate can be obtained by a combination of the basic gates we are already familiar with. For example, if we observe that the XOR function corresponds to $Z = X \oplus Y = (X + Y) \cdot \overline{(X \cdot Y)}$, we can realize the XOR gate by means of the circuit shown in Figure 12.30.

Figure 12.30 Realization of an XOR gate

Common IC logic gate configurations and device numbers are given at the end of the chapter. These devices are typically available in both of the two more common device families, TTL and CMOS. The devices listed at the end of the chapter are available in CMOS technology under the numbers MM54HCXX or MM74HCXX. The same logic gate ICs are also available as TTL devices.

DRILL EXERCISES

12.9 Show that one can obtain an OR gate using NAND gates only. [*Hint:* Use three NAND gates.]

12.10 Show that one can obtain an AND gate using NOR gates only. [*Hint:* Use three NOR gates.]

12.11 Prepare a step-by-step truth table for the following logic expressions:

a. $\overline{(X + Y + Z)} + (X \cdot Y \cdot Z) \cdot \overline{X}$
b. $\overline{X} \cdot Y \cdot Z + Y \cdot (Z + W)$
c. $(X \cdot \overline{Y} + Z \cdot \overline{W}) \cdot (W \cdot X + \overline{Z} \cdot Y)$

[*Hint:* Your truth table must have 2^n entries, where n is the number of logic variables.]

12.12 Implement the logic functions of Drill Exercise 12.11 using NAND and NOR gates only. [*Hint:* Use De Morgan's theorems and the fact that $\overline{\overline{f}} = f$.]

12.13 Implement the logic functions of Drill Exercise 12.11 using AND, OR, and NOT gates only.

12.14 Show that the XOR function can also be expressed as $Z = X \cdot \overline{Y} + Y \cdot \overline{X}$. Realize the corresponding function using NOT, AND, and OR gates. [*Hint:* Use truth tables for the logic function Z (as defined in the exercise) and for the XOR function.]

12.4 KARNAUGH MAPS AND LOGIC DESIGN

In examining the design of logic functions by means of logic gates, we have discovered that more than one solution is usually available for the implementation of a given logic expression. It should also be clear by now that some combinations of gates can implement a given function more efficiently than others. How can we be assured of having chosen the most efficient realization? Fortunately, there is a procedure that utilizes a map describing all possible combinations of the variables present in the logic function of interest. This map is called a **Karnaugh map,** after its inventor. Figure 12.31 depicts the appearance of Karnaugh maps for two-, three-, and four-variable expressions in two different forms. As can be seen, the row and column assignments for two or more variables are arranged so that all adjacent terms change by only one bit. For example, in the two-variable map, the columns next to column 01 are columns 00 and 10. Also note that each map consists of 2^N **cells,** where N is the number of logic variables.

Each cell in a Karnaugh map contains a **minterm,** that is, a product of the N variables that appear in our logic expression (in either uncomplemented or complemented form). For example, for the case of three variables ($N = 3$), there are $2^3 = 8$ such combinations, or minterms: $\overline{X} \cdot \overline{Y} \cdot \overline{Z}, \overline{X} \cdot \overline{Y} \cdot Z, \overline{X} \cdot Y \cdot \overline{Z}, \overline{X} \cdot Y \cdot Z, X \cdot \overline{Y} \cdot \overline{Z}, X \cdot \overline{Y} \cdot Z, X \cdot Y \cdot \overline{Z}$, and $X \cdot Y \cdot Z$. The content of each cell—that is, the minterm—is the product of the variables appearing at the corresponding

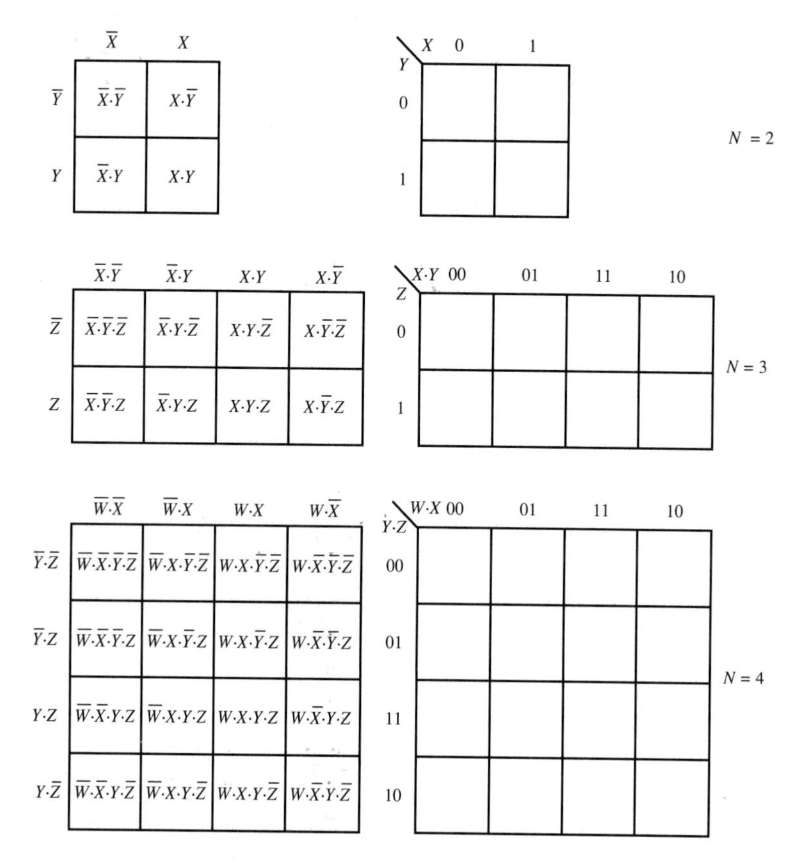

Figure 12.31 Two-, three-, and four-variable Karnaugh maps

vertical and horizontal coordinates. For example, in the three-variable map, $X \cdot Y \cdot \overline{Z}$ appears at the intersection of $X \cdot Y$ and $\overline{Z}$. The map is filled by placing a value of 1 for any combination of variables for which the desired output is a 1. For example, consider the function of three variables for which we desire to have an output of 1 whenever the variables X, Y, and Z have the following values:

$X = 0$	$Y = 1$	$Z = 0$
$X = 0$	$Y = 1$	$Z = 1$
$X = 1$	$Y = 1$	$Z = 0$
$X = 1$	$Y = 1$	$Z = 1$

The same truth table is shown in Figure 12.32 together with the corresponding Karnaugh map.

	$\overline{X}\overline{Y}$	$\overline{X}Y$	XY	$X\overline{Y}$
$\overline{Z}$	0	1	1	0
Z	0	1	1	0

Karnaugh map

X	Y	Z	Desired Function
0	0	0	0
0	0	1	0
0	1	0	1
0	1	1	1
1	0	0	0
1	0	1	0
1	1	0	1
1	1	1	1

Truth table

Figure 12.32 Truth table and Karnaugh map representations of a logic function

The Karnaugh map provides an immediate view of the values of the function in graphical form. Further, the arrangement of the cells in the Karnaugh map is such that any two adjacent cells contain minterms that vary in only one variable. This property, as will be verified shortly, is quite useful in the design of logic functions by means of logic gates, especially if we consider the map to be continuously wrapping around itself, as if the top and bottom, and right and left, edges were touching each other. For the three-variable map given in Figure 12.31, for example, the cell $X \cdot \overline{Y} \cdot \overline{Z}$ is adjacent to $\overline{X} \cdot \overline{Y} \cdot \overline{Z}$ if we "roll" the map so that the right edge touches the left. Note that these two cells differ only in the variable X, a property we earlier claimed adjacent cells have.[1]

W	X	Y	Z	Desired Function
0	0	0	0	1
0	0	0	1	1
0	0	1	0	0
0	0	1	1	0
0	1	0	0	0
0	1	0	1	1
0	1	1	0	1
0	1	1	1	0
1	0	0	0	0
1	0	0	1	1
1	0	1	0	1
1	0	1	1	0
1	1	0	0	0
1	1	0	1	0
1	1	1	0	0
1	1	1	1	1

Truth table for four-variable expression

	$\overline{W} \cdot \overline{X}$	$\overline{W} \cdot X$	$W \cdot X$	$W \cdot \overline{X}$
$\overline{Y} \cdot \overline{Z}$	1	0	0	0
$\overline{Y} \cdot Z$	1	1	0	1
$Y \cdot Z$	0	0	1	0
$Y \cdot \overline{Z}$	0	1	0	1

Figure 12.33 Karnaugh map for a four-variable expression

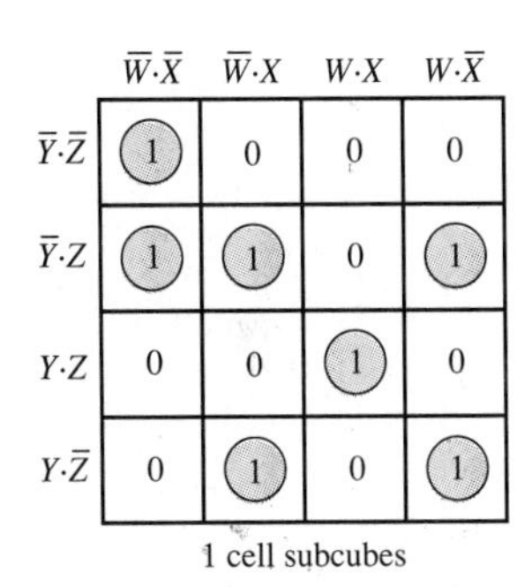

Shown in Figure 12.33 is a more complex, four-variable logic function, which will serve as an example in explaining how Karnaugh maps can be used directly to implement a logic function. First, we define a subcube as a set of 2^m adjacent cells, for $m = 1, 2, 3, \ldots, N$. Thus, a subcube can consist of 1, 2, 4, 8, 16, 32, ... cells. All possible subcubes for the four-variable map of Figure 12.33 are shown in Figure 12.34. Note that there are no four-cell subcubes in this particular case. Note also that there is some overlap between subcubes. Examples of four-cell and eight-cell subcubes are shown in Figure 12.35 for an arbitrary expression.

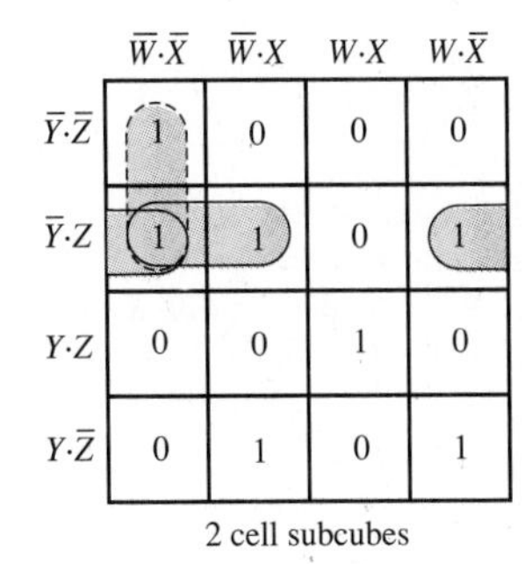

Figure 12.34 One- and two-cell subcubes for the Karnaugh map of Figure 12.33

[1] A useful rule to remember is that in a two-variable map there are two minterms adjacent to any given minterm; in a three-variable map, three minterms are adjacent to any given minterm; in a four-variable map, the number is four, and so on.

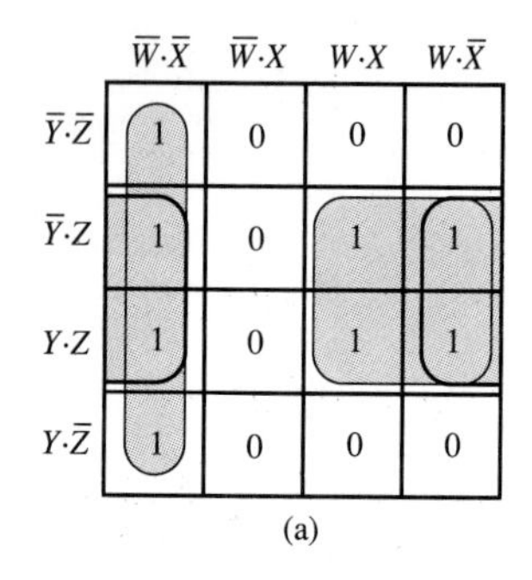

(a)

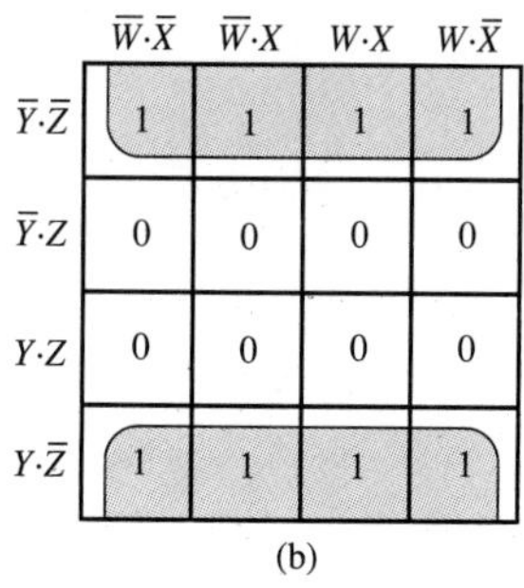

(b)

Figure 12.35 Four- and eight-cell subcubes for an arbitrary logic function

	$\overline{W}\cdot\overline{X}$	$\overline{W}\cdot X$	$W\cdot X$	$W\cdot\overline{X}$
$\overline{Y}\cdot\overline{Z}$	0	0	0	0
$\overline{Y}\cdot Z$	1	1	1	1
$Y\cdot Z$	0	0	0	0
$Y\cdot\overline{Z}$	0	0	0	0

Figure 12.36 Karnaugh map for the function $\overline{W}\cdot X\cdot\overline{Y}\cdot Z+\overline{W}\cdot\overline{X}\cdot\overline{Y}\cdot Z+W\cdot\overline{X}\cdot\overline{Y}\cdot Z+W\cdot X\cdot\overline{Y}\cdot Z$

In general, one tries to find the largest possible subcubes to cover all of the "1" entries in the map. How do maps and subcubes help in the realization of logic functions, then? The use of maps and subcubes in minimizing logic expressions is best explained by considering the following rule of Boolean algebra:

$$Y\cdot X+Y\cdot\overline{X}=Y$$

where the variable Y could represent a product of logic variables (for example, we could similarly write $(Z\cdot W)\cdot X+(Z\cdot W)\cdot\overline{X}=Z\cdot W$ with $Y=Z\cdot W$). This rule is easily proven by factoring Y:

$$Y\cdot(X+\overline{X})$$

and observing that $X+\overline{X}=1$, always. Then it should be clear that the variable X need not appear in the expression at all. Let us apply this rule to a more complex logic expression, to verify that it can also apply to this case. Consider the logic expression

$$\overline{W}\cdot X\cdot\overline{Y}\cdot Z+\overline{W}\cdot\overline{X}\cdot\overline{Y}\cdot Z+W\cdot\overline{X}\cdot\overline{Y}\cdot Z+W\cdot X\cdot\overline{Y}\cdot Z$$

and factor it as follows:

$$\begin{aligned}\overline{W}\cdot Z\cdot\overline{Y}\cdot(X+\overline{X})+W\cdot\overline{Y}\cdot Z\cdot(\overline{X}+X)&=\overline{W}\cdot Z\cdot\overline{Y}+W\cdot\overline{Y}\cdot Z\\&=\overline{Y}\cdot Z\cdot(\overline{W}+W)=\overline{Y}\cdot Z\end{aligned}$$

Quite a simplification! If we consider, now, a map in which we place a 1 in the cells corresponding to the minterms $\overline{W}\cdot X\cdot\overline{Y}\cdot Z$, $\overline{W}\cdot\overline{X}\cdot\overline{Y}\cdot Z$, $W\cdot\overline{X}\cdot\overline{Y}\cdot Z$, and $W\cdot X\cdot\overline{Y}\cdot Z$, forming the previous expression, we obtain the Karnaugh map of Figure 12.36. It can easily be verified that the map of Figure 12.36 shows a single four-cell subcube corresponding to the term $\overline{Y}\cdot Z$.

We have not established formal rules yet, but it definitely appears that the map method for simplifying Boolean expressions is a convenient tool. In effect, the map has performed the algebraic simplification automatically! We can see that in any subcube, one or more of the variables present will appear in both complemented *and* uncomplemented form in all their combinations with the other variables. These variables can be eliminated. As an illustration, in the *eight-cell* subcube case of Figure 12.37, the full-blown expression would be:

$$\begin{aligned}&\overline{W}\cdot\overline{X}\cdot\overline{Y}\cdot\overline{Z}+\overline{W}\cdot X\cdot\overline{Y}\cdot\overline{Z}+W\cdot X\cdot\overline{Y}\cdot\overline{Z}+W\cdot\overline{X}\cdot\overline{Y}\cdot\overline{Z}\\&+\overline{W}\cdot\overline{X}\cdot Y\cdot\overline{Z}+\overline{W}\cdot X\cdot Y\cdot\overline{Z}+W\cdot X\cdot Y\cdot\overline{Z}+W\cdot\overline{X}\cdot Y\cdot\overline{Z}\end{aligned}$$

However, if we consider the eight-cell subcube, we note that the three variables X, W, and Z appear both in complemented and uncomplemented form in all their combinations with the other variables and thus can be removed from the expression. This reduces the seemingly unwieldy expression simply to $\overline{Y}$! In logic design terms, a simple inverter is sufficient to implement the expression.

The example just shown is a particularly simple one, but it illustrates how simple it can be to determine the minimal expression for a logic function. It should be apparent that the larger a subcube, the greater the simplification that will result. For subcubes that do not intersect, as in the previous example, the solution can be found easily, and is unique.

Sum-of-Products Realizations

Although not explicitly stated, the logic functions of the preceding section were all in sum-of-products form. As you know, it is also possible to realize logic functions in product-of-sums form. This section discusses the implementation of logic functions in sum-of-products form and gives a set of design rules. The next section will do the same for product-of-sums form logical expressions. The following rules are a useful aid in determining the minimal sum-of-products expression:

1. Begin with isolated cells. These must be used as they are, since no simplification is possible.
2. Find all cells that are adjacent to only one other cell, forming two-cell subcubes.
3. Find cells that form four-cell subcubes, eight-cell subcubes, and so forth.
4. The minimal expression is formed by the collection of the *smallest number of maximal subcubes.*

The following examples illustrate the application of these principles to a variety of problems.

	$\overline{W}\cdot\overline{X}$	$\overline{W}\cdot X$	$W\cdot X$	$W\cdot\overline{X}$
$\overline{Y}\cdot\overline{Z}$	1	1	1	1
$\overline{Y}\cdot Z$	1	1	1	1
$Y\cdot Z$	0	0	0	0
$Y\cdot\overline{Z}$	0	0	0	0

Figure 12.37

Example 12.14

Design the logic circuit that implements the truth table shown. Use only two-input gates.

A	B	C	D	y
0	0	0	0	1
0	0	0	1	1
0	0	1	0	1
0	0	1	1	0
0	1	0	0	0
0	1	0	1	1
0	1	1	0	0
0	1	1	1	0
1	0	0	0	1
1	0	0	1	1
1	0	1	0	0
1	0	1	1	1
1	1	0	0	0
1	1	0	1	1
1	1	1	0	0
1	1	1	1	1

Solution:

A common way of solving problems such as this is to use a Karnaugh map to simplify and implement the function, with the aim of minimizing the complexity of the resulting circuit. The function y may be implemented by covering the Karnaugh map shown in Figure 12.38. The map covering shown in the figure leads to the expression

$$y = A \cdot D + \overline{C} \cdot D + \overline{A} \cdot \overline{B} \cdot \overline{D} + \overline{B} \cdot \overline{C}$$

with the corresponding implementation using two-input gates shown in Figure 12.39.

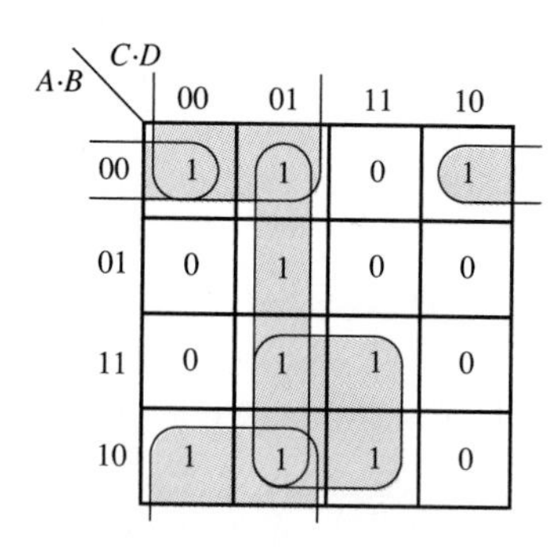

Figure 12.38 Karnaugh map for Example 12.14

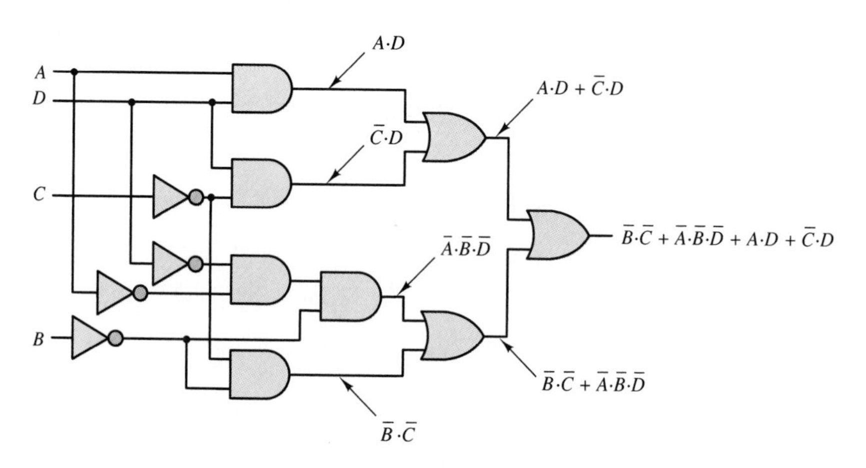

Figure 12.39 Logic circuit realization corresponding to Karnaugh map of Figure 12.38.

Example 12.15

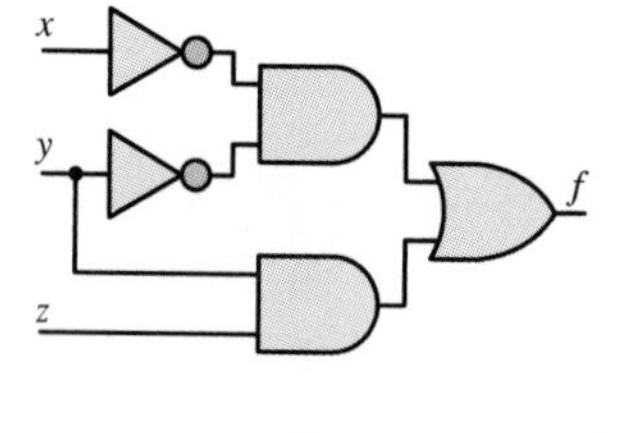

Figure 12.40

Construct a truth table and derive the minimum sum-of-products equation from the circuit of Figure 12.40.

Solution:

The first step is to make a truth table. The second step is to use a Karnaugh map to simplify and implement the function. The truth table may be obtained by writing the function corresponding to the circuit:

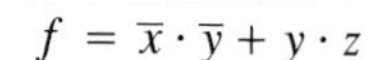

$$f = \overline{x} \cdot \overline{y} + y \cdot z$$

and filling the table accordingly, as follows:

x	y	z	f
0	0	0	1
0	0	1	1
0	1	0	0
0	1	1	1
1	0	0	0
1	0	1	0
1	1	0	0
1	1	1	1

Correspondingly, we can select the appropriate entries in a three-variable Karnaugh map:

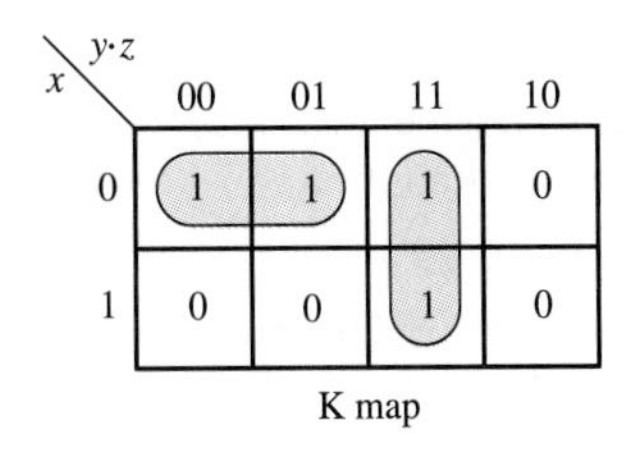

K map

The Karnaugh map verifies that the minimum sum-of-products expression is:

$$f = \bar{x} \cdot \bar{y} + y \cdot z$$

Example 12.16

Realize the following function using only two-input NAND gates:

$$f = (\bar{x} + \bar{y}) \cdot (\bar{z} + y)$$

Solution:

NAND gates naturally implement products, or serve as inverters, and thus there is no direct solution to this problem, unless the expression is manipulated into a different form. This can be done by applying De Morgan's laws:

1. $\bar{x} + \bar{y} = \overline{x \cdot y}$
2. $\bar{z} + y = \overline{z \cdot \bar{y}}$

Thus, the function f can be rewritten as $f = \overline{(x \cdot y)} \cdot \overline{(z \cdot \bar{y})}$ and consists only of product terms. The implementation with NAND gates requires the use of five gates, as shown in Figure 12.41.

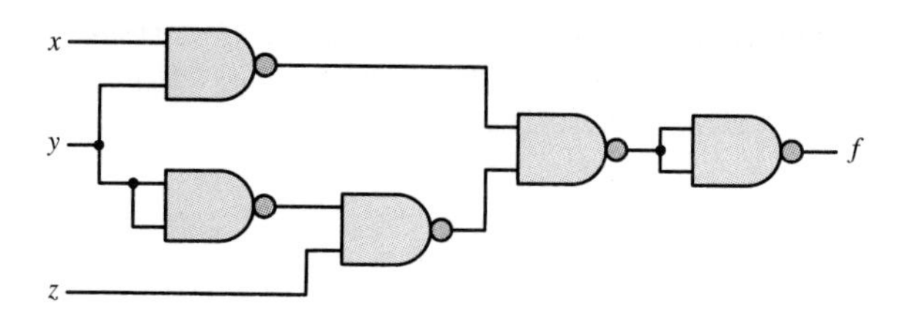

Figure 12.41

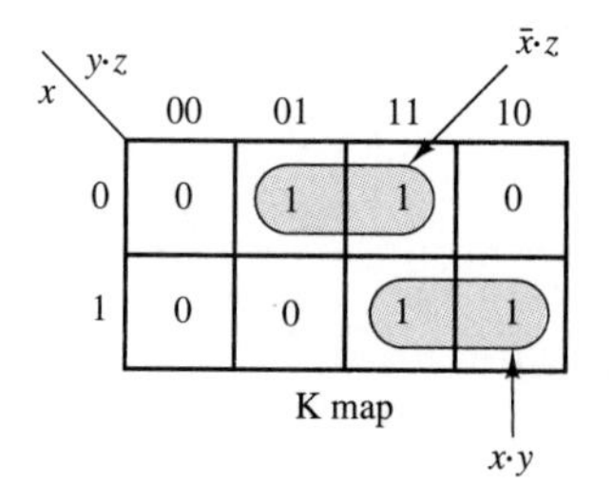

Figure 12.42

Example 12.17

Prove that $x \cdot y + \overline{x} \cdot z + y \cdot z = x \cdot y + \overline{x} \cdot z$, thus showing that $y \cdot z$ is a redundant term.

Solution:

One possible technique for simplifying a logic expression consists of using a Karnaugh map. In this case, the Karnaugh map is shown in Figure 12.42. The Karnaugh map immediately reveals that there is no need to circle the $y \cdot z$ term, thus making it redundant. Clearly, $f = x \cdot y + \overline{x} \cdot z$ is the minimum sum-of-products expression.

Example 12.18

Simplify the circuit of Figure 12.43 by reducing the number of gates to the minimum achievable number.

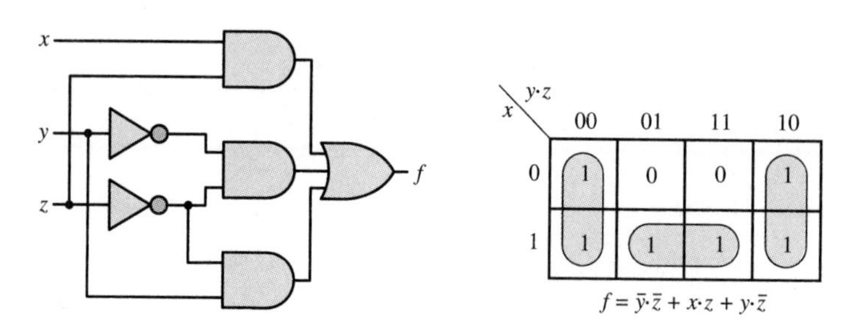

Figure 12.43 Figure 12.44

Solution:

To solve this problem, we need first to obtain a logic expression corresponding to the function of Figure 12.43. The second step is to use a Karnaugh map to simplify the function. The logic function corresponding to the logic circuit shown is

$$f = \overline{y} \cdot \overline{z} + x \cdot z + y \cdot \overline{z}$$

and the corresponding Karnaugh map is shown in Figure 12.44.

Inspection of the Karnaugh map, though, reveals that the map has been covered inefficiently, since two four-cell subcubes could have been used in place of three two-cell subcubes. Figure 12.45 depicts the alternative covering, which leads to the much simpler function

$$f = x + \bar{z}$$

The corresponding gate implementation is also shown in Figure 12.45.

This example emphasizes the importance of reducing logical expressions as much as possible prior to implementing a circuit. The Karnaugh map allows this simplification to be performed automatically, without having to explicitly reduce the logical expression.

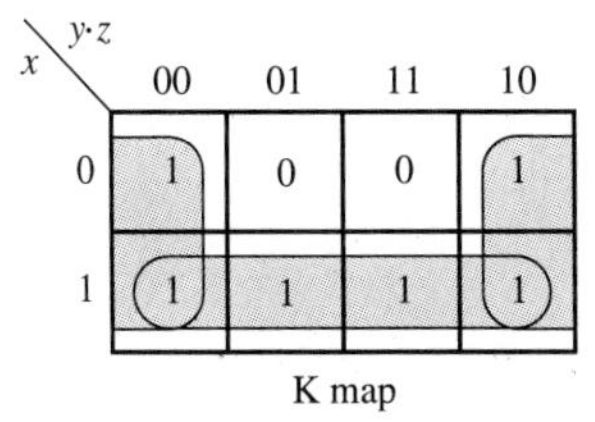

Figure 12.45

Product-of-Sums Realizations

Thus far, we have exclusively worked with sum-of-products expressions, that is, logic functions of the form $A \cdot B + C \cdot D$. We know, however, that De Morgan's laws state that there is an equivalent form that appears as a product of sums, for example, $(W + Y) \cdot (Y + Z)$. The two forms are completely equivalent, logically, but one of the two forms may lead to a realization involving a smaller number of gates. When using Karnaugh maps, we may obtain the product-of-sums form very simply by following these rules:

1. Solve for the 0s exactly as for the 1s in sum-of-products expressions.
2. Complement the resulting expression.

The same principles stated earlier apply in covering the map with subcubes and determining the minimal expression. The following examples illustrate how one form may result in a more efficient solution than the other.

Example 12.19

This example illustrates the design of a logic function using both sum-of-products and product-of-sums implementations, thus showing that it may be possible to realize some savings by using one implementation rather than the other.

1. Realize the function f by a Karnaugh map using 0s.
2. Realize the function f by a Karnaugh map using 1s.

x	y	z	f
0	0	0	0
0	0	1	1
0	1	0	1
0	1	1	1
1	0	0	1
1	0	1	1
1	1	0	0
1	1	1	0

Solution:

1. Using 0s, we obtain the Karnaugh map of Figure 12.46, leading to the product-of-sums expression

$$f = (x + y + z) \cdot (\overline{x} + \overline{y})$$

which requires five gates.

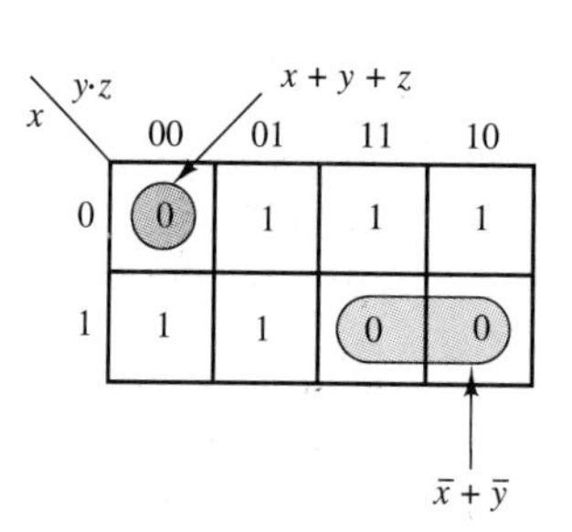

Figure 12.46

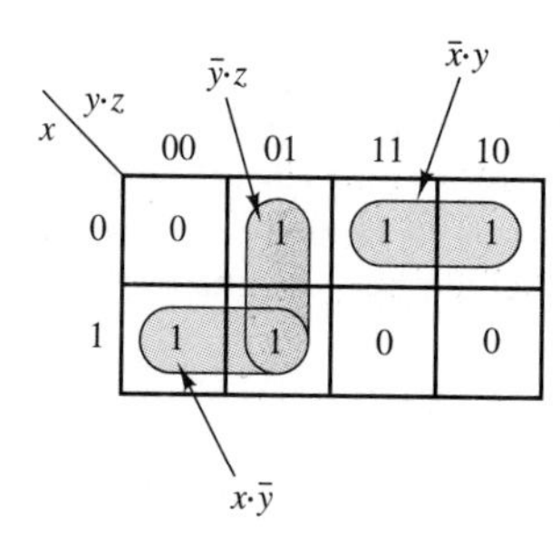

Figure 12.47

2. If 1s are used, as shown in Figure 12.47, a sum-of-products expression is obtained, of the form

$$f = \overline{x} \cdot y + x \cdot \overline{y} + \overline{y} \cdot z$$

which requires seven gates.

Example 12.20

Realize the logic function defined by the truth table shown as the simplest product-of-sums form:

x	y	z	f
0	0	0	1
0	0	1	0
0	1	0	1
0	1	1	0
1	0	0	1
1	0	1	0
1	1	0	0
1	1	1	0

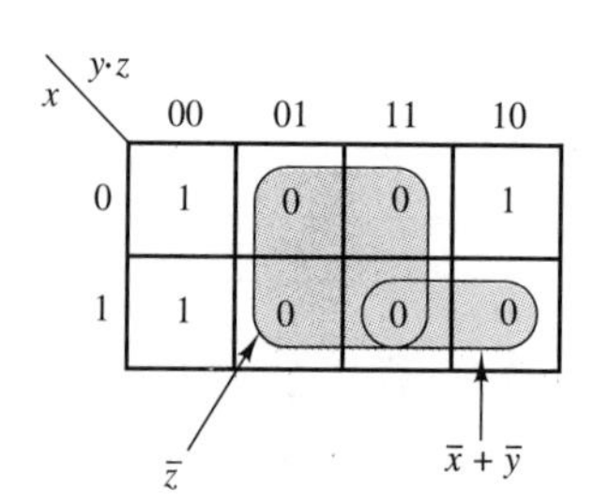

Figure 12.48

Solution:

To realize this function in product-of-sums form, the 0s are used in the map of Figure 12.48. The resulting expression is therefore given by

$$f = \overline{z} \cdot (\overline{x} + \overline{y})$$

Example 12.21 Safety Circuit for Operation of a Stamping Press

In this example, the techniques illustrated in the preceding example will be applied to a practical situation. To operate a stamping press, an operator must press two buttons (b_1 and b_2) one meter apart from each other and away from the press (this ensures that the operator's hands cannot be caught in the press). When the buttons are pressed, the logical variables b_1 and b_2 are equal to 1. Thus, we can define a new variable $A = b_1 \cdot b_2$; when $A = 1$, the operator's hands are safely away from the press. In addition to the safety requirement, however, other conditions must be satisfied before the operator can activate the press. The press is designed to operate on one of two workpieces, part I and part II, but not both. Thus, acceptable logic states for the press to be operated are "part I is in the press, but not part II" and "part II is in the press, but not part I." If we denote the presence of part I in the press by the logical variable $B = 1$ and the presence of part II by the logical variable $C = 1$, we can then impose additional requirements on the operation of the press. For example, a robot used to place either part in the press could activate a pair of switches (corresponding to logical variables B and C) indicating which part, if any, is in the press. Finally, in order for the press to be operable, it must be "ready," meaning that it has to have completed any previous stamping operation. Let the logical variable $D = 1$ represent the ready condition. We have now represented the operation of the press in terms of four logical variables, summarized in the truth table of Table 12.12. Note that only two combinations of the logical variables will result in operation of the press: $ABCD = 1011$ and $ABCD = 1101$. You should verify that these two conditions correspond to the desired operation of the press. Using a Karnaugh map, realize the logic circuitry required to implement the truth table shown.

Table 12.12 **Conditions for operation of stamping press**

(A) $b_1 \cdot b_2$	(B) Part I is in press	(C) Part II is in press	(D) Press is operable	Press operation 1 = pressing; 0 = not pressing
0	0	0	0	0
0	0	0	1	0
0	0	1	0	0
0	0	1	1	0
0	1	0	0	0
0	1	0	1	0
0	1	1	0	0
0	1	1	1	0
1	0	0	0	0
1	0	0	1	0
1	0	1	0	0
1	0	1	1	1
1	1	0	0	0
1	1	0	1	1
1	1	1	0	0
1	1	1	1	0

↑ Both buttons (b_1, b_2) must be pressed for this to be a 1.

Solution:

Table 12.12 can be converted to a Karnaugh map, as shown in Figure 12.49. Since there are many more 0s than 1s in the table, the use of 0s in covering the map will lead to greater

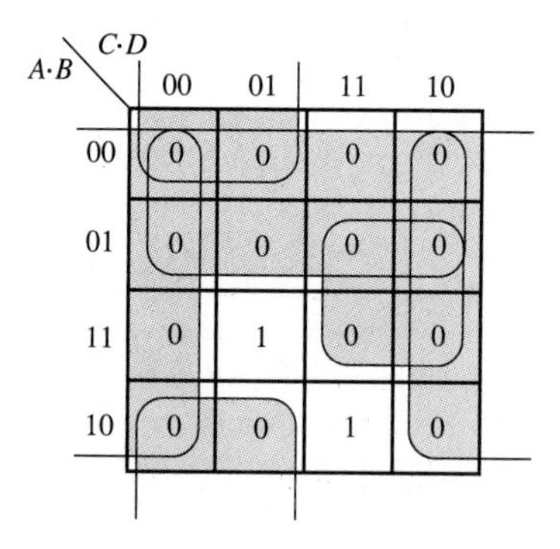

Figure 12.49

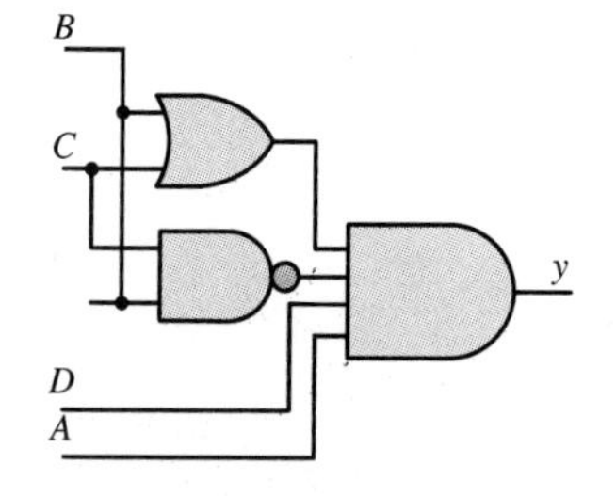

Figure 12.50

simplification. This will result in a product-of-sums expression. The four subcubes shown in Figure 12.49 yield the equation

$$A \cdot D \cdot (C + B) \cdot (\overline{C} + \overline{B})$$

By De Morgan's law, this equation is equivalent to

$$A \cdot D \cdot (C + B) \cdot \overline{(C \cdot B)}$$

which can be realized by the circuit of Figure 12.50.

For the purpose of comparison, the corresponding sum-of-products circuit is shown in Figure 12.51. Note that this circuit employs a greater number of gates and will therefore lead to a more expensive design.

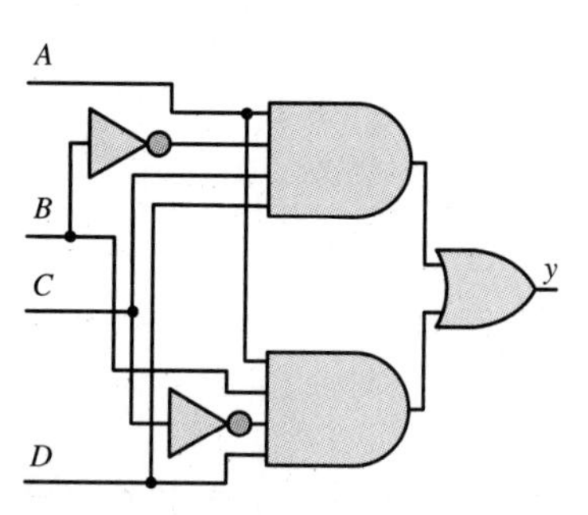

Figure 12.51

Don't Care Conditions

Another simplification technique may be employed whenever the value of the logic function to be implemented can be either a 1 or a 0. This condition may result from the specification of the problem and is not uncommon. Whenever it does not matter whether a position in the map is filled by a 1 or a 0, we use a so-called **don't care** entry, denoted by an **x**. Then the don't care can be used as either a 1 or a 0, depending on which results in a greater simplification (i.e., helps in forming the smallest number of maximal subcubes). The following examples illustrate the use of don't cares.

Example 12.22

When dealing with complex logic expressions involving several variables, it is often convenient to use a more compact form rather than write out the entire function. For example, one can specify that a certain logic function be "true" (i.e., logic 1) for certain combinations, which are then expressed as binary numbers. In this example, we wish to find a minimum sum-of-products expression for the function

$$f(a,b,c,d) = \bar{a}\bar{b}\bar{c}d + \bar{a}\bar{b}c\bar{d} + \bar{a}\bar{b}cd + \bar{a}b\bar{c}d + a\bar{b}cd + ab\bar{c}\bar{d}$$

Note that this could also be written as

$$f(a,b,c,d) = 0001 + 0010 + 0011 + 0101 + 1011 + 1100$$

if each uncomplemented variable is represented by a 1 and each complemented variable by a 0. Further, the binary numbers in the resulting expression may be converted to their decimal counterpart to obtain

$$f(a,b,c,d) = (1 + 2 + 3 + 5 + 11 + 12)_{10} = \sum(1,2,3,5,11,12)_{10}$$

In this example we also assume that the following combinations may be treated as don't cares:

$$f(a,b,c,d) = \sum(4,6,10,14)_{10}$$

Note that the **x**'s never occur, and so they may be assigned a 1 or a 0, whichever will best simplify the expression.

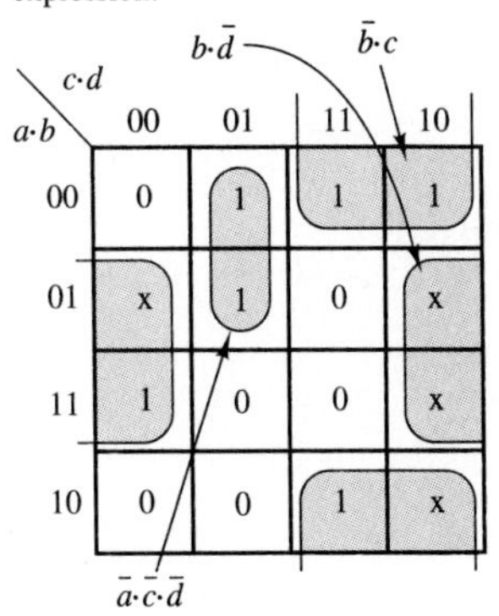

Figure 12.52

Solution:

The inputs that cannot occur may be treated as don't cares. If we use the symbol **x** to denote these inputs, the Karnaugh map corresponding to the problem specifications is as shown in Figure 12.52. If all **x** are given a value of 1, the following expression can be obtained by the map covering of Figure 12.52:

$$f = b\bar{d} + \bar{b}c + \bar{a}\cdot\bar{c}\cdot d$$

Example 12.23

Find a minimum sum-of-products expression for a function $f(a, b, c)$ whose output will be 1 for the combinations $(a, b, c)_{10} = 0, 2, 3$ and don't care for the combinations $(a, b, c)_{10} = 4, 5, 6$.

Solution:

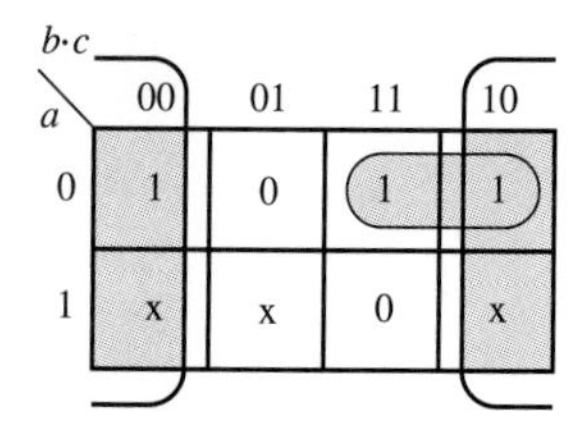

Figure 12.53

The key is to use the don't care terms *if* they will simplify the expression. The simplest function that can be obtained with the proper use of the don't care terms is shown in Figure 12.53. Thus the function f is

$$f = \overline{a} \cdot b + \overline{c}$$

Note that the don't care output for the combination $(a, b, c)_{10} = 5$ is not used. You should verify that using this additional don't care entry does not lead to any further simplification.

Example 12.24

Find a minimum sum-of-products expression for a function $f(a, b, c, d)$ whose output will be 1 for the combinations $(a, b, c, d)_{10} = 0, 3, 6, 9$ and don't care for the combinations $(a, b, c, d)_{10} = 10, 11, 13, 14, 15$.

Solution:

The Karnaugh map for the given function is shown in Figure 12.54(a). We know that for minimum sum-of-products, all the 1s have to be covered. With reference to Figure 12.54(a), the simplest function can be obtained by covering the map as shown in Figure 12.54(b). From this figure, it can be seen that the function is

$$f = \overline{a} \cdot \overline{b} \cdot \overline{c} \cdot \overline{d} + b \cdot c \cdot \overline{d} + a \cdot d + \overline{b} \cdot c \cdot d$$

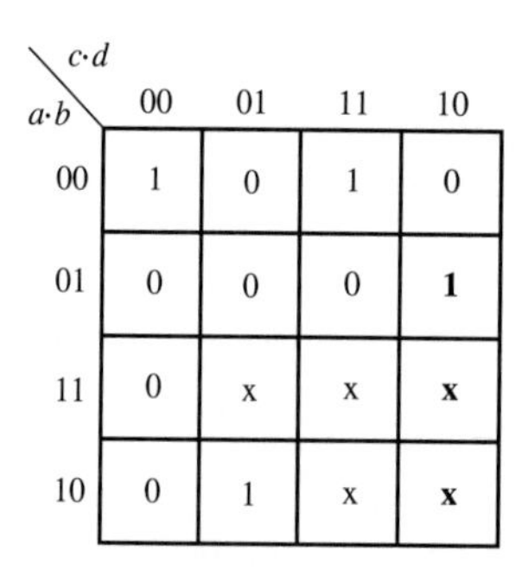

Figure 12.54(a)

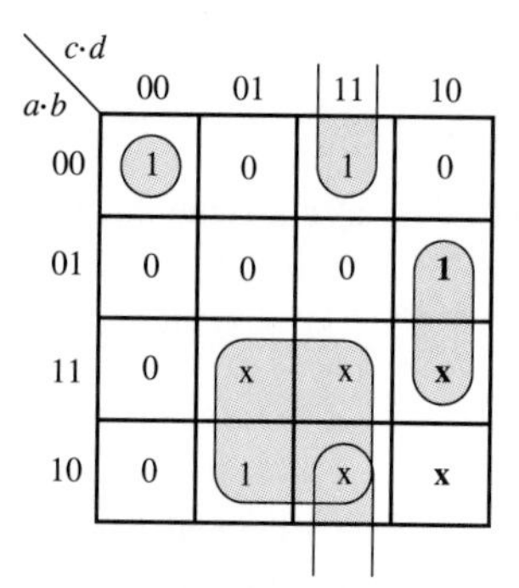

Figure 12.54(b)

DRILL EXERCISES

12.15 Simplify the following expression to show that it corresponds to the function $\overline{Z}$:

$$\overline{W}\cdot\overline{X}\cdot\overline{Y}\cdot\overline{Z}+\overline{W}\cdot X\cdot\overline{Y}\cdot\overline{Z}+W\cdot X\cdot\overline{Y}\cdot\overline{Z}+W\cdot\overline{X}\cdot\overline{Y}\cdot\overline{Z}+\overline{W}\cdot\overline{X}\cdot Y\cdot\overline{Z}$$
$$+\overline{W}\cdot X\cdot Y\cdot\overline{Z}+W\cdot X\cdot Y\cdot\overline{Z}+W\cdot\overline{X}\cdot Y\cdot\overline{Z}$$

12.16 Simplify the following expression, using a Karnaugh map:

$$\overline{W}\cdot\overline{X}\cdot\overline{Y}\cdot\overline{Z}+\overline{W}\cdot\overline{X}\cdot Y\cdot\overline{Z}+W\cdot X\cdot\overline{Y}\cdot\overline{Z}+W\cdot\overline{X}\cdot\overline{Y}\cdot\overline{Z}+W\cdot\overline{X}\cdot Y\cdot\overline{Z}$$
$$+W\cdot X\cdot Y\cdot\overline{Z}$$

12.17 Simplify the following expression, using a Karnaugh map:

$$\overline{W}\cdot\overline{X}\cdot\overline{Y}\cdot\overline{Z}+\overline{W}\cdot\overline{X}\cdot Y\cdot\overline{Z}+W\cdot X\cdot\overline{Y}\cdot\overline{Z}+W\cdot\overline{X}\cdot\overline{Y}\cdot\overline{Z}$$
$$+W\cdot\overline{X}\cdot Y\cdot\overline{Z}+\overline{W}\cdot X\cdot\overline{Y}\cdot\overline{Z}$$

12.18 The function y of Example 12.14 can be obtained with fewer gates if we use gates with three or four inputs. Find the minimum number of gates needed to obtain this function.

12.19 Verify that the product-of-sums expression for Example 12.19 can be realized with fewer gates.

12.20 Would a sum-of-products realization for Example 12.20 require fewer gates?

12.21 Prove that the circuit of Figure 12.51 can also be obtained from the sum of products.

12.22 In Example 12.22, assign a value of 0 to the don't care terms and derive the corresponding minimal expression. Is the new function simpler than the one obtained in Example 12.22?

12.23 In Example 12.23, assign a value of 0 to the don't care terms and derive the corresponding minimal expression. Is the new function simpler than the one obtained in Example 12.23?

12.24 In Example 12.23, assign a value of 1 to all don't care terms and derive the corresponding minimal expression. Is the new function simpler than the one obtained in Example 12.23?

12.25 In Example 12.24, assign a value of 0 to all don't care terms and derive the corresponding minimal expression. Is the new function simpler than the one obtained in Example 12.24?

12.26 In Example 12.24, assign a value of 1 to all don't care terms and derive the corresponding minimal expression. Is the new function simpler than the one obtained in Example 12.24?

12.5 COMBINATIONAL LOGIC MODULES

The basic logic gates described in the previous section are used to implement more advanced functions and are often combined to form logic modules, which, thanks to modern technology, are available in compact integrated circuit (IC) packages.

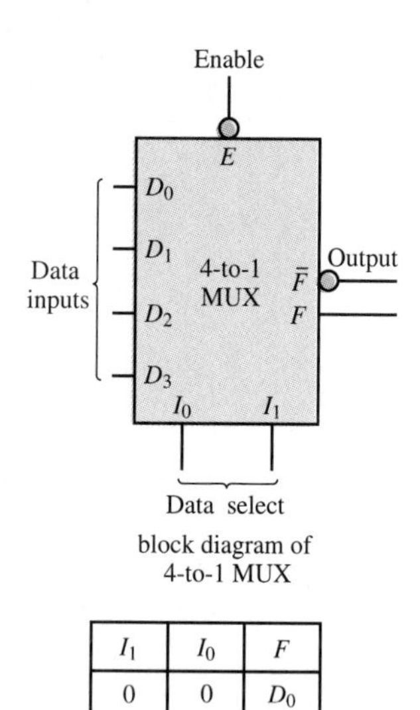

block diagram of 4-to-1 MUX

I_1	I_0	F
0	0	D_0
0	1	D_1
1	0	D_2
1	1	D_3

Truth table of 4-to-1 MUX

Figure 12.55 4-to-1 MUX

In this section and the next, we discuss a few of the more common **combinational logic modules**, illustrating how these can be used to implement advanced logic functions.

Multiplexers

Multiplexers, or **data selectors,** are combinational logic circuits that permit the selection of one of many inputs. A typical multiplexer (MUX) has 2^n **data lines,** n **address lines,** and one output. In addition, other control inputs (e.g., enables) may exist. Standard, commercially available MUXs allow for n up to 4; however, two or more MUXs can be combined if a greater range is needed. The MUX allows for one of 2^n inputs to be selected as the data output; the selection of which input is to appear at the output is made by way of the address lines. Figure 12.55 depicts the block diagram of a four-input MUX. The input data lines are labeled D_0, D_1, D_2, and D_3; the **data select**, or address, **lines** are labeled I_0 and I_1; and the output is available in both complemented and uncomplemented form, and is thus labeled F, or $\overline{F}$. Finally, an **enable** input, labeled E, is also provided, as a means of enabling or disabling the MUX: if $E = 1$, the MUX is disabled; if $E = 0$, it is enabled. The negative logic (MUX off when $E = 1$ and on when $E = 0$) is represented by the small "bubble" at the enable input, which represents a complement operation (just as at the output of NAND and NOR gates). The enable input is useful whenever one is interested in a cascade of MUXs; this would be of interest if we needed to select a line from a large number, say $2^8 = 256$. Then two 4-input MUXs could be used to provide the data selection of 1 of 8.

The material described in the previous sections is quite adequate to describe the internal workings of a multiplexer. Figure 12.56 shows the internal construction of a 4-to-1 MUX using exclusively NAND gates (inverters are also used, but the reader will recall that a NAND gate can act as an inverter if properly connected).

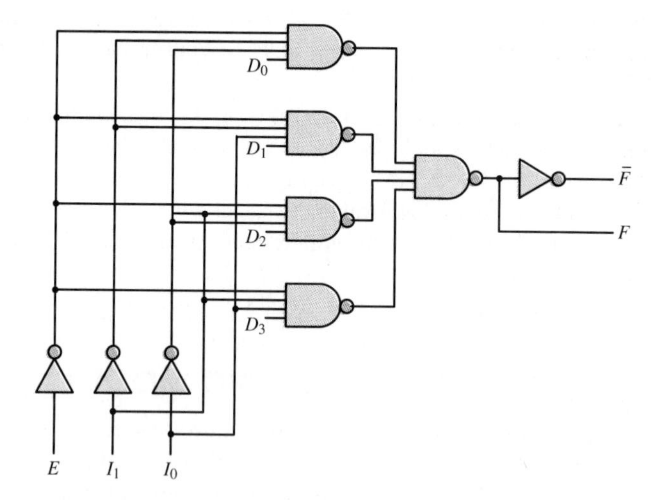

Figure 12.56 Internal structure of the 4-to-1 MUX

In the design of digital systems (for example, microcomputers), a single line is often required to carry two or more different digital signals. However, only one signal at a time can be placed on the line. A MUX will allow us to select, at different instants, the signal we wish to place on this single line. This property is shown here for a 4-to-1 MUX. Figure 12.57 depicts the functional diagram of a 4-to-1 MUX, showing four data lines, D_0 through D_3, and two select lines, I_0 and I_1.

The data selector function of a MUX is best understood in terms of Table 12.13. In this truth table, the **x**'s represent don't care entries. As can be seen from the truth table, the output selects one of the data lines depending on the values of I_1 and I_0, assuming that I_0 is the least significant bit. As an example, $I_1 I_0 = 10$ selects D_2, which means that the output, F, will select the value of the data line D_2. Therefore $F = 1$ if $D_2 = 1$ and $F = 0$ if $D_2 = 0$.

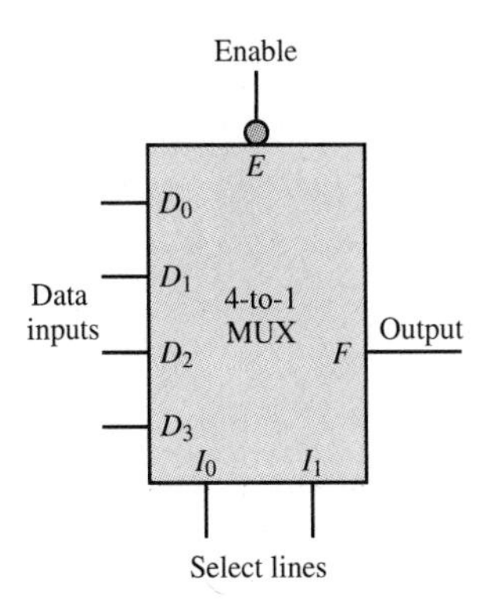

Figure 12.57 Functional diagram of four-input MUX

Table 12.13

I_1	I_0	D_3	D_2	D_1	D_0	F
0	0	x	x	x	0	0
0	0	x	x	x	1	1
0	1	x	x	0	x	0
0	1	x	x	1	x	1
1	0	x	0	x	x	0
1	0	x	1	x	x	1
1	1	0	x	x	x	0
1	1	1	x	x	x	1

Read-Only Memory (ROM)

Another common technique for implementing logic functions uses a **read-only memory,** or **ROM.** As the name implies, a ROM is a logic circuit that holds in storage ("memory") information—in the form of binary numbers—that cannot be altered but can be "read" by a logic circuit. A ROM is an array of memory cells, each of which can store either a 1 or a 0. The array consists of $2^m \times n$ cells, where n is the number of bits in each word stored in ROM. To access the information stored in ROM, m address lines are required. When an address is selected, in a fashion similar to the operation of the MUX, the binary word corresponding to the address selected appears at the output, which consists of n bits, that is, the same number of bits as the stored words. In some sense, a ROM can be thought of as a MUX that has an output consisting of a word instead of a single bit.

Figure 12.58 depicts the conceptual arrangement of a ROM with $n = 4$ and $m = 2$. The ROM table has been filled with arbitrary 4-bit words, just for the purpose of illustration. In Figure 12.58, if one were to select an enable input of 0 (i.e., on) and values for the address lines of $I_0 = 0$ and $I_1 = 1$, the output word would be $W_2 = 0110$, so that $b_0 = 0$, $b_1 = 1$, $b_2 = 1$, $b_3 = 0$. Depending on the content of the ROM and the number of address and output lines, one could implement an arbitrary logic function.

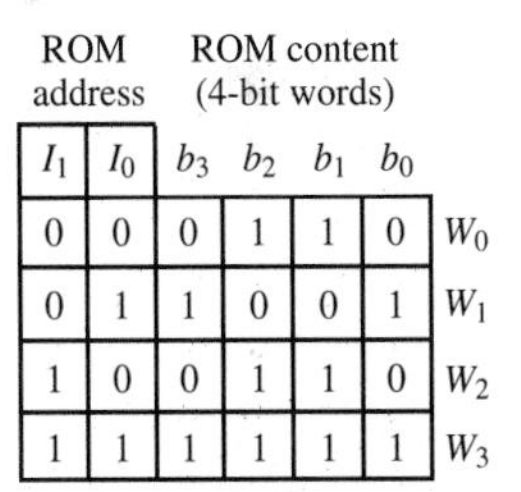

ROM address		ROM content (4-bit words)				
I_1	I_0	b_3	b_2	b_1	b_0	
0	0	0	1	1	0	W_0
0	1	1	0	0	1	W_1
1	0	0	1	1	0	W_2
1	1	1	1	1	1	W_3

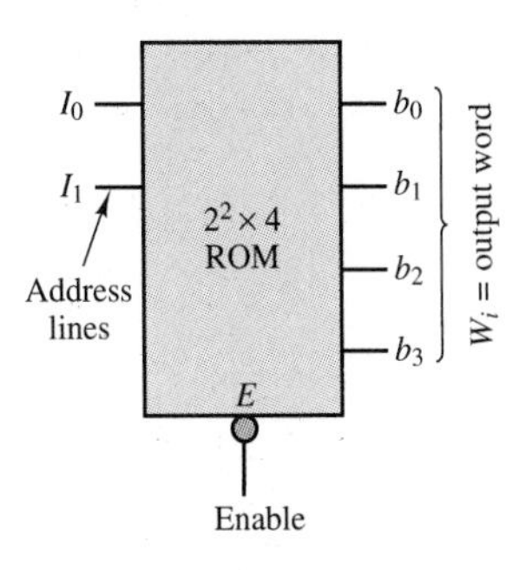

Figure 12.58 Read-only memory

Unfortunately, the data stored in read-only memories must be entered during fabrication and cannot be altered later. A much more convenient type of read-only memory is the **erasable programmable read-only memory (EPROM),** the content of which can be easily programmed and stored and may be changed if needed. EPROMs find use in many practical applications, because of their flexibility in content and ease of programming. The following example illustrates the use of an EPROM to perform the linearization of a nonlinear function.

Example 12.25 EPROM-Based Lookup Table

One of the most common applications of EPROMs is the *arithmetic lookup table.* A lookup table is similar in concept to the familiar multiplication table and is used to store precomputed values of certain functions, eliminating the need for actually computing the function. A practical application of this concept is present in every automobile manufactured in the United States since the early 1980s, as part of the exhaust emission control system. In order for the catalytic converter to minimize the emissions of exhaust gases (especially hydrocarbons, oxides of nitrogen, and carbon monoxide), it is necessary to maintain the *air-to-fuel ratio* (A/F) as close as possible to the stoichiometric value, that is, 14.7 parts of air for each part of fuel. Most modern-day engines are equipped with fuel injection systems that are capable of delivering accurate amounts of fuel to each individual cylinder; thus, the task of maintaining an accurate A/F amounts to measuring the mass of air that is aspirated into each cylinder and computing the corresponding mass of fuel. Many automobiles are equipped with a *mass airflow sensor,* capable of measuring the mass of air drawn into each cylinder during each engine cycle. Let the output of the mass airflow sensor be denoted by the variable M_A, and let this variable represent the mass of air (in g) actually entering a cylinder during a particular stroke. It is then desired to compute the mass of fuel, M_F (also expressed in g), required to achieve and A/F of 14.7. This computation is simply:

$$M_F = \frac{M_A}{14.7}$$

Although the above computation is a simple division, its actual calculation in a low-cost digital computer (such as would be used on an automobile) is rather complicated. It would be much simpler to tabulate a number of values of M_A, to precompute the variable M_F, and then to store the result of this computation into an EPROM. If the EPROM address were made to correspond to the tabulated values of air mass, and the content at each address to the corresponding fuel mass (according to the precomputed values of the expression $M_F = \frac{M_A}{14.7}$), it would not be necessary to perform the division by 14.7. For each measurement of air mass into one cylinder, an EPROM address is specified and the corresponding content is read. The content at the specific address is the mass of fuel required by that particular cylinder.

In practice, the fuel mass needs to be converted into a time interval corresponding to the duration of time during which the fuel injector is open. This final conversion factor can also be accounted for in the table. Suppose, for example, that the fuel injector is capable of injecting K_F g of fuel per second; then the time duration, T_F, during which the injector should be open in order to inject M_F g of fuel into the cylinder is given by:

$$T_F = \frac{M_F}{K_F} \text{ s}$$

Therefore, the complete expression to be precomputed and stored in the EPROM is:

$$T_F = \frac{M_A}{14.7 \times K_F} \text{ s}$$

Figure 12.59 illustrates this process graphically.

To provide a numerical illustration, consider a hypothetical engine capable of aspirating air in the range $0 < M_A < 0.51$ g and equipped with fuel injectors capable of injecting at the rate of 1.36 g/s. Thus, the relationship between T_F and M_A is:

$$T_F = 50 \times M_A \text{ ms} = 0.05 M_A \text{ s}$$

If the digital value of M_A is expressed in dg (decigrams, or tenths of g), the lookup table of Figure 12.60 can be implemented, illustrating the conversion capabilities provided by the EPROM. Note that in order to represent the quantities of interest in an appropriate binary format compatible with the 8-bit EPROM, the units of air mass and of time have been scaled.

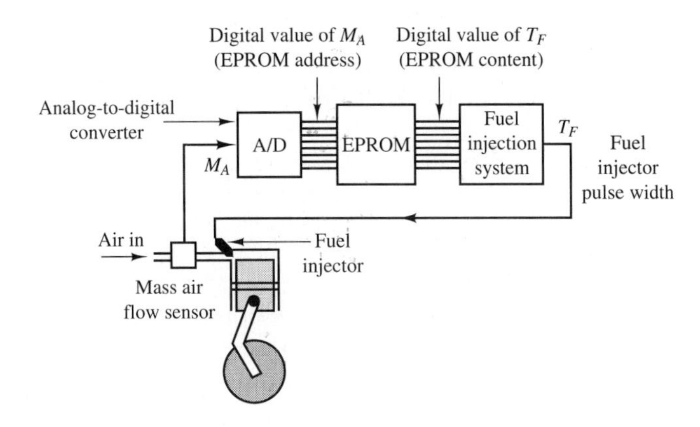

Figure 12.59 Use of EPROM lookup table in automotive fuel injection system

M_A(g) $\times 10^{-2}$	Address (digital value of M_A)	Content (digital value of T_F)	T_F(ms) $\times 10^{-1}$
0	00000000	00000000	0
1	00000001	00000101	5
2	00000010	00001010	10
3	00000011	00001111	15
4	00000100	00010100	20
5	00000101	00011001	25
⋮	⋮	⋮	⋮
51	00110011	11111111	255

Figure 12.60 Lookup table for automotive fuel injection application

Decoders and Read and Write Memory

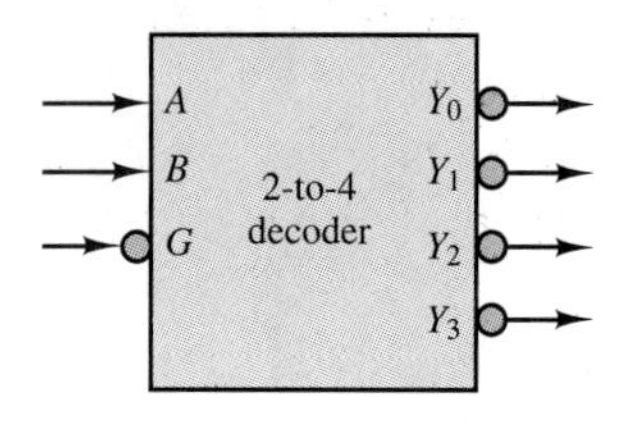

Inputs			Outputs			
Enable	Select					
$\overline{G}$	A	B	Y_0	Y_1	Y_2	Y_3
1	x	x	1	1	1	1
0	0	0	0	1	1	1
0	0	1	1	0	1	1
0	1	0	1	1	0	1
0	1	1	1	1	1	0

Figure 12.61 2-to-4 decoder

Decoders, which are commonly used for applications such as address decoding or memory expansion, are combinational logic circuits as well. Our reason for introducing decoders is to show some of the internal organization of semiconductor memory devices. An important application of decoders in the organization of a memory system will be discussed in Chapter 13.

Figure 12.61 shows the truth table for a 2-to-4 decoder. The decoder has an enable input, $\overline{G}$, and select inputs, B and A. It also has four outputs, Y_0 through Y_3. When the enable input is logic 1, all decoder outputs are forced to logic 1 regardless of the select inputs.

This simple description of decoders permits a brief discussion of the internal organization of an **SRAM (static random-access** or **read and write memory).** SRAM is internally organized to provide memory with high speed (i.e., short access time), a large bit capacity, and low cost. The memory array in this memory device has a column length equal to the number of words, W, and a row length equal to the number of bits per word, N. To select a word, an n-to-W decoder is needed. Since the address inputs to the decoder select only one of the decoder's outputs, the decoder selects one word in the memory array. Figure 12.62 shows the internal organization of a typical SRAM.

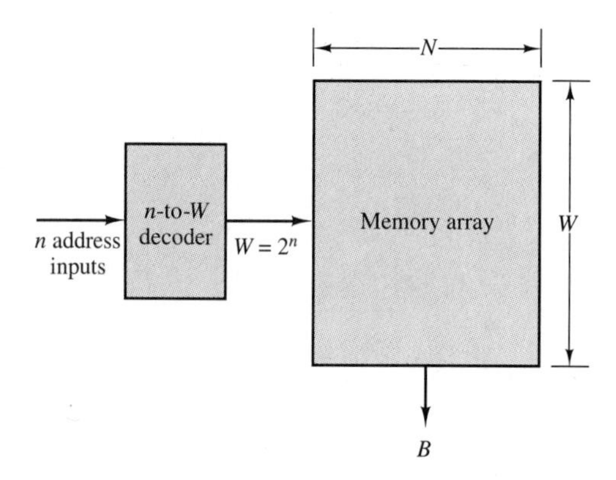

Figure 12.62 Internal organization of SRAM

Thus, to choose the desired word from the memory array, the proper address inputs are required. As an example, if the number of words in the memory array is 8, a 3-to-8 decoder is needed. Data sheets for 2-to-4 and 3-to-8 decoders from a CMOS family data book are provided at the end of the chapter.

DRILL EXERCISES

12.27 Which combination of the control lines will select the data line D_3 for a 4-to-1 MUX?

12.28 Show that an 8-to-1 MUX with eight data inputs (D_0 through D_7) and three control lines (I_0 through I_2) can be used as a data selector. Which combination of the control lines will select the data line D_5?

12.29 Which combination of the control lines will select the data line D_4 for an 8-to-1 MUX?

12.30 How many address inputs do you need if the number of words in a memory array is 16?

CONCLUSION

- Digital logic circuits are at the basis of digital computers. Such circuits operate strictly on binary signals according to the laws of Boolean algebra.
- Combinational logic circuits can implement arbitrary Boolean logic functions.
- Combinational logic circuits include all of the logic gates—AND, OR, NAND, NOT, and XOR—as well as logic modules such as multiplexers and read-only memory.

KEY TERMS

ANSWERS TO DRILL EXERCISES

12.1 (a) 100111; (b) 111011; (c) 1000000000; (d) 0.011100; (e) 0.11001; (f) 0.110011; (g) 100000000.11; (h) 10000001.1001; (i) 1000000000000.11101

12.2 (a) 13; (b) 27; (c) 23; (d) 0.6875; (e) 0.203125; (f) 0.2128906 0.2128906255; (g) 59.6875; (h) 91.203125; (i) 22.340820312

12.3 (a) 20.75_{10}; (b) 74_{10}; (c) 1.5_{10}; (d) 21_{10}; (e) 100000_2; (f) 1000000_2; (g) 110010.11_2; (h) 100011.11111_2

12.4 4,096

12.5 39 mV

12.6 (a) 111110000011; (b) 1111001001; (c) 10100110; (d) 1AE; (e) B9; (f) 6ED

12.7 (a) 00010111; (b) 01101001; (c) 0100010

12.8 (a) 00010111; (b) 01101001; (c) 0100010

12.16 $W \cdot \overline{Z} + \overline{X} \cdot \overline{Z}$

12.17 $\overline{Y} \cdot \overline{Z} + \overline{X} \cdot \overline{Z}$

12.18 Nine gates

12.20 No

12.22 $f = a \cdot b \cdot \overline{c} \cdot \overline{d} + \overline{a} \cdot \overline{c} \cdot d + \overline{a} \cdot \overline{b} \cdot c + \overline{b} \cdot c \cdot d$

12.23 $f = \overline{a} \cdot b + \overline{a} \cdot \overline{c}$

12.24 $f = \overline{a} \cdot b + a \cdot \overline{b} + \overline{c}$

12.25 $f = \overline{a} \cdot \overline{b} \cdot \overline{c} \cdot \overline{d} + \overline{a} \cdot \overline{b} \cdot c \cdot d + a \cdot \overline{b} \cdot \overline{c} \cdot d + \overline{a} \cdot b \cdot c \cdot \overline{d}$

12.26 $f = \overline{a} \cdot \overline{b} \cdot \overline{c} \cdot \overline{d} + b \cdot c \cdot \overline{d} + a \cdot d + \overline{b} \cdot c \cdot d + a \cdot c$

12.27 $I_1 I_0 = 11$

12.28 For the first part, use the same method as in Drill Exercise 12.27, but for an 8-to-1 MUX. For the second part, $I_2 I_1 I_0 = 101$.

12.29 $I_2 I_1 I_0 = 100$

12.30 4

DATA SHEETS FOR SELECTED LOGIC MODULES

Common IC Logic Gates

TTL Name	CMOS Name	Description
SN74LS04	MM74HC04	Hex inverter
SN74LS08	MM74HC08	Quad 2-input AND gate
SN74LS32	MM74HC32	Quad 2-input OR gate
SN74LS00	MM74HC00	Quad 2-input NAND gate
SN74LS02	MM74HC02	Quad 2-input NOR gate
SN74LS86	MM74HC86	Quad 2-input exclusive OR gate
SN74LS266	MM74HC266	Quad 2-input exclusive NOR gate
SN74LS10	MM74HC10	Triple 3-input NAND gate
SN74LS27	MM74HC27	Triple 3-input NOR gate
SN74LS20	MM74HC20	Dual 4-input NAND gate
SN74ALS25	CD4002C	Dual 4-input NOR gate
SN74LS30	MM74HC30	8-input NAND gate

MM54HC138/MM74HC138 3-to-8 Line Decoder

Features

- Typical propagation delay: 20 ns
- Wide power supply range: 2 V–6 V
- Low quiescent current: 80 μA maximum (74HC series)
- Low input current: 1 μA maximum
- Fanout of 10 LS-TTL loads

Connection Diagram

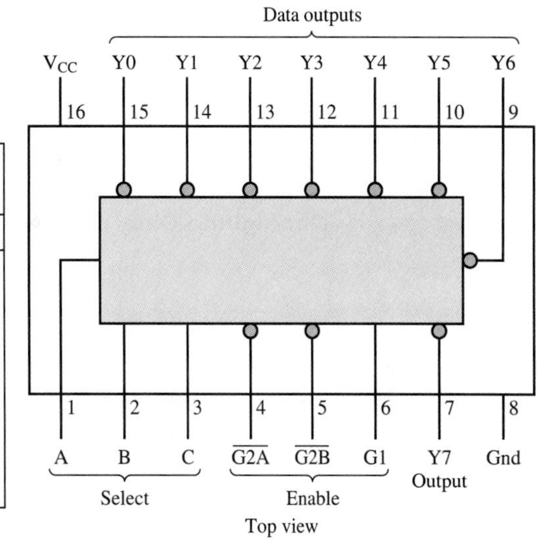

Truth Table

Inputs					Outputs							
Enable		Select										
G1	$\overline{G2}$*	C	B	A	Y0	Y1	Y2	Y3	Y4	Y5	Y6	Y7
X	H	X	X	X	H	H	H	H	H	H	H	H
L	X	X	X	X	H	H	H	H	H	H	H	H
H	L	L	L	L	L	H	H	H	H	H	H	H
H	L	L	L	H	H	L	H	H	H	H	H	H
H	L	L	H	L	H	H	L	H	H	H	H	H
H	L	L	H	H	H	H	H	L	H	H	H	H
H	L	H	L	L	H	H	H	H	L	H	H	H
H	L	H	L	H	H	H	H	H	H	L	H	H
H	L	H	H	L	H	H	H	H	H	H	L	H
H	L	H	H	H	H	H	H	H	H	H	H	L

*$\overline{G2}$ = G2A + G2B
H = high level, L = low level, X = don't care

Logic Diagram

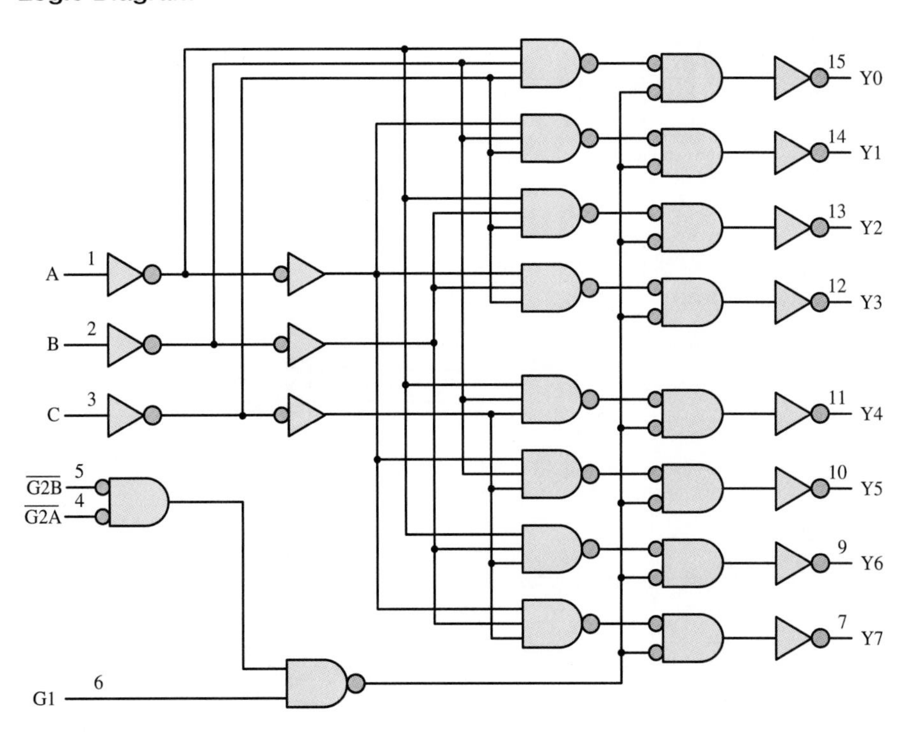

MM54HC139/MM74HC139 Dual 2-To-4 Line Decoder

Features

- Typical propagation delays—
  Select to outputs (4 delays): 18 ns
  Select to output (5 delays): 28 ns
  Enable to output: 20 ns
- Low power: 40 μW quiescent supply power
- Fanout of 10 LS-TTL devices
- Input current maximum 1 μA, typical 10 pA

Connection Diagram

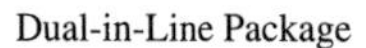

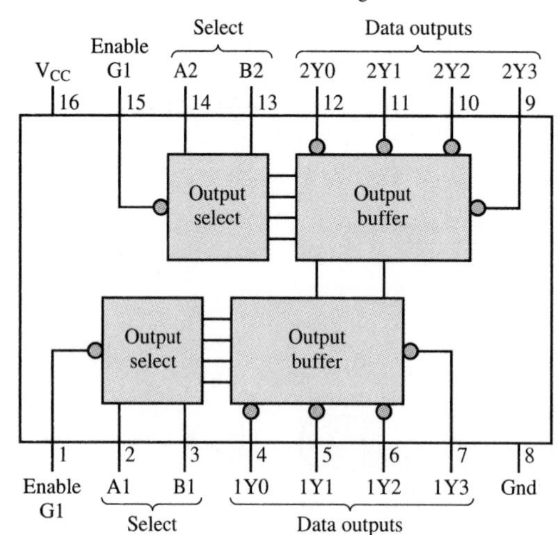

Truth Table

HC139

Inputs			Outputs			
Enable	Select					
G	B	A	Y0	Y1	Y2	Y3
H	X	X	H	H	H	H
L	L	L	L	H	H	H
L	L	H	H	L	H	H
L	H	L	H	H	L	H
L	H	H	H	H	H	L

H = high level, L = low level, X = don't care

Logic Diagram

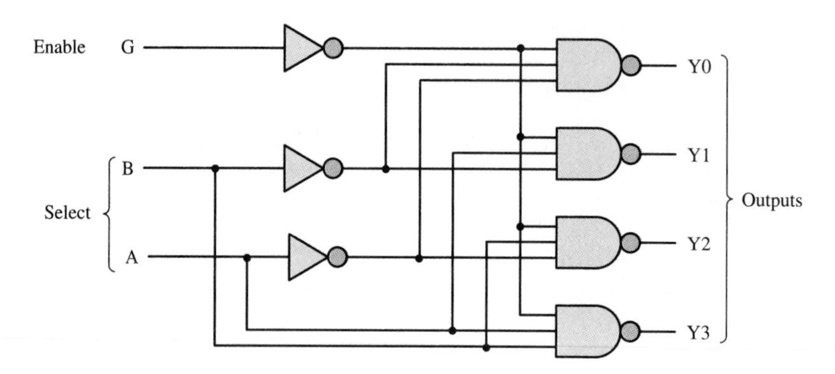

MM54HC153/MM74HC153 Dual 4-Input Multiplexer

Features

- Typical propagation delay: 24 ns
- Wide power supply range: 2 V–6 V
- Low quiescent current: 80 μA maximum (74HC series)
- Low input current: 1 μA maximum
- Fanout of 10 LS-TTL loads

Truth Table

Select Inputs		Data Inputs				Strobe	Output
B	A	C0	C1	C2	C3	G	Y
X	X	X	X	X	X	H	L
L	L	L	X	X	X	L	L
L	L	H	X	X	X	L	H
L	H	X	L	X	X	L	L
L	H	X	H	X	X	L	H
H	L	X	X	L	X	L	L
H	L	X	X	H	X	L	H
H	H	X	X	X	L	L	L
H	H	X	X	X	H	L	H

Select inputs A and B are common to both sections.
H = high level, L = low level, X = don't care

Connection Diagram

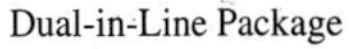

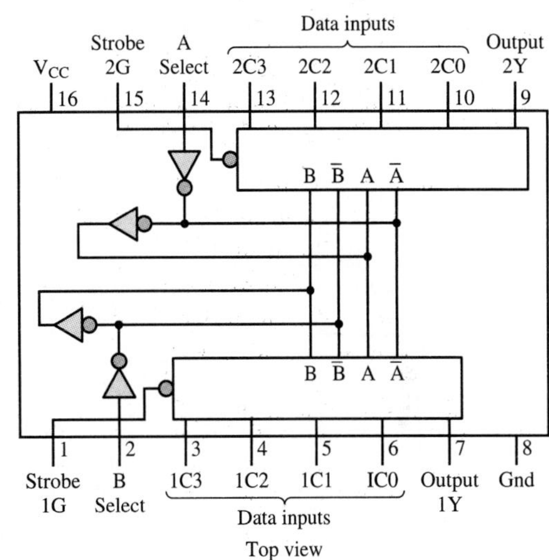

NMC98C64 8K x 8 CMOS Electrically Erasable PROM

Features

- Single 5-V power supply
- Low CMOS power
  - — Active, 10 mA typical
  - — Standby, 100 μA typical
  - — Quiescent, 100 μA typical
- Simple byte write and page write
  - — On-chip address and data latches
  - — Self-timed cycle, auto erase before write
  - — Page write up to 32 bytes per page
  - — Ready/$\overline{\text{Busy}}$ open drain status output and $\overline{\text{DATA}}$ polling verification
  - — Write protection
- Fast write time
  - — Byte or page write, 10 ms max
  - — Entire chip write in 2.6 seconds
  - — Page data load, 300 μs typical
- Fast access time: 200 ns/250 ns/350 ns
- CMOS and TTL compatible level inputs/outputs

Block and Connection Diagrams

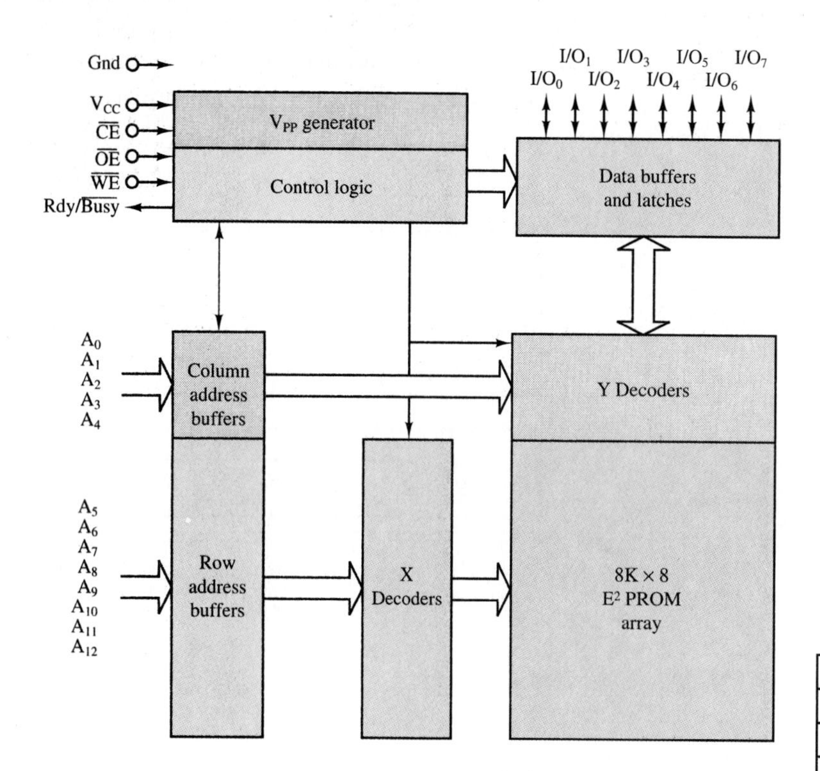

Pin	Name	Pin	Name
1	Ready/Bsy	28	V_{CC}
2	A_{12}	27	$\overline{WE}$
3	A_7	26	NC
4	A_6	25	A_8
5	A_5	24	A_9
6	A_4	23	A_{11}
7	A_3	22	$\overline{OE}$
8	A_2	21	A_{10}
9	A_1	20	$\overline{CE}$
10	A_0	19	I/O_7
11	I/O_0	18	I/O_6
12	I/O_1	17	I/O_5
13	I/O_2	16	I/O_4
14	V_{SS}	15	I/O_3

NMC98C64
8K × 8

A_0–A_4	Column Addresses
A_5–A_{12}	Row Addresses
I/O_0–I/O_7	Data Inputs/Outputs
$\overline{CE}$	Chip Enable
$\overline{OE}$	Output Enable
$\overline{WE}$	Write Enable
Rdy/$\overline{Busy}$	Device Ready/Busy
NC	No Connect

V61C16 Family High Performance Low Power 2K x 8 Bit CMOS Static RAM

Features

- High Speed
  - Maximum access time of 25/35/45/55/70 ns
  - Equal access and cycle times
- Low Power
  - 200 mW typical operating
  - 0.5 μW typical standby
  - 0.1 μW typical data retention
- Battery backup
  - 2 volt data retention (L version)
- Six transistor CMOS memory cell
- Fully static operation
  - No clock or refresh required
- Pin Compatible with standard 16K static RAMS and EPROMS in 300 and 600 mil DIP
- TTL compatible

Functional Block Diagram

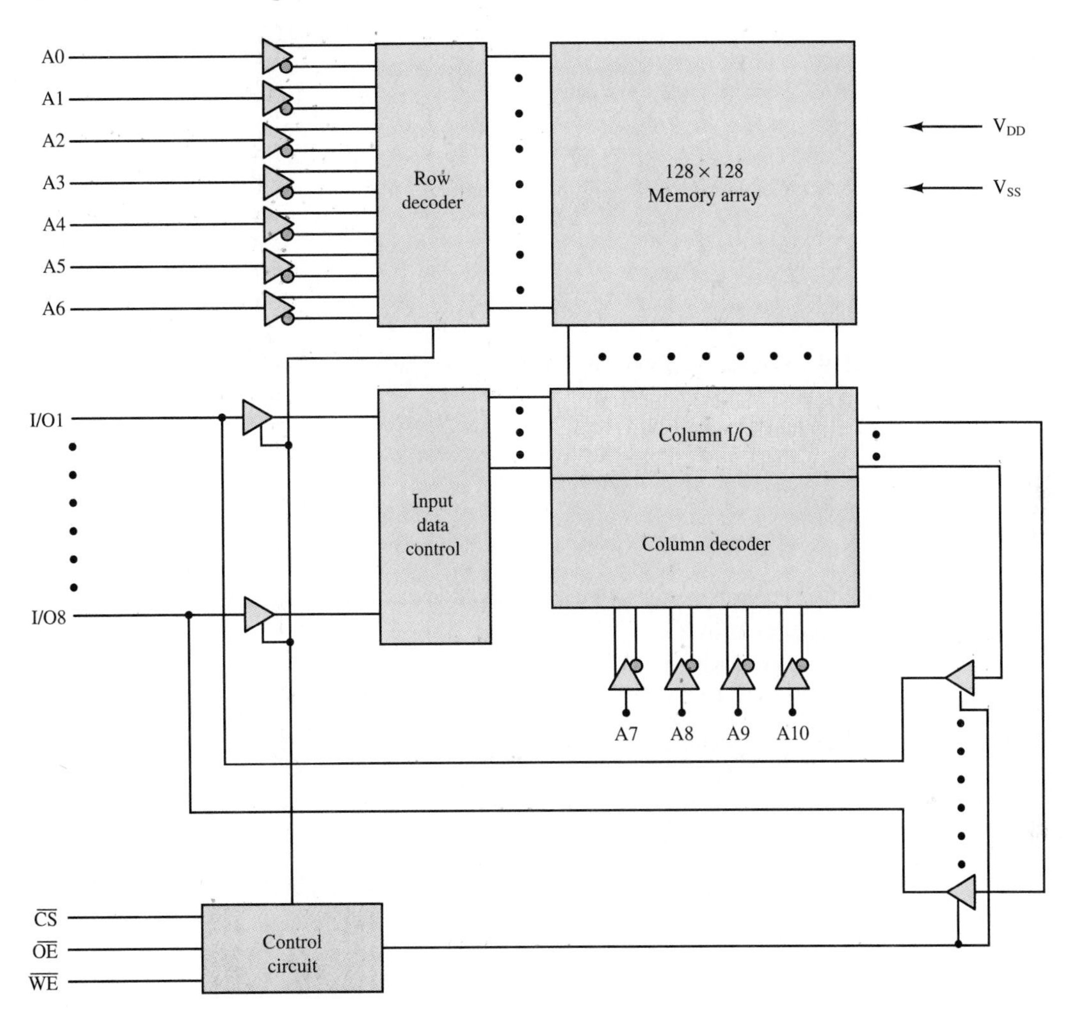

HOMEWORK PROBLEMS

12.1 Convert the following base 10 numbers to hex and binary:

a. 401
b. 273
c. 15
d. 38
e. 56

12.2 Convert the following hex numbers to base 10 and binary:

a. A
b. 66
c. 47
d. 21
e. 13

12.3 Convert the following base 10 numbers to binary:

a. 271.25
b. 53.375
c. 37.32
d. 54.27

12.4 Convert the following binary numbers to hex and base 10:

a. 1111
b. 1001101
c. 1100101
d. 1011100
e. 11101
f. 101000

12.5 Perform the following additions all in the binary system:

a. 11001011 + 101111
b. 10011001 + 1111011
c. 11101001 + 10011011

12.6 Perform the following subtractions all in the binary system:

a. 10001011 − 1101111
b. 10101001 − 111011
c. 11000011 − 10111011

12.7 Assuming that the most significant bit is the sign bit, find the decimal value of the following sign-magnitude form eight-bit binary numbers:

a. 11111000
b. 10011111
c. 01111001

12.8 Find the sign-magnitude form binary representation of the following decimal numbers:

a. 126
b. −126
c. 108
d. −98

12.9 Find the two's complement of the following binary numbers:

a. 1111
b. 1001101
c. 1011100
d. 11101

12.10 Use a truth table to prove that $B = AB + \overline{A}B$.

12.11 Use truth tables to prove that $BC + B\overline{C} + \overline{B}A = A + B$.

12.12 Using the method of proof by perfect induction, show that

$$(X + Y) \cdot (\overline{X} + X \cdot Y) = Y$$

12.13 Using De Morgan's theorems and the rules of Boolean algebra, simplify the following logic function:

$$F(X, Y, Z) = \overline{X} \cdot \overline{Y} \cdot \overline{Z} + \overline{X} \cdot Y \cdot Z + X \cdot (\overline{Y + Z})$$

12.14 Simplify the expression $f(A, B, C) = ABC + \overline{A}CD + \overline{B}CD$.

12.15 Simplify the logic function $F(A, B, C) = \overline{A} \cdot B \cdot \overline{C} + \overline{A} \cdot B \cdot C + A \cdot B \cdot \overline{C} + A \cdot B \cdot C$ using Boolean algebra.

12.16 Find the logic function defined by the truth table given in Figure P12.1.

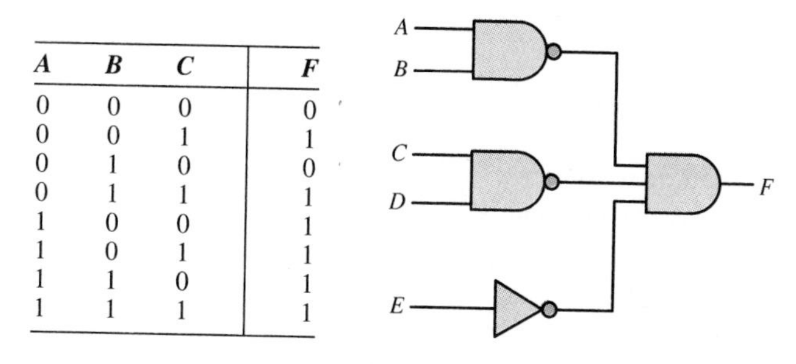

A	B	C	F
0	0	0	0
0	0	1	1
0	1	0	0
0	1	1	1
1	0	0	1
1	0	1	1
1	1	0	1
1	1	1	1

Figure P12.1 **Figure P12.2**

12.17 Determine the Boolean function describing the operation of the circuit shown in Figure P12.2.

12.18 Use a truth table to show when the output of the circuit of Figure P12.3 is 1.

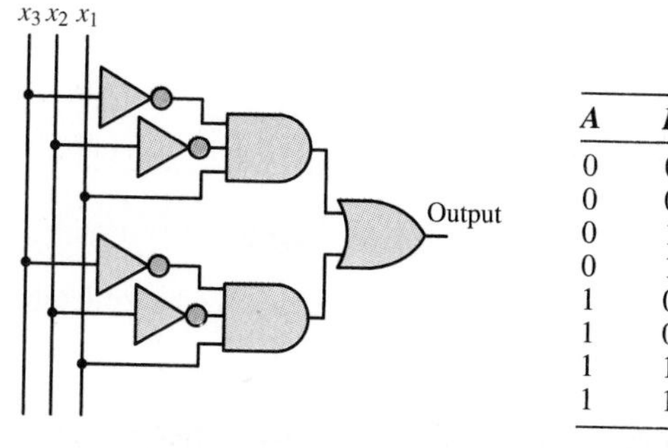

A	B	C	F
0	0	0	1
0	0	1	0
0	1	0	0
0	1	1	0
1	0	0	1
1	0	1	0
1	1	0	1
1	1	1	1

Figure P12.3 **Figure P12.4**

12.19 Find the logic function corresponding to the truth table of Figure P12.4 in the simplest sum-of-products form.

12.20 Find the minimum expression for the output of the logic circuit shown in Figure P12.5.

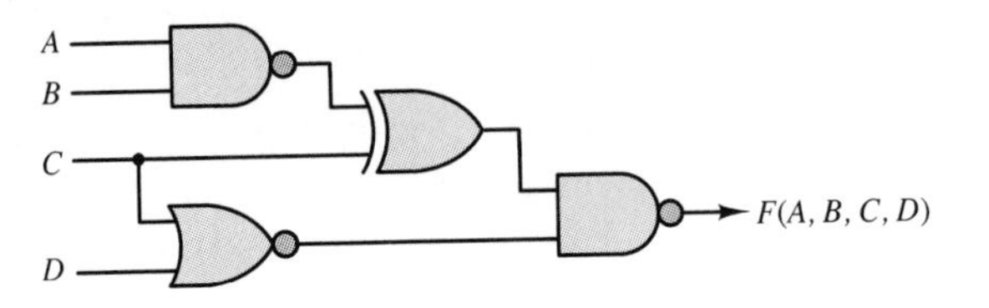

Figure P12.5

12.21 Use a Karnaugh map to minimize the function $f(A, B, C) = ABC + AB\overline{C} + A\overline{BC}$.

12.22 Determine the minimum expression for the following logic function, simplifying the expression:

$$f(A, B, C) = (A + B) \cdot A \cdot B + \overline{A} \cdot C + A \cdot \overline{B} \cdot C + \overline{B} \cdot \overline{C}$$

12.23

a. Build the Karnaugh map for the logic function defined by the truth table of Figure P12.6.
b. What is the minimum expression for this function?
c. Realize F using AND, OR, and NOT gates.

A	*B*	*C*	*D*	*F*
0	0	0	0	1
0	0	0	1	1
0	0	1	0	1
0	0	1	1	0
0	1	0	0	0
0	1	0	1	0
0	1	1	0	1
0	1	1	1	1
1	0	0	0	1
1	0	0	1	1
1	0	1	0	1
1	0	1	1	1
1	1	0	0	0
1	1	0	1	0
1	1	1	0	0
1	1	1	1	0

Figure P12.6

A	*B*	*C*	*f*(*A*, *B*, *C*)
0	0	0	0
0	0	1	1
0	1	0	1
0	1	1	0
1	0	0	1
1	0	1	0
1	1	0	0
1	1	1	1

Figure P12.7

12.24 Fill in the Karnaugh map for the function defined by the truth table of Figure P12.7, and find the minimum expression for the function.

12.25 A function, F, is defined such that it equals 1 when a 4-bit input code is equivalent to any of the decimal numbers 3, 6, 9, 12 or 15. F is 0 for input codes 0, 2, 8 and 10. Other input values cannot occur. Use a Karnaugh map to determine a minimal expression for this function. Design and sketch a circuit to implement this function using only AND and NOT gates.

12.26 The function described in Figure P12.8 can be constructed using only two gates. Design the circuit.

Input *A*	*B*	*C*	Output *F*
0	0	0	0
0	0	1	0
0	1	0	0
0	1	1	1
1	0	0	0
1	0	1	0
1	1	0	1
1	1	1	x

Figure P12.8

12.27 Design a logic circuit which will produce the one's complement of an 8-bit signed binary number.

12.28 Construct the Karnaugh map for the logic function defined by the truth table of Figure P12.9, and find the minimum expression for the function.

12.29 Modify the circuit for Problem 12.27 so that it produces the two's complement of the 8-bit signed binary input.

12.30 Find the minimum output expression for the circuit of Figure P12.10.

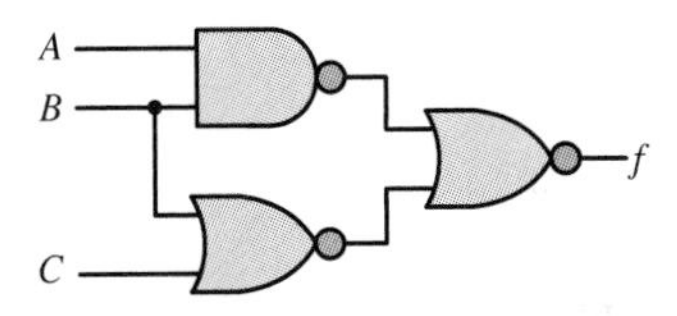

Figure P12.10

A	B	C	D	F
0	0	0	0	1
0	0	0	1	0
0	0	1	0	1
0	0	1	1	0
0	1	0	0	0
0	1	0	1	0
0	1	1	0	0
0	1	1	1	1
1	0	0	0	1
1	0	0	1	0
1	0	1	0	1
1	0	1	1	0
1	1	0	0	1
1	1	0	1	1
1	1	1	0	1
1	1	1	1	0

Figure P12.9

12.31 Design a combinational logic circuit which will add two 4-bit binary numbers.

12.32 Minimize the expression described in the truth table of Figure P12.11 and draw the circuit.

A	B	C	F
0	0	0	1
0	0	1	1
0	1	0	0
0	1	1	1
1	0	0	1
1	0	1	1
1	1	0	1
1	1	1	0

Figure P12.11

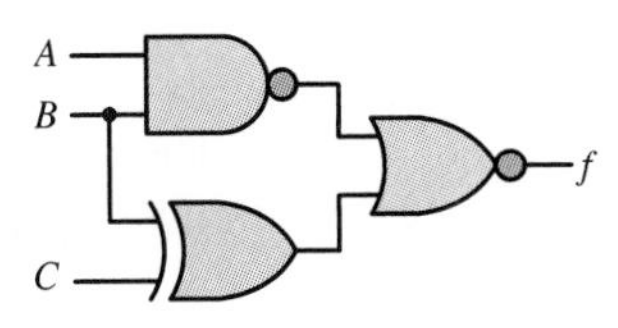

Figure P12.12

12.33 Find the minimum expression for the output of the logic circuit of Figure P12.12.

12.34 The objective of this problem is to design a combinational logic circuit which will aid in determination of the acceptability of emergency blood transfusions. It is known that human blood can be categorized into four types—A, B, AB, and O. Persons with type A blood can donate to both A and AB types, and can receive blood from both A and O types. Persons with type B blood can donate to both B and AB, and can receive from both B and O types. Persons with type AB blood can donate only to type AB, but can receive from any type. Persons with type O blood can donate to any type, but can receive only from type O. Make appropriate variable assignments and design a circuit which will approve or disapprove any particular transfusion based on these conditions.

12.35 Find the minimum expression for the logic function at the output of the logic circuit of Figure P12.13.

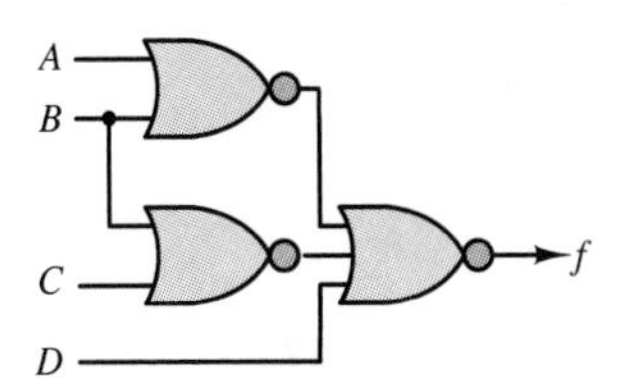

Figure P12.13

12.36 Design a combinational logic circuit which will accept a 4-bit binary number and:

If the number is even, divide it by 2_{10} and produce the binary result.

If the number is odd, multiply it by 2_{10} and produce the binary result.

12.37

a. Fill in the Karnaugh map for the logic function defined by the truth table of Figure P12.14.
b. What is the minimum expression for the function?
c. Realize the function using a 1-of-8 multiplexer.

A	B	C	D	f(A, B, C, D)
0	0	0	0	1
0	0	0	1	0
0	0	1	0	1
0	0	1	1	1
0	1	0	0	0
0	1	0	1	1
0	1	1	0	0
0	1	1	1	0
1	0	0	0	0
1	0	0	1	1
1	0	1	0	0
1	0	1	1	0
1	1	0	0	1
1	1	0	1	0
1	1	1	0	1
1	1	1	1	1

Figure P12.14

A	B	C	f(A, B, C)
0	0	0	1
0	0	1	1
0	1	0	0
0	1	1	1
1	0	0	1
1	0	1	1
1	1	0	1
1	1	1	0

Figure P12.15

12.38

a. Fill in the Karnaugh map for the function defined in the truth table of Figure P12.15.
b. What is the minimum expression for the function?
c. Draw the circuit, using AND, OR, and NOT gates.

12.39

a. Fill in the Karnaugh map for the logic function defined by the truth table of Figure P12.16.
b. What is the minimum expression for the function?

A	B	C	D	F
0	0	0	0	1
0	0	0	1	0
0	0	1	0	1
0	0	1	1	0
0	1	0	0	0
0	1	0	1	0
0	1	1	0	0
0	1	1	1	1
1	0	0	0	1
1	0	0	1	0
1	0	1	0	1
1	0	1	1	0
1	1	0	0	1
1	1	0	1	1
1	1	1	0	1
1	1	1	1	0

Figure P12.16

A	B	C	D	F
0	0	0	0	1
0	0	0	1	1
0	0	1	0	1
0	0	1	1	1
0	1	0	0	0
0	1	0	1	1
0	1	1	0	0
0	1	1	1	1
1	0	0	0	1
1	0	0	1	1
1	0	1	0	0
1	0	1	1	0
1	1	0	0	0
1	1	0	1	1
1	1	1	0	1
1	1	1	1	0

Figure P12.17

12.40

a. Fill in the Karnaugh map for the logic function defined by the truth table of Figure P12.17.
b. What is the minimum expression for the function?
c. Realize the function, using only NAND gates.

12.41 Design a circuit with a four-bit input representing the binary number $A_3A_2A_1A_0$. The output should be 1 if the input value is divisible by 3. Assume that the circuit is to be used only for the digits 0 through 9 (thus, values for 10 to 15 can be don't cares).

a. Draw the Karnaugh map and truth table for the function.
b. Determine the minimum expression for the function.
c. Draw the circuit, using only AND, OR, and NOT gates.

12.42 Find the simplified sum-of-products representation of the function from the Karnaugh map shown in Figure P12.18. Note that **x** is the don't care term.

12.43 Can the circuit for Problem 12.36 be simplified if it is known that the input represents a BCD (binary-coded decimal) number, i.e., it can never be greater than 10_{10}? If not, explain why not. Otherwise, design the simplified circuit.

12.44 Find the simplified sum-of-products representation of the function from the Karnaugh map shown in Figure P12.19.

12.45

a. Fill in the truth table for the multiplexer circuit shown in Figure P12.20.
b. What binary function is performed by these multiplexers?

$C \cdot D$ \ $A \cdot B$	00	01	11	10
00	0	1	0	0
01	1	1	0	0
11	0	x	1	0
10	0	0	1	0

Figure P12.18

$C \cdot D$ \ $A \cdot B$	00	01	11	10
00	0	1	x	0
01	0	1	x	0
11	0	1	0	1
10	x	x	1	0

Figure P12.19

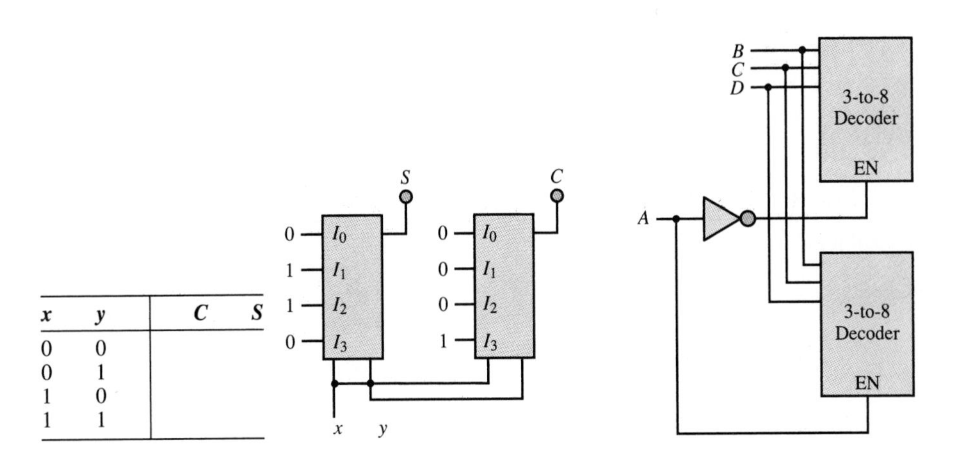

x	y	C	S
0	0		
0	1		
1	0		
1	1		

Figure P12.20 **Figure P12.21**

12.46 The circuit of Figure P12.21 can operate as a 4-to-16 decoder. Terminal EN denotes the enable input. Describe the operation of the 4-to-16 decoder. What is the role of logic variable A?

12.47 Show that the circuit given in Figure P12.22 converts 4-bit binary numbers to 4-bit Gray code.

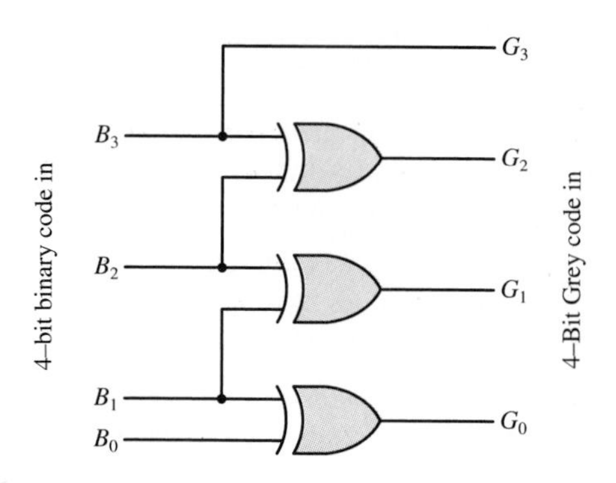

Figure P12.22

12.48 Suppose one of your classmates claims that the following Boolean expressions represent the conversion from 4-bit Gray code to 4-bit binary numbers:

$$
\begin{aligned}
B_3 &= G_3 \\
B_2 &= G_3 \oplus G_2 \\
B_1 &= G_3 \oplus G_2 \oplus G_1 \\
B_0 &= G_3 \oplus G_2 \oplus G_1 \oplus G_0
\end{aligned}
$$

a. Show that your classmate's claim is correct.
b. Draw the circuit which implements the conversion.

12.49 Select the proper inputs for a 4-input multiplexer to implement the function $f(A, B, C) = \bar{A}B\bar{C} + A\overline{BC} + AC$. Assume the inputs I_0, I_1, I_2, and I_3 correspond to $\overline{AB}$, $\bar{A}B$, $A\bar{B}$, and AB, respectively, and that each input may be 0, 1, $\bar{C}$, or C.

12.50 Select the proper inputs for an 8-bit multiplexer to implement the function $f(A, B, C, D) = \sum(2, 5, 6, 8, 9, 10, 11, 13, 14)_{10}$. Assume the inputs I_0 through I_7 correspond to $\overline{ABC}$, $A\overline{BC}$, $\bar{A}B\bar{C}$, $AB\bar{C}$, $\overline{AB}C$, $A\bar{B}C$, $\bar{A}BC$, and ABC, respectively, and that each input may be 0, 1, $\bar{D}$, or D.

CHAPTER

13

Digital Systems

The first half of Chapter 13 continues the analysis of digital circuits that was begun in Chapter 12 by focusing on sequential logic circuits, such as flip-flops, counters, and shift registers. The second half of the chapter is devoted to an overview of the basic functions of microprocessors and microcomputers. During the last decade, microcomputers have become a standard tool in the analysis of engineering data, in the design of experiments, and in the control of plants and processes. No longer a specialized electronic device to be used only by appropriately trained computer engineers, today's microcomputer—perhaps more commonly represented by the ubiquitous *personal computer*—is a basic tool in the engineering profession. The common thread in its application in various engineering fields is its use in digital data acquisition instruments and digital controllers.

Modern microcomputers are relatively easy to program, have significant computing power and excellent memory storage capabilities, and can be readily interfaced with other instruments and electronic devices, such as transducers, printers, and other computers. The basic functions performed by the microcomputer in a typical digital data acquisition or control application are easily described: input signals (often analog, sometimes already in digital form) are acquired by the computer and processed by means of suitable software to produce the desired result (i.e., they undergo some kind of mathematical manipulation), which is then outputted to either a display or a storage device, or is used in controlling a process, a plant, or an experiment. The objective of this chapter is to describe these various processes, with the aim of giving the reader enough background information to understand the notation used in data books and instruction manuals.

Upon completing this chapter you should be able to:

- Analyze sequential circuits including *RS, D,* and *JK* flip-flops.
- Understand the operation of binary, decade, and ring counters.
- Design simple sequential circuits using state transition diagrams.
- Understand the basic architecture of microprocessors and microcomputers.

13.1 SEQUENTIAL LOGIC MODULES

The discussion of logic devices in Chapter 12 focused on the general family of combinational logic devices. The feature that distinguishes combinational logic devices from the other major family—**sequential logic devices**—is that combinational logic circuits provide outputs that are based on a combination of present inputs only. On the other hand, sequential logic circuits depend on present and past input values. Because of this "memory" property, sequential circuits can store information; this capability opens a whole new area of application for digital logic circuits.

Latches and Flip-Flops

The basic information-storage device in a digital circuit is called a **flip-flop.** There are many different varieties of flip-flops; however, all flip-flops share the following characteristics:

1. A flip-flop is a **bistable device;** that is, it can remain in one of two stable states (0 and 1) until appropriate conditions cause it to change state. Thus, a flip-flop can serve as a memory element.
2. A flip-flop has two outputs, one of which is the complement of the other.

RS Flip-Flop

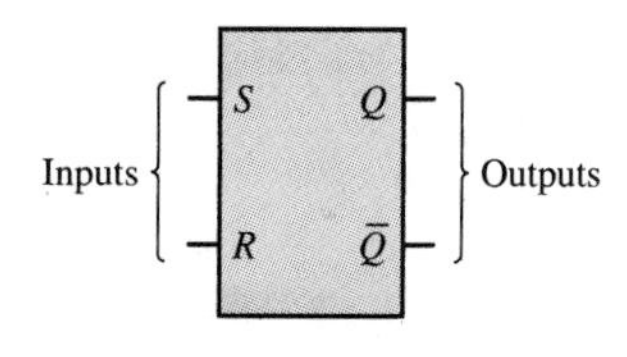

Figure 13.1 *RS* flip-flop

It is customary to depict flip-flops by their block diagram and a name—such as Q or X—representing the output variable. Figure 13.1 represents the so-called ***RS* flip-flop,** which has two inputs, denoted by S and R, and two outputs, Q and $\overline{Q}$. The value at Q is called the *state* of the flip-flop. If $Q = 1$, we refer to the device as *being in the 1 state.* Thus, we need define only one of the two outputs of the flip-flop. The two inputs, R and S, are used to change the state of the flip-flop, according to the following rules:

1. When $R = S = 0$, the flip-flop remains in its present state (whether 1 or 0).
2. When $S = 1$ and $R = 0$, the flip-flop is *set* to the 1 state (thus, the letter S, for **set**).
3. When $S = 0$ and $R = 1$, the flip-flop is *reset* to the 0 state (thus, the letter R, for **reset**).
4. It is not permitted for both S and R to be equal to 1. (This would correspond to requiring the flip-flop to set and reset at the same time.)

The rules just described are easily remembered by noting that 1s on the S and R inputs correspond to the set and reset commands, respectively.

A convenient means of describing the series of transitions that occur as the signals sent to the flip-flop inputs change is the **timing diagram.** A timing diagram is a graph of the inputs and outputs of the RS flip-flop (or any other logic device) depicting the transitions that occur over time. In effect, one could also represent these transitions in tabular form; however, the timing diagram provides a convenient visual representation of the evolution of the state of the flip-flop. Figure 13.2 depicts a table of transitions for an RS flip-flop Q, as well as the corresponding timing diagram.

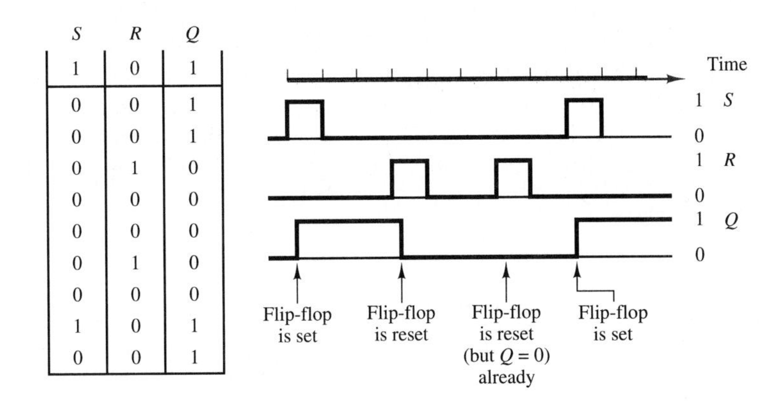

S	R	Q
1	0	1
0	0	1
0	0	1
0	1	0
0	0	0
0	0	0
0	1	0
0	0	0
1	0	1
0	0	1

Figure 13.2 Timing diagram for the RS flip-flop

It is important to note that the RS flip-flop is **level-sensitive.** This means that the set and reset operations are completed only after the R and S inputs have reached the appropriate levels. Thus, in Figure 13.2 we show the transitions in the Q output as occurring with a small delay relative to the transitions in the R and S inputs.

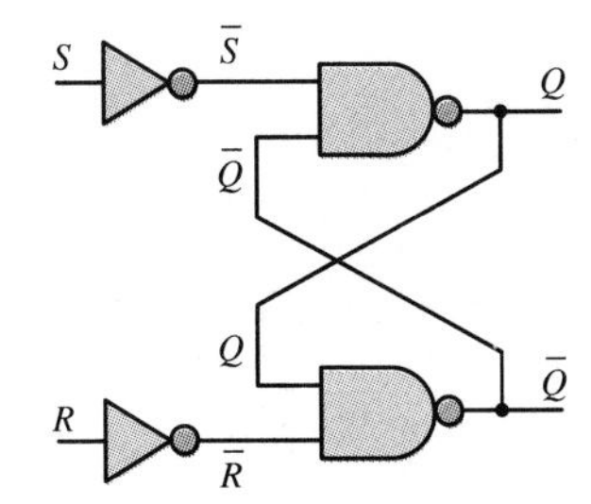

Figure 13.3 Logic gate implementation of the RS flip-flop

It is instructive to illustrate how an RS flip-flop can be constructed using simple logic gates. For example, Figure 13.3 depicts a realization of such a circuit consisting of four gates: two inverters and two NAND gates (actually, the same result could be achieved with four NAND gates). Consider the case in which the circuit is in the initial state $Q = 0$ (and therefore $\overline{Q} = 1$). If the input $S = 1$ is applied, the top NOT gate will see inputs $\overline{Q} = 1$ and $\overline{S} = 0$, so that

$Q = (\overline{\overline{S} \cdot \overline{Q}}) = (\overline{0 \cdot 1}) = 1$—that is, the flip-flop is set. Note that when Q is set to 1, $\overline{Q}$ becomes 0. This, however, does not affect the state of the Q output, since replacing $\overline{Q}$ with 0 in the expression

$$Q = (\overline{\overline{S} \cdot \overline{Q}})$$

does not change the result:

$$Q = (\overline{0 \cdot 0}) = 1$$

Thus, the cross-coupled feedback from outputs Q and $\overline{Q}$ to the input of the NAND gates is such that the set condition sustains itself. It is straightforward to show (by symmetry) that a 1 input on the R line causes the device to reset (i.e., causes $Q = 0$) and that this condition is also self-sustaining.

Example 13.1

What is the output of an *RS* flip-flop for the following series of inputs?

R	0	0	0	1	0	0	0
S	1	0	1	0	0	1	0

Solution:

On the basis of the rules stated earlier in this section, the state of the flip-flop can be described by means of the following truth table:

R	*S*	*Q*
0	1	1
0	0	1
0	1	1
1	0	0
0	0	0
0	1	1
0	0	1

A timing diagram can also be drawn, marking each transition:

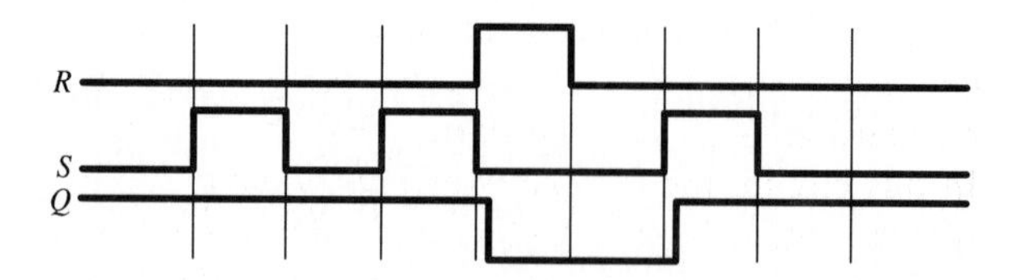

An extension of the *RS* flip-flop includes an additional enable input that is *gated* into each of the other two inputs. Figure 13.4 depicts an *RS* flip-flop consisting of two NOR gates. In addition, an enable input is connected through two AND gates to the *RS* flip-flop, so that an input to the *R* or *S* line will be effective only when the enable input is 1. Thus, any transitions will be controlled by the enable input, which acts as a synchronizing signal. The enable signal may consist of a **clock,** in which case the flip-flop is said to be **clocked** and its operation is said to be **synchronous.**

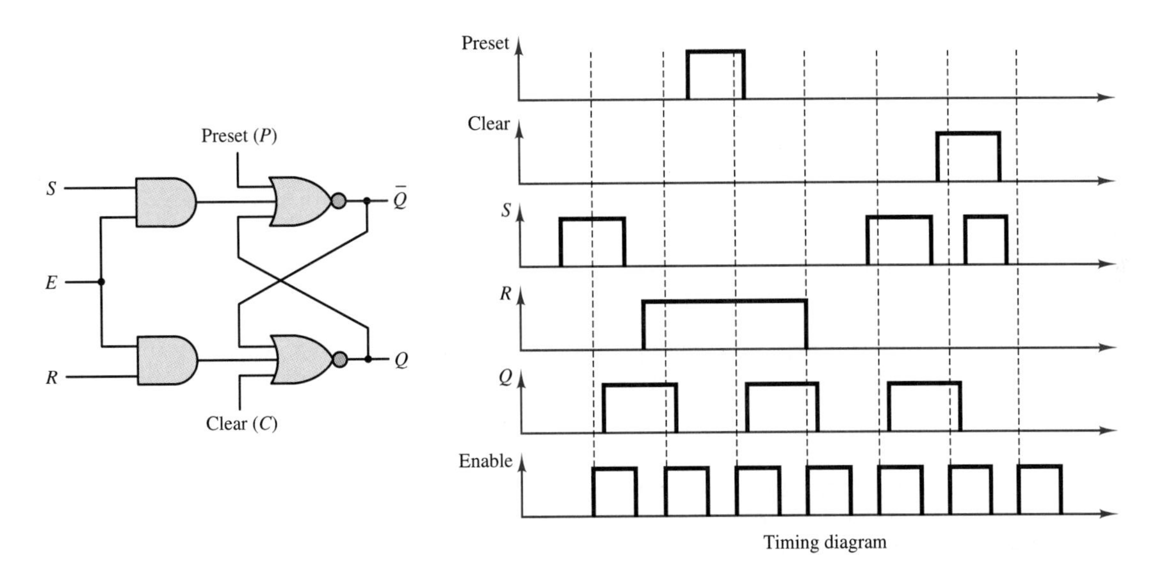

Figure 13.4 *RS* flip-flop with enable, preset, and clear lines

The same circuit of Figure 13.4 can be used to illustrate two additional features of flip-flops: the **preset** and **clear** functions, denoted by the inputs *P* and *C*, respectively. When *P* and *C* are 0, they do not affect the operation of the flip-flop. Setting $P = 1$ corresponds to setting $S = 1$, and therefore causes the flip-flop to go into the 1 state. Thus, the term *preset:* this function allows the user to preset the flip-flop to 1 at any time. When *C* is 1, the flip-flop is reset, or *cleared* (i.e., *Q* is made equal to 0). Note that these direct inputs are, in general, asynchronous; therefore, they allow the user to preset or clear the flip-flop at any time. A set of timing waveforms illustrating the function of the enable, preset, and clear inputs is also shown in Figure 13.4. Note how transitions occur only when the enable input goes high (unless the preset or clear inputs are used to overide the *RS* inputs).

Another extension of the *RS* flip-flop, called the **data latch,** is shown in Figure 13.5. In this circuit, the *R* input is always equal to the inverted *S* input, so that whenever the enable input is high, the flip-flop is set. This device has the dual advantage of avoiding the potential conflict that might arise if both *R* and *S* were high and reducing the number of input connections by eliminating the reset input. This circuit is called a data latch because once the enable input goes low,

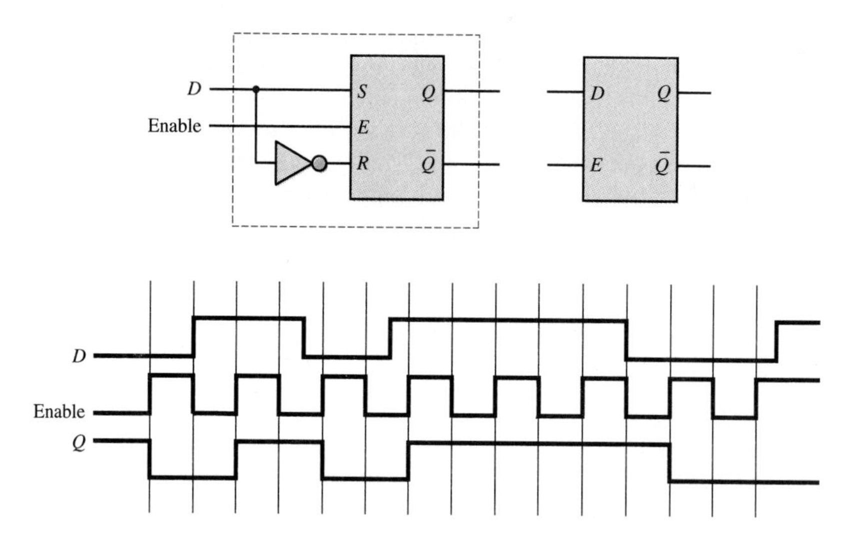

Figure 13.5 Data latch

the flip-flop is latched to the previous value of the input. Thus, this device can serve as a basic memory element.

D Flip-Flop

The ***D* flip-flop** is an extension of the data latch that utilizes two *RS* flip-flops, as shown in Figure 13.6. In this circuit, a clock is connected to the enable input of each flip-flop. Since Q_1 sees an inverted clock signal, the latch is enabled when the clock waveform goes low. However, since Q_2 is disabled when the clock is low, the output of the *D* flip-flop will not switch to the 1 state until the clock goes high, enabling the second latch and transferring the state of Q_1 to Q_2. It is important to note that the *D* flip-flop changes state only on the positive edge of the clock waveform: Q_1 is set on the negative edge of the clock, and Q_2 (and therefore Q) is set on the positive edge of the clock, as shown in the timing diagram of Figure 13.6. This type of device is said to be **edge-triggered.** This feature is indicated by the "knife edge" drawn next to the CLK input in the device symbol. The particular device described here is said to be positive edge–triggered, or **leading edge–triggered,** since the final output of the flip-flop is set on a positive-going clock transition.

On the basis of the rules stated in this section, the state of the *D* flip-flop can be described by means of the following truth table:

D	CLK	*Q*
0	↑	0
1	↑	1

where the symbol ↑ indicates the occurrence of a positive transition.

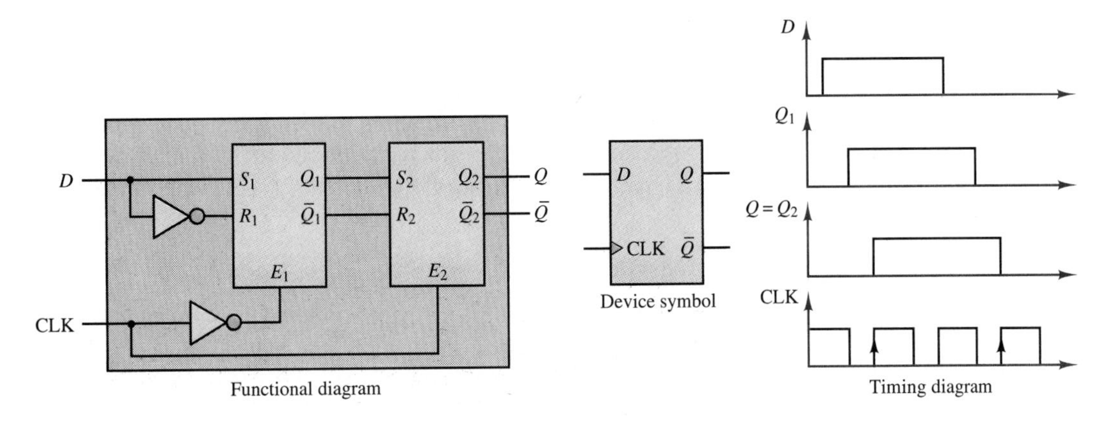

Figure 13.6 *D* flip-flop

JK Flip-Flop

Another very common type of flip-flop is the ***JK* flip-flop,** shown in Figure 13.7. The *JK* flip-flop operates according to the following rules:

- When *J* and *K* are both low, no change occurs in the state of the flip-flop.
- When $J = 0$ and $K = \downarrow$, the flip-flop is reset to 0.
- When $J = \downarrow$ and $K = 0$, the flip-flop is set to 1.
- When both *J* and *K* are high, the flip-flop will toggle between states at every transition of the clock input.

The symbol $\downarrow$ denotes a negative transition.

Note that, functionally, the operation of the *JK* flip-flop can also be explained in terms of two *RS* flip-flops. When the clock waveform goes high, the "master" flip-flop is enabled; the "slave" receives the state of the master upon a negative clock transition. The "bubble" at the clock input signifies that the device is negative or **trailing edge–triggered.** This behavior is similar to that of an *RS*

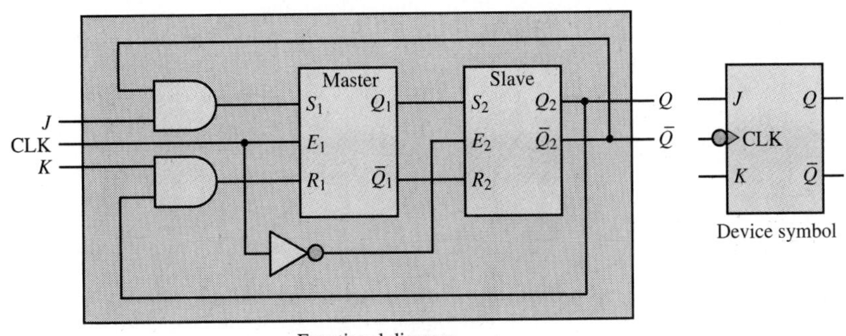

Figure 13.7 *JK* flip-flop

flip-flop, except for the $J = 1$, $K = 1$ condition, which corresponds to a toggle mode rather than to a disallowed combination of inputs.

Figure 13.8 depicts the truth table for the *JK* flip-flop. It is important to note that when both inputs are 0 the flip-flop remains in its previous state at the occurrence of a clock transition; when either input is high and the other is low, the *JK* flip-flop behaves like the *RS* flip-flop, whereas if both inputs are high, the output "toggles" between states every time the clock waveform undergoes a negative transition.

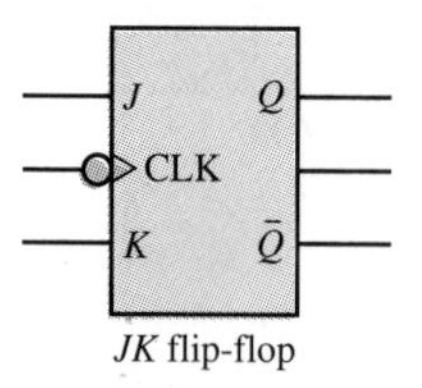

J_n	K_n	Q_{n+1}
0	0	Q_n
0	1	0 (reset)
1	0	1 (set)
1	1	$\bar{Q}_n$ (toggle)

Figure 13.8 Truth table for the *JK* flip-flop

Example 13.2 The *T* Flip-Flop

The ***T* flip-flop** is a *JK* flip-flop with the two inputs tied together (see Figure 13.9). The single-input *T* flip-flop follows the transition rule $Q_{k+1} = Q_k$ if $T = 0$ and $Q_{k+1} = \overline{Q}_k$ if $T = 1$. As shown in the associated timing diagram, the *T* flip-flop toggles between the high and low states at a frequency half that of the clock when the *T* input is held high. It is easy to verify that since the *T* input is always high, the flip-flop acts so as to divide the input clock frequency by 2, as shown in the timing diagram.

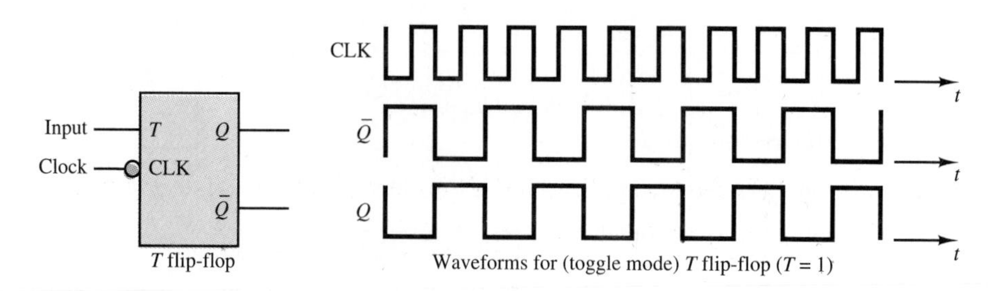

Figure 13.9

Example 13.3

If the following inputs are applied (in the order shown) to a *JK* flip-flop, what is the output after each transition, and what is the final output ($Q_{\text{initial}} = 1$)?

Sequence:

J	0	1	0	1	0	0	1
K	0	1	1	0	0	1	1

Solution:

Assume that the flip-flop is initially in the 1 state, that is, $Q_{\text{init}} = 1$. Then the output transition table is as follows:

J	*K*	Output (*Q*)
0	0	1
1	1	0
0	1	0
1	0	1
0	0	1
0	1	0
1	1	1

The timing diagram corresponding to these input/output combinations is as follows. Each line corresponds to a clock transition.

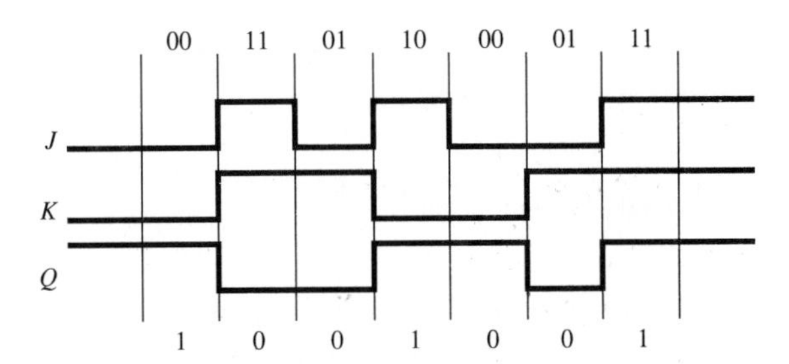

Digital Counters

One of the more immediate applications of flip-flops is in the design of **counters.** A counter is a sequential logic device that can take one of *N* possible states, stepping through these states in a sequential fashion. When the counter has reached its last state, it resets to zero and is ready to start counting again. For example, a three-bit **binary up counter** would have $2^3 = 8$ possible states, and might appear as shown in the functional block of Figure 13.10. The input clock waveform causes the counter to step through the eight states, making one transition for each clock pulse. We shall shortly see that a string of *JK* flip-flops can accomplish this task exactly. The device shown in Figure 13.10 also displays a reset input, which forces the counter output to equal 0: $b_2 b_1 b_0 = 000$.

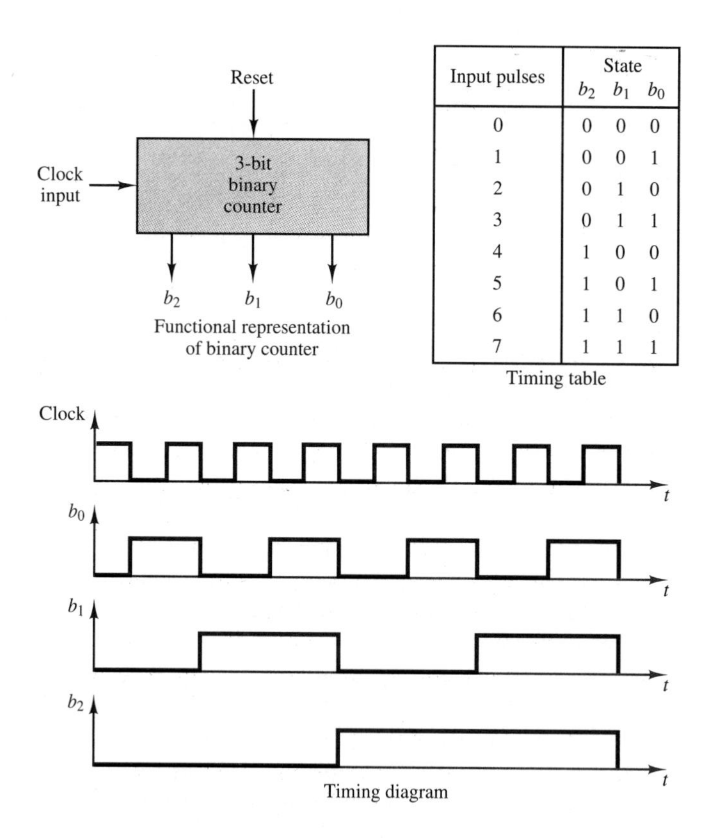

Input pulses	State b_2	b_1	b_0
0	0	0	0
1	0	0	1
2	0	1	0
3	0	1	1
4	1	0	0
5	1	0	1
6	1	1	0
7	1	1	1

Timing table

Figure 13.10 Binary up counter

Although binary counters are very useful in many applications, one is often interested in a **decade counter,** that is, a counter that counts from 0 to 9 and then resets. A four-bit binary counter can easily be configured in principle to provide this function by means of simple logic that resets the counter when it has reached the count $1001_2 = 9_{10}$. As shown in Figure 13.11, if we connect bits b_3 and b_1 to a four-input AND gate, along with $\overline{b}_2$ and $\overline{b}_0$, the output of the AND gate can be used to reset the counter after a count of 10. Additional logic can provide a "carry" bit whenever a reset condition is reached, which could be passed along to another decade counter, enabling counts up to 99. Decade counters can be cascaded so as to represent decimal digits in succession.

Although the decade counter of Figure 13.11 is attractive because of its simplicity, this configuration would never be used in practice, because of the presence

Input pulses	b_3	b_2	b_1	b_0
0	0	0	0	0
1	0	0	0	1
2	0	0	1	0
3	0	0	1	1
4	0	1	0	0
5	0	1	0	1
6	0	1	1	0
7	0	1	1	1
8	1	0	0	0
9	1	0	0	1

Figure 13.11 Decade counter

of **propagation delays.** These delays are caused by the finite response time of the individual transistors in each logic device and cannot be guaranteed to be identical for each gate and flip-flop. Thus, if the reset signal—which is presumed to be applied at exactly the same time to each of the four *JK* flip-flops in the four-bit binary counter—does not cause the *JK* flip-flops to reset at exactly the same time on account of different propagation delays, then the binary word appearing at the output of the counter will change from 1001 to some other number, and the output of the four-input NAND gate will no longer be high. In such a condition, the flip-flops that have not already reset will then not be able to reset, and the counting sequence will be irreparably compromised.

What can be done to obviate this problem? The answer is to use a systematic approach to the design of sequential circuits making use of **state transition diagrams.** This topic will be discussed in the next section.

A simple implementation of the binary counter we have described in terms of its functional behavior is shown in Figure 13.12. The figure depicts a three-bit binary **ripple counter,** which is obtained from a cascade of three *JK* flip-flops. The transition table shown in the figure illustrates how the Q output of each stage becomes the clock input to the next stage, while each flip-flop is held in the toggle

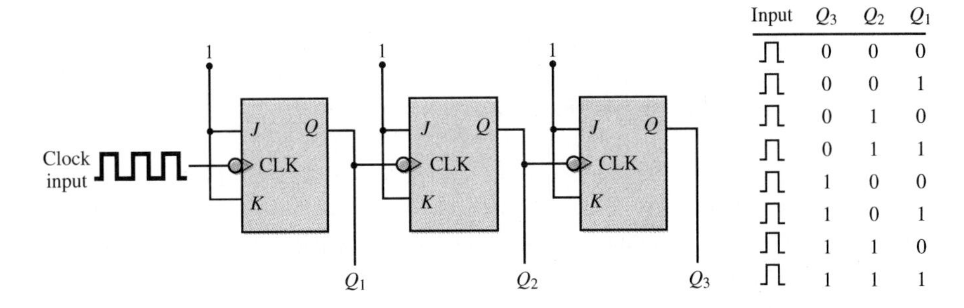

Input	Q_3	Q_2	Q_1
⎍	0	0	0
⎍	0	0	1
⎍	0	1	0
⎍	0	1	1
⎍	1	0	0
⎍	1	0	1
⎍	1	1	0
⎍	1	1	1

Figure 13.12 Ripple counter

mode. The output transitions assume that the clock, CLK, is a simple square wave (all *JK*s are negative edge–triggered).

This 3-bit ripple counter can easily be configured as a divide-by-8 mechanism, simply by adding an AND gate. To divide the input clock rate by 8, one output pulse should be generated for every eight clock pulses. If one were to output a pulse every time a binary 111 combination occurs, a simple AND gate would suffice to generate the required condition. This solution is shown in Figure 13.13. Note that the square wave is also included as an input to the AND gate; this ensures that the output is only as wide as the input signal. This application of ripple counters is further illustrated in the following example.

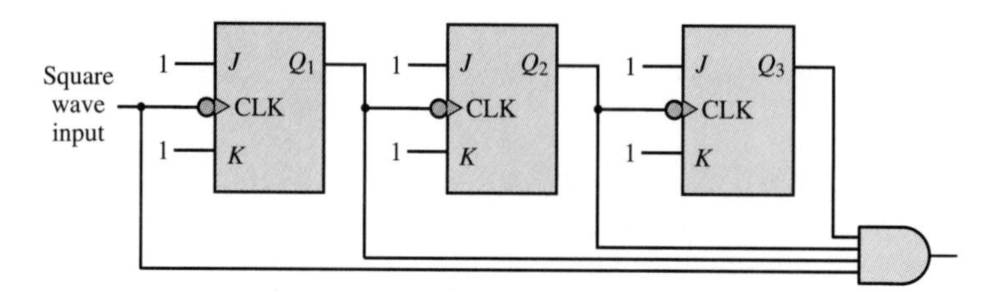

Figure 13.13 Divide-by-8 circuit

Example 13.4 Divider Circuit

A binary ripple counter provides a means of dividing the fixed output rate of a clock by powers of 2. For example, the circuit of Figure 13.14 is a divide-by-2 or divide-by-4 counter. Draw the timing diagrams for the clock input, Q_0, and Q_1 to demonstrate these functions (assume positive edge–triggered flip-flops).

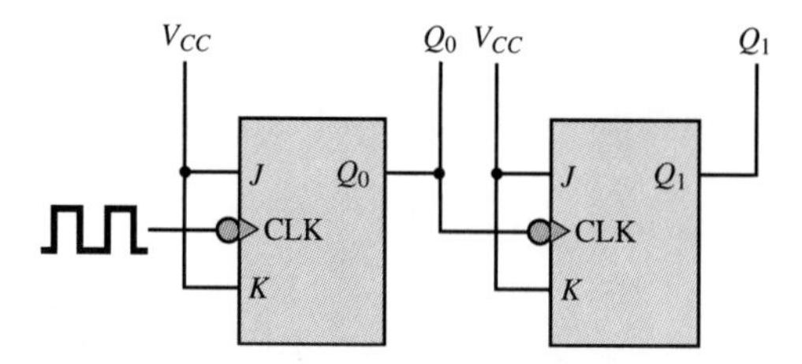

Figure 13.14

Solution:

To determine the operation of the divider circuit, a timing diagram is generated according to the rules for a positive edge–triggered *JK* flip-flop. From the accompanying diagram, it can be seen that Q_0 is switching at half the input clock frequency and Q_1 at half again the rate of Q_0, or one-fourth the original clock frequency.

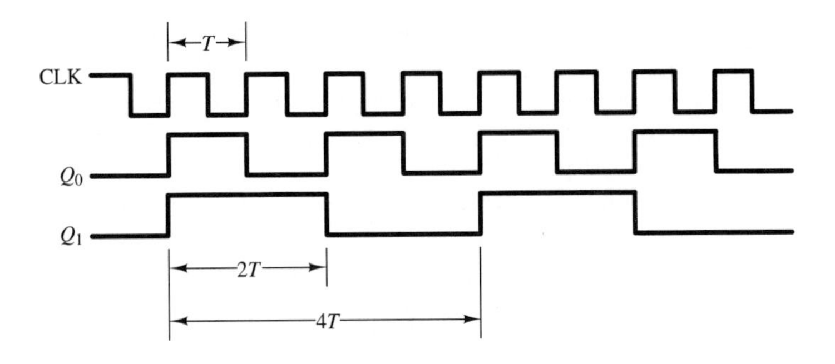

A slightly more complex version of the binary counter is the so-called **synchronous counter,** in which the input clock drives all of the flip-flops simultaneously. Figure 13.15 depicts a three-bit synchronous counter. In this figure, we have chosen to represent each flip-flop as a T flip-flop. The clocks to all the flip-flops are incremented simultaneously. The reader should verify that Q_0 toggles to 1 first and then Q_1 toggles to 1, and that the AND gate ensures that Q_2 will toggle only after Q_0 and Q_1 have both reached the 1 state ($Q_0 \cdot Q_1 = 1$).

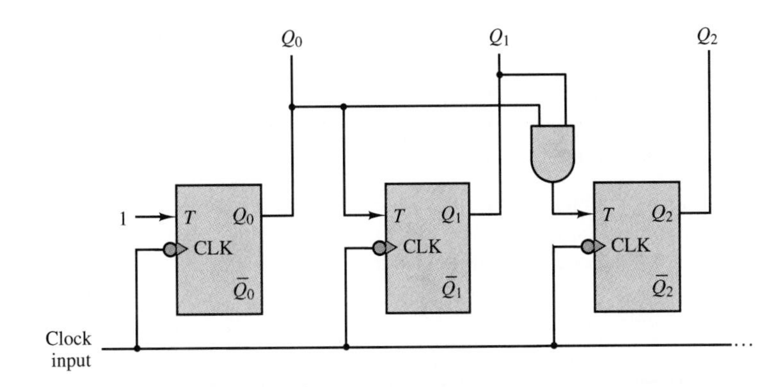

Figure 13.15 Three-bit synchronous counter

Other common counters are the **ring counter,** illustrated in Example 13.5, and the **up-down counter,** which has an additional select input that determines whether the counter counts up or down.

Example 13.5

The circuit shown in Figure 13.16 is called a *ring counter.* Analyze its operation by means of a timing diagram.

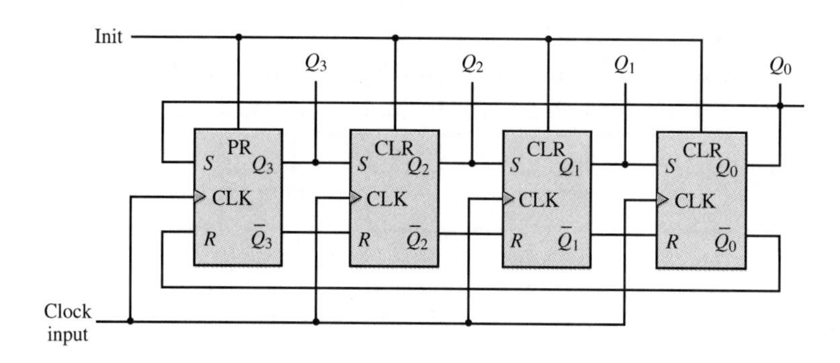

Figure 13.16 Ring counter

Solution:

The output from the first flip-flop (after initializing by means of a positive transition in the Init line) is $Q = 1, \overline{Q} = 0$. The complete sequence is as follows:

CLK	Q_3	Q_2	Q_1	Q_0
↑	1	0	0	0
↑	0	1	0	0
↑	0	0	1	0
↑	0	0	0	1
↑	1	0	0	0
↑	0	1	0	0
↑	0	0	1	0

This is called a ring counter because the logic 1 will propagate down the flip-flops and then rotate back to the first flip-flop in a circular fashion. This is how a shift register (introduced in the next section) would shift a value to the right at the rate of one bit per clock pulse.

Example 13.6 Measurement of Angular Position

Another type of angular position encoder, besides the angular encoder discussed in the previous chapter in Example 12.2, is the slotted encoder shown in Figure 13.17. This encoder can be used in conjunction with a pair of counters and a high-frequency clock to determine the speed of rotation of the slotted wheel. As shown in Figure 13.18, a clock of known frequency is connected to a counter while another counter records the number of slot pulses detected by an optical slot detector as the wheel rotates. Dividing the counter values, one could obtain the speed of the rotating wheel in radians per second. For example, assume a clocking frequency of 1.2 kHz. If both counters are started at zero and at some instant the timer counter reads 2,850 and the encoder counter reads 3,050, then the speed of the rotating encoder is found to be:

$$1{,}200\ \frac{\text{cycles}}{\text{second}} \cdot \frac{2{,}850\text{ slots}}{3{,}050\text{ cycles}} = 1{,}121.3\ \frac{\text{slots}}{\text{second}}$$

and

$$1{,}121.3 \text{ slots per second} \times 1^\circ \text{ per slot} \times 2\pi/360 \text{ rad/degree} = 19.6 \text{ rad/s}$$

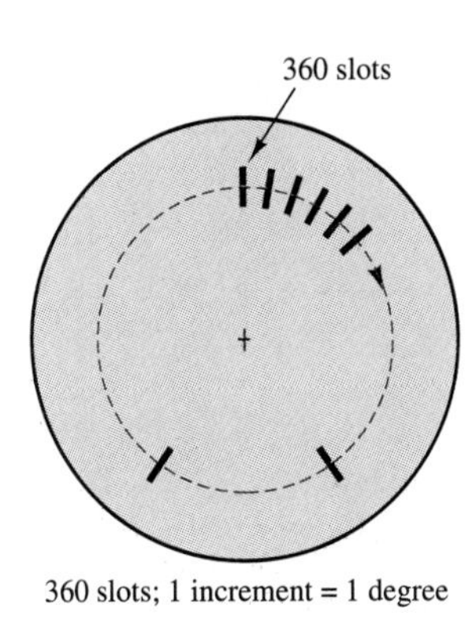

Figure 13.17

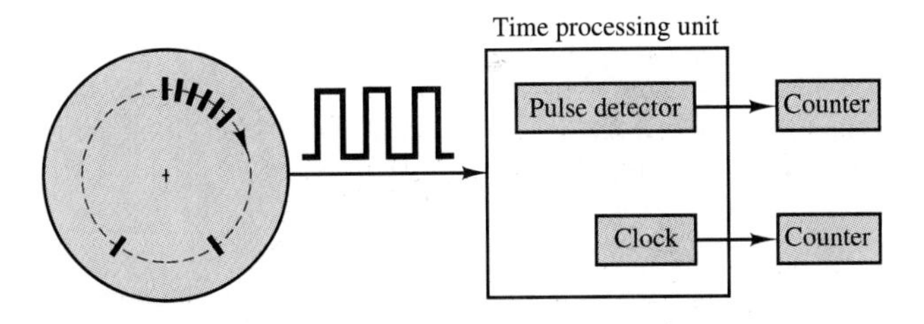

Figure 13.18 Calculating the speed of rotation of the slotted wheel

If this encoder is connected to a rotating shaft, it is possible to measure the angular position and velocity of the shaft. Such shaft encoders are used in measuring the speed of rotation of electric motors, machine tools, engines, and other rotating machinery. A typical application of the slotted encoder is to compute the ignition and injection timing in an automotive engine. In an automotive engine, information related to speed is obtained from the camshaft and the flywheel, which have known reference points. The reference points determine the timing for the ignition firing points and fuel injection pulses, and are identified by special slot patterns on the camshaft and crankshaft. Two methods are used to detect the special slots (reference points): *period measurement with additional transition detection (PMA),* and *period measurement with missing transition detection (PMM).* In the PMA method, an additional slot (reference point) determines a known reference position on the crankshaft or camshaft. In the PMM method, the reference position is determined by the absence of a slot. Figure 13.19 illustrates a typical PMA pulse sequence, showing the presence of an additional pulse. The additional slot may be used to determine the timing for the ignition pulses relative to a known position of the crankshaft. Figure 13.20 depicts a typical PMM pulse sequence. Because the period of the pulses is known, the additional slot or the missing slot can be easily detected and used as a reference position. How would you implement these pulse sequences using ring counters?

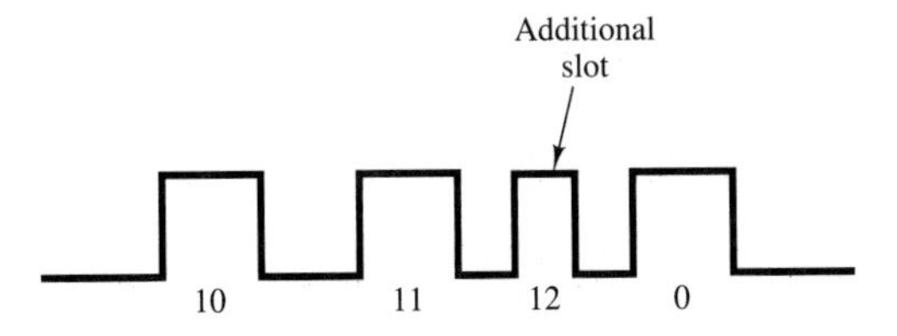

Figure 13.19 PMA pulse sequence

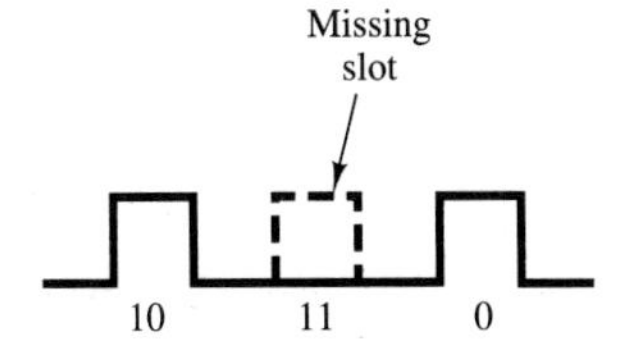

Figure 13.20 PMA pulse sequence

Registers

A register consists of a cascade of flip-flops that can store binary data, one bit in each flip-flop. The simplest type of register is the parallel input–parallel output register shown in Figure 13.21. In this register, the "load" input pulse, which acts on all clocks simultaneously, causes the parallel inputs $b_0b_1b_2b_3$ to be transferred to the respective flip-flops. The D flip-flop employed in this register allows the transfer from b_n to Q_n to occur very directly. Thus, D flip-flops are very commonly used in this type of application. The binary word $b_3b_2b_1b_0$ is now "stored," each bit being represented by the state of a flip-flop. Until the "load" input is applied again and a new word appears at the parallel inputs, the register will preserve the stored word.

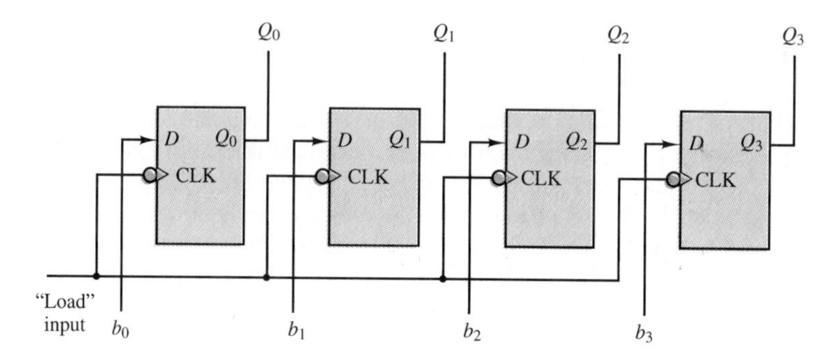

Figure 13.21 Four-bit parallel register

The construction of the parallel register presumes that the N-bit word to be stored is available in parallel form. However, it is often true that a binary word will arrive in serial form, that is, one bit at a time. A register that can accommodate this type of logic signal is called a **shift register.** Figure 13.22 illustrates how the same basic structure of the parallel register applies to the shift register, except that the input is now applied to the first flip-flop and shifted along at each clock pulse. Note that this type of register provides both a serial and a parallel output.

Data sheets for some common flip-flops and counters are shown at the end of the chapter.

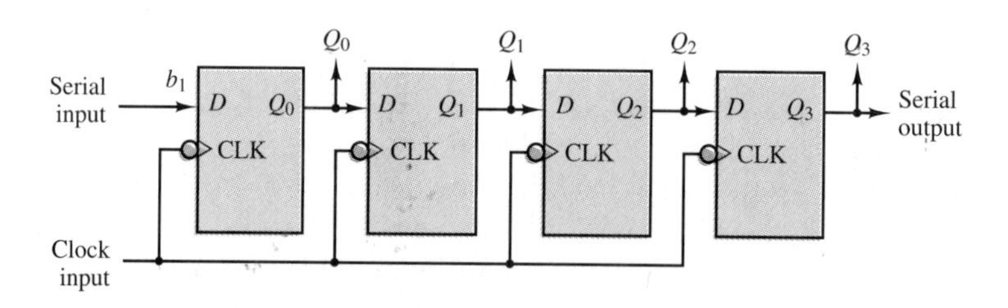

Figure 13.22 Four-bit shift register

Example 13.7 Seven-Segment Displays

A **seven-segment display** is a very convenient device for displaying digital data. The display is shown in Figure 13.23. Operation of a seven-segment display requires a decoder circuit to light the proper combinations of segments corresponding to the desired decimal digit.

A typical BCD to seven-segment decoder function block is shown in Figure 13.24, where the lowercase letters correspond to the segments shown in Figure 13.23. The decoder features four data inputs (A, B, C, D), which are used to light the appropriate segment(s). The outputs of the decoder are connected to the seven-segment display. The decoder will light up the appropriate segments corresponding to the incoming value. A BCD to seven-segment decoder function is similar to the 2-to-4 decoder function described in Chapter 12 and shown in Figure 12.61.

This display, with the appropriate decoder driver, is capable of displaying values ranging from 0 to 9.

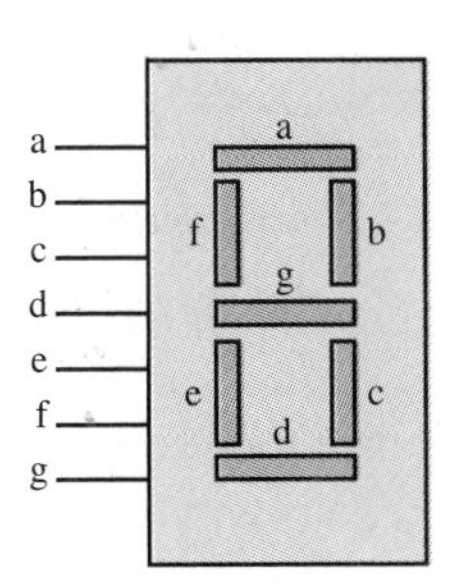

Figure 13.23 Seven-segment display

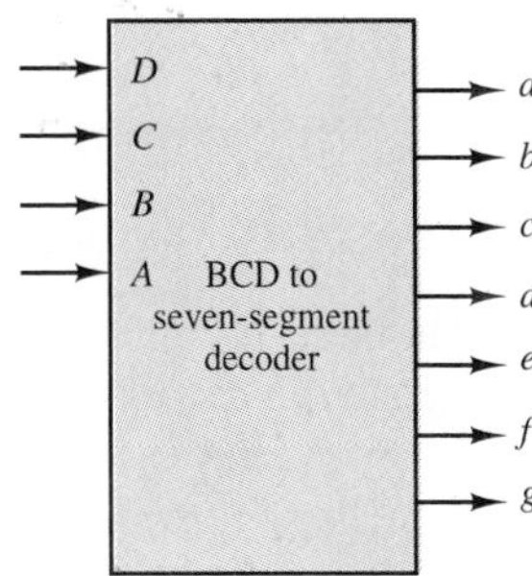

Figure 13.24

DRILL EXERCISES

13.1 The circuit shown in Figure 13.25 also serves as an *RS* flip-flop and requires only two NOR gates. Analyze the circuit to prove that it operates as an *RS* flip-flop. [*Hint:* Use a truth table with two variables, *S* and *R*.]

13.2 Derive the detailed truth table and draw a timing diagram for the *JK* flip-flop, using the two–flip-flop model of Figure 13.7.

13.3 The speed of the rotating encoder of Example 13.6 is found to be 9,425 rad/s. The encoder timer reads 10 and the clock counter reads 300. Assuming that both the timer counter and the encoder counter started at zero, find the clock frequency.

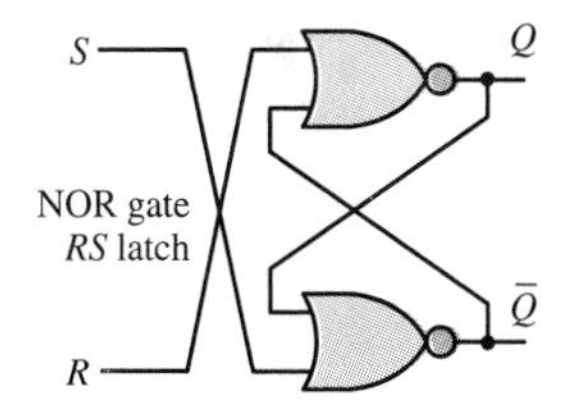

Figure 13.25

13.2 SEQUENTIAL LOGIC DESIGN

The design of sequential circuits, just like the design of combinational circuits, can be carried out by means of a systematic procedure. You will recall how the Karnaugh map, introduced in Chapter 12, allowed us to formalize the design procedures for an arbitrary combinational circuit. The equivalent of a Karnaugh map for a sequential circuit is the **state diagram,** with its associated **state transition table.** To illustrate these concepts, it is best to proceed with an example. Consider

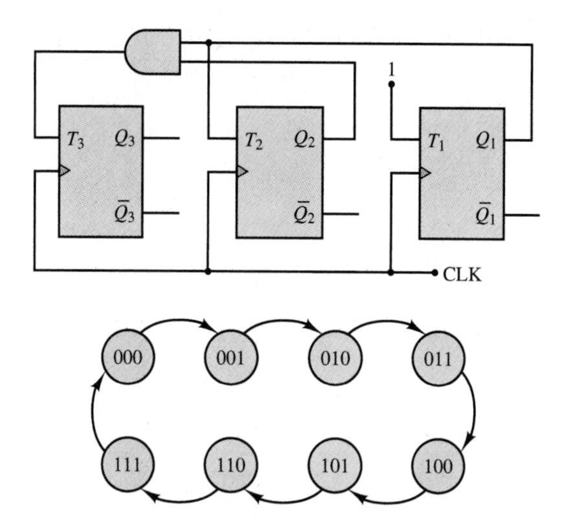

Figure 13.26 Three-bit binary counter and state diagram

the three-bit binary counter of Figure 13.26, which is made up of three T flip-flops. You can easily verify that the input equations for this counter are $T_1 = 1$, $T_2 = q_1$, and $T_3 = q_1 q_2$. Knowing the inputs, we can determine the three outputs from these relationships at any time. The outputs Q_1, Q_2, and Q_3 form the **state** of the machine. It is straightforward to show that as the clock goes through a series of cycles, the counter will go through the transitions shown in Table 13.1, where we indicate the current state by lowercase q and the next state by an uppercase Q. Note that the state diagram of Figure 13.26 provides information regarding the sequence of states assumed by the counter in graphical form. In a state diagram, each state is denoted by a circle called a **node,** and the transition from one state to another is indicated by a **directed edge,** that is, a line with a directional arrow. The analysis of sequential circuits consists of determining either their transition table or their state diagram.

Table 13.1 **State transition table for three-bit binary counter**

Current state			*Input*			*Next state*		
q_3	q_2	q_1	T_3	T_2	T_1	Q_3	Q_2	Q_1
0	0	0	0	0	1	0	0	1
0	0	1	0	1	1	0	1	0
0	1	0	0	0	1	0	1	1
0	1	1	1	1	1	1	0	0
1	0	0	0	0	1	1	0	1
1	0	1	0	1	1	1	1	0
1	1	0	0	0	1	1	1	1
1	1	1	1	1	1	0	0	0

The reverse of this analysis process is the design process. How can one systematically arrive at the design of a sequential circuit, such as a counter, by employing state transition tables and state diagrams? The design procedure will be explained in this section.

The initial specification for a logic circuit is usually in the form of either a transition table or a state diagram. The design will differ depending on the type of flip-flop used. Therefore one must first choose a flip-flop and define its behavior in the form of an excitation table. Truth tables and excitation tables for the *RS*, *D*, and *JK* flip-flops are given in Tables 13.2, 13.3, and 13.4.

Table 13.2 **Truth table and excitation table for *RS* flip-flop**

Truth table for RS flip-flop				*Excitation table for RS flip-flop*			
S	*R*	Q_t	Q_{t+1}	Q_t	Q_{t+1}	*S*	*R*
0	0	0	0	0	0	0	d[2]
0	0	1	1	0	1	1	0
0	1	0	0	1	0	0	1
0	1	1	0	1	1	d	0
1	0	0	1				
1	0	1	1				
1	1	X[1]	X				
1	1	X	X				

[1]An X indicates that this combination of inputs is not allowed.
[2]A "d" denotes a don't care entry.

Table 13.3 **Truth table and excitation table for *D* flip-flop**

Truth table for D flip-flop			*Excitation table for D flip-flop*		
D	Q_t	Q_{t+1}	Q_t	Q_{t+1}	*D*
0	0	0	0	0	0
0	1	0	0	1	1
1	0	1	1	0	0
1	1	1	1	1	1

Table 13.4 **Truth table and excitation table for *JK* flip-flop**

Truth table for JK flip-flop				*Excitation table for JK flip-flop*			
J	*K*	Q_t	Q_{t+1}	Q_t	Q_{t+1}	*J*	*K*
0	0	0	0	0	0	0	d
0	0	1	1	0	1	1	d
0	1	0	0	1	0	d	1
0	1	1	0	1	1	d	0
1	0	0	1				
1	0	1	1				
1	1	0	1				
1	1	1	0				

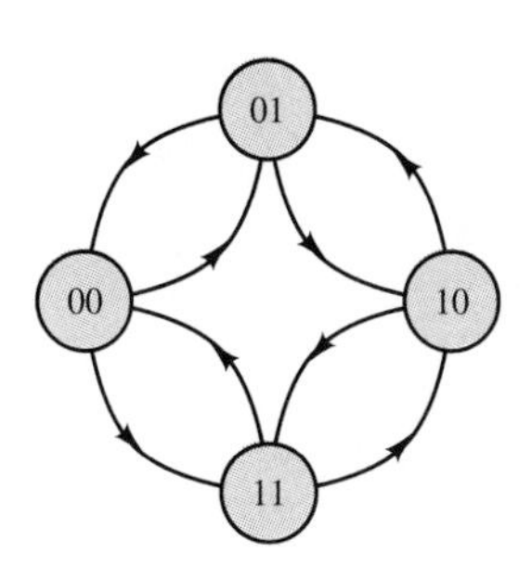

Figure 13.27 State diagram of a modulo-4 up-down counter

The use of excitation tables will now be demonstrated through an example. Let us design a **modulo-4 binary up-down counter,** that is, a counter that can change state counting up or down in the binary sequence from 0 to 3. For example, if the current state of the counter is 2, an input of 1 will cause the counter to change state "up" to 3, while an input of 0 will cause the counter to count "down" to 1. The state diagram for this counter is given in Figure 13.27. We choose two *RS* flip-flops for the implementation (the number of flip-flops must be sufficient to cover

Table 13.5 State transition table for modulo-4 up-down counter

Input x	Current state q_1	Current state q_2	Next state Q_1	Next state Q_2	S_1	R_1	S_2	R_2	Output y
0	0	0	1	1	1	0	1	0	1
0	0	1	0	0	0	*d*	0	1	0
0	1	0	0	1	0	1	1	0	1
0	1	1	1	0	*d*	0	0	1	0
1	0	0	0	1	0	d	1	0	1
1	0	1	1	0	1	0	0	1	0
1	1	0	1	1	*d*	0	1	0	1
1	1	1	0	0	0	1	0	1	0

all the necessary states—two flip-flops are sufficient for a four-state machine) and begin constructing Table 13.5 by listing the possible inputs, denoted by the variable x, and their effect on the counter. Since the counter can have four states and there are two inputs, we must look at eight possible combinations. The first five columns of Table 13.5 describe the behavior of the counter for all possible inputs and present states; the behavior of the counter consists of determining the next state, denoted by Q_1Q_2, given the input, x, and the current state, q_1q_2. Note that the first five columns of Table 13.5 contain exactly the same information that is given in the diagram of Figure 13.27. Now we can refer to the excitation table of the *RS* flip-flop to see what R and S inputs are required to obtain the desired counter function. For example, if $q_1 = 1$ and we wish to have $Q_1 = 0$, we must have $S_1 = 0$ and $R_1 = 1$ (we are resetting the first flip-flop). An entire state transition is handled by considering each flip-flop independently; for example, if we desire a transition from $q_1q_2 = 10$ to $Q_1Q_2 = 01$, we must have $S_1 = 0$ and $R_1 = 1$, as already stated, and $S_2 = 1$ and $R_2 = 0$. Repeating this analysis for each possible transition, one can then fill the next four columns of Table 13.5 with the values shown, where "d" represents a don't care condition.

So far, we have been able to determine the desired inputs for each flip-flop based on the counter input and on the desired state transition. Now we need to design a logic circuit that will cause the flip-flop inputs to be as stated in Table 13.5 in response to the input, x. This is a rather simple combinational logic problem, illustrated by the Karnaugh maps of Figure 13.28. From the Karnaugh maps we obtain the following expressions:

$$
\begin{aligned}
S_1 &= \overline{x}\,\overline{q}_1\overline{q}_2 + x\overline{q}_1\overline{q}_2 = (\overline{x}\,\overline{q}_2 + xq_2)\overline{q}_1 \\
R_1 &= \overline{x}q_1\overline{q}_2 + xq_1q_2 = (\overline{x}\,\overline{q}_2 + xq_2)q_1 \\
S_2 &= \overline{q}_2 \\
R_2 &= q_2
\end{aligned}
$$

which allow us to complete the design, as shown in Figure 13.29.

The procedure outlined in this section can be applied to more complex sequential circuits using the same basic steps. More advanced problems are explored in the homework problems.

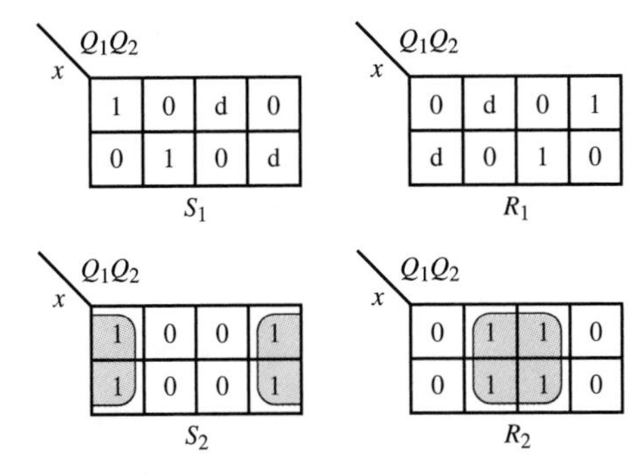

Figure 13.28 Karnaugh maps for flip-flop inputs in modulo-4 counter

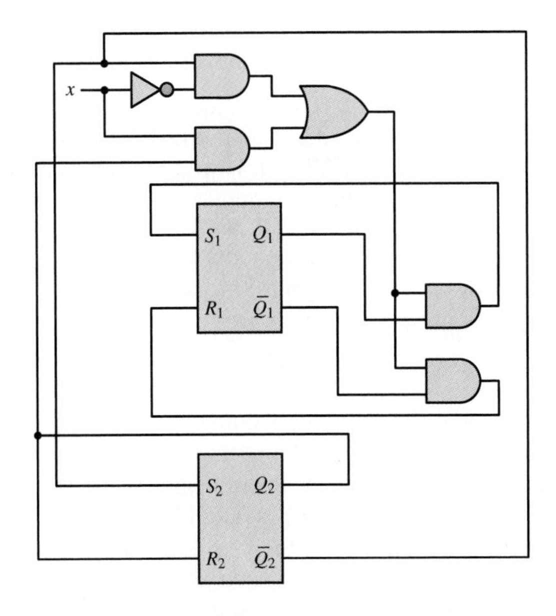

Figure 13.29 Implementation of modulo-4 counter

13.3 MICROCOMPUTERS

To bring the broad range of applicability of microcomputers in engineering into perspective, it will be useful to stop for a moment to consider the possible application of microcomputer systems to different fields. The following list—by no means exhaustive—provides a few suggestions; it would be a useful exercise to imagine other likely applications in your own discipline.

Civil engineering	Measurement of stresses and vibration in structures
Chemical engineering	Process control
Industrial engineering	Control of manufacturing processes
Material and metallurgical engineering	Measurement of material properties
Marine engineering	Instrumentation to determine ship location, ship propulsion control
Aerospace engineering	Instrumentation for flight control and navigation
Mechanical engineering	Mechanical measurements, robotics, control of machine tools
Nuclear engineering	Radiation measurement, reactor instrumentation
Biomedical engineering	Measurement of physiological functions (e.g., electrocardiography and electroencephalography), control of experiments

The massive presence of microcomputers in engineering laboratories and in plants and production facilities can be explained by considering the numerous advantages the computer can afford over conventional instrumentation and control technologies. Consider, for example, the following points:

- A single microcomputer can perform computations and send signals from many different sensors measuring different parameters to many different display, storage, or control devices, under control of a single software program.
- The microcomputer is easily reprogrammed for any changes or adjustments to the measurement or control procedures, or in the computations.
- A permanent record of the activities performed by the microcomputer can be easily stored and retained.

It should be evident that the microcomputer can perform repetitive tasks, or tasks that require great accuracy and repeatability, far better than could be expected of human operators and analog instruments. What, then, does constitute a **digital data acquisition and control system?** Figure 13.30 depicts the basic blocks that form such a system. In the figure, the user of the microcomputer system is shown to interact with the microcomputer by means of software, often called **application software.** Application software is a collection of programs written either in **high-level languages,** such as C, C^{++}, or Unix shell, or in **assembly language** (a programming language very close to the internal code used by the microcomputer). The particular application software used may be commercially available or may be provided by the user; a combination of these two cases is the norm.

The signals that originate from real-world sensors—signals related to temperatures, vibration, or flow, for example—are interfaced to the microcomputer by means of specialized circuitry that converts analog signals to digital form and times the flow of information into the microcomputer using a clock reference, which may be internal to the microcomputer or externally provided. The heart of the signal interface unit is the *analog-to-digital converter,* or *ADC,* which will be

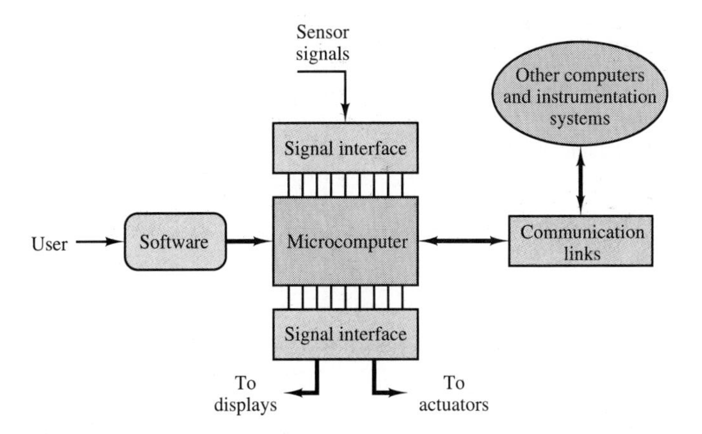

Figure 13.30 Structure of a digital data acquisition and control system

discussed in some detail in Chapter 14. Not all sensor signals are analog, though. For example, the position of a switch or an on-off valve might be of interest; signals of this type are binary in nature, and the signal interface unit can route such signals directly to the microcomputer. Once the sensor data has been acquired and converted to digital form, the microcomputer can perform computations on the data and either display or store the results, or generate command outputs to actuators through another signal interface. Actuators are devices that can generate a physical output (e.g., force, heat, flow, pressure) from an electrical input. Some actuators can be controlled directly by means of a digital signal (e.g., an on-off valve), but some require an analog input voltage or current, which can be obtained from the digital signal generated by the microcomputer by means of a *digital-to-analog converter,* or *DAC.*

In addition to the program control exercised by the user, the microcomputer may also respond to inputs originating from other computer and instrumentation systems through appropriate **communication links,** which also permit communication in the reverse direction. Thus, a microcomputer system dedicated to a complex task may consist of several microcomputers tied over a **communication network.**

The present chapter will describe the basic architecture and operation of a microcomputer, while Chapter 14 will explore instrumentation-related issues.

13.4 MICROCOMPUTER ARCHITECTURE

Prior to delving into a description of how microcomputers interface with external devices (such as sensors and actuators) and communicate with the outside world, it will be useful to discuss the general architecture of a microcomputer, in order to establish a precise nomenclature and paint a clear picture of the major functions required in the operation of a typical microcomputer.

The general structure of a microcomputer is shown in Figure 13.31. It should be noted immediately that each of the blocks that are part of the microcomputer is connected with the **CPU bus,** which is the physical wire connection allowing

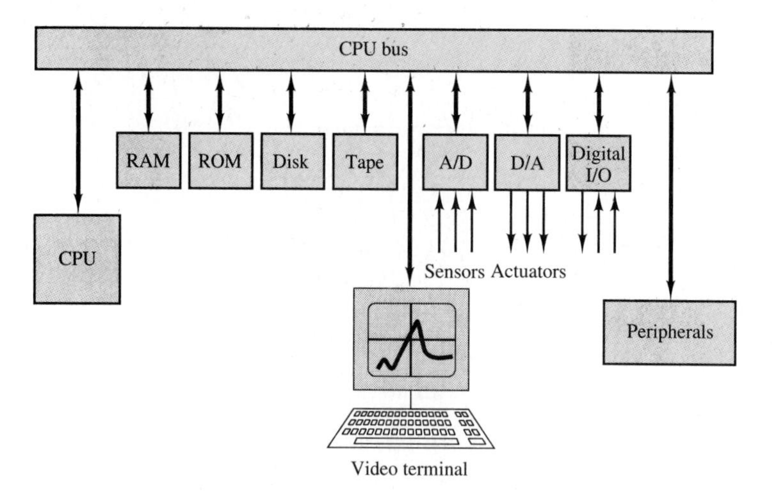

Figure 13.31 Microcomputer architecture

each of the subsystems to communicate with the others. In effect, the CPU bus is simply a set of wires; note, however, that since only one set of signals can travel over the data bus at any one time, it is extremely important that the transmission of data between different parts of the microcomputer (e.g., from the A/D unit to memory) be managed properly, to prevent interference with other functions (e.g., the display of unwanted data on a video terminal). As will be explained shortly, the task of managing the operation of the CPU bus resides within the **central processing unit,** or **CPU.** The CPU has the task of managing the flow of data and coordinating the different functions of the microcomputer, in addition to performing the data processing—in effect, the CPU is the heart and brains of the microcomputer. Some of the major functions of the CPU will be discussed in more detail shortly.

One of the important features of the microcomputer is its ability to store data. This is made possible by different types of **memory** elements: **read-only memory,** or **ROM; read and write memory (random-access memory),** or **RAM;** and **mass storage memory,** such as hard drive or floppy disk, tape, or optical drives. ROM is *nonvolatile memory:* it will retain its data whether the operating power is on or off. ROM memory contains software programs that are used frequently by the microcomputer; one example is the *bootstrap program,* which is necessary to first start up the computer when power is turned on. RAM is memory that can be accessed very rapidly by the CPU; data can be either read from or written into RAM very rapidly. RAM is therefore used primarily during the execution of programs to store partial or permanent results, as well as to store all of the software currently in use by the computer. The main difference between RAM and mass storage memory, such as a hard drive or a tape drive, is therefore in the speed of access: RAM can be accessed in tens of nanoseconds, whereas a hard drive requires access time on the order of microseconds, and tape drive, on the order of seconds. Another important distinction between RAM and mass memory is that the latter is far less expensive for an equivalent storage capability, the price typically being lower for longer access time.

A video terminal enables the user to enter programs and to display the data acquired by the microcomputer. The video terminal is one of many **peripherals** that enable the computer to communicate information to the outside world. Among these peripherals are printers, and devices that enable communication between computers, such as *modems* (a modem enables the computer to send and receive data over a telephone line). Finally, Figure 13.31 depicts real-time input/output (I/O) devices, such as analog-to-digital and digital-to-analog converters and digital I/O devices. These are the devices that allow the microcomputer to read signals from external sensors, to output signals to actuators, and to exchange data with other computers.

13.5 INTRODUCTION TO MICROPROCESSORS

This section provides an overview of the structure of a general-purpose microprocessor. The section builds on the foundation developed in Chapter 12 and requires no prior knowledge of digital computers. You may wish to explore this subject in greater depth, either by taking formal courses or through independent reading.

As explained earlier, the transfer of data from one location of the microprocessor to another (e.g., from the data registers to the ALU) occurs via a bus, which is a collection of data lines carrying individual bits that can form a word. Depending on the microprocessor structure, a word can be 8 bits, 16 bits, or 32 bits in length. An 8-bit word is referred to as a *byte*. The bits of a word can travel in parallel from one location to another. The width of the bus (number of data lines) is an important parameter in determining the processing speed and power of the microprocessor.

Figure 13.32 shows a simplified representation of a microprocessor's architecture. The two major functional units of a microprocessor are the **control unit** and the **arithmetic/logic unit (ALU).** In addition, a microprocessor has a program counter, an instruction register, general-purpose registers, temporary registers, and a stack pointer. These components are briefly discussed here.

The control unit of a microprocessor controls and synchronizes all data transfers and manipulations in the microprocessor. The control unit uses inputs from a master clock to derive timing and control signals associated with each instruction. The ALU is where all computations take place; it is capable of performing the following operations on binary data:

- Binary addition and subtraction
- Logical AND, OR, and XOR
- Complement
- Rotation left or right

The ALU is made up of the logic devices studied in Chapters 12 and 13 and has the function of performing arithmetic operations (such as the two's complement arithmetic described in Chapter 12) and Boolean logic operations.

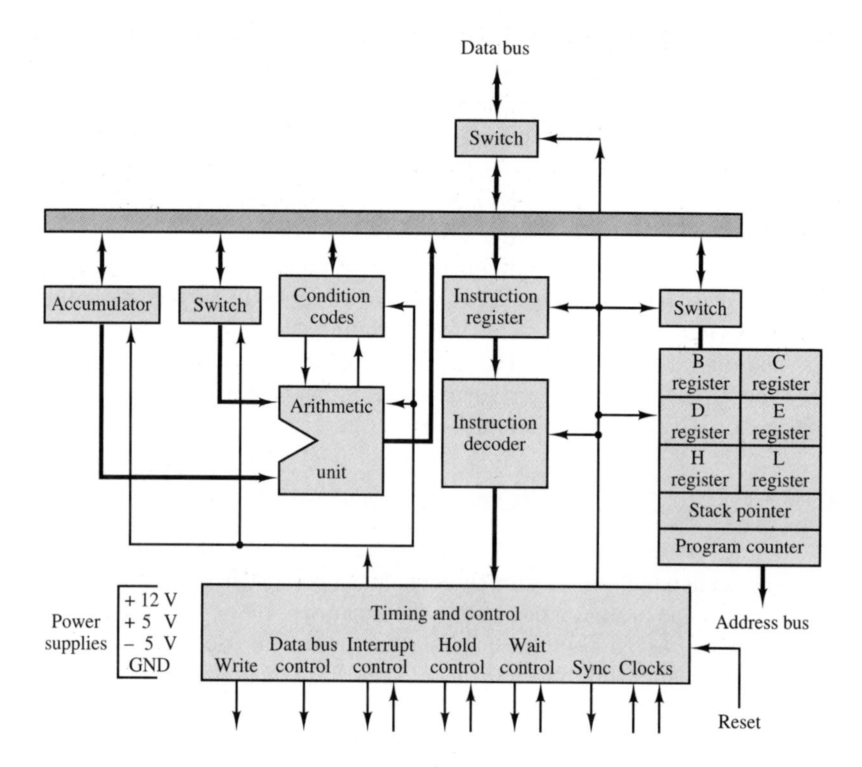

Figure 13.32 Example of microprocessor architecture

The **internal registers** include a **program counter,** the **instruction register, general-purpose registers,** and **temporary registers.** The sequence of instructions constituting a program (often called **microinstructions)** is generated according to a programming language, which is in principle similar to FORTRAN, BASIC, or Pascal. To keep track of which instruction is to be executed next, the control unit has a dedicated register called the program counter (PC). The program counter holds the address of either the next instruction to be executed or the address of a multiword instruction. When the control unit requests that the memory transfer the data to the microprocessor, the data is transferred into the microprocessor through a data bus latch, and then into the instruction register. Registers inside the microprocessor are interconnected by an internal data bus. Temporary registers are used by the control unit to hold some information temporarily until the information is transferred to another register or used in a computation. To allow greater flexibility for the microprocessor, instructions can carry out transfers between general-purpose registers. A microprocessor saves its register content in a storage register called the **stack** during subroutine calls and interrupts (an **interrupt** is a subroutine call initiated by external hardware). The stack consists of a group of specifically allocated read/write memory cells; a **stack pointer** is required to address a location or register in the stack.

Interrupts are an important part of the operation of a microprocessor. An interrupt is an electrical signal that is generated outside of the CPU and is connected to an input on the CPU. The interrupt causes the CPU to temporarily stop the execution of the program that is being executed to perform some operations on data coming from an external device. When an interrupt occurs, the processor automatically jumps to a designated program location and executes the interrupt service subroutine. When the interrupt subroutine is done, the computer returns to its place in the original program and resumes its operation. Example 13.8 illustrates the use of interrupts in a common application.

Example 13.8

In modern automotive instrumentation, a microprocessor performs all of the signal-processing operations for several measurements. A block diagram for such instrumentation is given in Figure 13.33. Depending on the technology used, the sensors' outputs can

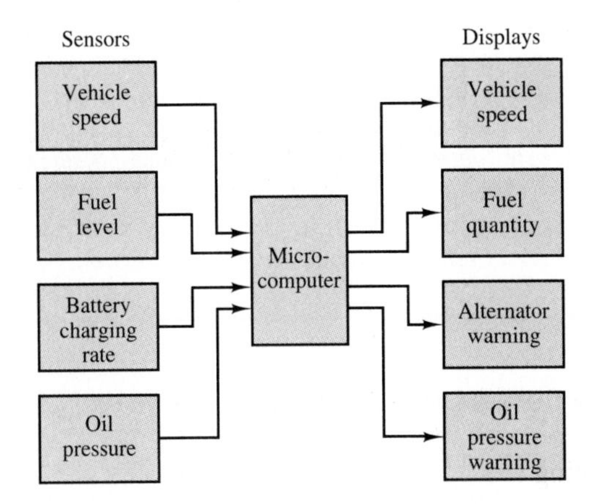

Figure 13.33 Automotive instrumentation

be either digital or analog. If the sensor signals are analog, they must be converted to digital format by means of an analog-to-digital converter (ADC), as shown in Figure 13.34. The analog-to-digital conversion process requires an amount of time that depends on the individual ADC, as will be explained in Chapter 14. After the conversion is completed, the ADC then signals the computer by changing the logic state on a separate line that sets

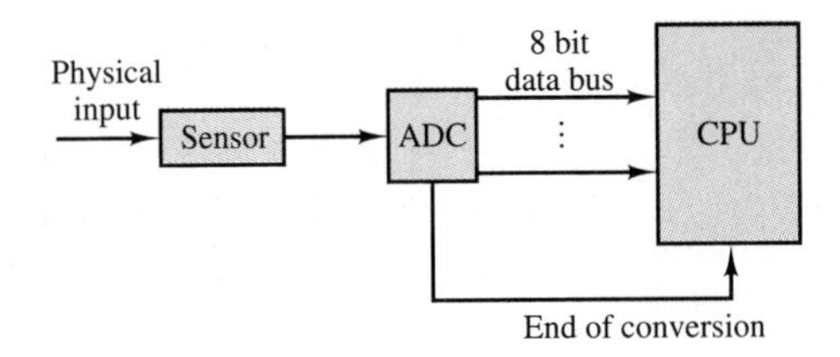

Figure 13.34 Sensor interface

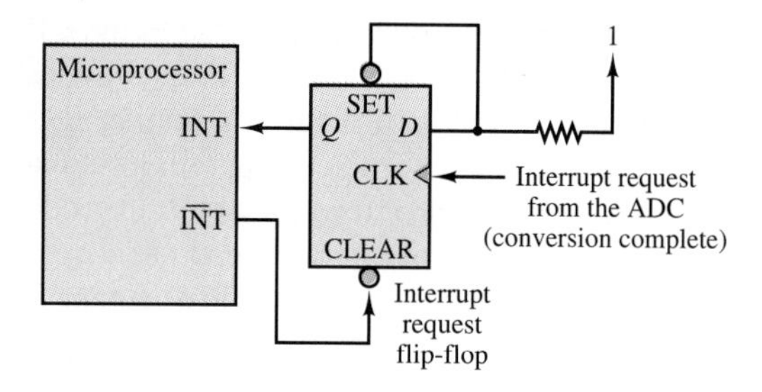

Figure 13.35 Interrupt request in a microprocessor

its *interrupt request* flip-flop. This flip-flop stores the ADC's interrupt request until it is acknowledged (see Figure 13.35).

When an interrupt occurs, the processor automatically jumps to a designated program location and executes the interrupt service subroutine. For the ADC, this would be a subroutine to read the conversion results and store them in some appropriate location, or to perform an operation on them. When the processor responds to the interrupt, the interrupt request flip-flop is cleared by a direct signal from the processor. To resume the execution of the program at the proper point upon completion of the ADC service subroutine, the program counter content is automatically saved before control is transferred to the service subroutine. The service subroutine saves in a stack the content of any registers it uses, and restores the registers' content before returning.

The interrupt may occur at any point in a program's execution, independent of the internal clock; it is therefore referred to as an *asynchronous* event.

The Intel 80X86 Microprocessor Family

In the remainder of this section, the properties and features of the Intel 80X86 microprocessor family will be explored. The reason for the choice of this device is that the 8086, 80186, 80286, 80386, 80486, and Pentium (80586) are commonly used processors found in personal computers (PCs); PCs are commonly employed in a variety of data acquisition and digital instrumentation applications in engineering plants and laboratories. We shall discuss the earliest (and simplest) version of this family of microprocessors, the 8086, in some detail, and then provide a summary of the characteristics of the more advanced versions.

The 8086 microprocessor was the first 16-bit processor from Intel Corporation. It is a 40-pin integrated circuit that was released in 1978. This processor is capable of addressing up to 1 Mbyte (2^{20} bytes) of memory and incorporates several hardware changes over previous products that make it function better and faster than its 8-bit predecessors, such as Intel's 8080 microprocessor. For example, the 8080 could address only up to 64 Kbytes of memory (1 Kbyte $= 2^{10}$ bytes), versus the 1 Mbyte of the 8086. The 8086 can operate with clock speeds up to 8 MHz, a substantial increase over the 4-MHz maximum rating for the 8080. The 8086 can also be used as an 8-bit processor for simple applications; because of the greater simplicity of the 8-bit configuration, this is the configuration used in the problems and examples throughout this chapter.

The 8086 package has 16 pins dedicated to connections to the bus system, to be described shortly. For the 8-bit operations discussed in this book, a derivative of the 8086 is actually being used: the Intel 8088. These two processors are identical, except that the 8088 has only eight pins for external bus connection. Although the 8086 has the advantage of being able to directly manipulate and transfer 16-bit words, the 8088 can in effect do the same by processing half of a 16-bit word, eight bits at a time. The two bytes then form a 16-bit word in the CPU, where operations proceed as they would in the 16-bit version. The drawback of this simplified design is a decrease in processing speed.

Internal Organization of the 8086/8088

Data Registers

The organization of the registers within the CPU is shown in Figure 13.36. The data registers hold information vital to instructions, both past and present. Three data registers are available: B, C, and D. These registers can be accessed as 16 bits (BX, CX, DX) or, separately, as 8 bits (BL, CL, DL for the lower 8 bits and BH, CH, DH for the upper 8 bits). Thus, the data register layout may look as follows:

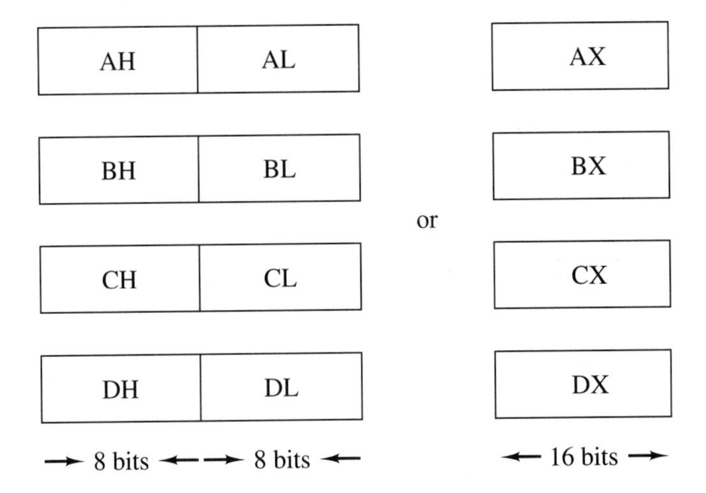

The **accumulator register (AX)** is also a data register, but it is used primarily for mathematical functions: results of mathematical operations appear in the accumulator and need to be transferred to other registers or into memory if they are to be saved. Also, the accumulator is used for I/O operations, storing data that is to be sent out to a port or holding information that has just been read. Thus, the AX register can be thought of as the main register, while the other data registers are used for temporary storage of variables and other data.

The reasons for using data registers are twofold. First, the registers are physically located on the microprocessor chip. Since off-chip operations (accessing memory in a floppy disk, reading from external input ports, or any other

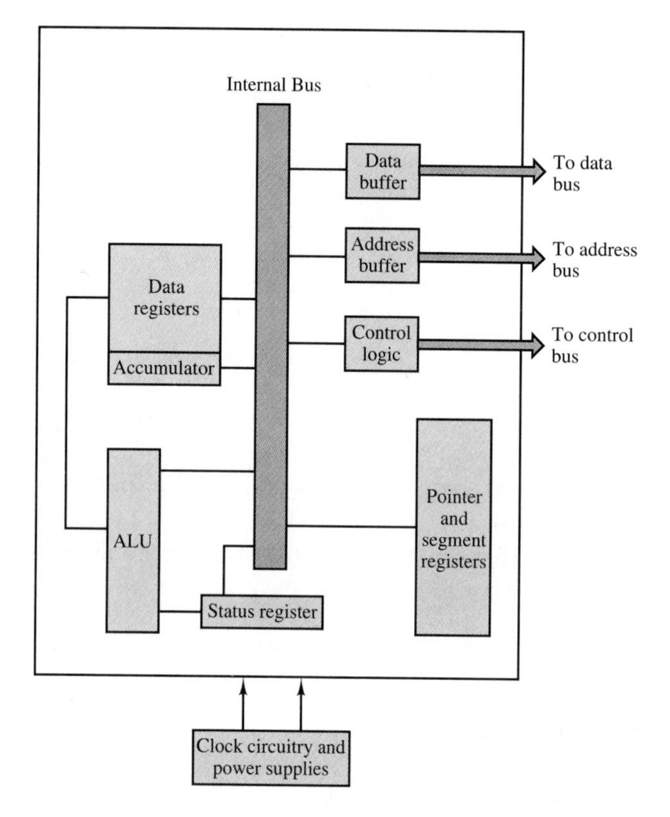

Figure 13.36 Register organization in the 8086/8088 microprocessor

operation that uses the external buses) are quite slow, using these registers will speed up access to the variables considerably. Second, some of these registers serve a dual purpose: many are dedicated to particular operations, much as the accumulator is dedicated to math functions.

ALU and Status Register

Next on the list is the ALU, or arithmetic/logic unit. This is the heart of the processor. It is used to decode instructions, add numbers, and perform comparisons. Also, the ALU does not only send the outcomes of some operations to the accumulator, but it can also detect conditions and set flags in the *status register.* This is the register used in acting on comparison statements, such as "if greater than . . . " and the like.

The status register is a 16-bit register; however, only 9 bits are used. The register looks as follows:

xxxxODITSZxAxPxC

Note that an "x" represents a bit that is not used, while the capital letters are bits that hold the following conditions:

O	Overflow tag
D	Direction
I	Interrupts enabled
T	Trap flag
S	Sign of operation
Z	Zero flag
A	Auxiliary carry
P	Parity flag
C	Carry flag

A detailed description of each of these flags is beyond the scope of this book. However, we shall give an example of how these flags are used. The easiest flag to understand is the zero flag, which is set when an arithmetic or logical operation results in accumulator contents equal to zero. Thus, when one compares two numbers and wishes to execute a *jump* to the location called *label* in the program, the operation that is in fact performed is the subtraction of the two numbers; a jump is executed if the zero flag is set (i.e., if $Z = 1$). You may compare this to a GO TO statement in FORTRAN or BASIC.

Pointer and Segment Registers

Next are the pointer and segment registers. These are registers that control the execution of the program and find the instructions to be executed. These hold the locations of instructions, the location of data, and the location of the *stack*.

The stack is simply an area of memory that is accessed using pop and push commands (described next); it provides for temporary storage of variables and values one does not wish to leave in the registers. By *pushing* the contents of a register onto the stack, the value is saved and the register can be used for another operation. Then, when the value is needed again, it is *popped* off the stack into its respective register. You may visualize the stack as a spring-loaded cylindrical container into which we may push coins at the top, to pop them out again when we need them.

The need for the pointer and segment registers to track the areas where the data and code are located is vital to accessing the information correctly. The architecture of the 8086 microprocessor dictates that memory be organized as 64-Kbyte blocks, since this is a 16-bit machine (recall that $2^{16} = 65{,}536$, or 64 Kbytes). Only one block is available to the CPU at any time, effectively limiting the amount of data and code that can be accessed at one time to 64 Kbytes. However, the 8086 can access up to 1 Mbyte of memory through a clever addressing scheme that shifts address information by 4 bits, effectively making the address locations 20 bits long, or up to 1 Mbyte. Thus, since one cannot directly access the entire 1-Mbyte range, one must "point" to the 64-K block that is to be accessed. This is the job of the pointer and segment registers. These hold the information that is vital to the CPU for loading memory data and finding the next instruction. For example, if a JMP statement will jump to code that is not within the current 64-K block, the registers must change to point to the block in which the code is located. The same applies to accessing data in different blocks. Control of these registers and their contents entails an extensive understanding of memory allocation procedures and protocol, and is beyond the scope of this book.

Bus Buffers

The final group of functions found inside the CPU is a group of buffers. As you can see from Figure 13.36, buffers are connected to three buses that exit the CPU. These are the system buses that enable access to input/output (I/O) ports and to RAM and ROM memory chips and provide information vital to the functioning of the rest of the computer. They are necessary to provide electrical isolation of subsystems that are not being used. Another function of the buffers is to drive the bus system connecting the CPU to a variety of external devices, taking into account the impedance of the bus. Thus, the bus buffers will condition the information through amplification and impedance isolation to ensure that loading does not occur.

External Support Circuitry

The CPU cannot function alone. It must be provided with power, and with a clock to drive its logic in an orderly fashion. The last functional block in Figure 13.36 is the circuitry required to accomplish these functions. For the 8086, the power supply will provide a 5-volt supply and two ground lines. The clock circuitry will vary with design; typically, one finds a crystal oscillator connected to a clock-generating IC. This IC is used to drive the oscillator as well as stabilize it, and to divide the oscillator's output down to the desired clock frequency for the CPU. Usually, the clock chip also provides outputs at various other frequencies for other applications, such as analog-to-digital conversion or communication.

Instruction Implementation in the 8086/8088

The basic instruction set in a microprocessor consists of a set of individual instructions, each of which consists of a simple sequence of elementary operations. Those sequences of operations that are performed routinely by the CPU are synthesized into instructions, which are then labeled with a *mnemonic,* that is, a simple abbreviation that suggests the basic function being performed. For example, JMP stands for the sequence of events that are necessary for the program to *jump* to a different point, MOV represents the operations required to *move* the content of a register to another register, and so forth. The essential instruction set for the 8086 is given in the accompanying box. The list is not complete, but it is representative of the complete instruction set.

In a microprocessor, an instruction (mnemonic) is decoded by an assembler into binary-coded information. For example, a "MOV AL,13H" mnemonic instruction (which means "move the value 13H into the lower 8 bits of the accumulator register") is converted into the hex code B013 by the assembler, where B0 is the MOV AL part and 13 is the number 13H, where the suffix "H" stands for hex. In binary form, this B013 becomes 1011 0000 0001 0011. This is the code the CPU will process through its hardware, either the ALU or operation-specific

Abbreviated 8086 Instruction Set

S = source, D = destination, AX = accumulator (AL = low byte, AH = high byte)

Memory access instructions		
MOV	D,S	Move byte/word from S to D
XCHG	D,S	Switch values in D and S
PUSH	S	Transfer S's value to the top of the stack
POP	D	Transfer the stack's top value to D
Mathematical instructions		
ADD	D,S	Add two numbers—results in D
SUB	D,S	Subtract S from D—result in D
MUL	S	Multiply AX by S—result in AX
DIV	S	Divide AX by S— result in AX
INC	D	Add 1 to D—result in D
DEC	D	Subtract 1 from D—result in D
NEG	D	Negate D (subtract D from 0)
Logical instructions		
AND	D,S	Bitwise logical AND of D and S—result in D
OR	D,S	Bitwise logical OR of D and S—result in D
XOR	D,S	Bitwise logical XOR of D and S—result in D
NOT	D	Invert the bits in D (form 1's complement)
TEST	D,S	Bitwise logical AND of D and S—result in Z flag
Compare and jump instructions		
CMP	D,S	Compare D and S—results in flags
JE	*Label*	Jump if equal to *Label*
JNE	*Label*	Jump if not equal to *Label*
JG	*Label*	Jump if greater than *Label*
JGE	*Label*	Jump if greater than or equal to *Label*
JL	*Label*	Jump if less than *Label*
JLE	*Label*	Jump is less than or equal to *Label*
JMP	*Label*	Jump to *Label*
Subroutine and interrupt instructions		
CALL	*Nambe*	Call the subroutine *Name*
RET		Return from subroutine entered with CALL
STI		Enable interrupt processing
INT	*type*	Generate an interrupt of *Type* in software
IRET		Return from an interrupt
I/O instructions		
IN	AX,*Port*	Read value of *Port* into *AX*
OUT	*Port*,AX	Write value in AX to *Port*
Miscellaneous instructions		
NOP		No operation—delay instructions

hardware (the 8086 microprocessor is capable of direct hardware multiplication and division, with a considerable increase in speed with respect to software operations). This sequence will run through the circuitry, producing outputs that accomplish the desired result. In this light, one can think of the 8086 as a very large logic circuit with 16 inputs capable of producing an enormous number of outputs. The operation is not quite as simple as this, but the general function of a microprocessor can be thought of as implementing large numbers of logic functions at very high speed.

Instructions are retrieved by the CPU using the pointer and segment registers, which tell the CPU where the next instruction is located. The binary form of the instruction is loaded into the instruction register (one can consider this register to be grouped with the segment and pointer registers), where it is then sent to the combinational circuitry to perform a specific function.

Example 13.9

In both large and small computers, it is desired to store the address of data in a main memory location, rather than have this address in an internal register. This procedure gives greater flexibility to the microprocessor. This example shows how this function is accomplished.

1. Assume that the desired data is 36H (the hexadecimal number 36), and use the 8086/8088 instruction set to write this data to I/O address A6H.

Solution:

The command to write to I/O ports is the OUT *Port,*AX command. Since this command uses the accumulator register and the 8088 system is an 8-bit system, the low byte of the accumulator must first be loaded with the value to be written. Thus, the code looks as follows:

```
MOV AL, 36H       ;Load accumulator with 36H
OUT 0A6H, AL      ;Write accumulator to I/O address A6H
```

Note that a zero must be written in front of the address, since it begins with a letter. This enables the compiler to distinguish it as a value rather than a label or constant.

2. Assume that an input device is wired to an 8088 system at the address E6H. Write the code necessary to read a byte value from this input device.

Solution:

The command to read from an I/O port is the IN AX, *Port* command. The values read are stored in the accumulator. Since the 8088 is an 8-bit system, you can load only a byte at a time into the accumulator.

To store the data in the low byte, the code is as follows:

```
IN AL, 0E6H
```

To store the data in the high byte of the register, the code is

IN AH, 0E6H

Note that a zero must be written in front of the address, since it begins with a letter. This enables the compiler to distinguish it as a value rather than a label or constant.

Example 13.10 Memory Access Command Examples

The microprocessor often has to access the memory, either to receive data from memory or to store data there. Two methods for transferring data to and from memory are **immediate addressing** and **indirect addressing.** In immediate addressing, the data transferred is part of the instruction. In indirect addressing, the address contained in the instruction specifies the address of the register where data is. In minicomputers and large computers, both of the registers involved in indirect addressing are usually external registers in RAM. In the case of microprocessors, however, the register that contains the actual address of the data is usually an internal register. Figure 13.37 illustrates the two addressing modes.

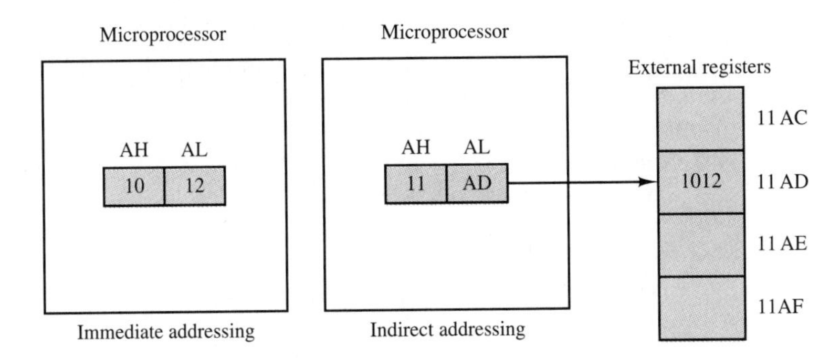

Figure 13.37 Addressing

For this example, the actual data is 1012H. As shown in Figure 13.37, the accumulator register holds the data for the case of immediate addressing. However, in the case of indirect addressing, the accumulator holds the address of the external register where the data is. Some commands and examples of memory access are given here.

Example of Immediate Addressing

Load register, immediate:

```
MOV AL, 12H     ;Load AL register with 12H
MOV AH, 10H     ;Load AH register with 10H
```

After the instructions are completed, the accumulator will hold the data (1012H).

Example of Indirect Addressing

Move the data to the accumulator:

```
LHLD 11ADH      ;Load data address into HL register pair
MOV AX, 1012H   ;Move data from memory location pointed to
                by HL to internal register AX
```

After the instructions are completed, the accumulator will hold the address of the register where the data resides.

8086-Based Microcomputer Architecture

After the preceding introduction to the functions and architecture of the 8086 microprocessor, it will not be too difficult to provide an overview of a typical computer, viewed in its functional blocks. With the CPU we described in the previous section as its heart, a microcomputer consists of *memory* for data and instruction storage, *I/O* ports for communication with the external world, and *peripherals* to extend the microprocessor's capabilities. All of these components are connected to the CPU via a bus structure. For the 8086, this bus is 16 bits wide (physically, 16 wires lying side by side), while it is 8 bits wide for the 8088. Either bus runs the length of the computer, enabling each device to communicate with its host. Control over the devices is entirely up to the CPU's instructions, and thus to the programmer. Figure 13.38 depicts the structure of a general-purpose computer based on the 8086 microprocessor.

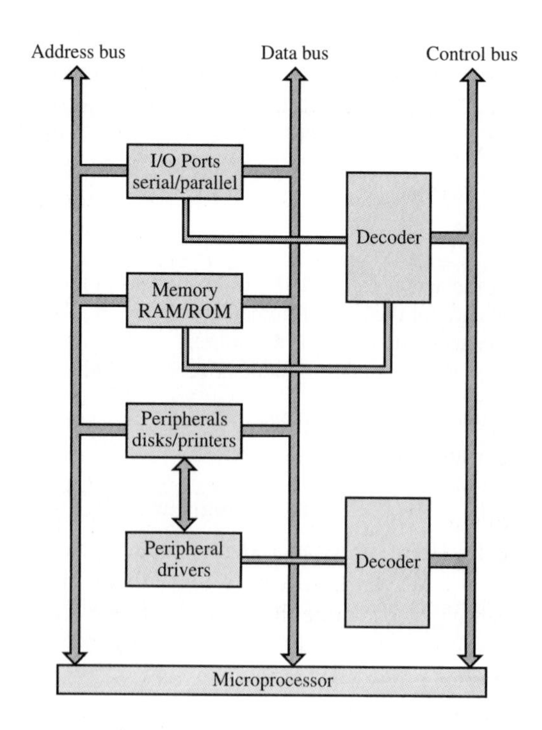

Figure 13.38 8086-based microcomputer

Bus System

In the block diagram of Figure 13.38, you can see three buses, each with a dedicated task. The **address bus,** which is a unidirectional bus, holds the address of the device the CPU wishes to communicate with. If one considers every address of each device a cell with a locked door, the values on the address bus would be the key to a particular door, and only that cell would be opened for access. The data bus, which is a bidirectional bus, plays its part by taking information to or from the device. Only information pointed to (or accessed by) the address bus will be available to the CPU through the data bus. The control bus carries the information concerning which operation is to be performed. For example, it selects whether data is to be read or written in memory or to I/O ports, and whether the peripherals being selected need to be controlled by drivers. The control bus carries the actual instruction, whereas the address bus carries the location information and the data bus contains either the result of the instruction or the command.

The two decoders shown are also an important part of the device selection process. Essentially, these decoders use the controller information to select a device, and they enable the device to become active. In the analogy of the locked cells, one can consider each device to be a building that contains cells. The decoders unlock the particular building and allow the address and data bus to pass through and unlock and react to the cell. All of this activity is controlled by the CPU and its instructions.

External Devices

The external devices shown in Figure 13.38 are divided into four groups, two of which are interconnected. The four are the I/O ports, memory, the peripherals, and the peripheral drivers. The **I/O ports** serve an important function in making the computer a practical tool for real-life applications. They enable one to connect sensors, alarms, converters, and any other instrument to the computer. Instructions can be written to control any instruments and monitor any sensors. In this fashion, one can interface the machine to the outside world to perform dangerous tasks or repetitive analysis, or to capture actions too fast for humans to perceive. The computer can also control these processes by sending instructions to instruments that can be computer-controlled.

Memory serves a basic purpose: storing instructions or data. The distinctions between read-and-write (RAM) and read-only memory (ROM) is important, since ROM is reserved for the storage of special instructions for use by the CPU. These are instructions that are vital to the CPU's life—instructions that must be available when the computer is turned on. RAM, on the other hand, is not so strictly controlled. This is the memory that allows programmers and users to temporarily store data, text, or measurements. The largest amount of memory the 8086 can access is 1 Mbyte; however, some of this is dedicated to program ROM, and so the user never has quite that much memory available.

The third and fourth devices, which are tied together, provide a connection with peripheral devices. Peripherals are devices that extend the computer's capabilities—for example, printers and disk drives. These, however, have additional hardware associated with them. While I/O ports and memories are directly accessible and controllable from the CPU, most peripherals are not. Disk drives must have their own controllers (or *drivers*), which spin the disk and read information from the magnetic surface; and printers must have their own drivers as well to select characters and advance the paper and ribbon. These drivers are under the control of the CPU and are a necessary addition to the peripheral device's operation.

With the knowledge of digital logic circuits from Chapter 12 and the material presented in this chapter, the following example is given to show how a memory system can be designed or expanded using auxiliary memory devices.

Example 13.11

This example illustrates the configuration of a memory system that is interfaced to an 8086 microprocessor. The memory system requires 8 Kbytes of ROM and 2 Kbytes of RAM. The ROM must start at location 0000H and must be immediately followed by the RAM. The following information is known for the processor: the system bus consists of a 16-bit address bus, A_0 through A_{15}, and an 8-bit bidirectional data bus, D_0 through D_7. Assume that the microprocessor has three control lines: $\overline{\text{WR}}$ (logic 0 indicates that the processor is writing), $\overline{\text{RD}}$ (logic 0 indicates that the processor is reading), and IO/$\overline{\text{M}}$ (logic 0 indicates that the address is that of the memory location as opposed to an I/O location).

Memory systems are designed to provide the amounts of ROM and RAM required for certain applications. The following design steps can be used:

1. Estimate the amount of ROM and RAM required for the given application.
2. Determine the address boundaries for ROM and RAM. Also, draw an initial memory map.
3. Select the types of memory devices that will be used for the design.
4. Determine the arrangement of memory devices needed to obtain the required word length and number of words.
5. Design the address-decoding logic.

These five steps will be applied to configure the required memory system. As the problem dictates, 8 Kbytes of ROM and 2 Kbytes of RAM are required. The next step is to draw the memory map for the system:

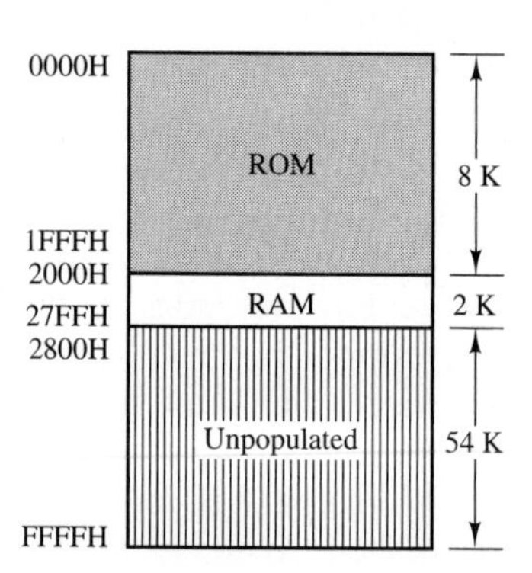

The storage requirements for this system can be met using only two devices: an NMC98C64 8-K × 8 EPROM, to provide the ROM; and a V61C16 2-K × 8 SRAM, to provide the RAM. Data sheets for these memory devices are given at the end of Chapter 12. Also, a 74ALS138 decoder is used for the decoding. Address-decoding logic is required to generate the chip enable signal for the EPROM and the chip select signal for the SRAM. Memory devices must be selected one at a time, to preclude bus contention (i.e., the simultaneous occurrence of two signals on the bus). The EPROM must be enabled only when an address in the range from 0000H to 1FFFH is generated and IO/$\overline{\text{M}}$ is logic 0. The SRAM chip must be selected only when the address is in the range 2000H through 27FFH and IO/$\overline{\text{M}}$ is logic 0.

The following address bit map shows which address bits are decoded by the memory device and which are decoded by the address-decoding logic:

		A_{15}–A_{12}	A_{11}–A_8	A_7–A_4	A_3–A_0
ROM	0000H	0 0 0 0	0 0 0 0	0 0 0 0	0 0 0 0
	1FFFH	0 0 0 1	1 1 1 1	1 1 1 1	1 1 1 1
		Selects first 8-K block	Selects location within 8-K ROM		
		A_{15}–A_{12}	A_{11}–A_8	A_7–A_4	A_3–A_0
RAM	2000H	0 0 1 0	0 0 0 0	0 0 0 0	0 0 0 0
	27FFH	0 0 1 0	0 1 1 1	1 1 1 1	1 1 1 1
		Selects second 2-K block	Selects location within 2-K RAM		

Using the memory map and address bit map, we can configure the required memory system as shown in Figure 13.39.

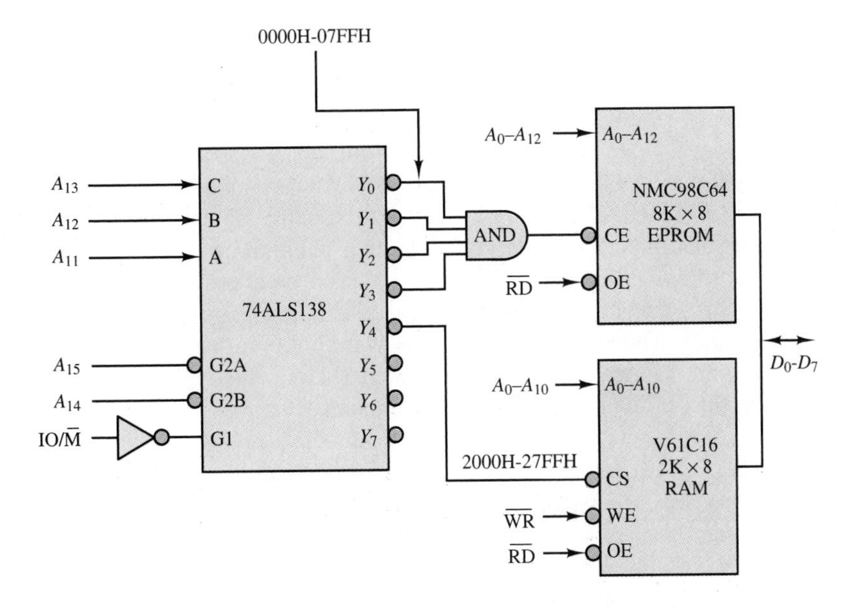

Figure 13.39 Memory system design

The implementation of Figure 13.39 has been designed to take advantage of the 74ALS138 8-to-1 decoder or its equivalent. The active high enable, G1, of the 74ALS138 is driven by the complement of $\mathrm{IO}/\overline{\mathrm{M}}$. Thus, the decoder is not enabled unless $\mathrm{IO}/\overline{\mathrm{M}}$ is logic 0. The active low enables, $\overline{\mathrm{G2A}}$ and $\overline{\mathrm{G2B}}$, are driven by address inputs A_{15} and A_{14}. This will ensure that the decoder is enabled only when both A_{15} and A_{14} are logic 0. The select inputs of the decoder, C, B, and A, are driven by A_{13}, A_{12}, and A_{11}. Therefore, each output of the decoder selects a 2-Kbyte block in the 64-Kbyte address space of the microprocessor. The first four 2-Kbyte blocks of the address space are occupied by the 2764A. The four-input AND will drive the chip enable low when any one of the first four 2-Kbyte blocks is selected. The output of the decoder that corresponds to the fifth 2-Kbyte block, 2000H–27FFH, drives the chip select of the V61C16. The memory device control inputs $\overline{\mathrm{WE}}$ and $\overline{\mathrm{OE}}$ are driven by $\overline{\mathrm{WR}}$ and $\overline{\mathrm{RD}}$, respectively. When an $\overline{\mathrm{RD}}$ signal occurs, only the selected memory device will have its output buffers enabled and will be able to drive the data bus.

Evolution of the 8086 into the 80X86 Family

In 1982, Intel introduced the 80186 and 80188. The 80186 microprocessor featured the processing power of the 8086 plus the support circuitry of 15 other chips. This "computer on a chip" contains two independent DMA (direct memory access) channels for high-speed peripherals such as disk drives, and a programmable interrupt controller. In terms of software, it provides the 8086 instruction set plus some additional instructions valuable to people who are designing high-level language translators or compilers.

The same year, Intel announced the 80286, the microprocessor at the heart of the IBM PC/AT. The 80286 is an enhanced 80186 microprocessor that provides special features necessary for memory management and protection. These features are important for multiuser applications, in which several users share the same computer, usually through a local area network. They are also important for multitasking applications, in which several programs run at the same time.

In recent years, Intel has introduced the 80386 and 80486 microprocessors. The 80386 is a 32-bit microprocessor with 32-bit separate address and data buses. It has 4 gigabytes (1 Gbyte $= 2^{30}$ bytes) of physical memory and 64 terabytes (1 Tbyte $= 2^{40}$ bytes) of virtual memory. The 80486 is an enhanced 80386 microprocessor that includes an integrated floating-point processor.

The objective of Sections 13.4 and 13.5 has been to provide an overview of the structure of a general-purpose microprocessor, and to motivate readers to pursue further study. The 8086 microprocessor was chosen for its simplicity and the fact that it is used in many personal computers (PCs), which are very commonly used in a variety of data acquisition applications.

DRILL EXERCISES

13.4 How many bits are in 256 bytes of data?

13.5 How many bits are in 4 Kbytes of data?

13.6 In Example 13.11, what is the output of the decoder that corresponds to the sixth 2-Kbyte block of the address?

13.7 Why are the address bits A_{15} and A_{14} connected to G2A and G2B in Example 13.11?

CONCLUSION

- Sequential logic circuits are digital logic circuits with memory capabilities; their operation is described by state transition tables and state diagrams. Counters and registers are the two principal classes of sequential circuits. Sequential circuits can be designed using a formal procedure analogous to the use of Karnaugh maps for combinational circuits.
- Digital systems play a prominent role in modern engineering. The microprocessor, in particular, has become an integral part of instrumentation and control systems. The microprocessor is very flexible in its application because it can be programmed to perform many different tasks. Depending on the computing power required, 8-, 16-, and 32-bit microprocessors are used for automating measurement and control functions in a variety of industrial applications.

KEY TERMS

ANSWERS TO DRILL EXERCISES

13.3 4.5 kHz

13.4 2,048

13.5 32,768

13.6 2800H–2FFFH

13.7 To ensure that the decoder is active only when both A_{15} and A_{14} are logic 0.

DEVICE DATA SHEETS

MM54HC74/MM74HC74 Dual *D* Flip-Flop with Preset and Clear

Features

- Typical propagation delay: 20 ns
- Wide power supply range: 2–6 V
- Low quiescent current: 40 μA maximum (74HC series)
- Low input current: 1 μA maximum
- Fanout of 10 LS-TTL loads

Truth table

Inputs				*Outputs*	
PR	CLR	CLK	*D*	*Q*	$\overline{Q}$
L	H	X	X	H	L
H	L	X	X	L	H
L	L	X	X	H*	H*
H	H	↑	H	H	L
H	H	↑	L	L	H
H	H	L	X	Q_0	$\overline{Q}_0$

Note: Q_0 = the level of Q before the indicated input conditions were established.
*This configuration is nonstable; that is, it will not persist when preset and clear inputs return to their inactive (high) level.

Connection and Logic Diagrams

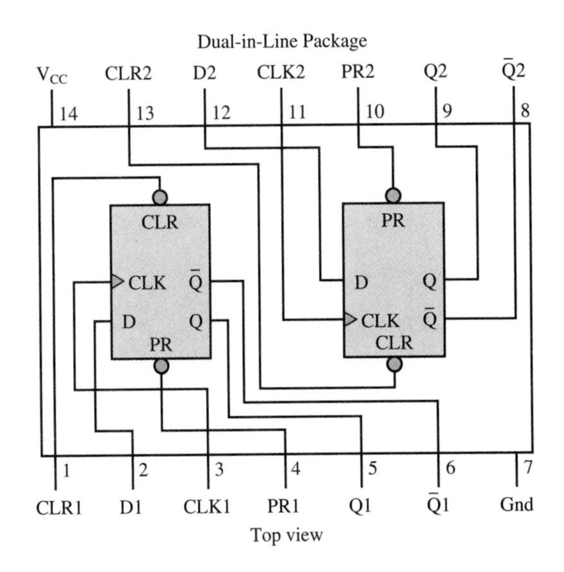

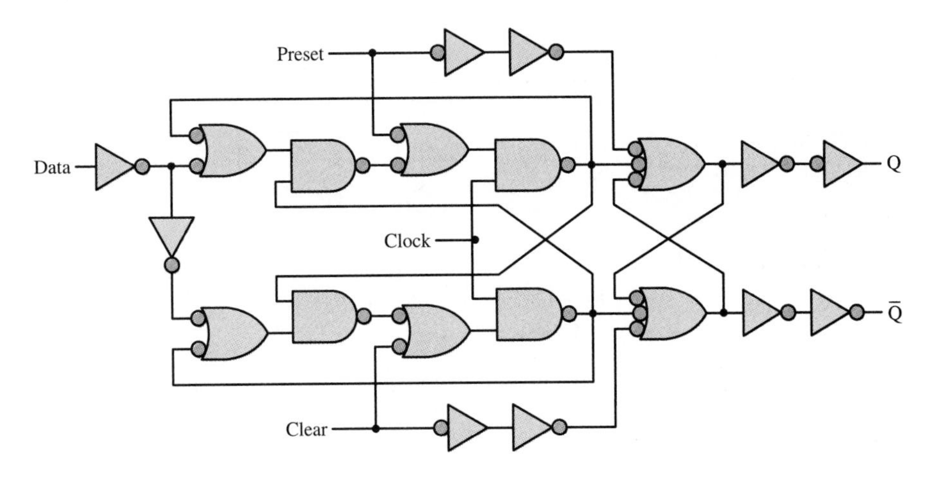

MM54HC73/MM74HC73 Dual *JK* Flip-Flops with Clear

Features

- Typical propagation delay: 16 ns
- Wide operating voltage range: 2–6 V
- Low input current: 1 μA maximum
- Low quiescent current: 40 μA (74HC series)
- High output drive: 10 LS-TTL loads

Connection and Logic Diagrams

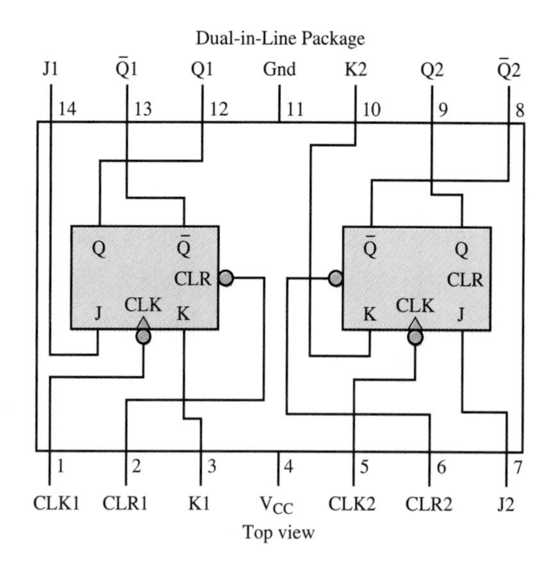

Truth table

Inputs				*Outputs*	
CLR	**CLK**	***J***	***K***	***Q***	$\overline{Q}$
L	X	X	X	L	H
H	↓	L	L	Q_0	$\overline{Q}_0$
H	↓	H	L	H	L
H	↓	L	H	L	H
H	↓	H	H	TOGGLE	
H	H	X	X	Q_0	$\overline{Q}_0$

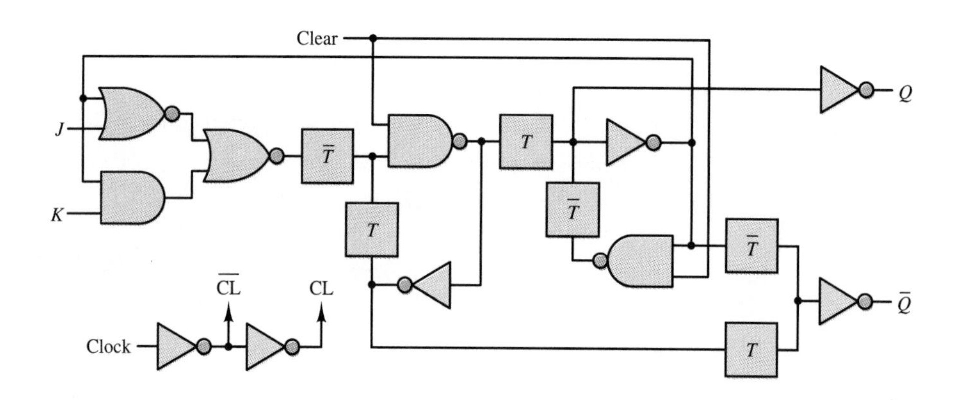

MM54HC76/MM74HC76 Dual *JK* Flip-Flops with Preset and Clear

Features

- Typical propagation delay: 16 ns
- Wide operating voltage range
- Low input current: 1 μA maximum
- Low quiescent current: 40 μA maximum (74HC series)
- High output drive: 10 LS-TTL loads

Connection and Logic Diagrams

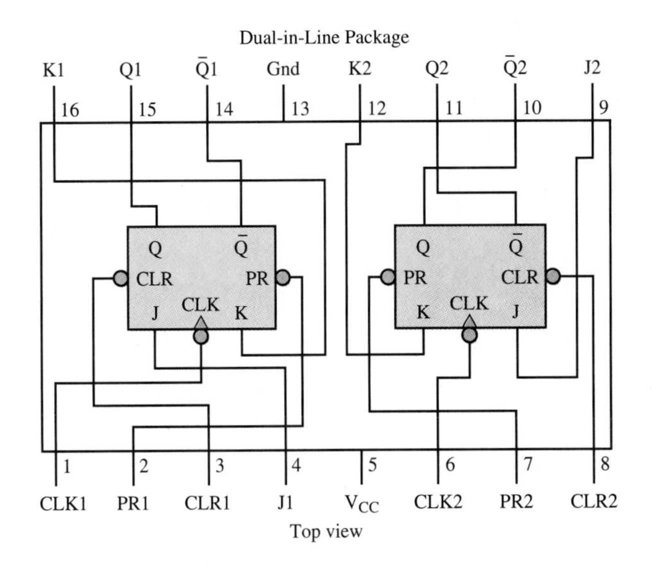

Truth table

Inputs					Outputs	
PR	**CLR**	**CLK**	***J***	***K***	***Q***	$\bar{Q}$
L	H	X	X	X	H	L
H	L	X	X	X	L	H
L	L	X	X	X	L*	L*
H	H	↓	L	L	Q_0	$\bar{Q}_0$
H	H	↓	H	L	H	L
H	H	↓	L	H	L	H
H	H	↓	H	H	TOGGLE	
H	H	H	X	X	Q_0	$\bar{Q}_0$

*This is an unstable condition and is not guaranteed.

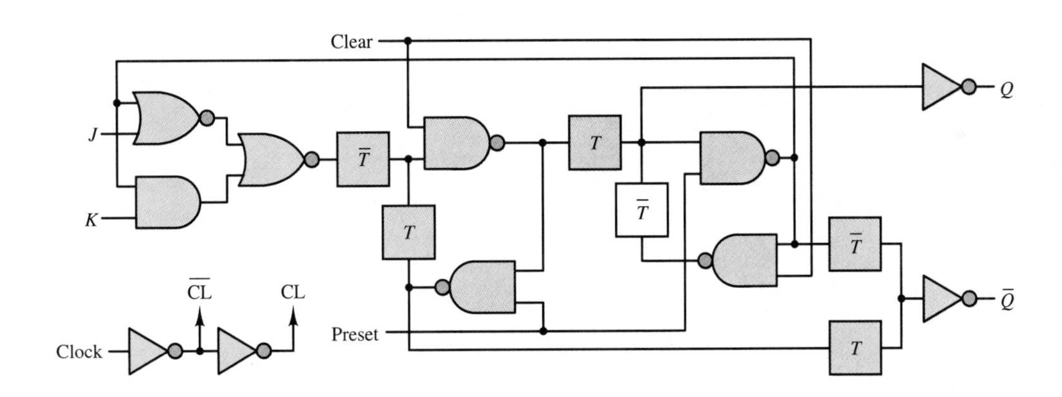

MM54HCT193/MM74HCT193 Synchronous Binary Up/Down Counters

Features

- Low quiescent supply current: 80 μA maximum (74HCT series)
- Low input current: 1 μA maximum
- TTL-compatible inputs

Connection and Logic Diagrams

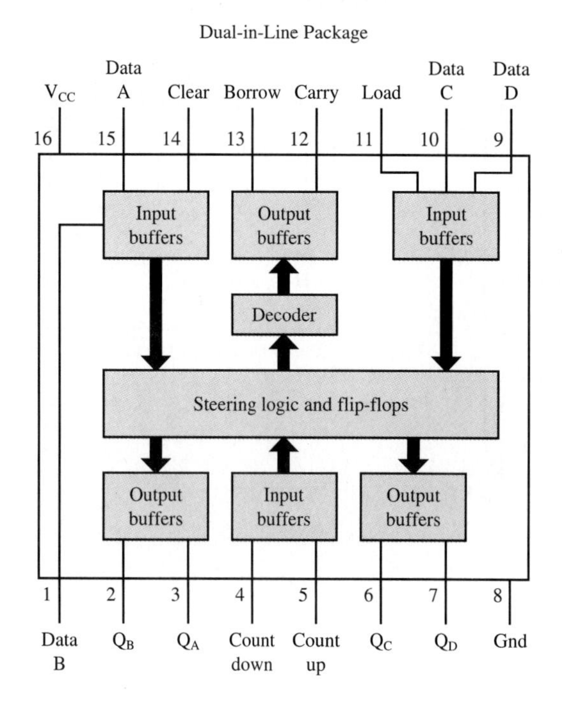

Truth table

Count		**Clear**	**Load**	**Function**
Up	**Down**			
↑	H	L	H	Count up
H	↑	L	H	Count down
X	X	H	X	Clear
X	X	L	L	Load

H = high level.
L = low level.
↑ = transition from low to high.
X = don't care.

HOMEWORK PROBLEMS

13.1 The input to the circuit of Figure P13.1 is a square wave having a period of 2 s, maximum value of 5 V, and minimum value of 0 V. Assume all flip-flops are initially in the RESET state.

a. Explain what the circuit does.
b. Sketch the timing diagram, including the input and all four outputs.

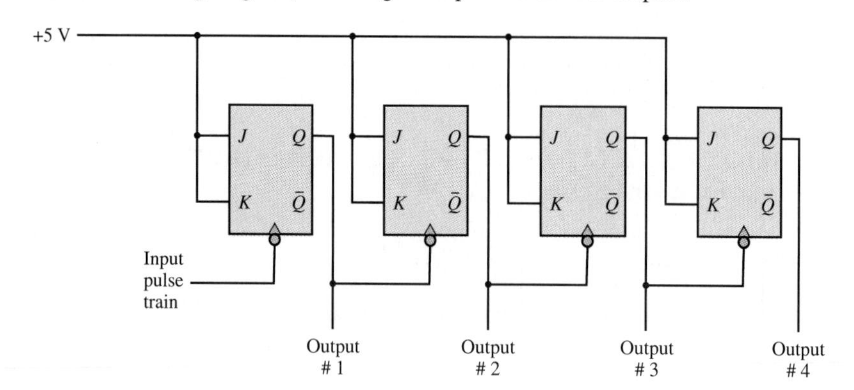

Figure P13.1

13.2 A binary pulse counter can be constructed by interconnecting T-type flip-flops in an appropriate manner. Assume it is desired to construct a counter which can count up to 100_{10}.

a. How many flip-flops would be required?
b. Sketch the circuit needed to implement this counter.

13.3 Explain what the circuit of Figure P13.2 does and how it works. Hint: This circuit is called a 2-bit synchronous binary up-down counter.

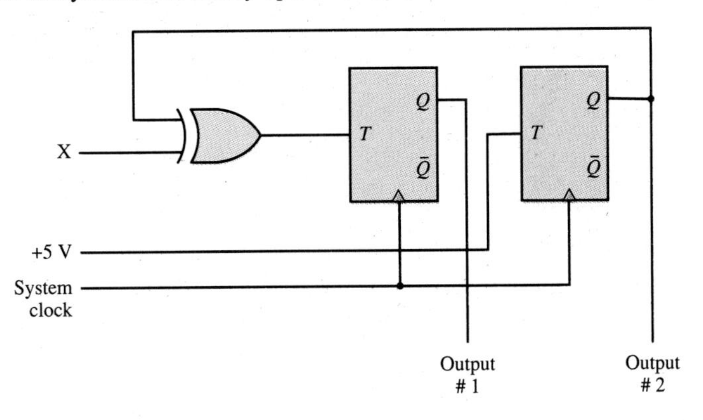

Figure P13.2

13.4 Suppose a circuit is constructed from 3 D-type flip-flops, with

$$D_0 = Q_2 \qquad D_1 = Q_2 \oplus Q_0 \qquad D_2 = Q_1$$

a. Draw the circuit diagram.
b. Assume the circuit starts with all flip-flops SET. Sketch a timing diagram which shows the outputs of all three flip-flops.

13.5 Suppose that you want to use a D flip-flop for a laboratory experiment. However, you have only T flip-flops. Assuming that you have all the logic gates available, make a D flip-flop using a T flip-flop and some logic gate(s).

13.6 Draw a timing diagram (four complete clock cycles) for A_0, A_1, and A_2 for the circuit of Figure P13.3. Assume that all initial values are 0. Note that all flip-flops are positive edge-triggered.

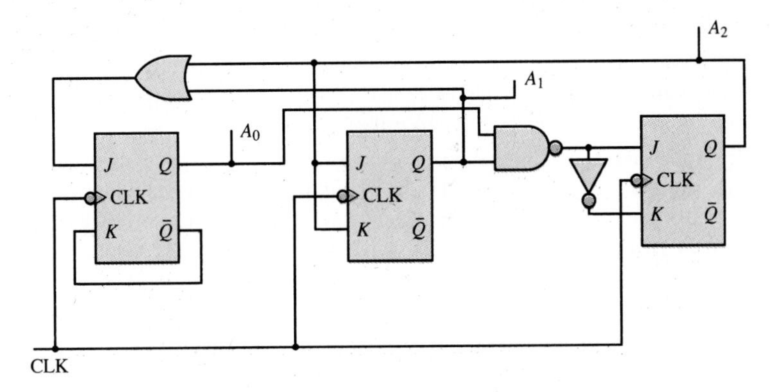

Figure P13.3

13.7 Assume that the slotted encoder shown in Figure P13.4 has a length of 1 meter and a total of 1,000 slots (i.e., there is one slot per millimeter). If a counter is incremented by 1 each time a slot goes past a sensor, design a digital counting system that determines the speed of the moving encoder (in meters per second).

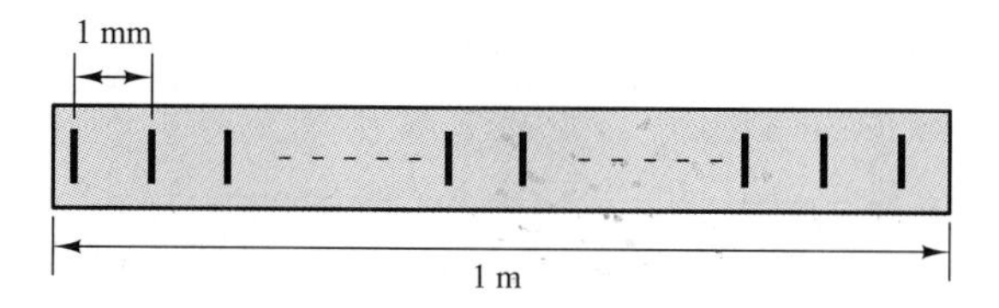

Figure P13.4

13.8 Find the output Q for the circuit of Figure P13.5.

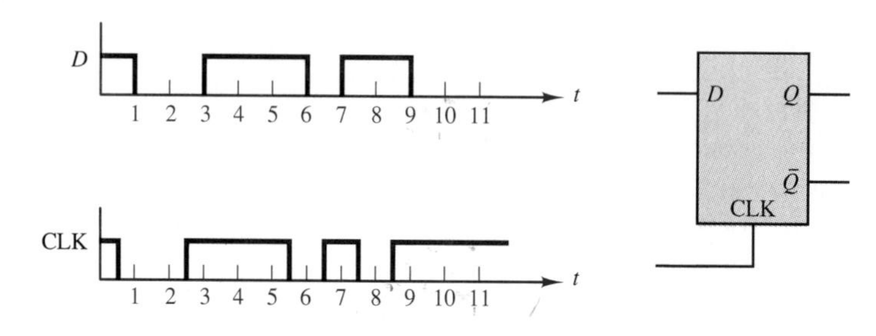

Figure P13.5

13.9 Describe how the ripple counter works. Why is it so named? What disadvantages can you think of for this counter?

13.10 Write the truth table for an *RS* flip-flop with enable (E), preset (P), and clear (C) lines.

13.11 A *JK* flip-flop is wired as shown in Figure P13.6 with a given input signal. Assuming that Q is at logic 0 initially and the trailing edge triggering is effective, sketch the output Q.

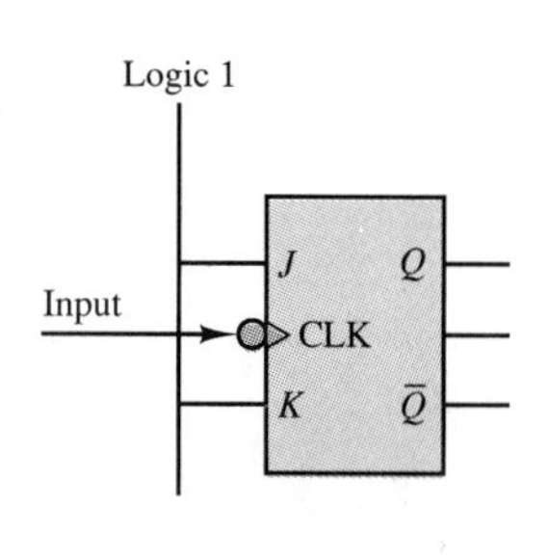

Figure P13.6

13.12 With reference to the *JK* flip-flop of Problem 13.11, assume that the output at the Q terminal is made to serve as the input to a second *JK* flip-flop wired exactly as the first. Sketch the Q output of the second flip-flop.

13.13 Assume that there is a flip-flop with the characteristic given in Figure P13.7, where A and B are the inputs to the flip-flop and Q is the next state output. Using necessary logic gates, make a T flip-flop from this flip-flop.

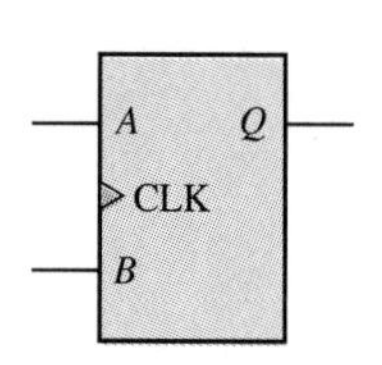

A	B	Q
0	0	$\bar{q}$
0	1	q
1	0	q
1	1	0

Figure P13.7

13.14 Suppose that two input devices are wired to an 8086 system, one at the address D1H, and the other at the address D2H. An output device is connected to the system at address D0H. Write the assembly language code necessary to read a byte from each input device, add the two values, and write the result to the output device. Assume no overflow occurs.

13.15 An input device is wired to an 8086 system at address F5H. A single-byte number is stored at memory address A1H. Write the necessary assembly language code to read a byte from the input device, multiply it by the number stored in memory location A1H, and write the result to address B0H. Assume no overflow occurs.

13.16 Two input devices are wired to an 8086 system, one at the address E5H, and the other at the address C2H. An output device is connected to the system at address B1H. Write the assembly language code necessary to read a byte from each input device, and write the number "1" to the output device if they are equal. Your program should write the number "0" to the output device if they are *not* equal.

13.17 A typical PC has 640 Kbytes of standard memory.

a. How many words is this?
b. How many nibbles is this?
c. How many bits is this?
d. If you could buy 256 Kbit memory chips for 20 cents each, how much would it cost you to expand the PC memory to 1 Mbyte?

13.18 Suppose a microprocessor has n registers.

a. How many control lines do you need to connect each register to all other registers?
b. How many control lines do you need if a bus is used?

13.19 Suppose it is desired to implement a 4K 16-bit memory.

a. How many bits are required for the memory address register?
b. How many bits are required for the memory data register?

13.20 What is the distinction between volatile and nonvolatile memory?

13.21 Draw a block diagram of a memory system that is interfaced to an 8086 microprocessor. The memory system requires 8 Kbytes of ROM and 8 Kbytes of RAM. The ROM must start at location 0000H and must be immediately followed by the RAM. The following information is known for the processor. The system bus consists of a 16-bit address bus, A_0 to A_{15}; and an 8-bit bidirectional data bus, D_0 to D_7. Assume that the microprocessor has three control lines: $\overline{WR}$ (logic 0 indicates that the processor is writing), $\overline{RD}$ (logic 0 indicates that the processor is reading), and $IO/\overline{M}$ (logic 0 indicates that the address is that of a memory location as opposed to an I/O location). The memory devices that are available are an 8K × 8 EPROM and an 8K × 8 RAM.

13.22 Suppose a particular magnetic tape can be formatted with eight tracks per centimeter of tape width. The recording density is 200 bits/cm, and the transport mechanism moves the tape past the read heads at a velocity of 25 cm/s. How many bytes/s can be read from a 2-cm-wide tape?

13.23 Draw a block diagram of a circuit that will interface two interrupts, INT0 and INT1, to the INT input of a CPU so that INT1 has the higher priority and INT0 has the lower. In other words, a signal on INT1 is to be able to interrupt the CPU even when the CPU is currently handling an interrupt generated by INT0, but not vice versa.

CHAPTER 14

Electronic Instrumentation and Measurements

This chapter will introduce measurement and instrumentation systems. It will summarize important concepts by building on the foundation provided in earlier chapters. The development of the chapter follows a logical thread, starting from the physical sensors and proceeding through wiring and grounding to signal conditioning and analog-to-digital conversion, and finally to digital data transmission.

The first section presents an overview of sensors commonly used in engineering measurements. Some sensing devices have already been covered in earlier chapters, and others will be discussed in later chapters; the main emphasis in this chapter will be on classifying physical sensors, and on providing additional detail on some sensors not presented elsewhere in this book—most notably, temperature transducers. The second section of the chapter describes the common signal connections and proper wiring and grounding techniques, with emphasis on noise sources and techniques for reducing undesired interference. Section 14.3 provides an essential introduction to digital signal conditioning, namely, a discussion of instrumentation amplifiers and active filters. The last three sections introduce analog-to-digital conversion, timing circuits, and digital data transmission, respectively.

Upon completing this chapter, you should be able to:

- Recognize the principal classes of sensors.
- Design proper circuit connections to minimize noise in floating, grounded, and differential-source circuits.
- Understand the concepts of shielding and grounding.
- Specify, analyze, and design instrumentation amplifiers and simple active filters.
- Understand the processes of analog-to-digital and digital-to-analog conversion, and specify the requirements of a data acquisition system.
- Design simple timing circuits using op-amps and integrated circuits.
- Understand the basic principles of digital data transmission.

14.1 MEASUREMENT SYSTEMS AND TRANSDUCERS

Measurement Systems

In virtually every engineering application there is a need for measuring some physical quantities, such as forces, stresses, temperatures, pressures, flows, or displacements. These measurements are performed by physical devices called **sensors** or **transducers,** which are capable of converting a physical quantity to a more readily manipulated electrical quantity. Most sensors, therefore, convert the change of a physical quantity (e.g., humidity, temperature) to a corresponding (usually proportional) change in an electrical quantity (e.g., voltage or current). Often, the direct output of the sensor requires additional manipulation before the electrical output is available in a useful form. For example, the change in resistance resulting from a change in the surface stresses of a material—the quantity measured by the resistance strain gauges described in Chapter 2—must first be converted to a change in voltage through a suitable circuit (the Wheatstone bridge) and then amplified from the millivolt to the volt level. The manipulations needed to produce the desired end result are referred to as *signal conditioning*. The wiring of the sensor to the signal conditioning circuitry requires significant attention to *grounding* and *shielding* procedures, to ensure that the resulting signal is as free from noise and interference as possible. Very often, the conditioned sensor signal is then converted to *digital* form and recorded in a computer for additional manipulation, or is displayed in some form. The apparatus used in manipulating a sensor output to produce a result that can be suitably displayed or stored is called a **measurement system.** Figure 14.1 depicts a typical computer-based measurement system in block diagram form.

Sensor Classification

There is no standard and universally accepted classification of sensors. Depending on one's viewpoint, sensors may be grouped according to their physical

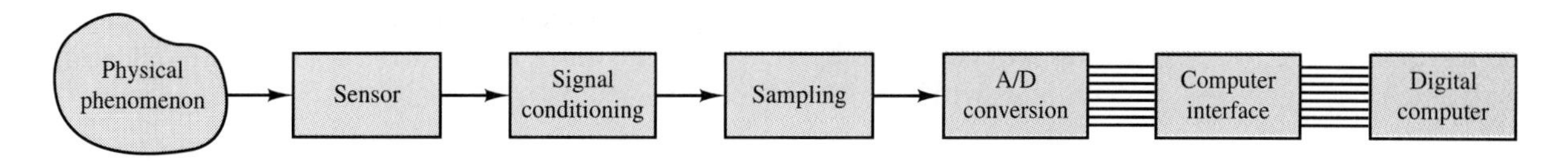

Figure 14.1 Measurement system

characteristics (e.g., electronic sensors, resistive sensors), or by the physical variable or quantity measured by the sensor (e.g., temperature, flow rate). Other classifications are also possible. Table 14.1 presents a partial classification of sensors grouped according to the variable sensed; we do not claim that the table is complete, but we can safely state that most of the engineering measurements of interest to the reader are likely to fall in the categories listed in Table 14.1. Also included in the table are section or example references to sensors described in this book.

Table 14.1 **Sensor classification**

Sensed variables	Sensors	Reference in this book
Motion and dimensional variables	Resistive potentiometers	Example 2.7
	Strain gauges	Examples 2.8, 2.13
	Differential transformers (LVDTs)	Example 15.1
	Variable-reluctance sensors	Examples 15.5, 15.6
	Capacitive sensors	Example 4.4
	Piezoelectric sensors	Example 11.9
	Electro-optical sensors	Examples 12.2, 13.6
	Moving-coil transducers	Section 15.5
	Seismic sensors	Example 6.9
Force, torque, and pressure	Strain gauges	Examples 2.8, 2.13
	Piezoelectric sensors	Example 11.9
	Capacitive sensors	Example 4.4
Flow	Pitot tube	
	Hot-wire anemometer	Section 14.1
	Differential pressure sensors	Section 14.1
	Turbine meters	Section 14.1
	Vortex shedding meters	
	Ultrasonic sensors	
	Electromagnetic sensors	
	Imaging systems	
Temperature	Thermocouples	Section 14.1
	Resistance thermometers (RTDs)	Section 14.1
	Semiconductor thermometers	
	Radiation detectors	
Liquid level	Motion transducers	
	Force transducers	
	Differential-pressure measurement devices	
Humidity	Semiconductor sensors	
Chemical composition	Gas analysis equipment	
	Solid-state gas sensors	

A sensor is usually accompanied by a set of specifications that indicate its overall effectiveness in measuring the desired physical variable. The following definitions will help the reader understand sensor data sheets:

Accuracy: Conformity of the measurement to the true value, usually in percent of full-scale reading

Error: Difference between measurement and true value, usually in percent of full-scale reading

Precision: Number of significant figures of the measurement

Resolution: Smallest measurable increment

Span: Linear operating range

Range: The range of measurable values

Linearity: Conformity to an ideal linear calibration curve, usually in percent of reading or of full-scale reading (whichever is greater)

Motion and Dimensional Measurements

The measurement of motion and dimension is perhaps the most commonly encountered engineering measurement. Measurements of interest include absolute position, relative position (displacement), velocity, acceleration, and jerk (the derivative of acceleration). These can be either translational or rotational measurements; usually, the same principle can be applied to obtain both kinds of measurements. These measurements are often based on changes in elementary properties, such as changes in the resistance of an element (e.g., strain gauges, potentiometers), in an electric field (e.g., capacitive sensors), or in a magnetic field (e.g., inductive, variable-reluctance, or eddy current sensors). Other mechanisms may be based on special materials (e.g., piezoelectric crystals), or on optical signals and imaging systems. Table 14.1 lists several examples of dimensional and motion measurement that can be found in this book.

Force, Torque, and Pressure Measurements

Another very common class of measurements is that of pressure and force, and the related measurement of torque. Perhaps the single most common family of force and pressure transducers comprises those based on strain gauges (e.g., load cells, diaphragm pressure transducers). Also very common are piezoelectric transducers. Capacitive transducers again find application in the measurement of pressure. Table 14.1 indicates where the reader can find examples of these measurements in this book.

Flow Measurements

In many engineering applications it is desirable to sense the flow rate of a fluid, whether compressible (gas) or incompressible (liquid). The measurement of fluid flow rate is a complex subject; in this section we simply summarize the concepts underlying some of the most common measurement techniques. Shown in Figure

14.2 are three different types of flow rate sensors. The sensor in Figure 14.2(a) is based on **differential pressure measurement** and on a **calibrated orifice:** the relationship between pressure across the orifice, $p_1 - p_2$, and flow rate through the orifice, q, is predetermined through the calibration; therefore, measuring the differential pressure is equivalent to measuring flow rate.

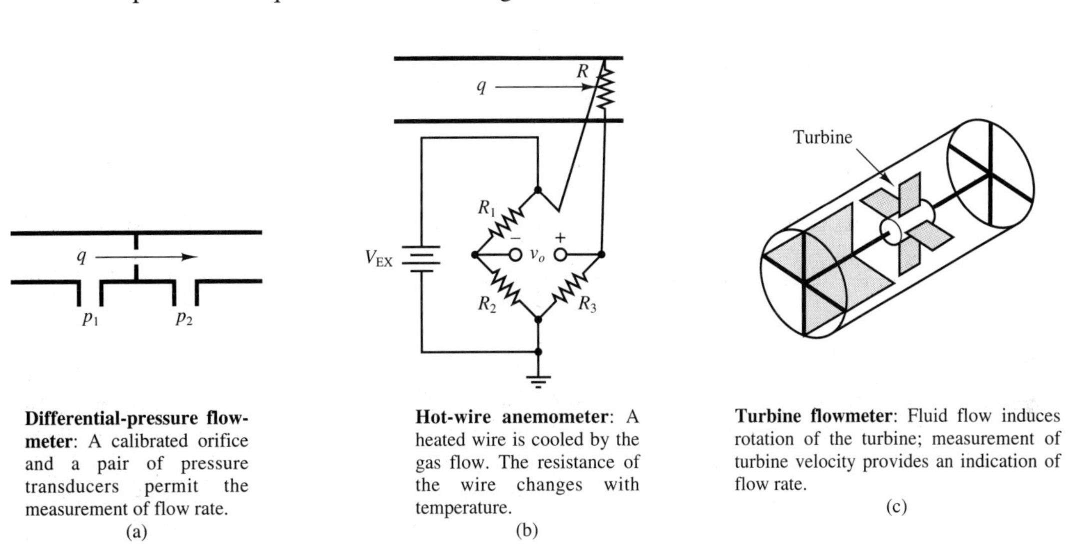

Differential-pressure flowmeter: A calibrated orifice and a pair of pressure transducers permit the measurement of flow rate.
(a)

Hot-wire anemometer: A heated wire is cooled by the gas flow. The resistance of the wire changes with temperature.
(b)

Turbine flowmeter: Fluid flow induces rotation of the turbine; measurement of turbine velocity provides an indication of flow rate.
(c)

Figure 14.2 Devices for the measurement of flow

The sensor in Figure 14.2(b) is called a **hot-wire anemometer,** because it is based on a heated wire that is cooled by the flow of a gas. The resistance of the wire changes with temperature, and a Wheatstone bridge circuit converts this change in resistance to a change in voltage. Also commonly used are **hot-film anemometers,** where a heated film is used in place of the more delicate wire. A very common application of the latter type of sensor is in automotive engines, where control of the air-to-fuel ratio depends on measurement of the engine intake mass airflow rate.

Figure 14.2(c) depicts a **turbine flowmeter,** in which the fluid flow causes a turbine to rotate; the velocity of rotation of the turbine (which can be measured by a noncontact sensor—e.g., a magnetic pickup) is related to the flow velocity.

Besides the techniques discussed in this chapter, many other techniques exist for measuring fluid flow, some of significant complexity.

Temperature Measurements

One of the most frequently measured physical quantities is temperature. The need to measure temperature arises in just about every field of engineering. This subsection is devoted to summarizing two common temperature sensors—the **thermocouple** and the **resistance temperature detector (RTD)**—and their related signal-conditioning needs.

Thermocouples

A thermocouple is formed by the junction of two dissimilar metals. This junction results on an open-circuit **thermoelectric voltage** due to the **Seebeck effect,** named after Thomas Seebeck, who discovered the phenomenon in 1821. Various types of thermocouples exist; they are usually classified according to the data of Table 14.2. The Seebeck coefficient shown in the table is specified at a given temperature because the output voltage of a thermocouple, v, has a nonlinear dependence on temperature. This dependence is typically expressed in terms of a polynomial of the following form:

$$T = a_0 + a_1 v + a_2 v^2 + a_3 v^3 + \cdots + a_n v^n \qquad \textbf{(14.1)}$$

Table 14.2 **Thermocouple data**

Type	Elements +/−	Seebeck coefficient (μV/°C)	Range (°C)	Range (mV)
E	Chromel/constantan	58.70 at 0°C	−270 to 1,000	−9.835 to 76.358
J	Iron/constantan	50.37 at 0°C	−210 to 1,200	−8.096 to 69.536
K	Chromel/alumel	39.48 at 0°C	−270 to 1,372	−6.548 to 54.874
R	Pt(10%)—Rh/Pt	10.19 at 600°C	−50 to 1,768	−0.236 to 18.698
T	Copper/constantan	38.74 at 0°C	−270 to 400	−6.258 to 20.869
S	Pt(13%)—Rh/Pt	11.35 at 600°C	−50 to 1,768	−0.226 to 21.108

For example, the coefficients of the J thermocouple in the range −100°C to +1,000°C are as follows:

$$a_0 = -0.048868252 \qquad a_1 = 19{,}873.14503 \qquad a_2 = -128{,}614.5353$$
$$a_3 = 11{,}569{,}199.78 \qquad a_4 = -264{,}917{,}531.4 \qquad a_5 = 2{,}018{,}441{,}314$$

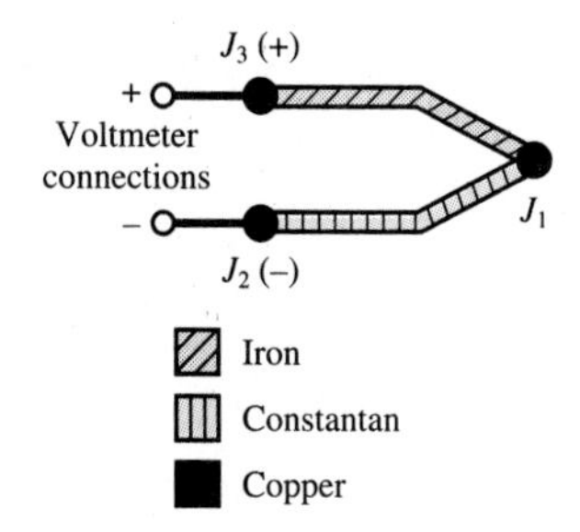

Figure 14.3 J thermocouple circuit

The use of a thermocouple requires special connections, because the junction of the thermocouple wires with other leads (such as voltmeter leads, for example), creates additional thermoelectric junctions that in effect act as additional thermocouples. For example, in the J thermocouple circuit of Figure 14.3, junction J_1 is exposed to the temperature to be measured, but junctions J_2 and J_3 also generate a thermoelectric voltage, which is dependent on the temperature at these junctions, that is, the temperature at the voltmeter connections. One would therefore have to know the voltages at these junctions, as well, in order to determine the actual thermoelectric voltage at J_1. To obviate this problem, a reference junction at known temperature can be employed; a traditional approach involves the use of a **cold junction,** so called because it consists of an ice bath, one of the easiest means of obtaining a known reference temperature. Figure 14.4 depicts a thermocouple measurement using an ice bath. The voltage measured in Figure 14.4 is dependent on the temperature difference $T_1 - T_{ref}$, where $T_{ref} = 0°C$. The connections to the voltmeter are made at an *isothermal block,* kept at a constant temperature; note that the same metal is used in both of the connections to the isothermal block. Thus (still assuming a J thermocouple), there is no difference

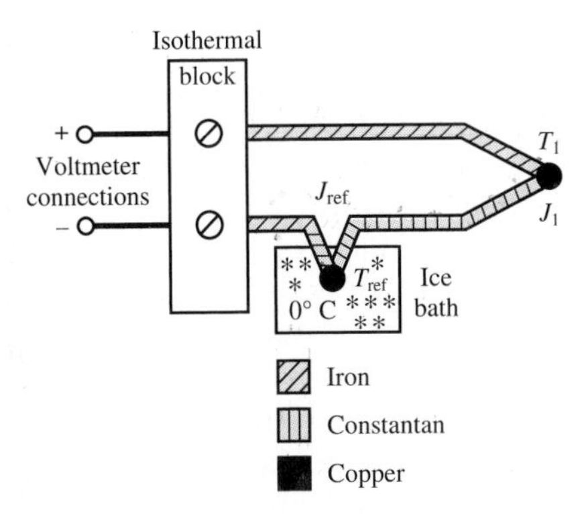

Figure 14.4 Cold-junction–compensated thermocouple circuit

between the thermoelectric voltages at the two copper-iron junctions; these will add to zero at the voltmeter. The voltmeter will therefore read a voltage proportional to $T_1 - T_{ref}$.

An ice bath is not always a practical solution. Other cold junction temperature compensation techniques employ an additional temperature sensor to determine the actual temperature of the junctions J_2 and J_3 of Figure 14.3.

Resistance Temperature Detectors (RTDs)

A resistance temperature detector (RTD) is a variable-resistance device whose resistance is a function of temperature. RTDs can be made with both positive and negative temperature coefficients and offer greater accuracy and stability than thermocouples. **Thermistors** are part of the RTD family. A characteristic of all RTDs is that they are *passive* devices, that is, they do not provide a useful output unless excited by an external source. The change in resistance in an RTD is usually converted to a change in voltage by forcing a current to flow through the device. An indirect result of this method is a **self-heating error,** caused by the i^2R heating of the device. Self-heating of an RTD is usually denoted by the amount of power that will raise the RTD temperature by 1°C. Reducing the excitation current can clearly help reduce self-heating, but it also reduces the output voltage.

The RTD resistance has a fairly linear dependence on temperature; a common definition of the **temperature coefficient** of an RTD is related to the change in resistance from 0° to 100°C. Let R_0 be the resistance of the device at 0°C and R_{100} the resistance at 100°C. Then the temperature coefficient, α, is defined to be

$$\alpha = \frac{R_{100} - R_0}{100 - 0} \frac{\Omega}{^\circ\mathrm{C}} \tag{14.2}$$

A more accurate representation of RTD temperature dependence can be obtained by using a nonlinear (cubic) equation and published tables of coefficients. As an example, a platinum RTD could be described either by the temperature coefficient $\alpha = 0.003911$, or by the equation

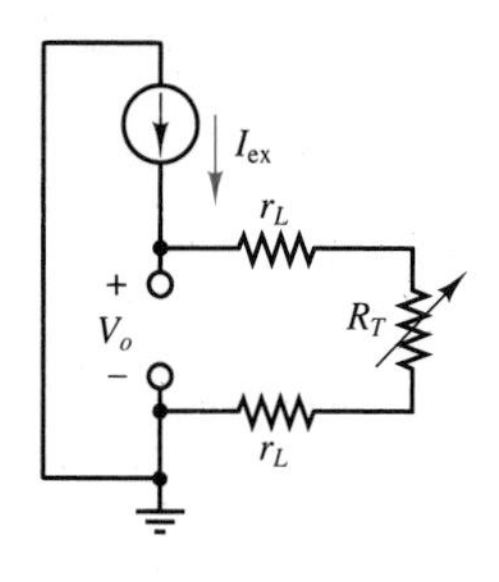

Figure 14.5 Effect of connection leads on RTD temperature measurement

$$R_T = R_0(1 + AT - BT^2 - CT^3) \qquad (14.3)$$
$$= R_0(1 + 3.6962 \times 10^{-3}T - 5.8495 \times 10^{-7}T^2 - 4.2325 \times 10^{-12}T^3)$$

where the coefficient C is equal to zero for temperatures above 0°C.

Because RTDs have fairly low resistance, they are sensitive to error introduced by the added resistance of the lead wires connected to them; Figure 14.5 depicts the effect of the lead resistances, r_L, on the RTD measurement. Note that the measured voltage includes the resistance of the RTD as well as the resistance of the leads. If the leads used are long (greater than 3 m is a good rule of thumb), then the measurement will have to be adjusted for this error. Two possible solutions to the lead problems are the *four-wire* RTD measurement circuit and the *three-wire* Wheatstone bridge circuit, shown in Figure 14.6(a) and (b), respectively. In the circuit of Figure 14.6(a), the resistance of the lead wires from the excitation, r_{L1} and r_{L4}, may be arbitrarily large, since the measurement is affected by the resistance of only the output lead wires, r_{L2} and r_{L3}, which can be kept small by making these leads short. The circuit of Figure 14.6(b) takes advantage of the properties of the Wheatstone bridge to cancel out the unwanted effect of the lead wires while still producing an output dependent on the change in temperature.

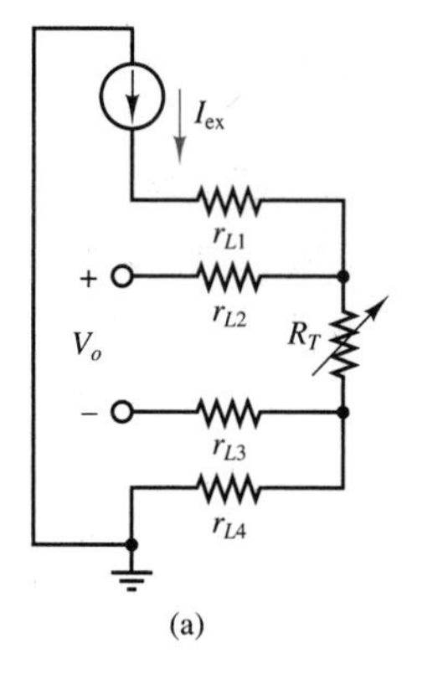

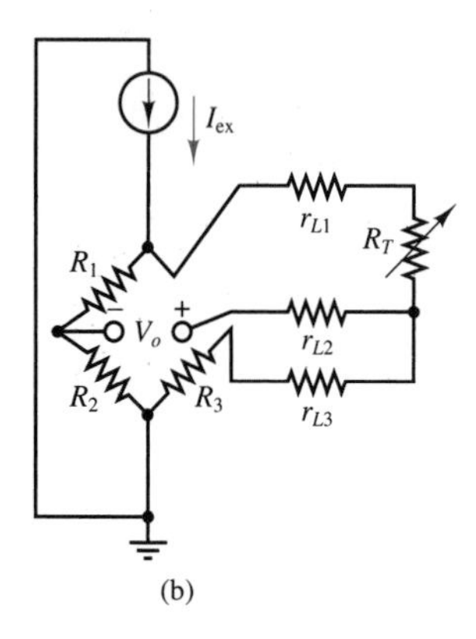

Figure 14.6 Four-wire RTD circuit (a) and three-wire Wheatstone bridge RTD circuit (b).

14.2 WIRING, GROUNDING, AND NOISE

The importance of proper circuit connections cannot be overemphasized. Unfortunately, this is a subject that is rarely taught in introductory electrical engineering courses. The present section summarizes some important considerations regarding signal source connections, various types of input configurations, noise sources and coupling mechanisms, and means of minimizing the influence of noise on a measurement.

Signal Sources and Measurement System Configurations

Before proper connection and wiring techniques can be presented, we must examine the difference between **grounded** and **floating signal sources.** Every sensor can be thought of as some kind of signal source; a general representation of the connection of a sensor to a measurement system is shown in Figure 14.7(a). The sensor is modeled as an ideal voltage source in series with a source resistance. Although this representation does not necessarily apply to all sensors, it will be adequate for the purposes of the present section. Figures 14.7(b) and (c) show two types of signal sources: grounded and floating. A grounded signal source is one in which a ground reference is established—for example, by connecting the *signal low* lead to a case or housing. A floating signal source is one in which neither signal lead is connected to ground; since ground potential is arbitrary, the signal source voltage levels (*signal low* and *signal high*) are at an unknown potential relative to the case ground. Thus, the signal is said to be *floating.* Whether a sensor can be characterized as a grounded or a floating signal source ultimately depends on the connection of the sensor to its case, but the choice of connection may depend on the nature of the source. For example, the thermocouple described in Section 12.1 is *intrinsically* a floating signal source, since the signal of interest is a difference between two voltages. The same thermocouple *could* become a grounded signal source if one of its two leads were directly connected to ground, but this is usually not a desirable arrangement for this particular sensor.

In analogy with a signal source, a measurement system can be either **ground-referenced** or **differential.** In a ground-referenced system, the signal low connection is tied to the instrument case ground; in a differential system, neither of the two signal connections is tied to ground. Thus, a differential measurement system is well suited to measuring the difference between two signal levels (such as the output of an ungrounded thermocouple).

One of the potential dangers in dealing with grounded signal sources is the introduction of **ground loops.** A ground loop is an undesired current path caused by the connection of two reference voltages to each other. This is illustrated in Figure 14.8, where a grounded signal source is shown connected to a ground-referenced measurement system. Notice that we have purposely denoted the signal source ground and the measurement system ground by two distinct symbols, to emphasize that these are not necessarily at the same potential—as also indicated by the voltage difference ΔV. Now, one might be tempted to tie the two grounds to each other, but this would only result in a current flowing from one ground to the other, through the small (but nonzero) resistance of the wire connecting the two. The net effect of this ground loop would be that the voltage measured by the instrument would include the unknown ground voltage difference ΔV, as shown in Figure 14.8. Since this latter voltage is unpredictable, you can see that ground loops can cause substantial errors in measuring systems. In addition, ground loops are the primary cause of conducted noise, as explained later in this section.

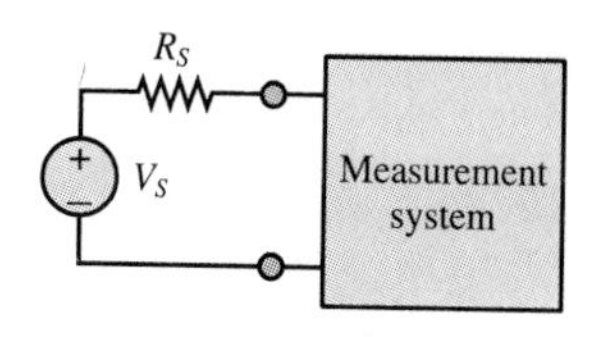

(a) Ideal signal source connected to measurement system

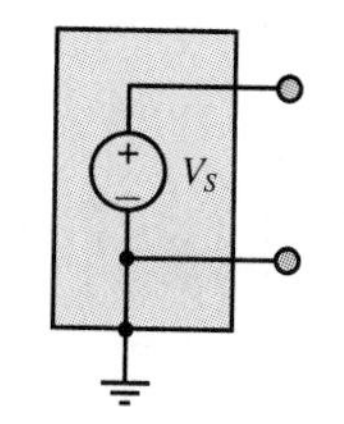

(b) Grounded signal source

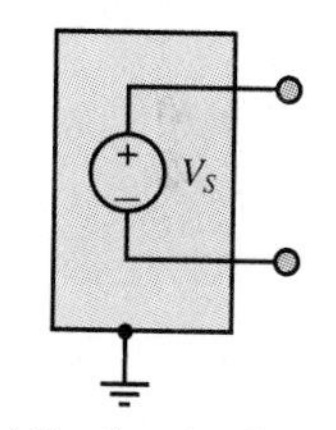

(c) Floating signal source

Figure 14.7 Measurement system and types of signal sources

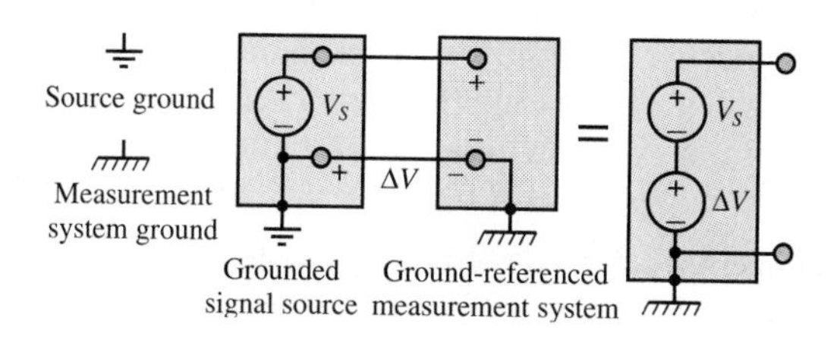

Figure 14.8 Ground loop in ground-referenced measurement system

A differential measurement system is often a way to avoid ground loop problems, because the signal source and measurement system grounds are not connected to each other, and especially because the signal low input of the measuring instrument is not connected to either instrument case ground. The connection of a grounded signal source and a differential measurement system is depicted in Figure 14.9.

If the signal source connected to the differential measurement system is floating, as shown in Figure 14.10, it is often a recommended procedure to reference the signal to the instrument ground by means of two identical resistors that

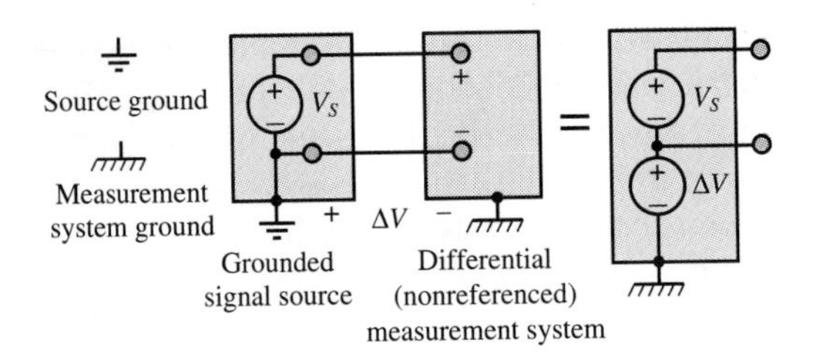

Figure 14.9 Differential (nonreferenced) measurement system

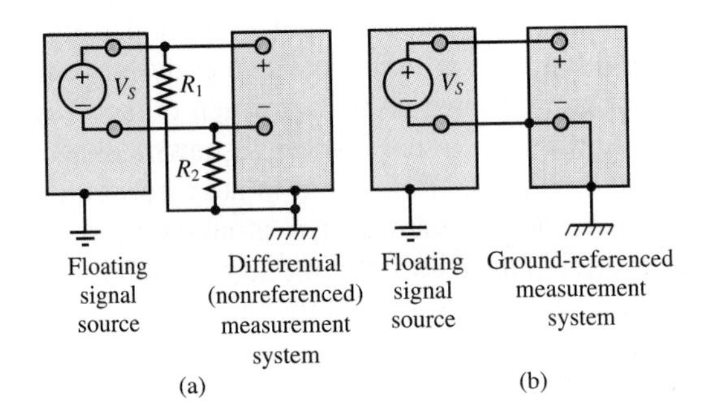

Figure 14.10 Measuring signals from a floating source: (a) differential input; (b) single-ended input

can provide a return path to ground for any currents present at the instrument. An example of such input currents would be the input bias currents inevitably present at the input of an operational or instrumentation amplifier.

The simple concepts illustrated in the preceding paragraphs and figures can assist the user and designer of instrumentation systems in making the best possible wiring connections for a given measurement.

Noise Sources and Coupling Mechanisms

Noise—meaning any undesirable signal interfering with a measurement—is an unavoidable element of all measurements. Figure 14.11 depicts a block diagram of the three essential stages of a noisy measurement: a **noise source,** a **noise coupling mechanism,** and a sensor or associated signal-conditioning circuit. Noise sources are always present, and are often impossible to eliminate completely; typical sources of noise in practical measurements are the electromagnetic fields caused by fluorescent light fixtures, video monitors, power supplies, switching circuits, and high-voltage (or current) circuits. Many other sources exist, of course, but often the simple sources in our everyday environment are the most difficult to defeat.

Figure 14.11 also indicates that various coupling mechanisms can exist between a noise source and an instrument. Noise coupling can be conductive; that is, noise currents may actually be conducted from the noise source to the

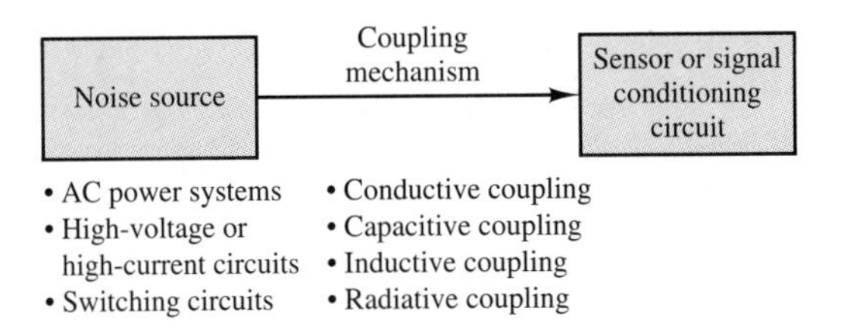

Figure 14.11 Noise sources and coupling mechanisms

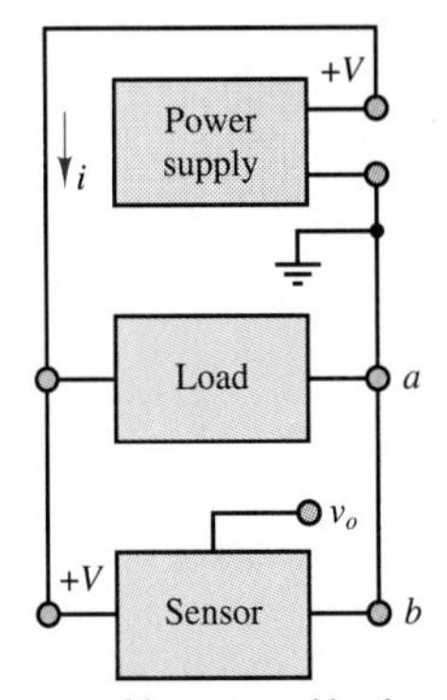

The ground loop created by the load circuit can cause a different ground potential between *a* and *b*.

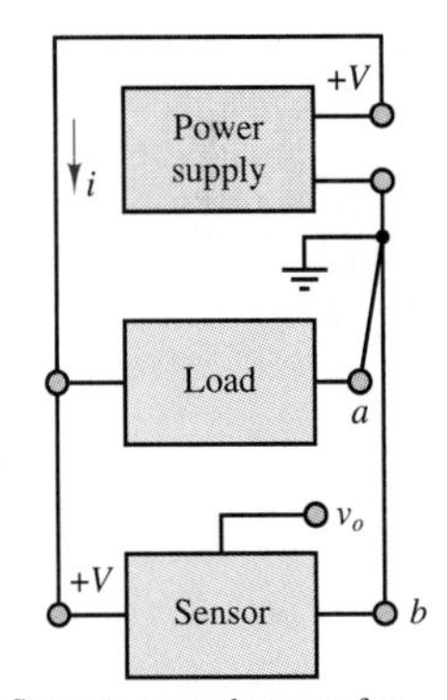

Separate ground returns for the load and the sensor circuit eliminate the ground loop.

Figure 14.12 Conductive coupling: ground loop and separate ground returns

instrument by physical wires. Noise can also be coupled capacitively, inductively, and radiatively.

Figure 14.12 illustrates how interference can be **conductively coupled** by way of a ground loop. In the figure, a power supply is connected to both a load and a sensor. We shall assume that the load may be switched on and off, and that it carries substantial currents. The top circuit contains a ground loop: the current i from the supply divides between the load and sensor; since the wire resistance is nonzero, a large current flowing through the load may cause the ground potential at point a to differ from the potential at point b. In this case, the measured sensor output is no longer v_o, but it is now equal to $v_o + v_{ba}$, where v_{ba} is the potential difference from point b to point a. Now, if the load is switched on and off and its current is therefore subject to large, abrupt changes, these changes will be manifested in the voltage v_{ba} and will appear as noise on the sensor output.

This problem can be cured simply and effectively by providing separate *ground returns* for the load and sensor, thus eliminating the ground loop.

The mechanism of **capacitive coupling** is rooted in electric fields that may be caused by sources of interference. The detailed electromagnetic analysis can be quite complex, but to understand the principle, refer to Figure 14.13(a), where a noise source is shown to generate an electric field. If a noise source conductor is sufficiently close to a conductor that is part of the measurement system, the two conductors (separated by air, a dielectric) will form a capacitor, through which any time-varying currents can flow. Figure 14.13(b) depicts an equivalent circuit in which the noise voltage V_N couples to the measurement circuit through an imaginary capacitor, representing the actual capacitance of the noise path.

The dual of capacitive coupling is **inductive coupling.** This form of noise coupling is due to the magnetic field generated by current flowing through a conductor. If the current is large, the magnetic fields can be significant, and the **mutual inductance** (see Chapters 5 and 15) between the noise source and the measurement circuit causes the noise to couple to the measurement circuit. Thus, inductive coupling, as shown in Figure 14.14, results when undesired (unplanned) magnetic coupling ties the noise source to the measurement circuit.

Noise Reduction

Various techniques exist for minimizing the effect of undesired interference, in addition to proper wiring and grounding procedures. The two most common methods are **shielding** and the use of **twisted-pair wire.** A shielded cable is shown

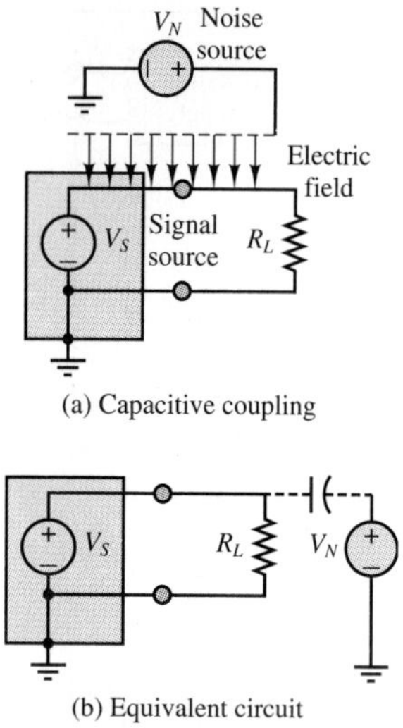

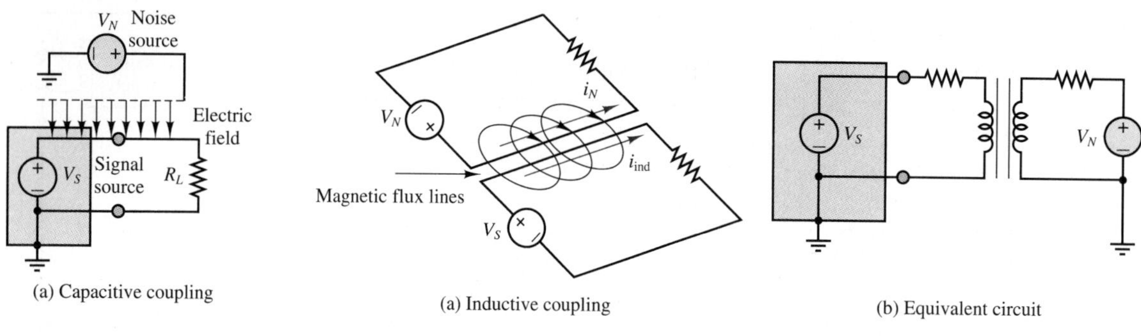

(a) Capacitive coupling

(a) Inductive coupling

(b) Equivalent circuit

(b) Equivalent circuit

Figure 14.13 Capacitive coupling and equivalent-circuit representaion

Figure 14.14 Inductive coupling and equivalent-circuit representation

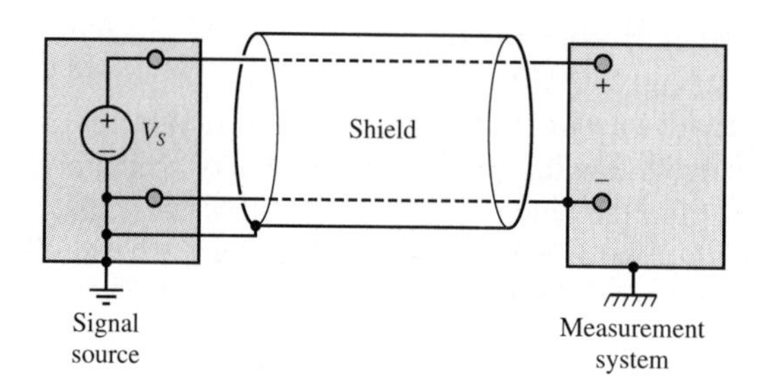

Figure 14.15 Shielding

in Figure 14.15. The shield is made of a copper braid or of foil and is usually grounded at the source end *but not at the instrument end,* because this would result in a ground loop. The shield can protect the signal from a significant amount of electromagnetic interference, especially at lower frequencies. Shielded cables with various numbers of conductors are available commercially. However, shielding cannot prevent inductive coupling. The simplest method for minimizing inductive coupling is the use of twisted-pair wire; the reason for using twisted pair is that untwisted wire can offer large loops that can couple a substantial amount of electromagnetic radiation (see Section 15.1). Twisting drastically reduces the loop area, and with it the interference. Twisted pair is available commercially.

14.3 SIGNAL CONDITIONING

A properly wired, grounded, and shielded sensor connection is a necessary first stage of any well-designed measurement system. The next stage consists of any **signal conditioning** that may be required to manipulate the sensor output into a form appropriate for the intended use. Very often, the sensor output is meant to be fed into a digital computer, as illustrated in Figure 14.1. In this case, it is important to condition the signal so that it is compatible with the process of data acquisition. Two of the most important signal-conditioning functions are *amplification* and *filtering*. Both are discussed in the present section.

Instrumentation Amplifiers

An **instrumentation amplifier (IA)** is a differential amplifier with very high input impedance, low bias current, and programmable gain that finds widespread application when low-level signals with large common-mode components are to be amplified in noisy environments. This situation occurs frequently when a low-level transducer signal needs to be preamplified, prior to further signal conditioning (e.g., filtering). Instrumentation amplifiers were briefly introduced in Chapter 11 (see Example 11.4), as an extension of the differential amplifier. You may recall that the IA introduced in Example 11.4 consisted of two stages, the first composed of two noninverting amplifiers, the second of a differential amplifier. Although the design in Chapter 11 is useful and is sometimes employed in practice, it suffers from a few drawbacks, most notably the requirement for very precisely matched resistors and source impedances to obtain the maximum possible cancellation of the common-mode signal. If the resistors are not matched exactly, the common-mode rejection ratio of the amplifier is significantly reduced, as the following will demonstrate.

The amplifier of Figure 14.16 has properly matched resistors ($R_2 = R'_2, R_F = R'_F$), except for resistors R and R', which differ by an amount ΔR such that $R' = R + \Delta R$. Let us compute the closed-loop gain for the amplifier. As shown in Example 11.4, the input-stage noninverting amplifiers have a closed-loop gain given by

$$A = \frac{v'_b}{v_b} = \frac{v'_a}{v_a} = 1 + \frac{2R_2}{R_1} \tag{14.4}$$

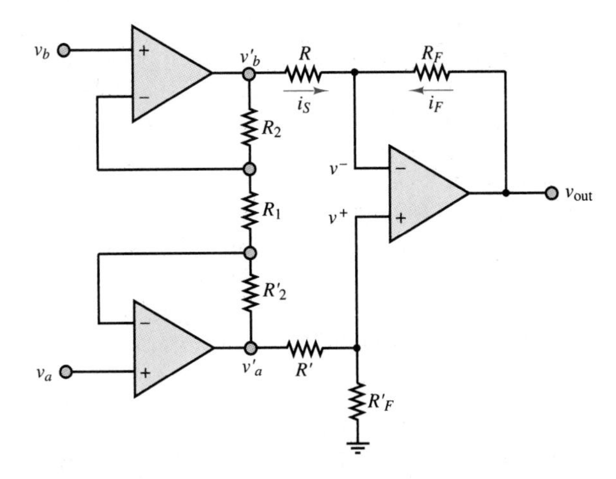

Figure 14.16 Discrete op-amp instrumentation amplifier

To compute the output voltage, we observe that the voltage at the noninverting terminal is

$$v^+ = \frac{R_F}{R_F + R + \Delta R} v_a' \tag{14.5}$$

and since the inverting-terminal voltage is $v^- = v^+$, the feedback current is given by

$$i_F = \frac{v_{\text{out}} - v^-}{R_F} = \frac{v_{\text{out}} - \frac{R_F}{R_F + R + \Delta R} v_a'}{R_F} \tag{14.6}$$

and the source current is

$$i_S = \frac{v_b' - v^-}{R} = \frac{v_b' - \frac{R_F}{R_F + R + \Delta R} v_a'}{R} \tag{14.7}$$

Applying KCL at the inverting node (under the usual assumption that the input current going into the op-amp is negligible), we set $i_F = -i_S$ and obtain the expression

$$\begin{aligned} \frac{v_{\text{out}}}{R_F} &= \frac{v_a'}{R_F + R + \Delta R} - \frac{v_b'}{R} + \frac{R_F}{R} \frac{v_a'}{R_F + R + \Delta R} \\ &= \left(1 + \frac{R_F}{R}\right) \frac{v_a'}{R_F + R + \Delta R} - \frac{v_b'}{R} \end{aligned}$$

so that the output voltage may be computed to be

$$v_{\text{out}} = R_F \left(\frac{R + R_F}{R}\right) \frac{v_a'}{R_F + R + \Delta R} - \frac{R_F}{R} v_b' \tag{14.8}$$

Note that if the term ΔR in the denominator were zero, the same result would be obtained as in Example 11.4: $v_{\text{out}} = (R_F/R)(v_a' - v_b')$; however, because of the resistor mismatch, there is a corresponding mismatch between the gains for the two differential signal components. Further—and more important—if the original signals, v_a and v_b, contained both differential-mode and common-mode components:

$$v_a = v_{a\,\text{dif}} + v_{\text{com}} \qquad v_b = v_{b\,\text{dif}} + v_{\text{com}} \tag{14.9}$$

such that

$$v_a' = A(v_{a\,\text{dif}} + v_{\text{com}}) \qquad v_b' = A(v_{b\,\text{dif}} + v_{\text{com}}) \tag{14.10}$$

then the common-mode components would not cancel out in the output of the amplifier, because of the gain mismatch, and the output of the amplifier would be given by

$$v_{\text{out}} = R_F \left(\frac{R + R_F}{R}\right) \frac{A(v_{a\,\text{dif}} + v_{\text{com}})}{R_F + R + \Delta R} - \frac{R_F}{R} A(v_{b\,\text{dif}} + v_{\text{com}}) \tag{14.11}$$

resulting in the following output voltage:

$$v_{\text{out}} = v_{\text{out, dif}} + v_{\text{out, com}} \tag{14.12}$$

with

$$v_{\text{out, dif}} = R_F \left(\frac{R + R_F}{R} \right) \frac{Av_{a\ \text{dif}}}{R_F + R + \Delta R} - \frac{R_F}{R} Av_{b\ \text{dif}} \tag{14.13}$$

and

$$\begin{aligned} v_{\text{out, com}} &= R_F \left(\frac{R + R_F}{R} \right) \frac{Av_{\text{com}}}{R_F + R + \Delta R} - \frac{R_F}{R} Av_{\text{com}} \\ &= \frac{R_F}{R} \left(\frac{R + R_F}{R_F + R + \Delta R} - 1 \right) Av_{\text{com}} \end{aligned} \tag{14.14}$$

The common-mode rejection ratio (CMRR; see Section 11.6) is given in units of dB by

$$\begin{aligned} \text{CMRR}_{\text{dB}} &= \left| \frac{A_{\text{dif}}}{A_{\text{com}}} \right| = 20 \log_{10} \left| \frac{A_{\text{dif}}}{v_{\text{out, com}}/v_{\text{com}}} \right| \\ &= 20 \log_{10} \left| \frac{A_{\text{dif}}}{\frac{R_F}{R} \left(\frac{R+R_F}{R_F+R+\Delta R} - 1 \right) A} \right| \end{aligned} \tag{14.15}$$

where A_{dif} is the *differential gain* (which is usually assumed equal to the nominal design value). Since the common-mode gain, $v_{\text{out, com}}/v_{\text{com}}$, should ideally be zero, the theoretical CMRR for the instrumentation amplifier with perfectly matched resistors is infinite. In fact, even a small mismatch in the resistors used would dramatically reduce the CMRR, as the drill exercises at the end of this subsection illustrate. Even with resistors having 1 percent tolerance, the maximum CMRR that could be attained for typical values of resistors and an overall gain of 1,000 would be only 60 dB. In many practical applications, a requirement for a CMRR of 100 or 120 dB is not uncommon, and these would demand resistors of 0.01 percent tolerance (see Drill Exercise 14.3). It should be evident, then, that the "discrete" design of the IA, employing three op-amps and discrete resistors, will not be adequate for the more demanding instrumentation applications.

Example 14.1

Compute the common-mode gain and CMRR for the amplifier of Figure 14.16 if $A = 10$, $R = 1\ \text{k}\Omega$, $R_F = 10\ \text{k}\Omega$, and $\Delta R = 2$ percent.

Solution:

To compute the gain of the common-mode signal component, we need to compute the ratio $v_{\text{out, com}}/v_{\text{com}}$. Using equation 14.14 and the numerical values given, we compute

$$\frac{v_{\text{out, com}}}{v_{\text{com}}} = 100 \left(\frac{11}{11.02} - 1 \right) = -0.1815$$

Now we can compute the CMRR in units of dB from equation 14.15 with $A_{\text{dif}} = A \times \frac{R_F}{R} = 100$:

$$\begin{aligned}
\text{CMRR}_{\text{dB}} &= \left|\frac{A_{\text{dif}}}{A_{\text{com}}}\right| = 20\log_{10}\left|\frac{A_{\text{dif}}}{v_{\text{out, com}}/v_{\text{com}}}\right| \\
&= 20\log_{10}\left|\frac{A_{\text{dif}}}{\frac{R_F}{R}\left(\frac{R+R_F}{R_F+R+\Delta R}-1\right)A}\right| \\
&= 20\log_{10}\left|\frac{100}{\frac{10}{1}\left(\frac{11}{11.02}-1\right)10}\right| = 54.82\text{ dB}
\end{aligned}$$

The general expression for the CMRR of the instrumentation amplifier of Figure 14.16, without assuming any of the resistors are matched, except for R_2 and R_2', is

$$\text{CMRR} = \left|\frac{A_{\text{dif}}}{A_{\text{com}}}\right| = \left|\frac{(R_F/R)(1+2R_2/R_1)}{\frac{R_F}{R}\left[\frac{R_F'}{R_F}\left(\frac{R_F+R}{R_F'+R'}\right)-1\right]}\right| \tag{14.16}$$

and it can easily be shown that the CMRR is infinite if the resistors are perfectly matched.

Example 14.1 illustrated some of the problems that are encountered in the design of instrumentation amplifiers using discrete components. Many of these problems can be dealt with very effectively if the entire instrumentation amplifier is designed into a single *monolithic integrated circuit,* where the resistors can be carefully matched by appropriate fabrication techniques and many other problems can also be avoided. The functional structure of an IC instrumentation amplifier is depicted in Figure 14.17. Specifications for a common IC instrumentation amplifier (and a more accurate circuit description) are shown in Figure 14.18. Among the features worth mentioning here are the programmable gains, which the user can set by suitably connecting one or more of the resistors labeled R_1 to the appropriate connection. Note that the user may also choose to connect additional resistors to control the amplifier gain, without adversely affecting the amplifier's performance, since R_1 requires no matching. In addition to the pin connection that permits programmable gains, two additional pins are provided, called **sense** and **reference.** These additional connections are provided to the user for the purpose of referencing the output voltage to a signal other than ground, by means of the reference terminal, or of further amplifying the output current (e.g., with a transistor stage), by connecting the sense terminal to the output of the current amplifier.

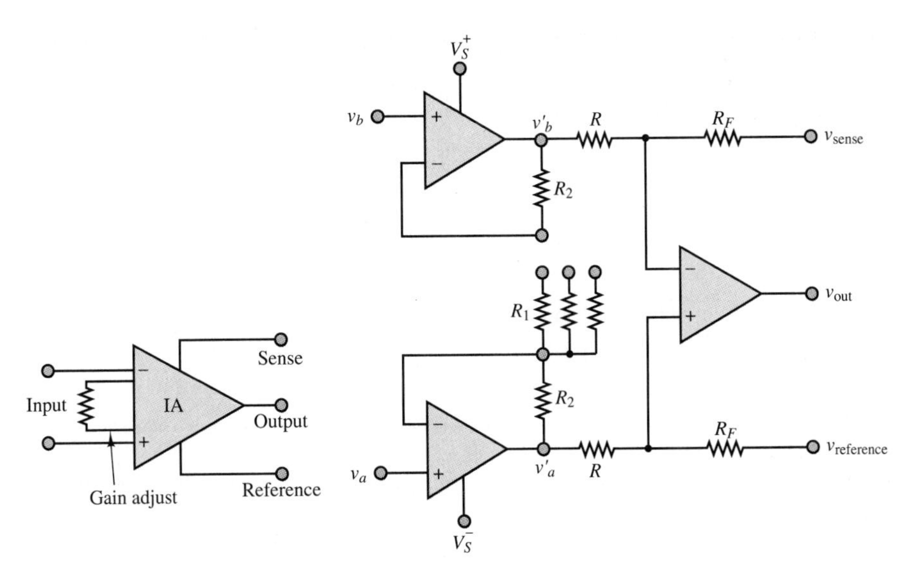

Figure 14.17 IC instrumentation amplifier

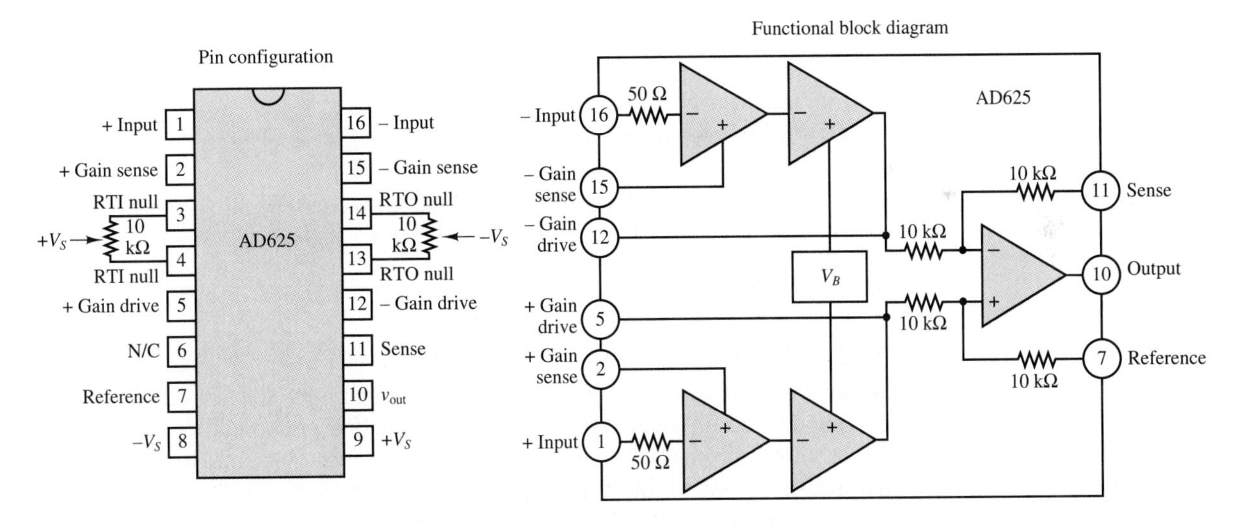

Figure 14.18 AD625 instrumentation amplifier data sheet

Example 14.2

Consider the IA of Figure 14.17, with the following resistor values: $R_F = R = 10\ \text{k}\Omega$, $R_2 = 20\ \text{k}\Omega$, and the following three choices for R_1: 80.2 Ω, 201 Ω, and 404 Ω. What are the possible gains that can be programmed with this amplifier?

Solution:

Recall that the gain of the input stage (for each of the differential inputs) is given by

$$A = 1 + \frac{2R_2}{R_1}$$

Thus, by connecting each of the three resistors, one can obtain the following gains:

$$A_1 = 1 + \frac{40{,}000}{80.2} = 500 \quad A_2 = 1 + \frac{40{,}000}{201} = 200 \quad A_3 = 1 + \frac{40{,}000}{404} = 100$$

It is also possible to obtain additional gains by connecting the resistors in parallel: $80.2 \parallel 201 = 57.3\ \Omega$ ($A_4 \approx 700$), $80.2 \parallel 404 = 66.9\ \Omega$ ($A_5 \approx 600$), $404 \parallel 201 = 134.2\ \Omega$ ($A_6 \approx 300$).

DRILL EXERCISES

14.1 Use the definition of the common-mode rejection ratio (CMRR) given in equation 14.16 to compute the CMRR (in dB) of the amplifier of Example 14.1 if $R_F/R = 100$ and $A = 10$, and if $\Delta R = 5\%$ of R. Assume $R = 1\ \text{k}\Omega$, $R_F = 100\ \text{k}\Omega$.

14.2 Repeat Drill Exercise 14.1 for a 1 percent variation in R.

14.3 Repeat Drill Exercise 14.1 for a 0.01 percent variation in R.

14.4 Calculate the mismatch in gains for the differential components for the 5 percent resistance mismatch of Drill Exercise 14.1.

14.5 Calculate the mismatch in gains for the differential components for the 1 percent resistance mismatch of Drill Exercise 14.2.

14.6 What value of resistance R_1 would permit a gain of 1,000 for the IA of Example 14.2?

Active Filters

The need to filter sensor signals that may be corrupted by noise or other interfering or undesired inputs has already been approached in two earlier chapters. In Chapter 6, simple passive filters made of resistors, capacitors, and inductors were analyzed. It was shown that three types of filter frequency response characteristics can be achieved with these simple circuits: low-pass, high-pass, and band-pass. In Chapter 11, the concept of active filters was introduced, to suggest that it may be desirable to exploit the properties of operational amplifiers to simplify filter design, to more easily match source and load impedances, and to eliminate the need for inductors. The aim of this section is to discuss more advanced active filter designs, which find widespread application in instrumentation circuits.

Figure 14.19 depicts the general characteristics of a low-pass active filter, indicating that within the pass-band of the filter, a certain deviation from the

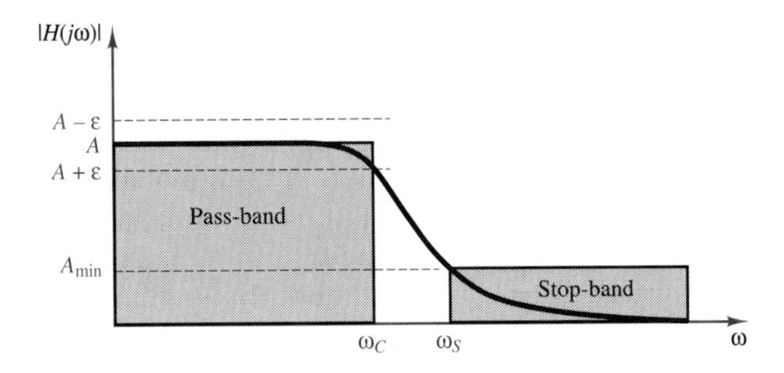

Figure 14.19 Prototype low-pass filter response

nominal filter gain, A, is accepted, as indicated by the minimum and maximum pass-band gains, $A + \varepsilon$ and $A - \varepsilon$. The width of the pass-band is indicated by the cutoff frequency, ω_C. On the other hand, the stop-band, starting at the frequency ω_S, does not allow a gain greater than $A_{\min}$. Different types of filter designs achieve different types of frequency responses, which are typically characterized either by having a particularly flat pass-band frequency response **(Butterworth filters),** or by a very rapid transition between pass-band and stop-band **(Chebyshev filters,** and **Cauer,** or **elliptical, filters),** or by some other characteristic, such as a very linear phase response **(Bessel filters).** Achieving each of these properties usually involves trade-offs; for example, a very flat pass-band response will usually result in a relatively slow transition from pass-band to stop-band.

In addition to selecting a filter from a certain family, it is also possible to select the *order* of the filter; this is equal to the order of the differential equation that describes the input-output relationship of a given filter. In general, the higher the order, the faster the transition from pass-band to stop-band (at the cost of greater phase shifts and amplitude distortion, however). Although the frequency response of Figure 14.19 pertains to a low-pass filter, similar definitions also apply to the other types of filters.

Butterworth filters are characterized by a *maximally flat* pass-band frequency response characteristic; their response is defined by a magnitude-squared function of frequency:

$$|H(j\omega)|^2 = \frac{H_0^2}{1 + \varepsilon^2 \omega^{2n}} \tag{14.17}$$

where $\varepsilon = 1$ for maximally flat response and n is the order of the filter. Figure 14.20 depicts the frequency response (normalized to $\omega_C = 1$) of first-, second-, third-, and fourth-order Butterworth low-pass filters. The **Butterworth polynomials,** given in Table 14.3 in factored form, permit the design of the filter by specifying the denominator as a polynomial in s. For $s = j\omega$, one obtains the frequency response of the filter. Examples 14.4 and 14.5 illustrate filter design procedures that make use of these tables.

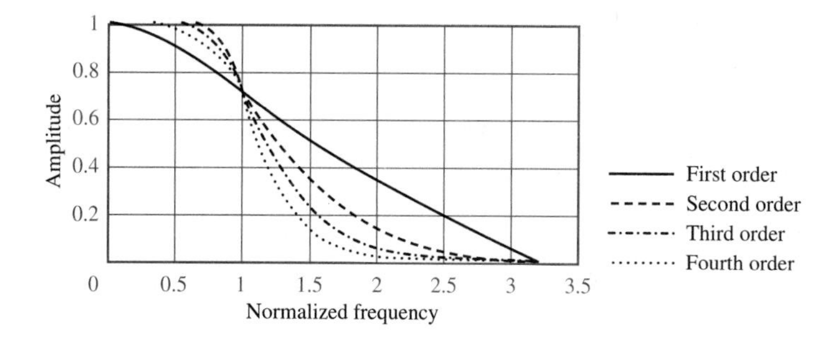

Figure 14.20 Butterworth low-pass filter frequency response

Table 14.3 **Butterworth polynomials in quadratic form**

Order, *n*	Quadratic factors
1	$(s + 1)$
2	$(s^2 + \sqrt{2}s + 1)$
3	$(s + 1)(s^2 + s + 1)$
4	$(s^2 + 0.7654s + 1)(s^2 + 1.8478s + 1)$
5	$(s + 1)(s^2 + 0.6180s + 1)(s^2 + 1.6180s + 1)$

Figure 14.21 depicts the normalized frequency response of first- to fourth-order low-pass Chebyshev filters ($n = 1$ to 4), for $\varepsilon = 1.06$. Note that a certain amount of ripple is allowed in the pass-band; the amplitude of the ripple is defined by the parameter ε, and is constant throughout the pass-band. Thus, these filters are also called **equiripple filters.** Cauer, or elliptical, filters are similar to Chebyshev filters, except for being characterized by equiripple both in the pass-band and in the stop-band. Design tables exist to select the appropriate order of Butterworth, Chebyshev, or Cauer filter for a specific application.

Three common configurations of second-order active filters, which can be used to implement **second-order** (or **quadratic**) **filter sections** using a single

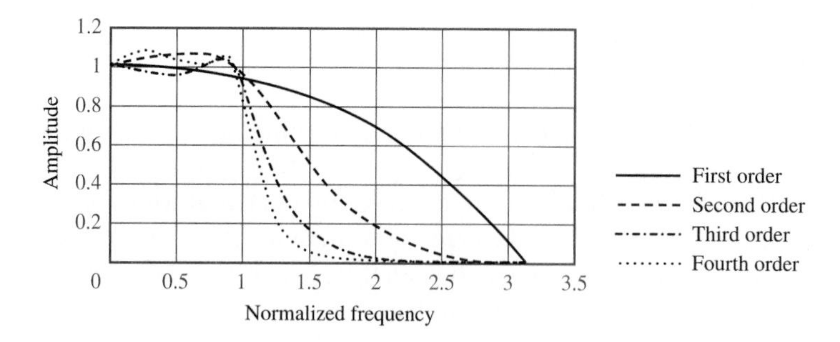

Figure 14.21 Chebyshev low-pass filter frequency response

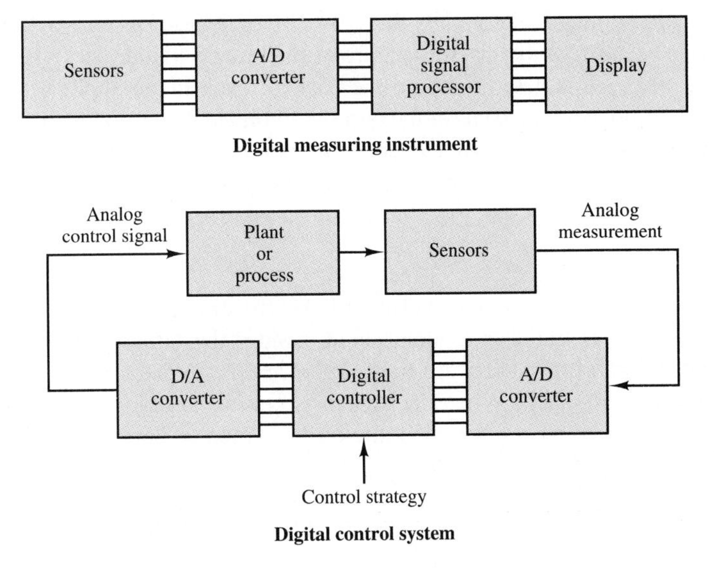

Figure 14.23 Block diagrams of a digital measuring instrument and a digital control system

general appearance of a digital measuring instrument and of a digital controller acting on a plant or process.

The objective of this section is to describe how the digital-to-analog (D/A) and analog-to-digital (A/D) conversion blocks of Figure 14.23 function. After illustrating discrete circuits that can implement simple A/D and D/A converters, we shall emphasize the use of ICs specially made for these tasks. Nowadays, it is uncommon (and impractical) to design such circuits using discrete components: the performance and ease of use of IC packages make them the preferred choice in virtually all applications.

Digital-to-Analog Converters

We discuss digital-to-analog conversion first because it is a necessary part of analog-to-digital conversion in many A/D conversion schemes. A **digital-to-analog converter (DAC)** will convert a binary word to an analog output voltage (or current). The binary word is represented in terms of 1s and 0s, where typically (but not necessarily), 1s correspond to a 5-volt level and 0s to a 0-volt signal. As an example, consider a four-bit binary word:

$$B = (b_3 b_2 b_1 b_0)_2 = (b_3 \cdot 2^3 + b_2 \cdot 2^2 + b_1 \cdot 2^1 + b_0 \cdot 2^0)_{10} \qquad \textbf{(14.21)}$$

The analog voltage corresponding to the digital word B would be

$$v_a = (8b_3 + 4b_2 + 2b_1 + b_0)\,\delta v \qquad \textbf{(14.22)}$$

where δv is the smallest *step size* by which v_a can increment. This least step size will occur whenever the least significant bit (LSB), b_0, changes from 0 to 1, and is the smallest increment the digital number can make. We shall also shortly see that the analog voltage obtained by the D/A conversion process has a "staircase" appearance because of the discrete nature of the binary signal.

The step size is determined on the basis of each given application, and is usually determined on the basis of the number of bits in the digital word to be converted to an analog voltage. We can see that, by extending the previous example for an n-bit word, the maximum value v_a can attain is

$$\begin{aligned} v_{a\,\max} &= (2^{n-1} + 2^{n-2} + \cdots + 2^1 + 2^0)\,\delta v \\ &= (2^n - 1)\,\delta v \end{aligned} \qquad \textbf{(14.23)}$$

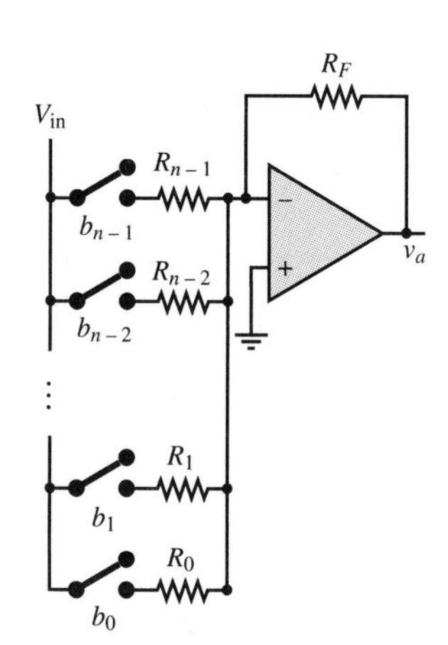

Figure 14.24 n-bit digital-to-analog converter (DAC)

It is relatively simple to construct a DAC by taking advantage of the summing amplifier illustrated in Chapter 11. Consider the circuit shown in Figure 14.24, where each bit in the word to be converted is represented by means of a 5-V source and a switch. When the switch is closed, the bit takes a value of 1 (5 V); when the switch is not closed, the bit has value 0. Thus, the output of the DAC is proportional to the word $b_{n-1}b_{n-2}\ldots b_1b_0$.

You will recall that a property of the summing amplifier is that the sum of the currents at the inverting node is zero, yielding the relationship

$$v_a = -\left(\frac{R_F}{R_i} \cdot b_i \cdot 5\right) \qquad i = 0, 1, \ldots, n-1 \qquad \textbf{(14.24)}$$

where R_i is the resistor associated with each bit and b_i is the decimal value of the ith bit (i.e., $b_0 = 2^0$, $b_1 = 2^1$, and so on). It is easy to verify that if we select

$$R_i = \frac{R_0}{2^i} \qquad \textbf{(14.25)}$$

we can obtain weighted gains for each bit so that

$$v_a = -\frac{R_F}{R_0}(2^{n-1}b_{n-1} + \cdots + 2^1 b_1 + 2^0 b_0) \cdot 5\ \text{V} \qquad \textbf{(14.26)}$$

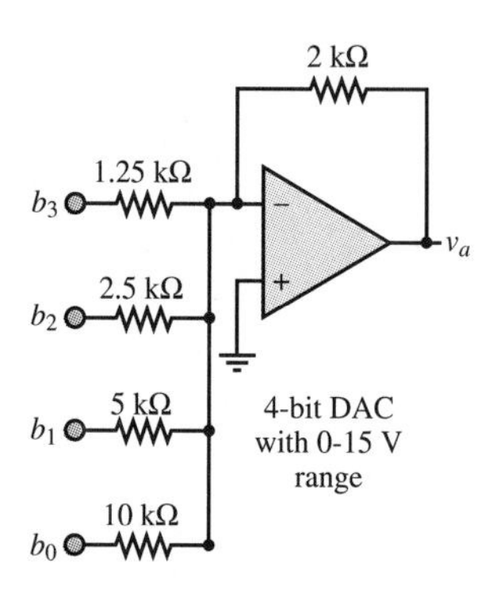

Figure 14.25 Four-bit DAC

and so that the analog output voltage is proportional to the decimal representation of the binary word. As an illustration, consider the case of a four-bit word; a reasonable choice for R_0 might be $R_0 = 10$ kΩ, yielding a resistor network consisting of 10-, 5-, 2.5-, and 1.25-kΩ resistors, as shown in Figure 14.25. The largest decimal value of a four-bit word is $2^4 - 1 = 15$, and so it is reasonable to divide this range into steps of 1 volt (i.e., $\delta v = 1$ V). Thus, the full-scale value of v_a is 15 V:

$$0 \le v_a \le 15\ \text{V}$$

and we select R_F according to the following expression:

$$R_F = \frac{\delta v\, R_0}{5} = 2\ \text{k}\Omega$$

The corresponding four-bit DAC is shown in Figure 14.25.

The DAC transfer characteristic is such that the analog output voltage, v_a, has a steplike appearance, because of the discrete nature of the binary signal. The coarseness of the "staircase" can be adjusted by selecting the number of bits in the binary representation.

Example 14.6

Assume that the maximum analog voltage ($v_{a\ max}$) of an eight-bit DAC is 12 volts. Find the smallest step size (δv) by which v_a can increment.

Solution:

On the basis of the discussion in this section, we can use the following formula:

$$\delta v = \frac{v_{a\ max}}{2^8 - 1}$$

Thus, applying the formula, we obtain:

$$\delta v = \frac{12}{2^8 - 1} = 47.1 \text{ mV}$$

The practical design of a DAC is generally not carried out in terms of discrete components, because of problems such as the accuracy required of the resistor value. Many of the problems associated with this approach can be solved by designing the complete DAC circuit in integrated circuit (IC) form. The specifications stated by the IC manufacturer include the resolution, that is, the minimum nonzero voltage; the full-scale accuracy; the output range; the output settling time; the power supply requirements; and the power dissipation. The following example illustrates the use of a common integrated circuit DAC.

Example 14.7

Define the following concepts for a DAC, and show how they can be used to find the minimum number of bits required.

1. Range
2. Resolution

Solution:

1. The maximum output voltage is produced when all bits are logic 1s; it can be found from the following equation:

$$V_{max} = V_{in}\frac{R_F}{R_0}(2^n - 1)$$

   where V_{in} is the voltage corresponding to logic 1 and n is the number of bits. The minimum output voltage is produced when all bits are logic 0. In this case, $V_{in} = 0$. Thus, the range of the DAC is from zero to V_{max}.
2. Resolution is the minimum nonzero voltage that can be produced; it can be found by using the following equation:

$$\delta v = \frac{V_{\text{max}}}{2^n - 1} = V_{\text{in}} \frac{R_F}{R_0}$$

where δv is the resolution of the DAC.

If both the range and the resolution are specified, the minimum number of bits required in a DAC can be found by using the following equation:

$$n = \frac{\log\left(\frac{V_{\text{max}} - V_{\text{min}}}{\delta v} + 1\right)}{\log 2}$$

Because n must be an integer, any fractional part should be rounded up to the next higher integer.

Example 14.8

A typical DAC one would use in conjunction with the 8086 microprocessor is the AD558. This is an IC that can be used in a "stand-alone" configuration (without a microprocessor) or with a microprocessor interfaced to it. Using the data supplied in the data sheets at the end of the chapter, answer the following questions:

1. If one were to set up the AD558 for an output swing of 0 to 10 volts, what would be the smallest voltage output increment attainable?
2. On what is the maximum operating frequency (the largest frequency on which the DAC can perform conversion) of the AD558 dependent? Determine the maximum frequency attainable if the converter is to be run at full-scale input.

Solution:

1. Since this DAC is an eight-bit device, the total number of digital increments one could expect is 256. Thus, the finest voltage steps one could expect at the output would be:

$$\frac{10}{255} = 39.2 \text{ mV}$$

This means that for every increment of one bit, the output would jump (in a stepwise fashion) by 39.2 mV.

2. The maximum frequency at which a DAC can run is dependent on the settling time. This is defined as the time it takes for the output to settle to within one half of the least significant bit of its final value. Thus, only one transition can be made per settling time. The settling time for the AD558 depends on the voltage range and is defined for a positive-going full-scale step to $\pm\frac{1}{2}$ LSB. The settling time is 1 μs, and the corresponding maximum conversion frequency is 1 MHz.

Analog-to-Digital Converters

You should by now have developed an appreciation for the reasons why it is convenient to process data in digital form. The device that makes conversion of analog signals to digital form is the **analog-to-digital converter (ADC),** and, just like the DAC, it is also available as a single IC package. This section will illustrate the essential features of four types of ADCs: the tracking ADC, which utilizes a DAC to perform the conversion; the integrating ADC; the flash ADC; and (later) the successive-approximation ADC. In addition to discussing analog-to-digital conversion, we shall also introduce the *sample-and-hold amplifier.*

Quantization

The process of converting an analog voltage (or current) to digital form requires that the analog signal be quantized and encoded in binary form. The process of **quantization** consists of subdividing the range of the signal into a finite number of intervals; usually, one employs $2^n - 1$ intervals, where n is the number of bits available for the corresponding binary word. Following this quantization, a binary word is assigned to each interval (i.e., to each range of voltages or currents); the binary word is then the digital representation of any voltage (current) that falls within that interval. You will note that the smaller the interval, the more accurate the digital representation is. However, some error is necessarily always present in the conversion process; this error is usually referred to as **quantization error.** Let v_a represent the analog voltage and v_d its quantized counterpart, as shown in Figure 14.26 for an analog voltage in the range 0–16 V. In the figure, the analog voltage v_a takes on a value of $v_d = 0$ whenever it is in the range 0–1 V; for $1 \leq v_a < 2$, the corresponding value is $v_d = 1$; for $2 \leq v_a < 3$, $v_d = 2$; and so on, until, for $15 \leq v_a < 16$, we have $v_d = 15$. You see that if we now represent the quantized voltage v_d by its binary counterpart, as shown in the table of Figure 14.26, each 1-volt analog interval corresponds to a unique binary word. In this example, a four-bit word is sufficient to represent the analog voltage, although the representation is not very accurate. As the number of bits increases, the quantized voltage is closer and closer to the original analog signal; however, the number of bits required to represent the quantized value increases.

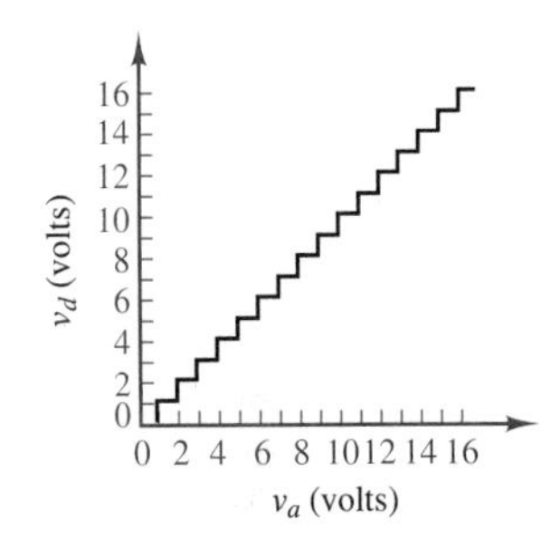

Quantized voltage v_d	Binary representation b_3	b_2	b_1	b_0
0	0	0	0	0
1	0	0	0	1
2	0	0	1	0
3	0	0	1	1
4	0	1	0	0
⋮		⋮		
14	1	1	1	0
15	1	1	1	1

Figure 14.26 A digital voltage representation of an analog voltage

Tracking ADC

Although not the most efficient in all applications, the **tracking ADC** is an easy starting point to illustrate the operation of an ADC, in that it is based on the DAC presented in the previous section. The tracking ADC, shown in Figure 14.27, compares the analog input signal with the output of a DAC; the comparator output determines whether the DAC output is larger or smaller than the analog input to be converted to binary form. If the DAC output is smaller, then the comparator output will cause an up-down counter (see Chapter 13) to count up, until it reaches a level close to the analog signal; if the DAC output is larger than the analog signal, then the counter is forced to count down. Note that the rate at which the up-down counter is incremented is determined by the external clock, and that

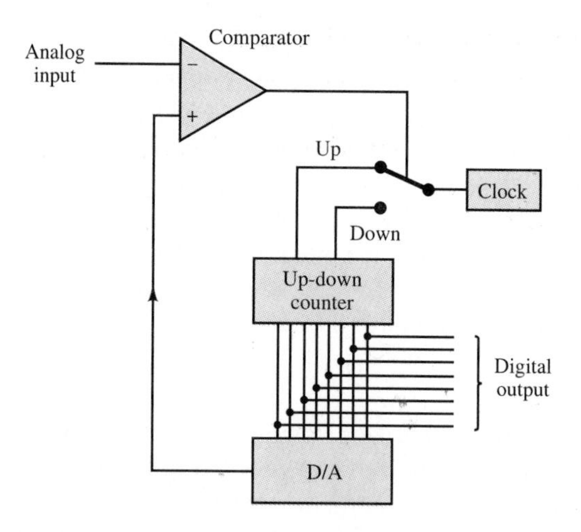

Figure 14.27 Tracking ADC

the binary counter output corresponds to the binary representation of the analog signal. A feature of the tracking ADC is that it follows ("tracks") the analog signal by changing one bit at a time.

Integrating ADC

The **integrating ADC** operates by charging and discharging a capacitor, according to the following principle: if one can ensure that the capacitor charges (discharges) linearly, then the time it will take for the capacitor to discharge is linearly related to the amplitude of the voltage that has charged the capacitor. In practice, to limit the time it takes to perform a conversion, the capacitor is not required to charge fully. Rather, a clock is used to allow the input (analog) voltage to charge the capacitor for a short period of time, determined by a fixed number of clock pulses. Then the capacitor is allowed to discharge through a known circuit, and the corresponding clock count is incremented until the capacitor is fully discharged. The latter condition is verified by a comparator, as shown in Figure 14.28. The clock count accumulated during the discharge time is proportional to the analog voltage.

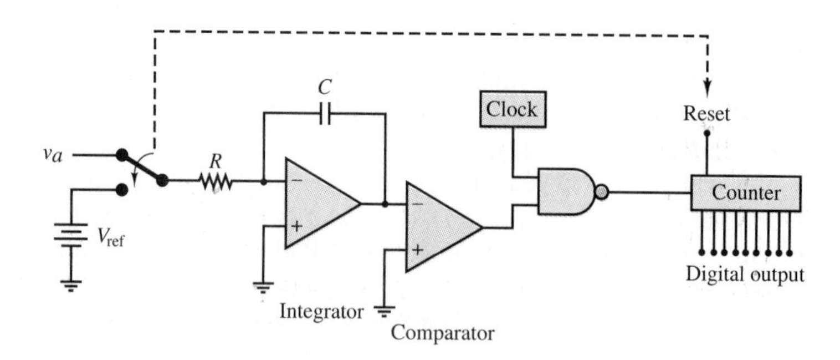

Figure 14.28 Integrating ADC

In the figure, the switch causes the counter to reset when it is connected to the reference voltage, V_{ref}. The reference voltage is used to provide a known, linear discharge characteristic through the capacitor (see the material on the op-amp integrator in Chapter 11). When the comparator detects that the output of the integrator is equal to zero, it switches state and disables the NAND gate, thus stopping the count. The binary counter output is now the digital counterpart of the voltage v_a.

Other common types of ADC are the so-called successive-approximation ADC and the flash ADC.

Flash ADC

The **flash ADC** is fully parallel and is used for high-speed conversion. A resistive divider network of 2^n resistors divides the known voltage range into that many equal increments. A network of $2^n - 1$ comparators then compares the unknown voltage with that array of test voltages. All comparators with inputs exceeding the unknown are "on"; all others are "off." This comparator code can be converted to conventional binary by a digital priority encoder circuit. For example, assume that the three-bit flash ADC of Figure 14.29 is set up with $V_{ref} = 8$ V. An input of 6.2 V is provided. If we number the comparators from the top of Figure 14.29, the state of each of the seven comparators is as given in Table 14.4.

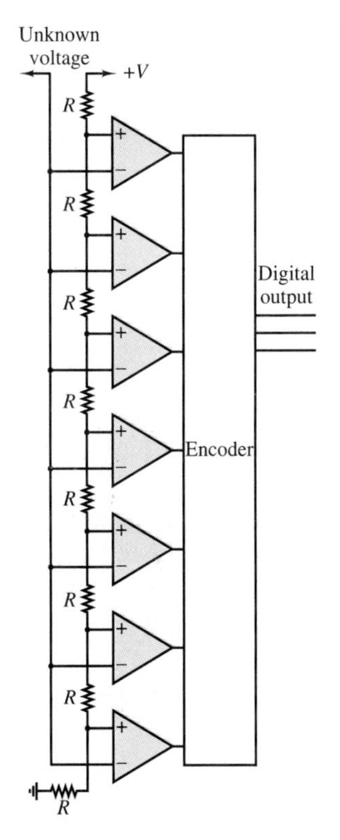

Figure 14.29 A three-bit flash ADC

Table 14.4 **State of comparators in a 3-bit flash ADC**

Comparator	Input on + line	Input on − line	Output
1	7 V	6.2 V	H
2	6 V	6.2 V	L
3	5 V	6.2 V	L
4	4 V	6.2 V	L
5	3 V	6.2 V	L
6	2 V	6.2 V	L
7	1 V	6.2 V	L

Example 14.9

Answer the following questions for the flash ADC:

1. How many parallel comparators do you need for a four-bit flash ADC?
2. State the advantage and the disadvantage of using a flash ADC.

Solution:

1. The flash ADC converter uses $2^n - 1$ parallel comparators to determine simultaneously all n bits of the digital output. Thus, 15 comparators are needed.
2. The flash converter has the advantage of high speed because of its parallel comparators. The disadvantage is that it is costly to use this technique for a large number of bits, because of the number of comparators required.

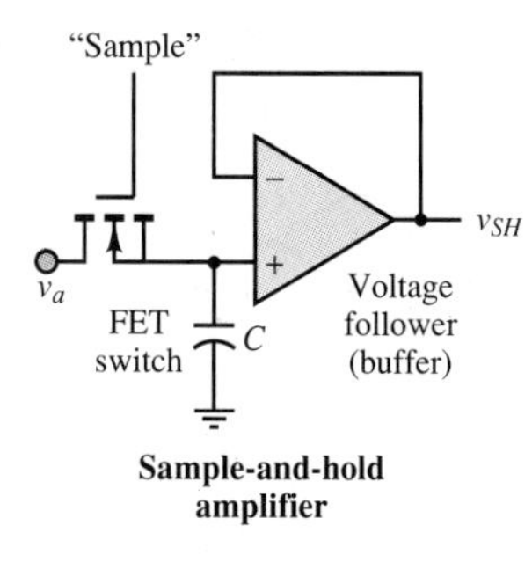

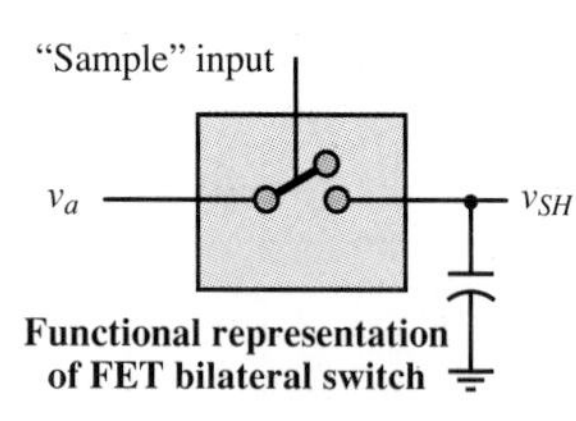

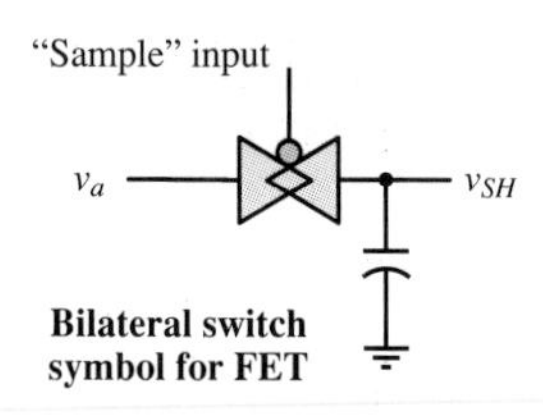

Figure 14.30 Description of the sample-and-hold process

In the preceding discussion, we explored a few different techniques for converting an analog voltage to its digital counterpart; these methods—and any others—require a certain amount of time to perform the A/D conversion. This is the ADC **conversion time,** and is usually quoted as one of the main specifications of an ADC device. A natural question at this point would be: If the analog voltage changes during the analog-to-digital conversion and the conversion process itself takes a finite time, how fast can the analog input signal change while still allowing the ADC to provide a meaningful digital representation of the analog input? To resolve the uncertainty generated by the finite ADC conversion time of any practical converter, it is necessary to use a sample-and-hold amplifier. The objective of such an amplifier is to "freeze" the value of the analog waveform for a time sufficient for the ADC to complete its task.

A typical sample-and-hold amplifier is shown in Figure 14.30. It operates as follows. A MOSFET analog switch (see Chapter 9) is used to "sample" the analog waveform. Recall that when a voltage pulse is provided to the sample input of the MOSFET switch (the gate), the MOSFET enters the ohmic region and in effect becomes nearly a short circuit for the duration of the sampling pulse. While the MOSFET conducts, the analog voltage, v_a, charges the "hold" capacitor, C, at a fast rate through the small "on" resistance of the MOSFET. The duration of the sampling pulse is sufficient to charge C to the voltage v_a. Because the MOSFET is virtually a short circuit for the duration of the sampling pulse, the charging (RC) time constant is very small, and the capacitor charges very quickly. When the sampling pulse is over, the MOSFET returns to its nonconducting state, and the capacitor holds the sampled voltage without discharging, thanks to the extremely high input impedance of the voltage-follower (buffer) stage. Thus, v_{SH} is the sampled-and-held value of v_a at any given sampling time.

Example 14.10

Using the data sheet for the AD582 sample-and-hold amplifier, answer the following questions:

1. What is the acquisition time?
2. How can the acquisition time be reduced?
3. What is the acquisition time for the AD582?

Solution:

1. The acquisition time, T, is the time required for the output of the sample-and-hold amplifier to reach its final value (within specified error bounds) after it has been switched from the hold mode to the sample mode. It includes the switch delay time, the slewing interval, and the amplifier settling time.
2. The acquisition time can be reduced by reducing the value of the hold capacitor, C_H.
3. From the data sheet, $T = 6\ \mu\text{s}$.

The appearance of the output of a typical sample-and-hold circuit is shown in Figure 14.31, together with the analog signal to be sampled. The time interval between samples, or **sampling interval,** $t_n - t_{n-1}$, allows the ADC to perform the conversion and make the digital version of the sampled signal available, say, to a computer or to another data acquisition and storage system. The sampling interval needs to be at least as long as the A/D conversion time, of course, but it is reasonable to ask how frequently one needs to sample a signal to preserve its fundamental properties (e.g., peaks and valleys, "ringing," fast transition). One might instinctively be tempted to respond that it is best to sample as frequently as possible, within the limitations of the ADC, so as to capture all the features of the analog signal. In fact, this is not necessarily the best strategy. How should we select the appropriate sampling frequency for a given application? Fortunately, an entire body of knowledge exists with regard to sampling theory, which enables the practicing engineer to select the best sampling rate for any given application. Given the scope of this chapter, we have chosen not to delve into the details of sampling theory, but, rather, to provide the student with a statement of the fundamental result: the **Nyquist sampling criterion.** The Nyquist criterion states that to prevent aliasing[1] when sampling a signal, *the sample rate should be selected to be at least twice the highest-frequency component present in the signal.* Thus, if we were sampling an audio signal (say, music), we would have to sample at a frequency of at least 40 kHz (twice the highest audible frequency, 20 kHz). In practice, it is advisable to select sampling frequencies substantially greater than the Nyquist rate; a good rule of thumb is 5 to 10 times greater. The following example illustrates how the designer might take the Nyquist criterion into account in designing a practical A/D conversion circuit.

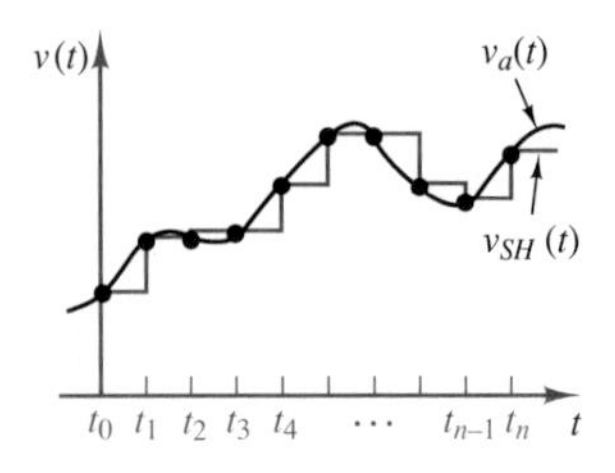

Figure 14.31 Sampled data

Example 14.11

A typical ADC one would use in conjunction with the 8086 microprocessor is the AD574. This is a successive-approximation converter. Using the data sheet supplied at the end of the chapter, answer the following questions:

1. What is the accuracy (in volts) the AD574 can provide if $V_{CC} = 15.0$ V and $0 \le V_{in} \le 15.0$ V?
2. On the basis of the data sheet, what is the highest-frequency signal you could convert using the AD574? (Assume that $V_{CC} = 15.0$ V.)
3. If the maximum conversion time available were 40 μs, what would be the highest-frequency signal you could expect to sample on the basis of the Nyquist criterion?

Solution:

1. According to the data sheet, the least significant bit (LSB) of this converter limits its accuracy, meaning that the output is accurate within ±1 bit. For the 0- to 15-V swing, this gives a voltage accuracy of

$$\frac{V_{max} - V_{min}}{2^n - 1} \quad \text{or} \quad \frac{15}{2^{12} - 1} \times (\pm 1 \text{ bit}) = \pm 3.66 \text{ mV}$$

[1]*Aliasing* is a form of signal distortion that occurs when an analog signal is sampled at an insufficient rate.

2. On the basis of the data sheet, the maximum conversion time is 35 μs. Therefore, the highest frequency of data conversion using the AD574 is

$$f_{\text{max}} = \frac{1}{35\ \mu\text{s}} = 28.57\ \text{kHz}$$

Thus, the highest signal frequency that could be represented, according to the Nyquist principle, is:

$$\frac{1}{2} f_{\text{max}} = \frac{28.57 \times 10^3}{2} = 14.285\ \text{kHz}$$

This is the maximum theoretical signal frequency that can be represented without distortion, according to the Nyquist principle.

3. Following the same procedure discussed in part 2,

$$\frac{1}{2} f_{\text{max}} = \left(\frac{1}{40 \times 10^{-6}}\right)\left(\frac{1}{2}\right) = 12.5\ \text{kHz}$$

Data Acquisition Systems

The structure of a data acquisition system, shown in Figure 14.32, can now be analyzed, at least qualitatively, since we have explored most of the basic building blocks. A typical data acquisition system often employs an *analog multiplexer,* to process several different input signals. A bank of bilateral analog MOSFET switches, such as the one we described together with the sample-and-hold amplifier, provides a simple and effective means of selecting which of the input signals should be sampled and converted to digital form. Control logic, employing standard gates and counters, is used to select the desired *channel* (input signal), and to trigger the sampling circuit and the ADC. When the A/D conversion is completed, the ADC sends an appropriate *end of conversion* signal to the control logic, thereby enabling the next channel to be sampled.

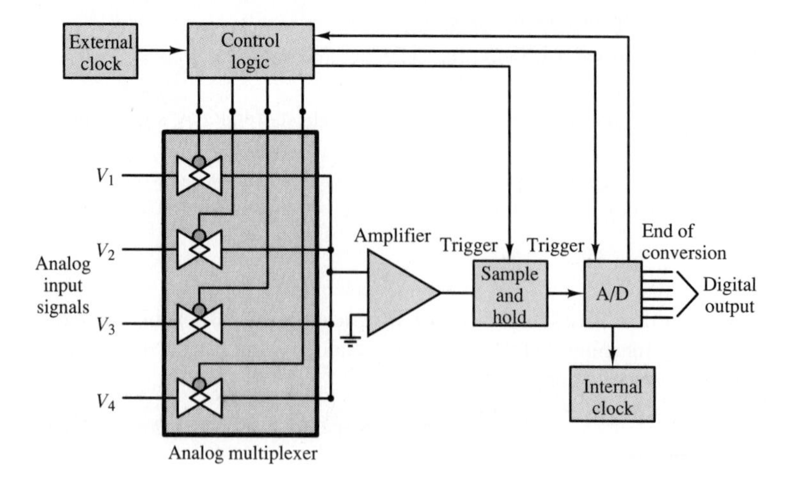

Figure 14.32 Data acquisition system

In the block diagram of Figure 14.32, four analog inputs are shown; if these were to be sampled at regular intervals, the sequence of events would appear as depicted in Figure 14.33. We notice, from a qualitative analysis of the figure, that the effective sampling rate for each channel is one fourth the actual external clock rate; thus, it is important to ensure that the sampling rate for each individual channel satisfies the Nyquist criterion. Further, although each sample is held for four consecutive cycles of the external clock, we must notice that the ADC can use only one cycle of the external clock to complete the conversion, since its services will be required by the next channel during the next clock cycle. Thus, the internal clock that times the ADC must be sufficiently fast to allow for a complete conversion of any sample within the design range. These and several other issues are discussed in the next example.

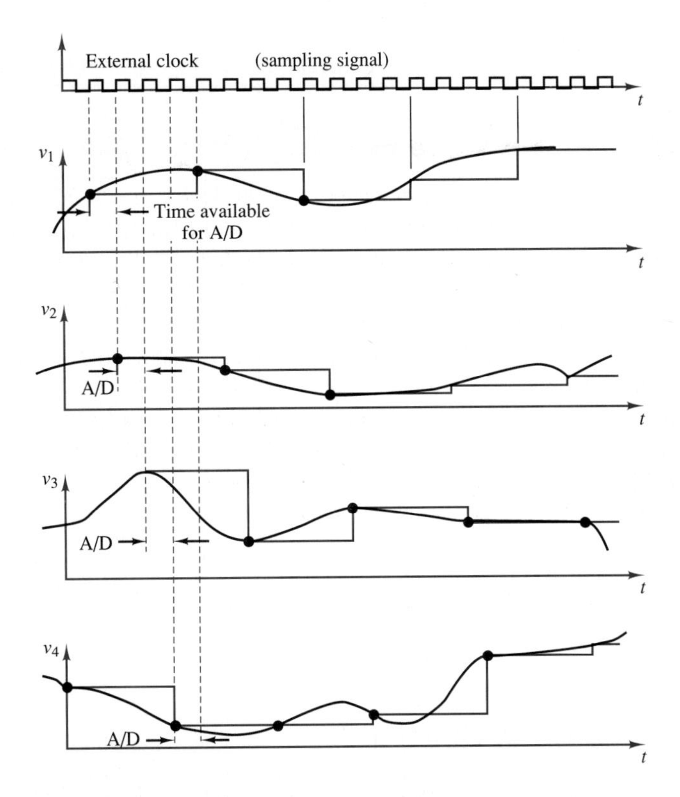

Figure 14.33 Multiplexed sampled data

Example 14.12 Data Acquisition System

This example discusses the internal structure of a typical data acquisition system, such as might be used in process monitoring and control, instrumentation, and test applications. The data acquisition system discussed in this example is the AT-MIO-16 from National Instruments. The AT-MIO-16 is a high-performance, multifunction analog, digital, and timing input/output (I/O) board for the IBM PC/AT and compatibles. It contains a 12-bit ADC with up to 16 analog inputs, two 12-bit DACs with voltage outputs, eight lines of transistor-transistor-logic (TTL)–compatible digital I/O, and three 16-bit counter/timer channels for timing I/O. If additional analog inputs are required, the AMUX-64T analog multiplexer can be used. By cascading up to four AMUX-64T's, 256 single-ended or 128 differential inputs can be obtained. The AT-MIO-16 also uses the RTSI bus (real-time system interface bus) to synchronize multiboard analog, digital, and counter/timer operations by communicating system-level timing signals between boards.

Figure 14.34 is a block diagram of the AT-MIO-16 circuitry. Its major functions are described next.

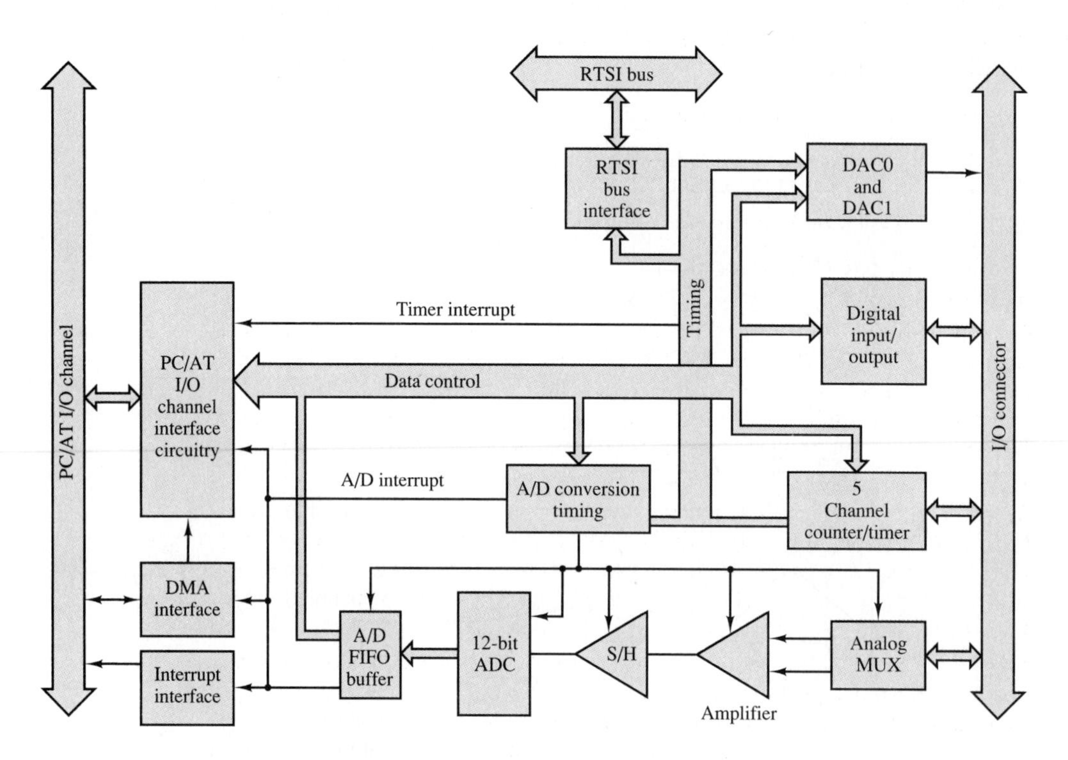

Figure 14.34 AT-MIO-16 block diagram

Analog Input: The AT-MIO-16 has two CMOS analog input multiplexers connected to 16 analog input channels. This data acquisition board has a software-programmable gain amplifier that can be used with voltage gains of 1, 10, 100, or 500 (the AT-MIO-16L) to accommodate low-level analog input signals, or with gains of 1, 2, 4, or 8 (the AT-MIO-16H) for high-level analog input signals. The AT-MIO-16 has a 12-bit ADC that gives an analog signal resolution of 4.88 mV with gain of 1 and an input range of ± 10 V. Finer resolutions up to 4.88 mV can be achieved by using gain and smaller input ranges. The board is available in three speeds: the AT-MIO-16(L/H)-9 contains a 9-μs ADC; the AT-MIO-16(L/H)-15 contains a 15-μs ADC; and the AT-MIO-16(L/H)-25 contains a 25-μs ADC. These conversions of the board have the following data acquisition sample rates on a single analog input channel:

Model number	Sampling rate	
	Typical case	Worst case
AT-MIO-16(L/H)-9	100 ksamples/s	91 ksamples/s
AT-MIO-16(L/H)-15	71 ksamples/s	59 ksamples/s
AT-MIO-16(L/H)-25	45 ksamples/s	37 ksamples/s

The timing of multiple A/D conversion is controlled either by the onboard counter/timer or by external timing signals. The onboard sample rate clock and sample counter control the onboard A/D timing. The AT-MIO-16 can generate both interrupts and DMA (direct memory access) requests on the PC/AT I/O channel. The interrupt can be generated when

1. An A/D conversion is available to be read from the A/D buffer.
2. The sample counter reaches its terminal count.
3. An error occurs.
4. One of the onboard timer clocks generates a pulse.

On the other hand, DMA requests can be generated whenever an A/D measurement is available from the A/D buffer.

Analog Output: The AT-MIO-16 has two double-buffered multiplying 12-bit DACs that are connected to two analog output channels. The resolution of the 12-bit DACs is 2.44 mV in the unipolar mode or 4.88 mV in the bipolar mode with the onboard 10-V reference. Finer resolutions can be achieved by using smaller voltages on the external reference. The analog output channels have an accuracy of ± 0.5 LSB and a differential linearity of ± 1 LSB. Voltage offset and gain error can be trimmed to zero.

Digital I/O: The AT-MIO-16 has eight digital I/O lines that are divided into two four-bit ports. The digital input circuitry has an eight-bit register that continuously reads the eight digital I/O lines, thus making read-back capability possible for the digital output ports, as well as reading incoming signals. The digital I/O lines are TTL-compatible.

Counter/Timer: The AT-MIO-16 uses the AM9513A counter/timer for time-related functions. The AM9513A contains five independent 16-bit counter/timers. A 1-MHz clock is the time baseline. Two of the AM9513A counter/timers are for multiple A/D conversion timing. The three remaining counters can be used for special data acquisition timing, such as expanding to a 32-bit sample counter or generating interrupts at user-programmable time intervals.

RTSI Bus Interface: The AT-MIO-16 is interfaced to the RTSI bus. You can send or receive the external analog input control signal; the waveform-generation timing signals; the output of counters 1, 2, and 5; the gate of counter 1; and the source of counter 5. You can send the RTSI bus the frequency output of the AM9513A.

PC/AT I/O Channel Interface: The PC/AT I/O channel interface circuitry includes address latches, address-decoding circuitry, data buffers, and interface timing and control signals.

I/O Connector: The I/O connector is a 50-pin male ribbon cable connector.

Software Support: The AT-MIO-16 also has software packages that control data acquisition functions on the PC-based data acquisition boards.

DRILL EXERCISES

14.10 Apply KCL at the inverting node of the summing amplifier of Figure 14.25 to show that equation 14.24 holds whenever $R_i = R_0/2^i$.

14.11 If the maximum analog voltage ($V_{a\,\max}$) of a 12-bit DAC is 15 volts, find the smallest step size (δv) by which v_a can increment.

14.12 Repeat Example 14.6 for the case of an eight-bit word with $R_0 = 10\ \text{k}\Omega$ and the same range of v_a. Find the value of δv and R_F. Assume that ideal resistor values are available.

14.13 For Figure 14.25, find $V_{\max}$ if $V_{\text{in}} = 4.5$ V.

14.14 For Figure 14.25, find the resolution if $V_{\text{ih}} = 3.8$ V.

14.15 Find the minimum number of bits required in a DAC if the range of the DAC is from 0.5 to 15 V and the resolution of the DAC is 20 mV.

14.16 In Example 14.11, if the maximum conversion time available to you were 50 μs, what would be the highest-frequency signal you could expect to sample on the basis of the Nyquist criterion?

14.5 COMPARATOR AND TIMING CIRCUITS

Timing and comparator circuits find frequent application in instrumentation systems. The aim of this section is to introduce the foundations that will permit the student to understand the operation of op-amp comparators and multivibrators, and of an integrated circuit timer.

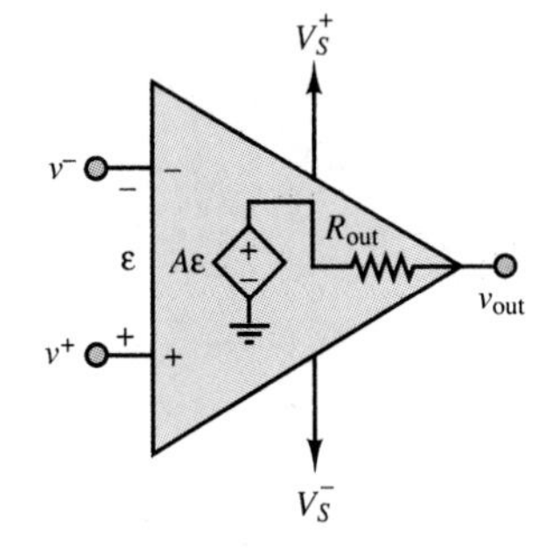

Figure 14.35 Op-amp in open-loop mode

The Op-Amp Comparator

The prototype of op-amp switching circuits is the op-amp comparator of Figure 14.35. This circuit, you will note, *does not employ feedback.* As a consequence of this,

$$v_{\text{out}} = A_{V(OL)}(v^+ - v^-) \tag{14.27}$$

Because of the large gain that characterizes the open-loop performance of the op-amp ($A_{V(OL)} > 10^5$), any small difference between input voltages, ε, will cause large outputs. In particular, for ε of the order of a few tens of microvolts, the op-amp will go into saturation at either extreme, according to the voltage supply values and the polarity of the voltage difference (recall the discussion of the op-amp voltage supply limitations in Section 11.6). For example, if ε were a 1-mV potential difference, the op-amp output would ideally be equal to 100 V, for an open-loop gain $A_{V(OL)} = 10^5$ (and in practice the op-amp would saturate at the voltage supply limits). Clearly, any difference between input voltages will cause the output to saturate toward either supply voltage, depending on the polarity of ε.

One can take advantage of this property to generate switching waveforms. Consider, for example, the circuit of Figure 14.36, in which a sinusoidal voltage source $v_{in}(t)$ of peak amplitude V is connected to the noninverting input. In this circuit, in which the inverting terminal has been connected to ground, the differential input voltage is given by

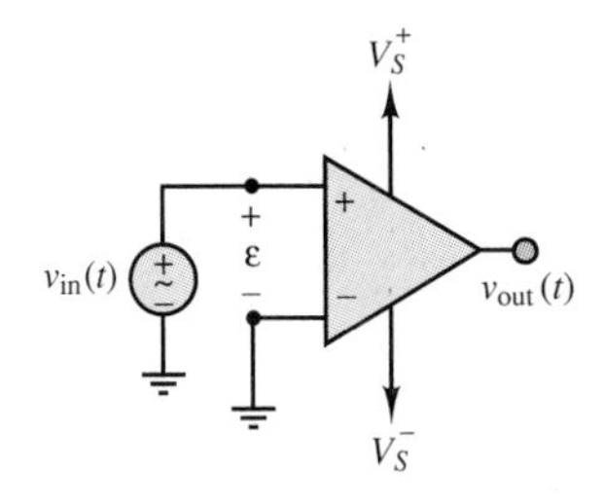

Figure 14.36 Noninverting op-amp comparator

$$\varepsilon = V \cos(\omega t) \tag{14.28}$$

and will be positive during the positive half-cycle of the sinusoid and negative during the negative half-cycle. Thus, the output will saturate toward V_S^+ or V_S^-, depending on the polarity of ε: the circuit is, in effect, *comparing* $v_{in}(t)$ and ground, producing a positive v_{out} when $v_{in}(t)$ is positive, and a negative v_{out} when $v_{in}(t)$ is negative, independent of the amplitude of $v_{in}(t)$ (provided, of course, that the peak amplitude of the sinusoidal input is at least 1 mV, or so). The circuit just described is therefore called a **comparator,** and in effect performs a binary decision, determining whether $v_{in}(t) > 0$ or $v_{in}(t) < 0$. The comparator is perhaps the simplest form of an *analog-to-digital converter,* that is, a circuit that converts a continuous waveform into discrete values. The comparator output consists of only two discrete levels: "greater than" and "less than" a reference voltage.

The input and output waveforms of the comparator are shown in Figure 14.37, where it is assumed that $V = 1$ V and that the saturation voltage corresponding to the $\pm$15-V supplies is approximately $\pm$13.5 V. This circuit will be termed a **noninverting comparator,** because a positive voltage differential, ε, gives rise to a positive output voltage. It should be evident that it is also possible to construct an inverting comparator by connecting the noninverting terminal to ground and connecting the input to the inverting terminal. Figure 14.38 depicts the waveforms for the **inverting comparator.** The analysis of any comparator circuit is greatly simplified if we observe that the output voltage is determined by the voltage difference present at the input terminals of the op-amp, according to the following relationship:

$$\begin{aligned} \varepsilon > 0 &\Rightarrow v_{out} = V_{sat}^+ \\ \varepsilon < 0 &\Rightarrow v_{out} = V_{sat}^- \end{aligned} \qquad \text{Operation of op-amp comparator} \tag{14.29}$$

where V_{sat} is the saturation voltage for the op-amp (somewhat lower than the supply voltage, as discussed in Chapter 11). Typical values of supply voltages for practical op-amps are $\pm$5 V to $\pm$24 V.

A simple modification of the comparator circuit just described consists of connecting a fixed reference voltage to one of the input terminals; the effect of the reference voltage is to raise or lower the voltage level at which the comparator will switch from one extreme to the other. Example 14.13 describes one such circuit.

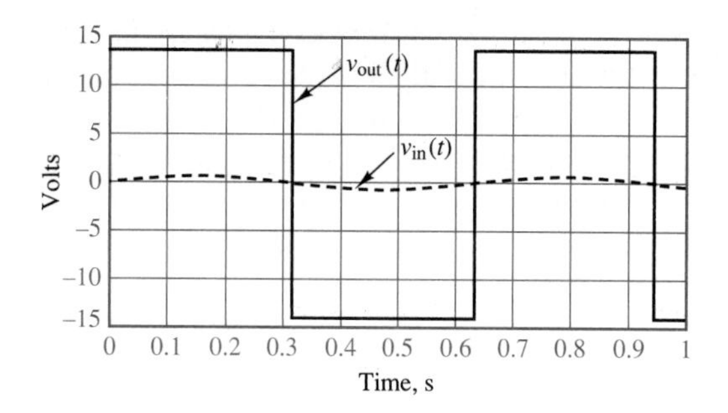

Figure 14.37 Input and output of noninverting comparator

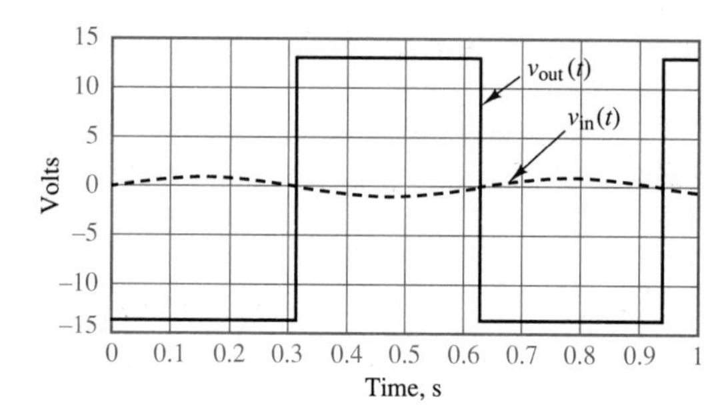

Figure 14.38 Input and output of inverting comparator

Example 14.13

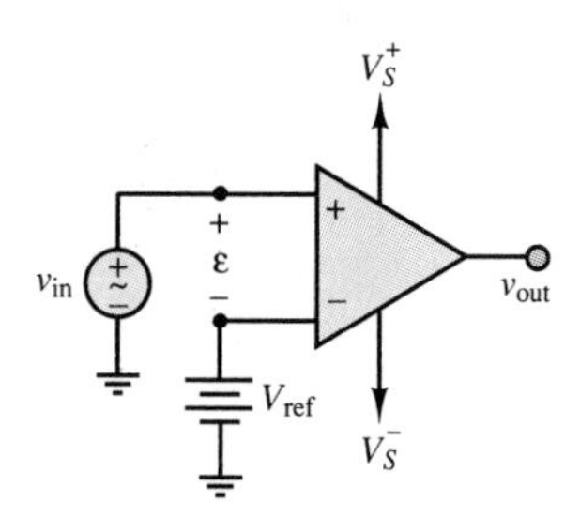

Figure 14.39

This example considers the effect of a reference DC voltage applied to one of the input terminals of a comparator. The circuit of Figure 14.39 depicts a noninverting comparator with a DC source connected to its inverting terminal. Let $v_{in}(t)$ be a 1-V peak sinusoid, and let $V_{ref} = 0.6$ V. Sketch the input and output waveforms of the comparator.

Solution:

The analysis of the circuit is best conducted by analyzing the differential voltage across the op-amp input terminals, ε. An expression for ε may be obtained as follows:

$$\varepsilon = v_{in} - V_{ref}$$

and it follows that, since the output will saturate to the positive extreme whenever ε is positive, the following conditions hold:

$$v_{in} > V_{ref} \Rightarrow v_{out} = V^+_{sat}$$
$$v_{in} < V_{ref} \Rightarrow v_{out} = V^-_{sat}$$

Thus, the comparator will switch levels whenever the sinusoidal voltage (1 V peak, in this example) exceeds 0.6 V, or falls below this level. Figure 14.40 depicts the relevant waveforms, also showing the 0.6-V level. Note that now the comparator output voltage is asymmetric—that is, it is no longer a symmetrical square wave. This type of circuit is sometimes called a **comparator with offset.**

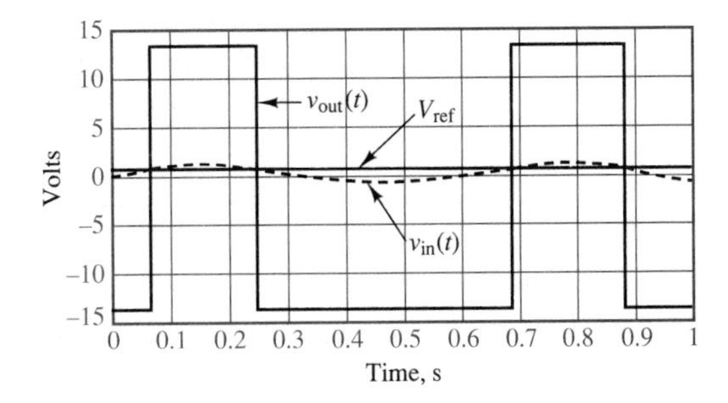

Figure 14.40 Waveforms of comparator with offset

Another useful interpretation of the op-amp comparator can be obtained by considering its **input-output transfer characteristic.** Figure 14.41 displays a plot of v_{out} versus v_{in} for a noninverting zero-reference (no offset) comparator. This circuit is often called a **zero-crossing comparator,** because the output voltage goes through a transition (V_{sat} to $-V_{sat}$, or vice versa) whenever the input voltage crosses the horizontal axis. You should be able to verify that Figure 14.42 displays the transfer characteristic for a comparator of the inverting type with a nonzero reference voltage.

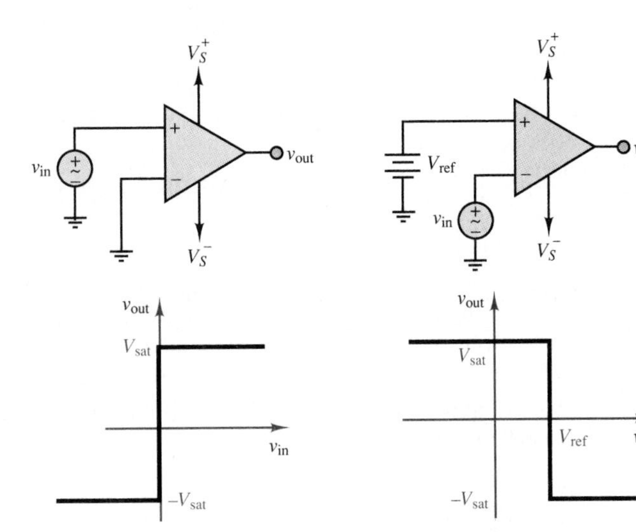

Figure 14.41 Transfer characteristic of zero-crossing comparator

Figure 14.42 Transfer characteristic of inverting comparator with offset

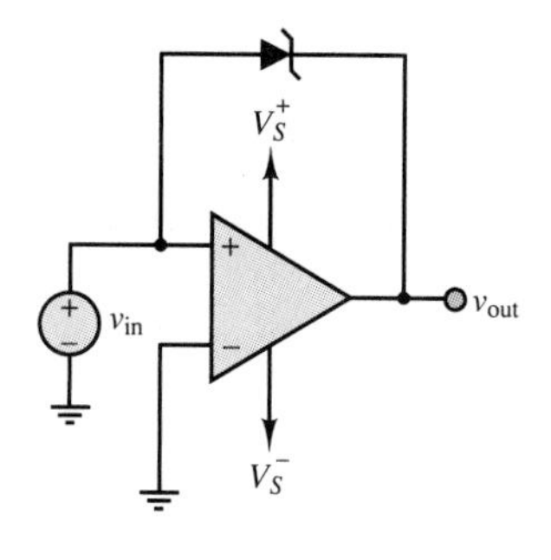

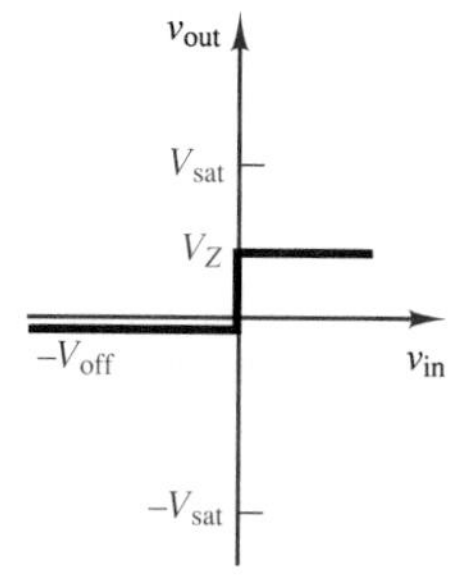

Figure 14.43 Level-clamped comparator

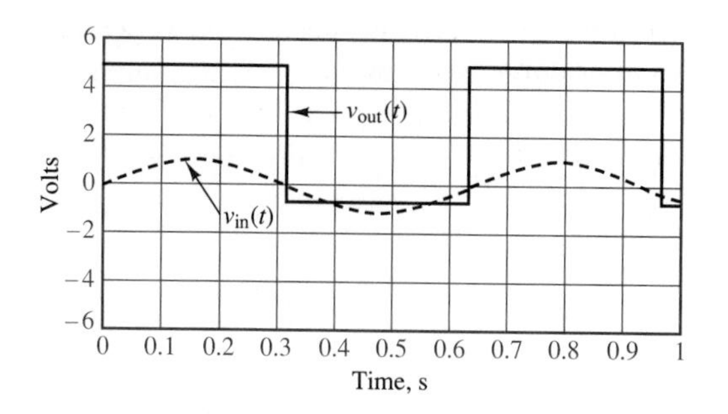

Figure 14.44 Zener-clamped comparator waveforms

Very often, in converting an analog signal to a binary representation, one would like to use voltage levels other than $\pm V_{sat}$. Commonly used voltage levels in this type of switching circuit are 0 V and 5 V, respectively. This modified voltage transfer characteristic can be obtained by connecting a Zener diode between the output of the op-amp and the noninverting input, in the configuration sometimes called a **level** or **Zener clamp.** The circuit shown in Figure 14.43 is based on the fact that a reversed-biased Zener diode will hold a constant voltage, V_Z, as was shown in Chapter 7. When the diode is forward-biased, on the other hand, the output voltage becomes the negative of the offset voltage, V_γ. An additional advantage of the level clamp is that it reduces the switching time. Input and output waveforms for a Zener-clamped comparator are shown in Figure 14.44, for the case of a sinusoidal $v_{in}(t)$ of peak amplitude 1 V and Zener voltage equal to 5 V.

Although the Zener-clamped circuit illustrates a specific issue of interest in the design of comparator circuits, namely, the need to establish desired reference output voltages other than the supply saturation voltages, this type of circuit is rarely employed in practice. Special-purpose integrated circuit (IC) packages are available that are designed specifically to serve as comparators. These can typically accept relatively large inputs and have provision for selecting the desired reference voltage levels (or, sometimes, are internally clamped to a specified voltage range). A representative product is the LM311, which provides an open-collector output, as shown in Figure 14.45. The open-collector output allows the user to connect the output transistor to any supply voltage of choice by means of an external pull-up resistor, thus completing the output circuit. The actual value of the resistor is not critical, since the transistor is operated in the saturation mode; values between a few hundred and a few thousand ohms are typical. In the remainder of the chapter it will be assumed, unless otherwise noted, that the comparator output voltage will switch between 0 V and 5 V.

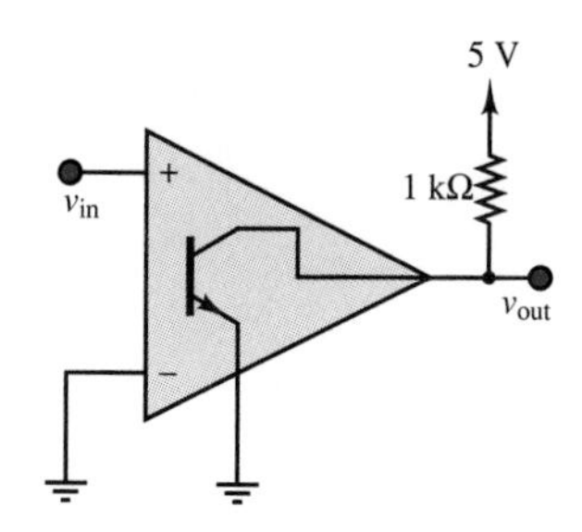

Figure 14.45 Open-collector comparator output with representative external supply connection

The Schmitt Trigger

One of the typical applications of the op-amp comparator is in detecting when an input voltage exceeds a present threshold level. The desired threshold is then represented by a DC reference, V_{ref}, connected to the noninverting input, and the input voltage source is connected to the inverting input, as in Figure 14.42. Under ideal conditions, for noise-free signals, and with an infinite slew rate for the op-amp, the operation of such a circuit would be as depicted in Figure 14.46. In practice, the presence of noise and the finite slew rate of practical op-amps will require special attention.

Two improvements of this circuit will be discussed in this section: how to improve the switching speed of the comparator, and how to design a circuit that can operate correctly even in the presence of noisy signals. If the input to the comparator is changing slowly, the comparator will not switch instantaneously, since its open-loop gain is not infinite and, more important, its slew rate further limits the switching speed. In fact, commercially available comparators have slew rates that are typically much lower than those of conventional op-amps. In this case, the comparator output would not switch very quickly at all. Further, in the presence of noisy inputs, a conventional comparator is inadequate, because the input signal could cross the reference voltage level repeatedly and cause multiple triggering. Figure 14.47 depicts the latter occurrence.

One very effective way of improving the performance of the comparator is by introducing positive feedback. As will be explained shortly, positive feedback can increase the switching speed of the comparator and provide noise immunity at the same time. Figure 14.48 depicts a comparator circuit in which the output has been tied back to the *noninverting* input (thus the terminology *positive* feedback) by means of a resistive voltage divider. The effect of this positive feedback connection is to provide a reference voltage at the noninverting input equal to a fraction of the comparator output voltage; since the comparator output is equal to either the positive or the negative saturation voltage, $\pm V_{\text{sat}}$, the reference voltage at the noninverting input can be either positive or negative.

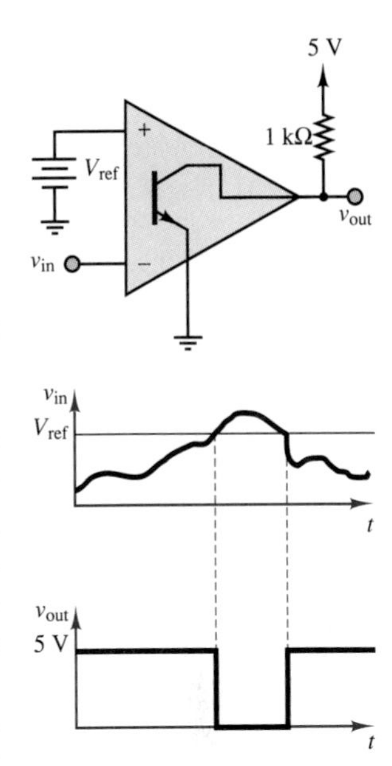

Figure 14.46 Waveforms for inverting comparator with offset

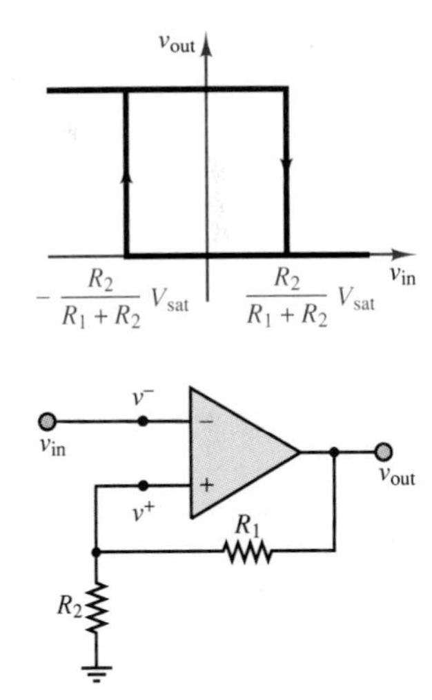

Figure 14.48 Transfer characteristic of the Schmitt trigger

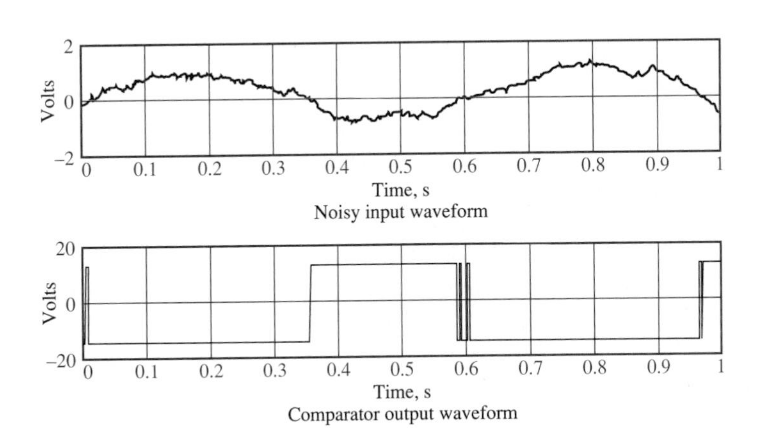

Figure 14.47 Comparator response to noisy inputs

Consider, first, the case when the comparator output is $v_{out} = +V_{sat}$. It follows that

$$v^+ = \frac{R_2}{R_2 + R_1} V_{sat} \tag{14.30}$$

and therefore the differential input voltage is

$$\varepsilon = v^+ - v^- = \frac{R_2}{R_2 + R_1} V_{sat} - v_{in} \tag{14.31}$$

For the comparator to switch from the positive to the negative saturation state, the differential voltage, ε, must then become negative; that is, the condition for the comparator to switch state becomes

$$v_{in} > \frac{R_2}{R_2 + R_1} V_{sat} \tag{14.32}$$

Since $\frac{R_2}{R_2+R_1} V_{sat}$ is a positive voltage, the comparator will not switch when the input voltage crosses the zero level, but it will switch when the input voltage exceeds some positive voltage, which can be determined by appropriate choice of R_1 and R_2.

Consider, now, the case when the comparator output is $v_{out} = -V_{sat}$. Then

$$v^+ = -\frac{R_2}{R_2 + R_1} V_{sat} \tag{14.33}$$

and therefore

$$\varepsilon = v^+ - v^- = -\frac{R_2}{R_2 + R_1} V_{sat} - v_{in} \tag{14.34}$$

For the comparator to switch from the negative to the positive saturation state, the differential voltage, ε, must then become positive; the condition for the comparator to switch state is now

$$v_{in} < -\frac{R_2}{R_2 + R_1} V_{sat} \tag{14.35}$$

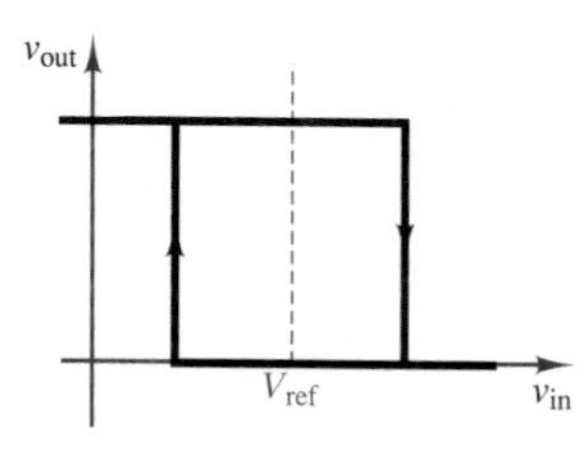

Thus, the comparator will not switch when the input voltage crosses the zero level (from the negative direction), but it will switch when the input voltage becomes more negative than a threshold voltage, determined by R_1 and R_2. Figure 14.48 depicts the effect of the different thresholds on the voltage transfer characteristic, showing the switching action by means of arrows.

The circuit just described finds frequent application and is called a **Schmitt trigger.**

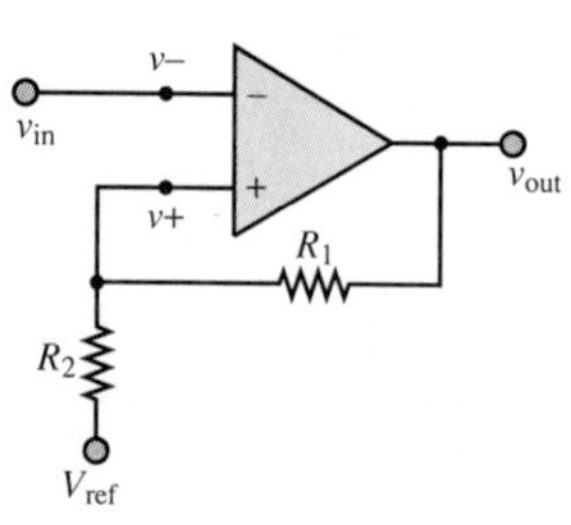

Figure 14.49 Schmitt trigger (general circuit)

If it is desired to switch about a voltage other than zero, a reference voltage can also be connected to the noninverting terminal, as shown in Figure 14.49. Now the expression for the noninverting terminal voltage is

$$v^+ = \frac{R_2}{R_2 + R_1} v_{out} + V_{ref} \frac{R_1}{R_2 + R_1} \tag{14.36}$$

and the switching levels for the Schmitt trigger are

$$v_{\text{in}} > \frac{R_2}{R_2 + R_1} V_{\text{sat}} + V_{\text{ref}} \frac{R_1}{R_2 + R_1} \tag{14.37}$$

for the positive-going transition, and

$$v_{\text{in}} < -\frac{R_2}{R_2 + R_1} V_{\text{sat}} + V_{\text{ref}} \frac{R_1}{R_2 + R_1} \tag{14.38}$$

for the negative-going transition. In effect, the Schmitt trigger provides a noise-rejection range equal to $\pm \frac{R_2}{R_2+R_1} V_{\text{sat}}$ within which the comparator cannot switch. Thus, if the noise amplitude is contained within this range, the Schmitt trigger will prevent multiple triggering. Figure 14.50 depicts the response of a Schmitt trigger with appropriate switching thresholds to a noisy waveform. Example 14.14 provides a numerical illustration of this process.

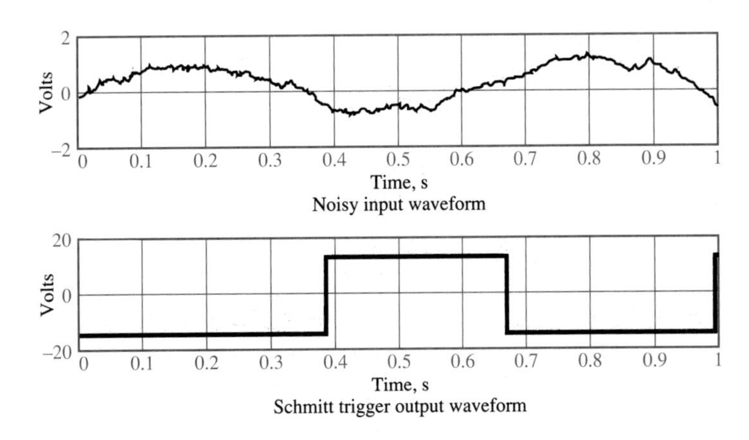

Figure 14.50 Schmitt trigger response to noisy waveforms

Example 14.14

In this example we analyze the practical design of a Schmitt trigger. Assume that the comparator has supply voltages equal to ±18 V and saturation voltages equal to ±16.5 V and it is desired to switch around the reference level $V_{\text{ref}} = 2$ V. Assume, further, that the noise corrupting the signal has a maximum excursion of ±100 mV. Find the required values of the resistors in the circuit of Figure 14.51.

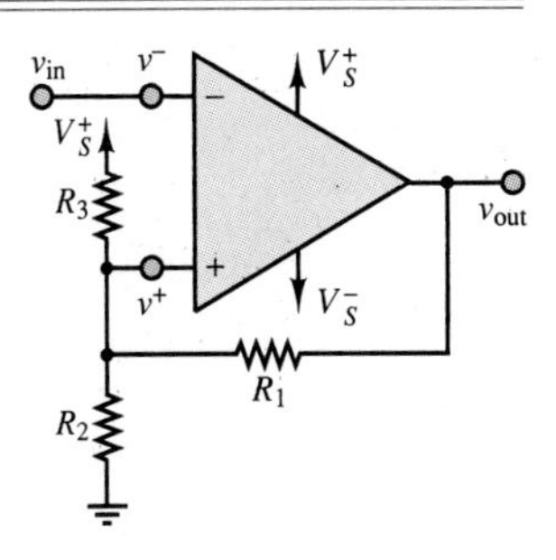

Figure 14.51

Solution:

Notice, first, that the reference voltage is obtained from the same (positive) voltage supply used for the op-amp. This is common practice and saves having a separate supply for the reference voltage. On the basis of the modified circuit, the expression for the noninverting-terminal voltage is

$$v^+ = \frac{R_2}{R_2 + R_1} v_{out} + \frac{R_2}{R_2 + R_3} V_S^+$$

Since the required noise protection level is ± 100 mV, R_1 and R_2 can be computed from

$$\frac{\Delta V}{2} = \frac{R_2}{R_2 + R_1} V_{sat} = \frac{R_2}{R_2 + R_1}(16.5) = 0.1$$

where $\Delta V = 200$ mV is the noise protection range shown in Figure 14.52. Thus,

$$\frac{R_2}{R_2 + R_1} = \frac{0.1}{16.5} = 6.06 \times 10^{-3}$$

For $R_1 = 100$ kΩ, R_2 can be calculated to be approximately 610 Ω. Since the required reference voltage is 2 V, we can find R_3 by solving the equation

$$\frac{R_2}{R_2 + R_3} V_S^+ = \frac{R_2}{R_2 + R_3}(18) = 2$$

to obtain $R_3 = 4{,}880\ \Omega$.

It is instructive at this point to calculate the actual thresholds at which the comparator will switch; these are given by the expression

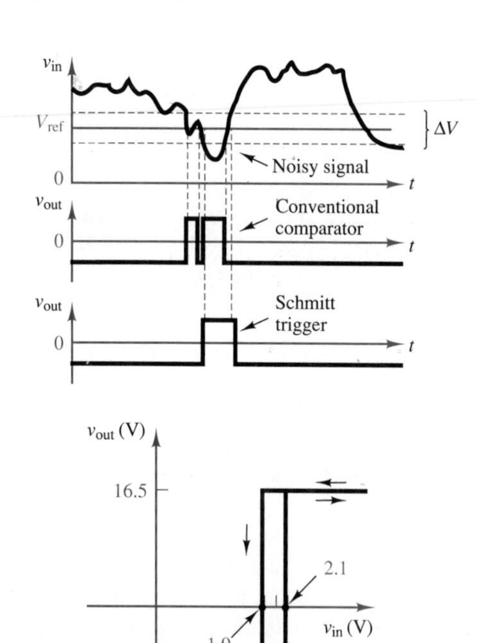

Figure 14.52

$$v^+ = \frac{R_2}{R_2 + R_1} v_{out} + \frac{R_2}{R_2 + R_3} V_S^+$$

where $v_{out} = \pm V_{sat}$. For a positive saturation voltage, the threshold will be

$$v^+ = \frac{R_2}{R_2 + R_1}(16.5) + 2 = 2.1 \text{ V}$$

while for the negative saturation voltage, the threshold is

$$v^+ = -\frac{R_2}{R_2 + R_1}(16.5) + 2 = 1.9 \text{ V}$$

as required. The transfer characteristic is shown in Figure 14.52.

The Op-Amp Astable Multivibrator

This section describes an op-amp circuit useful in the generation of timing, or clock, waveforms. In the previous discussion, it was shown how it is possible to utilize positive feedback in a Schmitt trigger circuit. A small fraction of the large output saturation voltage was used to delay the switching of a comparator to make it less sensitive to random fluctuations in the input signal. A very similar circuit can be employed to generate a square-wave signal of fixed period and amplitude. This circuit is called an **astable multivibrator,** because it periodically switches between two states without ever reaching a stable state. The circuit is shown in Figure 14.53 in a somewhat simplified form.

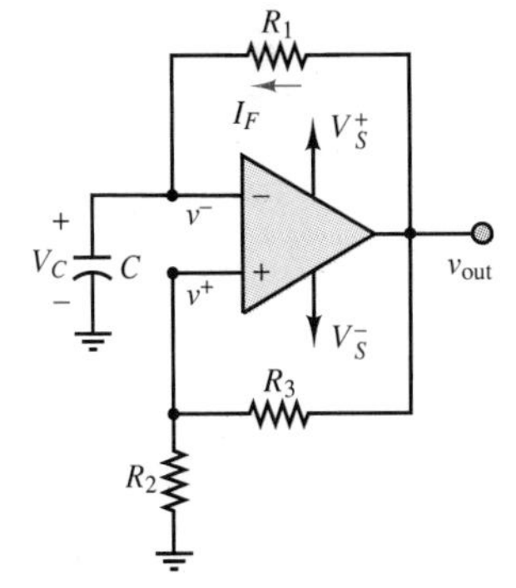

Figure 14.53 Simplified op-amp astable multivibrator

The analysis of the circuit is similar to that of the Schmitt trigger, except for the presence of both negative and positive feedback connections. We start by postulating that the op-amp output is saturated at the positive supply saturation voltage. One can easily verify that, in this case, the noninverting terminal voltage is

$$v^+ = \frac{R_2}{R_2 + R_3} v_{out} \tag{14.39}$$

and that the voltage at the inverting input is equal to the voltage across the capacitor, v_C. One can write an expression for the inverting terminal voltage, v^-, by noting that the capacitor charges at a rate determined by the feedback resistance and by the capacitance, according to the time constant:

$$\tau = R_1 C \tag{14.40}$$

Since the current flowing into the inverting terminal is negligible, the inverting input voltage is in effect given by

$$v^-(t) = v_C(t) = V_{sat}(1 - e^{-t/\tau}) \tag{14.41}$$

A sketch of the inverting input voltage as a function of time is shown in Figure 14.54.

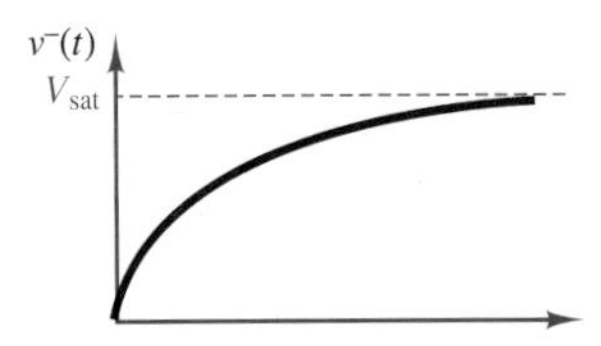

Figure 14.54

The behavior of this op-amp multivibrator is best understood if the differential voltage, $\varepsilon = v^+ - v^-$, is considered. Assuming that the capacitor starts charging at $t = 0$ from zero initial charge and that v_{out} is saturated at $+V_{sat}$, one can easily show that this condition will be sustained for as long as ε remains positive. However, as the capacitor continues to charge, v^- will eventually grow larger than v^+, and ε will become negative. Since

$$v^+ = \frac{R_2}{R_2 + R_3} V_{sat} \tag{14.42}$$

ε becomes negative as soon as the inverting-terminal voltage exceeds a threshold voltage:

$$v^- > \frac{R_2}{R_2 + R_3} V_{sat} \tag{14.43}$$

When the differential input voltage switches from a positive to a negative value, *the op-amp is forced to switch to the opposite extreme,* $-V_{sat}$. But when v_{out} switches to the negative saturation voltage, we see a sudden sign reversal in v^+:

$$v^+ = -\frac{R_2}{R_2 + R_3} V_{sat} \tag{14.44}$$

Further, the capacitor, which has been charging toward $+V_{sat}$, now sees a negative voltage, $-V_{sat}$. This condition may be analyzed by resorting to the transient analysis methods of Chapter 6. The reversal of the output voltage causes the capacitor to discharge toward the new value of v_{out}, $-V_{sat}$, according to the function

$$v_C(t) = [v_C(t_0) + V_{sat}]e^{-(t-t_0)/\tau} - V_{sat} \tag{14.45}$$

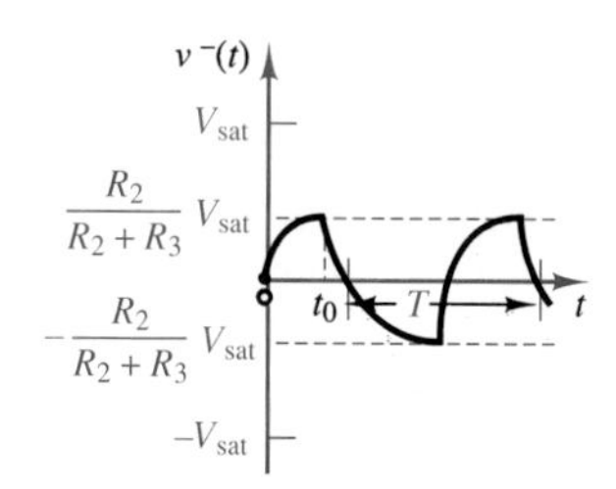

Figure 14.55 Astable multivibrator inverting-terminal voltage

where t_0 is the time at which the output voltage changes from $+V_{sat}$ to $-V_{sat}$. The discharging continues as long as the differential input voltage ε remains negative, since this guarantees that $v_{out} = -V_{sat}$. But this condition cannot last indefinitely, because at some point, as the capacitor discharges, $v_C(t) = v^-(t)$ will become more negative than $v^+(t)$ and ε will become positive again, causing the output of the op-amp to switch to $+V_{sat}$. This condition will occur when

$$v^- < -\frac{R_2}{R_2 + R_3} V_{sat} \tag{14.46}$$

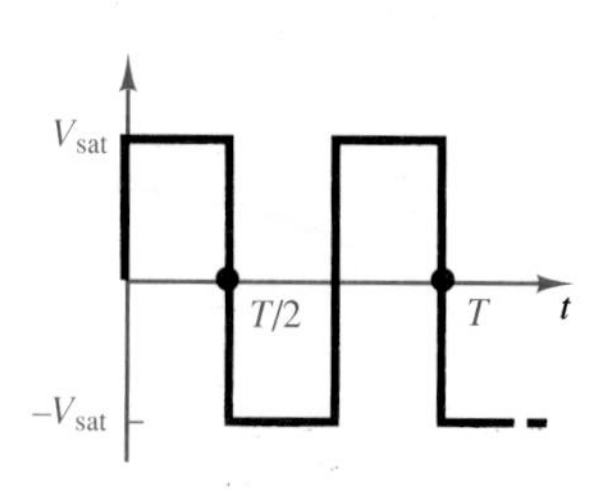

Figure 14.56 Astable multivibrator output waveform

Figure 14.55 graphically summarizes the operation of the astable multivibrator. The corresponding op-amp output is depicted in Figure 14.56. The period, T, of the waveform is determined by the charging and discharging cycle of the capacitor. It should be apparent that the introduction of a variable resistor R_1 gives rise to a circuit that can generate a variable-frequency square wave, since the frequency is directly determined by the time constant $\tau = R_1 C$. It can be shown that the period of the square-wave waveform resulting from the astable multivibrator is given by

$$T = 2R_1 C \log_e \left(\frac{2R_2}{R_3} + 1 \right) \tag{14.47}$$

The Op-Amp Monostable Multivibrator (One-Shot)

An extension of the astable multivibrator is the **one-shot,** or **monostable multivibrator.** One-shots are available in integrated circuit form but can also be constructed from more general-purpose op-amps. In this section, we first look at the op-amp one-shot, taking special note of the analogy with the astable multivibrator. Figure 14.57 depicts an op-amp monostable multivibrator. Its operation is summarized in the associated timing diagram, showing that if a negative voltage pulse is applied at the input, v_{in}, the output of the op-amp will switch from a stable state, $+V_{sat}$, to the unstable state, $-V_{sat}$, for a period of time T, which is determined by a charging time constant, after which the output will resume its stable value, $+V_{sat}$. In the analysis of the one-shot circuit, it will be assumed that the diode is an offset diode, with offset voltage V_{off}, and that $v_{out} = +V_{sat}$ for $t < t_0$.

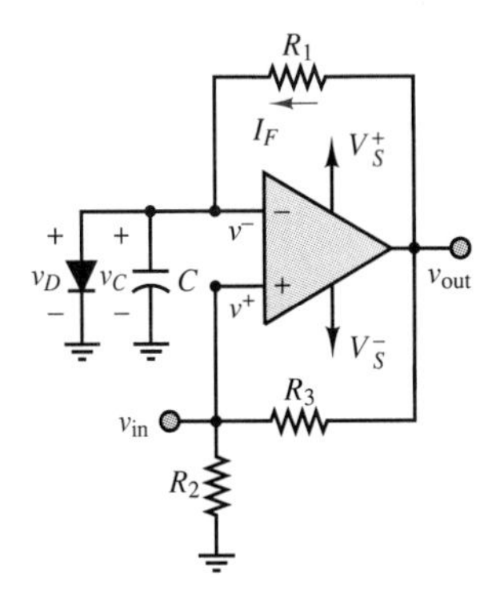

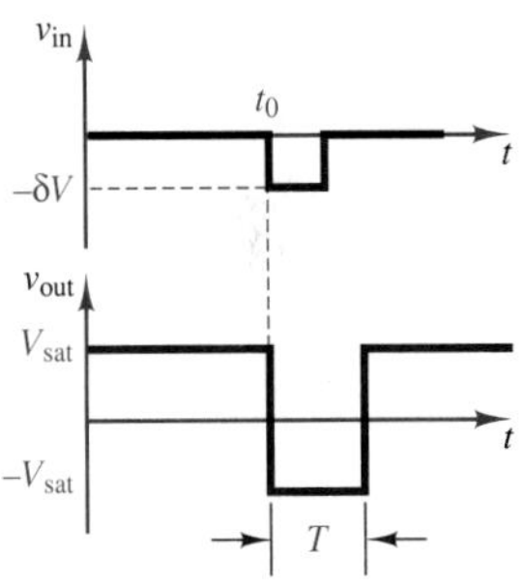

Figure 14.57 Op-amp monostable multivibrator and typical waveforms

For $t < t_0$, before the negative input pulse is applied, the input terminal voltages are given by

$$v^+ = V_{sat}\frac{R_2}{R_2 + R_3} \tag{14.48}$$

and

$$v^- = V_\gamma \tag{14.49}$$

Provided that $v^+ > v^-$, the op-amp will remain in its positive saturated state, then, until some external condition causes $\varepsilon = v^+ - v^-$ to become negative. This condition is brought about at $t = t_0$ by the "trigger" pulse v_{in}, which briefly lowers the noninverting-terminal voltage by δV volts. If δV is sufficiently large, the following condition will be satisfied:

$$v^+ = V_{sat}\frac{R_2}{R_2 + R_3} - \delta V < V_\gamma \tag{14.50}$$

and the op-amp will switch to the negative saturation state, as indicated in Figure 14.57. The noninverting-terminal voltage may be expressed as a function of time for $t > t_0$, considering that v_{out} has switched to $-V_{sat}$ at $t = t_0$ and the diode now acts as an open circuit, since v^- is negative. With the diode out of the picture, then, the R_1C circuit will discharge, causing v^- to drop from its initial voltage (V_γ) toward $-V_{sat}$.

The circuit of Figure 14.58 depicts the equivalent switching circuit, illustrating that the capacitor discharges from the initial value of $v_C(t_0) = V_\gamma$ toward the final value, $-V_{sat}$. Recalling the analysis of transients in Chapter 6, we know that the capacitor voltage is given by the expression

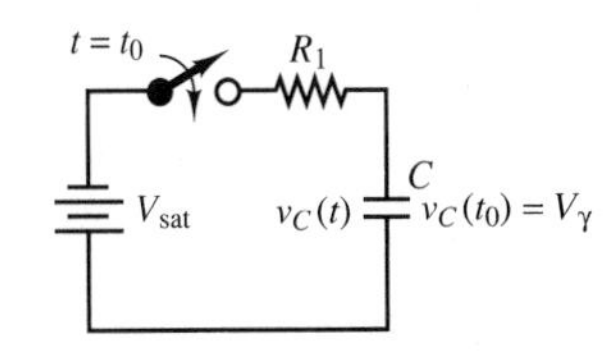

Figure 14.58 Equivalent charging circuit for monostable multivibrator

$$v_C(t) = [v_C(t_0) + V_{sat}]e^{-(t-t_0)/R_1C} - V_{sat} \tag{14.51}$$

As v^- becomes more negative, it eventually becomes smaller than v^+. Thus, when the condition

$$v_C(t) = [v_C(0) + V_{sat}]e^{-(t-t_0)/R_1C} - V_{sat} = v^+ \tag{14.52}$$

is met, the comparator will switch back to the positive saturation state, $+V_{sat}$, and remain in that state until a new trigger pulse is provided. The value of the noninverting-terminal voltage during the time the output is in the negative saturated state is determined by the positive feedback circuit:

$$v^+ = -V_{\text{sat}} \frac{R_2}{R_2 + R_3} \qquad \textbf{(14.53)}$$

Clearly, the duration of the output pulse provided by the one-shot is determined by the time constant of the R_1C circuit, as well as by the value of the resistive voltage divider in the positive feedback network, as illustrated in Example 14.15.

Example 14.15

The op-amp monostable multivibrator of Figure 14.57 is assembled with the following component values: $R_1 = 20\ \text{k}\Omega$, $R_2 = 670\ \Omega$, $V_\gamma = 0.6\ \text{V}$, $V_{\text{sat}} = 16\ \text{V}$, $R_3 = 100\ \text{k}\Omega$, and $C = 4.7\ \mu\text{F}$. Find the duration of the output pulse.

Solution:

Assume, without loss of generality, that $t_0 = 0$. Then the expression for the capacitor voltage (equal to the noninverting-terminal voltage) is

$$v_C(t) = [v_C(0) + V_{\text{sat}}]e^{-t/R_1C} - V_{\text{sat}}$$

The time at which $v^- = v^+$ may be found by setting this expression equal to the noninverting-terminal voltage:

$$v^+ = -V_{\text{sat}} \frac{R_2}{R_2 + R_3} = -0.1065\ \text{V}$$

and thus obtaining the following relationship:

$$-0.1065 = 16.6e^{-10.64t} - 16$$

or

$$t = -\frac{1}{10.64} \log_e \frac{15.8935}{16.6} = 4.09\ \text{ms}$$

Thus, the duration of the output pulse is 4.09 ms.

Monostable multivibrators are usually employed in IC package form. An IC one-shot can generate voltage pulses when triggered by a **rising** or a **falling edge,** that is, by a transition in either direction in the input voltage. Thus, a one-shot IC offers the flexibility of external selection of the type of transition that will cause a pulse to be generated: a rising edge (from low voltage to high, typically zero volts to some threshold level), or a falling edge (high-to-low transition). Various input connections are usually provided for selecting the preferred triggering mode, and the time constant is usually set by selection of an external RC circuit. The output pulse that may be generated by the one-shot can also occur as a positive or a negative transition. Figure 14.59 shows the response of a one-shot to a triggering signal for the four conditions that may be attained with a typical one-shot.

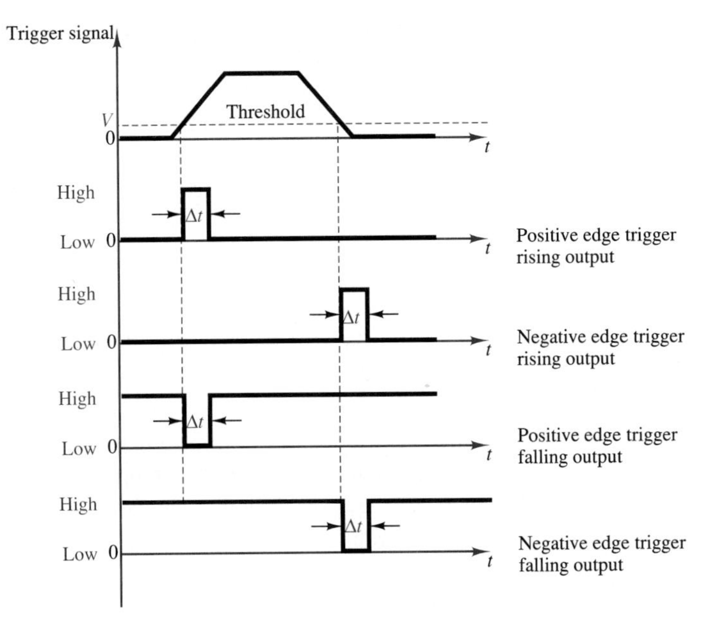

Figure 14.59 IC monostable multivibrator waveforms

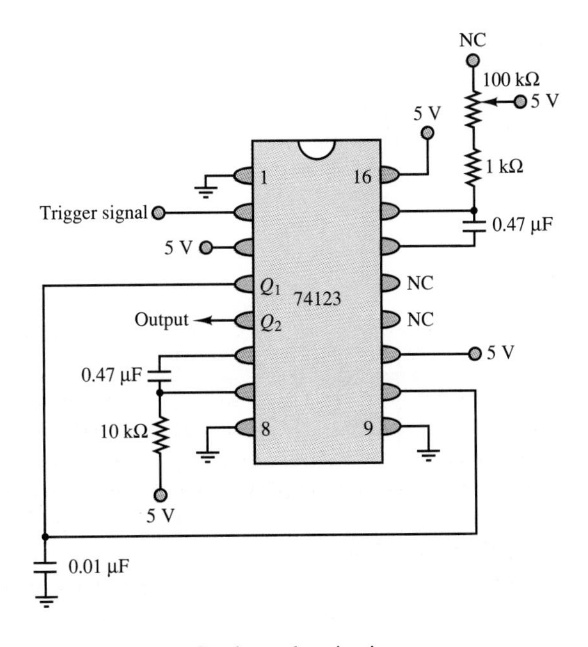

Figure 14.60 Dual one-shot circuit

A typical IC one-shot circuit based on the 74123 (see the data sheet at the end of the chapter) is displayed in Figure 14.60. The 74123 is a **dual one-shot,** meaning that the package contains two monostable multivibrators, which can be used independently. The outputs of the one-shot are indicated by the symbols Q_1, $\overline{Q}_1$, Q_2, and $\overline{Q}_2$, where the bar indicates the complement of the output. For example, if Q_1 corresponds to a positive-going output pulse, $\overline{Q}_1$ indicates a negative-going output pulse, of equal duration.

Timer ICs: The NE555

The discussion of op-amp–based timing circuits presented in the previous sections served the purpose of introducing a large family of integrated circuits that can provide flexible timing waveforms. These fall—for our purposes—into one of two classes: pulse generators, and clock waveform generators. Chapters 12 and 13 delve into a more detailed analysis of digital timing circuits, a family to which the circuits of the previous sections belong. This section will now introduce a multipurpose integrated circuit that can perform both the monostable and astable multivibrator functions. The main advantage of the integrated circuit implementation of these circuits (as opposed to the discrete op-amp version previously discussed) lies in the greater accuracy and repeatability one can obtain with ICs, their ease of application, and the flexibility provided in the integrated circuit packages. The NE555 is a timer circuit capable of producing accurate time delays (pulses) or oscillation. In the time-delay, or monostable, mode, the time delay or pulse duration is controlled by an external *RC* network. In the astable, or clock generator, mode, the frequency is controlled by two external resistors and one capacitor. Figure 14.61 depicts typical circuits for monostable and astable operation of the NE555. Note that the threshold level and the trigger level can also be externally controlled. For the monostable circuit, the pulse width can be computed from the following equation:

$$T = 1.1 R_1 C \qquad \textbf{(14.54)}$$

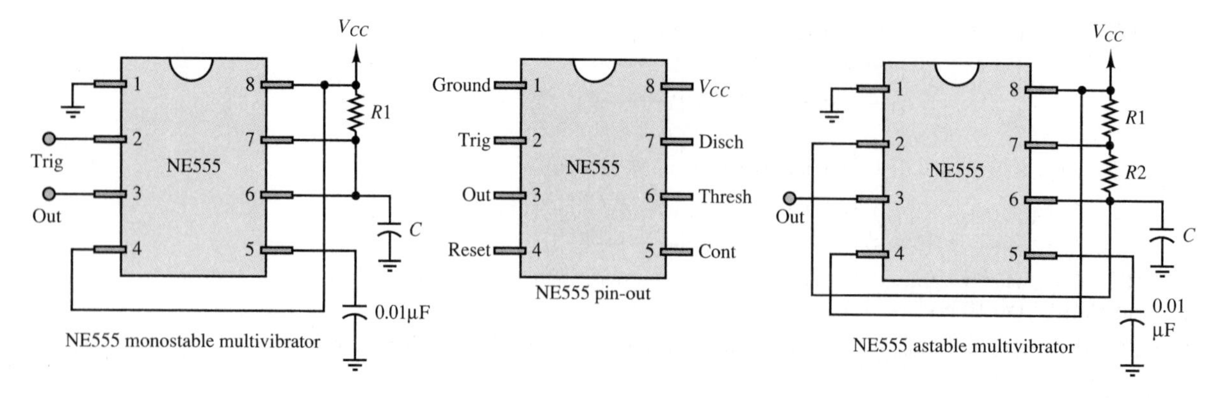

Figure 14.61 NE555 timer

For the astable circuit, the positive pulse width can be computed from the following equation:

$$T_+ = 0.69(R_1 + R_2)C \tag{14.55}$$

and the negative pulse width can be computed from

$$T_- = 0.69R_2C \tag{14.56}$$

The use of the NE555 timer is illustrated in Example 14.16.

Example 14.16

The period of a square wave generated by the monostable multivibrator of Figure 14.61 is 0.421 ms. Compute the values of R_1 and C.

Solution:

As was noted, the period of the square wave generated by the monostable multivibrator of the NE555 can be computed from the following equation:

$$T = 1.1R_1C$$

By choosing $C = 1\ \mu\text{F}$, we can obtain a value for R_1 from the following equation:

$$0.421\text{ ms} = 1.1R_1 \times 10^{-6}$$

Thus, $R_1 = 382.73\ \Omega$. Note that this answer is not unique.

DRILL EXERCISES

14.17 Verify that the transfer characteristic of Figure 14.42 is correct.

14.18 For the comparator circuit shown in Figure 14.62, sketch the waveforms $v_{out}(t)$ and $v_S(t)$ if $v_S(t) = 0.1 \cos(\omega t)$ and $V_{ref} = 50$ mV.

14.19 Derive the expressions for the switching thresholds of the Schmitt trigger of Figure 14.48.

14.20 Explain why positive feedback increases the switching speed of a comparator.

14.21 Compute the period of the square wave generated by the multivibrator of Figure 14.53 if $C = 1\ \mu\text{F}$, $R_1 = 10\ \text{k}\Omega$, $R_3 = 100\ \text{k}\Omega$, and $R_2 = 1\ \text{k}\Omega$.

14.22 Compute the value of R_3 that would increase the duration of the one-shot pulse in Example 14.15 to 10 ms.

14.23 Compute the value of C that would increase the duration of the one-shot pulse in Example 14.15 to 10 ms.

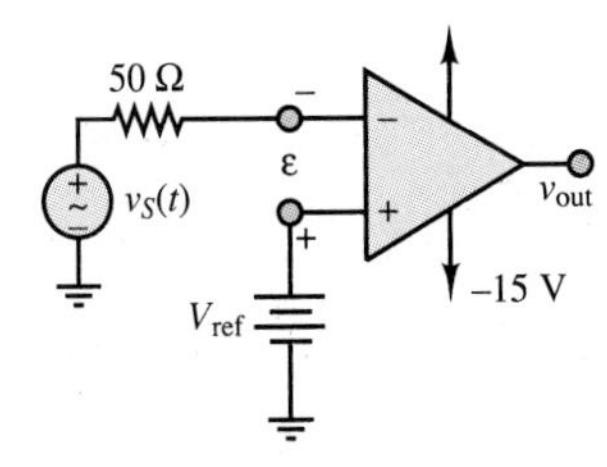

Figure 14.62

14.6 DATA TRANSMISSION IN DIGITAL INSTRUMENTS

One of the necessary aspects of data acquisition and control systems is the ability to transmit and receive data. Often, a microcomputer-based data acquisition system is interfaced to other digital devices, such as digital instruments or other microcomputers. In these cases it is necessary to transfer data directly in digital form. In fact, it is usually preferable to transmit data that is already in digital form, rather than analog voltages or currents. Among the chief reasons for the choice of digital over analog is that digital data is less sensitive to noise and interference than analog signals: in receiving a binary signal transmitted over a data line, the only decision to be made is whether the value of a bit is 0 or 1. Compared with the difficulty in obtaining a precise measurement of an analog voltage or current, either of which could be corrupted by noise or interference in a variety of ways, the probability of making an error in discerning between binary 0s and 1s is very small. Further, as will be shown shortly, digital data is often coded in such a way that many transmission errors may be detected and corrected for. Finally, storage and processing of digital data are much more readily accomplished than would be the case with analog signals. This section explores a few of the methods that are commonly employed in transmitting digital data; both parallel and serial interfaces are considered.

Digital signals in a microcomputer are carried by a bus, consisting of a set of parallel wires each carrying one bit of information. In addition to the signal-carrying wires, there are also control lines that determine under what conditions transmission may occur. A typical computer data bus consists of eight parallel wires and therefore enables the transmission of one byte; digital data is encoded in binary according to one of a few standard codes, such as the BCD code described in Chapter 12, or the ASCII code, which is summarized in Table 14.5. This bus configuration is usually associated with **parallel transmission,** whereby all of the bits are transmitted simultaneously, along with some control bits.

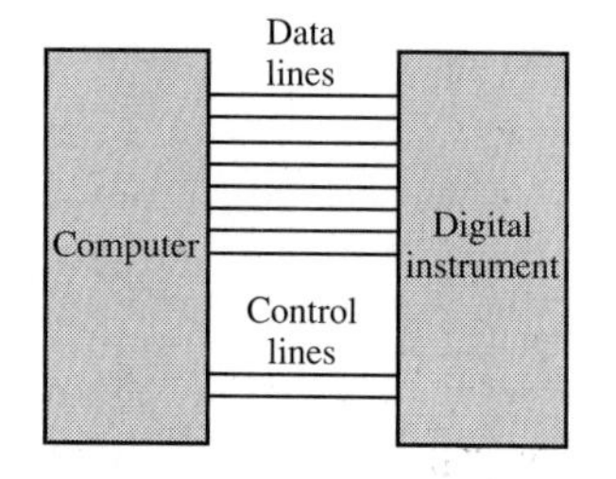

Figure 14.63 Parallel data transmission

Figure 14.63 depicts the general appearance of a parallel connection. Parallel data transmission can take place in one of two modes: **synchronous** or **asynchronous.** In synchronous transmission, a timing clock pulse is transmitted along with the data over a control line. The arrival of the clock pulse indicates that valid data has also arrived. While parallel synchronous transmission can be very fast, it requires the added complexity of a synchronizing clock, and is typically employed only for internal computer data transmission. Further, this type of communication can take place only over short distances (approximately 4 m). Asynchronous data transmission, on the other hand, does not take place at a fixed clock rate, but requires a **handshake protocol** between sending and receiving ends. The handshake protocol consists of the transmission of *data ready* and *acknowledge* signals over two separate control wires. Whenever the sending device is ready to transmit data, it sends a pulse over the *data ready* line. When this signal reaches the receiver, and if the receiver is ready to receive the data, an *acknowledge* pulse is sent back, indicating that the transmission may occur; at this point, the parallel data is transmitted.

Table 14.5 **ASCII code**

Graphic or control	ASCII (hex)	Graphic or control	ASCII (hex)	Graphic or control	ASCII (hex)
NUL	00	+	2B	V	56
SOH	01	,	2C	W	57
STX	02	-	2D	X	58
ETX	03	.	2E	Y	59
EOT	04	/	2F	Z	5A
ENQ	05	0	30	[	5B
ACK	06	1	31	\	5C
BEL	07	2	32	]	5D
BS	08	3	33	↑	5E
HT	09	4	34	←	5F
LF	0A	5	35	`	60
VT	0B	6	36	a	61
FF	0C	7	37	b	62
CR	0D	8	38	c	63
SO	0E	9	39	d	64
SI	0F	:	3A	e	65
DLE	10	;	3B	f	66
DC1	11	<	3C	g	67
DC2	12	=	3D	h	68
DC3	13	>	3E	i	69
DC4	14	?	3F	j	6A
NAK	15	@	40	k	6B
SYN	16	A	41	l	6C
ETB	17	B	42	m	6D
CAN	18	C	43	n	6E
EM	19	D	44	o	6F
SUB	1A	E	45	p	70
ESC	1B	F	46	q	71
FS	1C	G	47	r	72
GS	1D	H	48	s	73
RS	1E	I	49	t	74
US	1F	J	4A	u	75
SP	20	K	4B	v	76
!	21	L	4C	w	77
”	22	M	4D	x	78
#	23	N	4E	y	79
$	24	O	4F	z	7A
%	25	P	50	{	7B
&	26	Q	51	|	7C
’	27	R	52	}	7D
(	28	S	53	~	7E
)	29	T	54	DEL	7F
*	2A	U	55		

Perhaps the most common parallel interface is based on the **IEEE 488 standard,** leading to the so-called IEEE 488 bus, also referred to as **GPIB** (for **general-purpose instrument bus**).

The IEEE 488 Bus

The IEEE 488 bus, shown in Figure 14.64, is an eight-bit parallel asynchronous interface that has found common application in digital instrumentation applications. The physical bus consists of 16 lines, of which 8 are used to carry the data, 3 for the handshaking protocol, and the rest to control the data flow. The bus permits connection of up to 15 instruments and data rates of up to 1 Mbyte/s. There is a limitation, however, in the maximum total length of the bus cable, which is 20 m. The signals transmitted are TTL-compatible and employ negative logic (see Chapter 12), whereby a logic 0 corresponds to a TTL high state (> 2 V) and a logic 1 to a TTL low state (< 0.8 V). Often, the eight-bit word transmitted over an IEEE 488 bus is coded in ASCII format (see Table 14.5), as illustrated in Example 14.17.

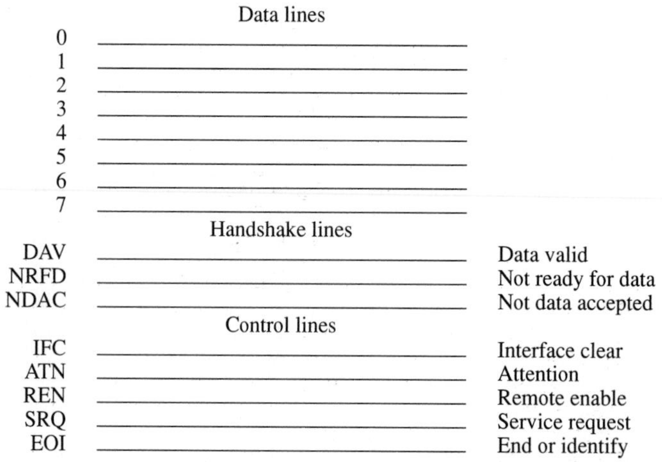

Data lines

0
1
2
3
4
5
6
7

Handshake lines

DAV — Data valid
NRFD — Not ready for data
NDAC — Not data accepted

Control lines

IFC — Interface clear
ATN — Attention
REN — Remote enable
SRQ — Service request
EOI — End or identify

Figure 14.64 IEEE 488 bus

Example 14.17

Determine the actual binary data sent by a digital voltmeter reading 3.405 V over an IEEE 488 bus if the data is encoded in ASCII format (see Table 14.5). Assume that the sequence is from most to least significant digit.

Solution:

From the ASCII conversion table (Table 14.5), we can operate the following conversions:

Control	ASCII (hex)
3	33
.	2E
4	34
0	30
5	35

Thus, the actual data sent, from most to least significant digit, is 33 2E 34 30 35 in hex code, or 110011 101110 110100 110000 110101 in binary.

In an IEEE 488 bus system, devices may play different roles and are typically classified as *controllers,* which manage the data flow; *talkers* (e.g., a digital voltmeter), which can only send data; *listeners* (e.g., a printer), which can only receive data; and *talkers/listeners* (e.g., a digital oscilloscope), which can receive as well as transmit data. The simplest system configuration might consist of just a talker and a listener. If more than two devices are present on the bus, a controller is necessary to determine when and how data transmission is to occur on the bus. For example, one of the key rules implemented by the controller is that only one talker can transmit at any one time; it is possible, however, for several listeners to be active on the bus simultaneously. If the data rates of the different listeners are different, the talker will have to transmit at the slowest rate, so that all of the listeners are assured of receiving the data correctly.

The set of rules by which the controller determines the order in which talking and listening are to take place is determined by a **protocol.** One aspect of the protocol is the handshake procedure, which enables the transmission of data. Since different devices (with different data rate capabilities) may be listening to the same talker, the handshake protocol must take into account these different capabilities. Let us discuss a typical handshake sequence that leads to transmission of data on an IEEE 488 bus. The three handshake lines used in the IEEE 488 have important characteristics that give the interface system wide flexibility, allowing interconnection of multiple devices that may operate at different speeds. The slowest active device controls the rate of data transfer, and more than one device can accept data simultaneously. The timing diagram of Figure 14.65 is used to illustrate the sequence in which the handshake and data transfer are performed:

1. All active listeners use the not ready for data (NRFD) line to indicate their state of readiness to accept a new piece of information. Nonreadiness to accept data is indicated if the NRFD line is held at zero volts. If even one active listener is not ready, the NRFD line of the entire bus is kept at zero volts and the active talker will not transmit the next byte. When all active listeners are ready and they have released the NRFD line, it now goes high.
2. The designated talker drives all eight data input/output lines, causing valid data to be placed on them.

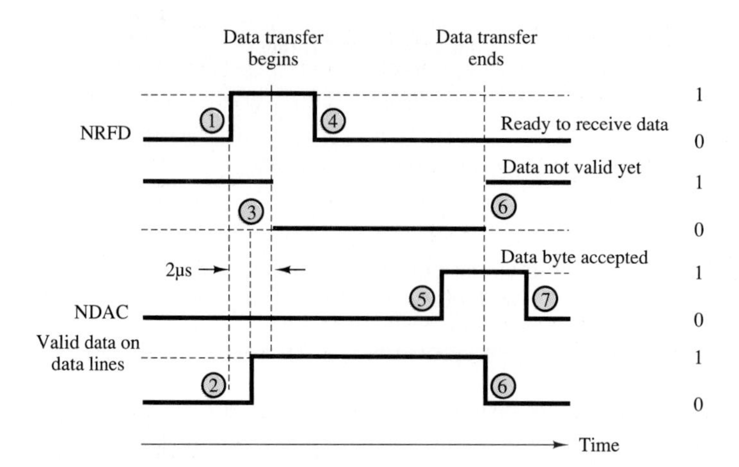

Figure 14.65 IEEE 488 data transmission protocol

3. Two microseconds after putting valid data on the data lines, the active talker pulls the data valid (DAV) line to zero volts and thereby signals the active listeners to read the information on the data bus. The 2-μs interval is required to allow the data put on the data lines to reach (settle to) valid logic levels.
4. After the DAV is asserted, the listeners respond by pulling the NRFD line back down to zero. This prevents any additional data transfers from being initiated. The listeners also begin accepting the data byte at their own rates.
5. When each listener has accepted the data, it releases the not data accepted (NDAC) line. Only when the last active listener has released its hold on the NDAC line will that line go to its high-voltage-level state.
6. (a) When the active talker sees that NDAC has come up to its high state, it stops driving the data line. (b) At the same time, the talker releases the DAV line, ending the data transfer. The talker may now put the next byte on the data bus.
7. The listeners pull down the NDAC line back to zero volts and put the byte "away."

Each of the instruments present on the data bus is distinguished by its own address, which is known to the controller; thus, the controller determines who the active talkers and listeners are on the bus by *addressing* them. To implement this and other functions, the controller uses the five control lines. Of these, ATN (attention) is used as a switch to indicate whether the controller is addressing or instructing the devices on the bus, or whether data transmission is taking place: when ATN is logic 1, the data lines contain either control information or addresses; with ATN = 1, only the controller is enabled to talk. When ATN = 0, only the devices that have been addressed can use the data lines. The IFC (interface clear) line is used to initialize the bus, or to clear it and reset it to a known condition in case of incorrect transmission. The REN (remote enable) line enables a remote instrument to be controlled by the bus; thus, any function that might normally be performed manually on the instrument (e.g., selecting a range or mode of operation) is now controlled by the bus via the data lines. The SRQ (service request) line is used by instruments on the bus whenever the instrument is ready to send or receive data; however, it is the controller who decides when to service the request. Finally, the EOI (end or identify) line can be used in two modes: when it is used by a talker, it signifies the end of a message; when it is used by the controller, it serves as a *polling* line, that is, a line used to interrogate the instrument about its data output.

Although it was mentioned earlier that the IEEE 488 bus can be used only over distances of up to 20 m, it is possible to extend its range of operation by connecting remote IEEE 488 bus systems over telephone communication lines. This can be accomplished by means of *bus extenders,* or by converting the parallel data to serial form (typically, in RS-232 format) and by transmitting the serial data over the phone lines by means of a modem. Serial communications and the RS-232 standard are discussed in the next section.

The RS-232 Standard

The primary reason why parallel transmission of data is not used exclusively is the limited distance range over which it is possible to transmit data on a parallel bus. Although there are techniques which permit extending the range for parallel transmission, these are complex and costly. Therefore, **serial transmission** is frequently used, whenever data is to be transmitted over a significant distance. Since serial data travels along one single path and is transmitted one bit at a time, the cabling costs for long distances are relatively low; further, the transmitting and receiving units are also limited to processing just one signal, and are also much simpler and less expensive. Two modes of operation exist for serial transmission: **simplex,** which corresponds to transmission in one direction only; and **duplex,** which permits transmission in either direction. Simplex transmission requires only one receiver and one transmitter, at each end of the link; on the other hand, duplex transmission can occur in one of two manners: **half-duplex** and **full-duplex.** In the former, although transmission can take place in both directions, it cannot occur simultaneously in both directions; in the latter case, both ends can simultaneously transmit and receive. Full-duplex transmission is usually implemented by means of four wires.

The data rate of a serial transmission line is measured in bits per second, since the data is transmitted one bit at a time. The unit of 1 bit/s is called a **baud;** thus, reference is often made to the baud rate of a serial transmission. The baud rate can be translated into a parallel transmission rate in words per second if the structure of the word is known; for example, if a word consists of 10 bits (start and stop bits plus an 8-bit data word) and the transmission takes place at 1,200 baud, 120 words are being transmitted every second. Typical data rates for serial transmission are standardized; the most common rates (familiar to the users of personal computer modem connections) are 300, 600, 1,200, and 2,400 baud. Baud rates can be as low as 50 baud or as high as 19,200 baud.

Like parallel transmission, serial transmission can also occur either synchronously or asynchronously. In the serial case, it is also true that asynchronous transmission is less costly but not as fast. A handshake protocol is also required for asynchronous serial transmission, as explained in the following. The most popular data-coding scheme for serial transmission is, once again, the ASCII code, consisting of a 7-bit word plus a **parity bit,** for a total of 8 bits per character. The role of the parity bit is to permit error detection in the event of erroneous reception (or transmission) of a bit. To see this, let us discuss the sequence of handshake events for asynchronous serial transmission and the use of parity bits to correct for errors. In serial asynchronous systems, handshaking is performed by using start and stop bits at the beginning and end of each character that is transmitted. The beginning of the transmission of a serial asynchronous word is announced by the "start" bit, which is always a 0 state bit. For the next five to eight successive bit times (depending on the code and the number of bits that specify the word length in that code), the line is switched to the 1 and 0 states required to represent the character being sent. Following the last bit of the data and the parity bit (which will be explained next), there is one bit or more in the 1 state, indicating "idle." The time period associated with this transmission is called the "stop" bit interval.

If noise pulses affect the transmission line, it is possible that a bit in the transmission could be misread. Thus, following the 5 to 8 transmitted data bits, there is a parity bit that is used for error detection. Here is how the parity bit works. If the transmitter keeps track of the number of 1s in the word being sent, it can send a parity bit, a 1 or a 0, to ensure that the total number of 1s sent is always even (even parity) or odd (odd parity). Similarly, the receiver can keep track of the 1s received to see whether there was an error with the transmission. If an error is detected, retransmission of the word can be requested.

Serial data transmission occurs most frequently according to the **RS-232 standard.** The RS-232 standard is based on the transmission of voltage pulses at a preselected baud rate; the voltage pulses are in the range -3 to -15 V for a logic 0 and in the range $+3$ to $+15$ V for a logic 1. It is important to note that this amounts to a negative logic convention and that the signals are *not* TTL-compatible. The distance over which such transmission can take place is up to approximately 17 m (50 ft). The RS-232 standard was designed to make the transmission of digital data compatible with existing telephone lines; since phone lines were originally designed to carry analog voice signals, it became necessary to establish some standard procedures to make digital transmission possible over them. The resulting standard describes the mechanical and electrical characteristics of the interface between *data terminal equipment* (DTE) and *data communication equipment* (DCE). DTE consists of computers, terminals, digital instruments, and related peripherals; DCE includes all of those devices that are used to encode digital data in a format that permits their transmission over telephone lines. Thus, the standard specifies how data should be presented by the DTE to the DCE so that digital data can be transmitted over voice lines.

A typical example of DCE is the **modem.** A modem converts digital data to audio signals that are suitable for transmission over a telephone line and is also capable of performing the reverse function, by converting the audio signals back to digital form. The term *modem* stands for *mod*ulate-*dem*odulate, because a modem modulates a sinusoidal carrier using digital pulses (for transmission) and demodulates the modulated sinusoidal signal to recover the digital pulses (at reception). Three methods are commonly used for converting digital pulses to an audio signal: **amplitude-shift keying, frequency-shift keying,** and **phase-shift keying,** depending on whether the amplitude, phase, or frequency of the sinusoid is modulated by the digital pulses. Figure 14.66 depicts the essential block of a data transmission system based on the RS-232 standard, as well as examples of digital data encoded for transmission over a voice line.

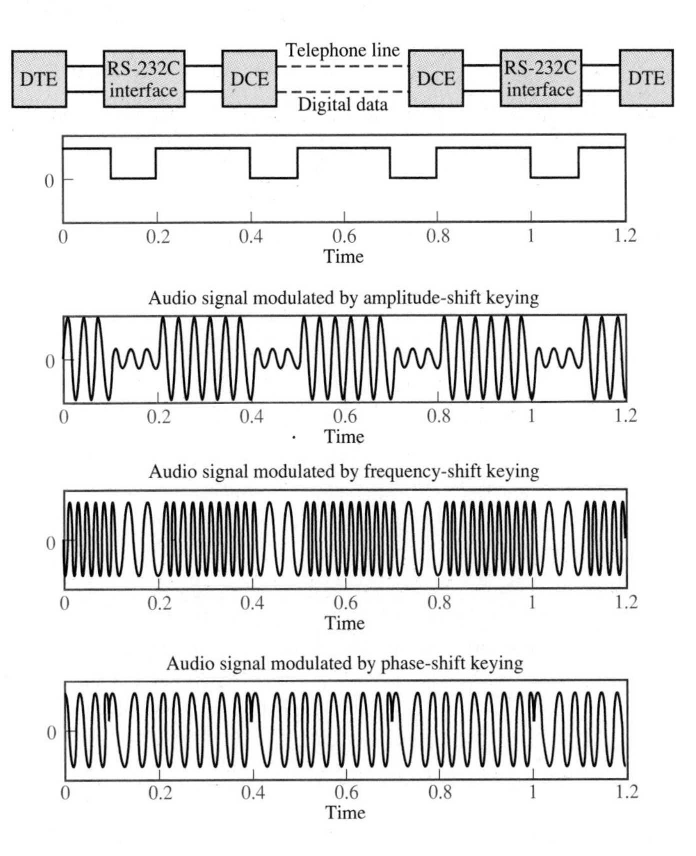

Figure 14.66 Digital data encoded for analog transmission

In addition to the function just described, however, the RS-232 standard also provides a very useful set of specifications for the direct transmission of digital data between computers and instruments. In other words, communication between digital terminal instruments may occur directly in digital form (i.e., without digital communication devices encoding the digital data in a form compatible with analog voice lines). Thus, this standard is also frequently used for direct digital communication.

The RS-232 standard can be summarized as follows:

- Data signals are encoded according to a negative logic convention using voltage levels of -3 to -15 V for logic 1 and +3 to +15 V for logic 0.
- Control signals use a positive logic convention (opposite to that of data signals).
- The maximum shunt capacitance of the load cannot exceed 2,500 pF; this, in effect, limits the maximum length of the cables used in the connection.
- The load resistance must be between 300 Ω and 3 kΩ.
- Three wires are used for data transmission. One wire each is used for receiving and transmitting data; the third wire is a signal return line (signal ground). In addition, there are 22 wires that can be used for a variety of control purposes between the DTE and DCE.
- The male part of the connector is assigned to the DTE and the female part to the DCE. Figure 14.67 labels each of the wires in the 25-pin

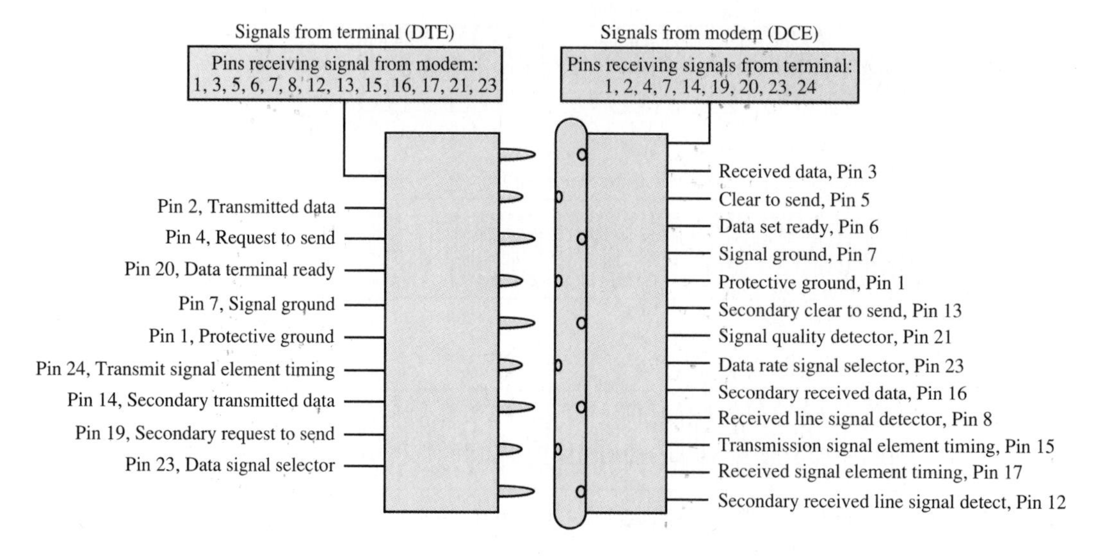

Figure 14.67 RS-232 connections

connector. Since each side of the connector has a *receive* and a *transmit* line, it has been decided by convention that the DCE transmits on the transmit line and receives on the receive line, while the DTE receives on the transmit line and transmits on the receive line.

- The baud rate is limited by the length of the cable; for a 17-m length, any rate from 50 baud to 19.2 kbaud is allowed. If a longer cable connection is desired, the maximum baud rate will decrease according to the length of the cable, and **line drivers** can be used to amplify the signals, which are transmitted over twisted-pair wires. Line drivers are simply signal amplifiers that are used directly on the digital signal, prior to encoding. For example, the signal generated by a DTE device (say a computer) may be transmitted over a distance of up to 3,300 m (at a rate of 600 baud) prior to being encoded by the DCE.
- The serial data can be encoded according to any code, although the ASCII code is by far the most popular.

DRILL EXERCISE

14.24 Determine the actual binary data sent by a digital voltmeter reading 15.06 V over an IEEE 488 bus if the data is encoded in ASCII format (see Table 14.5). Assume that the sequence is from most to least significant digit.

CONCLUSION

- Measurements and instrumentation are among the most important areas of electrical engineering because virtually all engineering disciplines require the ability to perform measurements of some kind.
- A measurement system consists of three essential elements: a sensor, signal-conditioning circuits, and recording or display devices. The last are often based on digital computers.
- Sensors are devices that convert a change in a physical variable into a corresponding change in an electrical variable—typically a voltage. A broad range of sensors exist to measure virtually all physical phenomena. Proper wiring, grounding, and shielding techniques are required to minimize undesired interference and noise.
- Often, sensor outputs need to be conditioned before further processing can take place. The most common signal-conditioning circuits are instrumentation amplifiers and active filters.

- If the conditioned sensor signals are to be recorded in digital form by a computer, it is necessary to perform an analog-to-digital conversion process; timing and comparator circuits are also often used in this context.
- Once the digital data corresponding to the measured quantity is available, the need for digital data transmission may arise. Standard transmission formats exist, of which the two most common are the IEEE 488 and the RS-232 standards.

KEY TERMS

Amplitude-Shift Keying *769*
Analog-to-Digital Converter (ADC) *737*
Astable Multivibrator *755*
Asynchronous Transmission *762*
Baud *767*
Bessel Filter *727*
Butterworth Filter *727*
Butterworth Polynomial *727*
Calibrated Orifice *713*
Capacitive Coupling *719*
Cauer (Elliptical) Filter *727*
Chebyshev Filter *727*
Cold Junction *714*
Comparator *747*
Comparator with Offset *749*
Conductive Coupling *719*
Constant-*K* (Sallen and Key) Filter *729*
Conversion Time *740*
Differential Measurement System *717*
Differential Pressure Measurement *713*
Digital-to-Analog Converter (DAC) *733*
Dual One-Shot *760*
Duplex Transmission *767*
Equiripple Filter *728*
Falling Edge *758*
Flash ADC *739*
Floating Signal Source *716*
Frequency-Shift Keying *769*
Full-Duplex Transmission *767*
General-Purpose Instrument Bus (GPIB) *764*
Ground Loop *717*
Ground-Referenced Measurement System *717*
Grounded Signal Source *716*
Half-Duplex Transmission *767*
Handshake Protocol *762*
Hot-Film Anemometer *713*
Hot-Wire Anemometer *713*
IEEE 488 Standard *764*
Inductive Coupling *719*
Input-Output Transfer Characteristic *749*
Instrumentation Amplifier (IA) *721*
Integrating ADC *738*
Inverting Comparator *747*
Level (Zener) Clamp *750*
Line Driver *771*
Measurement System *710*
Modem *769*
Mutual Inductance *719*

ANSWERS TO DRILL EXERCISES

14.1 66 dB

14.2 80 dB

14.3 120 dB

14.4 −6.1 dB

14.5 −20.1 dB

14.6 40 Ω

14.7 $n = 7$

14.8 42.1 dB

14.9 $R_1 = R_2 = 1$ kΩ; $C_1 = C_2 = 100$ μF; $K = 2$

14.11 3.66 mV

14.12 $\delta v = 47.1$ mV; $R_F = 94.2$ Ω

14.13 −13.5 V

14.14 0.76 V

14.15 10

14.16 $f_{max} = 10$ kHz

14.21 63 μs

14.22 9,300 Ω

14.23 11.5 μF

14.24 31 35 2E 30 36

DEVICE DATA SHEETS

AD625 Instrumentation Amplifier

SPECIFICATIONS (typical @ $V_S = \pm 15$ V, $R_L = 2$ kΩ, and $T_A = +25°$C unless otherwise specified)

Model	*AD625A/J/S* Min	Typ	Max	*AD625B/K* Min	Typ	Max	*AD625C* Min	Typ	Max	Units
Gain										
Gain equation		$\frac{2R_F}{R_G} + 1$			$\frac{2R_F}{R_G} + 1$			$\frac{2R_F}{R_G} + 1$		
Gain range	1		10,000	1		10,000	1		10,000	
Gain error		±0.035	±0.05		±0.02	±0.03		±0.01	±0.02	%
Nonlinearity, gain = 1–256			±0.005			±0.002			±0.001	%
gain > 256			±0.01			±0.008			±0.005	%
Gain vs. temp. gain < 1,000			5			5			5	ppm/°C
Input										
Input impedance										
Differential resistance		1			1			1		GΩ
Differential capacitance		4			4			4		pF
Common-mode resistance		1			1			1		GΩ
Common-mode capacitance		4			4			4		pF
Input voltage range										
Differ. input linear (V_D)			±10			±10			±10	V
Common-mode linear (V_{cm})		$12\text{ V} - \left(\frac{G}{2} \times V_D\right)$			$12\text{ V} - \left(\frac{G}{2} \times V_D\right)$			$12\text{ V} - \left(\frac{G}{2} \times V_D\right)$		
Common-mode rejection ratio										
DC to 60 Hz with 1-kΩ source imbalance										
$G = 1$	70	75		75	85		80	90		dB
$G = 10$	90	95		95	105		100	115		dB
$G = 100$	100	105		105	115		110	125		dB
$G = 1,000$	110	115		115	125		120	140		dB
Output Rating		±10 V @ 5 mA			±10 V @ 5 mA			±10 V @ 5 mA		
Dynamic response										
Small Signal −3 dB										
$G = 1$ ($R_F = 20$ kΩ)		650			650			650		kHz
$G = 10$		400			400			400		kHz
$G = 100$		150			150			150		kHz
$G = 1,000$		25			25			25		kHz
Slew rate		5.0			5.0			5.0		V/μs
Setting time to 0.01%, 20-V step										
$G = 1$ to 200		15			15			15		μs
$G = 500$		35			35			35		μs
$G = 1,000$		75			75			75		μs
Power supply										
Power supply range		±5 to ± 18			±5 to ± 18			±5 to ± 18		V
Quiescent current		3.5	5		3.5	5		3.5	5	mA

AD582 Low-Cost Sample-and-Hold Amplifier

Features

- Suitable for 12-bit applications
- High sample/hold current ratio: 10^7
- Low acquisition time: 6 μs to 0.1%
- Low charge transfer: < 2 pC
- High input impedance in sample-and-hold modes
- Connect in any op-amp configuration
- Differential logic inputs
- MIL-STD-883 compliant versions available

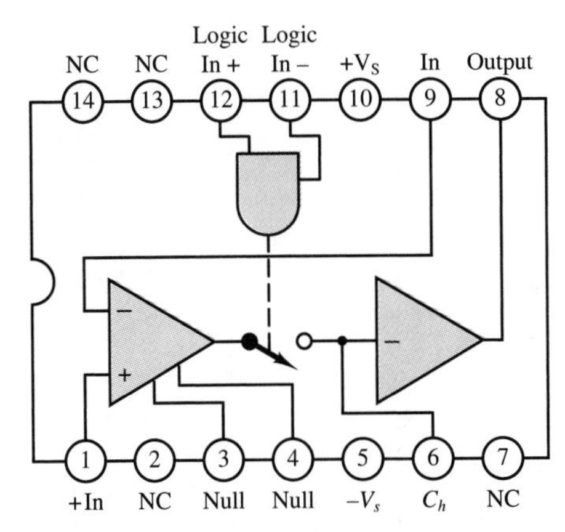

AD582—Specifications

Model	AD582K
Sample/hold characteristics	
Acquisition time, 10 V step to 0.1%, C_H = 100 pF	6 μs
Acquisition time, 10 V step to 0.01%, C_H = 1,000 pF	25 μs
Aperture delay, 20 V p-p input, hold 0 V	200 ns
Aperture jitter, 20 V p-p input, hold 0 V	15 ns
Settling time, 20 V p-p input, hold 0 V, to 0.01%	0.5 μs
Droop current, steady state, $\pm 10V_{out}$	100 pA max
Droop current, T_{min} to T_{max}	1 nA
Charge transfer	5 pC max (1.5 pC typ)
Sample to hold offset	0.5 mV
Feedthrough capacitance 20 V p-p, 10 kHz input	0.05 pF

AD582—Specifications (continued)

Model	AD582K
Transfer characteristics	
Open loop gain	
$V_{out} = 20$ V p-p, $R_L = 2$ kΩ	25 kΩ min (50 kΩ typ)
Common mode rejection	
$V_{CM} = 20$ V p-p	60 dB min (70 dB typ)
Small signal gain bandwidth	
$V_{out} = 100$ mV p-p, $C_H = 100$ pF	1.5 MHz
Full power bandwidth	
$V_{out} = 20$ V p-p, $C_H = 100$ pF	70 kHz
Slew rate	
$V_{out} = 20$ V p-p, $C_H = 100$ pF	3 V/μs
Output resistance	
Hold mode, $I_{out} = \pm 5$ mA	12 Ω
Linearity	
$V_{out} = 20$ V p-p, $R_L = 2$ kΩ	±0.01%
Output short circuit current	±25 mA
Analog input characteristics	
Offset voltage	6 mV max (2 mV typ)
Offset voltage, T_{min} to T_{max}	4 mV
Bias current	3 μA max (1.5 μA typ)
Offset current	300 nA max (75 nA typ)
Offset current, T_{min} to T_{max}	100 nA
Input capacitance, $f = 1$ MHz	2 pF
Input resistance, sample or hold	
20 V p-p input, $A = +1$	30 MΩ
Absolute max diff input voltage	30 V
Absolute max input voltage, either input	$\pm V_S$
Digital input characteristics	
+Logic input voltage	
Hold mode, T_{min} to T_{max}, −logic @ 0 V	+2 V min
Sample mode, T_{min} to T_{max}, −logic @ 0 V	+0.8 V max
+Logic input current	
Hold mode, +logic @ +5 V, −logic @ 0 V	1.5 μA
Sample mode, +logic @ 0 V, −logic @ 0 V	1 nA
−Logic input current	
Hold mode, +logic @ +5 V, −logic @ 0 V	24 μA
Sample mode, +logic @ 0 V, −logic @ 0 V	4 μA
Absolute max diff input voltage, +L to −L	+15 V/−6 V
Absolute max input voltage, either input	$\pm V_S$

AD558 DACPORT Low Cost, Complete μP-Compatible 8-Bit DAC

Features

- Complete 8-bit DAC
- Voltage output—2 calibrated ranges
- Internal precision band-gap reference
- Single-supply operation: +5 V to +15 V
- Full microprocessor interface
- Fast: 1-μs Voltage settling to $\pm\frac{1}{2}$ LSB

- Low power: 75 mW
- No user trims
- Guaranteed monotonic over temperature
- All errors specified T_{min} to T_{max}
- Small 16-pin DIP and 20-pin PLCC packages
- Single laser-wafer-trimmed chip for hybrids
- Low cost
- MIL-STD-883 compliant version available

Functional Block Diagram

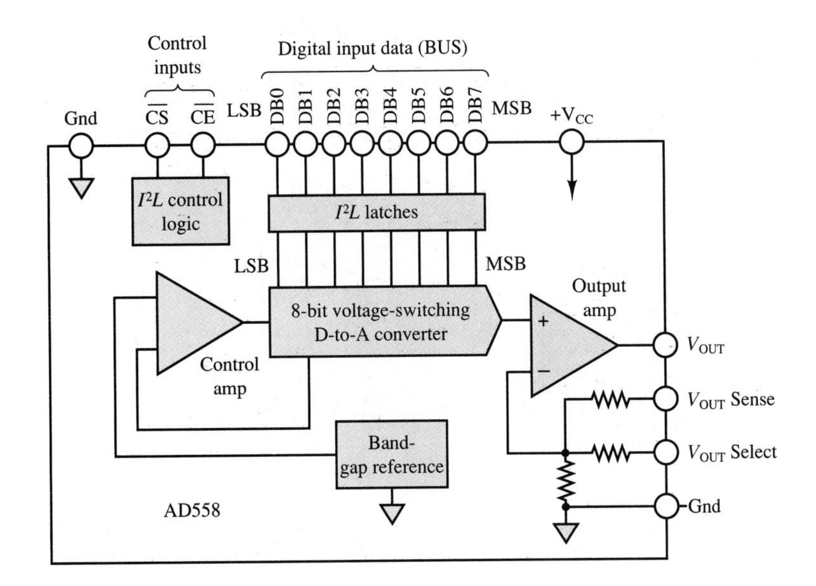

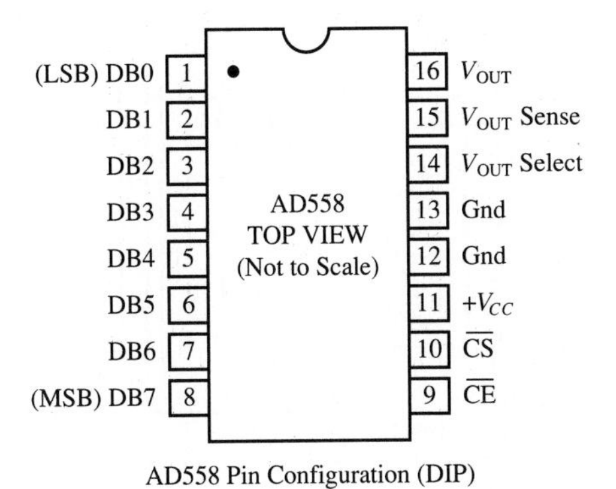

AD558 Pin Configuration (DIP)

AD574A Complete 12-bit A/D Converter

Features

- Complete 12-bit A/D converter with reference and clock
- 8- and 16-bit microprocessor bus interface
- Guaranteed linearity over temperature 0 to +70°C for AD574AJ, K, L; −55°C to +120°C for AD574AS, T, U
- No missing codes over temperature
- 35 μs maximum conversion time
- Buried Zener reference for long-term stability and low gain T.C. 10 ppm/°C max AD574AL; 12.5 ppm/°C max AD574AU
- Ceramic DIP, plastic DIP, or PLCC package
- Accuracy: ±1 bit

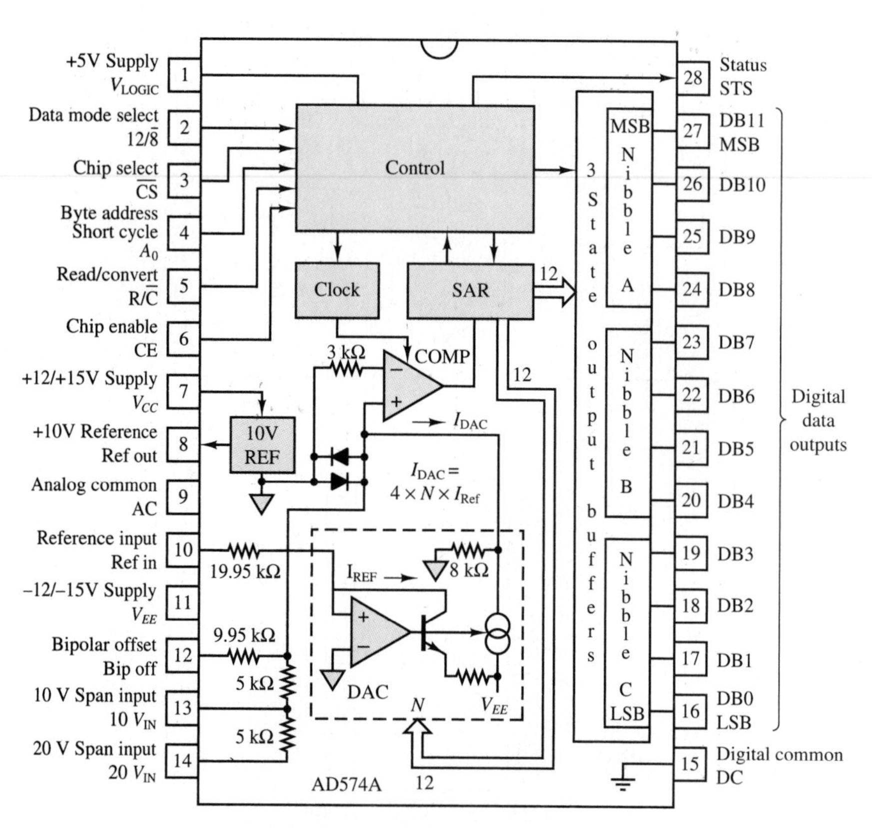

AD574A Block Diagram and Pin Configuration

LM311 High Performance Voltage Comparator

Typical Comparator Design Configuration

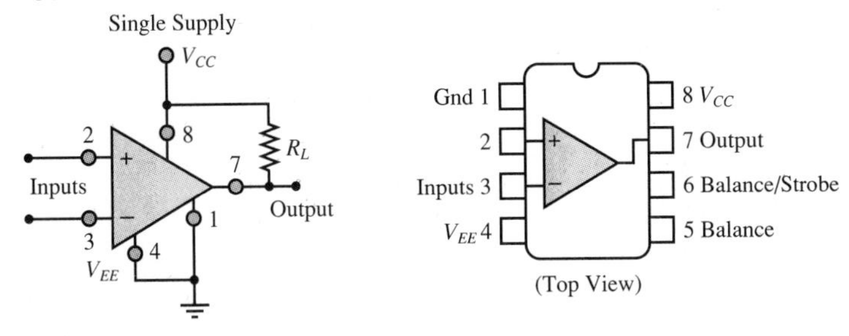

LM555/LM555C Timer

Features

- Direct replacement for SE555/NE555
- Timing from microseconds through hours
- Operates in both astable and monostable modes
- Adjustable duty cycle
- Output can source or sink 200 mA
- Output and supply TTL compatible
- Temperature stability better than 0.005% per °C
- Normally on and normally off output

Applications

- Precision timing
- Pulse generation
- Sequential timing
- Time delay generation
- Pulse width modulation
- Pulse position modulation
- Linear ramp generator

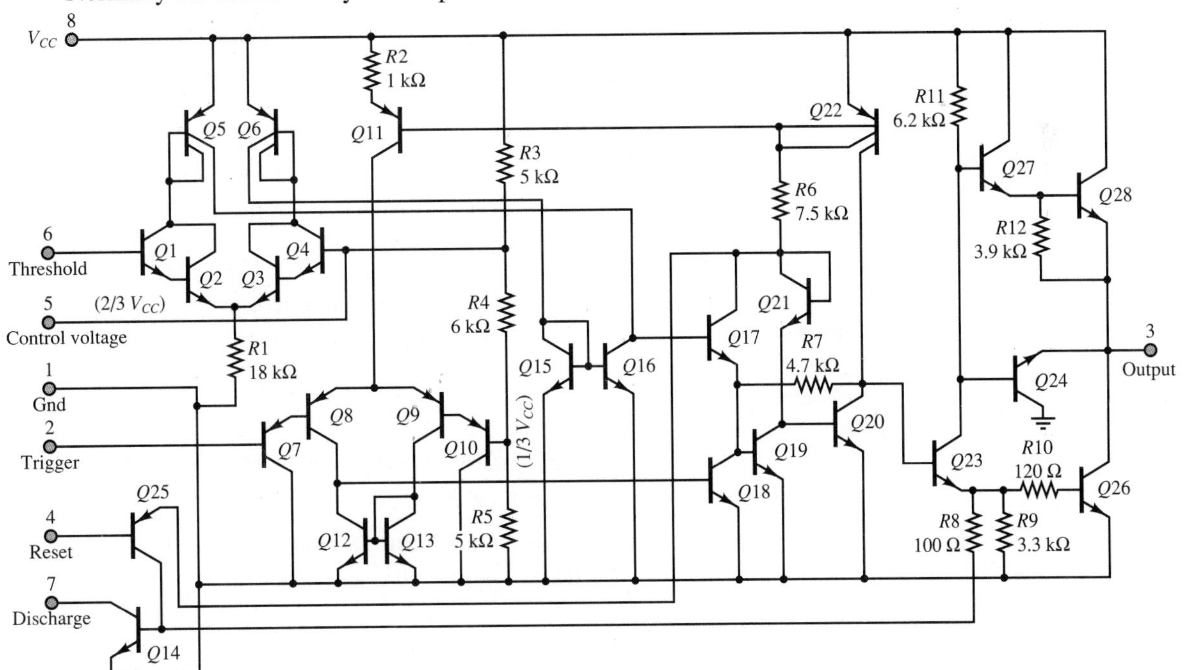

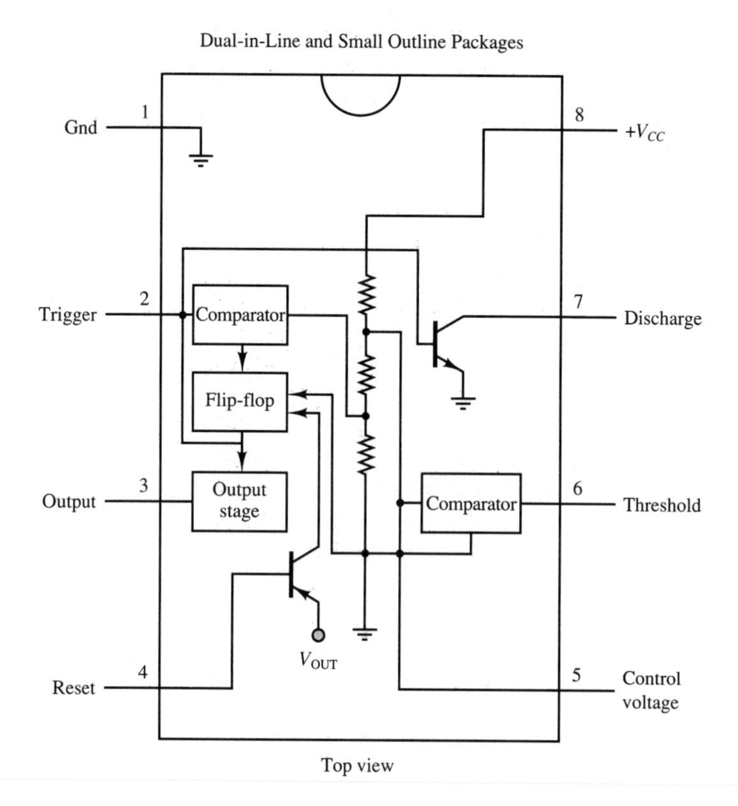

DM54LS123/DM74LS123 Dual Retriggerable One-Shot with Clear and Complementary Outputs

Features

- DC triggered from active-high transition or active-low transition inputs
- Retriggerable to 100% duty cycle
- Compensated for V_{CC} and temperature variations
- Triggerable from CLEAR input
- DTL, TTL compatible
- Input clamp diodes

Connection diagram

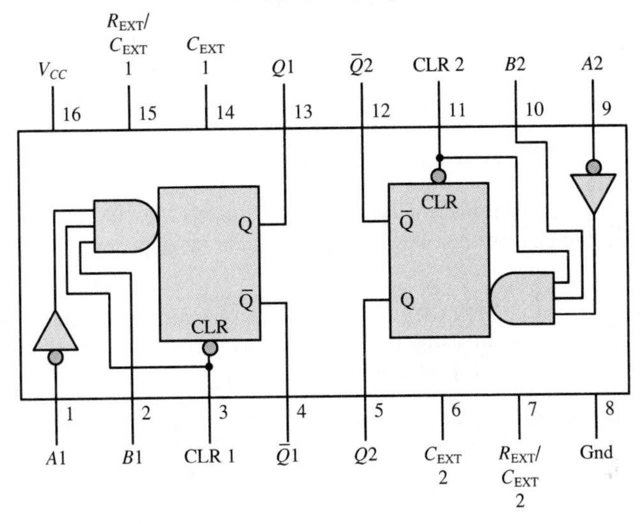

Function table

Inputs			Outputs	
CLEAR	*A*	*B*	*Q*	$\bar{Q}$
L	X	X	L	H
X	H	X	L	H
X	X	L	L	H
H	L	↑	⊓	⊔
H	↓	H	⊓	⊔
↑	L	H	⊓	⊔

H = High logic level
L = Low logic level
X = Can be either low or high
↑ = Positive going transition
↓ = Negative going transition
⊓ = A positive pulse
⊔ = A negative pulse

HOMEWORK PROBLEMS

14.1 Most motorcycles have engine speed tachometers, as well as speedometers, as part of their instrumentation. What differences, if any, are there between the two in terms of transducers?

14.2 Explain the differences between the engineering specifications you would write for a transducer to measure the frequency of an audible sound wave and a transducer to measure the frequency of a visible light wave.

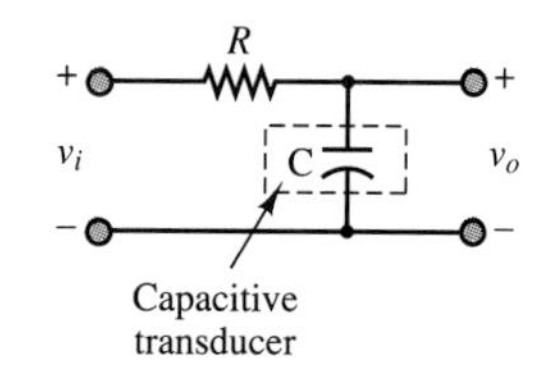

Figure P14.1

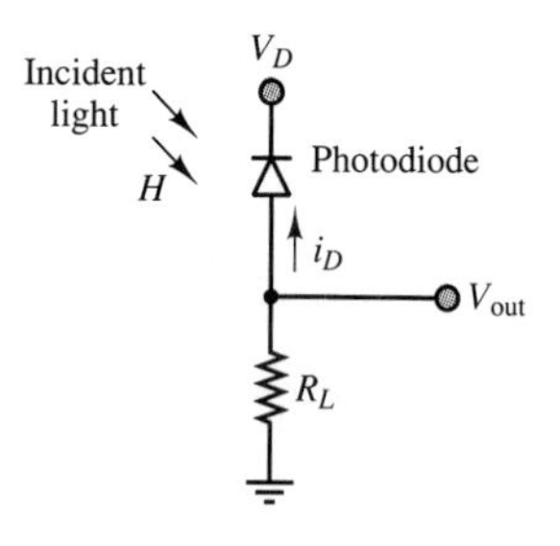

Figure P14.2

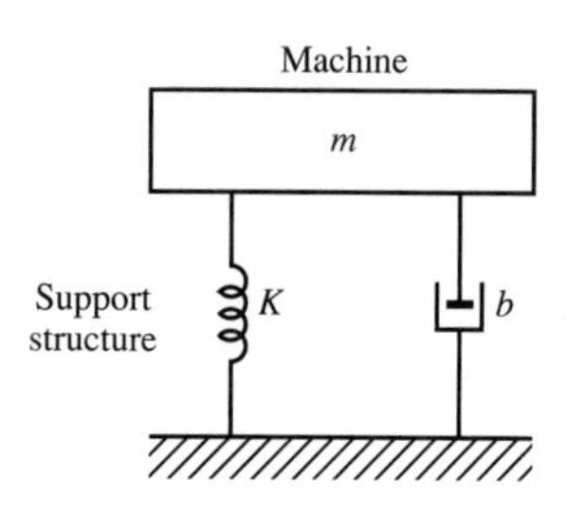

Figure P14.3

14.3 A measurement of interest in the summer is the temperature-humidity index, consisting of the sum of the temperature and the percentage relative humidity. How would you measure this? Sketch a simple schematic diagram.

14.4 Consider a capacitive displacement transducer as shown in Figure P14.1. Its capacitance is determined by the equation

$$C = \frac{0.255A}{d} \text{ Farads}$$

where A = cross-sectional area of the transducer plate (in^2), and d = air-gap length (in). Determine the change in voltage (Δv_0) when the air gap changes from 0.01 in to 0.015 in.

14.5 The circuit of Figure P14.2 may be used for operation of a photodiode. The voltage V_D is a reverse-bias voltage large enough to make diode current, i_D, proportional to the incident light intensity, H. Under this condition, $i_D/H = 0.5\ \mu\text{A-m}^2/\text{W}$.

a. Show that the output voltage, V_{out}, varies linearly with H.
b. If $H = 1{,}500\ \text{W/m}^2$, $V_D = 7.5$ V, and an output voltage of 1 V is desired, determine an appropriate value for R_L.

14.6 G is a material constant equal to 0.055 V-m/N for quartz in compressive stress and 0.22 V-m/N for polyvinylidene fluoride in axial stress.

a. A force sensor uses a piezoelectric quartz crystal as the sensing element. The quartz element is 0.25 in thick and has a rectangular cross section of 0.09 in^2. The sensing element is compressed and the output voltage measured across the thickness. What is the output of the sensor in volts per newton?
b. A polyvinylidene fluoride film is used as a piezoelectric load sensor. The film is 30 μm thick, 1.5 cm wide, and 2.5 cm in the axial direction. It is stretched by the load in the axial direction, and the output voltage is measured across the thickness. What is the output of the sensor in volts per newton?

14.7 Let b be the damping constant of the mounting structure of a machine as pictured in Figure P14.3. It must be determined experimentally. First, the spring constant, K, is determined by measuring the resultant displacement under a static load. The mass, m, is directly measured. Finally, the damping ratio, ξ, is measured using an impact test. The damping constant is given by $b = 2\xi\sqrt{Km}$. If the allowable levels of error in the measurements of K, m, and ξ are ± 5 percent, ± 2 percent, and ± 10 percent respectively, estimate a percentage error limit for b.

14.8 The quality control system in a plant that makes acoustical ceiling tile uses a proximity sensor to measure the thickness of the wet pulp layer every 2 feet along the sheet, and the roller speed is adjusted based on the last 20 measurements. Briefly, the speed is adjusted unless the probability that the mean thickness lies within $\pm 2\%$ of the sample mean exceeds 0.99.

A typical set of measurements (in mm) is as follows:

$$8.2, 9.8, 9.92, 10.1, 9.98, 10.2, 10.2, 10.16, 10.0, 9.94,$$

$$9.9, 9.8, 10.1, 10.0, 10.2, 10.3, 9.94, 10.14, 10.22, 9.8$$

Would the speed of the rollers be adjusted based on these measurements?

14.9 Discuss and contrast the following terms:

a. Measurement accuracy.
b. Instrument accuracy.
c. Measurement error.
d. Precision.

14.10 Four sets of measurements were taken on the same response variable of a process using four different sensors. The true value of the response was known to be constant. The four sets of data are shown in Figure P14.4. Rank these data sets (and hence the sensors) with respect to:

a. Precision.
b. Accuracy.

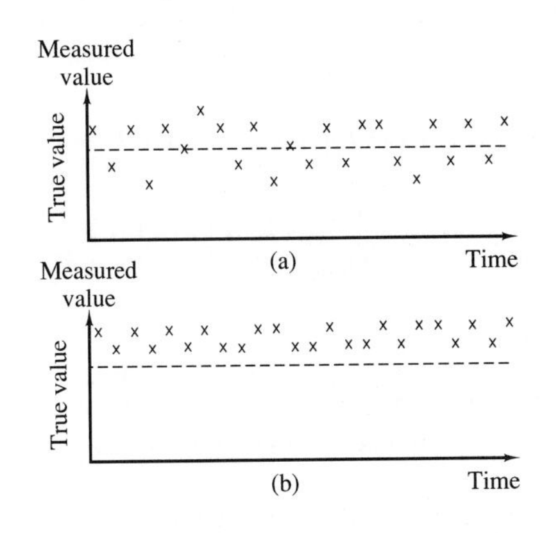

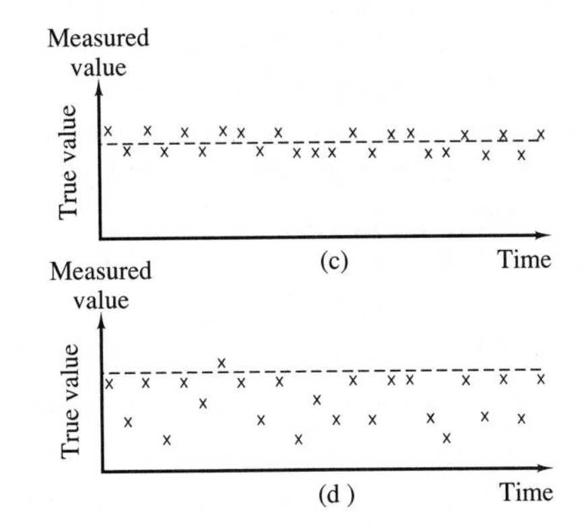

Figure P14.4

14.11 For the instrumentation amplifier of Figure P14.5, find the gain of the input stage if $R_1 = 1\ \text{k}\Omega$ and $R_2 = 5\ \text{k}\Omega$.

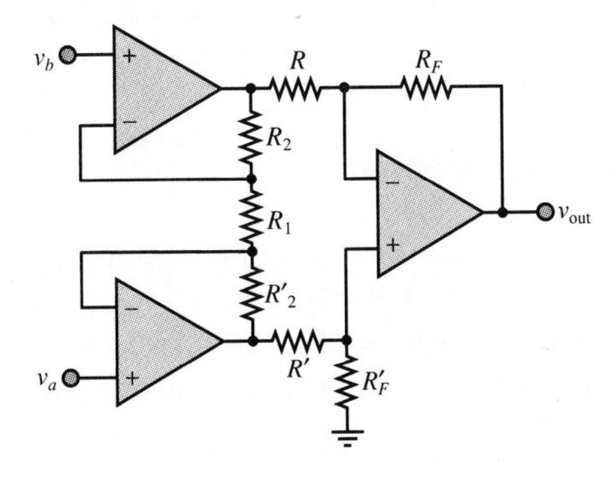

Figure P14.5

14.12 Consider again the instrumentation amplifier of Figure P14.5. Let $R_1 = 1\ \text{k}\Omega$. What value of R_2 should be used to make the gain of the input stage equal 50?

14.13 Again consider the instrumentation amplifier of Figure P14.5. Let $R_2 = 10\ \text{k}\Omega$. What value of R_1 will yield an input-stage gain of 16?

14.14 For the IA of Figure 14.16, find the gain of the input stage if $R_1 = 1\ \text{k}\Omega$ and $R_2 = 10\ \text{k}\Omega$.

14.15 For the IA of Figure 14.16, find the gain of the input stage if $R_1 = 1.5\ \text{k}\Omega$ and $R_2 = 80\ \text{k}\Omega$.

14.16 Find the differential gain for the IA of Figure 14.16 if $R_2 = 5\ \text{k}\Omega, R_1 = R' = R = 1\ \text{k}\Omega$, and $R_F = 10\ \text{k}\Omega$.

14.17 Suppose, for the circuit of Figure P14.5, that $R_F = 200\ \text{k}\Omega$, $R = 1\ \text{k}\Omega$, and $\Delta R = 2\%$ of R. Calculate the common-mode rejection ratio (CMRR) of the instrumentation amplifier. Express your result in dB.

14.18 Given the instrumentation amplifier of Figure P14.5, with the component values of Problem 14.17, calculate the mismatch in gains for the differential components. Express your result in dB.

14.19 Given $R_F = 10\ \text{k}\Omega$ and $R_1 = 2\ \text{k}\Omega$ for the IA of Figure 14.16, find R and R_2 so that a differential gain of 900 can be achieved.

14.20 Replace the cutoff frequency specification of Example 14.3 with $\omega_C = 10$ rad/s and determine the order of the filter required to achieve 40 dB attenuation at $\omega_S = 24$ rad/s.

14.21 The circuit of Figure P14.6 represents a low-pass filter with gain.

a. Derive the relationship between output amplitude and input amplitude.
b. Derive the relationship between output phase angle and input phase angle.

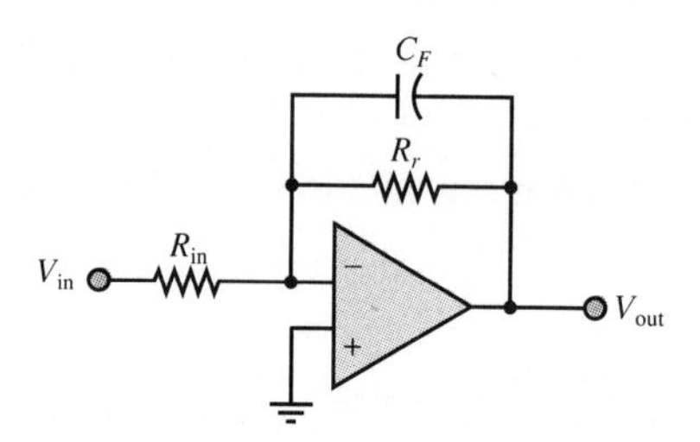

Figure P14.6

14.22 Consider again the circuit of Figure P14.6. Let $R_{in} = 20\ \text{k}\Omega, R_F = 100\ \text{k}\Omega$, and $C_F = 100$ pF. Determine an expression for $v_{out}(t)$ if $v_{in}(t) = 2\sin(2{,}000\pi t)$ V.

14.23 Derive the frequency response of the low-pass filter of Figure 14.22.

14.24 Derive the frequency response of the high-pass filter of Figure 14.22.

14.25 Derive the frequency response of the band-pass filter of Figure 14.22.

14.26 Consider again the circuit of Figure P14.6. Let $C_F = 100$ pF. Determine appropriate values for R_{in} and R_F if it is desired to construct a filter having a cutoff frequency of 20 kHz and a gain magnitude of 5.

14.27 Design a second-order Butterworth high-pass filter with a 10-kHz cutoff frequency, a DC gain of 10, $Q = 5$, and $V_S = \pm 15$ V.

14.28 Design a second-order Butterworth high-pass filter with a 25-kHz cutoff frequency, a DC gain of 15, $Q = 10$, and $V_S = \pm 15$ V.

14.29 The circuit shown in Figure P14.7 is claimed to exhibit a second-order Butterworth low-pass voltage gain characteristic. Derive the characteristic and verify the claim.

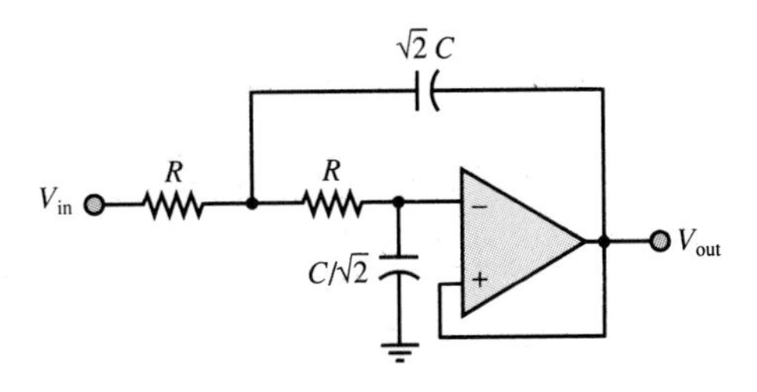

Figure P14.7

14.30 Design a second-order Butterworth low-pass filter with a 15-kHz cutoff frequency, a DC gain of 15, $Q = 5$, and $V_S = \pm 15$ V.

14.31 Design a band-pass filter with a low cutoff frequency of 200 Hz, a high cutoff frequency of 1 kHz, and a pass-band gain of 4. Calculate the value of Q for the filter. Also, draw the approximate frequency response of this filter.

14.32 Using the circuit of Figure P14.7, design a second-order low-pass Butterworth filter with a cutoff frequency of 10 Hz.

14.34 The circuit shown in Figure P14.8 exhibits low-pass, high-pass, and band-pass voltage gain characteristics, depending on whether the output is taken at node 1, node 2, or node 3. Find the transfer functions relating each of these outputs to V_{in}, and determine which is which.

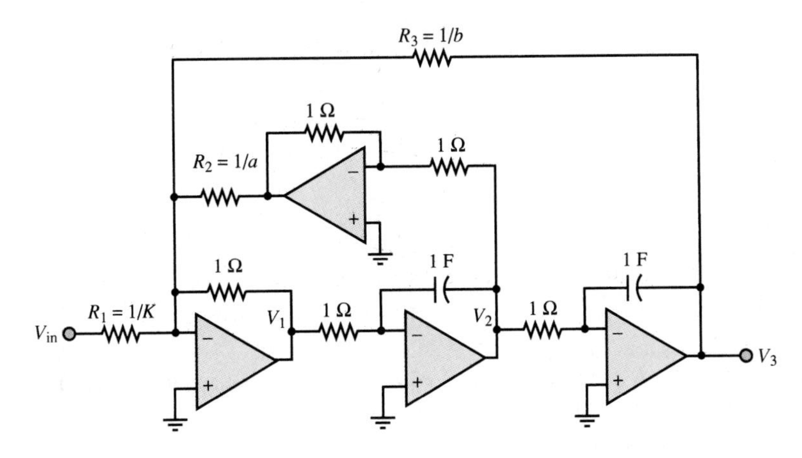

Figure P14.8

14.35 The filter shown in Figure P14.9 is called an *infinite-gain multiple-feedback filter.* Derive the following expression for the filter's frequency response:

$$H(j\omega) = \frac{-(1/R_3R_2C_1C_2)R_3/R_1}{(j\omega)^2 + \left(\frac{1}{R_1C_1} + \frac{1}{R_2C_1} + \frac{1}{R_3C_1}\right)j\omega + \frac{1}{R_3R_2C_1C_2}}$$

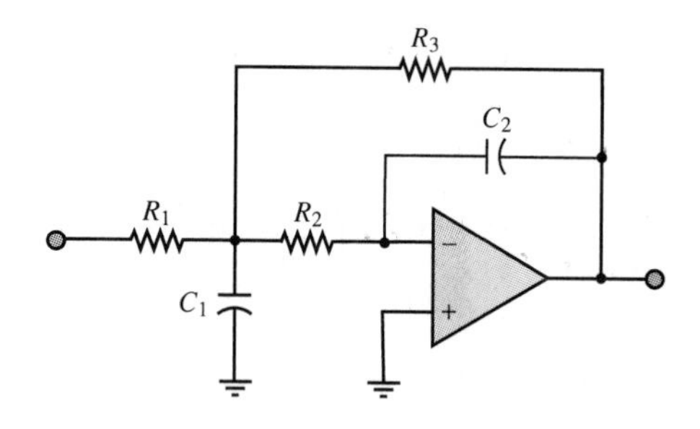

Figure P14.9

14.36 The filter shown in Figure P14.10 is a Sallen and Key band-pass filter circuit, where K is the DC gain of the filter. Derive the following expression for the filter's frequency response:

$$H(j\omega) = \frac{j\omega K/R_1C_1}{(j\omega)^2 + j\omega\left(\frac{1}{R_1C_1} + \frac{1}{R_3C_2} + \frac{1}{R_3C_1} + \frac{1-K}{R_2C_1}\right) + \frac{R_1+R_2}{R_1R_2R_3C_1C_2}}$$

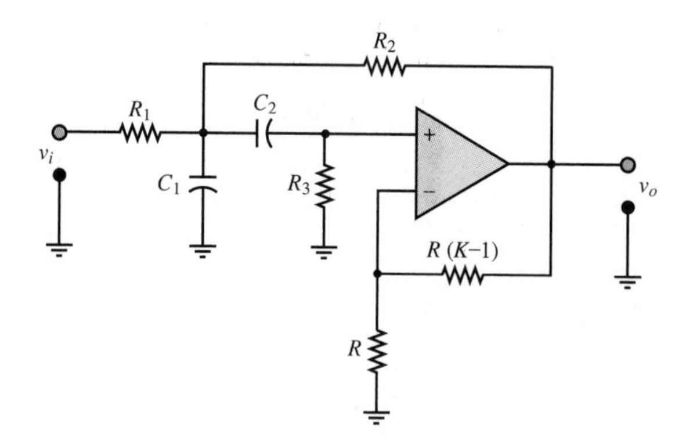

Figure P14.10

14.37 Show that the expression for Q in the filter of Problem 14.35 is given by

$$\frac{1}{Q} = \sqrt{R_2R_3\frac{C_2}{C_1}}\left(\frac{1}{R_1} + \frac{1}{R_2} + \frac{1}{R_3}\right)$$

14.38 List two advantages of digital signal processing over analog signal processing.

14.39 Discuss the role of a multiplexer in a data acquisition system.

14.40 The circuit shown in Figure P14.11 represents a sample-and-hold circuit, such as might be used in a successive-approximation ADC. Assume that the JFET is turned *on* when V_G is high, and *off* when V_G is low. Explain the operation of the circuit.

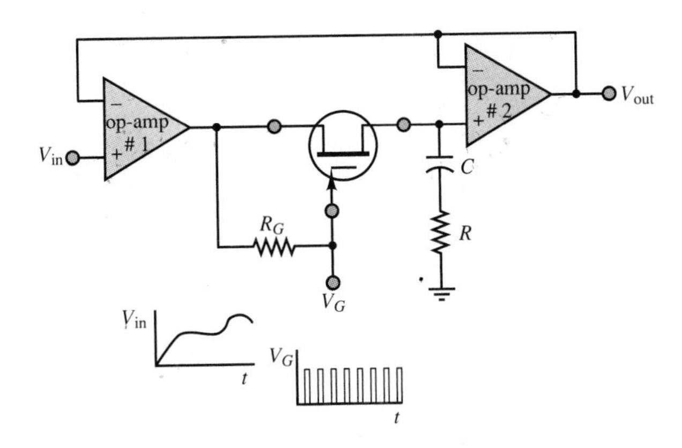

Figure P14.11

14.41 For the circuit shown in Figure P14.11, let V_{in} be a 1 kHz sinusoidal signal with 0° phase angle, 0 V DC offset, and 20 V peak-to-peak amplitude. Let V_G be a rectangular pulse train, with pulse width 10 μs, and period 100 μs, with the leading edge of the first pulse at $t = 0$.

a. Sketch V_{out} if the RC circuit has a time constant equal to 20 μs.
b. Sketch V_{out} if the RC circuit has a time constant equal to 1 ms.

14.42 The unsigned decimal number 12_{10} is inputted to a four-bit DAC. Given that $R_F = R_0/15$, logic 0 corresponds to 0 V, and logic 1 corresponds to 4.5 V,

a. What is the output of the DAC?
b. What is the maximum voltage that can be outputted from the DAC?
c. What is the resolution over the range 0 to 4.5 V?
d. Find the number of bits required in the DAC if an improved resolution of 20 mV is desired.

14.43 The unsigned decimal number 215_{10} is inputted to an eight-bit DAC. Given that $R_F = R_0/255$, logic 0 corresponds to 0 V, and logic 1 corresponds to 10 V,

a. What is the output of the DAC?
b. What is the maximum voltage that can be outputted from the DAC?
c. What is the resolution over the range 0 to 10 V?
d. Find the number of bits required in the DAC if an improved resolution of 3 mV is desired.

14.44 The circuit shown in Figure P14.12 represents a simple 4-bit digital-to-analog converter. Each switch is controlled by the corresponding bit of the digital number—if the bit is 1 the switch is up; if the bit is 0 the switch is down. Let the digital number be represented by $b_3b_2b_1b_0$. Determine an expression relating v_o to the binary input bits.

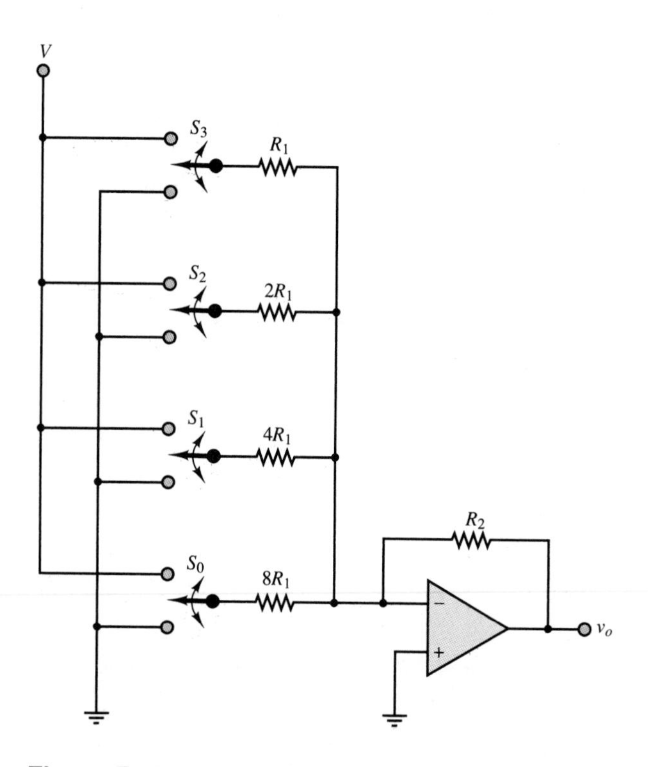

Figure P14.12

14.45 The unsigned decimal number 98_{10} is inputted to an eight-bit DAC. Given that $R_F = R_0/255$, logic 0 corresponds to 0 V and logic 1 corresponds to 4.5 V,

a. What is the output of the DAC?
b. What is the maximum voltage that can be outputted from the DAC?
c. What is the resolution over the range 0 to 4.5 V?
d. Find the number of bits required in the DAC if an improved resolution of 0.5 mV is desired.

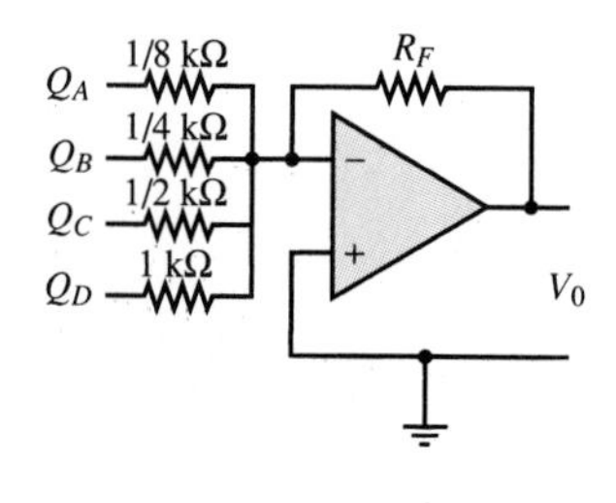

Figure P14.13

14.46 For the DAC circuit shown in Figure P14.13 (using an ideal op-amp), what value of R_F will give an output range of $-10 \leq V_0 \leq 0$ V? Assume that logic 0 = 0 V and logic 1 = 5 V.

14.47 Explain how to redesign the circuit of Figure P14.12 so that the overall circuit is a "noninverting" device.

14.48 The circuit of Figure P14.14 has been suggested as a means of implementing the switches needed for the digital-to-analog converter of Figure P14.12. Explain how the circuit works.

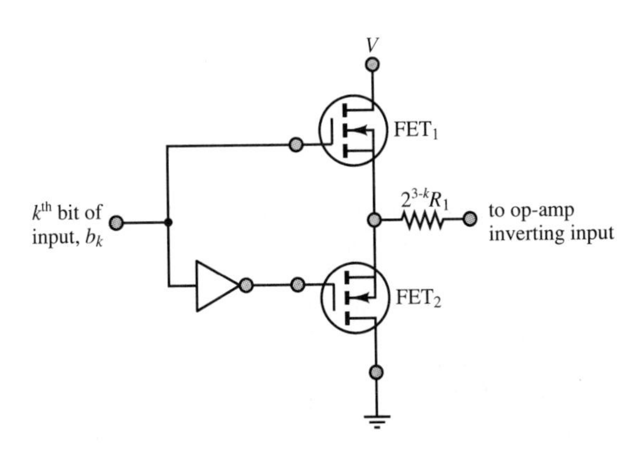

Figure P14.14

14.49 The unsigned decimal number 345_{10} is inputted to a 12-bit DAC. Given that $R_F = R_0/4{,}095$, logic 1 corresponds to 10 V, and logic 0 to 0 V,

a. What is the output of the DAC?
b. What is the maximum voltage that can be outputted from the DAC?
c. What is the resolution over the range 0 to 10 V?
d. Find the number of bits required in the DAC if an improved resolution of 0.5 mV is desired.

14.50 For the DAC circuit shown in Figure P14.13 (using an ideal op-amp), what value of R_F will give an output range of $-15 \le V_0 \le 0$ V?

14.51 Using the model of Figure P14.12, design a 4-bit digital-to-analog converter whose output is given by

$$v_o = -\frac{1}{10}(8b_3 + 4b_2 + 2b_1 + b_0)V$$

14.52 A data acquisition system uses a DAC with a range of ± 15 V and a resolution of 0.01 V. How many bits must be present in the DAC?

14.53 A data acquisition system uses a DAC with a range of ± 10 V and a resolution of 0.04 V. How many bits must be present in the DAC?

14.54 A data acquisition system uses a DAC with a range of -10 to $+15$ V and a resolution of 0.004 V. How many bits must be present in the DAC?

14.55 A DAC is to be used to deliver velocity commands to a motor. The maximum velocity is to be 2,500 rev/min, and the minimum nonzero velocity is to be 1 rev/min. How many bits are required in the DAC? What will the resolution be?

14.56 Assume the full-scale value of the analog input voltage to a particular analog-to-digital converter is 10 V.

a. If this is a 3-bit device, what is the resolution of the output?
b. If this is an 8-bit device, what is its resolution?
c. Make a general comment about the relationship between the number of bits and the resolution of an ADC.

14.57 The voltage range of feedback signal from a process is -5 V to $+15$ V, and a resolution of 0.05 percent of the voltage range is required. How many bits are required for the DAC?

14.58 Eight channels of analog information are being used by a computer to close eight control loops. Assume that all analog signals have identical frequency content and are multiplexed into a single ADC. The ADC requires 100 μs per conversion. The closed-loop software requires 500 μs of computation and output time for four of the loops, and for the other four it requires 250 μs. What is the maximum frequency content that the analog signal can have according to the Nyquist criterion?

14.59 A rotary potentiometer is to be used as a remote rotational displacement sensor. The maximum displacement to be measured is 180°, and the potentiometer is rated for 10 V and 270° of rotation.

a. What voltage increment must be resolved by an ADC to resolve an angular displacement of 0.5°? How many bits would be required in the ADC for full-range detection?
b. The ADC requires a 10-V input voltage for full-scale binary output. If an amplifier is placed between the potentiometer and the ADC, what amplifier gain should be used to take advantage of the full range of the ADC?

14.60 Suppose it is desired to digitize a 250-kHz analog signal to 10 bits using a successive-approximation ADC. Estimate the maximum permissible conversion time for the ADC.

14.61 A torque sensor has been mounted on a farm tractor engine. The voltage produced by the torque sensor is to be sampled by an ADC. The rotational speed of the crankshaft is 800 rev/min. Because of speed fluctuation caused by the reciprocating action of the engine, frequency content is present in the torque signal at twice the shaft rotation frequency. What is the minimum sampling period that can be used to ensure that the Nyquist criterion is satisfied?

14.62 The output voltage of an aircraft altimeter is to be sampled using an ADC. The sensor outputs 0 V at 0 m altitude and outputs 10 V at 10,000 m altitude. If the allowable error in sensing is 10 m, find the minimum number of bits required for the ADC. Assume the accuracy is ± 1 bit.

14.63 Consider a circuit that generates interrupts at fixed time intervals. Such a device is called a *real-time clock* and is used in control applications to establish the sample period as T seconds for control algorithms. Show how this can be done with a square wave (clock) that has a period equal to the desired time interval between interrupts.

14.64 What is the minimum number of bits required to digitize an analog signal with a resolution of:

a. 5%
b. 2%
c. 1%

14.65 Design a Schmitt trigger to operate in the presence of noise with peak amplitude = ± 150 mV. The circuit is to switch around the reference value -1 V. Assume an op-amp with ± 10-V supplies ($V_{sat} = 8.5$ V).

14.66 In the circuit of Figure P14.15, $R_1 = 100\ \Omega, R_2 = 56\ \text{k}\Omega, R_i = R_1 \| R_2$, and v_{in} is a 1-V peak-to-peak sine wave. Assuming that the supply voltages are ± 15 V, determine the threshold voltages (positive and negative v^+) and draw the output waveform.

14.67 The circuit in Figure P14.16 shows how a Schmitt trigger might be constructed with an op-amp. Explain the operation of this circuit.

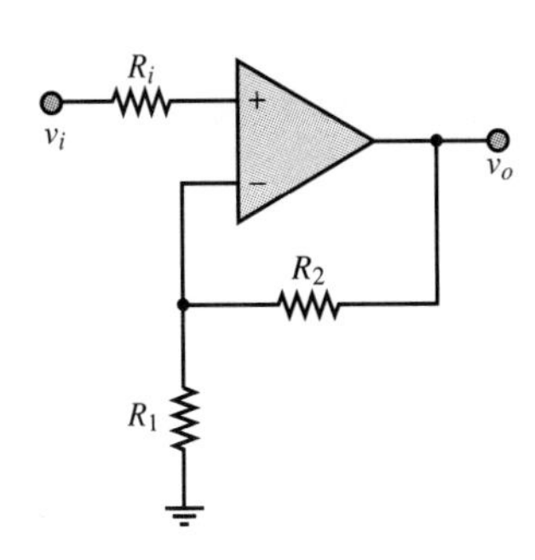

Figure P14.15

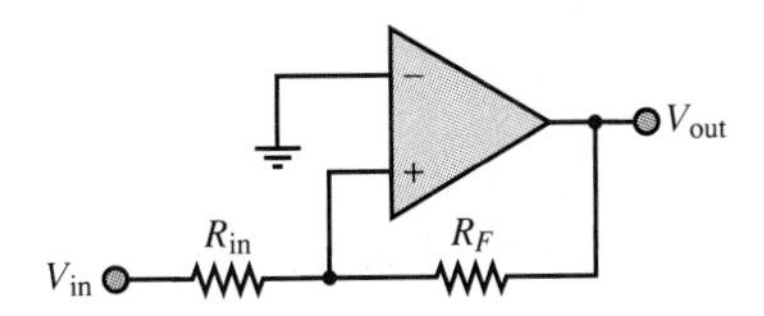

Figure P14.16

14.68 Consider again the circuit of Figure P14.16. Let the op-amp be an LM741 with ± 15 V bias supplies, and suppose R_F is chosen to be 104 kΩ. Assume V_{in} is a 1-kHz sinusoidal signal with 1V amplitude.

a. Determine the appropriate value for R_{in} if the output is to be high whenever $|V_{in}| \geq 0.25$ V.
b. Sketch the input and output waveforms.

14.69 For the circuit shown in Figure P14.17,

a. Draw the output waveform for v_{in} a 4-V peak-to-peak sine wave at 100 Hz and $V_{ref} = 2$ V.
b. Draw the output waveform for v_{in} a 4-V peak-to-peak sine wave at 100 Hz and $V_{ref} = -2$ V.

Note that the diodes placed at the input ensure that the differential voltage does not exceed the diode offset voltage.

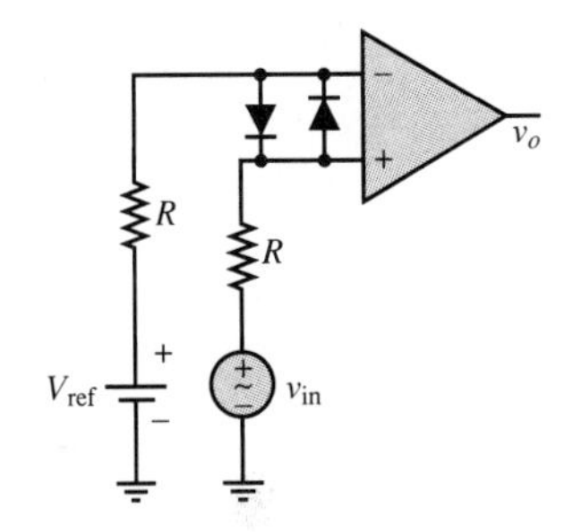

Figure P14.17

14.70 Figure P14.18 shows a simple *go-no go* detector application of a comparator.

a. Explain how the circuit works.
b. Design a circuit (i.e., choose proper values for the resistors) such that the green LED will turn on when V_{in} exceeds 5 V, and the red LED will be on whenever V_{in} is less than 5 V. Assume only 15 V supplies are available.

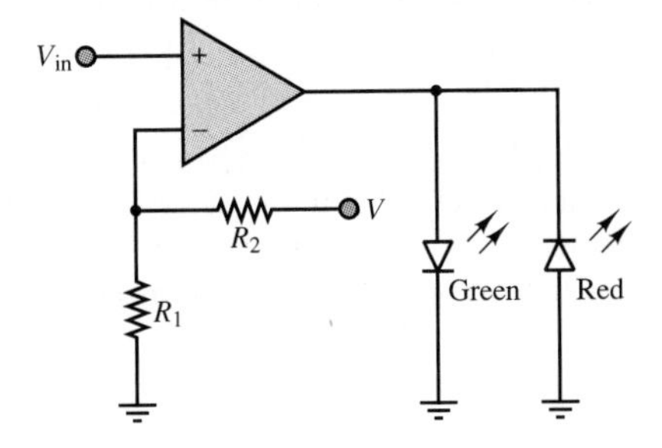

Figure P14.18

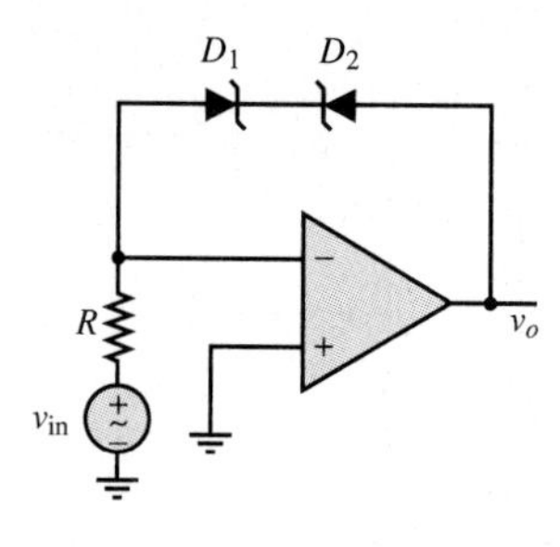

Figure P14.19

14.71 For the circuit of Figure P14.19, v_{in} is a 100-mV peak sine wave at 5 kHz, $R = 10\ \text{k}\Omega$, and D_1 and D_2 are 6.2-V Zener diodes. Draw the output voltage waveform.

14.72 Show that the period of oscillation of an op-amp astable multivibrator is given by the expression

$$T = 2R_1C \log_e\left(\frac{2R_2}{R_3} + 1\right)$$

14.73 Use the data sheets for the 74123 monostable multivibrator to analyze the connection shown in Figure 14.60. Draw a timing diagram indicating the approximate duration of each pulse, assuming that the trigger signal consists of a positive-going transition.

14.74 In the monostable multivibrator of Figure 14.61, $R_1 = 10\ \text{k}\Omega$ and the output pulse width $T = 10$ ms. Determine the value of C.

14.75 An ASCII (hex) encoded message is given below. Decode the message.

41 53 43 49 49 20 64 65 63 6F 64 69 6E 67 20 69 73 20 65 61 73 79 21

14.76 An ASCII (binary) encoded message is given below. Decode the messsage. Hint: Follow a line-by-line sequence, *not* column-by-column.

1010100 1101000 1101001 1110011 0100000 1101001 1110011 0100000 1100001 0100000 1110100
1101001 1101101 1100101 0101101 1100011 1101111 1101110 1110011 1110101 1101101 1101001
1101110 1100111 0100000 1110000 1110010 1101111 1100010 1101100 1100101 1101101 0101110

14.77 Express the following decimal numbers in ASCII form:

a. 12
b. 345.2
c. 43.5

14.78 Express the following words in ASCII form:

a. Digital
b. Computer
c. Ascii
d. ASCII

14.79 Explain why data transmission over long distances is usually done via a serial scheme rather than parallel.

14.80 A certain automated data-logging instrument has 16K-words of on-board memory. The device samples the variable of interest once every five minutes. How often must data be downloaded and the memory cleared in order to avoid losing any data?

14.81 Explain why three wires are required for the handshaking technique employed by IEEE 488 bus systems.

14.82 A CD-ROM can hold 650 Mbytes of information. Suppose the CD-ROMs are packed 50 per box. The manufacturer ships 100 boxes via commercial airliner from Los Angeles to New York. The distance between the two cities is 2,500 miles by air, and the airliner flies at a speed of 400 mi/hr. What is the data transmission rate between the two cities in bits/s?

PART III

ELECTROMECHANICS

CHAPTER

15

Principles of Electromechanics

The objective of this chapter is to introduce the fundamental notions of electromechanical energy conversion, leading to an understanding of the operation of various electromechanical transducers. The chapter also serves as an introduction to the material on electric machines to be presented in Chapters 16 and 17. The foundations for the material introduced in this chapter will be found in the circuit analysis chapters (1–6). In addition, the material on power electronics (Chapter 10) is also relevant, especially with reference to Chapters 16 and 17.

The subject of electromechanical energy conversion is one that should be of particular interest to the non–electrical engineer, because it forms one of the important points of contact between electrical engineering and other engineering disciplines. Electromechanical transducers are commonly used in the design of industrial and aerospace control systems and in biomedical applications, and they form the basis of many common appliances. In the course of our exploration of electromechanics, we shall illustrate the operation of practical devices, such as loudspeakers, relays, solenoids, sensors for the measurement of position and velocity, and other devices of practical interest.

Upon completion of the chapter, you should be able to:

- Analyze simple magnetic circuits, to determine electrical and mechanical performance and energy requirements.
- Size a relay or solenoid for a given application.
- Describe the energy-conversion process in electromechanical systems.
- Perform a simplified linear analysis of electromechanical transducers.

15.1 ELECTRICITY AND MAGNETISM

The notion that the phenomena of electricity and magnetism are interconnected was first proposed in the early 1800s by H. C. Oersted, a Danish physicist. Oersted showed that an electric current produces magnetic effects (more specifically, a magnetic field). Soon after, the French scientist André Marie Ampère expressed this relationship by means of a precise formulation, known as *Ampère's law*. A few years later, the English scientist Faraday illustrated how the converse of Ampère's law also holds true, that is, that a magnetic field can generate an electric field; in short, *Faraday's law* states that a changing magnetic field gives rise to a voltage. We shall undertake a more careful examination of both Ampère's and Faraday's laws in the course of this chapter.

As will be explained in the next few sections, the magnetic field forms a necessary connection between electrical and mechanical energy. Ampère's and Faraday's laws will formally illustrate the relationship between electric and magnetic fields, but it should already be evident from your own individual experience that the magnetic field can also convert magnetic energy to mechanical energy (for example, by lifting a piece of iron with a magnet). In effect, the devices we commonly refer to as *electromechanical* should more properly be referred to as electro*magneto*mechanical, since they almost invariably operate through a conversion from electrical to mechanical energy (or vice versa) by means of a magnetic field. Chapters 15 through 17 are concerned with the use of electricity and magnetic materials for the purpose of converting electrical energy to mechanical, and back.

The Magnetic Field and Faraday's Law

The quantities used to quantify the strength of a magnetic field are the **magnetic flux,** ϕ, in units of **webers** (Wb); and the **magnetic flux density, B,** in units of webers per square meter (Wb/m^2), or **teslas** (T). The latter quantity, as well as the associated **magnetic field intensity, H** (in units of amperes per meter, or A/m) are vectors.[1] Thus, the density of the magnetic flux and its intensity are in general described in vector form, in terms of the components present in each spatial direction (e.g., on the x, y, and z axes). In discussing magnetic flux density and field intensity in this chapter and the next, we shall almost always assume that the field is a *scalar field*, that is, that it lies in a single spatial direction. This will simplify many explanations.

[1] We'll use the boldface symbols **B** and **H** to denote the vector forms of B and H; the standard typeface will represent the scalar flux density or field intensity in a given direction.

It is customary to represent the magnetic field by means of the familiar *lines of force* (a concept also due to Faraday); we visualize the strength of a magnetic field by observing the density of these lines in space. You probably know from a previous course in physics that such lines are closed in a magnetic field, that is, that they form continuous loops exiting at a magnetic north pole (by definition) and entering at a magnetic south pole. The relative strengths of the magnetic fields generated by two magnets could be depicted as shown in Figure 15.1.

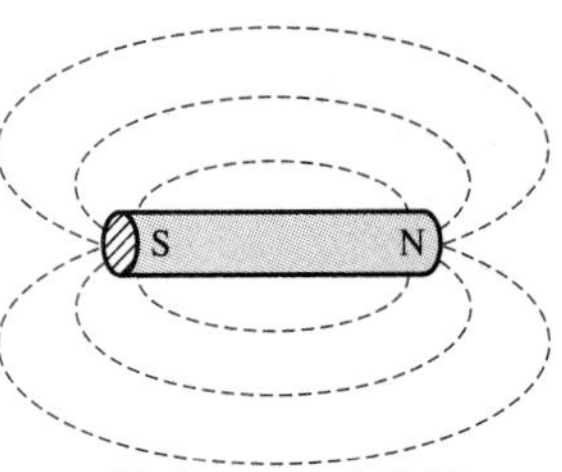

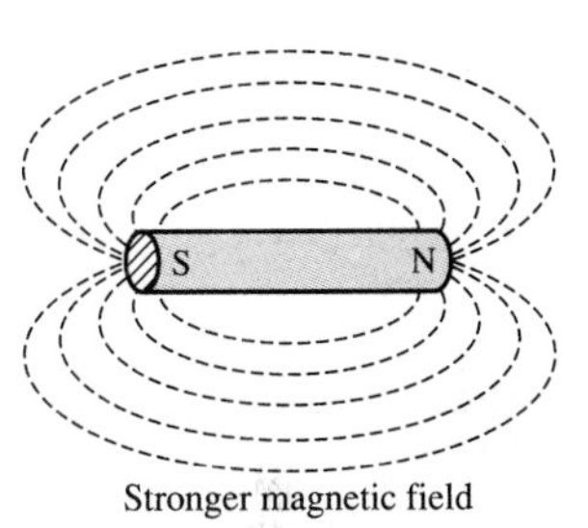

Figure 15.1 Lines of force in a magnetic field

Magnetic fields are generated by electric charge in motion, and their effect is measured by the force they exert on a moving charge. As you may recall from previous physics courses, the vector force **f** exerted on a charge of q moving at velocity **u** in the presence of a magnetic field with flux density **B** is given by the equation

$$\mathbf{f} = q\mathbf{u} \times \mathbf{B} \tag{15.1}$$

where the symbol $\times$ denotes the (vector) cross product. If the charge is moving at a velocity **u** in a direction that makes an angle θ with the magnetic field, then the magnitude of the force is given by

$$f = quB \sin\theta \tag{15.2}$$

and the direction of this force is at right angles with the plane formed by the vectors **B** and **u**. This relationship is depicted in Figure 15.2.

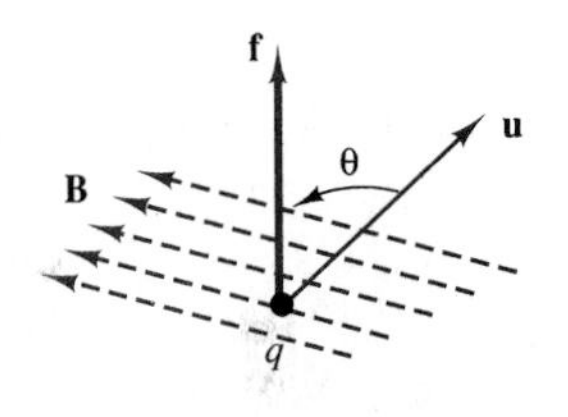

Figure 15.2 Charge moving in a constant magnetic field

The magnetic flux, ϕ, is then defined as the integral of the flux density over some surface area. For the simplified (but often useful) case of magnetic flux lines perpendicular to a cross-sectional area A, we can see that the flux is given by the following integral:

$$\phi = \int_A B\, dA \tag{15.3}$$

in webers (Wb), where the subscript A indicates that the integral is evaluated over the surface A. Furthermore, if the flux were to be uniform over the cross-sectional area A (a simplification that will be useful), the preceding integral could be approximated by the following expression:

$$\phi = B \cdot A \tag{15.4}$$

Figure 15.3 illustrates this idea, by showing hypothetical magnetic flux lines traversing a surface, delimited in the figure by a thin conducting wire.

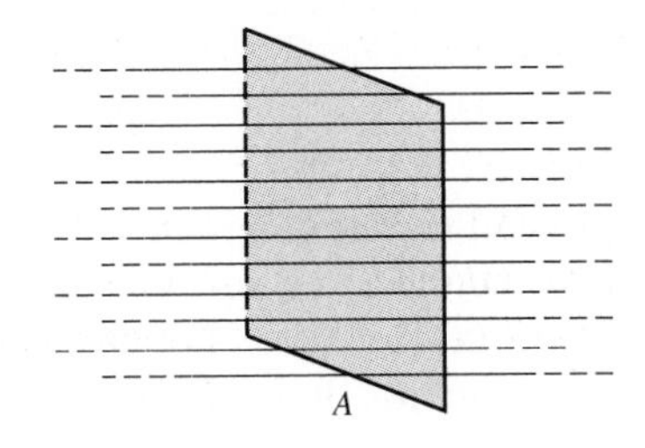

Figure 15.3 Magnetic flux lines crossing a surface

Faraday's law states that if the imaginary surface A were bounded by a conductor—for example, the thin wire of Figure 15.3—then a *changing* magnetic field would induce a voltage, and therefore a current, in the conductor. More precisely, Faraday's law states that a time-varying flux causes an induced **electromotive force,** or **emf,** e, as follows:

$$e = -\frac{d\phi}{dt} \qquad \textbf{(15.5)}$$

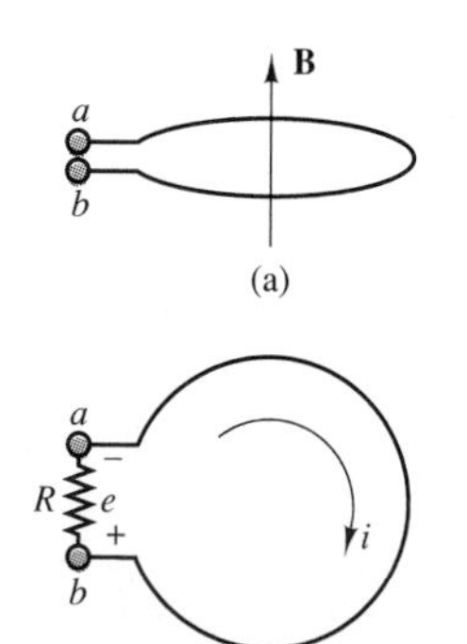

Current generating a magnetic flux opposing the increase in flux due to **B**

(b)

Figure 15.4 Flux direction

A little discussion is necessary at this point to explain the meaning of the minus sign in equation 15.5. Consider the one-turn coil of Figure 15.4, which forms a circular cross-sectional area, in the presence of a magnetic field with flux density **B** oriented in a direction perpendicular to the plane of the coil. If the magnetic field, and therefore the flux within the coil, is constant, no voltage will exist across terminals a and b; if, however, the flux were increasing and terminals a and b were connected—for example, by means of a resistor, as indicated in Figure 15.4(b)—current would flow in the coil in such a way that *the magnetic flux generated by the current would oppose the increasing flux.* Thus, the flux induced by such a current would be in the direction opposite to that of the original flux density vector, **B**. This principle is known as **Lenz's law.** The reaction flux would then point downward in Figure 15.4(a), or into the page in Figure 15.4(b). Now, by virtue of the **right-hand rule,** this reaction flux would induce a current flowing clockwise in Figure 15.4(b), that is, a current that flows out of terminal b and into terminal a. The resulting voltage across the hypothetical resistor R would then be negative. If, on the other hand, the original flux were decreasing, current would be induced in the coil so as to reestablish the initial flux; but this would mean that the current would have to generate a flux in the upward direction in Figure 15.4(a) (or out of the page in Figure 15.4(b)). Thus, the resulting voltage would change sign.

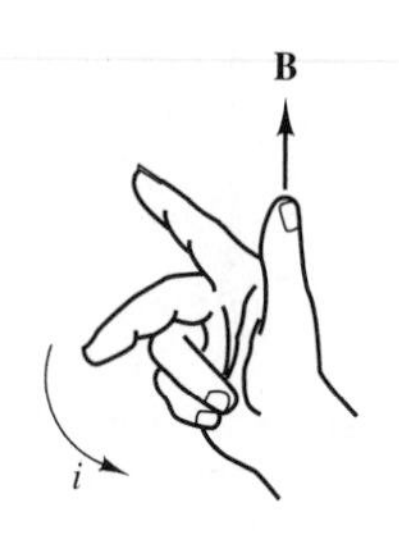

Right-hand rule

The polarity of the induced voltage can usually be determined from physical considerations; therefore the minus sign in equation 15.5 is usually left out. We will use this convention throughout the chapter.

In practical applications, the size of the voltages induced by the changing magnetic field can be significantly increased if the conducting wire is coiled many times around, so as to multiply the area crossed by the magnetic flux lines many times over. For an N-turn coil with cross-sectional area A, for example, we have the emf

$$e = N\frac{d\phi}{dt} \qquad \textbf{(15.6)}$$

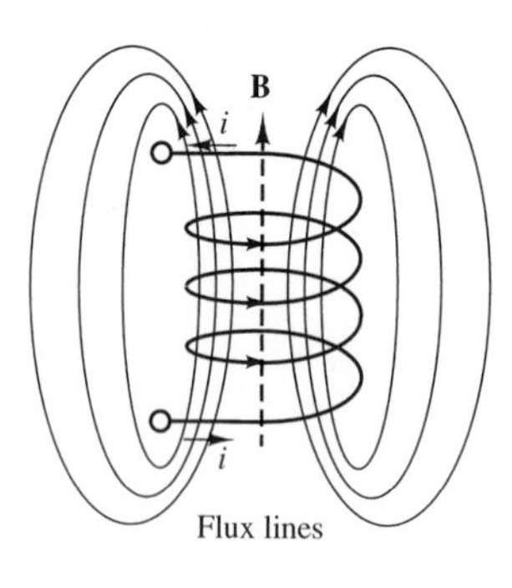

Figure 15.5 Concept of flux linkage

Figure 15.5 shows an N-turn coil *linking* a certain amount of magnetic flux; you can see that if N is very large and the coil is tightly wound (as is usually the case in the construction of practical devices), it is not unreasonable to presume that each turn of the coil links the same flux. It is convenient, in practice, to define the **flux linkage,** λ, as

$$\lambda = N\phi \qquad \textbf{(15.7)}$$

so that

$$e = \frac{d\lambda}{dt} \quad \textbf{(15.8)}$$

Note that equation 15.8, relating the derivative of the flux linkage to the induced emf, is analogous to the equation describing current as the derivative of charge:

$$i = \frac{dq}{dt} \quad \textbf{(15.9)}$$

In other words, flux linkage can be viewed as the dual of charge in a circuit analysis sense, provided that we are aware of the simplifying assumptions just stated in the preceding paragraphs, namely, a uniform magnetic field perpendicular to the area delimited by a tightly wound coil. These assumptions are not at all unreasonable when applied to the inductor coils commonly employed in electric circuits.

What, then, are the physical mechanisms that can cause magnetic flux to change, and therefore to induce an electromotive force? Two such mechanisms are possible. The first consists of physically moving a permanent magnet in the vicinity of a coil—for example, so as to create a time-varying flux. The second requires that we first produce a magnetic field by means of an electric current (how this can be accomplished is discussed later in this section) and then vary the current, thus varying the associated magnetic field. The latter method is more practical in many circumstances, since it does not require the use of permanent magnets and allows variation of field strength by varying the applied current; however, the former method is conceptually simpler to visualize. The voltages induced by a moving magnetic field are called **motional voltages;** those generated by a time-varying magnetic field are termed **transformer voltages.** We shall be interested in both in this chapter, for different applications.

In the analysis of linear circuits in Chapter 4, we implicitly assumed that the relationship between flux linkage and current was a linear one:

$$\lambda = Li \quad \textbf{(15.10)}$$

so that the effect of a time-varying current was to induce a transformer voltage across an inductor coil, according to the expression

$$v = L\frac{di}{dt} \quad \textbf{(15.11)}$$

This is, in fact, the defining equation for the ideal **self-inductance,** L. In addition to self-inductance, however, it is also important to consider the **magnetic coupling** that can occur between neighboring circuits. Self-inductance measures the voltage induced in a circuit by the magnetic field generated by a current flowing in the same circuit. It is also possible that a second circuit in the vicinity of the first may experience an induced voltage as a consequence of the magnetic field generated in the first circuit. As we shall see in Section 15.4, this principle underlies the operation of all transformers.

Self and Mutual Inductance

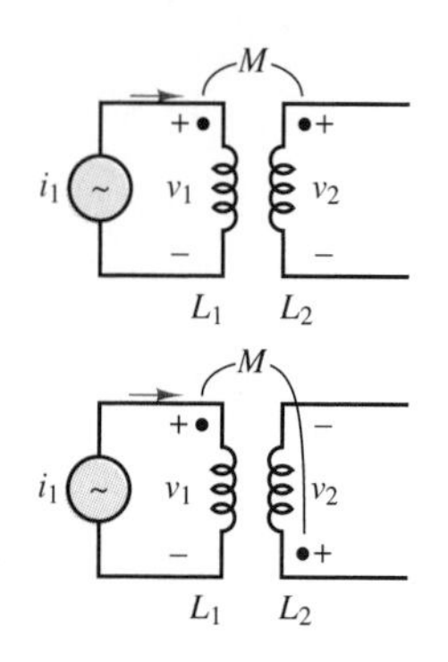

Figure 15.6 Mutual inductance

Figure 15.6 depicts a pair of coils, one of which, L_1, is excited by a current, i_1, and therefore develops a magnetic field and a resulting induced voltage, v_1. The second coil, L_2, is not energized by a current, but links some of the flux generated by the current i_1 around L_1 because of its close proximity to the first coil. The magnetic coupling between the coils established by virtue of their proximity is described by a quantity called **mutual inductance** and defined by the symbol M. The mutual inductance is defined by the equation

$$v_2 = M\frac{di_1}{dt} \qquad \textbf{(15.12)}$$

The dots shown in the two figures indicate the polarity of the coupling between the coils. If the dots are at the same end of the coils, the voltage induced in coil 2 by a current in coil 1 has the same polarity as the voltage induced by the same current in coil 1; otherwise, the voltages are in opposition, as shown in the lower part of Figure 15.6. Thus, the presence of such dots indicates that magnetic coupling is present between two coils. It should also be pointed out that if a current (and therefore a magnetic field) were present in the second coil, an additional voltage would be induced across coil 1. The voltage induced across a coil is, in general, equal to the sum of the voltages induced by self-inductance and mutual inductance.

Example 15.1 Linear Variable Differential Transformer (LVDT)

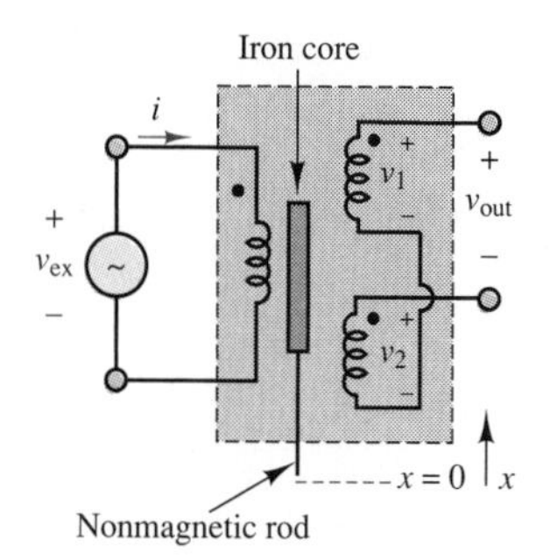

Figure 15.7 Linear variable differential transformer

The linear variable differential transformer (LVDT) is a displacement transducer based on the mutual inductance concept just discussed. Figure 15.7 shows a simplified representation of an LVDT, which consists of a primary coil, subject to AC excitation (v_{ex}), and of a pair of identical secondary coils, which are connected so as to result in the output voltage

$$v_{out} = v_1 - v_2$$

The ferromagnetic core between the primary and secondary coils can be displaced in proportion to some external motion, x, and determines the magnetic coupling between primary and secondary coils. Intuitively, as the core is displaced upward, greater coupling will occur between the primary coil and the top secondary coil, thus inducing a greater voltage in the top secondary coil. Hence, $v_{out} > 0$ for positive displacements. The converse is true for negative displacements. More formally, if the primary coil has resistance R_p and self-inductance L_p, we can write

$$iR_p + L_p\frac{di}{dt} = v_{ex}$$

and the voltages induced in the secondary coils are given by

$$v_1 = M_1\frac{di}{dt}$$

$$v_2 = M_2 \frac{di}{dt}$$

so that

$$v_{out} = (M_1 - M_2) \frac{di}{dt}$$

where M_1 and M_2 are the mutual inductances between the primary and the respective secondary coils. It should be apparent that each of the mutual inductances is dependent on the position of the iron core. For example, with the core at the *null position*, $M_1 = M_2$ and $v_{out} = 0$. The LVDT is typically designed so that $M_1 - M_2$ is linearly related to the displacement of the core, x.

Because the excitation is by necessity an AC signal (why?), the output voltage is actually given by the difference of two sinusoidal voltages at the same frequency, and is therefore itself a sinusoid, whose amplitude and phase depend on the displacement, x. Thus, v_{out} is an *amplitude-modulated (AM)* signal, similar to the one discussed in Example 4.19. To recover a signal proportional to the actual displacement, it is therefore necessary to use a demodulator circuit, such as the one discussed in Example 7.11.

In practical electromagnetic circuits, the self-inductance of a circuit is not necessarily constant; in particular, the inductance parameter, L, is not constant, in general, but depends on the strength of the magnetic field intensity, so that it will not be possible to use such a simple relationship as $v = L\,di/dt$, with L constant. If we revisit the definition of the transformer voltage,

$$e = N \frac{d\phi}{dt} \tag{15.13}$$

we see that in an inductor coil, the inductance is given by

$$L = \frac{N\phi}{i} = \frac{\lambda}{i} \tag{15.14}$$

This expression implies that the relationship between current and flux in a magnetic structure is linear (the inductance being the slope of the line). In fact, the properties of ferromagnetic materials are such that the flux-current relationship is nonlinear, as we shall see in Section 15.3, so that the simple linear inductance parameter used in electric circuit analysis is not adequate to represent the behavior of the magnetic circuits of the present chapter. In any practical situation, the relationship between the flux linkage, λ, and the current is nonlinear, and might be described by a curve similar to that shown in Figure 15.8. Whenever the i-λ curve is not a straight line, it is more convenient to analyze the magnetic system in terms of energy calculations, since the corresponding circuit equation would be nonlinear.

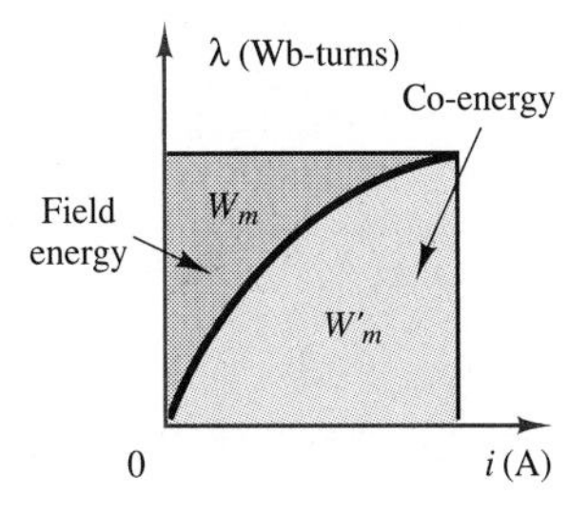

Figure 15.8

In a magnetic system, the energy stored in the magnetic field is equal to the integral of the instantaneous power, which is the product of voltage and current, just as in a conventional electrical circuit:

$$W_m = \int ei\, dt \tag{15.15}$$

However, in this case, the voltage corresponds to the induced emf, according to Faraday's law:

$$e = \frac{d\lambda}{dt} = N\frac{d\phi}{dt} \tag{15.16}$$

and is therefore related to the rate of change of the magnetic flux. The energy stored in the magnetic field could therefore be expressed in terms of the current by the integral

$$W_m = \int ei\, dt = \int \frac{d\lambda}{dt} i\, dt = \int i\, d\lambda \tag{15.17}$$

It should be straightforward to recognize that this energy is equal to the area above the λ-i curve of Figure 15.8. From the same figure, it is also possible to define a fictitious (but sometimes useful) quantity called **co-energy,** equal to the area under the curve and identified by the symbol W'_m. From the figure, it is also possible to see that the co-energy can be expressed in terms of the stored energy by means of the following relationship:

$$W'_m = i\lambda - W_m \tag{15.18}$$

Example 15.2 illustrates the calculation of energy, co-energy, and induced voltage using the concepts developed in these paragraphs.

Example 15.2

An iron-core inductor has the following empirically determined characteristic:

$$i = \lambda + 0.5\lambda^2$$

1. Determine the energy, the co-energy, and the incremental inductance for $\lambda = 0.5\text{V} \cdot \text{s}$.
2. Given that the coil resistance is 1 Ω and that the current through the coil is

$$i(t) = 0.625 + 0.01 \sin 400t$$

determine the voltage across the terminals of the inductor.

Solution:

1. The energy and co-energy can be determined by integrating the function

$$i = \lambda + 0.5\lambda^2$$

over λ:

$$W_m = \int_0^\lambda i(\lambda')\, d\lambda' = \frac{\lambda^2}{2} + 0.5\frac{\lambda^3}{3}$$

This function can be evaluated at $\lambda = 0.5$ to obtain

$$W_m = \frac{\lambda^2}{2} + 0.5\frac{\lambda^3}{3}\bigg|_{\lambda=0.5} = 0.1458 \text{ J}$$

and the corresponding co-energy is given by the expression

$$W'_m = i\lambda - W_m$$

To compute the current, i, required to generate a flux linkage $\lambda = 0.5$, we can use the following expression:

$$i = \lambda + 0.5\lambda^2 = 0.625 \text{ A}$$

and from the result, the co-energy may be computed as follows:

$$W'_m = (0.625)(0.5) - 0.1458 = 0.1667 \text{ J}$$

Further, since the *incremental inductance* (i.e., the inductance in the neighborhood of a particular point on the λ-i curve) is the reciprocal of the slope of the λ-i curve, it can be computed by differentiating the $i(\lambda)$ expression and evaluating it at $\lambda = 0.5$. Let the incremental inductance be L_Δ; then:

$$L_\Delta = \left(\frac{di}{d\lambda}\right)^{-1} = \frac{1}{1+\lambda} = 0.667 \text{ H}$$

Thus, for small variations of the current about the value $i = 0.625$ A, the nonlinear inductor will behave as if it were a linear inductor with incremental inductance $L_\Delta = 0.667$ H.

2. If the current consists of a small sinusoidal oscillation around the average current $i = 0.625$ A, it becomes possible to compute the corresponding voltage across the inductor using the well-known linear circuit laws:

$$\begin{aligned} v &= iR + L_\Delta \frac{di}{dt} \\ &= 0.625 + 0.01\sin(400t) \times 1 + (0.667)(4)\cos(400t) \\ &= 0.625 + 0.01\sin(400t) + 2.668\cos(400t) \\ &= 0.625 + 2.668\sin(400t + 89.8^\circ) \end{aligned}$$

The calculation of the energy stored in the magnetic field around a magnetic structure will be particularly useful later in the chapter, when the discussion turns to practical electromechanical transducers and it will be necessary to actually compute the forces generated in magnetic structures.

Ampère's Law

As explained in the previous section, Faraday's law is one of two fundamental laws relating electricity to magnetism. The second relationship, which forms a counterpart to Faraday's law, is **Ampère's law.** Qualitatively, Ampère's law states that the magnetic field intensity, **H**, in the vicinity of a conductor is related to the current carried by the conductor; thus Ampère's law establishes a dual relationship with Faraday's law.

In the previous section, we described the magnetic field in terms of its flux density, **B**, and flux ϕ. To explain Ampère's law and the behavior of magnetic materials, we need to define a relationship between the magnetic field intensity, **H,** and the flux density, **B**. These quantities are related by:

$$\begin{aligned}\mathbf{B} &= \mu\mathbf{H} \\ &= \mu_r\mu_0\mathbf{H} \qquad \text{Wb/m}^2 \text{ or T}\end{aligned} \tag{15.19}$$

where the parameter μ is a scalar constant for a particular physical medium (at least, for the applications we consider here) and is called the **permeability** of the medium. The permeability of a material can be factored as the product of the permeability of free space, $\mu_0 = 4\pi \times 10^{-7}$ H/m, times the relative permeability, μ_r, which varies greatly according to the medium. For example, for air and for most electrical conductors and insulators, μ_r is equal to 1. For ferromagnetic materials, the value of μ_r can take values in the hundreds or thousands. The size of μ_r represents a measure of the magnetic properties of the material. A consequence of Ampère's law is that, the larger the value of μ, the smaller the current required to produce a large flux density in an electromagnetic structure. Consequently, many electromechanical devices make use of ferromagnetic materials, called iron cores, to enhance their magnetic properties. Table 15.1 gives approximate values of μ_r for some common materials.

Table 15.1 **Relative permeabilities for common materials**

Material	μ_r
Air	1
Permalloy	100,000
Cast steel	1,000
Sheet steel	4,000
Iron	5,195

Conversely, the reason for introducing the magnetic field intensity is that it is independent of the properties of the materials employed in the construction of magnetic circuits. Thus, a given magnetic field intensity, **H**, will give rise to different flux densities in different materials. It will therefore be useful to define *sources* of magnetic energy in terms of the magnetic field intensity, so that different magnetic structures and materials can then be evaluated or compared for a given source. In analogy with electromotive force, this "source" will be termed **magnetomotive force (mmf).** As stated earlier, both the magnetic flux density and field intensity are vector quantities; however, for ease of analysis, scalar fields will be chosen by appropriately selecting the orientation of the fields, wherever possible.

Ampère's law states that the integral of the vector magnetic field intensity, **H,** around a closed path is equal to the total current linked by the closed path, *i*:

$$\oint \mathbf{H} \cdot d\mathbf{l} = \sum i \tag{15.20}$$

where $d\mathbf{l}$ is an increment in the direction of the closed path. If the path is in the same direction as the direction of the magnetic field, we can use scalar quantities to state that

$$\oint \mathbf{H} \cdot d\mathbf{l} = \sum i \tag{15.21}$$

Illustration of Ampère's law. By the right-hand rule, the current, *i*, generates a magnetic field intensity, **H**, in the direction shown.

Figure 15.9 illustrates the case of a wire carrying a current i, and of a circular path of radius r surrounding the wire. In this simple case, you can see that the magnetic field intensity, **H**, is determined by the familiar right-hand rule. This rule states that if the direction of the current i points in the direction of the thumb of one's right hand, the resulting magnetic field encircles the conductor in the direction in which the other four fingers would encircle it. Thus, in the case of Figure 15.9, the closed-path integral becomes equal to $H \cdot (2\pi r)$, since the path and the magnetic field are in the same direction, and therefore the magnitude of the magnetic field intensity is given by

$$H = \frac{i}{2\pi r} \tag{15.22}$$

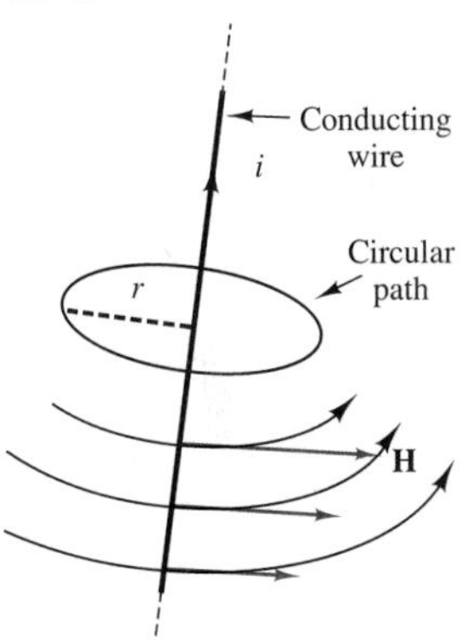

Figure 15.9 Illustration of Ampère's law

Now, the magnetic field intensity is unaffected by the material surrounding the conductor, but the flux density depends on the material properties, since $B = \mu H$. Thus, the density of flux lines around the conductor would be far greater in the presence of a magnetic material than if the conductor were surrounded by air. The field generated by a single conducting wire is not very strong; however, if we arrange the wire into a tightly wound coil with many turns, we can greatly increase the strength of the magnetic field. For such a coil, with N turns, one can verify visually that the lines of force associated with the magnetic field link all of the turns of the conducting coil, so that we have effectively increased the current linked by the flux lines N-fold. The product $N \cdot i$ is a useful quantity in electromagnetic circuits, and is called the magnetomotive force,[2] $\mathcal{F}$ (often abbreviated mmf), in analogy with the electromotive force defined earlier:

$$\mathcal{F} = Ni \qquad \text{ampere-turns } (A \cdot t) \tag{15.23}$$

[2]Note that, although dimensionally equal to amperes, the units of magnetomotive force are ampere-turns.

Figure 15.10 illustrates the magnetic flux lines in the vicinity of a coil. The magnetic field generated by the coil can be made to generate a much greater flux density if the coil encloses a magnetic material. The most common ferromagnetic materials are steel and iron; in addition to these, many alloys and oxides of iron—as well as nickel—and some artificial ceramic materials called **ferrites** also exhibit magnetic properties. Winding a coil around a ferromagnetic material accomplishes two useful tasks at once: it forces the magnetic flux to be concentrated near the coil and—if the shape of the magnetic material is appropriate—completely confines the flux within the magnetic material, thus forcing the closed path for the flux lines to be almost entirely enclosed within the ferromagnetic material. Typical arrangements are the iron-core inductor and the toroid of Figure 15.11. The flux densities for these inductors are given by the expressions

$$B = \frac{\mu N i}{l} \qquad \text{Flux density for tightly wound circular coil} \tag{15.24}$$

$$B = \frac{\mu N i}{2\pi r_2} \qquad \text{Flux density for toroidal coil} \tag{15.25}$$

Intuitively, the presence of a high-permeability material near a source of magnetic flux causes the flux to preferentially concentrate in the high-μ material, rather than in air, much as a conducting path concentrates the current produced by an electric field in an electric circuit. In the course of this chapter, we shall

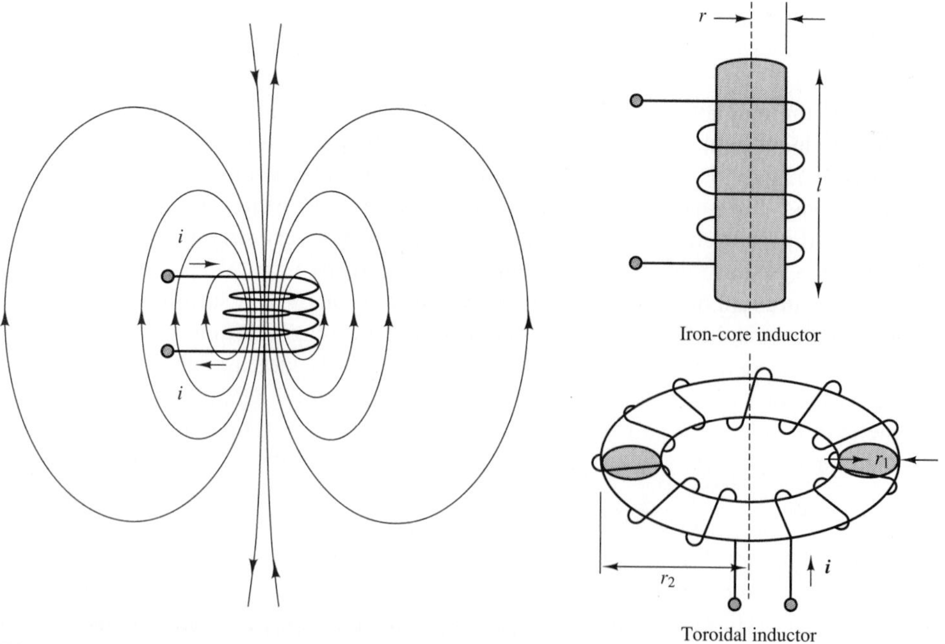

Figure 15.10 Magnetic field in the vicinity of a current-carrying coil

Figure 15.11 Practical inductors

continue to develop this analogy between electric circuits and magnetic circuits. Figure 15.12 depicts an example of a simple electromagnetic structure, which, as we shall see shortly, forms the basis of the practical transformer.

Table 15.2 summarizes the variables introduced thus far in the discussion of electricity and magnetism.

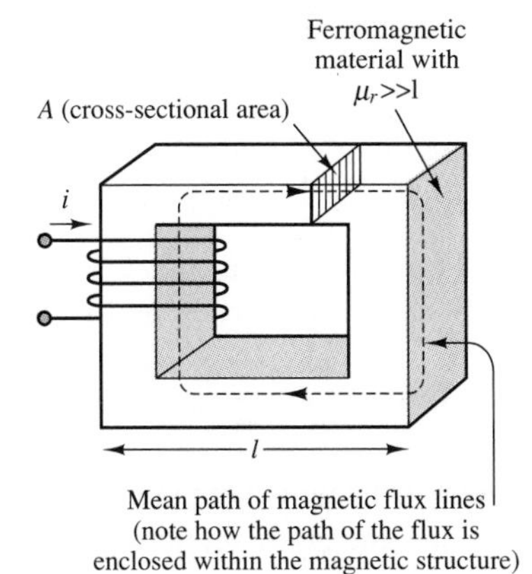

Figure 15.12 A simple electromagnetic structure

Table 15.2 Magnetic variables and units

Variable	Symbol	Units
Current	I	A
Magnetic flux density	B	$Wb/m^2 = T$
Magnetic flux	ϕ	Wb
Magnetic field intensity	H	A/m
Electromotive force	e	V
Magnetomotive force	$\mathcal{F}$	A · t
Flux linkage	λ	Wb · t

DRILL EXERCISES

15.1 A coil having 100 turns is immersed in a magnetic field that is varying uniformly from 80 mWb to 30 mWb in 2 seconds. Find the induced voltage in the coil.

15.2 The magnitude of **H** at a radius of 0.5 m from a long linear conductor is $1 \text{ A} \cdot \text{m}^{-1}$. Find the current in the wire.

15.3 The relation between the flux linkages and the current for a magnetic material is given by $\lambda = 6i/(2i + 1)$ Wb · t. Determine the energy stored in the magnetic field for $\lambda = 2$ Wb · t.

15.4 Verify that for the linear case, where the flux is proportional to the mmf, the energy stored in the magnetic field is $\frac{1}{2}Li^2$.

15.2 MAGNETIC CIRCUITS

It is possible to analyze the operation of electromagnetic devices such as the one depicted in Figure 15.12 by means of magnetic equivalent circuits, similar in many respects to the equivalent electrical circuits of the earlier chapters. Before we can present this technique, however, we need to make a few simplifying approximations. The first of these approximations assumes that there exists a **mean path** for the magnetic flux, and that the corresponding mean flux density is approximately constant over the cross-sectional area of the magnetic structure. Thus, a coil wound around a core with cross-sectional area A will have flux density

$$B = \frac{\phi}{A} \tag{15.26}$$

where A is assumed to be perpendicular to the direction of the flux lines. Figure 15.12 illustrates such a mean path and the cross-sectional area, A. Knowing the flux density, we obtain the field intensity:

$$H = \frac{B}{\mu} = \frac{\phi}{A\mu} \tag{15.27}$$

But then, knowing the field intensity, we can relate the mmf of the coil, $\mathcal{F}$, to the product of the magnetic field intensity, H, and the length of the magnetic (mean) path, l, for one leg of the structure:

$$\mathcal{F} = N \cdot i = H \cdot l \tag{15.28}$$

In summary, the mmf is equal to the magnetic flux times the length of the magnetic path, divided by the permeability of the material times the cross-sectional area:

$$\mathcal{F} = \phi \frac{1}{\mu A} \tag{15.29}$$

A review of this formula reveals that the magnetomotive force, $\mathcal{F}$, may be viewed as being analogous to the voltage source in a series electrical circuit, and that the flux, ϕ, is then equivalent to the electrical current in a series circuit and the term $l/\mu A$ to the *magnetic resistance* of one leg of the magnetic circuit. You will note that the term $l/\mu A$ is very similar to the term describing the resistance of a cylindrical conductor of length l and cross-sectional area A, where the permeability, μ, is analogous to the conductivity, σ. The term $l/\mu A$ occurs frequently enough to be assigned the name of **reluctance,** and the symbol $\mathcal{R}$.

In summary, when an N-turn coil carrying a current i is wound around a magnetic core such as the one indicated in Figure 15.12, the mmf, $\mathcal{F}$, generated by the coil produces a flux, ϕ, that is *mostly* concentrated within the core and is assumed to be uniform across the cross section. Within this simplified picture, then, the analysis of a magnetic circuit is analogous to that of resistive electrical circuits. This analogy is illustrated in Table 15.3 and in the examples in this section.

Table 15.3 **Analogy between electric and magnetic circuits**

Electrical quantity	Magnetic quantity
Electrical field intensity, E, V/m	Magnetic field intensity, H, $A \cdot t/m$
Voltage, v, V	Magnetomotive force, $\mathcal{F}$, $A \cdot t$
Current, i, A	Magnetic flux, ϕ, Wb
Current density, J, A/m^2	Magnetic flux density, B, Wb/m^2
Resistance, R, Ω	Reluctance, $\mathcal{R} = l/\mu A$, $A \cdot t/Wb$
Conductivity, σ, $1/\Omega \cdot m$	Permeability, μ, $Wb/A \cdot m$

The usefulness of the magnetic circuit analogy can be emphasized by analyzing a magnetic core similar to that of Figure 15.12, but with a slightly modified geometry. Figure 15.13 depicts the magnetic structure and its equivalent circuit analogy. In the figure, we see that the mmf, $\mathcal{F} = Ni$, excites the magnetic circuit, which is composed of four legs: two of mean path length l_1 and cross-sectional area $A_1 = d_1 w$, and the other two of mean length l_2 and cross section $A_2 = d_2 w$. Thus, the reluctance encountered by the flux in its path around the magnetic core is given by the quantity $\mathcal{R}_{\text{series}}$, with

$$\mathcal{R}_{\text{series}} = 2\mathcal{R}_1 + 2\mathcal{R}_2$$

and

$$\mathcal{R}_1 = \frac{l_1}{\mu A_1} \qquad \mathcal{R}_2 = \frac{l_2}{\mu A_2} \tag{15.30}$$

It is important at this stage to review the assumptions and simplifications made in analyzing the magnetic structure of Figure 15.13:

1. All of the magnetic flux is linked by all of the turns of the coil.
2. The flux is confined exclusively within the magnetic core.
3. The density of the flux is uniform across the cross-sectional area of the core.

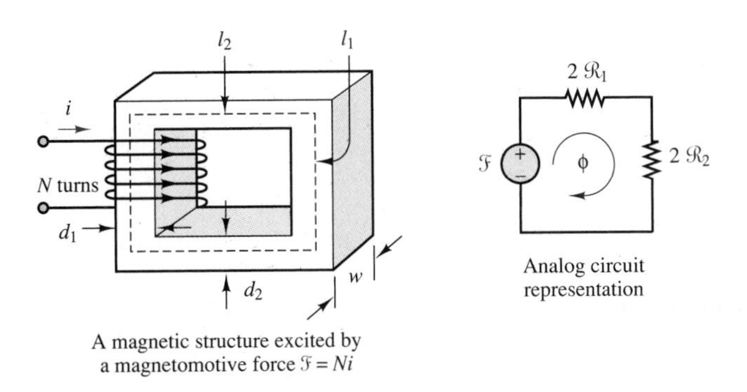

Figure 15.13 Analogy between magnetic and electric circuits

You can probably see intuitively that the first of these assumptions might not hold true near the ends of the coil, but that it might be more reasonable if the coil is tightly wound. The second assumption is equivalent to stating that the relative permeability of the core is infinitely higher than that of air (presuming that this is the medium surrounding the core): if this were the case, the flux would indeed be confined within the core. It is worthwhile to note that we make a similar assumption when we treat wires in electric circuits as perfect conductors: the conductivity of copper is substantially greater than that of free space, by a factor of approximately 10^{15}. In the case of magnetic materials, however, even for the best alloys, we have a relative permeability only on the order of 10^3 to 10^4. Thus, an approximation that is quite appropriate for electric circuits is not nearly as good in the case of magnetic circuits. Some of the flux in a structure such as those of Figures 15.12 and 15.13 would thus not be confined within the core (this is usually referred to as **leakage flux**). Finally, the assumption that the flux is uniform across the core cannot hold for a finite-permeability medium, but it is very helpful in giving an approximate *mean* behavior of the magnetic circuit.

The magnetic circuit analogy is therefore far from being exact. However, short of employing the tools of electromagnetic field theory and of vector calculus, or advanced numerical simulation software, it is the most convenient tool at the engineer's disposal for the analysis of magnetic structures. In the remainder of this chapter, the approximate analysis based on the electric circuit analogy will be used to obtain approximate solutions to problems involving a variety of useful magnetic circuits, many of which you are already familiar with. Among these will be the loudspeaker, solenoids, automotive fuel injectors, sensors for the measurement of linear and angular velocity and position, and other interesting applications.

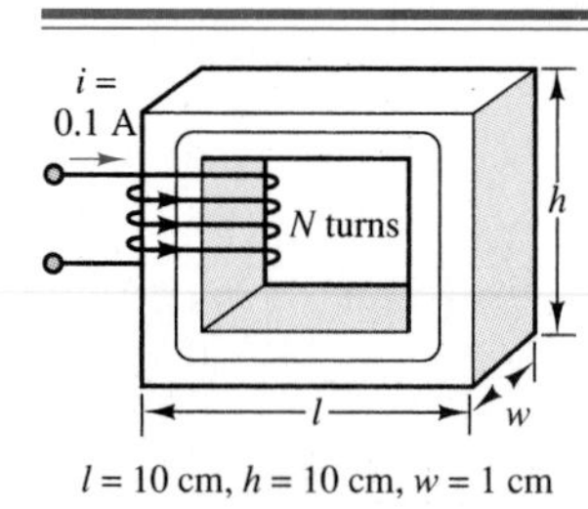

l = 10 cm, h = 10 cm, w = 1 cm

Figure 15.14

Example 15.3

Figure 15.14 shows a simple magnetic circuit. Calculate the flux, ϕ, the flux density, B, and the field intensity H, in the magnetic strucure. Assume that $\mu_r = 1{,}000, N = 500$, and $i = 100$ mA. Also, assume that the structure's cross section is square, with dimension w.

Solution:

First we find the magnetomotive force, which is the source of the flux:

$$\mathcal{F} = \text{mmf} = Ni = (500)(0.1) = 50 \text{ A} \cdot \text{t}$$

Next, we estimate the mean path of the flux. Note that we assume that the flux is completely contained inside the structure (or, equivalently, that the permeability of the structure is significantly greater than that of free space). According to this assumption, then, the mean path would run through the geometric center of the structure, as shown in Figure 15.15.

The total circuit path length is computed to be

$$l_c = 9 \text{ cm} + 9 \text{ cm} + 9 \text{ cm} + 9 \text{ cm} = 36 \text{ cm}$$

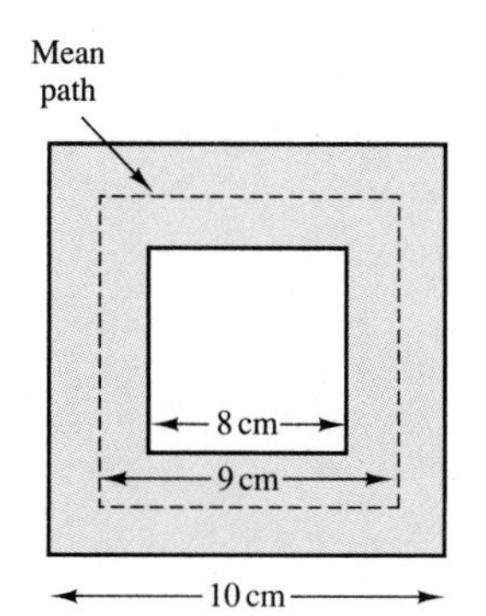

Figure 15.15

Since the cross section is 1 cm × 1 cm in all parts of the structure, we can use the total length of the path to find the reluctance, $\mathcal{R}$, of the magnetic circuit, according to the following expression:

$$\mathcal{R} = \frac{l_c}{\mu A} = \frac{0.36}{(1{,}000)(4\pi \times 10^{-7})(0.01)(0.01)} = 2.865 \times 10^6 \text{A} \cdot \text{t/Wb}$$

The corresponding equivalent magnetic circuit is shown in Figure 15.16.

A consequence of the simplifying approximations made in this section is that the mmf is linearly related to the flux:

$$\mathcal{F} = \phi\mathcal{R}$$

so that we can compute the magnetic flux in the structure to be

$$\phi = \frac{\mathcal{F}}{\mathcal{R}} = \frac{50 \text{ A} \cdot \text{t}}{2.865 \times 10^6 \text{ A} \cdot \text{t/Wb}}$$

$$= 1.75 \times 10^{-5} \text{ Wb}$$

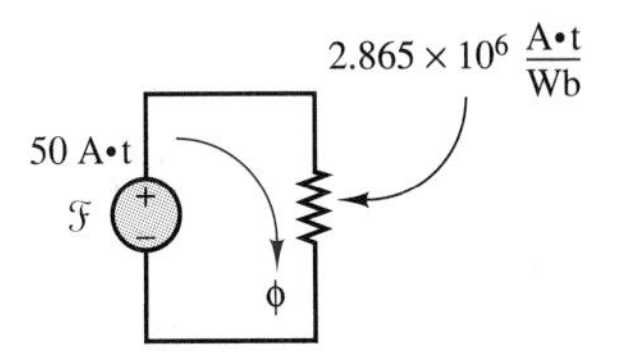

Figure 15.16

The corresponding flux density is then computed as follows:

$$B = \frac{\phi}{A} = \frac{17.5 \times 10^{-6} \text{ Wb}}{0.0001 \text{ m}^2}$$

$$= 0.175 \text{ Wb/m}^2$$

and the magnetic field intensity is therefore given by

$$H = \frac{B}{\mu} = \frac{0.175 \text{ Wb/m}^2}{1{,}000(4\pi \times 10^{-7}) \text{ H/m}}$$

$$= 139.0 \text{ A} \cdot \text{t/m}$$

These calculations are useful, in practice, in determining practical parameters in the design or application of electromagnetic structures. For example, determining the magnetomotive force required to generate a certain amount of flux is necessary to determine the current that must be provided. Since, in practice, any coil has a finite resistance, this current will determine the required excitation voltage. As will be shown in a later section, these calculations can also be related to the determination of energy and force in electromagnetic transducers.

It will be useful, before proceeding further, to consider the analysis of the same simple magnetic structure when an **air gap** is present. Air gaps are very common in magnetic structures; in rotating machines, for example, air gaps are necessary to allow for free rotation of the inner core of the machine. The magnetic circuit of Figure 15.17 differs from the circuit analyzed in Example 15.3 simply because of the presence of an air gap; the effect of the gap is to break the continuity of the high-permeability path for the flux, adding a high-reluctance component to the equivalent circuit. The situation is analogous to adding a very large series resistance to a series electrical circuit. It should be evident from Figure 15.17 that the basic concept of reluctance still applies, although now two different permeabilities must be taken into account.

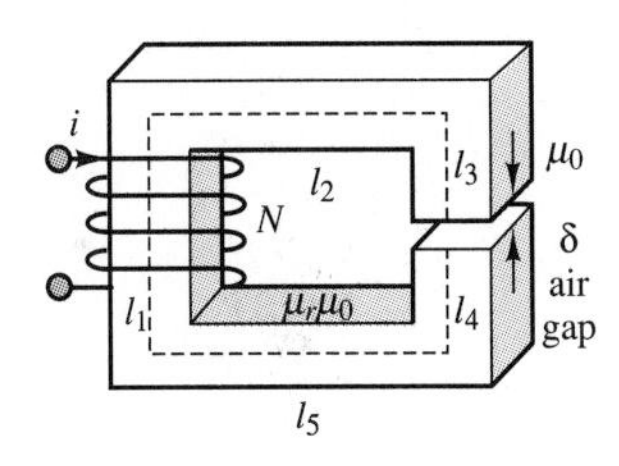

Figure 15.17 Magnetic circuit with air gap

The equivalent circuit for the structure of Figure 15.17 may be drawn as shown in Figure 15.18, where $\mathcal{R}_n$ is the reluctance of path l_n, for $n = 1, 2, \ldots, 5$, and $\mathcal{R}_g$ is the reluctance of the air gap. The reluctances can be expressed as follows, if we assume that the magnetic structure has a uniform cross-sectional area, A:

$$\mathcal{R}_1 = \frac{l_1}{\mu_r\mu_0 A} \qquad \mathcal{R}_2 = \frac{l_2}{\mu_r\mu_0 A}$$

$$\mathcal{R}_3 = \frac{l_3}{\mu_r\mu_0 A} \qquad \mathcal{R}_4 = \frac{l_4}{\mu_r\mu_0 A} \tag{15.31}$$

$$\mathcal{R}_5 = \frac{l_5}{\mu_r\mu_0 A} \qquad \mathcal{R}_g = \frac{\delta}{\mu_0 A_g}$$

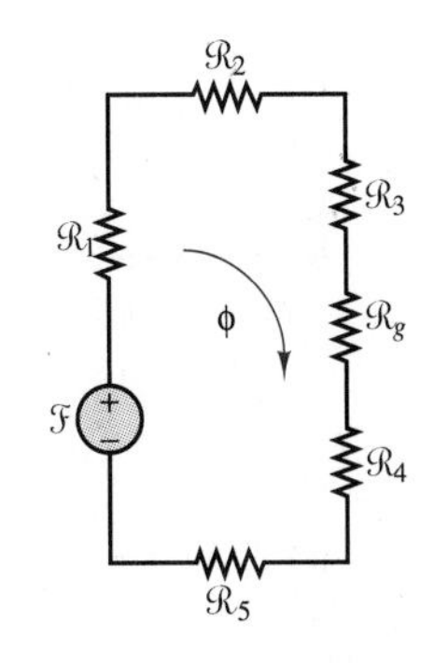

Figure 15.18 Equivalent representation of magnetic circuit with an air gap

Note that in computing $\mathcal{R}_g$, the length of the gap is given by δ and the permeability is given by μ_0, as expected, but A_g is different from the cross-sectional area, A, of the structure. The reason is that the flux lines exhibit a phenomenon known as **fringing** as they cross an air gap. The flux lines actually *bow out* of the gap defined by the cross section, A, not being contained by the high-permeability material any longer. Thus, it is customary to define an area A_g that is greater than A, to account for this phenomenon. Example 15.4 describes in more detail the procedure for finding A_g and also discusses the phenomenon of fringing.

Example 15.4 Magnetic Structures with Air Gaps

In practical applications, magnetic circuits such as the ones discussed in this section are often used to generate forces or displacements as a result of the application of an electrical current. For example, the magnetic structure of Figure 15.19 might be used to displace a conducting bar in order to *make* or *break* an electrical contact. The techniques introduced in this section may be applied to understand the behavior of practical devices, such as solenoids and relays, which will be described shortly; and to carry out approximate calculations—for example, to compute the current required to energize a contact. This example illustrates the use of these techniques in a practical magnetic structure. The structure includes two air gaps that the flux must cross to establish a closed path. Assume that the relative permeability of the magnetic material is $\mu_r = 10{,}000$. Find:

1. The reluctance of each of the air gaps.
2. The total reluctance of the magnetic circuit.
3. The flux density in the bottom plate of the structure.

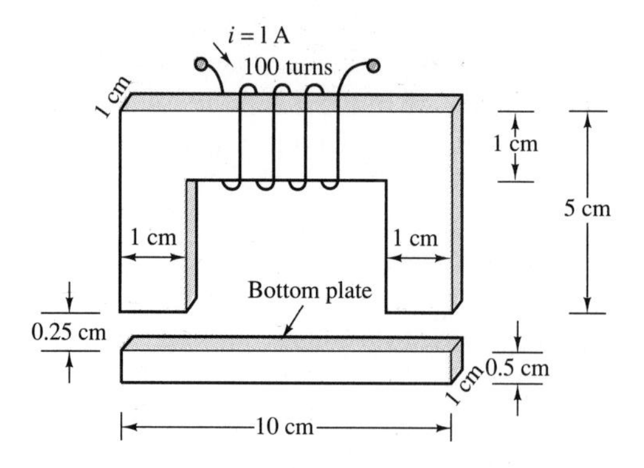

Figure 15.19 Electromagnetic structure with air gaps

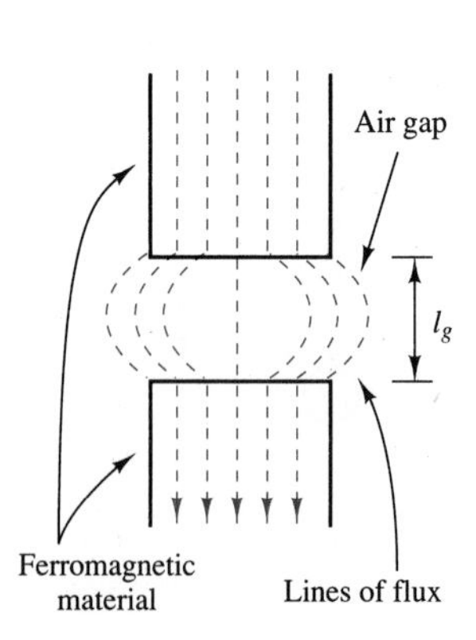

Figure 15.20 Fringing effects in air gap

Solution:

1. To find the reluctance of the air gaps, we must determine the effective cross-sectional area of the gaps. These air gaps will display the phenomenon known as *fringing*, which has been briefly discussed in this section; Figure 15.20 depicts the fringing of the flux lines across each gap.

   Notice how the flux lines tend to spread out in the air gap while in the magnetic structure they remain straight and completely contained. The assumption of constant flux density clearly breaks down in this case, but it is possible to partially compensate for this inaccuracy by increasing the effective area of the gap. A rule of thumb for this situation is to add the length of the air gap, l_g, to the gap's physical dimensions in both directions. For the structure in our example, the effective area in each gap will be found by the relation

$$A_g = (1 + l_g)(1 + l_g) = (1 + 0.25)(1 + 0.25) = 1.5625 \text{ cm}^2$$

The reluctance of the air gap is then found according to the usual expression, where now $\mu = \mu_0$:

$$\begin{aligned}\mathcal{R}_g &= \frac{l_g}{\mu_0 A_g} \\ &= \frac{0.0025 \text{ m}}{(4\pi \times 10^{-7} \text{ Wb/A}\cdot\text{m})(0.15625 \times 10^{-3} \text{ m}^2)} \\ &= 1.27 \times 10^7 \text{ A}\cdot\text{t/Wb}\end{aligned}$$

2. To find the total reluctance of the magnetic circuit, we must find the mean path followed by the flux. Figure 15.21 shows the mean path. The path indicates eight separate lengths that must be accounted for in finding the reluctance. The total length is given by

$$l_c = l_1 + l_2 + l_3 + l_4 + l_5 + l_6 + l_g + l_g$$

The total reluctance in this path is given by

$$\mathcal{R} = \mathcal{R}_1 + \mathcal{R}_2 + \mathcal{R}_3 + \mathcal{R}_4 + \mathcal{R}_5 + \mathcal{R}_6 + \mathcal{R}_g + \mathcal{R}_g$$

(where $\mathcal{R}_g$ has already been calculated in part 1). We also take note that the reluctance varies in proportion to the length of the path, and since l_5 and l_6 are small compared with the total length of the path, we choose to ignore these contributions. Then:

$$\mathcal{R} \approx \mathcal{R}_1 + \mathcal{R}_2 + \mathcal{R}_3 + \mathcal{R}_4 + 2\mathcal{R}_g$$

Now, to find the respective reluctances, we compute

$$\mathcal{R}_1 = \frac{l_1}{\mu A_1} = \frac{0.09}{(10{,}000)(4\pi \times 10^{-7})(0.0001)} = 71.6 \times 10^3 \text{ A}\cdot\text{t/Wb}$$

$$\begin{aligned}\mathcal{R}_3 = \mathcal{R}_2 &= \frac{l_2}{\mu A_2} = \frac{0.045}{(10{,}000)(4\pi \times 10^{-7})(0.0001)} \\ &= 35.8 \times 10^3 \text{ A}\cdot\text{t/Wb}\end{aligned}$$

$$\mathcal{R}_4 = \frac{l_4}{\mu A_4} = \frac{0.09}{(10{,}000)(4\pi \times 10^{-7})(0.00005)} = 143.2 \times 10^3 \text{ A}\cdot\text{t/Wb}$$

and

$$\begin{aligned}\mathcal{R} &= 71.6 \times 10^3 + 2 \times 35.8 \times 10^3 + 143.2 \times 10^3 + 2 \times 1.27 \times 10^7 \\ &= 25.686 \times 10^6 \text{ A}\cdot\text{t/Wb}\end{aligned}$$

You can see that the air gaps, in spite of their small physical dimensions, contribute the greater part of the reluctance of the magnetic circuit.

3. To find the flux density, we must first determine the flux in the magnetic circuit from the expression

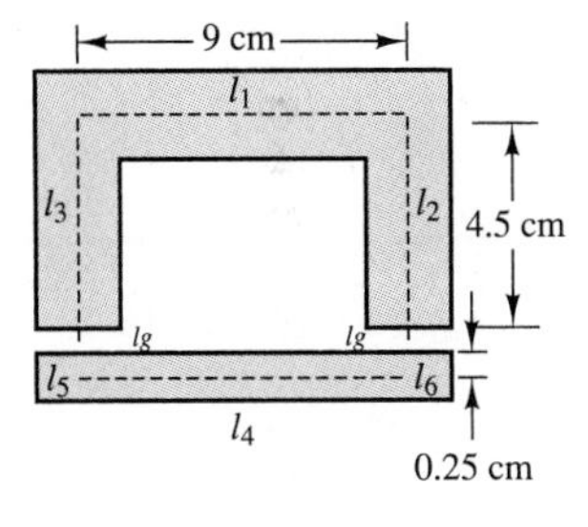

Figure 15.21

$$\phi = \frac{\mathcal{F}}{\mathcal{R}}$$

and since the mmf is given by

$$\mathcal{F} = Ni = (100)(1) = 100 \text{ A} \cdot \text{t}$$

the flux can be computed to be

$$\phi = \frac{100 \text{ A} \cdot \text{t}}{25.686 \times 10^6 \text{ A} \cdot \text{t/Wb}} = 3.89 \times 10^{-6} \text{ Wb}$$

Finally,

$$B = \frac{\phi}{A_4} = \frac{3.89 \times 10^{-6} \text{ Wb}}{0.00005 \text{ m}^2}$$
$$= 0.0778 \text{ Wb/m}^2$$

These calculations are essential in determining the specifications of electromechanical devices for a given application, since they can be used to determine the geometry and current required to accomplish a design objective.

Example 15.5 Magnetic Reluctance Position Sensor

A simple magnetic structure, very similar to those examined in the previous examples, finds very common application in the so-called variable-reluctance position sensor, which, in turn, finds widespread application in a variety of configurations for the measurement of linear and angular position and velocity. Figure 15.22 depicts one particular configuration that is used in many applications. In this structure, a permanent magnet with a coil of wire wound around it forms the sensor; a steel disk (typically connected to a rotating shaft) has a number of tabs that pass between the pole pieces of the sensor. The area of the tab is assumed equal to the area of the cross section of the pole pieces and is equal to a^2. The reason for the name *variable-reluctance sensor* is that the reluctance of the magnetic structure is variable, depending on whether or not a ferromagnetic tab lies between the pole pieces of the magnet.

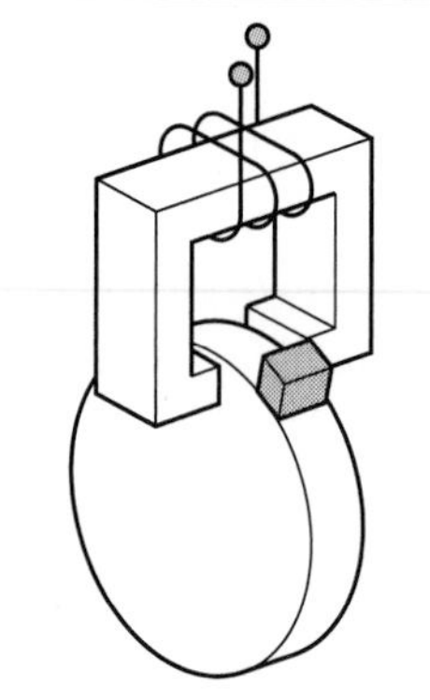

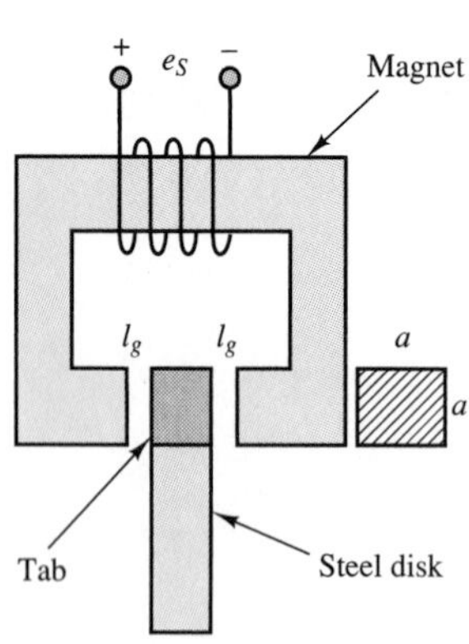

Figure 15.22 Variable-reluctance position sensor

The principle of operation of the sensor is that an electromotive force, e_S, is induced across the coil by the change in magnetic flux caused by the passage of the tab between the pole pieces when the disk is in motion. As the tab enters the volume between the pole pieces, the flux will increase, because of the lower reluctance of the configuration, until it reaches a maximum when the tab is centered between the poles of the magnet. Figure 15.23 depicts the approximate shape of the resulting voltage, which, according to Faraday's law, is given by

$$e_S = -\frac{d\phi}{dt}$$

The rate of change of flux is dictated by the geometry of the tab and of the pole pieces, and by the speed of rotation of the disk. It is important to note that, since the flux is changing only if the disk is rotating, this sensor cannot detect the static position of the disk.

One common application of this concept is in the measurement of the speed of rotation of rotating machines, including electric motors and internal combustion engines. In these applications, use is made of a *60-tooth wheel*, which permits the conversion of the speed rotation directly to units of revolutions per minute. The output of a variable-reluctance position sensor magnetically coupled to a rotating disk equipped with 60 tabs (teeth) is processed through a comparator or Schmitt trigger circuit (see Chapter 14). The voltage waveform generated by the sensor is nearly sinusoidal when the teeth are closely spaced, and it is characterized by one sinusoidal cycle for each tooth on the disk. If a negative zero-crossing detector (see Chapter 14) is employed, the trigger circuit will generate a pulse corresponding to the passage of each tooth, as shown in Figure 15.24. If the time between any two pulses is measured by means of a high-frequency clock, the speed of the engine can be directly determined in units of rev/min by means of a digital counter (see Chapter 13).

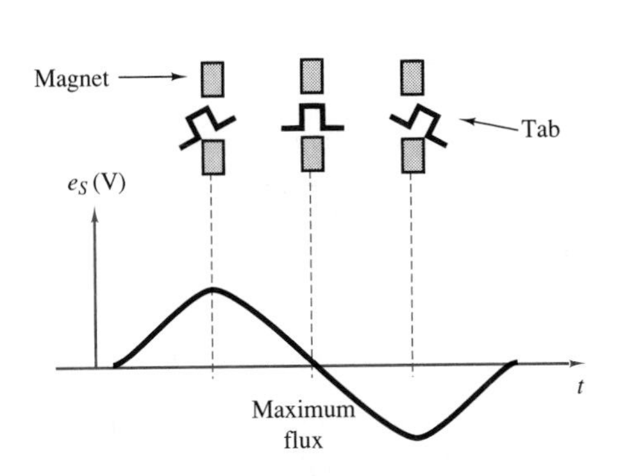

Figure 15.23 Variable-reluctance position sensor waveform

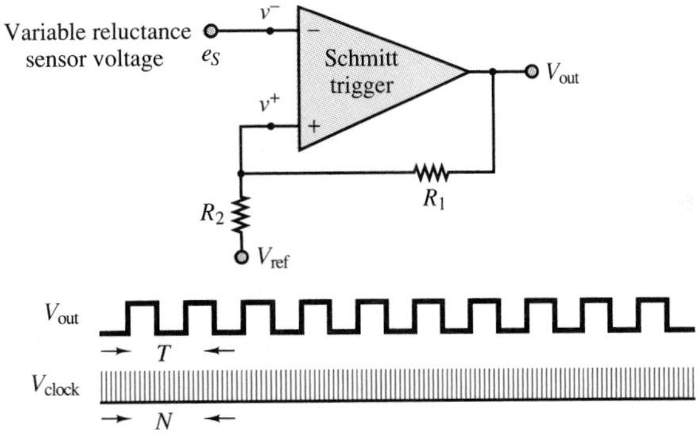

Figure 15.24 Signal processing for a 60-tooth wheel RPM sensor

Example 15.6

This example illustrates the calculation of the voltage induced in the magnetic reluctance sensor of Example 15.5 by a rotating toothed wheel. In particular, we will find an approximate expression for the reluctance and the induced voltage for the position sensor shown in Figure 15.25, and show that the induced voltage is speed-dependent. It will be assumed that the reluctance of the core and fringing at the air gaps are both negligible.

Solution:

From the geometry of Example 15.5, the equivalent reluctance of the magnetic structure is twice that of one gap, since the permeability of the tab and magnetic structure are assumed infinite (i.e., they have negligible reluctance). When the tab and the poles are aligned, the angle θ is zero, as shown in Figure 15.25, and the area of the air gap is maximum. For angles greater than $2\theta_0$, the magnetic length of the air gaps is so large that the magnetic field may reasonably be taken as zero.

To model the reluctance of the gaps, we assume the following simplified expression, where the area of overlap of the tab with the magnetic poles is assumed proportional to the angular displacement:

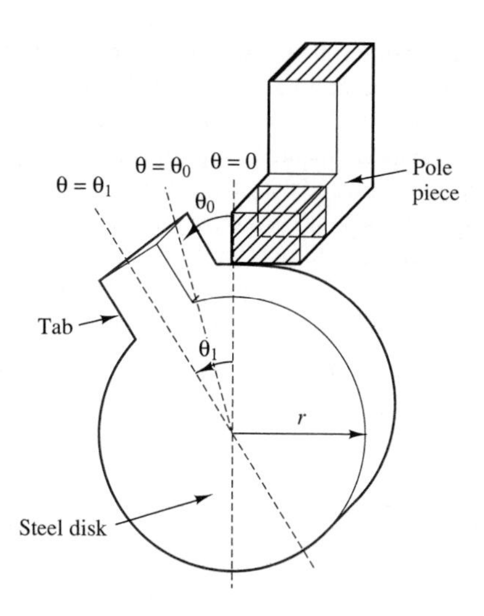

Figure 15.25 Reluctance sensor for measurement of angular position

$$\mathcal{R} = \frac{2l_g}{\mu_0 A} = \frac{2l_g}{\mu_0 ar(\theta_1 - \theta)} \qquad \text{for} \qquad 0 < \theta < \theta_1$$

Naturally, this is an approximation; however, the approximation captures the essential idea of this transducer, namely, that the reluctance will decrease with increasing overlap area until it reaches a minimum, and then it will increase as the overlap area decreases. For $\theta = \theta_1$, that is, with the tab outside the magnetic pole pieces, we have $\mathcal{R}_{max} \to \infty$. For $\theta = 0$, that is, with the tab perfectly aligned with the pole pieces, we have $\mathcal{R}_{min} = 2l_g/\mu_0 ar\theta_1$. The flux ϕ may therefore be computed as follows:

$$\phi = \frac{Ni}{\mathcal{R}} = \frac{Ni\mu_0 ar(\theta_1 - \theta)}{2l_g}$$

The induced voltage e_S is found by

$$e_S = -\frac{d\phi}{dt} = -\frac{d\phi}{d\theta}\frac{d\theta}{dt} = \frac{Ni\mu_0 ar}{2l_g} \times \omega$$

where $\omega = d\theta/dt$ is the rotational speed of the steel disk. It should be evident that the induced voltage is speed-dependent. For $a = 1$ cm, $r = 10$ cm, $l_g = 0.1$ cm, $N = 100$ turns, $i = 10$ mA, $\theta_1 = 6° \approx 0.1$ rad, and $\omega = 400$ rad/s (approximately 3,800 rev/min), we have

$$\mathcal{R}_{max} = \frac{2 \times 0.1 \times 10^{-2}}{4\pi \times 10^{-7} \times 1 \times 10^{-2} \times 10 \times 10^{-2} \times 0.1} = 1.59 \times 10^7 \text{ A} \cdot \text{t/Wb}$$

$$e_{S\,\text{peak}} = \frac{1{,}000 \times 10 \times 10^{-3} \times 4\pi \times 10^{-7} \times 1 \times 10^{-2} \times 10^{-1}}{2 \times 0.1 \times 10^{-2}} \times 400$$

$$= 2.5 \text{ mV}$$

That is, the peak amplitude of e_S will be 2.5 mV.

Example 15.7

Figure 15.26 shows a coil of wire wrapped around an iron core. If the flux in the core is given by the equation $\phi = 4\sin(377t)$ mWb, and if the coil has 100 turns, what is the voltage produced at the terminals of the coil? What polarity is the voltage during the time when the flux is decreasing in the figure? (Neglect the flux leakage.)

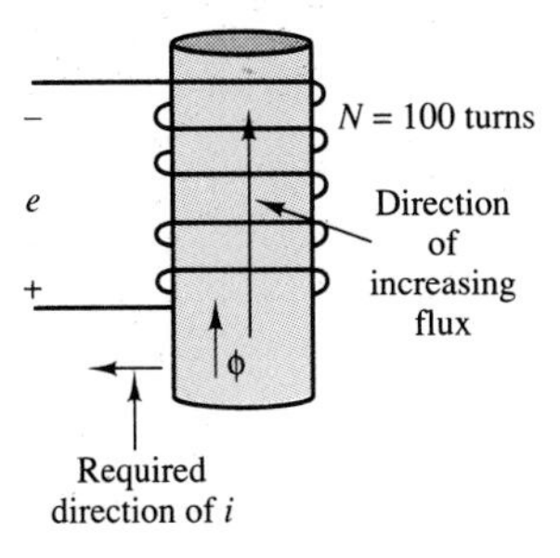

Figure 15.26

Solution:

The direction of the voltage while the flux is decreasing in the reference direction must be positive to negative, as shown in the figure. The magnitude of the induced voltage is given by

$$e = N\frac{d\phi}{dt} = 100 \times \frac{d}{dt}(4 \times 10^{-3} \sin 377t)$$

$$= 150.8\cos(377t) = 150.8\sin(377t + 90°)$$

DRILL EXERCISES

15.5 If $\mathcal{R} = 2\mathcal{R}_g$ in Example 15.4, calculate ϕ and B.

15.6 Determine the equivalent reluctance of the structure of Figure 15.27 as seen by the "source" if μ_r for the structure is 1,000, $l = 5$ cm, and all of the legs are 1 cm on a side.

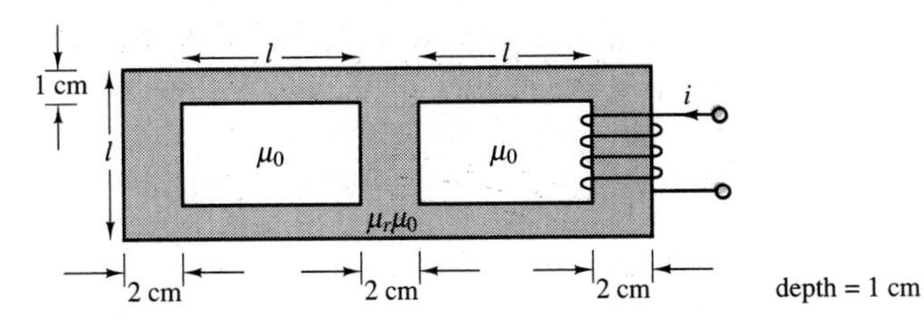

Figure 15.27

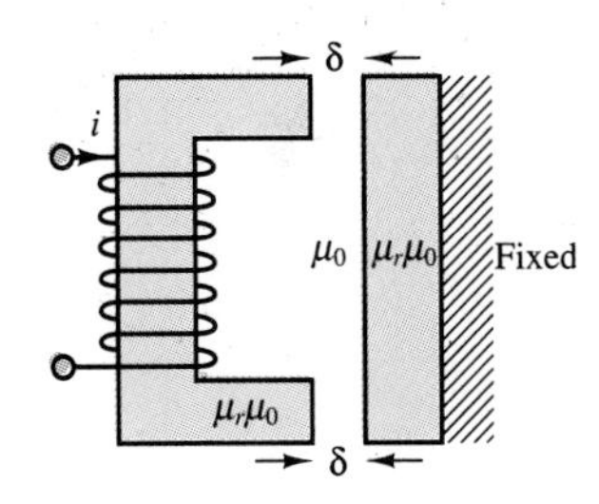

Figure 15.28

15.7 Find the equivalent reluctance of the magnetic circuit shown in Figure 15.28 if μ_r of the structure is infinite, $\delta = 2$ mm, and the physical cross section of the core is 1 cm^2. Do not neglect fringing.

15.8 Find the equivalent magnetic circuit of the structure of Figure 15.29 if μ_r is infinite. Give expressions for each of the circuit values if the physical cross-sectional area of each of the legs is given by

$$A = l \times w$$

Do not neglect fringing.

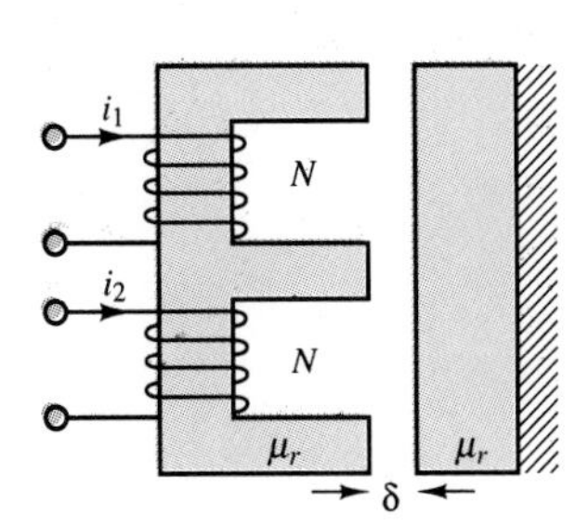

Figure 15.29

15.3 MAGNETIC MATERIALS AND *B-H* CURVES

In the analysis of magnetic circuits presented in the previous sections, the relative permeability, μ_r, was treated as a constant. In fact, the relationship between the magnetic flux density, **B**, and the associated field intensity, **H**,

$$\mathbf{B} = \mu \mathbf{H} \tag{15.32}$$

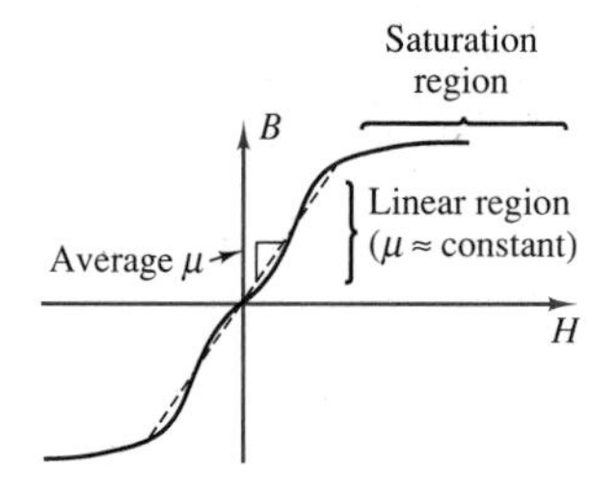

Figure 15.30 Permeability and magnetic saturation effects

is characterized by the fact that the relative permeability of magnetic materials is not a constant, but is a function of the magnetic field intensity. In effect, all magnetic materials exhibit a phenomenon called **saturation,** whereby the flux density increases in proportion to the field intensity until it cannot do so any longer. Figure 15.30 illustrates the general behavior of all magnetic materials. You will note that since the *B-H* curve shown in the figure is nonlinear, the value of μ (which is the slope of the curve) depends on the intensity of the magnetic field.

To understand the reasons for the saturation of a magnetic material, we need to briefly review the mechanism of magnetization. The basic idea behind magnetic materials is that the spin of electrons constitutes motion of charge, and therefore leads to magnetic effects, as explained in the introductory section of this chapter. In most materials, the electron spins cancel out, on the whole, and no net effect remains. In ferromagnetic materials, on the other hand, atoms can align so that the electron spins cause a net magnetic effect. In such materials, there exist small regions with strong magnetic properties (called **magnetic domains**), the effects of which are neutralized in unmagnetized material by other, similar regions that are oriented differently, in a random pattern. When the material is magnetized, the magnetic domains tend to align with each other, to a degree that is determined by the intensity of the applied magnetic field.

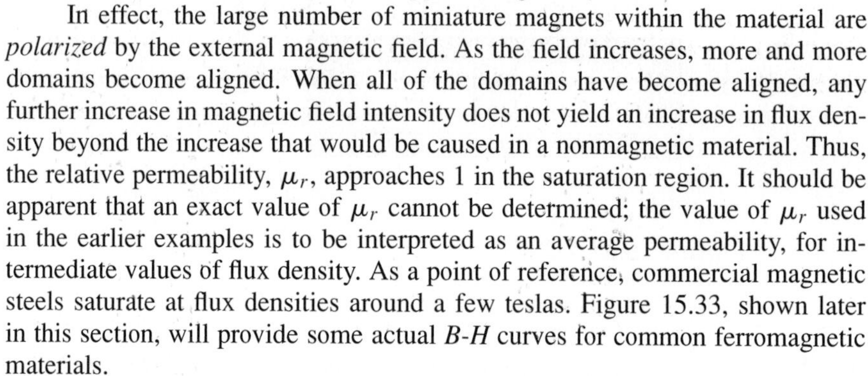

In effect, the large number of miniature magnets within the material are *polarized* by the external magnetic field. As the field increases, more and more domains become aligned. When all of the domains have become aligned, any further increase in magnetic field intensity does not yield an increase in flux density beyond the increase that would be caused in a nonmagnetic material. Thus, the relative permeability, μ_r, approaches 1 in the saturation region. It should be apparent that an exact value of μ_r cannot be determined; the value of μ_r used in the earlier examples is to be interpreted as an average permeability, for intermediate values of flux density. As a point of reference, commercial magnetic steels saturate at flux densities around a few teslas. Figure 15.33, shown later in this section, will provide some actual *B-H* curves for common ferromagnetic materials.

The phenomenon of saturation carries some interesting implications with regard to the operation of magnetic circuits: the results of the previous section would seem to imply that an increase in the mmf (that is, an increase in the current driving the coil) would lead to a proportional increase in the magnetic flux. This is true in the *linear region* of Figure 15.30; however, as the material reaches saturation, further increases in the driving current (or, equivalently, in the mmf) do not yield further increases in the magnetic flux.

There are two more features that cause magnetic materials to further deviate from the ideal model of the linear *B-H* relationship: **eddy currents** and **hysteresis.**

The first phenomenon consists of currents that are caused by any time-varying flux in the core material. As you know, a time-varying flux will induce a voltage, and therefore a current. When this happens inside the magnetic core, the induced voltage will cause "eddy" currents (the terminology should be self-explanatory) in the core, which depend on the resistivity of the core. Figure 15.31 illustrates the phenomenon of eddy currents. The effect of these currents is to dissipate energy in the form of heat. Eddy currents are reduced by selecting high-resistivity core materials, or by *laminating* the core, introducing tiny, discontinuous air gaps between core layers (see Figure 15.31). Lamination of the core reduces eddy currents greatly without affecting the magnetic properties of the core.

It is beyond the scope of this chapter to quantify the losses caused by induced eddy currents, but it will be important in Chapters 16 and 17 to be aware of this source of energy loss.

Hysteresis is another loss mechanism in magnetic materials; it displays a rather complex behavior, related to the magnetization properties of a material. The curve of Figure 15.32 reveals that the *B-H* curve for a magnetic material during magnetization (as *H* is increased) is displaced with respect to the curve that is measured when the material is demagnetized. To understand the hysteresis process, consider a core that has been energized for some time, with a field intensity of $H_1 A \cdot$ t/m. As the current required to sustain the mmf corresponding to H_1 is decreased, we follow the hysteresis curve from the point α to the point β. When the mmf is exactly zero, the material displays the **remanent** (or **residual**) **magnetization** B_r. To bring the flux density to zero, we must further decrease the mmf (i.e., produce a negative current), until the field intensity reaches the value $-H_0$ (point γ on the curve). As the mmf is made more negative, the curve eventually reaches the point α'. If the excitation current to the coil is now increased, the magnetization curve will follow the path $\alpha' = \beta' = \gamma' = \alpha$, eventually returning to the original point in the *B-H* plane, but via a different path.

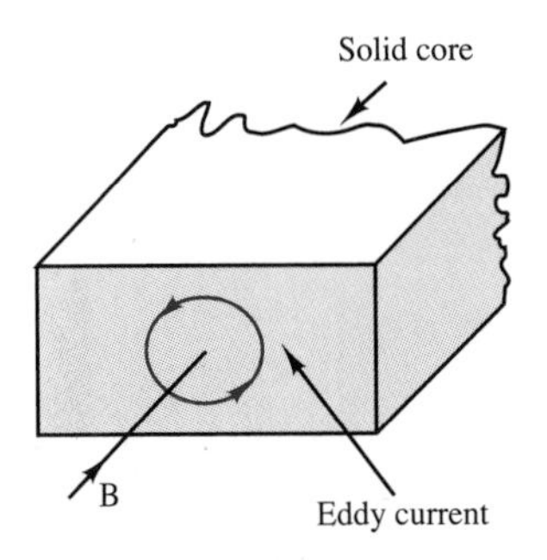

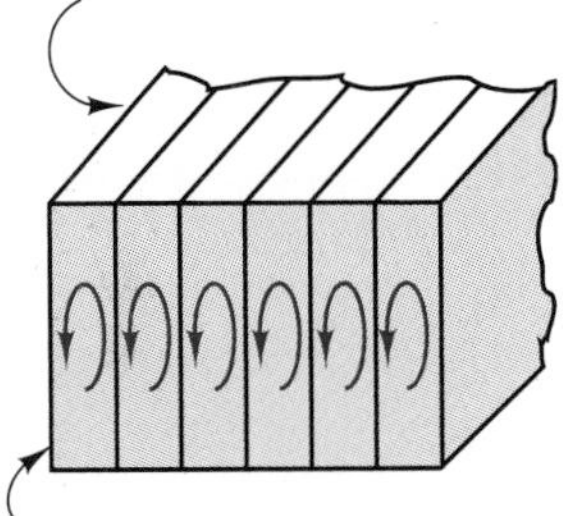

Figure 15.31 Eddy currents in magnetic structures

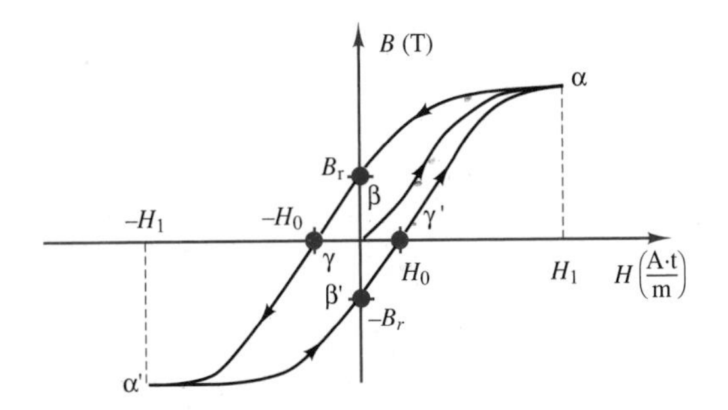

Figure 15.32 Hysteresis in magnetization curves

The result of this process, by which an *excess magnetomotive force* is required to magnetize or demagnetize the material, is a net energy loss. It is difficult to evaluate this loss exactly; however, it can be shown that it is related to the area between the curves of Figure 15.32. There are experimental techniques that enable the approximate measurement of these losses.

Figures 15.33(a)–(c) depict magnetization curves for three very common ferromagnetic materials: cast iron, cast steel, and sheet steel. These curves will be useful in solving some of the homework problems.

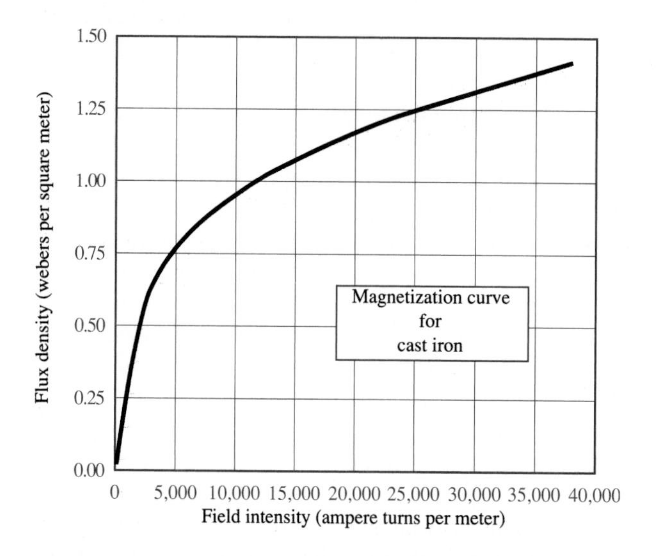

Figure 15.33(a) Magnetization curve for cast iron

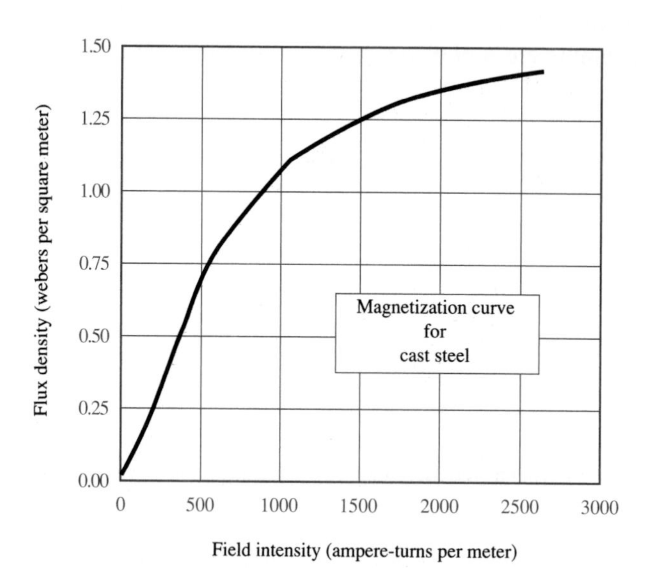

Figure 15.33(b) Magnetization curve for cast steel

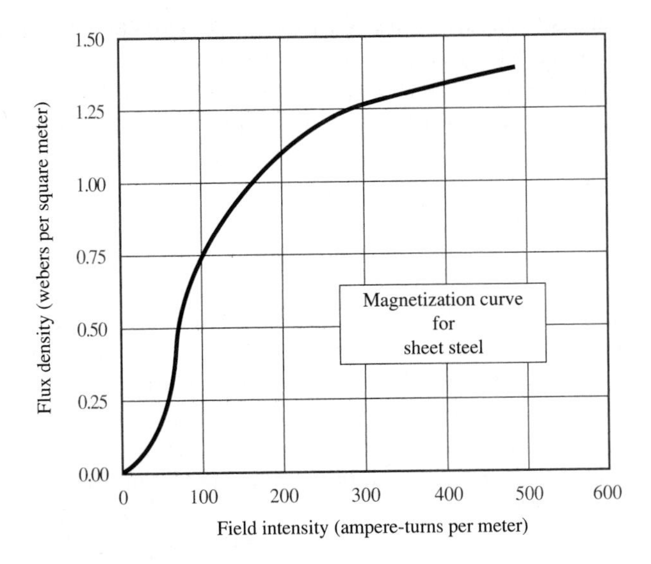

Figure 15.33(c) Magnetization curve for sheet steel

15.4 TRANSFORMERS

One of the more common magnetic structures in everyday applications is the transformer. The ideal transformer was introduced in Chapter 5 as a device that can step an AC voltage up or down by a fixed ratio, with a corresponding decrease or increase in current. The structure of a simple magnetic transformer is shown in Figure 15.34, which illustrates that a transformer is very similar to the magnetic circuits described earlier in this chapter. Coil L_1 represents the input side of the transformer, while coil L_2 is the output coil; both coils are wound around the same magnetic structure, which we show here to be similar to the "square doughnut" of the earlier examples.

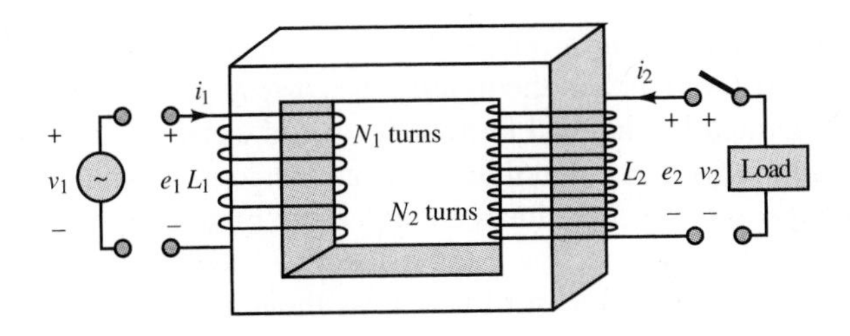

Figure 15.34 Structure of a transformer

The ideal transformer operates on the basis of the same set of assumptions we made in earlier sections: the flux is confined to the core, the flux links all turns of both coils, and the permeability of the core is infinite. The last assumption is equivalent to stating that an arbitrarily small mmf is sufficient to establish a flux in the core. In addition, we assume that the ideal transformer coils offer negligible resistance to current flow. Using this rather idealized picture, we can now resort to the principles developed earlier in this chapter to re-derive the already familiar results for the ideal transformer.

In essence, the operation of a transformer requires a time-varying current; if a time-varying voltage is applied to the primary side of the transformer, a corresponding current will flow in L_1; this current acts as an mmf and causes a (time-varying) flux in the structure. But the existence of a changing flux will induce an emf across the secondary coil! Without the need for a direct electrical connection, the transformer can couple a source voltage at the primary to the load; the coupling occurs by means of the magnetic field acting on both coils. Thus, a transformer operates by converting electric energy to magnetic, and then back to electric. The following derivation illustrates this viewpoint in the ideal case (no loss of energy), and compares the result with the definition of the ideal transformer in Chapter 5.

If a time-varying voltage source is connected to the input side, then by virtue of Faraday's law, a corresponding time-varying flux $d\phi/dt$ is established in coil L_1:

$$e_1 = N_1 \frac{d\phi}{dt} = v_1 \tag{15.33}$$

But since the flux thus produced also links coil L_2, an emf is induced across the output coil as well:

$$e_2 = N_2 \frac{d\phi}{dt} = v_2 \tag{15.34}$$

This induced emf can be measured as the voltage v_2 at the output terminals, and one can readily see that the ratio of the open-circuit output voltage to input terminal voltage is

$$\frac{v_2}{v_1} = \frac{N_2}{N_1} \tag{15.35}$$

If a load current i_2 is now required by the connection of a load to the output circuit (by closing the switch in the figure), the corresponding mmf is $\mathcal{F}_2 = N_2 i_2$. This mmf, generated by the load current i_2, would cause the flux in the core to change; however, this is not possible, since a change in ϕ would cause a corresponding change in the voltage induced across the input coil. But this voltage is determined (fixed) by the source v_1 (and is therefore $d\phi/dt$), so that the input coil is forced to generate a **counter mmf** to oppose the mmf of the output coil; this is accomplished as the input coil draws a current i_1 from the source v_1 such that

$$i_1 N_1 = i_2 N_2 \tag{15.36}$$

or

$$\frac{i_2}{i_1} = \frac{N_1}{N_2} = \alpha \tag{15.37}$$

where α is the ratio of primary to secondary turns (the transformer ratio) and N_1 and N_2 are the primary and secondary turns, respectively. If there were any net difference between the input and output mmf, flux balance required by the input voltage source would not be satisfied. Thus, the two mmf's must be equal. As you can easily verify, these results are the same as in Chapter 5; in particular, the ideal transformer does not dissipate any power, since

$$v_1 i_1 = v_2 i_2 \tag{15.38}$$

Note the distinction we have made between the induced voltages (emf's), e, and the terminal voltages, v. In general, these are not the same.

The results obtained for the ideal case do not completely represent the physical nature of transformers. A number of loss mechanisms need to be included in a practical transformer model, to account for the effects of leakage flux, for various magnetic core losses (e.g., hysteresis), and for the unavoidable resistance of the wires that form the coils.

Commercial transformer ratings are usually given on the so-called nameplate, which indicates the normal operating conditions. The nameplate includes the following parameters:

Primary-to-secondary voltage ratio
Design frequency of operation
(Apparent) rated output power

For example, a typical nameplate might read 480:240 V, 60 Hz, 2 kVA. The voltage ratio can be used to determine the turns ratio, while the rated output power represents the continuous power level that can be sustained without overheating. It is important that this power be rated as the apparent power in kVA, rather than real power in kW, since a load with low power factor would still draw current and therefore operate near rated power. Another important performance characteristic of a transformer is its **power efficiency,** defined by:

$$\text{Power efficiency} = \eta = \frac{\text{Output power}}{\text{Input power}} \tag{15.39}$$

The following examples illustrate the use of the nameplate ratings and the calculation of efficiency in a practical transformer, in addition to demonstrating the application of the circuit models.

Example 15.8

A transformer has the following nameplate ratings:

120 V/480 V
48 kVA
60 Hz

If the transformer's primary side is the 480-V side, find

1. The turns ratio $\alpha = N_1/N_2$.
2. The current for which the device is rated.

Solution:

1. From the ratio of primary to secondary voltages, the turns ratio is found to be

$$\alpha = \frac{N_1}{N_2} = \frac{480}{120} = 4$$

2. The kVA rating of the transformer can be used to find both primary current and secondary current: the primary current is 48 kVA/480 V = 100 A, while the secondary current is 48 kVA/120 V = 400 A. These calculations do not include the effects of the unavoidable losses that are always present in a transformer.

Example 15.9 Impedance Transformer

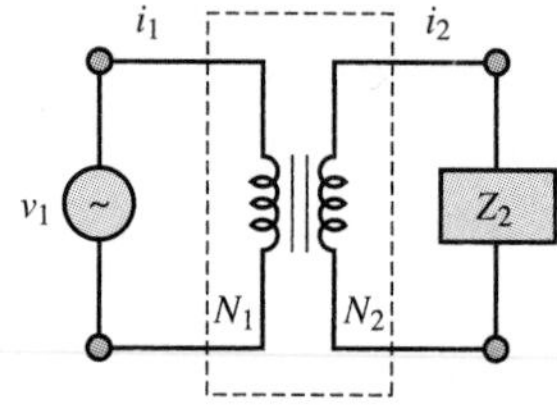

Figure 15.35 Ideal transformer

The ideal transformer of Figure 15.35 has a load Z_2 across its secondary output terminals. Find the equivalent reflected impedance from secondary to primary.

Solution:

For any value of load impedance Z_2, the secondary impedance looking into the secondary terminals from the load is

$$Z_1 = \frac{V_1}{I_1}$$

Since any changes in load at the secondary will be reflected as a change in current at the primary, it is convenient to simplify the transformer to a single equivalent circuit. From equation 15.35, we have

$$V_1 = \alpha V_2 \qquad I_1 = \frac{I_2}{\alpha}$$

Thus,

$$Z_1 = \frac{\alpha V_2}{I_2/\alpha} = \alpha^2 \frac{V_2}{I_2}$$

But V_2/I_2 is the secondary impedance, Z_2; therefore,

$$Z_1 = \alpha^2 Z_2$$

or

$$\frac{Z_1}{Z_2} = \alpha^2 = \left(\frac{N_1}{N_2}\right)^2$$

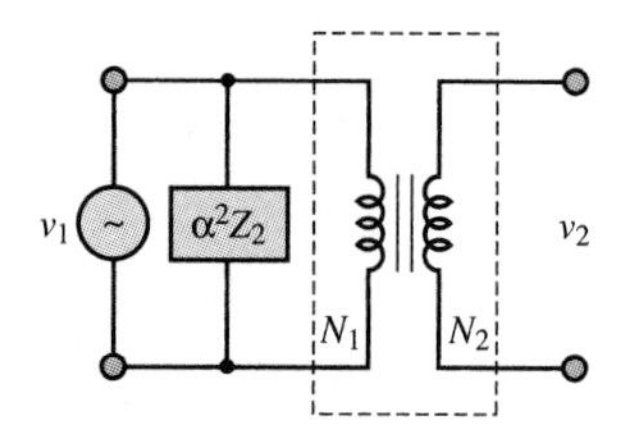

Figure 15.36

Figure 15.36 shows the impedance looking into the input terminals from the source when the secondary impedance has been reflected back to the primary.

DRILL EXERCISES

15.9 The high-voltage side of a transformer has 500 turns, and the low-voltage side has 100 turns. When the transformer is connected as a step-down transformer, the load current is 12 A. Calculate: (a) the turns ratio α; (b) the primary current.

15.10 Calculate the turns ratio if the transformer in Drill Exercise 15.9 is used as a step-up transformer.

15.11 The output of a transformer under certain conditions is 12 kW. The copper losses are 189 W and the core losses are 52 W. Calculate the efficiency of this transformer.

15.12 The output impedance of a servo amplifier is 250 Ω. The servomotor that the amplifier must drive has an impedance of 2.5 Ω. Calculate the turns ratio of the transformer required to match these impedances.

15.5 ELECTROMECHANICAL ENERGY CONVERSION

From the material developed thus far, it should be apparent that electromagnetomechanical devices are capable of converting mechanical forces and displacements to electromagnetic energy, and that the converse is also possible. The objective of this section is to formalize the basic principles of energy conversion in electromagnetomechanical systems, and to illustrate its usefulness and potential for application by presenting several examples of **energy transducers.** A transducer is a device that can convert electrical to mechanical energy (in this case, it is often called an **actuator**), or vice versa (in which case it is called a sensor).

Several physical mechanisms permit conversion of electrical to mechanical energy and back, the principal phenomena being the **piezoelectric effect,** consisting of the generation of a change in electric field in the presence of strain in certain crystals (e.g., quartz), and **electrostriction** and **magnetostriction,** in which changes in the dimension of certain materials lead to a change in their electrical (or magnetic) properties. Although these effects lead to some interesting applications, this chapter is concerned only with transducers in which electrical energy is converted to mechanical energy through the coupling of a magnetic field. It is important to note that all rotating machines (motors and generators) fit the basic definition of electromechanical transducers we have just given.

Forces in Magnetic Structures

It should be apparent by now that it is possible to convert mechanical forces to electrical signals, and vice versa, by means of the coupling provided by energy stored in the magnetic field. In this subsection, we discuss the computation of mechanical forces and of the corresponding electromagnetic quantities of interest; these calculations are of great practical importance in the design and application of electromechanical actuators. For example, a problem of interest is the computation of the current required to generate a given force in an electromechanical structure. This is the kind of application that is likely to be encountered by the engineer in the selection of an electromechanical device for a given task.

As already seen in this chapter, an electromechanical system includes an electrical system and a mechanical system, in addition to means through which the two can interact. The principal focus of this chapter has been the coupling that occurs through an electromagnetic field common to both the electrical and the mechanical system; to understand electromechanical energy conversion, it will be important to understand the various energy storage and loss mechanisms in the electromagnetic field. Figure 15.37 illustrates the coupling between the electrical and mechanical systems. In the mechanical system, energy loss can occur because of the heat developed as a consequence of *friction*, while in the electrical system, analogous losses are incurred because of *resistance*. Loss mechanisms are also present in the magnetic coupling medium, since *eddy current losses* and *hysteresis losses* are unavoidable in ferromagnetic materials. Either system can supply energy, and either system can store energy. Thus, the figure depicts the flow of energy from the electrical to the mechanical system, accounting for these various losses. The same flow could be reversed if mechanical energy were converted to electrical form.

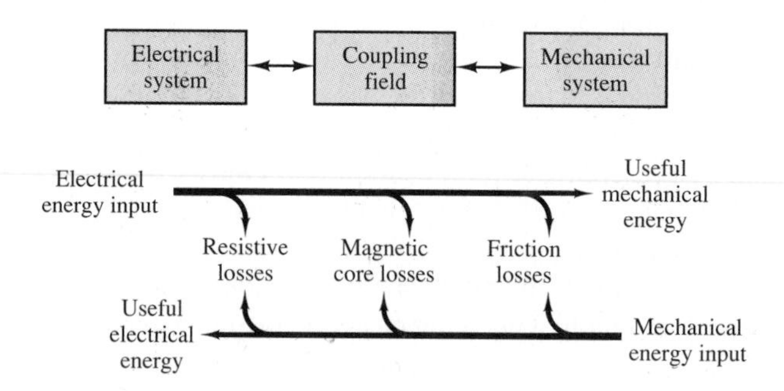

Figure 15.37

Moving-Iron Transducers

One important class of electromagnetomechanical transducers is that of **moving-iron transducers.** The aim of this section is to derive an expression for the magnetic forces generated by such transducers and to illustrate the application of these calculations to simple, yet common devices such as electromagnets, solenoids, and relays. The simplest example of a moving-iron transducer is the electromagnet of Figure 15.38, in which the U-shaped element is fixed and the bar is movable. In the following paragraphs, we shall derive a relationship between the current applied to the coil, the displacement of the movable bar, and the magnetic force acting in the air gap.

The principle that will be applied throughout the section is that in order for a mass to be displaced, some work needs to be done; this work corresponds to a

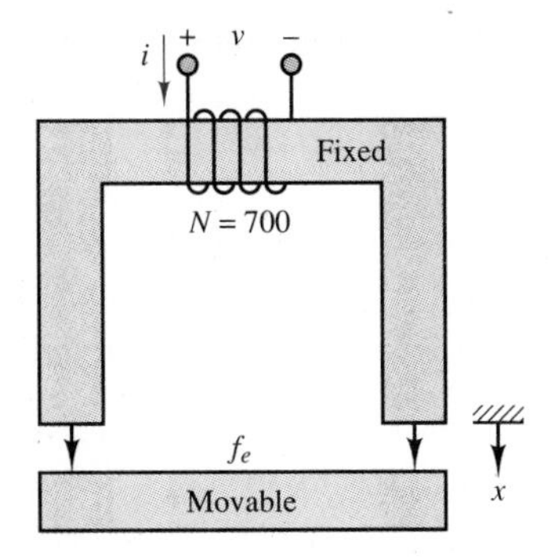

Figure 15.38

change in the energy stored in the electromagnetic field, which causes the mass to be displaced. With reference to Figure 15.38, let f_e represent the magnetic force acting on the bar and x the displacement of the bar, in the direction shown. Then the net work into the electromagnetic field, W_m, is equal to the sum of the work done by the electrical circuit plus the work done by the mechanical system. Therefore, for an incremental amount of work, we can write

$$dW_m = ei\,dt - f_e\,dx \tag{15.40}$$

where e is the electromotive force across the coil and the negative sign is due to the sign convention indicated in Figure 15.38. Recalling that the emf e is equal to the derivative of the flux linkage, we can further expand equation 15.40 to obtain

$$dW_m = ei\,dt - f_e\,dx = i\frac{d\lambda}{dt}\,dt - f_e\,dx = i\,d\lambda - f_e\,dx \tag{15.41}$$

or

$$f_e\,dx = i\,d\lambda - dW_m \tag{15.42}$$

Now we must observe that the flux in the magnetic structure of Figure 15.38 depends on two variables, which are in effect independent: the current flowing through the coil, and the displacement of the bar. Each of these variables can cause the magnetic flux to change. Similarly, the energy stored in the electromagnetic field is also dependent on both current and displacement. Thus we can rewrite equation 15.42 as follows:

$$f_e = i\left(\frac{\partial\lambda}{\partial i}\,di + \frac{\partial\lambda}{\partial x}\,dx\right) - \left(\frac{\partial W_m}{\partial i}\,di + \frac{\partial W_m}{\partial x}\,dx\right) \tag{15.43}$$

Since i and x are independent variables, we can write

$$f_e = i\frac{\partial\lambda}{\partial x} - \frac{\partial W_m}{\partial x} \qquad \text{and} \qquad 0 = i\frac{\partial\lambda}{\partial i} - \frac{\partial W_m}{\partial i} \tag{15.44}$$

From the first of the expressions in equation 15.44, we obtain the relationship

$$f_e = \frac{\partial}{\partial x}(i\lambda - W_m) = \frac{\partial}{\partial x}(W_c) \qquad \textbf{(15.45)}$$

where the term W_c corresponds to W'_m, defined as the co-energy in equation 15.18. Finally, we observe that the force acting to *pull* the bar toward the electromagnet structure, which we will call f, is of opposite sign relative to f_e, and therefore we can write

$$f = -f_e = -\frac{\partial}{\partial x}(W_c) = -\frac{\partial W_m}{\partial x} \qquad \textbf{(15.46)}$$

Equation 15.46 includes a very important assumption: that the energy is equal to the co-energy. If you make reference to Figure 15.8, you will realize that in general this is not true. Energy and co-energy are equal only if the λ-i relationship is linear. Thus, the useful result of equation 15.46, stating that the magnetic force acting on the moving iron is proportional to the rate of change of stored energy with displacement, applies only for linear magnetic structures.

Thus, in order to determine the forces present in a magnetic structure, it will be necessary to compute the energy stored in the magnetic field. In order to simplify the analysis, it will be assumed hereafter that the structures analyzed are magnetically linear. This is, of course, only an approximation, in that it neglects a number of practical aspects of electromechanical systems (for example, the nonlinear λ-i curves described earlier, and the core losses typical of magnetic materials), but it permits relatively simple analysis of many useful magnetic structures. Thus, although the analysis method presented in this section is only approximate, it will serve the purpose of providing a feeling for the direction and the magnitude of the forces and currents present in electromechanical devices. On the basis of a linear approximation, it can be shown that the stored energy in a magnetic structure is given by the expression

$$W_m = \frac{\phi\mathcal{F}}{2} \qquad \textbf{(15.47)}$$

and since the flux and the mmf are related by the expression

$$\phi = \frac{Ni}{\mathcal{R}} = \frac{\mathcal{F}}{\mathcal{R}} \qquad \textbf{(15.48)}$$

the stored energy can be related to the reluctance of the structure according to

$$W_m = \frac{\phi^2\mathcal{R}(x)}{2} \qquad \textbf{(15.49)}$$

where the reluctance has been explicitly shown to be a function of displacement, as is the case in a moving-iron transducer. Finally, then, we shall use the following approximate expression to compute the magnetic force acting on the moving iron:

$$f = -\frac{dW_m}{dx} = -\frac{\phi^2}{2}\frac{d\mathcal{R}(x)}{dx} \qquad \textbf{(15.50)}$$

The following examples illustrate the application of this approximate technique for the computation of forces and currents (the two problems of practical engineering interest to the user of such electromechanical systems) in some common devices.

Example 15.10 An Electromagnet

An electromagnet is used to support a solid piece of steel as shown in Figure 15.38. A force of 8,900 N is required to support the weight. The cross-sectional area of the magnet core (the fixed part) is 0.01 m^2. Determine the minimum current that can keep the weight from falling for $x = 1.5$ mm. Assume negligible reluctance for the steel parts, and negligible fringing in the air gap.

Solution:

We have already shown that in magnetic structures with air gaps, the reluctance is mostly due to the air gaps. This explains the assumption that the reluctance of the structure is negligible. For the structure of Figure 15.38, the reluctance is therefore given by

$$\mathcal{R}(x) = \frac{l}{\mu_0 A}$$

where $A = 0.01$ m^2 and $l = 2x$, and therefore

$$\mathcal{R}(x) = \frac{2x}{4\pi \times 10^{-7} \times 0.01} = \frac{x}{1.2566 \times 10^{-8}}$$

The magnitude of the force in the air gap is given by the expression

$$|f| = \frac{\phi^2}{2}\frac{d\mathcal{R}(x)}{dx} = \frac{N^2 i^2}{2\mathcal{R}^2}\frac{d\mathcal{R}(x)}{dx}$$

$$= \frac{i^2}{2}\frac{N^2}{\mathcal{R}^2}\frac{d\mathcal{R}}{dx} = \frac{i^2}{2}(700)^2\frac{6.2832 \times 10^{-9}}{x^2} = 8{,}900 \text{ N}$$

from which the current can be computed:

$$i^2 = 2 \times \frac{8{,}900(1.5 \times 10^{-3})^2}{(700)^2(6.2832 \times 10^{-9})} = 6.504 \text{ A}$$

or

$$i = 2.55 \text{ A}$$

You should recognize the practical importance of these calculations in determining approximate current requirements and force-generation capabilities of electromechanical transducers.

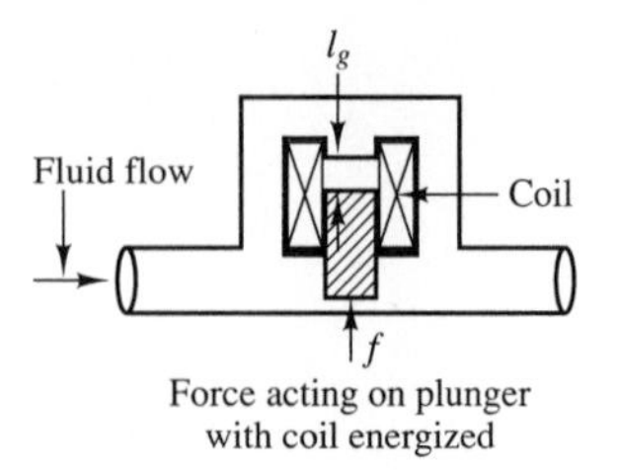

Figure 15.39 Application of the solenoid as a valve

One of the more common practical applications of the concepts discussed in this section is the **solenoid.** Solenoids find application in a variety of electrically controlled valves. The action of a solenoid valve is such that when it is energized, the plunger moves in such a direction as to permit the flow of a fluid through a conduit, as shown schematically in Figure 15.39.

The following example illustrates the calculations involved in the determination of forces and currents in a solenoid.

Example 15.11 A Solenoid

The magnetic structure shown in Figure 15.40 is a simplified representation of a solenoid. The presence of a flux in the air gap tends to pull the magnetic structures together, and serves the purpose of actuating the motion of the plunger.

1. Develop a general expression for the force exerted on the plunger as a function of the position, x.
2. Determine the mmf (and therefore the current) required to pull the plunger all the way into the field ($x = a$) if the spring constant is 1 N/m.

Assume that $a = 1$ cm, $l_g = 1$ mm, and μ_r is infinite. Neglect fringing, but assume that at $x = 0$, the plunger is still in the gap by the infinitesimal displacement ε.

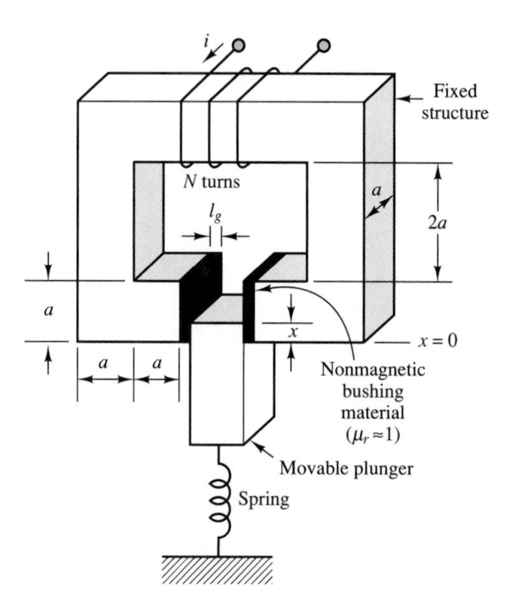

Figure 15.40 A solenoid

Solution:

1. As explained in this section, the force in the air gap can be expressed by

$$f_g = -\frac{1}{2}\phi^2\frac{d\mathcal{R}_g}{dx}$$

Since most of the reluctance is due to the air gap, we approximate the total reluctance by

$$\mathcal{R} = 2\mathcal{R}_g$$

where the gap reluctance is the reluctance of the nonmagnetic bushing. As shown in Figure 15.41, the area of this gap is variable, depending on the position of the plunger:

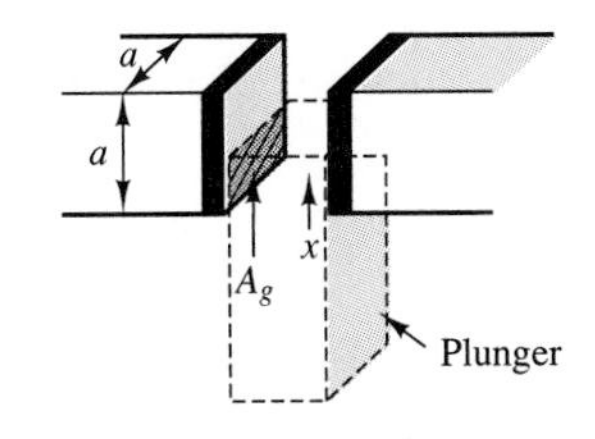

Figure 15.41

$$\mathcal{R}_g = \frac{l_g}{\mu_0 A_g} = \frac{l_g}{\mu_0 ax}$$

$$\mathcal{R} = 2\mathcal{R}_g = \frac{2l_g}{\mu_0 ax}$$

The magnetic flux in the structure can then be found by the expression

$$\phi = \frac{\mathcal{F}}{\mathcal{R}} = \frac{\mu_0 Niax}{2l_g}$$

Now,

$$\frac{d\mathcal{R}_g}{dx} = \frac{d}{dx}\left(\frac{l_g}{\mu_0 ax}\right) = \frac{-l_g}{\mu_0 ax^2}$$

As discussed earlier, the force in the air gap is given by

$$f_g = -\frac{1}{2}\phi^2\left(\frac{d\mathcal{R}_g}{dx}\right)$$

Since we have two air gaps, we need to consider the force contribution due to both to compute the force:

$$f_g = -\frac{1}{2}\phi^2\left(2\frac{d\mathcal{R}_g}{dx}\right) = -\frac{1}{2}\left(\frac{\mu_0 Niax}{2l_g}\right)^2\left(-2\frac{l_g}{\mu_0 ax^2}\right)$$

$$= \frac{1}{4}\frac{\mu_0 a}{l_g}(Ni)^2$$

Thus, we see that the force has no dependence on position—that is, it is constant through the gap.

2. To compute the mmf required to sustain the force computed in part 1 at the position $x = 1$ cm, we evaluate the expression

$$f_g = kx$$

at $x = 0.01$, to obtain

$$f_g = 1\text{ (N/m)}(0.01\text{ m}) = 0.01\text{ N}$$

and compute the required mmf:

$$\text{mmf} = \mathcal{F} = Ni = \left(\frac{4 f_g l_g}{\mu_0 a}\right)^{1/2} = \left[\frac{4(0.01)(0.001)}{4\pi \times 10^{-7}(0.01)}\right]^{1/2}$$

$$Ni = 56.42 \text{ A} \cdot \text{t}$$

This calculation permits the determination of two important variables: the current and number of turns required to produce the given force.

Another electromechanical device that finds common application in industrial practice is the **relay.** The relay is essentially an electromechanical switch that permits the opening and closing of electrical contacts by means of an electromagnetic structure similar to those discussed earlier in this section.

A relay such as would be used to start a high-voltage single-phase motor is shown in Figure 15.42. The magnetic structure has dimensions equal to 1 cm on all sides, and the transverse dimension is 8 cm. The relay works as follows. When the push button is pressed, an electrical current flows through the coil and generates a field in the magnetic structure. The resulting force draws the movable part toward the fixed part, causing an electrical contact to be made. The advantage of the relay is that a relatively low-level current can be used to control the opening and closing of a circuit that can carry large currents. In this particular example, the relay is energized by a 120-VAC contact, establishing a connection in a 240-VAC circuit. Such relay circuits are commonly employed to remotely switch large industrial loads.

Circuit symbols for relays are shown in Figure 15.43. An example of the calculations that would typically be required in determining the mechanical and electrical characteristics of a simple relay are given in Example 15.12.

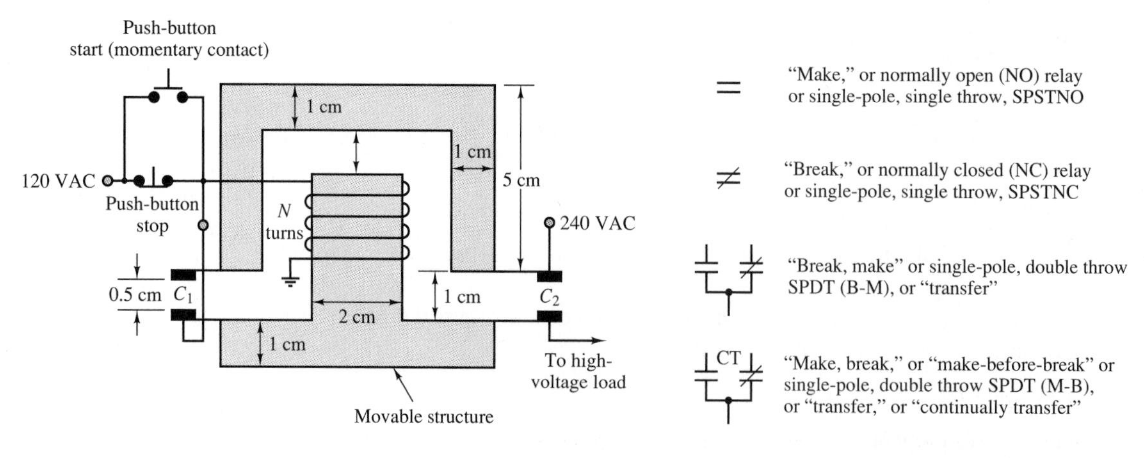

Figure 15.42 A relay

Figure 15.43 Circuit symbols for relays

Example 15.12 The Relay

For the simplified relay shown in Figure 15.44, determine the minimum amount of current required to keep the ferromagnetic plate at a distance of 0.5 cm from the electromagnet (which consists of a 10,000-turn coil) when the torque is 10 N · m. Assume that the ferromagnetic material is infinitely permeable and leakage and fringing effects are negligible.

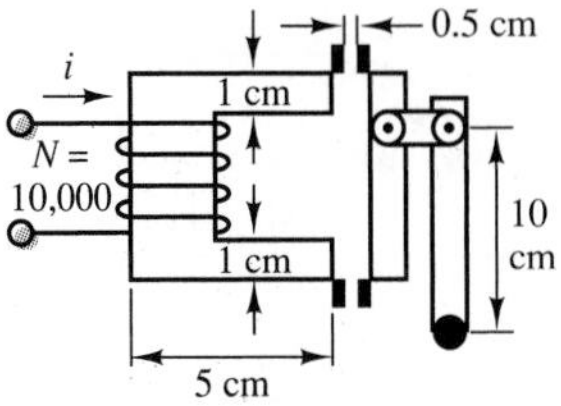

Figure 15.44

Solution:

The force on the plate corresponding to the stated torque at a radius of 10 cm is

$$f' = \frac{10}{0.1} = 100 \text{ N}$$

To balance this force, the magnitude of the force developed by the electromagnet must be equal to 100 N. The total reluctance in the magnetic structure is due primarily to the two air gaps. Thus, the total reluctance is

$$\mathcal{R} = \frac{2x}{4\pi \times 10^{-7} \times 1 \times 10^{-4}}$$

The inductance can be found to be

$$L = \frac{N^2}{\mathcal{R}} = \frac{10{,}000^2}{1.59 \times 10^{10} x} = \frac{0.629 \times 10^{-2}}{x}$$

and the stored magnetic energy can be expressed (see Drill Exercise 15.4) as

$$W_m = \frac{1}{2} L i^2 = \frac{1}{2}\frac{\lambda^2}{L}$$

Note that L is a function of λ and x and represents the effective inductance of the entire magnetic circuit. From $f = -\partial W_m(\lambda, x)/\partial x$, the developed force is

$$f = \frac{1}{2}\frac{\lambda^2}{2L^2}\frac{\partial L}{\partial x} = \frac{1}{2} i^2 \frac{\partial L}{\partial x}$$

$$= \frac{1}{2} i^2 \frac{dL}{dx} = -\frac{1}{2} i^2 \frac{0.629 \times 10^{-2}}{x^2}$$

For $x = 0.5$ cm,

$$f = -125.8 i^2$$

and since $f = -f'$, we have

$$125.8 i^2 = 100$$

corresponding to:

$$i = 0.892 \text{ A}$$

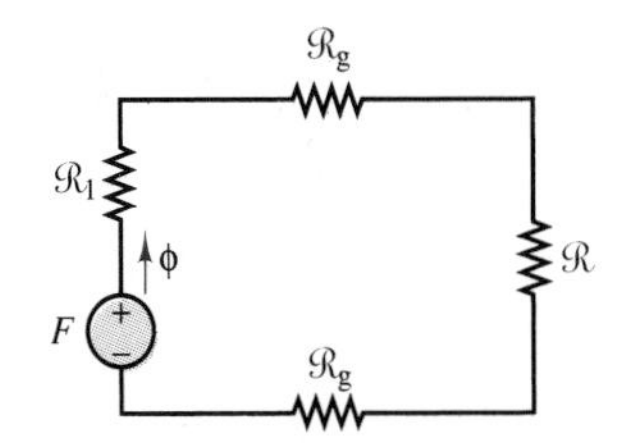

Figure 15.45

The resulting magnetic circuit is depicted in Figure 15.45. The circuit assumes that the reluctance of the air gaps is far greater than that of the magnetic material.

Example 15.13 A Practical Solenoid Circuit

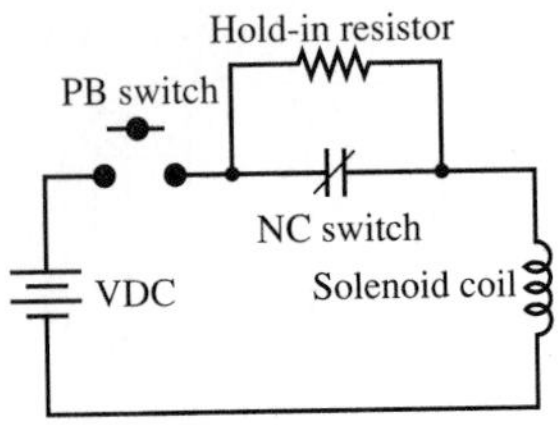

A common method to reduce the coil current required by a solenoid employs a normally closed (NC) switch in parallel with a *hold-in resistor*. In Figure 15.46, when the push button (PB) closes the circuit, full voltage is applied to the solenoid coil, bypassing the resistor through the NC switch. As the solenoid approaches the end of its stroke, a mechanical connection (it could also be electrical if the switch used is a relay) opens the switch, connecting the resistor in series with the coil. The additional voltage drop across the resistor is such that the solenoid voltage is reduced to the point where the voltage is just high enough to hold in, reducing the hold-in current to a minimum value, and thus preventing potential overheating.

Figure 15.46

Moving-Coil Transducers

Another important class of electromagnetomechanical transducers is that of **moving-coil transducers.** This class of transducers includes a number of common devices, such as microphones, loudspeakers, and all electric motors and generators. The aim of this section is to explain the relationship between a fixed magnetic field, the emf across the moving coil, and the forces and motions of the moving element of the transducer.

The basic principle of operation of electromechanical transducers was presented in Section 15.1, where we stated that a magnetic field exerts a force on a charge moving through it. The equation describing this effect is

$$\mathbf{f} = q\mathbf{u} \times \mathbf{B} \tag{15.51}$$

which is a vector equation, as explained earlier. In order to correctly interpret equation 15.51, we must recall the right-hand rule and apply it to the transducer, illustrated in Figure 15.47, depicting a structure consisting of a sliding bar which makes contact with a fixed conducting frame. Although this structure does not represent a practical actuator, it will be a useful aid in explaining the operation of moving-coil transducers such as motors and generators. In Figure 15.47, and in all similar figures in this section, a small cross represents the "tail" of an arrow pointing into the page, while a dot represents an arrow pointing out of the page; this convention will be useful in visualizing three-dimensional pictures.

Motor Action

A moving-coil transducer can act as a motor when an externally supplied current flowing through the electrically conducting part of the transducer is converted into a force that can cause the moving part of the transducer to be displaced. Such a current would flow, for example, if the support of Figure 15.47 were made of conducting material, so that the conductor and the right-hand side of the support "rail" were to form a loop (in effect, a 1-turn coil). In order to understand the effects of this current flow in the conductor, one must consider the fact that a charge moving at a velocity u' (along the conductor and perpendicular to the velocity of

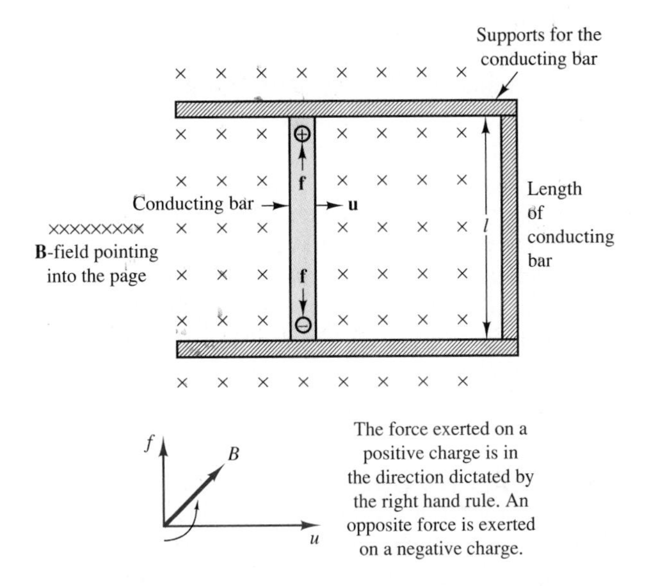

Figure 15.47 A simple electromechanical motion transducer

the conducting bar, as shown in Figure 15.48) corresponds to a current $i = \frac{dq}{dt}$ along the length l of the conductor. This fact can be explained by considering the current i along a differential element dl and writing

$$i\,dl = \frac{dq}{dt} \cdot u' dt \tag{15.52}$$

since the differential element dl would be traversed by the current in time dt at a velocity u'. Thus we can write

$$i\,dl = dq\,u' \tag{15.53}$$

or

$$il = qu' \tag{15.54}$$

for the geometry of Figure 15.48. From Section 15.1, the force developed by a charge moving in a magnetic field is, in general, given by

$$\mathbf{f} = q\mathbf{u} \times \mathbf{B} \tag{15.55}$$

For the term $q\mathbf{u}'$ we can substitute $i\mathbf{l}$, to obtain

$$\mathbf{f}' = i\mathbf{l} \times \mathbf{B} \tag{15.56}$$

Using the right-hand rule, we determine that the force $\mathbf{f}'$ generated by the current i is in the direction that would push the conducting bar to the left. The magnitude of this force is $f' = Bli$ if the magnetic field and the direction of the current are perpendicular. If they are not, then we must consider the angle γ formed by $\mathbf{B}$ and $\mathbf{l}$; in the more general case,

$$f' = Bli \sin\gamma \tag{15.57}$$

The phenomenon we have just described is sometimes referred to as the "Bli law."

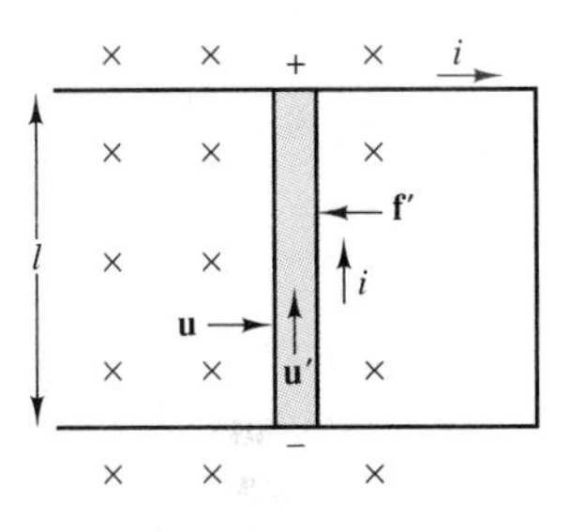

Figure 15.48

Generator Action

The other mode of operation of a moving-coil transducer occurs when an external force causes the coil (i.e., the moving bar, in Figure 15.47) to be displaced. This external force is converted to an emf across the coil, as will be explained in the following paragraphs.

It is important to observe that since positive and negative charges are forced in opposite directions in the transducer of Figure 15.47, a potential difference will appear across the conducting bar; this potential difference is the electromotive force, or emf. The emf must be equal to the force exerted by the magnetic field. In short, the electric force per unit charge (or electric field) e/l must equal the magnetic force per unit charge $f/q = Bu$. Thus, the relationship

$$e = Blu \tag{15.58}$$

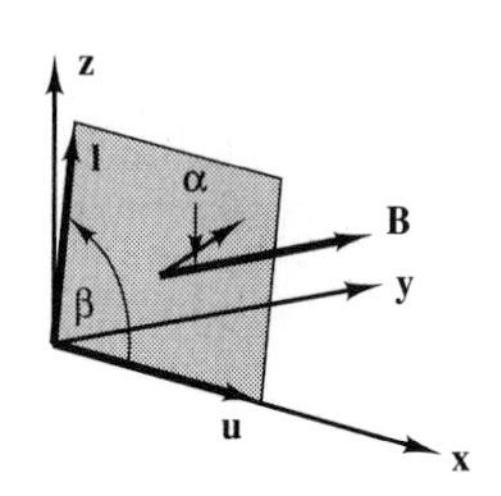

Figure 15.49

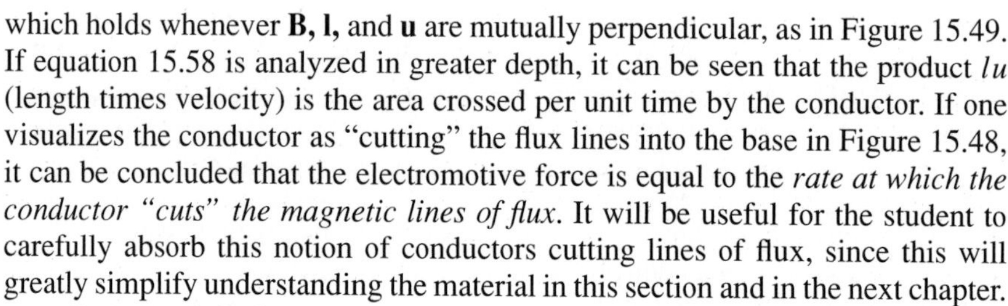

which holds whenever **B, l,** and **u** are mutually perpendicular, as in Figure 15.49. If equation 15.58 is analyzed in greater depth, it can be seen that the product lu (length times velocity) is the area crossed per unit time by the conductor. If one visualizes the conductor as "cutting" the flux lines into the base in Figure 15.48, it can be concluded that the electromotive force is equal to the *rate at which the conductor "cuts" the magnetic lines of flux*. It will be useful for the student to carefully absorb this notion of conductors cutting lines of flux, since this will greatly simplify understanding the material in this section and in the next chapter.

In general, **B**, **l**, and **u** are not necessarily perpendicular. In this case one needs to consider the angles formed by the magnetic field with the normal to the plane containing **l** and **u**, and the angle between **l** and **u**. The former is the angle α of Figure 15.49, the latter the angle β in the same figure. It should be apparent that the optimum values of α and β are 0° and 90°, respectively. Thus, most practical devices are constructed with these values of α and β. Unless otherwise noted, it will be tacitly assumed that this is the case. The "Blu law" just illustrated explains how a moving conductor in a magnetic field can generate an electromotive force.

To summarize the electromechanical energy conversion that takes place in the simple device of Figure 15.47, we must note now that the presence of a current in the loop formed by the conductor and the rail requires that the conductor move to the right at a velocity u (Blu law), thus cutting the lines of flux and generating the emf that gives rise to the current i. On the other hand, the same current causes a force f' to be exerted on the conductor (Bli law) in the direction opposite to the movement of the conductor. Thus, it is necessary that an *externally applied force* f_{ext} exist to cause the conductor to move to the right with a velocity u. The external force must overcome the force f'. This is the basis of electromechanical energy conversion.

An additional observation we must make at this point is that the current i flowing around a closed loop generates a magnetic field, as explained in Section 15.1. Since this additional field is generated by a one-turn coil in our illustration, it is reasonable to assume that it is negligible with respect to the field already present (perhaps established by a permanent magnet). Finally, we must consider that this coil links a certain amount of flux, which changes as the conductor moves from left to right. The area crossed by the moving conductor in time dt is

$$dA = lu\,dt \tag{15.59}$$

so that if the flux density, B, is uniform, the rate of change of the flux linked by the one-turn coil is

$$\frac{d\phi}{dt} = B\frac{dA}{dt} = Blu \tag{15.60}$$

In other words, *the rate of change* of the flux linked by the conducting loop is equal to the emf generated in the conductor. The student should realize that this statement simply confirms Faraday's law.

It was briefly mentioned that the Blu and Bli laws indicate that, thanks to the coupling action of the magnetic field, a conversion of mechanical to electrical energy—or the converse—is possible. The simple structures of Figures 15.47 and 15.48 can, again, serve as an illustration of this energy-conversion process, although we have not yet indicated how these idealized structures can be converted into a practical device. In this section we shall begin to introduce some physical considerations. Before we proceed any further, we should try to compute the power—electrical and mechanical—that is generated (or is required) by our ideal transducer. The electrical power is given by

$$P_E = ei = Blui \qquad \text{(W)} \tag{15.61}$$

while the mechanical power required, say, to move the conductor from left to right is given by the product of force and velocity:

$$P_M = f_{\text{ext}}u = Bliu \qquad \text{(W)} \tag{15.62}$$

The principle of conservation of energy thus states that in this ideal (lossless) transducer we can convert a given amount of electrical energy into mechanical energy, or vice versa. Once again we can utilize the same structure of Figure 15.47 to illustrate this reversible action. If the closed path containing the moving conductor is now formed from a closed circuit containing a resistance R and a battery, V_B, as shown in Figure 15.50, the externally applied force, f_{ext}, generates a positive current i into the battery provided that the emf is greater than V_B. When $e = Blu > V_B$, the ideal transducer acts as a *generator*. For any given set of values of B, l, R, and V_B, there will exist a velocity u for which the current i is positive. If the velocity is lower than this value—i.e., if $e = Blu < V_B$—then the current i is negative, and the conductor is forced to move to the right. In this case the battery acts as a source of energy and the transducer acts as a *motor* (i.e., electrical energy drives the mechanical motion).

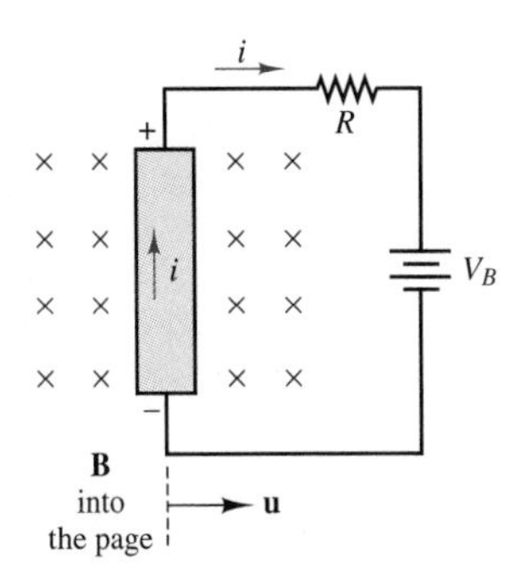

Figure 15.50 Motor and generator action in an ideal transducer

In practical transducers, we must be concerned with the inertia, friction, and elastic forces that are invariably present on the mechanical side of the transducer. Similarly, on the electrical side we must account for the inductance of the circuit, its resistance, and possibly some capacitance. Consider the structure of

Figure 15.51. In the figure, the conducting bar has been placed on a surface with coefficient of sliding friction d; it has a mass m and is attached to a fixed structure by means of a spring with spring constant k. The equivalent circuit representing the coil inductance and resistance is also shown.

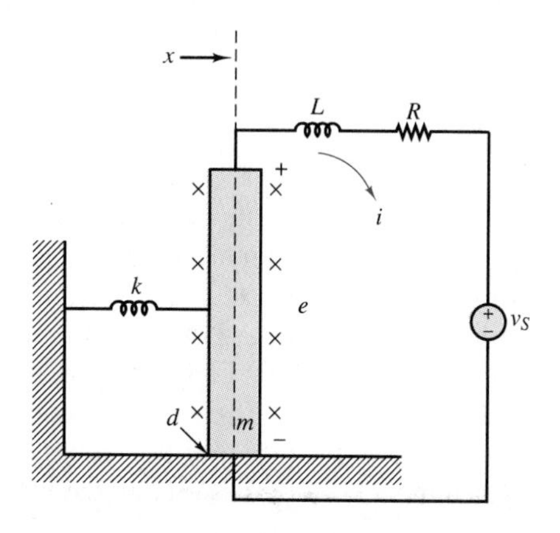

Figure 15.51 A more realistic representation of the transducer of Figure 15.50

If we recognize that $u = \frac{dx}{dt}$ in the figure, we can write the equation of motion for the conductor as:

$$f + m\frac{du}{dt} + du + \frac{l}{k}\int u\,dt = f' = Bli \qquad (15.63)$$

where the Bli term represents the driving input that causes the mass to move. The driving input in this case is provided by the electrical energy source, v_S; thus the transducer acts as a motor, and f is the net force acting on the mass of the conductor. On the electrical side, the circuit equation is:

$$v_S - L\frac{di}{dt} - Ri = e = Blu \qquad (15.64)$$

Equations 15.63 and 15.64 could then be solved by knowing the excitation voltage, v_S, and the physical parameters of the mechanical and electrical circuits. For example, if the excitation voltage were sinusoidal, with

$$v_S(t) = V_S \cos \omega t \qquad (15.65)$$

and the field density were constant:

$$B = B_0$$

we could postulate sinusoidal solutions for the transducer velocity, u, and current, i:

$$u = U \cos(\omega t + \theta_u) \qquad i = I \cos(\omega t + \theta_i) \qquad (15.66)$$

and use phasor notation to solve for the unknowns $(U, I, \theta_u, \theta_i)$.

The results obtained in the present section apply directly to transducers that are based on translational (linear) motion. These basic principles of electromechanical energy conversion and the analysis methods developed in the section will be applied to practical transducers in the next two examples.

Example 15.14

The device shown in Figure 15.52 is called a **seismic transducer** and can be used to measure the displacement, velocity, or acceleration of a body. The permanent magnet of mass m is supported on the case by a spring, k, and there is some viscous damping, d, between the magnet and the case; the coil is fixed to the case. You may assume that the coil has length l and resistance and inductance R_{coil} and L_{coil}, respectively; the magnet exerts a magnetic field B. Find the transfer function between the output voltage, v_{out}, and the acceleration of the body, $a(t)$. Note that $x(t)$ is not equal to zero when the system is at rest. We shall ignore this offset displacement.

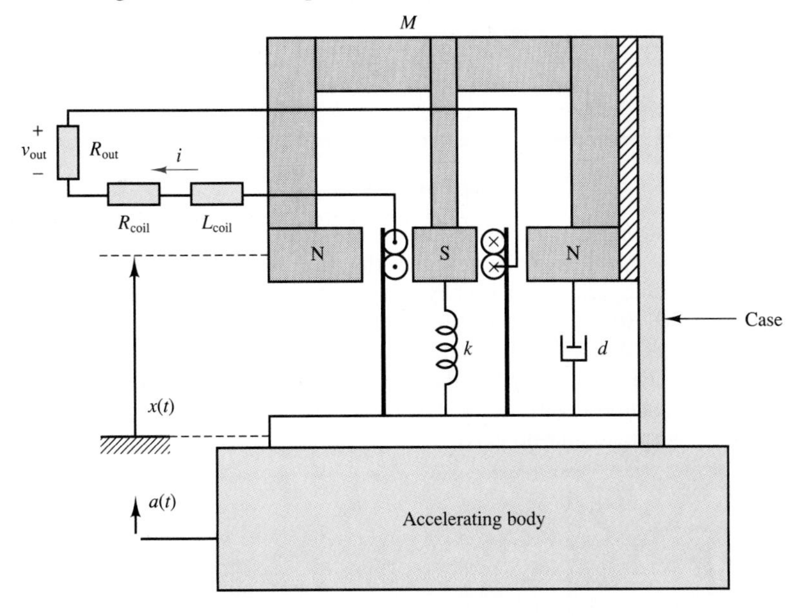

Figure 15.52 An electromagnetomechanical seismic transducer

Solution:

First we write the differential equation describing the electrical system:

$$L\frac{di}{dt} + (R_{\text{coil}} + R_{\text{out}})i + Bl\frac{dx}{dt} = 0$$

Also note that $v_{\text{out}} = -R_{\text{out}}i$.

Next, we write the differential equation describing the mechanical system:

$$M\left(a + \frac{d^2x}{dt^2}\right) + d\frac{dx}{dt} + kx = Bli$$

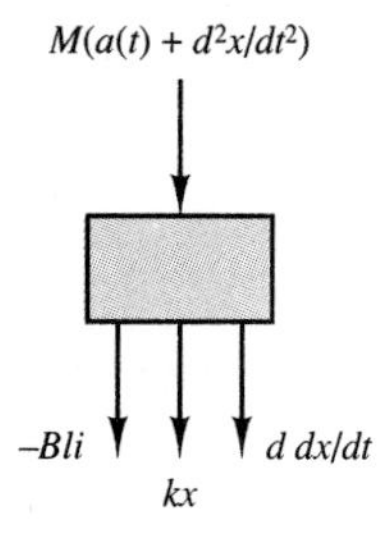

Free-body diagram for permanent magnet

The magnet experiences an inertial force due to the acceleration of the supporting body, $a(t)$, and to its own relative acceleration, $\frac{d^2x}{dt^2}$; thus, we can sketch a free-body diagram and apply Newton's second law to the permanent magnet, as shown in the sketch.

Finally, using the Laplace Transform, we determine the transfer function from $A(s)$ to $V_{\text{out}}(s)$. Let $R = R_{\text{coil}} + R_{\text{out}}$. Then

$$(Ls + R)I(s) + BlsX(s) = 0$$

$$BlI(s) - (Ms^2 + Ds + K)X(s) = MA(s)$$

Since we need the transfer function from A to V_{out}, we use the expression

$$V_{\text{out}}(s) = -R_{\text{out}}I(s)$$

and, after some algebra, find that

$$I(s) = \frac{MBlsA(s)}{(Ls + R)(Ms^2 + Ds + K) + B^2l^2s}$$

or

$$\frac{V_{\text{out}}(s)}{A(s)} = \frac{-MBsR_{\text{out}}}{(Ls + R)(Ms^2 + Ds + K) + B^2l^2s}$$

Example 15.15 The Loudspeaker

A loudspeaker, shown in Figure 15.53, uses a permanent magnet and a moving coil to produce the vibrational motion that generates the pressure waves we perceive as sound. Vibration of the loudspeaker is caused by changes in the input current to a coil; the coil is, in turn, coupled to a magnetic structure that can produce time-varying forces on the speaker diaphragm. A simplified model for the mechanics of the speaker is also shown in Figure 15.53. The force exerted on the coil is also exerted on the mass of the speaker diaphragm, as shown in Figure 15.54, which depicts a free-body diagram of the forces acting on the loudspeaker diaphragm.

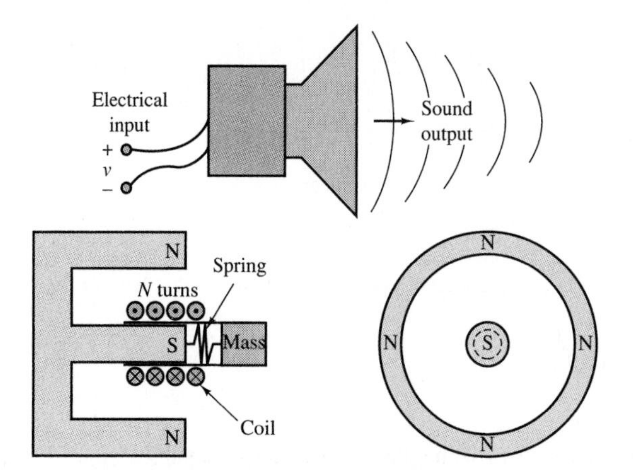

Figure 15.53 Loudspeaker

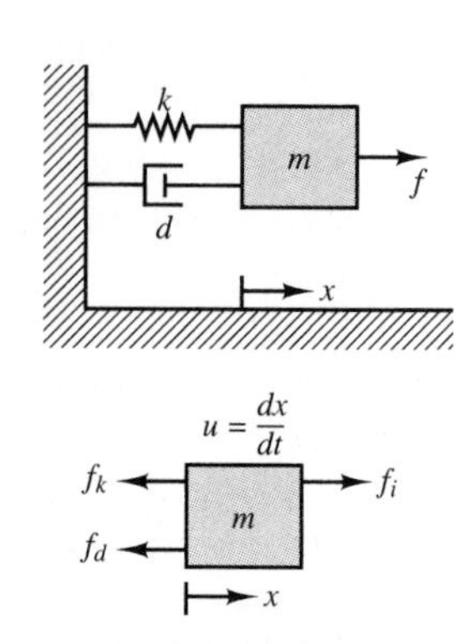

Figure 15.54 Forces acting on loudspeaker diaphragm

The force exerted on the mass, f_i, is the magnetic force due to current flow in the coil. The electrical circuit that describes the coil is shown in Figure 15.55, where L represents the inductance of the coil, R represents the resistance of the windings, and e is the emf induced by the coil moving through the magnetic field.

Determine the frequency response, $\frac{\mathbf{U}}{\mathbf{V}}(j\omega)$, of the speaker using phasor analysis if the model parameters are:

$$L \approx 0 \text{ H} \qquad R = 8\ \Omega \qquad m = 0.001 \text{ kg} \qquad d = 22.75$$

$$k \approx 500{,}000 \text{ N/m} \qquad N = 47 \qquad B = 1\ T$$

Radius of coil = 5 cm

Solution:

To determine the frequency response of the loudspeaker, we need to write the fundamental equations that describe the two subsystems that make up the loudspeaker. The electrical subsystem is described by the usual KVL relationship, applied to the circuit of Figure 15.55:

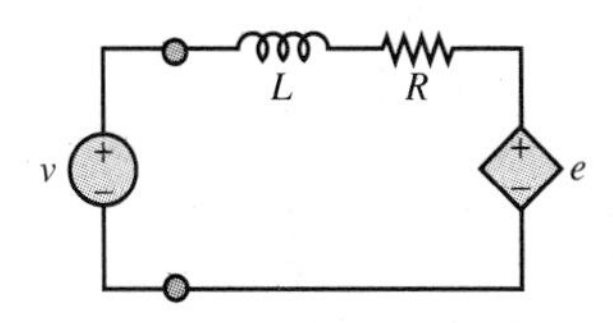

Figure 15.55 Model of transducer electrical side

$$v = L\frac{di}{dt} + Ri + e$$

where e is the emf generated by the motion of the coil in the magnetic field. Next, according to Newton's law, we can write a force balance equation to describe the dynamics of the mechanical subsystem:

$$m\frac{du}{dt} = f_i - f_d - f_k = f_i - du - kx$$

Now, the coupling between the electrical and mechanical systems is expressed in each of the two preceding equations by the terms e and f_i:

$$e = Blu$$

$$f_i = Bli$$

Since we desire the frequency response, we use phasor techniques to represent the electrical and mechanical subsystem equations:

$$\mathbf{V}(j\omega) = j\omega L\mathbf{I}(j\omega) + R\mathbf{I}(j\omega) + Bl\mathbf{U}(j\omega) \qquad \text{Electrical equation}$$

$$(j\omega m + d)\mathbf{U}(j\omega) + \frac{K}{j\omega}\mathbf{U}(j\omega) = Bl\mathbf{I}(j\omega) \qquad \text{Mechanical equation}$$

Having assumed that the inductance of the coil is negligible, we are able to simplify the electrical equation and to solve for $\mathbf{I}(j\omega)$:

$$\mathbf{I}(j\omega) = \frac{\mathbf{V}(j\omega) - Bl\mathbf{U}(j\omega)}{R}$$

Substituting this equivalence into the mechanical equation and accounting for the length of the coil, $l = 2\pi Nr$, the final expression for the frequency response of the loudspeaker is then given by:

$$\frac{\mathbf{U}}{\mathbf{V}}(j\omega) = \frac{2\pi NBr}{Rm} \times \frac{j\omega}{(j\omega)^2 + j\omega\left(\frac{d+\frac{(2\pi)^2B^2N^2r^2}{R}}{m}\right) + \frac{k}{m}}$$

or, numerically,

$$\frac{\mathbf{U}}{\mathbf{V}}(j\omega) = \frac{2\pi \times 47 \times 1 \times 0.05}{8(0.001)}$$

$$\times \frac{j\omega}{(j\omega)^2 + j\omega\left(\frac{22.75+\frac{(2\pi)^2(1)^2(47)^2(0.05)^2}{8}}{0.001}\right) + \frac{500{,}000}{0.001}}$$

$$\approx \frac{1{,}845\, j\omega}{(j\omega + 13.8 \times 10^3)(j\omega + 36.2 \times 10^3)}$$

$$= \frac{0.051\left(\frac{j\omega}{13.8\times10^3}\right)}{\left(1 + \frac{j\omega}{36.2 \times 10^3}\right)\left(1 + \frac{j\omega}{13.8 \times 10^3}\right)}$$

This frequency response shows that the speaker has a lower cutoff frequency $f_{cl} = \frac{13{,}800}{2\pi} \approx 2{,}200$ Hz and an upper cutoff frequency of $f_{ch} = \frac{36{,}000}{2\pi} \approx 5{,}800$ Hz. In practice, a loudspeaker with such a frequency response would be useful only as a mid-range speaker.

The methods introduced in this section will be applied in Chapters 16 and 17 to analyze rotating transducers, that is, electric motors and generators.

DRILL EXERCISES

15.13 The flux density of the earth's magnetic field is about 50 μT. Estimate the current required in a conductor of length 10 cm and mass 10 g to counteract the force of gravity if the wire is oriented in the optimum direction.

15.14 In Example 15.15, we examined the frequency response of a loudspeaker. However, over a period of time, permanent magnets may become demagnetized. Find the frequency response of the same loudspeaker if the permanent magnet has lost its strength to a point where $B = 0.95$ T.

15.15 In Example 15.11, a solenoid is used to exert force on a spring. Estimate the position of the plunger if the number of turns in the solenoid winding is 1,000 and the current going into the winding is 40 mA.

15.16 For the circuit in Figure 15.47, the conducting bar is moving with a velocity of 6 m/s. The flux density is 0.5 Wb/m², and $l = 1.0$ m. Find the magnitude of the resulting induced voltage.

CONCLUSION

- Magnetic fields form a coupling mechanism between electrical and mechanical systems, permitting the conversion of electrical energy to mechanical energy, and vice versa. The basic laws that govern such electromechanical energy conversion are Faraday's law, stating that a changing magnetic field can induce a voltage; and Ampère's law, stating that a current flowing through a conductor generates a magnetic field.
- The two fundamental variables in the analysis of magnetic structures are the magnetomotive force and the magnetic flux; if some simplifying approximations are made, these quantities are linearly related through the reluctance parameter, much in the same way as voltage and current are related through resistance according to Ohm's law. This simplified analysis permits approximate calculations of required forces and currents to be conducted with relative ease in magnetic structures.
- Magnetic materials are characterized by a number of nonideal properties, which should be considered in the detailed analysis of a magnetic structure. The most important phenomena are saturation, eddy currents, and hysteresis.
- Electromechanical transducers, which convert electrical signals to mechanical forces, or mechanical motion to electrical signals, can be analyzed according to the techniques presented in this chapter. Examples of such transducers are electromagnets, position and velocity sensors, relays, solenoids, and loudspeakers.

KEY TERMS

Accelerometer *839*
Actuator *825*
Air Gap *811*
Ampère's Law *804*
Co-energy *802*
Counter mmf *822*
Eddy Current *818*
Electromotive Force (emf) *798*
Electrostriction *825*
Energy Transducer *825*
Faraday's Law *798*
Ferrites *806*
Flux Linkage *798*
Fringing *812*
Hysteresis *818*
Leakage Flux *810*
Lenz's Law *798*
Magnetic Coupling *799*
Magnetic Domains *818*
Magnetic Field Intensity *796*
Magnetic Flux *796*
Magnetic Flux Density *796*
Magnetomotive Force (mmf) *804*
Magnetostriction *825*
Mean Path *807*
Motional Voltages *799*
Moving-Coil Transducer *834*
Moving-Iron Transducer *826*
Mutual Inductance *800*
Permeability *804*
Piezoelectric Effect *825*
Power Efficiency *823*
Relay *832*
Reluctance *808*
Remanent (Residual) Magnetization *819*
Right-Hand Rule *798*
Saturation *818*
Self-Inductance *799*
Solenoid *830*
Tesla (T) *796*
Transformer Voltages *799*
Weber (Wb) *796*

ANSWERS TO DRILL EXERCISES

15.1 $e = -2.5$ V

15.2 $I = \pi$ A

15.3 $W_m = 0.648$ J

15.5 $\phi = 3.94 \times 10^{-6}$ Wb; $B = 0.0788$ Wb/m^2

15.6 $\mathcal{R}_{eq} = 1.41 \times 10^6$ A · t/Wb

15.7 $\mathcal{R}_{eq} = 22 \times 10^6$ A · t/Wb

15.8 $\mathcal{R}_g = \mathcal{R}_1 = \mathcal{R}_2 = \mathcal{R}_3 = \frac{\delta}{\mu_0(l+\delta)(w+\delta)}; \mathcal{F}_1 = Ni_1; \mathcal{F}_2 = Ni_2$

15.9 $\alpha = 5; I_1 = \frac{I_2}{\alpha} = 2.4$ A

15.10 $\alpha = 0.2$

15.11 $\eta = 98\%$

15.12 $\alpha = 10$

15.13 $i = 196 \times 10^2$ A

15.14 $\frac{\mathbf{U}}{\mathbf{V}}(j\omega) = \frac{0.056(j\omega/15{,}950)}{(1+j\omega/15{,}950)(1+j\omega/31{,}347)}$

15.15 $x = 0.5$ cm

15.16 3 V

EIT EXAM REVIEW: ELECTRIC AND MAGNETIC FIELDS

The aim of this section is to test knowledge of the fundamental relationships relating electric and magnetic fields. Most of the useful relations may be found in Chapter 15. For a more complete review, you may wish to consult a physics text.

DRILL EXERCISES

15.17 Which of the following is a true characteristic of magnetic flux lines?

a. They cross each other.
b. They begin and end on electric charges.
c. They are parabolic.
d. They are continuous.
e. None of the above.

Solution:

As discussed in Chapter 15, magnetic flux lines are continuous. Thus, d is the correct answer.

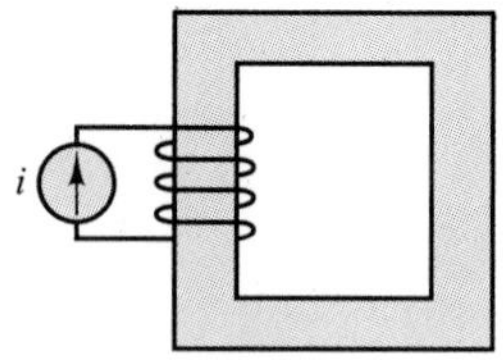

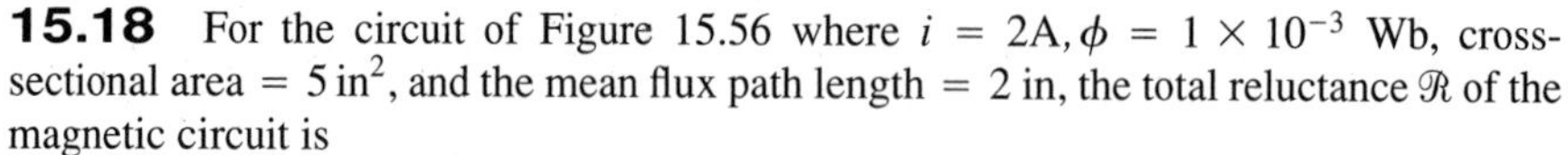

15.18 For the circuit of Figure 15.56 where $i = 2\text{A}, \phi = 1 \times 10^{-3}$ Wb, cross-sectional area $= 5\ \text{in}^2$, and the mean flux path length $= 2$ in, the total reluctance $\mathcal{R}$ of the magnetic circuit is

a. $1 \times 10^5\ \frac{\text{A·t·in}^2}{\text{Wb}}$
b. $2 \times 10^5\ \frac{\text{A·t·in}^2}{\text{Wb}}$
c. $1.5 \times 10^5\ \frac{\text{A·t·in}^2}{\text{Wb}}$
d. $3.5 \times 10^4\ \frac{\text{A·t}}{\text{Wb·in}}$
e. $2 \times 10^5\ \frac{\text{A·t}}{\text{Wb·in}^2}$

Figure 15.56

Solution:

From Chapter 15, the relationship between magnetomotive force and flux is

$$\mathcal{F} = \phi\mathcal{R}$$

Also, the magnetomotive force is related to the current i by

$$\mathcal{F} = Ni = 100 \times 2 = 200 \text{ A} \cdot \text{t}$$

which means that the reluctance is

$$\mathcal{R} = \frac{\mathcal{F}}{\phi} = \frac{200}{1 \times 10^{-3}} = 200{,}000 \ \frac{\text{A} \cdot \text{t} \cdot \text{in}^2}{\text{Wb}}$$

Therefore, the answer is b.

Additional examples illustrating the concept of magnetic circuits may be found in this chapter in Examples 15.2, 15.3, 15.4, 15.6, 15.7, 15.8, 15.11, 15.12, and 15.15. In addition, the homework problems marked EIT will provide additional practice material. Answers to these problems will be found in Appendix B.

HOMEWORK PROBLEMS

15.1 An iron-core inductor has the following characteristic:

$$i = \lambda + 0.5\lambda^2$$

a. Determine the energy, co-energy, and incremental inductance for $\lambda = 0.5$ V · s.
b. Given that the coil resistance is 1 Ω and that

$$i(t) = 0.625 + 0.01 \sin 400t \text{ A}$$

determine the voltage across the terminals on the inductor.

15.2 For the electromagnet of Figure P15.1:

a. Find the flux density in the core.
b. Sketch the magnetic flux lines and indicate their direction.
c. Indicate the north and south poles of the magnet.

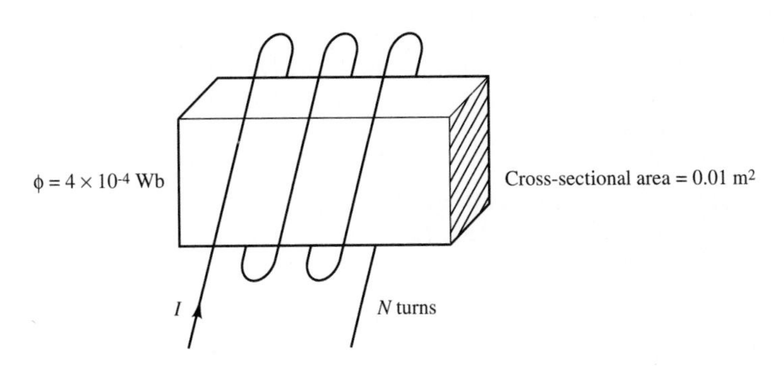

Figure P15.1

15.3 An iron-core inductor has the characteristic shown in Figure P15.2:

a. Determine the energy and the incremental inductance for $i = 1.0$ A.
b. Given that the coil resistance is 2 Ω and that $i(t) = 0.5 \sin 2\pi t$, determine the voltage across the terminals of the inductor.

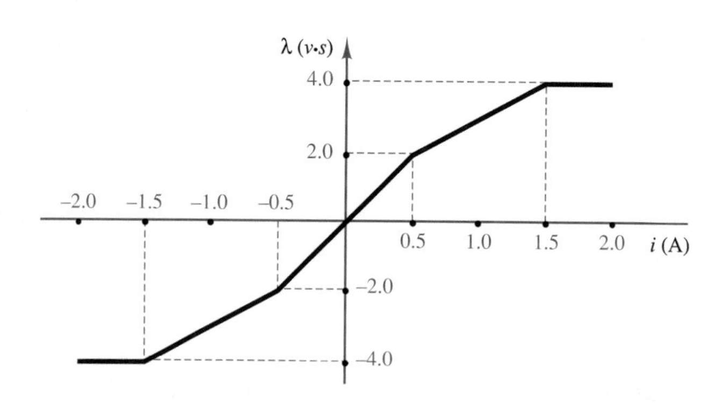

Figure P15.2

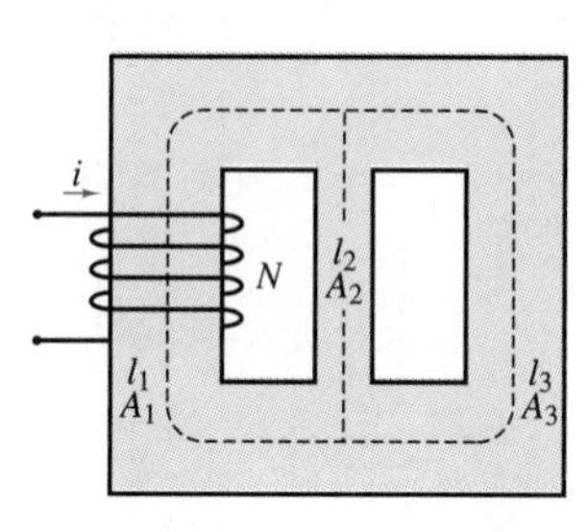

$N = 100$ turns $A_2 = 25$ cm²
$l_1 = 30$ cm $l_3 = 30$ cm
$A_1 = 100$ cm² $A_3 = 100$ cm²
$l_2 = 10$ cm

Figure P15.3

15.4

a. Find the reluctance of a magnetic circuit if a magnetic flux $\phi = 4.2 \times 10^{-4}$ Wb is established by an impressed mmf of 400 A · t.
b. Find the magnetizing force, H, in SI units if the magnetic circuit is 6 inches in length.

15.5 For the circuit shown in Figure P15.3:

a. Determine the reluctance values and show the magnetic circuit, assuming that $\mu = 3{,}000\mu_0$.
b. Determine the inductance of the device.
c. The inductance of the device can be modified by cutting an air gap in the magnetic structure. If a gap of 0.1 mm is cut in the arm of length l_3, what is the new value of inductance?
d. As the gap is increased in size (length), what is the limiting value of inductance? Neglect leakage flux and fringing effects.

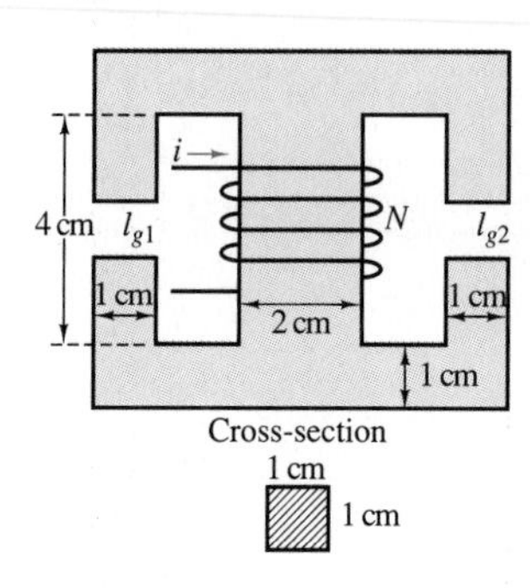

Figure P15.4

15.6 The magnetic circuit shown in Figure P15.4 has two parallel paths. Find the flux and flux density in each of the legs of the magnetic circuit. Neglect fringing at the air gaps and any leakage fields. $N = 1{,}000$ turns, $i = 0.2$ A, $l_{g1} = 0.02$ cm, and $l_{g2} = 0.04$ cm. Assume the reluctance of the magnetic core to be negligible.

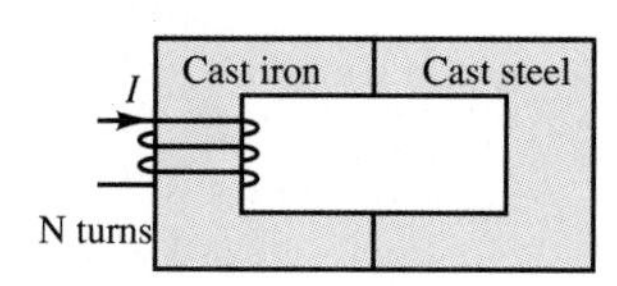

Figure P15.5

15.7 Find the current necessary to establish a flux of $\phi = 3 \times 10^{-4}$ Wb in the series magnetic circuit of Figure P15.5. Here, $l_{iron} = l_{steel} = 0.3$ m, Area(throughout) $= 5 \times 10^{-4}$ m², and $N = 100$ turns.

15.8 A cylindrical solenoid is shown in Figure P15.6. The plunger may move freely along its axis. The air gap between the shell and the plunger is uniform and equal to 1 mm, and the diameter, d, is 25 mm. If the exciting coil carries a current of 7.5 A, find the force acting on the plunger when $x = 2$ mm. Assume $N = 200$ turns, and neglect the reluctance of the steel shell.

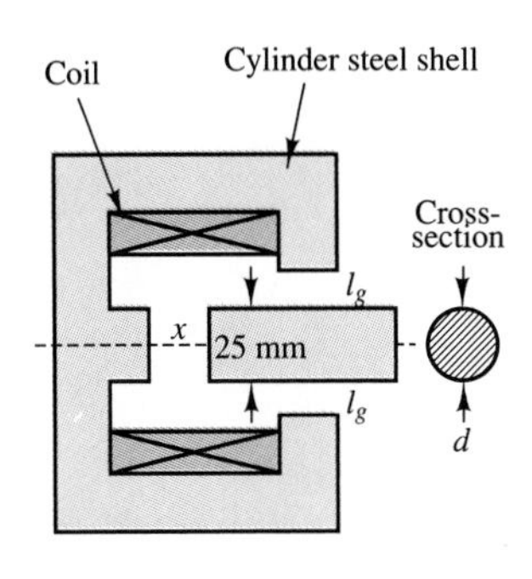

Figure P15.6

15.9

a. Find the current, I, required to establish a flux $\phi = 2.4 \times 10^{-4}$ Wb in the magnetic circuit of Figure P15.7. Here, Area(throughout) $= 2 \times 10^{-4}$ m^2, $l_{ab} = l_{ef} = 0.05$ m, $l_{af} = l_{be} = 0.02$ m, $l_{bc} = l_{dc}$, and the material is sheet steel.
b. Compare the mmf drop across the air gap to that across the rest of the magnetic circuit. Discuss your results using the value of μ for each material.

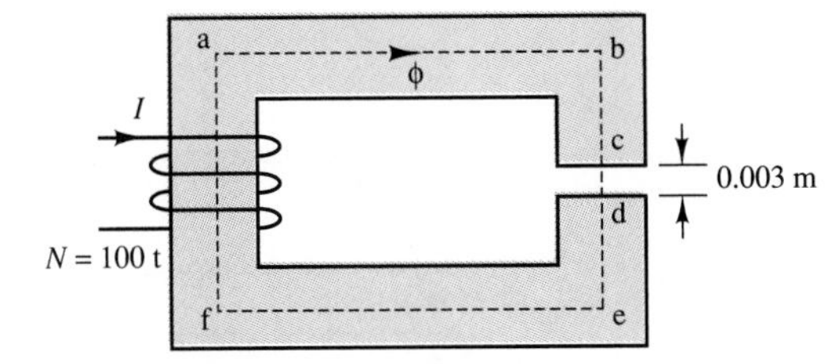

Figure P15.7

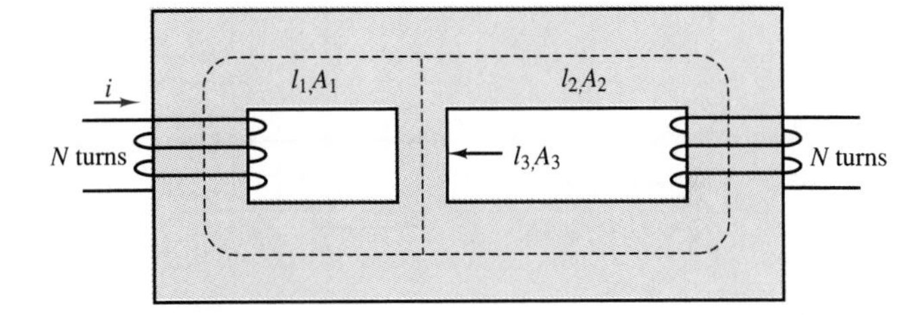

Figure P15.8

15.10 For the transformer shown in Figure P15.8, $N = 1{,}000$ turns, $l_1 = 16$ cm, $A_1 = 4$ cm^2, $l_2 = 22$ cm, $A_2 = 4$ cm^2, $l_3 = 5$ cm, and $A_3 = 2$ cm^2. The relative permeability of the material is $\mu_r = 1{,}500$.

a. Construct the equivalent magnetic circuit, and find the reluctance associated with each part of the circuit.
b. Determine the self-inductance and mutual inductance for the pair of coils (i.e., L_{11}, L_{22}, and $M = L_{12} = L_{21}$).

15.11 Find the magnetic flux, ϕ, established in the series magnetic circuit of Figure P15.9.

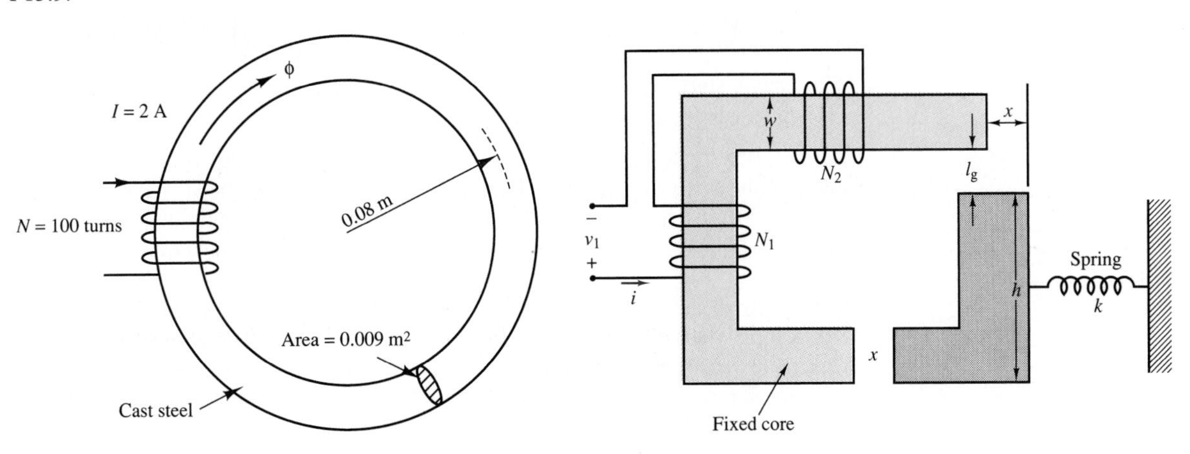

Figure P15.9

Figure P15.10

15.12 The double-excited electromechanical system shown in Figure P15.10 moves horizontally. Assuming that resistance, magnetic leakage, and fringing are negligible, the permeability of the core is very large, and the cross section of the structure is $w \times w$, find

a. The reluctance of the magnetic circuit.
b. The magnetic energy stored in the air gap.
c. The force on the movable part as a function of its position.

15.13 For the series-parallel magnetic circuit of Figure P15.11, find the value of I required to establish a flux in the gap of $\phi = 2 \times 10^{-4}$ Wb. Here, $l_{ab} = l_{bg} = l_{gh} = l_{ha} = 0.2$ m, $l_{bc} = l_{fg} = 0.1$ m, $l_{cd} = l_{ef} = 0.099$ m, and the material is sheet steel. Assume $\mu_r = 4{,}000$.

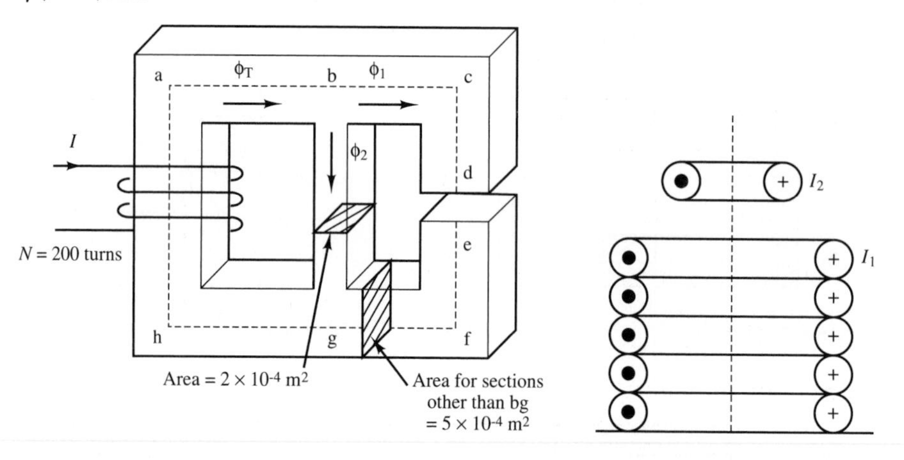

Figure P15.11

Figure P15.12

15.14 A single loop of wire carrying current I_2 is placed near the end of a solenoid having N turns and carrying current I_1, as shown in Figure P15.12. The solenoid is fastened to a horizontal surface, but the single coil is free to move. With the currents directed as shown, is there a resultant force on the single coil? If so, in what direction? Why?

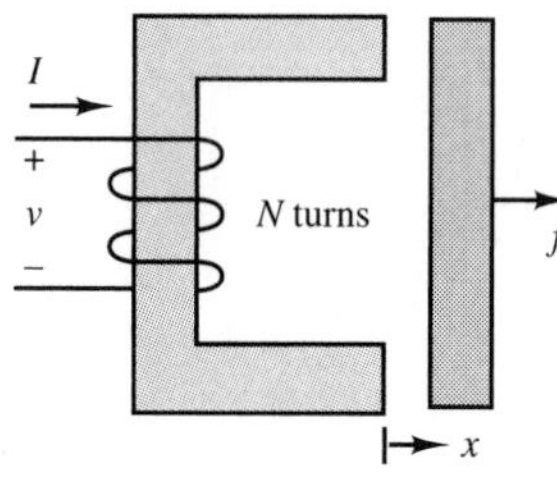
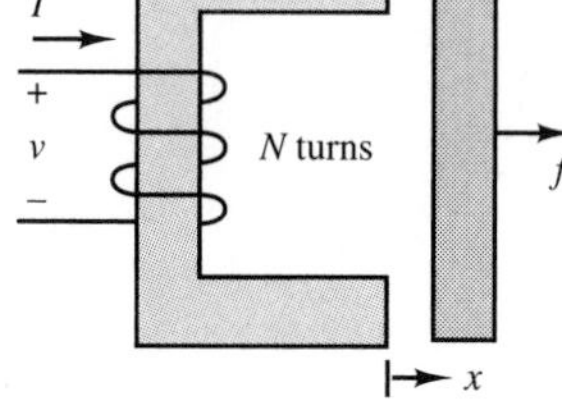

Figure P15.13

15.15 The electromagnet of Figure P15.13 has reluctance given by $\mathcal{R}(x) = 7 \times 10^8(0.002 + x)$ H^{-1}, where x is the length of the variable gap in meters. The coil has 980 turns and 30 Ω resistance. For an applied voltage of 120 VDC, find:

a. The energy stored in the magnetic field for $x = 0.005$ m.
b. The magnetic force for $x = 0.005$ m.

15.16 Determine the force, $\mathcal{F}$, between the faces of the poles (stationary coil and plunger) of the solenoid pictured in Figure P15.14 when it is energized. When energized, the plunger is drawn into the coil and comes to rest with only a negligible air gap separating the two. The flux density in the cast steel pathway is 1.1 T. The diameter of the plunger is 10 mm.

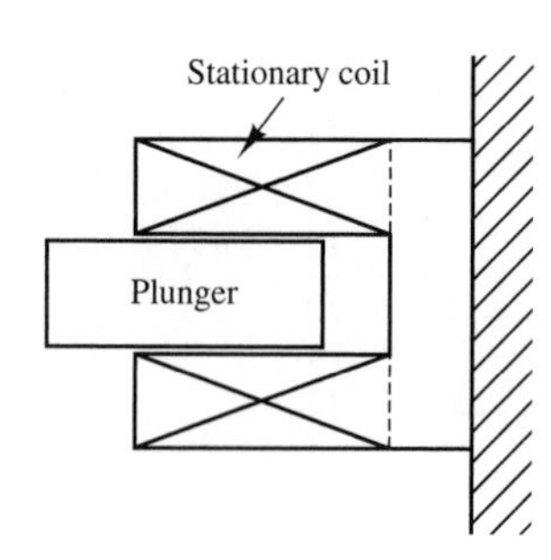

Figure P15.14

15.17 An electromagnet is used to support a solid piece of steel as shown in Example 15.10. A force of 10,000 N is required to support the weight. The cross-sectional area of the magnetic core (the fixed part) is 0.01 m^2. The coil has 1,000 turns. Determine the minimum current that can keep the weight from falling for $x = 1.0$ mm. Assume negligible reluctance for the steel parts and negligible fringing in the air gaps.

15.18 The armature, frame, and core of a 12-VDC control relay are made of sheet steel. The average length of the magnetic circuit is 12 cm when the relay is energized, and the average cross section of the magnetic circuit is 0.60 cm^2. The coil is wound with 250 turns and carries 50 mA. Determine:

a. The flux density, $\mathcal{B}$, in the magnetic circuit of the relay when the coil is energized.
b. The force, $\mathcal{F}$, exerted on the armature to close it when the coil is energized.

15.19 Refer to the actuator of Figure P15.15. The entire device is made of sheet steel. The coil has 2,000 turns. The armature is stationary so that the length of the air gaps, $g = 10$ mm, is fixed. A direct current passing through the coil produces a flux density of 1.2 T in the gaps. Determine:

a. The coil current.
b. The energy stored in the air gaps.
c. The energy stored in the steel.

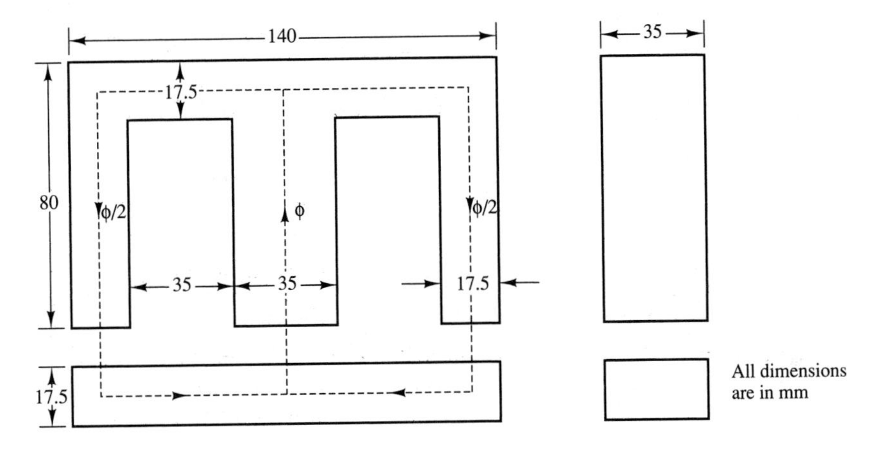

Figure P15.15

15.20 A core is shown in Figure P15.16, with $\mu_r = 2{,}000$ and $N = 100$. Find:

a. The current needed to produce a flux density of 0.4 Wb/m^2 in the center leg.
b. The current needed to produce a flux density of 0.8 Wb/m^2 in the center leg.

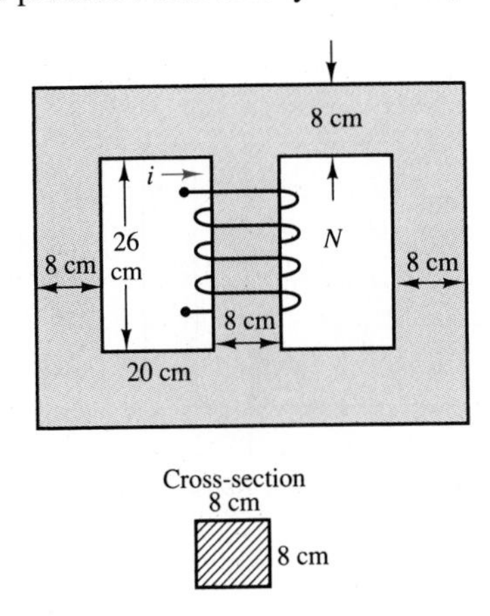

Figure P15.16

15.21 A transformer is delivering power to a 300-Ω resistive load. To achieve the desired power transfer, the turns ratio is chosen so that the resistive load referred to the primary is 7,500 Ω. The parameter values, *referred to the secondary winding,* are:

$$r_1 = 20\ \Omega \quad L_1 = 1.0\ \text{mH} \quad L_m = 25\ \text{mH}$$
$$r_2 = 20\ \Omega \quad L_2 = 1.0\ \text{mH}$$

Core losses are negligible.

a. Determine the turns ratio.
b. Determine the input voltage, current, and power and the efficiency when this transformer is delivering 12 W to the 300-Ω load at a frequency $f = 10{,}000/2\pi$ Hz.

15.22 Derive and sketch the frequency response of the loudspeaker of Example 15.15 for (1) $k = 50{,}000\frac{N}{m}$ and (2) $k = 5 \times 10^6 \frac{N}{m}$. Describe qualitatively how the loudspeaker frequency response changes as the spring stiffness, k, increases and decreases. What will the frequency response be in the limit as k approaches zero? What kind of speaker would this condition correspond to?

15.23 A 220/20-V transformer has 50 turns on its low-voltage side. Calculate

a. The number of turns on its high side.
b. The turns ratio α when it is used as a step-down transformer.
c. The turns ratio α when it is used as a step-up transformer.

15.24 The high-voltage side of a transformer has 750 turns, and the low-voltage side 50 turns. When the high side is connected to a rated voltage of 120 V, 60 Hz, a rated load of 40 A is connected to the low side. Calculate

a. The turns ratio.
b. The secondary voltage (assuming no internal transformer impedance voltage drops).
c. The resistance of the load.

15.25 A transformer is to be used to match an 8-Ω loudspeaker to a 500-Ω audio line. What is the turns ratio of the transformer, and what are the voltages at the primary and secondary terminals when 10 W of audio power is delivered to the speaker? Assume that the speaker is a resistive load and the transformer is ideal.

15.26 An electromechanical system is shown in Figure P15.17. Find the differential equations describing the system after the switch is closed. Neglect fringing and leakage and the friction.

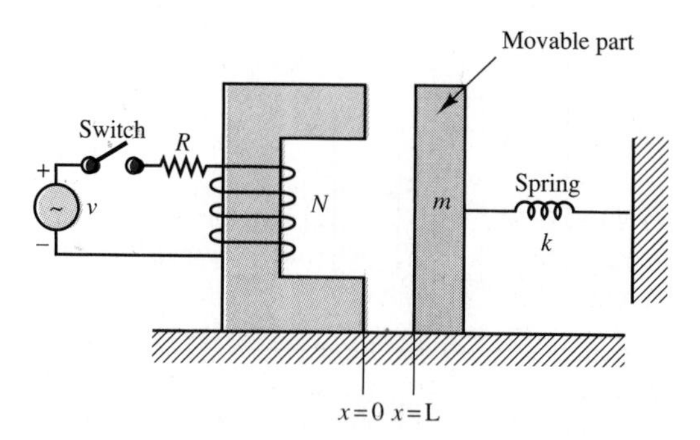

Figure P15.17

15.27 On the basis of Example 15.15, obtain an equivalent electrical circuit for the loudspeaker. [*Hint:* Eliminate velocity U and find the relation of $V = f(I)$ first.]

15.28 A practical LVDT is typically connected to a resistive load. Derive the LVDT equations in the presence of a resistive load, R_L, connected across the output terminals, using the results of Example 15.1.

15.29 A relay is shown in Figure P15.18. Find the differential equations describing the system.

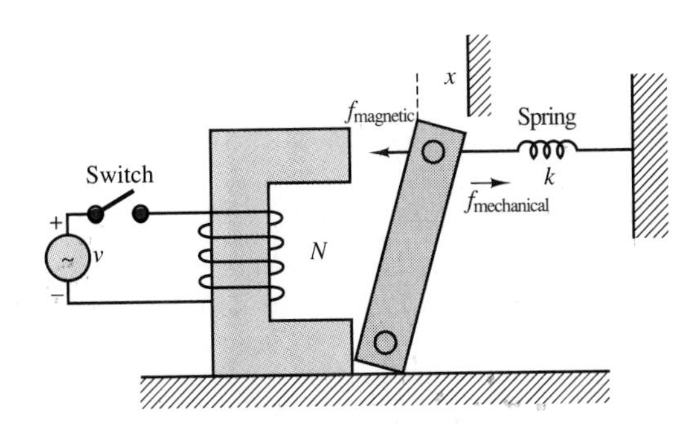

Figure P15.18

15.30 On the basis of the equations of Example 15.1, and of the results of Problem 15.28, derive the frequency response of the LVDT, and determine the range of frequencies for which the device will have maximum sensitivity for a given excitation. [*Hint:* Compute dv_{out}/dv_{ex}, and set the derivative equal to zero to determine the maximum sensitivity.]

15.31 The model of an electromechanical conversion system is shown in Figure P15.19. Find the differential equations describing the system.

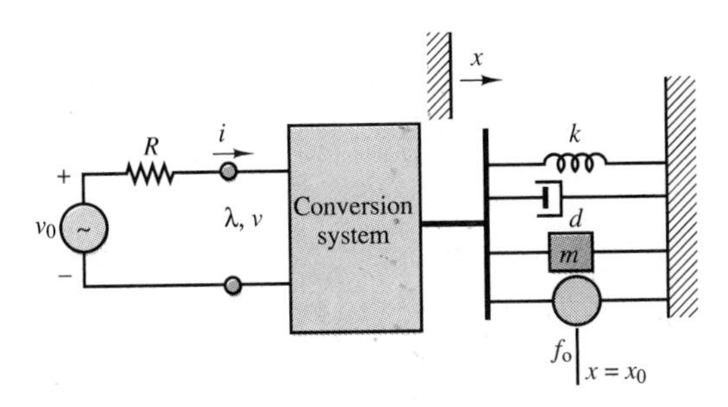

Figure P15.19

15.32 From the result of Problem 15.27, find the maximum mechanical impedance Z_m and the angular frequency where Z_m = max. [*Hint:* The mechanical parameters of Problem 15.27 are defined as the mechanical impedance.]

15.33 The air gap inductor shown in Figure P15.20 has a height, h, of 2.5 cm, width, w, of 2.5 cm, air-gap dimension $l_g = 0.1$ cm, and $N = 400$ turns, and it is excited by a DC current, $I = 5$ A. Determine the force of attraction between sides of the gap, neglecting the reluctance of the core, leakage, and fringing.

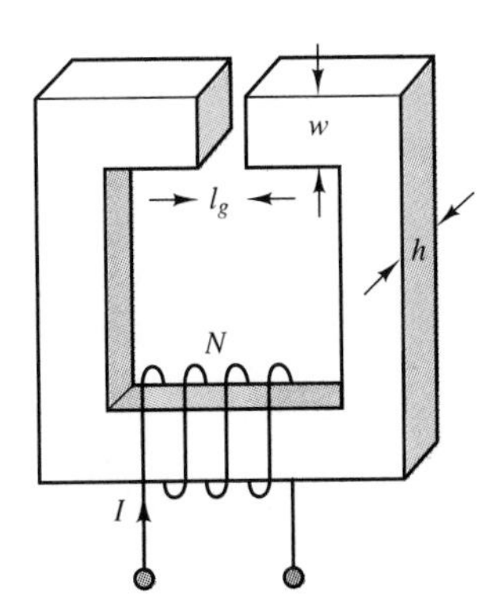

Figure P15.20

15.34 The high-voltage side of a step-down transformer has 800 turns, and the low-voltage side has 100 turns. A voltage of 240 VAC is applied to the high side, and the load impedance is 3 Ω (low side). Find

a. The secondary voltage and current.
b. The primary current.
c. The primary input impedance from the ratio of primary voltage and current.
d. The primary input impedance.

15.35 Calculate the transformer ratio of the transformer in Problem 15.34 when it is used as a step-up transformer.

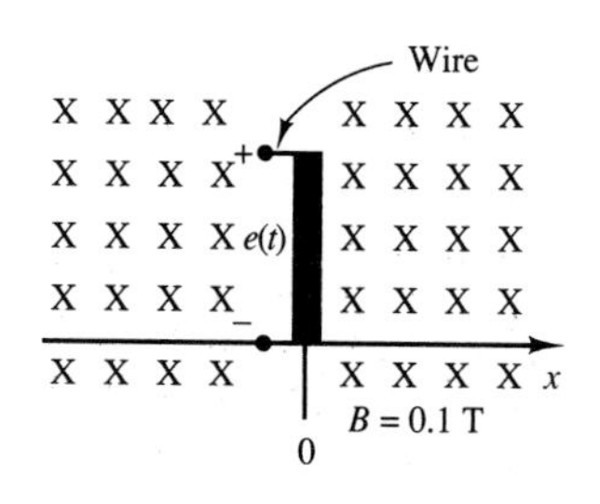

Figure P15.21

15.36 A 2,300/240-V, 60-Hz, 4.6-kVA transformer is designed to have an induced emf of 2.5 V/turn. Assuming an ideal transformer, find

a. The number of high-side turns, N_h, and low-side turns, N_l.
b. The rated current of the high-voltage side, I_h.
c. The transformer ratio when the device is used as a step-up transformer.

15.37 A wire of length 20 cm vibrates in one direction in a constant magnetic field with a flux density of 0.1 T; see Figure P15.21. The position of the wire as a function of time is given by $x(t) = 0.1 \sin 10t$ m. Find the induced emf across the length of the wire as a function of time.

15.38 The wire of Problem 15.37 induces a time-varying emf of

$$e_1(t) = 0.02 \cos 10t$$

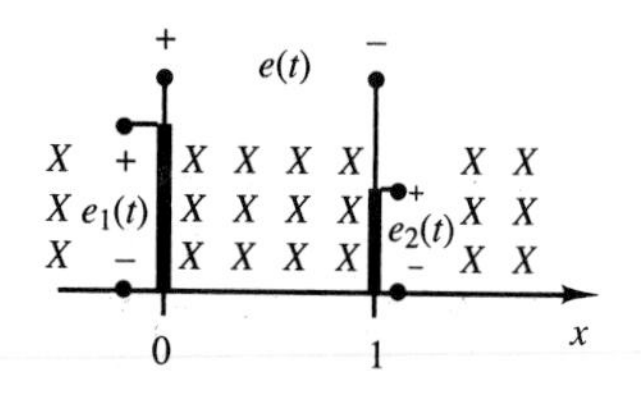

Figure P15.22

A second wire is placed in the same magnetic field but has a length of 0.1 m, as shown in Figure P15.22. The position of this wire is given by $x(t) = 1 - 0.1 \sin 10t$. Find the induced emf $e(t)$ defined by the difference in emf's $e_1(t)$ and $e_2(t)$.

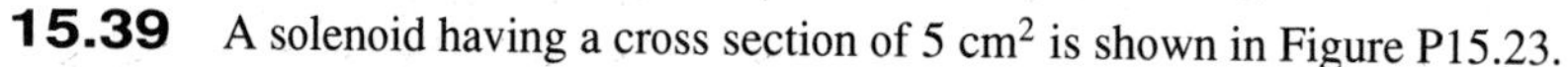

15.39 A solenoid having a cross section of 5 cm^2 is shown in Figure P15.23.

a. Calculate the force exerted on the plunger when the distance x is 2 cm and the current in the coil (where $N = 100$ turns) is 5 A. Assume that the fringing and leakage effects are negligible. The relative permeabilities of the magnetic material and the nonmagnetic sleeve are 2,000 and 1.
b. Develop a set of defferential equations governing the behavior of the solenoid.

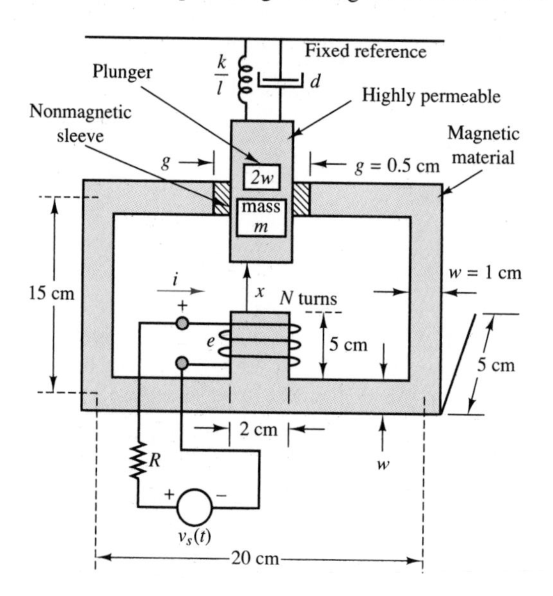

Figure P15.23

15.40 A conducting bar shown in Figure 14.47 in the text, is carrying 4 A of current in the presence of a magnetic field; $B = 0.3 \text{ Wb/m}^2$. Find the magnitude and direction of the force induced on the conducting bar.

15.41 A wire, shown in Figure P15.24, is moving in the presence of a magnetic field, with $B = 0.4 \text{ Wb/m}^2$. Find the magnitude and direction of the induced voltage in the wire.

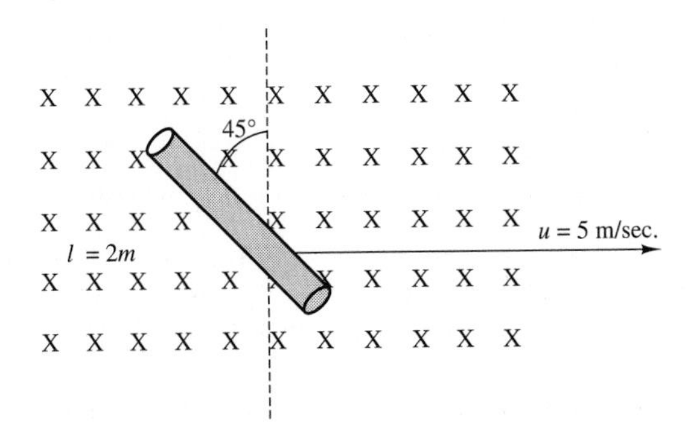

Figure P15.24

CHAPTER 16

Introduction to Electric Machines

The objective of this chapter is to introduce the basic operation of rotating electric machines. The operation of the three major classes of electric machines—DC, synchronous, and induction—will first be described as intuitively as possible, building on the material presented in Chapter 15. The second part of the chapter will be devoted to a discussion of the applications and selection criteria for the different classes of machines.

The emphasis of this chapter will be on explaining the properties of each type of machine, with its advantages and disadvantages with regard to other types; and on classifying these machines in terms of their performance characteristics and preferred field of application. Chapter 17 will be devoted to a survey of special-purpose electric machines—many of which find common application in industry—such as stepper motors, brushless DC motors, and single-phase induction motors. Selected examples and application notes will discuss some current issues of interest.

By the end of this chapter, you should be able to:

- Describe the principles of operation of DC and AC motors and generators.
- Interpret the nameplate data of an electric machine.

- Interpret the torque-speed characteristic of an electric machine.
- Specify the requirements of a machine given an application.

16.1 ROTATING ELECTRIC MACHINES

The range of sizes and power ratings and the different physical features of rotating machines are such that the task of explaining the operation of rotating machines in a single chapter may appear formidable at first. Some features of rotating machines, however, are common to all such devices. This introductory section is aimed at explaining the common properties of all rotating electric machines. We begin our discussion with reference to Figure 16.1, in which a hypothetical rotating machine is depicted in a cross-sectional view. In the figure, a box with a cross inscribed in it indicates current flowing into the page, while a dot represents current out of the plane of the page.

In Figure 16.1, we identify a **stator,** of cylindrical shape, and a **rotor,** which, as the name indicates, rotates inside the stator, separated from the latter by means of an air gap. The rotor and stator each consist of a magnetic core, some electrical insulation, and the windings necessary to establish a magnetic flux (unless this is created by a permanent magnet). The rotor is mounted on a bearing-supported shaft, which can be connected to *mechanical loads* (if the machine is a motor) or to a *prime mover* (if the machine is a generator) by means of belts, pulleys, chains, or other mechanical couplings. The windings carry the electric currents that generate the magnetic fields and flow to the electrical loads, and also provide the closed loops in which voltages will be induced (by virtue of Faraday's law, as discussed in the previous chapter).

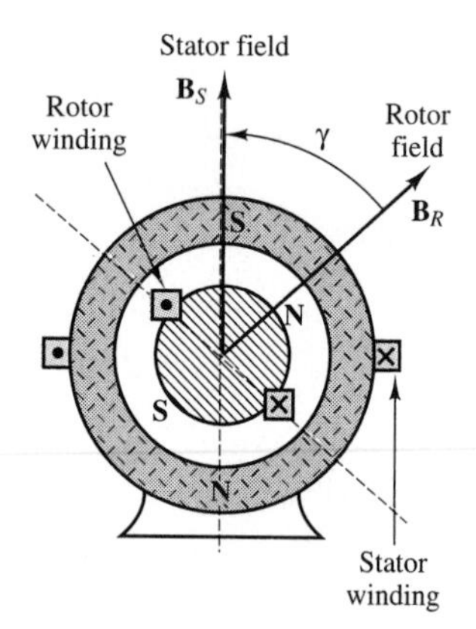

Figure 16.1 A rotating electric machine

Basic Classification of Electric Machines

An immediate distinction can be made between different types of windings characterized by the nature of the current they carry. If the current serves the sole purpose of providing a magnetic field and is independent of the load, it is called a *magnetizing,* or excitation, current, and the winding is termed a **field winding.** Field currents are nearly always DC and are of relatively low power, since their only purpose is to magnetize the core (recall the important role of high-permeability cores in generating large magnetic fluxes from relatively small currents). On the other hand, if the winding carries only the load current, it is called an **armature.** In DC and AC synchronous machines, separate windings exist to carry field and armature currents. In the induction motor, the magnetizing and load currents flow in the same winding, called the *input winding,* or *primary;* the output winding is then called the *secondary.* As we shall see, this terminology, which is reminiscent of transformers, is particularly appropriate for induction motors, which bear a significant analogy to the operation of the transformers studied in Chapters 5 and 15. Table 16.1 characterizes the principal machines in terms of their field and armature configuration.

It is also useful to classify electric machines in terms of their energy-conversion characteristics. A machine acts as a **generator** if it converts mechanical energy from a prime mover—e.g., an internal combustion engine—to electrical form. Examples of generators are the large machines used in power-generating plants, or the common automotive alternator. A machine is classified as a **motor** if it converts electrical energy to mechanical form. The latter class of machines is probably of more direct interest to you, because of its widespread application in engineering practice. Electric motors are used to provide forces and torques to generate motion in countless industrial applications. Machine tools, robots, punches, presses, mills, and propulsion systems for electric vehicles are but a few examples of the application of electric machines in engineering.

Note that in Figure 16.1 we have explicitly shown the direction of two magnetic fields: that of the rotor, $\mathbf{B}_R$, and that of the stator, $\mathbf{B}_S$. Although these fields are generated by different means in different machines (e.g., permanent magnets, AC currents, DC currents), the presence of these fields is what causes a rotating machine to turn and enables the generation of electric power. In particular, we see that in Figure 16.1 the north pole of the rotor field will seek to align itself with the south pole of the stator field. It is this magnetic attraction force that permits the generation of torque in an electric motor; conversely, a generator exploits the laws of electromagnetic induction to convert a changing magnetic field to an electric current.

To simplify the discussion in later sections, we shall presently introduce some basic concepts that apply to all rotating electric machines. Referring to Figure 16.2, we note that for all machines the force on a wire is given by the expression

$$\mathbf{f} = i_w \mathbf{l} \times \mathbf{B} \tag{16.1}$$

where i_w is the current in the wire, $\mathbf{l}$ is a vector along the direction of the wire, and $\times$ denotes the cross product of two vectors. Then the torque for a multiturn coil becomes

$$T = KBi_w \sin\alpha \tag{16.2}$$

where

B = magnetic flux density caused by the stator field
K = constant depending on coil geometry
α = angle between $\mathbf{B}$ and the normal to the plane of the coil

In the hypothetical machine of Figure 16.2, there are two magnetic fields: one generated within the stator, the other within the rotor windings. Either (but not both) of these fields could be generated either by a current or by a permanent magnet. Thus, we could replace the permanent-magnet stator of Figure 16.2 with a suitably arranged winding to generate a stator field in the same direction. If the stator were made of a toroidal coil of radius R (see Chapter 15), then the magnetic field of the stator would generate a flux density B, where

$$B = \mu H = \mu \frac{Ni}{2\pi R} \tag{16.3}$$

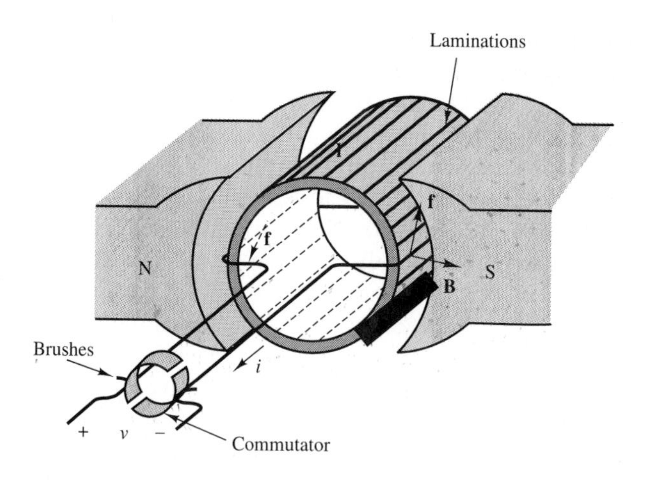

Figure 16.2 Stator and rotor fields and the force acting on a rotating machine

and where N is the number of turns and i is the coil current. The direction of the torque is always the direction determined by the rotor and stator fields as they seek to align to each other (i.e., counterclockwise in the diagram of Figure 16.1).

It is important to note that Figure 16.2 is merely a general indication of the major features and characteristics of rotating machines. A variety of configurations exist, depending on whether each of the fields is generated by a current in a coil or by a permanent magnet, and on whether the load and magnetizing currents are direct or alternating. The type of excitation (AC or DC) provided to the windings permits a first classification of electric machines (see Table 16.1). According to this classification, one can define the following types of machines:

- *Direct-current machines:* DC current in both stator and rotor
- *Synchronous machines:* AC current in one winding, DC in the other
- *Induction machines:* AC current in both

In most industrial applications, the induction machine is the preferred choice, because of the simplicity of its construction. However, the analysis of the performance of an induction machine is rather complex. On the other hand, DC

Table 16.1 **Configurations of the three types of electric machines**

Machine type	Winding	Winding type	Location	Current
DC	Input and output	Armature	Rotor	AC (winding) DC (at brushes)
	Magnetizing	Field	Stator	DC
Synchronous	Input and output	Armature	Stator	AC
	Magnetizing	Field	Rotor	DC
Induction	Input	Primary	Stator	AC
	Output	Secondary	Rotor	AC

machines are quite complex in their construction but can be analyzed relatively simply with the analytical tools we have already acquired. Therefore, the progression of this chapter will be as follows. We start with a section that discusses the physical construction of DC machines, both motors and generators. Then we continue with a discussion of synchronous machines, in which one of the currents is now alternating, since these can easily be understood as an extension of DC machines. Finally, we consider the case where both rotor and stator currents are alternating, and analyze the induction machine.

Other Characteristics of Electric Machines

As already stated earlier in this chapter, electric machines are **energy-conversion devices,** and we are therefore interested in their energy-conversion **efficiency.** Typical applications of electric machines as motors or generators must take into consideration the energy losses associated with these devices. Figure 16.3 represents the various loss mechanisms you must consider in analyzing the efficiency of an electric machine for the case of direct-current machines. It is important for you to keep in mind this conceptual flow of energy when analyzing electric machines. The sources of loss in a rotating machine can be separated into three fundamental groups: electrical (I^2R) losses, core losses, and mechanical losses.

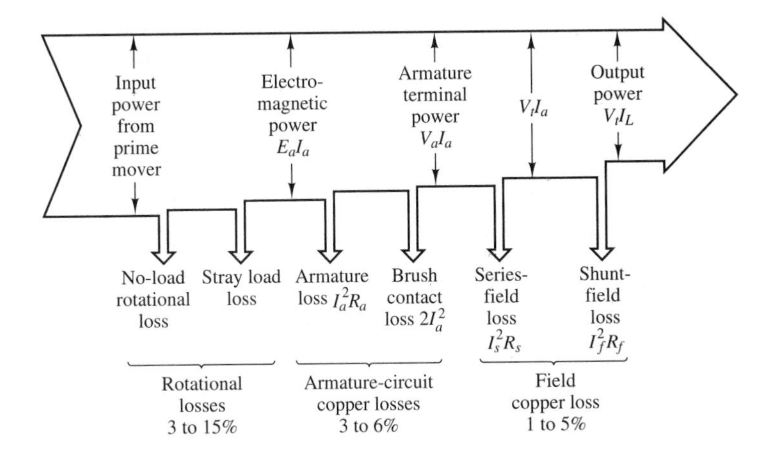

Figure 16.3(a) Generator losses, direct current

I^2R losses are usually computed on the basis of the DC resistance of the windings at 75°C; in practice, these losses vary with operating conditions. The difference between the nominal and actual I^2R loss is usually lumped under the category of *stray-load loss*. In direct-current machines, it is also necessary to account for the *brush contact loss* associated with slip rings and commutators.

Mechanical losses are due to *friction* (mostly in the bearings) and *windage,* that is, the air drag force that opposes the motion of the rotor. In addition, if

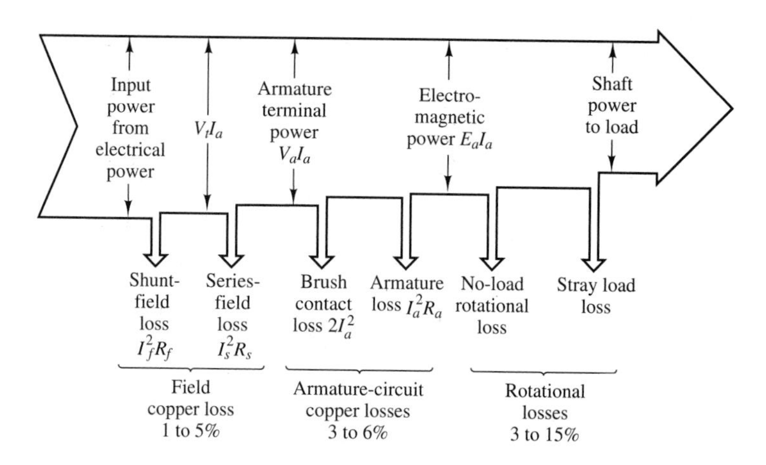

Figure 16.3(b) Motor losses, direct current

external devices (e.g., blowers) are required to circulate air through the machine for cooling purposes, the energy expended by these devices is also included in the mechanical losses.

Open-circuit core losses consist of hysteresis and eddy current losses, with only the excitation winding energized (see Chapter 15 for a discussion of hysteresis and eddy currents). Often these losses are summed with friction and windage losses to give rise to the *no-load rotational loss*. The latter quantity is useful if one simply wishes to compute efficiency. Since open-circuit core losses do not account for the changes in flux density caused by the presence of load currents, an additional magnetic loss is incurred that is not accounted for in this term. *Stray-load losses* are used to lump the effects of nonideal current distribution in the windings and of the additional core losses just mentioned. Stray-load losses are difficult to determine exactly and are often assumed to be equal to 1.0 percent of the output power for DC machines; these losses can be determined by experiment in synchronous and induction machines.

The performance of an electric machine can be quantified in a number of ways. In the case of an electric motor, it is usually portrayed in the form of a graphical **torque-speed characteristic.** The torque-speed characteristic of a motor describes how the torque supplied by the machine varies as a function of the speed of rotation of the motor for steady speeds. As we shall see in later sections, the torque-speed curves vary in shape with the type of motor (DC, induction, synchronous) and are very useful in determining the performance of the motor when connected to a mechanical load. Figure 16.4 depicts the torque-speed curve of a hypothetical motor. We shall presently describe the essential elements of such a graphical representation of motor performance, and we shall later return to analyze the typical performance curve of each type of motor we encounter in our discussion. It is quite likely that in most engineering applications, the engineer is required to make a decision regarding the performance characteristics of the

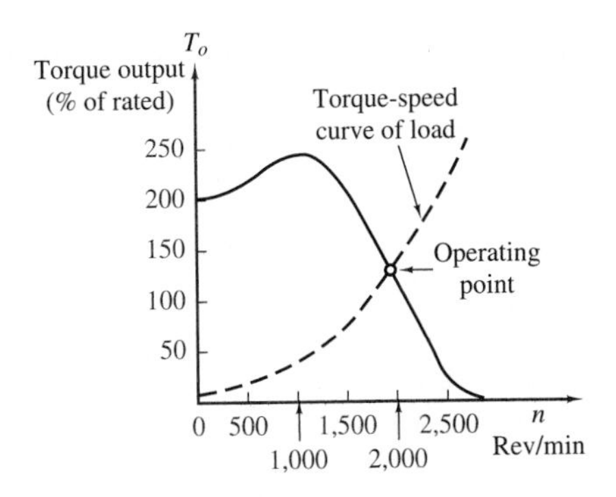

Figure 16.4 Torque-speed curve for an electric motor

motor best suited to a specified task. In this context, the torque-speed curve of a machine is a very useful piece of information.

The first feature we note of the torque-speed characteristic is that it bears a strong resemblance to the *i-v* characteristics used in earlier chapters to represent the behavior of electrical sources. It should be clear that, according to this torque-speed curve, the motor is not an ideal source of torque (if it were, the curve would appear as a horizontal line across the speed range). One can readily see, for example, that the hypothetical motor represented by the curves of Figure 16.4 would produce maximum torque in the range of speeds between approximately 800 and 1,400 rev/min. What determines the actual speed of the motor (and therefore its output torque and power) is the torque-speed characteristic of the load connected to it, much as a resistive load determines the current drawn from a voltage source. In the figure, we display the torque-speed curve of a load, represented by the dashed line; the operating point of the motor-load pair is determined by the intersection of the two curves.

Another important observation pertains to the fact that the motor of Figure 16.4 produces a nonzero torque at zero speed. This fact implies that as soon as electric power is connected to the motor, the latter is capable of supplying a certain amount of torque; this zero-speed torque is called the **starting torque.** If the load the motor is connected to requires less than the starting torque the motor can provide, then the motor can accelerate the load, until the motor speed and torque settle to a stable value, at the operating point. The motor-load pair of Figure 16.4 would behave in the manner just described. However, there may well be circumstances in which a motor might not be able to provide a sufficient starting torque to overcome the static load torque that opposes its motion. Thus, we see that a torque-speed characteristic can offer valuable insight into the operation of a motor. As we proceed to discuss each type of machine in greater detail, we shall devote some time to the discussion of its torque-speed curve.

The most common means of conveying information regarding electric machines is the *nameplate*. Typical information conveyed by the nameplate is:

1. Type of device (e.g., DC motor, alternator)
2. Manufacturer
3. Rated voltage and frequency
4. Rated current and volt-amperes
5. Rated speed and horsepower

The **rated voltage** is the terminal voltage for which the machine was designed, and which will provide the desired magnetic flux. Operation at higher voltages will increase magnetic core losses, because of excessive core saturation. The **rated current** and **rated volt-amperes** are an indication of the typical current and power levels at the terminal that will not cause undue overheating due to copper losses (I^2R losses) in the windings. These ratings are not absolutely precise, but they give an indication of the range of excitations for which the motor will perform without overheating. Peak power operation in a motor may exceed rated torque (horsepower) or currents by a substantial factor (up to as much as 6 or 7 times the rated value); however, continuous operation of the motor above the rated performance will cause the machine to overheat, and possibly to sustain damage. Thus, it is important to consider both peak and continuous power requirements when selecting a motor for a specific application. An analogous discussion is valid for the speed rating: while an electric machine may operate above rated speed for limited periods of time, the large centrifugal forces generated at high rotational speeds will eventually cause undesirable mechanical stresses, especially in the rotor windings, leading eventually even to self-destruction.

Another important feature of electric machines is the **regulation** of the machine speed or voltage, depending on whether it is used as a motor or as a generator, respectively. Regulation is the ability to maintain speed or voltage constant in the face of load variations. The ability to closely regulate speed in a motor or voltage in a generator is an important feature of electric machines; regulation is often improved by means of feedback control mechanisms, some of which will be briefly introduced in this chapter. We shall take the following definitions as being adequate for the intended purpose of this chapter:

$$\text{Speed regulation} = \frac{\text{Speed at no load} - \text{Speed at rated load}}{\text{Speed at rated load}} \tag{16.4}$$

$$\text{Voltage regulation} = \frac{\text{Voltage at no load} - \text{Voltage at rated load}}{\text{Voltage at rated load}} \tag{16.5}$$

Please note that the rated value is usually taken to be the nameplate value, and that the meaning of *load* changes depending on whether the machine is a motor, in which case the load is mechanical, or a generator, in which case the load is electrical.

Example 16.1 Regulation

The speed of a shunt DC motor drops from 1,800 rev/min at no load to 1,760 rev/min at rated load. Find the percent speed regulation.

Solution:

The percent speed regulation, SR%, is defined as follows:

$$\text{SR\%} = \frac{n_{nl} - n_{rl}}{n_{rl}} \times 100 = \frac{\omega_{nl} - \omega_{rl}}{\omega_{rl}} \times 100$$

where

n_{nl} = speed at no load, rev/min
ω_{nl} = speed at no load, rad/s
n_{rl} = speed at rated load, rev/min
ω_{rl} = speed at rated load, rad/s

The speed at no load is $n_{nl} = 1,800$ rev/min. The speed at rated load is $n_{rl} = 1{,}760$ rev/min. Thus,

$$\text{SR\%} = \frac{1{,}800 - 1{,}760}{1{,}760} \times 100 = 2.27\%$$

Example 16.2 Nameplate

The nameplate of a typical induction motor is shown in the table below. The model number (sometimes abbreviated as MOD) uniquely identifies the motor to the manufacturer. It may be a style number, a model number, an identification number, or an instruction sheet reference number.

MODEL	19308 J-X		
TYPE	CJ4B	FRAME	324TS
VOLTS	230/460	°C AMB.	40
		INS. CL.	B
FRT. BRG	210SF	EXT. BRG	312SF
SERV FACT	1.0	OPER INSTR	C-517
PHASE 3	Hz 60	CODE G	WDGS 1
H.P.	40		
R.P.M.	3,565		
AMPS	106/53		
NEMA NOM.	EFF		
NOM. P.F.			
DUTY	CONT.	NEMA DESIGN	B

The term *frame* (sometimes abbreviated as FR) refers principally to the physical size of the machine, as well as to certain construction features.

Ambient temperature (abbreviated as AMB, or MAX. AMB) refers to the maximum ambient temperature in which the motor is capable of operating. Operation of the motor in a higher ambient temperature may result in shortened motor life and reduced torque.

Insulation class (abbreviated as INS. CL.) refers to the type of insulation used in the motor. Most often used are class A (105°C) and class B (130°C).

The duty (DUTY), or time rating, denotes the length of time the motor is expected to be able to carry the rated load under usual service conditions. "CONT." means that the machine can be operated continuously.

The "CODE" letter sets the limits of starting kVA per horsepower for the machine. There are 19 levels, denoted by the letters A through V, excluding I, O, and Q.

Service factor (abbreviated as SERV FACT) is a term defined by NEMA (the National Electrical Manufacturers Association) as follows: "The service factor of a general-purpose alternating-current motor is a multiplier which, when applied to the rated horsepower, indicates a permissible horsepower loading which may be carried under the conditions specified for the service factor."

The voltage figure given on the nameplate refers to the voltage of the supply circuit to which the motor should be connected. Sometimes two voltages are given, for example, 230/460. In this case, the machine is intended for use on either a 230-V or a 460-V circuit. Special instructions will be provided for connecting the motor for each of the voltages.

The term "BRG" indicates the nature of the bearings supporting the motor shaft.

Example 16.3 Torque-Speed Curves

A variable-torque variable-speed motor has a torque output that varies directly with speed; hence, the horsepower output varies directly with the speed. Motors with this characteristic are commonly used with fans, blowers, and centrifugal pumps. Figure 16.5 shows typical torque-speed curves for this type of motor. Superimposed on the motor torque-speed curve is the torque-speed curve for a typical fan where the input power to the fan varies as the cube of the fan speed. Point *A* is the desired operating point, which could be determined graphically by plotting the load line and the motor torque-speed curve on the same graph, as illustrated in Figure 16.5.

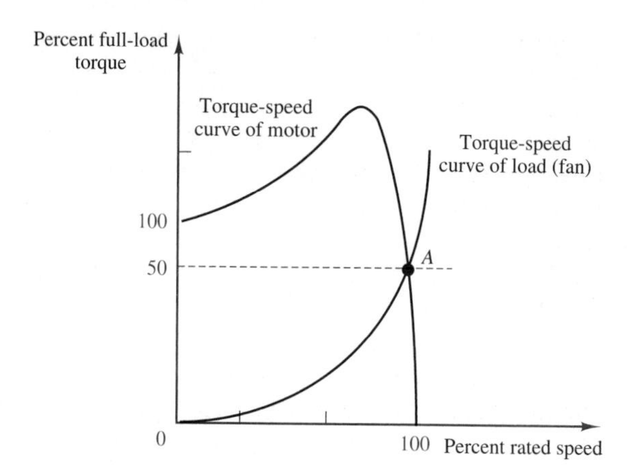

Figure 16.5

Basic Operation of All Rotating Machines

We have already seen in Chapter 15 how the magnetic field in electromechanical devices provides a form of coupling between electrical and mechanical systems. Intuitively, one can identify two aspects of this coupling, both of which play a role in the operation of electric machines: (1) magnetic attraction and repulsion forces generate mechanical torque, and (2) the magnetic field can induce a voltage in the machine windings (coils) by virtue of Faraday's law. Thus, we may think of the operation of an electric machine in terms of either a motor or a generator, depending on whether the input power is electrical and mechanical power is produced (motor action), or the input power is mechanical and the output power is electrical (generator action). Figure 16.6 illustrates the two cases graphically.

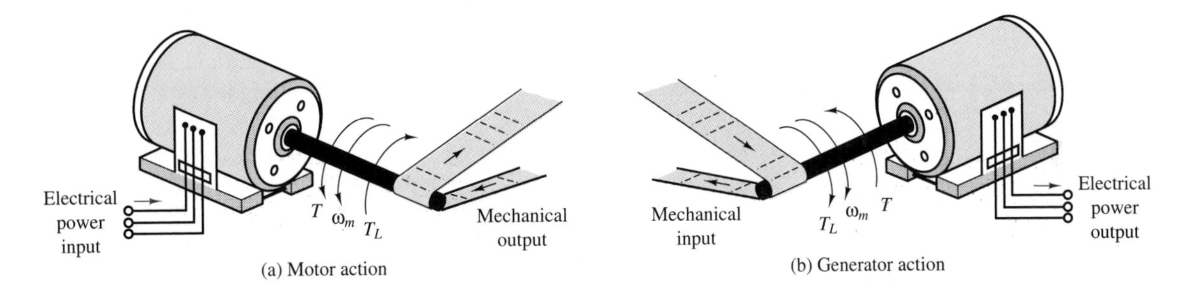

Figure 16.6 Generator and motor action in an electric machine

The coupling magnetic field performs a dual role, which may be explained as follows. When a current i flows through conductors placed in a magnetic field, a force is produced on each conductor, according to equation 16.1. If these conductors are attached to a cylindrical structure, a torque is generated, and if the structure is free to rotate, then it will rotate at an angular velocity ω_m. As the conductors rotate, however, they move through a magnetic field and cut through flux lines, thus generating an electromotive force in opposition to the excitation. This emf is also called "counter" emf; it opposes the source of the current i. If, on the other hand, the rotating element of the machine is driven by a prime mover (for example, an internal combustion engine), then an emf is generated across the coil that is rotating in the magnetic field (the armature). If a load is connected to the armature, a current i will flow to the load, and this current flow will in turn cause a reaction torque on the armature that opposes the torque imposed by the prime mover.

You see, then, that for energy conversion to take place, two elements are required: (1) a coupling field, $\mathbf{B}$, usually generated in the field winding; and (2) an armature winding that supports the load current, i, and the emf, e.

Magnetic Poles in Electric Machines

Before discussing the actual construction of a rotating machine, we should spend a few paragraphs to illustrate the significance of **magnetic poles** in an electric machine. In an electric machine, torque is developed as a consequence of magnetic forces of attraction and repulsion between magnetic poles on the stator and on the rotor; these poles produce a torque that accelerates the rotor and a reaction torque on the stator. Naturally, we would like a construction such that the torque generated as a consequence of the magnetic forces is continuous and in a constant direction. This can be accomplished if the number of rotor poles is equal to the number of stator poles. It is also important to observe that the number of poles must be even, since there have to be equal numbers of north and south poles.

Figure 16.7 depicts a two-pole machine in which the stator poles are constructed in such a way as to project closer to the rotor than to the stator structure. This type of construction is rather common, and poles constructed in this fashion are called **salient poles.** Note that the rotor could also be constructed to have salient poles.

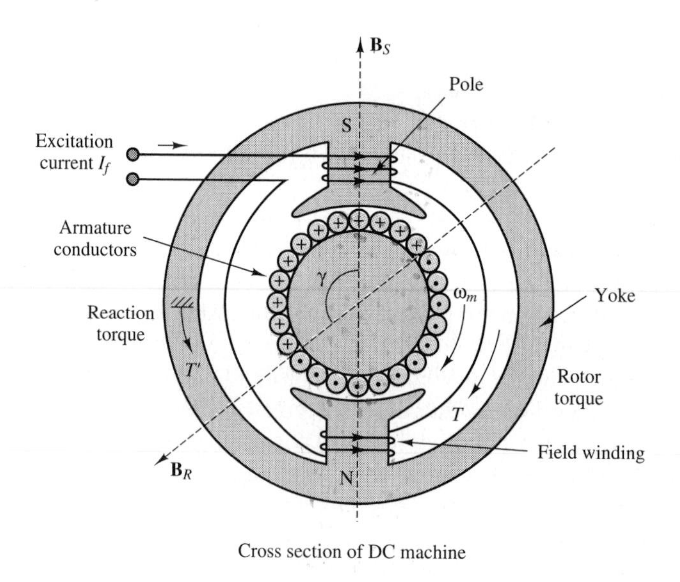

Figure 16.7 A two-pole machine with salient stator poles

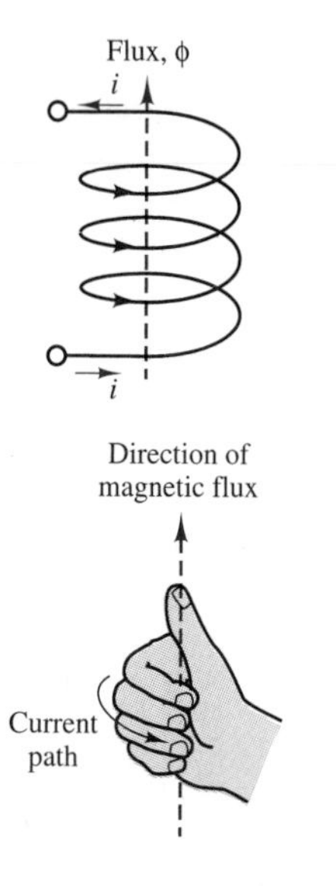

Figure 16.8 Right-hand rule

To understand magnetic polarity, we need to consider the direction of the magnetic field in a coil carrying current. Figure 16.8 shows how the *right-hand rule* can be employed to determine the direction of the magnetic flux. If one were to grasp the coil with the right hand, with the fingers curling in the direction of current flow, then the thumb would be pointing in the direction of the magnetic flux. As you may recall from the discussion in Chapter 15, magnetic flux is by convention viewed as entering the south pole and exiting from the north pole. Thus, to determine whether a magnetic pole is north or south, we must consider the direction of the flux. Figure 16.9 shows a cross section of a coil wound around a pair of salient rotor poles. In this case, one can readily identify the direction of the magnetic flux and therefore the magnetic polarity of the poles by applying the right-hand rule, as illustrated in the figure.

Often, however, the coil windings are not arranged as simply as in the case of salient poles. In many machines, the windings are embedded in slots cut into the stator or rotor, so that the situation is similar to that of the stator depicted in Figure 16.10. This figure is a cross section in which the wire connections between "crosses" and "dots" have been cut away. In Figure 16.10, the dashed line indicates the axis of the stator flux according to the right-hand rule, indicating that the slotted stator in effect behaves like a pole pair. The north and south poles indicated in the figure are a consequence of the fact that the flux exits the bottom part of the structure (thus, the north pole indicated in the figure) and enters the top half of the structure (thus, the south pole). In particular, if you consider that the windings are arranged so that the current entering the right-hand side of the stator (to the right of the dashed line) flows through the back end of the stator and then flows outward from the left-hand side of the stator slots (left of the dashed line), you can visualize the windings in the slots as behaving in a manner similar to the coils of Figure 16.9, where the flux axis of Figure 16.10 corresponds to the flux axis of each of the coils of Figure 16.9. The actual circuit that permits current flow is completed by the front and back ends of the stator, where the wires are connected according to the pattern *a*-*a*′, *b*-*b*′, *c*-*c*′, as depicted in the figure.

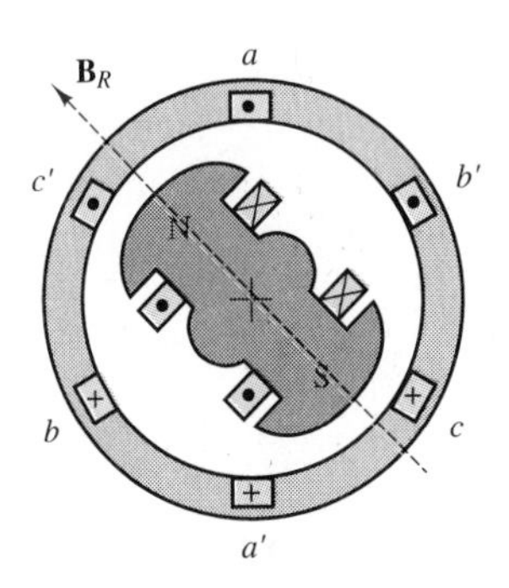

Figure 16.9 Magnetic field in a salient rotor winding

Another important consideration that facilitates understanding the operation of electric machines pertains to the use of AC currents. It should be apparent by now that if the current flowing into the slotted stator is alternating, the direction of the flux will also alternate, so that in effect the two poles will reverse polarity every time the current reverses direction, that is, every half-cycle of the sinusoidal current. Further—since the magnetic flux is approximately proportional to the current in the coil—as the amplitude of the current oscillates in a sinusoidal fashion, so will the flux density in the structure. Thus, *the magnetic field developed in the stator changes both spatially and in time.*

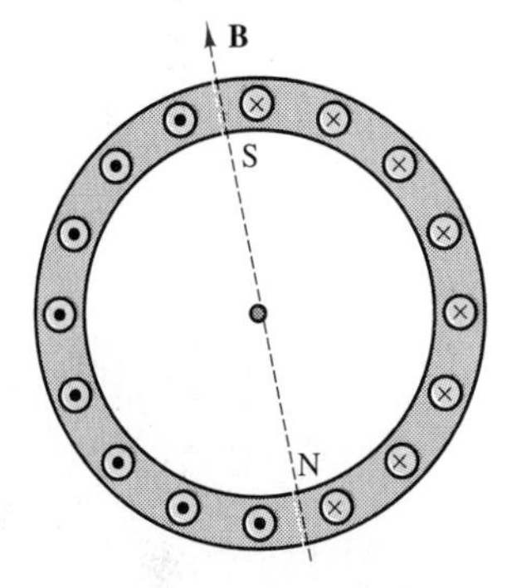

Figure 16.10 Magnetic field of stator

This property is typical of AC machines, where a *rotating magnetic field* is established by energizing the coil with an alternating current. As we shall see in the next section, the principles underlying the operation of DC and AC machines are quite different: in a direct-current machine, there is no rotating field, but a mechanical switching arrangement (the *commutator*) makes it possible for the rotor and stator magnetic fields to always align at right angles to each other.

DRILL EXERCISES

16.1 The percent speed regulation of a motor is 10 percent. If the full-load speed is 50π rad/s, find (a) the no-load speed in rad/s, and (b) the no-load speed in rev/min.

16.2 The percent voltage regulation for a 250-V generator is 10 percent. Find the no-load voltage of the generator.

16.3 The nameplate of a three-phase induction motor indicates the following values:

H.P. = 10	Volt = 220 V
R.P.M. = 1,750	Service factor = 1.15
Temperature rise = 60°C	Amp = 30 A

Find the rated torque, rated volt-amperes, and maximum continuous output power.

16.4 A motor having the characteristics shown in Figure 16.4 is to drive a load; the load has a linear torque-speed curve and requires 150 percent of rated torque at 1,500 rev/min. Find the operating point for this motor-load pair.

16.2 DIRECT-CURRENT MACHINES

As explained in the introductory section, direct-current (DC) machines are easier to analyze than their AC counterparts, although their actual construction is made rather complex by the need to have a commutator, which reverses the direction of currents and fluxes to produce a net torque. The objective of this section is to describe the major construction features and the operation of direct-current machines, as well as to develop simple circuit models that are useful in analyzing the performance of this class of machines.

Physical Structure of DC Machines

Figure 16.11 A DC machine

A representative DC machine was depicted in Figure 16.7, with the magnetic poles clearly identified, for both the stator and the rotor. Figure 16.11 is a photograph of the same type of machine. Note the salient pole construction of the stator and the slotted rotor. As previously stated, the torque developed by the machine is a consequence of the magnetic forces between stator and rotor poles. This torque is maximum when the angle γ between the rotor and stator poles is 90°. Also, as you can see from the figure, in a DC machine the armature is usually on the rotor, and the field winding is on the stator.

To keep this torque angle constant as the rotor spins on its shaft, a mechanical switch, called a **commutator,** is configured so the current distribution in the rotor winding remains constant and therefore the rotor poles are consistently at 90° with respect to the fixed stator poles. In a DC machine, the magnetizing current is DC, so that there is no spatial alternation of the stator poles due to time-varying currents. To understand the operation of the commutator, consider the simplified diagram of Figure 16.12. In the figure, the brushes are fixed, and the rotor revolves at an angular velocity ω_m; the instantaneous position of the rotor is given by the expression $\theta = \omega_m t - \gamma$.

The commutator is fixed to the rotor and is made up in this example of six segments that are made of electrically conducting material but are insulated from each other. Further, the rotor windings are configured so that they form six coils, connected to the commutator segments as shown in Figure 16.12.

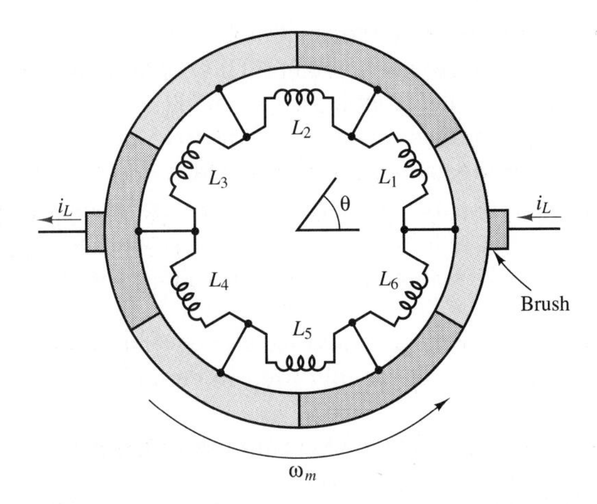

Figure 16.12 Rotor winding and commutator

As the commutator rotates counterclockwise, the rotor magnetic field rotates with it up to $\theta = 30°$. At that point, the direction of the current changes in coils L_3 and L_6 as the brushes make contact with the next segment. Now the direction of the magnetic field is $-30°$. As the commutator continues to rotate, the direction of the rotor field will again change from $-30°$ to $+30°$, and it will switch again when the brushes switch to the next pair of segments. In this machine, then, the torque angle, γ, is not always 90°, but can vary by as much as $\pm 30°$; the actual torque produced by the machine would fluctuate by as much as ± 14 percent, since the torque is proportional to $\sin\gamma$. As the number of segments increases, the torque fluctuation produced by the commutation is greatly reduced. In a practical machine, for example, one might have as many as 60 segments, and the variation of γ from 90° would be only $\pm 3°$, with a torque fluctuation of less than 1 percent. Thus, the DC machine can produce a nearly constant torque (as a motor) or voltage (as a generator).

Configuration of DC Machines

In DC machines, the field excitation that provides the magnetizing current is occasionally provided by an external source, in which case the machine is said to be **separately excited** (Figure 16.13(a)). More often, the field excitation is derived from the armature voltage and the machine is said to be **self-excited.** The latter configuration does not require the use of a separate source for the field excitation and is therefore frequently preferred. If a machine is in the separately excited configuration, an additional source, V_f, is required. In the self-excited case, one method used to provide the field excitation is to connect the field in parallel with the armature; since the field winding typically has significantly higher resistance than the armature circuit (remember that it is the armature that carries the load current), this will not draw excessive current from the armature. Further, a series

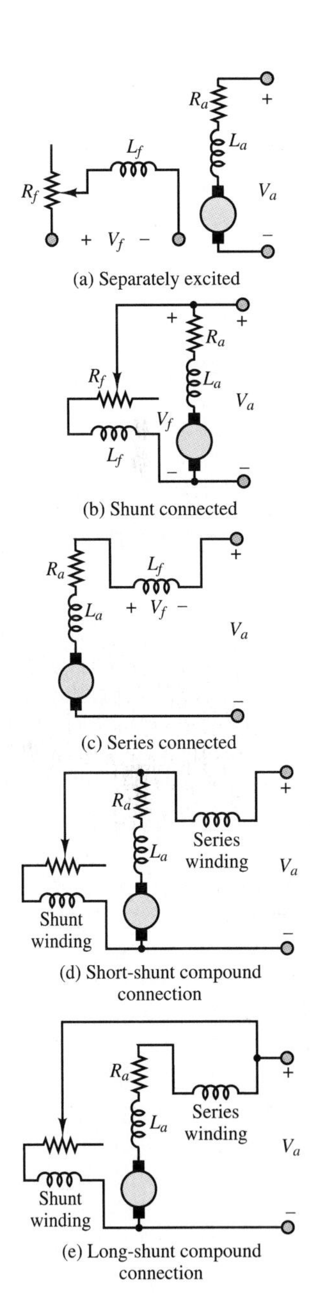

Figure 16.13

resistor can be added to the field circuit to provide the means for adjusting the field current independent of the armature voltage. This configuration is called a **shunt-connected** machine and is depicted in Figure 16.13(b). Another method for self-exciting a DC machine consists of connecting the field in series with the armature, leading to the **series-connected** machine, depicted in Figure 16.13(c); in this case, the field winding will support the entire armature current, and thus the field coil must have low resistance (and therefore relatively few turns). This configuration is rarely used for generators, since the generated voltage and the load voltage must always differ by the voltage drop across the field coil, which varies with the load current. Thus, a series generator would have poor (large) regulation. However, series-connected motors are commonly used in certain applications, as will be discussed in a later section.

The third type of DC machine is the **compound-connected** machine, which consists of a combination of the shunt and series configurations. Figures 16.13(d) and (e) show the two types of connections, called the **short shunt** and the **long shunt,** respectively. Each of these configurations may be connected so that the series part of the field adds to the shunt part (**cumulative compounding**) or so that it subtracts (**differential compounding**).

DC Machine Models

As stated earlier, it is relatively easy to develop a simple model of a DC machine, which is well suited to performance analysis, without the need to resort to the details of the construction of the machine itself. This section will illustrate the development of such models in two steps. First, steady-state models relating field and armature currents and voltages to speed and torque are introduced; second, the differential equations describing the dynamic behavior of DC machines are derived.

When a field excitation is established, a magnetic flux, ϕ, is generated by the field current, I_f. From equation 16.2, we know that the torque acting on the rotor is proportional to the product of the magnetic field and the current in the load-carrying wire; the latter current is the armature current, I_a (i_w, in equation 16.2). Assuming that, by virtue of the commutator, the torque angle, γ, is kept very close to 90°, and therefore $\sin\gamma = 1$, we obtain the following expression for the torque (in units of N-m) in a DC machine:

$$T = k_T \phi I_a \qquad \text{for } \gamma = 90° \tag{16.6}$$

You may recall that this is simply a consequence of the *Bli* law of Chapter 15. The mechanical power generated (or absorbed) is equal to the product of the machine torque and the mechanical speed of rotation, ω_m (in rad/s), and is therefore given by

$$P_m = \omega_m T = \omega_m k_T \phi I_a \tag{16.7}$$

Recall now that the rotation of the armature conductors in the field generated by the field excitation causes a **back emf,** E_b, in a direction that opposes the rotation of the armature. According to the *Blu* law (see Chapter 15), then, this back emf is given by the expression

$$E_b = k_a \phi \omega_m \tag{16.8}$$

where k_a is called the **armature constant** and is related to the geometry and magnetic properties of the structure. The voltage E_b represents a countervoltage (opposing the DC excitation) in the case of a motor, and the generated voltage in the case of a generator. Thus, the electric power dissipated (or generated) by the machine is given by the product of the back emf and the armature current:

$$P_e = E_b I_a \tag{16.9}$$

The constants k_T and k_a in equations 16.6 and 16.8 are related to geometry factors, such as the dimension of the rotor and the number of turns in the armature winding; and to properties of materials, such as the permeability of the magnetic materials. Note that in the ideal energy-conversion case, $P_m = P_e$, and therefore $k_a = k_T$. We shall in general assume such ideal conversion of electrical to mechanical energy (or vice versa) and will therefore treat the two constants as being identical: $k_a = k_T$. The constant k_a is given by

$$k_a = \frac{pN}{2\pi M} \tag{16.10}$$

where

p = number of magnetic poles

N = number of conductors per coil

M = number of parallel paths in armature winding

An important observation concerning the units of angular speed must be made at this point. The equality (under the no-loss assumption) between the constants k_a and k_T in equations 16.6 and 16.8 results from the choice of consistent units, namely, volts and amperes for the electrical quantities, and newton-meters and radians per second for the mechanical quantities. You should be aware that it is fairly common practice to refer to the speed of rotation of an electric machine in units of revolutions per minute (rev/min).[1] In this book, we shall uniformly use the symbol n to denote angular speed in rev/min; the following relationship should be committed to memory:

$$n \text{ (rev/min)} = \frac{60}{2\pi} \omega_m \text{ (rad/s)} \tag{16.11}$$

If the speed is expressed in rev/min, the armature constant changes as follows:

$$E_b = k'_a \phi n \tag{16.12}$$

[1]Note that the abbreviation RPM, although certainly familiar to the reader, is not a standard unit, and its use should be discouraged.

where

$$k'_a = \frac{pN}{60M} \tag{16.13}$$

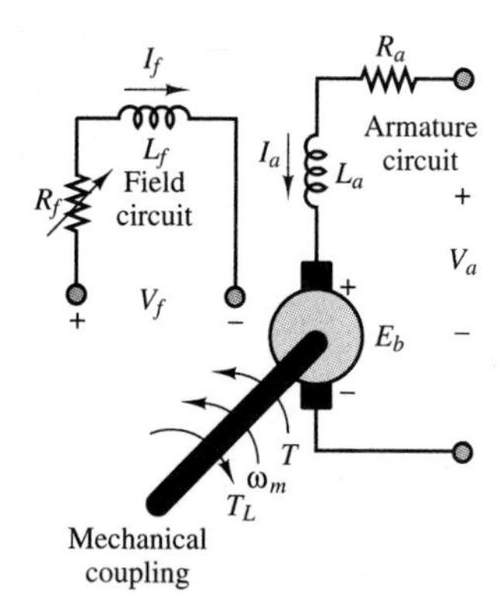

(a) Motor reference direction

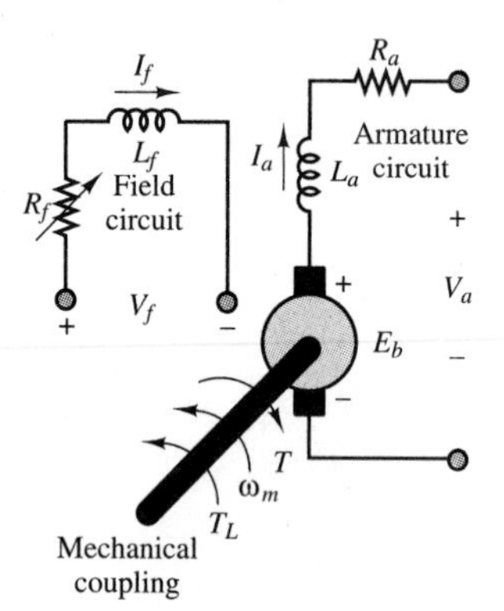

(b) Generator reference direction

Figure 16.14 Electrical circuit model of a separately excited DC machine

Having introduced the basic equations relating torque, speed, voltages, and currents in electric machines, we may now consider the interaction of these quantities in a DC machine at steady state, that is, operating at constant speed and field excitation. Figure 16.14 depicts the electrical circuit model of a separately excited DC machine, illustrating both motor and generator action. It is very important to note the reference direction of armature current flow, and of the developed torque, in order to make a distinction between the two modes of operation. The field excitation is shown as a voltage, V_f, generating the field current, I_f, that flows through a variable resistor, R_f, and through the field coil, L_f. The variable resistor permits adjustment of the field excitation. The armature circuit, on the other hand, consists of a voltage source representing the back emf, E_b, the armature resistance, R_a, and the armature voltage, V_a. This model is appropriate both for motor and for generator action. When $V_a < E_b$, the machine acts as a generator (I_a flows out of the machine). When $V_a > E_b$, the machine acts as a motor (I_a flows into the machine). Thus, according to the circuit model of Figure 16.14, the operation of a DC machine at steady state (i.e., with the inductors in the circuit replaced by short circuits) is described by the following equations:

$$\begin{aligned} I_f &= \frac{V_f}{R_f} \quad \text{and} \quad V_a = R_a I_a + E_b \quad \text{(motor action)} \\ I_f &= \frac{V_f}{R_f} \quad \text{and} \quad V_a = -R_a I_a + E_b \quad \text{(generator action)} \end{aligned} \tag{16.14}$$

Equation pair 16.14 together with equations 16.6 and 16.8 may be used to determine the steady-state operating condition of a DC machine.

The circuit model of Figure 16.14 permits the derivation of a simple set of differential equations that describe the *dynamic* analysis of a DC machine. The dynamic equations describing the behavior of a separately excited DC machine are as follows:

$$V_a(t) = I_a(t)R_a + L_a\frac{dI_a(t)}{dt} + E_b(t) \quad \text{(armature circuit)} \tag{16.15a}$$

$$V_f(t) = I_f(t)R_f + L_f\frac{dI_f(t)}{dt} \quad \text{(field circuit)} \tag{16.15b}$$

These equations can be related to the operation of the machine in the presence of a load. If we assume that the motor is rigidly connected to an inertial load with moment of inertia J and that the friction losses in the load are represented by a viscous friction coefficient, b, then the torque developed by the machine (in the motor mode of operation) can be written as follows:

$$T(t) = T_L + b\omega_m(t) + J\frac{d\omega_m(t)}{dt} \tag{16.16}$$

where T_L is the load torque. T_L is typically either constant or some function of speed, ω_m, in a motor. In the case of a generator, the load torque is replaced by the torque supplied by a prime mover, and the machine torque, $T(t)$, opposes the motion of the prime mover, as shown in Figure 16.14. Since the machine torque is related to the armature and field currents by equation 16.6, equations 16.16 and 16.17 are coupled to each other; this coupling may be expressed as follows:

$$T(t) = k_a \phi I_a(t) \tag{16.17}$$

or

$$k_a \phi I_a(t) = T_L + b\omega_m(t) + J\frac{d\omega_m(t)}{dt} \tag{16.18}$$

The dynamic equations described in this section apply to any DC machine. In the case of a *separately excited* machine, a further simplification is possible, since the flux is established by virtue of a separate field excitation, and therefore

$$\phi = \frac{N_f}{\mathcal{R}} I_f = k_f I_f \tag{16.19}$$

where N_f is the number of turns in the field coil, $\mathcal{R}$ is the reluctance of the structure, and I_f is the field current.

Magnetization Curve (Open-Circuit Characteristic)

To analyze the performance of a DC machine, it would be useful to obtain an open-circuit characteristic capable of predicting the voltage generated when the machine is driven at a constant speed ω_m by a prime mover. The common arrangement is to drive the machine at rated speed by means of a prime mover (or an electric motor). Then, with no load connected to the armature terminals, the armature voltage is recorded as the field current is increased from zero to some value sufficient to produce an armature voltage greater than the rated voltage. Since the load terminals are open-circuited, $I_a = 0$ and $E_b = V_a$; and since $k_a \phi = E_b/\omega_m$, the magnetization curve makes it possible to determine the value of $k_a \phi$ corresponding to a given field current, I_f, for the rated speed.

Figure 16.15 depicts a typical magnetization curve. Note that the armature voltage is nonzero even when no field current is present. This phenomenon is due to the *residual magnetization* of the iron core. The dashed lines in Figure 16.15 are called **field resistance curves** and are a plot of the voltage that appears across the field winding plus rheostat (variable resistor; see Figure 16.14) versus the field current, for various values of field winding plus rheostat resistance. Thus, the slope of the line is equal to the total field circuit resistance, R_f.

Analysis of Direct-Current Generators

The operation of a DC generator may be readily understood with reference to the magnetization curve of Figure 16.15. As soon as the armature is connected across the shunt circuit consisting of the field winding and the rheostat, a current will

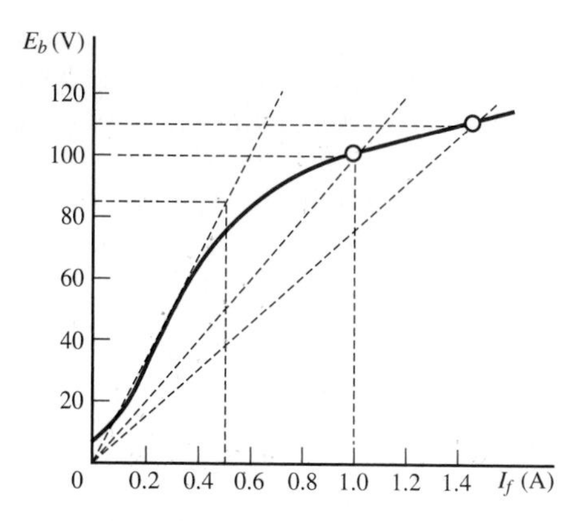

Figure 16.15 DC machine magnetization curve

flow through the winding, and this will in turn act to increase the emf across the armature. This **buildup** process continues until the two curves meet, that is, until the current flowing through the field winding is exactly that required to induce the emf. By changing the rheostat setting, the operating point at the intersection of the two curves can be displaced, as shown in Figure 16.15, and the generator can therefore be made to supply different voltages. The following examples illustrate the operation of the separately excited DC generator.

Example 16.4 Separately Excited DC Generator

A separately excited DC generator is rated at 100 V, 100 A, and 1,000 rev/min. The magnetization curve is shown in Figure 16.15. Some parameters of interest for the generator are as follows:

$$R_a = 0.14\ \Omega \qquad V_f = 100\ \text{V} \qquad R_f = 100\ \Omega$$

1. If the prime mover is driving the generator at 800 rev/min, what is the no-load terminal voltage, V_a?
2. If a 1-Ω load is connected to the generator, what is the generated voltage?

Solution:

1. The field current in the machine is

$$I_f = \frac{V_f}{R_f} = \frac{100}{100} = 1\ \text{A}$$

From the magnetization curve, it can be seen that this field current will produce 100 V at a speed of 1,000 rev/min. Since this generator is actually running at 800 rev/min, the induced emf may be found by assuming a linear relationship between speed and emf. This approximation is reasonable, provided that the departure from the nominal operating condition is small. Let n_0 and E_{b0} be the nominal speed and emf, respectively (i.e., 1,000 rev/min and 100 V); then,

$$\frac{E_b}{E_{b0}} = \frac{n}{n_0}$$

and therefore

$$E_b = \frac{n}{n_0} E_{b0} = \frac{800}{1{,}000} \times 100 = 80 \text{ V}$$

The open-circuit (output) terminal voltage of the generator is equal to the emf from the circuit model of Figure 16.14; therefore:

$$V_a = E_b = 80 \text{ V}$$

2. When a load resistance is connected to the circuit (the practical situation), the terminal (or load) voltage is no longer equal to E_b, since there will be a voltage drop across the armature winding resistance. The armature (or load) current may be determined from the expression

$$I_a = I_L = \frac{E_b}{R_a + R_L} = \frac{80}{0.14 + 1} = 70.2 \text{ A}$$

where $R_L = 1\ \Omega$ is the load resistance. The terminal (load) voltage is therefore given by

$$V_L = I_L R_L = 70.2 \times 1 = 70.2 \text{ V}$$

Example 16.5

A 1,000-kW, 2,000-V, 3,600-rev/min separately excited DC generator has an armature circuit resistance of 0.1 Ω. The flux per pole (ϕ) is 0.5 Wb. Find

1. The induced voltage.
2. The machine constant.
3. The torque developed at the rated conditions.

Solution:

1. The armature current may be found by observing that the rated power is equal to the product of the terminal (load) voltage and the armature (load) current; thus,

$$I_a = \frac{P_{\text{rated}}}{V_L} = \frac{1{,}000 \times 10^3}{2{,}000} = 500 \text{ A}$$

The generated voltage is equal to the sum of the terminal voltage and the voltage drop across the armature resistance (see Figure 16.14):

$$E_b = V_a + I_a R_a = 2{,}000 + 500 \times 0.1 = 2{,}050 \text{ V}$$

2. The speed of rotation of the machine in units of rad/s is

$$\omega_m = \frac{2\pi n}{60} = \frac{2\pi \times 3{,}600}{60} = 377 \text{ rad/s}$$

Thus, the machine constant is found to be

$$k_a = \frac{E_b}{\phi \omega_m} = \frac{2{,}050}{0.5 \times 377} = 10.876 \frac{\text{V-s}}{\text{Wb-rad}}$$

3. The torque developed is found from equation 16.6:

$$T = k_a \phi I_a = 10.875 \times 0.5 \times 500 = 2{,}718.9 \text{ N-m}$$

Since the compound-connected generator contains both a shunt and a series field winding, it is the most general configuration, and the most useful for developing a circuit model that is as general as possible. In the following discussion, we shall consider the so-called short-shunt, compound-connected generator, in which the flux produced by the series winding adds to that of the shunt winding. Figure 16.16 depicts the equivalent circuit for the compound generator; circuit models for the shunt generator and for the rarely used series generator can be obtained by removing the shunt or series field winding element, respectively. In the circuit of Figure 16.16, the generator armature has been replaced by a voltage source corresponding to the induced emf and a series resistance, R_a, corresponding to the resistance of the armature windings. The equations describing the DC generator at steady state (i.e., with the inductors acting as short circuits) are:

$$E_b = k_a \phi \omega_m \text{ V} \tag{16.20}$$

$$T = \frac{P}{\omega_m} = \frac{E_b I_a}{\omega_m} = k_a \phi I_a \text{ N-m} \tag{16.21}$$

$$V_L = E_b - I_a R_a - I_S R_S \tag{16.22}$$

$$I_a = I_S + I_f \tag{16.23}$$

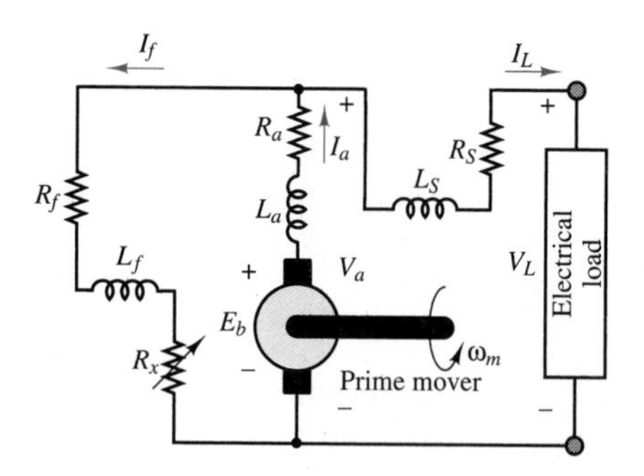

Figure 16.16 Compound generator circuit model

Note that in the circuit of Figure 16.16, the load and armature voltages are not equal, in general, because of the presence of a series field winding, represented by the resistor R_S and by the inductor L_S where the subscript "S" stands for "series." The expression for the armature emf is dependent on the air-gap flux, ϕ, to which the series and shunt windings in the compound generator both contribute, according to the expression

$$\phi = \phi_{\text{sh}} \pm \phi_S = \phi_{\text{sh}} \pm k_S I_a \tag{16.24}$$

Example 16.6

In many practical cases, it is not actually necessary to know the armature constant and the flux separately, but it is sufficient to know the value of the product $k_a\phi$. For example, suppose that the armature resistance of a DC machine is known and that, given a known field excitation, the armature current, load voltage, and speed of the machine can be measured. Then, the product $k_a\phi$ may be determined from equation 16.20, as follows:

$$k_a\phi = \frac{E_b}{\omega_m} = \frac{V_L + I_a(R_a + R_S)}{\omega_m}$$

where V_L, I_a, and ω_m are measured quantities for given operating conditions.

DRILL EXERCISES

16.5 A 24-coil, 2-pole DC generator has 16 turns per coil in its armature winding. The field excitation is 0.05 Wb per pole, and the armature angular velocity is 180 rad/s. Find the machine constant and the total induced voltage.

16.6 A 1,000-kW, 1,000-V, 2,400 rev/min separately excited DC generator has an armature circuit resistance of 0.04 Ω. The flux per pole is 0.4 Wb. Find: (a) the induced voltage; (b) the machine constant; and (c) the torque developed at the rated conditions.

16.7 A 100-kW, 250-V shunt generator has a field circuit resistance of 50 Ω and an armature circuit resistance of 0.05 Ω. Find: (a) the full-load line current flowing to the load; (b) the field current; (c) the armature current; and (d) the full-load generator voltage.

16.3 DIRECT-CURRENT MOTORS

DC motors are widely used in applications requiring accurate speed control—for example, in servo systems. Having developed a circuit model and analysis methods for the DC generator, we can rather straightforwardly extend these results to DC motors, since these are in effect DC generators with the roles of input and output reversed. Once again, we shall analyze the motor by means of both its magnetization curve and a circuit model. It will be useful to begin our discussion by referring to the schematic diagram of a cumulatively compounded motor, as shown in Figure 16.17. The choice of the compound-connected motor is the most convenient, since its model can be used to represent either a series or a shunt motor with minor modifications.

The equations that govern the behavior of the DC motor follow and are similar to those used for the generator. Note that the only differences between these equations and those that describe the DC generator appear in the last two equations in the group, where the source voltage is equal to the *sum* of the emf and the voltage drop across the series field resistance and armature resistance, and where the source current now equals the *sum* of the field shunt and armature series currents.

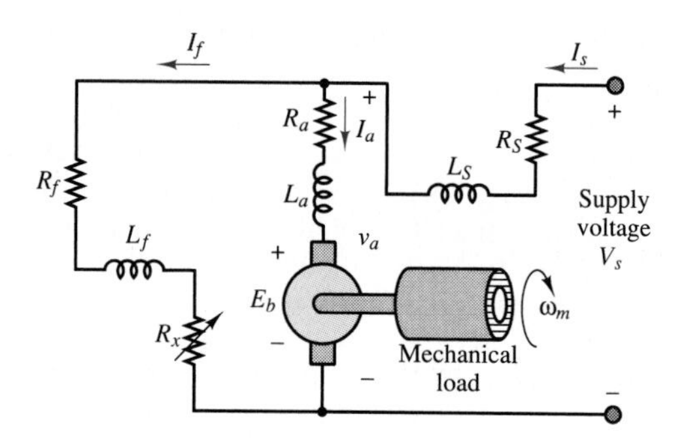

Figure 16.17 Equivalent circuit of a cumulatively compounded motor

$$E_b = k_a \phi \omega_m \tag{16.25}$$

$$T = k_a \phi I_a \tag{16.26}$$

$$V_s = E_b + I_a R_a + I_s R_S \tag{16.27}$$

$$I_s = I_f + I_a \tag{16.28}$$

Note that in these equations we have replaced the symbols V_L and I_L, used in the generator circuit model to represent the generator load current and voltage, with the symbols V_s and I_s, indicating the presence of an external source.

Speed-Torque and Dynamic Characteristics of DC Motors

The Shunt Motor

In a shunt motor (similar to the configuration of Figure 16.17, but with the series field short-circuited), the armature current is found by dividing the net voltage across the armature circuit (source voltage minus back emf) by the armature resistance:

$$I_a = \frac{V_s - k_a \phi \omega_m}{R_a} \tag{16.29}$$

An expression for the armature current may also be obtained from equation 16.26, as follows:

$$I_a = \frac{T}{k_a \phi} \tag{16.30}$$

It is then possible to relate the torque requirements to the speed of the motor by substituting equation 16.29 in equation 16.30:

$$\frac{T}{k_a \phi} = \frac{V_s - k_a \phi \omega_m}{R_a} \tag{16.31}$$

Equation 16.31 describes the steady-state torque-speed characteristic of the shunt motor. To understand this performance equation, we observe that if V_s, k_a, ϕ, and R_a are fixed in equation 16.31 (the flux is essentially constant in the shunt motor for a fixed V_s), then the speed of the motor is directly related to the armature current. Now consider the case where the load applied to the motor is suddenly increased, causing the speed of the motor to drop. As the speed decreases, the armature current increases, according to equation 16.29. The excess armature current causes the motor to develop additional torque, according to equation 16.30, until a new equilibrium is reached between the higher armature current and developed torque and the lower speed of rotation. The equilibrium point is dictated by the balance of mechanical and electrical power, in accordance with the relation

$$E_b I_a = T\omega_m \tag{16.32}$$

Thus, the shunt DC motor will adjust to variations in load by changing its speed to preserve this power balance. The torque-speed curves for the shunt motor may be obtained by rewriting the equation relating the speed to the armature current:

$$\omega_m = \frac{V_s - I_a R_a}{k_a \phi} = \frac{V_s}{k_a \phi} - \frac{R_a T}{(k_a \phi)^2} \tag{16.33}$$

To interpret equation 16.33, one can start by considering the motor operating at rated speed and torque. As the load torque is reduced, the armature current will also decrease, causing the speed to increase in accordance with equation 16.33. The increase in speed depends on the extent of the voltage drop across the armature resistance, $I_a R_a$. The change in speed will be on the same order of magnitude as this drop; it typically takes values around 10 percent. This corresponds to a relatively good speed regulation, which is an attractive feature of the shunt DC motor (recall the discussion of regulation in Section 16.1). Normalized torque and speed vs. power curves for the shunt motor are shown in Figure 16.18. Note that, over a reasonably broad range of powers, up to rated value, the curve is relatively flat, indicating that the DC shunt motor acts as a reasonably constant-speed motor.

The dynamic behavior of the shunt motor is described by equations 16.15 through 16.18, with the additional relation

$$I_a(t) = I_s(t) - I_f(t) \tag{16.34}$$

Compound Motors

It is interesting to compare the performance of the shunt motor with that of the compound-connected motor; the comparison is easily made if we recall that a series field resistance appears in series with the armature resistance and that the flux is due to the contributions of both series and shunt fields. Thus, the speed equation becomes

$$\omega_m = \frac{V_s - I_a(R_a + R_S)}{k_a(\phi_{sh} \pm \phi_S)} \tag{16.35}$$

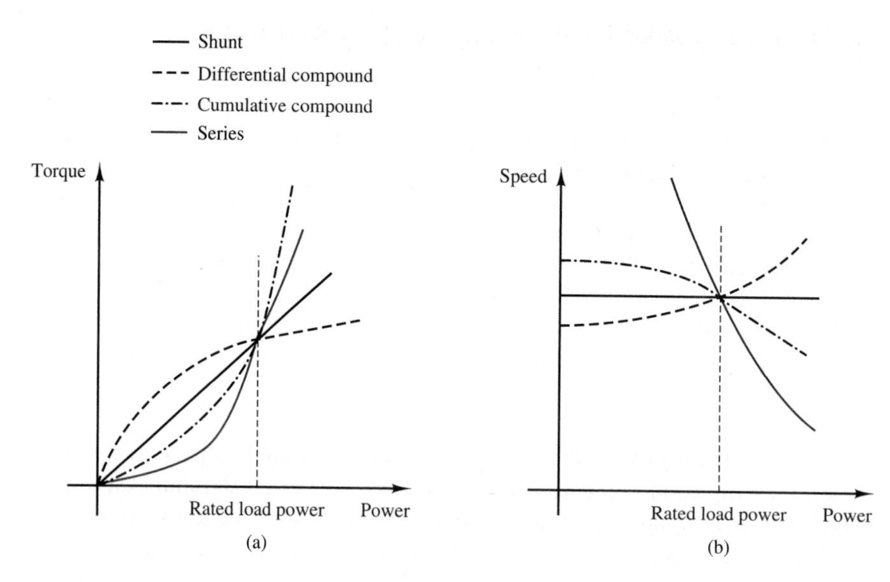

Figure 16.18 DC motor operating characteristics

where

+ in the denominator is for a cumulatively compounded motor.

− in the denominator is for a differentially compounded motor.

ϕ_{sh} is the flux set up by the shunt field winding, assuming that it is constant.

ϕ_S is the flux set up by the series field winding, $\phi_S = k_S I_a$.

For the cumulatively compound motor, two effects are apparent: the flux is increased by the presence of a series component, ϕ_S; and the voltage drop due to I_a in the numerator term is increased by an amount proportional to the resistance of the series field winding, R_S. As a consequence, when the load to the motor is reduced, the numerator increases more dramatically than in the case of the shunt motor, because of the corresponding decrease in armature current, while at the same time the series flux decreases. Each of these effects causes the speed to increase; therefore, it stands to reason that the speed regulation of the compound-connected motor is poorer than that of the shunt motor. Normalized torque and speed vs. power curves for the compound motor (both differential and cumulative connections) are shown in Figure 16.18.

The differential equation describing the behavior of a compound motor differs from that for the shunt motor in having additional terms due to the series field component:

$$\begin{aligned} V_s &= E_b(t) + I_a(t)R_a + L_a\frac{dI_a(t)}{dt} + I_s(t)R_S + L_S\frac{dI_s(t)}{dt} \\ &= V_a(t) + I_s(t)R_S + L_S\frac{dI_s(t)}{dt} \end{aligned} \qquad \textbf{(16.36)}$$

The differential equation for the field circuit can be written as

$$V_a = I_f(t)(R_f + R_x) + L_f \frac{dI_f(t)}{dt} \tag{16.37}$$

We also have the following basic relations:

$$I_a(t) = I_s(t) - I_f(t) \tag{16.38}$$

and

$$E_b(t) = k_a I_a(t)\omega_m(t) \qquad \text{and} \qquad T(t) = k_a \phi I_a(t) \tag{16.39}$$

Series Motors

The series motor (see Figure 16.13(c)) behaves somewhat differently from the shunt and compound motors because the flux is established solely by virtue of the series current flowing through the armature. It is relatively simple to derive an expression for the emf and torque equations for the series motor if we approximate the relationship between flux and armature current by assuming that the motor operates in the linear region of its magnetization curve. Then we can write

$$\phi = k_S I_a \tag{16.40}$$

and the emf and torque equations become

$$E_b = k_a \omega_m \phi = k_a \omega_m k_S I_a \tag{16.41}$$

$$T = k_a \phi I_a = k_a k_S I_a^2 \tag{16.42}$$

The circuit equation for the series motor becomes

$$V_s = E_b + I_a(R_a + R_S) = (k_a \omega_m k_S + R_T) I_a \tag{16.43}$$

where R_a is the armature resistance, R_S is the series field winding resistance, and R_T is the total series resistance. From equation 16.43, we can solve for I_a and substitute in the torque expression (equation 16.42) to obtain the following torque-speed relationship:

$$T = k_a k_S \frac{V_s^2}{(k_a \omega_m k_S + R_T)^2} \tag{16.44}$$

which indicates the inverse squared relationship between torque and speed in the series motor. This expression describes a behavior that can, under certain conditions, become unstable. Since the speed increases when the load torque is reduced, one can readily see that if one were to disconnect the load altogether, the speed would tend to increase to dangerous values. To prevent excessive speeds, series motors are always mechanically coupled to the load. This feature is not necessarily a drawback, though, because series motors can develop very high torque at low speeds, and therefore can serve very well for traction-type loads (e.g., conveyor belts or vehicle propulsion systems). Torque and speed vs. power curves for the series motor are also shown in Figure 16.18.

The differential equation for the armature circuit of the motor can be given as

$$V_s = I_a(t)(R_a + R_S) + L_a\frac{dI_a(t)}{dt} + L_S\frac{dI_a(t)}{dt} + E_b$$
$$= I_a(t)(R_a + R_S) + L_a\frac{dI_a(t)}{dt} + L_S\frac{dI_a(t)}{dt} + k_a k_S I_a \omega_m \quad (16.45)$$

Permanent-Magnet DC Motors

Permanent-magnet (PM) DC motors have become increasingly common in applications requiring relatively low torques and efficient use of space. The construction of PM direct-current motors differs from that of the motors considered thus far in that the magnetic field of the stator is produced by suitably located poles made of magnetic materials. Thus, the basic principle of operation, including the idea of commutation, is unchanged with respect to the wound-stator DC motor. What changes is that there is no need to provide a field excitation, whether separately or by means of the self-excitation techniques discussed in the preceding sections. Therefore, the PM motor is intrinsically simpler than its wound-stator counterpart.

The equations that describe the operation of the PM motor follow. The torque produced is related to the armature current by a torque constant, k_{PM}, which is determined by the geometry of the motor:

$$T = k_{TPM} I_a \quad (16.46)$$

As in the conventional DC motor, the rotation of the rotor produces the usual counter or back emf, E_b, which is linearly related to speed by a voltage constant, k_{aPM}:

$$E_b = k_{aPM}\omega_m \quad (16.47)$$

The equivalent circuit of the PM motor is particularly simple, since we need not model the effects of a field winding. Figure 16.19 shows the circuit model and the torque-speed curve of a PM motor.

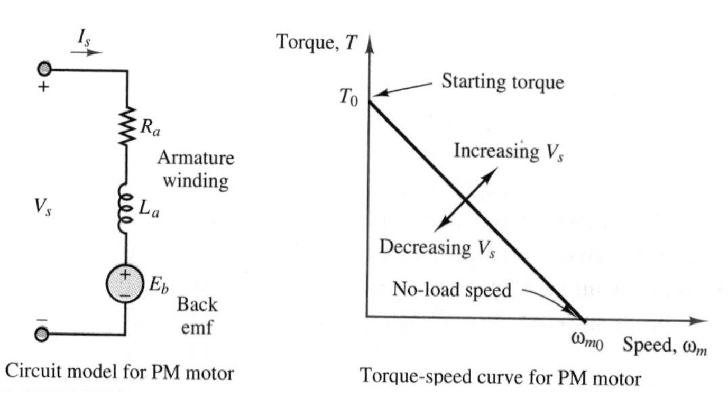

Figure 16.19 Circuit model and torque-speed curve of PM motor

We can use the circuit model of Figure 16.19 to predict the torque-speed curve shown in the same figure, as follows. From the circuit model, for a constant speed (and therefore constant current), we may consider the inductor a short circuit and write the equation

$$\begin{aligned} V_s &= I_a R_a + E_b \\ &= I_a R_a + k_{aPM}\omega_m \\ &= \frac{T}{k_{TPM}} R_a + k_{aPM}\omega_m \end{aligned} \qquad (16.48)$$

thus obtaining the equations relating speed and torque:

$$\omega_m = \frac{V_s}{k_{aPM}} - \frac{T R_a}{k_{aPM} k_{TPM}} \qquad (16.49)$$

and

$$T = \frac{V_s}{R_a} k_{TPM} - \frac{\omega_m}{R_a} k_{aPM} k_{TPM} \qquad (16.50)$$

From these equations, one can extract the stall torque, T_0, that is, the zero-speed torque:

$$T_0 = \frac{V_s}{R_a} k_{TPM} \qquad (16.51)$$

and the no-load speed, ω_{m0}:

$$\omega_{m0} = \frac{V_s}{k_{aPM}} \qquad (16.52)$$

Under dynamic conditions, assuming an inertia plus viscous friction load, the torque produced by the motor can be expressed as

$$T = k_{TPM} I_a(t) = T_{\text{load}}(t) + d\omega_m(t) + J\frac{d\omega_m(t)}{dt} \qquad (16.53)$$

The differential equation for the armature circuit of the motor can be given as

$$\begin{aligned} V_s &= I_a(t) R_a + L_a \frac{dI_a(t)}{dt} + E_b \\ &= I_a(t) R_a + L_a \frac{dI_a(t)}{dt} + k_{aPM}\omega_m(t) \end{aligned} \qquad (16.54)$$

The fact that the air-gap flux is constant in a permanent-magnet DC motor makes its characteristics somewhat different from those of the wound DC motor. A direct comparison of PM and wound-field DC motors reveals the following advantages and disadvantages of each configuration:

1. PM motors are smaller and lighter than wound motors for a given power rating. Further, their efficiency is greater because there are no field winding losses.

2. An additional advantage of PM motors is their essentially linear speed-torque characteristic, which makes analysis (and control) much easier. Reversal of rotation is also accomplished easily, by reversing the polarity of the source.
3. A major disadvantage of PM motors is that they can become demagnetized by exposure to excessive magnetic fields, application of excessive voltage, or operation at excessively high or low temperatures.
4. A less obvious drawback of PM motors is that their performance is subject to greater variability from motor to motor than is the case for wound motors, because of variations in the magnetic materials.

In summary, four basic types of DC motors are commonly used. Their principal operating characteristics are summarized as follows, and their general torque and speed versus power characteristics are depicted in Figure 16.18, assuming motors with identical voltage, power, and speed ratings.

Shunt wound motor: Field connected in parallel with the armature. With constant armature voltage and field excitation, the motor has good speed regulation (flat speed-torque characteristic).

Compound wound motor: Field winding has both series and shunt components. This motor offers better starting torque than the shunt motor, but worse speed regulation.

Series wound motor: The field winding is in series with the armature. The motor has very high starting torque and poor speed regulation. It is useful for low-speed, high-torque applications.

Permanent-magnet motor: Field windings are replaced by permanent magnets. The motor has adequate starting torque, with speed regulation somewhat worse than that of the compound wound motor.

Example 16.7 DC Shunt Motor

A four-pole, 3-hp, 240-VDC shunt motor has 1,000 conductors in the armature (i.e., $N = 1{,}000$, $M = 4$, as defined in equation 16.10). The load current is 30 A. The armature resistance is 0.6 Ω and the flux is 20 mWb. The field winding current is 1.4 A. Find the speed and torque.

Solution:

The output power at full load in watts may be computed using the conversion factor 1 hp = 746 W:

$$P_{\text{out}} = 3 \times 746 = 2{,}238 \text{ W}$$

The armature current may then be computed as the difference between the source current and the field current:

$$I_a = I_s - I_f = 30 - 1.4 = 28.6 \text{ A}$$

Thus, the no-load voltage is

$$E_b = V_s - I_a R_a = 240 - 28.6 \times 0.6 = 222.84 \text{ V}$$

and the armature constant is found from equation 16.10 to be

$$k_a = \frac{pN}{2\pi M} = \frac{4 \times 1{,}000}{2\pi \times 4} = \frac{500}{\pi} = 159.15$$

The speed may then be computed as follows:

$$\omega_m = \frac{E_a}{k_a \phi} = \frac{222.84}{159.15 \times 20 \times 10^{-3}} = 70 \text{ rad/s}$$

The torque developed by the motor is found as the ratio of output power to speed of rotation:

$$T = \frac{P_{\text{out}}}{\omega_m} = \frac{2{,}238}{70} = 31.97 \text{ N-m}$$

Example 16.8 DC Shunt Motor

A DC machine is used as a shunt motor. The magnetization curve for the machine used is shown in Figure 16.20. The size of the motor and its characteristics are typical of the small motors used in hand-held tools, such as power ratchets and screwdrivers. The motor armature is known to draw 8 A at full load and to have a speed of 120 rev/min. The armature winding resistance is estimated to be 0.2 Ω, and the battery voltage is 7.2 V. The field winding has 200 turns. The circuit configuration is shown in Figure 16.21.

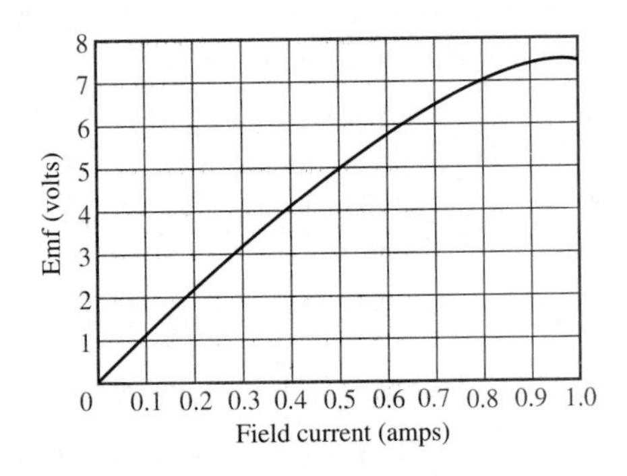

Figure 16.20 Magnetization curve for a small DC motor

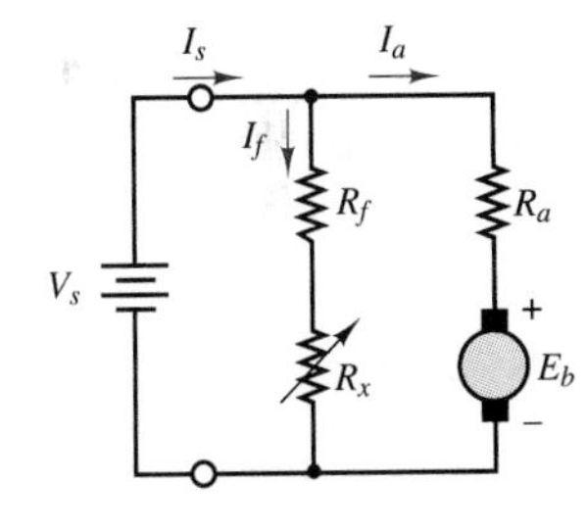

Figure 16.21 Shunt motor configuration

1. Estimate the current, I_f, required for full-load operation.
2. Determine the no-load speed.
3. Plot the speed-torque curve in the range from no-load torque to rated torque.
4. What is the power output at rated load?

Solution:

1. To find the field current, we must find the generated emf since R_f is not known. Writing KVL around the armature circuit, we obtain

$$V_s = E_b + I_a R_a$$
$$E_b = V_s - I_a R_a = 7.2 - 8(0.2) = 5.6 \text{ V}$$

Having found the back emf, we can find the field current from the magnetization curve. At $E_b = 5.6$ V, we find that the field current and field resistance are

$$I_f = 0.6 \text{ A} \quad \text{and} \quad R_f = \frac{7.2}{0.6} = 12 \ \Omega$$

2. To obtain the no-load speed, we use the equations

$$E_b = k_a \phi \frac{2\pi n}{60} \qquad T = k_a \phi I_a$$

leading to

$$V_s = I_a R_a + E_b = I_a R_a + k_a \phi \frac{2\pi}{60} n$$

or

$$n = \frac{V_s - I_a R_a}{k_a \phi (2\pi/60)}$$

At no load, and assuming no mechanical losses, the torque is zero, and we see that the current I_a must also be zero in the torque equation ($T = k_a \phi I_a$). Thus, the motor speed at no load is given by

$$n_{\text{no-load}} = \frac{V_s}{k_a \phi (2\pi/60)}$$

We can obtain an expression for $k_a \phi$ knowing that, at full load,

$$E_b = 5.6 \text{ V} = k_a \phi \frac{2\pi n}{60}$$

so that, for constant field excitation,

$$k_a \phi = E_b \left(\frac{60}{2\pi n} \right) = 5.6 \left(\frac{60}{2\pi(120)} \right) = 0.44563 \frac{\text{V} \cdot \text{s}}{\text{rad}}$$

Finally, we may solve for the no-load speed, in rev/min:

$$n_{\text{no-load}} = \frac{V_s}{k_a \phi (2\pi/60)} = \frac{7.2}{(0.44563)(2\pi/60)}$$
$$= 154.3 \text{ rev/min}$$

3. The torque at rated speed may be found as follows:

$$T_{\text{full-load}} = k_a \phi I_a = (0.44563)(8) = 3.565 \text{ N-m}$$

Now we have the two points necessary to construct the torque-speed curve for this motor, which is shown in Figure 16.22.

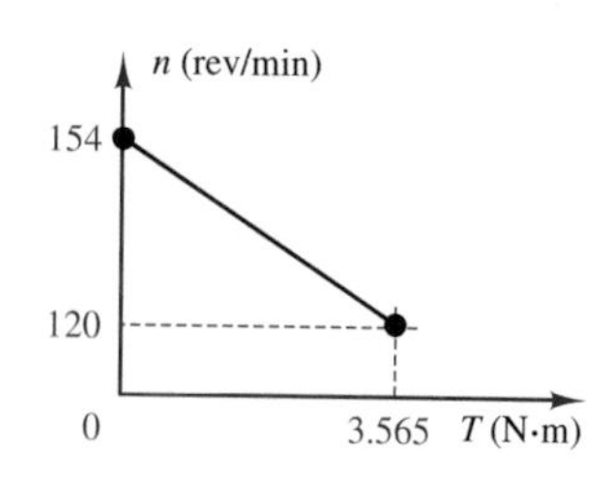

Figure 16.22

4. The power is related to the torque by the frequency of the shaft:

$$P = T\omega_m = (3.565)\left(\frac{120}{60} \right) 2\pi = 44.8 \text{ W}$$

or, equivalently,

$$P = 44.8 \text{ W} \times \frac{1 \text{ hp}}{746 \text{ W}} = 0.06 \text{ hp}$$

Example 16.9 DC Series Motor

A 10-hp, 115-VDC series motor draws 40 A when operating at a full-load speed of 1,800 rev/min. If the supply current changes to 60 A, find the torque developed by the motor when it operates within its linear region.

Solution:

Within the linear region of operation, the flux per pole is directly proportional to the current in the field winding. That is,

$$\phi = k_S I_a$$

The full-load speed is

$$n = 1{,}800 \text{ rev/min}$$

or

$$\omega_m = \frac{2\pi n}{60} = 60\pi \text{ rad/s}$$

Rated output power is

$$P_{\text{out}} = 10 \times 746 = 7{,}460 \text{ W}$$

and full-load torque is

$$T_{40\text{A}} = \frac{7{,}460}{60\pi} = 39.58 \text{ N-m}$$

Thus, the machine constant may be computed from the torque equation for the series motor:

$$T = k_a k_S I_a^2 = K I_a^2$$

Thus, at full load,

$$K = k_a k_S = \frac{39.58}{40^2} = 0.0247$$

and we can compute the torque developed for a 60-A supply current to be

$$T_{60\text{A}} = 0.0247 \times 60^2 = 88.92 \text{ N-m}$$

Example 16.10 DC Motor Dynamic Response

The aim of this example is to illustrate the computation of a permanent-magnet DC motor dynamic response. Assume, for simplicity, that the mechanical friction parameter, b, in equation 16.17 is negligible and that the two torque constants for the PM motor are equal, that is, $k_{TPM} = k_{aPM} = k_{PM}$. Find the dynamic equations of a DC motor.

Solution:

To determine the dynamic response of the motor, two balance equations are needed; the first describes the mechanical behavior of the motor:

$$T = T_{\text{load}} + J\frac{d\omega_m(t)}{dt} \tag{16.55}$$

while the second is related to the electrical performance:

$$V_L = E_b + R_a I_a(t) + L_a\frac{dI_a(t)}{dt} \tag{16.56}$$

These two equations are related to each other by the characteristic equations for the permanent-magnet DC motor:

$$E_b = k_{PM}\omega_m(t) \tag{16.57}$$

and

$$T = k_{PM} I_a(t) \tag{16.58}$$

When these two relations are substituted in the mechanical and electrical balance equations, the following differential equations are obtained:

$$k_{PM} I_a(t) = T_{\text{load}} + J\frac{d\omega_m(t)}{dt} \tag{16.59}$$

and

$$V_L = k_{PM}\omega_m(t) + R_a I_a(t) + L_a\frac{dI_a(t)}{dt} \tag{16.60}$$

Solving for $\omega_m(t)$ in equation 16.59 and substituting in equation 16.60, we obtain a second-order differential equation in the armature current:

$$k_{PM} I_a(t) = T_{\text{load}} + \frac{J}{k_{PM}}\left[\frac{dV_L}{dt} - R_a\frac{dI_a(t)}{dt} - L_a\frac{d^2I_a(t)}{dt^2}\right] \tag{16.61}$$

A similar equation in $\omega_m(t)$ may also be obtained by substituting equation 16.60 in equation 16.59:

$$V_L = k_{PM}\omega_m(t) + \frac{R_a}{k_{PM}}\left[T_{\text{load}} + J\frac{d\omega_m(t)}{dt}\right] + \frac{L_a}{k_{PM}}\left[\frac{dT_{\text{load}}}{dt} + J\frac{d^2\omega_m(t)}{dt^2}\right] \tag{16.62}$$

The two differential equations may be manipulated to obtain

$$\frac{L_a J}{k_{PM}}\frac{d^2I_a(t)}{dt^2} + \frac{R_a J}{k_{PM}}\frac{dI_a(t)}{dt} + k_{PM} I_a(t) = \frac{J}{k_{PM}}\frac{dV_L}{dt} + T_{\text{load}} \tag{16.63}$$

and

$$\frac{L_a J}{k_{PM}}\frac{d^2\omega_m(t)}{dt^2} + \frac{R_a J}{k_{PM}}\frac{d\omega_m(t)}{dt} + k_{PM}\omega_m(t) = V_L - \frac{R_a}{k_{PM}}T_{\text{load}} - \frac{L_a}{k_{PM}}\frac{dT_{\text{load}}}{dt} \tag{16.64}$$

If we define the constants

$$\tau_m = \frac{R_a J}{k_{PM}^2} \quad \text{and} \quad \tau_a = \frac{L_a}{R_a} \tag{16.65}$$

the differential equations may be simplified as follows:

$$k_{PM}\tau_a\tau_m \frac{d^2 I_a(t)}{dt^2} + k_{PM}\tau_m \frac{dI_a(t)}{dt} + k_{PM} I_a(t) = \frac{J}{k_{PM}}\frac{dV_L}{dt} + T_{\text{load}} \tag{16.66}$$

and

$$k_{PM}\tau_a\tau_m \frac{d^2 \omega_m(t)}{dt^2} + k_{PM}\tau_m \frac{d\omega_m(t)}{dt} + k_{PM}\omega_m(t) = V_L - \frac{R_a}{k_{PM}} T_{\text{load}} - \frac{L_a}{k_{PM}}\frac{dT_{\text{load}}}{dt} \tag{16.67}$$

Speed Control of DC Motors

DC motor drives were briefly discussed in Chapter 10. Which type of DC drive is best suited for a given application depends in part on the characteristics of the load. Thus, it will be necessary to delve into a brief discussion of the different types of loads that are commonly encountered. Typical applications of various types of DC motors will also be discussed. *Constant-torque loads* are quite common, and are characterized by a need for constant torque over the entire speed range. This need is usually due to friction; the load will demand increasing horsepower at higher speeds, since power is the product of speed and torque. Thus, the power required will increase linearly with speed. This type of loading is characteristic of conveyors, extruders, and surface winders.

Another type of load is one that requires *constant horsepower* over the speed range of the motor. Since torque is inversely proportional to speed with constant horsepower, this type of load will require higher torque at low speeds. Examples of constant-horsepower loads are machine tool spindles (e.g., lathes). This type of application requires very high starting torques.

Variable-torque loads are also common. In this case, the load torque is related to the speed in some fashion, either linearly or geometrically. For some loads, for example, torque is proportional to the speed (and thus horsepower is proportional to speed squared); examples of loads of this type are positive displacement pumps. More common than the linear relationship is the squared-speed dependence of inertial loads such as centrifugal pumps, some fans, and all loads in which a flywheel is used for energy storage.

To select the appropriate motor and adjustable speed drive for a given application, we need to examine how each method for speed adjustment operates on a DC motor. Armature voltage control serves to smoothly adjust speed from 0 to 100 percent of the nameplate rated value (i.e., base speed), provided that the field excitation is also equal to the rated value. Within this range, it is possible to fully control motor speed for a constant-torque load, thus providing a linear increase in horsepower, as shown in Figure 16.23. Field weakening allows for increases in speed of up to several times the base speed; however, field control changes the characteristics of the DC motor from constant torque to constant horsepower, and therefore the torque output drops with speed, as shown in Figure 16.23. Operation above base speed requires special provision for field control, in addition to the circuitry required for armature voltage control, and is therefore more complex and costly.

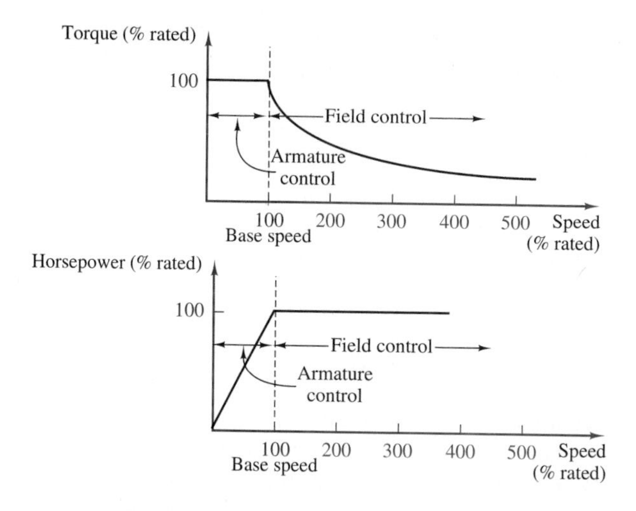

Figure 16.23 Speed adjustment in DC motors

DRILL EXERCISES

16.8 A series motor draws a current of 25 A and develops a torque of 100 N-m Find: (a) the torque when the current rises to 30 A if the field is unsaturated; and (b) the torque when the current rises to 30 A and the increase in current produces a 10 percent increase in flux.

16.9 A 200-V DC shunt motor draws 10 A at 1,800 rev/min. The armature circuit resistance is 0.15 Ω and the field winding resistance is 350 Ω. What is the torque developed by the motor?

16.10 Describe the cause-and-effect behavior of the speed control method of changing armature voltage for a shunt DC motor.

16.4 AC MACHINES

From the previous sections, it should be apparent that it is possible to obtain a wide range of performance characteristics from DC machines, as both motors and generators. A logical question at this point should be, Would it not be more convenient in some cases to take advantage of the single- or multiphase AC power that is available virtually everywhere than to expend energy and use additional hardware to rectify and regulate the DC supplies required by direct-current motors? The answer to this very obvious question is certainly a resounding yes. In fact, the AC induction motor is the workhorse of many industrial applications, and synchronous generators are used almost exclusively for the generation of electric power worldwide. Thus, it is appropriate to devote a significant portion of this chapter to the study of AC machines, and of induction motors in particular. The objective of this section is to explain the basic operation of both synchronous and induction machines, and to outline their performance characteristics. In doing so, we shall also point out the relative advantages and disadvantages of these machines in comparison with direct-current machines.

Rotating Magnetic Fields

As mentioned in Section 16.1, the fundamental principle of operation of AC machines is the generation of a rotating magnetic field, which causes the rotor to turn at a speed that depends on the speed of rotation of the magnetic field. We shall now explain how a rotating magnetic field can be generated in the stator and air gap of an AC machine by means of AC currents.

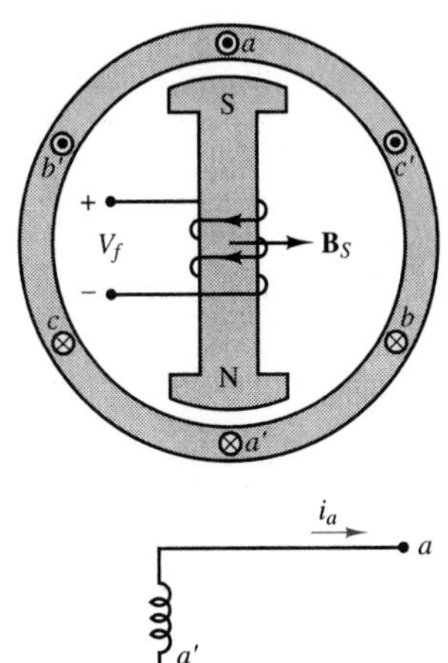

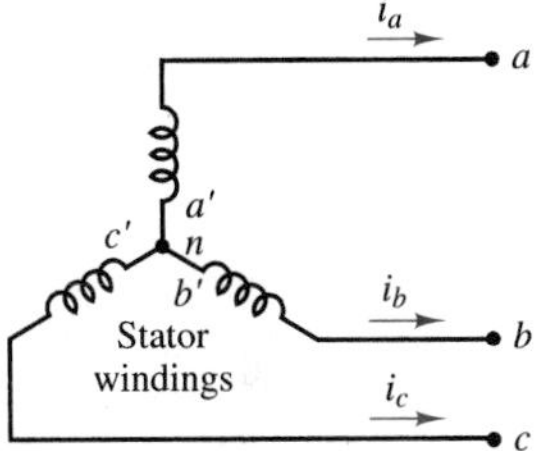

Figure 16.24 Two-pole three-phase stator

Consider the stator shown in Figure 16.24, which supports windings a-a', b-b' and c-c'. The coils are geometrically spaced 120° apart, and a three-phase voltage is applied to the coils. As you may recall from the discussion of AC power in Chapter 5, the currents generated by a three-phase source are also spaced by 120°, as illustrated in Figure 16.25. The phase voltages referenced the neutral terminal, would then be given by the expressions

$$v_a = A\cos(\omega_e t)$$

$$v_b = A\cos\left(\omega_e t - \frac{2\pi}{3}\right)$$

$$v_c = A\cos\left(\omega_e t + \frac{2\pi}{3}\right)$$

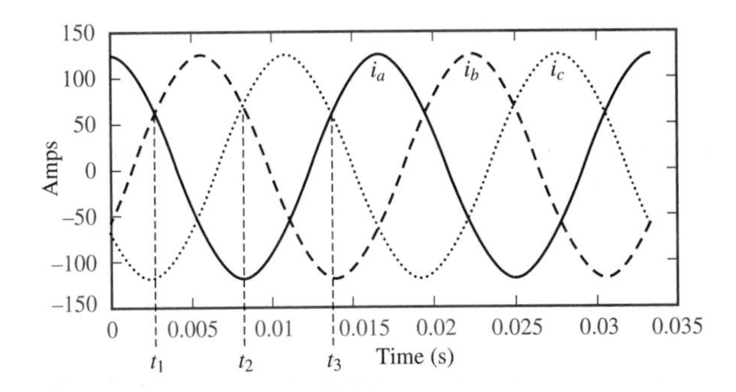

Figure 16.25 Three-phase stator winding currents

where ω_e is the frequency of the AC supply, or line frequency. The coils in each winding are arranged in such a way that the flux distribution generated by any one winding is approximately sinusoidal. Such a flux distribution may be obtained by appropriately arranging groups of coils for each winding over the stator surface. Since the coils are spaced 120° apart, the flux distribution resulting from the sum of the contributions of the three windings is the sum of the fluxes due to the separate windings, as shown in Figure 16.26. Thus, the flux in a three-phase machine rotates in space according to the vector diagram of Figure 16.27, and is constant in amplitude. A stationary observer on the machine's stator would see a sinusoidally varying flux distribution as shown in Figure 16.26.

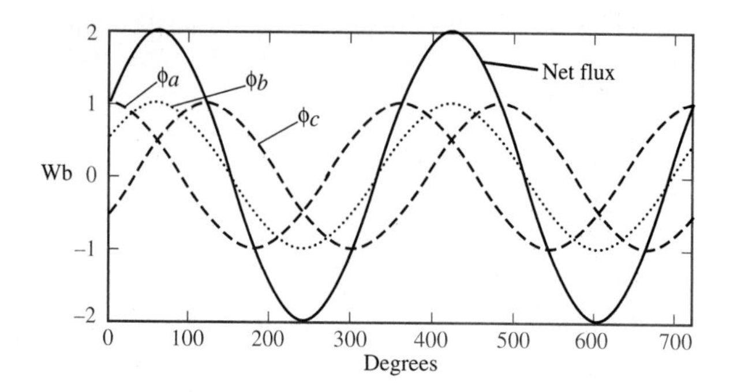

Figure 16.26 Flux distribution in a three-phase stator winding as a function of angle of rotation

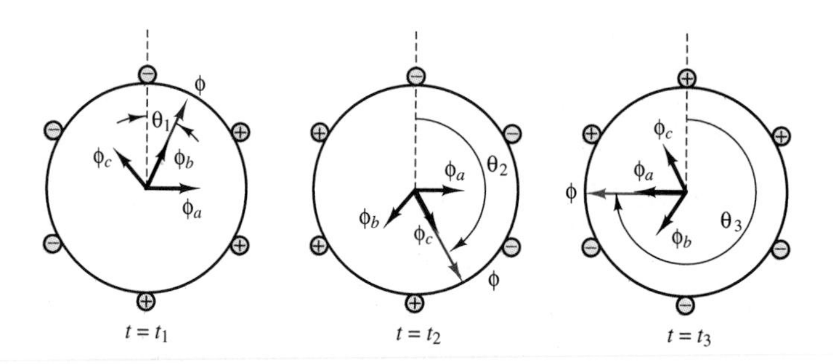

Figure 16.27 Rotating flux in a three-phase machine

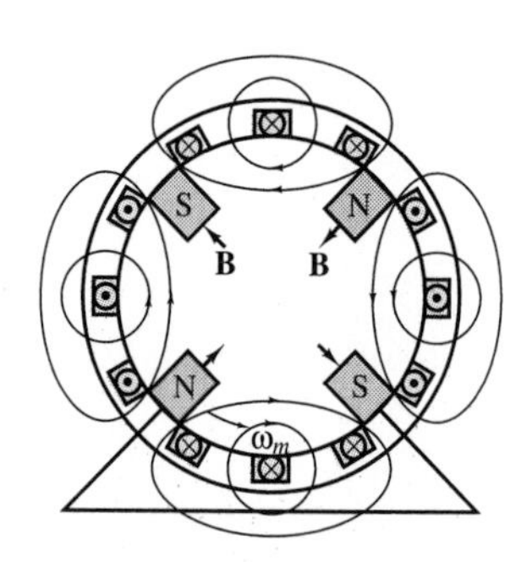

Figure 16.28 Four-pole stator

Since the resultant flux of Figure 16.26 is generated by the currents of Figure 16.25, the speed of rotation of the flux must be related to the frequency of the sinusoidal phase currents. In the case of the stator of Figure 16.24, the number of magnetic poles resulting from the winding configuration is two; however, it is also possible to configure the windings so that they have more poles. For example, Figure 16.28 depicts a simplified view of a four-pole stator.

In general, the speed of the rotating magnetic field is determined by the frequency of the excitation current, f, and by the number of poles present in the stator, p, according to the equation

$$n_s = \frac{120f}{p} \text{ rev/min}$$

or **(16.68)**

$$\omega_s = \frac{2\pi n_S}{60} = \frac{2\pi \times 2f}{p}$$

where n_s (or ω_s) is usually called the **synchronous speed.**

Now, the structure of the windings in the preceding discussion is the same whether the AC machine is a motor or a generator; the distinction between the two depends on the direction of power flow. In a generator, the electromagnetic torque is a reaction torque that opposes rotation of the machine; this is the torque against which the prime mover does work. In a motor, on the other hand, the rotational (motional) voltage generated in the armature opposes the applied voltage; this voltage is the counter (or back) emf. Thus, the description of the rotating magnetic field given thus far applies to both motor and generator action in AC machines.

It is worthwhile to devote a few paragraphs to an additional discussion of magnetic poles in AC machines, since an intuitive understanding of how a rotating magnetic field is generated is essential. The motion and associated electromagnetic torque of an electric machine are the result of two magnetic fields that are trying to align with each other so that the south pole of one field attracts the north pole of the other. Figure 16.29 illustrates this action by analogy with two permanent magnets, one of which is allowed to rotate about its center of mass.

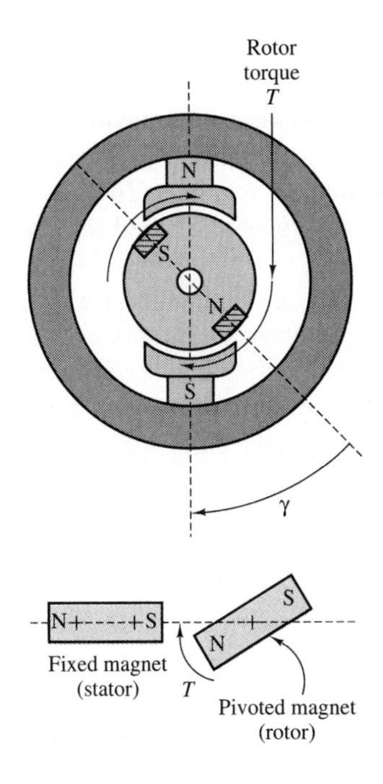

Figure 16.29 Alignment action of poles

As described a few paragraphs earlier, the stator magnetic field rotates in an AC machine, and therefore the rotor cannot "catch up" with the stator field and is in constant pursuit of it. The speed of rotation of the rotor will therefore depend on the number of magnetic poles present in the stator and in the rotor. The magnitude of the torque produced in the machine is a function of the angle γ between the stator and rotor magnetic fields; precise expressions for this torque depend on how the magnetic fields are generated and will be given separately for the two cases of synchronous and induction machines. What is common to all rotating machines is that the number of stator and rotor poles must be identical if any torque is to be generated. Further, the number of poles must be even, since for each north pole there must be a corresponding south pole.

One important desired feature in an electric machine is an ability to generate a constant electromagnetic torque. With a constant-torque machine, one can avoid torque pulsations that could lead to undesired mechanical vibration in the motor itself and in other mechanical components attached to the motor (e.g., mechanical loads, such as spindles or belt drives). A constant torque may not always be achieved, although it will be shown that it is possible to accomplish this goal when the excitation currents are multiphase. A general rule of thumb, in this respect, is that it is desirable, insofar as possible, to produce a constant flux per pole.

16.5 THE ALTERNATOR (SYNCHRONOUS GENERATOR)

One of the most common AC machines is the **synchronous generator,** or **alternator.** In this machine, the field winding is on the rotor, and the connection is made by means of brushes, in an arrangement similar to that of the DC machines studied earlier. The rotor field is obtained by means of a DC current provided to the rotor winding, or by permanent magnets. The rotor is then connected to a mechanical source of power and rotates at a speed that we will consider constant to simplify the analysis.

Figure 16.30 depicts a two-pole three-phase synchronous machine. Figure 16.31 depicts a four-pole three-phase alternator, in which the rotor poles are generated by means of a wound salient pole configuration and the stator poles are the result of windings embedded in the stator according to the simplified arrangement shown in the figure, where each of the pairs a/a', b/b', and so on, contributes to the generation of the magnetic poles, as follows. The group a/a', b/b', c/c' produces a

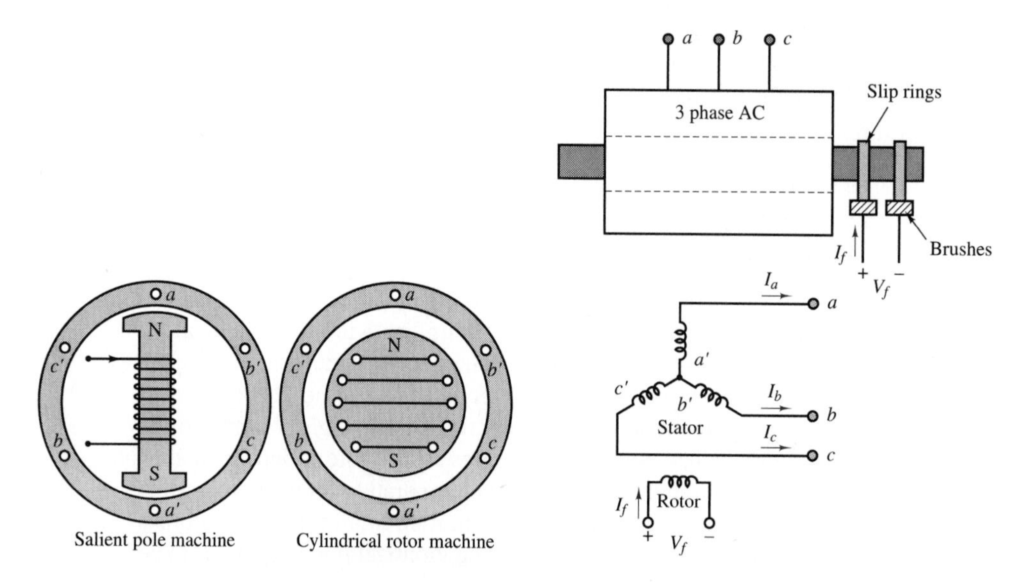

Figure 16.30 Two-pole synchronous machine

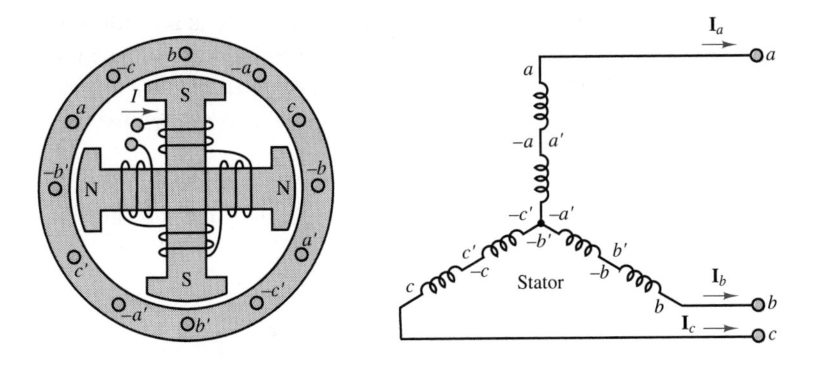

Figure 16.31 Four-pole three-phase alternator

sinusoidally distributed flux (see Figure 16.26) corresponding to one of the pole pairs, while the group $-a/-a'$, $-b/-b'$, $-c/-c'$ contributes the other pole pair. The connections of the coils making up the windings are also shown in Figure 16.31. Note that the coils form a wye connection (see Chapter 5). The resulting flux distribution is such that the flux completes two sinusoidal cycles around the circumference of the air gap. Note also that each arm of the three-phase wye connection has been divided into two coils, wound in different locations, according to the schematic stator diagram of Figure 16.31. One could then envision analogous configurations with greater numbers of poles, obtained in the same fashion, that is, by dividing each arm of a wye connection into more windings.

The arrangement shown in Figure 16.31 requires that a further distinction be made between mechanical degrees, θ_m, and electrical degrees, θ_e. In the four-pole alternator, the flux will see two complete cycles during one rotation of the rotor, and therefore the voltage that is generated in the coils will also oscillate at twice the frequency of rotation. In general, the electrical degrees (or radians) are related to the mechanical degrees by the expression

$$\theta_e = \frac{p}{2}\theta_m \tag{16.69}$$

where p is the number of poles. In effect, the voltage across a coil of the machine goes through one cycle every time a pair of poles moves past the coil. Thus, the frequency of the voltage generated by a synchronous generator is

$$f = \frac{p}{2}\frac{n}{60} \text{ Hz} \tag{16.70}$$

where n is the mechanical speed in rev/min. Alternatively, if the speed is expressed in rad/s, we have

$$\omega_e = \frac{p}{2}\omega_m \tag{16.71}$$

where ω_m is the mechanical speed of rotation in rad/s. The number of poles employed in a synchronous generator is then determined by two factors: the frequency desired of the generated voltage (e.g., 60 Hz, if the generator is used to produce AC power), and the speed of rotation of the prime mover. In the latter respect, there is a significant difference, for example, between the speed of rotation of a steam turbine generator and that of a hydroelectric generator, the former being much greater.

A common application of the alternator is in automotive battery-charging systems—in which, however, the generated AC voltage is rectified to provide the DC current required for charging the battery.

16.6 THE SYNCHRONOUS MOTOR

Synchronous motors are virtually identical to synchronous generators with regard to their construction, except for an additional winding for helping start the motor and minimizing motor speed over- and undershoots. The principle of operation is, of course, the opposite: an AC excitation provided to the armature generates a magnetic field in the air gap between stator and rotor, resulting in a mechanical torque. To generate the rotor magnetic field, some DC current must be provided to the field windings; this is often accomplished by means of an **exciter,** which consists of a small DC generator propelled by the motor itself, and therefore mechanically connected to it. It was mentioned earlier that to obtain a constant torque in an electric motor, it is necessary to keep the rotor and stator magnetic fields constant relative to each other. This means that the electromagnetically rotating field in the stator and the mechanically rotating rotor field should be aligned at all times. The only condition for which this can occur is if both fields are rotating at the synchronous speed, $n_s = 120f/p$. Thus, synchronous motors are by their very nature constant-speed motors.

For a non–salient pole (cylindrical-rotor) synchronous machine, the torque can be written in terms of the AC stator current, $i_S(t)$, and of the DC rotor current, I_f:

$$T = ki_S(t)I_f \sin(\gamma) \qquad \textbf{(16.72)}$$

where γ is the angle between the stator and rotor fields (see Figure 16.29). Let the angular speed of rotation be

$$\omega_m = \frac{d\theta_m}{dt} \text{ rad/s} \qquad \textbf{(16.73)}$$

where $\omega_m = 2\pi n/60$, and let ω_e be the electrical frequency of $i_S(t)$, where $i_S(t) = \sqrt{2}I_S \sin(\omega_e t)$. Then the torque may be expressed as follows:

$$T = k\sqrt{2}I_S \sin(\omega_e t)I_f \sin(\gamma) \qquad \textbf{(16.74)}$$

where k is a machine constant, I_S is the rms value of the stator current, and I_f is the DC rotor current. Now, the rotor angle γ can be expressed as a function of time by

$$\gamma = \gamma_0 + \omega_m t \qquad \textbf{(16.75)}$$

where γ_0 is the angular position of the rotor at $t = 0$; the torque expression then becomes

$$\begin{aligned} T &= k\sqrt{2}I_S I_f \sin(\omega_e t)\sin(\omega_m t + \gamma_0) \\ &= k\frac{\sqrt{2}}{2}I_S I_f \cos[(\omega_m - \omega_e)t - \gamma_0] - \cos[(\omega_m + \omega_e)t + \gamma_0] \end{aligned} \qquad \textbf{(16.76)}$$

It is a straightforward matter to show that the average value of this torque, $\langle T \rangle$, is different from zero only if $\omega_m = \pm\omega_e$, that is, only if the motor is turning at the synchronous speed. The resulting average torque is then given by

$$\langle T \rangle = k\sqrt{2}I_S I_f \cos(\gamma_0) \qquad \textbf{(16.77)}$$

Note that equation 16.76 corresponds to the sum of an average torque plus a fluctuating component at twice the original electrical (or mechanical) frequency. The fluctuating component results because, in the foregoing derivation, a single-phase current was assumed. The use of multiphase currents reduces the torque fluctuation to zero and permits the generation of a constant torque.

A per-phase circuit model describing the synchronous motor is shown in Figure 16.32, where the rotor circuit is represented by a field winding equivalent resistance and inductance, R_f and L_f, respectively, and the stator circuit is represented by equivalent stator winding inductance and resistance, L_S and R_S, respectively, and by the induced emf, E_b. From the exact equivalent circuit as given in Figure 16.32, we have

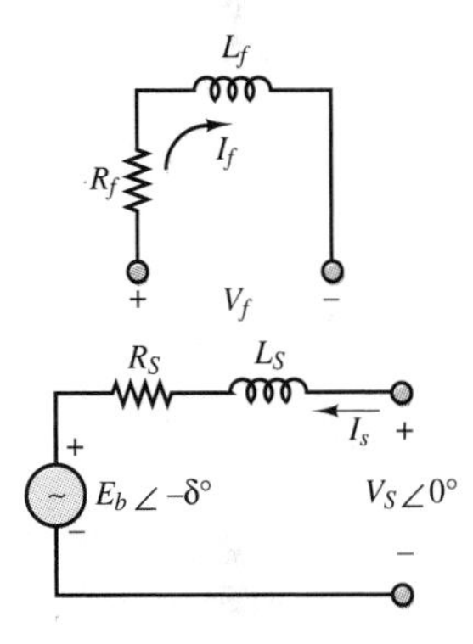

Figure 16.32 Per-phase circuit model

$$V_S = E_b + I_S(R_S + jX_S) \qquad \textbf{(16.78)}$$

where X_S is known as the synchronous reactance and includes magnetizing reactance.

The motor power is

$$P_{\text{out}} = \omega_S T = |V_S||I_S|\cos(\theta) \qquad \textbf{(16.79)}$$

for each phase, where T is the developed torque and θ is the angle between V_S and I_S.

When the phase winding resistance R_S is neglected, the circuit model of a synchronous machine can be redrawn as shown in Figure 16.33. The input power (per phase) is equal to the output power in this circuit, since no power is dissipated in the circuit; that is:

$$P_\phi = P_{\text{in}} = P_{\text{out}} = |\mathbf{V}_S||\mathbf{I}_S|\cos(\theta) \qquad \textbf{(16.80)}$$

Also by inspection of Figure 16.33, we have

$$d = |\mathbf{E}_b|\sin(\delta) = |\mathbf{I}_S|X_S\cos(\theta) \qquad \textbf{(16.81)}$$

Then

$$|\mathbf{E}_b||\mathbf{V}_S|\sin(\delta) = |\mathbf{V}_S||\mathbf{I}_S|X_S\cos(\theta) = X_S P_\phi \qquad \textbf{(16.82)}$$

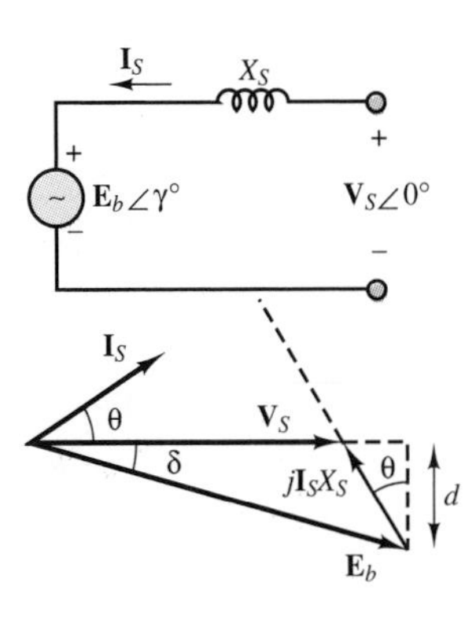

Figure 16.33

The total power of a three-phase synchronous machine is then given by

$$P = (3)\frac{|\mathbf{V}_S||\mathbf{E}_b|}{X_S}\sin(\delta) \qquad \textbf{(16.83)}$$

Because of the dependence of the power upon the angle δ, this angle has come to be called the **power angle.** If δ is zero, the synchronous machine cannot develop useful power. The developed power has its maximum value at δ equal to 90°. If we assume that $|\mathbf{E}_b|$ and $|\mathbf{V}_S|$ are constant, we can draw the curve shown in Figure 16.34, relating the power and power angle in a synchronous machine.

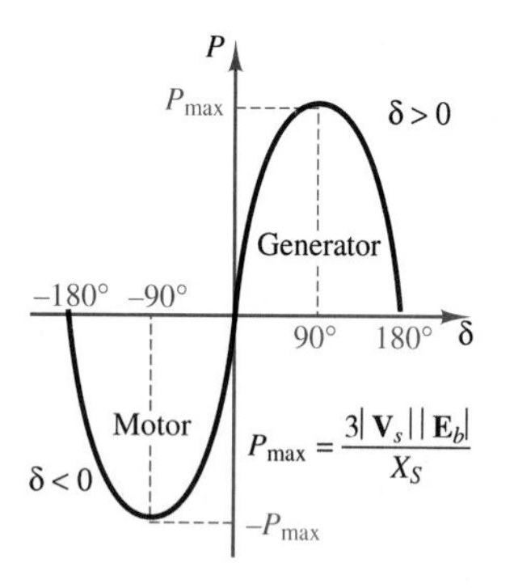

Figure 16.34 Power versus power angle for a synchronous machine

A synchronous generator is usually operated at a power angle varying from 15° to 25°. For synchronous motors and small loads, δ is close to 0°, and the motor torque is just sufficient to overcome its own windage and friction losses; as the load increases, the rotor field falls further out of phase with the stator field (although the two are still rotating at the same speed), until δ reaches a maximum at 90°. If the load torque exceeds the maximum torque, which is produced for $\delta = 90°$, the motor is forced to slow down below synchronous speed. This condition is undesirable, and provisions are usually made to shut the motor down automatically whenever synchronism is lost. The maximum torque is called the **pull-out torque** and is an important measure of the performance of the synchronous motor.

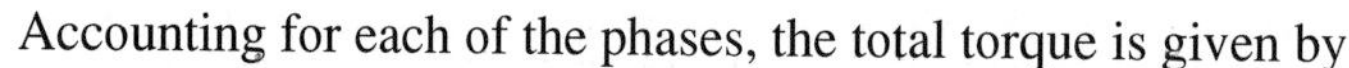

Accounting for each of the phases, the total torque is given by

$$T = \frac{m}{\omega_s}|\mathbf{V}_S||\mathbf{I}_S|\cos(\theta) \qquad \textbf{(16.84)}$$

where m is the number of phases. From Figure 16.33, we have $E_b\sin(\delta) = X_S I_S\cos(\theta)$. Therefore, for a three-phase machine, the developed torque is

$$T = \frac{P}{\omega_s} = \frac{3}{\omega_s}\frac{|\mathbf{V}_S||\mathbf{E}_b|}{X_S}\sin(\delta) \qquad \text{N-m} \qquad \textbf{(16.85)}$$

Typically, analysis of multiphase motors is performed on a per-phase basis, as illustrated in the examples that follow.

Example 16.11

A 460-V, three-phase, wye-connected synchronous motor has a full-load stator winding current of 12.5 A. The synchronous impedance of the motor is $1.0 + j12\ \Omega$. Find:

1. The kVA rating of the motor.
2. The induced voltage and the power angle of the rotor when the motor is fully loaded at 0.707 power factor lagging.

Solution:

The circuit model for the motor is shown in Figure 16.35.

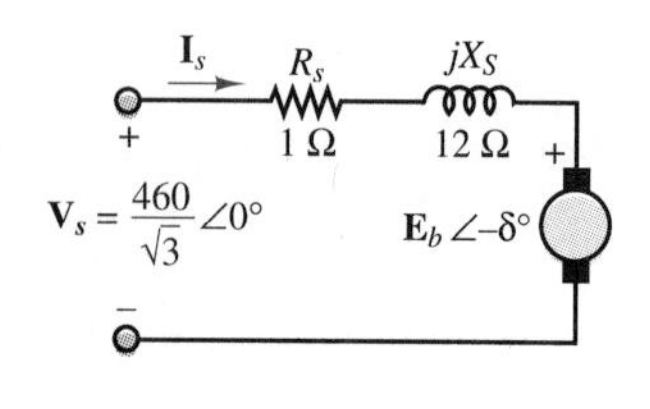

Figure 16.35

1. The per-phase current in the wye-connected stator winding is

$$I_S = |\mathbf{I}_S| = 12.5\ \text{A}$$

The per-phase voltage is

$$V_S = |\mathbf{V}_S| = \frac{460}{\sqrt{3}} = 265.58 \text{ V}$$

The kVA rating of the motor is expressed in terms of the apparent power, S (see Chapter 5):

$$S = 3V_S I_S = 3 \times 265.58 \times 12.5 = 9{,}959 \text{ W}$$

2. From the equivalent circuit, we have

$$\begin{aligned}\mathbf{E}_b &= \mathbf{V}_S - \mathbf{I}_S(R_S + jX_S)\\ &= 265.58 - (12.5\angle -45°) \times (1 + j12) = 179.31\angle -32.83°\end{aligned}$$

The induced line voltage is defined to be

$$V_{\text{line}} = \sqrt{3}E_b = \sqrt{3} \times 179.31 = 310.57 \text{ V}$$

From the expression for $\mathbf{E}_b$, we can find the power angle:

$$\delta = -32.83°$$

The minus sign indicates that the machine is in the motor mode.

Example 16.12

A 208-V, 45-kVA, 0.8 power factor leading, Δ-connected 60-Hz synchronous machine has a synchronous reactance of 2.5 Ω and a negligible armature resistance. Its friction and windage losses are 1.5 kW, and the core losses are 1.0 kW. Initially, the shaft is supplying a 15-hp load. Find $\mathbf{I}_S$, $\mathbf{I}_{\text{line}}$, and $\mathbf{E}_b$.

Solution:

The output power of the motor is 15 hp; that is:

$$P_{\text{out}} = 15 \text{ hp} \times 0.746 \text{ kW/hp} = 11.19 \text{ kW}$$

The electric power supplied to the machine is

$$\begin{aligned}P_{\text{in}} &= P_{\text{out}} + P_{\text{mech}} + P_{\text{core-loss}} + P_{\text{elec-loss}}\\ &= 11.19 \text{ kW} + 1.5 \text{ kW} + 1.0 \text{ kW} + 0 \text{ kW} = 13.69 \text{ kW}\end{aligned}$$

As discussed in Chapter 5, the resulting line current is

$$I_{\text{line}} = \frac{P_{\text{in}}}{\sqrt{3}V \cos\theta} = \frac{13{,}690}{\sqrt{3} \times 208 \times 0.8} = 47.5 \text{ A}$$

Because of the Δ connection, the armature current is

$$\mathbf{I}_S = \frac{1}{\sqrt{3}}\mathbf{I}_{\text{line}} = 27.4\angle 36.87° \text{A}$$

The emf may be found from the equivalent circuit and KVL:

$$\begin{aligned}\mathbf{E}_b &= \mathbf{V}_S - jX_S\mathbf{I}_S\\ &= 208\angle 0° - j2.5(27.4\angle 36.87°) = 255\angle -12.4° \text{ V}\end{aligned}$$

The power angle is

$$\delta = -12.4^\circ$$

Synchronous motors are not very commonly used in practice, for various reasons, among which are that they are essentially required to operate at constant speed (unless a variable-frequency AC supply is available) and that they are not self-starting. Further, separate AC and DC supplies are required. It will be seen shortly that the induction motor overcomes most of these drawbacks.

DRILL EXERCISES

16.11 A synchronous generator has a multipolar construction that permits changing its synchronous speed. If only two poles are energized, at 50 Hz, the speed is 3,000 rev/min. If the number of poles is progressively increased to 4, 6, 8, 10, and 12, find the synchronous speed for each configuration.

16.12 Draw the complete equivalent circuit of a synchronous generator and its phasor diagram.

16.13 Find an expression for the maximum pull-out torque of the synchronous motor.

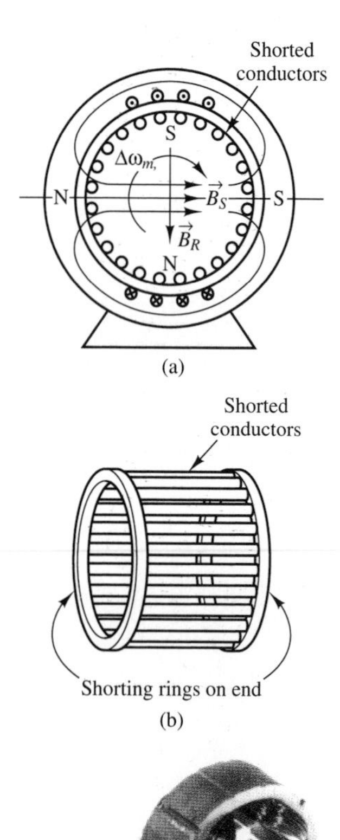

Figure 16.36 (a) Squirrel-cage induction motor; (b) conductors in rotor; (c) photo of squirrel-cage induction motor

16.7 THE INDUCTION MOTOR

The induction motor is the most widely used electric machine, because of its relative simplicity of construction. The stator winding of an induction machine is similar to that of a synchronous machine; thus, the description of the three-phase winding of Figure 16.24 also applies to induction machines. The primary advantage of the induction machine, which is almost exclusively used as a motor (its performance as a generator is not very good), is that no separate excitation is required for the rotor. The rotor typically consists of one of two arrangements: a **squirrel cage,** or a **wound rotor.** The former contains conducting bars short-circuited at the end and embedded within it; the latter consists of a multiphase winding similar to that used for the stator, but electrically short-circuited.

In either case, the induction motor operates by virtue of currents induced from the stator field in the rotor. In this respect, its operation is similar to that of a transformer, in that currents in the stator (which acts as a primary coil) induce currents in the rotor (acting as a secondary coil). In most induction motors, no external electrical connection is required for the rotor, thus permitting a simple, rugged construction, without the need for slip rings or brushes. Unlike the synchronous motor, the induction motor does not operate at synchronous speed, but at a somewhat lower speed, which is dependent on the load. Figure 16.36 illustrates the appearance of a squirrel-cage induction motor. The following discussion will focus mainly on this very common configuration.

You are by now acquainted with the notion of a rotating stator magnetic field. Imagine now that a squirrel-cage rotor is inserted in a stator in which such a rotating magnetic field is present. The stator field will induce voltages in the cage conductors, and if the stator field is generated by a three-phase source, the resulting rotor currents—which circulate in the bars of the squirrel cage, with the conducting path completed by the shorting rings at the end of the cage—are also three-phase, and are determined by the magnitude of the induced voltages and by the impedance of the rotor. Since the rotor currents are induced by the stator field, the number of poles and the speed of rotation of the induced magnetic field are the same as those of the stator field, *if the rotor is at rest*. Thus, when a stator field is initially applied, the rotor field is synchronous with it, and the fields are stationary with respect to each other. Thus, according to the earlier discussion, a *starting torque* is generated.

If the starting torque is sufficient to cause the rotor to start spinning, the rotor will accelerate up to its operating speed. However, an induction motor can never reach synchronous speed; if it did, the rotor would appear to be stationary with respect to the rotating stator field, since it would be rotating at the same speed. But in the absence of relative motion between the stator and rotor fields, no voltage would be induced in the rotor. Thus, an induction motor is limited to speeds somewhere below the synchronous speed, n_s. Let the speed of rotation of the rotor be n; then, the rotor is losing ground with respect to the rotation of the stator field at a speed $(n_s - n)$. In effect, this is equivalent to backward motion of the rotor at the **slip speed,** defined by $(n_s - n)$. The **slip,** s, is usually defined as a fraction of n_s:

$$s = \frac{n_s - n}{n_s} \tag{16.86}$$

which leads to the following expression for the rotor speed:

$$n = n_s(1 - s) \tag{16.87}$$

The slip, s, is a function of the load, and the amount of slip in a given motor is dependent on its construction and rotor type (squirrel cage or wound rotor). Since there is a relative motion between the stator and rotor fields, voltages will be induced in the rotor at a frequency called the **slip frequency,** related to the relative speed of the two fields. This gives rise to an interesting phenomenon: the rotor field travels relative to the rotor at the slip speed sn_s, but the rotor is mechanically traveling at the speed $(1 - s)n_s$, so that the net effect is that the rotor field travels at the speed

$$sn_s + (1 - s)n_s = n_s \tag{16.88}$$

that is, at synchronous speed. The fact that the rotor field rotates at synchronous speed—although the rotor itself does not—is extremely important, because it means that the stator and rotor fields will continue to be stationary with respect to each other, and therefore a net torque can be produced.

As in the case of DC and synchronous motors, important characteristics of induction motors are the starting torque, the maximum torque, and the torque-speed curve. These will be discussed shortly, after some analysis of the induction motor is performed in the next few paragraphs.

Example 16.13

A four-pole, three-phase, 230-V, 60-Hz induction motor runs at a speed of 1,725 rev/min under full-load conditions. Find the slip and the frequency of the voltage induced in the rotor at the rated speed.

Solution:

The synchronous speed of the motor is

$$n_s = \frac{120f}{p} = \frac{120 \times 60}{4} = 1{,}800 \text{ rev/min}$$

The slip is

$$s = \frac{n_s - n}{n_s} = \frac{1{,}800 - 1{,}725}{1{,}800} = 0.0417$$

The rotor frequency, f_R, is

$$f_R = sf = 0.0417 \times 60 = 2.5 \text{ Hz}$$

The induction motor can be described relatively simply by means of an equivalent circuit, which is essentially that of a rotating transformer. (See Chapter 15 for a circuit model of the transformer.) Figure 16.37 depicts such a circuit model, where:

R_S = stator resistance per phase, R_R = rotor resistance per phase
X_S = stator reactance per phase, X_R = rotor reactance per phase
X_m = magnetizing (mutual) reactance
R_C = equivalent core-loss resistance
E_S = per-phase induced voltage in stator windings
E_R = per-phase induced voltage in rotor windings

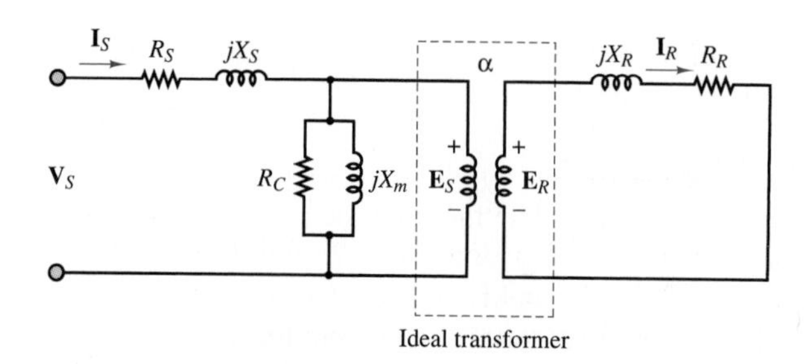

Figure 16.37 Circuit model for induction machine

The primary internal stator voltage, $\mathbf{E}_S$, is coupled to the secondary rotor voltage, $\mathbf{E}_R$, by an ideal transformer with an effective turns ratio α. For the rotor circuit, the induced voltage at any slip will be

$$\mathbf{E}_R = s\mathbf{E}_{R0} \tag{16.89}$$

where $\mathbf{E}_{R0}$ is the induced rotor voltage at the condition in which the rotor is stationary. Also, $X_R = \omega_R L_R = 2\pi f_R L_R = 2\pi s f L_R = sX_{R0}$, where $X_{R0} = 2\pi f L_R$ is the reactance when the rotor is stationary. The rotor current is given by the expression

$$\mathbf{I}_R = \frac{\mathbf{E}_R}{R_R + jX_R} = \frac{s\mathbf{E}_{R0}}{R_R + jsX_{R0}} = \frac{\mathbf{E}_{R0}}{R_R/s + jX_{R0}} \tag{16.90}$$

The resulting rotor equivalent circuit is shown in Figure 16.38.

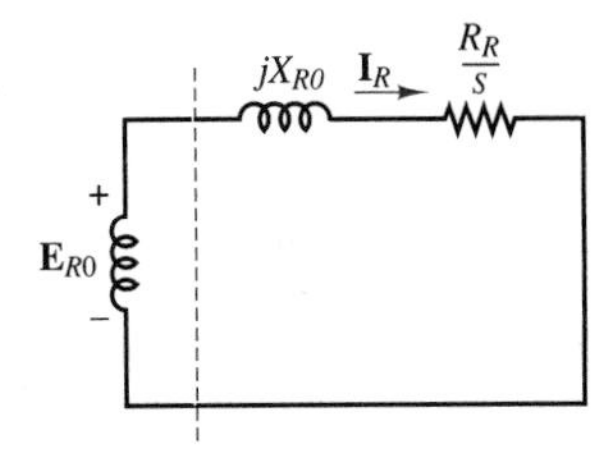

Figure 16.38 Rotor circuit

The voltages, currents, and impedances on the secondary (rotor) side can be reflected to the primary (stator) by means of the effective turns ratio. When this transformation is effected, the transformed rotor voltage is given by

$$\mathbf{E}_2 = \mathbf{E}'_R = \alpha\mathbf{E}_{R0} \tag{16.91}$$

The transformed (reflected) rotor current is

$$\mathbf{I}_2 = \frac{\mathbf{I}_R}{\alpha} \tag{16.92}$$

The transformed rotor resistance can be defined as

$$R_2 = \alpha^2 R_R \tag{16.93}$$

and the transformed rotor reactance can be defined by

$$X_2 = \alpha^2 X_{R0} \tag{16.94}$$

The final per-phase equivalent circuit of the induction motor is shown in Figure 16.39.

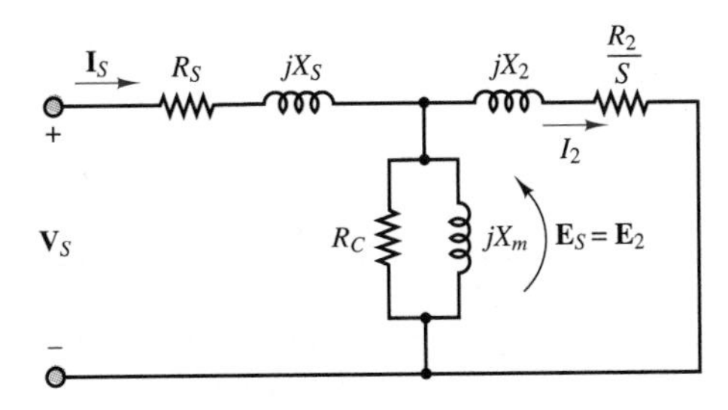

Figure 16.39 Equivalent circuit of an induction machine

The following examples illustrate the use of the circuit model in determining the performance of the induction motor.

Example 16.14

A 460-V, 25-hp, 60-Hz, four-pole, wye-connected induction machine has the following parameters (per phase), referred to the stator circuit:

$$R_S = 0.641\ \Omega \qquad R_2 = 0.332\ \Omega$$
$$X_S = 1.106\ \Omega \qquad X_2 = 0.464\ \Omega \qquad X_m = 26.3\ \Omega$$

For $s = 0.022$ at the rated voltage and frequency and an output power of 14 hp, find

1. The speed.
2. The stator current.
3. The power factor.
4. The output torque.

You may neglect core losses ($R_C = 0$).

Solution:

1. The per-phase equivalent circuit is shown in Figure 16.39. The synchronous speed is found to be

$$n_s = \frac{120f}{p} = \frac{120 \times 60}{4} = 1{,}800 \text{ rev/min}$$

or

$$\omega_s = 1{,}800 \times \frac{2\pi}{60} = 188.5 \text{ rad/s}$$

The rotor mechanical speed is

$$n = (1 - s)n_s = 1{,}760 \text{ rev/min}$$

or

$$\omega_m = (1 - s)\omega_s = 184.4 \text{ rad/s}$$

2. The reflected rotor impedance is found from the parameters of the per-phase circuit to be

$$Z_2 = \frac{R_2}{s} + jX_2 = \frac{0.332}{0.022} + j0.464$$
$$= 15.09 + j0.464\ \Omega$$

The combined magnetization plus rotor impedance is therefore equal to

$$Z = \frac{1}{1/jX_m + 1/Z_2} = \frac{1}{-j0.038 + 0.0662\angle -1.76^\circ} = 12.94\angle 31.1^\circ$$

and the total impedance is

$$Z_{\text{total}} = Z_S + Z = 0.641 + j1.106 + 11.08 + j6.68$$
$$= 11.72 + j7.79 = 14.07\angle 33.6^\circ\ \Omega$$

Finally, the stator current is given by

$$I_S = \frac{V_S}{Z_{\text{total}}} = \frac{460/\sqrt{3}\angle 0^\circ}{14.07\angle 33.6^\circ} = 18.88\angle -33.6^\circ \text{ A}$$

3. The power factor is

$$\text{pf} = \cos 33.6^\circ = 0.883 \qquad \text{lagging}$$

4. The output power, P_{out}, is

$$P_{\text{out}} = 14 \times 746 = 10.444 \text{ kW}$$

and the output torque is

$$T = \frac{P_{\text{out}}}{\omega_m} = \frac{10{,}444 \text{ W}}{184.4 \text{ rad/s}} = 56.64 \text{ N-m}$$

Example 16.15

A 500-V, three-phase, 50-Hz, eight-pole, wye-connected induction motor has the following equivalent-circuit parameters:

$$R_S = 0.13\ \Omega \qquad R'_R = 0.32\ \Omega \qquad X_S = 0.6\ \Omega \qquad X'_R = 1.48\ \Omega$$

The magnetic branch admittance is

$$Y_m = G_C + jB_m = 0.004 - j0.05\ \Omega^{-1}$$

referred to the primary side, where G_C is the conductance and B_m the susceptance associated with the core loss and mutual inductance, respectively. The full-load slip is 5 percent. Find:

1. The full-load electromagnetic torque
2. The stator input current
3. The power factor

using the approximate equivalent circuit. The effective stator-to-rotor turns ratio per phase is 1/1.57. Neglect mechanical losses.

Solution:

The approximate equivalent circuit of the three-phase induction motor on a per-phase basis is shown in Figure 16.40. The parameters of the model are calculated as follows:

$$R_2 = 0.32 \times \left(\frac{1}{1.57}\right)^2 = 0.13\ \Omega$$

$$X_2 = 1.48 \times \left(\frac{1}{1.57}\right)^2 = 0.6\ \Omega$$

$$Z = R_S + \frac{R_2}{S} + j(X_S + X_2)$$

$$= 0.13 + \frac{0.13}{0.05} + j(0.6 + 0.6) = 2.73 + j1.2\ \Omega$$

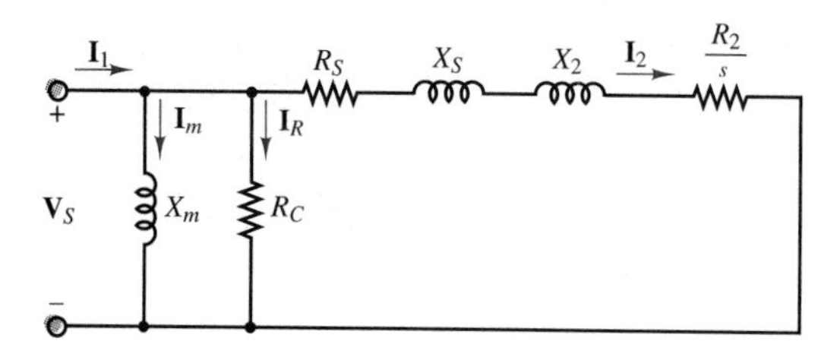

Figure 16.40

Using the approximate circuit,

$$\mathbf{I}_2 = \frac{\mathbf{V}_S}{Z} = \frac{(500/\sqrt{3})\angle 0^\circ}{2.73 + j1.2} = 88.8 - j39 \text{ A}$$

$$\mathbf{I}_R = \mathbf{V}_S G_S = 288.7 \times 0.004 = 1.15 \text{ A}$$

$$\mathbf{I}_m = -j\mathbf{V}_S B_m = 288.7 \times (-j0.05) = -j14.4 \text{ A}$$

$$\mathbf{I}_1 = \mathbf{I}_2 + \mathbf{I}_R + \mathbf{I}_m = 89.95 - j53.4 \text{ A}$$

$$\text{Input power factor} = \frac{89.95}{105} = 0.856 = 85.6\% \text{ lagging}$$

$$\text{Torque} = \frac{3P}{\omega_S} = \frac{3I_2^2 R_2/s}{4\pi f/p} = 935 \text{ N-m}$$

Performance of Induction Motors

The performance of induction motors can be described by torque-speed curves similar to those already used for DC motors. Figure 16.41 depicts an induction motor torque-speed curve, with five torque ratings marked a through e. Point a

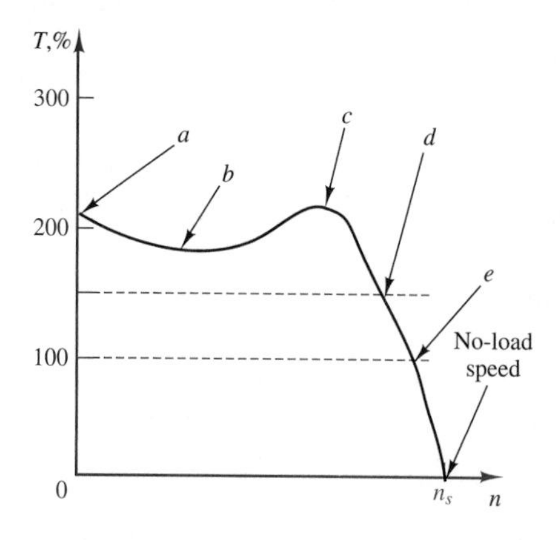

Figure 16.41 Performance curve for induction motor

is the *starting torque,* also called **breakaway torque,** and is the torque available with the rotor "locked," that is, in a stationary position. At this condition, the frequency of the voltage induced in the rotor is highest, since it is equal to the frequency of rotation of the stator field; consequently, the inductive reactance of the rotor is greatest. As the rotor accelerates, the torque drops off, reaching a maximum value called the **pull-up torque** (point b); this typically occurs somewhere between 25 and 40 percent of synchronous speed. As the rotor speed continues to increase, the rotor reactance decreases further (since the frequency of the induced voltage is determined by the relative speed of rotation of the rotor with respect to the stator field). The torque becomes a maximum when the rotor inductive reactance is equal to the rotor resistance; maximum torque is also called **breakdown torque** (point c). Beyond this point, the torque drops off, until it is zero at synchronous speed, as discussed earlier. Also marked on the curve are the *150 percent torque* (point d), and the *rated torque* (point e).

A general formula for the computation of the induction motor steady-state torque-speed characteristic is

$$T = \frac{1}{\omega_e} \frac{mV_S^2 R_R/s}{\left[\left(R_S + \frac{R_R}{s}\right)^2 + (X_S + X_R)^2\right]} \qquad \textbf{(16.95)}$$

where m is the number of phases.

Different construction arrangements permit the design of induction motors with different torque-speed curves, thus permitting the user to select the motor that best suits a given application. Figure 16.42 depicts the four basic classifications, classes A, B, C, and D, as defined by NEMA. The determining features in the classification are the locked-rotor torque and current, the breakdown torque, the pull-up torque, and the percent slip. Class A motors have a higher breakdown torque than class B motors, and a slip of 5 percent or less. Motors in this class are often designed for a specific application. Class B motors are general-purpose motors; this is the most commonly used type of induction motor, with typical values of slip of 3 to 5 percent. Class C motors have a high starting torque for a given starting current, and a low slip. These motors are typically used in applications demanding high starting torque but having relatively normal running loads, once running speed has been reached. Class D motors are characterized by high starting torque, high slip, low starting current, and low full-load speed. A typical value of slip is around 13 percent.

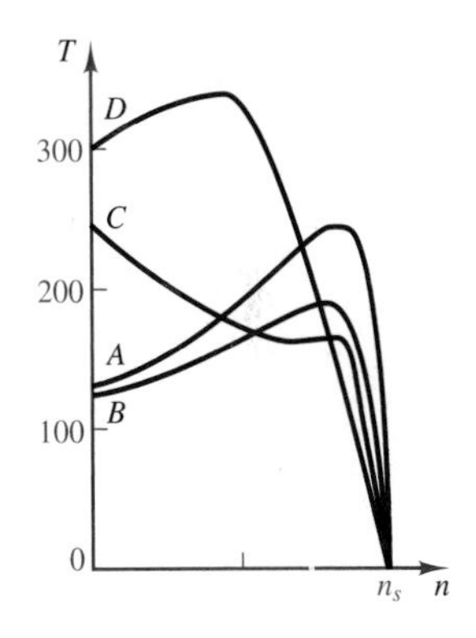

Figure 16.42 Induction motor classification

Factors that should be considered in the selection of an AC motor for a given application are the *speed range,* both minimum and maximum, and the speed variation. For example, it is important to determine whether constant speed is required; what variation might be allowed, either in speed or in torque; or whether variable-speed operation is required, in which case a variable-speed drive will be needed. The torque requirements are obviously important as well. The starting and running torque should be considered; they depend on the type of load. Starting torque can vary from a small percentage of full-load to several times full-load torque. Furthermore, the excess torque available at start-up determines the *acceleration characteristics* of the motor. Similarly, *deceleration characteristics* should be considered, to determine whether external braking might be required.

Another factor to be considered is the *duty cycle* of the motor. The duty cycle, which depends on the nature of the application, is an important consideration when the motor is used in repetitive, noncontinuous operation, such as is encountered in some types of machine tools. If the motor operates at zero or reduced load for periods of time, the duty cycle—that is, the percentage of the time the motor is loaded—is an important selection criterion. Last, but by no means least, are the *heating properties* of a motor. Motor temperature is determined by internal losses and by ventilation; motors operating at a reduced speed may not generate sufficient cooling, and forced ventilation may be required.

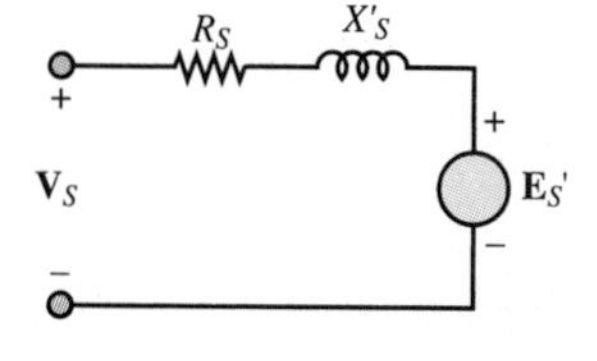

Figure 16.43 Simplified induction motor dynamic model

Thus far, we have not considered the dynamic characteristics of induction motors. Among the integral-horsepower induction motors (i.e.: motors with horsepower rating greater than one), the most common dynamic problems are associated with starting and stopping and with the ability of the motor to continue operation during supply system transient disturbances. Dynamic analysis methods for induction motors depend to a considerable extent on the nature and complexity of the problem and the associated precision requirements. When the electrical transients in the motor are to be included as well as the motional transients, and especially when the motor is an important element in a large network, the simple transient equivalent circuit of Figure 16.43 provides a good starting approximation. In the circuit model of Figure 16.43, X_S' is called the *transient reactance*. The voltage E_S' is called the *voltage behind the transient reactance* and is assumed to be equal to the initial value of the induced voltage, at the start of the transient. R_S is the stator resistance. The dynamic analysis problem consists of selecting a sufficiently simple but reasonably realistic representation that will not unduly complicate the dynamic analysis, particularly through the introduction of nonlinearities.

It should be remarked that the basic equations of the induction machine, as derived from first principles, are quite nonlinear. Thus, an accurate dynamic analysis of the induction motor, without any linearizing approximations, requires the use of computer simulation.

DRILL EXERCISES

16.14 A three-phase induction motor has six poles. (a) If the line frequency is 60 Hz, calculate the speed of the magnetic field in rev/min. (b) Repeat the calculation if the frequency is changed to 50 Hz.

16.15 A four-pole induction motor operating at a frequency of 60 Hz has a full-load slip of 4 percent. Find the frequency of the voltage induced in the rotor (a) at the instant of starting and (b) at full load.

16.16 A four-pole, 1,746-rev/min, 220-V, 3-phase, 60-Hz, 10-hp, Y-connected induction machine has the following parameters: $R_S = 0.4\,\Omega, R_2 = 0.14\,\Omega, X_m = 16\,\Omega, X_S = 0.35\,\Omega, X_2 = 0.35\,\Omega, R_C = 0$. Using Figure 16.39, find: (a) the stator current; (b) the rotor current; (c) the motor power factor; and (d) the total stator power input.

CONCLUSION

The principles developed in Chapter 16 can be applied to rotating electric machines, to explain how mechanical energy can be converted to electrical energy, and vice versa. The former function is performed by electric generators, while the latter is provided by electric motors.

- Electric machines are described in terms of their mechanical characteristics, their torque-speed curves, and their electrical characteristics, including current and voltage requirements. Losses and efficiency are an important part of the operation of electric machines, and it should be recognized that there will be electrical losses (due to the resistance of the windings), mechanical losses (friction and windage), and magnetic core losses (eddy currents, hysteresis). The main mechanical components of an electric machine are the stator, rotor, and air gap. Electrically, the important parameters are the armature (load current–carrying) circuit, and the field (magnetizing) circuit. Magnetic fields establish the coupling between the electrical and mechanical systems.
- Electric machines are broadly classified into DC and AC machines; the former use DC excitation for both the field and armature circuits, while the latter may be further subdivided into two classes: synchronous machines, and induction motors. AC synchronous machines are characterized by DC field and AC armature excitation. Induction machines (of the squirrel-cage type), on the other hand, do not require a field excitation, since this is provided by electromagnetic induction. Typically, DC machines have the armature winding on the rotor, while AC machines have it on the stator.
- The performance of electric machines can be approximately predicted with the use of circuit models, or of performance curves. The selection of a particular machine for a given application is driven by many factors, including the availability of suitable electrical supplies (or prime movers), the type of load, and various other concerns, of which heat dissipation and thermal characteristics are probably the most important.

KEY TERMS

Armature *856*
Armature Constant *871*
Back emf *870*
Breakaway Torque *907*
Breakdown Torque *907*
Buildup *874*
Commutator *868*
Compound-Connected DC Machine *870*
Cumulative Compounding *870*
Differential Compounding *870*
Efficiency *859*
Energy-Conversion Device *859*
Exciter *896*
Field Resistance Curve *873*
Field Winding *856*
Generator *857*
Long Shunt Connection *870*
Magnetic Poles *865*
Motor *857*
Power Angle *898*
Pull-Out Torque *898*
Pull-Up Torque *907*
Rated Current *862*
Rated Volt-Amperes *862*
Rated Voltage *862*
Regulation *862*

ANSWERS TO DRILL EXERCISES

16.1 (a) $\omega = 55\pi$ rad/s; (b) $n = 1{,}650$ rev/min

16.2 275 V

16.3 $T = 40.7$ N-m; volt-amperes = 11,431 VA; $P_{\max} = 11.5$ hp

16.4 170% of rated torque; 1,700 rev/min

16.5 $k_a = 5.1$; $E_b = 45.9$ V

16.6 (a) $E_b = 1{,}040$ V; (b) $k_a = 10.34$; (c) $T = 4{,}138$ N-m

16.7 (a) 400 A; (b) 5 A; (c) 405 A; (d) 270.25 V

16.8 (a) 144 N-m; (b) 132 N-m

16.9 $T = \dfrac{P}{\omega_m} = 9.93$ N-m

16.10 Increasing the armature voltage leads to an increase in armature current. Consequently, the motor torque increases until it exceeds the load torque, causing the speed to increase as well. The corresponding increase in back emf, however, causes the armature current to drop and the motor torque to decrease until a balance condition is reached between motor and load torque and the motor runs at constant speed.

16.11 1,500 rev/min; 1,000 rev/min; 750 rev/min; 600 rev/min; 500 rev/min

16.13 $T_{\max} = \dfrac{3V_S E_b}{\omega_m X_S}$

16.14 (a) $n = 1{,}200$ rev/min; (b) $n = 1{,}000$ rev/min

16.15 (a) $f_R = 60$ Hz; (b) $f_R = 2.4$ Hz

16.16 (a) $25.92\angle -22.45°$ A; (b) $24.39\angle -6.51°$ A; (c) 0.9243; (d) 8,476 W

EIT EXAM REVIEW: ELECTRICAL MACHINERY

The nonelectrical engineer will be frequently exposed to electrical machinery. This subject is covered in Chapters 16 and 17. The most popular electric machines in engineering applications are the DC motor and the AC induction motor. The former is discussed in the first half of Chapter 16, while the latter is discussed in Chapter 16 (three-phase motors) and in Chapter 17 (single-phase motors).

DRILL EXERCISES

16.17 A four-pole synchronous motor operating from a 60-Hz supply will have a synchronous speed, in rev/min, of

a. 900
b. 1,800
c. 1,200
d. 3,600
e. 4,800

Solution:

$$\text{Frequency} = (\text{no. of poles}/2) \times (\text{rev/min}/60)$$

Therefore,

$$60 = (4/2) \times (\text{rev/min}/60)$$

or $\text{rev/min} = 1,800$. Thus, b is the answer.

16.18 The armature resistance of a 55-hp, 525-V, DC shunt wound motor is 0.4 Ω. The full-load armature current of this motor is 80 A. If the initial starting current is 175 percent of the full-load value, the resistance of the starting coil in Ω should be nearest to

a. 6.6125
b. 3.75
c. 3.35
d. 4.15
e. 2.75

Solution:

$$I_{\text{start}}(\text{starting current}) = 1.75 \times 80 = 140 \text{ A}$$

Therefore, the total resistance is

$$R = \frac{V}{I_S} = \frac{525}{140} = 3.75 \ \Omega$$

Thus, the resistance of the starter is $3.75 - 0.4 = 3.35 \ \Omega$. The correct answer is c.

16.19 The speed of an AC electric motor

a. Is independent of the frequency.
b. Is directly proportional to the square of the frequency.
c. Varies directly with the number of poles.
d. Varies inversely with the number of poles.
e. None of the above.

Solution:

The speed varies inversely with the number of poles and directly with the frequency. Thus, d is the correct answer.

In addition to the sample problems given above, the following examples can assist you in your review of electric machines: 16.1, 16.4, 16.5, 16.6, 16.7, 16.9, 16.11, 16.13, 16.14. Additional review material may be found in the homework problems marked EIT. The answers to these problems will be found in Appendix B.

HOMEWORK PROBLEMS

16.1 Calculate the force exerted by each conductor, 6 in. long, on the armature of a DC motor when it carries a current of 90 A and lies in a field the density of which is 5.2×10^{-4} Wb per square in.

EIT 16.2 In a DC machine, the air-gap flux density is 4 W_b/m^2. The area of the pole face is 2 cm × 4 cm. Find the flux per pole in the machine.

16.3 A 220-V shunt motor has an armature resistance of 0.32 ohm and a field resistance of 110 ohms. At no load the armature current is 6 A and the speed is 1,800 rpm. Assume that the flux does not vary with load and calculate:

a. The speed of the motor when the line current is 62 A (assume a 2-volt brush drop).
b. The speed regulation of the motor.

16.4 A 120-V, 10-A shunt generator has an armature resistance of 0.6 Ω. The shunt field current is 2 A. Determine the voltage regulation of the generator.

16.5 A 50-hp, 550-volt shunt motor has an armature resistance, including brushes, of 0.36 ohm. When operating at rated load and speed, the armature takes 75 amp. What resistance should be inserted in the armature circuit to obtain a 20 percent speed reduction when the motor is developing 70 percent of rated torque? Assume that there is no flux change.

16.6 A 20-kW, 230-V separately excited generator has an armature resistance of 0.2 Ω and a load current of 100 A. Find:

a. The generated voltage when the terminal voltage is 230 V.
b. The output power.

EIT 16.7 A 10-kW, 120-VDC series generator has an armature resistance of 0.1 Ω and a series field resistance of 0.05 Ω. Assuming that it is delivering rated current at rated speed, find (a) the armature current and (b) the generated voltage.

16.8 The armature resistance of a 30-kW, 440-V shunt generator is 0.1 Ω. Its shunt field resistance is 200 Ω. Find

a. The power developed at rated load.
b. The load, field, and armature currents.
c. The electrical power loss.

16.9 A four-pole, 450-kW, 4.6-kV shunt generator has armature and field resistances of 2 and 333 Ω. The generator is operating at the rated speed of 3,600 rev/min. Find the no-load voltage of the generator and terminal voltage at half load.

16.10 A shunt DC motor has a shunt field resistance of 400 Ω and an armature resistance of 0.2 Ω. The motor nameplate rating values are 440 V, 1,200 rev/min, 100 hp, and full-load efficiency of 90 percent. Find:

a. The motor line current.
b. The field and armature currents.
c. The counter emf at rated speed.
d. The output torque.

EIT 16.11 A 30-kW, 240-V generator is running at half load at 1,800 rev/min with efficiency of 85 percent. Find the total losses and input power.

16.12 A 240-volt series motor has an armature resistance of 0.42 ohm and a series-field resistance of 0.18 ohm. If the speed is 500 rev/min when the current is 36 amp, what will be the motor speed when the load reduces the line current to 21 amp? (Assume a 3-volt brush drop and that the flux is proportional to the current.)

16.13 A 220-VDC shunt motor has an armature resistance of 0.2 Ω and a rated armature current of 50 A. Find

a. The voltage generated in the armature.
b. The power developed.

16.14 A 550-volt series motor takes 112 A and operates at 820 rev/min when the load is 75 hp. If the effective armature-circuit resistance is 0.15 Ω, calculate the horsepower output of the motor when the current drops to 84 A, assuming that the flux is reduced by 15 percent.

16.15 A 200-VDC shunt motor has the following parameters:

$$R_a = 0.1\ \Omega \qquad R_f = 100\ \Omega$$

When running at 1,100 rev/min with no load connected to the shaft, the motor draws 4 A from the line. Find E and the rotational losses at 1,100 rev/min (assuming that the stray-load losses can be neglected).

EIT 16.16 A 230-VDC shunt motor has the following parameters:

$$R_a = 0.5\ \Omega \qquad R_f = 75\ \Omega$$
$$P_{\text{rot}} = 500\ \text{W (at 1,120 rev/min)}$$

When loaded, the motor draws 46 A from the line. Find

a. The speed, P_{dev}, and T_{sh}.
b. If $L_f = 25$ H, $L_a = 0.008$ H, and the terminal voltage has a 115-V change, find $i_a(t)$ and $\omega_m(t)$.

16.17 A 200-VDC shunt motor with an armature resistance of 0.1 Ω and a field resistance of 100 Ω draws a line current of 5 A when running with no load at 955 rev/min. Determine the motor speed, the motor efficiency, the total losses (i.e., rotational and I^2R losses), and the load torque (i.e., T_{sh}) that will result when the motor draws 40 A from the line. Assume rotational power losses are proportional to the square of shaft speed.

16.18 A self-excited DC shunt generator is delivering 20 A to a 100-V line when it is driven at 200 rad/s. The magnetization characteristic is shown in Figure P16.1. It is known that $R_a = 1.0\ \Omega$ and $R_f = 100\ \Omega$. When the generator is disconnected from the line, the drive motor speeds up to 220 rad/s. What is the terminal voltage?

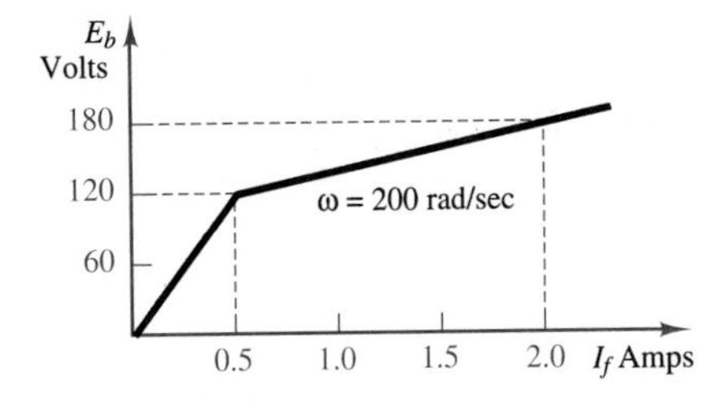

Figure P16.1

16.19 A 50-hp, 230-volt shunt motor has a field resistance of 17.7 Ω and operates at full load when the line current is 181 A at 1,350 rev/min. To increase the speed of the motor to 1,600 rev/min, a resistance of 5.3 Ω is "cut in" via the field rheostat; the line current then increases to 190 A. Calculate:

a. The power loss in the field and its percentage of the total power input for the 1,350 rev/min speed.
b. The power losses in the field and the field rheostat for the 1,600 rev/min speed.
c. The percent losses in the field and in the field rheostat at 1,600 rev/min.

16.20 A 10-hp, 230-volt shunt-wound motor has rated speed of 1,000 rev/min and full-load efficiency of 86 percent. Armature circuit resistance is 0.26 Ω; field-circuit resistance is 225 Ω. If this motor is operating under rated load and the field flux is very quickly reduced to 50 percent of its normal value, what will be the effect upon counter emf, armature current and torque? What effect will this change have upon the operation of the motor, and what will be its speed when stable operating conditions have been regained?

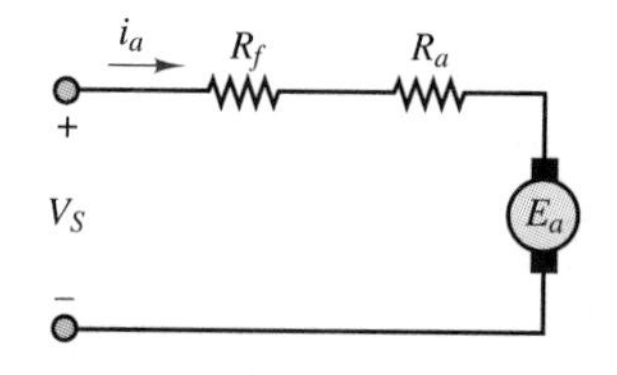

Figure P16.2

EIT 16.21 The machine of example 16.8 is being used in a series connection. That is, the field coil is connected in series with the armature, as shown in Figure P16.2. The machine is to be operated under the same conditions as in the previous example, that is, $n = 120$ rev/min, $I_a = 8$ A. In the operating region, $\phi = kI_f$, and $k = 200$. The armature resistance is 0.2 Ω, and the resistance of the field winding is negligible.

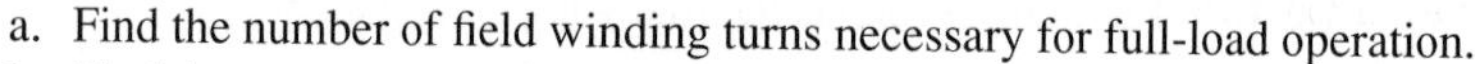
a. Find the number of field winding turns necessary for full-load operation.
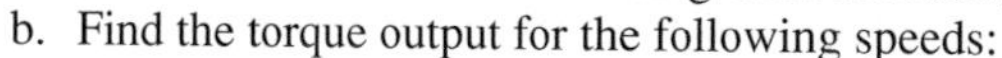
b. Find the torque output for the following speeds:
   1. $n' = 2n$
   2. $n' = 3n$
   3. $n' = \frac{n}{2}$
   4. $n' = \frac{n}{4}$

c. Plot the speed-torque characteristic for the conditions of part b.

16.22 A non-salient pole, Y-connected, three-phase, two-pole synchronous machine has a synchronous reactance of 7 Ω and negligible resistance and rotational losses. One point on the open-circuit characteristic is given by $V_o = 400$ V (phase voltage) for a field current of 3.32 A. The machine is to be operated as a motor, with a terminal voltage of 400 V (phase voltage). The armature current is 50 A, with power factor 0.85, leading. Determine E_b, field current, torque developed, and power angle δ.

16.23 A factory load of 900 kW at 0.6 power factor lagging is to be increased by the addition of a synchronous motor that takes 450 kW. At what power factor must this motor operate, and what must be its KVA input if the overall power factor is to be 0.9 lagging?

16.24 A non-salient pole, Y-connected, three-phase, two-pole synchronous generator is connected to a 400-V (line to line), 60-Hz, three-phase line. The stator impedance is $0.5 + j1.6$ Ω (per phase). The generator is delivering rated current (36 A) at unity power factor to the line. Determine the power angle for this load and the value of E_b for this condition. Sketch the phasor diagram, showing $\mathbf{E}_b$, $\mathbf{I}_S$, and $\mathbf{V}_S$.

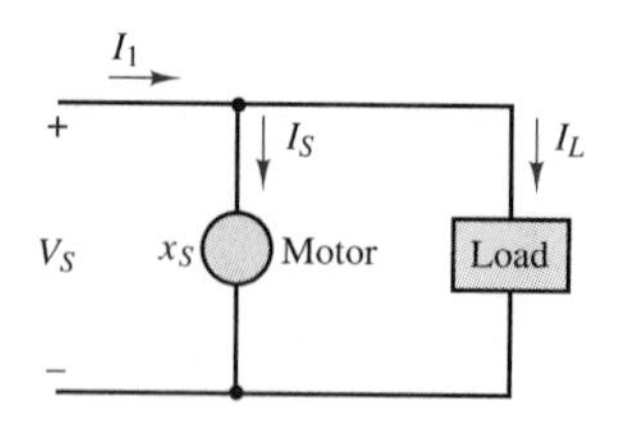

Figure P16.3

EIT 16.25 A non-salient pole, three-phase, two-pole synchronous motor is connected in parallel with a three-phase, Y-connected load so that the per-phase equivalent circuit is as shown in Figure P16.3. The parallel combination is connected to a 220-V (line to line), 60-Hz, three-phase line. The load current $\mathbf{I}_L$ is 25 A at a power factor of 0.866 inductive. The motor has $X_S = 2$ Ω and is operating with $I_f = 1$ A and $T = 50$ N-m at a power angle of $-30°$. (Neglect all losses for the motor.)
Find $\mathbf{I}_S$, P_{in} (to the motor), the overall power factor (i.e., angle between $\mathbf{I}_1$ and $\mathbf{V}_S$), and the total power drawn from the line.

16.26 An automotive alternator is rated 500 V-A and 20 V. It delivers its rated V-A at a power factor of 0.85. The resistance per phase is 0.05 Ω, and the field takes 2 A at 12 V. If friction and windage loss is 25 W and core loss is 30 W, calculate the percent efficiency under rated conditions.

16.27 A four-pole, three-phase, Y-connected, non–salient pole synchronous motor has a synchronous reactance of 10 Ω. This motor is connected to a $230\sqrt{3}$ V (line to line), 60-Hz, three-phase line and is driving a load such that $T_{\text{shaft}} = 30$ N-m. The line current is 15 A, leading the phase voltage. Assuming that all losses can be neglected, determine the power angle δ and E for this condition. If the load is removed, what is the line current, and is it leading or lagging the voltage?

16.28 A 10-hp, 230-V, 60 Hz, three-phase wye-connected synchronous motor delivers full load at a power factor of 0.8 leading. The synchronous reactance is 6 Ω, the rotational loss is 230 W, and the field loss is 50 W. Find

a. The armature current.
b. The motor efficiency.
c. The power angle.

Neglect the stator winding resistance.

16.29 The circuit of Figure P16.4 represents a voltage regulator for a car alternator. Briefly, explain the function of Q, D, Z, and SCR. Note that unlike other alternators, a car alternator is *not* driven at constant speed.

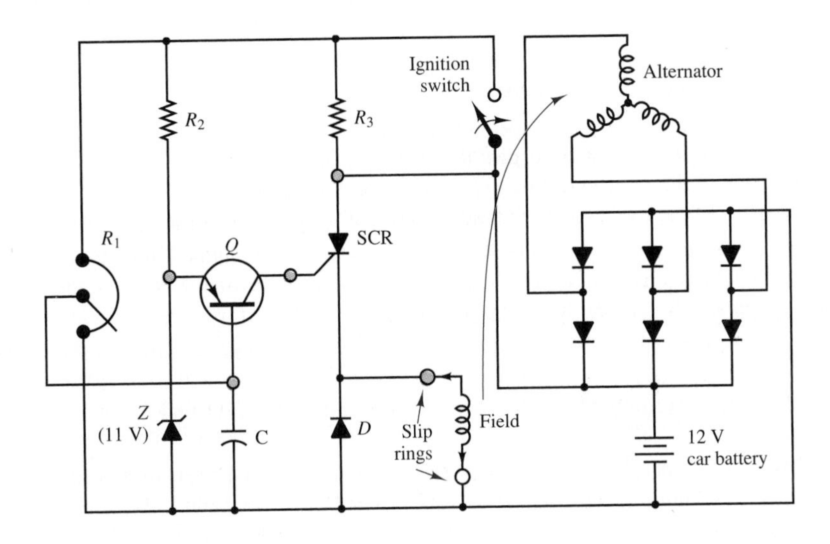

Figure P16.4

16.30 It has been determined by test that the synchronous reactance, X_s, and armature resistance, r_a, of a 2,300-V, 500-KVA, three-phase synchronous generator are 8.0 Ω and 0.1 Ω respectively. If the machine is operating at rated load and voltage at a power factor of 0.867 lagging, find the generated voltage per phase and the torque angle.

EIT 16.31 A 2,000-hp, unity power factor, three-phase, Y-connected, 2,300-V, 30-pole, 60-Hz synchronous motor has a synchronous reactance of 1.95 Ω per phase. Neglect all losses. Find the maximum power and torque.

16.32 A 1,200-V, three-phase, wye-connected synchronous motor takes 110 kW (exclusive of field winding loss) when operated under a certain load at 1,200 rev/min. The back emf of the motor is 2,000 V. The synchronous reactance is 10 Ω per phase, with negligible winding resistance. Find the line current and the torque developed by the motor.

16.33 The per-phase impedance of a 600-V, three-phase, Y-connected synchronous motor is $5 + j50\ \Omega$. The motor takes 24 kW at a leading power factor of 0.707. Determine the induced voltage and the power angle of the motor.

16.34 A 74.6-kW, three-phase, 440-V (line to line), four-pole, 60-Hz induction motor has the following (per-phase) parameters referred to the stator circuit:

$$R_S = 0.06\ \Omega \quad X_S = 0.3\ \Omega \quad X_m = 5\ \Omega \quad R_R = 0.08\ \Omega \quad X_R = 0.3\ \Omega$$

The no-load power input is 3,240 W at a current of 45 A. Determine the line current, the input power, the developed torque, the shaft torque, and the efficiency at $s = 0.02$.

EIT 16.35 A 60-Hz, four-pole, Y-connected induction motor is connected to a 400-V (line to line), three-phase, 60-Hz line. The equivalent circuit parameters are:

$$R_S = 0.2\ \Omega \quad R_R = 0.1\ \Omega$$
$$X_S = 0.5\ \Omega \quad X_R = 0.2\ \Omega$$
$$X_m = 20\ \Omega$$

When the machine is running at 1,755 rev/min, the total rotational and stray-load losses are 800 W. Determine the slip, input current, total input power, mechanical power developed, shaft torque, and efficiency.

16.36 A three-phase, 60-Hz induction motor has eight poles and operates with a slip of 0.05 for a certain load. Determine

a. The speed of the rotor with respect to the stator.
b. The speed of the rotor with respect to the stator magnetic field.
c. The speed of the rotor magnetic field with respect to the rotor.
d. The speed of the rotor magnetic field with respect to the stator magnetic field.

16.37 A three-phase, two-pole, 400-V (per phase), 60-Hz induction motor develops 37 kW (total) of mechanical power (P_m) at a certain speed. The rotational loss at this speed is 800 W (total). (Stray-load loss is negligible.)

a. If the total power transferred to the rotor is 40 kW, determine the slip and the output torque.
b. If the total power into the motor (P_{in}) is 45 kW and R_S is 0.5 Ω find I_S and the power factor.

16.38 The name-plate speed of a 25-Hz induction motor is 720 rev/min. If the speed at no load is 745 rev/min, calculate:

a. The slip
b. The percent regulation

16.39 The name plate of a squirrel-cage induction motor has the following information: 25 hp, 220 volts, three-phase, 60 Hz, 830 rev/min, 8 poles, 64A line current. If the motor draws 20,800 watts when operating at full load, calculate:

a. Slip
b. Percent regulation if the no-load speed is 895 rpm
c. Power factor
d. Torque
e. Efficiency

16.40 A squirrel-cage motor develops 80 lb-ft of starting torque when the 50 percent tap on an autotransformer is used to start the machine by a reduced voltage method. What starting torque will be developed if:

a. The 65 percent tap is used
b. The 80 percent tap is used

EIT 16.41 A 60-Hz, four-pole, Y-connected induction motor is connected to a 200-V (line to line), three-phase, 60 Hz line. The equivalent circuit parameters are:

$R_S = 0.48\ \Omega$ Rotational loss torque $= 3.5$ N-m

$X_S = 0.8\ \Omega$ $R_R = 0.42\ \Omega$ (referred to the stator)

$X_m = 30\ \Omega$ $X_R = 0.8\ \Omega$ (referred to the stator)

The motor is operating at slip $s = 0.04$. Determine the input current, input power, mechanical power, and shaft torque (assuming that stray-load losses are negligible).

16.42

a. A three-phase, 220-V, 60-Hz induction motor runs at 1,140 rev/min. Determine the number of poles (for minimum slip), the slip, and the frequency of the rotor currents.
b. To reduce the starting current, a three-phase squirrel-cage induction motor is started by reducing the line voltage to $V_s/2$. By what factor are the starting torque and the starting current reduced?

16.43 The power rating of a motor can be modified to account for different ambient temperature, according to the following table:

Ambient temperature	30°C	35°C	40°C	45°C	50°C	55°C
Variation of rated power	+8%	+5%	0	−5%	−12.5%	−25%

A motor with $P_e = 10$ kW is rated up to 85°C. Find the actual power for each of the following conditions:

a. Ambient temperature is 50°C.
b. Ambient temperature is 25°C.

16.44 The speed-torque characteristic of an induction motor has been empirically determined as follows:

Speed	1,470	1,440	1,410	1,300	1,100	900	750	350	0	rev/min
Torque	3	6	9	13	15	13	11	7	5	N-m

The motor will drive a load requiring a starting torque of 4 N-m and increase linearly with speed to 8 N-m at 1,500 rev/min.

a. Find the steady-state operating point of the motor.
b. Equation 16.95 predicts that the motor speed can be regulated in the face of changes in load torque by adjusting the stator voltage. Find the change in voltage required to maintain the speed at the operating point of part a if the load torque increases to 10 N-m.

EIT 16.45 An induction motor, operating from a 60-Hz line, develops 10 hp at 1,745 rev/min. Find the speed and the horsepower if the load torque is reduced by one half.

16.46 A blocked-rotor test was performed on a 5-hp, 220-V, four-pole, 60 Hz, three-phase induction motor. The following data were obtained: $V = 48$ V, $I = 18$ A, $P = 610$ W. Calculate:

a. The equivalent stator resistance per phase, R_S,
b. The equivalent rotor resistance per phase, R_R, and
c. The equivalent blocked-rotor reactance per phase, X_R.

16.47 Calculate the starting torque of the motor of Problem 16.46 when it is started at:

a. 220 V
b. 110 V

The starting torque equation is:

$$T = \frac{m}{\omega_e} \cdot V_S^2 \cdot \frac{R_R}{\left[(R_R + R_S)^2 + (X_R + X_S)^2\right]}$$

16.48 Find the speed of the rotating field of a six-pole, three-phase motor connected to (a) a 60-Hz line and (b) a 50-Hz line, in rev/min and rad/s.

16.49 A six-pole, three-phase, 440-V, 60-Hz induction motor has the following model impedances:

$$R_S = 0.8\ \Omega \quad X_S = 0.7\ \Omega$$
$$R_R = 0.3\ \Omega \quad X_R = 0.7\ \Omega$$
$$X_m = 35\ \Omega$$

Calculate the input current and power factor of the motor for a speed of 1,200 rev/min.

EIT 16.50 An eight-pole, three-phase, 220-V, 60-Hz induction motor has the following model impedances:

$$R_S = 0.78\ \Omega \quad X_S = 0.56\ \Omega \quad X_m = 32\ \Omega$$
$$R_R = 0.28\ \Omega \quad X_R = 0.84\ \Omega$$

Find the input current and power factor of this motor for $s = 0.02$.

16.51 A nameplate is given in Example 16.2. Find the rated torque, rated volt amperes, and maximum continuous output power for this motor.

16.52 A 3-phase induction motor, at rated voltage and frequency, has a starting torque of 140 percent and a maximum torque of 210 percent of full-load torque. Neglect stator resistance and rotational losses and assume constant rotor resistance. Determine:

a. The slip at full load.
b. The slip at maximum torque.
c. The rotor current at starting as a percent of full-load rotor current.

16.53 The system shown in Figure P16.5 was widely used to convert balanced supply frequency voltages to other frequencies. The 3-phase wound rotor induction machine is rigidly coupled to the 3-phase synchronous motor. The terminals of the 3-phase rotor winding of the induction machine are brought out to slip rings. The induction machine is driven by the synchronous motor at the proper speed and in the proper direction so that 3-phase 180-Hz voltages are available at the slip rings. The induction machine has a 6-pole stator winding.

a. How many poles must the rotor winding of the induction machine have?
b. If the stator field in the induction machine rotates in a counterclockwise direction, what must the direction of rotation of its rotor be?

c. What must the speed in rpm be?
d. How many poles must the synchronous motor have?

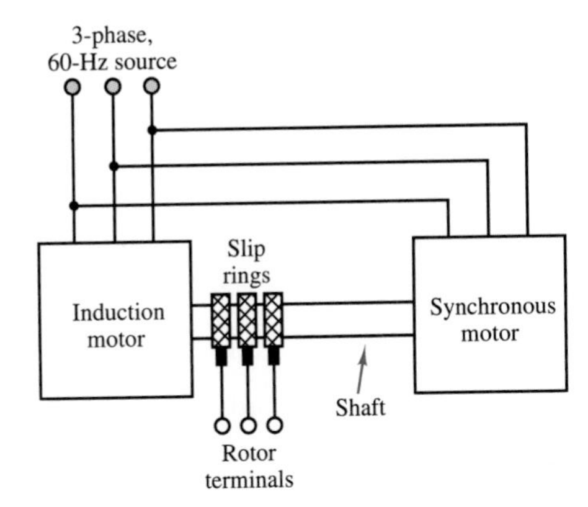

Figure P16.5

16.54 In the system shown in Figure P16.5, the synchronous motor has four poles and drives the interconnecting shaft in a clockwise direction. The induction machine has 20 poles, and its stator windings are connected to the lines to produce a *counter*clockwise rotating field.

a. At what speed does the motor run?
b. What is the frequency of the rotor voltages in the induction machine?

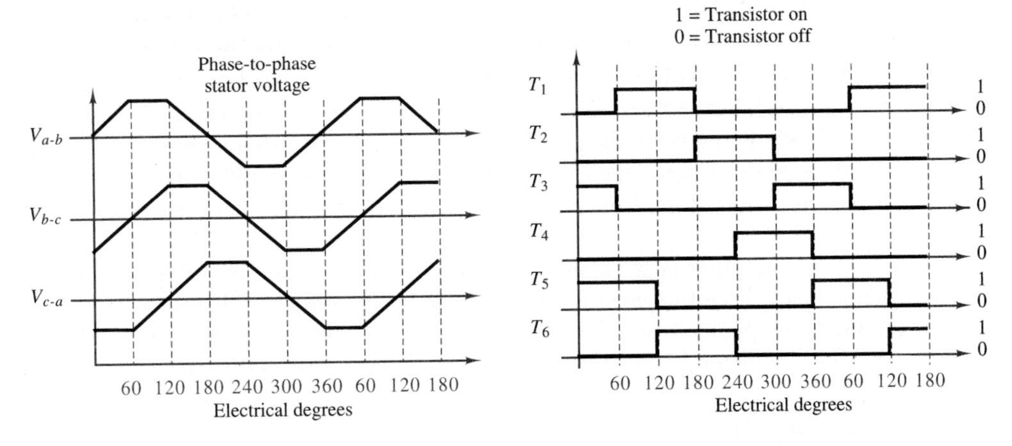

Figure 17.4 Phase voltages and transistor (SCR) switching sequence for the brushless DC motor drive of Figure 17.3

addition of the three phase voltages of Figure 17.4 leads to a constant voltage. The brushless DC motor is therefore similar to a standard permanent-magnet DC motor, and can be described by the following simplified equations:

$$V = k_a \omega_m + R_w I \quad \textbf{(17.1)}$$

$$T = k_T I \quad \textbf{(17.2)}$$

where

$$k_a = k_T$$

and where

V = motor voltage

k_a = armature constant

ω_m = mechanical speed

R_w = winding resistance

T = motor torque

k_T = torque constant

I = motor (armature) current

The speed and torque of a brushless DC motor can therefore be controlled with any variable-speed DC supply, such as one of the supplies briefly discussed in Chapter 10. Further, since the brushless motor has intrinsically higher torque and lower inertia than its DC counterpart, its response speed is superior to that obtained from traditional DC motors.

One important difference between the conventional DC and the brushless motor, however, is due to the coarseness of the electronic switching compared with the mechanical switching of the brush-type DC motor (recall the discussion of torque ripple due to the commutation effect in DC motors in Chapter 16). In practice, one cannot obtain the exact trapezoidal emf of Figure 17.4 by means of the transistor switching circuit of Figure 17.3, and a voltage ripple results as a consequence, leading to a torque ripple in the motor. Additional phase windings on the stator could solve the problem, at the expense of further complexity in the drive electronics, since the switching sequence would be more complex. Thus, brushless motors suffer from an inherent trade-off between torque ripple and drive complexity.

Among other applications, brushless DC motors find use in the design of servo loops in control systems—for example, in computer disk drives, and in propulsion systems for electric vehicles. The comparisons between the conventional DC motor and the brushless DC motor are summarized in the following table:

Conventional DC motors

Advantages
1. Controllability over a wide range of speeds
2. Capability of rapid acceleration and deceleration
3. Convenient control of shaft speed and position by servo amplifiers

Disadvantages
1. Commutation (through brushes) causes wear, electrical noise, and sparking

Brushless DC motors

Advantages
1. Controllability over a wide range of speeds
2. Capability of rapid acceleration and deceleration
3. Convenient control of shaft speed and position
4. No mechanical wear or sparking problem due to commutation
5. Better heat dissipation capabilities

Disadvantages
1. Need for more complex power electronics than the brush-type DC motor for equivalent power rating and control range

Example 17.1

This example explains a sinusoidal torque-generation scheme for brushless motors, in contrast to the system that produces the trapezoidal waveforms of Figure 17.4. The torque-generation scheme uses a magnet and field coil arrangement that produces a torque function resulting from the sine-cosine relationship between the currents in two coils, as shown in Figure 17.5.

The torques produced by the respective coils are

$$\text{Coil 1: } T_1 = I_{m1} k_T \sin\theta$$

$$\text{Coil 2: } T_2 = I_{m2} k_T \cos\theta$$

where T_1 and T_2 are instantaneous torque values and k_T is the torque constant of each winding. The current in each coil is controlled as a function of shaft angle by the switching electronics, as follows:

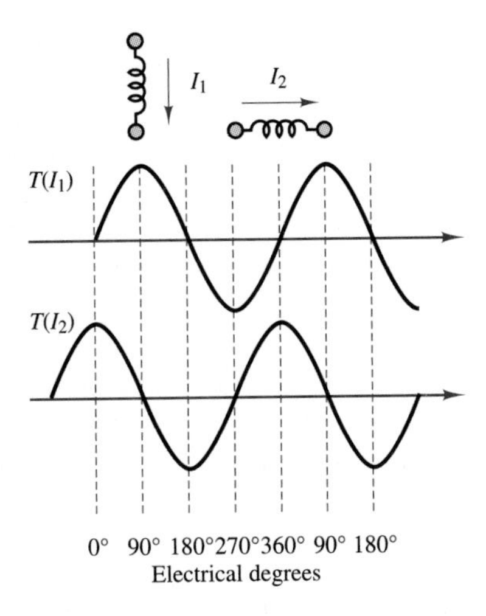

Figure 17.5 Sinusoidal torque-generation circuit and current waveforms for a brushless DC motor

$$
\begin{aligned}
I_{m1} &= I_m \sin\theta \\
I_{m2} &= I_m \cos\theta
\end{aligned}
$$

where I_m is the amplitude of the desired motor current. Thus, the total output torque is

$$
\begin{aligned}
T &= T_1 + T_2 \\
&= I_m k_T \sin\theta \times \sin\theta + I_m k_T \cos\theta \times \cos\theta \\
&= k_T I_m (\sin^2\theta + \cos^2\theta) = k_T I_m
\end{aligned}
$$

Thus, the torque generated is directly proportional to the supply current amplitude:

$$T = k_T I_m$$

Example 17.2

A three-phase brushless DC motor is running at 500 rev/min. Four phases conduct in one cycle. The motor parameters are

$$k_T = 0.037 \text{ N-m/A (per phase)} \qquad I = 3.6 \text{ A (per phase)}$$

Find the torque generated by the motor.

Solution:

Since four phases conduct during each cycle, the torque is given by

$$T = 4k_T I = 4 \times 0.037 \times 3.6 = 0.53 \text{ N-m}$$

DRILL EXERCISES

17.1 List four features of the brushless DC motor that differ from a conventional shunt-type DC motor.

17.2 Repeat Example 17.2 for $I = 7.2$ A.

17.3 Convert the result of Drill Exercise 17.2 to oz-in.

17.4 Verify that the transistor and SCR switching circuits of Figure 17.3 perform the same function.

(a) Permanent-magnet stepping motor

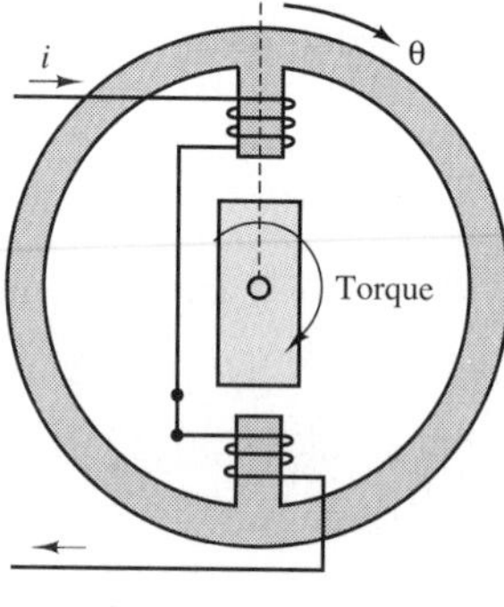

(b) Variable-reluctance stepping motor

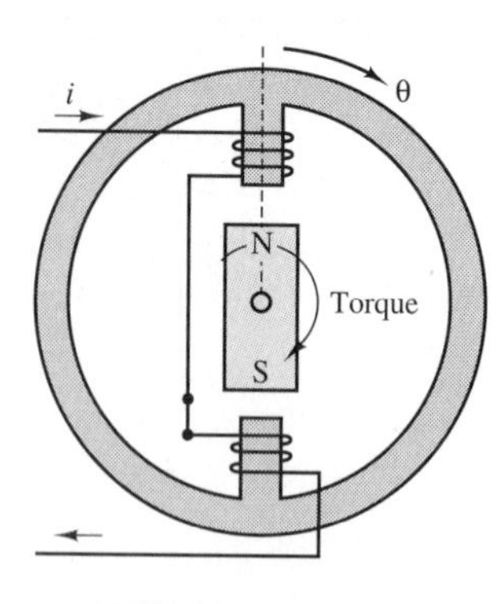

(c) Hybrid stepping motor

Figure 17.6 Stepping motor configurations

17.2 STEPPING MOTORS

Stepping, or **stepper, motors** are motors that convert digital information to mechanical motion. The principles of operation of stepping motors have been known since the 1920s; however, their application has seen a dramatic rise with the increased use of digital computers. Stepping motors, as the name suggests, rotate in distinct steps, and their position can be controlled by means of logic signals. Typical applications of stepping motors are line printers, positioning of heads in magnetic disk drives, and any other situation where continuous or stepwise displacements are required.

Stepping motors can generally be classified in one of three categories: variable-reluctance, permanent-magnet, and hybrid types. It will soon be shown that the principles of operation of each of these devices bear a definite resemblance to those of devices already encountered in this book. Stepping motors have a number of special features that make them particularly useful in practical applications. Perhaps the most important feature of a stepping motor is that the angle of rotation of the motor is directly proportional to the number of input pulses; further, the angle error per step is very small and does not accumulate. Stepping motors are also capable of rapid response to starting, stopping, and reversing commands, and can be driven directly by digital signals. Another important feature is a self-holding capability that makes it possible for the rotor to be held in the stopped position without the use of brakes. Finally, a wide range of rotating speeds—proportional to the frequency of the pulse signal—may be attained in these motors.

Figure 17.6 depicts the general appearance of three types of stepping motors. The **permanent-magnet-rotor stepping motor,** Figure 17.6(a), permits a nonzero holding torque when the motor is not energized. Depending on the construction of the motor, it is typically possible to obtain step angles of 7.5, 11.25, 15, 18, 45, or 90 degrees. The angle of rotation is determined by the number of stator poles, as will be illustrated shortly in an example. The **variable-reluctance stepping motor,** Figure 17.6(b), has an iron multipole rotor and a laminated wound stator, and rotates when the teeth on the rotor are attracted to the electromagnetically energized stator teeth. The rotor inertia of a variable-reluctance stepping motor is low, and the response is very quick, but the allowable load inertia is small. When the windings are not energized, the static torque of this type of motor is zero. Generally, the step angle of the variable-reluctance stepping motor is 15 degrees.

The **hybrid stepping motor,** Figure 17.6(c), is characterized by multi-toothed stator and rotor, the rotor having an axially magnetized concentric magnet around its shaft. It can be seen that this configuration is a mixture of the variable-reluctance and permanent-magnet types. This type of motor generally has high accuracy and high torque and can be configured to provide a step angle as small as 1.8 degrees.

For any of these configurations, the principle of operation is essentially the same: when the coils are energized, magnetic poles are generated in the stator, and the rotor will align in accordance with the direction of the magnetic field developed in the stator. By reversing the phase of the currents in the coils, or by energizing only some of the coils (this is possible in motors with more than two stator poles), the alignment of the stator magnetic field can take one of a discrete number of positions; if the currents in the coils are pulsed in the appropriate sequence, the rotor will advance in a step-by-step fashion. Thus, this type of motor can be very useful whenever precise incremental motion must be attained. As mentioned earlier, typical applications are printer wheels, computer disk drives, and plotters. Other applications are found in the control of the position of valves (e.g., control of the throttle valve in an engine, or of a hydraulic valve in a fluid power system), and in drug-dispensing apparatus for clinical applications.

The following examples illustrate the operation of a four-pole, two-phase permanent-magnet stepping motor, and of a similar motor of the variable-reluctance type. The operation of these motors is representative of all stepping motors.

Example 17.3

The PM stepper motor shown in Figure 17.7 can be operated by applying either or both of the currents i_1 and i_2, and by selecting the polarity of the currents. The currents are provided by a bipolar source (i.e., both positive and negative currents can be provided).

Initially, it will be assumed that the rotor is at rest; this condition can be attained with $i_1 > 0$ and $i_2 = 0$. The simplest mode of operation is the full-step, single-phase sequence, in which current is supplied to only one of the two coils at any given time while the other coil is deenergized. The current sequence shown in Table 17.1 will then cause the rotor to advance in 90° increments. Thus, if current is supplied to the coils with the appropriate switching sequence, the motor will rotate by a 90° step with every current pulse.

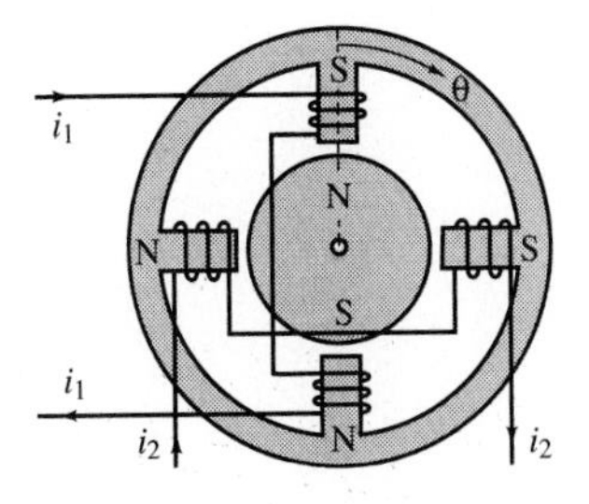

Figure 17.7 Two-phase four-pole PM stepper motor

Table 17.1 **Full-step, single-phase sequence**

i_1	i_2	θ
+	0	0°
0	+	90°
−	0	180°
0	−	270°
+	0	0°

It is also possible to obtain intermediate steps by supplying current to both coils. In this case, the magnetic stator poles will cause the rotor to align halfway between the stator teeth, also in increments of 90°, but shifted by 45° with respect to the single-phase stepping sequence. Table 17.2 illustrates the stepping sequence for this full-step, two-phase mode of operation.

Table 17.2 Full-step, two-phase sequence

i_1	i_2	θ
+	+	45°
−	+	135°
−	−	225°
+	−	315°
+	+	45°

Finally, the two sequences may be combined, since the current commands required by each are distinct, to obtain the half-step sequence described in Table 17.3. It should be apparent that for this particular motor, it is not possible to obtain finer increments in position than 45°. However, by increasing the number of windings and stator teeth, greater resolution can be obtained.

Table 17.3 Half-step sequence

i_1	i_2	θ
+	0	0°
+	+	45°
0	+	90°
−	+	135°
−	0	180°
−	−	225°
0	−	270°
+	−	315°
+	0	0°

Example 17.4

Variable-reluctance (VR) stepper motors are characterized by the presence of salient poles in both the rotor and the stator, as shown in Figure 17.8, where a two-pole rotor is depicted.

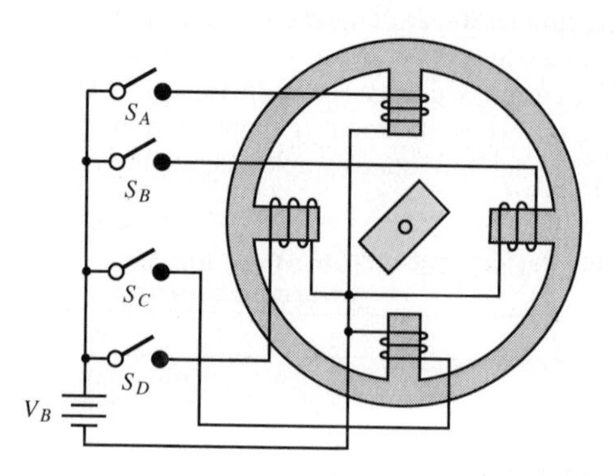

Figure 17.8 Two-phase four-pole VR stepping motor

The operation of the VR motor is slightly different from that of the PM type, because the rotor is not magnetically polarized and therefore it is not necessary to have a bipolar supply to cause the rotor to rotate through a full revolution. Figure 17.8 depicts an ideal arrangement in which a DC source and four switches (controlled by a logic circuit, such as one of the circuits discussed in Chapters 12 and 13) are used to step the motor through its sequence of positions.

Figure 17.9 illustrates how the rotor would be positioned for three different positions of the switches; you should be able to infer the complete switching sequence for the motor. Note that, just as in the preceding example, the smallest increment in position that can be attained by this motor is 45°.

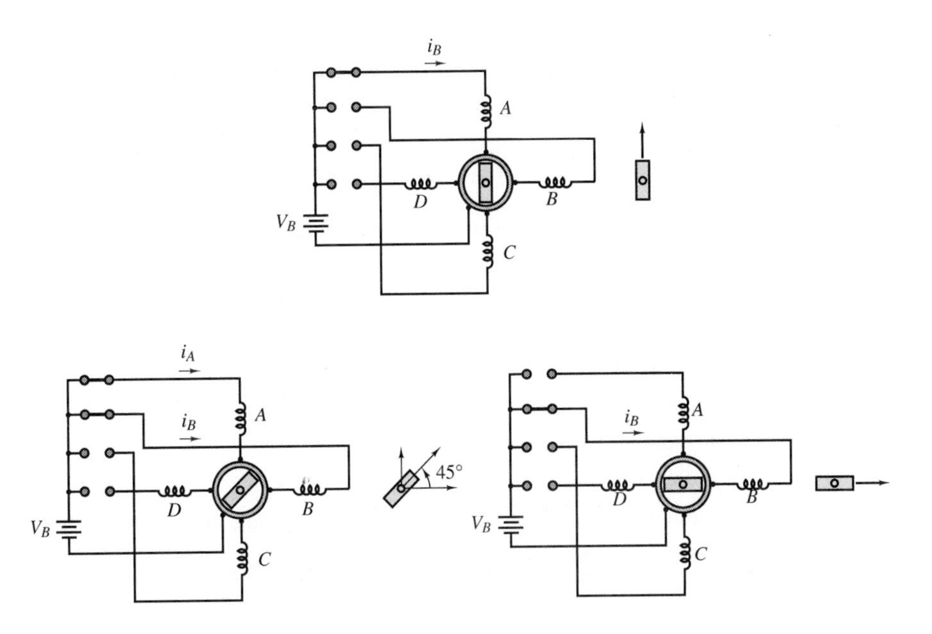

Figure 17.9 Two-phase four-pole VR motor positioning sequence

Example 17.5

The step angle of a VR stepping motor is determined by the number of teeth on the rotor and the stator, as well as the number of phases. The following equations give the relationship between the number of steps per revolution, the number of phases, the number of teeth in each phase, and the step angle (resolution):

$$N = tm$$

where

N = number of steps per revolution

t = number of teeth per phase

m = number of phases

and

$$\Delta\theta = \frac{360^\circ}{tm}$$

where $\Delta\theta$ = step angle, in degrees, or resolution.

For example, if a VR stepping motor has 12 teeth on the rotor and three phases, we can compute the steps per revolution as follows:

$$N = tm = 12 \times 3 = 36$$

It also follows that the step angle is given by

$$\Delta\theta = \frac{360^\circ}{tm} = \frac{360^\circ}{36} = 10^\circ$$

Example 17.6

The torque equation of a stepping motor may be written as

$$T = 0.314 \times 10^{-6} \frac{tL(r + g/2)\mathcal{F}^2}{g} \text{ N-m}$$

where

T = torque, N-m
t = number of teeth per phase
L = axial length of rotor, m
g = rotor-to-stator radial air gap, m
r = rotor radius, m
$\mathcal{F}$ = magnetomotive force developed across the two air gaps in series through which a line of flux must pass in one phase

For a stepping motor having the following data:

t = 16 teeth per phase (48 steps per revolution, for three-phase excitation)
$g = 6.35 \times 10^{-5}$ m
$r = 1.29 \times 10^{-2}$ m
$\mathcal{F}$ = 720 A-t
$L = 6.35 \times 10^{-3}$ m

we may compute the torque T using the preceding relationship, as follows:

$$T = \frac{0.314 \times 10^{-6} \times 16 \times 6.35 \times 10^{-3}(1.29 \times 10^{-2} + \frac{1}{2} \times 6.35 \times 10^{-5}) \times 720^2}{6.35 \times 10^{-5}}$$

$$= 3.37 \text{ N-m}$$

From the preceding examples, you should now have a feeling for the operation of variable-reluctance and PM stepping motors. The hybrid configuration is characterized by multitooth rotors that are made of magnetic materials, thus providing a variable-reluctance geometry in conjunction with a permanent-magnet rotor.

An ideal torque-speed characteristic for a stepper motor is shown in Figure 17.10. Two distinct modes of operation are marked on the curve: the **locked-step mode,** and the **slewing mode.** In the first mode, the rotor comes to rest (or at least decelerates) between steps; this is the mode commonly used to achieve a given rotor position. In the locked-step mode, the rotor can be started, stopped, and reversed. The slewing mode, on the other hand, does not allow stopping or reversal of the rotor, although the rotor still advances in synchronism with the stepping sequence, as described in the preceding examples. This second mode could be used, for example, in rewinding or fast-forwarding a tape drive.

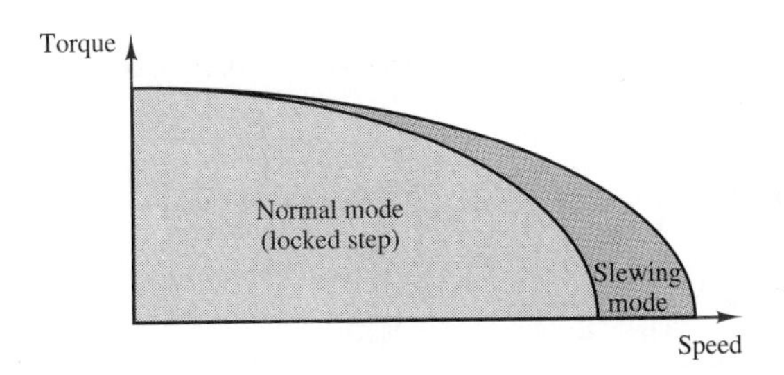

Figure 17.10 Ideal torque-speed characteristic of a stepping motor

The power supply, or driver, required by a stepping motor is shown in block diagram form in Figure 17.11; it includes a DC power supply, to provide the required current to drive the motor, in addition to logic and switching circuits to provide the appropriate inputs at the right time. One of the important considerations in driving a stepping motor is the excitation mode, which can be one-phase or two-phase. The driver is the circuit that arranges, distributes, and amplifies pulse trains from the logic circuit determining the stepping sequence; the driver excites each winding of the stepping motor at specified times. In the **one-phase excitation mode,** current is supplied to one phase at a time, with the advantages of low power consumption and good step-angle accuracy. Input signal pulses and the change in the condition of each phase excitation are shown in Figure 17.12. In the **two-phase excitation mode,** current is simultaneously provided to two phases.

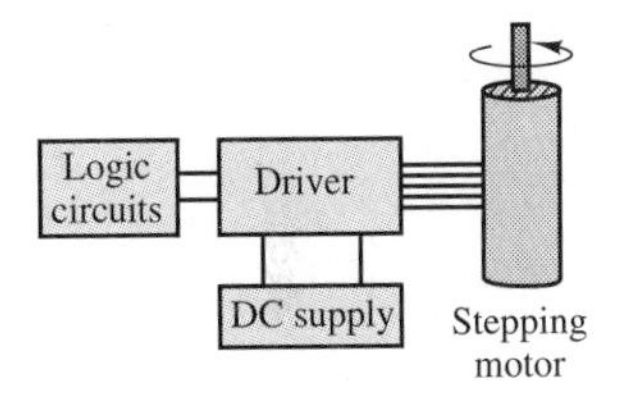

Figure 17.11 Power supply for stepping motor

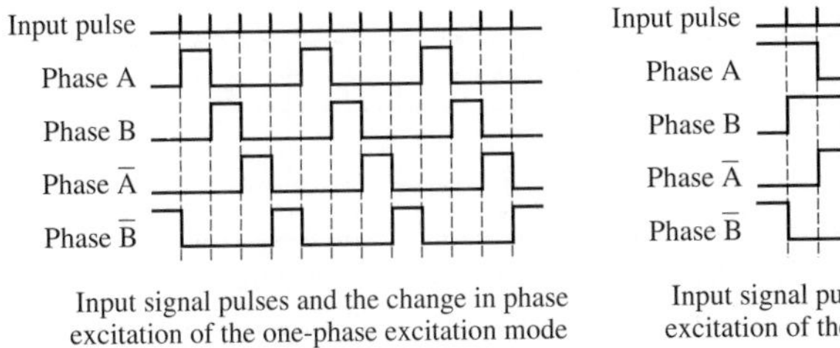

Figure 17.12 One- and two-phase excitation waveforms for stepper motors

Input signal pulses and the change in the condition of each phase excitation are also shown in Figure 17.12.

In addition to the classification of the excitation by phase, stepping motor drives are also classified according to whether the drive supplies are unipolar or bipolar, that is, whether they can supply current in one or both directions. Unipolar excitation is clearly simpler, although in the case of the two-phase excitation mode, only half of the motor windings are used, with an obvious decrease in performance. Figure 17.13 shows a circuit diagram of the unipolar drive and the sequence of phase excitation.

When a bipolar drive is used, motor windings are used effectively, because of the bidirectional exciting current; when operated in this mode, a stepping motor can generate a large output torque at low speed compared with the unipolar drive.

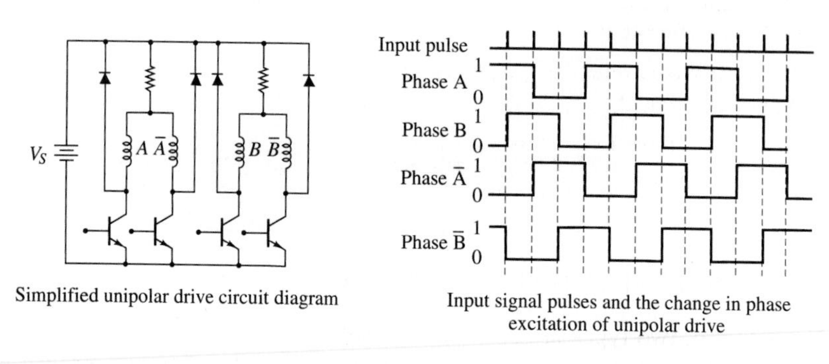

Simplified unipolar drive circuit diagram

Input signal pulses and the change in phase excitation of unipolar drive

Figure 17.13 Unipolar drive for stepper motor

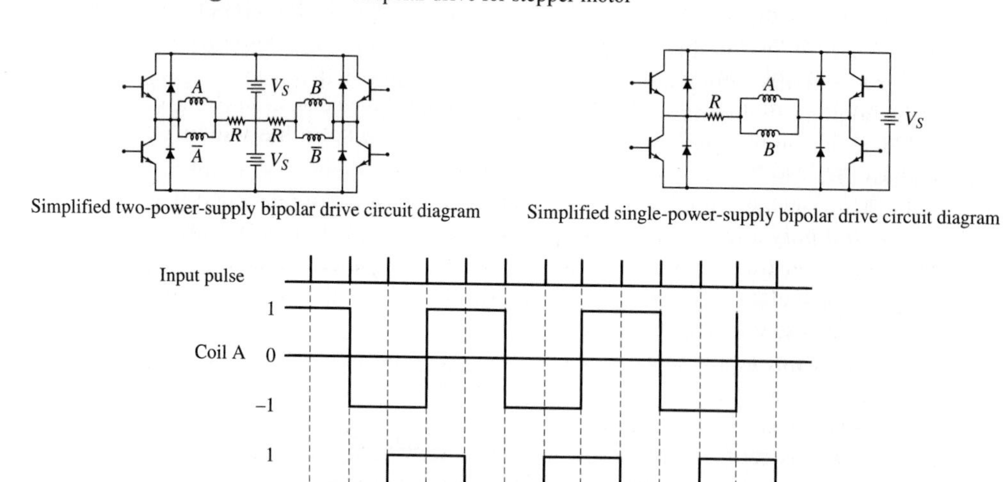

Simplified two-power-supply bipolar drive circuit diagram

Simplified single-power-supply bipolar drive circuit diagram

Input pulse signals and the change in phase excitation of bipolar drive

Figure 17.14 Bipolar drive for stepper motors

Figure 17.14 shows two versions of the bipolar drive. The first requires two power supplies, one for each polarity, while the second requires only one power supply but needs four switching transistors per phase to reverse the polarity.

DRILL EXERCISES

17.5 Explain why the term *variable-reluctance* is used to describe the class of stepping motors so named.

17.6 Determine the smallest increment in angular position that can be obtained with a PM stepper motor with six stator teeth and three-phase current excitation.

17.7 Express the stepping sequence of the variable-reluctance stepping motor of Example 17.4 as a four-digit binary sequence.

17.8 Derive the excitation waveforms corresponding to the direction of rotation opposite to that caused by the stepping sequence shown in Figure 17.12.

17.9 For Example 17.6, express the torque in units of lb-in.

17.3 SINGLE-PHASE AC MOTORS

In Chapter 16, two types of AC machines were discussed: synchronous, and induction. In the discussion of these devices, especially in their motor applications, three-phase excitation was assumed; however, in many practical applications—and especially in small household appliances and small industrial motors—three-phase sources are not readily available, and it would be desirable to use single-phase excitation. Unfortunately, single-phase power does not lend itself to the generation of a rotating magnetic field: single-phase currents in the winding of an AC machine lead to a magnetic field that pulsates in amplitude but does not rotate in space. Thus, it would not be possible to use the AC machines described in Chapter 16 if only single-phase power were available. The aim of this section is to discuss the construction and the operating and performance characteristics of single-phase AC motors. The discussion will focus mainly on the *universal motor* and on **single-phase induction motors.**

Fractional-horsepower (as opposed to **integral-horsepower**) **motors** represent by far the major share of the industrial production of electric motors. Many fractional-horsepower motors are designed for single-phase use, since single-phase AC power is readily available practically anywhere. Many applications are related to household appliances: refrigerator compressors, air conditioners, fans, electric tools, washer and dryer motors, and others. For the rest of this chapter, we shall examine qualitatively the principle of operation of single-phase motors and look at a few applications. The variety of designs for practical single-phase motors is such that it would not be possible to present the detailed principles of operation for all common types. However, it is hoped that the introduction provided in this chapter will help you in decoding the manufacturer's specifications for a given motor, and in making a preliminary selection for a given application.

Example 17.7

A small motor, as defined by the American Standards Association (ASA) and the National Electrical Manufacturers Association (NEMA; see Chapter 16) is a "motor built in a frame smaller than that having a continuous rating of 1 hp, open type, at 1700 to 1800 rev/ min." Small motors are generally considered *fractional-horsepower motors*. However, since the determination is based on frame size and on a given speed range, the classification of a motor is not always obvious. Let us give two examples.

1. Consider a $\frac{3}{4}$-hp, 1,200-rev/min motor. This motor is not considered a fractional-horsepower motor, because of its frame size. If the same frame size were used for an 1,800-rev/min motor, it would produce a rating of more than 1 hp. Thus, it is considered an integral-horsepower motor of

$$0.75 \text{ hp} \times \frac{1{,}800}{1{,}200} = 1.125 \text{ hp}$$

   In other words, since the motor is capable of integral-horsepower performance at speeds of 1,700 to 1,800 rev/min, it is classified as an integral-horsepower motor.

2. Consider now a 1.25-hp, 3,600-rev/min motor. This motor is classified as a fractional-horsepower motor, in spite of the fact that its power output is actually greater than 1 hp. If the same motor were used at a speed of 1,800 rev/min, it would produce a rating of less than 1 hp:

$$1.25 \times \frac{1{,}800}{3{,}600} = 0.625 \text{ hp}$$

Thus, we see once again that some attention must be paid to the speed of operation of the motor in determining its classification. The term *fractional horsepower* relates more to the physical size of the machine than to the actual power output rating.

The Universal Motor

If it were possible to operate a DC motor from a single-phase AC supply, a wide range of simple applications would become readily available. Recall that the direction of the torque produced by a DC machine is determined by the direction of current flow in the armature conductors and by the polarity of the field; torque is developed in a DC machine because the commutator arrangement permits the field and armature currents to remain in phase, thus producing torque in a constant direction. A similar result can be obtained by using an AC supply, and by connecting the armature and field windings in series, as shown in Figure 17.15. A series DC motor connected in this configuration can therefore operate on a single-phase AC supply, and is referred to as a **universal motor.** An additional consideration is that, because of the AC excitation, it is necessary to reduce AC core losses by laminating the stator; thus, the universal motor differs from the series DC motor

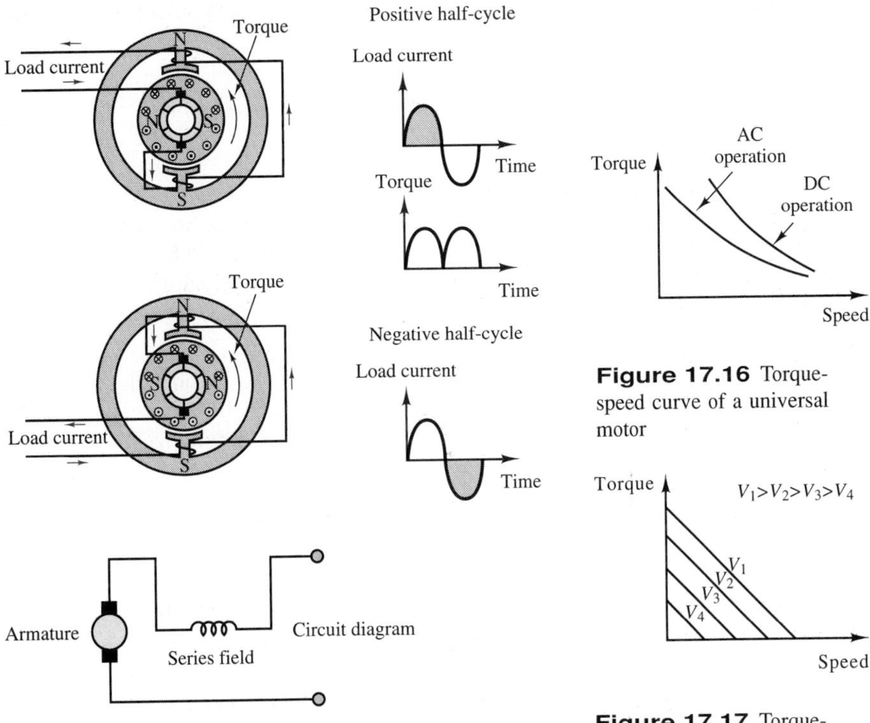

Figure 17.15 Operation and circuit diagram of a universal motor

Figure 17.16 Torque-speed curve of a universal motor

Figure 17.17 Torque-speed characteristics of a universal motor

discussed in Chapter 16 in its construction features. Typical torque-speed curves for AC and DC operation of a universal motor are shown in Figure 17.16. As shown in Figure 17.15, the load current is sinusoidal and therefore reverses direction each half-cycle; however, the torque generated by the motor is always in the same direction, resulting in a pulsating torque, with nonzero average value.

As in the case of a DC series motor, the best method for controlling the speed of a universal motor is to change its (rms) input voltage. The higher the rms input voltage, the greater the resulting speed of the motor. Approximate torque-speed characteristics of a universal motor as a function of voltage are shown in Figure 17.17.

Example 17.8

A 120-V, 60-Hz, two-pole universal motor operates at a speed of 800 rev/min at full load and draws a current of 17.85 A at a lagging power factor of 0.912. The parameters of the windings are:

Series field winding:

$$R_f = 0.65\ \Omega \qquad X_f = 1.2\ \Omega$$

Armature winding:

$$R_a = 1.36\ \Omega \qquad X_a = 1.6\ \Omega$$

Find:

1. The generated voltage.
2. The power output.
3. The shaft torque.
4. The efficiency of the motor (assume that the rotational loss is 80 W).

Solution:

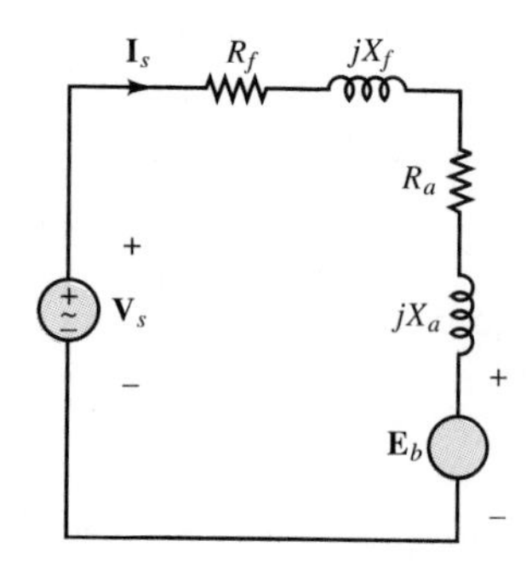

Figure 17.18 Equivalent circuit of a universal motor

Using the circuit model for the series motor discussed in Chapter 16, we can obtain the circuit shown in Figure 17.18. Note that now voltages and currents are phasor quantities.

1. The impedance angle may be found from the power factor:

$$\cos^{-1}\theta = 0.912 \qquad \theta = -24.22^\circ$$

and therefore the expression for the back emf can be evaluated as follows:

$$\begin{aligned}\mathbf{E}_b &= \mathbf{V}_S - \mathbf{I}_a(R_f + R_a + jX_f + jX_a)\\ &= 120 - (17.85\angle -24.22^\circ)(0.65 + 1.36 + j1.2 + j1.6)\\ &= 73.56\angle -24.8^\circ\ \text{V}\end{aligned}$$

2. The total power developed by the motor is given by the product of the induced emf and the series current:

$$P_{\text{total}} = E_b \times I_S = 73.56 \times 17.85 = 1{,}313.15\ \text{W}$$

The useful output power is the difference between the total power developed by the motor and the rotational losses:

$$P_{\text{out}} = P_{\text{total}} - P_{\text{rot}} = 1{,}313.15 - 80 = 1{,}233.15\ \text{W}$$

3. The angular speed in units of rad/s is

$$\omega_m = \frac{2\pi \times 800}{60} = 83.776\ \text{rad/s}$$

and therefore the output torque, T_{out}, is

$$T_{\text{out}} = \frac{P_{\text{out}}}{\omega_m} = \frac{1{,}233.15}{83.776} = 14.72\ \text{N-m}$$

4. The input power (or power supplied to the motor) is found by computing the phasor product of the supply voltage and current:

$$P_{\text{in}} = V_S I_S \cos\theta = 120 \times 17.85 \times 0.912 = 1{,}953.5\ \text{W}$$

Knowing both the input and the output power, we can compute the efficiency of the motor according to the following expression:

$$\text{Efficiency} = \frac{1{,}233.15}{1{,}953.5} = 0.6312 = 63.12\%$$

Example 17.9

Compute an expression for the average torque of a universal motor based on the circuit diagram of Figure 17.18. Assume that the magnetic flux is such that the motor operates in the linear region of its magnetization curve.

Solution:

The instantaneous magnetic flux ϕ of an AC series motor produced by the motor current $i_S(t)$ in the field is

$$\phi = k_S i_S(t)$$

The instantaneous torque is given by

$$T(t) = k_T \phi(t) i_S(t)$$

Thus, the time-average torque, T_{ave}, over one period of the supply waveform, τ, is

$$T_{\text{ave}} = \frac{1}{\tau\omega}\int_0^{\tau\omega} k_T k_S I_S^2 dt$$

When operated on AC supplies, the universal motor develops a pulsating torque, one pulse for each half-cycle, as shown in Figure 17.15. The average value of this torque, based on the preceding expression, may then be computed by integrating over one period of rotation. Let α denote mechanical degrees; then,

$$T_{\text{ave}} = \frac{1}{\tau/\omega}\int_0^{\tau/\omega} k_T k_S I_S^2 d\alpha = k_T k_S I_S^2 = K I_S^2$$

where $K = k_T k_S$ and I_S is the rms value of the armature current, i_S.

Single-Phase Induction Motors

A typical single-phase induction motor bears close resemblance to the polyphase squirrel-cage induction motor discussed in Chapter 16, the major difference being in the configuration of the stator winding. A simplified schematic diagram of such a motor, with a single winding, is shown in Figure 17.19; the winding is typically distributed around the stator so as to produce an approximately sinusoidal mmf.

Assume that the mmf for a practical motor can be generated so as to approximate the following function:

$$\mathscr{F} = F_{\max}\cos(\omega t)\cos(\theta_m)$$

This function can be written as the sum of two components, as follows:

$$\mathscr{F}^+ = \tfrac{1}{2}F_{\max}\cos(\theta_m - \omega t)$$

$$\mathscr{F}^- = \tfrac{1}{2}F_{\max}\cos(\theta_m + \omega t)$$

These two components may be interpreted as representing two mmf waves traveling in opposite directions around the stator. Each of these mmf's produces torque

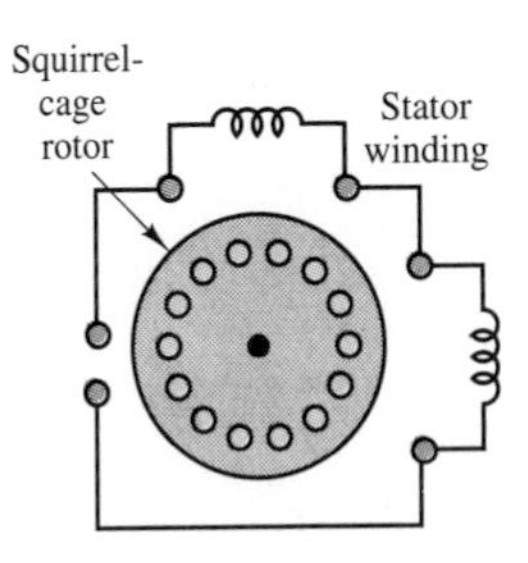

Figure 17.19 Single-phase induction motor

according to the induction principles described in Chapter 16; however, the two components are equal and opposite, and no net torque results if the rotor is at rest. The resulting mmf is pulsating (i.e., changing in amplitude), but not rotating in space, as it would be in a polyphase stator. If the rotor is made to turn in either direction, however, the two mmf's will not be equal any longer, because the motion of the rotor will induce an additional mmf, which will add to one of the two mmf's and subtract from the other. Thus, a net torque will be established, causing the motor to continue its rotation in the same direction in which it was started. In particular, if the rotor is started in the forward direction, the forward mmf, $\mathcal{F}^+$, will be greater than the backward mmf, and the motor will continue to rotate in the forward direction.

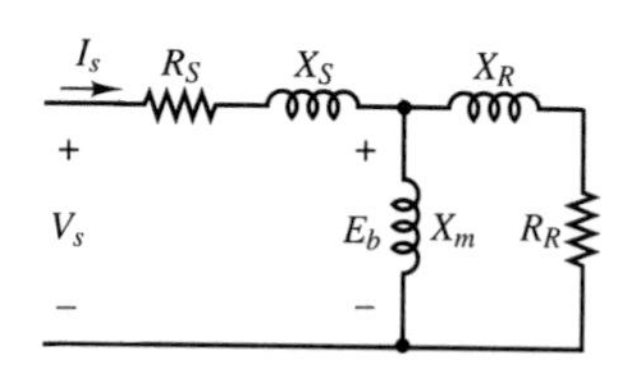

Figure 17.20 Circuit model for single-phase induction motor with rotor at standstill

Figure 17.20 depicts an equivalent circuit for the single-phase induction motor *with stationary rotor*, where the parameters in the circuit are defined as follows:

V_s = supply voltage

R_S = resistance of stator winding

X_S = leakage reactance of stator winding

X_m = magnetizing reactance of stator winding

X_R = leakage reactance of rotor referred to stator at standstill

R_R = leakage resistance of rotor referred to stator at standstill

E_b = voltage induced in the stator winding by the (stationary) pulsating flux in air gap

Figure 17.21 depicts the equivalent circuit for the same motor with the rotor rotating with slip s. Note that the circuit is asymmetrical, because of the different air-gap flux forward and backward components, E_f and E_b, respectively. The factors of 0.5 come from the resolution of the pulsating stator mmf into forward and backward components. Note further that the reflected rotor impedance is asymmetrical because of the presence of the slip parameter in the expression for the

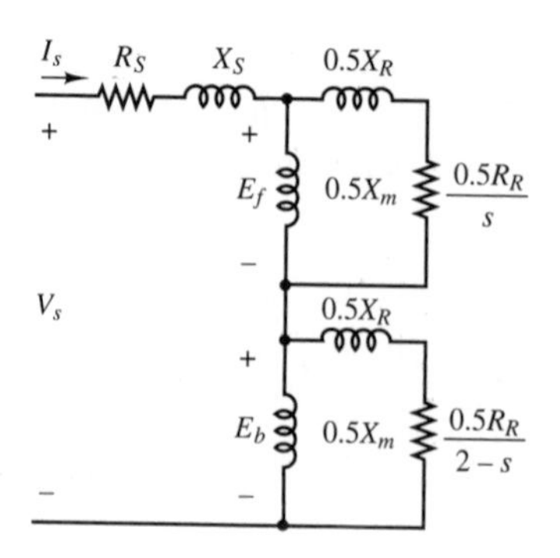

Figure 17.21 Circuit model for single-phase induction motor with rotor in motion

reflected rotor resistance. Further, the circuit model also confirms that the forward induced voltage, E_f, must be greater than the backward voltage, E_b, since the slip is always less than 1.

It can be shown that the torque components in the forward and backward directions are given by the expressions

$$T_f = \frac{P_f}{\omega_s} \tag{17.3}$$

and

$$T_b = \frac{P_b}{\omega_s} \tag{17.4}$$

where ω_s is the synchronous speed and

$$P_f = I_s^2 R_f \tag{17.5}$$

Here, R_f is the resistive component of the forward field impedance; also,

$$P_b = I_s^2 R_b \tag{17.6}$$

where R_b is the resistive component of the backward field impedance. Since the torque produced by the backward field is in the opposite direction to that produced by the forward field, the net torque will consist of the difference between the two:

$$T = T_f - T_b = \frac{I_s^2(R_f - R_b)}{\omega_s} \tag{17.7}$$

The mechanical power developed by the motor is

$$\begin{aligned} P_{\text{mech}} &= T\omega_m = T\omega_s(1 - s) = (P_f - P_b)(1 - s) \\ &= I_s^2(R_f - R_b)(1 - s) \end{aligned} \tag{17.8}$$

Example 17.10

A 115-V, 60-Hz, four-pole, single-phase induction motor is rotating at a speed of 1,710 rev/min. Find the slip of the field in

1. The direction of rotation.
2. The backward direction.

Solution:

1. First, we find the synchronous speed of the motor:

$$n_s = \frac{120f}{p} = \frac{120 \times 60}{4} = 1{,}800 \text{ rev/min}$$

The slip in the direction of rotation can now be computed, since we know the mechanical speed of rotation of the motor:

$$s_f = \frac{n_s - n_m}{n_s} = \frac{1{,}800 - 1{,}710}{1{,}800} = 0.05 = 5\%$$

2. The slip in the backward direction is simply computed as:

$$s_b = 2 - s_f = 2 - 0.05 = 1.95$$

Example 17.11

A four-pole single-phase induction motor is rated at $\frac{1}{4}$ hp, 110 V, 60 Hz. The parameters of the circuit model are:

$$R_S = 1.5\ \Omega \qquad X_S = 2\ \Omega$$

$$R_R = 3\ \Omega \qquad X_R = 2\ \Omega$$

$$X_m = 50\ \Omega \qquad s = 0.05$$

The motor is operating at rated voltage and rated frequency. Find the input current and electromagnetic torque.

Solution:

Using the equivalent circuit shown in Figure 17.21, we find that for the given numerical values, the impedance seen by the emf, E_b, is much smaller than that seen by E_f. This observation permits the following approximation:

$$0.5Z_b = (0.5)\frac{jX_m[R_R/(2-s) + jX_R]}{R_R/(2-s) + j(X_m + X_R)} \approx 0.5\left(\frac{R_R}{2-s} + jX_R\right)$$

This approximation is valid for values of slip less than 0.15. The corresponding equivalent circuit is shown in Figure 17.22.

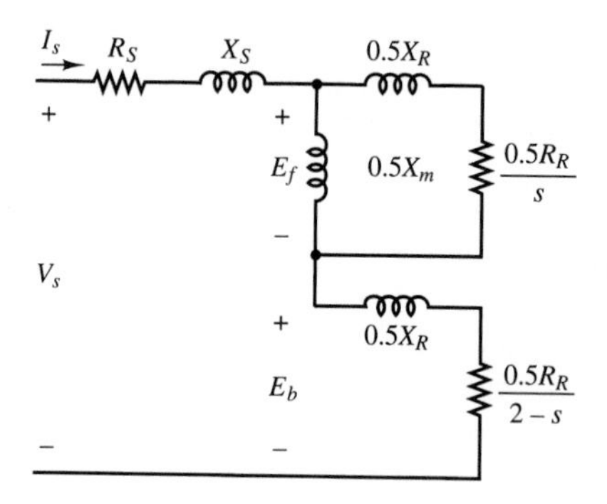

Figure 17.22 Approximate circuit model for single-phase induction motor

Using the approximate circuit, we find that

$$0.5Z_b = 0.5(1.538 + j2)\ \Omega = 0.5R_b + j0.5X_b$$

and, for the forward direction of rotation,

$$0.5Z_f = \frac{j25(30 + j1)}{30 + j26} = 18.9\angle 51^\circ$$

$$= 11.9 + j14.69 = 0.5R_f + j0.5X_f$$

Let $Z_S = R_S + jX_S$; then,

$$Z_S = 1.5 + j2\ \Omega$$

and the total impedance is

$$Z = Z_S + 0.5Z_f + 0.5Z_b = 14.169 + j17.69 = 22.66\angle 51.3^\circ\ \Omega$$

The stator current may then be found as follows:

$$I_s = \frac{V_s}{Z} = \frac{110}{22.66\angle 51.3^\circ} = 4.85\angle -51.3^\circ \text{A}$$

The power factor is pf $= \cos(51.3^\circ) = 0.625$. Such a low value for the line power factor is typical of single-phase motors.

The forward field air-gap power is

$$P_f = I_s^2 0.5R_f = (4.85)^2 \times 11.9 = 279.9\ \text{W}$$

The backward field air-gap power is

$$P_b = I_s^2 0.5R_b = (4.85)^2 \times 0.769 = 18.1\ \text{W}$$

The synchronous speed is

$$\omega_s = \frac{4\pi f}{p} = 188.5\ \text{rad/s}$$

The torque is

$$T = \frac{P_f - P_b}{\omega_s} = 1.39\ \text{N-m}$$

Example 17.12

The machine in Example 17.11 is operated at rated voltage and frequency and with slip $s = 0.05$. The combined mechanical and core losses are 30 W. Find

1. The output torque.
2. The output power.
3. The efficiency.

Solution:

For the same operating conditions as in Example 17.11, the mechanical power is

$$P_{\text{mech}} = (1 - s)(P_f - P_b) = 0.95(279.9 - 18.1) = 248.71\ \text{W}$$

Subtract the losses to find the output power:

$$P_{\text{out}} = P_{\text{mech}} - P_{\text{loss}} = 248.71 - 30 = 218.71\ \text{W}$$

The shaft speed is

$$\omega_m = (1 - s)\omega_s = 179\ \text{rad/s}$$

The output torque is

$$T_{\text{out}} = \frac{P_{\text{out}}}{\omega_m} = \frac{218.71}{179} = 1.22 \text{ N-m}$$

The electrical losses are

$$P_{R_S} = I_s^2 R_S = (4.85)^2 \times 1.5 = 35.3 \text{ W}$$

$$P_{R_R(f)} = sP_f = 0.05 \times 279.9 = 14 \text{ W}$$

$$P_{R_R(b)} = (2 - s)P_b = 1.95 \times 18.1 = 35.3 \text{ W}$$

and the total loss is

$$P_{\text{loss}} = P_{R_S} + P_{R_R(f)} + P_{R_R(b)} + P_{\text{rot}} = 114.6 \text{ W}$$

Finally, the efficiency may be computed to be:

$$\text{Efficiency} = 1 - \frac{\sum \text{losses}}{P_{\text{out}} + \sum \text{losses}} = 1 - \frac{114.6}{218.71 + 114.6} = 0.656 = 65.6\%$$

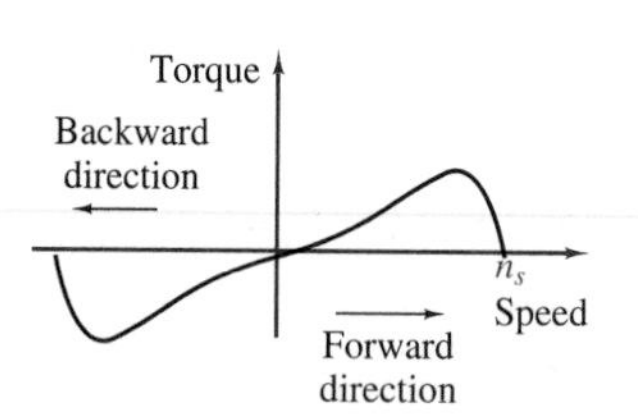

Figure 17.23 Torque-speed curve of a single-phase induction motor

The equations and circuit models in the preceding examples suggest that a single-phase induction motor is capable of sustaining a torque, and of reaching its operating speed, once it is started by external means. However, because the magnetic field in a single-phase winding is stationary, a single-phase motor is not self-starting. The speed-torque characteristic of a typical single-phase induction motor shown in Figure 17.23 clearly shows that the starting torque for this motor is zero. The curve also shows that the motor can operate in either direction, depending on the direction of the initial starting torque, which must be provided by separate means.

Classification of Single-Phase Induction Motors

Thus far, we have not mentioned how the initial starting torque can be provided to a single-phase motor. In practice, single-phase motors are classified by their starting and running characteristics, and several methods exist to provide nonzero starting torque. The aim of this section is to classify single-phase motors by describing their configuration on the basis of the method of starting. For each class of motor, a torque-speed characteristic will also be described.

Split-Phase Motors

Split-phase motors are constructed with two separate stator windings, called **main** and **auxiliary windings;** the axes of the two windings are actually at 90° with respect to each other, as shown in Figure 17.24. The auxiliary winding current is designed to be out of phase with the main winding current, as a result of different reactances of the two windings. Different winding reactances can be attained by having a different ratio of resistance to inductance—for example, by increasing the resistance of the auxiliary winding. In particular, the auxiliary

The rated torque is

$$T = \frac{K \times \text{power}}{\text{rev/min}}$$

where the constant K is given by

$$K = 0.97376 \text{ when } T \text{ is expressed in meter-kilograms}$$
$$K = 9.549 \text{ when } T \text{ is expressed in newton-meters}$$
$$T = 9.549 \times \frac{248.7}{1{,}725} = 1.377 \text{ N-m}$$

Summary of Single-Phase Motor Characteristics

Four basic classes of single-phase motors are commonly used:

1. Single-phase induction motors are used for the larger home and small business tasks, such as furnace oil burner pumps, or hot water or hot air circulators. Refrigerator compressors, lathes, and bench-mounted circular saws are also powered with induction motors.
2. Shaded-pole motors are used in the smaller sizes for quiet, low-cost applications. The size range is from $\frac{1}{30}$ hp (24.9 W) to $\frac{1}{2}$ hp (373 W), particularly for fans and similar drives in which the starting torque is low.
3. Universal motors will operate on any household AC frequency or on DC without modification or adjustment. They can develop very high speed while loaded, and very high power for their size. Vacuum cleaners, sewing machines, kitchen food mixers, portable electric drills, portable circular saws, and home motion-picture projectors are examples of applications of universal motors.
4. The capacitor-type motor finds its widest field of application at low speeds (below 900 rev/min) and in ratings from $\frac{3}{4}$ hp (0.5595 kW) to 3 hp (2.238 kW) at all speeds, especially in fan drives.

DRILL EXERCISES

17.10 Prove that the mmf for the single-phase induction motor can be expressed as the sum of two components traveling in opposite directions.

17.11 For the circuit shown in Figure 17.29, draw the starting phasor diagram relating $\mathbf{V}, \mathbf{I}, \mathbf{I}_{\text{main}}$, and $\mathbf{I}_{\text{aux}}$.

17.12

a. What is the zero-speed torque for a single-phase induction motor?
b. If external torque is applied to the machine, what is its terminal speed?

17.13 For the motor of Example 17.14, calculate the rated torque in meter-kilograms.

17.4 MOTOR SELECTION AND APPLICATION

The objective of this section is to outline the selection process of a motor for application to an electrical drive, and to summarize the characteristics of the most common drive motors, with emphasis on fractional-horsepower applications. An electrical motor should satisfy a set of precise requirements to be considered for a specific application. These include:

1. Starting characteristics (torque and current)
2. Acceleration characteristics (dependent on the load)
3. Efficiency at rated load
4. Overload capability
5. Electrical and thermal safety
6. Cost

These requirements suggest that the specific details of the application should be clear in the designer's mind. For example, the nature of the load, of the available electrical supplies, and of the ambient conditions should be carefully investigated before the motor selection process is initiated. Once the application environment is known, it is usually possible to narrow the selection of a drive motor to a few choices. In this section we shall provide some insight into the motor selection process; in addition, Tables 17.6 and 17.7, found near the end of the section, summarize the principal characteristics of fractional-horsepower drive motors and will serve as a useful reference to the reader.

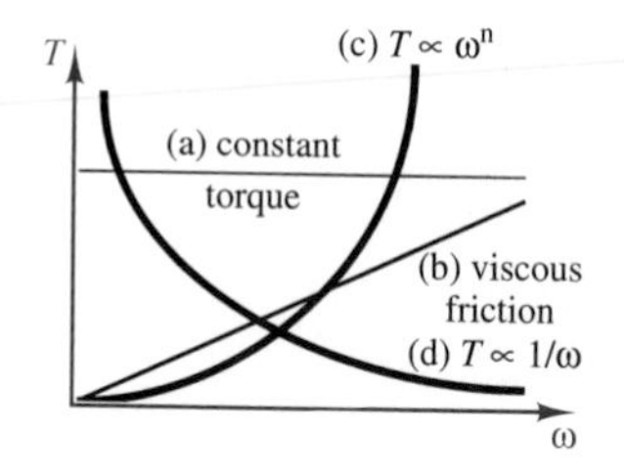

Figure 17.33 Typical load torque-speed curves

Motor Performance Calculations

To better understand the motor selection process, it is important to review some of the basic ideas underlying the motion of the motor and of the load. For rotational systems, the relationships of Table 17.4 hold. Figure 17.33 summarizes the various types of load profiles that are likely to be encountered in practical applications. These include constant-torque loads; viscous friction-type loads, where

Table 17.4 **Equations of motion and definitions of variables**

Equations of motion	Definition of terms
$\omega_2 = \omega_1 + \alpha t$	ω_1 = initial velocity (rad-s^{-1})
$\theta = \omega_1 t + \frac{1}{2}(\alpha t^2)$	ω_2 = final velocity (rad-s^{-1})
$\omega_2^2 = \omega_1^2 + 2\alpha\theta$	α = acceleration (rad-s^{-2})
$P = T\omega$	θ = angular displacement (rad)
$T = Fr = J\alpha$	n = speed of rotation (rev/min)
$W = T\theta$	P = output power (W)
$\omega = \frac{2\pi n}{60} = 0.105n$	W = work (J)
$J = mk^2$	F = force (N)
	T = torque (N-m)
	J = polar moment of inertia (kg-m^2)
	r = radius of arm (m)
	k = radius of gyration (m)

torque is proportional to speed; loads in which the torque is proportional to a power of speed (e.g., fans, pumps); and constant-power loads, where torque is inversely proportional to speed.

Reflected Load Inertia Calculations

To calculate the motor requirements, one must compute the required torque referenced to the motor output shaft. Since gearing systems are often employed, the inertias of all rotating components must be referred back to the motor shaft. Using the terminology of Table 17.4, we then conclude that the reflected load torque at the motor shaft, T_r, is related to the load torque, T_L, by the relationship

$$T_r = \frac{\omega_L}{\omega_r} T_L \quad (17.9)$$

where $\omega_r = \omega_m$ is the motor shaft speed, and the ratio of load speed to motor speed is equal to the gear ratio:

$$\frac{\omega_L}{\omega_r} = \frac{n_L}{n_r} \quad (17.10)$$

If we equate the kinetic energy on the motor side to that on the load side, we can also derive an expression for the reflected load inertia:

$$J_L \omega_L^2 = J_r \omega_r^2 \quad (17.11)$$

or

$$J_r = J_L \left(\frac{n_L}{n_r} \right)^2 \quad (17.12)$$

Thus, the reflected inertia seen by the motor at the shaft is equal to the load inertia times the square of the gear ratio. Note that this is a mechanical impedance transformation similar to that used in the case of transformers. For all practical purposes, one can think of a gearing system as a mechanical transformer that, in the ideal case, conserves power. Under this ideal gearing assumption, it can be shown that the acceleration of the load is given by the following expression:

$$\alpha = \frac{T_{m\,\text{peak}}}{\frac{\omega_r}{\omega_L} \left[J_m + \frac{J_L}{\left(\frac{\omega_r}{\omega_L} \right)^2} \right]} \quad (17.13)$$

where the numerator on the right-hand side is the peak torque the motor can produce and J_m is the motor inertia. If one wished to determine what gear ratio were required to obtain maximum acceleration of the load, equation 17.13 would be differentiated and set equal to zero, to obtain the result

$$\frac{\omega_r}{\omega_L} = \sqrt{\frac{J_L}{J_m}} \quad (17.14)$$

This expression implies that the load inertia should be made equivalent to the motor inertia by appropriate gearing, in order to obtain the best acceleration. Substituting equation 17.14 in 17.13, we can show that the maximum acceleration is given by

$$\alpha_{\max} = \frac{1}{2}\frac{T_{m\,\text{peak}}\omega_L}{J_m\omega_r} \tag{17.15}$$

It should be apparent that equations 17.9 through 17.15 are very useful in sizing a motor and in determining whether any gearing will be necessary to achieve the desired performance.

Torque Definitions

Although the definition of various torques was introduced in Chapter 16, it will be useful to briefly review these definitions in light of the preceding subsection. In sizing a motor, it is important to ensure that the motor is capable of overcoming the static friction at start-up, to accelerate to the desired speed in an acceptable fashion, and to handle any overloads that may occur. The following definitions will help in the analysis:

1. **Locked-rotor,** or **static, torque:** The minimum torque the motor will develop at rest for all angular positions under rated conditions.
2. **Breakdown torque:** The maximum torque a motor will develop under rated conditions without an abrupt drop in speed.
3. **Full-load torque:** The torque necessary to produce rated power output at full-load speed.
4. **Acceleration torque:** At any specified speed, acceleration torque, i.e., the torque available for acceleration, is $T_{\text{acc}} = T_m - T_L - T_F$, where T_m is the motor torque, T_L is the load torque, and T_F is the frictional load torque.

Clearly, in order for the motor to accelerate to full-speed operation, the motor torque at standstill must exceed the total static-load torque. When the motor torque-speed curve intersects the load torque-speed curve, then a balanced operating condition has been reached.

Acceleration Calculations

The equation that defines the acceleration characteristics of a motor-load pair is

$$T_m - T_L = J\frac{d\omega}{dt} \tag{17.16}$$

where T_L is the total load torque. From this equation we can calculate the time required to accelerate the load from a speed ω_1 to a speed ω_2 as follows:

$$t = J\int_{\omega_1}^{\omega_2}\frac{d\omega}{T_m - T_L} \quad \text{(s)} \tag{17.17}$$

or, in units of rev/min,

$$t = \frac{2\pi J_T}{60T}(n_1 - n_2) \text{ (s)} \quad \textbf{(17.18)}$$

where T is the net torque (motor torque minus load torque) and J_T is the total system inertia (motor inertia plus reflected load inertia).

Efficiency Calculations

The efficiency of a motor is the ratio of the mechanical power output to the electrical power input, that is, the effectiveness of the electromechanical energy conversion. We have already discussed the sources of loss in Chapter 16 and classified them as electrical losses, magnetic losses, and mechanical losses; refer to Section 16.1 for these definitions. The efficiency of a motor, η, is defined by:

$$\eta = 1 - \left(\frac{P_{\text{loss}}}{P_{\text{input}}}\right) = 1 - \left(\frac{P_{\text{loss}}}{VI}\right) \quad \textbf{(17.19)}$$

Thermal Calculations

The calculation of the temperature rise and thermal dissipation in a motor can be quite complex and depends very much on the motor construction. For the purpose of illustration, we briefly discuss only the thermal characteristics of a DC motor and perform some thermal calculations for this type of machine.

Thermal dissipation is one of the most important limiting factors in the operation of DC machines. We assume that most power losses take place in the armature (a reasonable assumption, since most of the electrical power flows through the armature circuit), and we use the thermal-electrical system analogy (see Example 6.16 in Chapter 6), where the thermal difference, $\Delta\theta°$, is given by

$$\Delta\theta° = I_a^2 R_a \times R_T \quad \textbf{(17.20)}$$

and where I_a is the armature current, R_a the armature resistance, and R_T the thermal resistance of the rotor. The thermal time constant of the motor is then defined to be the time (in seconds) taken by the armature to reach 63% of the temperature rise corresponding to a given constant power dissipation. Now, the maximum continuous torque the motor can develop is related to the power dissipation, because the motor torque is proportional to the armature current:

$$T_{\text{max}} = K_T I_{a\,\text{max}} = K_T\sqrt{\frac{P_{\text{diss}}}{R_{\text{max}}}} = K_T\sqrt{\frac{\Delta\theta°}{R_T R_{\text{max}}}} \quad \textbf{(17.21)}$$

where P_{diss} is the dissipated power and R_{max} the rotor resistance at the maximum temperature, R_T is the rotor thermal resistance at ambient temperature, and $\Delta\theta$ is the temperature rise. The temperature rise of copper can be determined from the known resistance of the wound rotor by computing the maximum temperature as follows:

$$\theta^{\circ}_{\text{max}} = \left[\frac{R_{\text{max}}}{R_T}\right](\theta^{\circ}_{\text{ambient}} + 235) - 235 \tag{17.22}$$

and by computing $\Delta\theta^{\circ} = \theta^{\circ}_{\text{max}} - \theta^{\circ}_{\text{ambient}}$, it is possible to use equation 17.21 to determine the maximum acceptable torque.

Conversely, to ensure that a given motor can operate within its thermal limits, one can calculate an average rms current requirement, I_{rms}, consisting of the acceleration, deceleration, and running current, and use it to compute the temperature as follows:

$$\Delta\theta^{\circ} = I^2_{\text{rms}} R_a R_T \tag{17.23}$$

Motor Selection

The range of electric motor applications is so broad that it is difficult to establish precise rules for motor selection. The differences between applications such as vehicle traction, robot motion, micromotors, disk drives, manufacturing machines, and pump systems, for example, are so many that it is virtually impossible to specify what the best motor would be, unless the application and its environment are clearly specified. The aim of this subsection is simply to outline a procedure that can help in narrowing down the choice of a suitable drive motor to a few most likely candidates.

The first step in selecting a motor is the analysis of the requirements imposed by the application; these can be divided into three groups: (1) motor requirements, (2) load requirements, and (3) control requirements. Table 17.5 summarizes the important considerations for each of these.

Table 17.5 Motor selection requirements

Motor requirements	Load requirements	Control requirements
Operating speed	Determine worst-case operating conditions	Available power (AC, DC)
Life span and maintenance	Dynamic acceleration, full-load and overload conditions	Motor operating voltage and current
Torque characteristics	Starting conditions	Open- or closed-loop
Mechanical aspects (size, weight, noise level, environment)	Transients	Forward/reverse operation
Applicable standards (e.g., radio frequency interference)	Need for gearing, selection of optimum gear ratio	Motoring and/or braking
Overload characteristics	Frictional characteristics	Torque, position, or speed control
Thermal dissipation characteristics		Accuracy of speed or position control
		Controller complexity and cost

On the basis of the requirements listed in Table 17.5, and with the assistance of the summary of motor performance characteristics supplied in Tables 17.6 and 17.7, one can undertake the task of selecting a motor for a specific application.

Motion Requirements

The first step in the drive selection process is to understand the application-driven specifications, that is, issues such as the type of motion, duty cycle, the required acceleration and gearing system, and the type of control that may be required (position, velocity, torque).

Motor Sizing

The second step in the drive selection process concerns the sizing of the motor itself. This is done by first calculating the maximum speed; next, the reflected inertia of the load and drive components is calculated, as discussed earlier in this section. From the inertia calculations, the peak torque required by the application can be calculated. The maximum speed and torque requirements thus obtained will narrow the field significantly. Next, one should determine the appropriate constants for each of the candidate motors; these include, in general, inertias, resistances (electrical and thermal), and torque and back emf constants. With these constants it becomes possible to determine that the motor can operate within its thermal specifications by calculating the **temperature rise** of the machine in operation. This, of course, can be a greater limitation during certain portions of the motion cycle—for example, during a hard acceleration.

Defining the Power Requirements

This step involves calculating peak voltage and current, to determine the supply requirements.

Choosing a Transmission

Although we have been assuming that the mechanical drive system was known beforehand, so that the reflected inertia and peak torque could be calculated, there are many issues that need to be investigated in establishing the drive system—for example, the effect that elastic couplings might have in creating mechanical resonances; noise and vibration characteristics; and backlash due to gearing system imperfections.

It should be apparent from this brief discussion that the process of selecting an electromechanical drive is quite complex, and that it requires a good understanding of many aspects of engineering, including heat transfer, kinematics, dynamics, electronics, systems, and, of course, electromechanics. We hope that this brief introduction will provide the motivation to pursue further studies in this exciting area of engineering.

Table 17.6 AC Motor application and performance characteristics

	Split-phase induction	Capacitor-start induction	Permanent split capacitor (PSC)	Capacitor-start capacitor-run
Applications	Domestic appliances, office machinery, industrial fractional-hp applications	High inertial loads—pumps, compressors, machine tools, conveyor belts	Direct-drive fans and blowers	High torque
Specifications				
Duty	Continuous	Continuous	Continuous	Continuous
Supply	AC	AC	AC	AC
Reversibility	Only at zero speed	During rotation	During rotation	During rotation
Speed	Nearly constant	Nearly constant	Controllable	Controllable
Start torque	130–200% of rated	Up to 300% of rated	50–100% of full load	> 200% of rated
Start current	High	Medium	Low	Medium to high
Advantages	General-purpose, low cost Good efficiency, starting torque, and acceleration Fairly constant speed	Greater torque than PSC Capacitor improves starting characteristics and reduces starting current Fairly efficient Frequent starting OK Quiet	Voltage control permits speed control Frequent starting OK Quiet High efficiency No centrifugal switch needed	Extremely quiet High starting torque and acceleration Use of two capacitors optimizes both efficiency and power factor Designed for continuous operation
Disadvantages	Special switching required for reversal during rotation Not suited to frequent stops and starts (overheats) Starting current 5–10 times run current Not suited for large inertial loads Auxiliary winding subject to damage if connected for more than a few seconds Cannot control speed Low power factor	Performance is very sensitive to correct choice of capacitor Cannot control speed	Can overheat at light loads Needs to run near rated load Low torque compared with other capacitor motors Not suitable for belt-driven loads Most expensive of capacitor motors Capacitor value is a compromise between start and run	Optimal value of run capacitor usually depends on load

Table 17.6 *continued*

Shaded-pole induction	Synchronous reluctance	Hysteresis synchronous	Permanent-magnet synchronous	Universal
Many consumer and industrial applications	Constant-speed applications; drive for various mechanisms	Electric clocks, synchronous servos, tape, turntable, disk drives; gyros	Combines best features of induction and permanent-magnet motors	Constant-speed applications; drive for various mechanisms
Continuous	Continuous	Continuous	Continuous	Intermittent
AC	AC or PWM AC	AC	AC or PWM	AC or DC
Not available	No	At zero speed	Yes	Not usually
Nearly constant	Constant	Constant	Great range	Varies with load
50–100% of rated	50–100% of pull-in torque	100% of rated	300% of rated	>175% of rated torque
Low	Low	Low	—	High
Simple and low-cost, suitable for mass production Reliable Quiet and low vibration Speed insensitive to load variations Speed can be controlled by series resistance	Constant-speed (synchronous) operation Speed dependent on supply frequency and number of poles Speed control possible with variable-frequency AC drives Can have squirrel-cage construction	Reliable Quiet and smooth Good efficiency Low surge currents High shaft power relative to size Constant speed	Good power-to-weight ratio Good cost and size per unit power Good starting torque and acceleration Linear torque-speed characteristics Fast transient response	Best power-to-weight ratio of any single-phase AC motor Capable of speeds greater than 3,600 rev/min High starting torque Adjustable speed over 4,000–10,000 rev/min
Low efficiency—will run quite warm Low start and running torques Low power factor	Will not achieve synchronism for high loads, and large load inertias Requires separate starting Expensive Low power factor Less efficient than induction motors	Voltage variations reduce efficiency Low supply voltage reduces torque output	Expensive Requires sophisticated drive	Poor speed and torque regulation in open-loop mode Cannot be reversed Brushes require maintenance Very noisy Moderate to low efficiency

Table 17.7 **Summary of DC motor types**

	Series-wound	Shunt-wound	Permanent-magnet	DC servo
Applications	Battery-powered vehicles, traction applications	Constant speed at a given control setting	Used in numerous products, such as cars, consumer electronics	Used where starts and stops must be made quickly and accurately
Specifications				
Duty	Continuous	Continuous	Continuous	Continuous
Power supply	AC or DC	DC	DC	DC
Reversibility	Usually unidirectional	At rest or rotating	At rest or rotating	At rest or rotating
Speed	Variable with load	Fairly constant	Adjustable	Variable with load
Start torque	>175% of rated torque	125–200% of rated	>175% of rated	High
Start current	High	Normal	High	High
Advantages	Good starting and acceleration torque Efficient Moderate speed control easily obtained with field diverter resistance	Earliest and most reliable of DC motor types Will operate from rectified AC Speed stability in open-loop mode is quite good Can control speed by varying either field current or armature voltage Brush life is generally good Good starting torque Efficient Fairly load-independent	High starting torque enables inertial loads Efficiency moderate to high Simple to reverse and dynamically brake High torques at low speed Linear torque-speed curve and ease of electronics control Use of rare earth magnets has reduced overall size and increased scope of applications Low cost Ideal for stop/start No electrical power required to generate stator field	Speeds from stall to 2,000–4,000 rev/min Substantial transient overload capability Low rotor inertia Rugged brushes and commutator to handle high powers Low electrical time constant High torque linearity with current input High torque/inertia ratio Fast response at all speeds due to rapid current rise in armature Excellent low-speed characteristics
Disadvantages	With no load on motor, speed can run away because torque-speed curve rises steeply Expensive	Brush life adversely affected by reversing when rotating Dynamic braking is severe on brushes Expensive	Cannot operate continuously at high torques, since serious overheating may result Overload may cause partial demagnetization Limited to 100 VDC because of field demagnetization Speed regulation good for constant voltage and moderate loads High voltage produces excess arcing of brushes and commutator Brush life decreases with higher commutator surface speeds Limited to small ratings	Expensive; requires directly mounted tachogenerator Requires closed-loop electrical control system Special design with particular characteristics—performance can vary widely

Table 17.7 *continued*

Printed-circuit motor	Shell armature	VR stepper	PM stepper
For incremental-motion application in competition with steppers; thin in relation to diameter	Used for servo applications and incremental motion control	Digital control of machine functions; positional control of printers, plotters, etc.	Digital control of machine functions; positional control of printers, plotters, etc.
Continuous	Continuous	Continuous	Continuous
DC	DC	Switched DC	Switched DC
At rest or rotating	At rest or rotating	At rest or rotating	At rest or rotating
Adjustable	Adjustable	Adjustable	Adjustable
Up to 500% of rated	>175% of rated	N/A	N/A
High	High	Normal	Normal
Suitable for intermittent duty (positioning servo) where torque smoothness required Speed control possible within a single revolution High torque/inertia ratio Armature has no "wound" windings Moderately efficient Speeds as low as 1 rev/min Fast response Excellent power/volume ratio Quiet and simple High torque linearity Reliable, high brush life with multiple brushes	Very low inertia and high acceleration Low inductance and thus low electrical time constant High torque/inertia ratio Maximum speed 10,000–15,000 rev/min	Steppers are excellent positioning drives, in both performance and cost Can replace mechanical devices, e.g., clutches, with improved reliability Accuracy Suitable for high-speed/high-performance operation Robust rotor Lower-cost stepper	Steppers are excellent positioning drives, in both performance and cost Can replace mechanical devices, e.g., clutches, with improved reliability Accuracy Compact low-weight, low-cost unit Detent torque up to 15% of holding torque available Ideally suited for μP control
Fragile construction with thin armature Windings arranged across a very large radius Radius factor contributes substantially to armature inertia	Low thermal time constant for armature and long thermal time constant for housing Armature can overheat Fragile construction Expensive; employ only where unique performance characteristic required	Use only for light-load applications Fairly low-torque inertial capability Prone to some resonance problems No detent torque when coils not energized Not efficient Fixed-angle devices Tend to "ring" on stopping; hence, some means of damping usually required When inertial loads are being driven, speed ramping usually required	Fairly low-torque speed capability Not efficient Fixed-angle devices Limited ability to handle large inertial loads Tend to "ring" on stopping; hence, some means of damping usually required When inertial loads are being driven, speed ramping usually required Rough low-speed operation without microstepping

Table 17.7 *continued*

	Hybrid stepper	Brushless DC	Switched-reluctance
Applications	Digital control of machine functions	Used in long-life applications: robotics, office machinery, tools	Domestic appliances and industrial drives; vehicle traction
Specifications			
Duty	Continuous	Continuous	Continuous
Power supply	Switched DC	3-phase	Switched DC
Reversibility	At rest or rotating	At rest or rotating	N/A
Speed	Adjustable	Adjustable	Controllable
Start torque	N/A	>175% of rated	Up to 200% of rated
Start current	Normal	High (rotor position feedback required)	Low (rotor position feedback required)
Advantages	Steppers are excellent positioning drives, in both performance and cost Can replace mechanical devices, e.g., clutches, with improved reliability Accuracy Combines best feature of VR and PM types High performance at reasonable cost High resolutions available, 1.8° steps being common	No brushes or commutator to limit life Reliable, relatively maintenance-free drive Long-life motors, operational-life rated power not limited to wear considerations Very low noise Speeds up to 60,000 rev/min not unusual Peak torque maintainable up to high speed	Simple low-inertia rotor Stator is simple to wind Torque-speed characteristics can be tailored to applications more easily than many other motors Most losses appear in stator, which is easily cooled Rotor losses very small; hence, no rotor cooling problems as with other motors Torque is independent of the polarity of the phase current High starting torques without the problem of surge currents
Disadvantages	Not efficient Fixed-angle devices Tend to "ring" on stopping; hence, some means of damping usually required When inertial loads are being driven, speed ramping usually required	Electronically complex compared with other DC systems Cannot be reversed by simple power source polarity Excess coasting; special circuits can be added at cost Copper losses are high, not very efficient in low-voltage systems Very expensive compared with PMDC Limited economically to small motor sizes	Cannot equal the power density or efficiency of PMDC in small frame sizes Large torque ripple Current ripple No smaller than an induction motor designed to the same specification Shaft position sensing required Cannot start from sinusoidal AC

CONCLUSION

- The most common engineering applications of electric motors make use of a number of special-purpose motors, with operating characteristics that may be derived from the fundamental principles presented in Chapters 15 and 16.
- The brushless DC motor is a PM synchronous motor in which the mechanical commutation of conventional DC motors is replaced by electronic commutation. Brushless DC motors can be made quite compact and find application in electric vehicle propulsion and motion control.
- Stepping motors—of the variable-reluctance, PM, or hybrid type—permit fine angular displacement control by moving in fixed, discrete steps. Typical applications are in robotics and control systems.
- The universal motor is very similar in construction to a series DC motor but can operate on AC supplies; its speed can be controlled by means of electronic circuitry of modest complexity. Thus, the universal motor finds common application in both variable- and fixed-speed appliances, such as power drills and vacuum cleaners, respectively.
- Squirrel-cage induction motors can be operated on a single-phase AC supply if a means is provided for establishing a starting torque. Various techniques are commonly employed, such as split-phase, capacitor-start, and shaded-pole construction. The different types are characterized by differing torque-speed characteristics that make the single-phase induction motor a very versatile device. This is the most commonly employed electric machine.

KEY TERMS

ANSWERS TO DRILL EXERCISES

17.2 $T = 1.07$ N-m

17.3 $T = 150.9$ oz-in

17.6 $\Delta\theta = 20°$

17.7

17.8

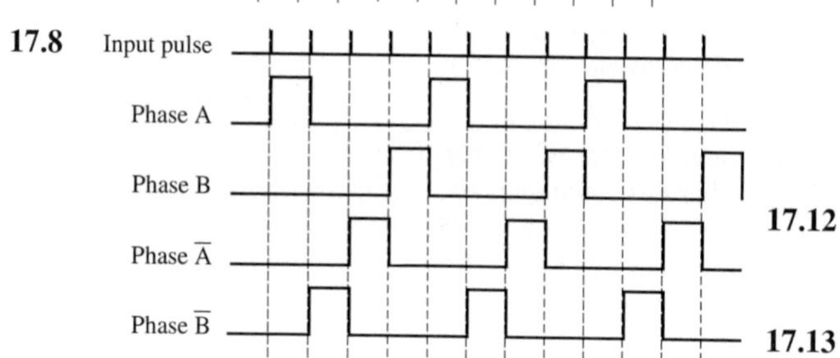

17.9 29.875 lb-in

17.11

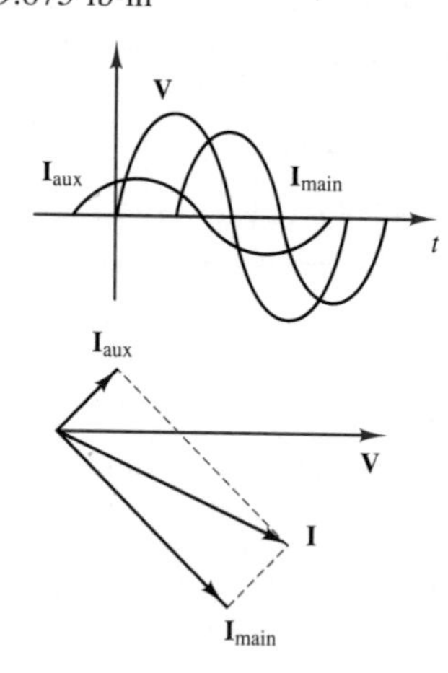

17.12 (a) Zero, without a start winding; (b) $n = (1 - s)n_s$, where $n_s = \frac{120f}{p}$ and s = slip (determined by shaft load).

17.13 0.14 m-kg

HOMEWORK PROBLEMS

17.1 It is found that $\lambda_m = 0.1$ V-s for a permanent magnet six-pole two-phase synchronous machine. Calculate the amplitude (peak value) of the open-circuit phase voltage measured when the rotor is turned at 60 rev/sec.

17.2 A four-pole two-phase brushless dc motor is driven by a mechanical source at $n = 3600$ rev/min. The open-circuit voltage across one of the phases is 50 V rms.

a. Calculate λ.
b. The mechanical source is removed and the following voltages are applied: $V_a = \sqrt{2}\,25\cos\theta$, $V_b = \sqrt{2}\,25\sin\theta$ where $\theta = \omega_e t$. Calculate the no-load rotor speed ω in rad/s.

17.3 Derive the dynamic equation for a stepping motor coupled to a load. The motor moment of inertia is J_m, the load moment of inertia is J_L, the viscous damping coefficient is D, and motor friction torque is T_f.

17.4 Sketch the rotor-stator configuration of a hybrid stepper motor capable of 18° steps. [*Hint:* The rotor will have five teeth.]

17.5 Use a binary counter and logic gates to implement the stepping motor binary sequence of Drill Exercise 17.7.

17.6 A two-phase permanent-magnet stepper motor has 50 rotor teeth. When the rotor is driven by an external mechanical source at $\omega = 100$ rad/s, the measured open circuit phase voltage is 25 V, peak-to-peak. Calculate λ. If $i_a = 1$A and $i_b = 0$, express the developed torque. Assume the winding resistance is 0.1Ω.

17.7 The schematic diagram of a four-phase, two-pole PM stepper motor is shown in Figure P17.1. The phase coils are excited in sequence by means of a logic circuit. Find

a. The logic schedule for full-stepping of this motor.
b. The displacement angle of the full step.

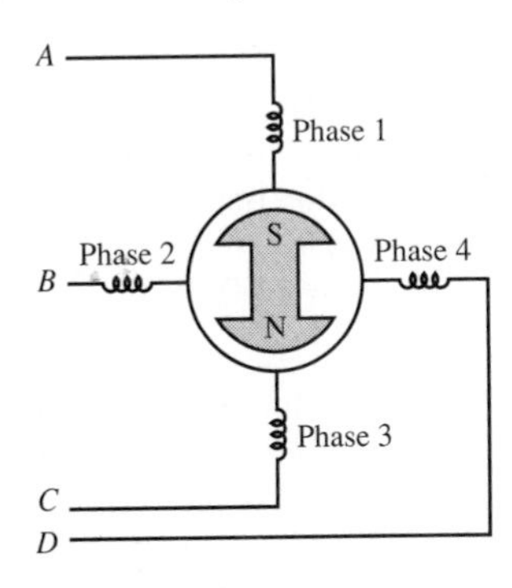

Figure P17.1

17.8 A PM stepper motor is designed to provide a full-step angle of 15°. Find the number of stator and rotor poles.

17.9 A bridge driver scheme for a two-phase stepping motor is shown in Figure P17.2. Find the excitation sequences of the bridge operation (fill in the blanks of the table).

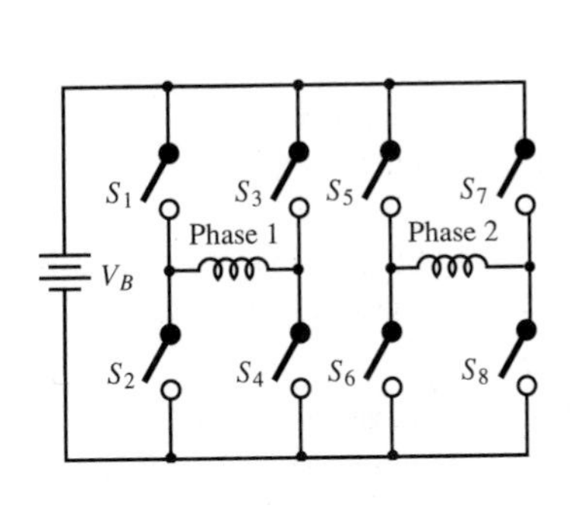

Clock state	Reset	1	2	3	4	5	6	7	8
S_1									
S_2									
S_3									
S_4									
S_5									
S_6									
S_7									
S_8									

Figure P17.2

17.10 A permanent-magnet stepper motor with a 15° step angle is used to directly drive a 0.100 in. lead screw. Determine:

a. The resolution of the stepper motor in steps/revolution,
b. The distance the lead screw travels (in inches) for each 15° step of the stepper motor,
c. The number of full 15° steps required to move the lead screw and the stepper motor shaft through 17.5 revolutions, and
d. The shaft speed (in rev/min) when the stepping frequency is 220 pps.

17.11 Determine whether the following motors are integral- or fractional-horse-power motors:

a. $\frac{3}{4}$ hp, 900 rev/min
b. $1\frac{1}{2}$ hp, 3,600 rev/min
c. $\frac{3}{4}$ hp, 1,800 rev/min
d. $1\frac{1}{2}$ hp, 6,000 rev/min

17.12 The spatial fluctuation of the stator mmf $\mathcal{F}_1$ is expressed as

$$\mathcal{F}_1 = F_{1(\text{peak})} \cos \theta$$

where θ is the electrical angle measured from the stator coil axis and $F_{1(\text{peak})}$ is the instantaneous value of the mmf wave at the coil axis and is proportional to the instantaneous stator current. If the stator current is a cosine function of time, the instantaneous value of the spatial peak of the pulsating mmf wave is

$$F_{1(\text{peak})} = F_{1(\text{max})} \cos \omega t$$

where $F_{1(\text{max})}$ is the peak value corresponding to maximum instantaneous current. Derive the expression for $\mathcal{F}_1$, and verify that for a single-phase winding, both forward and backward components are present.

17.13 A 200-V, 60-Hz, 10-hp single-phase induction motor operates at an efficiency of 0.86 and a power factor of 0.9. What capacitor should be placed in parallel with the motor so that the feeder supplying the motor will operate at unity power factor?

17.14 A 230-V, 50-Hz, two-pole single-phase induction motor is designed to run at 3 percent slip, Find the slip in the opposite direction of rotation. What is the speed of the motor in the normal direction of rotation?

17.15 Determine the amount of time (in seconds) it will take for a stepper motor with a 15° step angle, operating in one-phase excitation mode, to rotate through 28 rev. when the pulse rate is 180 pps. Note: $t = \theta/\omega$.

17.16 A $\frac{1}{4}$-hp, 110-V, 60-Hz, four-pole capacitor-start motor has the following parameters:

$$R_S = 2.02\ \Omega \quad X_S = 2.8\ \Omega$$
$$R_R = 4.12\ \Omega \quad X_R = 2.12\ \Omega$$
$$X_m = 66.8\ \Omega \quad s = 0.05$$

Find

a. The stator current.
b. The mechanical power.
c. The rotor speed.

17.17 A $\frac{1}{4}$-hp, four-pole, 110-V, 60-Hz single-phase induction motor has the following data:

$$R_S = 1.86\ \Omega \quad X_S = 2.56\ \Omega$$
$$R_R = 3.56\ \Omega \quad X_R = 2.56\ \Omega$$
$$X_m = 53.5\ \Omega \quad s = 0.05$$

Find the mechanical power output.

17.18 A one-phase, 115-V, 60-Hz, four-pole induction motor has the following parameters:

$$R_S = 0.5\ \Omega \quad X_S = 0.4\ \Omega$$
$$R_R = 0.25\ \Omega \quad X_R = 0.4\ \Omega$$
$$X_m = 35\ \Omega$$

Find the input current and developed torque when the motor speed is 1,730 rev/min.

17.19 The no-load test of a single-phase induction motor is made by running the motor without load at rated voltage and rated frequency. Derive the equivalent circuit of a single-phase induction motor for the no-load test. [*Hint:* The no-load slip is very small.]

17.20 Derive the equivalent circuit of a single-phase induction motor for the locked-rotor test. Neglect the magnetizing current.

17.21 The design for a $\frac{1}{8}$-hp, two-pole, 115-V universal motor gives the effective resistances of the armature and series field as 4 Ω and 6 Ω, respectively. The output torque is 0.17 N-m when the motor is drawing rated current of 1.5 A (rms) at a power factor of 0.88 at rated speed. Find:

a. The full-load efficiency.
b. The rated speed.
c. The full-load copper losses.
d. The combined windage, friction, and iron losses.
e. The motor speed when the rms current is 0.5 A, neglecting phase differences and saturation.

17.22 A 240-V, 60-Hz, two-pole universal motor operates at a speed of 12,000 rev/min on full load and draws a current of 6.5 A at 0.94 power factor lagging. The series field-winding impedance is $4.55 + j3.2$ ohms, and the armature circuit impedance is $6.15 + j9.4$ ohms. Find

a. The back emf of the motor.
b. The mechanical power developed by the motor.
c. The power output if the rotational loss is 65 W.
d. The efficiency of the motor.

17.23 A single-phase motor is drawing 20 A from a 400-V, 50-Hz supply. The power factor is 0.8 lagging. What value of capacitor connected across the circuit will be necessary to raise the power factor to unity?

17.24 A three-phase induction motor is required to operate from a single-phase source. One possible connection is shown in Figure P17.3. Will the motor work? Explain why or why not.

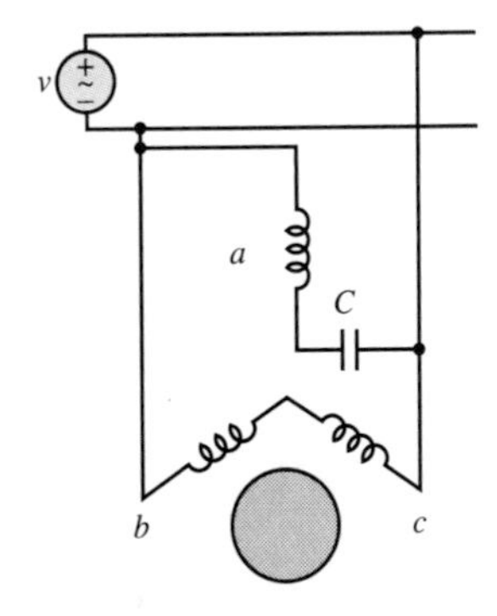

Figure P17.3

17.25 What type of motor would you select to perform the following tasks? Justify your selection.

a. Vacuum cleaner
b. Refrigerator
c. Air conditioner compressor
d. Air conditioner fan
e. Variable-speed sewing machine
f. Clock
g. Electric drill
h. Tape drive
i. *X-Y* plotter

17.26 In performing a brake-load test upon a 1/4-hp capacitor-start motor with its output adjusted to rated value, the following data were obtained: $E = 115$ volts; $I = 3.8$ amp; $P = 310$ W; rev/min $= 1725$. Calculate:

a. Efficiency
b. Power factor
c. Torque in pound-inches

17.27 A 5 hp, 1150 rev/min shunt motor has its speed controlled by means of a tapped field resistor as shown in Figure P17.4. With the tap at position 3, determine the speed of the motor and the torque available at the maximum permissible load.

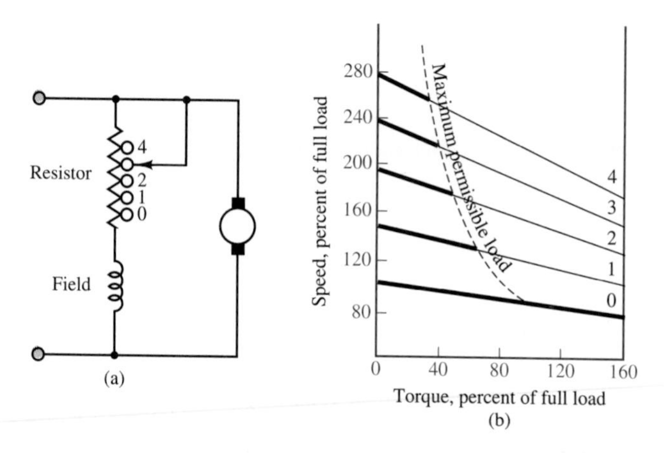

Figure P17.4

17.28 Which single-phase motor would you choose for the following applications?

a. Inexpensive analog electric clock.
b. Bathroom ventilator fan.
c. Escalator which must start under all load conditions.
d. Kitchen blender.
e. Table model circular saw operating at about 3500 rev/min.
f. Hand-held circular saw operating at 15,000 rev/min.
g. Water pump.

17.29 The power required to drive a fan varies as the cube of the speed. If a motor driving a shaft-mounted fan is loaded to 100 percent of its horsepower rating on the top speed connection, what is the horsepower output in percent of rating:

a. At a speed reduction of 20 percent?
b. At a speed reduction of 30 percent?
c. At a speed reduction of 50 percent?

17.30 An industrial plant has a load of 800 kW at a power factor of 0.8 lagging. It is desired to purchase a synchronous motor of sufficient capacity to deliver a load of 200 kW and also serve to correct the overall plant power factor to 0.92. Assuming that the synchronous motor has an efficiency of 91 percent, determine its KVA input rating and the power factor at which it will operate.

APPENDIX

A

Linear Algebra and Complex Numbers

I. SOLVING SIMULTANEOUS LINEAR EQUATIONS, CRAMER'S RULE AND MATRIX EQUATION

The solution of simultaneous equations, such as those that are often seen in circuit theory, may be obtained relatively easily by using Cramer's rule. This method applies to 2×2 or higher systems of equations. Cramer's rule requires the use of the concept of determinant. The method of determinants is valuable because it is systematic, general, and useful in solving complicated problems. A determinant is defined as a square array of numbers having a numerical value, such as

$$det = \begin{vmatrix} a_{11} & a_{12} \\ a_{21} & a_{22} \end{vmatrix} \tag{A.1}$$

In this case the determinant is a 2×2 array, with two rows and two columns and its determinant defined as

$$det = a_{11}a_{22} - a_{12}a_{21} \tag{A.2}$$

A third-order, or 3 × 3, determinant such as

$$det = \begin{vmatrix} a_{11} & a_{12} & a_{13} \\ a_{21} & a_{22} & a_{23} \\ a_{31} & a_{32} & a_{33} \end{vmatrix} \tag{A.3}$$

has three rows and three columns and its determinant is defined as

$$det = a_{11}(a_{22}a_{33} - a_{23}a_{32}) - a_{12}(a_{21}a_{33} - a_{23}a_{31}) + a_{13}(a_{21}a_{32} - a_{22}a_{31}) \tag{A.4}$$

For higher-order determinants, you may refer to a linear algebra book. To illustrate Cramer's method, a set of two equations in general form will be solved here. A set of two linear simultaneous algebraic equations in two unknowns can be written in the form

$$\begin{aligned} a_{11}x_1 + a_{12}x_2 &= b_1 \\ a_{21}x_1 + a_{22}x_2 &= b_2 \end{aligned} \tag{A.5}$$

where x_1 and x_2 are the two unknowns to be solved for. The coefficients a_{11}, a_{12}, a_{21} and a_{22} are known quantities. The two quantities on the right-hand side, b_1 and b_2, are also known (these are typically the source currents and voltages in a circuit problem). The set of equations can be arranged in matrix form, as shown in equation A.6.

$$\begin{bmatrix} a_{11} & a_{12} \\ a_{21} & a_{22} \end{bmatrix} \begin{bmatrix} x_1 \\ x_2 \end{bmatrix} = \begin{bmatrix} b_1 \\ b_2 \end{bmatrix} \tag{A.6}$$

In equation A.6, a coefficient matrix multiplied by a vector of unknown variables is equated to a solution vector. Cramer's rule can then be applied to find x_1 and x_2 using the following formulas:

$$\begin{aligned} x_1 &= \frac{\begin{vmatrix} b_1 & a_{12} \\ b_2 & a_{22} \end{vmatrix}}{\begin{vmatrix} a_{11} & a_{12} \\ a_{21} & a_{22} \end{vmatrix}} \\ x_2 &= \frac{\begin{vmatrix} a_{11} & b_1 \\ a_{21} & b_2 \end{vmatrix}}{\begin{vmatrix} a_{11} & a_{12} \\ a_{21} & a_{22} \end{vmatrix}} \end{aligned} \tag{A.7}$$

Thus, the solution is given by the ratio of two determinants: the denominator is the determinant of the matrix of coefficients, while the numerator is the determinant of the same matrix with the solution vector ($\begin{bmatrix} b_1 \\ b_2 \end{bmatrix}$ in this case) substituted in place of the column of the coefficient matrix corresponding to the desired variable (i.e., first column for x_1, second column for x_2, etc.). In a circuit analysis problem, the coefficient matrix is formed by the resistance (or conductance) values, the vector of unknowns is composed of the mesh currents (or node voltages), and the solution vector contains the source currents or voltages.

In practice, many calculations involve solving higher-order systems of linear equations. Therefore, a variety of computer software packages are often used to solve higher-order systems of linear equations.

DRILL EXERCISES

A.1 Use Cramer's rule to solve the system

$$5v_1 + 4v_2 = 6$$
$$3v_1 + 2v_2 = 4$$

A.2 Use Cramer's rule to solve the system

$$i_1 + 2i_2 + i_3 = 6$$
$$i_1 + i_2 - 2i_3 = 1$$
$$i_1 - i_2 + i_3 = 0$$

A.3 Convert the following system of linear equations into a matrix equation as shown in equation A.6 and find matrices A and b.

$$2i_1 - 2i_2 + 3i_3 = -10$$
$$-3i_1 + 3i_2 - 2i_3 + i_4 = -2$$
$$5i_1 - i_2 + 4i_3 - 4i_4 = 4$$
$$i_1 - 4i_2 + i_3 + 2i_4 = 0$$

II. INTRODUCTION TO COMPLEX ALGEBRA

From your earliest training in arithmetic, you have dealt with real numbers such as 4, -2, $\frac{5}{9}$, π, e, etc., which may be used to measure distances in one direction or another from a fixed point. However, a number such as x that satisfies the equation

$$x^2 + 9 = 0 \tag{A.8}$$

is not a real number. Imaginary numbers were introduced to solve equations such as equation A.8. Imaginary numbers add a new dimension to our number system. To deal with imaginary numbers, a new element, j, is added to the number system having the property

$$j^2 = -1$$

or **(A.9)**

$$j = \sqrt{-1}$$

Thus, we have $j^3 = -j$, $j^4 = 1$, $j^5 = j$, etc. Using equation A.9, we can see that the solutions to equation A.8 are $\pm j3$. In mathematics, the symbol i is used for the imaginary unit, but this might be confused with current in electrical engineering. Therefore, the symbol j is used in this book.

A complex number (indicated in boldface notation) is an expression of the form

$$\mathbf{A} = a + jb \tag{A.10}$$

where a and b are real numbers. The complex number $\mathbf{A}$ has a real part, a, and an imaginary part, b, which can be expressed as

$$a = \operatorname{Re}\mathbf{A}$$
$$b = \operatorname{Im}\mathbf{A} \tag{A.11}$$

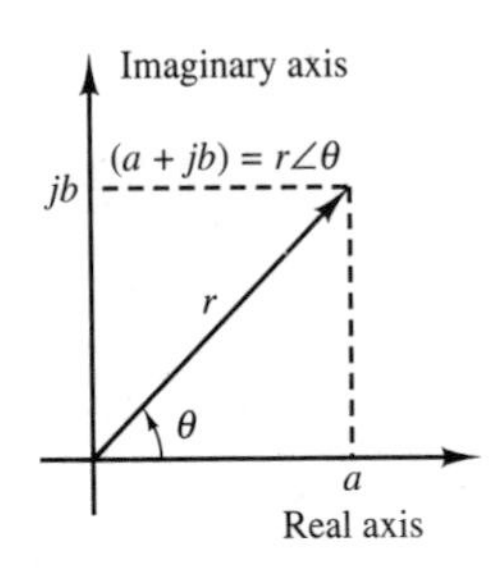

Figure A.1 Polar form representation of complex numbers

It is important to note that a and b are both real numbers. The complex number $a+jb$ can be represented on a rectangular coordinate plane, called the *complex plane*, by interpreting it as a point (a, b). That is, the horizontal coordinate is a in real axis and the vertical coordinate is b in imaginary axis, as shown in Figure A.1. The complex number $\mathbf{A} = a + jb$ can also be uniquely located in the complex plane by specifying the distance r along a straight line from the origin and the angle θ, which this line makes with the real axis, as shown in Figure A.1. From the right triangle of Figure A.1, we can see that:

$$\begin{aligned} r &= \sqrt{a^2 + b^2} \\ \theta &= \tan^{-1}\left(\frac{b}{a}\right) \\ a &= r\cos\theta \\ b &= r\sin\theta \end{aligned} \tag{A.12}$$

Then, we can represent a complex number by the expression:

$$\mathbf{A} = re^{j\theta} = r\angle\theta \tag{A.13}$$

which is called the polar form of the complex number. The number r is called the magnitude (or amplitude) and the number θ is called the angle (or argument). The two numbers are usually denoted by: $r = |\mathbf{A}|$ and $\theta = \arg \mathbf{A} = \angle\mathbf{A}$.

Given a complex number $\mathbf{A} = a + jb$, the complex conjugate of $\mathbf{A}$, denoted by the symbol $\mathbf{A}^*$, is defined by the following equalities:

$$\begin{aligned} \operatorname{Re}\mathbf{A}^* &= \operatorname{Re}\mathbf{A} \\ \operatorname{Im}\mathbf{A}^* &= -\operatorname{Im}\mathbf{A} \end{aligned} \tag{A.14}$$

That is, the sign of the imaginary part is reversed in the complex conjugate.

Finally, we should remark that two complex numbers are equal *if and only if* the real parts are equal and the imaginary parts are equal. This is equivalent to stating that two complex numbers are equal only if their magnitudes are equal and their arguments are equal.

The following examples and drill exercises should help clarify these explanations.

Example A.1

Convert the complex number $\mathbf{A} = 3 + j4$ to its polar form.

Solution:

$$r = \sqrt{3^2 + 4^2} = 5 \quad \theta = \tan^{-1}\left(\frac{4}{3}\right) = 53.13°$$

$$\mathbf{A} = 5\angle 53.13°$$

Example A.2

Convert the number $\mathbf{A} = 4\angle -60°$ to its complex form.

Solution:

$$a = 4\cos(-60°) = 4\cos(60°) = 2$$

$$b = 4\sin(-60°) = -4\sin(60°) = -2\sqrt{3}$$

Thus, $\mathbf{A} = 2 - j2\sqrt{3}$.

Addition and *subtraction* of complex numbers take place according to the following rules:

$$\begin{aligned}(a_1 + jb_1) + (a_2 + jb_2) &= (a_1 + a_2) + j(b_1 + b_2)\\ (a_1 + jb_1) - (a_2 + jb_2) &= (a_1 - a_2) + j(b_1 - b_2)\end{aligned} \tag{A.15}$$

Multiplication of complex numbers in polar form follows the law of exponents. That is, the magnitude of the product is the product of the individual magnitudes, and the angle of the product is the sum of the individual angles, as shown below.

$$(\mathbf{A})(\mathbf{B}) = (Ae^{j\theta})(Be^{j\phi}) = ABe^{j(\theta+\phi)} = AB\angle(\theta + \phi) \tag{A.16}$$

If the numbers are given in rectangular form and the product is desired in rectangular form, it may be more convenient to perform the multiplication directly, using the rule that $j^2 = -1$, as illustrated in equation A.17.

$$\begin{aligned}(a_1 + jb_1)(a_2 + jb_2) &= a_1a_2 + ja_1b_2 + ja_2b_1 + j^2b_1b_2\\ &= (a_1a_2 + j^2b_1b_2) + j(a_1b_2 + a_2b_1)\\ &= (a_1a_2 - b_1b_2) + j(a_1b_2 + a_2b_1)\end{aligned} \tag{A.17}$$

Division of complex numbers in polar form follows the law of exponents. That is, the magnitude of the quotient is the quotient of the magnitudes, and the angle of the quotient is the difference of the angles, as shown in equation A.18.

$$\frac{\mathbf{A}}{\mathbf{B}} = \frac{Ae^{j\theta}}{Be^{j\phi}} = \frac{A\angle\theta}{B\angle\phi} = \frac{\mathbf{A}}{\mathbf{B}}\angle(\theta - \phi) \tag{A.18}$$

Division in the rectangular form can be accomplished by multiplying the numerator and denominator by the complex conjugate of the denominator. Multiplying the denominator by its complex conjugate converts the denominator to a real number and simplifies division. This is shown in Example A.4. Powers and roots of a complex number in polar form follow the laws of exponents, as shown in equations A.19 and A.20.

$$\mathbf{A}^n = (Ae^{j\theta})^n = A^ne^{jn\theta} = A^n\angle n\theta \tag{A.19}$$

$$\begin{aligned}\mathbf{A}^{1/n} &= (Ae^{j\theta})^{1/n} = A^{1/n}e^{j1/n\theta}\\ &= \sqrt[n]{A}\angle\left(\frac{\theta + k2\pi}{n}\right) \qquad k = 0, \pm1, \pm2, \ldots\end{aligned} \tag{A.20}$$

Example A.3

Perform the following operations given that $\mathbf{A} = 2 + j3$ and $\mathbf{B} = 5 - j4$.

(a) $\mathbf{A} + \mathbf{B}$ (b) $\mathbf{A} - \mathbf{B}$ (c) $2\mathbf{A} + 3\mathbf{B}$

Solution:

$$A + B = (2 + 5) + j(3 + (-4)) = 7 - j$$
$$A - B = (2 - 5) + j(3 - (-4)) = -3 + j7$$

For part c, $2A = 4 + j6$ and $3B = 15 - j12$. Thus, $2A + 3B = (4 + 15) + j(6 + (-12)) = 19 - j6$

Example A.4

Perform the following operations both in rectangular and polar form, given that $A = 3 + j3$ and $B = 1 + j\sqrt{3}$.

(a) AB (b) $A \div B$

Solution:

a. In rectangular form:

$$\begin{aligned} AB &= (3 + j3)(1 + j\sqrt{3}) = 3 + j3\sqrt{3} + j3 + j^2 3\sqrt{3} \\ &= (3 + j^2 3\sqrt{3}) + j(3 + j3\sqrt{3}) \\ &= (3 - 3\sqrt{3}) + j(3 + j3\sqrt{3}) \end{aligned}$$

To obtain the answer in polar form, we need to convert A and B to their polar forms:

$$A = 3\sqrt{2}e^{j45^\circ} = 3\sqrt{2}\angle 45^\circ$$
$$B = \sqrt{4}e^{j60^\circ} = 2\angle 60^\circ$$

Then,

$$AB = (3\sqrt{2}e^{j45^\circ})(\sqrt{4}e^{j60^\circ}) = 6\sqrt{2}\angle 105^\circ$$

b. To find $A \div B$ in rectangular form, we can multiply A and B by B^*.

$$\frac{A}{B} = \frac{3 + j3}{1 + j\sqrt{3}} \frac{1 - j\sqrt{3}}{1 - j\sqrt{3}}$$

Then,

$$\frac{A}{B} = \frac{(3 + 3\sqrt{3}) + j(3 - 3\sqrt{3})}{4}$$

In polar form, the same operation may be performed as follows:

$$A \div B = \frac{3\sqrt{2}\angle 45^\circ}{2\angle 60^\circ} = \frac{3\sqrt{2}}{2}\angle(45^\circ - 60^\circ) = \frac{3\sqrt{2}}{2}\angle -15^\circ$$

Euler's Identity

This formula extends the usual definition of the exponential function to allow for complex numbers as arguments. Euler's identity states that

$$e^{j\theta} = \cos\theta + j\sin\theta \tag{A.21}$$

All the standard trigonometry formulas in the complex plane are direct consequences of Euler's identity. The two important formulas are:

$$\cos\theta = \frac{e^{j\theta} + e^{-j\theta}}{2} \qquad \sin\theta = \frac{e^{j\theta} - e^{-j\theta}}{2j} \tag{A.22}$$

Example A.5

Using Euler's formula, show that

$$\cos\theta = \frac{e^{j\theta} + e^{-j\theta}}{2}$$

Solution:

Using Euler's formula

$$e^{j\theta} = \cos\theta + j\sin\theta$$

Extending the above formula, we can obtain

$$e^{-j\theta} = \cos(-\theta) + j\sin(-\theta) = \cos\theta - j\sin\theta$$

Thus,

$$\cos\theta = \frac{e^{j\theta} + e^{-j\theta}}{2}$$

DRILL EXERCISES

A.4 In a certain AC circuit, $\boldsymbol{V} = \boldsymbol{ZI}$ where $\boldsymbol{Z} = 7.75\angle 90^\circ$ and $\boldsymbol{I} = 2\angle -45^\circ$. Find $\boldsymbol{V}$.

A.5 In a certain AC circuit, $\boldsymbol{V} = \boldsymbol{ZI}$ where $\boldsymbol{Z} = 5\angle 82^\circ$ and $\boldsymbol{V} = 30\angle 45^\circ$. Find $\boldsymbol{I}$.

A.6 Show that the polar form of $\boldsymbol{AB}$ in Example A.4 is equivalent to its rectangular form.

A.7 Show that the polar form of $\boldsymbol{A} \div \boldsymbol{B}$ in Example A.4 is equivalent to its rectangular form.

A.8 Using Euler's formula, show that $\sin\theta = \frac{e^{j\theta} - e^{-j\theta}}{2j}$.

Answers to Selected Problems

CHAPTER 2

2.1
a. 5.184×10^6 J
b. 864×10^3 J

2.2
a. 48,500 C
b. 489.9 kJ

2.7 3 A

2.10
a. 20 W
b. −36 W

2.13 −1200 W

2.17 2.88 Ω, 1.92 Ω; 1.152 Ω.

2.21 $R = 18$ kΩ, $v = 16$ V, $v_1 = 2$ V, $I = 0.5$ mA.

2.25 20/3 A

2.28
a. 12 Ω
b. 0.5 A
c. 3 W
d. −2 V
e. 1 W

2.37
a. $v_{\text{out}}(x) = \dfrac{e^x}{2203}$
b. $x = 9.08$ cm

2.42
a. $r_B = 0.061$ Ω
b. $r_B = 84.1$ Ω

2.44
a. $i \approx 1$ mA
b. $r_a = 9.28$ Ω

2.48

	With meter in circuit	Without meter in circuit
a	8.61 mA	8.92 mA
b	39.6 mA	47.2 mA
c	61.9 mA	82.6 mA
d	65.6 mA	89.3 mA

2.50 39.98 kΩ

CHAPTER 3

3.8 $V_a = 5.36$ V, $V_b = 7.89$ V.

3.9 25 mW

3.12 $I_1 = 0.1923$ A, $I_2 = 0.865$ A, $v_{10\ \Omega} = 6.727$ V (+ref. at left)

3.21 $v_{1\ \Omega} = 1.655$ V (+ref. at bottom)

3.23 $V_a = 5.36$ V, $V_b = 7.895$ V.

3.35 $R_T = 500\ \Omega$, $v_{OC} = 0$ V.

3.37 $R_T = 10\ \Omega$, $v_{OC} = 2$ V.

3.39
a. 4.44 W
b. 22.2 W
c. $v_{OC} = 10$ V and $R_T = 5\ \Omega$.
d. $\frac{20}{9}$ W
e. No, No.

3.41
a. $R_T = 999\ \Omega$, $v_T = 12$ mV
b. $32.04 \times 10^{-9} = 32$ nW
c. 64 nW
d. 144.1 mW

3.48 2.33 V

3.50 $I = 14$ mA, $V_{out} = 11$ V.

3.53
a. $\approx 0°$
b. $V_{out} \approx 3.5$ V
c. A few points from the curve are (T °C, V_{out} V): $(-20, 4.1), (-10, 3.5), (0, 3), (10, 2.7), (20, 2.6)$.

CHAPTER 4

4.1 1.155 V.

4.6 6.40 V.

4.11
a. 10 W.
b. 10 W.

4.19 For $0 < t < 0.1$ s,
$i_L = 0.5t + 0.01e^{-t/0.02}$ A
For $0.1 < t < 0.2$ s,
$i_L = 0.49663t - 0.009596$ A
For $0.2 < t < 0.3$ s,
$i_L = 0.5t - 0.01e^{(t-0.3)/0.02} - 0.0102$ A
For $0.3 < t < 0.4$ s
$i_L = 0.1298$ A

4.23 $i_C = 500 \times 10^{-6} \times 3{,}750$ for $0 < t < 4$ ms
$i_C = 500 \times 10^{-6} \times (-3{,}750)$ for $4 < t < 8$ ms
$i_L(t) = 18{,}750t^2$ for $0 < t < 4$ ms
$i_L(t) = 300t - 18{,}750t^2 - 0.6$ A for $4 < t < 8$ ms

4.29 $\frac{dv_c(t)}{dt} + 1667v_c(t) = 1667v_S(t)$

4.32 $v_{out}(t) = \frac{I_m R}{\sqrt{1 + (\omega RC)^2}} \cos(\omega t + \tan^{-1}(-\omega RC))$ V

4.34
a. $v_{out} = 203 \sin 377t - 662 \cos 377t$ V
b. $i_R = v_{out}/40$
c. $i_L = \frac{1}{L} \int_{-\infty}^{t} v_{out}(\tau)\, d\tau$

4.37 a. $\mathbf{I}(\omega) = 0.5\angle 45°$ A

4.41
a. $4 - j4;\ 2 + j8;\ -5 - j2.$
b. and c. $1.25\angle 36.87°;\ 0.485\angle 165.96°;\ 0.1857\angle -158.2°.$

4.44
a. $1 + j5$
b. $C = 510.1\ \mu$F
c. $26\ \Omega$

4.46 $R_2 = \frac{\omega^2 R_1 L_1^2}{R_1^2 + (\omega L_1)^2}$ $L_2 = \frac{R_1^2 L_1}{R_1^2 + (\omega L_1)^2}$

4.53 $V_L(t) = 14.4 \cos(1000t + 53.13°)$ V

4.57
a. $Z_T = 500 + j10.01\ \Omega$, $\mathbf{V}_T = 10\angle 0°$ V
b. $Z_T = 500 - j10.01\ \Omega$, $\mathbf{V}_T = 10\angle 0°$ V

4.59 $i(t) = 0.2357 \cos(2t + 45°)$ A

4.63 $Z_T = 2\ \Omega$

4.68 $V_T = 1.414\angle 45°$ V

CHAPTER 5

5.3 5.76 kW

5.6
a. 800 W
b. 800 W
c. 478.6 W
d. 1700 W

5.9
a. 0.56
b. 56.25°
c. $5.56 \pm j8.31\ \Omega$
d. $5.56\ \Omega$

5.12
a. 0.848 leading
b. 0.9925 lagging
c. 0.08716 leading
d. 0.9487 lagging

5.16 a. $S = 228\angle -40°$ VA
b. $P_L = 174.66$ W
c. $Z_L = 63.16\angle -40°\ \Omega$

5.21 a. 4,072.3 W
b. 407.23 W
c. $4641.4\angle 15.26°$ VA
d. 0.9578 lagging
e. 0.9647 lagging

5.25 $S = 2{,}070.3\angle 28.97°$ VA

5.28 a. 717.4 W
b. 0.847 lagging
c. 19 μF

5.34 a. 869.57 A
b. 400 kW
c. 320 kW
d. 280 kW
e. 0.75

5.39 $119.94\angle 180.3°$ V

5.44 $11.448\angle 135.18°$ A

5.47 $\mathbf{I}_R = 22$ A; $\mathbf{I}_W = 22\angle 120°$ A;
$\mathbf{I}_B = 22\angle 240°$ A; $\mathbf{I}_N = 0$ A

5.50 24,387 W

5.54 Wye circuit:

$S = 345.6 + j259.2$ VA;

$P = 345.6$ W

Delta circuit:

$S = 1037.6 + j778.2$ VA;

$P = 1037.6$ W

5.58 $\mathbf{I}_A = 5.3\angle -14°$ A; $\mathbf{I}_B = 5.3\angle 106°$ A
$\mathbf{I}_C = 5.3\angle -106°$ A; $\mathbf{I}_N = 2.56\angle -30°$ A
$P = 2262.8$ W

CHAPTER 6

6.3 a. $$\left|\frac{\mathbf{V}_{\text{out}}}{\mathbf{V}_{\text{in}}}(j\omega)\right| = \frac{1}{\sqrt{4 + 0.01\omega^2}}$$
$$\phi(j\omega) = -\arctan(0.05\omega)$$

6.7 a. $$\left|\frac{\mathbf{V}_{\text{out}}}{\mathbf{V}_{\text{in}}}\right| = \frac{100000}{\sqrt{100000)^2 + 0.01\omega^2}}$$
$$\phi(\omega) = -\arctan\left(\frac{0.1\omega}{100000}\right)$$

6.13 a. $$\left|\frac{\mathbf{V}_{\text{o}}}{\mathbf{V}_{\text{i}}}\right| = \frac{\omega RC}{\sqrt{(1 - \omega^2 LC)^2 + (\omega RC)^2}}$$
b. $$\omega_{\text{center}} = \frac{1}{\sqrt{LC}} = 1000 \text{ rad/s}$$

6.17 a. $$Z_{ab} = \frac{j\omega L}{1 - \omega^2 LC}$$
b. $$\omega = \frac{1}{\sqrt{LC}} = 10^6 \text{ rad/s}$$

6.22 $$\frac{\mathbf{V}_{\text{out}}(\omega)}{\mathbf{V}_S(\omega)} = \frac{R_L}{R_S + \dfrac{j\omega L}{1 - \omega^2 LC} + R_L}$$

6.33 a. $v_c(0^-) = v_c(0^+) = 0$ V
b. $\tau = 48$ s
c. $v_c(t) = -8e^{-1/48t} + 8$ V $\quad t > 0$
d. $v_c(0) = 0$ V; $v_c(\tau) = 5.06$ V;
$v_c(2\tau) = 6.9$ V; $v_c(5\tau) = 7.95$ V;
$v_c(10\tau) = 8.0$ V

6.35 a. $v_c(0^-) = 11.67$ V
b. $V_c(t) = -11.09e^{-3/70t} + 11.67$ V $\quad t > 0$

6.38 $\tau = 0.923$ ms; $\tau' = 1.333$ ms.

6.44 a. $\tau = 5.005$ ms
b. $\tau = 5$ μs

6.46 a. $t_{\text{ready}} = 6.9$ s
b. $W = 42.27$ mJ
c. $W = 31.6$ mJ

6.50 $$\frac{d^2 v_c}{dt^2} + 100\frac{dv_c}{dt} + 100{,}000 v_c = 0$$

6.53 $v(t) = -12e^{-2t} - 12te^{-2t} + 12$ V for $t > 0$

6.56 $v(t) = 18e^{-t} - 3e^{-6t}$ V $\quad t > 0$

6.60 A 10 μF capacitor in series with the following three capacitors in parallel: 5 μF, 0.5 μF, 0.5 μF.

CHAPTER 7

7.7 $V_{\text{in}} \geq 3$ V

7.11 $V_{\text{o}} = 4$ V

7.14 a. $i_{DQ} \approx 4.3$ mA, $v_{DQ} \approx 0.7$ V;
$P_D = 3.01$ mW
b. $i_{DQ} \approx 0.85$ mA, $v_{DQ} \approx 0.58$ V;
$P_D = 0.493$ mW

7.18 a. $i_D = 1$ mA for $t < 10$ ms
$i_D = 0$ mA for $10 \text{ ms} < t < 20$ ms
b. $i_D = 0.94$ mA for $t < 10$ ms
$i_D = 0$ mA for $10 \text{ ms} < t < 20$ ms
c. $i_D = 0.85$ mA for $t < 10$ ms
$i_D = 0$ mA for $10 \text{ ms} < t < 20$ ms

7.20 a. Source voltage is a 16.97-V peak sinusoid. Load voltage is a 15.77-V half-wave rectified sinusoid. D1 and D4 are on during the positive half cycle of the source voltage; D2 and D3 are on during the negative half cycle of the source voltage.
b. The load voltage waveform is a periodic exponential charging-discharging waveform going from 1.96 to 15.77 V and back.
c. Now the amplitude of the "ripple" is greatly reduced, going from 13.34 V to 15.77 V.

7.27 a. ≈ 0.61 V
b. 5 V
c. 0.63 V
d. 0.63 V

7.33 a. 3.66 Ω
b. 3.7 mA.

7.36 For $0 < v_S < V_2 = 12$ V
$v_L = 0.3125 v_S + 7.5$ V
For $12 < v_S < 15$,
$v_L = 0.833 v_S + 7.5$
For $15 < v_S < 20$
$v_L = 0.4545 v_S + 6.818$
For $-20 < v_S < 0$
$v_L = 7.5 + 0.3125 v_S$

7.39 For Figure P7.36a
a. $I_{SW} = 0; I_S = I_B = 0.31$ A
b. $I_S = 13$ A, $I_B = -0.96$ A; $I_{SW} = 13.96$ A
c. The battery will discharge quickly.
For Figure P7.36b
a. $I_{SW} = 0; I_S = I_B = 0.25$ A
b. $I_S = I_{SW} = 13$ A, $I_B = 0$
c. The diode prevents current from flowing out of the battery.

CHAPTER 8

8.2 $V_1 = 2.7$ V; $I_E = \frac{2}{3}$ mA; $I_B = 13$ μA
$I_C = 0.654$ mA; $\beta = 50.3$; $V_3 = 8.73$ V

8.6 a. $V_{CE} = 6.65$ V
b. $I_C = 5.9$ mA
c. $P \approx 39$ mW

8.11 $\frac{I_C}{I_B} = 96 < \beta$

8.15 a. 170; 165; 143.
b. $V_{CE} = 6.29$ V

8.20 $V_{BB} > 1.56$ V

8.23 $A_V = -2.725$

8.29 $V_T = 2$ V; $k = 62.5 \times 10^{-6} \frac{\text{A}}{\text{V}^2}$

8.33 $i_D = 1.395$ mA

8.38 $V_T = 0; K = 4.17 \times 10^{-5}$ mA/V²

8.42 a. $r_{DS} = 250$ Ω
b. $v_{GS} = 0$ V

CHAPTER 9

9.2 a. $I_B = 0.142$ mA; $I_C = 21.3$ mA; $V_{CE} = 5.78$ V.
b. $h_{ie} = 4.93$ kΩ; $h_{fe} \approx \beta = 150$.

9.6 $R_2 = 10$ kΩ; $R_1 = 20$ kΩ; $R_E = \frac{V_E}{I_E} =$ 5 kΩ; $R_C = \frac{5}{I_C} = 5$ kΩ.

9.10 a. $R_E = 1.81$ kΩ
b. $R_C = 8.27$ kΩ
d. $A_V = -4.15$

9.13 a. $P_{max} = 0.02694 + 0.5 = 527$ mW
b. 3.69 ms

9.15 a. $A_i = 9{,}300$
b. $R_{in} = 27.7$ kΩ

9.33 $5{,}833\ \Omega \le R_B \le 18{,}333\ \Omega$

9.37 This circuit performs the function of a 2-input NAND gate.

9.40 The two transistors at the top are cut off and the two at the bottom are on.

CHAPTER 10

10.2 $R_S = \frac{V_Z - V_{BE}}{I} = \frac{V_Z - 0.6}{I}$

10.3 $V_{out} = V_Z + V_\gamma$

10.6 a. $v_L(t) = (10-0.7)\sin\omega t$ (positive half cycle)
b. $< v_L > \approx 2.96$ V

CHAPTER 11

11.3 $v = 6\cos 2t$ V

11.6 $v_o = -12$ V

11.7 $i = 2$ A

11.14 a. $V_{OC(a)} = \dfrac{\left(1 + \dfrac{|\Delta R|}{R_o}\right)v_S}{2}$;

$R_{T(a)} = \dfrac{R_o}{2}\left(1 - \left(\dfrac{|\Delta R|}{R_o}\right)^2\right)$

b. $V_{out} = 2K\Delta T$

11.17 $A_v = 9.4$

11.23 $A_v = -\dfrac{1}{10}$

11.26 a. $\dfrac{V_{out}}{V_{in}}(j\omega) = -\dfrac{j\omega R_2 C}{j\omega R_1 C + 1}$
b. -0.04 dB
c. Gain $= -0.9988$, phase $= -177.14°$
d. $\omega > 130.8$ rad/s

11.30 a. $\dfrac{\mathbf{V}_{out}}{\mathbf{V}_{in}}(j\omega) = -\dfrac{C_1}{C_2}\dfrac{1 + j\omega R_2 C_2}{1 + j\omega R_1 C_1}$
b. 6.0 dB
c. Gain $= 1.9977$, phase $= -177.85°$

11.35 a. $\dfrac{V_{out}}{V_{in}} = -\dfrac{1 + j\omega R_2 C}{j\omega R_1 C}$
b. $.04dB$
c. Gain $= -1.0008$; Phase $= 177.71°$
d. $\omega > 196.5$ rad/s

11.39 $\dfrac{d^2x}{dt^2} - 4{,}000x(t) = 16{,}000f(t)$

11.45 a. $\dfrac{\mathbf{V}_{out}}{(\mathbf{V}_1 - \mathbf{V}_2)} = \dfrac{R_1/R_2}{1 + j\omega C R_1}$
b. 5.12 dB (gain is $-$ 5.12 dB)
c. $|H(\omega = 2{,}000)| = 0.4472$
$\angle H(\omega = 2{,}000) = -63.4°$
d. $\omega < 328.7$ rad/s
e. 0.4455 dB(gain is $-$ 0.4455 dB)

11.48 a. $R_i = r_o + (A_{OL} + 1)R_{in}$
b. $A_{OL} = 2 \times 10^{12}\ \Omega$

11.52 a. $A_{CL} = \left(\dfrac{R_4 + R_3}{R_3}\right)\left(\dfrac{R_2}{R_1 + R_2}\right)\left(\dfrac{1}{\dfrac{R_4 + R_3}{R_3 A_{OL}} + 1}\right)$
b. $v_o(t) = A_{CL}v_{in}(t) = 4.68\cos 27.6t$

CHAPTER 12

12.2 a. 10, 1010
b. 102, 1100110
c. 71, 1000111
d. 33, 100001
e. 19, 10011

12.6 a. 11100
b. 1101110
c. 1000

12.13 $F(X, Y, Z) = \overline{YZ} + XYZ$

12.16 $F = A + C$

12.20 $F(A, B, C, D) = \left(\overline{CD}\right)\left(\left(\overline{A} + \overline{B}\right)\overline{C} + ABC\right)$

12.24 $f = \overline{BC} + A\overline{B} + \overline{A}BC$

12.30 $\overline{A} + \overline{B}$

12.35 $f = B\overline{D} + AC\overline{D}$

12.42 $F = \overline{ACD} + \overline{A}B\overline{C} + ABC$

12.46 The circuit operates as a 4 of 16 decoder.

CHAPTER 13

13.7 A 10 kHz clock increments a 16-bit binary counter. The count is held by a latch, and then converted to BCD for use with seven-segment displays.

13.9 This is briefly discussed in the digital counters section, on page 669.

13.18 a. $n(n-1)$
b. $2n$

13.20 "Static" means that memory contents do not have to be refreshed. "Nonvolatile" means that the information in the memory is not lost when the power is off.

CHAPTER 14

14.14 $A = 21$

14.16 $A_{diff} = 110$

14.20 $n = 2$

14.24 $\frac{v_{out}(j\omega)}{v_s(j\omega)} = \frac{(j\omega)^2/K}{(j\omega)^2 + j\omega K_1 - K_2}$

$K = \frac{R_B}{R_A + R_B}$

$K_1 = \frac{1}{KR_1C_1} - \frac{1}{R_2C_2} - \frac{1}{R_1C_1} - \frac{1}{R_2C_1}$

$K_2 = \frac{1}{R_1R_2C_1C_2}$

14.28 $C_1 = C_2 = 1\ \mu F; R_1 = 1.8\ \Omega, R_2 = 23\ \Omega.$

14.38 1. Digital signals are less subject to noise.
2. Digital signals are directly compatible with digital computers, and can therefore be easily stored on a disk, or exchanged between computers.

14.42 a. $v_a = -3.6$ V
b. $(v_a)_{max} = -4.5$ V
c. $\delta v 0.3$ V
d. $n = 8$

14.45 a. $v_a = -1.729$ V
b. $(v_a)_{max} = -4.5$ V
c. $\delta v = 17.6$ mV
d. $n = 14.$

14.49 a. $v_a = -0.8425$ V
b. $(v_a)_{max} = -10$ V
c. $\delta v = 2.44$ mV
d. Choose a 15-bit ADC.

14.54 $n = 13$

14.58 $f_{max} = 104.15$ Hz

14.62 $n = 9$

14.65 Since the required noise protection level is ± 150 mV, $R_1 = 100\ k\Omega$, $R_2 = 1.8\ k\Omega$. $R_3 = 16.2\ k\Omega$.

14.74 $C = 0.91\ \mu F$

CHAPTER 15

15.3 a. $W_m = 1.25$ J
$L_\Delta = 2$ H
b. $v_L(t) = \sin(2\pi t) + 4\pi\cos(2\pi t)$

15.8 $f = 173.5$ N

15.12 a. $R = \frac{x}{\mu_0 w^2} + \frac{l_g}{\mu_0 w(w - x)}$
b. $W_m = \frac{(N_1 + N_2)^2 i^2}{2R}$
c. $f = \frac{i^2}{2}\frac{(N_1 + N_2)^2 \mu_0 w^2}{x^2}$

15.17 $i = 1.784$ A

15.21 a. $N = 5$
b. $\mathbf{V}_1 = 353.6\angle -0.84°$ V
$\mathbf{I}_1 = 0.066\angle -50.9°$ A
$P_{in} = 14.98$ W
efficiency $= \eta = 80.1\%$

15.26 $v = i\,R + \frac{N^2\mu_0 A}{2x}\frac{di}{dt}$

$m\frac{d^2x}{dt^2} + kx = \frac{N^2 i^2 \mu A}{4x^2}.$

15.30 $\frac{V_{out}(s)}{V_{ex}(s)} = \frac{Ms}{(L_p s + R_p)2(L - M)s + 1 + 2\frac{R_s}{R_L}}$

Let $s = j\omega$ and find the maximum.

15.35 $\alpha = 1/8$

15.39 a. $f = -1.55$ N

CHAPTER 16

16.2 $\phi = 0.32$ mWb

16.7 a. $I_a = 83.33$ A
b. $V_a = 124.17$ V

16.11 $P_{loss} = 2.647$ kW; $P_{in} = 17.647$ kW.

16.16 a. $i_a = 42.93$ A, $\omega_m = 106.3$ rad/sec
$P_{dev} = 8.952$ kW
$P_o = 8{,}452$ W
$T_{sh} = 72.1$ N-m

16.21 a. $N_{series} = \frac{120}{8} = 15$ turns
b. (1) $T_X = 1.13$ NM
(2) $T_X = 0.55$ N-m
(3) $T_X = 9.54$ N-m
(4) $T_X = 20.53$ N-m

16.25 $\mathbf{I}_s = 54.23\angle 24.16°$ A
$P_{in\ motor} = 18.85$ kW
$P_{in\ total} = 27.10$ kW
$pf = 0.991$ leading

16.31 $P_{max} = 3.096$ MW
$T_{max} = 123.2$ kN-m

16.35 $s = 0.025$; $\mathbf{I}_s = 54.6\angle -19.98°$ A;
$P_{in} = 35.6$ kW; $P_{sh} = 32.17$ kW;
$T_{sh} = 175$ N-m; efficiency $= 0.904$

16.41 $I_s = 11.02\angle -26.2°$ A; $P_{in} = 3{,}426$ W;
$P_m = 3{,}121$ W; $T_{sh} = 17.24$ N-m

16.45 If the load torque is reduced by one half the speed will increase and the horsepower delivered will reduce by approximately one half.

16.50 $I_s = 9.39\angle -27.15°$ A
$pf = 0.8898$ lagging

CHAPTER 17

17.3
$$v = Ri + L\frac{di}{dt} + K_E\omega$$
$$T = K_T i = (J_m + J_L)\frac{d\omega}{dt} + D\omega + T_F + T_L$$

17.8 The motor will require 24 stator teeth and 2 rotor teeth.

17.12
$$\begin{aligned} F_1 &= F_{1max}\cos(\omega t)\cos\theta \\ &= \frac{1}{2}F_{1max}\cos\theta\cos(\omega t) - \frac{1}{2}F_{1max}\cos\theta\sin(\omega t) \\ &\quad + \frac{1}{2}F_{1max}\cos\theta\cos(\omega t) + \frac{1}{2}F_{1max}\cos\theta\sin(\omega t) \\ &= F_{CW} + F_{CCW} \end{aligned}$$

17.17 $P_{mech} = 201.68$ W

17.21
a. efficiency = 61.43%
b. 5,238.1 rev/min
c. 22.5 W
d. 36.05 W
e. 6,091.9 rev/min

17.24 It will work. The *b* and *c* windings will produce a magnetic field similar to a single phase machine, that is, two components rotating in opposite directions and the *a* winding would act as a starting winding. The phase shift provided by the capacitor is needed to provide a nonzero starting torque.

Index

G

H

I

O

P